PREFACE

This solutions guide is a supplement to *Precalculus Functions and Graphs: A Graphing Approach,* Second Edition and *Precalculus with Limits: A Graphing Approach,* Second Edition by Roland E. Larson, Robert P. Hostetler, and Bruce H. Edwards. All references to chapters, theorems, and definitions apply to this text.

Part I of this guide contains solutions to the *even-numbered* exercises in the text, with all essential algebraic steps included. Part II contains Chapter Project solutions, and Part III contains Focus on Concepts solutions.

We have made every effort to see that the solutions are correct. However, we would appreciate hearing about any errors or other suggestions for improvement.

Bruce H. Edwards
University of Florida
Department of Mathematics
Gainesville, Florida 32611
be@math.ufl.edu

Dianna L. Zook
Indiana University
Purdue University at Fort Wayne
Fort Wayne, Indiana 46805

Emily J. Keaton
Peabody, MA

Solutions to Even-Numbered Exercises

PRECALCULUS FUNCTIONS AND GRAPHS: A Graphing Approach

and

PRECALCULUS WITH LIMITS: A Graphing Approach

SECOND EDITION

Larson / Hostetler / Edwards

Bruce H. Edwards
University of Florida

Dianna L. Zook
Indiana University
Purdue University at Fort Wayne

Emily J. Keaton

HOUGHTON MIFFLIN COMPANY Boston New York

Sponsoring Editor: Christine B. Hoag
Senior Associate Editor: Maureen Brooks
Managing Editor: Catherine B. Cantin
Assistant Editor: Carolyn Johnson
Supervising Editor: Karen Carter
Associate Project Editor: Rachel D'Angelo Wimberly
Editorial Assistant: Caroline Lipscomb
Production Supervisor: Lisa Merrill
Art Supervisor: Gary Crespo
Associate Marketing Manager: Ros Kane
Marketing Assistant: Kate Burden Thomas

Printed in the United States of America.

International Standard Book Number: 0-669-41730-0

123456789-VG 01 00 99 98 97

CONTENTS

PART I
Solutions to Even-Numbered Exercises

CHAPTER P
Prerequisites

CHAPTER P
Prerequisites

Section P.1 The Cartesian Plane

Solutions to Even-Numbered Exercises

2.

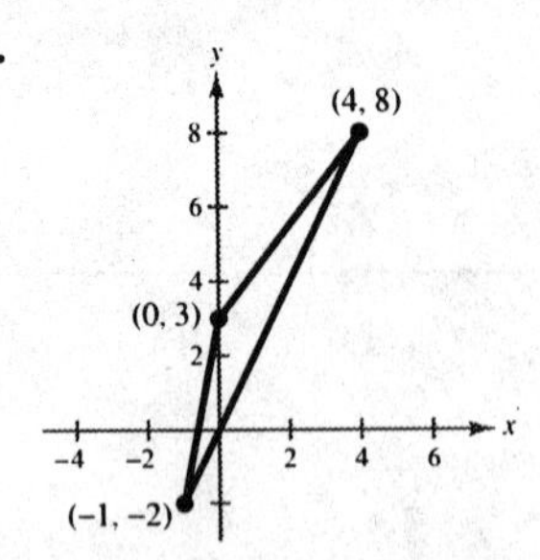

4.

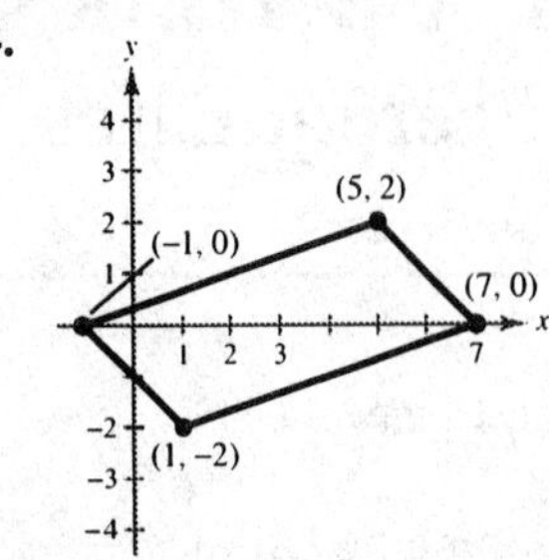

6. $A: \left(\frac{3}{2}, -4\right)$; $B: (0, -2)$; $C: \left(-3, \frac{5}{2}\right)$; $D: (-6, 0)$

8. $A: (-4, 0)$; $B: (-5, -5)$; $C: (3.5, -2.5)$; $D: (2, 0)$

10. $(4, -8)$

12. $(-12, 0)$

14. No, it is not true that the scales on the x- and y-axis must be the same. The scales depend on the magnitude of the coordinates.

16. $x < 0$ and $y < 0$ in Quadrant III.

18. $x > 2$ and $y = 3$ in Quadrant I.

20. $x > 4$ in Quadrants I and IV.

22. If $(-x, y)$ is in Quadrant IV, then (x, y) must be in Quadrant III.

24. If $xy < 0$, then x and y have opposite signs. This happens in Quadrants II and IV.

26.

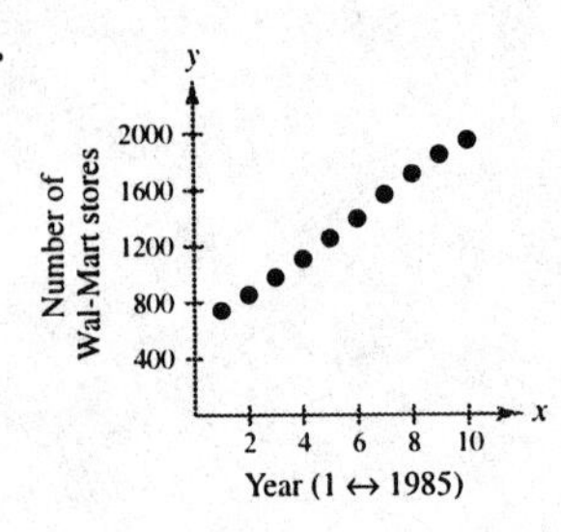

28.
$$(-3 + 6, 6 - 3) = (3, 3)$$
$$(-5 + 6, 3 - 3) = (1, 0)$$
$$(-3 + 6, 0 - 3) = (3, -3)$$
$$(-1 + 6, 3 - 3) = (5, 0)$$

30. $y = 2 - \frac{1}{2}x^2$

x	-2	-1	$-\frac{1}{2}$	0	$\frac{1}{2}$	1	2
y	0	$\frac{3}{2}$	$\frac{15}{8}$	2	$\frac{15}{8}$	$\frac{3}{2}$	0

32. (a) Cost during Super Bowl XV (1981) = \$275,000
Cost during Super Bowl V (1971) = \$75,000
Increase = \$275,000 − \$75,000 = \$200,000

(b) Cost during Super Bowl XXV (1991) = \$800,000
Increase = \$800,000 − \$275,000 = \$525,000

34. Approximate highest price = \$13.70
Approximate price paid in 1993 = \$12.80

$$\text{Percent drop} = \frac{13.70 - 12.80}{13.70} \approx 0.07 \text{ or } 7\%$$

36. Minimum wage in 1990 = \$3.80
Minimum wage in 1994 = \$4.25

$$\text{Percent increase} = \frac{\$4.25 - 3.80}{3.80} \approx 0.118 \text{ or } 11.8\%$$

38. No, there are many variables that will affect the final exam score.

40. (a) $(1, 0), (13, 5)$

$d = \sqrt{(13-1)^2 + (5-0)^2} = \sqrt{12^2 + 5^2}$
$= \sqrt{169} = 13$

$(13, 5), (13, 0)$
$d = |5 - 0| = |5| = 5$

$(1, 0), (13, 0)$
$d = |1 - 13| = |-12| = 12$

(b) $5^2 + 12^2 = 25 + 144 = 169 = 13^2$

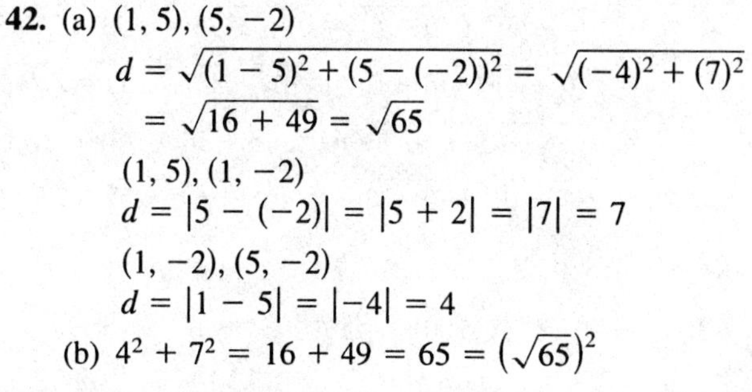

42. (a) $(1, 5), (5, -2)$

$d = \sqrt{(1-5)^2 + (5-(-2))^2} = \sqrt{(-4)^2 + (7)^2}$
$= \sqrt{16 + 49} = \sqrt{65}$

$(1, 5), (1, -2)$
$d = |5 - (-2)| = |5 + 2| = |7| = 7$

$(1, -2), (5, -2)$
$d = |1 - 5| = |-4| = 4$

(b) $4^2 + 7^2 = 16 + 49 = 65 = \left(\sqrt{65}\right)^2$

44. $d = |1 - 8| = |-7| = 7$

46. $d = |-4 - 6| = |-10| = 10$

48. (a)

(b) $d = \sqrt{(1-6)^2 + (12-0)^2}$
$= \sqrt{25 + 144} = 13$

(c) $\left(\frac{1+6}{2}, \frac{12+0}{2}\right) = \left(\frac{7}{2}, 6\right)$

50. (a)

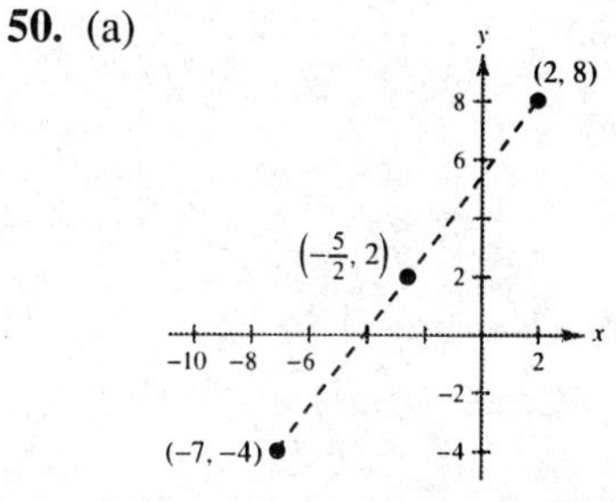

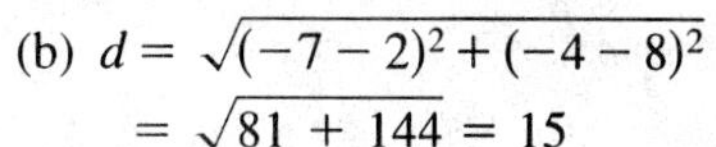

(b) $d = \sqrt{(-7-2)^2 + (-4-8)^2}$
$= \sqrt{81 + 144} = 15$

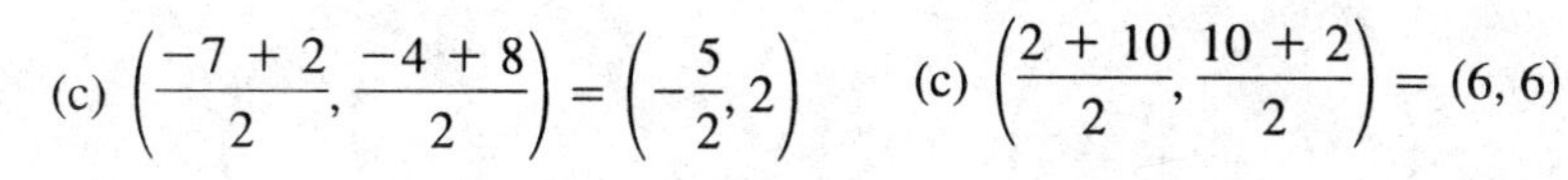

(c) $\left(\frac{-7+2}{2}, \frac{-4+8}{2}\right) = \left(-\frac{5}{2}, 2\right)$

52. (a)

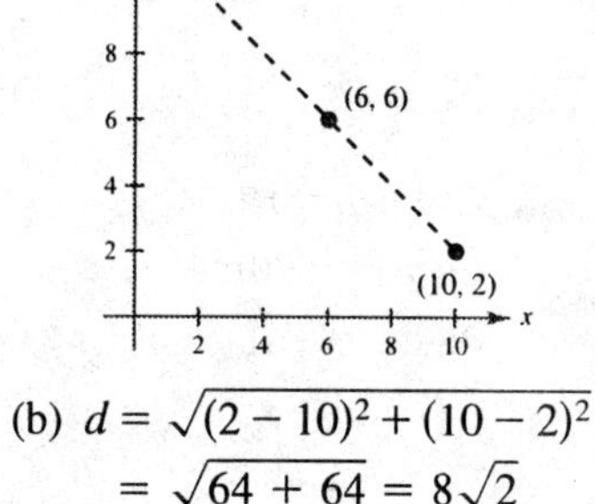

(b) $d = \sqrt{(2-10)^2 + (10-2)^2}$
$= \sqrt{64 + 64} = 8\sqrt{2}$

(c) $\left(\frac{2+10}{2}, \frac{10+2}{2}\right) = (6, 6)$

54. (a)

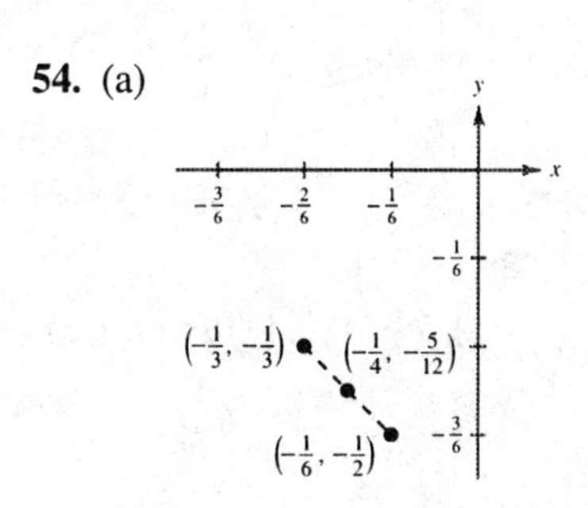

(b) $d = \sqrt{\left(-\frac{1}{3} + \frac{1}{6}\right)^2 + \left(-\frac{1}{3} + \frac{1}{2}\right)^2} = \sqrt{\frac{1}{36} + \frac{1}{36}} = \frac{\sqrt{2}}{6}$

(c) $\left(\frac{(-1/3) - (1/6)}{2}, \frac{(-1/3) - (1/2)}{2}\right) = \left(\frac{-1/2}{2}, \frac{-5/6}{2}\right) = \left(-\frac{1}{4}, -\frac{5}{12}\right)$

56. (a)

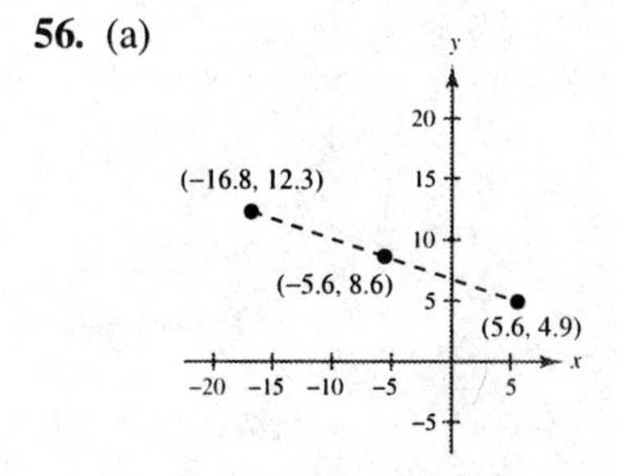

(b) $d = \sqrt{(-16.8 - 5.6)^2 + (12.3 - 4.9)^2}$
$= \sqrt{501.76 + 54.76} = \sqrt{556.52}$

(c) $\left(\frac{-16.8 + 5.6}{2}, \frac{12.3 + 4.9}{2}\right) = (-5.6, 8.6)$

58. (a)

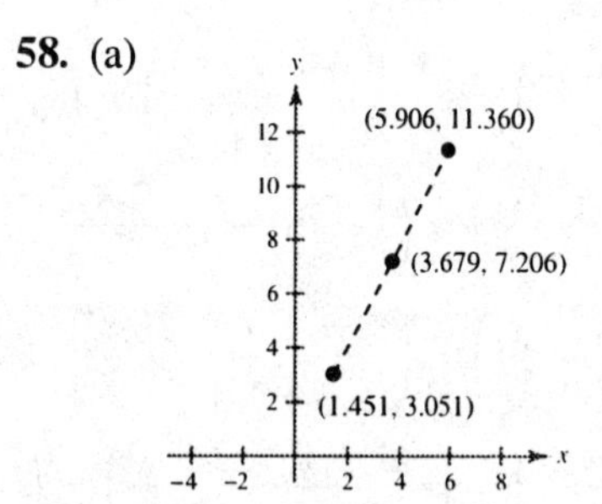

(b) $d = \sqrt{(1.451 - 5.906)^2 + (3.051 - 11.360)^2}$
$\approx \sqrt{88.887}$

(c) $\left(\frac{1.451 + 5.906}{2}, \frac{3.051 + 11.360}{2}\right) \approx (3.679, 7.206)$

60. $\left(\frac{1991 + 1995}{2}, \frac{\$4{,}200{,}000 + \$5{,}650{,}000}{2}\right)$
$= (1993, \$4{,}925{,}000)$

In 1993 the sales were $4,925,000.

62. $d_1 = \sqrt{(1-3)^2 + (-3-2)^2} = \sqrt{4 + 25} = \sqrt{29}$
$d_2 = \sqrt{(3+2)^2 + (2-4)^2} = \sqrt{25 + 4} = \sqrt{29}$
$d_3 = \sqrt{(1+2)^2 + (-3-4)^2} = \sqrt{9 + 49} = \sqrt{58}$
$d_1 = d_2$

64. $d_1 = \sqrt{(4-0)^2 + (0-6)^2} = \sqrt{16+36} = \sqrt{52} = 2\sqrt{13}$

$d_2 = \sqrt{(0+4)^2 + (6-0)^2} = \sqrt{16+36} = \sqrt{52} = 2\sqrt{13}$

$d_3 = \sqrt{(-4-0)^2 + (0+6)^2} = \sqrt{16+36} = \sqrt{52} = 2\sqrt{13}$

$d_4 = \sqrt{(4-0)^2 + (0+6)^2} = \sqrt{16+36} = \sqrt{52} = 2\sqrt{13}$

$d_1 = d_2 = d_3 = d_4$

66. $d_1 = \sqrt{(0-3)^2 + (1-7)^2} = \sqrt{9+36} = \sqrt{45} = 3\sqrt{5}$

$d_2 = \sqrt{(3-4)^2 + (7-4)^2} = \sqrt{1+9} = \sqrt{10}$

$d_3 = \sqrt{(4-1)^2 + (4+2)^2} = \sqrt{9+36} = \sqrt{45} = 3\sqrt{5}$

$d_4 = \sqrt{(0-1)^2 + (1+2)^2} = \sqrt{1+9} = \sqrt{10}$

Opposite sides have equal lengths of $3\sqrt{5}$ and $\sqrt{10}$.

68. $(x-0)^2 + (y-0)^2 = 5^2$

$x^2 + y^2 = 25$

70. $(x-0)^2 + \left(y - \frac{1}{3}\right)^2 = \left(\frac{1}{3}\right)^2$

$x^2 + \left(y - \frac{1}{3}\right)^2 = \frac{1}{9}$

72. $r = \sqrt{(3-(-1))^2 + (-2-1)^2} = \sqrt{16+9} = 5$

$(x-3)^2 + (y+2)^2 = 5^2 = 25$

74. Center $= \left(\dfrac{-4+4}{2}, \dfrac{-1+1}{2}\right) = (0, 0)$

$r = \sqrt{(4-0)^2 + (1-0)^2} = \sqrt{17}$

$x^2 + y^2 = 17$

76. $x^2 + y^2 = 16$

Center: $(0, 0)$

Radius: $\sqrt{16} = 4$

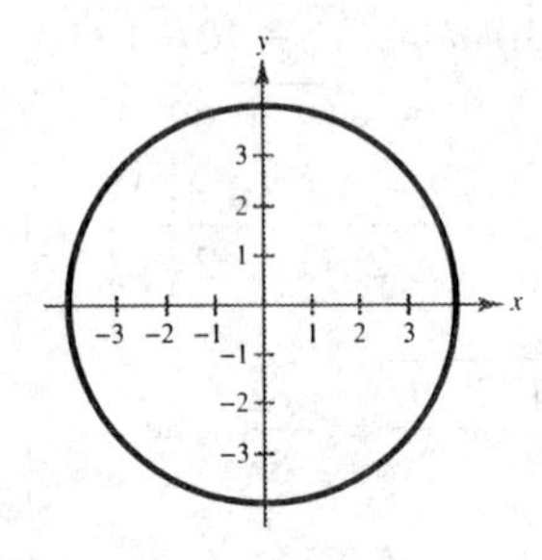

78. $x^2 + (y-1)^2 = 4$

Center: $(0, 1)$

Radius: $\sqrt{4} = 2$

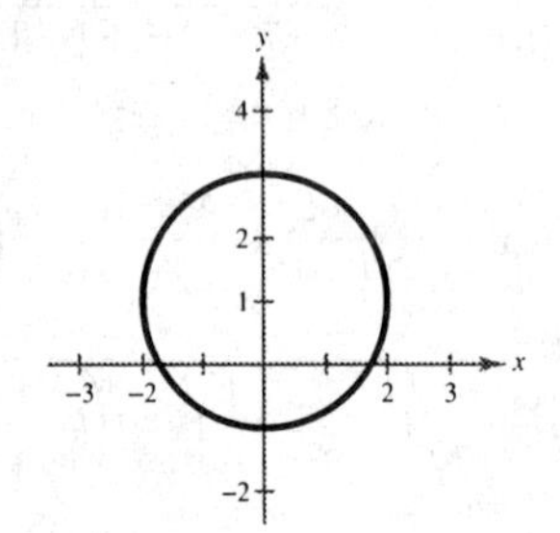

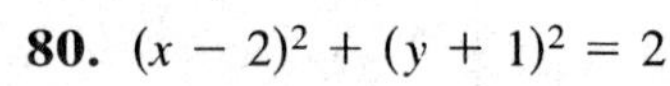

80. $(x-2)^2 + (y+1)^2 = 2$

Center: $(2, -1)$

Radius: $\sqrt{2}$

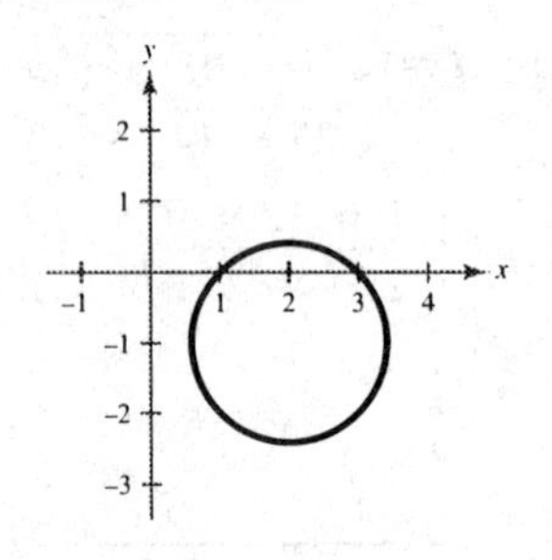

82. (a) $(x_2, y_2) = (2x_m - x_1, 2y_m - y_1)$

$= (2 \cdot 4 - 1, 2(-1) - (-2)) = (7, 0)$

(b) $(x_2, y_2) = (2x_m - x_1, 2y_m - y_1)$

$= (2 \cdot 2 - (-5), 2 \cdot 4 - 11) = (9, -3)$

84. (a) $\left(\dfrac{3x_1 + x_2}{4}, \dfrac{3y_1 + y_2}{4}\right) = \left(\dfrac{3 \cdot 1 + 4}{4}, \dfrac{3(-2) - 1}{4}\right) = \left(\dfrac{7}{4}, -\dfrac{7}{4}\right)$

$\left(\dfrac{x_1 + x_2}{2}, \dfrac{y_1 + y_2}{2}\right) = \left(\dfrac{1+4}{2}, \dfrac{-2-1}{2}\right) = \left(\dfrac{5}{2}, -\dfrac{3}{2}\right)$

$\left(\dfrac{x_1 + 3x_2}{4}, \dfrac{y_1 + 3y_2}{4}\right) = \left(\dfrac{1 + 3 \cdot 4}{4}, \dfrac{-2 + 3(-1)}{4}\right) = \left(\dfrac{13}{4}, -\dfrac{5}{4}\right)$

(b) $\left(\dfrac{3x_1 + x_2}{4}, \dfrac{3y_1 + y_2}{4}\right) = \left(\dfrac{3(-2) + 0}{4}, \dfrac{3(-3) + 0}{4}\right) = \left(-\dfrac{3}{2}, -\dfrac{9}{4}\right)$

$\left(\dfrac{x_1 + x_2}{2}, \dfrac{y_1 + y_2}{2}\right) = \left(\dfrac{-2+0}{2}, \dfrac{-3+0}{2}\right) = \left(-1, -\dfrac{3}{2}\right)$

$\left(\dfrac{x_1 + 3x_2}{4}, \dfrac{y_1 + 3y_2}{4}\right) = \left(\dfrac{-2+0}{4}, \dfrac{-3+0}{4}\right) = \left(-\dfrac{1}{2}, -\dfrac{3}{4}\right)$

86. Let $(0, 0)$ represent the coordinates of the point of departure and let $(100, 150)$ represent the coordinates of the destination.

$d = \sqrt{(0-100)^2 + (0-150)^2}$

$= \sqrt{10{,}000 + 22{,}500}$

$= \sqrt{325{,}000}$

$= 50\sqrt{13} \approx 180.28$ miles

88. The midpoint of the diagonal connecting $(0, 0)$ and $(a + b, c)$ is $\big((a+b)/2, c/2\big)$. The midpoint of the diagonal connecting $(a, 0)$ and (b, c) is $\big((a+b)/2, c/2\big)$. Thus, the diagonals bisect each other.

Section P.2 Graphs and Graphing Utilities

Solutions to Even-Numbered Exercises

2. $y = x^2 - 3x + 2$

(a) $(2, 0)$: $(2)^2 - 3(2) + 2 \stackrel{?}{=} 0$

$4 - 6 + 2 \stackrel{?}{=} 0$

$0 = 0$

Yes, the point *is* on the graph.

(b) $(-2, 8)$: $(-2)^2 - 3(-2) + 2 \stackrel{?}{=} 8$

$4 + 6 + 2 \stackrel{?}{=} 8$

$12 \neq 8$

No, the point *is not* on the graph.

4. $y = \frac{1}{3}x^3 - 2x^2$

(a) $\left(2, -\frac{16}{3}\right)$: $\frac{1}{3}(2)^3 - 2(2)^2 \stackrel{?}{=} -\frac{16}{3}$

$\frac{1}{3} \cdot 8 - 2 \cdot 4 \stackrel{?}{=} -\frac{16}{3}$

$\frac{8}{3} - 8 \stackrel{?}{=} -\frac{16}{3}$

$\frac{8}{3} - \frac{24}{3} \stackrel{?}{=} -\frac{16}{3}$

$-\frac{16}{3} = -\frac{16}{3}$

Yes, the point *is* on the graph.

(b) $(-3, 9)$: $\frac{1}{3}(-3)^3 - 2(-3)^2 \stackrel{?}{=} 9$

$\frac{1}{3}(-27) - 2(9) \stackrel{?}{=} 9$

$-9 - 18 \stackrel{?}{=} 9$

$-27 \neq 9$

No, the point *is not* on the graph.

6. $x^2 + y^2 = 20$

(a) $(3, -2)$: $3^2 + (-2)^2 \stackrel{?}{=} 20$

$9 + 4 \stackrel{?}{=} 20$

$13 \neq 20$

No, the point *is not* on the graph.

(b) $(-4, 2)$: $(-4)^2 + 2^2 \stackrel{?}{=} 20$

$16 + 4 \stackrel{?}{=} 20$

$20 = 20$

Yes, the point *is* on the graph.

8. $y = \dfrac{1}{x^2 + 1}$

(a) $(0, 0)$: $\dfrac{1}{0^2 + 1} \stackrel{?}{=} 0$

$\dfrac{1}{1} \stackrel{?}{=} 0$

$1 \neq 0$

No, the point *is not* on the graph.

(b) $(3, 0.1)$: $\dfrac{1}{3^2 + 1} \stackrel{?}{=} 0.1$

$\dfrac{1}{9 + 1} \stackrel{?}{=} 0.1$

$\dfrac{1}{10} \stackrel{?}{=} 0.1$

$0.1 = 0.1$

Yes, the point *is* on the graph.

10. $y = \frac{3}{2}x - 1$

x	-2	0	$\frac{2}{3}$	1	2
y	-4	-1	0	$\frac{1}{2}$	2

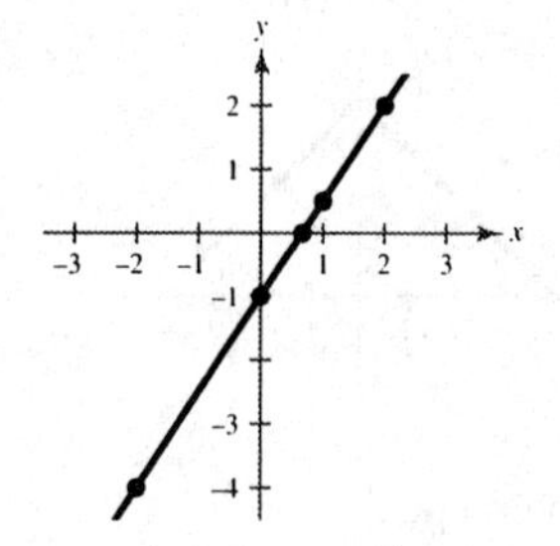

12. $y = 4 - x^2$

x	-2	-1	0	1	2
y	0	3	4	3	0

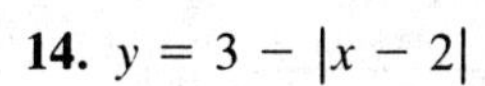

14. $y = 3 - |x - 2|$

x	0	1	2	3	4
y	1	2	3	2	1

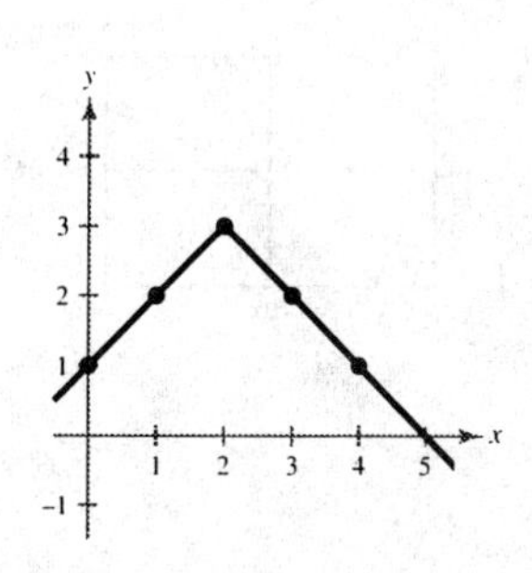

16. $y = \dfrac{6x}{x^{-2} + 1}$

x	-2	-1	0	1	2
y	$-\frac{48}{5}$	-3	Undefined	3	$\frac{48}{5}$

Note: hole at (0, 0)

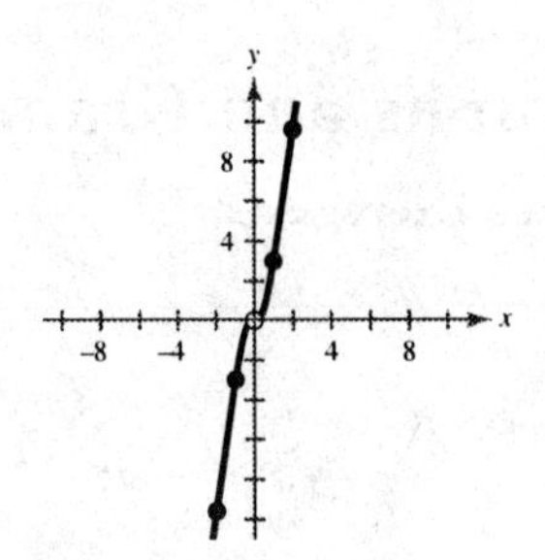

18.

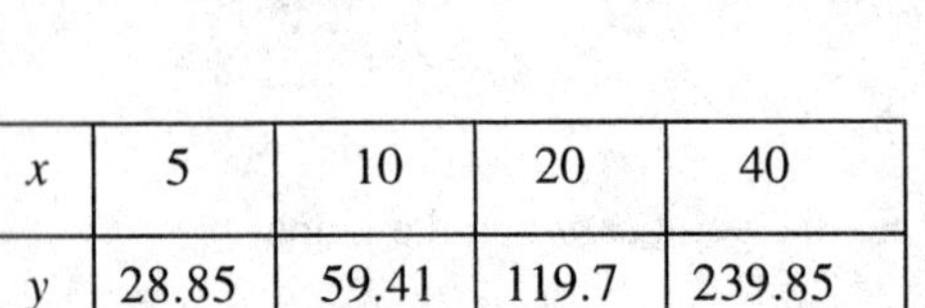

x	5	10	20	40
y	28.85	59.41	119.7	239.85

The value of y approaches infinity. No, y is positive for positive values of x.

20.

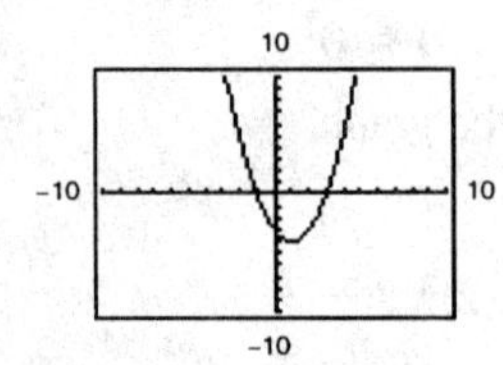

Intercepts: $(-1, 0), (0, -3), (3, 0)$

22.

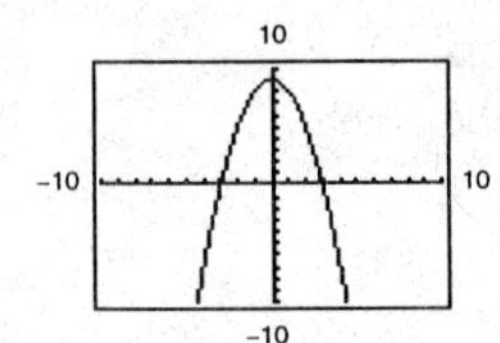

Intercepts: $(-3, 0), (0, 9), (3, 0)$

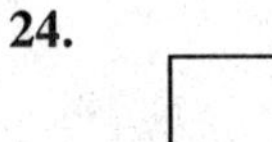

24.

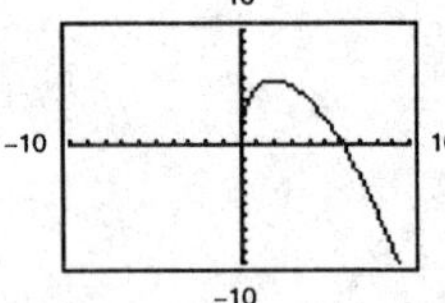

Intercepts: $(0, 0), (6, 0)$

26.

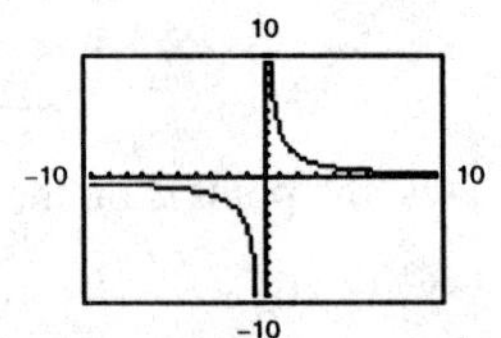

No intercepts

28. $y = x^2 - 2x$ is a parabola. Matches (c).

30. $y = 2\sqrt{x}$ passes through the origin. Matches (e).

32. $y = |x| - 3$ involves an absolute value. Matches (b).

34. No symmetry

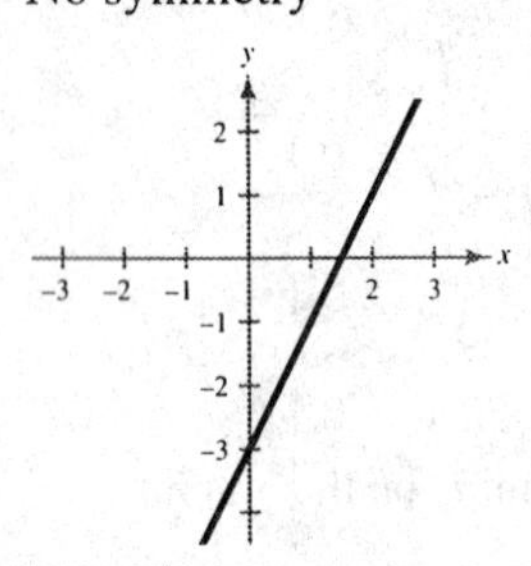

36. y-axis symmetry

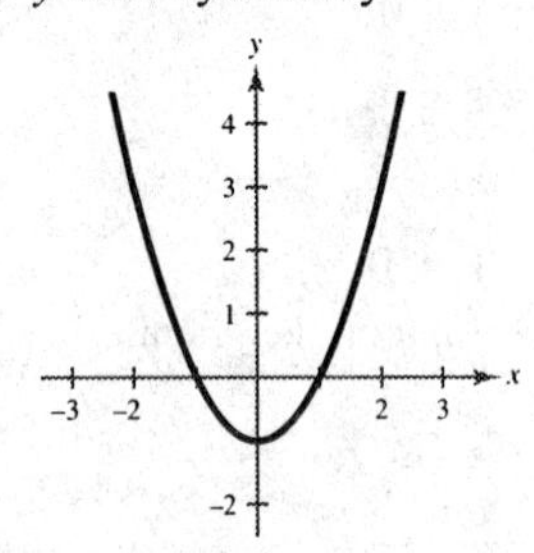

38. No symmetry

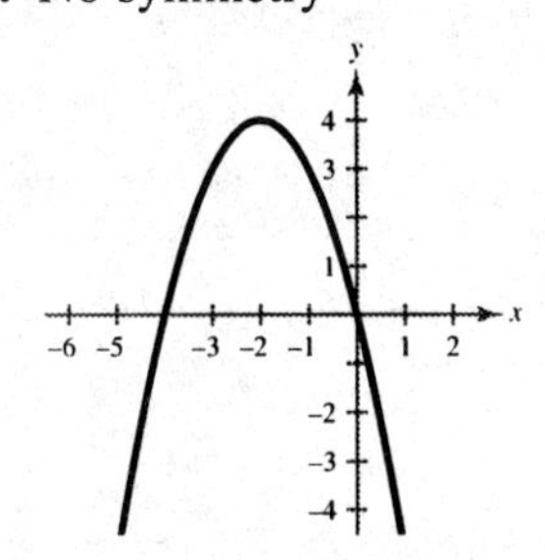

40. No symmetry

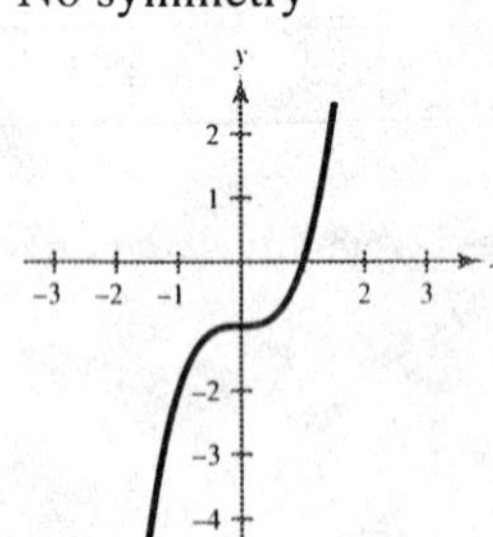

42. No symmetry

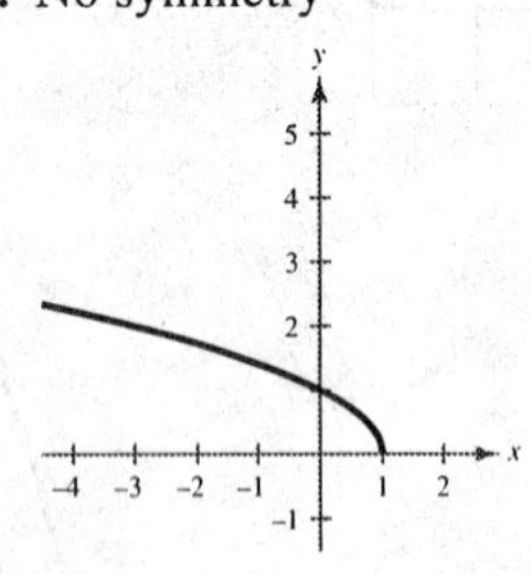

44. y-axis symmetry

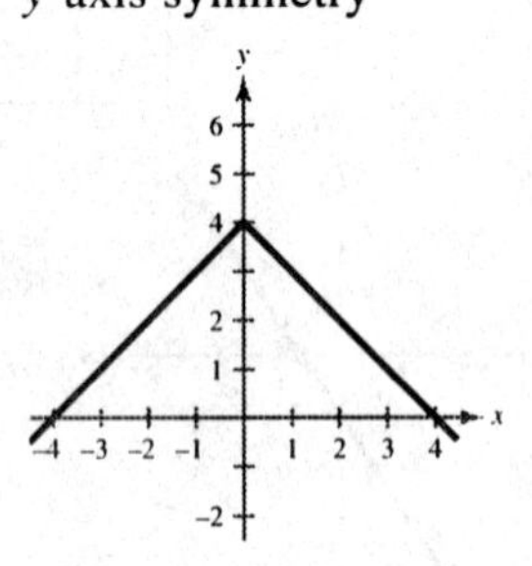

46. x-axis symmetry

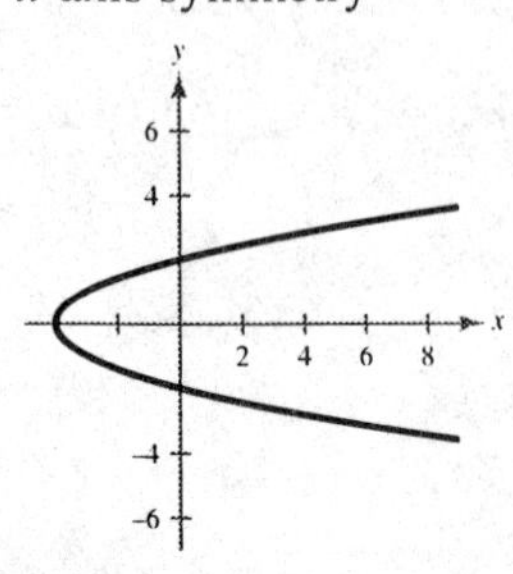

48. Intercepts: $(0, -1), (\frac{3}{2}, 0)$

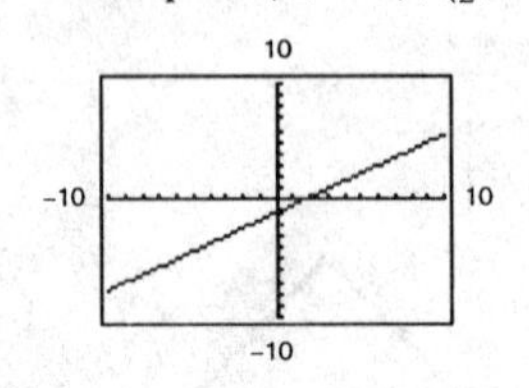

50. Intercepts: $(-4, 0), (2, 0), (0, -4)$

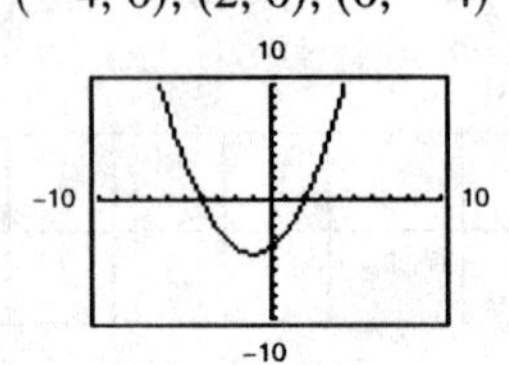

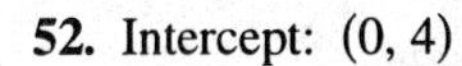

52. Intercept: (0, 4)

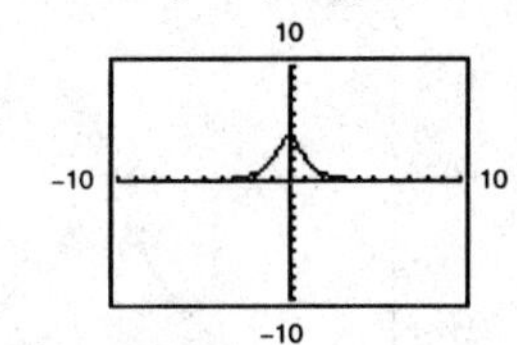

54. Intercepts: $(-1, 0), (0, 1)$

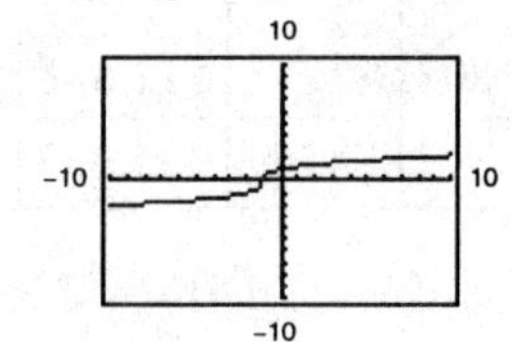

56. $y = -3x + 50$

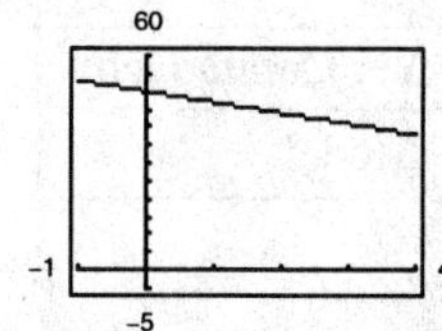

The specified setting gives a more complete graph. (The y-intercept is visible.)

58. $y = 4(x + 5)\sqrt{4 - x}$

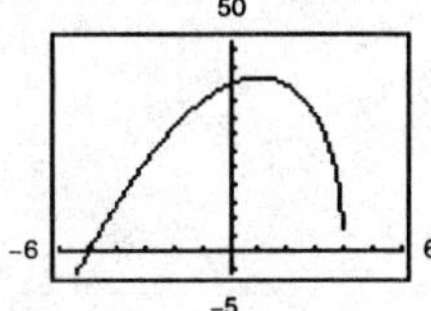

The specified setting gives a more complete graph.

60. $y = x^3 - 3x^2 + 4$

Range/Window

Xmin = -3
Xmax = 5
Xscl = 1
Ymin = -3
Ymax = 5
Yscl = 1

62. $y = 8\sqrt[3]{x - 6}$

Range/Window

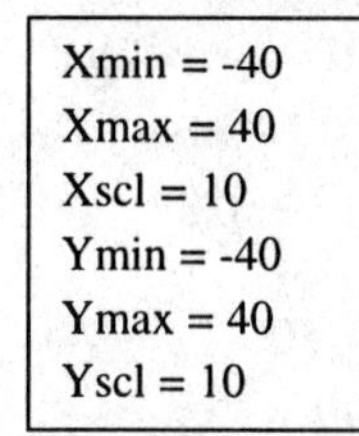

Xmin = -40
Xmax = 40
Xscl = 10
Ymin = -40
Ymax = 40
Yscl = 10

64.

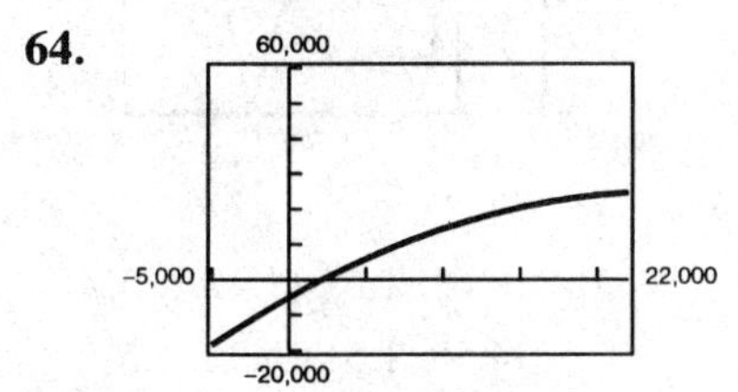

24,000 / -5,000 / 22,000 / -5,000

Xmin = -5000
Xmax = 22000
Xscl = 5000
Ymin = -5000
Ymax = 24000
Yscl = 5000

This viewing rectangle would be selected.

66. $y_1 = 2 + \sqrt{16 - (x - 1)^2}$

$y_2 = 2 - \sqrt{16 - (x - 1)^2}$

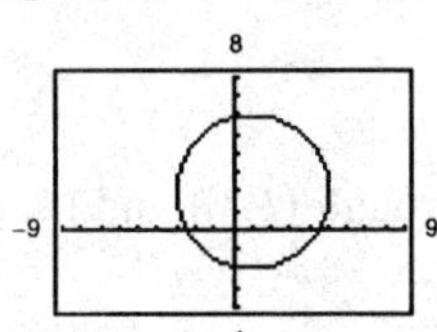

A circle is bounded by their graphs.

68. $y_1 = 1 + \sqrt{25 - (x - 3)^2}$

$y_2 = 1 - \sqrt{25 - (x - 3)^2}$

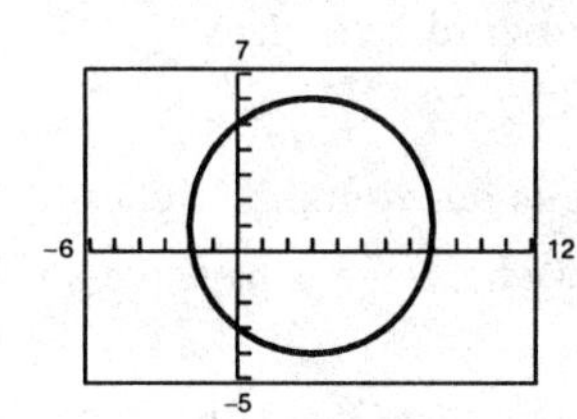

70. $y_1 = \frac{1}{2}x + (x + 1)$

$y_2 = \frac{3}{2}x + 1$

Graphing these with a graphing utility shows that their graphs are identical. The Associative Property is illustrated.

72. $y_1 = (x - 3) \cdot \dfrac{1}{x - 3}$

$y_2 = 1$

Graphing these with a graphing utility shows that their graphs are identical. The Multiplicative Inverse Property is illustrated. (Except for hole at $x = 3$.)

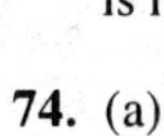

74. (a)

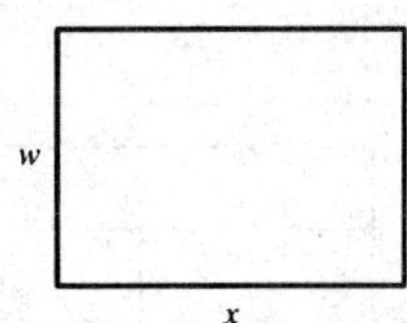

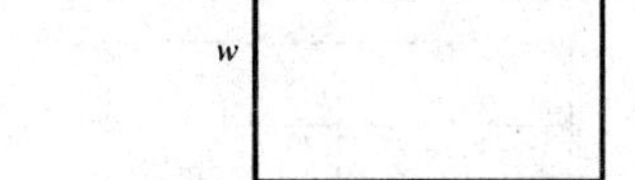

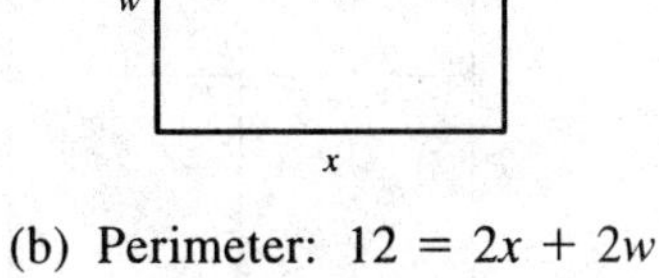

(b) Perimeter: $12 = 2x + 2w$

$12 = 2(x + w)$

$6 = x + w$

Thus, $w = 6 - x$.

Area $= xw = x(6 - x)$

(c)

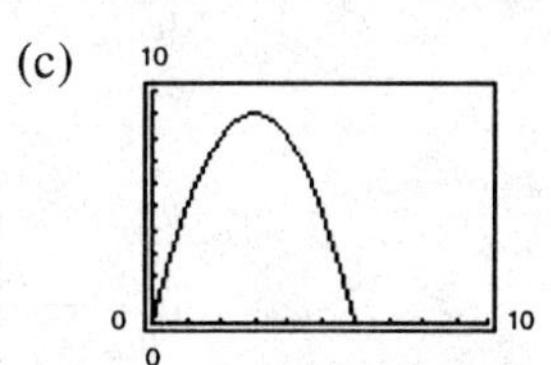

(d) The maximum area corresponds to the highest point on the graph, which appears to be (3, 9). Thus, $x = 3$ and $w = 6 - x = 6 - 3 = 3$.

76. (a)

Year	1950	1960	1970	1980	1990	1994
Per Capita Debt	\$1688	\$1572	\$1807	\$3981	\$12,848	\$15,750
Model	\$1570	\$1416	\$1972	\$4769	\$11,337	\$15,362

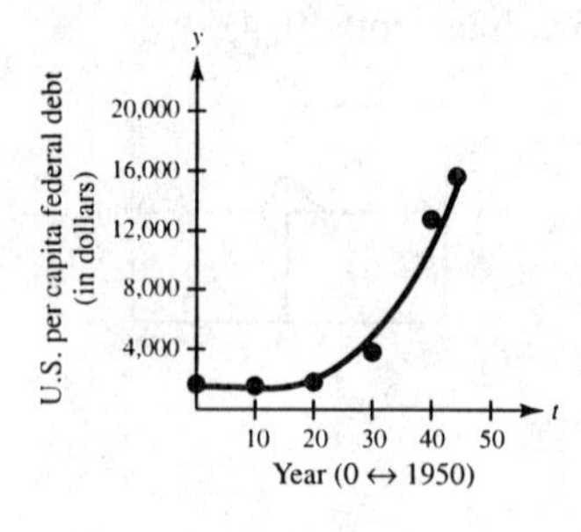

(b) When $t = 48$, $y = 0.255(48)^3 - 4.096(48)^2 + 1570.417 \approx \$20,334$.

(c) When $t = 50$, $y = 0.255(50)^3 - 4.096(50)^2 + 1570.417 \approx \$23,205$.

78. $y = (x - a)(x - b)$

$(-2, 0)$: $0 = (x + 2)(x - 6) \implies a = -2$

$(5, 0)$: $0 = (x + 2)(x - 5) \implies b = 5$

80. $y = \dfrac{10,770}{x^2} - 0.37$

When $x = 50$,

$$y = \frac{10,770}{50^2} - 0.37 = \frac{10,770}{2500} - 0.37 \approx 3.9 \text{ ohms.}$$

82.

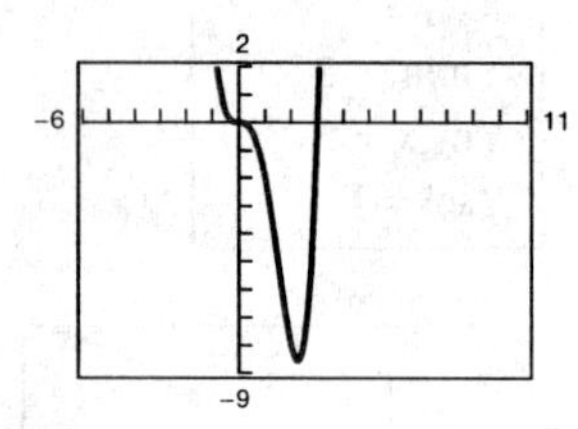

(a) $(2.25, -8.54)$

(b) $(-1.63, 20)$, $(3.48, 20)$

84.

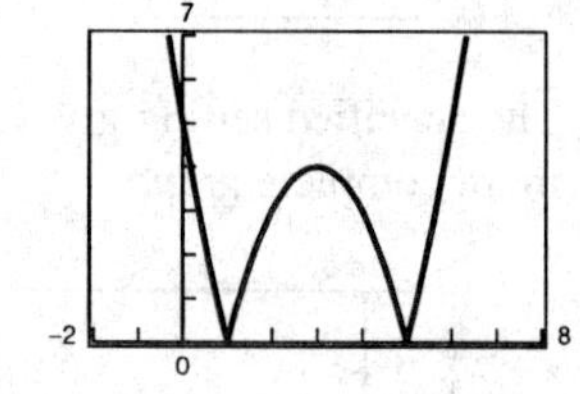

(a) $(2, 3)$

(b) $(0.65, 1.5)$, $(1.42, 1.5)$ $(4.58, 1.5)$, $(5.35, 1.5)$

Section P.3 Lines in the Plane

Solutions to Even-Numbered Exercises

2. (a) $m = 0$. The line is horizontal. Matches L_2.

(b) $m = -\frac{3}{4}$. Because the slope is negative, the line falls. Matches L_1.

(c) $m = 1$. Because the slope is positive, the line rises. Matches L_3.

4. The line appears to go through $(1, 0)$ and $(4, 8)$.

$$\text{Slope} = \frac{y_2 - y_1}{x_2 - x_1} = \frac{8 - 0}{4 - 1} = \frac{8}{3}$$

6. The line appears to go through $(0, 7)$ and $(7, 0)$.

$$\text{Slope} = \frac{y_2 - y_1}{x_2 - x_1} = \frac{0 - 7}{7 - 0} = -1$$

8. The line appears to go through $(0, 1)$ and $(8, 7)$.

$$\text{Slope} = \frac{y_2 - y_1}{x_2 - x_1} = \frac{7 - 1}{8 - 0} = \frac{3}{4}$$

10.

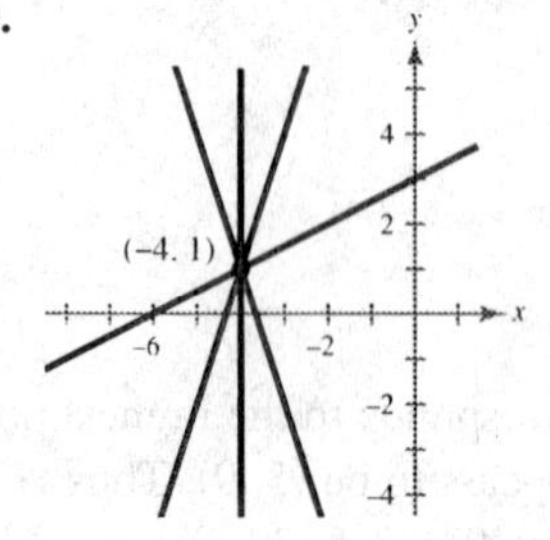

12. $\text{Slope} = \dfrac{-4 - 4}{4 - 2} = -4$

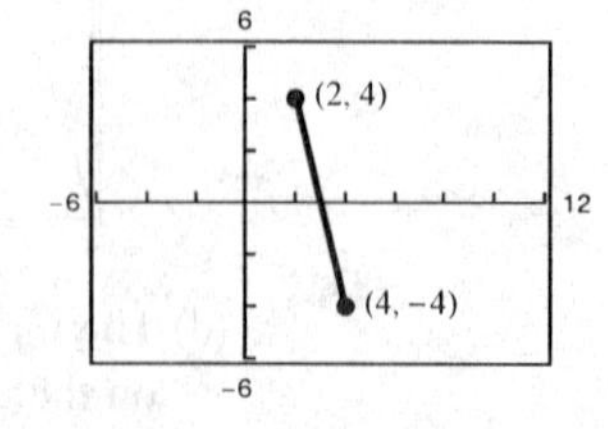

14. $\text{Slope} = \dfrac{0 - (-10)}{-4 - 0} = -\dfrac{5}{2}$

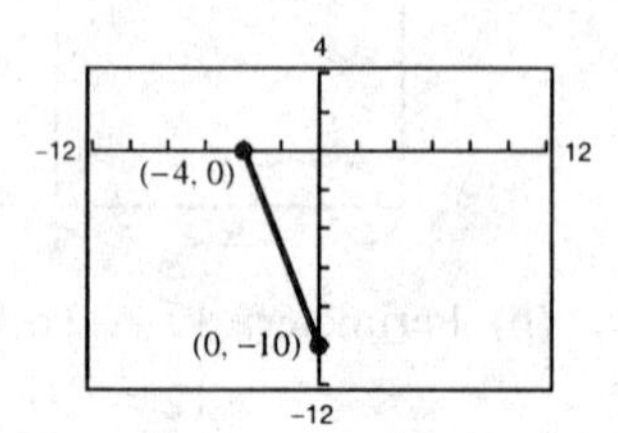

16. Slope $= \dfrac{-\frac{1}{4} - \frac{3}{4}}{\frac{5}{4} - \frac{7}{8}} = \dfrac{-1}{\frac{3}{8}} = -\dfrac{8}{3}$

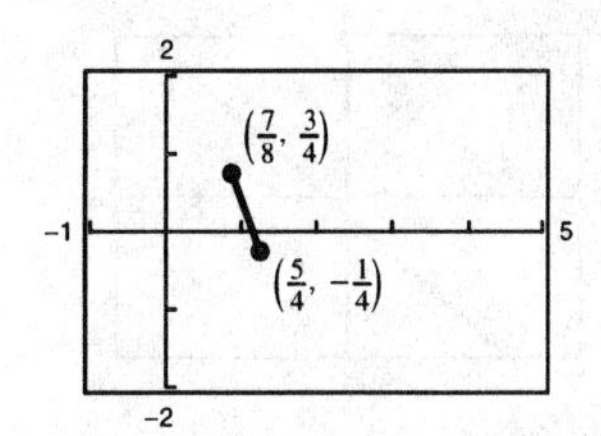

18. Because m is undefined, x does not change. Three other points are: $(-4, 0)$, $(-4, 3)$, $(-4, 5)$.

20. Because $m = -1$, y decreases by 1 for every one unit increase in x. Three other points are: $(0, 4)$, $(9, -5)$, $(11, -7)$.

22. Because $m = 0$, y does not change. Three other points are: $(-4, -1)$, $(-2, -1)$, $(0, -1)$.

24. L_1: $(-2, -1)$, $(1, 5)$

$$m_1 = \frac{5 - (-1)}{1 - (-2)} = \frac{6}{3} = 2$$

L_2: $(1, 3)$, $(5, -5)$

$$m_2 = \frac{-5 - 3}{5 - 1} = \frac{-8}{4} = -2$$

The lines are neither parallel nor perpendicular.

26. L_1: $(4, 8)$, $(-4, 2)$

$$m_1 = \frac{2 - 8}{-4 - 4} = \frac{-6}{-8} = \frac{3}{4}$$

L_2: $(3, -5)$, $\left(-1, \dfrac{1}{3}\right)$

$$m_2 = \frac{(1/3) - (-5)}{-1 - 3} = \frac{16/3}{-4} = -\frac{4}{3}$$

The lines are perpendicular.

28. No, the slopes of two perpendicular lines have opposite signs (assume that neither line is vertical nor horizontal).

30. (a) $m = 400$. The revenues are increasing \$400 per day.

(b) $m = 100$. The revenues are increasing \$100 per day.

(c) $m = 0$. There is no change in revenue. (Revenue remains constant.)

32. (a)

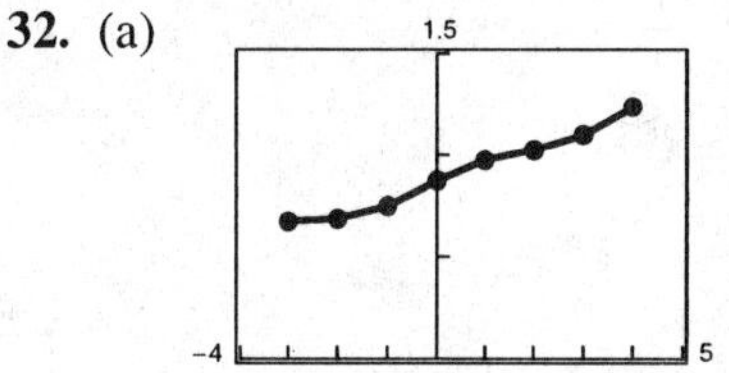

(b) The steepest portion of the graph is between 1993 and 1994.

34.

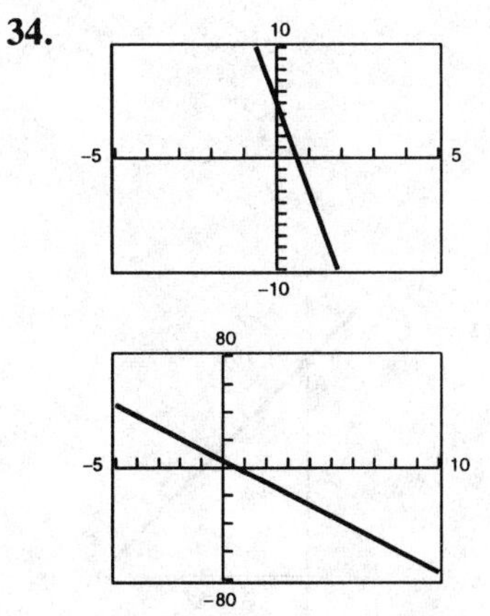

The first setting shows the x- and y-intercepts more clearly.

36. $2x + 3y - 9 = 0$

$$3y = -2x + 9$$

$$y = -\frac{2}{3}x + 3$$

Slope: $m = -\dfrac{2}{3}$

y-intercept: $(0, 3)$

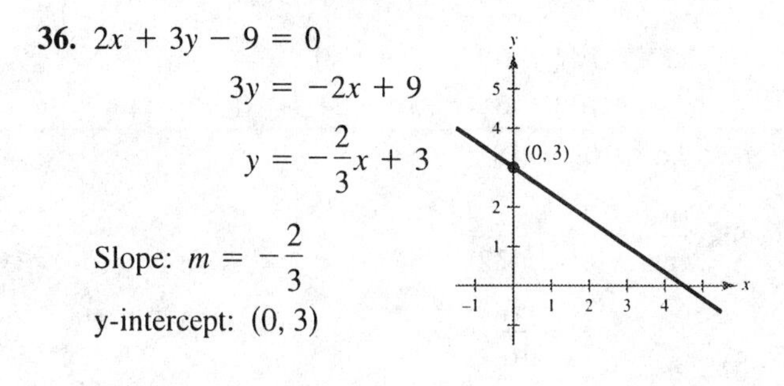

38. $3y + 5 = 0$

$$3y = -5$$

$$y = -\frac{5}{3}$$

Slope: $m = 0$

y-intercept: $\left(0, -\dfrac{5}{3}\right)$

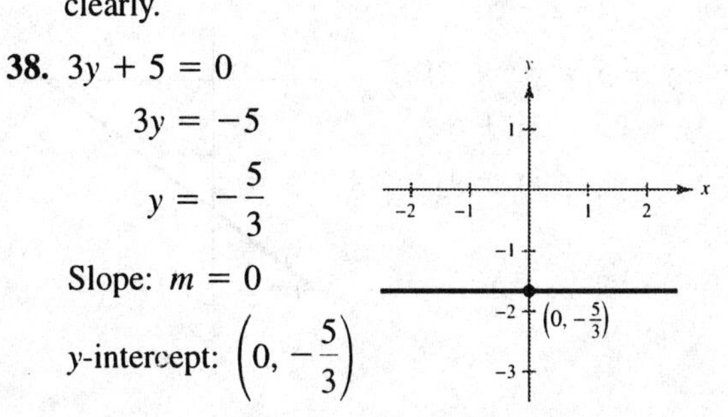

40. $x - y - 10 = 0$

$$x - 10 = y$$

Slope: $m = 1$

y-intercept: $(0, -10)$

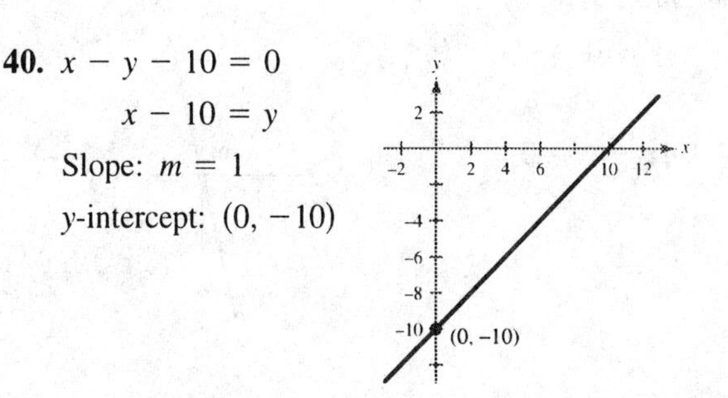

42. $(4, 3), (-4, -4)$

$$y - 3 = \frac{-4 - 3}{-4 - 4}(x - 4)$$
$$y - 3 = \frac{7}{8}(x - 4)$$
$$8y - 24 = 7x - 28$$
$$7x - 8y - 4 = 0$$

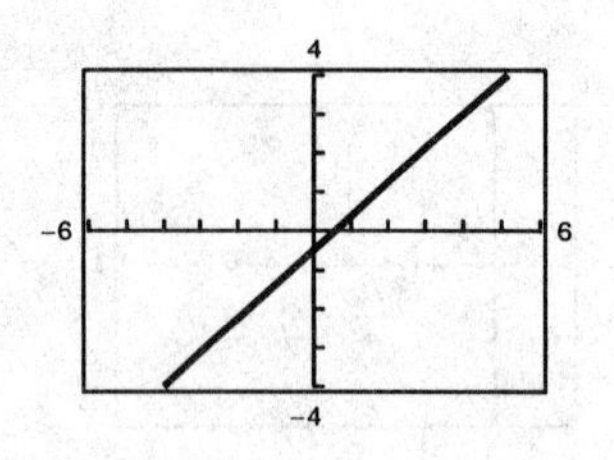

44. $(-1, 4), (6, 4)$

$$y - 4 = \frac{4 - 4}{6 - (-1)}(x + 1)$$
$$y - 4 = 0(x + 1)$$
$$y - 4 = 0$$

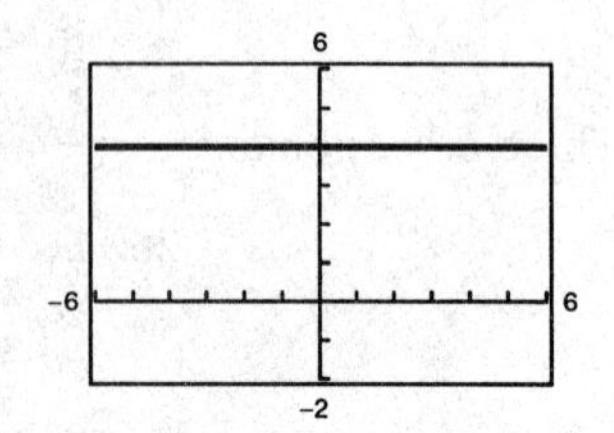

46. $(1, 1), \left(6, -\frac{2}{3}\right)$

$$y - 1 = \frac{-\frac{2}{3} - 1}{6 - 1}(x - 1)$$
$$y - 1 = -\frac{1}{3}(x - 1)$$
$$y - 1 = -\frac{1}{3}x + \frac{1}{3}$$
$$3y - 3 = -x + 1$$
$$x + 3y - 4 = 0$$

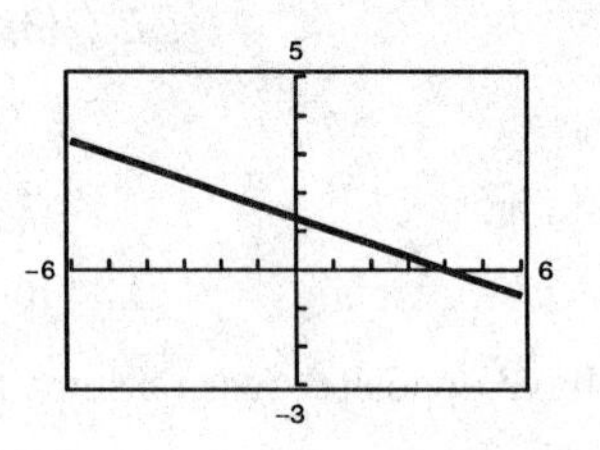

48. $(-8, 0.6), (2, -2.4)$

$$y - 0.6 = \frac{-2.4 - 0.6}{2 - (-8)}(x + 8)$$
$$y - 0.6 = -\frac{3}{10}(x + 8)$$
$$10y - 6 = -3(x + 8)$$
$$10y - 6 = -3x - 24$$
$$3x + 10y + 18 = 0$$

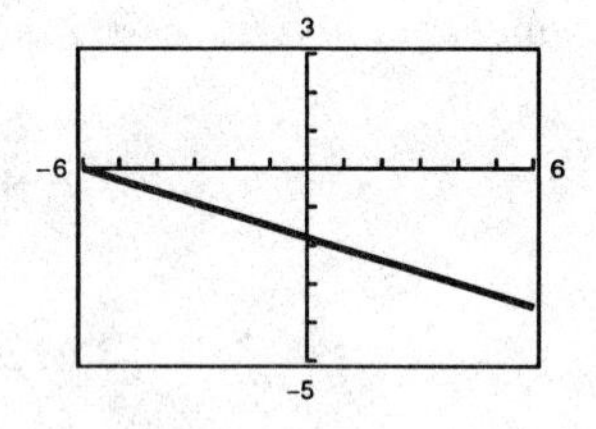

50. $\frac{\text{rise}}{\text{run}} = \frac{3}{4} = \frac{x}{\frac{1}{2}(32)}$

$$\frac{3}{4} = \frac{x}{16}$$
$$4x = 48$$
$$x = 12$$

The maximum height in the attic is 12 feet.

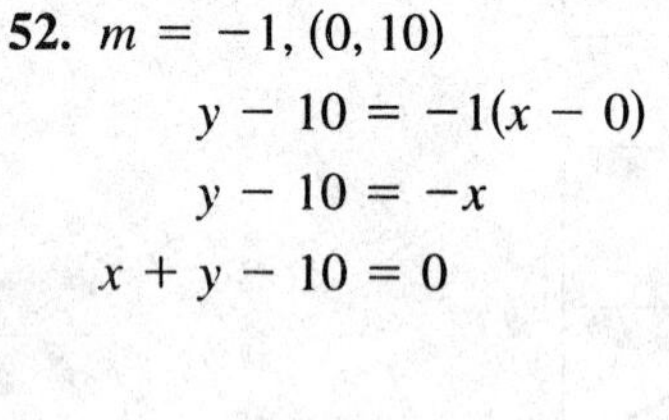

52. $m = -1, (0, 10)$

$$y - 10 = -1(x - 0)$$
$$y - 10 = -x$$
$$x + y - 10 = 0$$

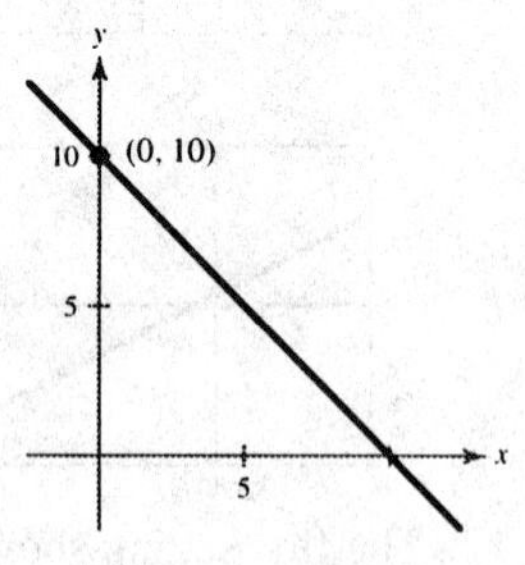

54. $m = 4, (0, 0)$

$$y - 0 = 4(x - 0)$$
$$y = 4x$$
$$4x - y = 0$$

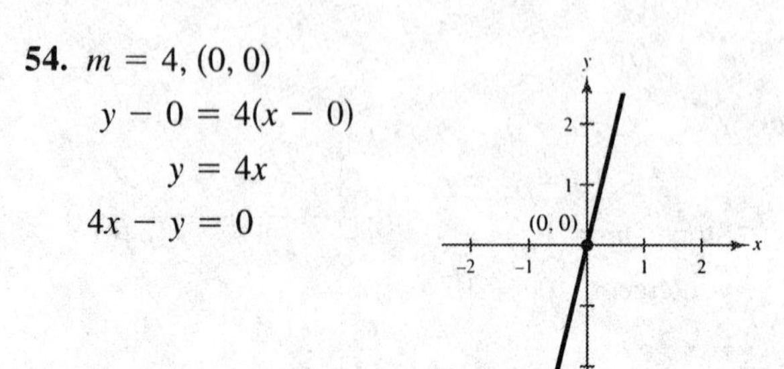

56. $m = \frac{3}{4}, (-2, -5)$

$$y + 5 = \frac{3}{4}(x + 2)$$
$$4y + 20 = 3x + 6$$
$$0 = 3x - 4y - 14$$

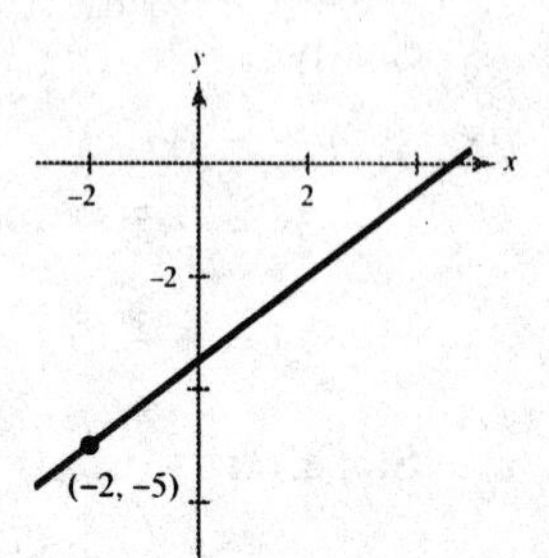

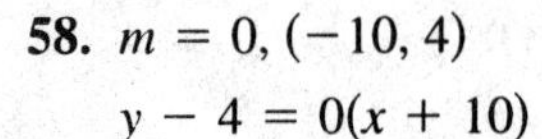

58. $m = 0, (-10, 4)$

$y - 4 = 0(x + 10)$

$y - 4 = 0$

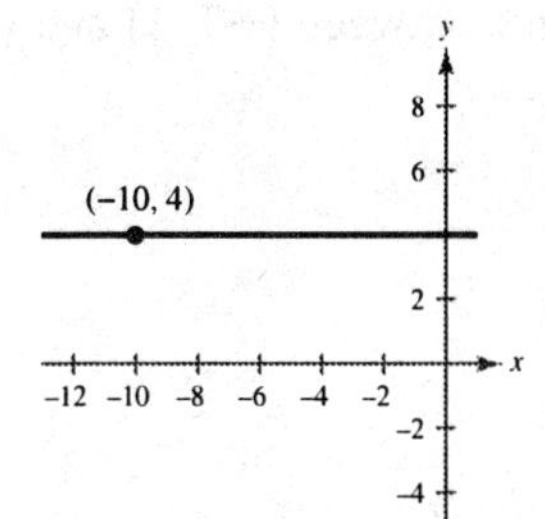

60. $m = -3, \left(-\frac{1}{2}, \frac{3}{2}\right)$

$y - \frac{3}{2} = -3\left(x + \frac{1}{2}\right)$

$y - \frac{3}{2} = -3x - \frac{3}{2}$

$3x + y = 0$

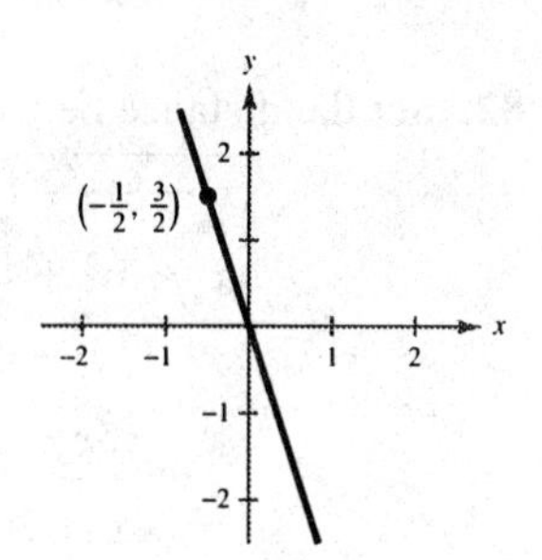

62. $\frac{x}{a} + \frac{y}{b} = 1$

$\frac{x}{-6} + \frac{y}{2} = 1$

$y = 2\left(1 + \frac{x}{6}\right)$

$y = \frac{x}{3} + 2$

a and b are the x- and y-intercepts.

64. $\frac{x}{a} + \frac{y}{b} = 1$

$\frac{x}{-1/6} + \frac{y}{-2/3} = 1$

$-6x - \frac{3}{2}y = 1$

$12x + 3y + 2 = 0$

66. $5x + 3y = 0$

$3y = -5x$

$y = -\frac{5}{3}x$

Slope: $m = -\frac{5}{3}$

(a) $m = -\frac{5}{3}, \left(\frac{7}{8}, \frac{3}{4}\right)$

$y - \frac{3}{4} = -\frac{5}{3}\left(x - \frac{7}{8}\right)$

$24y - 18 = -40\left(x - \frac{7}{8}\right)$

$24y - 18 = -40x + 35$

$40x + 24y - 53 = 0$

(b) $m = \frac{3}{5}, \left(\frac{7}{8}, \frac{3}{4}\right)$

$y - \frac{3}{4} = \frac{3}{5}\left(x - \frac{7}{8}\right)$

$40y - 30 = 24\left(x - \frac{7}{8}\right)$

$40y - 30 = 24x - 21$

$24x - 40y + 9 = 0$

68. $x = 4$

m is undefined.

(a) $(2, 5)$, m is undefined.

$x = 2$

$x - 2 = 0$

(b) $(2, 5)$, $m = 0$

$y = 5$

$y - 5 = 0$

70. $L_1 = y = \frac{2}{3}x;\ L_2 = y = -\frac{3}{2}x;\ L_3 = y = \frac{2}{3}x + 2$

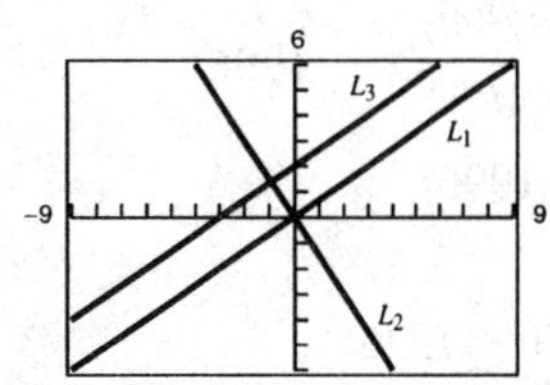

L_1 is parallel to L_3. L_2 is perpendicular to L_1 and L_3.

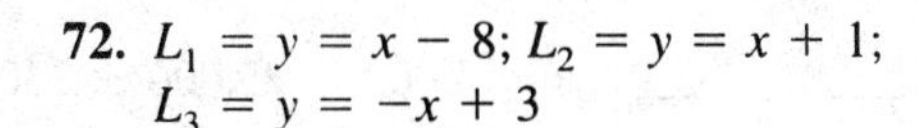

72. $L_1 = y = x - 8;\ L_2 = y = x + 1;$
$L_3 = y = -x + 3$

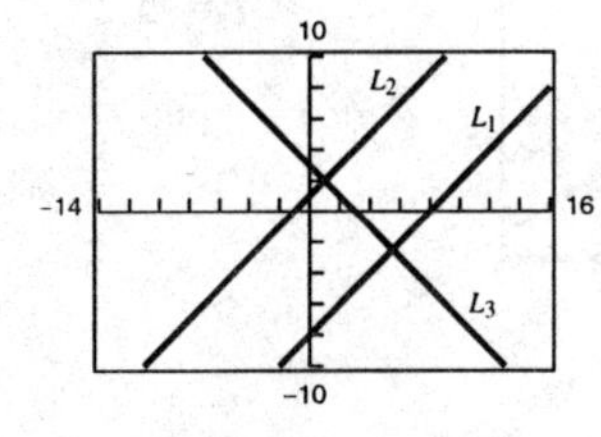

L_1 is parallel to L_2. L_3 is perpendicular to L_1 and L_2.

74. One point on the line is $(6, 156)$, and the slope is $m = 4.5$.

$V - 156 = 4.5(t - 6)$

$V - 156 = 4.5t - 27$

$V = 4.5t + 129$

76. One point on the line is $(6, 245{,}000)$, and the slope is -5600.

$V - 245{,}000 = -5600(t - 6)$

$V = 278{,}600 - 5600t$

78. The y-intercept is 1250 and the slope is 1.5, which represents the increase in hourly wage per unit produced. Matches graph (c).

80. The y-intercept is 600 and the slope is -100 which represents the decrease in the value of the word processor each year. Matches graph (d).

82. Set the distance between $\left(3, \frac{5}{2}\right)$ and (x, y) equal to the distance between $(-7, 1)$ and (x, y).

$$\sqrt{(x-3)^2 + \left(y - \frac{5}{2}\right)^2} = \sqrt{[x-(-7)]^2 + (y-1)^2}$$
$$(x-3)^2 + \left(y - \frac{5}{2}\right)^2 = (x+7)^2 + (y-1)^2$$
$$x^2 - 6x + 9 + y^2 - 5y + \frac{25}{4} = x^2 + 14x + 49 + y^2 - 2y + 1$$
$$-6x - 5y + \frac{61}{4} = 14x - 2y + 50$$
$$-24x - 20y + 61 = 56x - 8y + 200$$
$$80x + 12y + 139 = 0$$

This line is the perpendicular bisector of the line segment connecting $\left(3, \frac{5}{2}\right)$ and $(-7, 1)$.

84. $F = \frac{9}{5}C + 32$

$F = 0°$: $0 = \frac{9}{5}C + 32$
$$-32 = \frac{9}{5}C$$
$$-17.9 \approx C$$

$C = 10°$: $F = \frac{9}{5}(10) + 32$
$$F = 18 + 32$$
$$F = 50$$

$F = 90°$: $90 = \frac{9}{5}C + 32$
$$58 = \frac{9}{5}C$$
$$32.2 \approx C$$

$C = -10°$: $F = \frac{9}{5}(-10) + 32$
$$F = -18 + 32$$
$$F = 14$$

$F = 68°$: $68 = \frac{9}{5}C + 32$
$$36 = \frac{9}{5}C$$
$$20 = C$$

$C = 177°$: $F = \frac{9}{5}(177) + 32$
$$F = 318.6 + 32$$
$$F = 350.6$$

C	−17.8°	−10°	10°	20°	32.2°	177°
F	0°	14°	50°	68°	90°	350.6°

86. (1994, 2546) and (1996, 2702)

$$y - 2702 = \frac{2702 - 2546}{1996 - 1994}(x - 1996)$$
$$y - 2702 = 78(x - 1996)$$

For $x = 2000$:

$$y - 2702 = 78(2000 - 1996)$$
$$y = 312 + 2702 = 3014$$

The college will have 3014 students in 2000.

88. (a) (0, 25,000) and (10, 2000)

$$V - 25{,}000 = \frac{25{,}000 - 2000}{0 - 10}t$$
$$V - 25{,}000 = -2300t$$
$$V = 25{,}000 - 2300t$$

(b)

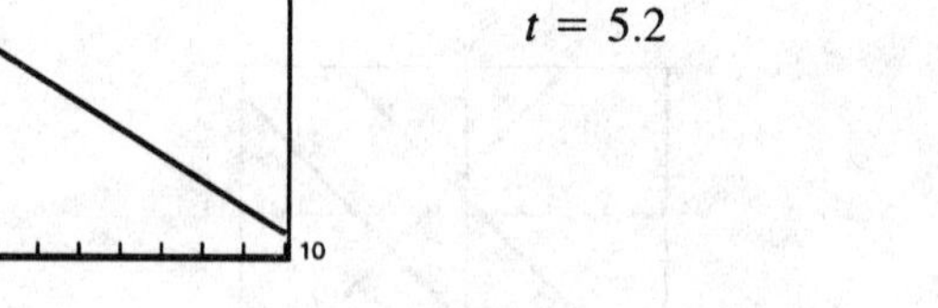

$$13{,}000 = 25{,}000 - 2300t$$
$$t = 5.2$$

90. (Hourly wage) = (Base pay) + (Piecework pay)

$$W = 11.5 + 0.75x$$

92. (a)

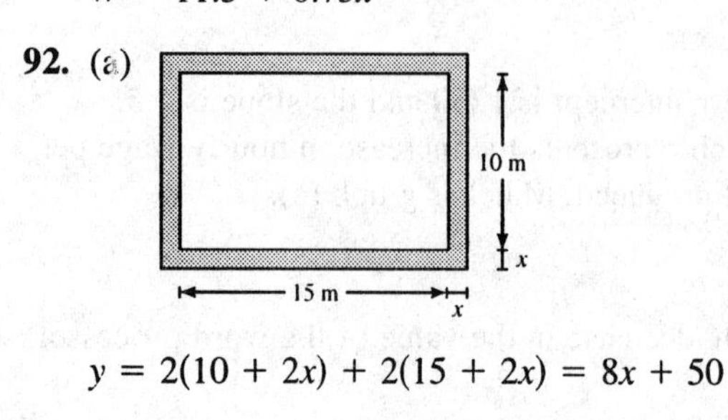

$$y = 2(10 + 2x) + 2(15 + 2x) = 8x + 50$$

(b)

150

0

0

10

(c) Slope is 8 meters For each additional 1-meter increase, y increases by 8 meters.

94. (a) $y = 4x + 19$

(b)

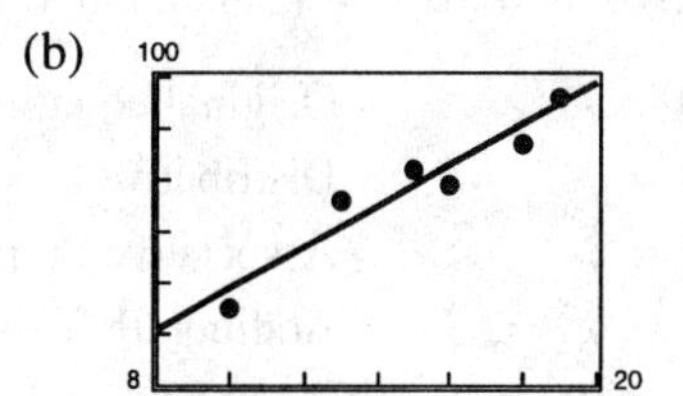

(c) $y = 4(17) + 19 = 87$

(d) Average increase in exam score for each one point increase in quiz scores

(e) Vertical shift four units upward

$y = 4x + 23$

96. (a) $12{,}000 - x$ is invested in the fund paying 8% interest.

(b) Model: $\begin{pmatrix}\text{Interest from}\\ 5\frac{1}{2}\%\text{ fund}\end{pmatrix} + \begin{pmatrix}\text{Interest from}\\ 8\%\text{ fund}\end{pmatrix} = \begin{pmatrix}\text{Annual}\\ \text{interest}\end{pmatrix}$

Labels: Interest from $5\frac{1}{2}\%$ fund $= 0.055x$
Interest from 8% fund $= 0.08(12{,}000 - x)$
Annual interest $= y$

Equation: $y = 0.055x + 0.08(12{,}000 - x)$

$y = 0.055x + 960 - 0.08x$

$y = -0.025x + 960$

(c)

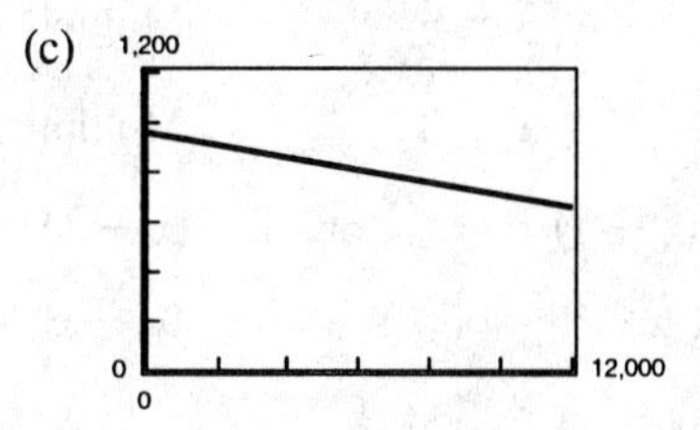

(d) As the amount invested at the lower interest increases, the annual interest decreases.

Section P.4 Solving Equations Algebraically and Graphically

Solutions to Even-Numbered Exercises

2. $3 + \dfrac{1}{x+2} = 4$

(a) $x = -1$

$3 + \dfrac{1}{-1+2} \stackrel{?}{=} 4$

$3 + \dfrac{1}{1} \stackrel{?}{=} 4$

$3 + 1 \stackrel{?}{=} 4$

$4 = 4$

$x = -1$ *is* a solution.

(b) $x = -2$

$3 + \dfrac{1}{-2+2} \stackrel{?}{=} 4$

$3 + \dfrac{1}{0} \stackrel{?}{=} 4$

Division by zero is undefined.

$x = -2$ *is not* a solution.

(c) $x = 0$

$3 + \dfrac{1}{0+2} \stackrel{?}{=} 4$

$3\frac{1}{2} \neq 4$

$x = 0$ *is not* a solution.

(d) $x = 5$

$3 + \dfrac{1}{5+2} \stackrel{?}{=} 4$

$3\frac{1}{7} \neq 4$

$x = 5$ *is not* a solution.

4. $\sqrt[3]{x-8} = 3$

(a) $x = 2$

$\sqrt[3]{2-8} \stackrel{?}{=} 3$

$\sqrt[3]{6} \neq 3$

$x = 2$ *is not* a solution.

(b) $x = -5$

$\sqrt[3]{-5-8} \stackrel{?}{=} 3$

$\sqrt[3]{-13} \neq 3$

$x = -5$ *is not* a solution.

(c) $x = 35$

$\sqrt[3]{35-8} \stackrel{?}{=} 3$

$\sqrt[3]{27} \stackrel{?}{=} 3$

$3 = 3$

$x = 35$ *is* a solution.

(d) $x = 8$

$\sqrt[3]{8-8} \stackrel{?}{=} 3$

$0 \neq 3$

$x = 8$ *is not* a solution.

6. $3(x + 2) = 5x + 4$ is *conditional.* There are real values of x for which the equation is not true (for example, $x = 0$).

8. $x^2 + 2(3x - 2) = x^2 + 6x - 4$ is an *identity* by simplification. It is true for all real values of x.

10.
$3(x - 4) + 10 = 7$ Original equation
$3x - 12 + 10 = 7$ Distributive Property
$3x - 2 = 7$ Associative Property
$3x - 2 + 2 = 7 + 2$ Addition Property of Equality
$3x = 9$ Additive Inverse Property
$\frac{3x}{3} = \frac{9}{3}$ Multiplicative Property of Equality
$x = 3$ Multiplicative Inverse Property

12. (a) $3(x - 1) = 4$ or $3(x - 1) = 4$
$x - 1 = \frac{4}{3}$ $3x - 3 = 4$
$x = \frac{4}{3} + 1$ $3x = 7$
$x = \frac{7}{3}$ $x = \frac{7}{3}$

(b) $\frac{3}{4}(z - 4) = 6$ or $\frac{3}{4}(z - 4) = 6$
$\left(\frac{4}{3}\right)\frac{3}{4}(z - 4) = \left(\frac{4}{3}\right)6$ $\frac{3}{4}z - 3 = 6$
$z - 4 = 8$ $\frac{3}{4}z = 9$
$z = 12$ $\frac{4}{3} \cdot \frac{3}{4}z = \frac{4}{3} \cdot 9$
$z = \frac{36}{3} = 12$

The first way is easier because the fraction is eliminated in the first step.

14.
$7x + 3 = 3x - 13$
$7x + 3 - 3 - 3x = 3x - 13 - 3 - 3x$
$4x = -16$
$x = -4$

16.
$2(13t - 15) + 3(t - 19) = 0$
$26t - 30 + 3t - 57 = 0$
$29t - 87 = 0$
$29t = 87$
$t = 3$

18.
$3(x + 3) = 5(1 - x) - 1$
$3x + 9 = 5 - 5x - 1$
$3x + 9 = 4 - 5x$
$8x = -5$
$x = -\frac{5}{8}$

20.
$\frac{x}{5} - \frac{x}{2} = 3$
$10\left(\frac{x}{5}\right) - 10\left(\frac{x}{2}\right) = 10(3)$
$2x - 5x = 30$
$-3x = 30$
$x = -10$

22.
$\frac{3x}{2} + \frac{1}{4}(x - 2) = 10$
$(4)\left(\frac{3x}{2}\right) + (4)\frac{1}{4}(x - 2) = (4)10$
$6x + (x - 2) = 40$
$7x - 2 = 40$
$7x = 42$
$x = 6$

24.
$0.60x + 0.40(100 - x) = 50$
$0.60x + 40 - 0.40x = 50$
$0.20x = 10$
$x = 50$

26.
$\frac{17 + y}{y} + \frac{32 + y}{y} = 100$
$(y)\frac{17 + y}{y} + (y)\frac{32 + y}{y} = 100(y)$
$17y + 32 + y = 100y$
$49 + 2y = 100y$
$49 = 98y$
$\frac{1}{2} = y$

28.
$\frac{15}{x} - 4 = \frac{6}{x} + 3$
$\frac{15}{x} - \frac{6}{x} = 7$
$\frac{9}{x} = 7$
$9 = 7x$
$\frac{9}{7} = x$

30.
$$\frac{1}{x-2} + \frac{3}{x+3} = \frac{4}{x^2+x-6}$$
$$(x^2+x-6)\frac{1}{x-2} + (x^2+x-6)\frac{3}{x+3} = (x^2+x-6)\frac{4}{x^2+x-6}$$
$$(x+3) + 3(x-2) = 4$$
$$x + 3 + 3x - 6 = 4$$
$$4x - 3 = 4$$
$$4x = 7$$
$$x = \frac{7}{4}$$

32.
$$\frac{2}{(x-4)(x-2)} = \frac{1}{x-4} + \frac{2}{x-2}$$
$$(x-4)(x-2)\frac{2}{(x-4)(x-2)} = (x-4)(x-2)\frac{1}{x-4} + (x-4)(x-2)\frac{2}{x-2}$$
$$2 = (x-2) + 2(x-4)$$
$$2 = x - 2 + 2x - 8$$
$$2 = 3x - 10$$
$$12 = 3x$$
$$4 = x$$

Check: $\dfrac{2}{(4-4)(4-2)} = \dfrac{1}{4-4} + \dfrac{2}{4-2}$
$$\frac{2}{0 \cdot 2} = \frac{1}{0} + \frac{2}{2}$$

Division by zero is undefined. Thus, $x = 4$ is not a solution, and the original equation has no solution.

34.
$$\frac{4}{u-1} + \frac{6}{3u+1} = \frac{15}{3u+1}$$
$$(u-1)(3u+1)\frac{4}{u-1} + (u-1)(3u+1)\frac{6}{3u+1} = (u-1)(3u+1)\frac{15}{3u+1}$$
$$4(3u+1) + 6(u-1) = 15(u-1)$$
$$12u + 4 + 6u - 6 = 15u - 15$$
$$18u - 2 = 15u - 15$$
$$3u = -13$$
$$u = -\frac{13}{3}$$

36.
$$\frac{6}{x} - \frac{2}{x+3} = \frac{3(x+5)}{x(x+3)}$$
$$x(x+3)\frac{6}{x} - x(x+3)\frac{2}{x+3} = x(x+3)\frac{3(x+5)}{x(x+3)}$$
$$6(x+3) - 2x = 3(x+5)$$
$$6x + 18 - 2x = 3x + 15$$
$$4x + 18 = 3x + 15$$
$$x = -3$$

Check: $\dfrac{6}{-3} - \dfrac{2}{-3+3} = \dfrac{3(-3+5)}{-3(-3+3)}$
$$-2 - \frac{2}{0} = \frac{-6}{-3(0)}$$

Division by zero is undefined. Thus, $x = -3$ is not a solution, and the original equation has no solution.

38.
$$3 = 2 + \frac{2}{z+2}$$
$$1 = \frac{2}{z+2}$$
$$z + 2 = 2$$
$$z = 0$$

40.
$$(x+1)^2 + 2(x-2) = (x+1)(x-2)$$
$$x^2 + 2x + 1 + 2x - 4 = x^2 - x - 2$$
$$5x = 1$$
$$x = \frac{1}{5}$$

42. $W_1x = W_2(L - x)$

$W_1 = 200$ pounds

$W_2 = 550$ pounds

$L = 5$ feet

$200x = 550(5 - x)$

$200x = 2750 - 550x$

$750x = 2750$

$x = 3\frac{2}{3}$

44. $y = 4 - x^2$

Let $y = 0$: $0 = 4 - x^2 \Rightarrow x = 2, -2$

$\Rightarrow (2, 0), (-2, 0)$ x-intercepts

Let $x = 0$: $y = 4 - 0^2 = 4 \Rightarrow (0, 4)$ y-intercept

46. $xy = 4 \Rightarrow y = 4/x$

Neither y nor x can be zero. Hence, there are no intercepts.

48. $x^2y - x^2 + 4y = 0$

Let $y = 0$: $-x^2 = 0 \Rightarrow x = 0$

$\Rightarrow (0, 0)$ x-intercept

Let $x = 0$: $4y = 0 \Rightarrow y = 0$

$\Rightarrow (0, 0)$ y-intercept

50. $f(x) = 3(x - 5) + 9$

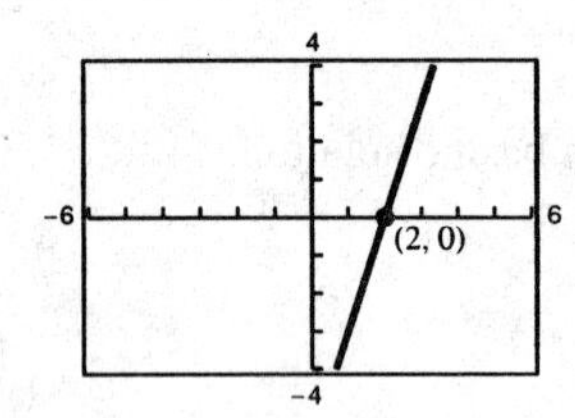

52. $f(x) = x^3 - 9x^2 + 18x$

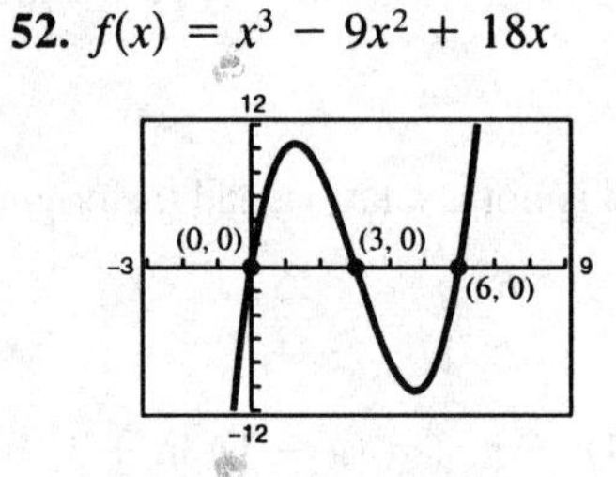

54. $f(x) = x - 3 - \dfrac{10}{x}$

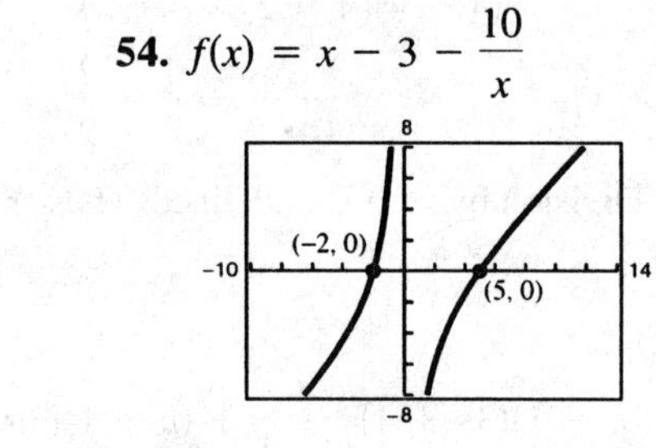

56.

$$0 = \tfrac{4}{3}x + 2$$
$$-\tfrac{4}{3}x = 2$$
$$\left(-\tfrac{3}{4}\right)\left(-\tfrac{4}{3}x\right) = \left(-\tfrac{3}{4}\right)2$$
$$x = -\tfrac{3}{2}$$

Intercept: $\left(-\frac{3}{2}, 0\right)$

The solution to $0 = \frac{4}{3}x + 2$ is the same as the x-intercept of $y = \frac{4}{3}x + 2$. They are both $x = -\frac{3}{2}$.

58.

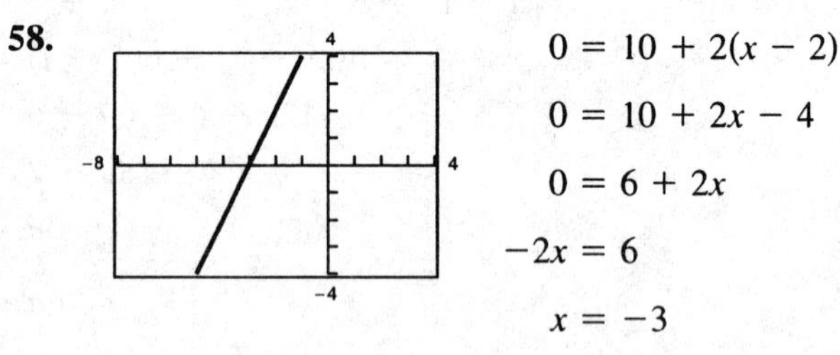

$$0 = 10 + 2(x - 2)$$
$$0 = 10 + 2x - 4$$
$$0 = 6 + 2x$$
$$-2x = 6$$
$$x = -3$$

Intercept: $(-3, 0)$

The solution to $0 = 10 + 2(x - 2)$ is the same as the x-intercept of $0 = 10 + 2(x - 2)$ They are both $x = -3$.

60. $0.60x + 0.40(100 - x) = 50$

$0.60x + 40 - 0.40x = 50$

$0.20x = 10$

$x = 50$

$f(x) = 0.60x + 0.40(100 - x) - 50 = 0$

$x = 50.0$

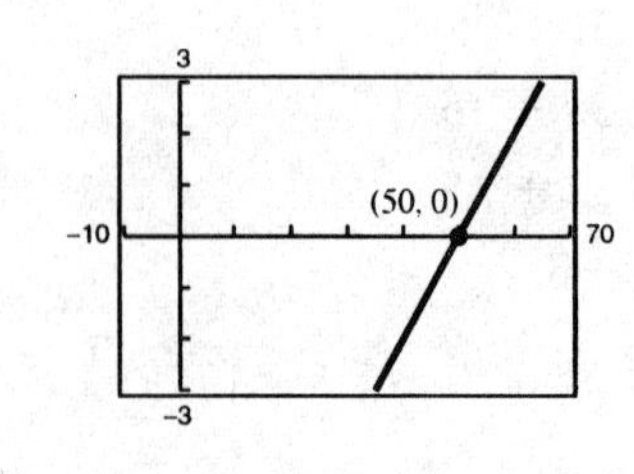

62. $(x + 1)^2 + 2(x - 2) = (x + 1)(x - 2)$

$$x^2 + 2x + 1 + 2x - 4 = x^2 - x - 2$$
$$5x = 1$$
$$x = \frac{1}{5}$$
$$f(x) = (x + 1)^2 + 2(x - 2) - (x + 1)(x - 2) = 0$$
$$x = 0.20$$

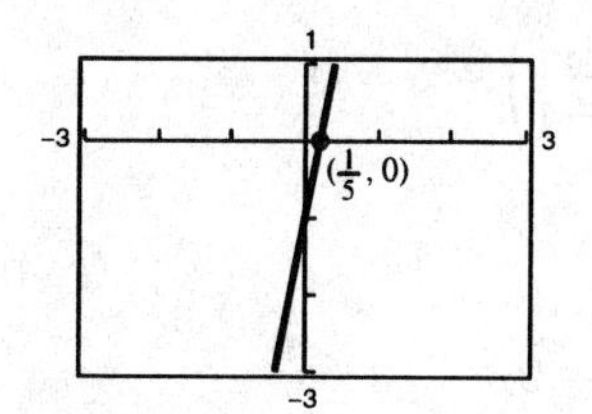

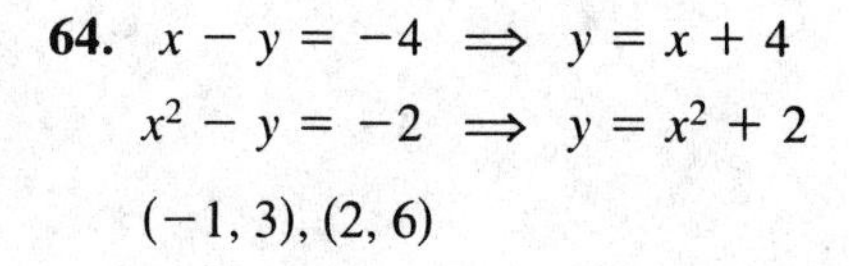

64. $x - y = -4 \Rightarrow y = x + 4$

$x^2 - y = -2 \Rightarrow y = x^2 + 2$

$(-1, 3), (2, 6)$

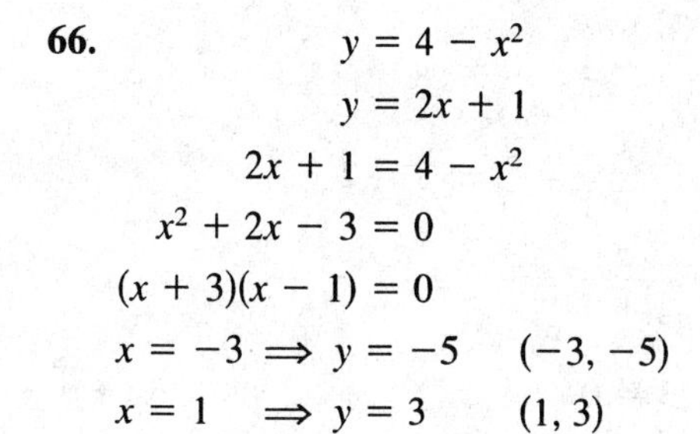

66.

$$y = 4 - x^2$$
$$y = 2x + 1$$
$$2x + 1 = 4 - x^2$$
$$x^2 + 2x - 3 = 0$$
$$(x + 3)(x - 1) = 0$$
$$x = -3 \Rightarrow y = -5 \quad (-3, -5)$$
$$x = 1 \Rightarrow y = 3 \quad (1, 3)$$

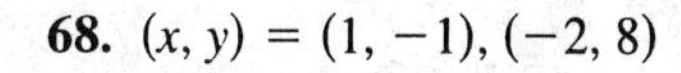

68. $(x, y) = (1, -1), (-2, 8)$

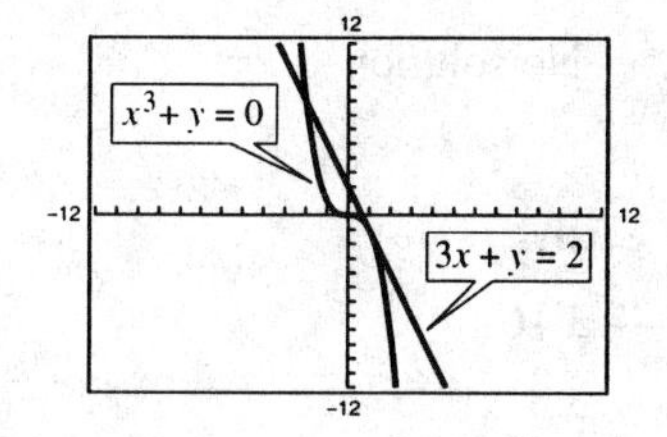

70. $(x, y) = (1.670, 1.660)$

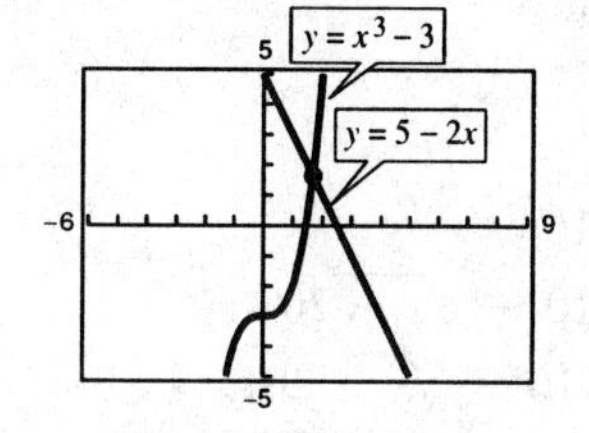

72. $(x, y) = (2.050, 32)$

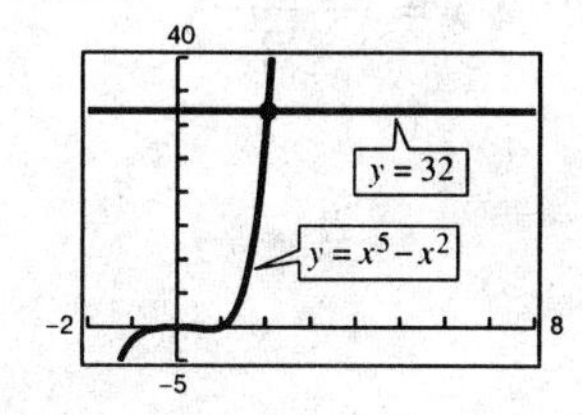

74. $(x, y) = (0, 0)\ (3, -3)$

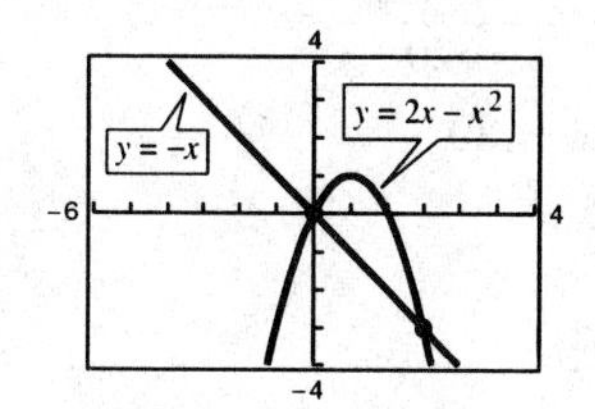

76. $11x^2 + 33x = 0$

$$11(x^2 + 3x) = 0$$
$$x(x + 3) = 0$$
$$x = 0$$
$$x + 3 = 0 \Rightarrow x = -3$$

78. $x^2 + 3x - \frac{3}{4} = 0$

$$x^2 + 3x + \left(\frac{3}{2}\right)^2 = \frac{3}{4} + \frac{9}{4}$$
$$\left(x + \frac{3}{2}\right)^2 = 3$$
$$x + \frac{3}{2} = \pm\sqrt{3}$$
$$x = -\frac{3}{2} \pm \sqrt{3}$$

80.

$$20x^3 - 125x = 0$$
$$5x(4x^2 - 25) = 0$$
$$5x(2x + 5)(2x - 5) = 0$$
$$5x = 0 \Rightarrow x = 0$$
$$2x + 5 = 0 \Rightarrow x = -\frac{5}{2}$$
$$2x - 5 = 0 \Rightarrow x = \frac{5}{2}$$

82.

$$4x^4 - 65x^2 + 16 = 0$$
$$(4x^2 - 1)(x^2 - 16) = 0$$
$$(2x + 1)(2x - 1)(x + 4)(x - 4) = 0$$
$$2x + 1 = 0 \Rightarrow x = -\frac{1}{2}$$
$$2x - 1 = 0 \Rightarrow x = \frac{1}{2}$$
$$x + 4 = 0 \Rightarrow x = -4$$
$$x - 4 = 0 \Rightarrow x = 4$$

84. $6\left(\frac{s}{s+1}\right)^2 + 5\left(\frac{s}{s+1}\right) - 6 = 0$

Let $u = s/(s+1)$.

$$6u^2 + 5u - 6 = 0$$
$$(3u - 2)(2u + 3) = 0$$
$$3u - 2 = 0 \Rightarrow u = \frac{2}{3}$$
$$2u + 3 = 0 \Rightarrow u = -\frac{3}{2}$$
$$\frac{s}{s+1} = \frac{2}{3} \Rightarrow s = 2$$
$$\frac{s}{s+1} = -\frac{3}{2} \Rightarrow s = -\frac{3}{5}$$

86.
$$3x^{1/3} + 2x^{2/3} = 5$$
$$2x^{2/3} + 3x^{1/3} - 5 = 0$$
$$(2x^{1/3} + 5)(x^{1/3} - 1) = 0$$
$$x^{1/3} = -\frac{5}{2} \Rightarrow x = \frac{-125}{8}$$
$$x^{1/3} = 1 \Rightarrow x = 1$$

88.
$$\sqrt{5 - x} - 3 = 0$$
$$\sqrt{5 - x} = 3$$
$$5 - x = 9$$
$$x = -4$$

90.
$$\sqrt[3]{3x + 1} - 5 = 0$$
$$\sqrt[3]{3x + 1} = 5$$
$$3x + 1 = 125$$
$$3x = 124$$
$$x = \frac{124}{3}$$

92.
$$\sqrt{x + 5} = \sqrt{x - 5}$$
$$x + 5 = x - 5$$
$$5 = -5$$

No solution

94.
$$\sqrt{x} + \sqrt{x - 20} = 10$$
$$\sqrt{x} = 10 - \sqrt{x - 20}$$
$$\left(\sqrt{x}\right)^2 = \left(10 - \sqrt{x - 20}\right)^2$$
$$x = 100 - 20\sqrt{x - 20} + x - 20$$
$$-80 = -20\sqrt{x - 20}$$
$$4 = \sqrt{x - 20}$$
$$16 = x - 20$$
$$36 = x$$

96.
$$(x^2 - x - 22)^{4/3} = 16$$
$$x^2 - x - 22 = \pm 16$$
$$x^2 - x - 22 = \pm 8$$
$$x^2 - x - 30 = 0 \Rightarrow x = -5, 6$$
$$x^2 - x - 14 = 0 \Rightarrow x = \frac{1 \pm \sqrt{57}}{2}$$

98.
$$\frac{4}{x} - \frac{5}{3} = \frac{x}{6}$$
$$(6x)\frac{4}{x} - (6x)\frac{5}{3} = (6x)\frac{x}{6}$$
$$24 - 10x = x^2$$
$$x^2 + 10x - 24 = 0$$
$$(x + 12)(x - 2) = 0$$
$$x + 12 = 0 \Rightarrow x = -12$$
$$x - 2 = 0 \Rightarrow x = 2$$

100.
$$\frac{x}{x^2 - 4} + \frac{1}{x + 2} = 3$$
$$(x + 2)(x - 2)\frac{x}{x^2 - 4} + (x + 2)(x - 2)\frac{1}{x + 2} = 3(x + 2)(x - 2)$$
$$x + x - 2 = 3x^2 - 12$$
$$3x^2 - 2x - 10 = 0$$

$a = 3,\ b = -2,\ c = -10$

$$x = \frac{-(-2) \pm \sqrt{(-2)^2 - 4(3)(-10)}}{2(3)} = \frac{2 \pm \sqrt{124}}{6} = \frac{2 \pm 2\sqrt{31}}{6} = \frac{1 \pm \sqrt{31}}{3}$$

102. $4x + 1 = \dfrac{3}{x}$

$(x)4x + (x)1 = (x)\dfrac{3}{x}$

$4x^2 + x = 3$

$4x^2 + x - 3 = 0$

$(4x - 3)(x + 1) = 0$

$4x - 3 = 0 \Rightarrow x = \dfrac{3}{4}$

$x + 1 = 0 \Rightarrow x = -1$

104. $\dfrac{x+1}{3} - \dfrac{x+1}{x+2} = 0$

$3(x + 2)\dfrac{x+1}{3} - 3(x + 2)\dfrac{x+1}{x+2} = 0$

$(x + 2)(x + 1) - 3(x + 1) = 0$

$x^2 + 3x + 2 - 3x - 3 = 0$

$x^2 - 1 = 0$

$(x + 1)(x - 1) = 0$

$x + 1 = 0 \Rightarrow x = -1$

$x - 1 = 0 \Rightarrow x = 1$

106. $|3x + 2| = 7$

$3x + 2 = 7 \Rightarrow x = \frac{5}{3}$

$-(3x + 2) = 7$

$-3x - 2 = 7 \Rightarrow x = -3$

108. $|x - 10| = x^2 - 10x$

First equation:

$x - 10 = x^2 - 10x$

$0 = x^2 - 11x + 10$

$0 = (x - 1)(x - 10)$

$0 = x - 1 \Rightarrow x = 1$, not a solution

$0 = x - 10 \Rightarrow x = 10$

Second equation:

$-(x - 10) = x^2 - 10x$

$0 = x^2 - 9x - 10$

$0 = (x - 10)(x + 1)$

$0 = x - 10 \Rightarrow x = 10$

$0 = x + 1 \Rightarrow x = -1$

110. $y = 2x^4 - 15x^3 + 18x^2$

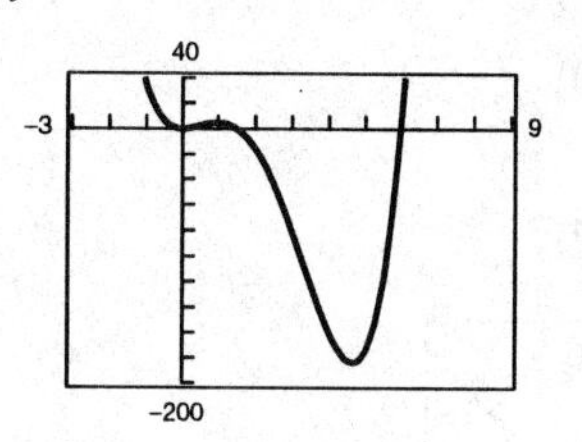

$0 = 2x^4 - 15x^3 + 18x^2$

$= x^2(2x^2 - 15x + 18)$

$= x^2(2x - 3)(x - 6)$

$0 = x^2 \Rightarrow x = 0$

$0 = 2x - 3 \Rightarrow x = \frac{3}{2}$

$0 = x - 6 \Rightarrow x = 6$

x-intercepts: $(0, 0)$, $\left(\frac{3}{2}, 0\right)$, $(6, 0)$

112. $y = x^4 - 29x^2 + 100$

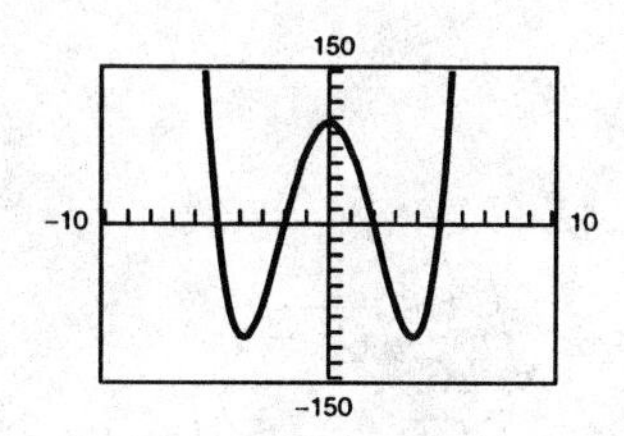

$0 = x^4 - 29x^2 + 100$

$= (x^2 - 4)(x^2 - 25)$

$= (x + 2)(x - 2)(x + 5)(x - 5)$

$0 = x + 2 \Rightarrow x = -2$

$0 = x - 2 \Rightarrow x = 2$

$0 = x + 5 \Rightarrow x = -5$

$0 = x - 5 \Rightarrow x = 5$

x-intercepts: $(-2, 0)$, $(2, 0)$, $(-5, 0)$, $(5, 0)$

114. $y = 2x - \sqrt{15 - 4x}$

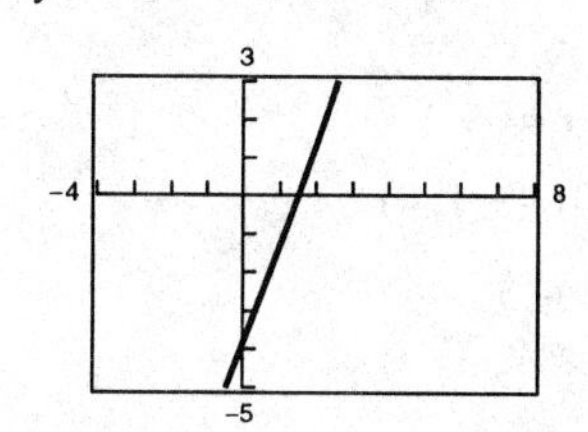

$0 = 2x - \sqrt{15 - 4x}$

$\sqrt{15 - 4x} = 2x$

$15 - 4x = 4x^2$

$0 = 4x^2 + 4x - 15$

$0 = (2x + 5)(2x - 3)$

$0 = 2x + 5 \Rightarrow x = -\frac{5}{2}$

$0 = 2x - 3 \Rightarrow x = \frac{3}{2}$

$x = -\frac{5}{2}$ is extraneous. The x-intercept is $\left(\frac{3}{2}, 0\right)$.

116. $y = 3\sqrt{x} - \dfrac{4}{\sqrt{x}} - 4$

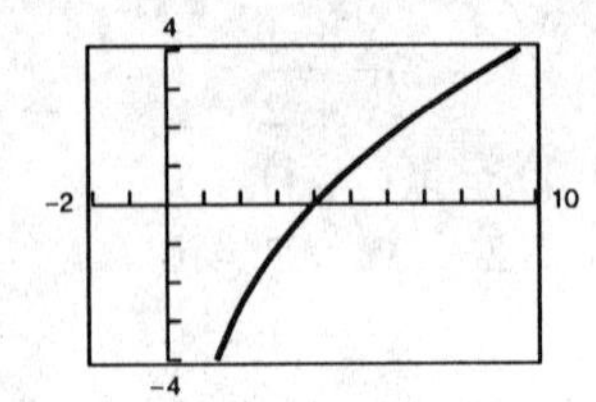

$$0 = 3\sqrt{x} - \frac{4}{\sqrt{x}} - 4$$
$$0 = \sqrt{x}\left(3\sqrt{x} - \frac{4}{\sqrt{x}} - 4\right)$$
$$0 = 3x - 4 - 4\sqrt{x}$$
$$0 = \left(3\sqrt{x} + 2\right)\left(\sqrt{x} - 2\right)$$
$$0 = 3\sqrt{x} + 2 \Rightarrow x = \frac{4}{9}, \text{ extraneous}$$
$$0 = \sqrt{x} - 2 \Rightarrow x = 4$$

x-intercept: $(4, 0)$

118. $y = x + \dfrac{9}{x + 1} - 5$

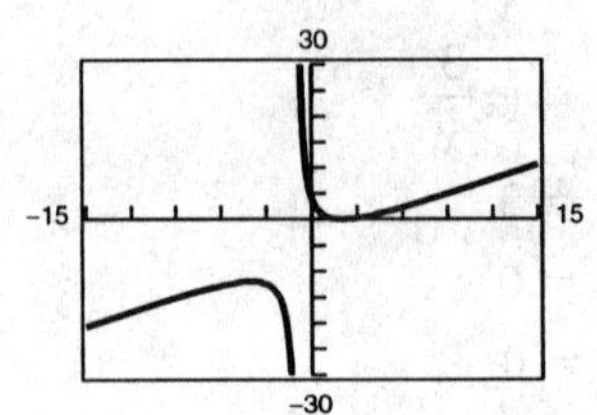

$$0 = x + \frac{9}{x + 1} - 5$$
$$0 = x(x + 1) + (x + 1)\frac{9}{x + 1} - 5(x + 1)$$
$$0 = x^2 + x + 9 - 5x - 5$$
$$0 = x^2 - 4x + 4$$
$$0 = (x - 2)(x - 2)$$
$$0 = x - 2 \Rightarrow x = 2$$

x-intercept: $(2, 0)$

120. $y = |x - 2| - 3$

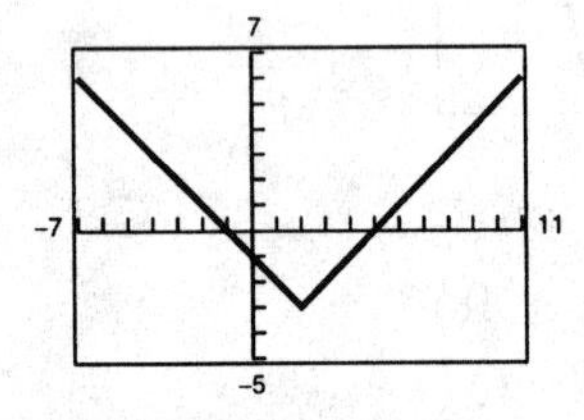

$$0 = |x - 2| - 3$$
$$3 = |x - 2|$$

First equation:

$$x - 2 = 3 \Rightarrow x = 5$$

Second equation:

$$-(x - 2) = 3$$
$$-x + 2 = 3 \Rightarrow x = -1$$

x-intercepts: $(5, 0), (-1, 0)$

122.

$$A = P\left(1 + \frac{r}{n}\right)^{nt}$$
$$\frac{A}{\left(1 + \frac{r}{n}\right)^{nt}} = P$$
$$A\left(1 + \frac{r}{n}\right)^{-nt} = P$$

124.

$$S = \frac{rL - a}{r - 1}$$
$$S(r - 1) = rL - a$$
$$Sr - S = rL - a$$
$$Sr - rL = S - a$$
$$r(S - L) = S - a$$
$$r = \frac{S - a}{S - L}$$

126.

$$i = \pm\sqrt{\frac{1}{LC}}\sqrt{Q^2 - q}$$
$$i^2 = \left(\pm\sqrt{\frac{1}{Lc}}\sqrt{Q^2 - q}\right)^2$$
$$i^2 = \frac{1}{LC}(Q^2 - q)$$
$$LCi^2 = Q^2 - q$$
$$LCi^2 + q = Q^2$$
$$\pm\sqrt{LCi^2 + q} = Q$$

128. (a) The graph indicates that there are 5 zeros.

(b) If $k = 25$, then there is one zero (≈ -4.25). If $k = 15$, then there are three zeros.

(c) No, the range of this function is $(-\infty, \infty)$ and hence, the graph must cross the x-axis at least once.

Section P.5 Solving Inequalities Algebraically and Graphically

Solutions to Even-Numbered Exercises

2. $x \geq 5$

Matches (a).

4. $0 \leq x \leq \frac{9}{2}$

Matches (b).

6. $-1 < \dfrac{3-x}{2} \leq 1$

(a) $x = 0$

$-1 \overset{?}{<} \dfrac{3-0}{2} \overset{?}{\leq} 1$

$-1 < \dfrac{3}{2} \nleq 1$

No, $x = 0$ is not a solution.

(b) $x = \sqrt{5}$

$-1 \overset{?}{<} \dfrac{3-\sqrt{5}}{2} \overset{?}{\leq} 1$

$-1 < 0.382 \leq 1$

Yes, $x = \sqrt{5}$ is a solution.

(c) $x = 1$

$-1 \overset{?}{<} \dfrac{3-1}{2} \overset{?}{\leq} 1$

$-1 < 1 \leq 1$

Yes, $x = 1$ is a solution.

(d) $x = 5$

$-1 \overset{?}{<} \dfrac{3-5}{2} \overset{?}{\leq} 1$

$-1 \nless -1 \leq 1$

No, $x = 5$ is not a solution.

8. $|2x - 3| < 15$

(a) $x = -6$

$|2(-6) - 3| \overset{?}{<} 15$

$15 \nless 15$

No, $x = -6$ is not a solution.

(b) $x = 0$

$|2(0) - 3| \overset{?}{<} 15$

$1 < 15$

Yes, $x = 0$ is a solution.

(c) $x = 12$

$|2(12) - 3| \overset{?}{<} 15$

$21 \nless 15$

No, $x = 12$ is not a solution.

(d) $x = 7$

$|2(7) - 3| \overset{?}{<} 15$

$11 < 15$

Yes, $x = 7$ is a solution.

10. $2x > 3 \Rightarrow x > \frac{3}{2}$

12. $2x + 7 < 3$

$2x < -4$

$x < -2$

14. $-8 \leq 1 - 3(x - 2) < 13$

$-8 \leq 1 - 3x + 6 < 13$

$-8 \leq -3x + 7 < 13$

$-15 \leq -3x < 6$

$5 \geq x > -2 \Rightarrow -2 < x \leq 5$

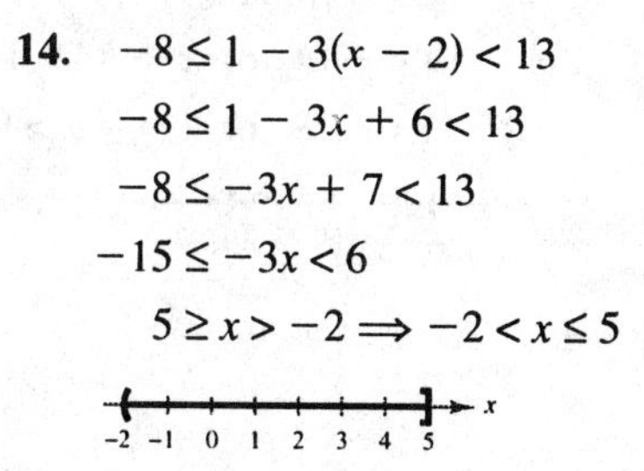

16. $0 \leq \dfrac{x+3}{2} < 5$

$0 \leq x + 3 < 10$

$-3 \leq x < 7$

18. $\dfrac{3}{4} > x + 1 > \dfrac{1}{4}$

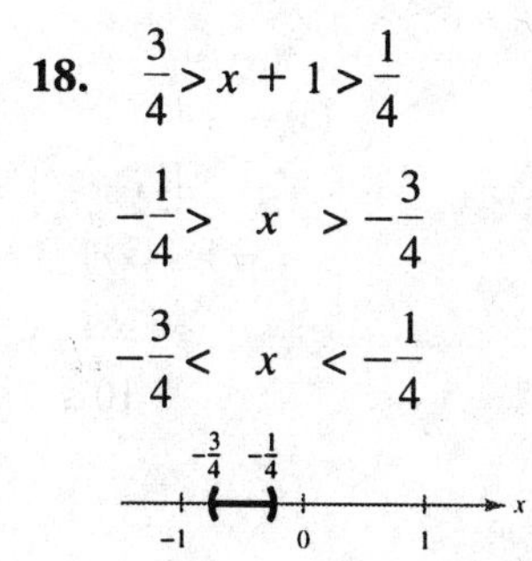

$-\dfrac{1}{4} > x > -\dfrac{3}{4}$

$-\dfrac{3}{4} < x < -\dfrac{1}{4}$

20. $3x - 1 \leq 5$

$3x \leq 6$

$x \leq 2$

22. $3(x + 1) < x + 7$

$3x + 3 < x + 7$

$2x < 4$

$x < 2$

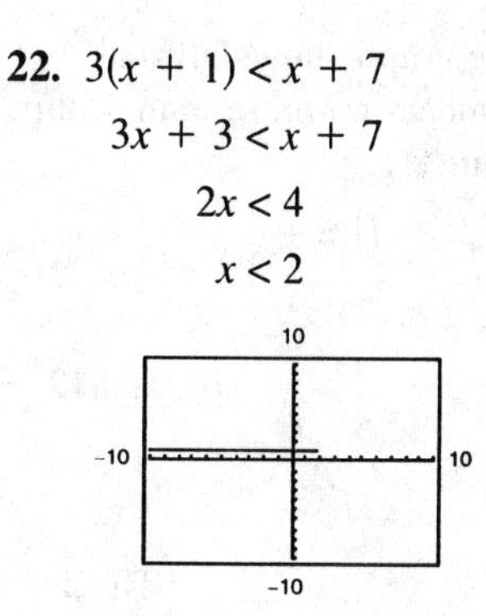

24. $-2 < 3x + 1 < 10$

$-3 < 3x < 9$

$-1 < x < 3$

26. $y = \frac{2}{3}x + 1$

(a) $y \le 5$

$\frac{2}{3}x + 1 \le 5$

$\frac{2}{3}x \le 4$

$x \le 6$

(b) $y \ge 0$

$\frac{2}{3}x + 1 \ge 0$

$\frac{2}{3}x \ge -1$

$x \ge -\frac{3}{2}$

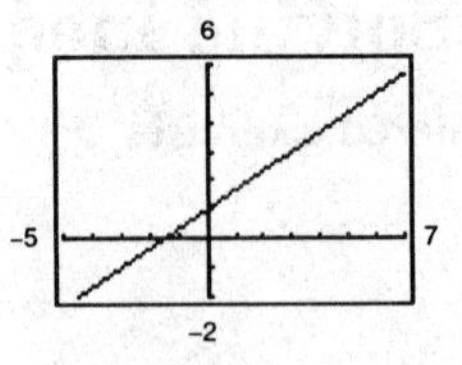

28. $y = -3x + 8$

(a) $-1 \le y \le 3$

$-1 \le -3x + 8 \le 3$

$-9 \le -3x \le -5$

$3 \ge x \ge \frac{5}{3}$

$\frac{5}{3} \le x \le 3$

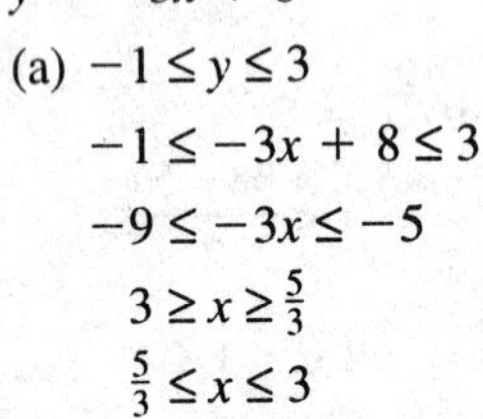

(b) $y \le 0$

$-3x + 8 \le 0$

$-3x \le -8$

$x \ge \frac{8}{3}$

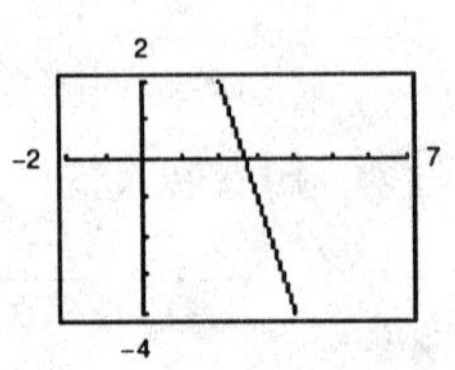

30. $\sqrt[4]{6x + 15}$

$6x + 15 \ge 10$

$6x \ge -15$

$x \ge -\frac{5}{2}$

$\left[-\frac{5}{2}, \infty\right)$

32. $|5x| > 10$

$5x < -10$ or $5x > 10$

$x < -2$ or $x > 2$

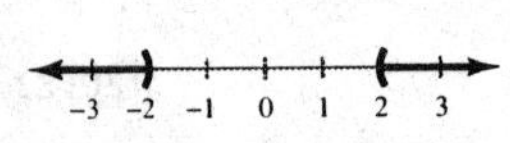

34. $|x - 7| < 6$

$-6 < x - 7 < 6$

$1 < x < 13$

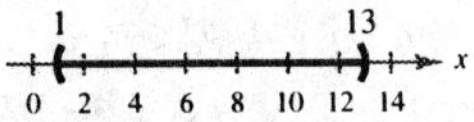

36. $|x + 14| + 3 > 17$

$|x + 14| > 14$

$x + 14 < -14$ or $x + 14 > 14$

$x < -28$ or $x > 0$

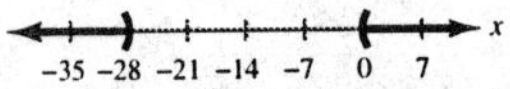

38. $|1 - 2x| < 5$

$-5 < 1 - 2x < 5$

$-6 < -2x < 4$

$3 > x > -2$

$-2 < x < 3$

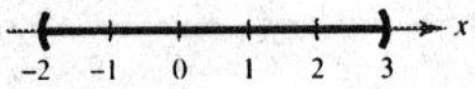

40. $3|4 - 5x| \le 9$

$|4 - 5x| \le 3$

$-3 \le 4 - 5x \le 3$

$-7 \le -5x \le -1$

$\frac{7}{5} \ge x \ge \frac{1}{5}$

$\frac{1}{5} \le x \le \frac{7}{5}$

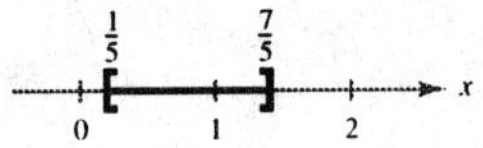

42. $y = \left|\frac{1}{2}x + 1\right|$

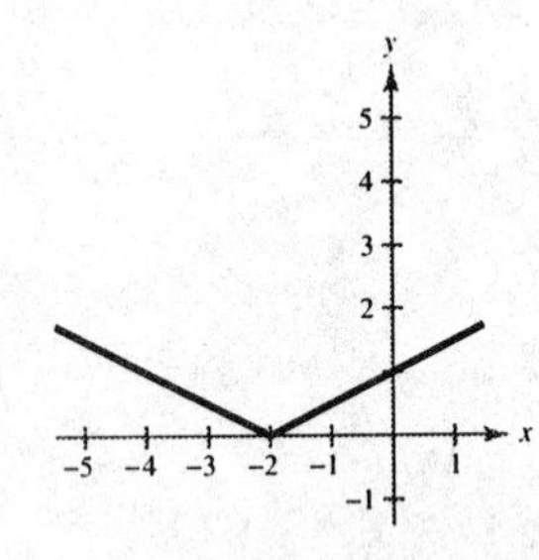

(a) $y \le 4$

$\left|\frac{1}{2}x + 1\right| \le 4$

$-4 \le \frac{1}{2}x + 1 \le 4$

$-5 \le \frac{1}{2}x \le 3$

$-10 \le x \le 6$

(b) $y \ge 1$

$\left|\frac{1}{2}x + 1\right| \ge 1$

$\frac{1}{2}x + 1 \le -1$ or $\frac{1}{2}x + 1 \ge 1$

$\frac{1}{2}x \le -2$ or $\frac{1}{2}x \ge 0$

$x \le -4$ or $x \ge 0$

44. The graph shows all real numbers no more than 3 units from 0.

$|x - 0| > 3$

$|x| > 3$

46. The graph shows all real numbers no more than 4 units from -1.

$|x + 1| \le 4$

48. All real numbers more than 5 units from -3

$|x + 3| > 5$

50. (a)

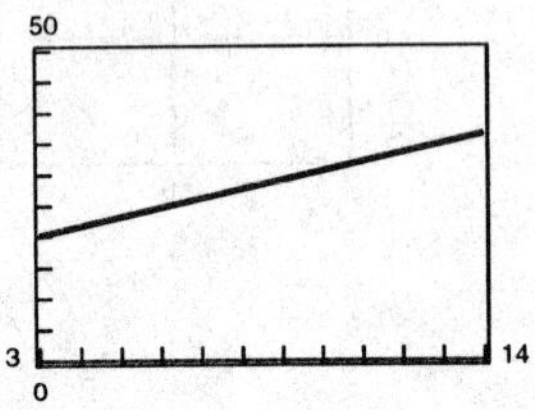

(b) $15.812 + 1.472t > 40$

$1.472t > 24.188$

$t > 16.432$

The average salary will exceed \$40,000 when $t \ge 17$.

52. $(x + 6)^2 \le 8$

$x^2 + 12x + 28 \le 0$

Zeros: $x = \dfrac{-12 \pm \sqrt{12^2 - 4(1)(28)}}{2(1)} = -6 \pm 2\sqrt{2}$

Critical numbers: $x = -6 + 2\sqrt{2}, x = -6 - 2\sqrt{2}$

Test intervals: $\left(-\infty, -6 - 2\sqrt{2}\right) \Rightarrow x^2 + 12x + 28 > 0$

$\left(-6, - 2\sqrt{2}, -6 + 2\sqrt{2}\right) \Rightarrow x^2 + 12x + 28 < 0$

$\left(-6 + 2\sqrt{2}, \infty\right) \Rightarrow x^2 + 12x + 28 > 0$

Solution interval: $\left[-6 - 2\sqrt{2}, -6 + 2\sqrt{2}\right]$

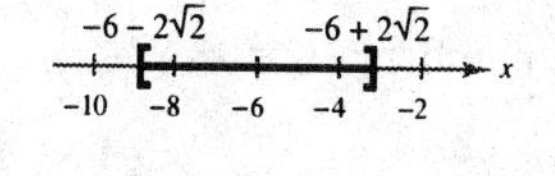

54. $x^2 - 6x + 9 < 16$

$x^2 - 6x - 7 < 0$

$(x + 1)(x - 7) < 0$

Critical numbers: $x = -1, x = 7$

Test intervals: $(-\infty, -1) \Rightarrow (x + 1)(x - 7) > 0$

$(-1, 7) \Rightarrow (x + 1)(x - 7) < 0$

$(7, \infty) \Rightarrow (x + 1)(x - 7) > 0$

Solution interval: $(-1, 7)$

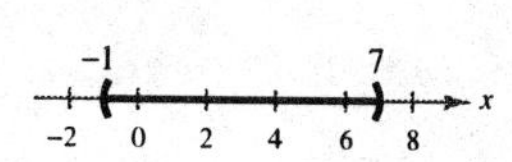

56. $4x^3 - 12x^2 > 0$

$4x^2(x - 3) > 0$

Critical numbers: $x = 0, x = 3$

Test intervals: $(-\infty, 0) \Rightarrow 4x^2(x - 3) < 0$

$(0, 3) \Rightarrow 4x^2(x - 3) < 0$

$(3, \infty) \Rightarrow 4x^2(x - 3) > 0$

Solution interval: $(3, \infty)$

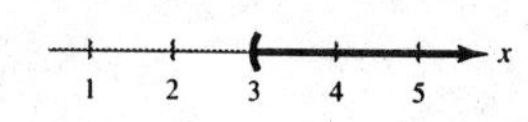

58. $x^4(x - 3) \le 0$

Critical numbers: $x = 0, x = 3$

Test intervals: $(-\infty, 0) \Rightarrow x^4(x - 3) < 0$

$(0, 3) \Rightarrow x^4(x - 3) < 0$

$(3, \infty) \Rightarrow x^4(x - 3) > 0$

Solution intervals: $(-\infty, 0] \cup [0, 3]$ or $(-\infty, 3]$

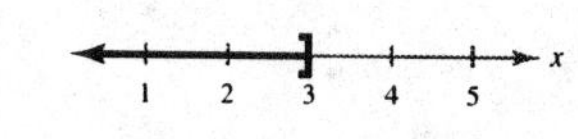

60. $y = \frac{1}{2}x^2 - 2x + 1$

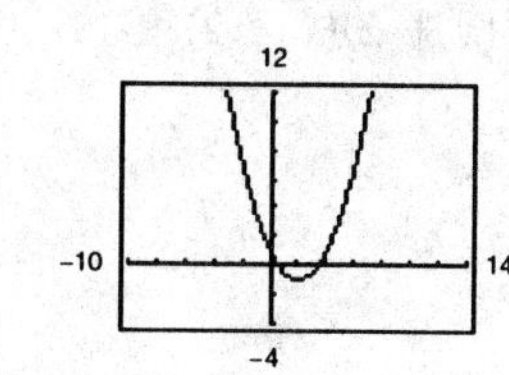

(a) $y \le 1$

$\frac{1}{2}x^2 - 2x + 1 \le 1$

$x^2 - 4x \le 0$

$x(x - 4) \le 0$

$y \le 1$ when $0 \le x \le 4$.

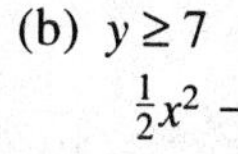

(b) $y \ge 7$

$\frac{1}{2}x^2 - 2x + 1 \ge 7$

$x^2 - 4x - 12 \ge 0$

$(x - 6)(x + 2) \ge 0$

$y \ge 7$ when $x \le -2$ or $x \ge 6$.

62. $y = x^3 - x^2 - 16x + 16$

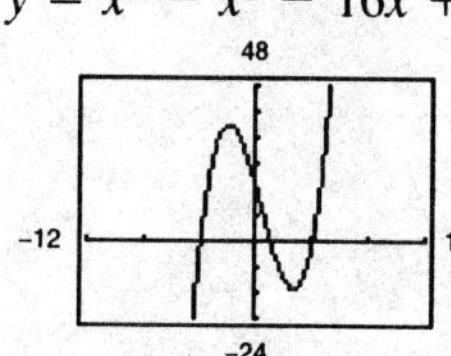

(a) $y \le 0$

$x^3 - x^2 - 16x + 16 \le 0$

$x^2(x - 1) - 16(x - 1) \le 0$

$(x - 1)(x^2 - 16) \le 0$

$y \le 0$ when $-\infty < x \le -4, 1 \le x \le 4$.

(b) $y \ge 36$

$x^3 - x^2 - 16x + 16 \ge 36$

$x^3 - x^2 - 16x - 20 \ge 0$

$(x + 2)(x - 5)(x + 2) \ge 0$

$y \ge 36$ when $x = -2, 5 \le x < \infty$.

64. $\dfrac{1}{x} - 4 < 0$

$\dfrac{1 - 4x}{x} < 0$

Critical numbers: $x = 0, x = \dfrac{1}{4}$

Test intervals: $(-\infty, 0) \Rightarrow \dfrac{1 - 4x}{x} < 0$

$\left(0, \dfrac{1}{4}\right) \Rightarrow \dfrac{1 - 4x}{x} > 0$

$\left(\dfrac{1}{4}, \infty\right) \Rightarrow \dfrac{1 - 4x}{x} < 0$

Solution interval: $(-\infty, 0) \cup \left(\dfrac{1}{4}, \infty\right)$

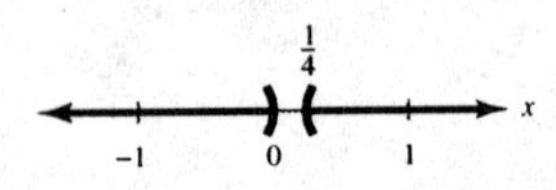

66. $\dfrac{x + 12}{x + 2} - 3 \geq 0$

$\dfrac{x + 12 - 3(x + 2)}{x + 2} \geq 0$

$\dfrac{6 - 2x}{x + 2} \geq 0$

Critical numbers: $x = -2, x = 3$

Test intervals: $(-\infty, -2) \Rightarrow \dfrac{6 - 2x}{x + 2} < 0$

$(-2, 3) \Rightarrow \dfrac{6 - 2x}{x + 2} > 0$

$(3, \infty) \Rightarrow \dfrac{6 - 2x}{x + 2} < 0$

Solution interval: $(-2, 3]$

68. $y = \dfrac{2(x - 2)}{x + 1}$

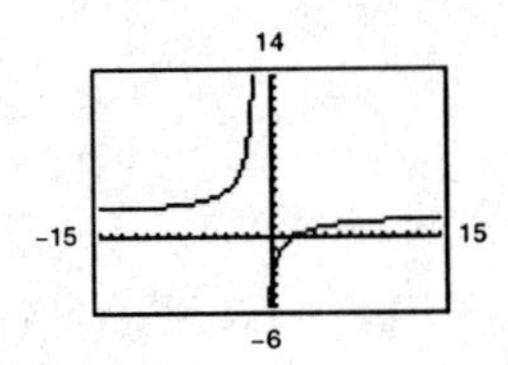

(a) $y \leq 0$

$\dfrac{2(x - 2)}{x + 1} \leq 0$

$y \leq 0$ when $-1 < x \leq 2$.

(b) $y \geq 8$

$\dfrac{2(x - 2)}{x + 1} \geq 8$

$\dfrac{2(x - 2) - 8(x + 1)}{x + 1} \geq 0$

$\dfrac{-6x - 12}{x + 1} \geq 0$

$\dfrac{-6(x + 2)}{x + 1} \geq 0$

$y \geq 8$ when $-2 \leq x < -1$.

70. $y = \dfrac{5x}{x^2 + 4}$

(a) $y \geq 1$

$\dfrac{5x}{x^2 + 4} \geq 1$

$\dfrac{5x - (x^2 + 4)}{(x^2 + 4)} \geq 0$

$\dfrac{(x - 4)(x - 1)}{x^2 + 4} \geq 0$

$y \geq 1$ when $1 \leq x \leq 4$.

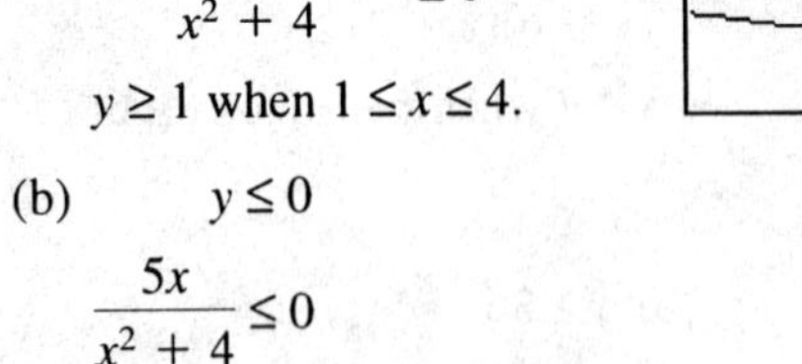

(b) $y \leq 0$

$\dfrac{5x}{x^2 + 4} \leq 0$

$y \leq 0$ when $-\infty < x \leq 0$.

72. $x^2 - 4 \geq 0$

$(x + 2)(x - 2) \geq 0$

Critical numbers: $x = -2, x = 2$

Test intervals: $(-\infty, -2) \Rightarrow (x + 2)(x - 2) > 0$

$(-2, 2) \Rightarrow (x + 2)(x - 2) < 0$

$(2, \infty) \Rightarrow (x + 2)(x - 2) > 0$

Domain: $(-\infty, -2] \cup [2, \infty)$

Section P.6 Exploring Data: Representing Data Graphically

Solutions to Even-Numbered Exercises

2. (a) 900 pounds

(b) (600, 1300)

4.

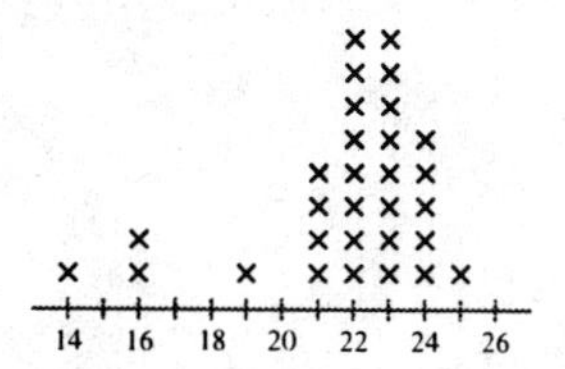

22 and 23 occurred with the greatest frequency.

6.

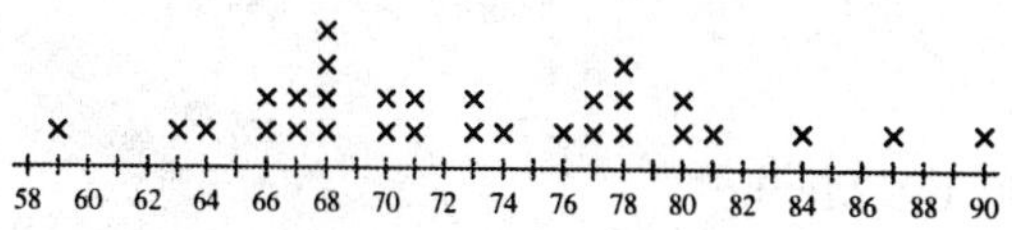

68 occurred with the greatest frequency.

8.

Leaves (Exam 2)	*Stem*	*Leaves (Exam 1)*
9	5	
8 8 8 8 7 7 6 6 4 3	6	
8 8 8 7 7 6 4 3 3 1 1 0 0	7	0 5 5 5 7 7 8 8 8
7 4 1 0 0	8	1 1 1 1 2 3 4 5 5 5 5 7 8 9 9 9
0	9	0 2 8
	10	0 0

Exam 2 scores are higher.

10.

Stem	*Leaves*
1	49
2	15 25 70
3	05 70 83 88
4	64 91 94
5	12 80
6	57 91 96
7	15 57 59 82
8	63 71 96
9	32 45 64 92
10	49 84
11	19 30 58
12	45 67 95
13	19 66
15	03
17	22 47
20	94
21	78
23	62
25	21
26	62
42	89
45	36
48	94

12. Corn: $\frac{43}{240} \times 100 = 17.9\%$

Soybeans: $\frac{21}{61} \times 100 = 34.4\%$

Wheat: $\frac{33}{66} \times 100 = 54.5\%$

14.

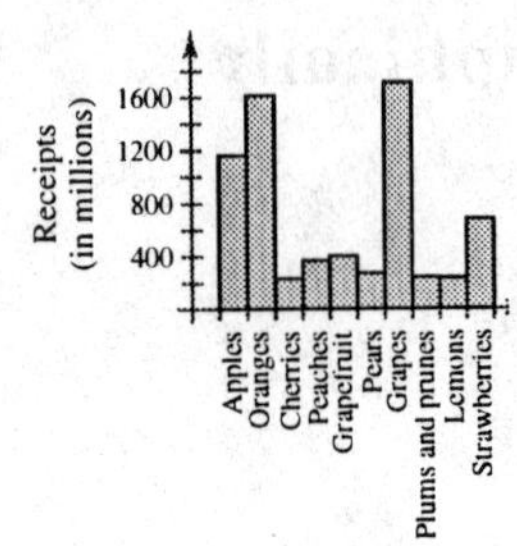

16.

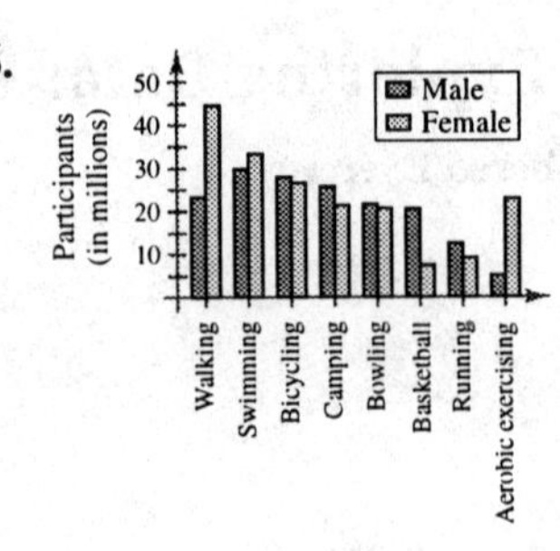

18. (a) Mining

(b) Greatest: 1987
Least: 1992, 1993

20.

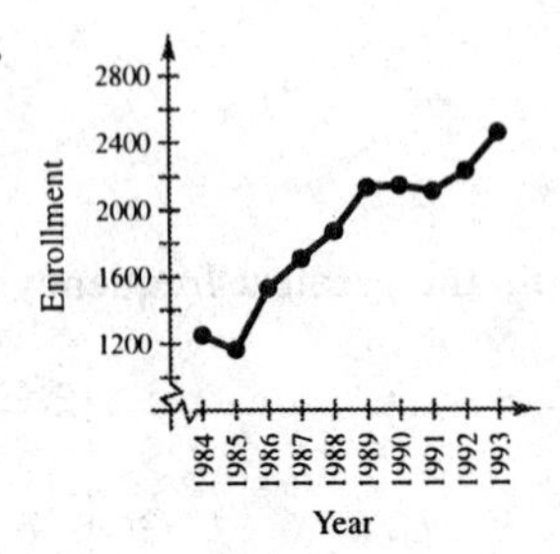

Imports have risen most of the period 1985–1993.

❑ Review Exercises for Chapter P

Solutions to Even-Numbered Exercises

2.

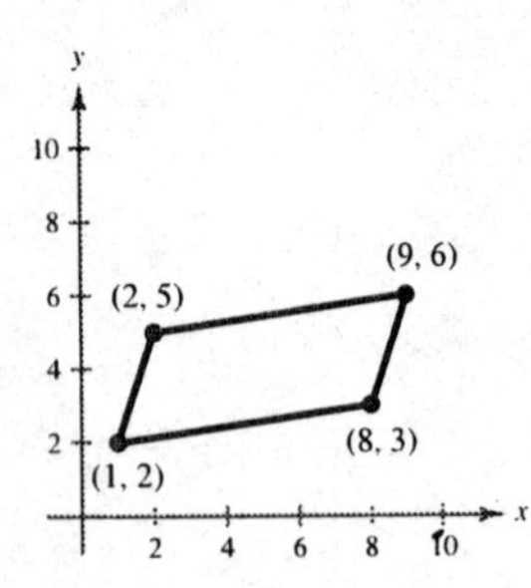

$$d_1 = \sqrt{(1-8)^2 + (2-3)^2} = \sqrt{49+1} = \sqrt{50} = 5\sqrt{2}$$
$$d_2 = \sqrt{(8-9)^2 + (3-6)^2} = \sqrt{1+9} = \sqrt{10}$$
$$d_3 = \sqrt{(9-2)^2 + (6-5)^2} = \sqrt{49+1} = \sqrt{50} = 5\sqrt{2}$$
$$d_4 = \sqrt{(1-2)^2 + (2-5)^2} = \sqrt{1+9} = \sqrt{10}$$

Opposite sides have equal lengths of $\sqrt{10}$ and $5\sqrt{2}$.

4. If $xy = 4$ then the coordinates have the same sign. This happens in Quadrants I and III.

6.

x	-1	0	1	2	3
y	4	0	-2	-2	0

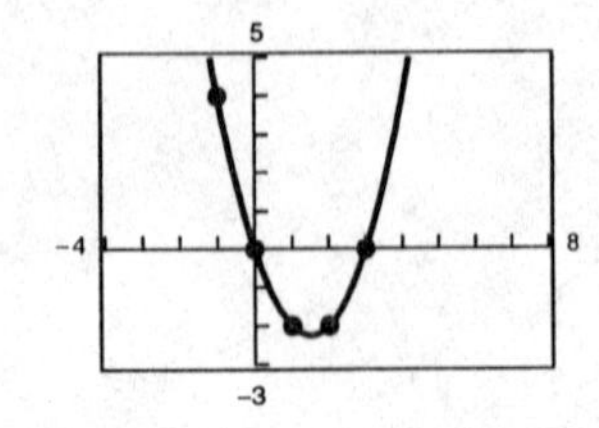

8. $x - 5 = 0$

$x = 5$ is a vertical line through $(5, 0)$.

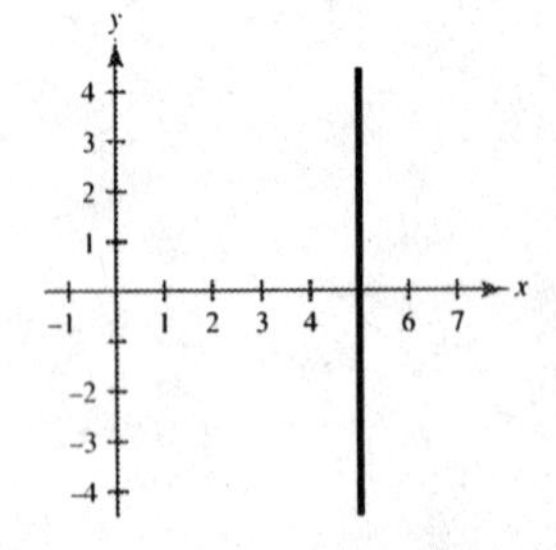

10.

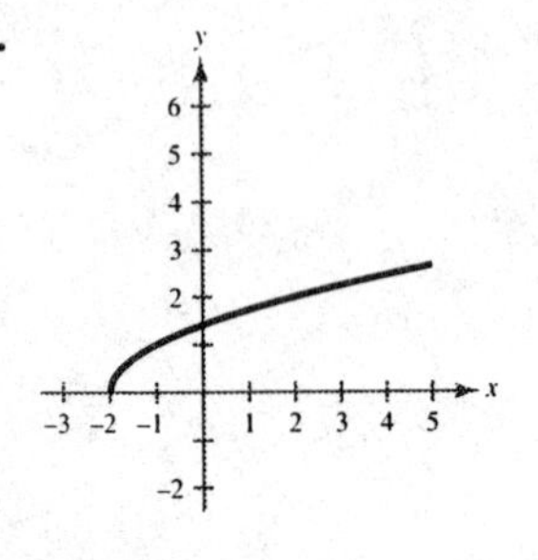

12.

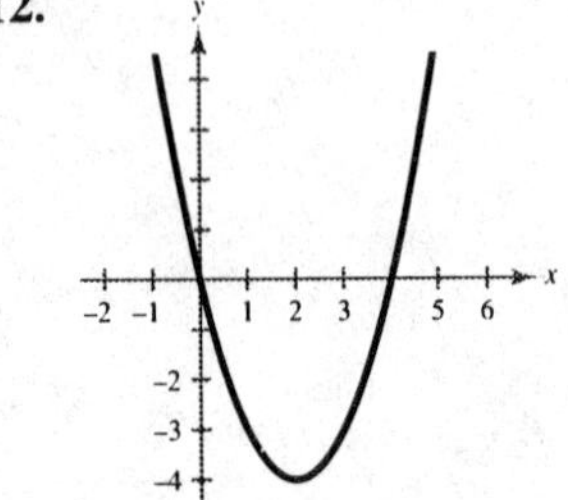

14.

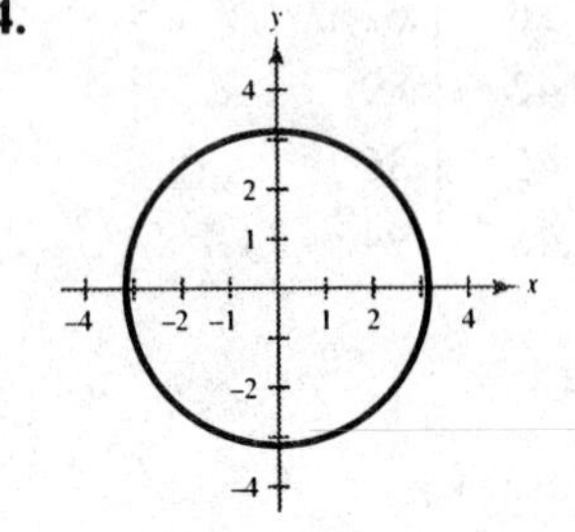

16.

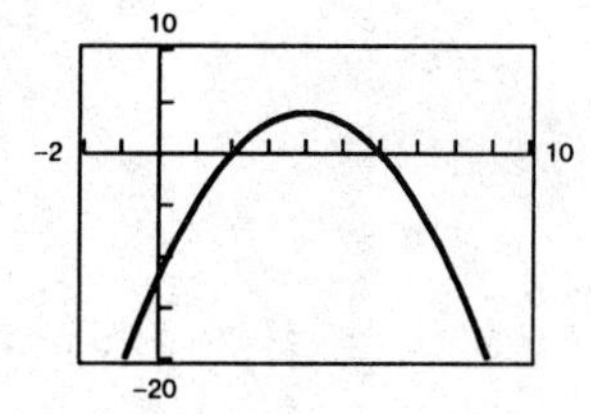

Intercepts: $(6, 0), (2, 0), (0, -12)$

18.

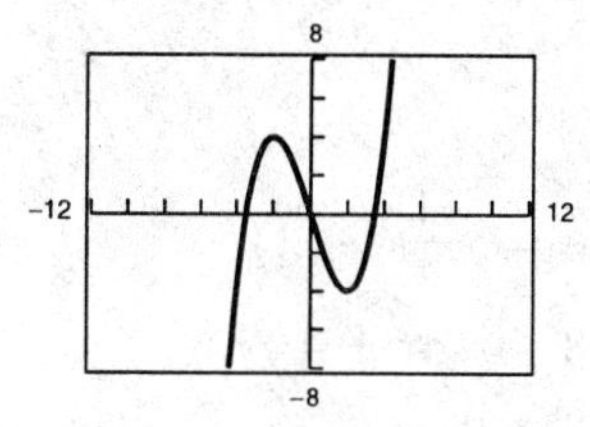

Intercepts: $(0, 0), \left(\pm 2\sqrt{3}, 0\right)$

20.

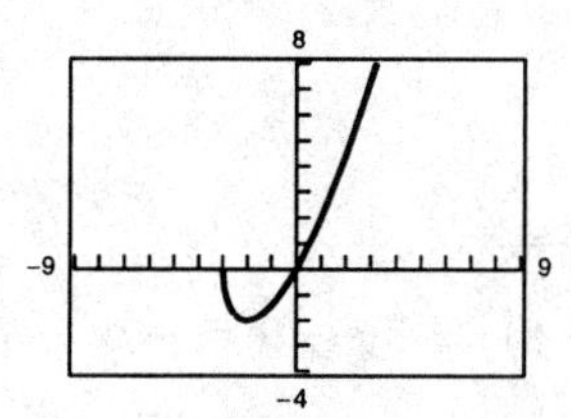

Intercepts: $(0, 0), (-3, 0)$

22.

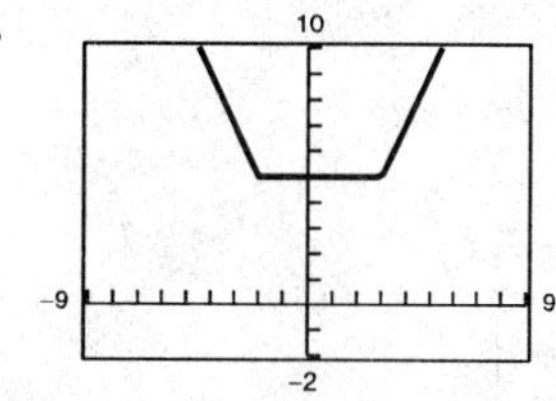

Intercept: $(0, 5)$

24. Midpoint is center of circle: $\left(\dfrac{0+4}{2}, \dfrac{0-6}{2}\right) = (2, -3)$

Radius is distance from center to one point: $r = \sqrt{(2-0)^2 + (-3-0)^2} = \sqrt{13}$

Equation: $(x-2)^2 + (y+3)^2 = 13$

26. (a)

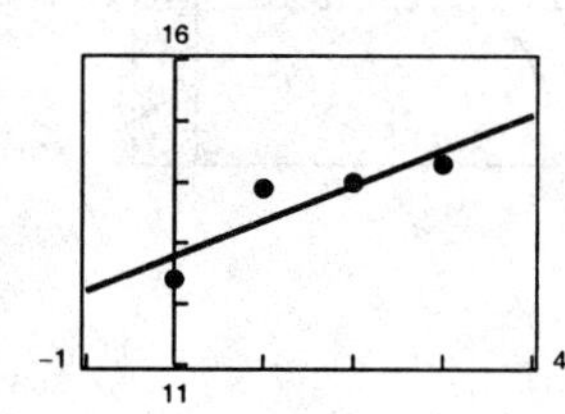

(b) $t = 8$ for 1998: $y = 0.58(8) + 12.78 = 17.42$

$t = 10$ for 2000: $y = 0.58(10) + 12.78 = 18.58$

28.

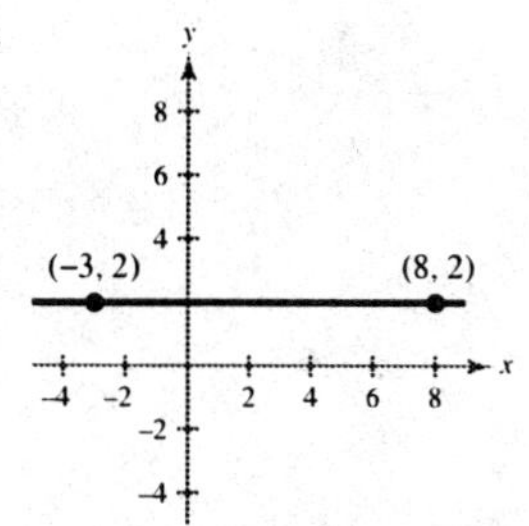

Slope $= \dfrac{2-2}{8-(-3)} = 0$

30.

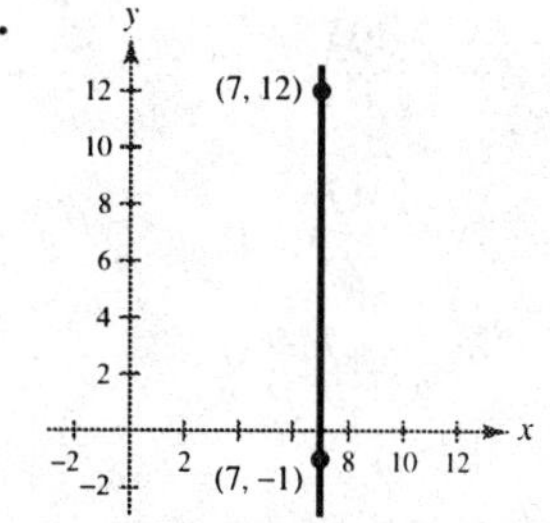

Slope $= \dfrac{12-(-1)}{7-7}$, undefined

32. $(-1, 2), (1, -1), (-5, 8)$

34. $(10, 0), (10, 2), (10, 5)$

36. Slope $= \dfrac{4-0}{-1-2} = -\dfrac{4}{3}$

Line: $y - 0 = -\dfrac{4}{3}(x-2)$

$3y + 4x - 8 = 0$

38. Slope $= \dfrac{2-(-10)}{-2-3} = -\dfrac{12}{5}$

Line: $y - (-10) = -\dfrac{12}{5}(x-3)$

$5y + 50 = -12x + 36$

$12x + 5y + 14 = 0$

40. Slope $= \dfrac{6-2}{1-4} = -\dfrac{4}{3}$

Line: $4 - 6 = -\dfrac{4}{3}(x - 1)$

$3y - 18 = -4x + 4$

$3y + 4x - 22 = 0$

42.

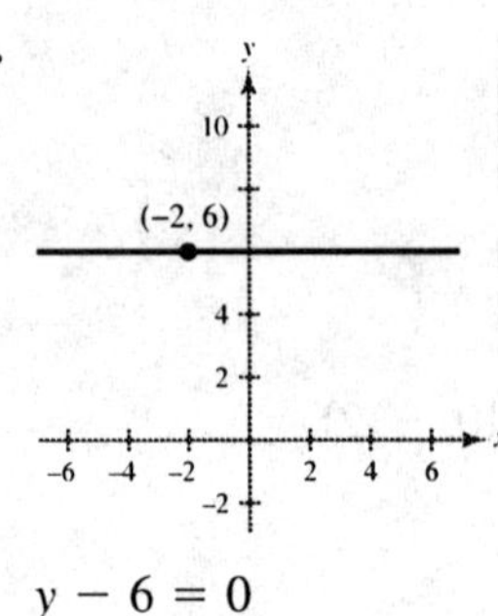

$y - 6 = 0$

44.

$x - 5 = 0$

46. Slope of given line $m = -\frac{2}{3}$

(a) $y - 3 = -\frac{2}{3}(x + 8) \Rightarrow 3y - 9 = -2x - 16$

$\Rightarrow 2x + 3y + 7 = 0$

(b) $y - 3 = \frac{3}{2}(x + 8) \Rightarrow 2y - 6 = 3x + 24$

$\Rightarrow 3x - 2y + 30 = 0$

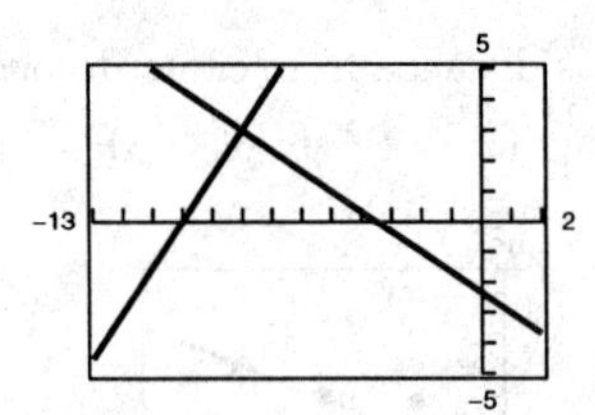

48. $y = 5.15(t - 6) + 72.95 = 5.15t + 42.05$

50.
$$\sqrt{(x-1)^2 + \left(y - \frac{7}{2}\right)^2} = \sqrt{(x-5)^2 + (y-0)^2}$$

$$(x^2 - 2x + 1) + \left(y^2 - 7y + \frac{49}{4}\right) = x^2 - 10x + 25 + y^2$$

$$8x - 7y - \frac{47}{4} = 0$$

$$32x - 28y - 47 = 0$$

52. $3(x - 2) + 2x = 2(x + 3)$

$3x - 6 + 2x = 2x + 6$

$3x = 12$

$x = 4$

Conditional statement

54. $\frac{1}{2}(x - 3) - 2(x + 1) = 5$

$\frac{1}{2}x - \frac{3}{2} - 2x - 2 = 5$

$-\frac{3}{2}x = \frac{17}{2}$

$x = -\frac{17}{3}$

56. $\dfrac{1}{x-2} = 3$

$1 = 3(x - 2)$

$1 = 3x - 6$

$7 = 3x$

$\dfrac{7}{3} = x$

58. $15 + x - 2x^2 = 0$

$(5 + 2x)(3 - x) = 0$

$5 + 2 = 0 \Rightarrow x = -\dfrac{5}{2}$

$3 - x = 0 \Rightarrow x = 3$

60. $16x^2 = 25$

$x^2 = \dfrac{25}{16}$

$x = \pm\sqrt{\dfrac{25}{16}} = \pm\dfrac{5}{4}$

62. $x^2 + 6x - 3 = 0$

$a = 1, b = 6, c = -3$

$$x = \frac{-6 \pm \sqrt{6^2 - 4(1)(-3)}}{2(1)} = \frac{-6 \pm \sqrt{48}}{2}$$

$$= -3 \pm 2\sqrt{3}$$

64. $4x^3 - 6x^2 = 0$

$$x^2(4x - 6) = 0$$
$$x^2 = 0 \Rightarrow x = 0$$
$$4x - 6 = 0 \Rightarrow x = \frac{3}{2}$$

66. $\dfrac{1}{(t+1)^2} = 1$

$$1 = (t+1)^2$$
$$0 = t^2 + 2t$$
$$0 = t(t+2)$$
$$0 = t \Rightarrow t = 0$$
$$0 = t + 2 \Rightarrow t = -2$$

68. $\sqrt{3x - 2} = 4 - x$

$$3x - 2 = (4 - x)^2$$
$$3x - 2 = 16 - 8x + x^2$$
$$0 = 18 - 11x + x^2$$
$$0 = (x - 9)(x - 2)$$
$$0 = x - 9 \Rightarrow x = 9, \text{ extraneous}$$
$$0 = x - 2 \Rightarrow x = 2$$

70. $5\sqrt{x} - \sqrt{x - 1} = 6$

$$5\sqrt{x} = 6 + \sqrt{x - 1}$$
$$25x = 36 + 12\sqrt{x - 1} + x - 1$$
$$24x - 35 = 12\sqrt{x - 1}$$
$$576x^2 - 1680x + 1225 = 144(x - 1)$$
$$576x^2 - 1824x + 1369 = 0$$
$$x = \frac{-(-1824) \pm \sqrt{(-1864)^2 - 4(576)(1369)}}{2(576)}$$
$$= \frac{1824 \pm \sqrt{172{,}800}}{1152} = \frac{1824 \pm 240\sqrt{3}}{1152}$$
$$x = \frac{38 + 5\sqrt{3}}{24}$$
$$x = \frac{38 - 5\sqrt{3}}{25}, \text{ extraneous}$$

72. $(x + 2)^{3/4} = 27$

$$x + 2 = 27^{3/4}$$
$$x + 2 = 81$$
$$x = 79$$

74. $|x^2 - 6| = x$

$$x^2 - 6 = x$$
$$x^2 - x - 6 = 0$$
$$(x - 3)(x + 2) = 0$$
$$x - 3 = 0 \Rightarrow x = 3$$
$$x + 2 = 0 \Rightarrow x = -2, \text{ extraneous}$$

or

$$-(x^2 - 6) = x$$
$$x^2 + x - 6 = 0$$
$$(x + 3)(x + 2) = 0$$
$$x - 2 = 0 \Rightarrow x = 2$$
$$x + 3 = 0 \Rightarrow x = -3, \text{ extraneous}$$

76. $y = 12x^3 - 84x^2 + 120x$

$0 = 12x^3 - 84x^2 + 120x$

$0 = 12x(x^2 - 7x + 10)$

$0 = 12x(x - 5)(x - 2)$

$0 = 12x \Rightarrow x = 0$

$0 = x - 5 \Rightarrow x = 5$

$0 = x - 2 \Rightarrow x = 2$

x-intercepts: $(0, 0)$, $(5, 0)$, $(2, 0)$

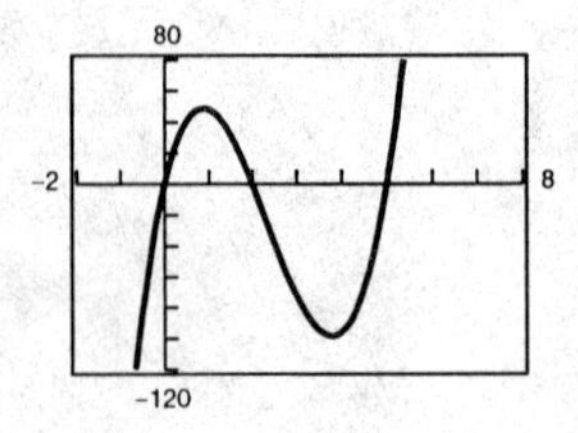

78. $y = \dfrac{4}{x - 3} - \dfrac{4}{x} - 1$

$0 = \dfrac{4}{x - 3} - \dfrac{4}{x} - 1$

$0 = 4x - 4(x - 3) - x(x - 3)$

$0 = 4x - 4x + 12 - x^2 + 3x$

$0 = x^2 - 3x - 12$

$$x = \frac{-(-3) \pm \sqrt{(-3)^2 - 4(1)(-12)}}{2(1)} = \frac{3 \pm \sqrt{57}}{2} \approx 5.275, -2.275$$

80. $y = |2x - 3| - 5$

Graphically, $x = -1, 4$.

Analytically, $|2x - 3| - 5 = 0$.

$2x - 3 = 5$ or $-(2x - 3) = 5$

$2x = 8$ $\quad$ $-2x = 2$

$x = 4$ $\quad$ $x = -1$

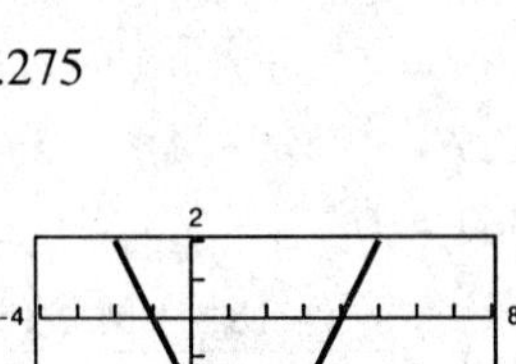

82. $Z = \sqrt{R^2 - X^2}$

$Z^2 = R^2 - X^2$

$X^2 = R^2 - Z^2$

$X = \pm\sqrt{R^2 - Z^2}$

84. $E = 2kw\left(\dfrac{v}{2}\right)^2$

$\dfrac{E}{2kw} = \left(\dfrac{v}{2}\right)^2$

$\pm\sqrt{\dfrac{E}{2kw}} = \dfrac{v}{2}$

$\pm\sqrt{\dfrac{E}{2kw}} = v$

$\pm\sqrt{\dfrac{4E}{2kw}} = v$

$\pm\sqrt{\dfrac{2E}{kw}} = v$

86. $x^2 - 2x \geq 3$

$x^2 - 2x - 3 \geq 0$

$(x - 3)(x + 1) \geq 0$

Test intervals: $(-\infty, -1)$, $(-1, 3)$, $(3, \infty)$

$x \geq 3$ or $x \leq -1$

$(-\infty, -1] \cup [3, \infty)$

88. $\dfrac{2}{x+1} \le \dfrac{3}{x-1}$

$\dfrac{2(x-1) - 3(x+1)}{(x+1)(x-1)} \le 0$

$\dfrac{2x - 2 - 3x - 3}{(x+1)(x-1)} \le 0$

$\dfrac{-(x+5)}{(x+1)(x-1)} \le 0$

Test Intervals: $(-\infty, -5) \Rightarrow \dfrac{-(x+5)}{(x+1)(x-1)} > 0$

$(-5, -1) \Rightarrow \dfrac{-(x+5)}{(x+1)(x-1)} < 0$

$(-1, 1) \Rightarrow \dfrac{-(x+5)}{(x+1)(x-1)} > 0$

$(1, -\infty) \Rightarrow \dfrac{-(x+5)}{(x+1)(x-1)} < 0$

Solution intervals: $[-5, -1) \cup (1, \infty)$

90. $|x| \le 4$

$-4 \le x \le 4$

$[-4, 4]$

92. $|x - 3| > 4$

$x - 3 > 4$ or $x - 3 < -4$

$x > 7$ $\qquad x < -1$

$(-\infty, -1) \cup (7, \infty)$

94. $2x^2 + x \ge 15$

$2x^2 + x - 15 \ge 0$

$(2x - 5)(x + 3) \ge 0$

Test intervals: $(-\infty, -3) \Rightarrow (2x - 5)(x + 3) > 0$

$\left(-3, \frac{5}{2}\right) \Rightarrow (2x - 5)(x + 3) < 0$

$\left(\frac{5}{2}, \infty\right) \Rightarrow (2x - 5)(x + 3) > 0$

Solution interval: $(-\infty, -3] \cup \left[\frac{5}{2}, \infty\right)$

96. $|x(x - 6)| < 5$

$x(x - 6) < 5$ or $x(x - 6) > -5$

$x^2 - 6x - 5 < 0$ $\qquad x^2 - 6x + 5 > 0$

Critical numbers: $-0.74, 6.74, 1, 5$

Test intervals: $(-\infty, -0.74) \Rightarrow |x(x - 6)| > 5$

$(-0.74, 1) \Rightarrow |x(x - 6)| < 5$

$(1, 5) \Rightarrow |x(x - 6)| > 5$

$(5, 6.74) \Rightarrow |x(x - 6)| < 5$

$(6.74, \infty) \Rightarrow |x(x - 6)| > 5$

Solution interval: $(-0.74, 1) \cup (5, 6.74)$

98. $M = 500x(20 - x)$

(a)

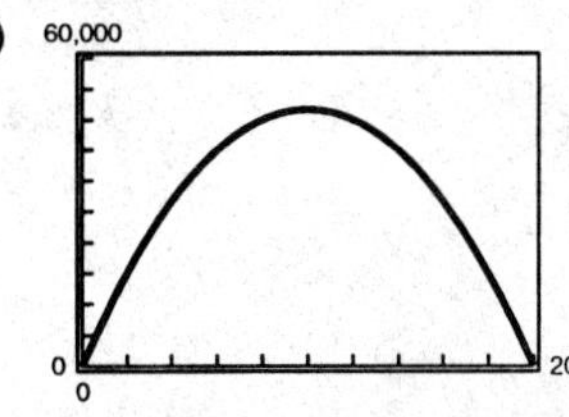

(b) $M = 0$ for $x = 0$ and $x = 20$

(c) $M = 50{,}000$ is greatest for $x = 10$

(d) $M = 500x(20 - x) < 40{,}000$

$x(20 - x) < 80$

$0 < x^2 - 20x + 80$

The zeros of $x^2 - 20x + 80$ are $\dfrac{20 \pm \sqrt{(-20)^2 - 4(80)}}{2} = 10 \pm \sqrt{20}.$

Intervals: $0 \le x < 10 - \sqrt{20}, \quad 10 + \sqrt{20} < x \le 20$

$0 \le x < 10 - 2\sqrt{5}, \quad 10 + 2\sqrt{5} < x \le 20$

CHAPTER 1
Functions and Their Graphs

CHAPTER 1
Functions and Their Graphs

Section 1.1 Functions

Solutions to Even-Numbered Exercises

2. No, it is not a function.
The domain value of -1 is matched with two output values.

4. Yes, it is a function.
Each domain value is matched with only one range value.

6. No, the table does not represent a function.
The input values of 0 and 1 are each matched with two different output values.

8. Yes, the table does represent a function.
Each input value is matched with only one output value.

10. (a) The element c in A is matched with two elements, 2 and 3 of B, so it is not a function.
(b) Each element of A is matched with exactly one element of B, so it does represent a function.
(c) This is not a function from A to B (it represents a function from B to A instead).

12. Reading from the graph, $f(1988)$ is approximately 22 million.

14. $x = y^2 \implies y = \pm\sqrt{x}$
Thus, y *is not* a function of x.

16. $x + y^2 = 4 \Rightarrow y = \pm\sqrt{4 - x}$

Thus, y *is not* a function of x.

18. $(x - 2)^2 + y^2 \Rightarrow y = \pm\sqrt{4 - (x - 2)^2}$

Thus, y *is not* a function of x.

20. $y = \sqrt{x + 5}$

This is a function of x.

22. $|y| = 4 - x \Rightarrow y = 4 - x$ or $y = -(4 - x)$

Thus, y *is not* a function of x.

24. $g(x) = x^2 - 2x$

(a) $g(2) = (2)^2 - 2(2) = 0$

(b) $g(-3) = (-3)^2 - 2(-3) = 15$

(c) $g(t + 1) = (t + 1)^2 - 2(t + 1) = t^2 - 1$

(d) $g(x + c) = (x + c)^2 - 2(x + c)$

$= x^2 + 2cx + c^2 - 2x - 2c$

26. $g(y) = 7 - 3y$

(a) $g(0) = 7 - 3(0) = 7$

(b) $g(\frac{7}{3}) = 7 - 3(\frac{7}{3}) = 0$

(c) $g(s + 2) = 7 - 3(s + 2)$

$= 7 - 3s - 6 = 1 - 3s$

28. $V(r) = \dfrac{4}{3}\pi r^3$

(a) $V(3) = \dfrac{4}{3}\pi(3)^3 = 36\pi$

(b) $V\left(\dfrac{3}{2}\right) = \dfrac{4}{3}\pi\left(\dfrac{3}{2}\right)^3 = \dfrac{4}{3} \cdot \dfrac{27}{8}\pi = \dfrac{9\pi}{2}$

(c) $V(2r) = \dfrac{4}{3}\pi(2r)^3 = \dfrac{32\pi r^3}{3}$

30. $f(x) = \sqrt{x + 8} + 2$

(a) $f(-8) = \sqrt{(-8) + 8} + 2 = 2$

(b) $f(1) = \sqrt{(1) + 8} + 2 = 5$

(c) $f(x - 8) = \sqrt{(x - 8) + 8} + 2 = \sqrt{x} + 2$

32. $q(t) = \dfrac{2t^2 + 3}{t^2}$

(a) $q(2) = \dfrac{2(2)^2 + 3}{(2)^2} = \dfrac{8 + 3}{4} = \dfrac{11}{4}$

(b) $q(0) = \dfrac{2(0)^2 + 3}{(0)^2}$ Division by zero is undefined.

(c) $q(-x) = \dfrac{2(-x)^2 + 3}{(-x)^2} = \dfrac{2x^2 + 3}{x^2}$

34. $f(x) = |x| + 4$

(a) $f(2) = |2| + 4 = 6$

(b) $f(-2) = |-2| + 4 = 6$

(c) $f(x^2) = |x^2| + 4 = x^2 + 4$

36. $f(x) = \begin{cases} x^2 + 2, & x \le 1 \\ 2x^2 + 2, & x > 1 \end{cases}$

(a) $f(-2) = (-2)^2 + 2 = 6$

(b) $f(1) = (1)^2 + 2 = 3$

(c) $f(2) = 2(2)^2 + 2 = 10$

38. $g(x) = \sqrt{x - 3}$

$g(3) = \sqrt{3 - 3} = 0$

$g(4) = \sqrt{4 - 3} = 1$

$g(5) = \sqrt{5 - 3} = \sqrt{2}$

$g(6) = \sqrt{6 - 3} = \sqrt{3}$

$g(7) = \sqrt{7 - 3} = 2$

x	3	4	5	6	7
$g(x)$	0	1	$\sqrt{2}$	$\sqrt{3}$	2

40. $f(s) = \dfrac{|s - 2|}{s - 2}$

$f(0) = \dfrac{|0 - 2|}{0 - 2} = \dfrac{2}{-2} = -1$

$f(1) = \dfrac{|1 - 2|}{1 - 2} = \dfrac{1}{-1} = -1$

$f\left(\dfrac{3}{2}\right) = \dfrac{|\frac{3}{2} - 2|}{\frac{3}{2} - 2} = \dfrac{\frac{1}{2}}{-\frac{1}{2}} = -1$

$f\left(\dfrac{5}{2}\right) = \dfrac{|\frac{5}{2} - 2|}{\frac{5}{2} - 2} = \dfrac{\frac{1}{2}}{\frac{1}{2}} = 1$

$f(4) = \dfrac{|4 - 2|}{4 - 2} = \dfrac{2}{2} = 1$

s	0	1	$\frac{3}{2}$	$\frac{5}{2}$	4
$f(s)$	-1	-1	-1	1	1

42. $h(x) = \begin{cases} 9 - x^2, & x < 3 \\ x - 3, & x \geq 3 \end{cases}$

$h(1) = 9 - (1)^2 = 8$

$h(2) = 9 - (2)^2 = 5$

$h(3) = (3) - 3 = 0$

$h(4) = (4) - 3 = 1$

$h(5) = (5) - 3 = 2$

s	1	2	3	4	5
$h(x)$	8	5	0	1	2

44. $f(x) = 0$

$\frac{3x - 4}{5} = 0$

$3x - 4 = 0$

$3x = 4$

$x = \frac{4}{3}$

46. $f(x) = 0$

$x^3 - x = 0$

$x(x^2 = 1) = 0$

$x(x + 1)(x - 1) = 0$

$x = 0$

$x + 1 = 0 \Rightarrow x = -1$

$x - 1 = 0 \Rightarrow x = 1$

48. $f(x) = g(x)$

$x^2 + 2x + 1 = 3x + 3$

$x^2 - x - 2 = 0$

$(x - 2)(x + 1) = 0$

$x - 2 = 0 \Rightarrow x = 2$

$x + 1 = 0 \Rightarrow x = -1$

50. $f(x) = g(x)$

$x^4 - 2x^2 = 2x^2$

$x^4 - 4x^2 = 0$

$x^2(x^2 - 4) = 0$

$x^2(x + 2)(x - 2) = 0$

$x^2 = 0 \Rightarrow x = 0$

$x + 2 = 0 \Rightarrow x = -2$

$x - 2 = 0 \Rightarrow x = 2$

52. $f(x) = 1 - 2x^2$

Because $f(x)$ is a polynomial, the domain is all real numbers x.

54. $s(y) = \frac{3y}{y + 5}$

$y + 5 \neq 0$

$y \neq -5$

The domain is all real numbers $y \neq 5$.

56. $f(t) = \sqrt[3]{t + 4}$

Because $f(t)$ is a cube root, the domain is all real numbers t.

58. $h(x) = \frac{10}{x^2 - 2x}$

$x^2 - 2x \neq 0$

$x(x - 2) \neq 0$

The domain is all real numbers $x \neq 0$ and $x \neq 2$.

60. $f(s) = \frac{\sqrt{s - 1}}{s - 4}$

$s - 1 \geq 0$ and $s - 4 \neq 0$

$s \geq 1$ and $s \neq 4$

The domain is $s \geq 1, s \neq 4$.

62. $f(x) = \frac{2x}{x^2 + 1}$

$\left\{\left(-2, -\frac{4}{5}\right), (-1, -1), (0, 0), (1, 1), \left(2, \frac{4}{5}\right)\right\}$

64. $f(x) = |x + 1|$

$\{(-2, 1), (-1, 0), (0, 1), (1, 2), (2, 3)\}$

66. An advantage to function notation is that it gives a name to the relationship so it can be easily referenced. When evaluating a function, you see both the input and output values.

68. By plotting the data, you can see that it represents a line, or $f(x) = cx$. Because $(0, 0)$ and $\left(1, \frac{1}{4}\right)$ are on the line, the slope is $\frac{1}{4}$. Thus, $f(x) = \frac{1}{4}x$.

70. By plotting the data, you can see that it represents $h(x) = c\sqrt{|x|}$. Because $\sqrt{|-4|} = 2$ and $\sqrt{|-1|} = 1$ but the corresponding y values are 6 and 3. Thus, $c = 3$ and $h(x) = 3\sqrt{|x|}$.

72.
$$\begin{aligned} f(x) &= 5x - x^2 \\ f(5+h) &= 5(5+h) - (5+h)^2 \\ &= 25 + 5h - (25 + 10h + h^2) \\ &= 25 + 5h - 25 - 10h - h^2 \\ &= -h^2 - 5h \\ f(5) &= 5(5) - (5)^2 \\ &= 25 - 25 = 0 \\ \frac{f(5+h) - f(5)}{h} &= \frac{-h^2 - 5h}{h} \\ &= \frac{-h(h+5)}{h} = -(h+5),\ h \neq 0 \end{aligned}$$

74.
$$\begin{aligned} f(x) &= 2x \\ f(x+c) &= 2(x+c) = 2x + 2c \\ f(x) &= 2x \\ \frac{f(x+c) - f(x)}{c} &= \frac{2x + 2c - 2x}{c} = 2 \end{aligned}$$

76.
$$\begin{aligned} f(t) &= \frac{1}{t} \\ f(1) &= \frac{1}{1} = 1 \end{aligned}$$
$$\frac{f(t) - f(1)}{t - 1} = \frac{\frac{1}{t} - 1}{t - 1} = \frac{\frac{1}{t} - \frac{t}{t}}{t - 1} = \frac{\frac{1 - t}{t}}{t - 1} = \frac{\left(-\frac{1}{t}\right)(t - 1)}{t - 1} = -\frac{1}{t},\ t \neq 1$$

78. $A = \frac{1}{2}bh$, in an equilateral triangle $b = s$ and:
$$s^2 = h^2 + \left(\frac{s}{2}\right)^2$$
$$h = \sqrt{s^2 - \left(\frac{s}{2}\right)^2}$$
$$h = \sqrt{\frac{4s^2}{4} - \frac{s^2}{4}} = \frac{\sqrt{3}s}{2}$$
$$A = \frac{1}{2}s \cdot \frac{\sqrt{3}s}{2} = \frac{\sqrt{3}s^2}{4}$$

80. (a)

Units x	*Price*	*Profit P*
102	$90 - 2(0.15)$	$102[90 - 2(0.15)] - 102(60) = 3029.40$
104	$90 - 4(0.15)$	$104[90 - 4(0.15)] - 104(60) = 3057.60$
106	$90 - 6(0.15)$	$106[90 - 6(0.15)] - 106(60) = 3084.60$
108	$90 - 8(0.15)$	$108[90 - 8(0.15)] - 108(60) = 3110.40$
110	$90 - 10(0.15)$	$110[90 - 10(0.15)] - 110(60) = 3135.00$
112	$90 - 12(0.15)$	$112[90 - 112(0.15)] - 12(60) = 3158.40$

(b)

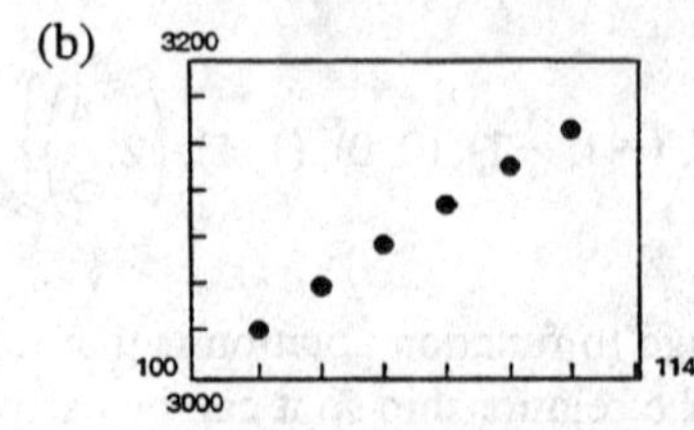

Yes, P is a function.

$$\begin{aligned} \text{Profit} &= \text{Revenue} - \text{Cost} \\ &= (\text{price per unit})(\text{number of units}) - (\text{cost})(\text{number of units}) \\ &= [90 - (x - 100)(0.15)]x - 60x,\ x > 100 \\ &= (90 - 0.15x + 15)x - 60x \\ &= (105 - 0.15x)x - 60x \\ &= 105x - 0.15x^2 - 60x \\ &= 45x - 0.15x^2,\ x > 100 \end{aligned}$$

82. $A = l \cdot w = (2x)y = 2xy$

But $y = \sqrt{36 - x^2}$, so $A = 2x\sqrt{36 - x^2},\ 0 < x < 6$.

84.

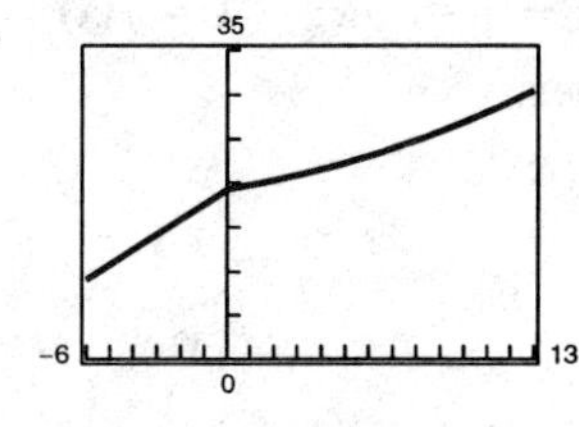

In 1978: $p(-2) = 19.247 + 1.694(-2) = 15.859$

The average mobile home price in 1978 was $15,859.

In 1988: $p(8) = 19.305 + 0.427(8) + 0.033(8)^2 = 24.833$

The average mobile home price in 1988 was $24,833.

In 1993: $p(13) = 19.305 + 0.427(13) + 0.033(13)^2 = 30.433$

The average mobile home price in 1993 was $30,433.

86. (a) $R = (\text{Rate})(\text{number of people}) = [8 - 0.05(n - 80)]n = [12 - 0.05n]n$

(b)

n	90	100	110	120	130	140	150
$R(n)$	$675	$700	$715	$720	$715	$700	$675

The revenue is maximum when $n = 120$.

(c)

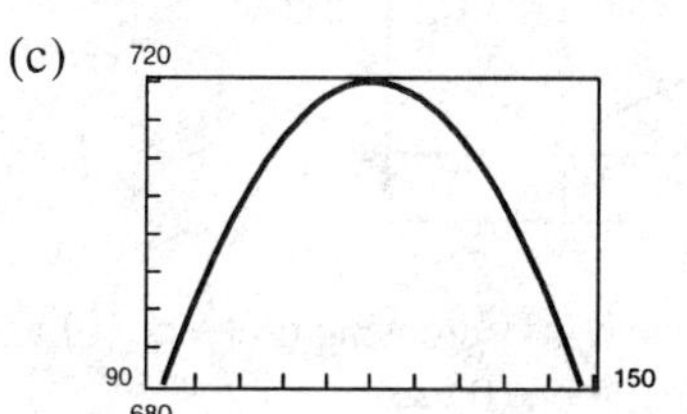

$n = 120$ yields a maximum revenue of $720.

Section 1.2 Graphs of Functions

Solutions to Even-Numbered Exercises

2. $f(x) = \frac{1}{2}|x - 2|$

Domain: $(-\infty, \infty)$

Range: $[0, \infty)$

4. $g(x) = \dfrac{|x - 1|}{x - 1}$

Domain: $x - 1 \neq 0$

$x \neq 1$

$(-\infty, 1), (1, \infty)$

Range: $-1, 1$

6. $f(x) = \sqrt{x - 1}$

Domain: $x - 1 \geq 0$

$x \geq 1$

$[1, \infty)$

Range: $[0, \infty)$

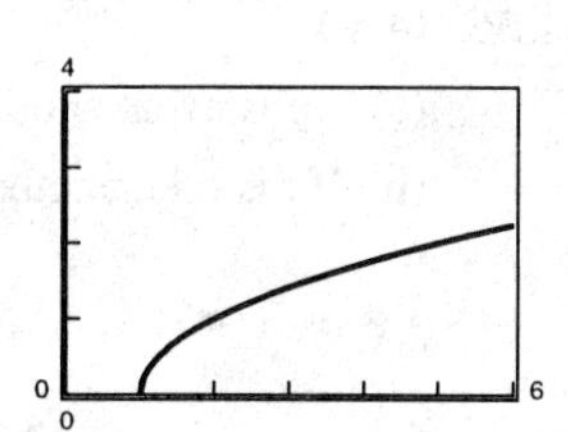

8. $h(t) = \sqrt{4 - t^2}$

$4 - t^2 \geq 0 \Rightarrow t^2 \leq 4$

Domain: $[-2, 2]$

Range: $[0, 2]$

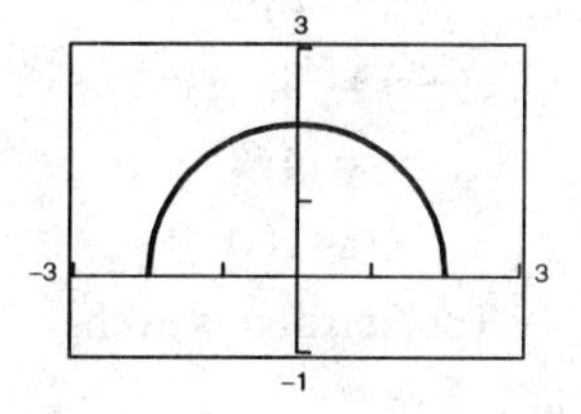

10. $y = \frac{1}{4}x^3$

A vertical line intersects the graph no more than once, so y is a function of x.

12. $x^2 + y^2 = 25$

A vertical line intersects the graph more than once, so y is not a function of x.

14. $x = |y + 2|$

A vertical line intersects the graph more than once, so y is not a function of x.

16. $f(x) = 6[x - (0.1x)^5]$

The third setting shows the most complete graph.

18. $f(x) = 10x\sqrt{400 - x^2}$

The third setting shows the most complete graph.

20. $f(x) = x^2 - 4x$

(a) The graph is decreasing on $(-\infty, 2)$ and increasing on $(2, \infty)$.

(b) $f(-x) = (-x)^2 - 4(-x) = x^2 + 4x$

$x^2 + 4x \neq f(x)$

$x^2 + 4x \neq -f(x)$

The function is neither odd nor even.

22. $f(x) = \sqrt{x^2 - 1}$

(a) The graph is decreasing on $(-\infty, -1)$ and increasing on $(1, \infty)$.

(b) $f(-x) = \sqrt{(-x)^2 - 1} = \sqrt{x^2 - 1} = f(x)$

The function is even.

24. No, for some values of y there corresponds more than one value of x.

26. $f(x) = x^{2/3}$

(a)

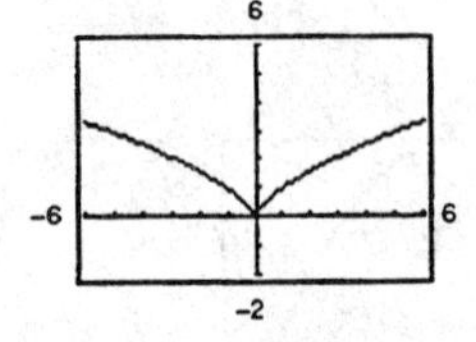

(b) The graph is decreasing on $(-\infty, 0)$ and increasing on $(0, \infty)$.

(c) $f(-x) = (-x)^{2/3} = x^{2/3} = f(x)$

The function is even.

28. $f(x) = |x + 1| + |x - 1|$

(a)

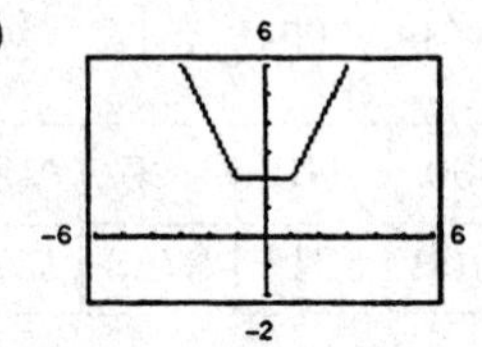

(b) The graph is decreasing on $(-\infty, -1)$, constant on $(-1, 1)$, and increasing on $(1, \infty)$.

(c) $f(-x) = |-x + 1| + |x - 1|$

$= |(-1)(x - 1)| + |(-1)(x + 1)|$

$= |x - 1| + |x + 1|$

$= f(x)$

The function is even.

30.
$$\begin{aligned} h(x) &= x^3 - 5 \\ h(-x) &= (-x)^3 - 5 \\ &= -x^3 - 5 \\ &\neq h(x) \\ &\neq -h(x) \end{aligned}$$

The function is neither odd nor even.

32.
$$\begin{aligned} f(x) &= x\sqrt{1 - x^2} \\ f(-x) &= -x\sqrt{1 - (-x)^2} \\ &= -x\sqrt{1 - x^2} \\ &= -f(x) \end{aligned}$$

The function is odd.

34.
$$\begin{aligned} g(s) &= 4s^{2/3} \\ g(-s) &= 4(-s)^{2/3} \\ &= 4s^{2/3} \\ &= g(s) \end{aligned}$$

The function is even.

36. $(4, 9)$

(a) If f is even, another point is $(-4, 9)$.

(b) If f is odd, another point is $(-4, -9)$.

38. $g(x) = x$

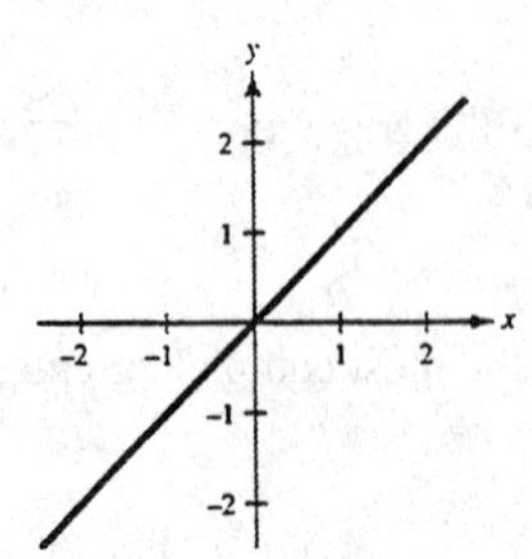

$g(-x) = -x = -g(x)$

The function is odd.

40. $h(x) = x^2 - 4$

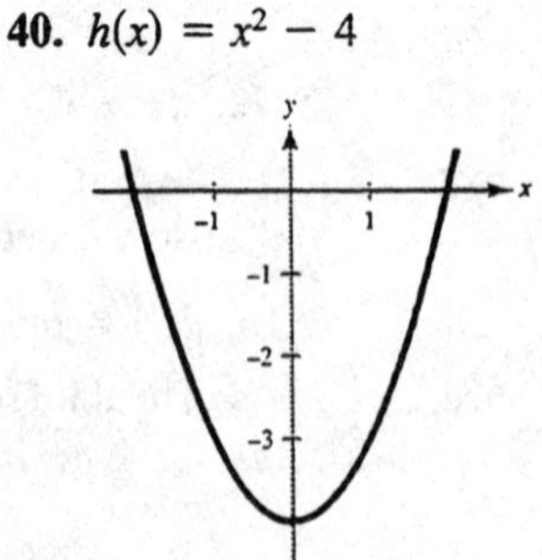

$h(-x) = (-x)^2 - 4$

$= x^2 - 4 = h(x)$

The function is even.

42. $f(x) = x^{3/2}$

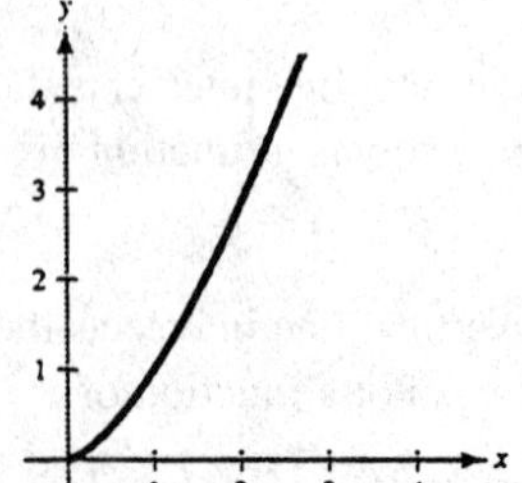

The graph is neither odd nor even.

44. $f(x) = |x + 2|$

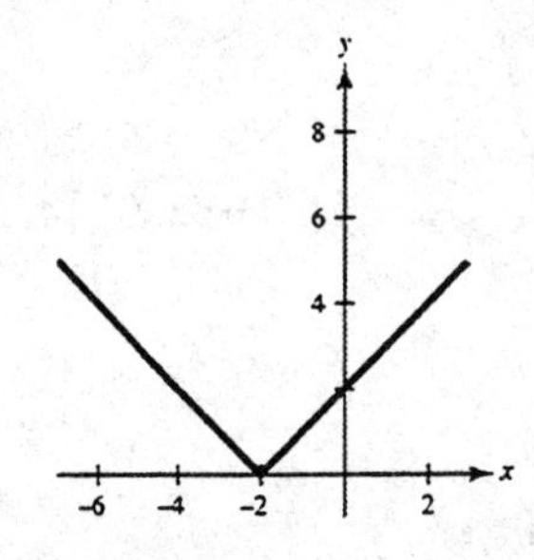

The graph is neither odd nor even.

46. $f(x) = \begin{cases} 2x + 1, & x \le -1 \\ x^2 - 2, & x > -1 \end{cases}$

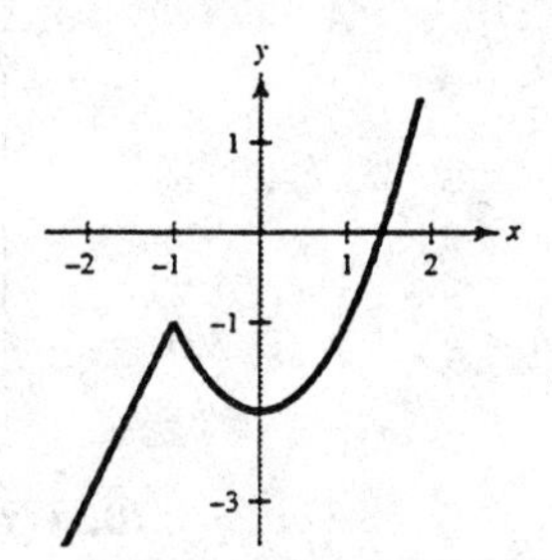

The graph is neither odd nor even.

48. $f(x) = \begin{cases} \sqrt{4 + x}, & x < 0 \\ \sqrt{4 - x}, & x \ge 0 \end{cases}$

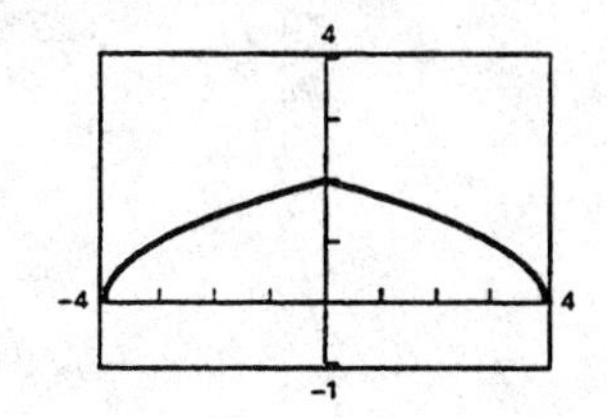

50. $f(x) = \begin{cases} 1 - (x - 1)^2, & x \le 2 \\ \sqrt{x - 2}, & x > 2 \end{cases}$

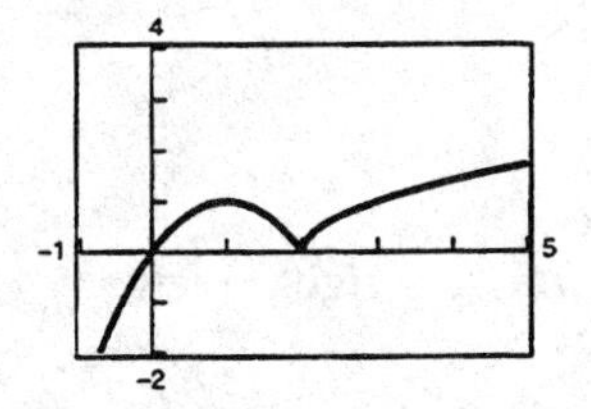

52.

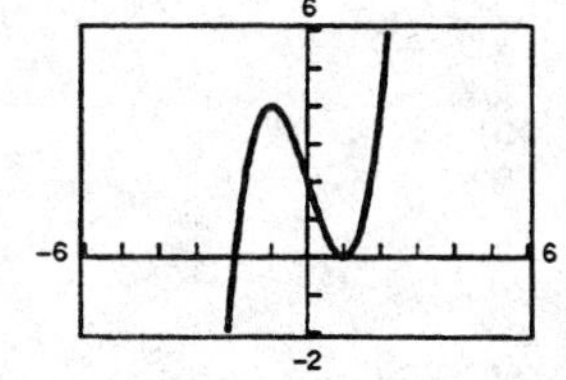

Relative minimum: $(1, 0)$
Relative maximum: $(-1, 4)$

54.

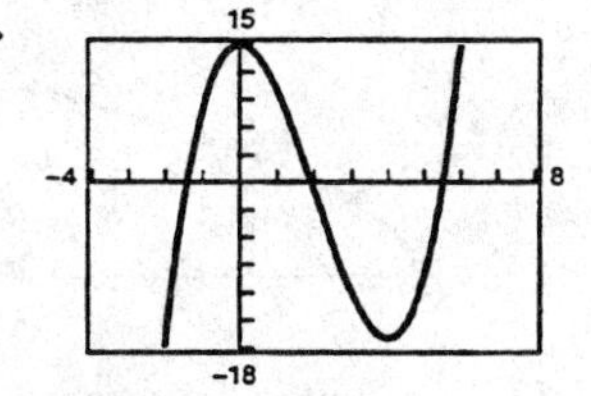

Relative minimum: $(4, -17)$
Relative maximum: $(0, 15)$

56.

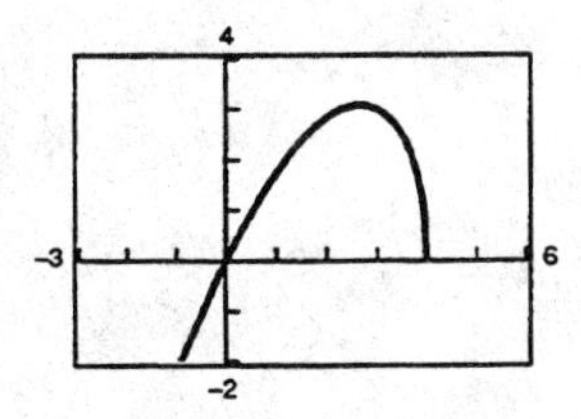

Maximum: $(2.67, 3.08)$

58. $p = 100 - 0.0001x$

$C = 350{,}000 + 30x$

$P = R - C = xp - C = x(100 - 0.0001x) - 350{,}000 - 30x$

$= x(100 - 0.0001x) - 350{,}000 - 30x$

$= -0.0001x^2 + 70x - 350{,}000$

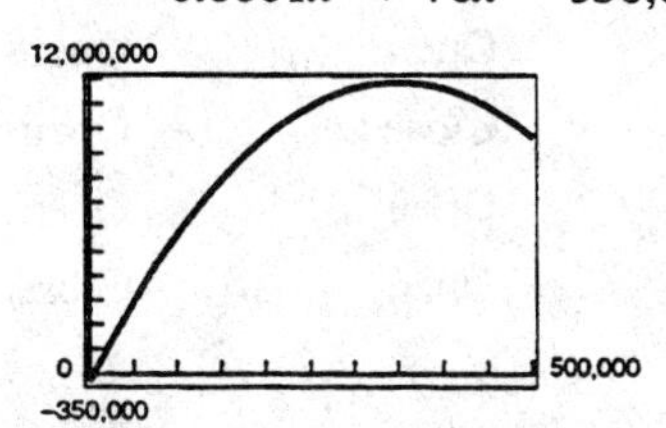

Maximum at 350,000 units

60. $g(x) = 2\left(\frac{1}{4}x - [\![\frac{1}{4}x]\!]\right)^2$

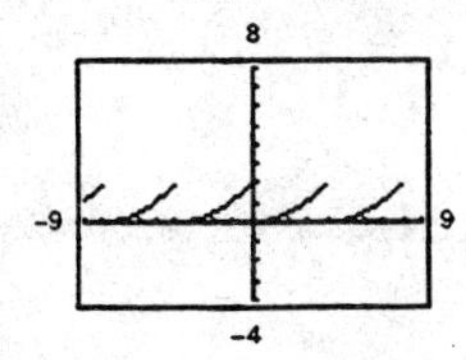

Domain: $(-\infty, \infty)$
Range: $[0, 2)$
Pattern: Sawtooth

62. *Model*: (Total cost) = (Flat rate) + (Rate per pound)

Labels: Total cost = C
Flat rate = 9.80
Rate per pound = $2.50[\![x]\!],\ x > 0$

Equation: $C = 9.80 + 2.50[\![x]\!],\ x > 0$

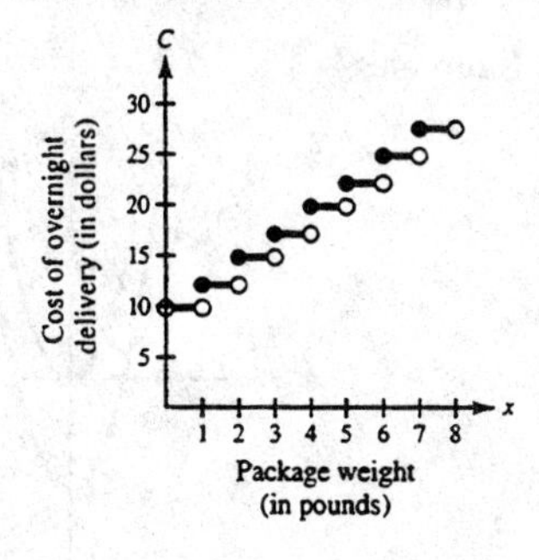

64. $f(x) = 4x + 2$

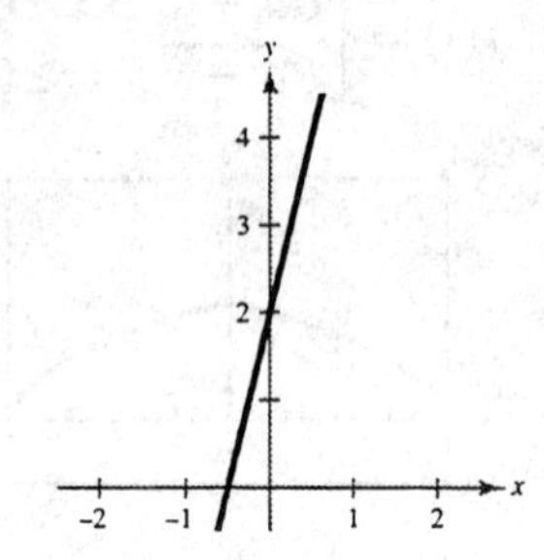

$$f(x) \geq 0$$
$$4x + 2 \geq 0$$
$$4x \geq -2$$
$$x \geq -\frac{1}{2}$$
$$\left[-\tfrac{1}{2}, \infty\right)$$

66. $f(x) = x^2 - 4x$

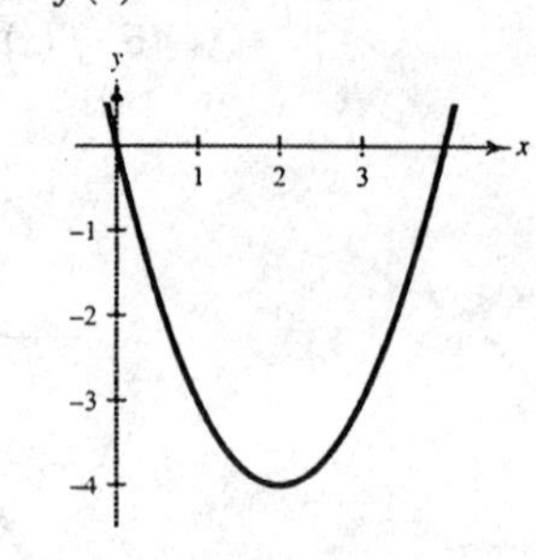

$$f(x) \geq 0$$
$$x^2 - 4x \geq 0$$
$$x(x - 4) \geq 0$$
$$(-\infty, 0], [4, \infty)$$

68. $f(x) = \sqrt{x + 2}$

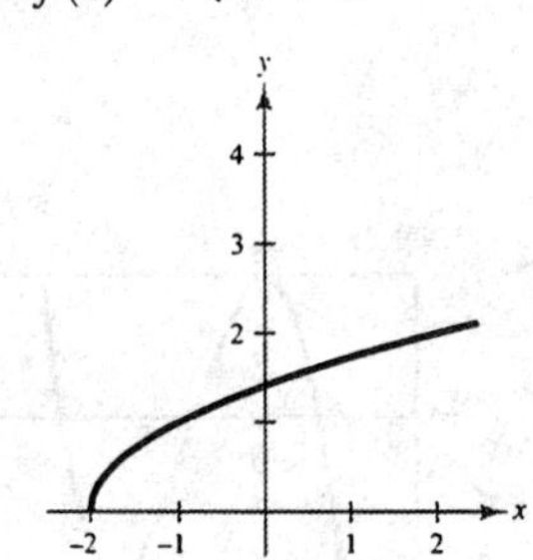

$$f(x) \geq 0$$
$$\sqrt{x + 2} \geq 0$$
$$x + 2 \geq 0$$
$$x \geq -2$$
$$[-2, \infty)$$

70. $f(x) = -(1 + |x|)$

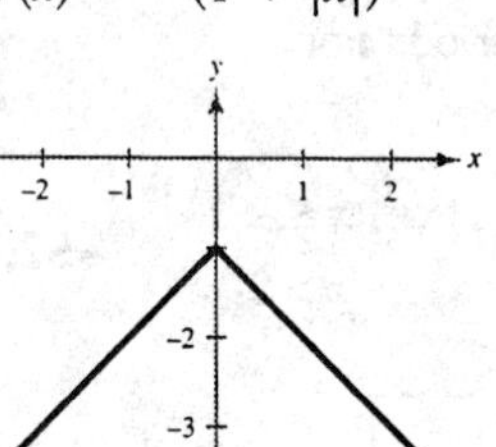

$f(x)$ is never greater than 0. ($f(x) < 0$ for all x.)

72. $h = \text{top} - \text{bottom}$
$= 3 - (4x - x^2)$
$= 3 - 4x + x^2$

74. $h = \text{top} - \text{bottom}$
$= 2 - \sqrt[3]{x}$

76. $L = \text{right} - \text{left}$
$= 2 - \sqrt[3]{2y}$

78.

Interval	*Intake Pipe*	*Drainpipe 1*	*Drainpipe 2*
[0, 5]	Open	Closed	Closed
[5, 10]	Open	Open	Closed
[10, 20]	Closed	Closed	Closed
[20, 30]	Closed	Closed	Open
[30, 40]	Open	Open	Open
[40, 45]	Open	Closed	Open
[45, 50]	Open	Open	Open
[50, 60]	Open	Open	Closed

80. $f(x) = a_{2n}x^{2n} + a_{2n-2}x^{2n-2} + \cdots + a_2x^2 + a_0$

$f(-x) = a_{2n}(-x)^{2n} + a_{2n-2}(-x)^{2n-2} + \cdots + a_2(-x)^2 + a_0$

$= a_{2n}x^{2n} + a_{2n-2}x^{2n-2} + \cdots + a_2x^2 + a_0 = f(x)$

$f(-x) = f(x)$; thus, $f(x)$ is even.

Section 1.3 Shifting, Reflecting, and Stretching Graphs

Solutions to Even-Numbered Exercises

2.

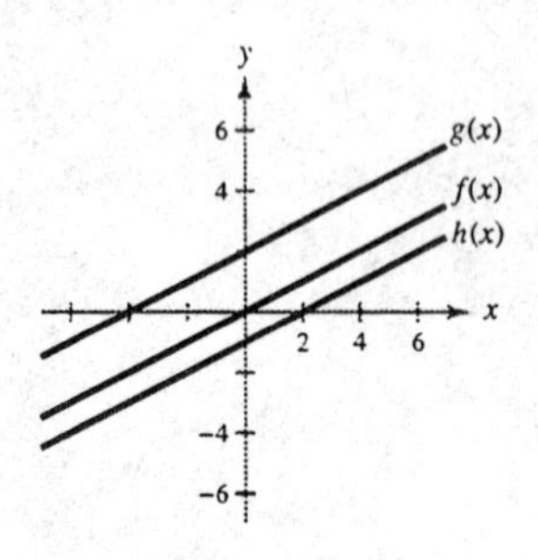

4.

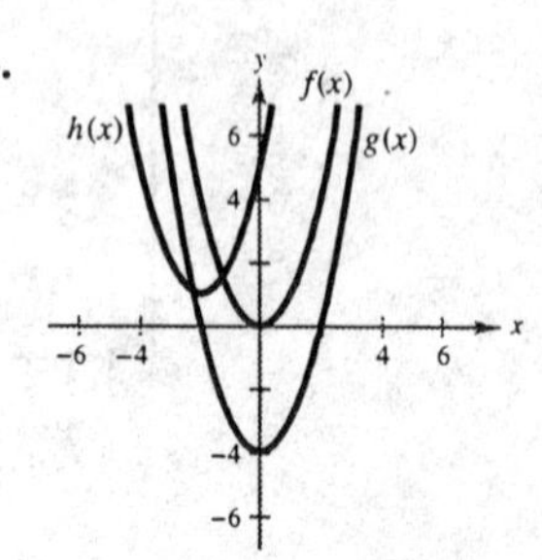

6.

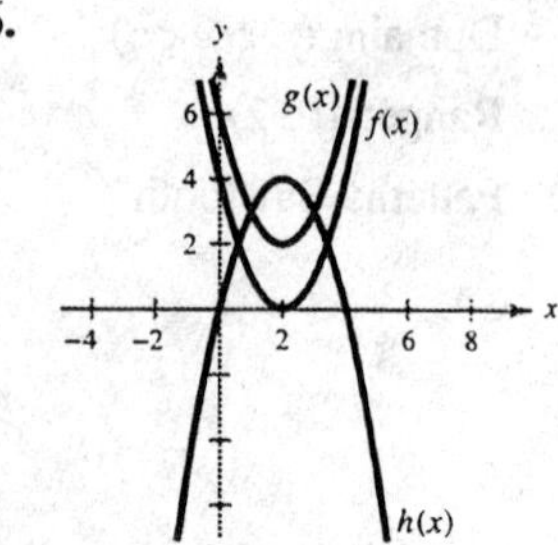

8.

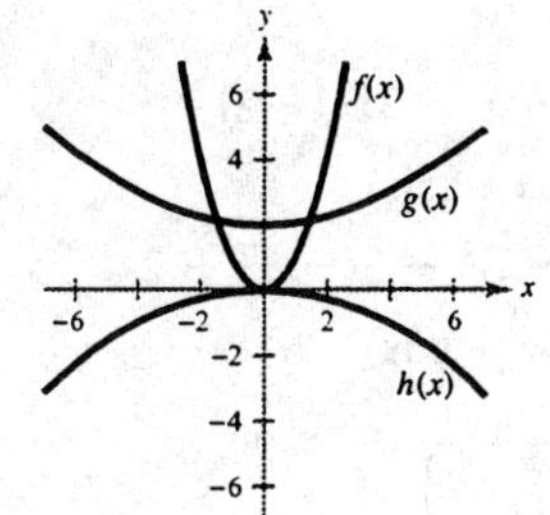

10.

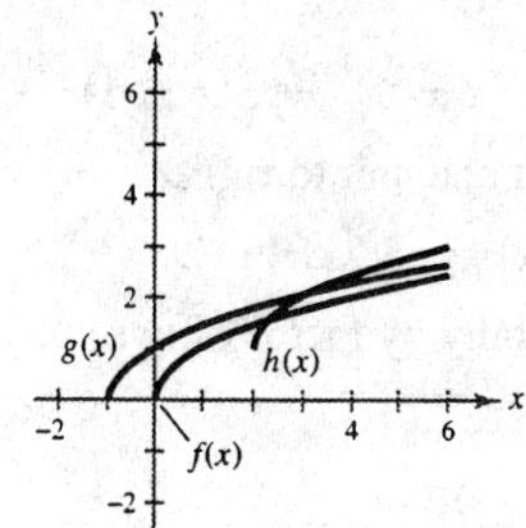

12. (a)

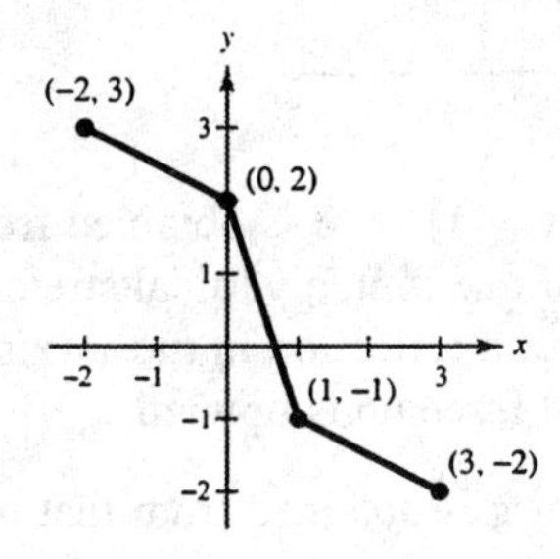

(b)

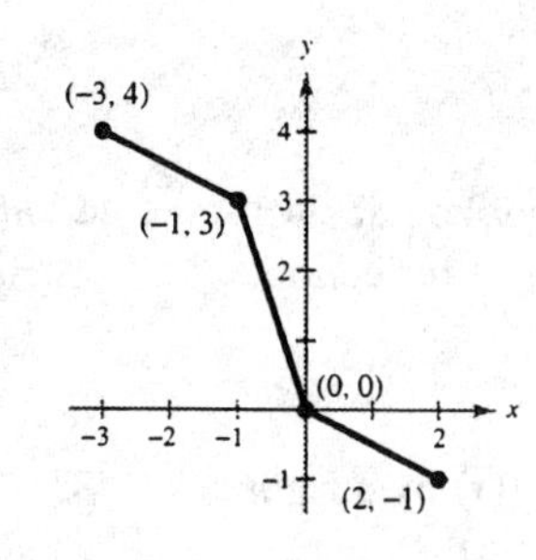

(c)

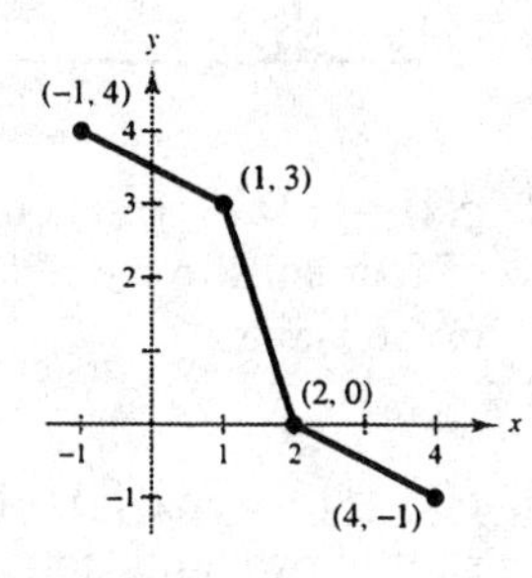

(d)

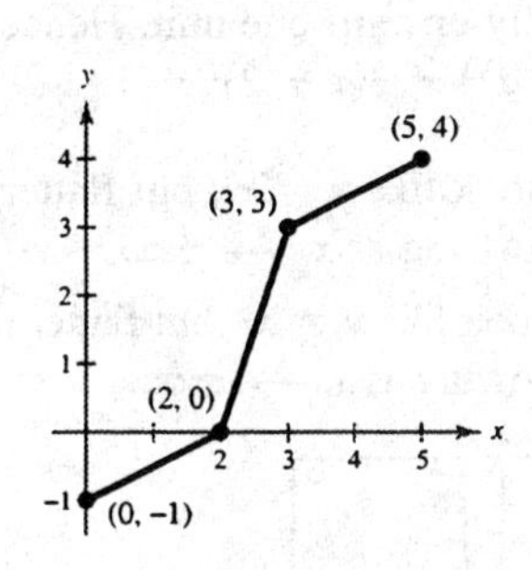

(e)

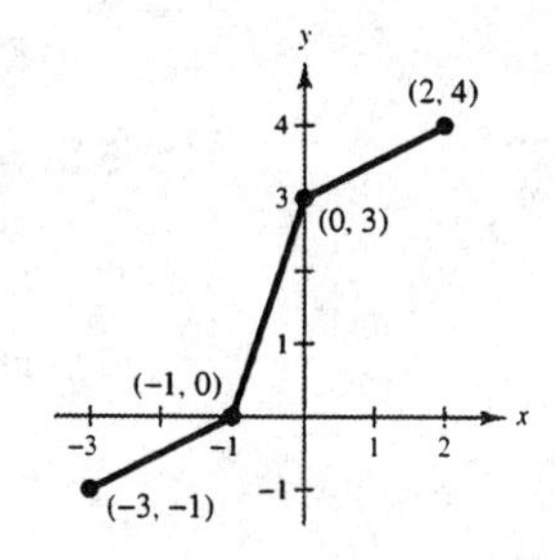

(f)

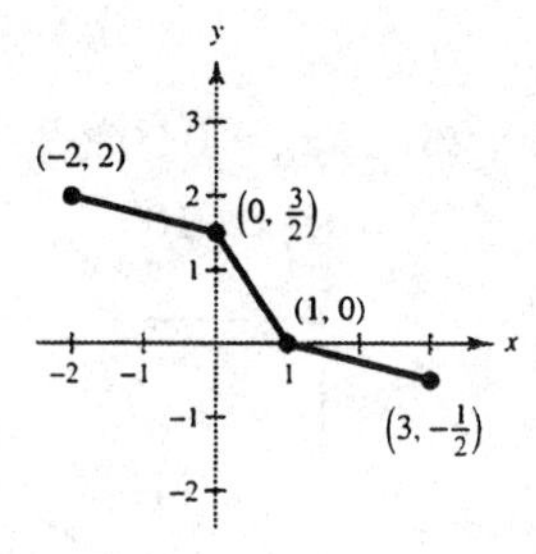

14. Vertical shrink of $y = x$

$y = \frac{1}{2}x$

16. Constant function

$y = 7$

18. Horizontal shift of $y = |x|$

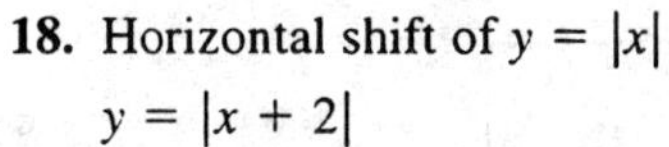

$y = |x + 2|$

20. Reflection in the x-axis of $y = x^2$ followed by vertical and horizontal shifts

$y = 1 - (x + 1)^2$

22. Horizontal and vertical shifts of $y = x^3$

$y = (x - 1)^3 + 1$

24. Reflection in the y-axis of $y = \sqrt{x}$ followed by a horizontal shift

$y = \sqrt{3 - x}$

26. $y = -\sqrt{x}$ is $f(x)$ reflected in the x-axis.

28. $y = \sqrt{x + 3}$ is $f(x)$ shifted left three units.

30. $y = \sqrt{-x}$ is $f(x)$ reflected in the y-axis.

32. $y = |x| - 3$ is $f(x) = |x|$ shifted down three units.

34. $y = |-x|$ is a reflection in the y-axis. In fact $y = |-x| = |x|$.

36. $y = \left|\frac{1}{2}x\right|$ is a horizontal stretch.

38. $f(x) = x^3 - 3x^2 + 2$

$g(x) = f(x - 1) = (x - 1)^3 - 3(x - 1)^2 + 2$

(horizontal shift one unit to right)

$h(x) = f(2x) = (2x)^3 - 3(2x)^2 + 2$

(shrink horizontally by factor of two)

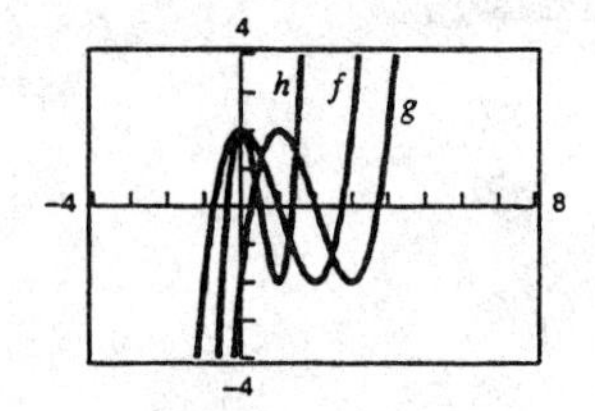

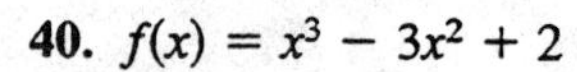

40. $f(x) = x^3 - 3x^2 + 2$

$g(x) = -f(x) = -(x^3 - 3x^2 + 2)$

(reflection in the x-axis)

$h(x) = f(-x) = (-x)^3 - 3(-x)^2 + 2$

(reflection in the y-axis)

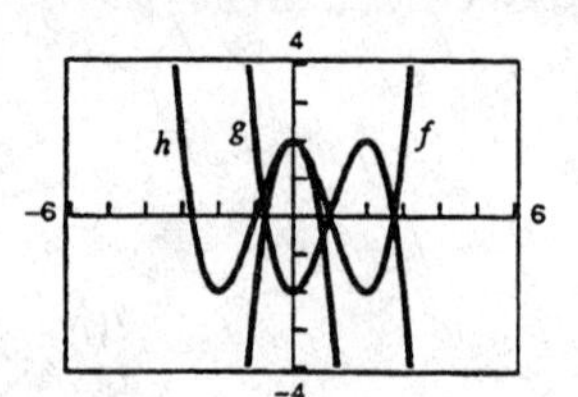

42. $g(x) = -(x - 4)^3$ is obtained by a horizontal shift of four units to the right, followed by a reflection in the x-axis.

44. $h(x) = -2(x - 1)^3 + 3$ is obtained from $f(x)$ by a right shift of one unit, a vertical stretch by a factor of two, a reflection in the x-axis, and a vertical shift three units upward.

46. $p(x) = [3(x - 2)]^3$ is obtained from $f(x)$ by a right shift of two units, followed by a vertical stretch.

48. The graph of g is obtained from that of f by first shifting horizontally two units to the right, and then vertically upward one unit. Hence, $g(x) = (x - 2)^3 - 3(x - 2)^2 + 1$.

50. (a) $H(x) = 0.002x^2 + 0.005x - 0.029$

$10 \le x \le 100$

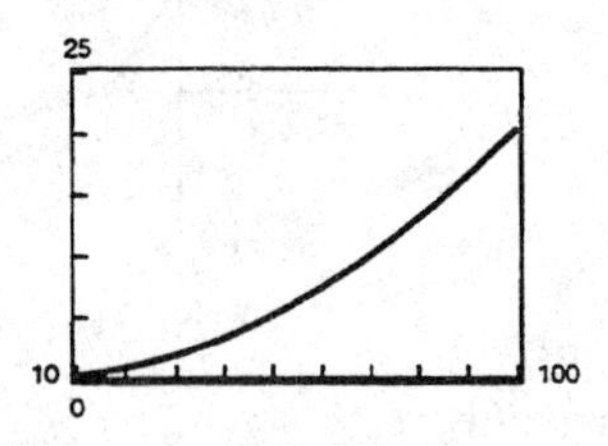

(b) $H(x) = 0.002(1.6x)^2 + 0.005(1.6x) - 0.029$

$= 0.00512x^2 + 0.008x - 0.029$

This is a horizontal shrink.

52. $y = x^7$ will look like $y = x^5$, but flatter in $(-1, 1)$, and faster growing as $x \to \pm\infty$.

$y = x^8$ will look like $y = x^6$, but flatter in $(-1, 1)$, and faster growing as $x \to \pm\infty$.

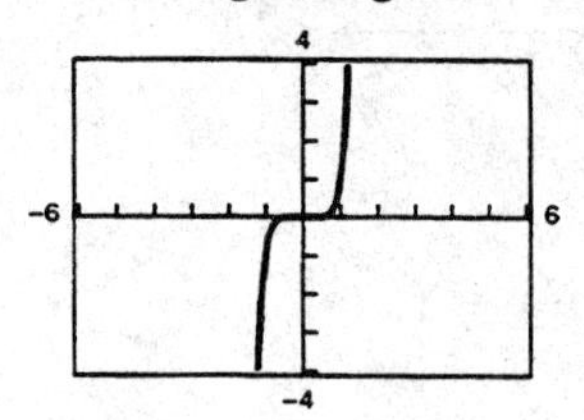

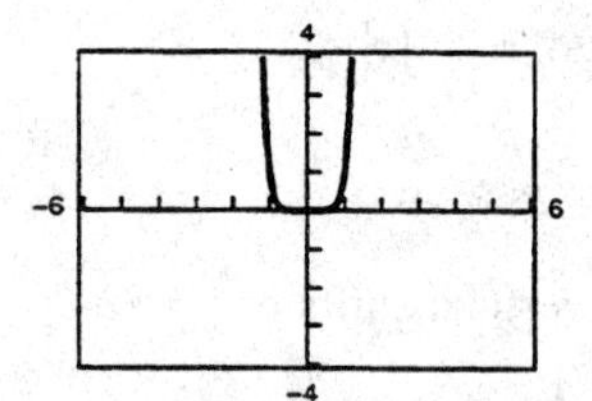

54. $y = (x + 1)^2$

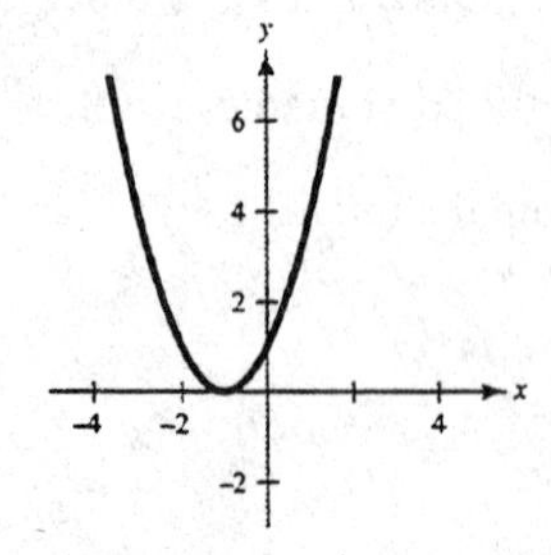

56. $f(x) = x^3(x - 6)^2$

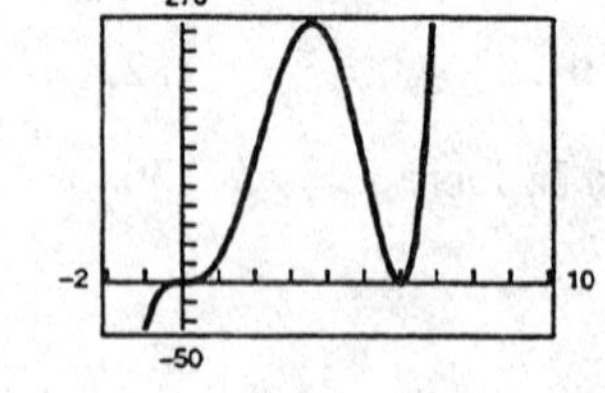

58. $f(x) = x^3(x - 6)^3$

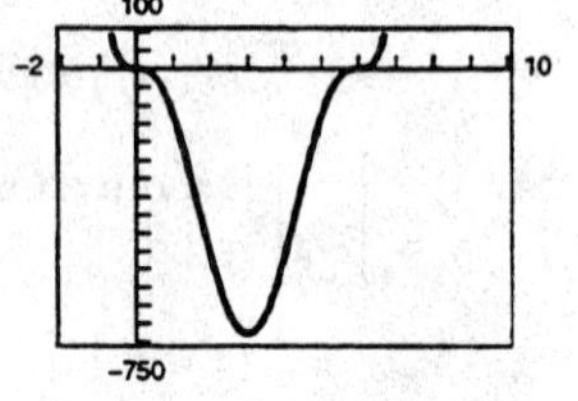

60. (a) 10,000

(b) 500,000

(c) 1

62. $f(x + 2)$

64. $f(x) - 1$

66. $\frac{1}{2}f(x)$

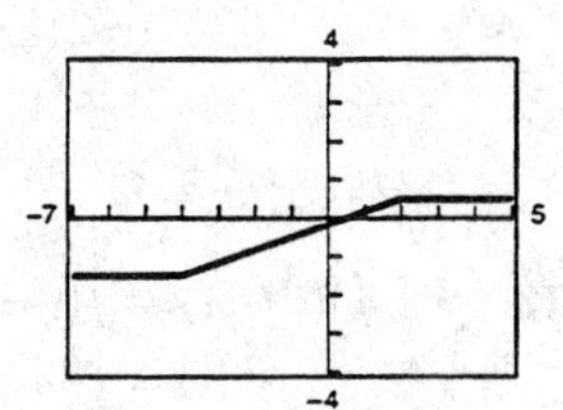

Section 1.4 Combinations of Functions

Solutions to Even-Numbered Exercises

2.

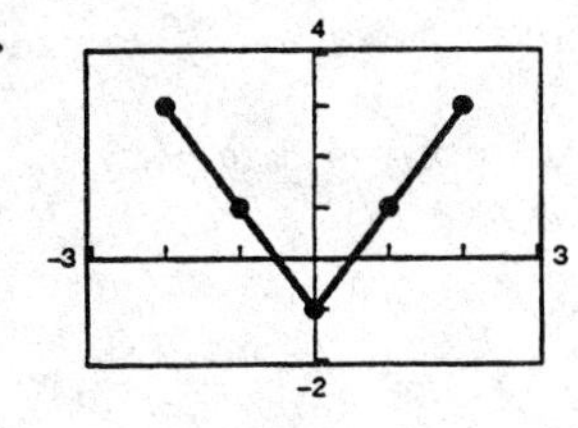

4.

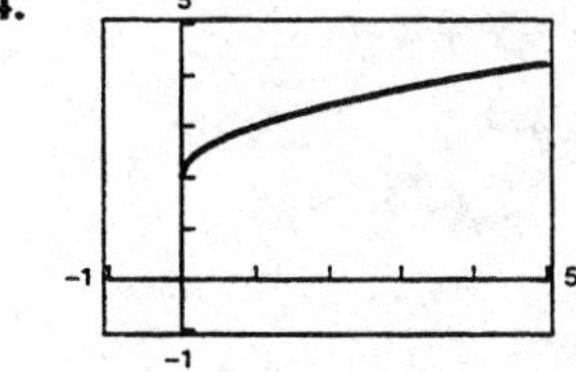

6. $f(x) = 2x - 5,\ g(x) = 1 - x$

(a) $(f + g)(x) = 2x - 5 + 1 - x$
$= x - 4$

(b) $(f - g)(x) = 2x - 5 - (1 - x)$
$= 2x - 5 - 1 + x$
$= 3x - 6$

(c) $(fg)(x) = (2x - 5)(1 - x)$
$= 2x - 2x^2 - 5 + 5x$
$= -2x^2 + 7x - 5$

(d) $\left(\dfrac{f}{g}\right)(x) = \dfrac{2x - 5}{1 - x}$

Domain: $1 - x \neq 0$
$x \neq 1$

8. $f(x) = 2x - 5,\ g(x) = 5$

(a) $(f + g)(x) = 2x - 5 + 5 = 2x$

(b) $(f - g)(x) = 2x - 5 - 6 = 2x - 10$

(c) $(fg)(x) = (2x - 5)(5) = 10x - 25$

(d) $\left(\dfrac{f}{g}\right)(x) = \dfrac{2x - 5}{5} = \dfrac{2}{5}x - 1$

Domain: $-\infty < x < \infty$

10. $f(x) = \sqrt{x^2 - 4},\ g(x) = \dfrac{x^2}{x^2 + 1}$

(a) $(f + g)(x) = \sqrt{x^2 - 4} + \dfrac{x^2}{x^2 + 1}$

(b) $(f - g)(x) = \sqrt{x^2 - 4} - \dfrac{x^2}{x^2 + 1}$

(c) $(fg)(x) = \left(\sqrt{x^2 - 4}\right)\left(\dfrac{x^2}{x^2 + 1}\right) = \dfrac{x^2\sqrt{x^2 - 4}}{x^2 + 1}$

(d) $\left(\dfrac{f}{g}\right)(x) = \sqrt{x^2 - 4} \div \dfrac{x^2}{x^2 + 1}$
$= \dfrac{(x^2 + 1)\sqrt{x^2 - 4}}{x^2}$

Domain: $x^2 - 4 \geq 0$
$x^2 \geq 4 \Rightarrow x \geq 2 \text{ or } x \leq -2$

Domain: $|x| \geq 2$

12. $f(x) = \dfrac{x}{x + 1},\ g(x) = x^3$

(a) $(f + g)(x) = \dfrac{x}{x + 1} + x^3 = \dfrac{x + x^4 + x^3}{x + 1}$

(b) $(f - g)(x) = \dfrac{x}{x + 1} - x^3 = \dfrac{x - x^4 - x^3}{x + 1}$

(c) $(fg)(x) = \dfrac{x}{x + 1} \cdot x^3 = \dfrac{x^4}{x + 1}$

(d) $\left(\dfrac{f}{g}\right)(x) = \dfrac{x}{x + 1} \div x^3$
$= \dfrac{x}{x + 1} \cdot \dfrac{1}{x^3} = \dfrac{1}{x^2(x + 1)}$

Domain: $x \neq 0, x \neq -1$

14. $(f - g)(-2) = f(-2) - g(-2)$
$= (-2)^2 + 1 - (-2 - 4)$
$= 4 + 1 - (-6)$
$= 11$

16. $(f + g)(1) = f(1) + g(1)$
$= (1)^2 + 1 + (1) - 4$
$= -1$

18. $(f + g)(t - 1) = f(t - 1) + g(t - 1)$
$= (t - 1)^2 + 1 + (t - 1) - 4$
$= t^2 - 2t + 1 + 1 + t - 1 - 4$
$= t^2 - t - 3$

20. $(fg)(-6) = f(-6) \cdot g(-6)$
$= [(-6)^2 + 1][(-6) - 4]$
$= (37)(-10)$
$= -370$

22. $\left(\frac{f}{g}\right)(0) = \frac{f(0)}{g(0)} = \frac{0^2 + 1}{0 - 4} = -\frac{1}{4}$

24. $(2f)(5) = 2 \cdot f(5) = 2(5^2 + 1) = 52$

26. $f(x) = \frac{1}{3}x,\ g(x) = -x + 4$
$(f + g)(x) = \frac{1}{3}x - x + 4 = -\frac{2}{3}x + 4$

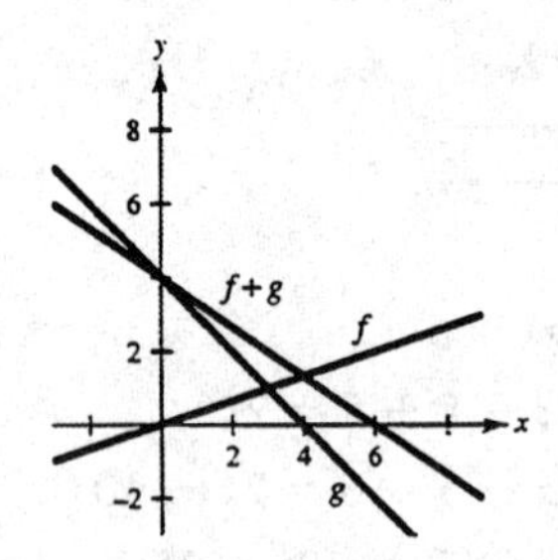

28. $f(x) = 4 - x^2,\ g(x) = x$
$(f + g)(x) = 4 - x^2 + x = 4 + x - x^2$

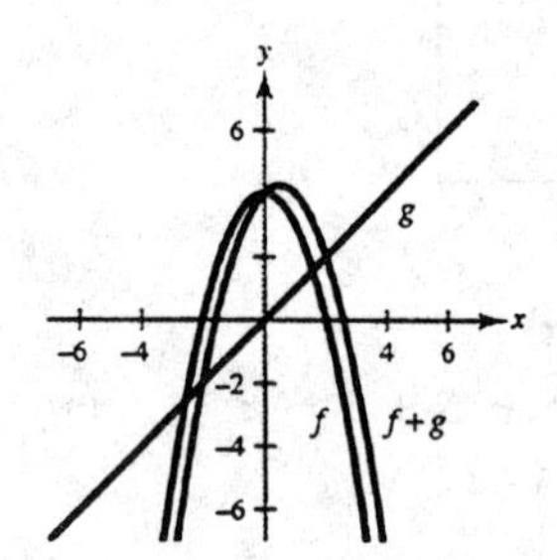

30. $f(x) = \frac{x}{2},\ g(x) = \sqrt{x}$
$(f + g)(x) = \frac{x}{2} + \sqrt{x}$

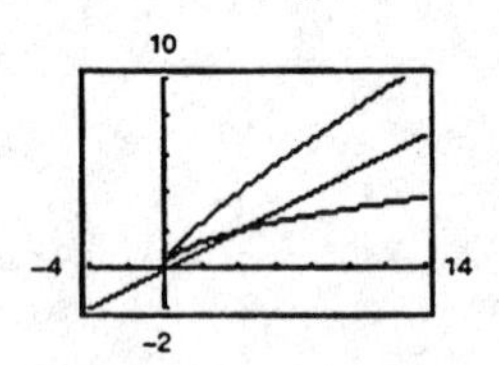

$g(x)$ contributes most to the magnitude of the sum for $0 \le x \le 2$. $f(x)$ also contributes most to the magnitude of the sum for $x > 5$.

32. (a) Total sales $= R_1 + R_2$
$= 480 - 8t - 0.8t^2 + 254 + 0.78t$
$= 734 - 7.22t - 0.8t^2$

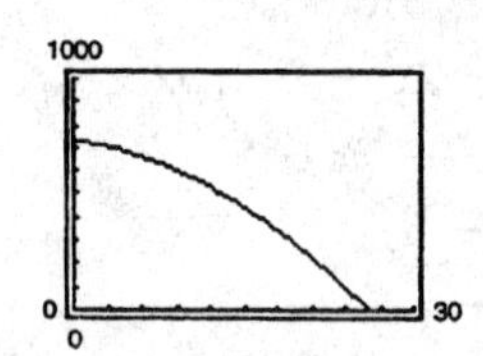

(b) Total sales have been decreasing.

34. $y_1 = 0.16t^2 - 2.43t + 13.96$
$y_2 = 0.17t + 0.38$
$y_3 = 0.04t + 0.44$
$y_1 + y_2 + y_3 = 0.16t^2 - 2.43t + 13.96 + 0.17t + 0.38 + 0.04t + 0.44$
$= 0.16t^2 - 2.22t + 14.78$

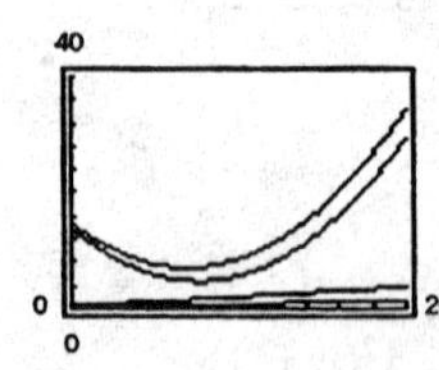

$(y_1 + y_2 + y_3)(15) = 0.16(15)^2 - 2.22(15) + 14.78$
$= 17.48$

The total variable costs per mile in 1995 was 17.49 cents.

36. $f(x) = \sqrt[3]{x-1},\ g(x) = x^3 + 1$

(a) $(f \circ g)(x) = f(g(x))$
$= f(x^3 + 1)$
$= \sqrt[3]{(x^3+1) - 1}$
$= \sqrt[3]{x^3} = x$

(b) $(g \circ f)(x) = g(f(x))$
$= g\left(\sqrt[3]{x-1}\right)$
$= \left(\sqrt[3]{x-1}\right)^3 + 1$
$= (x - 1) + 1 = x$

(c) $(f \circ f)(x) = f(f(x))$
$= f\left(\sqrt[3]{x-1}\right)$
$= \sqrt[3]{\sqrt[3]{x-1} - 1}$

38. $f(x) = x^3,\ g(x) = \dfrac{1}{x}$

(a) $(f \circ g)(x) = f(g(x))$
$= f\left(\dfrac{1}{x}\right)$
$= \left(\dfrac{1}{x}\right)^3 = \dfrac{1}{x^3}$

(b) $(g \circ f)(x) = g(f(x))$
$= g(x^3)$
$= \dfrac{1}{x^3}$

(c) $(f \circ f)(x) = f(f(x))$
$= f(x^3)$
$= (x^3)^3 = x^9$

40. (a) $(f \circ g)(x) = f(x^3 - 2) = \sqrt[3]{(x^3 - 2) + 1} = \sqrt[3]{x^3 - 1}$
$(g \circ f)(x) = g\left(\sqrt[3]{x+1}\right) = \left[\sqrt[3]{x+1}\right]^3 - 2 = x + 1 - 2 = x - 1$

(b) $f \circ g \neq g \circ f$

42. (a) $(f \circ g)(x) = (g \circ f)(x) = \sqrt{\sqrt{x}} = x^{1/4} = \sqrt[4]{x}$

(b) Equal

44. (a) $(f \circ g)(x) = f(x + 6) = |x + 6|$
$(g \circ f)(x) = g(|x|) = |x| + 6$

(b) $f \circ g \neq g \circ f$

46. (a) $(f - g)(1) = f(1) - g(1)$
$= 2 - 3 = -1$

(b) $(fg)(4) = f(4) \cdot g(4)$
$= 4 \cdot 0 = 0$

48. (a) $(f \circ g)(1) = f(g(1))$
$= f(3) = 2$

(b) $(g \circ f)(3) = g(f(3))$
$= g(2) = 2$

50. $h(x) = (1 - x)^3$

One possibility: Let $g(x) = 1 - x$ and $f(x) = x^3$.

$(f \circ g)(x) = f(1 - x) = (1 - x)^3 = h(x)$

52. $h(x) = \sqrt{9 - x}$

One possibility: Let $g(x) = 9 - x$ and $f(x) = \sqrt{x}$.

$(f \circ g)(x) = f(9 - x) = \sqrt{9 - x} = h(x)$

54. $h(x) = \dfrac{4}{(5x + 2)^2}$

One possibility:

Let $g(x) = 5x + 2$ and $f(x) = \dfrac{4}{x^2}$.

$(f \circ g)(x) = f(5x + 2) = \dfrac{4}{(5x + 2)^2}$

56. $h(x) = (x + 3)^{3/2}$

One possibility:

Let $g(x) = x + 3$ and $f(x) = x^{3/2}$.

$(f \circ g) = f(x + 3)$
$= (x + 3)^{3/2} = h(x)$

58. (a) Domain of f: all $x \neq 0$

(b) Domain of g: all x

(c) Domain of $(f \circ g)$: $x \neq -3$

60. $f(x) = 2x + 3,\ g(x) = \dfrac{x}{2}$

(a) Domain of f: all real numbers

(b) Domain of g: all real numbers

(c) $(f \circ g)(x) = f\left(\dfrac{x}{2}\right) = 2\left(\dfrac{x}{2}\right) + 3 = x + 3$

Domain: all real numbers

62.

$$f(x) = 1 - x^2$$
$$f(x+h) = 1 - (x+h)^2 = 1 - (x^2 + 2hx + h^2) = 1 - x^2 - 2hx - h^2$$
$$\frac{f(x+h) - f(x)}{h} = \frac{1 - x^2 - 2hx - h^2 - (1 - x^2)}{h} = \frac{-2hx - h^2}{h} = -2x - h,\ h \neq 0$$

64.

$$f(x) = \sqrt{2x+1}$$
$$f(x+h) = \sqrt{2(x+h)+1}$$
$$\frac{f(x+h) - f(x)}{h} = \frac{\sqrt{2(x+h)+1} - \sqrt{2x+1}}{h}$$
$$= \frac{\sqrt{2(x+h)+1} - \sqrt{2x+1}}{h} \cdot \frac{\sqrt{2(x+h)+1} + \sqrt{2x+1}}{\sqrt{2(x+h)+1} + \sqrt{2x+1}}$$
$$= \frac{[2(x+h)+1] - (2x+1)}{h\left(\sqrt{2(x+h)+1} + \sqrt{2x+1}\right)}$$
$$= \frac{2x + 2h + 1 - 2x - 1}{h\left(\sqrt{2(x+h)+1} + \sqrt{2x+1}\right)}$$
$$= \frac{2}{\left(\sqrt{2(x+h)+1} + \sqrt{2x+1}\right)},\ h \neq 0$$

66. (a) $r(x) = \dfrac{x}{2}$

(b) $A(r) = \pi r^2$

(c) $(A \circ r)(x) = A(r(x))$

$$= A\left(\frac{x}{2}\right) = \pi\left(\frac{x}{2}\right)^2 = \frac{1}{4}\pi x^2$$

$A \circ r$ represents the area of the circular base of the tank with radius $x/2$.

68. $x = 150 \text{ miles} - (450 \text{ mph})(t \text{ hours})$

$y = 200 \text{ miles} - (450 \text{ mph})(t \text{ hours})$

$$s = \sqrt{x^2 + y^2} = \sqrt{(150 - 450t)^2 + (200 - 450t)^2} = 50\sqrt{162t^2 - 126t + 25}$$

70. (a) $R = p - 1200$

(b) $S = 0.92p$

(c) $(R \circ S)(p) = 0.92p - 1200$

$(S \circ R)(p) = 0.92(p - 1200)$

(d) $(R \circ S)(18{,}400) = 15{,}728$

$(S \circ R)(18{,}400) = 15{,}824$

The discount first yields a lower cost.

72. Let $f(x)$ be even and $g(x)$ be odd. Define $h(x) = f(x)g(x)$, then

$$h(-x) = f(-x)g(-x) = [f(x)][-g(x)] = -f(x)g(x) = -h(x).$$

Thus, h is odd.

74. $f(x) = \frac{1}{2}[f(x) + f(-x)] + \frac{1}{2}[f(x) - f(-x)]$

$= g(x) + h(x)$, where $g(x)$ even and $h(x)$ odd by Exercise 73.

76. (a)

n	1	2	3	4	5	6	7	8	9	10
$R_n(0)$	0.3333	0.6667	0.8819	0.9965	1.0524	1.0787	1.0908	1.0963	1.0988	1.1000

(b)

n	1	2	3	4	5	6	7	8	9	10
$R_n(0)$	3	3.4641	3.5305	3.5398	3.5412	3.5414	3.5414	3.5414	3.5414	3.5414

(c) Approaches a constant.

78. 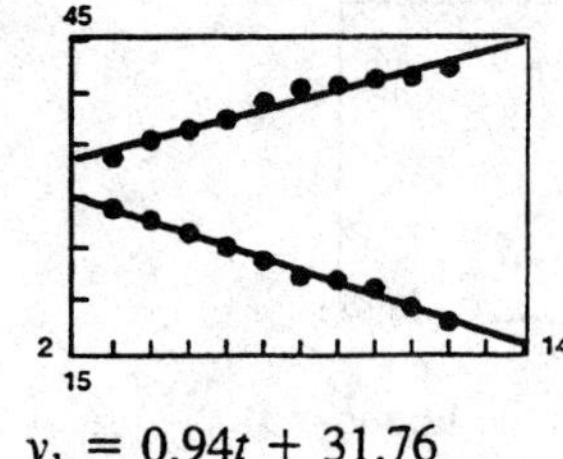

$y_1 = 0.94t + 31.76$

$y_2 = -1.19t + 32.32$

The difference between circulations is increasing.

80. $(-8, 3)$

82. $d = \sqrt{(5-0)^2 + (4-8)^2}$
$= \sqrt{25 + 16}$
$= \sqrt{41}$

Section 1.5 Inverse Functions

Solutions to Even-Numbered Exercises

2. The inverse is a line through (0, 6) and (6, 0). Matches graph (b).

4. The inverse is a third-degree equation through (0, 0). Matches graph (d).

6.
$$f(x) = \tfrac{1}{5}x$$
$$f^{-1}(x) = 5x$$
$$f(f^{-1}(x)) = f(5x) = \tfrac{1}{5}(5x) = x$$
$$f^{-1}(f(x)) = f^{-1}\left(\tfrac{1}{5}x\right) = 5\left(\tfrac{1}{5}x\right) = x$$

8.
$$f(x) = x - 5$$
$$f^{-1}(x) = x + 5$$
$$f(f^{-1}(x)) = f(x+5) = (x+5) - 5 = x$$
$$f^{-1}(f(x)) = f^{-1}(x-5) = (x-5) + 5 = x$$

10.
$$f(x) = x^5$$
$$f^{-1}(x) = \sqrt[5]{x}$$
$$f(f^{-1}(x)) = f\left(\sqrt[5]{x}\right) = \left(\sqrt[5]{x}\right)^5 = x$$
$$f^{-1}(f(x)) = f^{-1}(x^5) = \sqrt[5]{x^5} = x$$

12. $f(x) = x - 5,\ g(x) = x + 5$
$$f(g(x)) = f(x+5) = (x+5) - 5 = x$$
$$g(f(x)) = g(x-5) = (x-5) + 5 = x$$

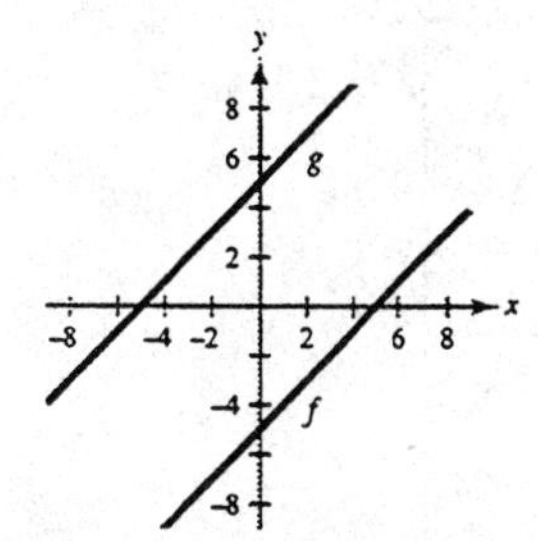

14. $f(x) = 3 - 4x,\ g(x) = \dfrac{3-x}{4}$
$$f(g(x)) = f\left(\frac{3-x}{4}\right) = 3 - 4\left(\frac{3-x}{4}\right) = 3 - (3 - x) = x$$
$$g(f(x)) = g(3 - 4x) = \frac{3 - (3 - 4x)}{4} = \frac{4x}{4} = x$$

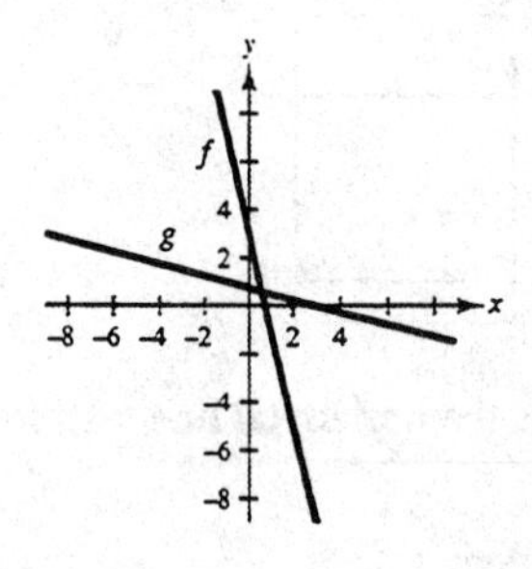

16. $f(x) = \dfrac{1}{x},\ g(x) = \dfrac{1}{x}$

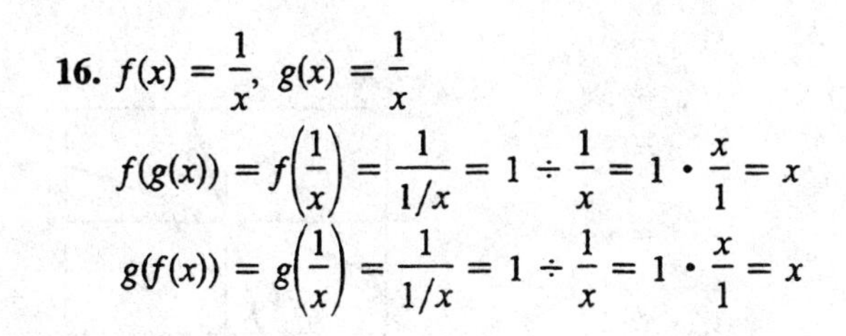

$$f(g(x)) = f\left(\frac{1}{x}\right) = \frac{1}{1/x} = 1 \div \frac{1}{x} = 1 \cdot \frac{x}{1} = x$$
$$g(f(x)) = g\left(\frac{1}{x}\right) = \frac{1}{1/x} = 1 \div \frac{1}{x} = 1 \cdot \frac{x}{1} = x$$

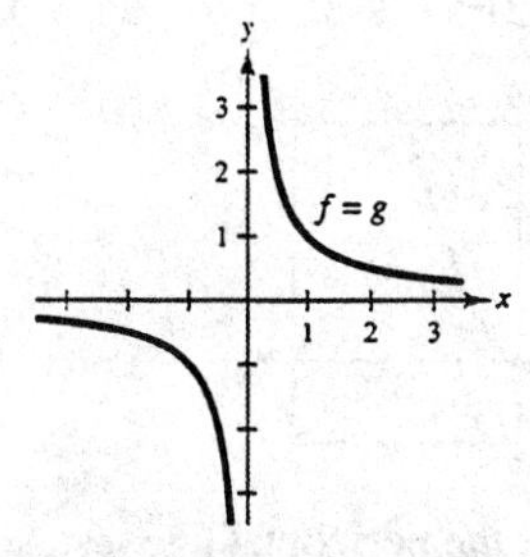

Reflections in the line $y = x$

18. $f(x) = 9 - x^2, x \geq 0$

$g(x) = \sqrt{9 - x}, x \leq 9$

$f(g(x)) = f(\sqrt{9 - x}) = 9 - (\sqrt{9 - x})^2 = 9 - (9 - x) = x$

$g(f(x)) = g(9 - x^2) = \sqrt{9 - (9 - x^2)} = \sqrt{x^2} = x$

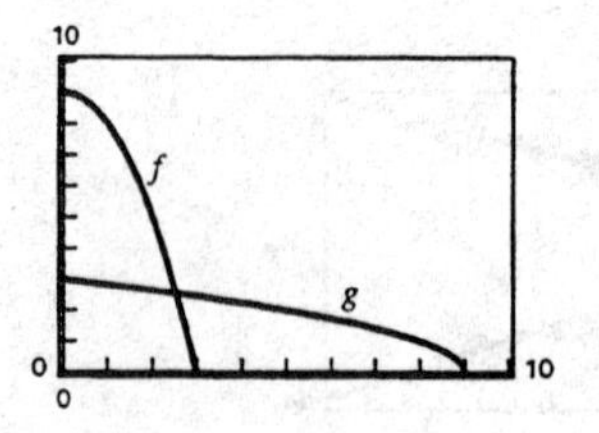

Reflections in the line $y = x$

20. $f(x) = \dfrac{1}{1 + x}, \; x \geq 0; \; g(x) = \dfrac{1 - x}{x}, \; 0 < x \leq 1$

$$f(g(x)) = f\left(\frac{1 - x}{x}\right) = \frac{1}{1 + \left(\frac{1 - x}{x}\right)} = \frac{1}{\frac{x}{x} + \frac{1 - x}{x}} = \frac{1}{\frac{1}{x}} = x$$

$$g(f(x)) = g\left(\frac{1}{1 + x}\right) = \frac{1 - \left(\frac{1}{1 + x}\right)}{\left(\frac{1}{1 + x}\right)} = \frac{\frac{1 + x}{1 + x} - \frac{1}{1 + x}}{\frac{1}{1 + x}} = \frac{\frac{x}{1 + x}}{\frac{1}{1 + x}} = \frac{x}{1 + x} \cdot \frac{x + 1}{1} = x$$

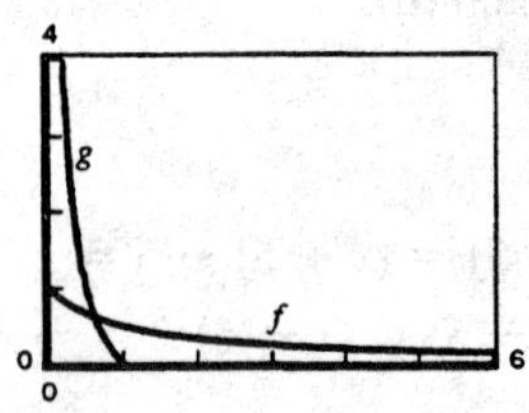

Reflections in the line $y = x$

22. No, because some horizontal lines intersect the graph twice, f does not have an inverse.

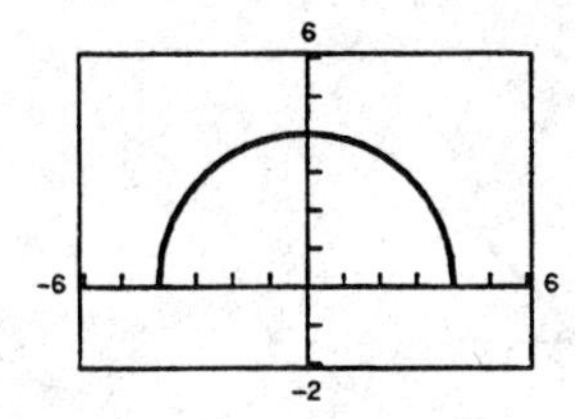

24. Yes, because no horizontal lines intersect the graph at more than one point, f has an inverse.

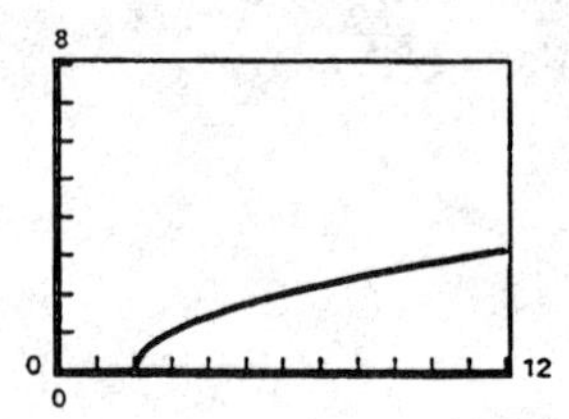

26. $f(x) = 10$

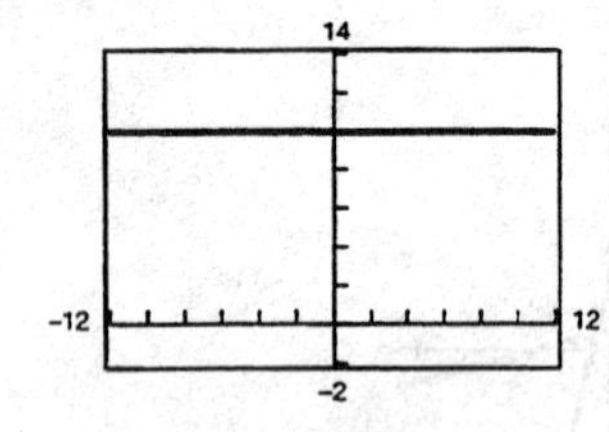

f does not pass the horizontal line test, so f has no inverse.

28. $g(x) = (x + 5)^3$

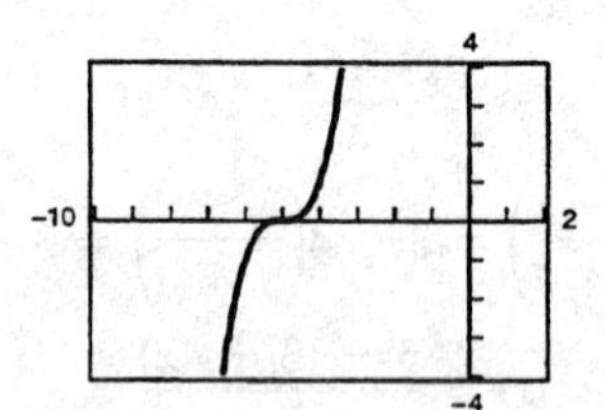

g passes the horizontal line test, so g has an inverse.

30. $f(x) = \frac{1}{8}(x + 2)^2 - 1$

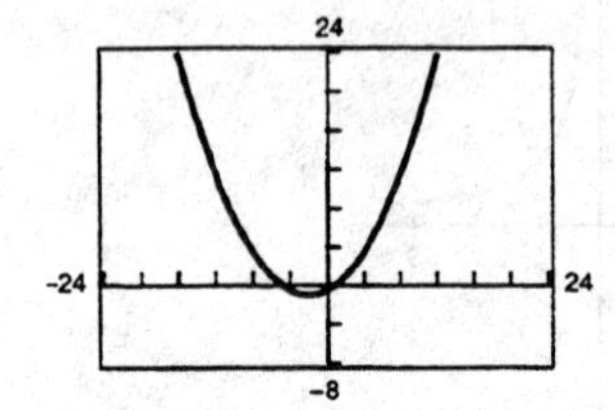

f does not pass the horizontal line test, so f has no inverse.

32. $f(x) = 3x$

$y = 3x$

$x = 3y$

$\dfrac{x}{3} = y$

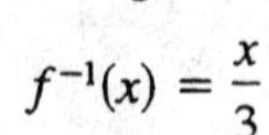

$f^{-1}(x) = \dfrac{x}{3}$

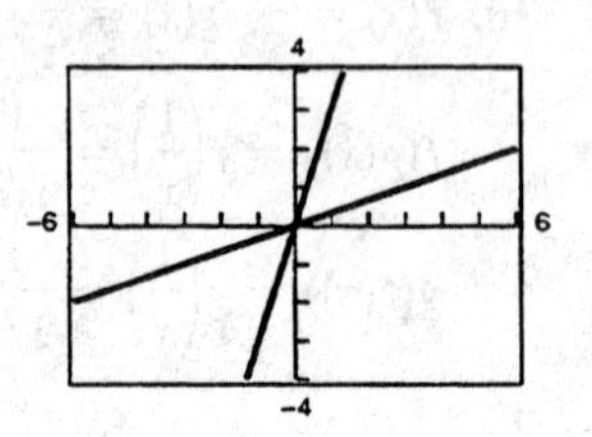

Reflections in the line $y = x$

34. $f(x) = x^3 + 1$

$$y = x^3 + 1$$
$$x = y^3 + 1$$
$$x - 1 = y^3$$
$$\sqrt[3]{x-1} = y$$
$$f^{-1}(x) = \sqrt[3]{x-1}$$

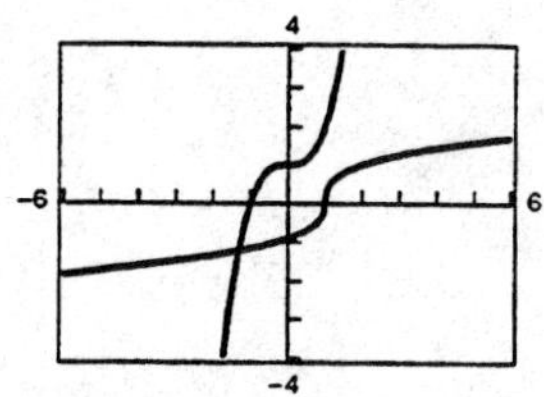

Reflections in the line $y = x$

36. $f(x) = x^2, x \geq 0$

$$y = x^2$$
$$x = y^2$$
$$\sqrt{x} = y$$
$$f^{-1}(x) = \sqrt{x}$$

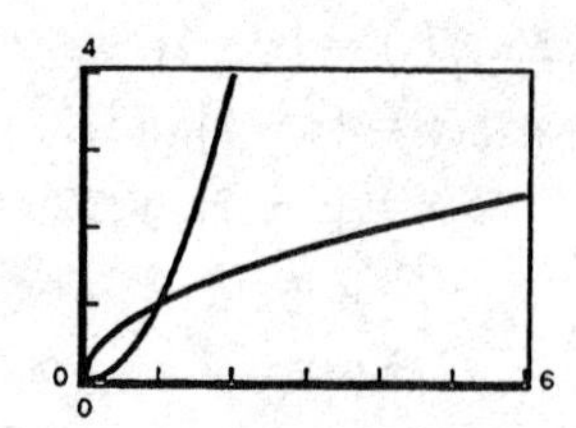

Reflections in the line $y = x$

38. $f(x) = \dfrac{4}{x}$

$$y = \frac{4}{x}$$
$$x = \frac{4}{y}$$
$$y = \frac{4}{x}$$
$$f^{-1}(x) = \frac{4}{x}$$

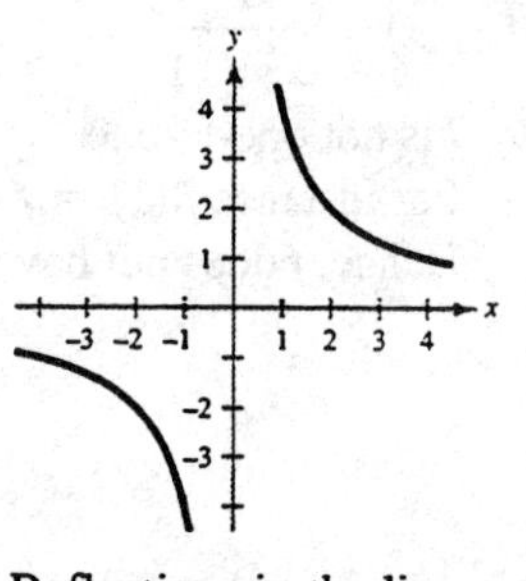

Reflections in the line $y = x$

40. $f(x) = x^{3/5}$

$$y = x^{3/5}$$
$$x = y^{3/5}$$
$$x^{5/3} = (y^{3/5})^{5/3}$$
$$x^{5/3} = y$$
$$f^{-1}(x) = x^{5/3}$$

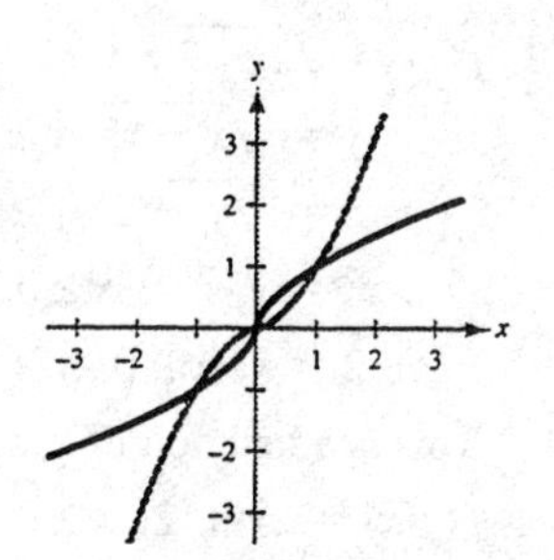

Reflections in the line $y = x$

42. $f(x) = \dfrac{1}{x^2}$

f is not one-to-one.
For instance, $f(1) = f(-1)$.
f does not have an inverse.

44. $f(x) = 3x + 5$

f is one-to-one.

$$y = 3x + 5$$
$$x = 3y + 5$$
$$x - 5 = 3y$$
$$\frac{x-5}{3} = y$$

This is a function of x, so f has an inverse.

$$f^{-1}(x) = \frac{x-5}{3}.$$

46. $f(x) = \dfrac{3x+4}{5}$

f is one-to-one.

$$y = \frac{3x+4}{5}$$
$$x = \frac{3y+4}{5}$$
$$5x = 3y + 4$$
$$5x - 4 = 3y$$
$$\frac{5x-4}{3} = y$$

This is a function of x, so f has an inverse.

$$f^{-1}(x) = \frac{5x-4}{3}$$

48. $q(x) = (x - 5)^2$

q is not one-to-one.
For instance, $q(0) = q(10)$.
Hence, q does not have an inverse.

50. $f(x) = |x - 2|, \ x \le 2 \Rightarrow y \ge 0$

$y = |x - 2|, \ x \le 2, \ y \ge 0$

$x = |y - 2|, \ y \le 2, \ x \ge 0$

$x = y - 2 \quad \text{or} \quad -x = y - 2$

$2 + x = y \quad \text{or} \quad 2 - x = y$

The portion that satisfies the conditions $y \le 2$ and $x \ge 0$ is $2 - x = y$. This is a function of x, so f has an inverse.

$f^{-1}(x) = 2 - x, \ x \ge 0$

52. $f(x) = \sqrt{x - 2} \Rightarrow x \ge 2, \ y \ge 0$

$y = \sqrt{x - 2}, \ x \ge 2, \ y \ge 0$

$x = \sqrt{y - 2}, \ y \ge 2, \ x \ge 0$

$x^2 = y - 2, \ x \ge 0, \ y \ge 2$

$x^2 + 2 = y, \ x \ge 0, \ y \ge 2$

This is a function of x, so f has an inverse.

$f^{-1}(x) = x^2 + 2, \ x \ge 0$

54. $f(x) = \dfrac{x^2}{x^2 + 1}$

f is not one-to-one.

For instance $f(1) = f(-1)$.

Hence, f does not have an inverse.

56. $f(x) = ax + b, \ a \ne 0$

$y = ax + b$

$x = ay + b$

$x - b = ay$

$\dfrac{x - b}{a} = y$

This is a function of x, so f has an inverse.

58. If we let $f(x) = 1 - x^4, \ x \ge 0$, then f has an inverse. [Note: we could also let $x \le 0$.]

$f(x) = 1 - x^4, \ x \ge 0 \Rightarrow y \le 1$

$y = 1 - x^4, \ x \ge 0, \ y \le 1$

$x = 1 - y^4, \ y \ge 0, \ x \le 1$

$y^4 = 1 - x, \ y \ge 0, \ x \le 1$

$y = \sqrt[4]{1 - x}, \ x \le 1, \ y \ge 0$

Thus, $f^{-1}(x) = \sqrt[4]{1 - x}, \ x \le 1$.

60. If we let $f(x) = |x - 2|, \ x \ge 2$, then f has an inverse. [Note: we could also let $x \le 2$.]

$f(x) = |x - 2|, \ x \ge 2$

$f(x) = x - 2$ when $x \ge 2$.

$y = x - 2, \ x \ge 2, \ y \ge 0$

$x = y - 2, \ x \ge 0, \ y \ge 2$

$x + 2 = y, \ x \ge 0, \ y \ge 2$

Thus, $f^{-1}(x) = x + 2, \ x \ge 0$.

62.

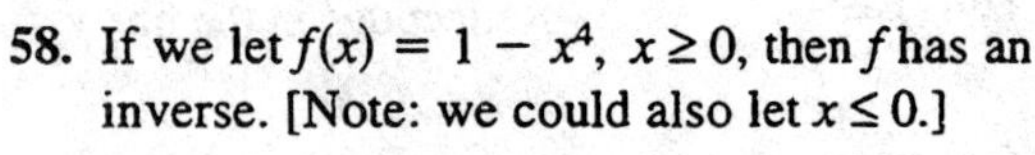

x	$f(x)$
4	−3
3	−2
−1	0
−2	6

x	$f^{-1}(x)$
−3	4
−2	3
0	−1
6	−2

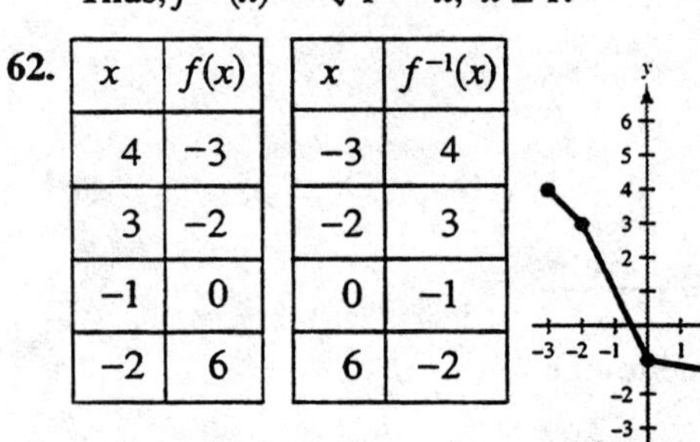

64. (a) and (b)

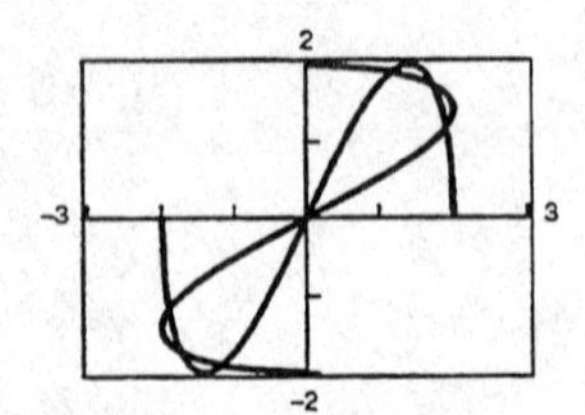

(c) Not an inverse function since it does not satisfy the Vertical Line Test.

66. (a) and (b)

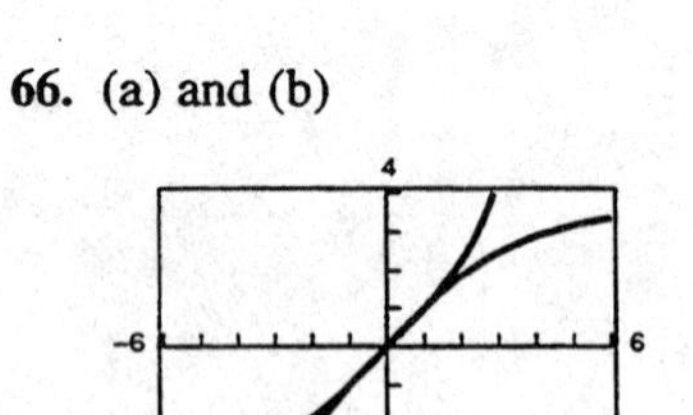

(c) Inverse function since it satisfies the Vertical

68. True

70. False, $f(x) = \dfrac{1}{x}$ has an inverse $f^{-1}(x) = \dfrac{1}{x}$.

In Exercises 72, 74, and 76, $f(x) = \frac{1}{8}x - 3$, $g(x) = x^3$, $f^{-1}(x) = 8(x + 3)$, $g^{-1}(x) = \sqrt[3]{x}$.

72. $$\begin{aligned}(g^{-1} \circ f^{-1})(-3) &= g^{-1}(f^{-1}(-3)) \\ &= g^{-1}(8(-3 + 3)) \\ &= g^{-1}(0) = \sqrt[3]{0} = 0\end{aligned}$$

74. $$\begin{aligned}(g^{-1} \circ g^{-1})(-4) &= g^{-1}(g^{-1}(-4)) \\ &= g^{-1}(\sqrt[3]{-4}) \\ &= \sqrt[3]{\sqrt[3]{-4}} = -\sqrt[9]{4}\end{aligned}$$

76. $$\begin{aligned}g^{-1} \circ f^{-1} &= g^{-1}(f^{-1}(x)) \\ &= g^{-1}(8(x + 3)) \\ &= \sqrt[3]{8(x + 3)} \\ &= 2\sqrt[3]{x + 3}\end{aligned}$$

In Exercises 78 and 80, $f(x) = x + 4$, $g(x) = 2x - 5$, $f^{-1}(x) = x - 4$, $g^{-1}(x) = \dfrac{x + 5}{2}$.

78. $$\begin{aligned}f^{-1} \circ g^{-1}(x) &= f^{-1}(g^{-1}(x)) \\ &= f^{-1}\left(\frac{x + 5}{2}\right) \\ &= \frac{x + 5}{2} - 4 \\ &= \frac{x + 5 - 8}{2} \\ &= \frac{x - 3}{2}\end{aligned}$$

80. $$\begin{aligned}(g \circ f)(x) &= g(f(x)) \\ &= g(x + 4) \\ &= 2(x + 4) - 5 \\ &= 2x + 8 - 5 \\ &= 2x + 3 \\ y &= 2x + 3 \\ x &= 2y + 3 \\ x - 3 &= 2y \\ \frac{x - 3}{2} &= y \\ (g \circ f)^{-1}(x) &= \frac{x - 3}{2}\end{aligned}$$

82. If f is one-to-one, then f^{-1} exists. If f is odd, then $f(-x) = -f(x)$. Consider $f(x) = y \Leftrightarrow f^{-1}(y) = x$. Then, $f^{-1}(-y) = f^{-1}(-f(x)) = f^{-1}(f(-x)) = -x = -f^{-1}(y)$.

84. (a) *Model*: $(\text{Total cost}) = \left(\begin{array}{c}\text{Cost of} \\ \text{first commodity}\end{array}\right) + \left(\begin{array}{c}\text{Cost of} \\ \text{second commodity}\end{array}\right)$

Labels: Total cost $= y$
Amount of first commodity $= x$
Amount of second commodity $= 50 - x$
Cost of first commodity $= 1.25x$
Cost of second commodity $= 1.60(50 - x)$

Equation: $y = 1.25x + 1.60(50 - x)$

(b) $$\begin{aligned}x &= 1.25y + 1.60(50 - y) \\ x &= 1.25y + 80 - 1.60y \\ x - 80 &= -0.35y \\ \frac{x - 80}{-0.35} &= y \\ y &= \frac{80 - x}{0.35}\end{aligned}$$

$x =$ total cost
$y =$ number of pounds of less expensive commodity

(c) To keep the number of pounds of less expensive commodity nonnegative, $x \le 80$.

(d) $\dfrac{80 - 73}{0.35} = y = 20$ pounds

86. If $f(x) = k(2 - x - x^3)$ has an inverse and $f^{-1}(3) = -2$, then $f(-2) = 3$. Thus,

$$\begin{aligned}f(-2) = k(2 - (-2) - (-2)^3) &= 3 \\ k(2 + 2 + 8) &= 3 \\ 12k &= 3 \\ k &= \tfrac{3}{12} = \tfrac{1}{4}.\end{aligned}$$

Thus, $k = \frac{1}{4}$.

Section 1.6 Exploring Daa: Linear Models and Scatter Plots

Solutions to Even-Numbered Exercises

2. $r \approx 0$ (no correlation)

4. $r > 0$ (positive correlation)

6. (a)

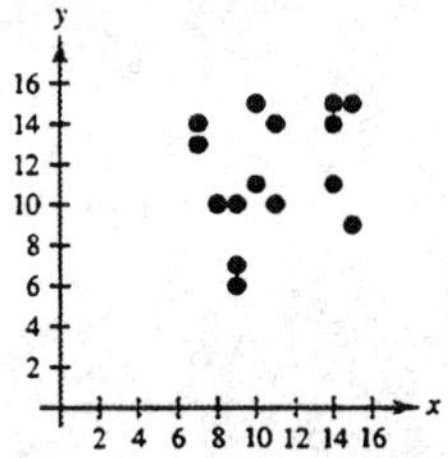

(b) No. Quiz scores are dependent on a several variables such as study time, class attendance, etc. These variables may change from the time of one quiz to the next.

8. (a) $y = -1.25x + 3$

(b)

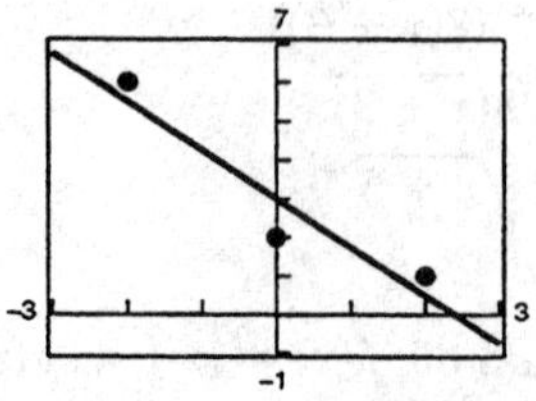

(c) The model appears valid.

10. (a) $y = -1.15x + 6.85$

(b)

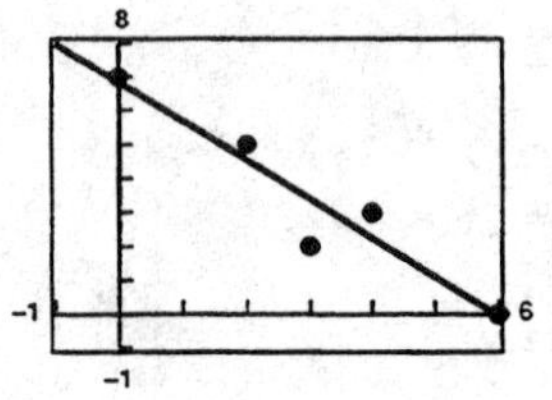

(c) The model appears valid.

12. (a) $x = 9.7t + 0.4$

(b)

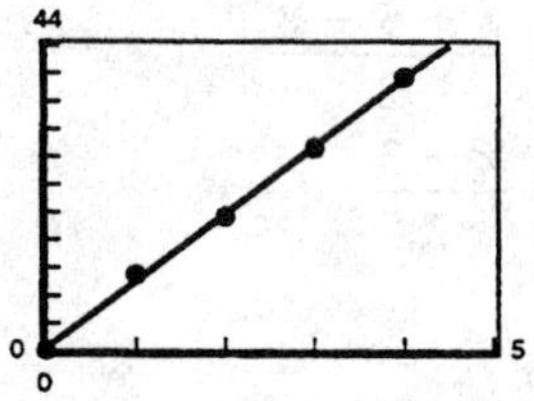

The model fits well.

(c) 24.65

14. (a) $P = 3.71t + 59.11$

(b)

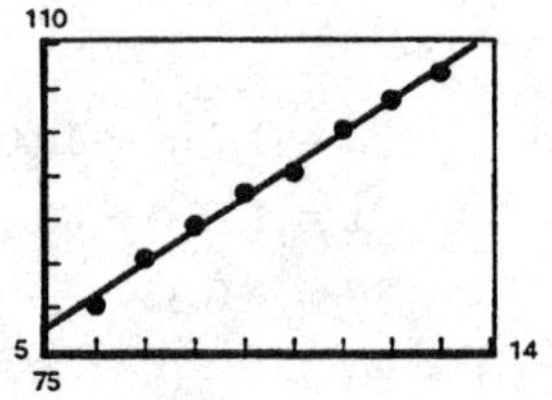

(c) The average increase per year in the price of homes

(d) $P(20) = 3.71(20) + 59.11$

$= 133.31$

Median price in year 2000 is $133,310.00

16. (a) $y = 2.62x + 2.66$

(b)

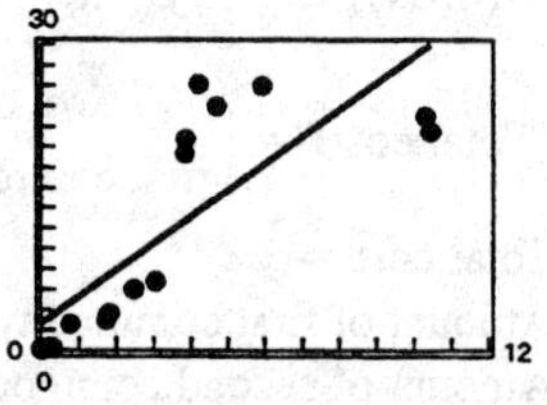

(c) Denmark, Japan, Finland, France, Italy

18. (a) $y = 0.77x + 9.30$

(b)

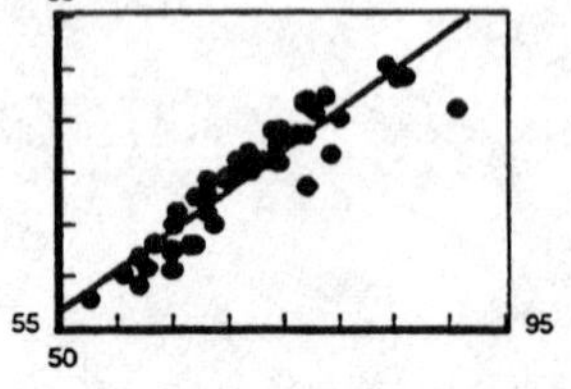

(c) Alabama, Mississippi, North Dakota

(d) On average, approximately 77% of the registered voters actually vote.

❑ Review Exercises for Chapter 1

Solutions to Even-Numbered Exercises

2. Yes, for each input value there is one, and only one, output value.

4. Yes, $y = 2x - 3$.

6. No, does not pass vertical line test.

8. $g(x) = x^{4/3}$

(a) $g(8) = 8^{4/3} = 2^4 = 16$

(b) $g(t + 1) = (t + 1)^{4/3}$

(c) $\dfrac{g(8) - g(1)}{8 - 1} = \dfrac{16 - 1}{8 - 1} = \dfrac{15}{7}$

10. $f(t) = \sqrt[4]{t}$

(a) $f(16) = \sqrt[4]{16} = 2$

(b) $f(t + 5) = \sqrt[4]{t + 5}$

(c) $\dfrac{f(16) - f(0)}{16} = \dfrac{2 - 0}{16} = \dfrac{1}{8}$

12. $f(x) = \begin{cases} x^2 + 2, & x < 0 \\ |x - 2|, & x \ge 0 \end{cases}$

(a) $f(-4) = (-4)^2 + 2 = 18$

(b) $f(0) = |0 - 2| = 2$

(c) $f(1) = |1 - 2| = 1$

14. $(-\infty, \infty)$

16. $x^2 + 8x = x(x + 8) \ge 0$

Domain: $(-\infty, 8] \cup [0, \infty)$

18. The domain of $h(t) = |t + 1|$ is all t.

20. Third setting

22.

24.

26.

28.

30.

32.

34. Vertical shift upward one unit.

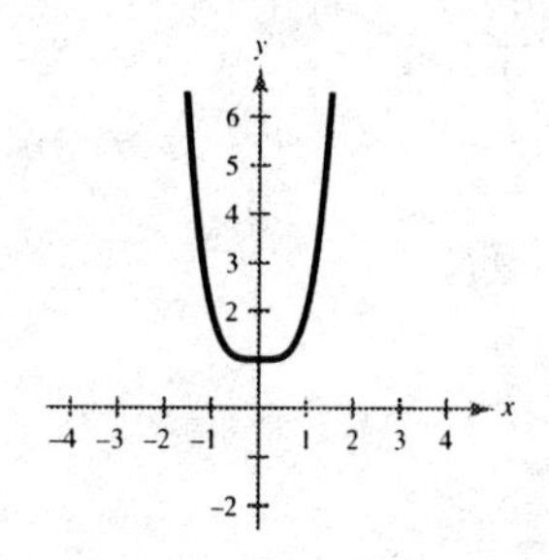

36. Reflection in the x-axis followed by a vertical shift upward 2 units.

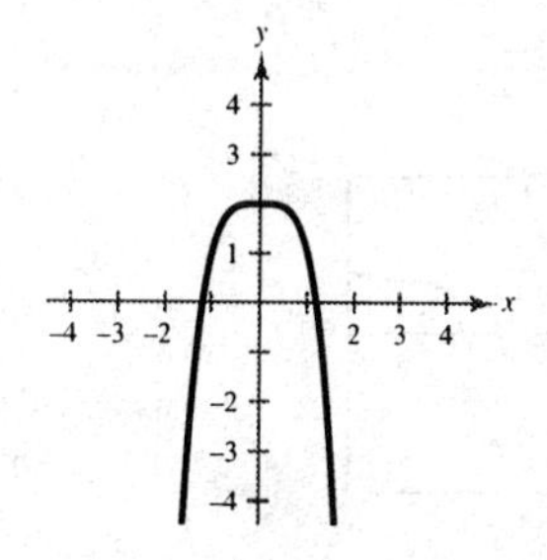

38. Horizontal shift 2 units to the right: $y = \sqrt{x - 2}$

40. Reflection in the x-axis followed by a vertical shift upward 2 units: $y = 2 - \sqrt{x}$

42. $f(x) = (x^2 - 4)^2$

(a) Increasing: $(-2, 0), (2, \infty)$
Decreasing: $(-\infty, -2), (0, 2)$

(b) Relative maximum: $(0, 16)$
Relative minimum: $(-2, 0), (2, 0)$

(c) The function is even: $f(-x) = f(x)$

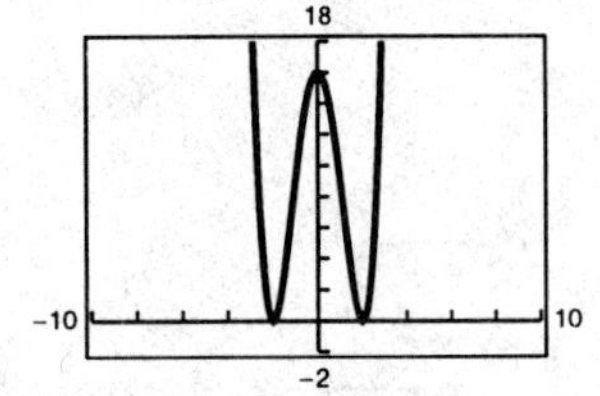

44. $g(x) = \sqrt[3]{x(x + 3)^2}$

(a) Increasing: $(-\infty, -3), (-1, \infty)$
Decreasing: $(-3, -1)$

(b) Relative maximum: $(-3, 0)$
Relative minimum: $(-1, -1.59)$

(c) Neither even nor odd.

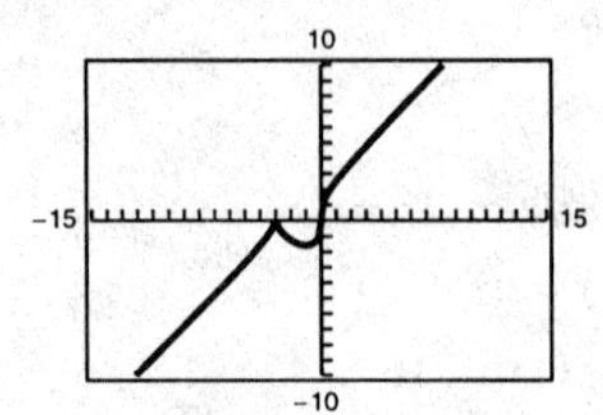

46.

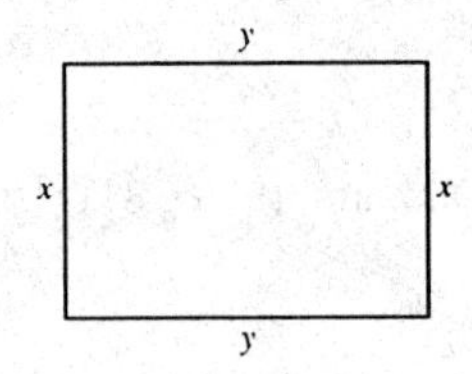

(a) $2x + 2y = 24$

$$y = 12 - x$$

$$A = xy = x(12 - x)$$

(b) Since x and y cannot be negative, we have $0 \le x \le 12$. The domain is $[0, 12]$.

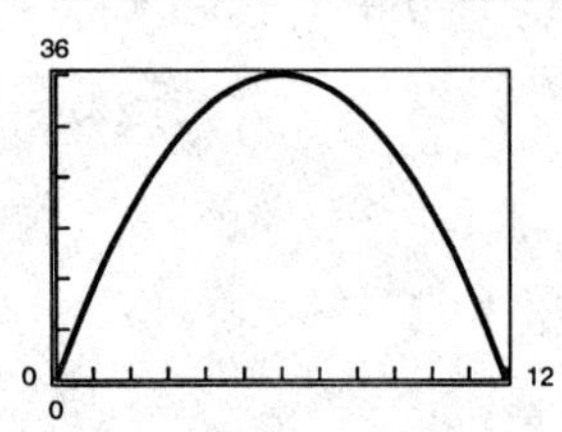

(c) The maximum area of 36 occurs when $x = 6$ and the rectangle is a 6×6 square.

48. (a)

$$y = 5x - 7$$

$$y + 7 = 5x$$

$$f^{-1}(x) = \frac{x + 7}{5}$$

(b)

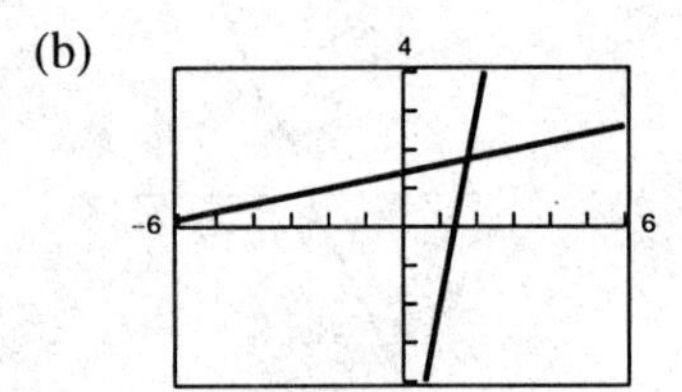

(c) $f^{-1}(f(x)) = f^{-1}(5x - 7)$

$$= \frac{5x - 7 + 7}{5}$$

$$= x$$

$$f(f^{-1}(x)) = f\left(\frac{x + 7}{5}\right)$$

$$= 5\left(\frac{x + 7}{5}\right) - 7$$

$$= x$$

50. (a)

$$y = x^3 + 2$$

$$y - 2 = x^3$$

$$f^{-1}(x) = \sqrt[3]{x - 2}$$

(b)

(c) $f^{-1}(f(x)) = f^{-1}(x^3 + 2)$

$$= \sqrt[3]{x^3 + 2 - 2}$$

$$= x$$

$$f(f^{-1}(x)) = f(\sqrt[3]{x - 2})$$

$$= (x - 2) + 2$$

$$= x$$

52. (a)
$$\begin{aligned} f(x) &= \sqrt[3]{x+1} \\ y &= \sqrt[3]{x+1} \\ x &= \sqrt[3]{y+1} \\ x^3 &= y+1 \\ y &= x^3-1 \\ f^{-1}(x) &= x^3-1 \end{aligned}$$

(b)

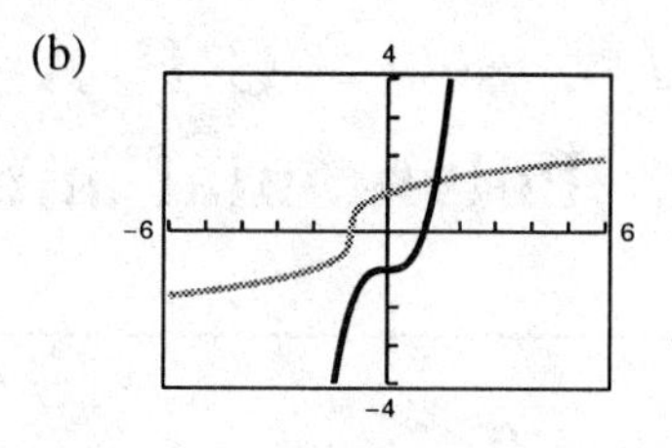

(c)
$$\begin{aligned} f^{-1}(f(x)) &= f^{-1}\left(\sqrt[3]{x+1}\right) \\ &= (x+1)-1 = x \\ f(f^{-1}(x)) &= f(x^3-1) \\ &= \sqrt[3]{x^3-1+1} = x \end{aligned}$$

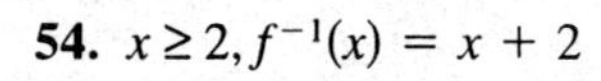

54. $x \geq 2, f^{-1}(x) = x + 2$

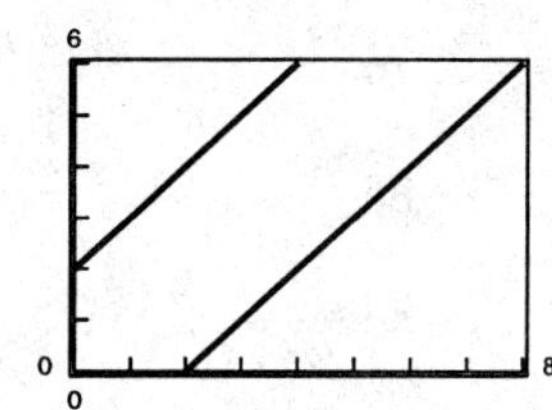

56.
$$\begin{aligned} f(x) &= x^{4/3},\ x \geq 0 \\ y &= x^{4/3} \\ x &= y^{4/3} \\ y &= x^{3/4} \\ f^{-1}(x) &= x^{3/4},\ x \geq 0 \end{aligned}$$

58.
$$\begin{aligned} (f+h)(5) &= f(5) + h(5) \\ &= -7 + 77 \\ &= 70 \end{aligned}$$

60. $\left(\dfrac{g}{h}\right)(1) = \dfrac{g(1)}{h(1)} = \dfrac{1}{5}$

62. $(g \circ f)(-2) = g(7) = \sqrt{7}$

64. $(h \circ f^{-1})(1) = h(f^{-1}(1)) = h(1) = 5$

$\left[\text{Note: } f^{-1}(x) = \dfrac{3-x}{2}\right]$

66. (a) $y = 8.22x + 24.47$

(b)

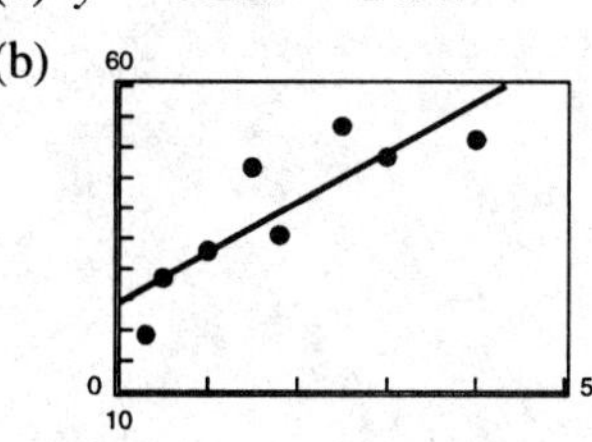

(c) The average increase per year in sales for each additional year of experience

(d) $65,600

CHAPTER 2
Polynomial and Rational Functions

CHAPTER 2
Polynomial and Rational Functions

Section 2.1 Quadratic Functions

Solutions to Even-Numbered Exercises

2. $f(x) = (x + 4)^2$ opens upward and has vertex $(-4, 0)$. Matches graph (c).

4. $f(x) = 3 - x^2$ opens downward and has vertex $(0, 3)$. Matches graph (h).

6. $f(x) = (x + 1)^2 - 2$ opens upward and has vertex $(-1, -2)$. Matches graph (a).

8. $f(x) = -(x - 4)^2$ opens downward and has vertex $(4, 0)$. Matches graph (d).

10. (a) $y = x^2 + 1$

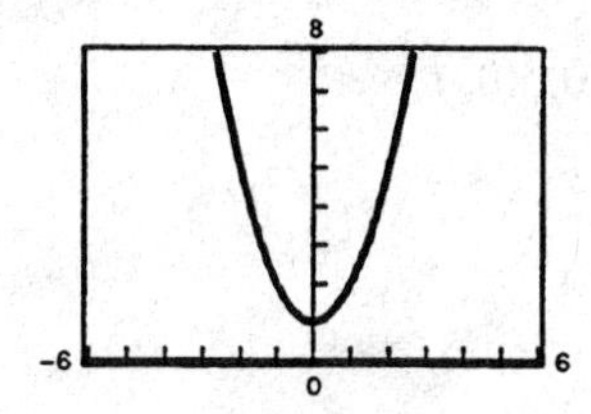

Vertical translation 1 unit upward

(b) $y = x^2 - 1$

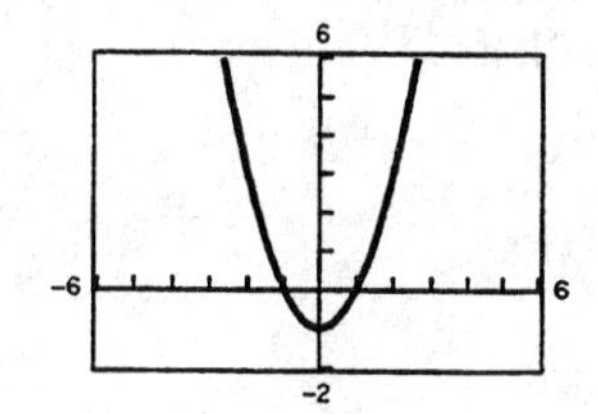

Vertical translation 1 unit downward

(c) $y = x^2 + 3$

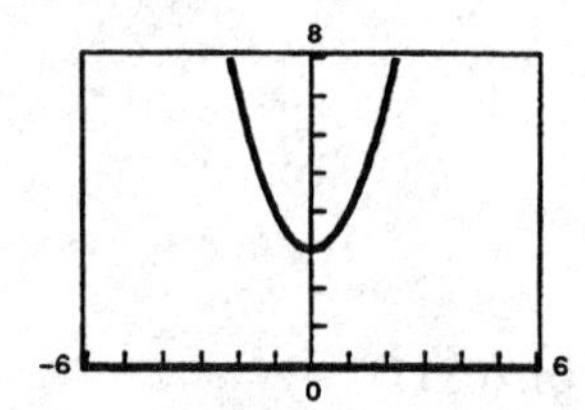

Vertical translation 3 units upward

(d) $y = x^2 - 3$

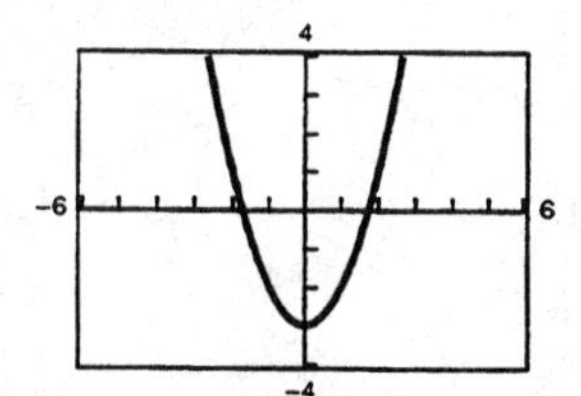

Vertical translation 3 units downward

12. (a) $y = -\frac{1}{2}(x - 2)^2 + 1$

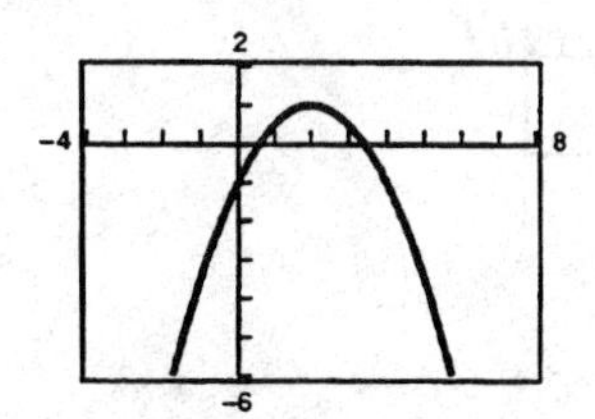

Horizontal translation 2 units to right, vertical shrink by $\frac{1}{2}$, reflection in the x-axis, and vertical translation 1 unit upward

(b)

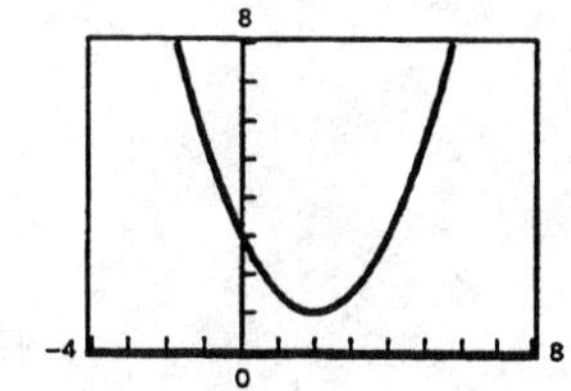

Horizontal translation 2 units to the right, vertical shrink by $\frac{1}{2}$, vertical translation 1 unit upward

14. $f(x) = \frac{1}{2}x^2 - 4$

Vertex: $(0, -4)$

Intercepts: $(\pm 2\sqrt{2}, 0), (0, -4)$

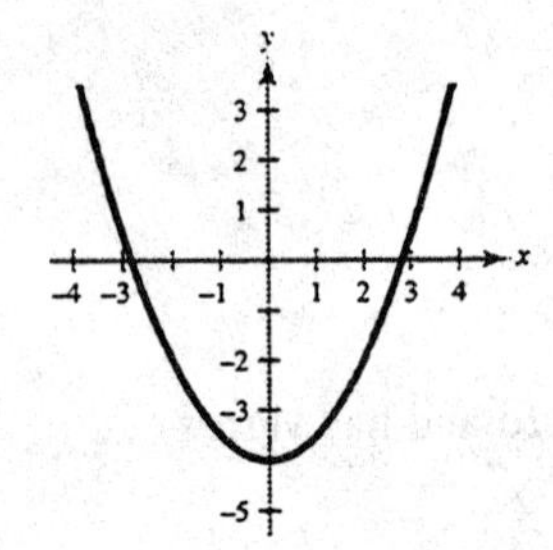

16. $g(x) = x^2 + 2x + 1 = (x + 1)^2$

Vertex: $(-1, 0)$

Intercepts: $(-1, 0), (0, 1)$

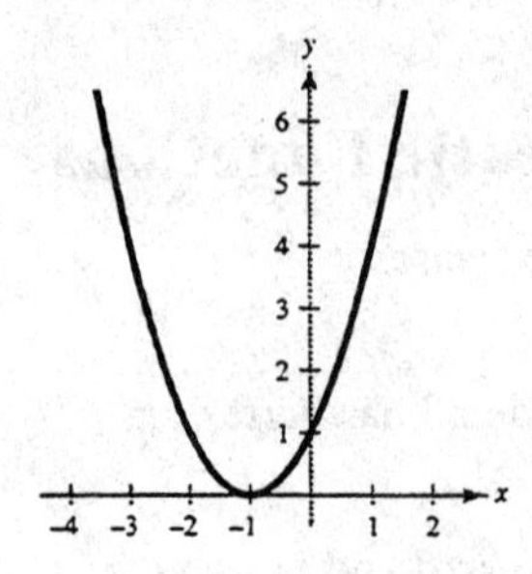

18. $f(x) = x^2 + 3x + \frac{1}{4} = \left(x + \frac{3}{2}\right)^2 - 2$

Vertex: $\left(-\frac{3}{2}, -2\right)$

Intercepts: $\left(-\frac{3}{2} \pm \sqrt{2}, 0\right), \left(0, \frac{1}{4}\right)$

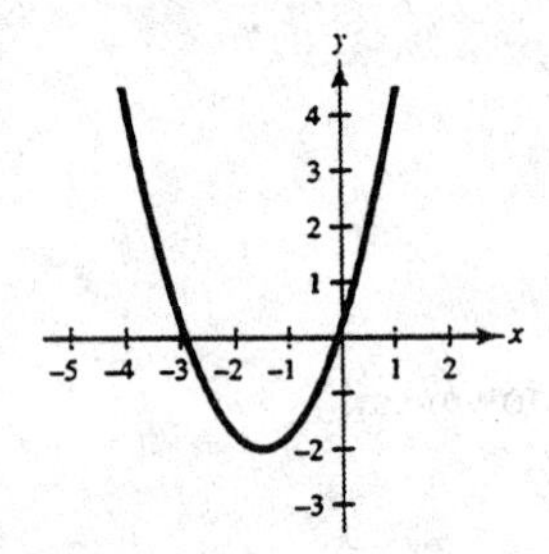

20. $f(x) = -x^2 - 4x + 1 = -1(x^2 + 4x - 1)$
$= -1[(x + 2)^2 - 5]$

Vertex: $(-2, 5)$

Intercepts: $\left(-2 \pm \sqrt{5}, 0\right), (0, 1)$

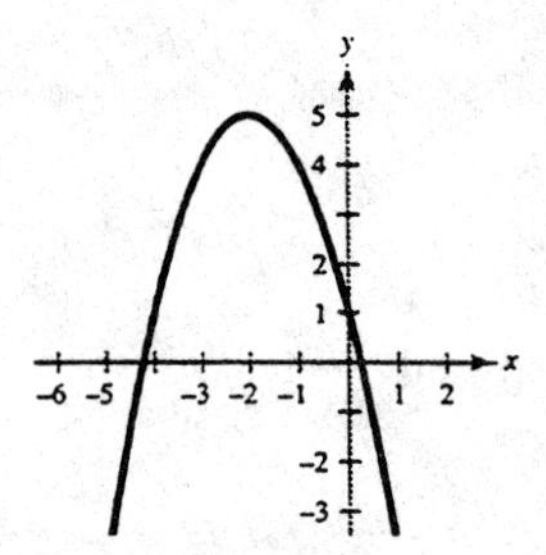

22. $f(x) = 2x^2 - x + 1$

$= 2\left(x^2 - \frac{1}{2}x\right) + 1$

$= 2\left(x - \frac{1}{4}\right)^2 - \frac{1}{8} + 1$

$= 2\left(x - \frac{1}{4}\right)^2 + \frac{7}{8}$

Vertex: $\left(\frac{1}{4}, \frac{7}{8}\right)$

Intercept: $(0, 1)$

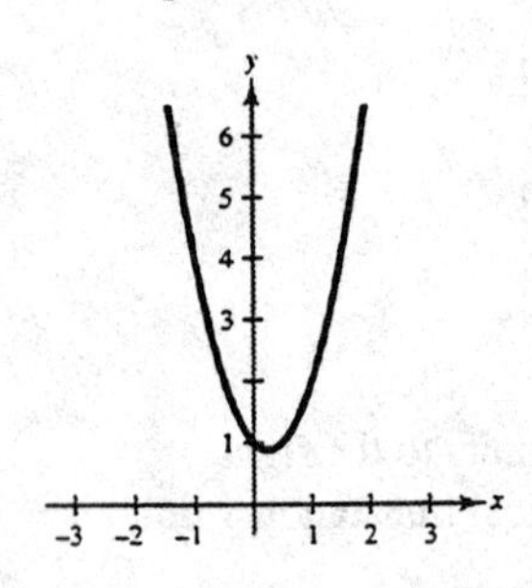

24. $g(x) = x^2 + 8x + 11$

$= (x + 4)^2 - 5$

Vertex: $(-4, -5)$

Intercepts: $\left(-4 \pm \sqrt{5}, 0\right), (0, 11)$

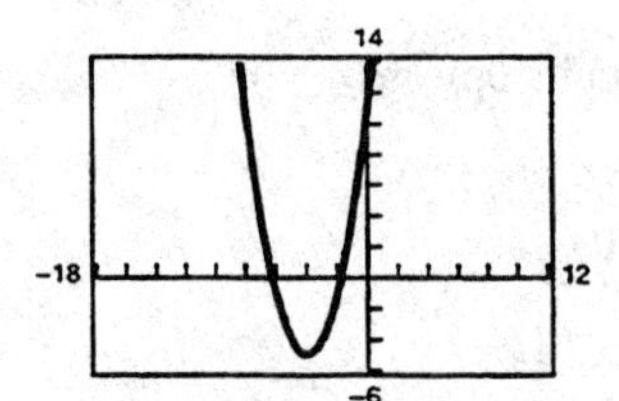

26. $g(x) = \frac{1}{2}(x^2 + 4x - 2)$

$= \frac{1}{2}(x^2 + 4x) - 1$

$= \frac{1}{2}(x + 2)^2 - 3$

Vertex: $(-2, -3)$

Intercepts: $\left(-2 \pm \sqrt{6}, 0\right), (0, -1)$

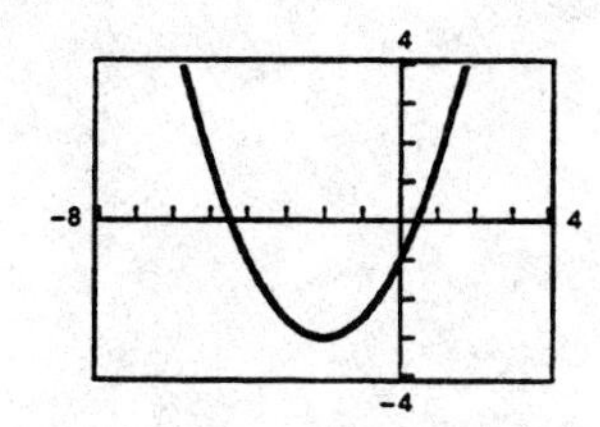

28. $(-2, -1)$ is the vertex.

$f(x) = a(x + 2)^2 - 1$

Since the graph passes through $(0, 3)$,

$3 = a(0 + 2)^2 - 1$

$3 = 4a - 1$

$4 = 4a$

$1 = a.$

Thus, $y = (x + 2)^2 - 1.$

30. $(2, 0)$ is the vertex.

$f(x) = a(x - 2)^2 + 0 = a(x - 2)^2$

Since the graph passes through $(3, 2)$,

$2 = a(3 - 2)^2$

$2 = a.$

Thus, $y = 2(x - 2)^2.$

32. $(4, -1)$ is the vertex.

$f(x) = a(x - 4)^2 - 1$

Since the graph passes through $(2, 3)$,

$3 = a(2 - 4)^2 - 1$

$3 = 4a - 1$

$4 = 4a$

$1 = a.$

Thus, $f(x) = (x - 4)^2 - 1.$

34. $(2, 3)$ is the vertex.

$f(x) = a(x - 2)^2 + 3$

Since the graph passes through $(0, 2)$,

$2 = a(0 - 2)^2 + 3$

$2 = 4a + 3$

$-1 = 4a$

$-\frac{1}{4} = a.$

Thus, $f(x) = -\frac{1}{4}(x - 2)^2 + 3.$

36. $(-2, -2)$ is the vertex.

$f(x) = a(x + 2)^2 - 2$

Since the graph passes through $(-1, 0)$,

$0 = a(-1 + 2)^2 - 2$

$0 = a - 2$

$2 = a.$

Thus, $f(x) = 2(x + 2)^2 - 2.$

38. $y = x^2 - 6x + 9$

x-intercept: $(3, 0)$

$0 = x^2 - 6x + 9$

$0 = (x - 3)^2$

$x = 3$

40. $y = 2x^2 + 5x - 3$

x-intercepts: $\left(\frac{1}{2}, 0\right), (-3, 0)$

$0 = 2x^2 + 5x - 3$

$0 = (2x - 1)(x + 3)$

$x = \frac{1}{2}, -3$

42. $y = x^2 - 9x + 18$

x-intercepts: $(3, 0), (6, 0)$

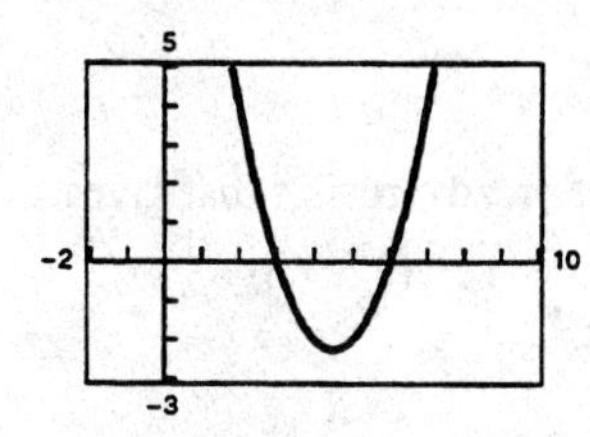

$0 = x^2 - 9x + 18$

$0 = (x - 6)(x - 3)$

$x = 3, 6$

44. $y = -\frac{1}{2}(x^2 - 6x - 7)$

x-intercepts: $(7, 0), (-1, 0)$

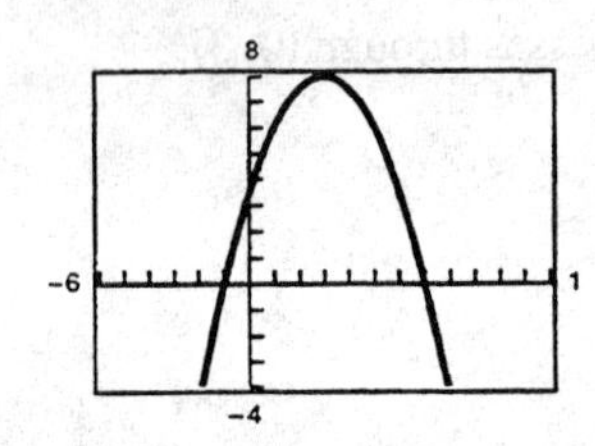

$0 = -\frac{1}{2}(x^2 - 6x - 7)$

$0 = x^2 - 6x - 7$

$0 = (x - 7)(x + 1)$

$x = 7, -1$

46. $f(x) = (x - 4)(x - 8)$

$= x^2 - 12x + 32$, opens upward

$g(x) = -f(x)$, opens downward

$g(x) = -x^2 + 12x - 32$

48. $f(x) = 2\left[x - \left(-\frac{5}{2}\right)\right](x - 2)$

$= 2\left(x + \frac{5}{2}\right)(x - 2)$

$= 2\left(x^2 + \frac{1}{2}x - 5\right)$

$= 2x^2 + x - 10$, opens upward

$g(x) = -f(x)$, opens downward

$g(x) = -2x^2 - x + 10$

50. Let x = first number and y = second number. Then, $x + y = S$, $y = S - x$. The product is $P(x) = xy = x(S - x)$.

$$P(x) = Sx - x^2$$
$$= -x^2 + Sx$$
$$= -\left(x^2 - Sx + \frac{S^2}{4} - \frac{S^2}{4}\right)$$
$$= -\left(x - \frac{S}{2}\right)^2 + \frac{S^2}{4}$$

The maximum value of the product occurs at the vertex of $P(x)$ and is $S^2/4$. This happens when $x = y = S/2$.

52. Let x = length of rectangle and y = width of rectangle.

$2x + 2y = 36$

$y = 18 - x$

(a) $A(x) = xy = x(18 - x)$

Domain: $0 < x < 18$

(b)

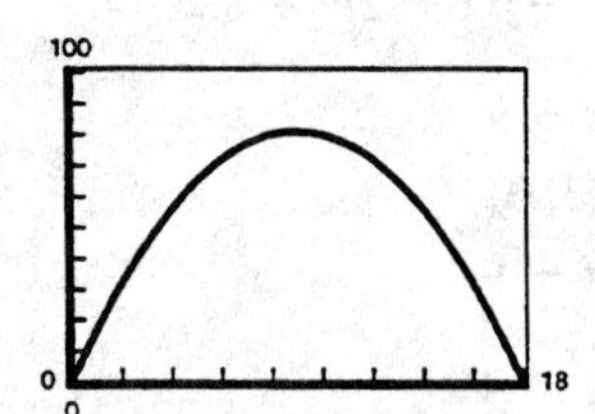

(c) The area is maximum (81 square meters) when $x = y = 9$ meters. The rectangle has dimensions 9 meters × 9 meters.

54. $P = 230 + 20x - 0.5x^2$

The vertex occurs at

$$x = -\frac{b}{2a} = -\frac{20}{2(-0.5)} = 20.$$

Because x is in hundreds of dollars, $20 \times 100 = 2000$ dollars is the amount spent on advertising that gives maximum profit.

56. (a)

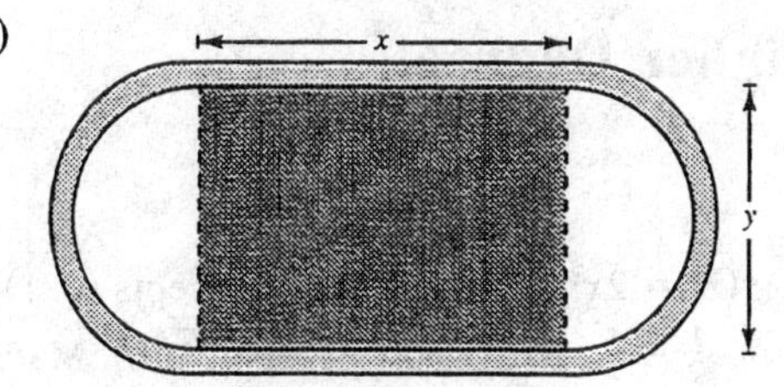

(b) Radius of semicircular ends of track: $r = \frac{1}{2}y$

distance around two semicircular parts of track:

$$d = 2\pi r = 2\pi\left(\frac{1}{2}y\right) = \pi y$$

(c) Distance traveled around track in one lap:

$$d = \pi y + 2x = 200$$
$$\pi y = 200 - 2x$$
$$y = \frac{200 - 2x}{\pi}$$

(d) and (e)

Area of rectangular region:

$$A = xy = x\left(\frac{200 - 2x}{\pi}\right)$$
$$= \frac{1}{\pi}(200x - 2x^2)$$
$$= -\frac{2}{\pi}(x^2 - 100x)$$
$$= -\frac{2}{\pi}(x^2 - 100x + 2500 - 2500)$$
$$= -\frac{2}{\pi}(x - 50)^2 + \frac{5000}{\pi}$$

The area is maximum when $x = 50$ and

$$y = \frac{200 - 2(50)}{\pi} = \frac{100}{\pi}.$$

(e)

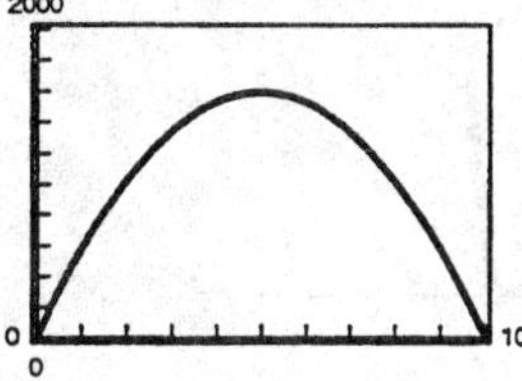

58. $y = -\frac{4}{9}x^2 + \frac{24}{9}x + 12$

The maximum height of the dive occurs at the vertex, $x = -\frac{b}{2a} = -\frac{\frac{24}{9}}{2\left(-\frac{4}{9}\right)} = 3.$

The height at $x = 3$ is $-\frac{4}{9}(3)^2 + \frac{24}{9}(3) + 12 = 16.$

The maximum height of the dive is 16 feet.

60. (a)

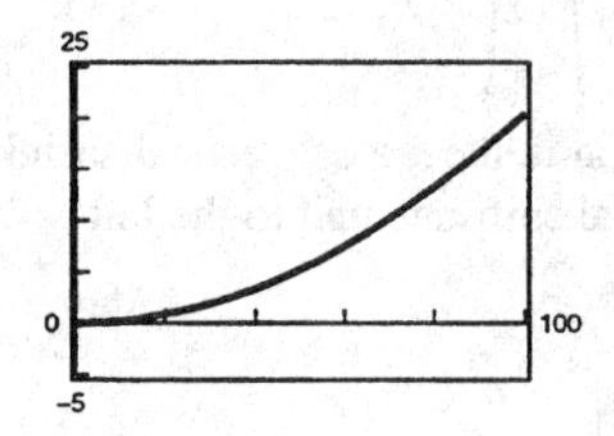

(b) $0.002s^2 + 0.005s - 0.029 = 10$

$$2s^2 + 5s - 29 = 10{,}000$$
$$2s^2 + 5s - 10{,}029 = 0$$

$a = 2, b = 5, c = -10{,}029$

$$s = \frac{-5 \pm \sqrt{5^2 - 4(2)(-10{,}029)}}{2(2)}$$
$$s = \frac{-5 \pm \sqrt{80{,}257}}{4}$$
$$s \approx -72.1, 69.6$$

The maximum speed if power is not to exceed 10 horsepower is 69.6 miles per hour.

62. (a) $y = -0.73t^2 + 19.46t - 58.77$

(b)

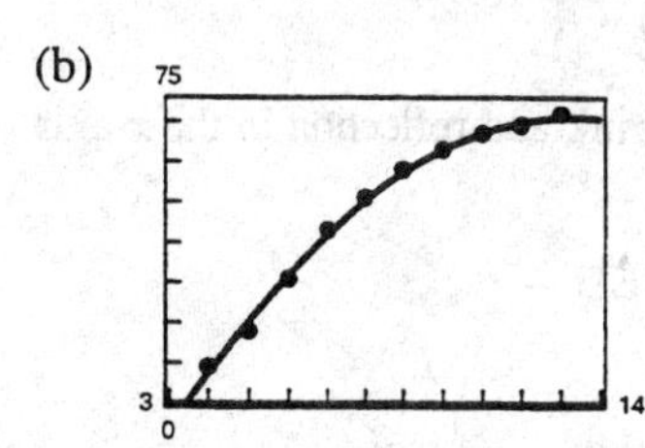

(c) No, the model would probably not give accurate predictions in 2000 because the model will begin to decrease and will eventually become negative.

Section 2.2 Polynomial Functions of Higher Degree

Solutions to Even-Numbered Exercises

2. $f(x) = x^2 - 4x$ is a parabola with intercepts $(0, 0)$ and $(4, 0)$ and opens upward. Matches graph (h).

4. $f(x) = 2x^3 - 3x + 1$ has intercepts $(0, 1)$, $(1, 0)$, $\left(-\frac{1}{2} - \frac{1}{2}\sqrt{3}, 0\right)$ and $\left(-\frac{1}{2} + \frac{1}{2}\sqrt{3}, 0\right)$. Matches graph (a).

6. $f(x) = -\frac{1}{3}x^3 + x^2 - \frac{4}{3}$ has y-intercept $\left(0, -\frac{4}{3}\right)$. Matches graph (d).

8. $f(x) = \frac{1}{5}x^5 - 2x^3 + \frac{9}{5}x$ has intercepts $(0, 0)$, $(1, 0)$, $(-1, 0)$, $(3, 0)$, $(-3, 0)$. Matches (b).

10. $y = x^5$

(a) $f(x) = (x + 1)^5$

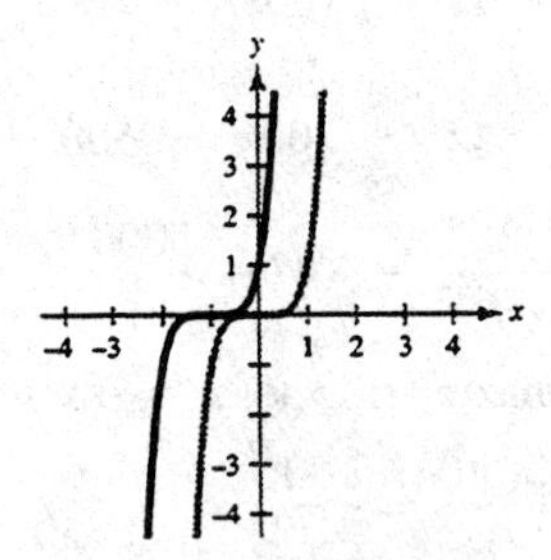

Horizontal shift one unit to the left

(b) $f(x) = x^5 + 1$

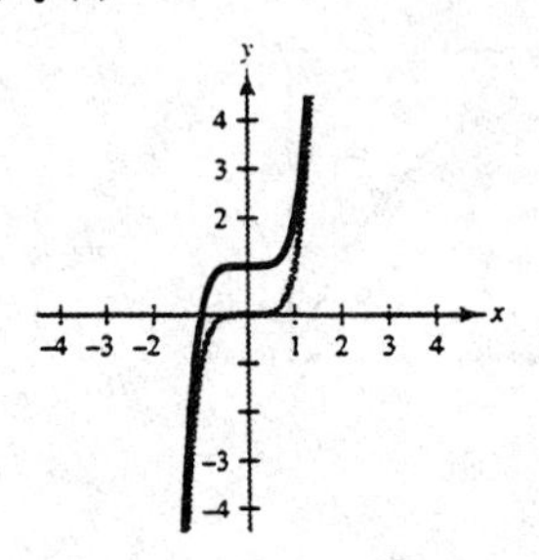

Vertical shift one unit upward

(c) $f(x) = 1 - \frac{1}{2}x^5$

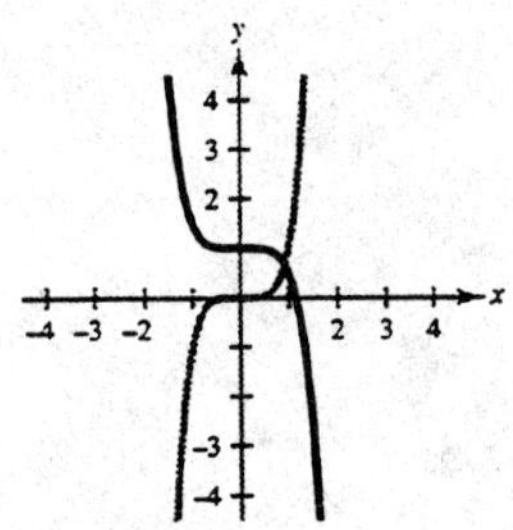

Reflection in the x-axis, vertical shrink and vertical shift one unit upward

(d) $f(x) = -\frac{1}{2}(x + 1)^5$

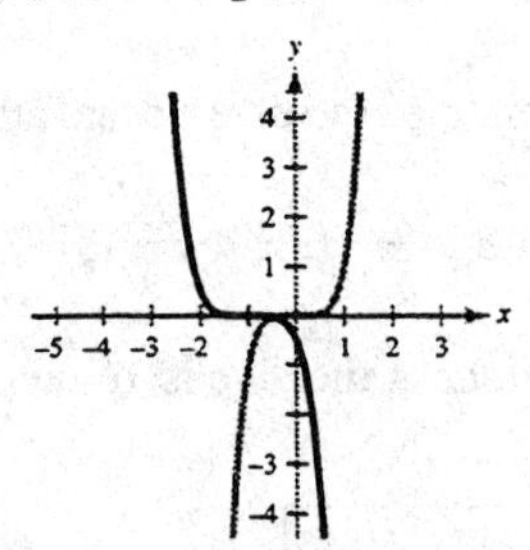

Reflection in the x-axis, vertical shrink and horizontal shift one unit to the left

12. $y = x^6$

(a) $f(x) = -\frac{1}{8}x^6$

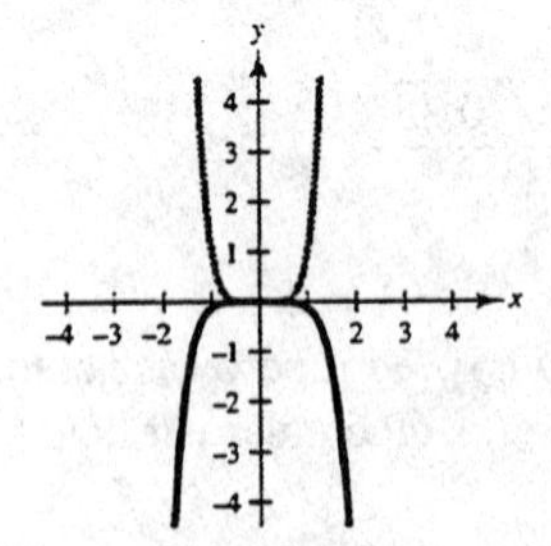

Vertical shrink and reflection in the x-axis

(b) $f(x) = x^6 - 4$

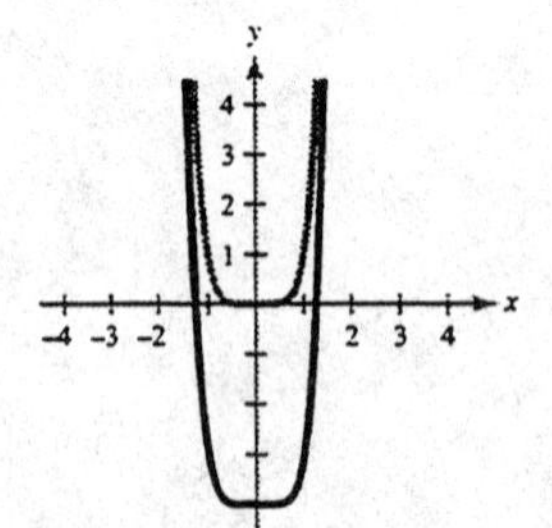

Vertical shift 4 units downward

—CONTINUED—

12. —CONTINUED—

(c) $f(x) = -\frac{1}{4}x^6 + 1$

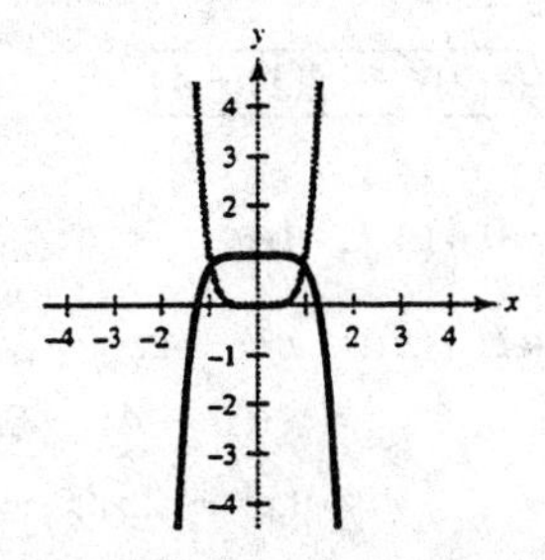

Vertical shrink, vertical shift upward one unit, and reflection in the x-axis

(d) $f(x) = (x + 2)^6 - 4$

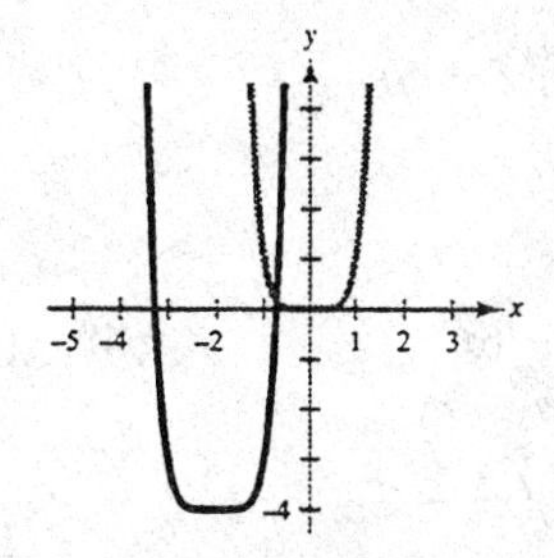

Horizontal shift two units to the left and vertical shift 4 units downward

14. $f(x) = -\frac{1}{3}(x^3 - 3x + 2), g(x) = -\frac{1}{3}x^3$

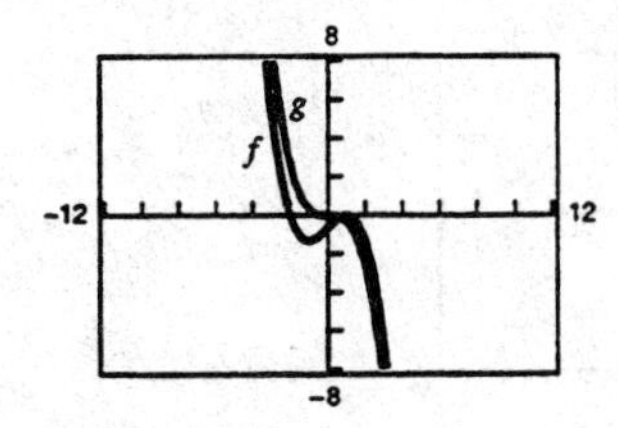

16. $f(x) = 3x^4 - 6x^2, g(x) = 3x^4$

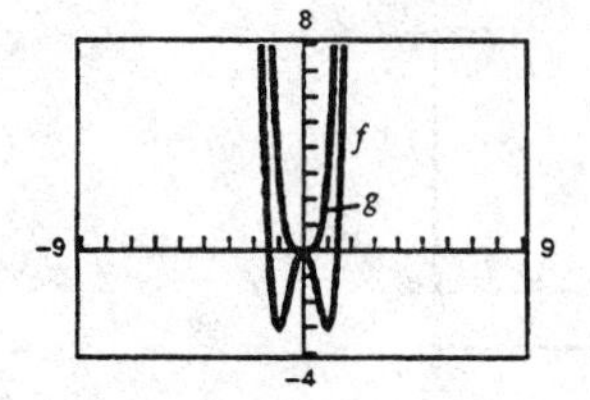

18. $f(x) = \frac{1}{3}x^3 + 5x$

Degree: 3

Leading coefficient: $\frac{1}{3}$

The degree is odd and the leading coefficient is positive. The graph falls to the left and rises to the right.

20. $f(x) = -2.1x^5 + 4x^3 - 2$

Degree: 5

Leading coefficient: -2.1

The degree is odd and the leading coefficient is negative. The graph rises to the left and falls to the right.

22. $h(x) = 1 - x^6$

Degree: 6

Leading coefficient: -1

The degree is even and the leading coefficient is negative. The graph falls to the left and right.

24. $f(x) = \dfrac{3x^4 - 2x + 5}{4}$

Degree: 4

Leading coefficient: $\frac{3}{4}$

The degree is even and the leading coefficient is positive. The graph rises to the left and right.

26. $f(s) = -\frac{7}{8}(s^3 + 5s^2 - 7s + 1)$

Degree: 3

Leading coefficient: $-\frac{7}{8}$

The degree is odd and the leading coefficient is negative. The graph rises to the left and falls to the right.

28. $f(x) = 49 - x^2$

$= (7 - x)(7 + x)$

$x = \pm 7$

30. $f(x) = x^2 + 10x + 25$

$= (x + 5)^2$

$x = -5$

32. $f(x) = \dfrac{1}{2}x^2 + \dfrac{5}{2}x - \dfrac{3}{2}$

$a = \dfrac{1}{2}, b = \dfrac{5}{2}, c = -\dfrac{3}{2}$

$$x = \frac{-\frac{5}{2} \pm \sqrt{\left(\frac{5}{2}\right)^2 - 4\left(\frac{1}{2}\right)\left(-\frac{3}{2}\right)}}{1} = -\frac{5}{2} \pm \sqrt{\frac{37}{4}}$$

$$x = -\frac{5}{2} + \frac{1}{2}\sqrt{37} = \frac{-5 + \sqrt{37}}{2}$$

$$x = -\frac{5}{2} - \frac{1}{2}\sqrt{37} = \frac{-5 - \sqrt{37}}{2}$$

34. $f(x) = x^4 - x^3 - 20x^2$
$= x^2(x^2 - x - 20)$
$= x^2(x + 4)(x - 5)$
$x = 0, -4, 5$

36. $g(x) = 5(x^2 - 2x - 1)$
$a = 1, b = -2, c = -1$
$$x = \frac{-(-2) \pm \sqrt{(-2)^2 - 4(1)(-1)}}{2}$$
$x = 1 \pm \sqrt{2} \approx -0.414, 2.414$

38. $f(x) = x^5 + x^3 - 6x$
$= x(x^4 + x^2 - 6)$
$= x(x^2 + 3)(x^2 - 2)$
$x = 0, \pm\sqrt{2} \approx 0, \pm 1.414$

40. $g(t) = t^5 - 6t^3 + 9t$
$= t(t^4 - 6t^2 + 9)$
$= t(t^2 - 3)^2$
$t = 0, \pm\sqrt{3} \approx 0, \pm 1.732$

42. $f(x) = x^3 - 4x^2 - 25x + 100$
$= x^2(x - 4) - 25(x - 4)$
$= (x^2 - 25)(x - 4)$
$= (x + 5)(x - 5)(x - 4)$
$x = \pm 5, 4$

44. $y = 4x^3 + 4x^2 - 7x + 2$

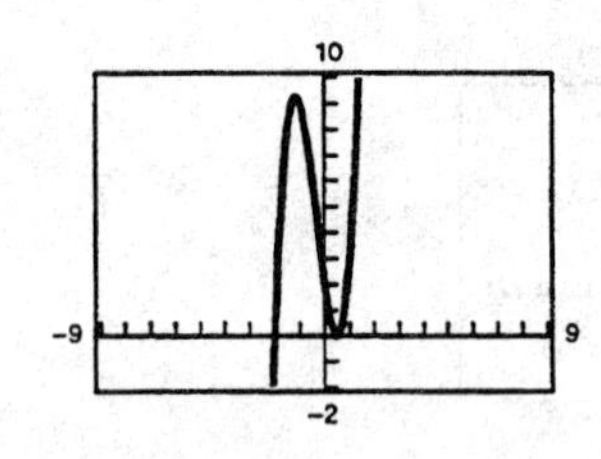

$0 = 4x^3 + 4x^2 - 7x + 2$
$= (2x - 1)(2x^2 + 3x - 2)$
$= (2x - 1)(2x - 1)(x + 2)$
$x = -2, \frac{1}{2}$
x-intercepts: $(-2, 0), \left(\frac{1}{2}, 0\right)$

46. $y = \frac{1}{4}x^3(x^2 - 9)$

$0 = \frac{1}{4}x^3(x^2 - 9)$
$x = 0, \pm 3$
x-intercepts: $(0, 0), (\pm 3, 0)$

48. $f(x) = (x - 0)(x - (-3))$
$= x(x + 3)$
$= x^2 + 3x$
Note: $f(x) = ax(x + 3)$ has zeros 0 and -3 for all real numbers a.

50. $f(x) = (x - (-4))(x - 5)$
$= (x + 4)(x - 5)$
$= x^2 - x - 20$
Note: $f(x) = a(x + 4)(x - 5)$ has zeros -4 and 5 for all real numbers a.

52. $f(x) = (x - 0)(x - 2(x - 5)$
$= x(x - 2)(x - 5)$
$= x(x^2 - 7x + 10)$
$= x^3 - 7x^2 + 10x$
Note: $f(x) = ax(x - 2)(x - 5)$ has zeros 0, 2, 5 for all real numbers a.

54. $f(x) = (x - (-2))(x - (-1))(x - 0)(x - 1)(x - 2)$
$= x(x + 2)(x + 1)(x - 1)(x - 2)$
$= x(x^2 - 4)(x^2 - 1)$
$= x(x^4 - 5x^2 + 4)$
$= x^5 - 5x^3 + 4x$
Note: $f(x) = ax(x + 2)(x + 1)(x - 1)(x - 2)$ has zeros $-2, -1, 0, 1, 2$ for all real numbers a.

56. $f(x) = (x - 2)\left[x - (4 + \sqrt{5})\right]\left[x - (4 - \sqrt{5})\right]$
$= (x - 2)\left[x - (4 + \sqrt{5})\right]\left[(x - 4) - \sqrt{5}\right]$
$= (x - 2)\left[(x - 4)^2 - 5\right]$
$= x(x - 4)^2 - 5x - 2(x - 4)^2 + 10$
$= x^3 - 8x^2 + 16x - 5x - 2x^2 + 16x - 32 + 10$
$= x^3 - 10x^2 + 27x - 22$
Note: $f(x) = a(x^3 - 10x^2 + 27x - 22)$ has these zeros for all real numbers x.

58. (a) $f(x) = 0.11x^3 - 2.07x^2 + 9.81x - 6.88$

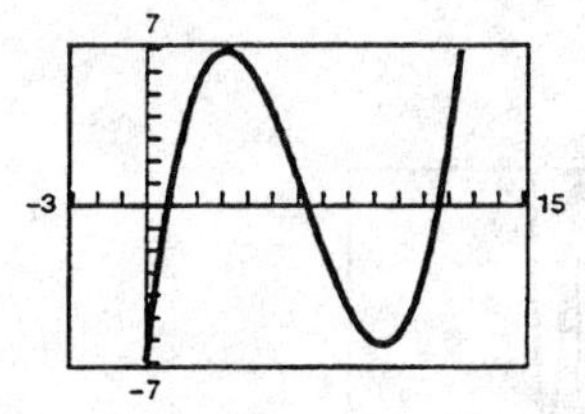

The function has three zeros. They are in the intervals (0, 1), (6, 7), and (11, 12).

(b) Approximate x-values: 0.845, 6.385, 11.588

60. (a) $h(x) = x^4 - 10x^2 + 3$

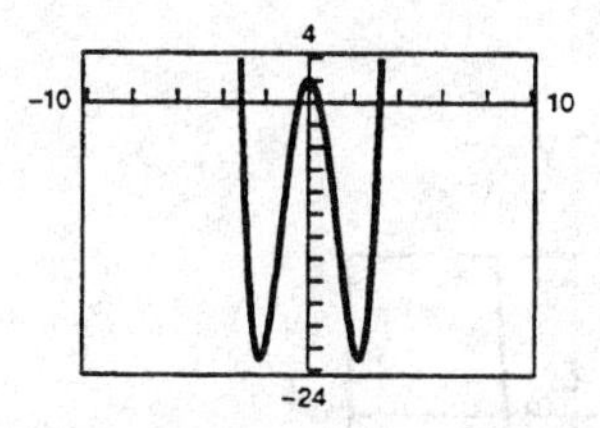

The function has four zeros. They are in the intervals $(-4, -3)$, $(-1, 0)$, $(0, 1)$, and $(3, 4)$.

(b) Approximate x values: -3.130, -0.452, 0.452, 3.130

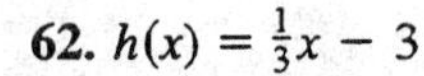

62. $h(x) = \frac{1}{3}x - 3$

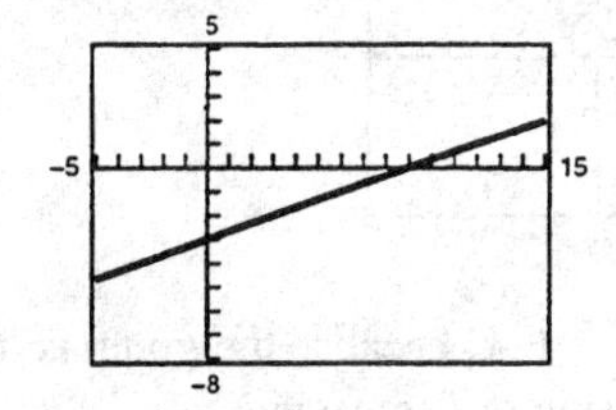

Xmin = -5
Xmax = 15
Xscl = 1
Ymin = -8
Ymax = 5
Yscl = 1

64. $g(x) = -x^2 + 10x - 16$

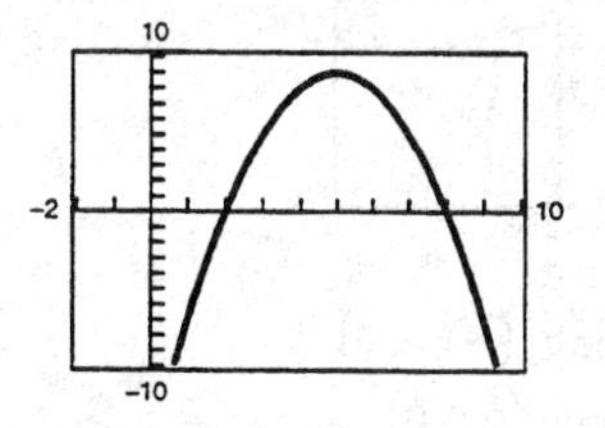

Xmin = -2
Xmax = 10
Xscl = 1
Ymin = -10
Ymax = 10
Yscl = 1

66. $h(x) = \frac{1}{3}x^3(x - 4)^2$

Two zeros: $x = 0, x = 4$

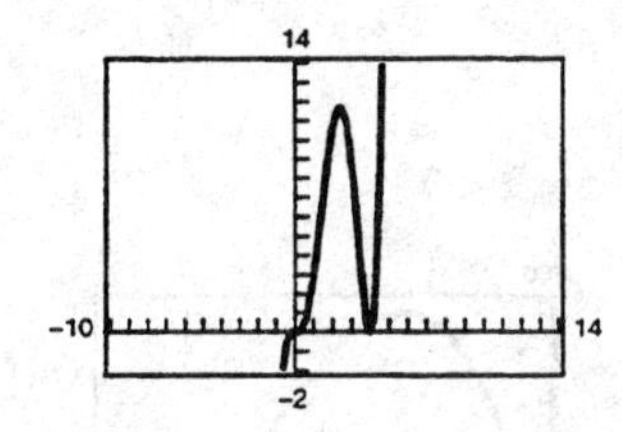

68. $g(x) = \frac{1}{10}(x + 1)^2(x - 3)^3$

Zeros: $-1, 3$

Right: moves up

Left: moves down

x	-2	-1	0	1	2
$f(x)$	-12.5	0	-2.7	-3.2	-0.9

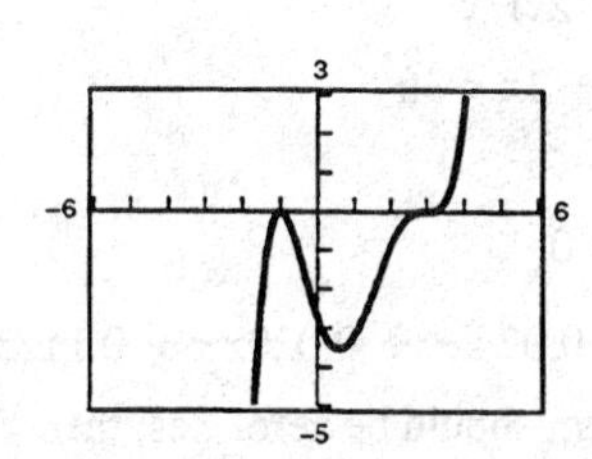

70. $f(x) = \frac{1}{4}x^4 - 2x^2$

Three zeros

y-axis symmetry

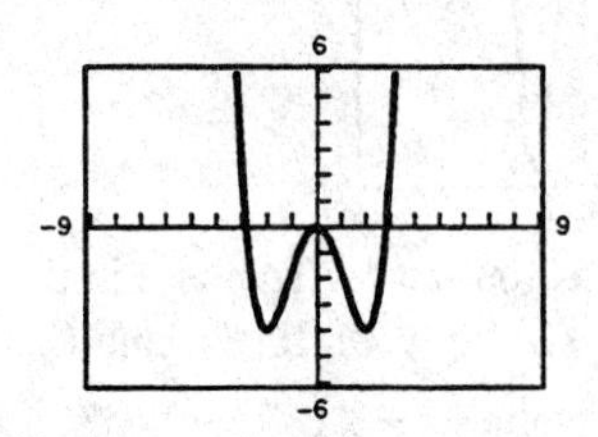

72. $h(x) = \frac{1}{5}(x + 2)^2(3x - 5)^2$

Two zeros

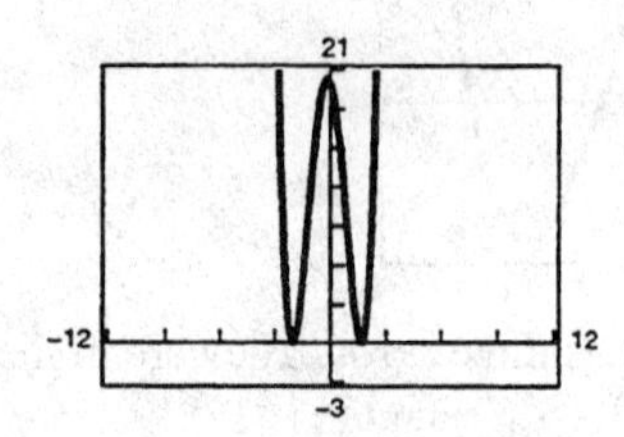

74. (a)

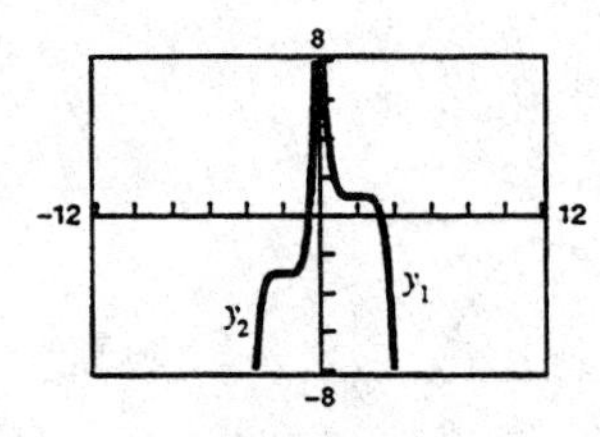

y_1 is decreasing; $\left(-\frac{1}{3} < 0\right)$

y_2 is increasing; $\left(\frac{3}{5} > 0\right)$

(b) Always increasing or decreasing, and determined by a.

(c)

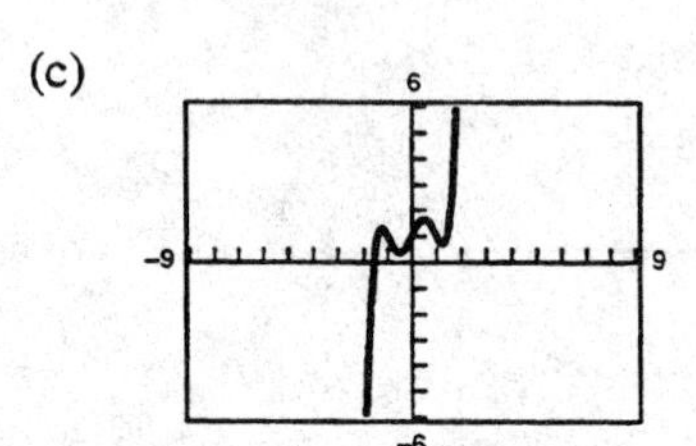

$H \neq a(x - h)^5 + k$, because the graph is not always increasing or decreasing

(d) Odd natural numbers

76. (a) and (b)

Height, x	Length and Width	Volume, V
1	$36 - 2(1)$	$1[36 - 2(1)]^2 = 1156$
2	$36 - 2(2)$	$2[36 - 2(2)]^2 = 2048$
3	$36 - 2(3)$	$3[36 - 2(3)]^2 = 2700$
4	$36 - 2(4)$	$4[36 - 2(4)]^2 = 3136$
5	$36 - 2(5)$	$5[36 - 2(5)]^2 = 3380$
6	$36 - 2(6)$	$6[36 - 2(6)]^2 = 3456$
7	$36 - 2(7)$	$7[36 - 2(7)]^2 = 3388$

(c) Volume = length × width × height

Because the box is made from a square, length = width. Thus:

$$\text{Volume} = (\text{length})^2 \times \text{height}$$
$$= (36 - 2x)^2 x$$

Domain: $0 < 36 - 2x < 36$

$$-36 < -2x < 0$$
$$18 > x > 0$$

(d) $5 < x < 7$

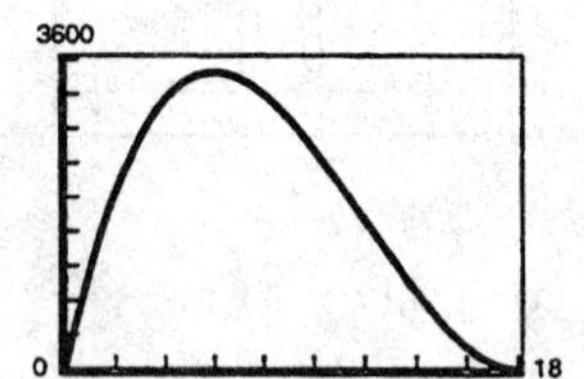

$x = 6$ when $V(x)$ is maximum.

78. (a) $y = 0.003x^4 - 0.024x^3 + 0.020x^2 + 0.113x$

(c) The constant term should be zero. Yes, the model has zero as its constant term.

(b)

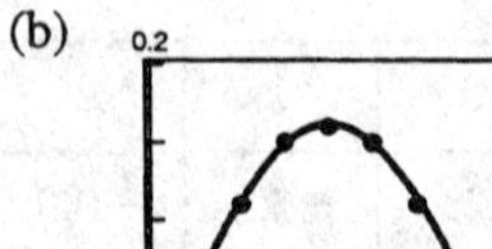

Section 2.3 Real Zeros of Polynomial Functions

Solutions to Even-Numbered Exercises

2. $y_2 = 3 + \dfrac{4}{x-3}$

$= \dfrac{3(x-3)+4}{x-3}$

$= \dfrac{3x-9+4}{x-3}$

$= \dfrac{3x-5}{x-3}$

$= y_1$

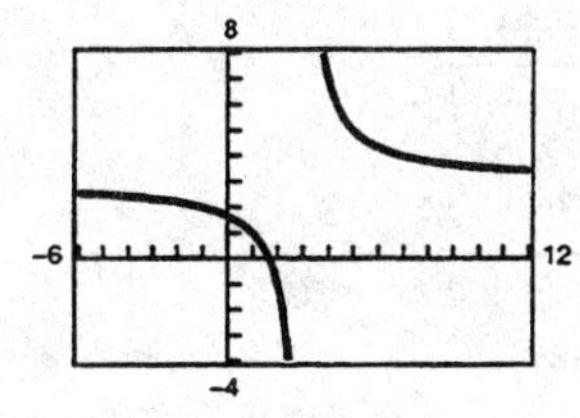

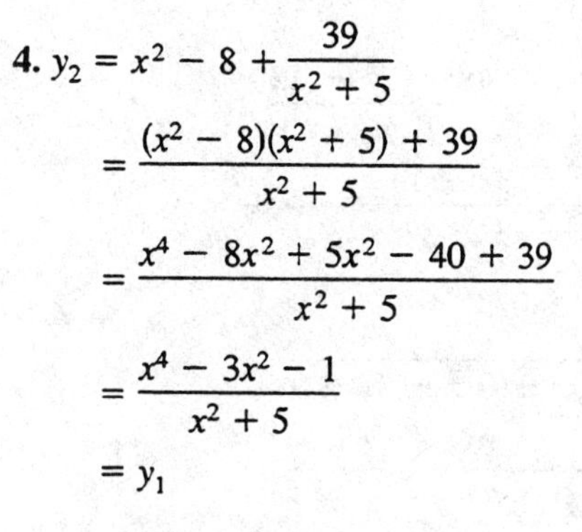

4. $y_2 = x^2 - 8 + \dfrac{39}{x^2+5}$

$= \dfrac{(x^2-8)(x^2+5)+39}{x^2+5}$

$= \dfrac{x^4 - 8x^2 + 5x^2 - 40 + 39}{x^2+5}$

$= \dfrac{x^4 - 3x^2 - 1}{x^2+5}$

$= y_1$

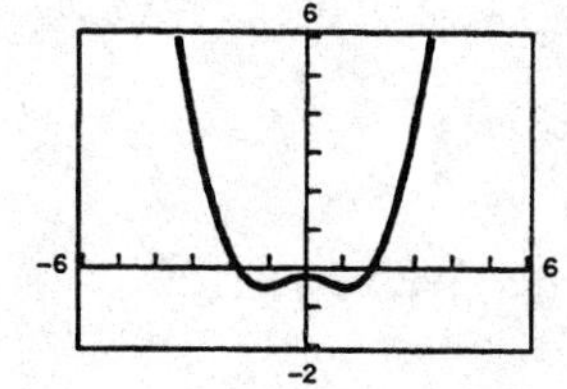

6. $y_2 = x - 3 + \dfrac{2(x+4)}{x^2+x+1}$

$= \dfrac{x(x^2+x+1) - 3(x^2+x+1) + 2(x+4)}{x^2+x+1}$

$= \dfrac{x^3 + x^2 + x - 3x^2 - 3x - 3 + 2x + 8}{x^2+x+1}$

$= \dfrac{x^3 - 2x^2 + 5}{x^2+x+1}$

$= y_1$

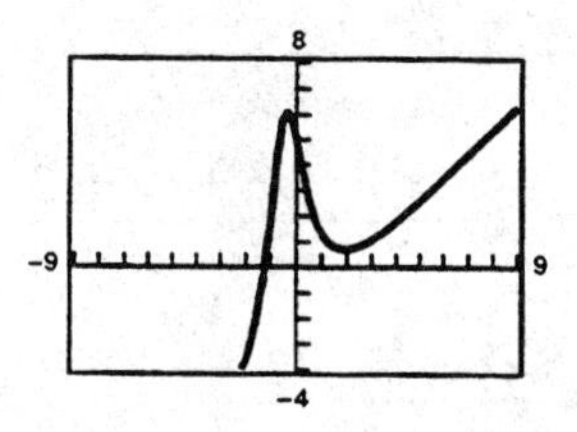

8.

$$\begin{array}{r}
5x + 3 \\
x - 4\overline{)\,5x^2 - 17x - 12} \\
\underline{5x^2 - 20x} \\
3x - 12 \\
\underline{3x - 12} \\
0
\end{array}$$

$\dfrac{5x^2 - 17x - 12}{x-4} = 5x + 3$

10.

$$\begin{array}{r}
2x^2 - 4x + 3 \\
3x - 2\overline{)\,6x^3 - 16x^2 + 17x - 6} \\
\underline{6x^3 - 4x^2} \\
-12x^2 + 17x \\
\underline{-12x^2 + 8x} \\
9x - 6 \\
\underline{9x - 6} \\
0
\end{array}$$

$\dfrac{6x^3 - 16x^2 + 17x - 6}{3x-2} = 2x^2 - 4x + 3$

12.
$$\begin{array}{r} x+4 \\ x^2-3\overline{)x^3+4x^2-3x-12} \\ \underline{x^3 \qquad -3x \qquad\quad} \\ 4x^2 \qquad -12 \\ \underline{4x^2 \qquad -12} \\ 0 \end{array}$$

$$\frac{x^3+4x^2-3x-12}{x^2-3} = x+4$$

14.
$$\begin{array}{r} 4 \\ 2x+1\overline{)8x-5} \\ \underline{8x+4} \\ -9 \end{array}$$

$$\frac{8x-5}{2x+1} = 4 - \frac{9}{2x+1}$$

16.
$$\begin{array}{r} x \qquad\qquad\quad \\ x^2+1\overline{)x^3+0x^2+0x-9} \\ \underline{x^3 \qquad\quad +x \qquad} \\ -x-9 \end{array}$$

$$\frac{x^3-9}{x^2+1} = x - \frac{x+9}{x^2+1}$$

18.
$$\begin{array}{r} x^2 \qquad\qquad\qquad\qquad\quad \\ x^3-1\overline{)x^5+0x^4+0x^3+0x^2+0x+7} \\ \underline{x^5 \qquad\qquad\qquad -x^2 \qquad\qquad} \\ x^2+7 \end{array}$$

$$\frac{x^5+7}{x^3-1} = x^2 + \frac{x^2+7}{x^3-1}$$

20. $(x-1)^3 = x^3-3x^2+3x-1$

$$\begin{array}{r} x+3 \\ x^3-3x^2+3x-1\overline{)x^4 \qquad\qquad\qquad\qquad} \\ \underline{x^4-3x^3+3x^2-x \qquad} \\ 3x^3-3x^2+x \qquad \\ \underline{3x^3-9x^2+9x-3} \\ 6x^2-8x+3 \end{array}$$

$$\frac{x^4}{(x-1)^3} = x+3+\frac{6x^2-8x+3}{(x-1)^3}$$

22.
$$\begin{array}{r} x^{2n}-x^n+3 \\ x^n-2\overline{)x^{3n}-3x^{2n}+5x^n-6} \\ \underline{x^{3n}-2x^{2n} \qquad\qquad\quad} \\ -x^{2n}+5x^n \qquad \\ \underline{-x^{2n}+2x^n \qquad} \\ 3x^n-6 \\ \underline{3x^n-6} \\ 0 \end{array}$$

$$\frac{x^{3n}-3x^{2n}+5x^n-6}{x^n-2} = x^{2n}-x^n+3$$

24.
$$\begin{array}{r|rrrr} 2 & 9 & -18 & -16 & 32 \\ & & 18 & 0 & -32 \\ \hline & 9 & 0 & -16 & 0 \end{array}$$

$$\frac{9x^3-18x^2-16x+32}{x-2} = 9x^2-16$$

26.
$$\begin{array}{r|rrrr} -2 & 5 & 0 & 6 & 8 \\ & & -10 & 20 & -52 \\ \hline & 5 & -10 & 26 & -44 \end{array}$$

$$\frac{5x^3+6x+8}{x+2} = 5x^2-10x+26-\frac{44}{x+2}$$

28.
$$\begin{array}{r|rrrr} -3 & 5 & 0 & 0 & 0 \\ & & -15 & 45 & -135 \\ \hline & 5 & -15 & 45 & -135 \end{array}$$

$$\frac{5x^3}{x+3} = 5x^2-15x+45-\frac{135}{x+3}$$

30.
$$\begin{array}{r|rrrrr} -2 & -3 & 0 & 0 & 0 & 0 \\ & & 6 & -12 & 24 & -48 \\ \hline & -3 & 6 & -12 & 24 & -48 \end{array}$$

$$\frac{-3x^4}{x+2} = -3x^3+6x^2-12x+24-\frac{48}{x+2}$$

32.
$$\begin{array}{r|rrrr} \frac{3}{2} & 3 & -4 & 0 & 5 \\ & & \frac{9}{2} & \frac{3}{4} & \frac{9}{8} \\ \hline & 3 & \frac{1}{2} & \frac{3}{4} & \frac{49}{8} \end{array}$$

$$\frac{3x^3-4x^2+5}{x-\frac{3}{2}} = 3x^2+\frac{1}{2}x+\frac{3}{4}+\frac{49}{8x-12}$$

34. You can check polynomial division by multiplying the quotient by the divisor. This should yield the original dividend if the multiplication was performed correctly.

36. $f(x) = 15x^4 + 10x^3 - 6x^2 + 14, \quad k = -\dfrac{2}{3}$

$$\begin{array}{r|rrrrr} -\frac{2}{3} & 15 & 10 & -6 & 0 & 14 \\ & & -10 & 0 & 4 & -\frac{8}{3} \\ \hline & 15 & 0 & -6 & 4 & \frac{34}{3} \end{array}$$

$$f(x) = \left(x + \frac{2}{3}\right)(15x^3 - 6x + 4) + \frac{34/3}{[x + (2/3)]}$$

$$f\left(-\frac{2}{3}\right) = \frac{34}{3}$$

38. $f(x) = 4x^3 - 6x^2 - 12x - 4, \quad k = 1 - \sqrt{3}$

$$\begin{array}{r|rrrr} 1-\sqrt{3} & 4 & -6 & -12 & -4 \\ & & 4-4\sqrt{3} & 10-2\sqrt{3} & 4 \\ \hline & 4 & -2-4\sqrt{3} & -2-2\sqrt{3} & 0 \end{array}$$

$$f(x) = \left(x - 1 + \sqrt{3}\right)\left[4x^2 - \left(2 + 4\sqrt{3}\right)x - \left(2 + 2\sqrt{3}\right)\right]$$

$$f\left(1 - \sqrt{3}\right) = 0$$

40. $f(x) = 0.4x^4 - 1.6x^3 + 0.7x^2 - 2$

(a)
$$\begin{array}{r|rrrrl} 1 & 0.4 & -1.6 & 0.7 & 0 & -2 \\ & & 0.4 & -1.2 & -0.5 & -0.5 \\ \hline & 0.4 & -1.2 & -0.5 & -0.5 & -2.5 = f(1) \end{array}$$

(b)
$$\begin{array}{r|rrrrl} -2 & 0.4 & -1.6 & 0.7 & 0 & -2 \\ & & -0.8 & 4.8 & -11 & 22 \\ \hline & 0.4 & -2.4 & 5.5 & -11 & 20 = f(-2) \end{array}$$

(c)
$$\begin{array}{r|rrrrl} 5 & 0.4 & -1.6 & 0.7 & 0 & -2 \\ & & 2.0 & 2.0 & 13.5 & 67.5 \\ \hline & 0.4 & 0.4 & 2.7 & 13.5 & 65.5 = f(5) \end{array}$$

(d)
$$\begin{array}{r|rrrrl} -10 & 0.4 & -1.6 & 0.7 & 0 & -2 \\ & & -4.0 & 56.0 & -567 & 5670 \\ \hline & 0.4 & -5.6 & 56.7 & -567 & 5668 = f(-10) \end{array}$$

42.
$$\begin{array}{r|rrrr} -4 & 1 & 0 & -28 & -48 \\ & & -4 & 16 & 48 \\ \hline & 1 & -4 & -12 & 0 \end{array}$$

$$\begin{aligned} x^3 - 28x - 48 &= (x + 4)(x^2 - 4x - 12) \\ &= (x + 4)(x - 6)(x + 2) \end{aligned}$$

Zeros: $-4, -2, 6$

44.
$$\begin{array}{r|rrrr} \frac{2}{3} & 48 & -80 & 41 & -6 \\ & & 32 & -32 & 6 \\ \hline & 48 & -48 & 9 & 0 \end{array}$$

$$\begin{aligned} 48x^3 - 80x^2 + 41x - 6 &= \left(x - \tfrac{2}{3}\right)(48x^2 - 48x + 9) \\ &= \left(x - \tfrac{2}{3}\right)(4x - 3)(12x - 3) \\ &= (3x - 2)(4x - 3)(4x - 1) \end{aligned}$$

Zeros: $\frac{2}{3}, \frac{3}{4}, \frac{1}{4}$

46.
$$\begin{array}{r|rrrr} \sqrt{2} & 1 & 2 & -2 & -4 \\ & & \sqrt{2} & 2\sqrt{2}+2 & 4 \\ \hline & 1 & 2+\sqrt{2} & 2\sqrt{2} & 0 \end{array}$$

$$\begin{aligned} x^3 + 2x^2 - 2x - 4 &= (x - \sqrt{2})\left[x^2 + (2 + \sqrt{2})x + 2\sqrt{2}\right] \\ &= (x - \sqrt{2})(x + 2)(x + \sqrt{2}) \end{aligned}$$

Zeros: $-2, -\sqrt{2}, \sqrt{2}$

48.
$$\begin{array}{r|rrrr} 2-\sqrt{5} & 1 & -1 & -13 & -3 \\ & & 2-\sqrt{5} & 7-3\sqrt{5} & 3 \\ \hline & 1 & 1-\sqrt{5} & -6-3\sqrt{5} & 0 \end{array}$$

$$\begin{aligned} x^3 - x^2 - 13x - 3 &= \left[x - (2 - \sqrt{5})\right]\left[x^2 + (1 - \sqrt{5})x - (6 + 3\sqrt{5})\right] \\ &= (x - 2 + \sqrt{5})(x - 2 - \sqrt{5})(x + 3) \end{aligned}$$

Zeros: $2 - \sqrt{5}, 2 + \sqrt{5}, -3$

50. $f(x) = (x - k)q(x) + r$

(a) $5 = (2 - k)q(2) + r.$

Let $k = 2$, $q(x) = x^2$, $r = 5$.

$f(x) = (x - 2)x^2 + 5 = x^3 - 2x^2 + 5$

(b) $1 = (-3 - k)q(-3) + r.$

Let $k = -3$, $q(x) = -x^2$, $r = 1$.

$f(x) = (x + 3)(-x^2) + 1 = -x^3 - 3x^2 + 1$

52. $f(x) = x^3 - 4x^2 - 4x + 16$

$p =$ factor of 16

$q =$ factor of 1

Possible rational zeros: $\pm 1, \pm 2, \pm 4, \pm 8, \pm 16$

Zeros shown on graph: $-2, 2, 4$

54. $f(x) = 6x^3 - 71x^2 - 13x + 12$

$p =$ factor of 12

$q =$ factor of 6

Possible rational zeros: $\pm 1, \pm 2, \pm 3, \pm 4, \pm 6, \pm 12,$
$\pm \frac{1}{2}, \pm \frac{1}{3}, \pm \frac{1}{6}, \pm \frac{2}{3}, \pm \frac{3}{2}, \pm \frac{4}{3}$

Zeros shown on graph: $-\frac{1}{2}, \frac{1}{3}, 12$

56. $f(x) = -3x^3 + 20x^2 - 36x + 16$

(a) Possible real zeros: $\pm 1, \pm 2, \pm 4, \pm 8, \pm 16, \pm \frac{1}{3},$
$\pm \frac{2}{3}, \pm \frac{4}{3}, \pm \frac{8}{3}, \pm \frac{16}{3}$

(b)

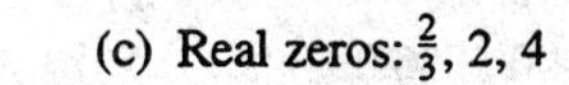

(c) Real zeros: $\frac{2}{3}, 2, 4$

58. $f(x) = 4x^3 - 12x^2 - x + 15$

(a) Possible real zeros: $\pm 1, \pm 3, \pm 5, \pm 15, \pm \frac{1}{2}, \pm \frac{3}{2},$
$\pm \frac{5}{2}, \pm \frac{15}{2}, \pm \frac{1}{4}, \pm \frac{3}{4}, \pm \frac{5}{4}, \pm \frac{15}{4}$

(b)

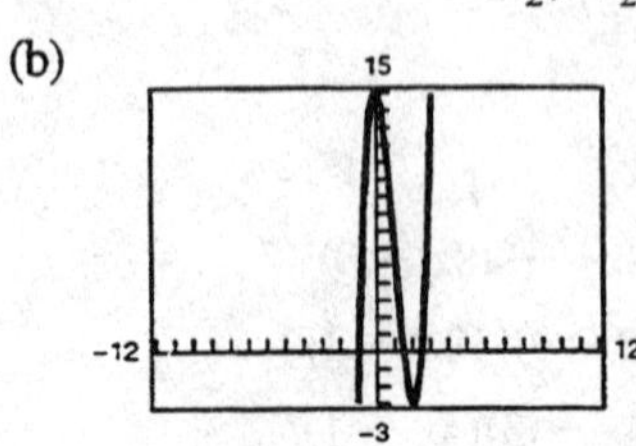

(c) Real zeros: $-1, \frac{3}{2}, \frac{5}{2}$

60. $f(x) = 4x^4 - 17x^2 + 4$

(a) Possible real zeros: $\pm 1, \pm 2, \pm 4, \pm \frac{1}{2}, \pm \frac{1}{4}$

(b)

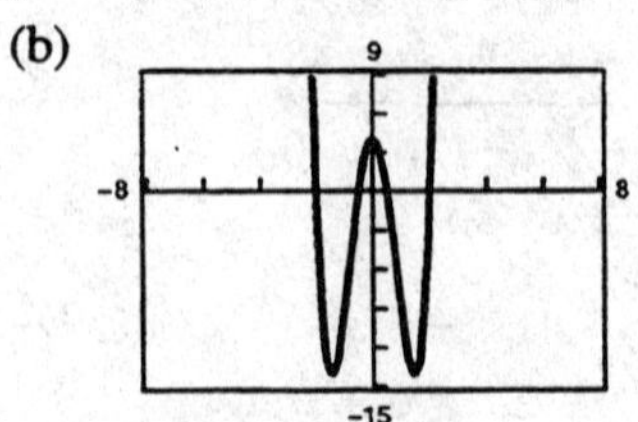

(c) Real zeros: $\pm 2, \pm \frac{1}{2}$

62. $f(x) = 6x^3 - x^2 - 13x + 8$

(a) $\pm\frac{1}{6}, \pm\frac{1}{3}, \pm\frac{1}{2}, \pm\frac{2}{3}, \pm 1, \pm\frac{4}{3}, \pm 2, \pm\frac{8}{3}, \pm 4, \pm 8$

(b)

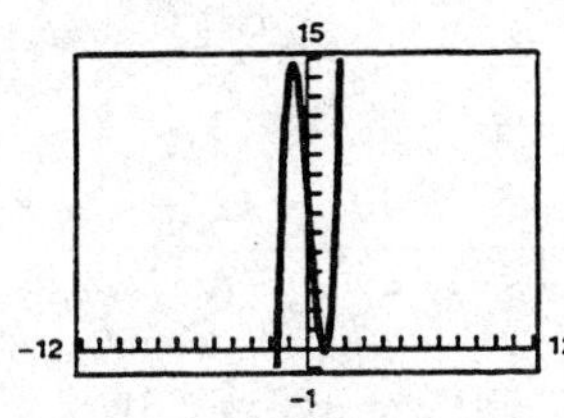

(c) Real zeros: $1, \dfrac{-5 \pm \sqrt{217}}{12}$

64. $f(x) = 2x^3 + 5x^2 - 21x - 10$

(a) $\pm\frac{1}{2}, \pm 1, \pm 2, \pm\frac{5}{2}, \pm 5, \pm 10$

(b)

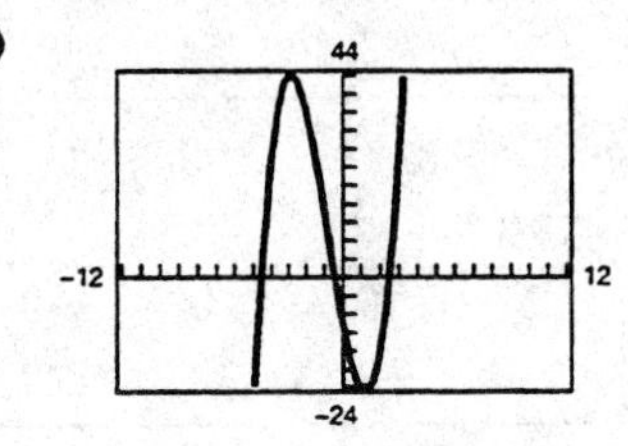

(c) Real zeros: $\dfrac{5}{2}, \dfrac{-5 \pm \sqrt{17}}{2}$

66. $x^4 - x^3 - 29x^2 - x - 30 = 0$. Using a graphing utility and synthetic division, $x = 6$ and $x = -5$ are rational zeros. Hence,
$(x - 6)(x + 5)(x^2 + 1) = 0 \Longrightarrow x = -5, 6.$

68. $2y^4 + 7y^3 - 26y^2 + 23y - 6 = 0$
Using a graphing utility and synthetic division, $1/2$, 1, and -6 are rational zeros. Hence,
$(y + 6)(y - 1)^2\left(y - \frac{1}{2}\right) = 0 \Longrightarrow y = -6, 1, \frac{1}{2}.$

70. $x^5 - x^4 - 3x^3 + 5x^2 - 2x = 0$

$x(x^4 - x^3 - 3x^2 + 5x - 2) = 0$

$$\begin{array}{r|rrrrr} 1 & 1 & -1 & -3 & 5 & -2 \\ & & 1 & 0 & -3 & 2 \\ \hline & 1 & 0 & -3 & 2 & 0 \end{array}$$

$$\begin{array}{r|rrrr} -2 & 1 & 0 & -3 & 2 \\ & & -2 & 4 & -2 \\ \hline & 1 & -2 & 1 & 0 \end{array}$$

$x(x - 1)(x + 2)(x^2 - 2x + 1) = 0$

$x(x - 1)(x + 2)(x - 1)(x - 1) = 0$

The real zeros are $-2, 0, 1$.

72. $P(t) = t^4 - 7t^2 + 12$

(a) $t = \pm 2, \pm\sqrt{3}$

(b) $P(t) = t^4 - 7t^2 + 12$
$= (t^2 - 3)(t^2 - 4)$
$= \left(t - \sqrt{3}\right)\left(t + \sqrt{3}\right)(t - 2)(t + 2)$

74. $g(x) = 6x^4 - 11x^3 - 51x^2 + 99x - 27$

(a) $x = \pm 3, \frac{3}{2}, \frac{1}{3}$

(b)
$$\begin{array}{r|rrrrr} 3 & 6 & -11 & -51 & 99 & -27 \\ & & 18 & 21 & -90 & 27 \\ \hline & 6 & 7 & -30 & 9 & 0 \end{array}$$

$$\begin{array}{r|rrrr} -3 & 6 & 7 & -30 & 9 \\ & & -18 & 33 & -9 \\ \hline & 6 & -11 & 3 & 0 \end{array}$$

$g(x) = (x - 3)(x + 3)(6x^2 - 11x + 3)$
$= (x - 3)(x + 3)(3x - 1)(2x - 3)$

76. $f(x) = 2x^3 - 3x^2 - 12x + 8$

(a)
$$\begin{array}{r|rrrr} 4 & 2 & -3 & -12 & 8 \\ & & 8 & 20 & 32 \\ \hline & 2 & 5 & 8 & 40 \end{array}$$

4 is an upper bound.

(b)
$$\begin{array}{r|rrrr} -3 & 2 & -3 & -12 & 8 \\ & & -6 & 27 & -45 \\ \hline & 2 & -9 & 15 & -37 \end{array}$$

-3 is a lower bound.

78. $f(x) = 2x^4 - 8x + 3$

(a)

3	2	0	0	−8	3
		6	18	54	138
	2	6	18	46	141

3 is an upper bound.

(b)

−4	2	0	0	−8	3
		−8	32	−128	544
	2	−8	32	−136	547

−4 is a lower bound.

80. $f(x) = \frac{1}{2}(2x^3 - 3x^2 - 23x + 12)$

Possible rational zeros: $\pm 1, \pm 2, \pm 3, \pm 4, \pm 6, \pm 12, \pm\frac{1}{2}, \pm\frac{3}{2}$

4	2	−3	−23	12
		8	20	−12
	2	5	−3	0

$f(x) = \frac{1}{2}(x - 4)(2x^2 + 5x - 3)$

$= \frac{1}{2}(x - 4)(2x - 1)(x + 3)$

Rational zeros: $-3, \frac{1}{2}, 4$

82. $f(x) = \frac{1}{6}(6z^3 + 11z^2 - 3z - 2)$

Possible rational zeros: $\pm 1, \pm 2, \pm\frac{1}{2}, \pm\frac{1}{3}, \pm\frac{2}{3}, \pm\frac{1}{6}$

−2	6	11	−3	−2
		−12	2	2
	6	−1	−1	0

$f(x) = \frac{1}{6}(x + 2)(6x^2 - x - 1)$

$= \frac{1}{6}(x + 2)(3x + 1)(2x - 1)$

Rational zeros: $-2, -\frac{1}{3}, \frac{1}{2}$

84. $f(x) = x^3 - 2$

$= \left(x - \sqrt[3]{2}\right)\left(x^2 + \sqrt[3]{2}x + \sqrt[3]{4}\right)$

Rational zeros: 0

Irrational zeros: 1 $\left(x = \sqrt[3]{2}\right)$

Matches (a).

86. $f(x) = x^3 - 2x$

$= x(x^2 - 2)$

$= x\left(x + \sqrt{2}\right)\left(x - \sqrt{2}\right)$

Rational zeros: 1 $(x = 0)$

Irrational zeros: 2 $\left(x = \pm\sqrt{2}\right)$

Matches (c).

88. $g(x) = 3f(x)$. This function would also have the same zeros as $f(x)$ because $g(x) = 3f(x) = 3(x - r_1)(x - r_2)(x - r_3)$.

The roots are r_1, r_2, and r_3

90. $g(x) = f(2x)$

$= (2x - r_1)(2x - r_2)(2x - r_3)$

The roots are $\frac{r_1}{2}, \frac{r_2}{2}$, and $\frac{r_3}{2}$.

92. $g(x) = f(-x)$

$= (-x - r_1)(-x - r_2)(-x - r_3)$

The roots are $-r_1, -r_2, -r_3$.

94. (a) Combined length and width:

$4x + y = 120 \Rightarrow y = 120 - 4x$

$\text{Volume} = l \cdot w \cdot h = x^2y$

$= x^2(120 - 4x)$

$= 4x^2(30 - x)$

(b)

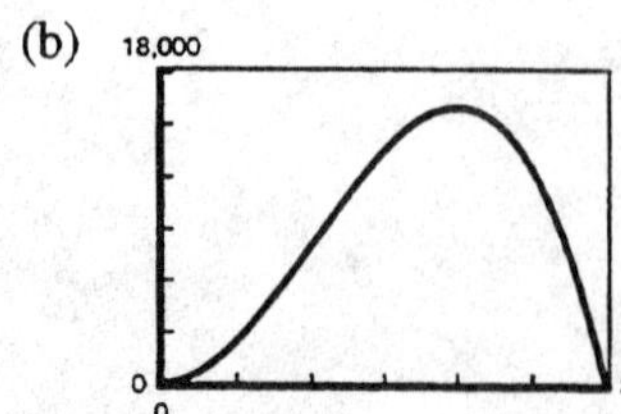

Dimensions with maximum volume: $20 \times 20 \times 40$

(c) $13{,}500 = 4x^2(30 - x)$

$4x^3 - 120x^2 + 13{,}500 = 0$

$x^3 - 30x^2 + 3375 = 0$

15	1	−30	0	3375
		15	−225	−3375
	1	−15	−225	0

$(x - 15)(x^2 - 15x - 225) = 0$

Using the Quadratic equation,

$$x = 15, \frac{15 \pm 15\sqrt{5}}{2}$$

The value of $\frac{15 - 15\sqrt{5}}{2}$ is not possible because it is negative.

96.
$$P = -45x^3 + 2500x^2 - 275{,}000$$
$$800{,}000 = -45x^3 + 2500x^2 - 275{,}000$$
$$0 = 45x^3 - 2500x^2 + 1{,}075{,}000$$
$$0 = 9x^3 - 500x^2 + 215{,}000$$

The zeros of this equation are $x \approx -18.0$, $x \approx 31.5$, and $x \approx 42.0$. Because $0 \le x \le 50$, disregard $x \approx -18.02$. The smaller remaining solution is $x \approx 31.5$, or \$315,000.

Section 2.4 Complex Numbers

Solutions to Even-Numbered Exercises

2. $a + bi = 13 + 4i$
$a = 13$
$b = 4$

4. $(a + 6) + 2bi = 6 - 5i$
$2b = -5$
$b = -\frac{5}{2}$
$a + 6 = 6$
$a = 0$

6. $3 + \sqrt{-16} = 3 + 4i$

8. $1 + \sqrt{-8} = 1 + 2\sqrt{2}i$

10. 45

12. $-4i^2 + 2i = -4(-1) + 2i$
$= 4 + 2i$

14. $(\sqrt{-4})^2 - 5 = -4 - 5 = -9$

16. $\sqrt{-0.0004} = 0.02i$

18. $(13 - 2i) + (-5 + 6i) = 8 + 4i$

20. $(3 + 2i) - (6 + 13i) = 3 + 2i - 6 - 13i = -3 - 11i$

22. $(8 + \sqrt{-18}) - (4 + 3\sqrt{2}i) = 8 + 3\sqrt{2}i - 4 - 3\sqrt{2}i = 4$

24. $22 + (-5 + 8i) + 10i = 17 + 18i$

26. $(1.6 + 3.2i) + (-5.8 + 4.3i) = -4.2 + 7.5i$

28. $\sqrt{-5} \cdot \sqrt{-10} = (\sqrt{5}i)(\sqrt{10}i)$
$= \sqrt{50}i^2 = 5\sqrt{2}(-1) = -5\sqrt{2}$

30. $(\sqrt{-75})^2 = (\sqrt{75}i)^2 = 75i^2 = -75$

32. $(6 - 2i)(2 - 3i) = 12 - 18i - 4i + 6i^2$
$= 12 - 22i - 6$
$= 6 - 22i$

34. $-8i(9 + 4i) = -72i - 32i^2$
$= 32 - 72i$

36. $(3 + \sqrt{-5})(7 - \sqrt{-10}) = (3 + \sqrt{5}i)(7 - \sqrt{10}i)$
$= 21 - 3\sqrt{10}i + 7\sqrt{5}i - \sqrt{50}i^2$
$= 21 + \sqrt{50} + 7\sqrt{5}i - 3\sqrt{10}i$
$= (21 + 5\sqrt{2}) + (7\sqrt{5} - 3\sqrt{10})i$

38. $(2 - 3i)^3 = (4 - 12i + 9i^2)(2 - 3i)$
$= (-5 - 12i)(2 - 3i)$
$= -10 + 15i - 24i + 36i^2$
$= -46 - 9i$

40. $(1 - 2i)^2 - (1 + 2i)^2 = 1 - 4i + 4i^2 - (1 + 4i + 4i^2)$
$= 1 - 4i + 4i^2 - 1 - 4i - 4i^2$
$= -8i$

42. False. If the complex number is real, the number equals its conjugate.

44. The complex conjugate of $9 - 12i$ is $9 + 12i$.
$(9 - 12i)(9 + 12i) = 81 - 144i^2$
$= 81 - (-144)$
$= 225$

46. The complex conjugate of $-4 + \sqrt{2}i$ is $-4 - \sqrt{2}i$.

$$(-4 + \sqrt{2}i)(-4 - \sqrt{2}i) = 16 - 2i^2 = 16 - (-2) = 18$$

48. The complex conjugate of $\sqrt{-15} = \sqrt{15}i$ is $-\sqrt{15}i$.

$$(\sqrt{15}i)(-\sqrt{15}i) = -15i^2 = -(-15) = 15$$

50. The complex conjugate of $1 + \sqrt{8}$ is $1 - \sqrt{8}$.

$$(1 + \sqrt{8})(1 - \sqrt{8}) = 1 - 8 = -7$$

52. $-\dfrac{10}{2i} \cdot \dfrac{-2i}{-2i} = \dfrac{20i}{-4i} = \dfrac{20i}{4} = 5i$

54. $\dfrac{3}{1-i} \cdot \dfrac{1+i}{1+i} = \dfrac{3+3i}{1-i^2} = \dfrac{3+3i}{2} = \dfrac{3}{2} + \dfrac{3}{2}i$

56. $\dfrac{8-7i}{1-2i} \cdot \dfrac{1+2i}{1+2i} = \dfrac{8+16i-7i-14i^2}{1-4i^2} = \dfrac{22+9i}{5} = \dfrac{22}{5} + \dfrac{9}{5}i$

58. $\dfrac{8+20i}{2i} \cdot \dfrac{-2i}{-2i} = \dfrac{-16i-40i^2}{-4i^2} = 10 - 4i$

60. $\dfrac{(2-3i)(5i)}{2+3i} \cdot \dfrac{2-3i}{2-3i} = \dfrac{(10i-15i^2)(2-3i)}{4-9i^2} = \dfrac{20i-30i^2-30i^2+45i^3}{13} = \dfrac{60+20i-45i}{13} = \dfrac{60-25i}{13} = \dfrac{60}{13} - \dfrac{25}{13}i$

62. $\dfrac{2i}{2+i} + \dfrac{5}{2-i} = \dfrac{2i(2-i)}{(2+i)(2-i)} + \dfrac{5(2+i)}{(2+i)(2-i)} = \dfrac{4i-2i^2+10+5i}{4-i^2} = \dfrac{12+9i}{5} = \dfrac{12}{5} + \dfrac{9}{5}i$

64. $\dfrac{1+i}{i} - \dfrac{3}{4-i} = \dfrac{(1+i)(4-i)-3i}{i(4-i)} = \dfrac{4-i+4i-i^2-3i}{4i-i^2} = \dfrac{5}{1+4i} \cdot \dfrac{1-4i}{1-4i} = \dfrac{5-20i}{1-16i^2} = \dfrac{5}{17} - \dfrac{20}{17}i$

66. (a) $i^{40} = i^4 \cdot i^4 \cdot i^4 \cdot i^4 \cdot i^4 \cdot i^4 \cdot i^4 \cdot i^4 \cdot i^4 \cdot i^4 = 1 \cdot 1 \cdot 1 \cdot 1 \cdot 1 \cdot 1 \cdot 1 \cdot 1 \cdot 1 \cdot 1 = 1$

(b) $i^{25} = i^4 \cdot i^4 \cdot i^4 \cdot i^4 \cdot i^4 \cdot i^4 \cdot i = 1 \cdot 1 \cdot 1 \cdot 1 \cdot 1 \cdot 1 \cdot i = i$

(c) $i^{50} = i^{25} \cdot i^{25} = i \cdot i = i^2 = -1$

(d) $i^{67} = i^{50} \cdot i^{17} = -1 \cdot i^4 \cdot i^4 \cdot i^4 \cdot i^4 \cdot i = -i$

68. $4i^2 - 2i^3 = -4 + 2i$

70. $(-i)^3 = (-1)(i^3) = (-1)(-i) = i$

72. $(\sqrt{-2})^6 = (\sqrt{2}i)^6 = 8i^6 - 8i^4i^2 = -8$

74. $\dfrac{1}{(2i)^3} = \dfrac{1}{8i^3} = \dfrac{1}{-8i} \cdot \dfrac{8i}{8i} = \dfrac{8i}{-64i^2} = \dfrac{1}{8}i$

76. $-1 - 2i$

78. $2 - 6i$

80.

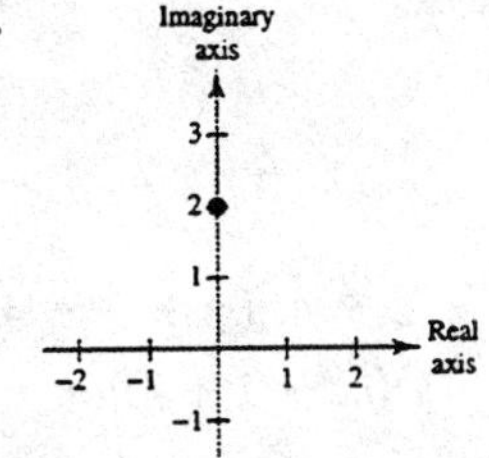

82.

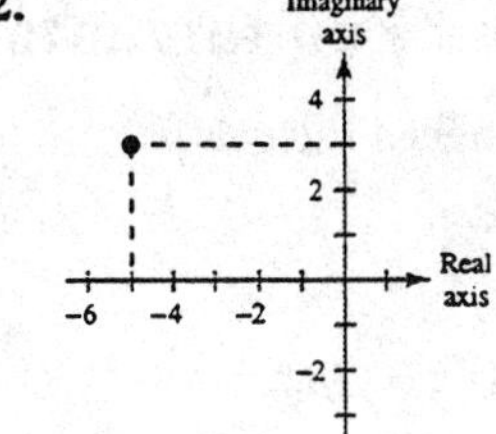

84. 2

$$2^2 + 2 = 6$$
$$6^2 + 2 = 38$$
$$38^2 + 2 = 1446$$
$$1446^2 + 2 = 2{,}090{,}918$$
$$4.4 \times 10^{12}$$

Not bounded. $c = 2$ is not in the Mandelbrot Set.

86. $-i$

$$(-i)^2 - i = -1 - i$$
$$(-1 - i)^2 - i = i$$
$$i^2 - i = -1 - i$$
$$(-1 - i)^2 - i = i$$
$$i^2 - i = -1 - i$$

Bounded. $c = -i$ is in the Mandelbrot Set.

88. -1

$$(-1)^2 - 1 = 0$$
$$0^2 - 1 = -1$$
$$(-1)^2 - 1 = 0$$
$$0^2 - 1 = -1$$
$$(-1)^2 - 1 = 0$$

Bounded. $c = -1$ is in the Mandelbrot Set.

90. $2^4 = 16,\ (-2)^4 = 16$

$$(2i)^4 = 2^4 i^4 = 16(1) = 16$$
$$(-2i)^4 = (-2)^4 i^4 = 16(1) = 16$$

92. $(a + bi)(a - bi) = a^2 - b^2i^2 + abi - abi$

$$= a^2 + b^2, \text{ a real number}$$

94. $(a_1 + b_1i) + (a_2 + b_2i) = (a_1 + a_2) + (b_1 + b_2)i$

Conjugate of sum $= (a_1 + a_2) - (b_1 + b_2)i$

Sum of conjugates $= (a_1 - b_1i) + (a_2 - b_2i)$

$= (a_1 + a_2) - (b_1 + b_2)i$

96. $(4 + 3x) + (8 - 6x - x^2) = -x^2 - 3x + 12$

98. $(2x - 5)^2 = 4x^2 - 20x + 25$

100.

$$V = \frac{4}{3}\pi a^2 b$$
$$\frac{3V}{4\pi b} = a^2$$
$$a = \sqrt{\frac{3V}{4\pi b}} = \frac{1}{2}\sqrt{\frac{3V}{\pi b}},\ (a > 0)$$

102. (Final concentration)(Amount) = (Solution 1 concentration)(Amount) + (Solution 2 concentration)(Amount)

$$(60\%)(5) = (50\%)(5 - x) + (100\%)x$$
$$3 = 0.5(5 - x) + x$$
$$3 = 2.5 + 0.5x$$
$$0.5 = 0.5x$$
$$x = 1 \text{ liter}$$

Section 2.5 The Fundamental Theorem of Algebra

Solutions to Even-Numbered Exercises

2. $g(x) = (x - 2)(x + 4)^3$

$= (x - 2)(x + 4)(x + 4)(x + 4)$

The four zeros are: $2, -4, -4, -4$

4. $h(m) = (m - 4)^2(m - 2 + 4i)(m - 2 - 4i)$

$= (m - 4)(m - 4)(m - 2 + 4i)(m - 2 - 4i)$

The four zeros are: $4, 4, 2 - 4i, 2 + 4i$

6. $f(x) = x^3 - 4x^2 - 4x + 16$

$= x^2(x - 4) - 4(x - 4)$

$= (x^2 - 4)(x - 4)$

$= (x + 2)(x - 2)(x - 4)$

The zeros are: $x = 2, -2$, and 4. This corresponds to the x-intercepts of $(-2, 0)$, $(2, 0)$, and $(4, 0)$ on the graph.

8. $f(x) = x^4 - 3x^2 - 4$

$= (x^2 - 4)(x^2 + 1)$

$= (x + 2)(x - 2)(x^2 + 1)$

The only real zeros are $x = -2, 2$. This corresponds to the x-intercepts of $(-2, 0)$ and $(2, 0)$ on the graph.

10. $f(x) = x^2 - x + 56$

Zeros: $x = \dfrac{1 \pm \sqrt{223}i}{2}$

$$f(x) = \left(x - \frac{1 - \sqrt{223}i}{2}\right)\left(x - \frac{1 + \sqrt{223}i}{2}\right)$$

12. $g(x) = x^2 + 10x + 23$

Zeros: $x = \dfrac{-10 \pm \sqrt{8}}{2} = -5 \pm \sqrt{2}$

$g(x) = (x + 5 + \sqrt{2})(x + 5 - \sqrt{2})$

14. $f(y) = y^4 - 625$

Zeros: $x = \pm 5, \pm 5i$

$f(y) = (y + 5)(y - 5)(y + 5i)(y - 5i)$

16. $h(x) = x^3 - 3x^2 + 4x - 2$

$$\begin{array}{r|rrrr} 1 & 1 & -3 & 4 & -2 \\ & & 1 & -2 & 2 \\ \hline & 1 & -2 & 2 & 0 \end{array}$$

Zeros: $x = 1, \dfrac{2 \pm \sqrt{4}i}{2} = 1 \pm i$

$h(x) = (x - 1)(x - 1 - i)(x - 1 + i)$

18. $f(x) = x^3 + 11x^2 + 39x + 29$

$$\begin{array}{r|rrrr} -1 & 1 & 11 & 39 & 29 \\ & & -1 & -10 & -29 \\ \hline & 1 & 10 & 29 & 0 \end{array}$$

Zeros: $x = -1, \dfrac{-10 \pm \sqrt{16}i}{2} = -5 \pm 2i$

$f(x) = (x + 1)(x + 5 + 2i)(x + 5 - 2i)$

20. $f(s) = 2s^3 - 5s^2 + 12s - 5$

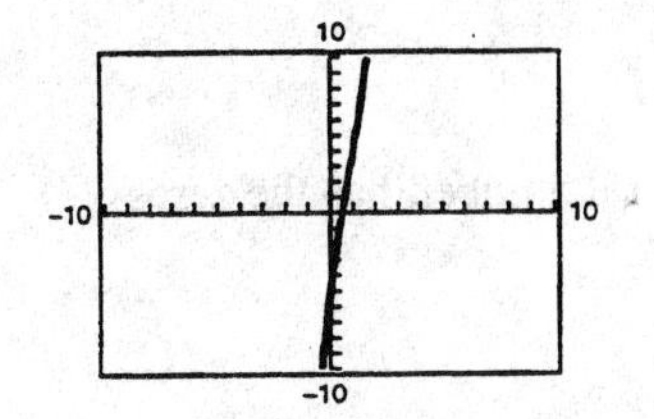

The graph reveals one zero at $x = \frac{1}{2}$.

$$\begin{array}{r|rrrr} \frac{1}{2} & 2 & -5 & 12 & -5 \\ & & 1 & -2 & 5 \\ \hline & 2 & -4 & 10 & 0 \end{array}$$

Zeros: $s = \dfrac{1}{2}, \dfrac{4 \pm \sqrt{64}i}{4} = 1 \pm 2i$

$f(s) = (2s - 1)(s - 1 + 2i)(s - 1 - 2i)$

22. $g(x) = 3x^3 - 4x^2 + 8x + 8$

$$\begin{array}{r|rrrr} -\frac{2}{3} & 3 & -4 & 8 & 8 \\ & & -2 & 4 & -8 \\ \hline & 3 & -6 & 12 & 0 \end{array}$$

Zeros: $x = -\dfrac{2}{3}, \dfrac{6 \pm \sqrt{108}i}{6} = 1 \pm \sqrt{3}i$

$g(x) = (3x + 2)(x - 1 + \sqrt{3}i)(x - 1 - \sqrt{3}i)$

24. $f(x) = x^4 + 29x^2 + 100$

$= (x^2 + 25)(x^2 + 4)$

Zeros: $x = \pm 2i, \pm 5i$

$f(x) = (x + 2i)(x - 2i)(x + 5i)(x - 5i)$

26. $h(x) = x^4 + 6x^3 + 10x^2 + 6x + 9$

$$\begin{array}{r|rrrrr} -3 & 1 & 6 & 10 & 6 & 9 \\ & & -3 & -9 & -3 & -9 \\ \hline -3 & 1 & 3 & 1 & 3 & 0 \\ & & -3 & 0 & -3 & \\ \hline & 1 & 0 & 1 & 0 & \end{array}$$

Zeros: $x = -3, \pm i$

$h(x) = (x + 3)^2(x + i)(x - i)$

28. $g(x) = x^5 - 8x^4 + 28x^3 - 56x^2 + 64x - 32$

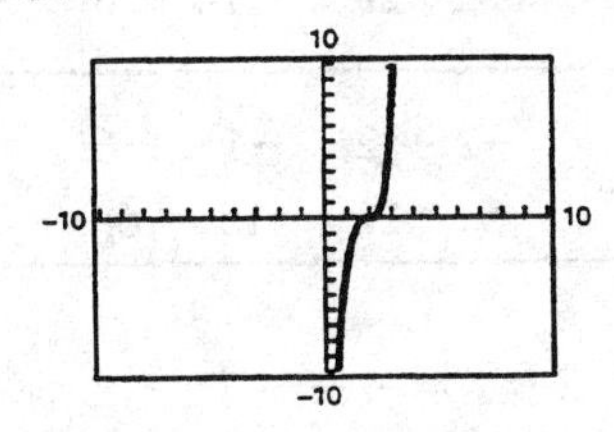

The graph reveals one zero at $x = 2$.

$$\begin{array}{r|rrrrrr} 2 & 1 & -8 & 28 & -56 & 64 & -32 \\ & & 2 & -12 & 32 & -48 & 32 \\ \hline 2 & 1 & -6 & 16 & -24 & 16 & 0 \\ & & 2 & -8 & 16 & -16 & \\ \hline 2 & 1 & -4 & 8 & -8 & 0 & \\ & & 2 & -4 & 8 & & \\ \hline & 1 & -2 & 4 & 0 & & \end{array}$$

Zeros: $x = 2, \dfrac{2 \pm \sqrt{12}i}{2} = 1 \pm \sqrt{3}i$

$g(x) = (x - 2)^3(x - 1 + \sqrt{3}i)(x - 1 - \sqrt{3}i)$

30. $f(x) = (x - 4)(x - 3i)(x + 3i)$

$= (x - 4)(x^2 + 9)$

$= x^3 - 4x^2 + 9x - 36$

Note: $f(x) = a(x^3 - 4x^2 + 9x - 36)$, where a is any real number, has the zeros 4, $3i$ and $-3i$.

32. $f(x) = (x - 6)(x - (-5 + 2i))(x - (-5 - 2i))$

$= (x - 6)(x^2 + 10x + 29)$

$= x^3 + 4x^2 - 31x - 174$

34. $f(x) = (x - 2)^3(x - 4i)(x + 4i)$
$= (x^3 - 6x^2 + 12x - 8)(x^2 + 16)$
$= x^5 - 6x^4 + 28x^3 - 104x^2 + 192x - 128$

Note: $f(x) = a(x^5 - 6x^4 + 28x^3 - 104x^2 + 192x - 128)$,where a is any real number, has the zeros $2, 2, 2, 4i, -4i$.

36. $f(x) = x^2(x - 4)(x - 1 - i)(x - 1 + i)$
$= (x^3 - 4x^2)(x^2 - 2x + 2)$
$= x^5 - 6x^4 + 10x^3 - 8x^2$

Note: $f(x) = a(x^5 - 6x^4 + 10x^3 - 8x^2)$, where a is any real number, has the zeros $0, 0, 4, 1 + i$.

38. $f(x) = x^4 - 2x^3 - 3x^2 + 12x - 18$

(a) $f(x) = (x^2 - 6)(x^2 - 2x + 3)$

(b) $f(x) = (x + \sqrt{6})(x - \sqrt{6})(x^2 - 2x + 3)$

(c) $f(x) = (x + \sqrt{6})(x - \sqrt{6})(x - 1 - \sqrt{2}i)(x - 1 + \sqrt{2}i)$

40. $f(x) = x^4 - 3x^3 - x^2 - 12x - 20$

(a) $f(x) = (x^2 + 4)(x^2 - 3x - 5)$

(b) $f(x) = (x^2 + 4)\left(x - \dfrac{3 + \sqrt{29}}{2}\right)\left(x - \dfrac{3 - \sqrt{29}}{2}\right)$

(c) $f(x) = (x + 2i)(x - 2i)\left(x - \dfrac{3 + \sqrt{29}}{2}\right)\left(x - \dfrac{3 - \sqrt{29}}{2}\right)$

42. $f(x) = x^3 + x^2 + 9x + 9$

Since $3i$ is a zero, so is $-3i$.

$$\begin{array}{r|rrrr} 3i & 1 & 1 & 9 & 9 \\ & & 3i & -9 + 3i & -9 \\ \hline -3i & 1 & 1 + 3i & 3i & 0 \\ & & -3i & -3i & \\ \hline & 1 & 1 & 0 & \end{array}$$

The zero of $x + 1$ is $x = -1$.

The zeros of f are $x = -1, \pm 3i$.

44. $g(x) = x^3 - 7x^2 - x + 87$

Since $5 + 2i$ is a zero, so is $5 - 2i$.

$$\begin{array}{r|rrrr} 5 + 2i & 1 & -7 & -1 & 87 \\ & & 5 + 2i & -14 + 6i & -87 \\ \hline 5 - 2i & 1 & -2 + 2i & -15 + 6i & 0 \\ & & 5 - 2i & 15 - 6i & \\ \hline & 1 & 3 & 0 & \end{array}$$

The zero of $x + 3$ is $x = -3$.

The zeros of f are $x = -3, 5 \pm 2i$.

46. $h(x) = 3x^3 - 4x^2 + 8x + 8$

Since $1 - \sqrt{3}i$ is a zero, so is $1 + \sqrt{3}i$.

$$\begin{array}{r|rrrr} 1 - \sqrt{3}i & 3 & -4 & 8 & 8 \\ & & 3 - 3\sqrt{3}i & -10 - 2\sqrt{3}i & -8 \\ \hline 1 + \sqrt{3}i & 3 & -1 - 3\sqrt{3}i & -2 - 2\sqrt{3}i & 0 \\ & & 3 + 3\sqrt{3}i & 2 + 2\sqrt{3}i & \\ \hline & 3 & 2 & 0 & \end{array}$$

The zero of $3x + 2$ is $x = -\frac{2}{3}$. The zeros of f are $x = \pm 3i, -\frac{2}{3}$.

48. $f(x) = x^3 + 4x^2 + 14x + 20$

Since $-1 - 3i$ is a zero, so is $-1 + 3i$.

$$\begin{array}{r|rrrr} -1-3i & 1 & 4 & 14 & 20 \\ & & -1-3i & -12-6i & -20 \\ \hline & 1 & 3-3i & 2-6i & 0 \end{array}$$

$$\begin{array}{r|rrr} -1+3i & 1 & 3-3i & 2-6i \\ & & -1+3i & -2+6i \\ \hline & 1 & 2 & 0 \end{array}$$

The zero of $x + 2$ is $x = -2$. The zeros of f are $x = -2, -1 \pm 3i$.

50. $f(x) = 25x^3 - 55x^2 - 54x - 18$

Since $\frac{1}{5}(-2 + \sqrt{2}i) = \frac{-2 + \sqrt{2}i}{5}$ is a zero, so is $\frac{-2 - \sqrt{2}i}{5}$.

$$\begin{array}{r|rrrr} \frac{-2+\sqrt{2}i}{5} & 25 & -55 & -54 & -18 \\ & & -10+5\sqrt{2}i & 24-15\sqrt{2}i & 18 \\ \hline \end{array}$$

$$\begin{array}{r|rrrr} \frac{-2-\sqrt{2}i}{5} & 25 & -65+5\sqrt{2}i & -30-15\sqrt{2}i & 0 \\ & & -10-5\sqrt{2}i & 30+15\sqrt{2}i & \\ \hline & 25 & -75 & 0 & \end{array}$$

The zero of $25x - 75$ is $x = 3$. The zeros of f are $x = 3, \frac{-2 \pm \sqrt{2}i}{5}$.

52. $f(x) = x^4 - 4x^2 + k$

(a) f has 4 real zeros if $0 < k < 4$.

(b) f has 2 real zeros each of multiplicity two if $k = 4$. $x^4 - 4k^2 + 4 = (x^2 - 2)^2$.

(c) f has 2 real zeros and 2 complex zeros if $k < 0$.

(d) f has 4 complex zeros if $k > 4$.

54. $-16t^2 + 48t = 64, \ 0 \le t \le 3$

$-16t^2 + 48t - 64 = 0$

$$t = \frac{-48 \pm \sqrt{1792}i}{-32}$$

Since this results in imaginary roots, the ball will never reach a height of 64 feet. Verifying this with a graphing utility, we can graph $y_1 = h = -16t^2 + 48t$ and $y_2 = 64$ which do not intersect.

56. $f(x) = (x - \sqrt{b}i)(x + \sqrt{b}i)$

$= x^2 + b$

58.

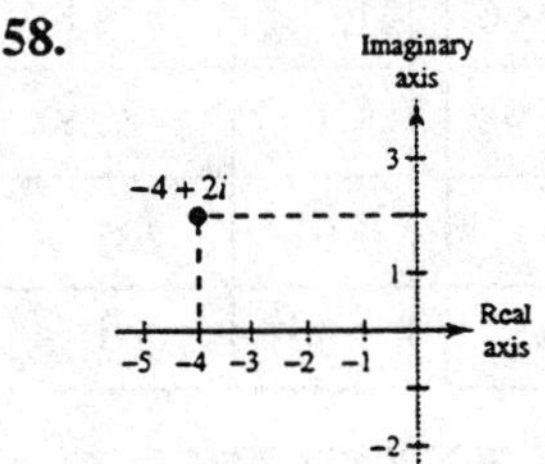

Section 2.6 Rational Functions and Asymptotes

Solutions to Even-Numbered Exercises

2. $f(x) = \dfrac{5x}{x-1}$

(a)

x	$f(x)$
0.5	−5
0.9	−45
0.99	−495
0.999	−4995

x	$f(x)$
1.5	15
1.1	55
1.01	505
1.001	5005

x	$f(x)$
5	6.25
10	$5.\overline{55}$
100	$5.\overline{05}$
1000	$5.\overline{005}$

(b) The zero of the denominator is $x = 1$, so $x = 1$ is a vertical asymptote. The degree of the numerator is equal to the degree of the denominator, so the line $y = \frac{5}{1} = 5$ is a horizontal asymptote.

(c) The domain is all real numbers except $x = 1$.

4. $f(x) = \dfrac{3}{|x-1|}$

(a)

x	$f(x)$
0.5	6
0.9	30
0.99	300
0.999	3000

x	$f(x)$
1.5	6
1.1	30
1.01	300
1.001	3000

x	$f(x)$
5	0.75
10	$0.\overline{33}$
100	$0.\overline{03}$
1000	$0.\overline{003}$

(b) The zero of the denominator is $x = 1$, so $x = 1$ is a vertical asymptote. Because the degree of the numerator is less than the degree of the denominator, the x-axis or $y = 0$ is a horizontal asymptote.

(c) The domain is all real numbers except $x = 1$.

6. $f(x) = \dfrac{4x}{x^2-1}$

(a)

x	$f(x)$
0.5	$-2.\overline{66}$
0.9	−18.95
0.99	−199
0.999	−1999

x	$f(x)$
1.5	4.8
1.1	20.95
1.01	201
1.001	2001

x	$f(x)$
5	$0.\overline{833}$
10	$0.\overline{40}$
100	0.04
1000	0.004

(b) The zeros of the denominator are $x = \pm 1$ so both $x = 1$ and $x = -1$ are vertical asymptotes. Because the degree of the numerator is less than the degree of the denominator, the x-axis or $y = 0$ is a horizontal asymptote.

(c) The domain is all real numbers except $x = \pm 1$.

8. $f(x) = \frac{1}{x - 3}$

Vertical asymptote: $x = 3$
Horizontal asymptote: $y = 0$
Matches graph (d).

10. $f(x) = \frac{1 - x}{x}$

Vertical asymptote: $x = 0$
Horizontal asymptote: $y = -1$
Matches graph (e).

12. $f(x) = -\frac{x + 2}{x + 4}$

Vertical asymptote: $x = -4$
Horizontal asymptote: $y = -1$
Matches graph (f).

14. $f(x) = \frac{4}{(x - 2)^3}$

Domain: all real numbers except $x = 2$
Vertical asymptote: $x = 2$
Horizontal asymptote: $y = 0$
[Degree of $p(x) <$ degree of $q(x)$]

16. $f(x) = \frac{1 - 5x}{1 + 2x} = \frac{-5x + 1}{2x + 1}$

Domain: all real numbers except $x = -\frac{1}{2}$
Vertical asymptote: $x = -\frac{1}{2}$
Horizontal asymptote: $y = -\frac{5}{2}$
[Degree of $p(x) =$ degree of $q(x)$]

18. $f(x) = \frac{2x^2}{x + 1}$

Domain: all real numbers except $x = -1$
Vertical asymptote: $x = -1$
Horizontal asymptote: None
[Degree of $p(x) >$ degree of $q(x)$]

20. $f(x) = \frac{3x^2 + x - 5}{x^2 + 1}$

Domain: All real numbers. The denominator has no real zeros. [Try the Quadratic Formula on the denominator.]
Vertical asymptote: None
Horizontal asymptote: $y = 3$
[Degree of $p(x) =$ degree of $q(x)$]

22. $f(x) = \frac{x^2(x - 3)}{x^2 - 3x}$, $g(x) = x$

(a) Domain of f: all real numbers except 0 and 3
Domain of g: all real numbers

(b) Because $x^2 - 3x$ is a common factor of both the numerator and the denominator of $f(x)$, neither $x = 0$ nor $x = 3$ is a vertical asymptote of f. Thus, f has no vertical asymptotes.

(c)

x	-1	0	1	2	3	3.5	4
$f(x)$	-1	Undef.	1	2	Undef.	3.5	4
$g(x)$	-1	0	1	2	3	3.5	4

(d) f and g differ only where f is undefined.

24. $f(x) = \frac{2x - 8}{x^2 - 9x + 20}$, $g(x) = \frac{2}{x - 5}$

(a) Domain of f: all real numbers except 4 and 5
Domain of g: all real numbers except 5

(b) Because $x - 4$ is a common factor of both the numerator and the denominator of f, $x = 4$ is not a vertical asymptote of f. The only vertical asymptote is $x = 5$.

(c)

x	0	1	2	3	4	5	6
$f(x)$	$-\frac{2}{5}$	$-\frac{1}{2}$	$-\frac{2}{3}$	-1	Undef.	Undef.	2
$g(x)$	$-\frac{2}{5}$	$-\frac{1}{2}$	$-\frac{2}{3}$	-1	-2	Undef.	2

(d) f and g differ only at $x = 4$ where f is undefined and g is defined.

26. $f(x) = \dfrac{1}{1 + x^2}$

28. $f(x) = \dfrac{-3x^2}{x(2x - 5)} = \dfrac{-3x^2}{2x^2 - 5x}$

30. $f(x) = 2 + \dfrac{1}{x - 3}$

(a) As $x \to \pm\infty, f(x) \to 2$.

(b) As $x \to \infty, f(x) \to 2$ but is greater than 2.

(c) As $x \to -\infty, f(x) \to 2$ but is less than 2.

32. $f(x) = \dfrac{2x - 1}{x^2 + 1}$

(a) As $x \to \pm\infty, f(x) \to 0$.

(b) As $x \to \infty, f(x) \to 0$ but is greater than 0.

(c) As $x \to -\infty, f(x) \to 0$ but is less than 0.

34. $h(x) = 4 + \dfrac{5}{x^2 + 2}$

There are no real zeros.

36. $g(x) = \dfrac{x^3 - 8}{x^2 + 1}$

The zero of g corresponds to the zero of the numerator and is $x = 2$.

38. (a) $C = \dfrac{25{,}000(15)}{100 - 15} \approx 4411.76$

The cost would be \$4411.76.

(b) $C = \dfrac{25{,}000(50)}{100 - 50} = 25{,}000$

The cost would be \$25,000.

(c) $C = \dfrac{25{,}000(90)}{100 - 90} = 225{,}000$

The cost would be \$225,000.

(d) No. The model is undefined for $p = 100$.

40. (a) Use data $\left(10, \frac{1}{7}\right), \left(20, \frac{1}{10}\right), \left(30, \frac{1}{14}\right), \left(40, \frac{1}{22}\right), \left(50, \frac{1}{40}\right)$. The least squares line for this data $(x, 1/y)$ is:

$$\frac{1}{y} = 0.164 - 0.0029x \Longrightarrow y = \frac{1}{0.164 - 0.0029x} = \frac{154{,}000}{25260 - 447x} = \frac{154{,}000}{3(8420 - 149x)}$$

(b)

x	10	20	30	40	50
y	7.4	9.4	13.0	20.9	52.9

(c) No, the function is negative for $x = 60$.

42. The moth will become satiated at the horizontal asymptote:

$y = \dfrac{1.568}{6.360} \approx 0.247$ mg

44. $P = \dfrac{0.5 + 0.6(n - 1)}{1 + 0.8(n - 1)} = \dfrac{0.6n - 0.1}{0.8n + 0.2}$

The percentage of correct responses is limited by a horizontal asymptote:

$P = \dfrac{0.6}{0.8} = 0.75 = 75\%$

46.

$$t^3 - 50t = 0$$
$$t(t^2 - 50) = 0$$
$$t\left(t - \sqrt{50}\right)\left(t + \sqrt{50}\right) = 0 \Longrightarrow t = 0, \pm\sqrt{50} = 0, \pm 5\sqrt{2}$$

48.

$$x(10 - x) = 25$$
$$x^2 - 10x + 25 = 0$$
$$(x - 5)^2 = 0 \Longrightarrow x = 5$$

Section 2.7 Graphs of Rational Functions

Solutions to Even-Numbered Exercises

2.

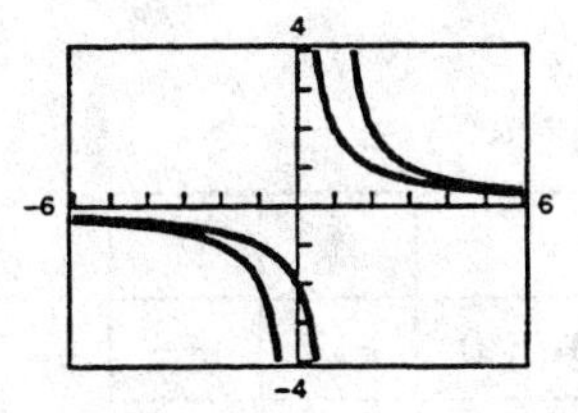

Horizontal shift one unit to the right

4.

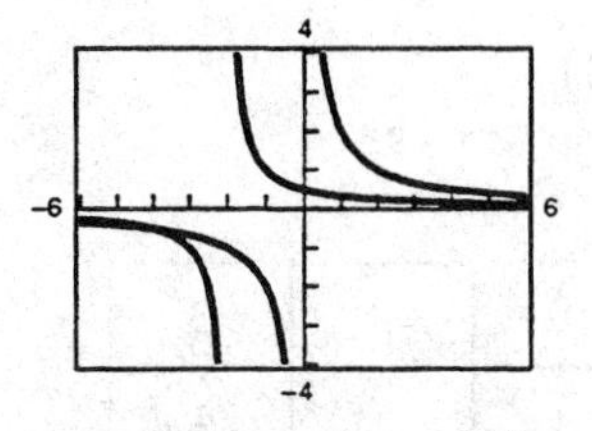

Horizontal shift two units to the left, and vertical shrink

6.

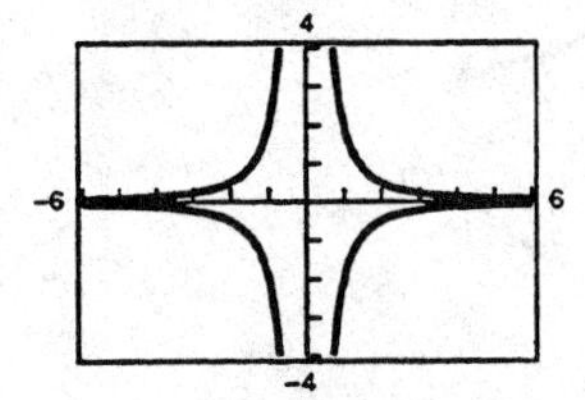

Reflection in the x-axis

8.

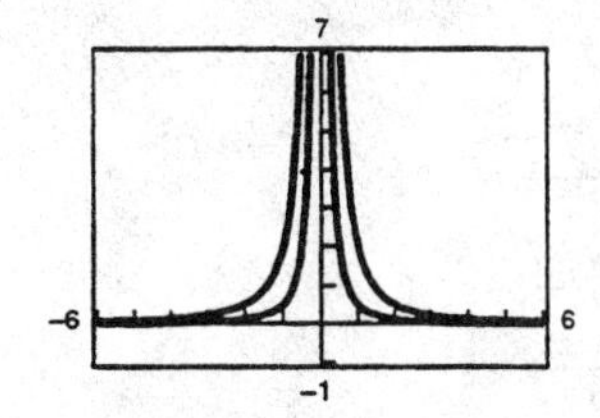

Each y-value is multiplied by $\frac{1}{4}$. Vertical shrink

10.

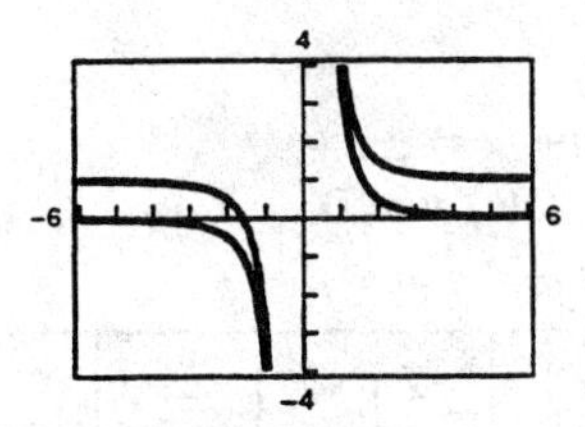

Vertical shift one unit upward

12.

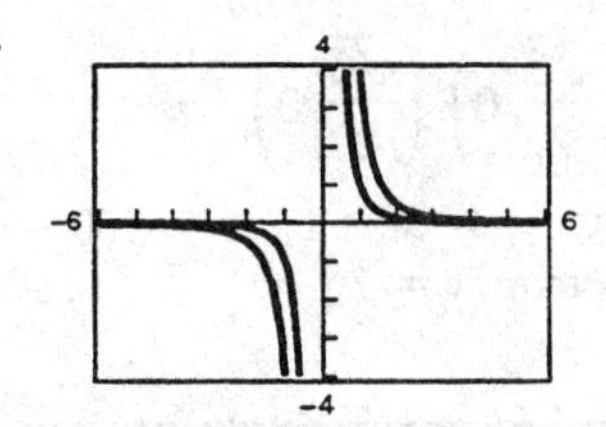

Each y-value is multiplied by $\frac{1}{4}$.Vertical shrink

14. $f(x) = \dfrac{1}{x - 3}$

y-intercept: $\left(0, -\dfrac{1}{3}\right)$

Vertical asymptote: $x = 3$

Horizontal asymptote: $y = 0$

x	0	1	2	4	5	6
y	$-\frac{1}{3}$	$-\frac{1}{2}$	-1	1	$\frac{1}{2}$	$\frac{1}{3}$

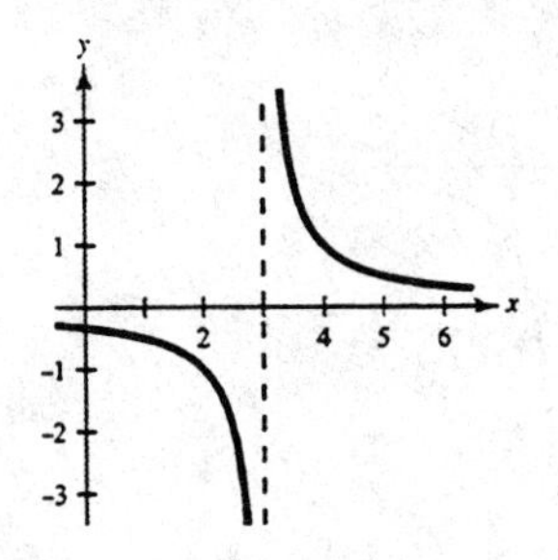

16. $g(x) = \dfrac{1}{3 - x} = \dfrac{-1}{x - 3}$

y-intercept: $\left(0, \dfrac{1}{3}\right)$

Vertical asymptote: $x = 3$

Horizontal asymptote: $y = 0$

x	0	1	2	4	5	6
y	$\frac{1}{3}$	$\frac{1}{2}$	1	-1	$-\frac{1}{2}$	$-\frac{1}{3}$

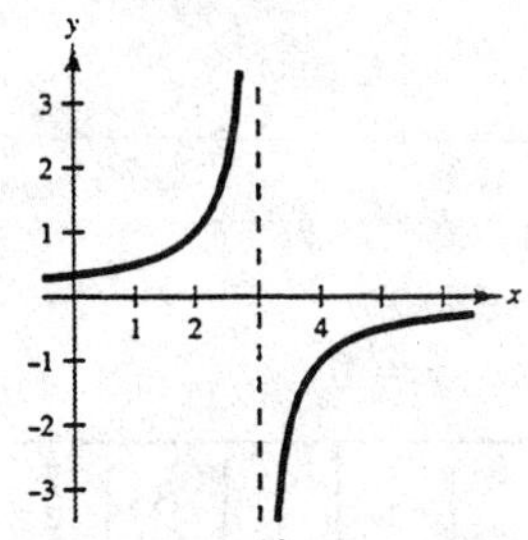

Note: This is the graph of $f(x) = \dfrac{1}{x - 3}$ (Exercise 14) reflected about the x-axis.

18. $P(x) = \dfrac{1 - 3x}{1 - x} = \dfrac{3x - 1}{x - 1}$

x-intercept: $\left(\dfrac{1}{3}, 0\right)$

y-intercept: $(0, 1)$

Vertical asymptote: $x = 1$

Horizontal asymptote: $y = 3$

x	-1	0	2	3
y	2	1	5	4

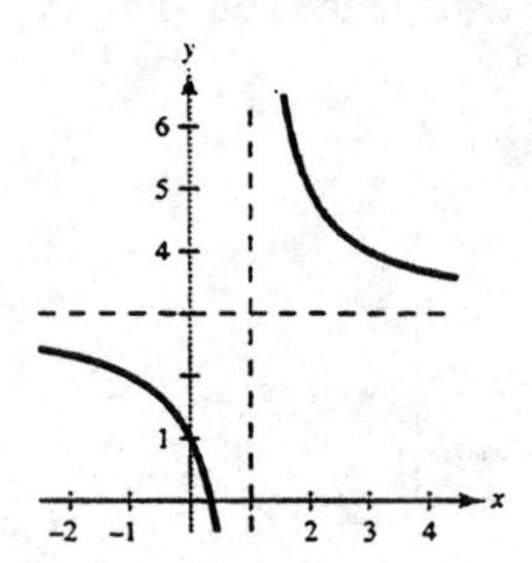

20. $f(t) = \dfrac{1 - 2t}{t} = -\dfrac{2t - 1}{t}$

t-intercept: $\left(\dfrac{1}{2}, 0\right)$

Vertical asymptote: $t = 0$

Horizontal asymptote: $y = -2$

x	-2	-1	$\frac{1}{2}$	1	2
y	$-\frac{5}{2}$	-3	0	-1	$-\frac{3}{2}$

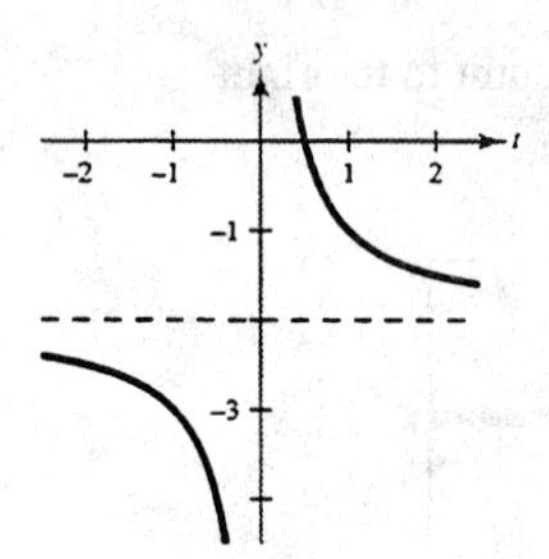

22. $f(x) = 2 - \dfrac{3}{x^2}$

x-intercepts: $\left(-\dfrac{\sqrt{6}}{2}, 0\right), \left(\dfrac{\sqrt{6}}{2}, 0\right)$

Vertical asymptote: $x = 0$

Horizontal asymptote: $y = 2$

y-axis symmetry

x	-2	-1	$-\frac{1}{2}$	$\frac{1}{2}$	1	2
y	$\frac{5}{4}$	-1	-10	-10	-1	$\frac{5}{4}$

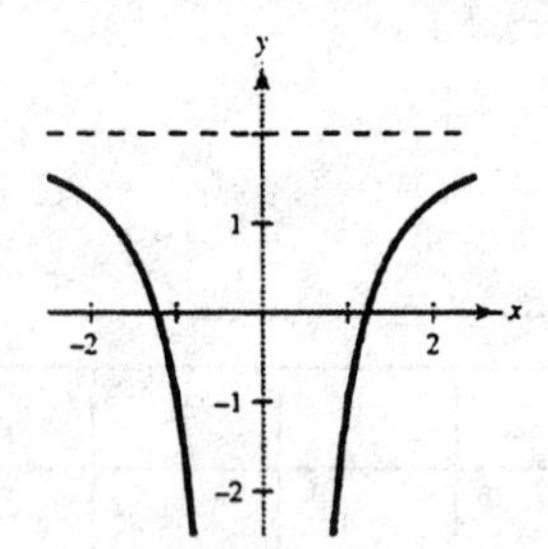

24. $g(x) = \dfrac{x}{x^2 - 9}$

Intercepts: $(0, 0)$

Vertical asymptote: $x = \pm 3$

Horizontal asymptote: $y = 0$

Origin symmetry

x	-5	-4	-2	0	2	4	5
y	$-\frac{5}{16}$	$-\frac{4}{7}$	$\frac{2}{5}$	0	$-\frac{2}{5}$	$\frac{4}{7}$	$\frac{5}{16}$

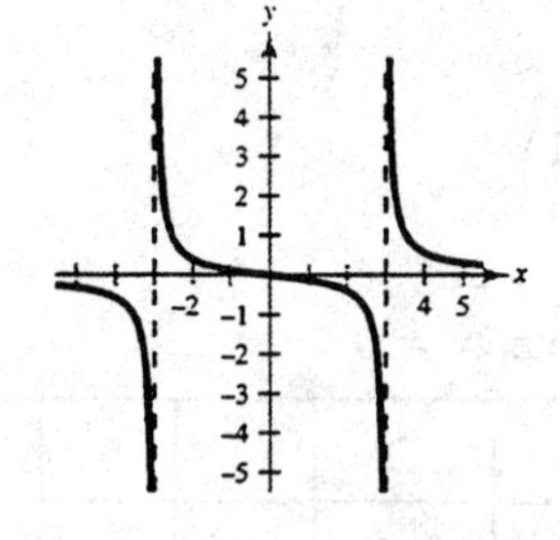

26. $f(x) = -\dfrac{1}{(x - 2)^2}$

y-intercept: $\left(0, -\dfrac{1}{4}\right)$

Vertical asymptote: $x = 2$

Horizontal asymptote: $y = 0$

x	0	$\frac{1}{2}$	1	$\frac{3}{2}$	$\frac{5}{2}$	3	$\frac{7}{2}$	4
y	$-\frac{1}{4}$	$-\frac{4}{9}$	-1	-4	-4	-1	$-\frac{4}{9}$	$-\frac{1}{4}$

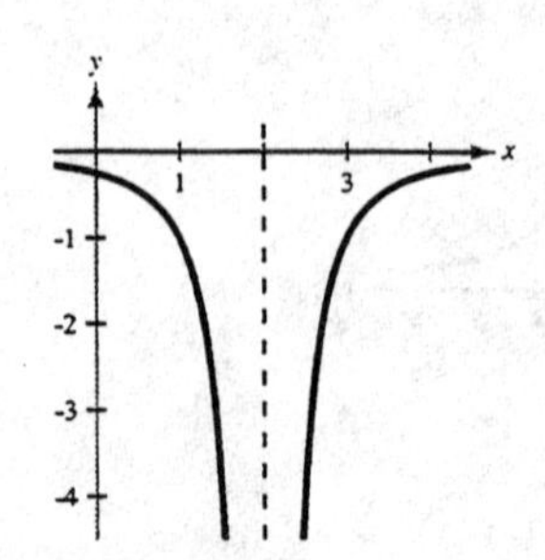

28. $h(x) = \dfrac{2}{x^2(x - 2)}$

Vertical asymptotes: $x = 0, x = 2$
Horizontal asymptote: $y = 0$

x	-2	-1	$\frac{1}{2}$	1	$\frac{3}{2}$	$\frac{5}{2}$	3
y	$-\frac{1}{8}$	$-\frac{2}{3}$	$-\frac{16}{5}$	-2	$-\frac{16}{9}$	$\frac{16}{25}$	$\frac{2}{9}$

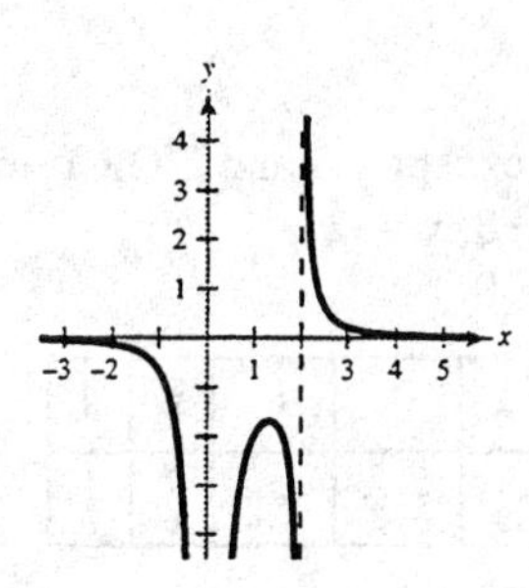

30. $f(x) = \dfrac{2x}{x^2 + x - 2} = \dfrac{2x}{(x + 2)(x - 1)}$

Intercept: $(0, 0)$
Vertical asymptotes: $x = -2, 1$
Horizontal asymptote: $y = 0$

x	-4	-3	-1	0	$\frac{1}{2}$	2	3
y	$-\frac{4}{5}$	$-\frac{3}{2}$	1	0	$-\frac{4}{5}$	1	$\frac{3}{5}$

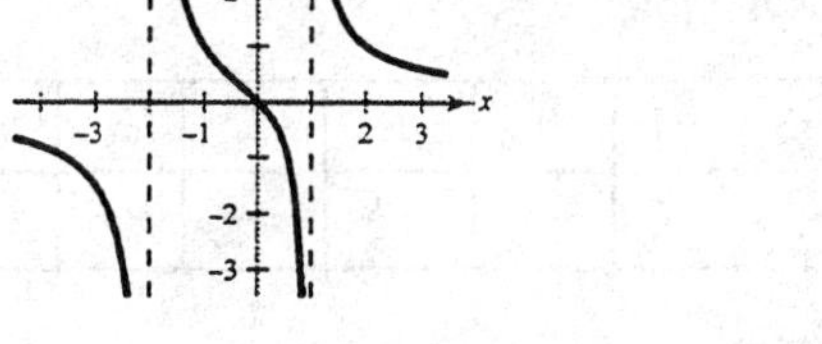

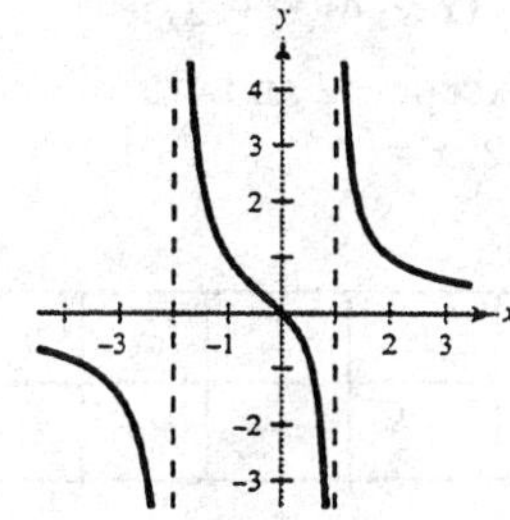

32. $f(x) = \dfrac{3 - x}{2 - x} = \dfrac{x - 3}{x - 2}$

x-intercept: $(3, 0)$
y-intercept: $\left(0, \dfrac{3}{2}\right)$
Vertical asymptote: $x = 2$
Horizontal asymptote: $y = 1$
Domain: all $x \neq 2$

x	0	1	3	4
y	$\frac{3}{2}$	2	0	$\frac{1}{2}$

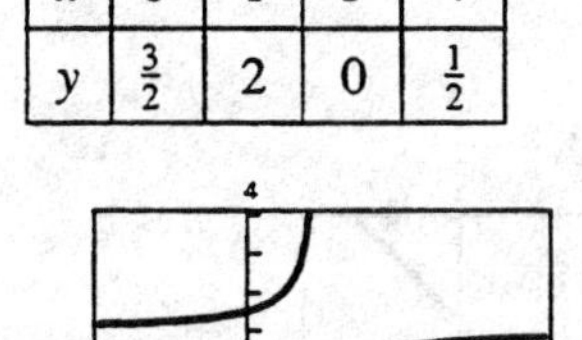

34. $h(x) = \dfrac{1}{x - 3} + 1 = \dfrac{1 + (x - 3)}{x - 3} = \dfrac{x - 2}{x - 3}$

x-intercept: $(2, 0)$
y-intercept: $\left(0, \dfrac{2}{3}\right)$
Vertical asymptote: $x = 3$
Horizontal asymptote: $y = 1$
Domain: all $x \neq 3$

x	0	1	2	4	5	6
y	$\frac{2}{3}$	$\frac{1}{2}$	0	2	$\frac{3}{2}$	$\frac{4}{3}$

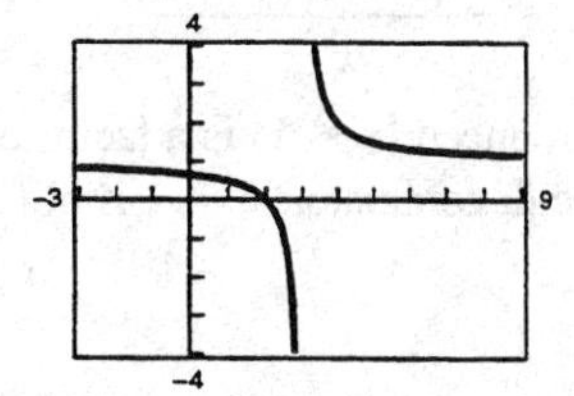

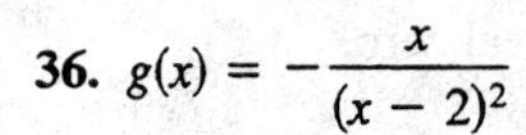

36. $g(x) = -\dfrac{x}{(x - 2)^2}$

Domain: all real numbers except 2 OR $(-\infty, 2) \cup (2, \infty)$
Vertical asymptote: $x = 2$
Horizontal asymptote: $y = 0$

x	-1	0	1	$\frac{3}{2}$	$\frac{5}{2}$	3	4
y	$\frac{1}{9}$	0	-1	-6	-10	-3	-1

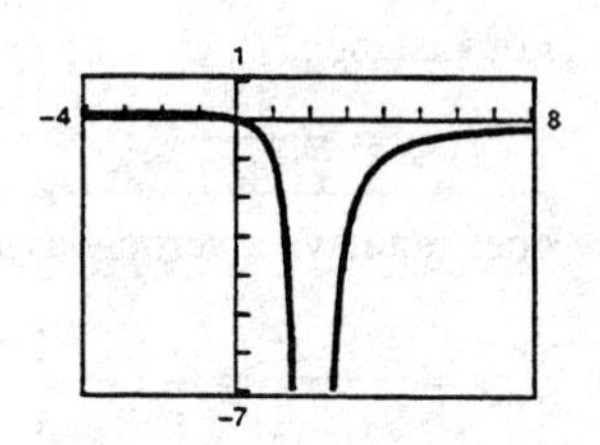

38. $f(x) = \dfrac{x+4}{x^2+x-6}$

Domain: all real numbers except -3 and 2 OR $(-\infty, -3) \cup (-3, 2) \cup (2, \infty)$
Vertical asymptotes: $x = -3, x = 2$
Horizontal asymptote: $y = 0$

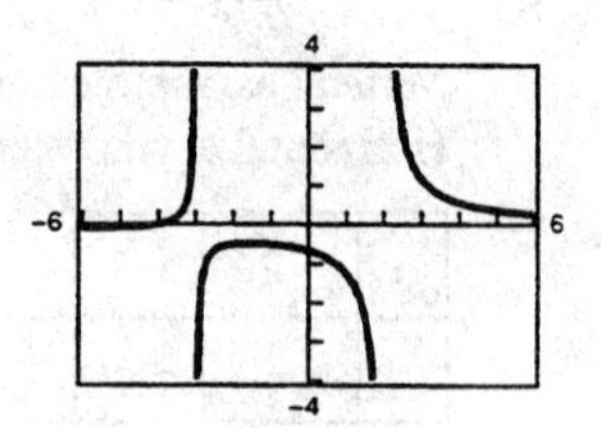

x	-6	-4	-2	-1	0	1	3	4
y	$-\frac{1}{12}$	0	$-\frac{1}{2}$	$-\frac{1}{2}$	$-\frac{2}{3}$	$-\frac{5}{4}$	$\frac{7}{6}$	$\frac{4}{7}$

40. $f(x) = 5\left(\dfrac{1}{x-4} - \dfrac{1}{x+2}\right) = \dfrac{30}{(x-4)(x+2)}$

Domain: all real numbers except -2 and 4
Vertical asymptotes: $x = -2, x = 4$
Horizontal asymptote: $y = 0$

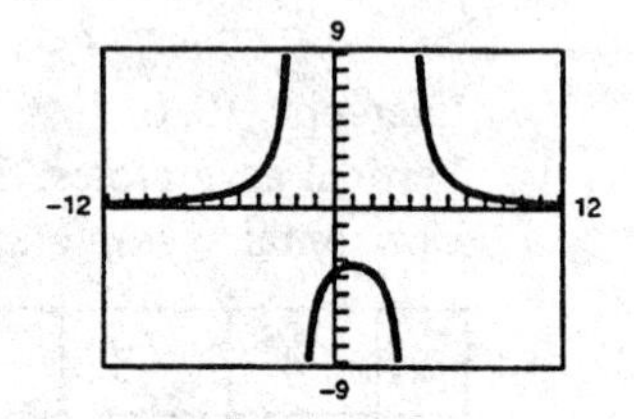

x	-4	-3	-1	0	1	2	3	5	6	7
y	$\frac{15}{8}$	$\frac{30}{7}$	-6	$-\frac{15}{4}$	$-\frac{10}{3}$	$-\frac{15}{4}$	-6	$\frac{30}{7}$	$\frac{15}{8}$	$\frac{10}{9}$

42. $g(x) = \dfrac{4|x-2|}{x+1}$

Horizontal asymptotes: $y = \pm 4$
Vertical asymptote: $x = -1$

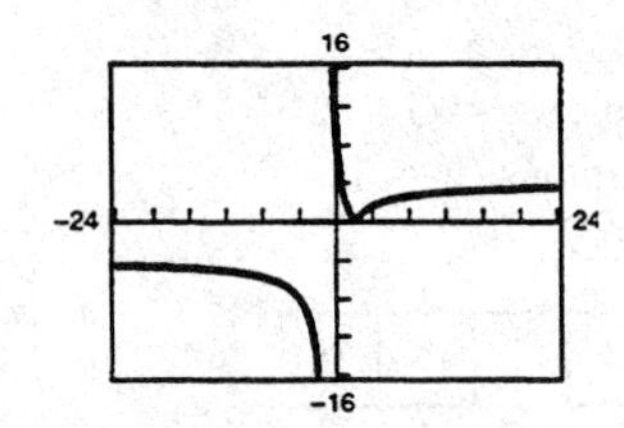

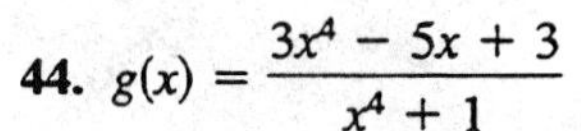

44. $g(x) = \dfrac{3x^4 - 5x + 3}{x^4 + 1}$

Horizontal asymptote: $y = 3$

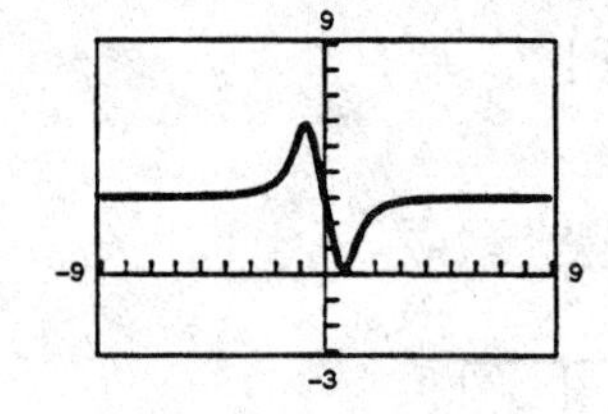

46. $g(x) = \dfrac{x^2+x-2}{x-1} = \dfrac{(x+2)(x-1)}{x-1}$

Since $g(x)$ is not reduced $(x-1)$ is a factor of both the numerator and the denominator, $x = 1$ is not a horizontal asymptote.

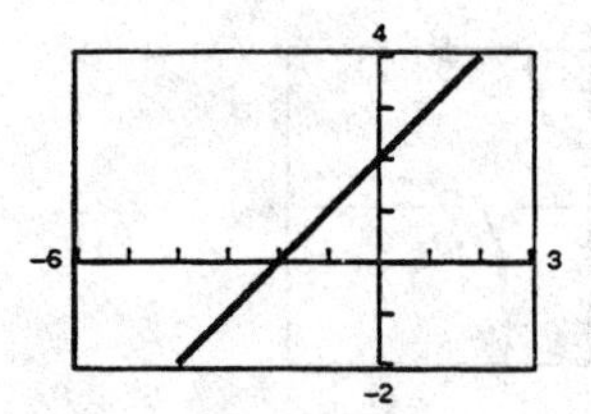

48. No, a rational function will only have a vertical asymptote if the denominator has a real zero. For instance,

$$f(x) = \frac{1}{x^2+1}$$

does not have a vertical asymptote because the denominator has no real zero.

50. $f(x) = \dfrac{1-x^2}{x} = -x + \dfrac{1}{x}$

x-intercepts: $(-1, 0), (1, 0)$
Vertical asymptote: $x = 0$
Slant asymptote: $y = -x$
Origin symmetry

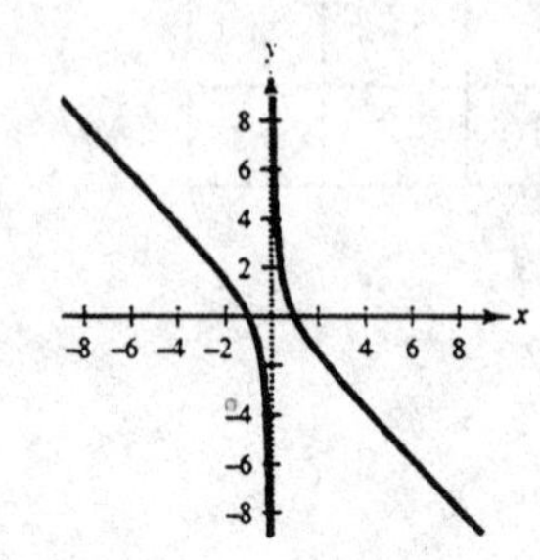

52. $h(x) = \dfrac{x^2}{x - 1} = x + 1 + \dfrac{1}{x - 1}$

Intercept: $(0, 0)$
Vertical asymptote: $x = 1$
Slant asymptote: $y = x + 1$

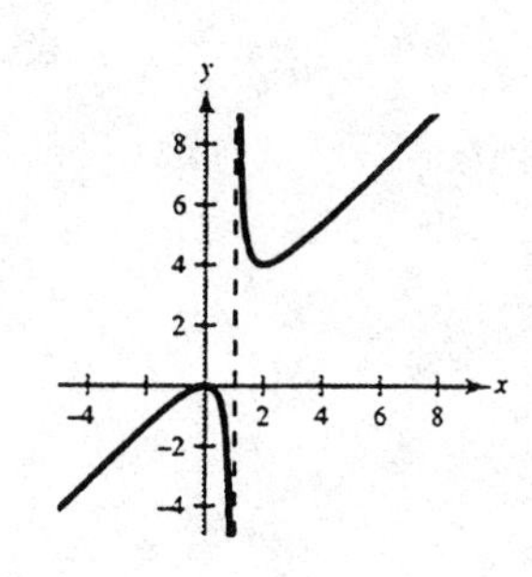

54. $g(x) = \dfrac{x^3}{2x^2 - 8} = \dfrac{1}{2}x + \dfrac{4x}{2x^2 - 8}$

Intercept: $(0, 0)$
Vertical asymptotes: $x = \pm 2$
Slant asymptote: $y = \dfrac{1}{2}x$
Origin symmetry

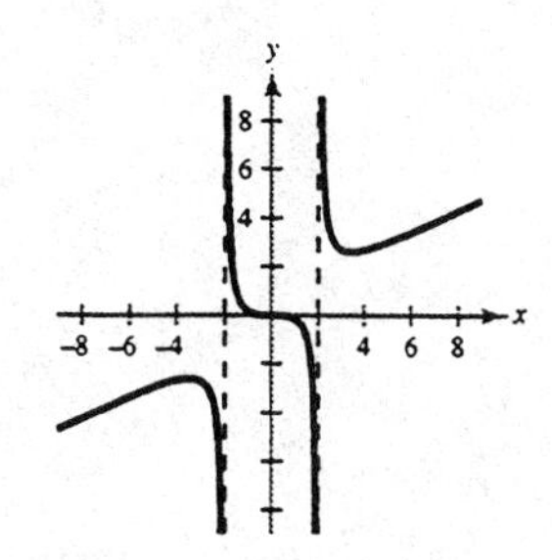

56. $f(x) = \dfrac{2x^2 - 5x + 5}{x - 2} = 2x - 1 + \dfrac{3}{x - 2}$

y-intercept: $\left(0, -\dfrac{5}{2}\right)$
Vertical asymptote: $x = 2$
Slant asymptote: $y = 2x - 1$

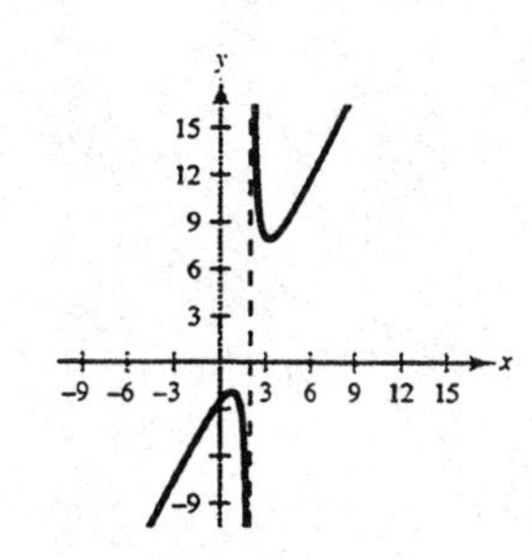

58. (a) x-intercept: $(0, 0)$

(b) $0 = \dfrac{2x}{x - 3}$

$0 = 2x$

$0 = x$

60. (a) x-intercepts: $(1, 0), (2, 0)$

(b) $0 = x - 3 + \dfrac{2}{x}$

$0 = x^2 - 3x + 2$

$0 = (x - 1)(x - 2)$

$x = 1, 2$

62.

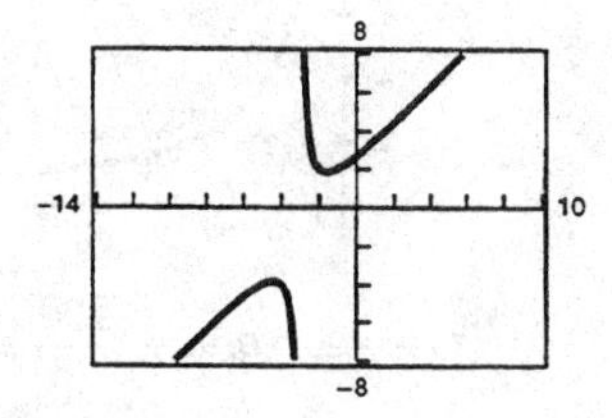

Domain: all $x \neq -3$
Vertical asymptote: $x = -3$
Slant asymptote: $y = x + 2$

64. $h(x) = \dfrac{12 - 2x - x^2}{2(4 + x)} = -\dfrac{1}{2}x + 1 + \dfrac{2}{4 + x}$

Domain: all real numbers except -4 OR $(-\infty, -4) \cup (-4, \infty)$
x-intercepts: $(-4.61, 0), (2.61, 0)$
y-intercept: $\left(0, \dfrac{3}{2}\right)$
Vertical asymptote: $x = -4$
Slant asymptote: $y = -\dfrac{1}{2}x + 1$

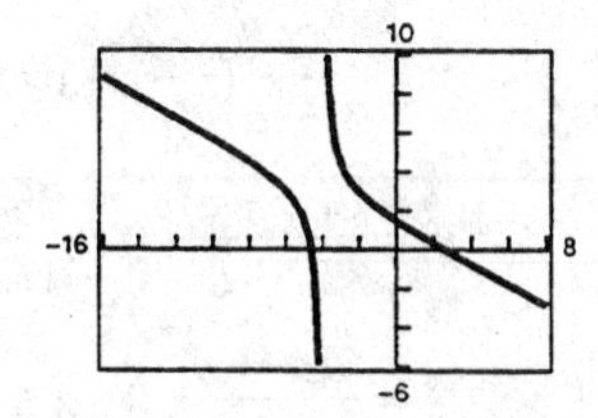

66. $y = 20\left(\frac{2}{x+1} - \frac{3}{x}\right)$

(a)

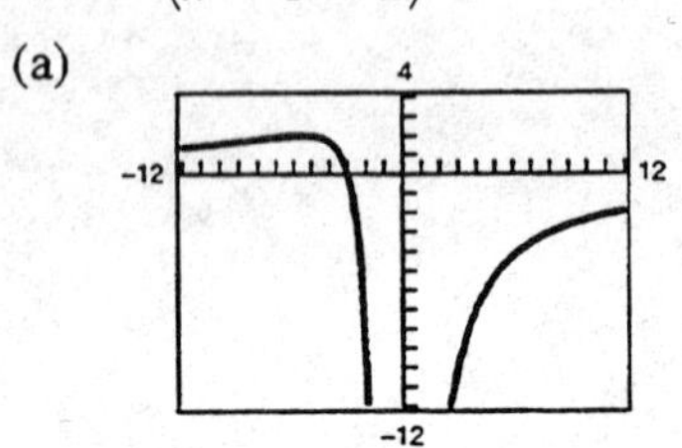

x-intercept: $(-3, 0)$

(b) $0 = 20\left(\frac{2}{x+1} - \frac{3}{x}\right)$

$\frac{3}{x} = \frac{2}{x+1}$

$2x = 3(x + 1)$

$2x = 3x + 3$

$-3 = x$

68. $y = x - \frac{9}{x}$

(a)

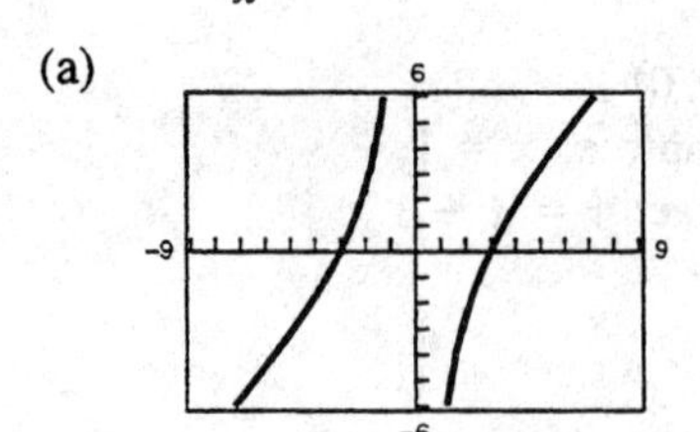

x-intercepts: $(-3, 0), (3, 0)$

(b) $0 = x - \frac{9}{x}$

$\frac{9}{x} = x$

$9 = x^2$

$\pm 3 = x$

70. (a) Area $= xy = 500$

$y = \frac{500}{x}$

(b) Domain: $x > 0$

(c)

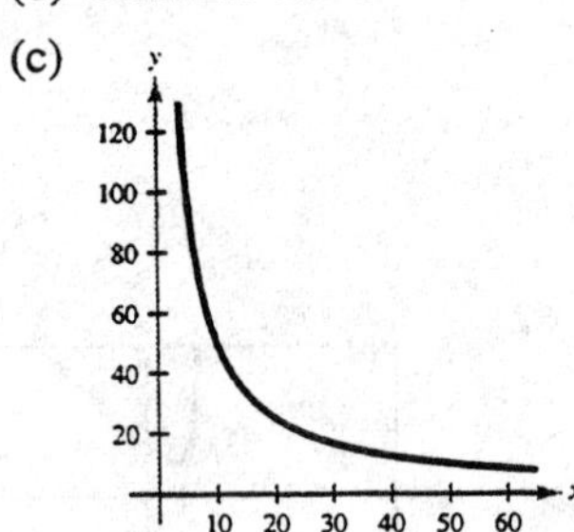

For $x = 30$, $y = \frac{500}{30} = 16\frac{2}{3}$ meters.

72. $C(x) = x + \frac{32}{x}$

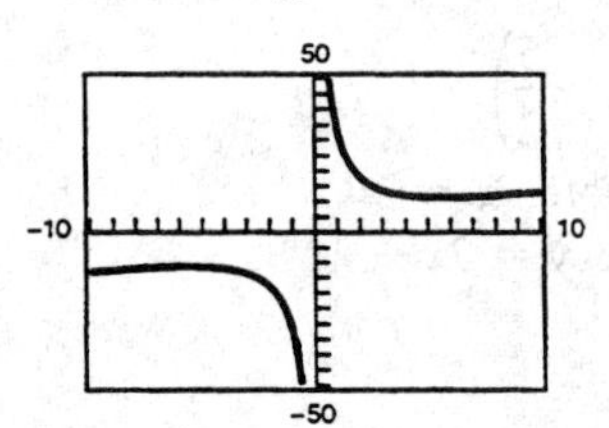

Local minimum: $(5.66, 11.31)$
Local maximum: $(-5.66, -11.31)$

74. (a) The line passes through the points $(a, 0)$ and $(3, 2)$ and has a slope of

$$m = \frac{2-0}{3-a} = \frac{2}{3-a}.$$

$y - 0 = \frac{2}{3-a}(x - a)$ by the point-slope form

$$y = \frac{2(x-a)}{3-a} = \frac{-2(a-x)}{-1(a-3)}$$

$$= \frac{2(a-x)}{a-3}, 0 \le x \le a$$

(b) The area of a triangle is $A = \frac{1}{2}bh$.

$b = a$

$h = y$ when $x = 0$, so $h = \frac{2(a-0)}{a-3} = \frac{2a}{a-3}.$

$$A = \frac{1}{2}a\left(\frac{2a}{a-3}\right) = \frac{a^2}{a-3}$$

(c) $A = \frac{a^2}{a-3} = a + 3 + \frac{9}{a-3}$

Vertical asymptote: $a = 3$
Slant asymptote: $A = a + 3$

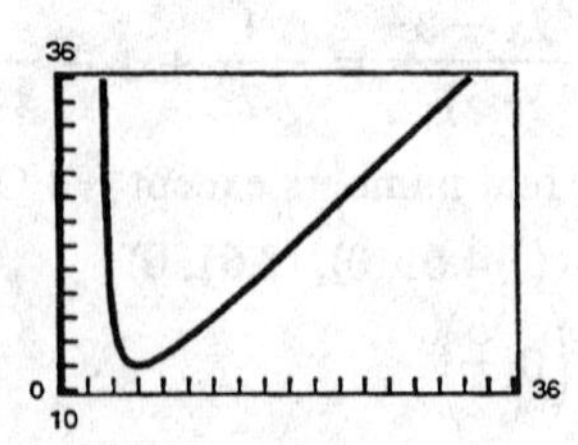

A is a minimum when $a = 6$ and $A = 12$.

76. $\overline{C} = \dfrac{C}{x} = \dfrac{0.2x^2 + 10x + 5}{x}$

x	0.5	1	2	3	4	5	6	7
$\overline{C}$	20.1	15.2	12.9	≈ 12.3	12.05	12	≈ 12.0	12.1

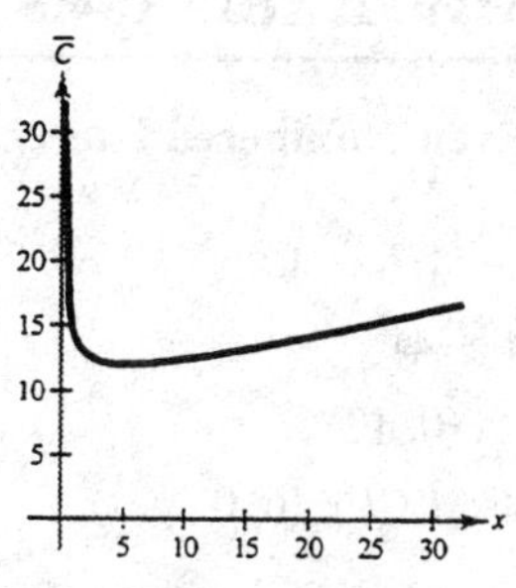

The minimum average cost occurs when $x = 5$.

78. (a) Rate × Time = Distance or $\dfrac{\text{Distance}}{\text{Rate}} = \text{Time}$

$\dfrac{100}{x} + \dfrac{100}{y} = \dfrac{200}{50} = 4$

$\dfrac{25}{x} + \dfrac{25}{y} = 1$

$25y + 25x = xy$

$25x = xy - 25y$

$25x = y(x - 25)$

$y = \dfrac{25x}{x - 25}$

(b) Vertical asymptote: $x = 25$
Horizontal asymptote: $y = 25$

(c)

x	30	35	40	45	50	55	60
y	150	87.5	66.7	56.3	50	45.8	42.9

(d) Yes, the results in the table are unexpected. You would expect the average speed for the round trip to be the average of the average speeds for the two parts of the trip.

(e) No, it is not possible to average 20 miles per hour in one direction and still average 50 miles per hour on the round trip. At 20 miles per hour you would use more time in one direction than is required for the round trip at an average speed of 50 miles per hour.

80. (a) $N_1 = -0.41t + 16.43$

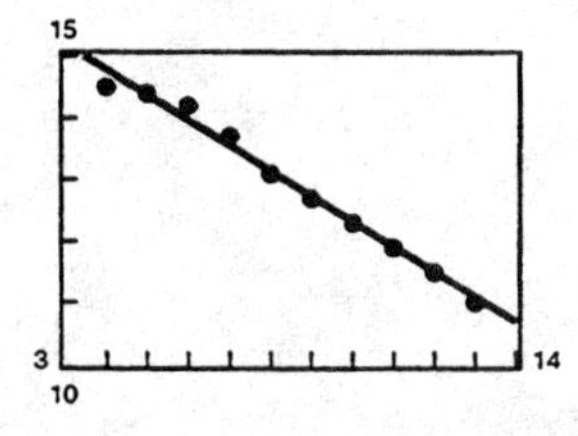

(b) $N_2 = \dfrac{1{,}197{,}488}{3015t + 67{,}787}$

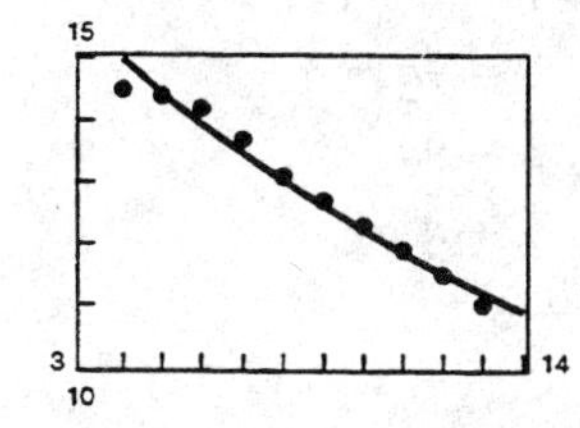

(c)

t	4	5	6	7	8
N_1	14.8	14.4	14.0	13.6	13.2
N_2	15.0	14.5	13.9	13.5	13.0

Both fit the data well. Either model will do.

(d)

t	9	10	11	12	13
N_1	12.7	12.3	11.9	11.5	11.1
N_2	12.6	12.2	11.9	11.5	11.2

82. $y = \dfrac{2(x - 3)}{(x + 1)}$

This has a vertical asymptote at $x = -1$, the zero of the denominator, a horizontal asymptote at $y = 2$ because the degree of the denominator equals the degree of the numerator, and has $x = 3$ as a zero of the function because it is the zero of the numerator.

Review Exercises for Chapter 2

Solutions to Even-Numbered Exercises

2. $f(x) = (x - 4)^2 - 4$

Vertex: $(4, -4)$

y-intercept: $(0, 12)$

x-intercepts: $(2, 0), (6, 0)$

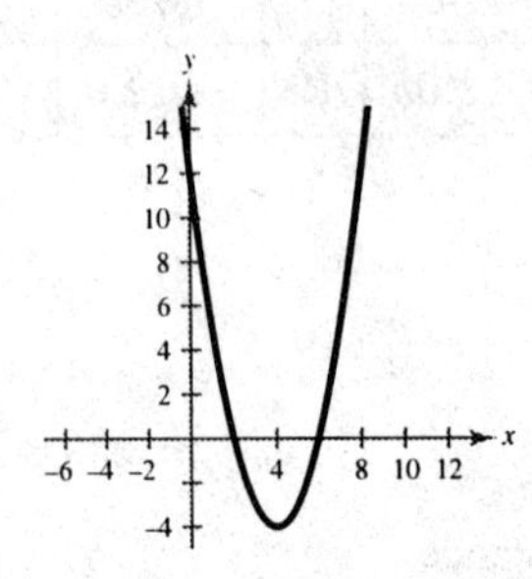

4. $f(x) = 3x^2 - 12x + 11$

$$= 3\left(x^2 - 4x + 4 - 4 + \frac{11}{3}\right)$$

$$= 3\left[(x - 2)^2 - \frac{1}{3}\right]$$

$$= 3(x - 2)^2 - 1$$

Vertex: $(2, -1)$

y-intercept: $(0, 11)$

x-intercepts: $x = \dfrac{12 \pm \sqrt{12}}{6} = 2 \pm \dfrac{1}{3}\sqrt{3}$

$\left(2 + \frac{1}{3}\sqrt{3}, 0\right), \left(2 - \frac{1}{3}\sqrt{3}, 0\right)$

6. Vertex: $(2, 3) \Longrightarrow f(x) = a(x - 2)^2 + 3$

Point: $(-1, 6) \Longrightarrow 6 = a(-1 - 2)^2 + 3$

$$6 = 9a + 3$$

$$3 = 9a$$

$$\tfrac{1}{3} = a$$

$f(x) = \frac{1}{3}(x - 2)^2 + 3$

8. (a) $y = x^2 - 4$

Vertical shift 4 units downward

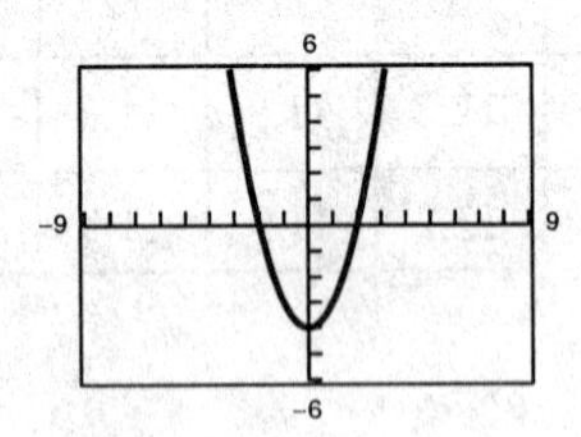

(b) $y = 4 - x^2$

Reflection in the x-axis and a vertical shift 4 units upward

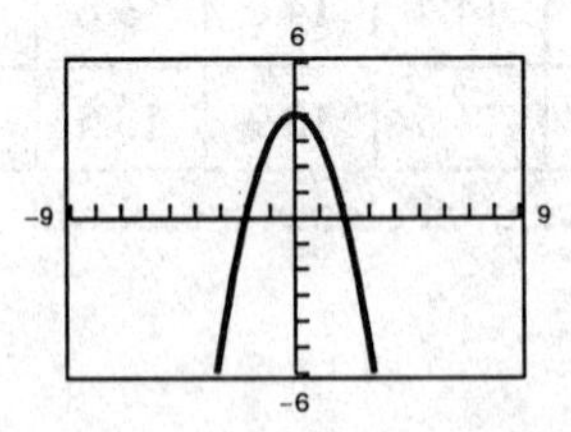

(c) $y = (x - 3)^2$

Horizontal shift 3 units to the right

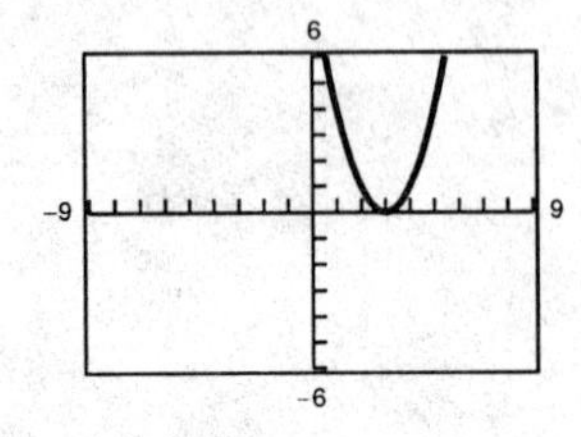

(d) $y = \frac{1}{2}x^2 - 1$

Vertical shrink and a vertical shift 1 unit downward

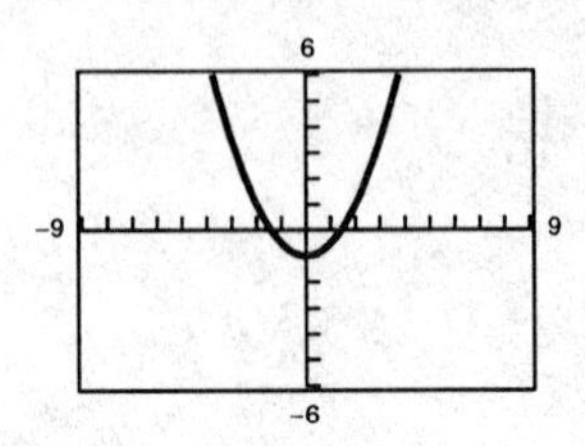

10. $f(x) = x^2 + 8x + 10$
$= x^2 + 8x + 16 - 16 + 10$
$= (x + 4)^2 - 6$

The minimum occurs at the vertex $(-4, -6)$.

12. $h(x) = 3 + 4x - x^2$
$= -(x^2 - 4x - 3)$
$= -(x^2 - 4x + 4 - 4 - 3)$
$= -[(x - 2)^2 - 7]$
$= -(x - 2)^2 + 7$

The maximum occurs at the vertex $(2, 7)$.

14. $h(x) = 4x^2 + 4x + 13$
$= 4\left(x^2 + x + \frac{1}{4} - \frac{1}{4} + \frac{13}{4}\right)$
$= 4\left[\left(x + \frac{1}{2}\right)^2 + 3\right]$
$= 4\left(x + \frac{1}{2}\right)^2 + 12$

The minimum occurs at the vertex $\left(-\frac{1}{2}, 12\right)$.

16. $f(x) = 4x^2 + 4x + 5$
$= 4\left(x^2 + x + \frac{1}{4} - \frac{1}{4} + \frac{5}{4}\right)$
$= 4\left[\left(x + \frac{1}{2}\right)^2 + 1\right]$
$= 4\left(x + \frac{1}{2}\right)^2 + 4$

The minimum occurs at the vertex $\left(-\frac{1}{2}, 4\right)$.

18. $P = 230 + 20x - \frac{1}{2}x^2$

Using the zoom and trace features, $x = 20$ yields a maximum profit. Or, completing the square,

$P = -\frac{1}{2}x^2 + 20x + 230$
$= -\frac{1}{2}(x^2 - 40x + 400) + 230 + 200$
$= -\frac{1}{2}(x - 20)^2 + 430.$

Hence, $x = 20$ (\$2000) yields a maximum profit (of 430).

20. $f(x) = \frac{1}{2}x^3 + 2x$

The degree is odd and the leading coefficient is positive. The graph falls to the left and rises to the right.

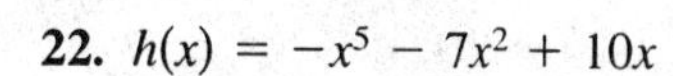

22. $h(x) = -x^5 - 7x^2 + 10x$

The degree is odd and the leading coefficient is negative. The graph rises to the left and falls to the right.

24.

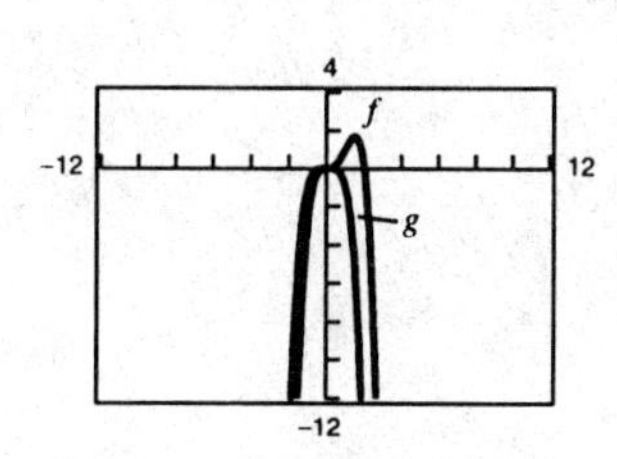

26.

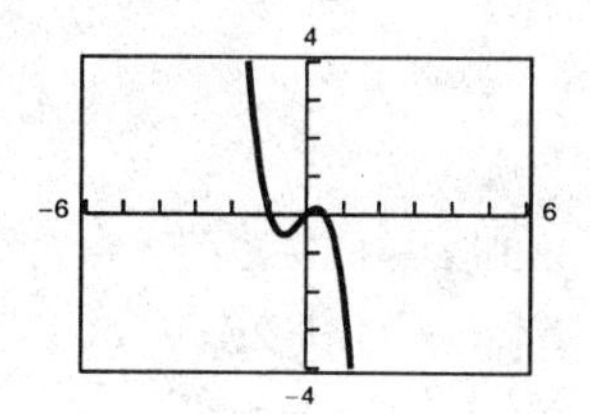

28.

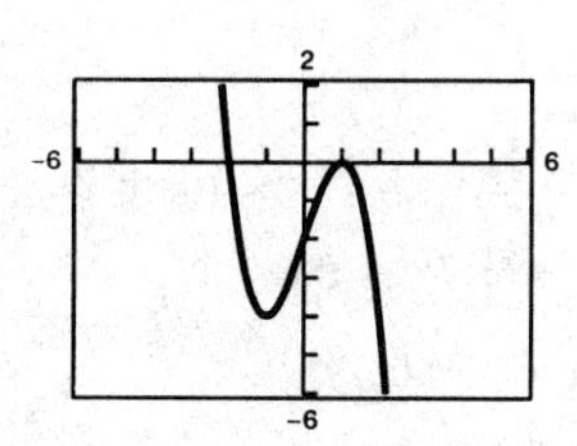

30.

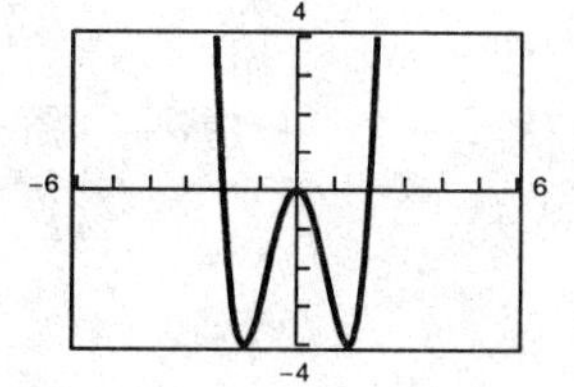

32. $y_1 = \dfrac{x^4 + 1}{x^2 + 2}$

$y_2 = x^2 - 2 + \dfrac{5}{x^2 + 2}$

$= \dfrac{x^2(x^2 + 2)}{x^2 + 2} - \dfrac{2(x^2 + 2)}{x^2 + 2} + \dfrac{5}{x^2 + 2}$

$= \dfrac{x^4 + 2x^2 - 2x^2 - 4 + 5}{x^2 + 2}$

$= \dfrac{x^4 + 1}{x^2 + 2} = y_1$

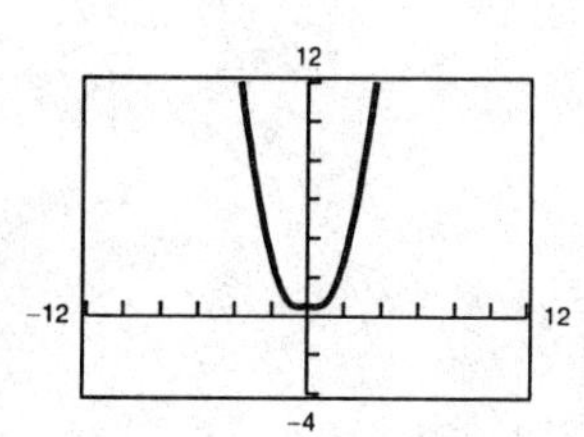

34.

$$\begin{array}{r} \frac{4}{3} \\ 3x - 2 \overline{) 4x + 7} \\ \underline{4x - \frac{8}{3}} \\ \frac{29}{3} \end{array}$$

$$\frac{4x + 7}{3x - 2} = \frac{4}{3} + \frac{29}{3(3x - 2)}$$

36.

$$\begin{array}{r} 3x^2 \quad + 3 \\ x^2 - 1 \overline{) 3x^4 + 0x^3 + 0x^2 + 0x + 0} \\ \underline{3x^4 \qquad - 3x^2} \\ 3x^2 \qquad + 0 \\ \underline{3x^2 \qquad - 3} \\ 3 \end{array}$$

$$\frac{3x^4}{x^2 - 1} = 3x^2 + 3 + \frac{3}{x^2 - 1}$$

38.

$$\begin{array}{c|cccc} \frac{1}{2} & 2 & 2 & -1 & 2 \\ & & 1 & \frac{3}{2} & \frac{1}{4} \\ \hline & 2 & 3 & \frac{1}{2} & \frac{9}{4} \end{array}$$

$$\frac{2x^3 + 2x^2 - x + 2}{x - (1/2)} = 2x^2 + 3x + \frac{1}{2} + \frac{9/4}{x - (1/2)}$$

40.

$$\begin{array}{c|cccc} 5 & 0.1 & 0.3 & 0 & -0.5 \\ & & 0.5 & 4 & 20 \\ \hline & 0.1 & 0.8 & 4 & 19.5 \end{array}$$

$$\frac{0.1x^3 + 0.3x^2 - 0.5}{x - 5} = 0.1x^2 + 0.8x + 4 + \frac{19.5}{x - 5}$$

42. $$\left(\frac{\sqrt{2}}{2} - \frac{\sqrt{2}}{2}i\right) - \left(\frac{\sqrt{2}}{2} + \frac{\sqrt{2}}{2}i\right) = \frac{\sqrt{2}}{2} - \frac{\sqrt{2}}{2}i - \frac{\sqrt{2}}{2} - \frac{\sqrt{2}}{2}i = -2\frac{\sqrt{2}}{2}i = -\sqrt{2}i$$

44.

$$\begin{aligned} i(6 + i)(3 - 2i) &= i(18 - 12i + 3i - 2i^2) \\ &= i(20 - 9i) \\ &= 20i - 9i^2 \\ &= 9 + 20i \end{aligned}$$

46.

$$\begin{aligned} \frac{3 + 2i}{5 + i} &= \frac{3 + 2i}{5 + i} \cdot \frac{5 - i}{5 - i} \\ &= \frac{15 + 7i - 2i^2}{25 - i^2} \\ &= \frac{17 + 7i}{26} \\ &= \frac{17}{26} + \frac{7i}{26} \end{aligned}$$

48.

$$\begin{aligned} f(x) &= (x - 2)(x + 3)[x - (1 - 2i)][x - (1 + 2i)] \\ &= (x^2 + x - 6)[x^2 - (1 + 2i)x - (1 - 2i)x + 5] \\ &= (x^2 + x - 6)(x^2 - 2x + 5) \\ &= x^4 - 2x^3 + 5x^2 + x^3 - 2x^2 + 5x - 6x^2 + 12x - 30 \\ &= x^4 - x^3 - 3x^2 + 17x - 30 \end{aligned}$$

Note: $f(x) = a(x^4 - x^3 - 3x^2 + 17x - 30)$, where a is any real number, has zeros $2, -3, 1 - 2i, 1 + 2i$.

50. $f(x) = 10x^3 + 21x^2 - x - 6$

$$\begin{array}{c|cccc} -2 & 10 & 21 & -1 & -6 \\ & & -20 & -2 & 6 \\ \hline & 10 & 1 & -3 & 0 \end{array}$$

Zeros: $-2, -\frac{3}{5}, \frac{1}{2}$

52. $f(x) = x^3 - 1.3x^2 - 1.7x + 0.6$

$$\begin{array}{c|cccc} 2 & 1 & -1.3 & -1.7 & 0.6 \\ & & 2 & 1.4 & -0.6 \\ \hline & 1 & 0.7 & -0.3 & 0 \end{array}$$

$$\begin{array}{c|ccc} -1 & 1 & 0.7 & -0.3 \\ & & -1 & 0.3 \\ \hline & 1 & -0.3 & 0 \end{array}$$

Thus, $f(x) = (x - 2)(x + 1)(x - 0.3)$ and the zeros of f are $x = 2, -1, 0.3$.

54. $f(x) = 5x^4 + 126x^2 + 25$

$f(x) = (5x^2 + 1)(x^2 + 25)$

$5x^2 + 1 = 0$

$x^2 = -\frac{1}{5}$

$x = \pm\frac{\sqrt{5}}{5}i$

$x^2 + 25 = 0$

$x^2 = -25$

$x = \pm 5i$

56. $g(x) = x^3 - 3x^2 + 3x + 2$

(a)

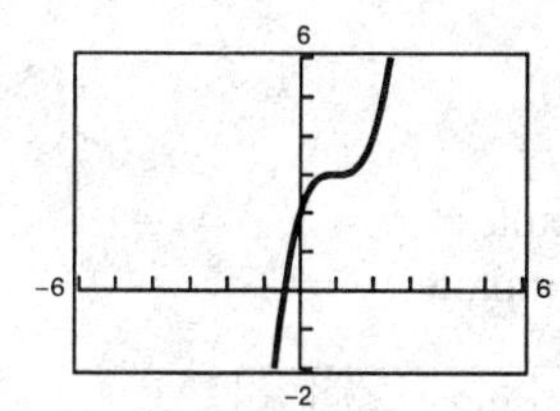

(b) One real zero because the graph has only one x-intercept.

(c) The zero is $x \approx -0.44$.

58. $f(x) = x^5 + 2x^3 - 3x - 20$

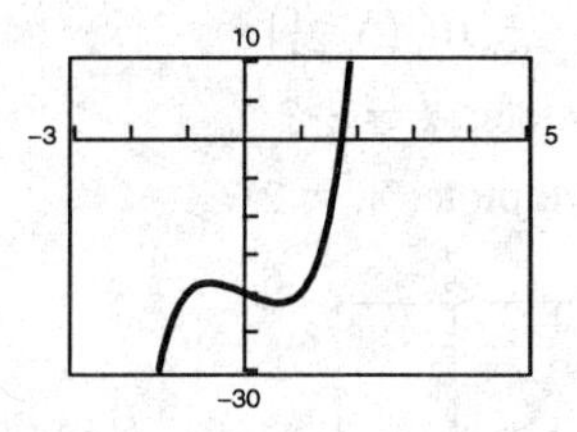

(b) One real zero because the graph has only one x-intercept.

(c) The zero is $x \approx 1.72$.

60. $30 = -0.00428x^2 + 1.442x - 3.136$

$0 = -0.00428x^2 + 1.442x - 33.136$

$$x = \frac{-1.442 \pm \sqrt{(1.442)^2 - 4(-0.00428)(-33.136)}}{2(-0.00428)}$$

$x \approx 24.8, 312.11$

The age of the bride is approximately 24.8 years when the age of the groom is 30 years.

62. $h(x) = \frac{x - 3}{x - 2}$

x-intercept: $(3, 0)$

y-intercept: $\left(0, \frac{3}{2}\right)$

Vertical asymptote: $x = 2$

Horizontal asymptote: $y = 1$

x	-1	0	1	3	4	5
y	$\frac{4}{3}$	$\frac{3}{2}$	2	0	$\frac{1}{2}$	$\frac{2}{3}$

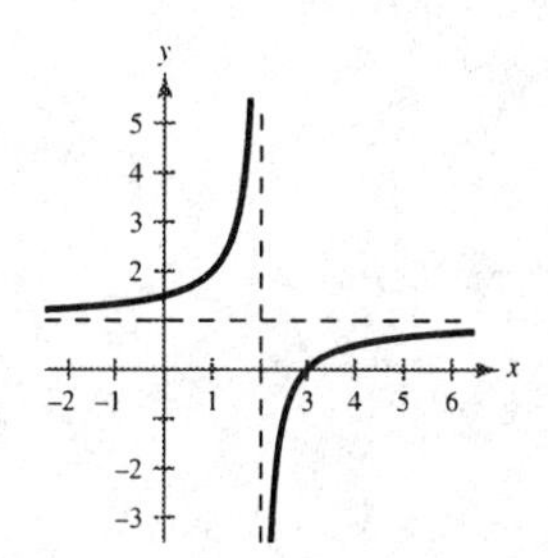

64. $f(x) = \frac{2x}{x^2 + 4}$

Intercept: $(0, 0)$

Origin symmetry

Horizontal asymptote: $y = 0$

x	-2	-1	0	1	2
y	$-\frac{1}{2}$	$-\frac{2}{5}$	0	$\frac{2}{5}$	$\frac{1}{2}$

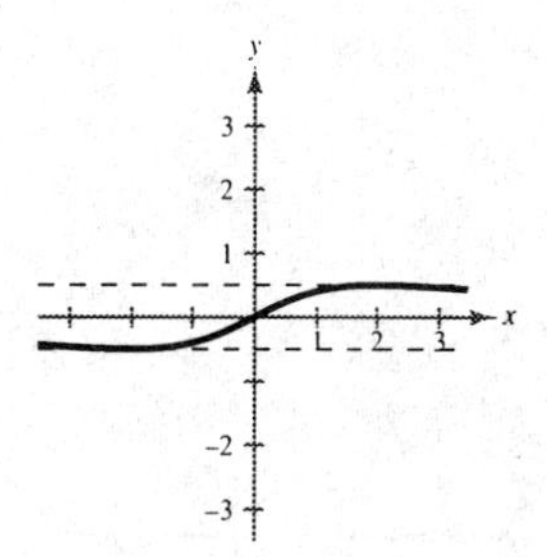

66. $h(x) = \frac{4}{(x - 1)^2}$

y-intercept: $(0, 4)$

Vertical asymptote: $x = 1$

Horizontal asymptote: $y = 0$

x	-2	-1	0	2	3	4
y	$\frac{4}{9}$	1	4	4	1	$\frac{4}{9}$

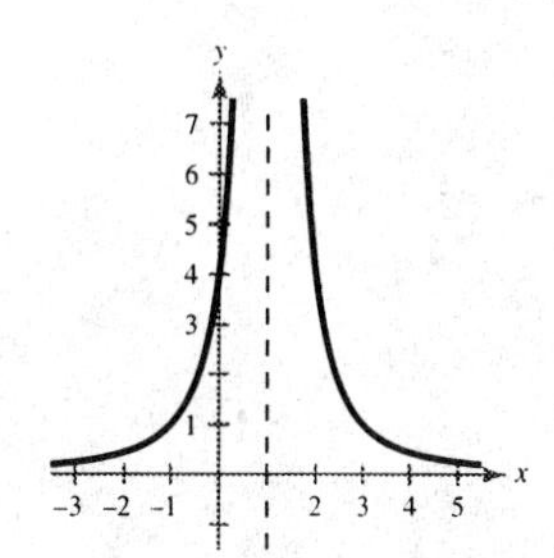

68. $y = \dfrac{2x^2}{x^2 - 4}$

Intercept: $(0, 0)$
y-axis symmetry
Vertical asymptotes: $x = 2, x = -2$
Horizontal asymptote: $y = 2$

x	± 5	± 4	± 3	± 1	0
y	$\frac{50}{21}$	$\frac{8}{3}$	$\frac{18}{5}$	$-\frac{2}{3}$	0

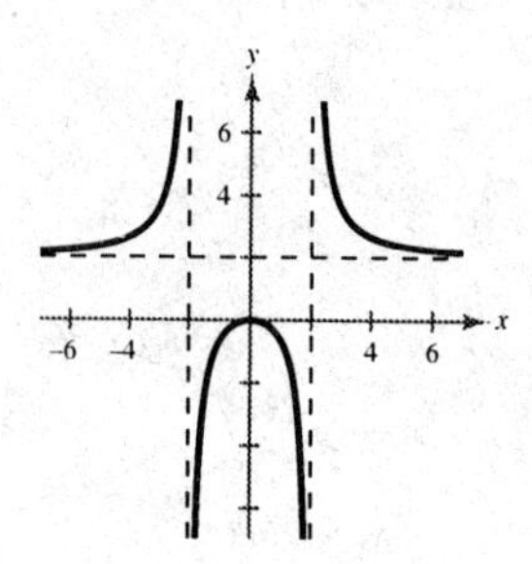

70. $y = \dfrac{5x}{x^2 - 4}$

Intercept: $(0, 0)$
Vertical asymptotes: $x = 2, x = -2$
Horizontal asymptote: $y = 0$

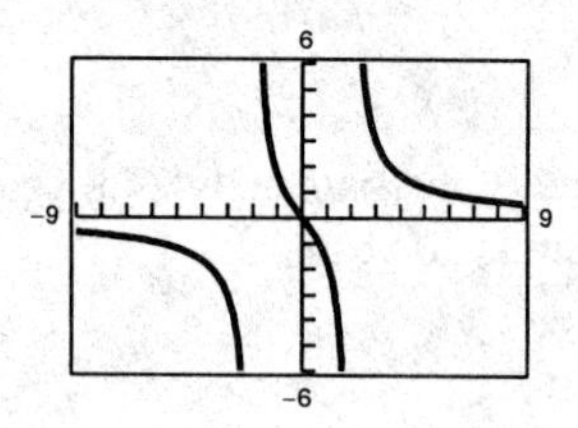

72. $y = \dfrac{1}{x + 3} + 2 = \dfrac{2x + 7}{x + 3}$

Intercepts: $(-3.5, 0), \left(0, 2\frac{1}{3}\right)$
Vertical asymptote: $x = -3$
Horizontal asymptote: $y = 2$

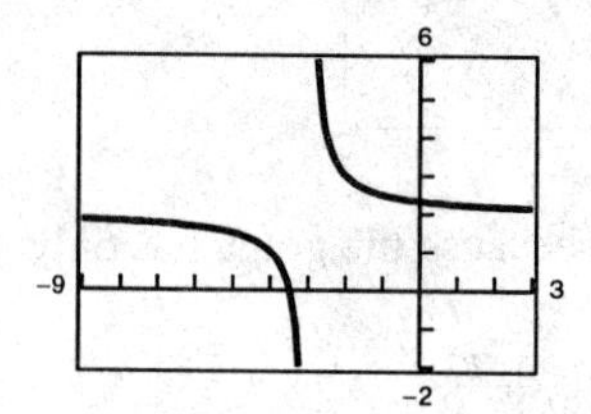

74. $f(x) = \dfrac{2x^2 - 10x + 1}{x - 5}$

This answer is not unique.

CHAPTER 3
Exponential and Logarithmic Functions

CHAPTER 3
Exponential and Logarithmic Functions

Section 3.1 Exponential Functions and Their Graphs

Solutions to Even-Numbered Exercises

2. $5000(2^{-1.5}) \approx 1767.767$

4. $8^{2\pi} \approx 472{,}369.379$

6. $\sqrt[3]{4395} \approx 16.380$

8. $e^{1/2} \approx 1.649$

10. $e^{3.2} \approx 24.533$

12. $g(x) = 2^{2x+6}$
$= 2^{2x} \cdot 2^{6}$
$= 64(2^{2x})$
$= 64(2^{2})^{x}$
$= 64(4^{x})$
$= h(x)$
Thus, $g(x) = h(x)$ but $g(x) \neq f(x)$.

14. $f(x) = 5^{-x} + 3$
$g(x) = 5^{3-x} = 5^{3} \cdot 5^{-x}$
$h(x) = -5^{x-3} = -(5^{x} \cdot 5^{-3})$
Thus, $f(x)$, $g(x)$ and $h(x)$ are all distinct.

16. $f(x) = \left(\frac{3}{2}\right)^{x}$

x	-2	-1	0	1	2
y	$\frac{4}{9}$	$\frac{2}{3}$	1	$\frac{3}{2}$	$\frac{9}{4}$

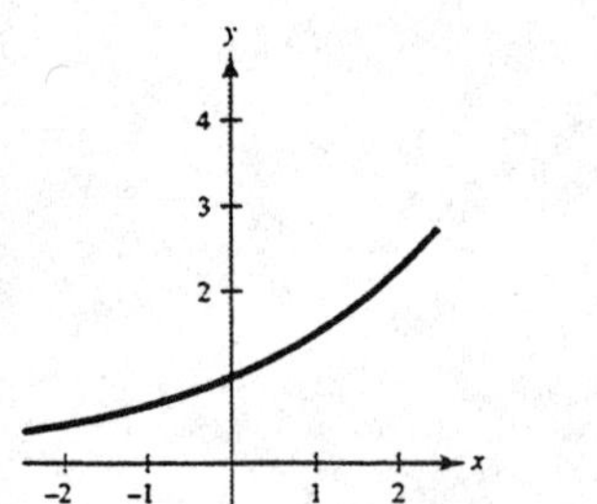

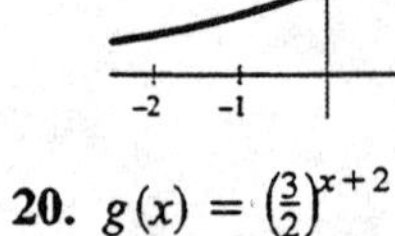

Asymptote: $y = 0$
Intercept: $(0, 1)$
Increasing

18. $h(x) = \left(\frac{3}{2}\right)^{-x}$

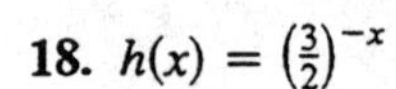

x	-2	-1	0	1	2
y	$\frac{9}{4}$	$\frac{3}{2}$	1	$\frac{2}{3}$	$\frac{4}{9}$

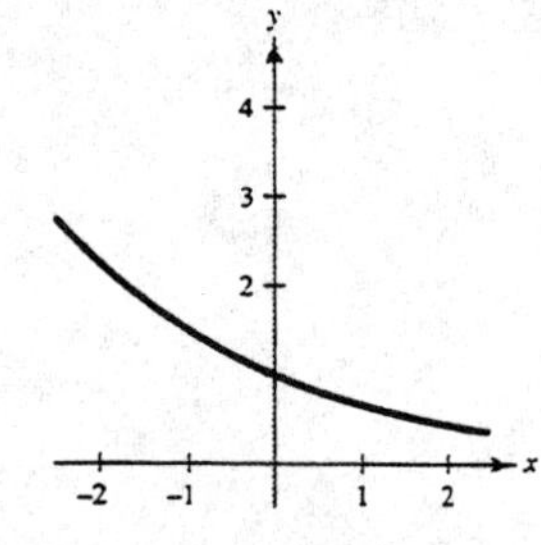

Asymptote: $y = 0$
Intercept: $(0, 1)$
Decreasing

20. $g(x) = \left(\frac{3}{2}\right)^{x+2}$

x	-4	-3	-2	-1	0
y	$\frac{4}{9}$	$\frac{2}{3}$	1	$\frac{3}{2}$	$\frac{9}{4}$

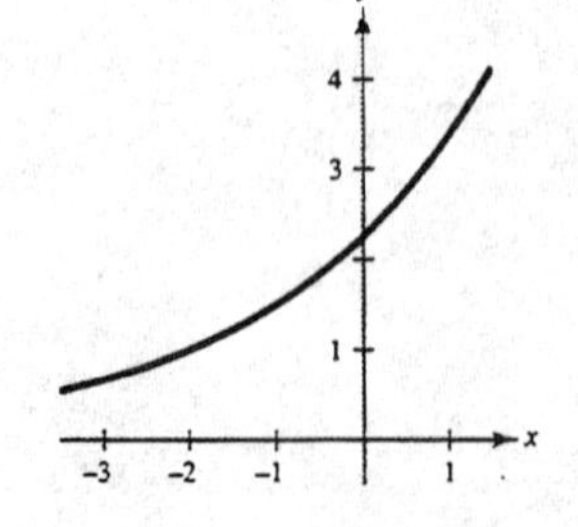

Asymptote: $y = 0$
Intercept: $\left(0, \frac{9}{4}\right)$
Increasing

22. $f(x) = \left(\frac{3}{2}\right)^{-x} + 2$

x	-2	-1	0	1	2
y	$\frac{17}{4}$	$\frac{7}{2}$	3	$\frac{8}{3}$	$\frac{22}{9}$

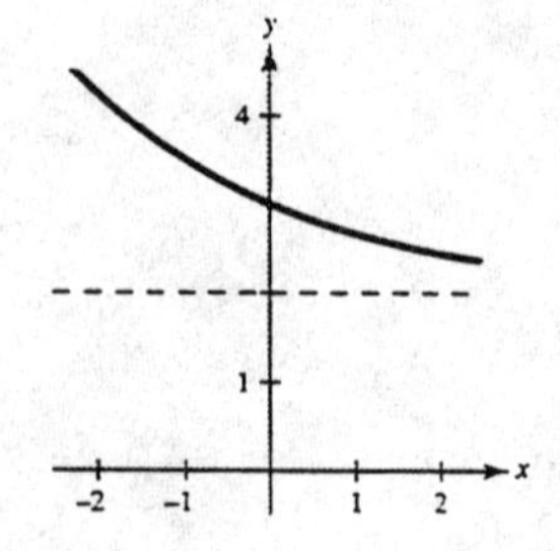

Asymptote: $y = 2$
Intercept: $(0, 3)$
Decreasing

24. $f(x) = -2^{x}$ is negative and decreasing. Matches graph (h).

26. $f(x) = -2^{-x}$ is negative and increasing. Matches graph (b).

28. $f(x) = 2^{x} + 1$ is increasing and has $(0, 2)$ intercept. Matches graph (f).

30. $f(x) = 2^{x-2}$ is increasing and has $\left(0, \frac{1}{4}\right)$ intercept. Matches graph (d).

32.

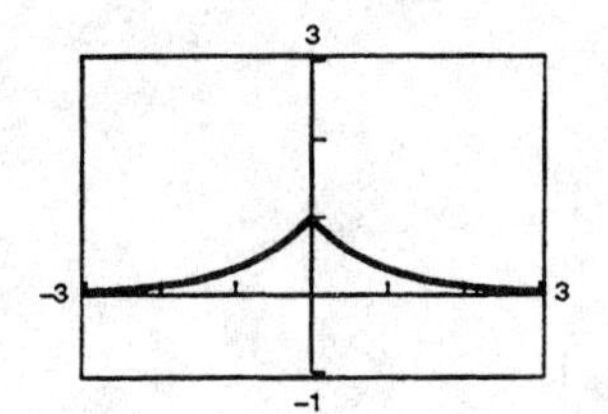

34.

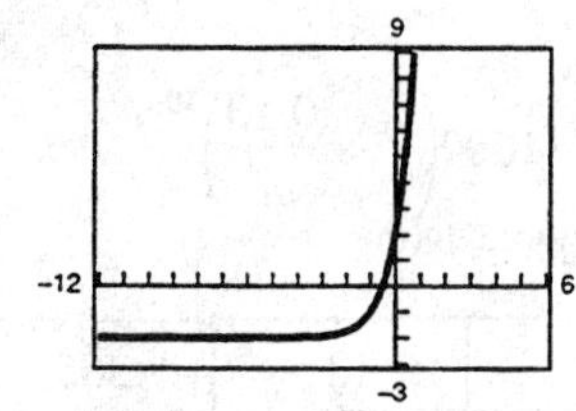

36.

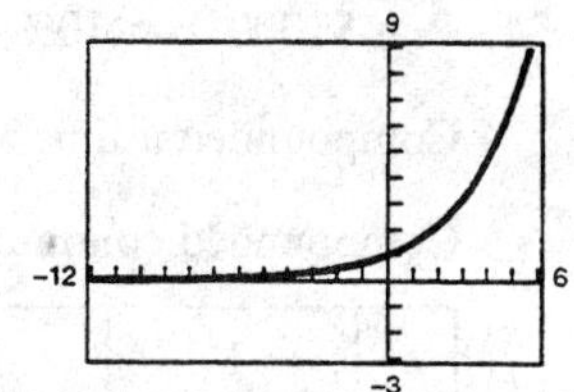

38.

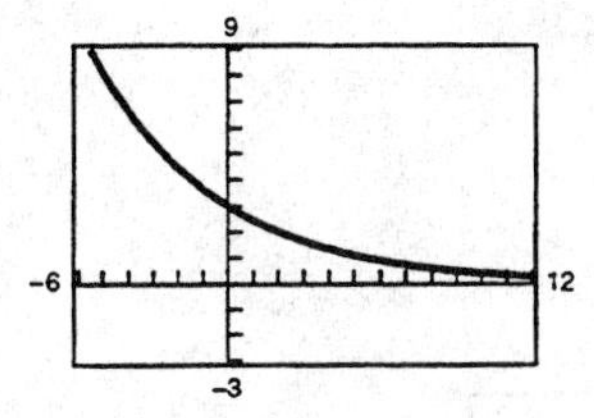

40.

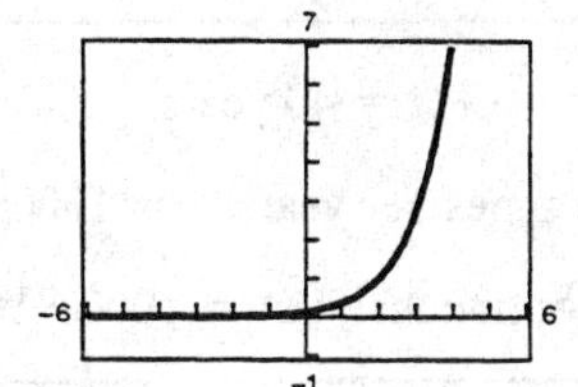

42. (a)

x	-1	-0.5	0	0.5	1
$f(x)$	2	1.4142	1	0.7071	0.5
$g(x)$	4	2	1	0.5	0.25

$\left(\frac{1}{4}\right)^x < \left(\frac{1}{2}\right)^x$ for $x > 0$

(b)

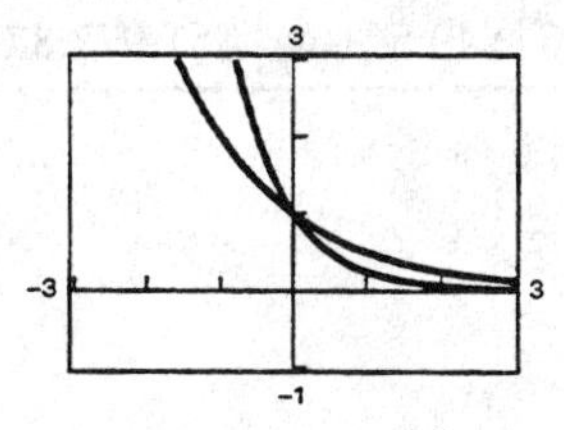

(i) $\left(\frac{1}{4}\right)^x < \left(\frac{1}{2}\right)^x$ for $x > 0$

(ii) $\left(\frac{1}{4}\right)^x > \left(\frac{1}{2}\right)^x$ for $x < 0$

44. (a) $f(x) = \dfrac{8}{1 + e^{-0.5x}}$

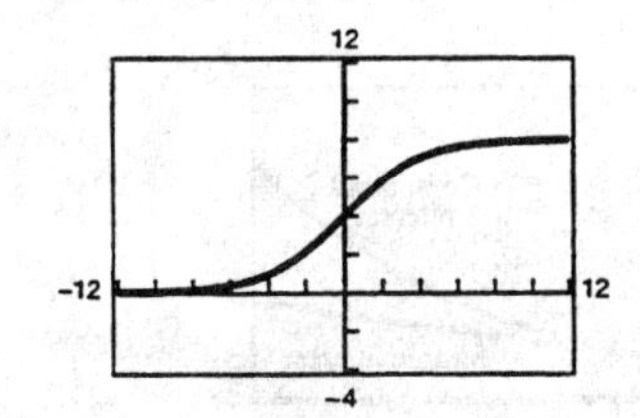

Horizontal asymptotes at $y = 0$ and $y = 8$

(b) $g(x) = \dfrac{8}{1 + e^{-0.5/x}}$

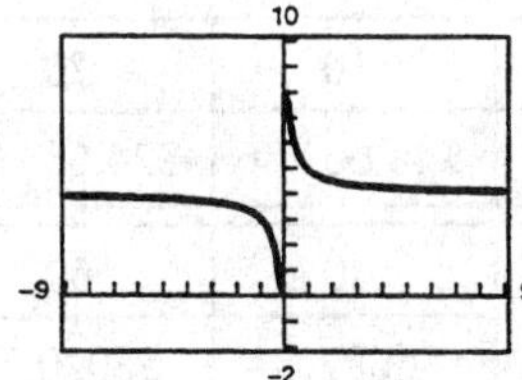

Horizontal asymptote at $y = 4$

46.

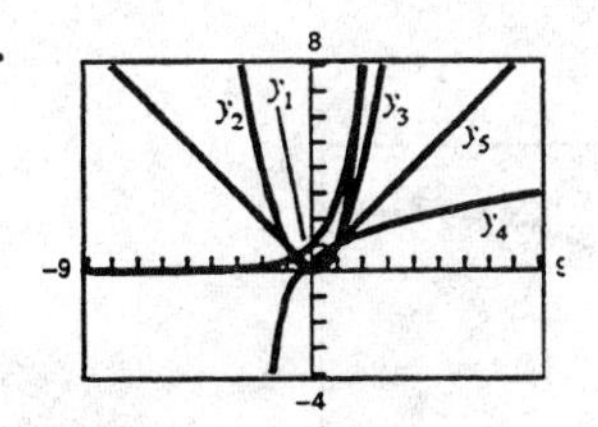

The function that increases at the fastest rate for "large" values of x is $y_1 = e^x$.

48. For a quantity that is described as growing exponentially, it usually implies rapid growth.

50. In Exercise 49, $f(x) = [1 + (0.5/x)^x]$ appears to approach $g(x) = e^{0.5}$ as x increases without bound. Therefore, the value of $[1 + (r/x)]^x$ approaches e^r as x increases without bound. For example, if $r = 1$,

x	1	10	100	200	500	1100	10,000
$\left[1 + \left(\frac{1}{x}\right)\right]^x$	2	2.5937	2.7048	2.7115	2.7156	2.7170	2.718

$e^1 \approx 2.718281828\ldots$

52. $A = Pe^{rt} = 25{,}000e^{(0.0875)(25)} \approx \$222{,}822.57$

54. $P = \$1000, r = 10\%, t = 10$ years

Compounded n times per year: $A = 1000\left(1 + \frac{0.10}{n}\right)^{10n}$

Compounded continuously: $A = 1000e^{0.10(10)}$

n	1	2	4	12	365	Continuous
A	\$2,593.74	\$2,653.30	\$2,685.06	\$2,707.04	\$2,717.90	\$2,718.28

56. $P = \$1000, r = 10\%, t = 40$ years

Compounded n times per year: $A = 1000\left(1 + \frac{0.10}{n}\right)^{40n}$

Compounded continuously: $A = 1000e^{0.10(40)}$

n	1	2	4	12	365	Continuous
A	\$45,259.26	\$49,561.44	\$51,977.87	\$53,700.66	\$54,568.25	\$54.598.15

58.

$$A = Pe^{rt}$$

$$100{,}000 = Pe^{0.12t}$$

$$\frac{100{,}000}{e^{0.12t}} = P$$

$$P = 100{,}000e^{-0.12t}$$

t	1	10	20	30	40	50
P	\$88,692.04	\$30,119.42	\$9071.80	\$2732.37	\$822.97	\$247.88

60.

t	1	10	20
P	\$93,240.01	\$48,661.86	\$24,663.01

t	30	40	50
P	\$12,248.11	\$6082.64	\$3020.75

62. (a)

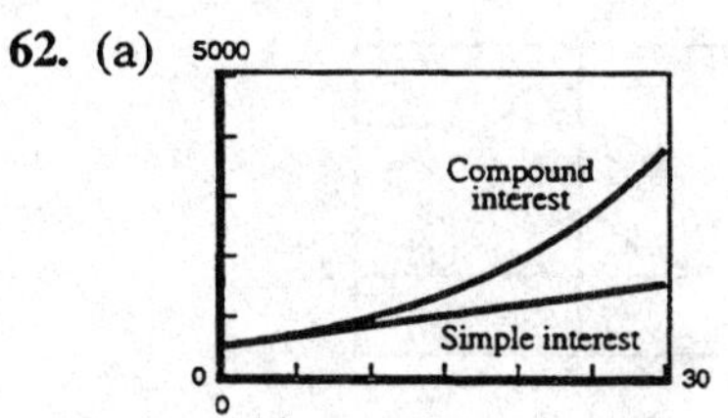

(b) $A = 500(1.07)^t$

$A = 500(0.07)t + 500$

64. $P(t) = 2500e^{0.0293t}$

$P(10) = 2500e^{0.0293(10)} \approx 3351$ (year 2000)

$P(20) = 2500e^{0.0293(20)} \approx 4492$ (year 2010)

66. (a) and (b)

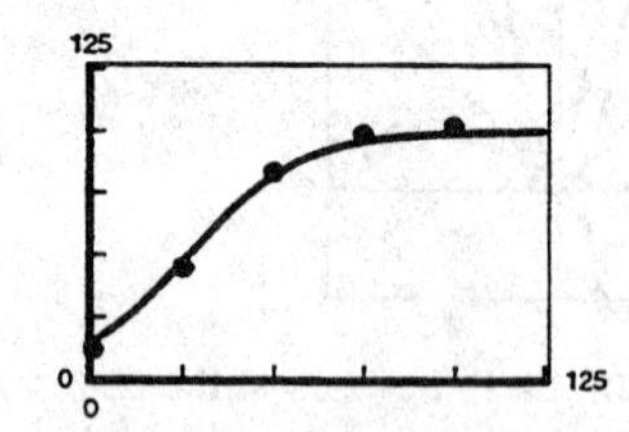

The model fits the data well.

(c)

x	0	25	50	75	100
y	15	47	82	96	99

(d) If $x = 36, y \approx 64.7\%$.

(e) If $y = 66.7\%, x \approx 36.9$.

68. $Q = 10\left(\frac{1}{2}\right)^{t/3730}$

(a) 10 units

(b) 7.85 units when $t = 2000$

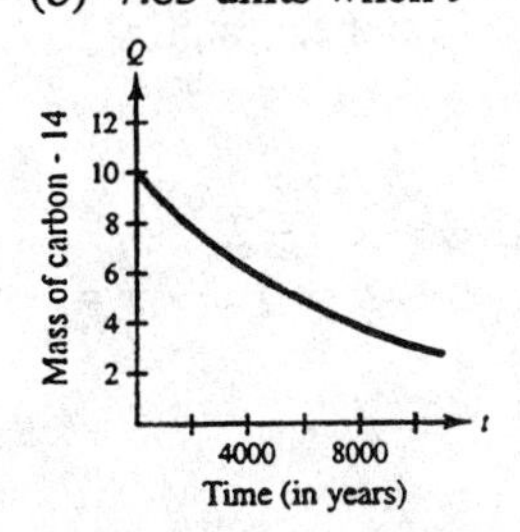

70. $C(t) = P(1.04)^t$

$= 23.95(1.04)^{10} \approx \$35.45$

72. False. e is an irrational number.

74. (c) and (d) are exponential functions.

(a) and (b) are polynomials.

76. $y_4 = 1 + \dfrac{x}{1!} + \dfrac{x^2}{2!} + \dfrac{x^3}{3!} + \dfrac{x^4}{4!}$

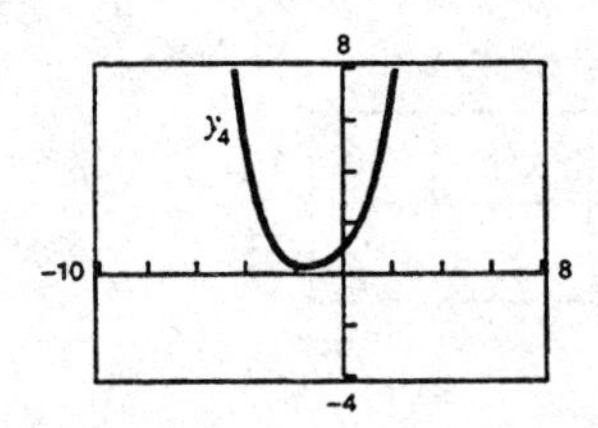

y_n gets closer to e^x as n increases.

Section 3.2 Logarithmic Functions and Their Graphs

Solutions to Even-Numbered Exercises

2. $\log_3 81 = 4 \Rightarrow 3^4 = 81$

4. $\log_{10} \frac{1}{1000} = -3 \Rightarrow 10^{-3} = \frac{1}{1000}$

6. $\log_{16} 8 = \frac{3}{4} \Rightarrow 16^{3/4} = 8$

8. $\ln 4 = 1.386\ldots \Rightarrow e^{1.386\ldots} = 4$

10. $8^2 = 64 \Rightarrow \log_8 64 = 2$

12. $9^{3/2} = 27 \Rightarrow \log_9 27 = \frac{3}{2}$

14. $10^{-3} = 0.001 \Rightarrow \log_{10} 0.001 = -3$

16. $e^0 = 1 \Rightarrow \ln 1 = 0$

18. $u^v = w \Rightarrow \log_u w = v$

20. $\log_2 \frac{1}{8} = \log_2 2^{-3} = -3$

22. $\log_{27} 9 = \log_{27} 27^{2/3} = \frac{2}{3}$

24. $\log_{10} 1000 = \log_{10} 10^3 = 3$

26. $\log_{10} 10 = \log_{10} 10^1 = 1$

28. $\ln 1 = 0$ because $e^0 = 1$.

30. $\log_a\left(\dfrac{1}{a}\right) = \log_a a^{-1} = -\log_a a = -1$

32. $\log_{10} \dfrac{4}{5} \approx -0.097$

34. $\log_{10} 12.5 \approx 1.097$

36. $\ln\sqrt{42} \approx 1.869$

38. $\ln(\sqrt{5} - 2) \approx -1.444$

40. $\ln 0.75 \approx -0.288$

42. $f(x) = 5^x,\ g(x) = \log_5 x$

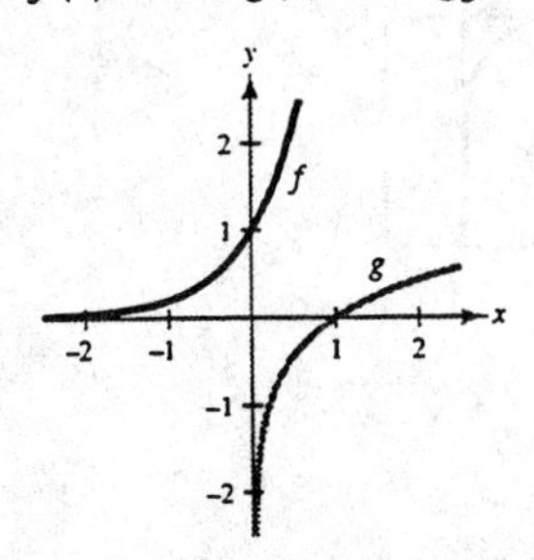

f and g are inverses. Their graphs are reflected about the line $y = x$.

44. $f(x) = 10^x,\ g(x) = \log_{10} x$

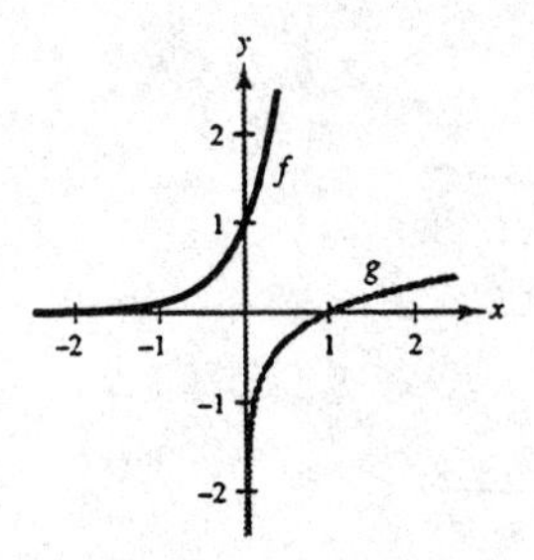

f and g are inverses. Their graphs are reflected about the line $y = x$.

46. $f(x) = -\log_3 x$

Asymptote: $x = 0$
Point on graph: $(1, 0)$
Matches graph (f).

48. $f(x) = \log_3(x - 1)$

Asymptote: $x = 1$
Point on graph: $(2, 0)$
Matches graph (e).

50. $f(x) = -\log_3(-x)$

Asymptote: $x = 0$
Point on graph: $(-1, 0)$
Matches graph (a).

52. $g(x) = \log_6 x$

Domain: $(0, \infty)$
Vertical asymptote: $x = 0$
x-intercept: $(1, 0)$

$y = \log_6 x \Rightarrow 6^y = x$

x	$\frac{1}{6}$	1	$\sqrt{6}$	36
y	-1	0	$\frac{1}{2}$	2

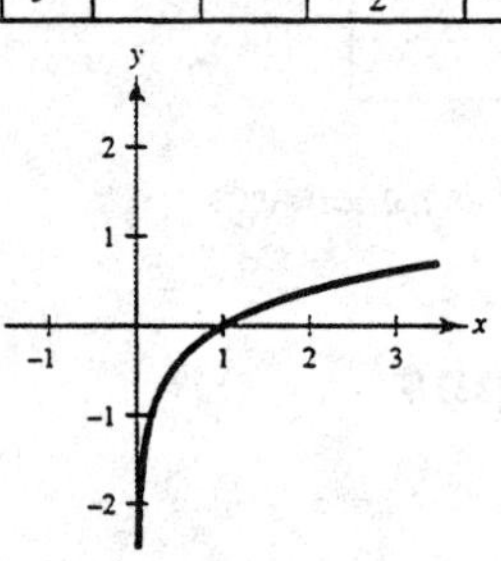

54. $f(x) = -\log_6(x + 2)$

Domain: $x + 2 > 0 \Rightarrow x > -2$
Vertical asymptote: $x + 2 = 0 \Rightarrow x = -2$
x-intercept: $(-1, 0)$

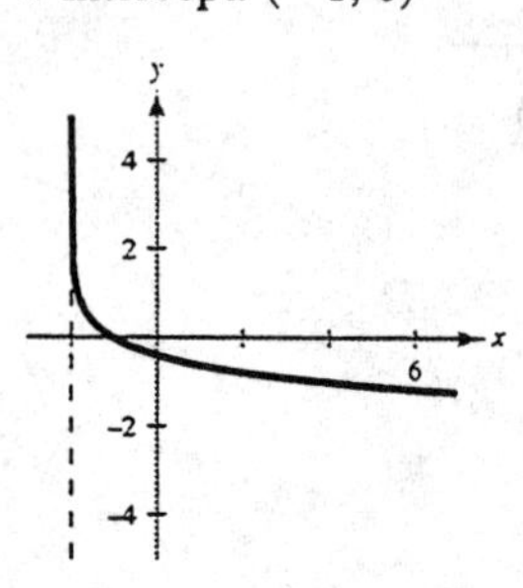

56. $y = \log_5(x - 1) + 4$

Domain: $x - 1 > 0 \Rightarrow x > 1$
The domain is $(1, \infty)$.
Vertical asymptote: $x - 1 = 0 \Rightarrow x = 1$
x-intercept: $\log_5(x - 1) + 4 = 0$

$$\log_5(x - 1) = -4$$
$$5^{-4} = x - 1$$
$$\frac{1}{625} = x - 1$$
$$\frac{626}{625} = x$$

The x-intercept is $\left(\frac{626}{625}, 0\right)$.

$y = \log_5(x - 1) + 4 \Rightarrow 5^{y-4} + 1 = x$

x	1.00032	1.0016	1.008	1.04	1.2
y	-1	0	1	2	3

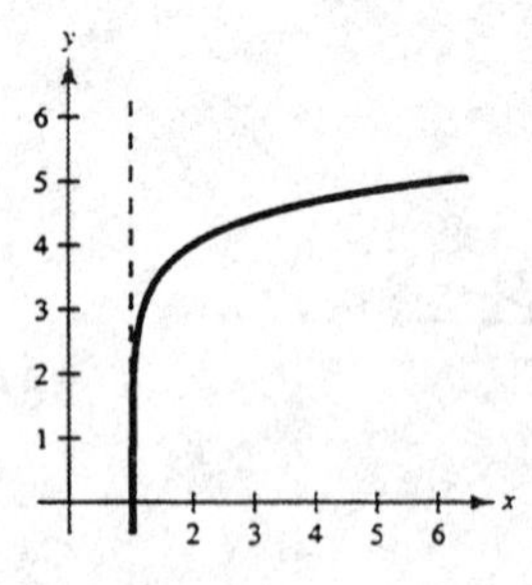

58. $y = \log_{10}(-x)$

Domain: $-x > 0 \Rightarrow x < 0$
The domain is $(-\infty, 0)$.
Vertical asymptote: $x = 0$
x-intercept: $\log_{10}(-x) = 0$

$$10^0 = -x$$
$$-1 = x$$

The x-intercept is $(-1, 0)$.

$y = \log_{10}(-x) \Rightarrow -10^y = x$

x	$-\frac{1}{100}$	$-\frac{1}{10}$	-1	-10
y	-2	-1	0	1

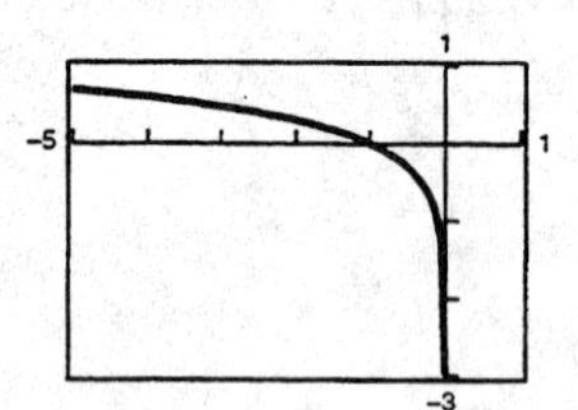

60. $h(x) = \ln(x + 1)$

Domain: $x + 1 > 0 \Rightarrow x > -1$

The domain is $(-1, \infty)$.

Vertical asymptote: $x + 1 = 0 \Rightarrow x = -1$

x-intercept: $\ln(x + 1) = 0$

$$e^0 = x + 1$$
$$1 = x + 1$$
$$0 = x$$

The x-intercept is $(0, 0)$.

$y = \ln(x + 1) \Rightarrow e^y - 1 = x$

x	-0.39	0	1.72	6.39	19.09
y	$-\frac{1}{2}$	0	1	2	3

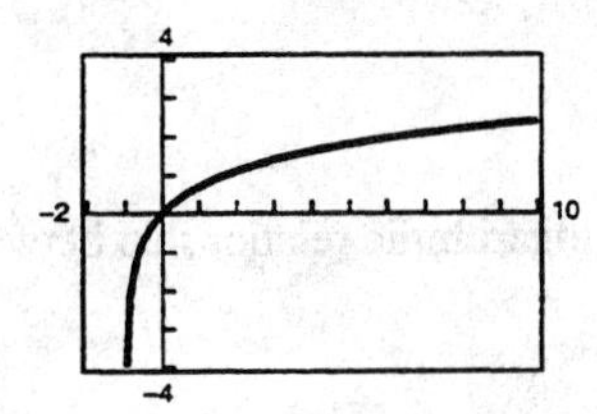

62. $f(x) = \ln(3 - x)$

Domain: $3 - x > 0 \Rightarrow x < 3$

The domain is $(-\infty, 3)$.

Vertical asymptote: $3 - x = 0 \Rightarrow x = 3$

x-intercept: $\ln(3 - x) = 0$

$$e^0 = 3 - x$$
$$1 = 3 - x$$
$$2 = x$$

The x-intercept is $(2, 0)$.

$y = \ln(3 - x) \Rightarrow x = 3 - e^y$

x	2.95	2.86	2.63	2	0.28
y	-3	-2	-1	0	1

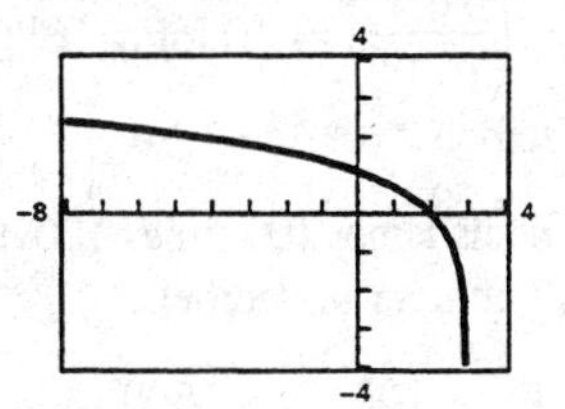

64. $g(x) = \dfrac{12 \ln x}{x}$

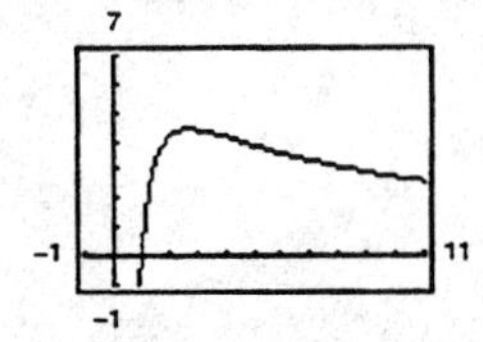

Domain: $(0, \infty)$

Increasing on $(0, 2.72)$.

Decreasing on $(2.72, \infty)$.

Relative maximum: $(2.72, 4.41)$

66. $f(x) = \dfrac{x}{\ln x}$

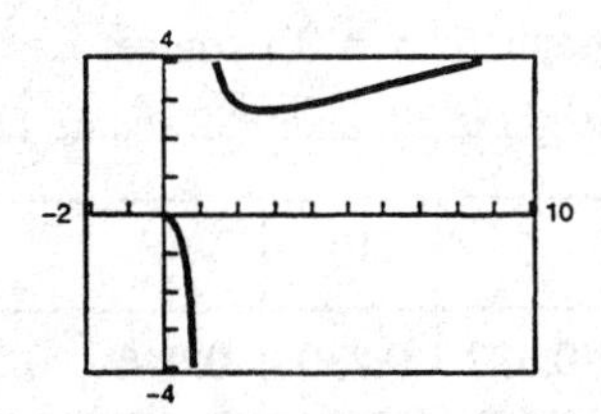

Domain: $(0, \infty)$

Increasing on $(0, 2.72)$.

Decreasing on $(2.72, \infty)$.

Relative maximum: $(2.72, 2.72)$

68. $f(t) = 80 - 17 \log_{10}(t + 1)$

(a) $f(0) = 80$

(b) $f(4) = 68.1$

(c) $f(10) = 62.3$

70.

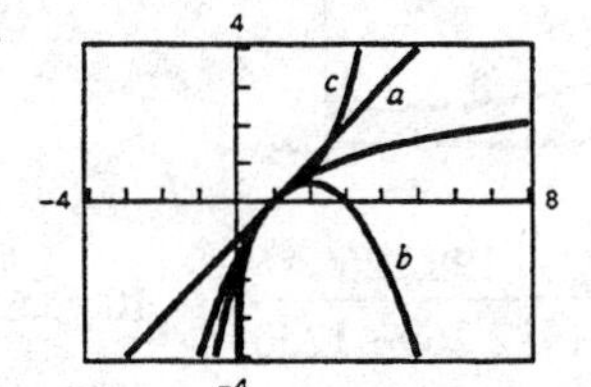

72. (a)

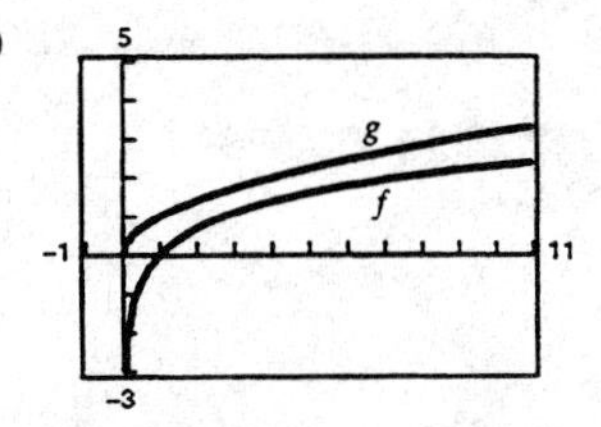

$g(x)$ is increasing more rapidly.

(b)

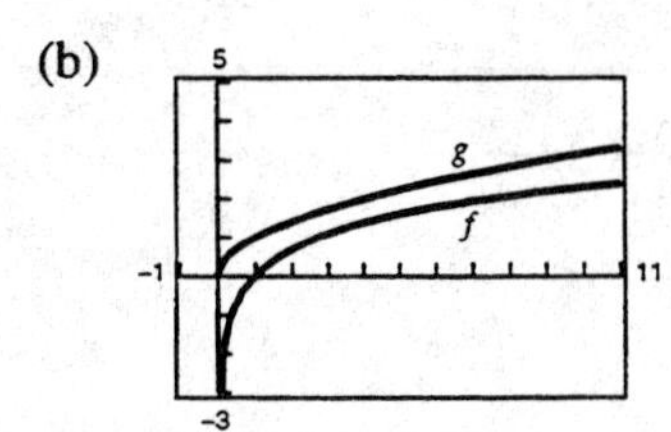

$g(x)$ is increasing more rapidly. As x increases without bound, $\sqrt[4]{x}$ eventually increases at a faster rate than $\ln x$.

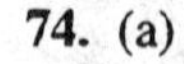

74. (a)

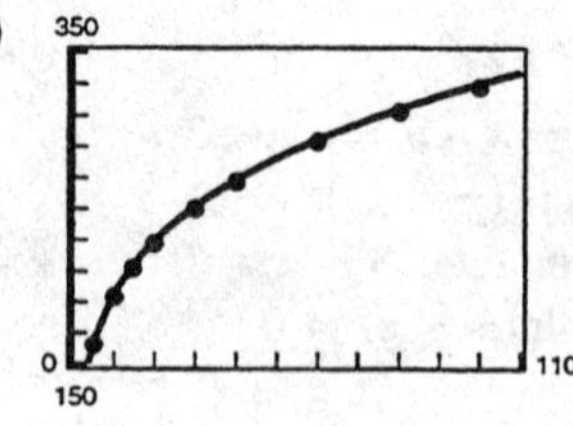

(b) $T > 300°$ F when $x > 67.3$ pounds per square inch.

76.

r	0.005	0.010	0.015
t	138.6 yr	69.3 yr	46.2 yr

r	0.020	0.025	0.030
t	34.7 yr	27.7 yr	23.1 yr

The doubling time decreases as r increases.

78. $\beta = 10 \log_{10}\left(\frac{I}{10^{-16}}\right)$

(a) $\beta = 10 \log_{10}\left(\frac{10^{-4}}{10^{-16}}\right) = 10 \log_{10}(10^{12}) = 10(12)$

$\beta = 120$ decibels

(b) $\beta = 10 \log_{10}\left(\frac{10^{-6}}{10^{-16}}\right) = 10 \log_{10}(10^{10}) = 10(10)$

$= 100$ decibels

(c) No, 120 decibels is not 100 times 100 decibels. The difference is due to the logarithmic relationship between intensity and number of decibels.

80. $y = 80.4 - 11 \ln x, \quad 100 \le x \le 1500$

(a) $\frac{450 \text{ cubic ft per minute}}{30 \text{ children}} = 15$ cubic feet per minute per child

(b) From the graph, for $y = 15$ you get $x \approx 382$ cubic feet.

(c) If ceiling height is 30, then 382 square feet of floor space is needed.

82. (a)

x	1	5	10	10^2	10^4	10^6
$f(x)$	0	0.322	0.230	0.046	0.00092	0.0000138

(b) As x increases without bound, $f(x)$ approaches 0.

(c)

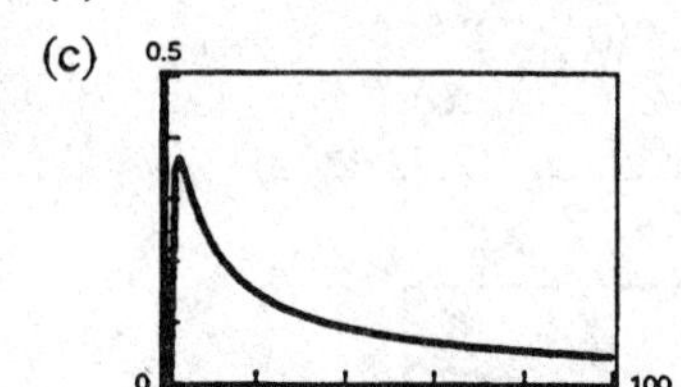

84. $t = 10.042 \ln\left(\frac{1982.26}{1982.26 - 1250}\right) \approx 10$ years

86. Total amount $= (1982.26)(10)(12) \approx \$237{,}871$

Interest $= \$237{,}871 - \$150{,}000 = \$87{,}871$

Section 3.3 Properties of Logarithms

Solutions to Even-Numbered Exercises

2. $f(x) = \ln x$

$g(x) = \dfrac{\log_{10} x}{\log_{10} e}$

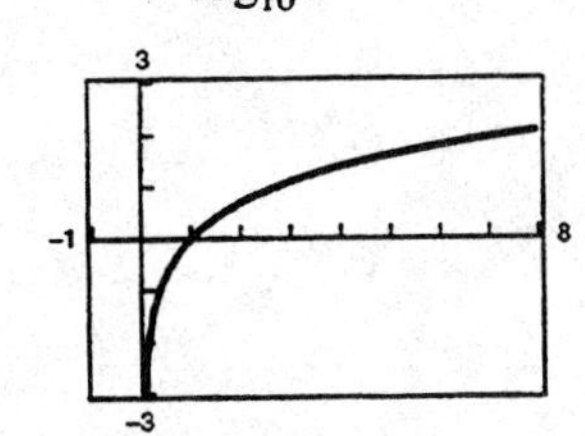

$f(x) = g(x)$

4. $\log_4 10 = \dfrac{\log_{10} 10}{\log_{10} 4} = \dfrac{1}{\log_{10} 4}$

6. $\ln 5 = \dfrac{\log_{10} 5}{\log_{10} e}$

8. $\log_4 10 = \dfrac{\ln 10}{\ln 4}$

10. $\log_{10} 5 = \dfrac{\ln 5}{\ln 10}$

12. $\log_7 4 = \dfrac{\log_{10} 4}{\log_{10} 7} = \dfrac{\ln 4}{\ln 7} \approx 0.712$

14. $\log_4 0.55 = \dfrac{\log_{10} 0.55}{\log_{10} 4} = \dfrac{\ln 0.55}{\ln 4} \approx -0.431$

16. $\log_{20} 125 = \dfrac{\log_{10} 125}{\log_{10} 20} = \dfrac{\ln 125}{\ln 20} \approx 1.612$

18. $\log_{1/3} 0.015 = \dfrac{\log_{10} 0.015}{\log_{10}(1/3)} = \dfrac{\ln 0.015}{\ln(1/3)} \approx 3.823$

20. $\log_{10} 10z = \log_{10} 10 + \log_{10} z$

22. $\log_{10} \dfrac{y}{2} = \log_{10} y - \log_{10} 2$

24. $\log_6 z^{-3} = -3 \log_6 z$

26. $\ln \sqrt[3]{t} = \ln t^{1/3} = \dfrac{1}{3} \ln t$

28. $\ln \dfrac{xy}{z} = \ln x + \ln y - \ln z$

30.
$$\begin{aligned} \ln\left(\frac{x^2 - 1}{x^3}\right) &= \ln(x^2 - 1) - \ln x^3 \\ &= \ln[(x + 1)(x - 1)] - \ln x^3 \\ &= \ln(x + 1) + \ln(x - 1) - 3 \ln x \end{aligned}$$

32.
$$\begin{aligned} \ln \sqrt{\frac{x^2}{y^3}} = \ln\left(\frac{x^2}{y^3}\right)^{1/2} &= \frac{1}{2} \ln\left(\frac{x^2}{y^3}\right) \\ &= \frac{1}{2}(\ln x^2 - \ln y^3) \\ &= \frac{1}{2}(2 \ln x - 3 \ln y) \end{aligned}$$

34.
$$\begin{aligned} \ln\left(\frac{x}{\sqrt{x^2 + 1}}\right) &= \ln x - \ln \sqrt{x^2 + 1} \\ &= \ln x - \ln(x^2 + 1)^{1/2} \\ &= \ln x - \frac{1}{2} \ln(x^2 + 1) \end{aligned}$$

36.
$$\begin{aligned} \ln \sqrt{x^2(x + 2)} &= \ln[x^2(x + 2)]^{1/2} \\ &= \ln[x(x + 2)^{1/2}] \\ &= \ln x + \ln(x + 2)^{1/2} \\ &= \ln x + \frac{1}{2} \ln(x + 2) \end{aligned}$$

38.
$$\begin{aligned} \log_b \frac{\sqrt{x}\, y^4}{z^4} &= \log_b \sqrt{x}\, y^4 - \log_b z^4 \\ &= \log_b x^{1/2} + \log_b y^4 - \log_b z^4 \\ &= \frac{1}{2} \log_b x + 4 \log_b y - 4 \log_b z \end{aligned}$$

40. $y_1 = \ln\left(\dfrac{\sqrt{x}}{x - 2}\right)$

$y_2 = \dfrac{1}{2} \ln x - \ln(x - 2)$

$y_1 = y_2$

42. $\ln y + \ln z = \ln yz$

44. $\log_5 8 - \log_5 t = \log_5 \dfrac{8}{t}$

46. $-4 \log_6 2x = \log_6(2x)^{-4} = \log_6 \dfrac{1}{16x^4}$

48. $\dfrac{3}{2} \log_7(z - 2) = \log_7(z - 2)^{3/2}$

50. $2 \ln 8 + 5 \ln z = \ln 8^2 + \ln z^5$
$= \ln 64z^5$

52. $3 \ln x + 2 \ln y - 4 \ln z = \ln x^3 + \ln y^2 - \ln z^4$
$= \ln x^3y^2 - \ln z^4$
$= \ln \dfrac{x^3y^2}{z^4}$

54. $4[\ln z + \ln(z + 5)] - 2 \ln(z - 5) = 4[\ln z(z + 5)] - \ln(z - 5)^2$
$= \ln[z(z + 5)]^4 - \ln(z - 5)^2$
$= \ln \dfrac{z^4(z + 5)^4}{(z - 5)^2}$

56. $2[\ln x - \ln(x + 1) - \ln(x - 1)] = 2\left[\ln \dfrac{x}{x + 1} - \ln(x - 1)\right]$
$= 2\left[\ln \dfrac{x}{(x + 1)(x - 1)}\right]$
$= 2\left[\ln \dfrac{x}{x^2 - 1}\right]$
$= \ln\left(\dfrac{x}{x^2 - 1}\right)^2$

58. $\frac{1}{2}[\ln(x + 1) + 2 \ln(x - 1)] + 3 \ln x = \frac{1}{2}[\ln(x + 1) + \ln(x - 1)^2] + \ln x^3$
$= \frac{1}{2}[\ln(x + 1)(x - 1)^2] + \ln x^3$
$= \ln[(x + 1)(x - 1)^2]^{1/2} + \ln x^3$
$= \ln[(x + 1)^{1/2}(x - 1)] + \ln x^3$
$= \ln\left[x^3(x - 1)\sqrt{x + 1}\right]$

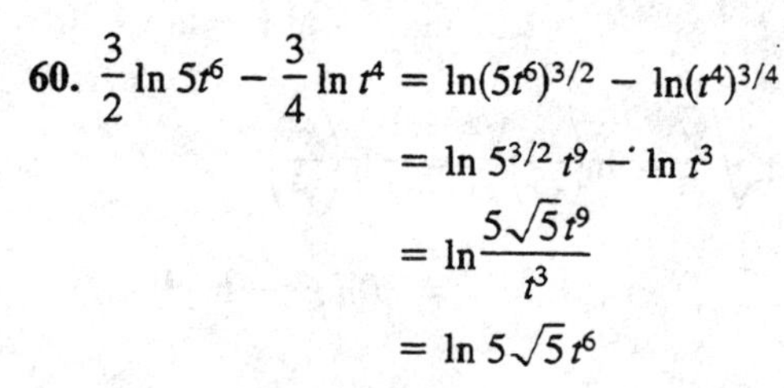

60. $\dfrac{3}{2} \ln 5t^6 - \dfrac{3}{4} \ln t^4 = \ln(5t^6)^{3/2} - \ln(t^4)^{3/4}$
$= \ln 5^{3/2} t^9 - \ln t^3$
$= \ln \dfrac{5\sqrt{5}t^9}{t^3}$
$= \ln 5\sqrt{5}t^6$

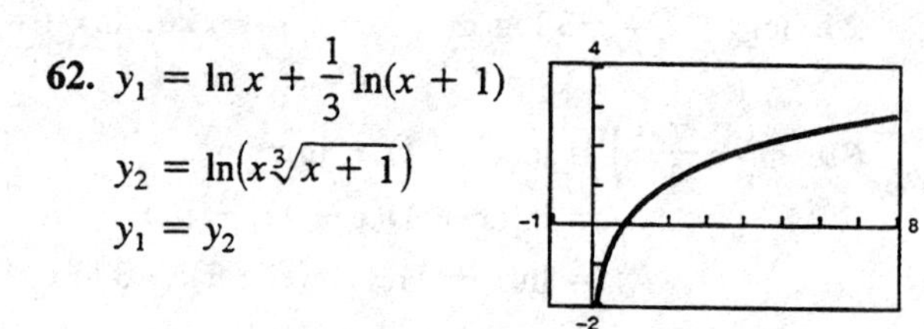

62. $y_1 = \ln x + \dfrac{1}{3} \ln(x + 1)$
$y_2 = \ln\left(x\sqrt[3]{x + 1}\right)$
$y_1 = y_2$

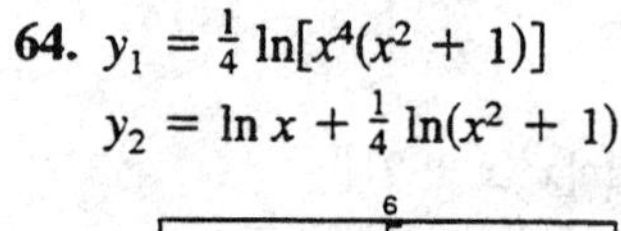

64. $y_1 = \frac{1}{4} \ln[x^4(x^2 + 1)]$
$y_2 = \ln x + \frac{1}{4} \ln(x^2 + 1)$

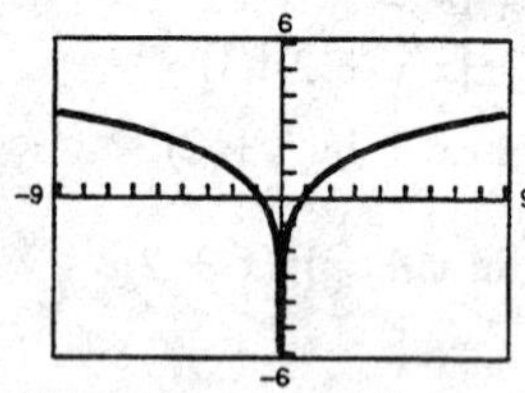

They are not equivalent. The domain of y_1 is all real numbers except 0. The domain of y_2 is $x > 0$.

66. $\ln 2 \approx 0.6931, \ln 3 \approx 1.0986, \ln 5 \approx 1.6094$

$\ln 2 \approx 0.6931$

$\ln 3 \approx 1.0986$

$\ln 4 = \ln 2 + \ln 2 \approx 0.6931 + 0.6931 = 1.3862$

$\ln 5 \approx 1.6094$

$\ln 6 = \ln 2 + \ln 3 \approx 0.6931 + 1.0986 = 1.7917$

$\ln 8 = \ln 2^3 = 3 \ln 2 \approx 3(0.6931) = 2.0793$

$\ln 9 = \ln 3^2 = 2 \ln 3 \approx 2(1.0986) = 2.1972$

$\ln 10 = \ln 5 + \ln 2 \approx 1.6094 + 0.6931 = 2.3025$

$\ln 12 = \ln 2^2 + \ln 3 = 2 \ln 2 + \ln 3 \approx 2(0.6931) + 1.0986 = 2.4848$

$\ln 15 = \ln 5 + \ln 3 \approx 1.6094 + 1.0986 = 2.7080$

$\ln 16 = \ln 2^4 = 4 \ln 2 \approx 4(0.6931) = 2.7724$

$\ln 18 = \ln 3^2 + \ln 2 = 2 \ln 3 + \ln 2 \approx 2(1.0986) + 0.6931 = 2.8903$

$\ln 20 = \ln 5 + \ln 2^2 = \ln 5 + 2 \ln 2 \approx 1.6094 + 2(0.6931) = 2.9956$

68. $\log_6 \sqrt[3]{6} = \log_6 6^{1/3} = \frac{1}{3} \log_6 6 = \frac{1}{3}(1) = \frac{1}{3}$

70. $\log_5 \frac{1}{125} = \log_5 5^{-3} = -3 \log_5 5 = -3(1) = -3$

72. $\log_2(-16)$ is undefined because -16 is not in the domain of $\log_2 x$.

74. $\log_4 2 + \log_4 32 = \log_4 4^{1/2} + \log_4 4^{5/2}$
$= \frac{1}{2} \log_4 4 + \frac{5}{2} \log_4 4$
$= \frac{1}{2}(1) + \frac{5}{2}(1)$
$= 3$

76. $3 \ln e^4 = (3)(4) \ln e$
$= 12(1) = 12$

78. $\ln 1 = 0$

80. $\ln \sqrt[4]{e^3} = \ln e^{3/4}$
$= \frac{3}{4} \ln e$
$= \frac{3}{4}(1) = \frac{3}{4}$

82. $\log_5\left(\frac{1}{15}\right) = \log_5 1 - \log_5 15 = 0 - (\log_5 3 + \log_5 5)$
$= -1 - \log_5 3$

84. $\log_2(4^2 \cdot 3^4) = \log_2 4^2 + \log_2 3^4$
$= 2 \log_2 4 + 4 \log_2 3$
$= 2 \log_2 2^2 + 4 \log_2 3$
$= 4 \log_2 2 + 4 \log_2 3$
$= 4 + 4 \log_2 3$

86. $\log_{10} \frac{9}{300} = \log_{10} \frac{3}{100}$
$= \log_{10} 3 - \log_{10} 100$
$= \log_{10} 3 - \log_{10} 10^2$
$= \log_{10} 3 - 2 \log_{10} 10$
$= \log_{10} 3 - 2$

88. $\ln \frac{6}{e^2} = \ln 6 - \ln e^2$
$= \ln 6 - 2 \ln e$
$= \ln 6 - 2$

90. $f(t) = 90 - 15 \log_{10}(t + 1), 0 \le t \le 12$

(a)

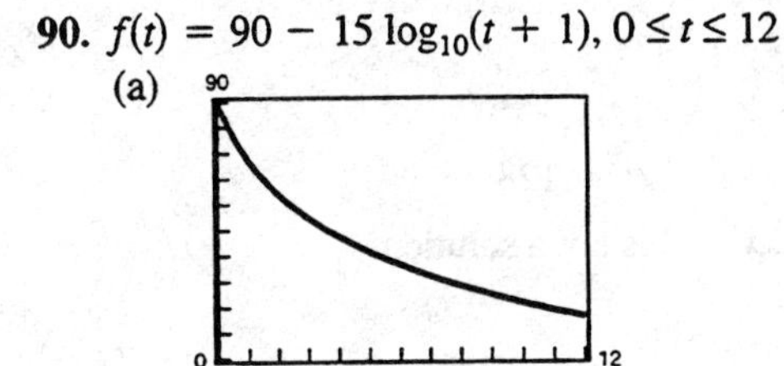

(b) When $t = 0, f(0) = 90$.

(c) $f(6) \approx 77$

(d) $f(12) \approx 73$

(e) $f(t) = 75$ when $t \approx 9$ months.

92. If $y = ab^x$, then $\ln y = \ln(ab^x) = \ln a + x \ln b$, which is linear. If $y = \dfrac{1}{cx + d}$, then $\dfrac{1}{y} = cx + d$.

94. $f(x) = \log_7 x = \dfrac{\log_{10} x}{\log_{10} 7} = \dfrac{\ln x}{\ln 7}$

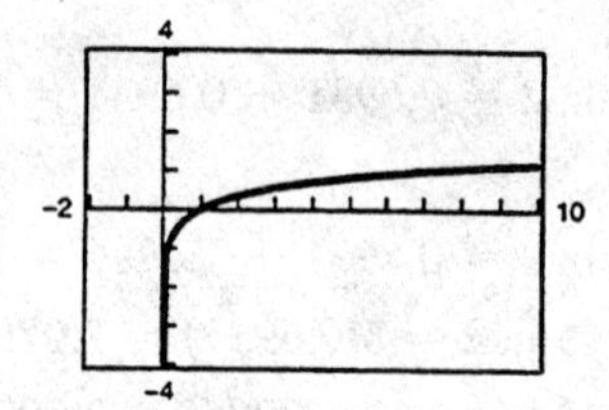

96. $f(t) = \log_5 \dfrac{t}{3} = \log_5 t - \log_5 3 = \dfrac{\ln t}{\ln 5} - \dfrac{\ln 3}{\ln 5}$

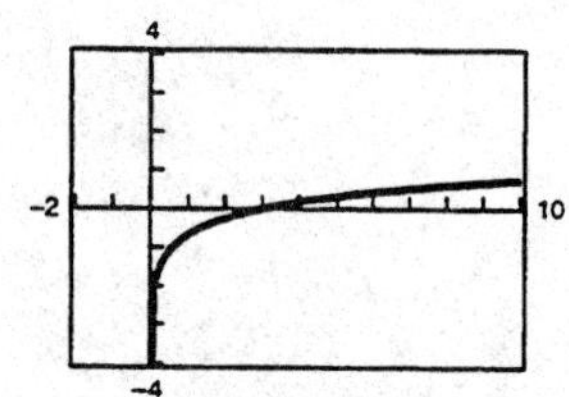

98. $f(ax) = f(a) + f(x),\ a > 0,\ x > 0$

True, because

$f(ax) = \ln ax = \ln a + \ln x$

$= f(a) + f(x)$.

100. $\sqrt{f(x)} = \frac{1}{2} f(x)$; False.

$\sqrt{f(x)} = \sqrt{\ln x}$ can't be simplified further.

$f(\sqrt{x}) = \ln \sqrt{x} = \ln x^{1/2} = \frac{1}{2} \ln x = \frac{1}{2} f(x)$

102. If $f(x) < 0$, then $0 < x < 1$.

True.

104. Let $x = \log_b u$, then $u = b^x$ and $u^n = b^{nx}$.

$\log_b u^n = \log_b b^{nx} = nx = n \log_b u$

106.
$$\left(\frac{2x^2}{3y}\right)^{-3} = \left(\frac{3y}{2x^2}\right)^3 = \frac{(3y)^3}{(2x^2)^3} = \frac{27y^3}{8x^6}$$

108.
$$xy(x^{-1} + y^{-1})^{-1} = \frac{xy}{x^{-1} + y^{-1}} = \frac{xy}{\frac{1}{x} + \frac{1}{y}} = \frac{xy}{\frac{y + x}{xy}} = \frac{(xy)^2}{x + y}$$

Section 3.4 Solving Exponential and Logarithmic Equations

Solutions to Even-Numbered Exercises

2. $2^{3x+1} = 32$

(a) $x = -1$

$2^{3(-1)+1} = 2^{-2} = \frac{1}{4}$

No, $x = -1$ is not a solution.

(b) $x = 2$

$2^{3(2)+1} = 2^7 = 128$

No, $x = 2$ is not a solution.

4. $5^{2x+3} = 812$

(a) $x = -1.5 + \log_5\sqrt{812}$

$$5^{2(-1.5+\log_5\sqrt{812})+3} = 5^{-3+2\log_5\sqrt{812}+3}$$
$$= 5^{2\log_5 812^{1/2}}$$
$$= 5^{\log_5 (812^{1/2})^2}$$
$$= 5^{\log_5 812} = 812$$

Yes, $x = -1.5 + \log_5\sqrt{812}$ is a solution.

(b) $x \approx 0.5813$

$$5^{2(0.5813)+3} = 5^{4.1626} \approx 812$$

Yes, $x \approx 0.5813$ is a solution.

(c) $x = \dfrac{1}{2}\left(-3 + \dfrac{\ln 812}{\ln 5}\right)$

$$5^{2[1/2(-3+(\ln 812/\ln 5))]+3} = 5^{-3+(\ln 812/\ln 5)+3}$$
$$= 5^{\ln 812/\ln 5}$$
$$= 5^{\log_5 812}$$
$$= 812$$

Yes, $x = \dfrac{1}{2}\left(-3 + \dfrac{\ln 812}{\ln 5}\right)$ is a solution.

6. $\ln(x - 1) = 3.8$

(a) $x = 1 + e^{3.8}$

$\ln(1 + e^{3.8} - 1) = \ln e^{3.8} = 3.8$

Yes, $x = 1 + e^{3.8}$ is a solution.

(b) $x \approx 45.7012$

$\ln(45.7012 - 1) = \ln(44.7012) \approx 3.8$

Yes, $x \approx 45.7012$ is a solution.

(c) $x = 1 + \ln 3.8$

$\ln(1 + \ln 3.8 - 1) = \ln(\ln 3.8) \approx 0.289$

No, $x = 1 + \ln 3.8$ is not a solution.

8. $f(x) = g(x)$

$$27^x = 9$$
$$27^x = 27^{2/3}$$
$$x = \tfrac{2}{3}$$

Point of intersection: $\left(\tfrac{2}{3}, 9\right)$

10.
$$f(x) = g(x)$$
$$\ln(x - 4) = 0$$
$$x - 4 = e^0$$
$$x - 4 = 1$$
$$x = 5$$

Point of intersection: $(5, 0)$

12. $3^x = 243$

$$3^x = 3^5$$
$$x = 5$$

14. $8^x = 4$

$$8^x = 8^{2/3}$$
$$x = \tfrac{2}{3}$$

16. $3^{x-1} = 27$

$$3^{x-1} = 3^3$$
$$x - 1 = 3$$
$$x = 4$$

18. $\log_5 5x = 2$

$$5^2 = 5x$$
$$5 = x$$

20. $\ln(2x - 1) = 0$

$$e^0 = 2x - 1$$
$$1 = 2x - 1$$
$$2 = 2x$$
$$1 = x$$

22. $\ln e^{2x-1} = 2x - 1$

24. $-1 + \ln e^{2x} = -1 + 2x = 2x - 1$

26. $-8 + e^{\ln x^3} = -8 + x^3 = x^3 - 8$

28. $4e^x = 91$

$$e^x = \tfrac{91}{4}$$
$$\ln e^x = \ln \tfrac{91}{4}$$
$$x = \ln \tfrac{91}{4} \approx 3.125$$

30. $-14 + 3e^x = 11$

$$3e^x = 25$$
$$e^x = \tfrac{25}{3}$$
$$\ln e^x = \ln \tfrac{25}{3}$$
$$x = \ln \tfrac{25}{3} \approx 2.120$$

32. $1000e^{-4x} = 75$

$$e^{-4x} = \tfrac{3}{40}$$
$$\ln e^{-4x} = \ln \tfrac{3}{40}$$
$$-4x = \ln \tfrac{3}{40}$$
$$x = -\tfrac{1}{4}\ln \tfrac{3}{40} \approx 0.648$$

34.
$$10^x = 570$$
$$\log_{10} 10^x = \log_{10} 570$$
$$x = \log_{10} 570 \approx 2.756$$

36.

x	1.6	1.7	1.8	1.9	2.0
$f(x)$	24.53	29.96	36.60	44.70	54.60

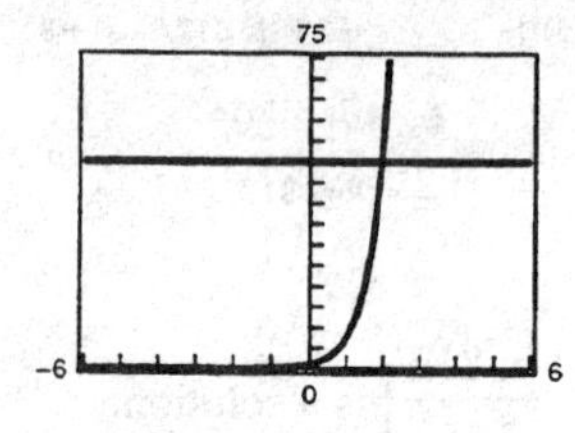

$x \approx 1.956$

38.

x	0	1	2	3	4
$f(x)$	200	292	352	381	393

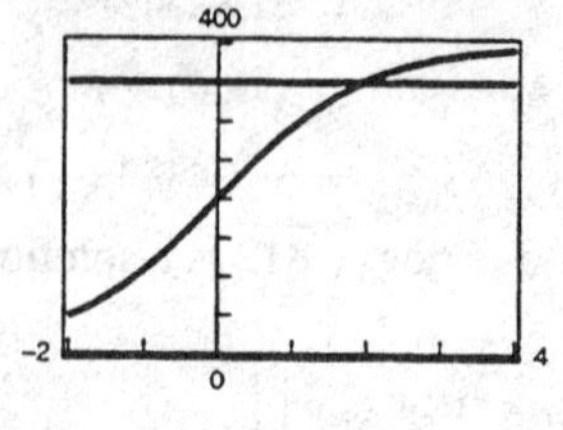

$x \approx 1.946$

40.
$$\begin{aligned} e^{2x} - 5e^x + 6 &= 0 \\ (e^x - 2)(e^x - 3) &= 0 \\ e^x = 2 \text{ or } e^x &= 3 \\ x = \ln 2 \approx 0.693 \text{ or } x &= \ln 3 \approx 1.099 \end{aligned}$$

42.
$$\begin{aligned} 6^{5x} &= 3000 \\ \ln 6^{5x} &= \ln 3000 \\ (5x) \ln 6 &= \ln 3000 \\ 5x &= \frac{\ln 3000}{\ln 6} \\ x &= \frac{\ln 3000}{5 \ln 6} \approx 0.894 \end{aligned}$$

44.
$$\begin{aligned} 4^{-3t} &= 0.10 \\ \ln 4^{-3t} &= \ln 0.10 \\ (-3t) \ln 4 &= \ln 0.10 \\ -3t &= \frac{\ln 0.10}{\ln 4} \\ t &= -\frac{\ln 0.10}{3 \ln 4} \approx 0.554 \end{aligned}$$

46.
$$\begin{aligned} \frac{3000}{2 + e^{2x}} &= 2 \\ 1500 &= 2 + e^{2x} \\ 1498 &= e^{2x} \\ \ln 1498 &= \ln e^{2x} \\ \ln 1498 &= 2x \\ \frac{\ln 1498}{2} &= x \approx 3.656 \end{aligned}$$

48.
$$\begin{aligned} 3(5^{x-1}) &= 21 \\ 5^{x-1} &= 7 \\ \ln 5^{x-1} &= \ln 7 \\ (x - 1) \ln 5 &= \ln 7 \\ x - 1 &= \frac{\ln 7}{\ln 5} \\ x &= \frac{\ln 7}{\ln 5} + 1 \approx 2.209 \end{aligned}$$

50. $\left(1 + \dfrac{0.065}{365}\right)^{365t} = 4 \Longrightarrow t = 21.330$

52. $f(x) = 3e^{3x/2} - 962$

The zero is $x \approx 3.847$.

300
-6
9
-1200

54. $h(t) = e^{0.125t} - 8$

The zero is $t \approx 16.636$.

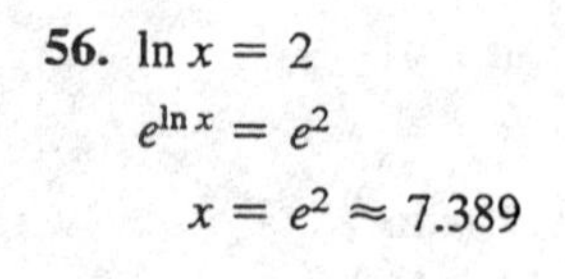

56.
$$\begin{aligned} \ln x &= 2 \\ e^{\ln x} &= e^2 \\ x &= e^2 \approx 7.389 \end{aligned}$$

58. $\ln(x+1)^2 = 2$

$$e^{\ln(x+1)^2} = e^2$$
$$(x+1)^2 = e^2$$
$$x+1 = e \text{ or } x+1 = -e$$
$$x = e - 1 \approx 1.718$$
or
$$x = -e - 1 \approx -3.718$$

60. $\log_4 x - \log_4(x-1) = \frac{1}{2}$

$$\log_4\left(\frac{x}{x-1}\right) = \frac{1}{2}$$
$$4^{\log_4(x/x-1)} = 4^{1/2}$$
$$\frac{x}{x-1} = 2$$
$$x = 2(x-1)$$
$$x = 2x - 2$$
$$2 = x$$

62.

x	4	5	6	7	8
$f(x)$	8.99	9.66	10.20	10.67	11.07

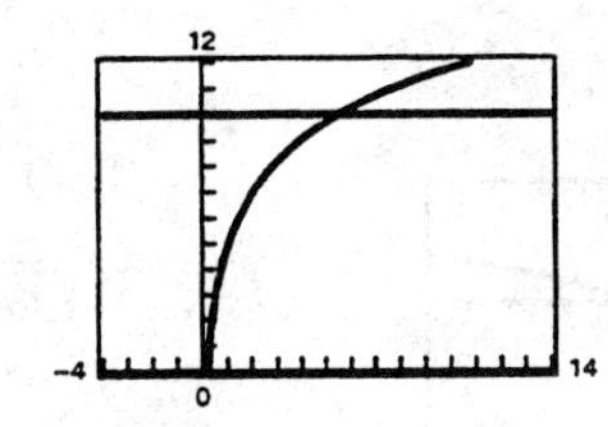

$x \approx 5.606$

64.

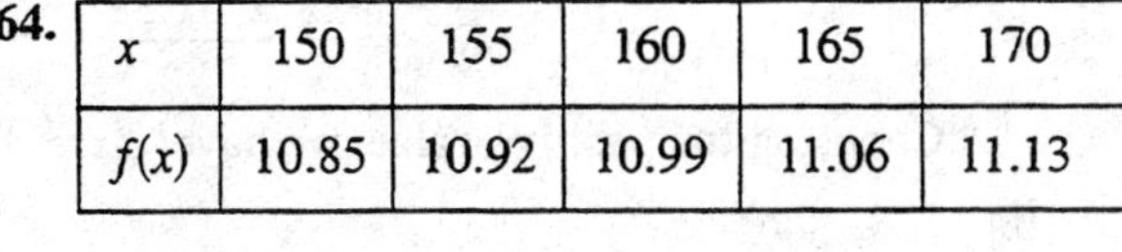

x	150	155	160	165	170
$f(x)$	10.85	10.92	10.99	11.06	11.13

$x \approx 160.489$

66. $\log_{10} x^2 = 6$

$$10^{\log_{10} x^2} = 10^6$$
$$x^2 = 10^6$$
$$x = \pm\sqrt{10^6} = \pm 1000$$

68. $\ln 4x = 1$

$$e^{\ln 4x} = e^1$$
$$4x = e$$
$$x = \frac{e}{4} \approx 0.680$$

70. $\ln x + \ln(x+3) = 1$

$$\ln[x(x+3)] = 1$$
$$e^{\ln[x(x+3)]} = e^1$$
$$x(x+3) = e^1$$
$$x^2 + 3x - e = 0$$
$$x = \frac{-3 \pm \sqrt{9+4e}}{2}$$

Using the positive value for x, we have $x = \frac{-3 + \sqrt{9+4e}}{2} \approx 0.729$.

72. $\log_2 x + \log_2(x+2) = \log_2(x+6)$

$$\log_2[x(x+2)] = \log_2(x+6)$$
$$x^2 + 2x = x + 6$$
$$x^2 + x - 6 = 0$$
$$x = \frac{-1 \pm \sqrt{1^2 - 4(1)(-6)}}{2}$$
$$= \frac{-1 \pm 5}{2} = -3, 2$$

Choosing the positive value of x, we have $x = 2$.

74. $\ln(x+1) - \ln(x-2) = \ln x^2$

$$\ln\left(\frac{x+1}{x-2}\right) = \ln x^2$$
$$\frac{x+1}{x-2} = x^2$$
$$x + 1 = x^3 - 2x^2$$
$$0 = x^3 - 2x^2 - x - 1$$

From the graph, we have $x \approx 2.547$.

76. $\log_{10} 8x - \log_{10}(1 + \sqrt{x}) = 2$

$$\log_{10}\frac{8x}{1 + \sqrt{x}} = 2$$

$$\frac{8x}{1 + \sqrt{x}} = 10^2$$

$$8x = 100 + 100\sqrt{x}$$

$$8x - 10\sqrt{x} - 100 = 0$$

$$2x - 25\sqrt{x} - 25 = 0$$

$$\sqrt{x} = \frac{25 \pm \sqrt{25^2 - 4(2)(-25)}}{4}$$

$$= \frac{25 \pm 5\sqrt{33}}{4}$$

Choosing the positive value, we have $\sqrt{x} \approx 13.431$ and $x \approx 180.384$.

78. $y_1 = 500$

$y_2 = 1500e^{-x/2}$

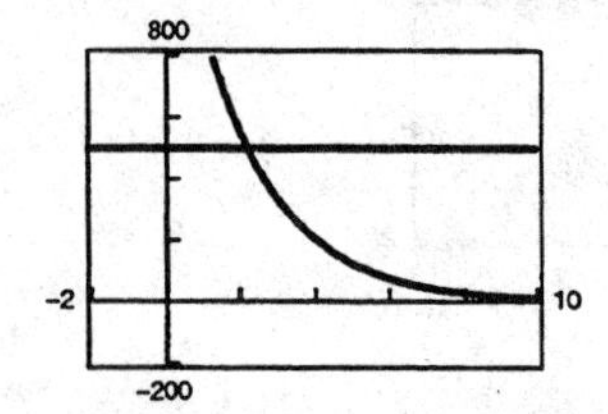

From the graph, we have $x \approx 2.197$.

80. $y_1 = 10$

$y_2 = 4 \ln(x - 2)$

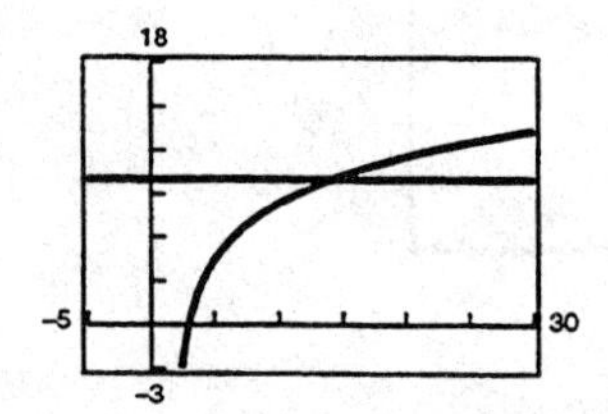

From the graph, we have $x \approx 14.182$.

82.

$$r = 0.12$$
$$A = Pe^{rt}$$
$$2000 = 1000e^{0.12t}$$
$$2 = e^{0.12t}$$
$$\ln 2 = \ln e^{0.12t}$$
$$\ln 2 = 10.12t$$
$$\frac{\ln 2}{0.12} = t$$
$$t \approx 5.8 \text{ years}$$

84. To find the length of time it takes for an investment P to double to $2P$, solve

$$2P = Pe^{rt}$$
$$2 = e^{rt}$$
$$\ln 2 = rt$$
$$\frac{\ln 2}{r} = t.$$

Thus, you can see that the time is not dependent on the size of the investment, but rather the

86.

$$r = 0.12$$
$$A = Pe^{rt}$$
$$3000 = 1000e^{0.12t}$$
$$3 = e^{0.12t}$$
$$\ln 3 = \ln e^{0.12t}$$
$$\ln 3 = 0.12t$$
$$\frac{\ln 3}{0.12} = t$$
$$t = 9.2 \text{ years}$$

88. $P = \dfrac{0.83}{1 + e^{-0.2n}}$

(a)

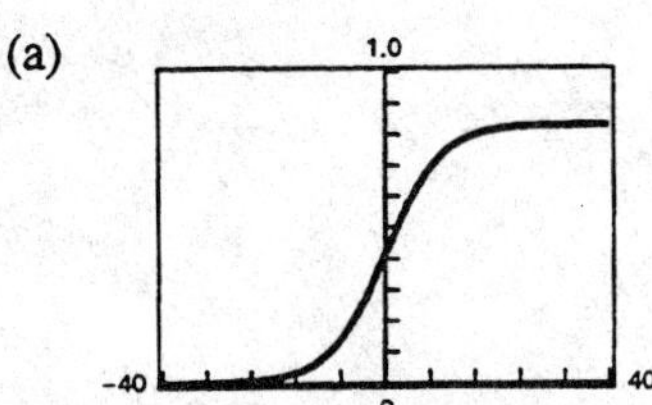

(b) Horizontal asymptotes: $y = 0, y = 0.83$
The upper asymptote, $y = 0.83$, indicates that the proportion of correct responses will approach 0.83 as the number of trials increases.

(c) When $P = 60\%$ or $P = 0.60$:

$$0.60 = \frac{0.83}{1 + e^{-0.2n}}$$

$$1 + e^{-0.2n} = \frac{0.83}{0.60}$$

$$e^{-0.2n} = \frac{0.83}{0.60} - 1$$

$$\ln e^{-0.2n} = \ln\left(\frac{0.83}{0.60} - 1\right)$$

$$-0.2n = \ln\left(\frac{0.83}{0.60} - 1\right)$$

$$n = -\frac{\ln\left(\frac{0.83}{0.60} - 1\right)}{0.2} \approx 5 \text{ trials}$$

90. $p = 5000\left(1 - \dfrac{4}{4 + e^{-0.002x}}\right)$

(a) When $p = \$600$:

$$600 = 5000\left(1 - \frac{4}{4 + e^{-0.002x}}\right)$$

$$0.12 = 1 - \frac{4}{4 + e^{-0.002x}}$$

$$\frac{4}{4 + e^{-0.002x}} = 0.88$$

$$4 = 3.52 + 0.88e^{-0.002x}$$

$$0.48 = 0.88e^{-0.002x}$$

$$\frac{6}{11} = e^{-0.002x}$$

$$\ln\frac{6}{11} = \ln e^{-0.002x}$$

$$\ln\frac{6}{11} = -0.002x$$

$$x = \frac{\ln(6/11)}{0.002} \approx 303 \text{ units}$$

(b) When $p = \$400$:

$$400 = 5000\left(1 - \frac{4}{4 + e^{-0.002x}}\right)$$

$$0.08 = 1 - \frac{4}{4 + e^{-0.002x}}$$

$$\frac{4}{4 + e^{-0.002x}} = 0.92$$

$$4 = 3.68 + 0.92e^{-0.002x}$$

$$0.32 = 0.92e^{-0.002x}$$

$$\frac{8}{23} = e^{-0.002x}$$

$$\ln\frac{8}{23} = \ln e^{-0.002x}$$

$$x = \frac{\ln(8/23)}{0.002} \approx 528 \text{ units}$$

92. $N = 68(10^{-0.04x})$

When $N = 21$:

$$21 = 68(10^{-0.04x})$$

$$\frac{21}{68} = 10^{-0.04x}$$

$$\log_{10}\frac{21}{68} = -0.04x$$

$$x = -\frac{\log(21/68)}{0.04} \approx 12.76 \text{ inches}$$

94. (a) $y = 29.08 - 75.96 \ln x$

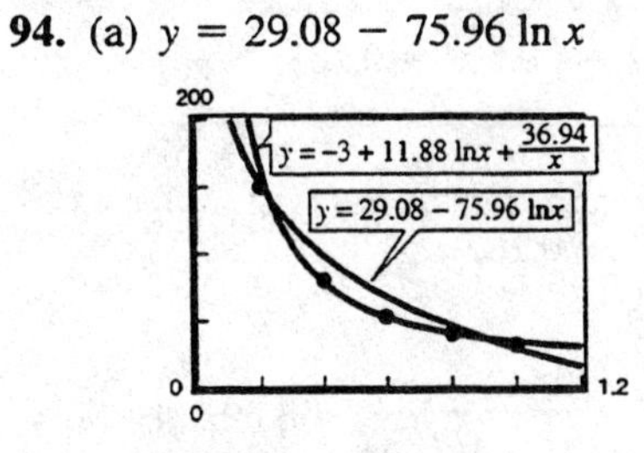

Answers will vary.

(b) 1.2 meters, 1.0 meters.
Answers will vary.

96. $2x + 5y - 10 = 0 \Rightarrow y = \frac{1}{5}(-2x + 10) = -\frac{2}{5}x + 2$
Matches (c).

98. $\dfrac{x}{2} + \dfrac{y}{4} = 1$
Intercepts: (2, 0), (0, 4)
Matches (e).

100. $x + 2 = 0 \Rightarrow x = -2$
Vertical line
Matches (a).

Section 3.5 Exponential and Logarithmic Models

Solutions to Even-Numbered Exercises

2. $y = 6e^{-x/4}$
This is an exponential decay model. Matches graph (c).

4. $y = \dfrac{12}{x + 4}$
This is a rational function.
Matches graph (d).

6. $y = \sqrt{x}$
This is a square root.
Matches graph (a).

8. Since $A = 20{,}000e^{0.105t}$, the time to double is given by $40{,}000 = 20{,}000e^{0.105t}$, and we have

$$t = \frac{\ln 2}{0.105} \approx 6.60 \text{ years.}$$

Amount after 10 years:

$$A = 20{,}000e^{0.105(10)} \approx \$57{,}153.02$$

10. Since $A = 10{,}000e^{rt}$ and $A = 20{,}000$ when $t = 5$, we have

$$20{,}000 = 10{,}000e^{5r}$$

$$r = \frac{\ln 2}{5} \approx 0.1386 \text{ or } 13.86\%.$$

Amount after 10 years:

$$A = 10{,}000e^{0.1386(10)} \approx \$39{,}988.23$$

12. Since $A = 600e^{rt}$ and $A = 19{,}205$ when $t = 10$, we have

$$19{,}205 = 600e^{10r}$$

$$r = \frac{\ln(19{,}205/600)}{10} \approx 0.3466 \text{ or } 34.66\%.$$

The time to double is given by

$$1200 = 600e^{0.3466t}$$

$$t = \frac{\ln 2}{0.3466} \approx 2 \text{ years.}$$

14. Since $A = Pe^{0.08t}$ and $A = 20{,}000$ when $t = 10$, we have

$$20{,}000 = Pe^{0.08(10)}$$

$$P = \frac{20{,}000}{e^{0.08(10)}} \approx \$8986.58.$$

The time to double is given by

$$t = \frac{\ln 2}{0.08} \approx 8.66 \text{ years.}$$

16.

$$A = P\left(1 + \frac{r}{n}\right)^{nt}$$

$$500{,}000 = P\left(1 + \frac{0.12}{12}\right)^{12(40)}$$

$$P = \$4214.16$$

18. $P = 1000$, $r = 10.5\% = 0.105$

(a) $n = 1$

$$t = \frac{\ln 2}{\ln(1 + 0.105)} \approx 6.94 \text{ years}$$

(b) $n = 12$

$$t = \frac{\ln 2}{12\ln\left(1 + \frac{0.105}{12}\right)} \approx 6.63 \text{ years}$$

(c) $n = 365$

$$t = \frac{\ln 2}{365\ln\left(1 + \frac{0.105}{365}\right)} \approx 6.602 \text{ years}$$

(d) Compounded continuously

$$t = \frac{\ln 2}{0.105} \approx 6.601 \text{ years}$$

20. 

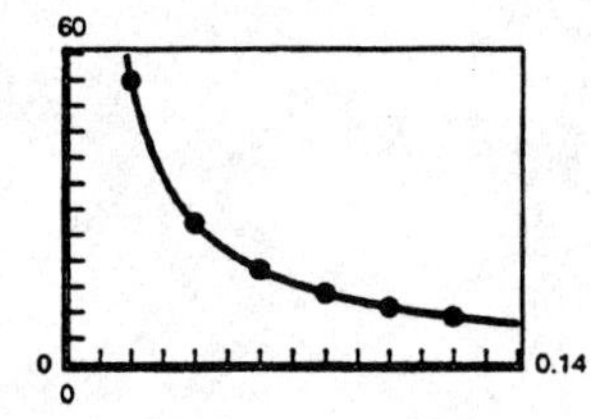

Using the power regression feature of a graphing utility, $t = 1.099r^{-1}$.

22.

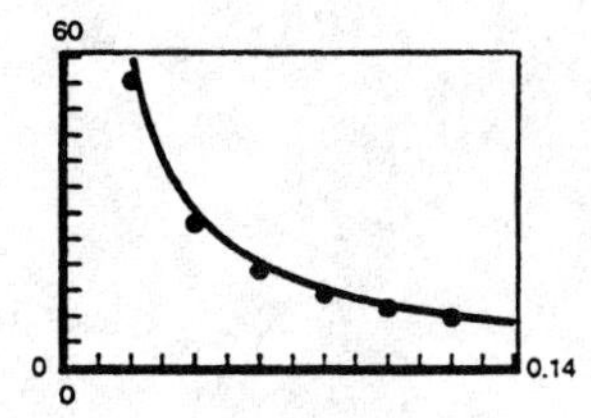

Using the power regression feature of a graphing utility, $t = 1.222r^{-1}$.

24.

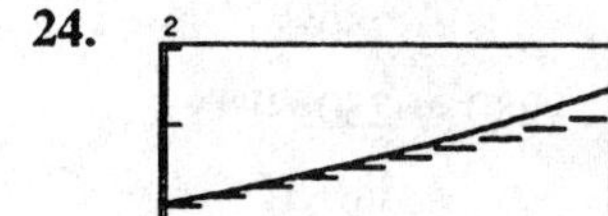

From the graph, $5\frac{1}{2}\%$ compounded daily grows faster than 6% simple interest.

26. 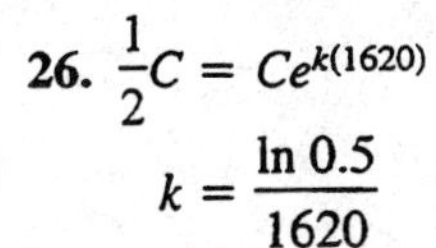

$$\frac{1}{2}C = Ce^{k(1620)}$$

$$k = \frac{\ln 0.5}{1620}$$

Given $y = 1.5$ grams after 1000 years, we have

$$1.5 = Ce^{[(\ln 0.5)/1620](1000)}$$

$$C \approx 2.30 \text{ grams.}$$

28.

$$\frac{1}{2}C = Ce^{k(24,360)}$$

$$k = \frac{\ln 0.5}{24,360}$$

Given $y = 0.4$ grams after 1000 years, we have

$$0.4 = Ce^{[(\ln 0.5)/24,360](1000)}$$

$$C \approx 0.41 \text{ grams.}$$

30.

$$y = ae^{bx}$$

$$\frac{1}{2} = ae^{b(0)} \implies a = \frac{1}{2}$$

$$5 = \frac{1}{2}e^{b(4)}$$

$$10 = e^{4b}$$

$$\ln 10 = 4b$$

$$\frac{\ln 10}{4} = b \implies b \approx 0.5756$$

Thus, $y = \frac{1}{2}e^{0.5756x}$.

32.

$$y = ae^{bx}$$

$$5 = ae^{b(0)} \implies a = 5$$

$$1 = 5e^{b(4)}$$

$$\frac{1}{5} = e^{4b}$$

$$\ln\frac{1}{5} = 4b$$

$$b = \frac{\ln(1/5)}{r} \implies b \approx -0.4024$$

Thus, $y = 5e^{-0.4024x}$.

34.

$$P = 240,360e^{0.012t}$$

$$275,000 = 240,360e^{0.012t}$$

$$\ln\frac{27,500}{24,036} = 0.012t$$

$$t = \frac{\ln(27,500/24,036)}{0.012} \approx 11$$

The population will reach 275,000 in 2001.

36. For 1960, we use $t = -30$.

$$100,250 = 140,500e^{k(-30)}$$

$$k = -\frac{\ln(100,250/140,500)}{30} \approx 0.0113$$

For 2000, we use $t = 10$.

$$P = 140,500e^{-[(\ln(100,250/140,500))/30](10)} \approx 157,232$$

38.

$$y = ae^{bt}$$

$$2.30 = ae^{b(0)} \implies a \approx 2.30$$

$$2.65 = 2.30e^{b(10)}$$

$$\frac{2.65}{2.30} = e^{10b}$$

$$\ln\left(\frac{2.65}{2.30}\right) = 10b \implies b \approx 0.0142$$

Thus, $y = 2.30e^{0.0142t}$. When $t = 20$, $y = 230e^{(0.0142)(20)} \approx 3.05$ million.

40.

$$y = ae^{bt}$$

$$9.17 = ae^{b(0)} \implies a = 9.17$$

$$8.57 = 9.17e^{b(10)}$$

$$\frac{8.57}{9.17} = e^{10b}$$

$$\ln\left(\frac{8.57}{9.17}\right) = 10b \implies b \approx -0.068$$

Thus, $y = 9.17e^{-0.0068t}$. When $t = 20$, $y = 9.17e^{-0.0068(20)} \approx 8.00$ million.

42. The constant b in the equation $y = ae^{bt}$ determines whether the population is increasing $(b > 0)$ or is decreasing $(b < 0)$.

44. $N = 250e^{kt}$

$280 = 250e^{k(10)}$

$k = \dfrac{\ln 1.12}{10}$

$N = 250e^{[(\ln 1.12)/10]t}$

$500 = 250e^{[(\ln 1.12)/10]t}$

$t = \dfrac{\ln 2}{(\ln 1.12)/10} \approx 61.16$ hours

46. $y = Ce^{kt}$

$\dfrac{1}{2}C = Ce^{5730k}$

$\ln \dfrac{1}{2} = 5730k$

$k = \dfrac{\ln(1/2)}{5730}$

The ancient charcoal has only 15% as much radioactive carbon.

$0.15C = Ce^{[(\ln 0.5)/5730]t}$

$\ln 0.15 = \dfrac{\ln 0.5}{5730}t$

$t = \dfrac{5730 \ln 0.15}{\ln 0.5} \approx 15{,}683$ years

48. (a) $V = mt + b$; $V(0) = 22{,}000 \Rightarrow b = 22{,}000$

$V(2) = 13{,}000 \Rightarrow 13{,}000 = 2m + 22{,}000 \Rightarrow m = -4500$

$V(t) = -4500t + 22{,}000$

(b) $V = ae^{kt}$; $V(0) = 22{,}000 \Rightarrow a = 22{,}000$

$V(2) = 13{,}000 \Rightarrow 13{,}000 = 22{,}000e^{2k}$

$\frac{13}{22} = e^{2k}$

$\ln \frac{13}{22} = 2k$

$k = \frac{1}{2} \ln \frac{13}{22} \approx -0.263$

$V = 22{,}000e^{-0.263t}$

(c)

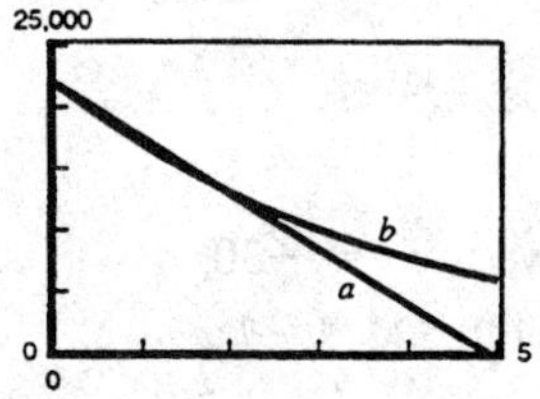

The exponential model depreciates faster in the first two years.

(d) Straight line: $V(1) = \$17{,}500$

$V(3) = \$8500$

Exponential: $V(1) = \$16{,}912$

$V(3) = \$9993$

(e) The negative slope means the car depreciates $4500 per year.

50. $S = \dfrac{500{,}000}{1 + 0.6e^{kt}}$

(a) $300{,}000 = \dfrac{500{,}000}{1 + 0.6e^{2k}}$

$1 + 0.6e^{2k} = \dfrac{5}{3}$

$0.6e^{2k} = \dfrac{2}{3}$

$e^{2k} = \dfrac{10}{9}$

$2k = \ln\left(\dfrac{10}{9}\right)$

$k = \dfrac{1}{2}\ln\left(\dfrac{10}{9}\right) \approx 0.053$

$S = \dfrac{500{,}000}{1 + 0.6e^{0.053t}}$

(b) When $t = 5$:

$S = \dfrac{500{,}000}{1 + 0.6e^{[0.5 \ln(10/9)](5)}} \approx \$280{,}771$

52.
$$y = ae^{bt}$$
$$632{,}000 = 742{,}000e^{b(2)}$$
$$\tfrac{632}{742} = e^{2b}$$
$$b = \tfrac{1}{2}\ln\left(\tfrac{632}{742}\right)$$
$$y = 742{,}000e^{0.5(3)\ln(632/742)} \approx \$583{,}275$$

54. $p(t) = \dfrac{1000}{1 + 9e^{-0.1656t}}$

(a)

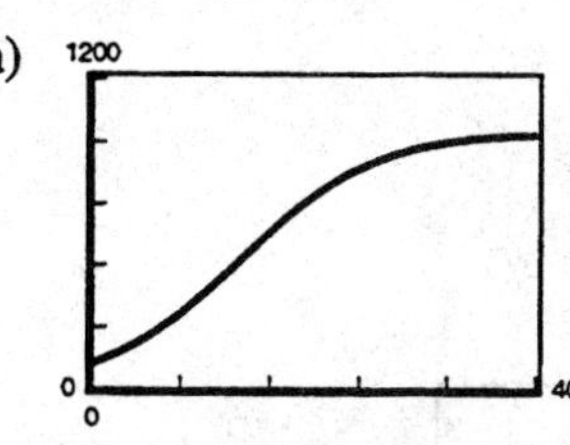

The horizontal asymptotes are $y = 0$ and $y = 1000$. The asymptote with the larger p-value, $y = 1000$, indicates that the population size will approach 1000 as time increases.

(b) $p(5) = \dfrac{1000}{1 + 9e^{-0.1656(5)}} \approx 203$ animals

(c)
$$500 = \frac{1000}{1 + 9e^{-0.1656t}}$$
$$1 + 9e^{-0.1656t} = 2$$
$$9e^{-0.1656t} = 1$$
$$e^{-0.1656t} = \frac{1}{9}$$
$$t = -\frac{\ln(1/9)}{0.1656} \approx 13 \text{ months}$$

56. $R = \log_{10}\dfrac{I}{I_0} = \log_{10} I$ since $I_0 = 1$.

(a) $8.6 = \log_{10} I$

$10^{8.6} = I \approx 398{,}107{,}171$

(b) $6.7 = \log_{10} I$

$10^{6.7} = I \approx 5{,}011{,}872$

58. $\beta(I) = 10\log_{10}\dfrac{I}{I_0}$ where $I_0 = 10^{-16}$ watt/cm^2.

(a) $\beta(10^{-13}) = 10\log_{10}\dfrac{10^{-13}}{10^{-16}} = 10\log_{10}10^3 = 30$ decibels

(b) $\beta(10^{-7.5}) = 10\log_{10}\dfrac{10^{-7.5}}{10^{-16}} = 10\log_{10}10^{8.5} = 85$ decibels

(c) $\beta(10^{-7}) = 10\log_{10}\dfrac{10^{-7}}{10^{-16}} = 10\log_{10}10^{9} = 90$ decibels

(d) $\beta(10^{-4.5}) = 10\log_{10}\dfrac{10^{-4.5}}{10^{-16}} = 10\log_{10}10^{11.5} = 115$ decibels

60.
$$\beta = 10\log_{10}\frac{I}{I_0}$$
$$10^{\beta/10} = \frac{I}{I_0}$$
$$I = I_0\,10^{\beta/10}$$
$$\%\text{ decrease} = \frac{I_0\,10^{8.8} - I_0\,10^{7.2}}{I_0\,10^{8.8}} \times 100 \approx 97\%$$

62. $\text{pH} = -\log_{10}[H^+] = -\log_{10}[11.3 \times 10^{-6}] \approx 4.95$

64.
$$3.2 = -\log_{10}[H^+]$$
$$19^{-3.2} = H^+$$
$$H^+ \approx 6.3 \times 10^{-4} \text{ moles per liter}$$

66.
$$\text{pH} - 1 = \log_{10}[H^+]$$
$$-(\text{pH} - 1) = \log_{10}[H^+]$$
$$10^{-(\text{pH}-1)} = [H^+]$$
$$10^{-\text{pH}+1} = [H^+]$$
$$10^{-\text{pH}} \cdot 10 = [H^+]$$

The hydrogen ion concentration is increased by a factor of 10.

68. $u = 120{,}000\left[\dfrac{0.095t}{1 - \left(1 + \dfrac{0.095}{12}\right)^{12t}} - 1\right]$

(a)

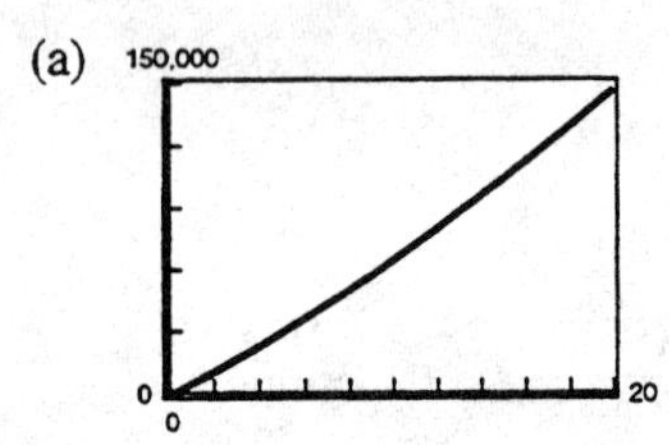

(b) From the graph, when $u = 120{,}000$, $t \approx 17$ years. Yes, if a person takes an approximately 30-year long mortgage the interest will be twice as much as the mortgage.

70. (a) $V = -995.57t + 11{,}018.10$

$V = 53.39t^2 - 1262.54t + 11{,}196.07$

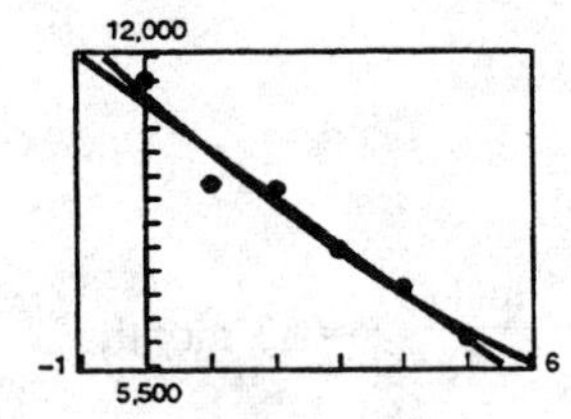

(b) Average depreciation per year

(c) No. To the right of the vertex it will increase.

(d) $V = 11{,}215.50(0.8888)^t$

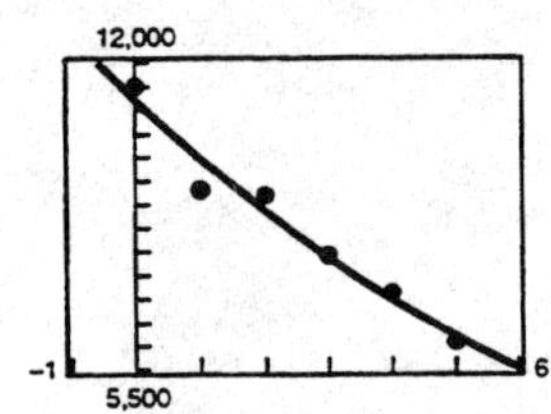

The model fits well.

(e) $V = \dfrac{10{,}000{,}000}{153t + 864}$

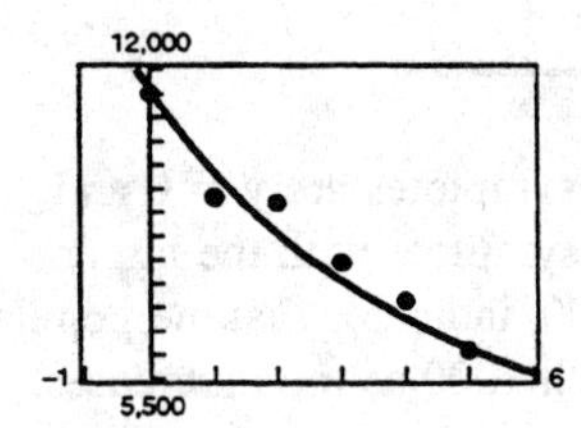

The model fits well.

(f) $V = 0$. As time increases, the value approaches 0.

72. Answers will vary.

74. -4

	4	4	−39	36
		−16	48	−36
	4	−12	9	0

$\dfrac{4x^3 + 4x^2 - 39x + 36}{x + 4} = 4x^2 - 12x + 9$

76. 4

	2	−8	3	−9
		8	0	12
	2	0	3	3

$\dfrac{2x^3 - 8x^2 + 3x - 9}{x - 4} = 2x^2 + 3 + \dfrac{3}{x - 4}$

Section 3.6 Exploring Data: Nonlinear Models

Solutions to Even-Numbered Exercises

2. Linear model

4. Exponential model

6. Logistics model

8. Linear model

10.

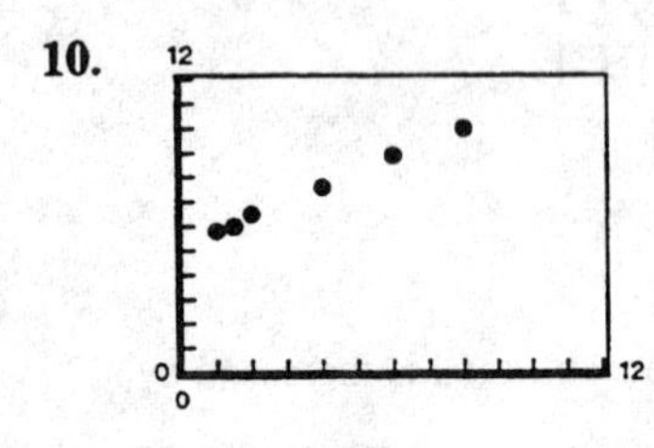

Linear model

12. Exponential model

14. Logarithmic model

16. $y = 5.544(1.3650)^x$

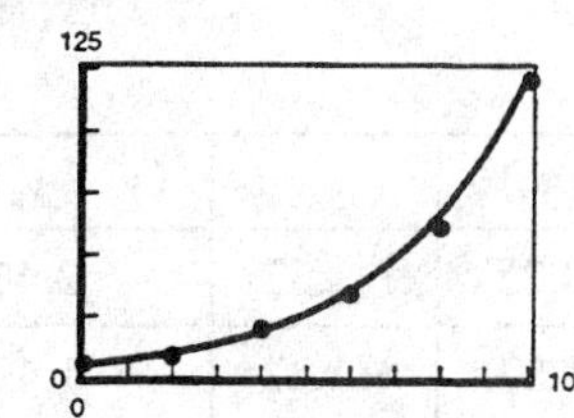

18. $y = 87.262(0.9438)^x$

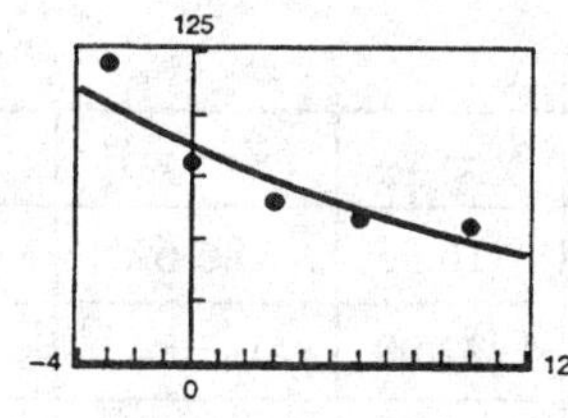

20. $y = 9.027 + 2.537 \ln x$

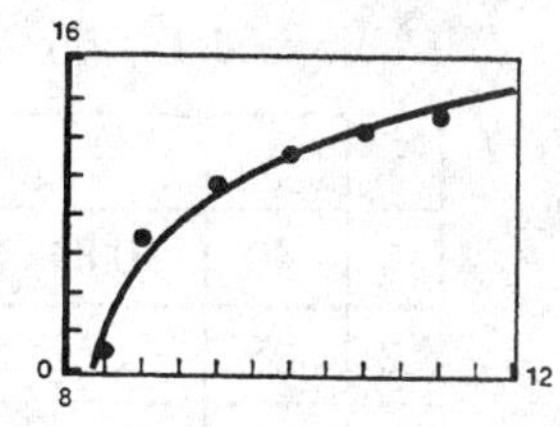

22. $y = 20.076 - 5.027 \ln x$

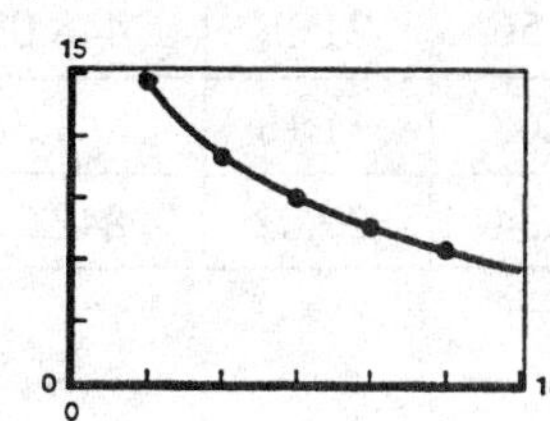

24. $y = 3.397x^{1.602}$

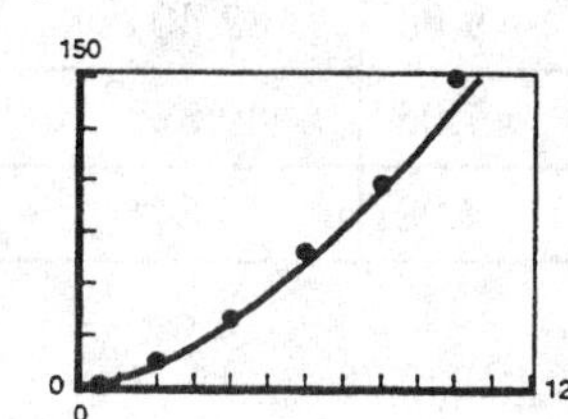

26. $y = 525.428x^{-0.226}$

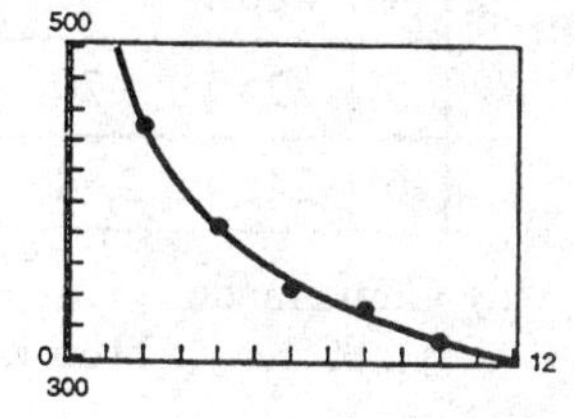

28. (a) $y = 19.826 + 9.848 \ln t$

(b)

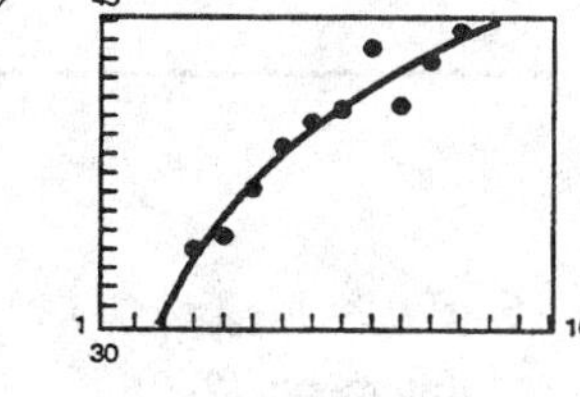
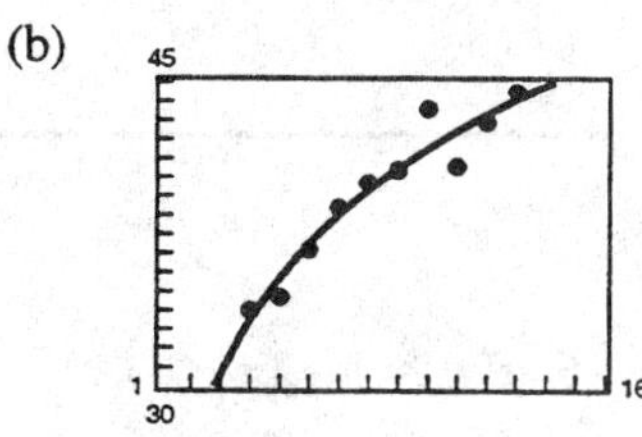

(c) For $t = 21$, $y = 49.8$

30. (a) $h = 0$ is not in the domain of the logarithmic function.

(b) $h = 0.863 - 6.447 \ln p$

(c)

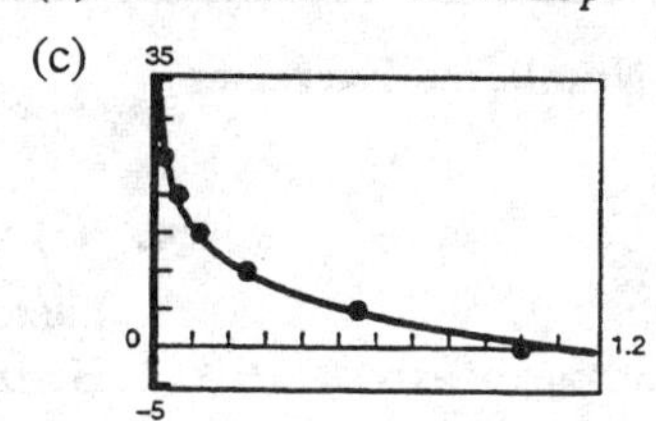

(d) If $p = 0.75$, then $h = 2.72$ kilometers.

(e) If $h = 13$, then $p = 0.15$ atmosphere.

32. (a) $y = 268.868(1.095)^x$

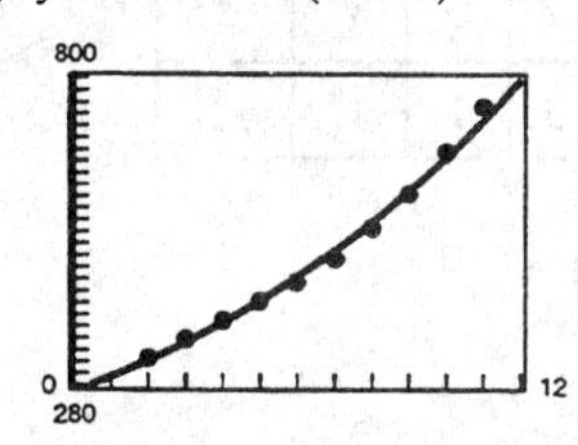

(b) $y = 93.979 + 233.307 \ln x$

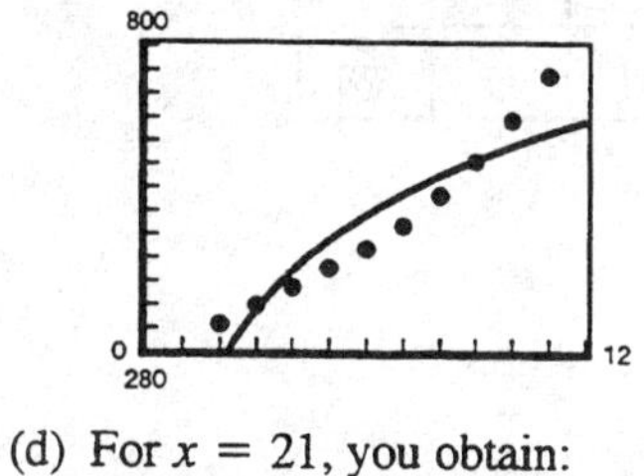

(c) Exponential is better unless the rate of growth of health costs is slowed. Then the logarithmic model would be better.

(d) For $x = 21$, you obtain:
Exponential: 1808.2
Logarithmic: 804.3

34. (a) $P = 0.039S^{3.000}$ (b) For $S = 45$, $P = 3553$ (c) Proportional to x^3

36. (a) $y_1 = -1.81x^3 + 14.58x^2 + 16.39x + 10.00$

$y_2 = 23.07 + 121.08 \ln x$

$y_3 = 38.38(1.4227)^x$

$y_4 = 41.57x^{1.0525}$

(b)

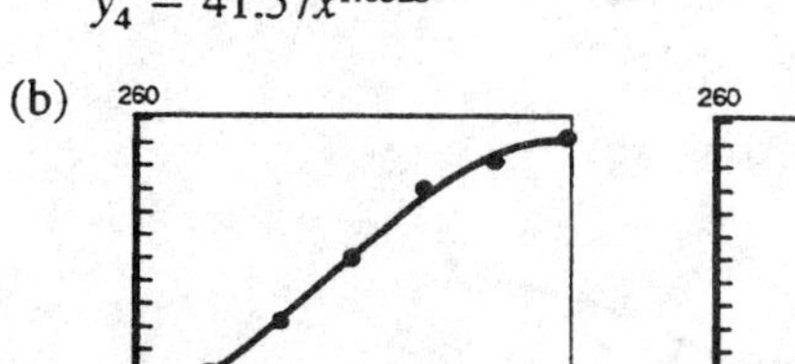

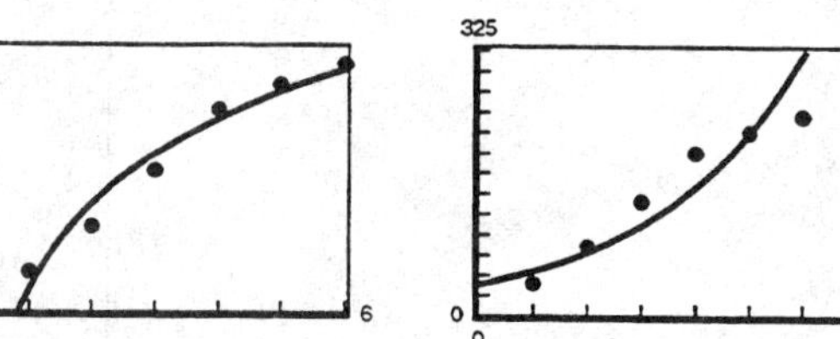

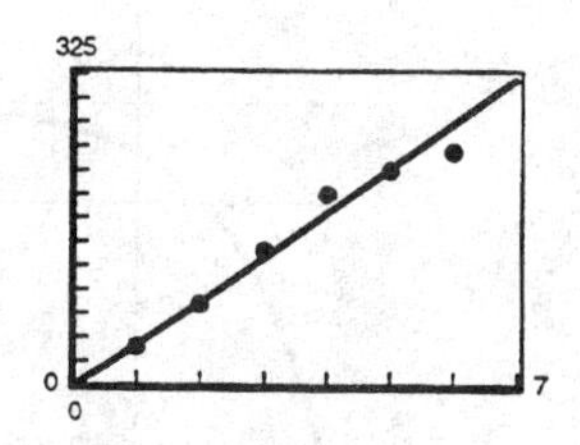

—CONTINUED—

36. —CONTINUED—

(c) Cubic model

x	y	$y - y_1$	$(y - y_1)^2$	$y - y_2$	$(y - y_2)^2$	$y - y_3$	$(y - y_3)^2$	$y - y_4$	$(y - y_4)^2$
1	40	0.84	0.71	16.93	286.62	−14.60	213.25	−1.57	2.46
2	85	−1.62	2.62	−22.00	483.84	7.32	53.52	−1.22	1.49
3	140	−1.52	2.31	−16.09	258.89	29.48	869.01	7.89	62.18
4	200	7.00	49.00	9.08	82.40	42.76	1828.56	21.17	448.03
5	225	5.20	27.04	7.06	49.83	1.30	1.68	−1.18	1.38
6	245	2.74	7.51	4.98	24.84	−73.26	5367.34	29.02	842.23

(d) Cubic model
y_1: 89.19; y_2: 1186.42; y_3: 8333.37; y_4: 1357.79

(e) The sums represent the sum of the squares of the errors.

❑ Review Exercises for Chapter 3

Solutions to Even-Numbered Exercises

2. $f(x) = 4^{-x}$
Intercept: (0,1)
Horizontal asymptote: x-axis
Decreasing on: $(-\infty, \infty)$
Matches graph (f).

4. $f(x) = 4^x + 1$
Intercept: (0, 2)
Horizontal asymptote: $y = 1$
Increasing on: $(-\infty, \infty)$
Matches graph (c).

6. $f(x) = \log_4(x - 1)$
Intercept: (2, 0)
Vertical asymptote: $x = 1$
Increasing on: $(1, \infty)$
Matches graph (d).

8. $f(x) = 0.3^x = \left(\frac{3}{10}\right)^x$

x	−2	−1	0	1	2
y	$\frac{100}{9}$	$\frac{10}{3}$	1	$\frac{3}{10}$	$\frac{9}{100}$

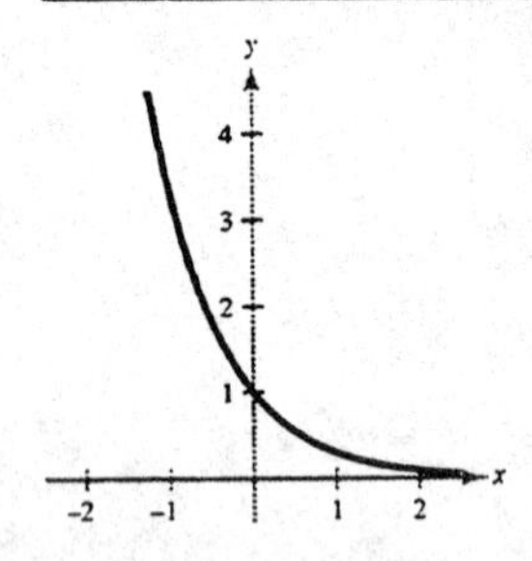

10. $g(x) = 0.3^{-x}$

x	−2	−1	0	1	2
y	0.09	0.3	1	$3\frac{1}{3}$	$11\frac{1}{9}$

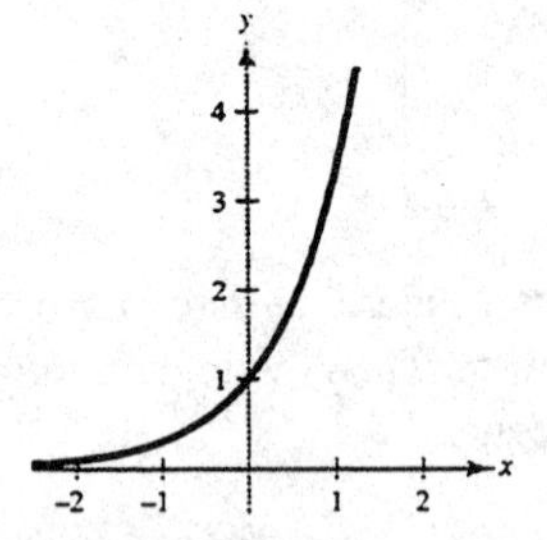

12. $h(x) = 2 - e^{-x/2}$

x	−2	−1	0	1	2
y	−0.72	0.35	1	1.39	1.63

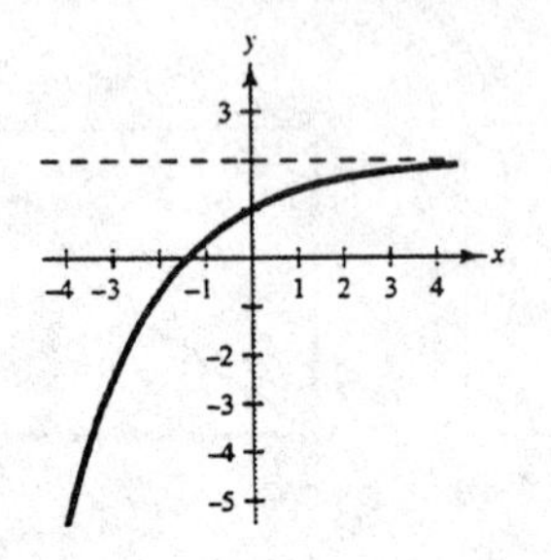

14. $s(t) = 4e^{-2/t}, t > 0$

t	$\frac{1}{2}$	1	2	3	4
y	0.07	0.54	1.47	2.05	2.43

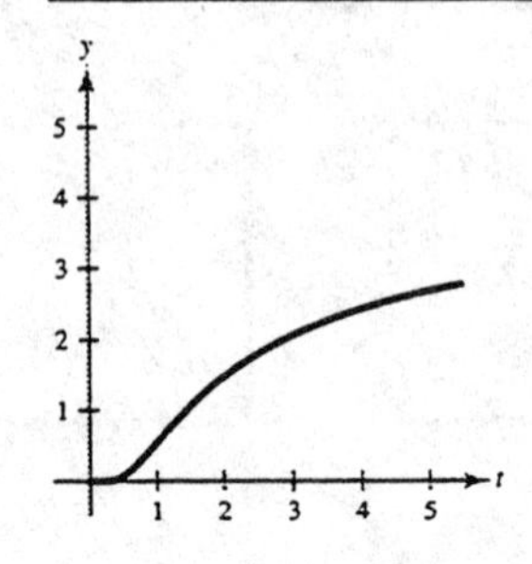

16.

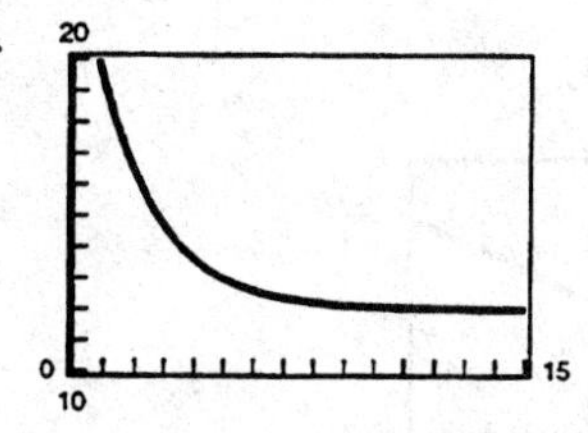

Asymptote: $y = 12$

18. 

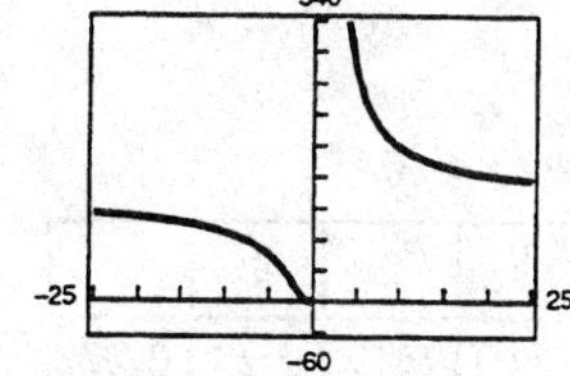

Asymptote: $y = 200$

20. $A = 2000\left(1 + \dfrac{0.12}{n}\right)^{30n}$ or $A = 2000e^{(0.12)(30)}$

n	1	2	4	12	365	Continuous
A	\$59,919.84	\$65,975.38	\$60,421.97	\$71,899.28	\$73,153.17	\$73,196.47

22. $200{,}000 = P\left(1 + \dfrac{0.10}{12}\right)^{12t}$

$$P = \frac{200{,}000}{\left(1 + \dfrac{0.10}{12}\right)^{12t}}$$

t	1	10	20	30	40	50
P	\$181.042.49	\$73,881.39	\$27,292.30	\$10,081.97	\$3724.35	\$1375.80

24. $A = Pe^{rt} = 50{,}000e^{(0.0875)(35)} \approx \$1{,}069{,}047.14$

26. $F(t) = 1 - e^{-t/3}$

(a) $F\left(\frac{1}{2}\right) \approx 0.154$

(b) $F(2) \approx 0.487$

(c) $F(5) \approx 0.811$

28. $C(t) = P(1.045)^t = 69.95(1.045)^{10} \approx \$108.63$

30. $g(x) = \log_5 x \Rightarrow 5^y = x$

Domain: $(0, \infty)$

Vertical asymptote: $x = 0$

x	$\frac{1}{25}$	$\frac{1}{5}$	1	5	25
y	-2	-1	0	1	2

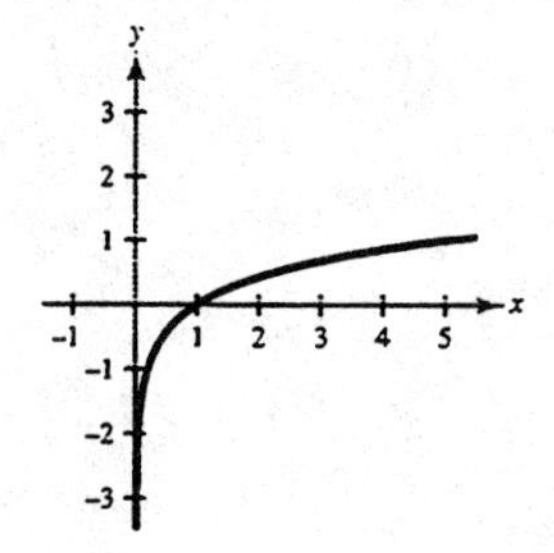

32. $f(x) = \ln(x - 3)$

Domain: $(3, \infty)$

Vertical asymptote: $x = 3$

x	3.5	4	4.5	5	5.5
y	-0.69	0	0.41	0.69	0.92

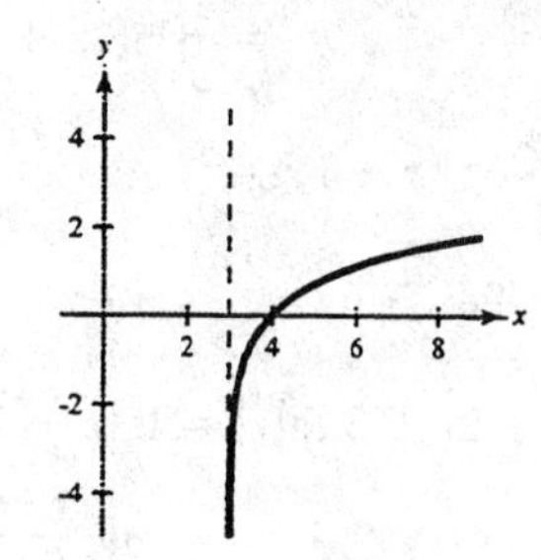

34. $f(x) = \frac{1}{4} \ln x$

Domain: $(0, \infty)$

Vertical asymptote: $x = 0$

x	$\frac{1}{2}$	1	$\frac{3}{2}$	2	$\frac{5}{2}$	3
y	−0.17	0	0.10	0.17	0.23	0.27

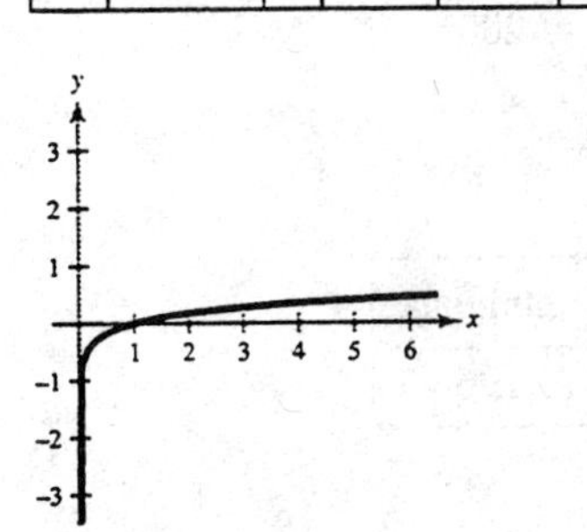

36. $y = \sqrt{x} \ln(x + 1)$

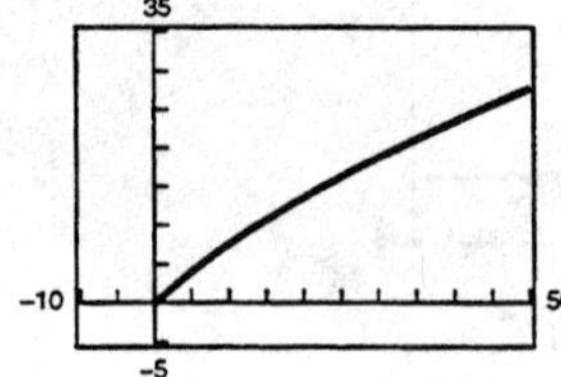

38. $25^{3/2} = 125$

$\log_{25} 125 = \frac{3}{2}$

40. $\log_9 3 = \log_9 9^{1/2} = \frac{1}{2}$

42. $\log_4 \dfrac{1}{16} = \log_4 (4^{-2}) = -2$

44. $\log_a \dfrac{1}{a} = \log_a a^{-1} = -1$

46. $\ln e^{-3} = -3 \ln e = -3$

48. $\log_{1/2} 5 = \dfrac{\log_{10} 5}{\log_{10}(1/2)} \approx -2.322$

$\log_{1/2} 5 = \dfrac{\ln 5}{\ln(1/2)} \approx -2.322$

50. $\log_3 0.28 = \dfrac{\log_{10} 0.28}{\log_{10} 3} \approx -1.159$

$\log_3 0.28 = \dfrac{\ln 0.28}{\ln 3} \approx -1.159$

52. $\log_7 \dfrac{\sqrt{x}}{4} = \log_7 \sqrt{x} - \log_7 4$

$= \log_7 x^{1/2} - \log^7 4$

$= \dfrac{1}{2} \log_7 x - \log_7 4$

54. $\ln\left|\dfrac{x - 1}{x + 1}\right| = \ln|x - 1| - \ln|x + 1|$

56. $\ln \sqrt[5]{\dfrac{4x^2 - 1}{4x^2 + 1}} = \dfrac{1}{5} \ln\left(\dfrac{4x^2 - 1}{4x^2 + 1}\right) = \dfrac{1}{5} \ln(4x^2 - 1) - \dfrac{1}{5} \ln(4x^2 + 1)$

$= \dfrac{1}{5} \ln[(2x + 1)(2x - 1)] - \dfrac{1}{5} \ln(4x^2 + 1)$

$= \dfrac{1}{5}[\ln(2x + 1) + \ln(2x - 1) - \ln(4x^2 + 1)]$

58. $\log_6 y - 2 \log_6 z = \log_6 y - \log_6 z^2$

$= \log_6 \dfrac{y}{z^2}$

60. $5 \ln|x - 2| - \ln|x + 2| - 3 \ln|x| = \ln|x - 2|^5 - \ln|x + 2| - \ln|x|^3$

$= \ln\left|\dfrac{(x - 2)^5}{(x + 2)x^3}\right|$

62. $3[\ln x - 2 \ln(x^2 + 1)] + 2 \ln 5 = \ln x^3 - \ln(x^2 + 1)^6 + \ln 5^2$

$= \ln \dfrac{25x^3}{(x^2 + 1)^6}$

64. $e^{x-1} = e^x \cdot e^{-1} = \dfrac{e^x}{e}$

True (by properties of exponents).

66. $\ln(x + y) = \ln(x \cdot y)$ False

$\ln(x \cdot y) = \ln x + \ln y \neq \ln(x + y)$

68. $\log\left(\dfrac{10}{x}\right) = \log 10 - \log x = 1 - \log x$

True

70. $\log_b\left(\dfrac{25}{9}\right) = \log_b 25 - \log_b 9 = 2\log_b 5 - 2\log_b 3$

$\approx 2(0.8271) - 2(0.5646) = 0.525$

72. $\log_b 30 = \log_b(2 \cdot 3 \cdot 5) = \log_b 2 + \log_b 3 + \log_b 5$

$\approx 0.3562 + 0.5646 + 0.8271 = 1.7479$

74. $s = 25 - \dfrac{13\ln(h/12)}{\ln 3},\ 2 \le h \le 15$

$s(10) \approx 27.16$ miles

76.
$$e^{3x} = 25$$
$$\ln e^{3x} = \ln 25$$
$$3x = \ln 25$$
$$x = \frac{\ln 25}{3} \approx 1.073$$

78.
$$14e^{3x+2} = 560$$
$$e^{3x+2} = 40$$
$$\ln e^{3x+2} = \ln 40$$
$$3x + 2 = \ln 40$$
$$x = \frac{(\ln 40) - 2}{3} \approx 0.563$$

80.
$$e^{2x} - 7e^x + 10 = 0$$
$$(e^x - 5)(e^x - 2) = 0$$
$$e^x = 5 \quad \text{or } e^x = 2$$
$$x = \ln 5 \quad \text{or } x = \ln 2$$
$$x \approx 1.609 \text{ or } x \approx 0.693$$

82.
$$2\ln 4x = 15$$
$$\ln 4x = \tfrac{15}{2}$$
$$4x = e^{7.5}$$
$$x = \tfrac{1}{4}e^{7.5} \approx 452.011$$

84.
$$\ln\sqrt{x + 1} = 2$$
$$\tfrac{1}{2}\ln(x + 1) = 2$$
$$\ln(x + 1) = 4$$
$$x + 1 = e^4$$
$$x = e^4 - 1 \approx 53.598$$

86.
$$\log(1 - x) = -1$$
$$1 - x = 10^{-1}$$
$$1 - \tfrac{1}{10} = x$$
$$x = \tfrac{9}{10}$$

88.
$$25e^{-0.3x} = 12$$
$$25e^{-0.3x} - 12 = 0$$

Graph $y_1 = 25e^{-0.3x} - 12$.

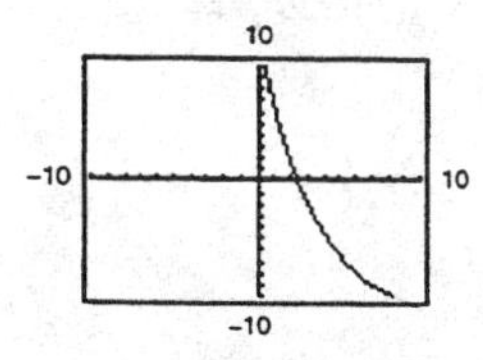

The x-intercept is at $x \approx 2.45$.

90. $6\log_{10}(x^2 + 1) - x = 0$

Graph

$y_1 = 6\log_{10}(x^2 + 1) - x.$

The x-intercepts are at $x = 0$, $x = 0.416$ and $x \approx 13.63$.

92.
$$y = ae^{bx}$$
$$\tfrac{1}{2} = ae^{b(0)} \implies a = \tfrac{1}{2}$$
$$5 = \tfrac{1}{2}e^{b(5)}$$
$$10 = e^{5b}$$
$$\ln 10 = 5b \implies b \approx 0.4605$$

Thus, $y = \tfrac{1}{2}e^{0.4605x}$.

94. $y = ae^{bx}$

$2 = ae^{b(0)} \Rightarrow a = 2$

$1 = 2e^{b(5)} \Rightarrow \frac{1}{2} = e^{5b} \Rightarrow 5b = \ln\frac{1}{2} \Rightarrow b = \frac{1}{5}\ln\frac{1}{2} = -\frac{1}{5}\ln 2 \approx -0.1386$

$y = 2e^{-0.1386x}$

96. $N = \dfrac{157}{1 + 5.4e^{-0.12t}}$

(a) When $N = 50$:

$$50 = \frac{157}{1 + 5.4e^{-0.12t}}$$

$$1 + 5.4e^{-0.12t} = \frac{157}{50}$$

$$5.4e^{-0.12t} = \frac{107}{50}$$

$$e^{-0.12t} = \frac{107}{270}$$

$$-0.12t = \ln\frac{107}{270}$$

$$t = \frac{\ln(107/270)}{-0.12} \approx 7.7 \text{ weeks}$$

(b) When $N = 75$:

$$75 = \frac{157}{1 + 5.4e^{-0.12t}}$$

$$1 + 5.4e^{-0.12t} = \frac{157}{75}$$

$$5.4e^{-0.12t} = \frac{82}{75}$$

$$e^{-0.12t} = \frac{82}{405}$$

$$-0.12^{t} = \ln\frac{82}{405}$$

$$t = \frac{\ln(82/405)}{-0.12} \approx 13.3 \text{ weeks}$$

98. $A = Pe^{rt} = 10{,}000e^{rt}$

(a) $20{,}000 = 10{,}000\, e^{r(5)}$

$$2 = e^{5r} \Rightarrow 5r = \ln 2 \Rightarrow r \approx \frac{\ln 2}{5} \approx 0.1386$$

$$= 13.86\%$$

(b) $A = 10{,}000e^{0.1386(1)} \approx \$11{,}486.65$

(c) $\dfrac{11{,}486.98}{10{,}000} = 1.1487 \Rightarrow 14.87\%$

100. $R = \log_{10}\left(\dfrac{I}{I_0}\right) = \log_{10}(I) \Rightarrow I = 10^R$

(a) $I = 10^{8.4}$

(b) $I = 10^{6.85}$

(c) $I = 10^{9.1}$

102. (a) $y = 2.29t + 2.34$

(b) $y = 1.54 + 8.37 \ln t$

(c)

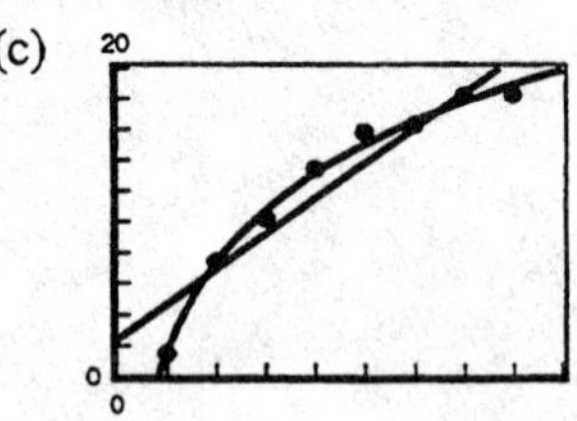

Logarithmic model is a better model.

CHAPTER 4
Trigonometric Functions

CHAPTER 4
Trigonometric Functions

Section 4.1 Radian and Degree Measure

Solutions to Even-Numbered Exercises

2. 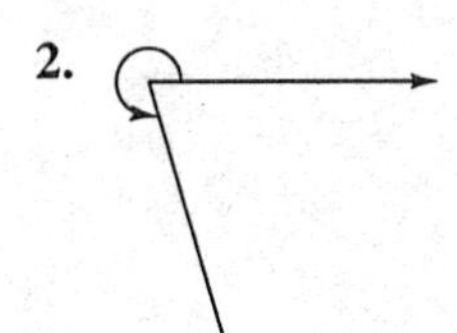

The angle shown is approximately 5 radians.

4. 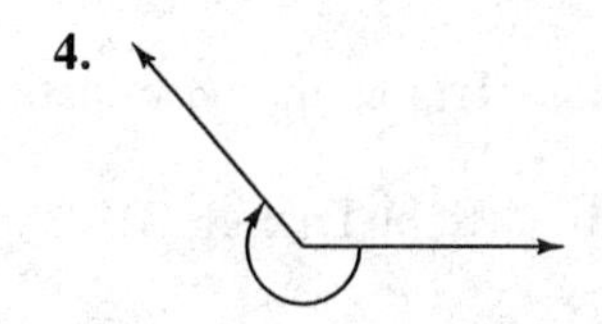

The angle shown is approximately -4 radians.

6. (a) Since $\pi < \frac{5\pi}{4} < \frac{3\pi}{2}$; $\frac{5\pi}{4}$ lies in Quadrant III.

(b) Since $\frac{3\pi}{2} < \frac{7\pi}{4} < 2\pi$; $\frac{7\pi}{4}$ lies in Quadrant IV.

8. (a) Since $-\frac{\pi}{2} < -1 < 0$; -1 lies in Quadrant IV.

(b) Since $-\pi < -2 < -\frac{\pi}{2}$; -2 lies in Quadrant III.

10. (a) Since $\frac{3\pi}{2} < 5.63 < 2\pi$; 5.63 lies in Quadrant IV.

(b) Since $-\pi < -2.25 < -\frac{\pi}{2}$; -2.25 lies in Quadrant III.

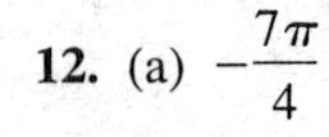

12. (a) $-\frac{7\pi}{4}$

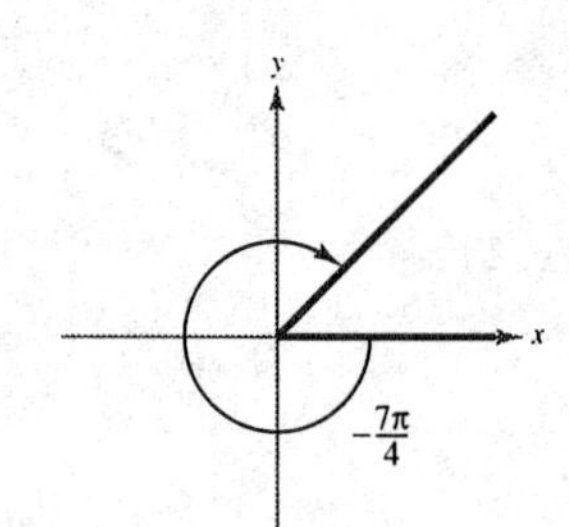

(b) $-\frac{5\pi}{2}$

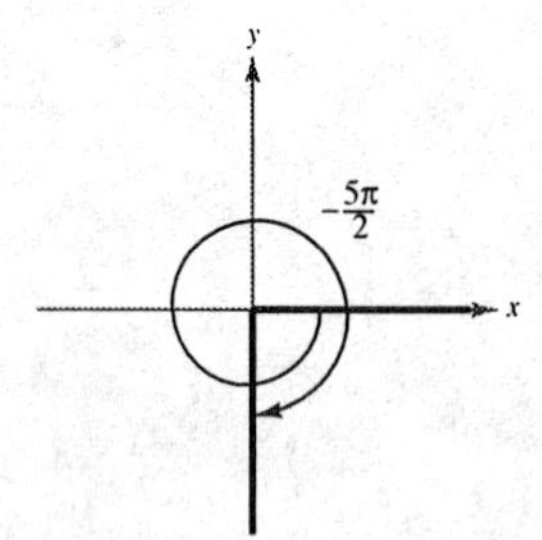

14. (a) 4

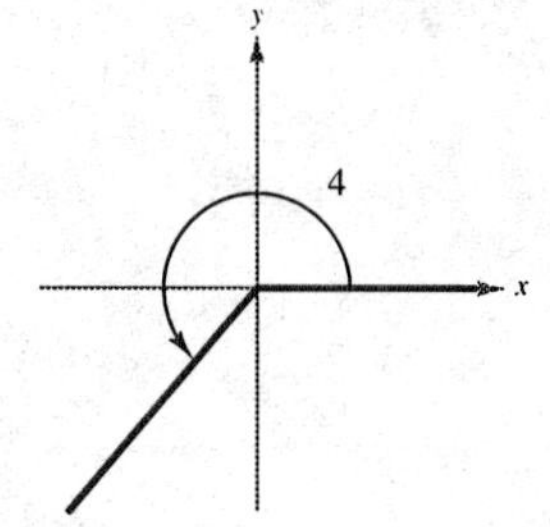

(b) -3

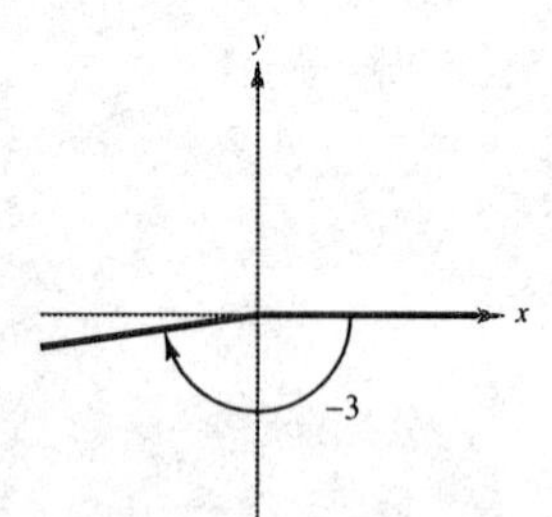

16. (a) $\frac{7\pi}{6} + 2\pi = \frac{19\pi}{6}$

$\frac{7\pi}{6} - 2\pi = -\frac{5\pi}{6}$

(b) $-\frac{11\pi}{6} + 2\pi = \frac{\pi}{6}$

$-\frac{11\pi}{6} - 2\pi = -\frac{23\pi}{6}$

18. (a) $\frac{8\pi}{9} + 2\pi = \frac{26\pi}{9}$

$\frac{8\pi}{9} - 2\pi = -\frac{10\pi}{9}$

(b) $\frac{8\pi}{45} + 2\pi = \frac{98\pi}{45}$

$\frac{8\pi}{45} - 2\pi = -\frac{82\pi}{45}$

20.

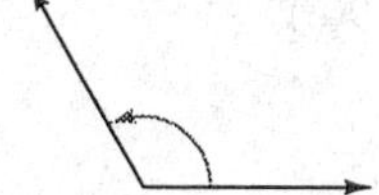

The angle shown is approximately 120°.

22.

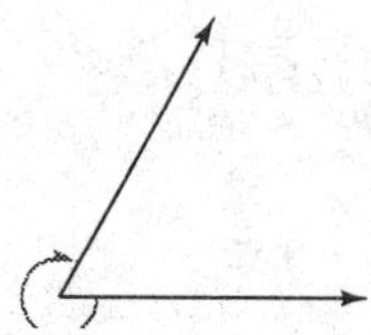

The angle shown is approximately $-300°$.

24. (a) Since $0° < 8.3° < 90°$; 8.3° lies in Quadrant I.

(b) Since $180° < 257°\,30' < 270°$; $257°\,30'$ lies in Quadrant III.

26. (a) Since $-270° < -260° < -180°$; $-260°$ lies in Quadrant II.

(b) Since $-90° < -3.4° < 0°$; $-3.4°$ lies in Quadrant IV.

28. (a) $-270°$

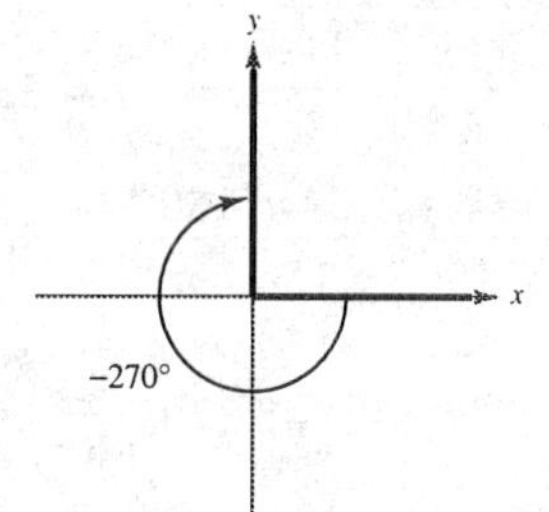

(b) $-120°$

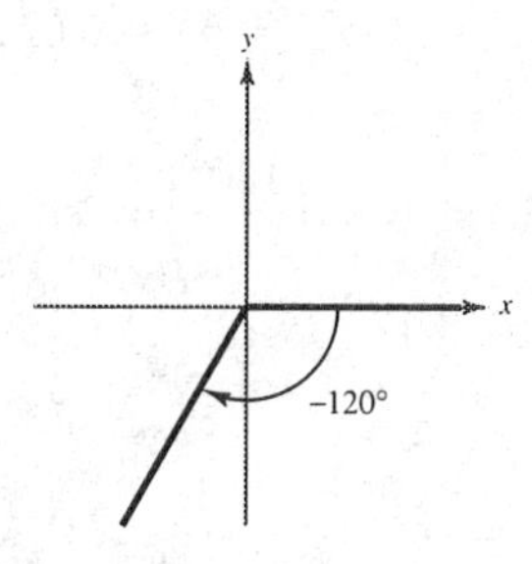

30. (a) 750°

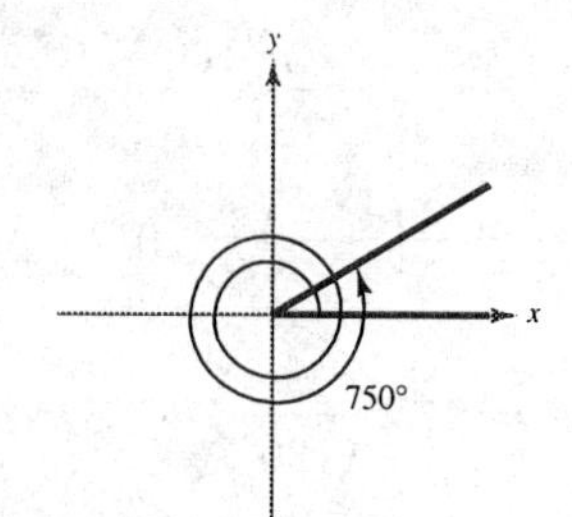

(b) $-600°$

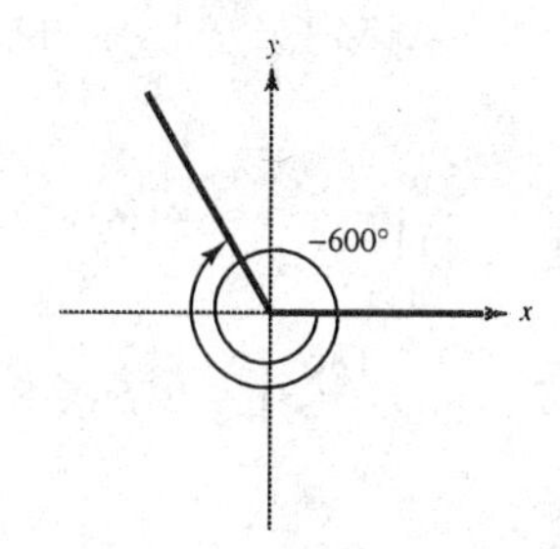

32. (a) $120° + 360° = 480°$

$120° - 360° = -240°$

(b) $-390° + 720° = 330°$

$-390° + 360° = -30°$

34. (a) $-420° + 720° = 300°$

$-420° + 360° = -60°$

(b) $230° - 360° = -130°$

$230° + 360° = 590°$

36. (a) $245°\,10' = 245° + \left(\frac{10}{60}\right)^{\circ}$

$\approx 245° + 0.167°$

$= 245.167°$

(b) $2°\,12' = 2° + \left(\frac{12}{60}\right)^{\circ}$

$= 2° + 0.2° = 2.2°$

38. (a) $-135°\,36'' = -135° - \left(\frac{36}{3600}\right)^{\circ}$

$= -135° - 0.01° = -135.01°$

(b) $-408°\,16'\,20'' = -\left(408° + \left(\frac{16}{60}\right)^{\circ} + \left(\frac{20}{3600}\right)^{\circ}\right)$

$\approx -(408° + 0.267° + 0.006°)$

$= -408.272°$

40. (a) $-345.12° = -(345° + (0.12)(60'))$

$= -(345° + 7' + 0.2(60''))$

$= -345°\,7'\,12''$

(b) $0.45 = 0.45\left(\frac{180}{\pi}\right)^{\circ}$

$\approx 25.7831°$

$= 25° + (0.7831)(60')$

$\approx 25° + 46' + (0.986)(60'')$

$\approx 25°\,46'\,59''$

42. (a) $-0.355 = -0.355\left(\frac{180}{\pi}\right)^{\circ}$

≈ -20.34

$= -(20° + (0.34)(60'))$

$= -(20° + 20' + 0.4(60''))$

$= -20°\,20'\,24''$

(b) $0.7865 = 0.7865\left(\frac{180}{\pi}\right)^{\circ}$

≈ 45.0631

$= 45° + (0.0631)(60')$

$= 45° + 3' + 0.786(60'')$

$\approx 45°\,3'\,47''$

44. (a) Complement: $90° - 79° = 11°$

Supplement: $180° - 79° = 101°$

(b) Complement: none $(150° > 90°)$

Supplement: $180° - 150° = 30°$

46. (a) Complement: $\frac{\pi}{2} - 1 \approx 0.57$

Supplement: $\pi - 1 \approx 2.14$

(b) Complement: none $\left(2 > \frac{\pi}{2}\right)$

Supplement: $\pi - 2 \approx 1.14$

48. (a) $315° = 315°\left(\frac{\pi}{180°}\right) = \frac{7\pi}{4}$

(b) $120° = 120°\left(\frac{\pi}{180°}\right) = \frac{2\pi}{3}$

50. (a) $-270° = -270°\left(\frac{\pi}{180°}\right) = -\frac{3\pi}{2}$

(b) $144° = 144°\left(\frac{\pi}{180°}\right) = \frac{4\pi}{5}$

52. (a) $-\frac{7\pi}{12} = -\frac{7\pi}{12}\left(\frac{180°}{\pi}\right) = -105°$

(b) $\frac{\pi}{9} = \frac{\pi}{9}\left(\frac{180°}{\pi}\right) = 20°$

54. (a) $\frac{11\pi}{6} = \frac{11\pi}{6}\left(\frac{180°}{\pi}\right) = 330°$

(b) $\frac{34\pi}{15} = \frac{34\pi}{15}\left(\frac{180°}{\pi}\right) = 408°$

56. $87.4° = 87.4°\left(\frac{\pi}{180°}\right) \approx 1.525$ radians

58. $-48.27° = -48.27°\left(\frac{\pi}{180°}\right) \approx -0.842$ radian

60. $0.54° = 0.54°\left(\frac{\pi}{180°}\right) \approx 0.009$ radian

62. $345° = 345°\left(\frac{\pi}{180°}\right) \approx 6.021$ radians

64. $\frac{5\pi}{11} = \frac{5\pi}{11}\left(\frac{180°}{\pi}\right) \approx 81.818°$

66. $6.5\pi = 6.5\pi\left(\frac{180°}{\pi}\right) = 1170°$

68. $4.8 = 4.8\left(\frac{180°}{\pi}\right) \approx 275.020°$

70. $-0.57 = -0.57\left(\frac{180°}{\pi}\right) \approx -32.659°$

72. $S = r\theta$

$31 = 12\theta$

$\theta = \frac{31}{12} = 2\frac{7}{12}$ radians

74. $S = r\theta$

$60 = 75\theta$

$\theta = \frac{60}{75} = \frac{4}{5}$ radian

Because the angle represented is clockwise, this angle is $-\frac{4}{5}$ radian.

76. $r = 16$ feet, $s = 10$ feet

$\theta = \frac{s}{r} = \frac{10}{16} = \frac{5}{8}$ radian

78. $r = 80$ kilometers, $s = 160$ kilometers

$\theta = \frac{s}{r} = \frac{160}{80} = 2$ radians

80. $r = 9$ feet, $\theta = 60° = \frac{\pi}{3}$

$s = r\theta = 9\left(\frac{\pi}{3}\right) = 3\pi$ feet

82. $r = 40$ centimeters, $\theta = \frac{3\pi}{4}$

$s = r\theta = 40\left(\frac{3\pi}{4}\right) = 30\pi$ centimeters

84. $r = 4000$ miles

$\theta = 47°\,36'\,32'' - 37°\,46'\,39'' = 9°\,49'\,53''$

≈ 0.1716 radian

$s = r\theta \approx (4000)(0.1716) \approx 686.4$ miles

86. $r = 4000$ miles

$\theta = 31°\,47' + 26°\,10' = 57°\,57'$

≈ 1.0114 radians

$s = r\theta \approx (4000)(1.0114) \approx 4045.7$ miles

88. $r = 6378$ kilometers, $s = 800$ kilometers

$$\theta = \frac{s}{r} = \frac{800}{6378} \approx 0.125 \text{ radian} = 0.125\left(\frac{180^\circ}{\pi}\right) \approx 7.19^\circ$$

90. $\theta = \dfrac{s}{r} = \dfrac{12}{5} = 2.4 \text{ radians} = 2.4\left(\dfrac{180^\circ}{\pi}\right) \approx 137.5^\circ$

92. Linear velocity for either pulley: $1700(2\pi) = 3400\pi$ in/min

(a) Angular speed of motor pulley: $\omega = \dfrac{v}{r} = \dfrac{3400\pi}{1} = 3400\pi$ rad/min

Angular speed of the saw arbor: $\omega = \dfrac{v}{r} = \dfrac{3400\pi}{2} = 1700\pi$ rad/min

(b) Revolutions per minute of the saw arbor: $\dfrac{1700\pi}{2\pi} = 850$ rev/min

94. If θ is constant, the length of the arc is proportional to the radius ($s = r\theta$).

96. speed $=$ (360 revolutions/minute)($2\pi(1.68)$ inches/revolution)

$= 1209.6\pi$ inches/minute

$= 20.16\pi$ inches/second

98. Let A be the area of a circular sector of radius r and central angle θ. Then

$$\frac{A}{\pi r^2} = \frac{\theta}{2\pi} \Rightarrow A = \frac{1}{2}r^2\theta.$$

100. Because $s = r\theta$, $\theta = \frac{12}{15}$. Hence, $A = \frac{1}{2}r^2\theta = \frac{1}{2}15^2\left(\frac{12}{15}\right) = 90 \text{ ft}^2$.

102. $A = \frac{1}{2}r^2\theta$, $s = r\theta$

(a) $\theta = 0.8 \Rightarrow A = \frac{1}{2}r^2(0.8) = 0.4r^2$ Domain: $r > 0$

$s = r\theta = r(0.8)$ Domain: $r > 0$

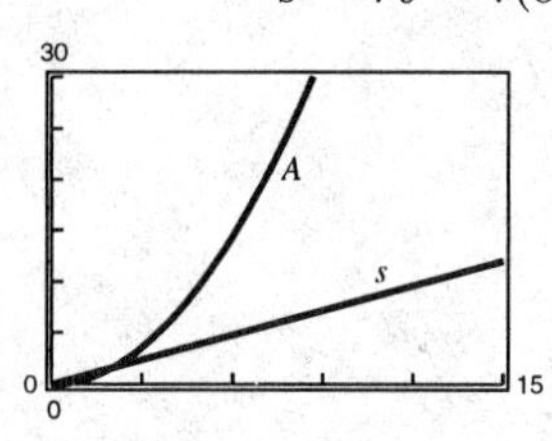

The area function changes more rapidly for $r > 1$ because it is quadratic and the arc length function in linear.

(b) $r = 10 \Rightarrow A = \frac{1}{2}(10^2)\theta = 50\,\theta$ Domain: $0 < \theta < 2\pi$

$s = r\theta = 10\,\theta$ Domain: $0 < \theta < 2\pi$

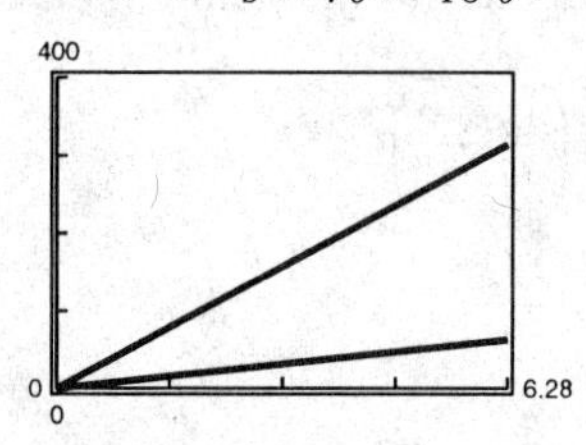

104. $\theta = 0.031^\circ = 5.41052 \times 10^{-4}$ radians

$s = r\theta = (4000)(5.41052 \times 10^{-4}) = 2.1642$ miles

Section 4.2 Trigonometric Functions: The Unit Circle

Solutions to Even-Numbered Exercises

2. $(x, y) = \left(\frac{12}{13}, \frac{5}{13}\right)$

$\sin t = y = \frac{5}{13}$

$\cos t = x = \frac{12}{13}$

$\tan t = \frac{y}{x} = \frac{5/13}{12/13} = \frac{5}{12}$

$\csc t = \frac{1}{y} = \frac{1}{5/13} = \frac{13}{5}$

$\sec t = \frac{1}{x} = \frac{1}{12/13} = \frac{13}{12}$

$\cot t = \frac{x}{y} = \frac{12/13}{5/13} = \frac{12}{5}$

4. $(x, y) = \left(-\frac{4}{5}, -\frac{3}{5}\right)$

$\sin t = y = -\frac{3}{5}$

$\cos t = x = -\frac{4}{5}$

$\tan t = \frac{y}{x} = \frac{-3/5}{-4/5} = \frac{3}{4}$

$\csc t = \frac{1}{y} = \frac{1}{-3/5} = -\frac{5}{3}$

$\sec t = \frac{1}{x} = \frac{1}{-4/5} = -\frac{5}{4}$

$\cot t = \frac{x}{y} = \frac{-4/5}{-3/5} = \frac{4}{3}$

6. $(x, y) = (-0.8668, 0.4987)$

$\sin t = 0.4987$

$\cos t = -0.8668$

$\tan t = \frac{y}{x} = -0.5753$

$\csc t = \frac{1}{y} = 2.0052$

$\sec t = \frac{1}{x} = -1.1537$

$\cot t = \frac{x}{y} = -1.7381$

8. $t = \frac{\pi}{3} \Rightarrow \left(\frac{1}{2}, \frac{\sqrt{3}}{2}\right)$

10. $t = \frac{5\pi}{4} \Rightarrow \left(-\frac{\sqrt{2}}{2}, -\frac{\sqrt{2}}{2}\right)$

12. $t = \frac{11\pi}{6} \Rightarrow \left(\frac{\sqrt{3}}{2}, -\frac{1}{2}\right)$

14. $t = \pi \Rightarrow (-1, 0)$

16. $t = -\frac{\pi}{4}$ corresponds to the point:

$(x, y) = \left(\frac{\sqrt{2}}{2}, -\frac{\sqrt{2}}{2}\right)$

$\sin\left(-\frac{\pi}{4}\right) = y = -\frac{\sqrt{2}}{2}$

$\cos\left(-\frac{\pi}{4}\right) = x = \frac{\sqrt{2}}{2}$

$\tan\left(-\frac{\pi}{4}\right) = \frac{y}{x} = \frac{-\sqrt{2}/2}{\sqrt{2}/2} = -1$

18. $t = \frac{\pi}{3}$ corresponds to the point:

$(x, y) = \left(\frac{1}{2}, \frac{\sqrt{3}}{2}\right)$

$\sin\frac{\pi}{3} = y = \frac{\sqrt{3}}{2}$

$\cos\frac{\pi}{3} = x = \frac{1}{2}$

$\tan\frac{\pi}{3} = \frac{y}{x} = \frac{\sqrt{3}/2}{1/2} = \sqrt{3}$

20. $t = -\frac{5\pi}{6}$ corresponds to the point:

$$(x, y) = \left(-\frac{\sqrt{3}}{2}, -\frac{1}{2}\right)$$

$$\sin\left(-\frac{5\pi}{6}\right) = y = -\frac{1}{2}$$

$$\cos\left(-\frac{5\pi}{6}\right) = x = -\frac{\sqrt{3}}{2}$$

$$\tan\left(-\frac{5\pi}{6}\right) = \frac{y}{x} = \frac{-1/2}{-\sqrt{3}/2} = \frac{1}{\sqrt{3}}$$

22. $t = \frac{2\pi}{3}$ corresponds to the point:

$$(x, y) = \left(-\frac{1}{2}, \frac{\sqrt{3}}{2}\right)$$

$$\sin\frac{2\pi}{3} = y = \frac{\sqrt{3}}{2}$$

$$\cos\frac{2\pi}{3} = x = -\frac{1}{2}$$

$$\tan\frac{2\pi}{3} = \frac{y}{x} = \frac{\sqrt{3}/2}{-1/2} = -\sqrt{3}$$

24. $t = \frac{7\pi}{4}$ corresponds to the point:

$$(x, y) = \left(\frac{\sqrt{2}}{2}, -\frac{\sqrt{2}}{2}\right)$$

$$\sin\frac{7\pi}{4} = y = -\frac{\sqrt{2}}{2}$$

$$\cos\frac{7\pi}{4} = x = \frac{\sqrt{2}}{2}$$

$$\tan\frac{7\pi}{4} = \frac{y}{x} = \frac{-\sqrt{2}/2}{\sqrt{2}/2} = -1$$

26. $t = -2\pi$ corresponds to the point: $(x, y) = (1, 0)$

$$\sin(-2\pi) = y = 0$$

$$\cos(-2\pi) = x = 1$$

$$\tan(-2\pi) = \frac{y}{x} = \frac{0}{1} = 0$$

28. $t = -\frac{2\pi}{3}$ corresponds to the point:

$$(x, y) = \left(-\frac{1}{2}, -\frac{\sqrt{3}}{2}\right)$$

$$\sin\left(-\frac{2\pi}{3}\right) = y = -\frac{\sqrt{3}}{2}$$

$$\cos\left(-\frac{2\pi}{3}\right) = x = -\frac{1}{2}$$

$$\tan\left(-\frac{2\pi}{3}\right) = \frac{y}{x} = \frac{-\sqrt{3}/2}{-1/2} = \sqrt{3}$$

$$\csc\left(-\frac{2\pi}{3}\right) = \frac{1}{y} = \frac{1}{-\sqrt{3}/2} = -\frac{2\sqrt{3}}{3}$$

$$\sec\left(-\frac{2\pi}{3}\right) = \frac{1}{x} = \frac{1}{-1/2} = -2$$

$$\cot\left(-\frac{2\pi}{3}\right) = \frac{x}{y} = \frac{-1/2}{-\sqrt{3}/2} = \frac{\sqrt{3}}{3}$$

30. $t = \frac{3\pi}{2}$ corresponds to the point: $(x, y) = (0, -1)$

$$\sin\frac{3\pi}{2} = y = -1$$

$$\cos\frac{3\pi}{2} = x = 0$$

$$\tan\frac{3\pi}{2} = \frac{y}{x} = \frac{-1}{0} \Rightarrow \text{undefined}$$

$$\csc\frac{3\pi}{2} = \frac{1}{y} = \frac{1}{-1} = -1$$

$$\sec\frac{3\pi}{2} = \frac{1}{x} = \frac{1}{0} \Rightarrow \text{undefined}$$

$$\cot\frac{3\pi}{2} = \frac{x}{y} = \frac{0}{-1} = 0$$

32. $t = -\frac{11\pi}{6}$ corresponds to the point:

$$(x, y) = \left(\frac{\sqrt{3}}{2}, \frac{1}{2}\right)$$

$$\sin\left(-\frac{11\pi}{6}\right) = y = \frac{1}{2}$$

$$\cos\left(-\frac{11\pi}{6}\right) = x = \frac{\sqrt{3}}{2}$$

$$\tan\left(-\frac{11\pi}{6}\right) = \frac{y}{x} = \frac{1/2}{\sqrt{3}/2} = \frac{\sqrt{3}}{3}$$

$$\csc\left(-\frac{11\pi}{6}\right) = \frac{1}{y} = \frac{1}{1/2} = 2$$

$$\sec\left(-\frac{11\pi}{6}\right) = \frac{1}{x} = \frac{1}{\sqrt{3}/2} = \frac{2\sqrt{3}}{3}$$

$$\cot\left(-\frac{11\pi}{6}\right) = \frac{x}{y} = \frac{\sqrt{3}/2}{1/2} = \sqrt{3}$$

34. Because $3\pi = 2\pi + \pi$:

$$\cos 3\pi = \cos(2\pi + \pi) = \cos \pi = -1$$

36. Because $\frac{9\pi}{4} = 2\pi + \frac{\pi}{4}$:

$$\sin\frac{9\pi}{4} = \sin\left(2\pi + \frac{\pi}{4}\right) = \sin\frac{\pi}{4} = \frac{\sqrt{2}}{2}$$

38. Because $-\frac{13\pi}{6} = -2\pi - \frac{\pi}{6}$:

$$\sin\left(-\frac{13\pi}{6}\right) = \sin\left(-2\pi - \frac{\pi}{6}\right) = \sin\left(-\frac{\pi}{6}\right) = -\sin\left(\frac{\pi}{6}\right) = -\frac{1}{2}$$

40. Because $-\frac{8\pi}{3} = -4\pi + \frac{4\pi}{3}$:

$$\cos\left(-\frac{8\pi}{3}\right) = \cos\left(-4\pi + \frac{4\pi}{3}\right) = \cos\frac{4\pi}{3} = -\frac{1}{2}$$

42. $\sin(-t) = \frac{2}{5}$

(a) $\sin t = -\sin(-t) = -\frac{2}{5}$

(b) $\csc t = \frac{1}{\sin(t)} = \frac{1}{-\sin(-t)} = -\frac{5}{2}$

44. $\cos t = -\frac{3}{4}$

(a) $\cos(-t) = \cos t = -\frac{3}{4}$

(b) $\sec(-t) = \frac{1}{\cos(-t)} = \frac{1}{\cos t} = -\frac{4}{3}$

46. $\cos t = \frac{4}{5}$

(a) $\cos(\pi - t) = -\cos t = -\frac{4}{5}$

(b) $\cos(t + \pi) = -\cos t = -\frac{4}{5}$

48. $\tan \pi = 0$

50. $\cot 1 = \frac{1}{\tan 1} \approx 0.6421$

52. $\csc 2.3 = \frac{1}{\sin 2.3} \approx 1.3410$

54. $\sec 1.8 = \frac{1}{\cos 1.8} \approx -4.4014$

56. $\sin(-0.9) \approx -0.7833$

58. (a) $\sin 0.75 = y \approx 0.7$

(b) $\cos 2.5 = x \approx -0.8$

60. (a) $\sin t = -0.75$

$t \approx 4.0$ or $t \approx 5.4$

(b) $\cos t = 0.75$

$t \approx 0.72$ or $t \approx 5.56$

62. $\sin 0.25 \approx 0.2474$

$\sin 0.75 \approx 0.6816$

$\sin 1 \approx 0.8415$

$\sin(0.25 + 0.75) = \sin 1 \neq \sin 0.25 + \sin 0.75$

because, $0.8415 \neq 0.2474 + 0.6816 = 0.9290$.

64. (a) The points (x_1, y_1) and (x_2, y_2) are symmetric about the origin.

(b) Because of the symmetry of the points, you can make the conjecture that $\sin(t_1 + \pi) = -\sin t_1$.

(c) Because of the symmetry of the points, you can make the conjecture that $\cos(t_1 + \pi) = -\cos t_1$.

66. $y(t) = \frac{1}{4}e^{-t} \cos 6t$

(a) When $t = 0$: $y(0) = \frac{1}{4}e^{-0} \cos 0 = 0.2500$ foot

(b) When $t = \frac{1}{4}$: $y\left(\frac{1}{4}\right) = \frac{1}{4}e^{-1/4} \cos\left(6 \cdot \frac{1}{4}\right) \approx 0.0138$ foot

(c) When $t = \frac{1}{2}$: $y\left(\frac{1}{2}\right) = \frac{1}{2}e^{-1/2} \cos\left(6 \cdot \frac{1}{2}\right) \approx -0.1501$ foot

68. $\cos \theta = x = \cos(-\theta)$

$\sin \theta = \dfrac{1}{x} = \sec(-\theta)$

$\sin \theta = y$

$\sin(-\theta) = -y = -\sin \theta$

$\sec \theta = \dfrac{1}{y}$

$\sec(-\theta) = -\dfrac{1}{y} = -\sec \theta$

$\tan \theta = \dfrac{y}{x}$

$\tan(-\theta) = \dfrac{-y}{x} = -\tan \theta$

$\cot \theta = \dfrac{x}{y}$

$\cot(-\theta) = \dfrac{x}{-y} = -\cot \theta$

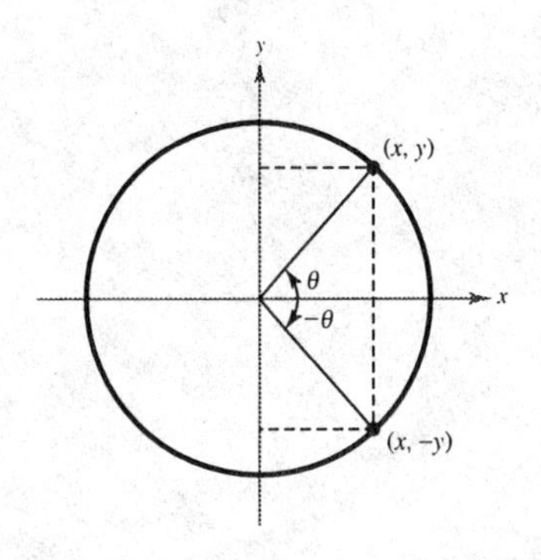

70. $f(t) = \sin t$ and $g(t) = \tan t$

Both f and g are odd functions.

$h(t) = f(t)g(t) = \sin t \tan t$

$h(-t) = \sin(-t) \tan(-t)$

$= (-\sin t)(-\tan t)$

$= \sin t \tan t = h(t)$

The function $h(t) = f(t)g(t)$ is even.

72. $f(x) = \frac{1}{4}x^3 + 1$

$$y = \frac{1}{4}x^3 + 1$$
$$x = \frac{1}{4}y^3 + 1$$
$$x - 1 = \frac{1}{4}y^3$$
$$4(x - 1) = y^3$$
$$y = \sqrt[3]{4(x - 1)}$$
$$f^{-1}(x) = \sqrt[3]{4(x - 1)}$$

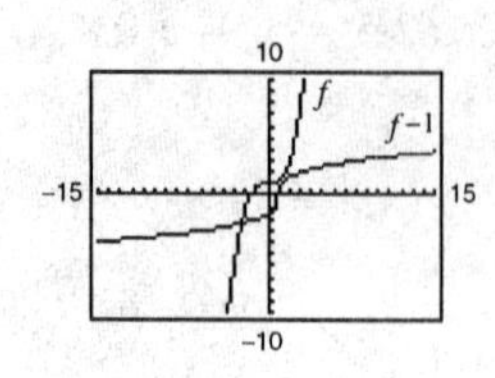

74. $f(x) = \dfrac{2x}{x + 1}, x > -1$

$$y = \frac{2x}{x + 1}, x > -1$$
$$x = \frac{2y}{y + 1}$$
$$xy + x = 2y$$
$$x = 2y - xy$$
$$x = y(2 - x)$$
$$\frac{x}{2 - x} = y, x < 2$$
$$f^{-1}(x) = \frac{x}{2 - x}, x < 2$$

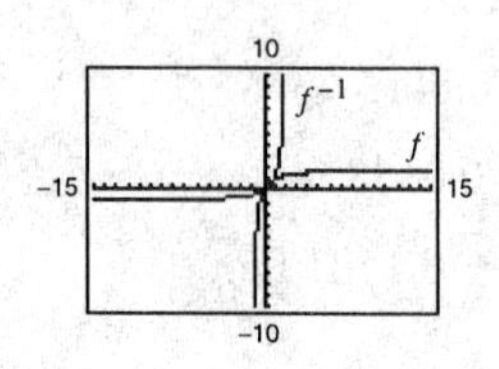

Section 4.3 Right Triangle Trigonometry

Solutions to Even-Numbered Exercises

2. $b = \sqrt{13^2 - 5^2} = \sqrt{169 - 25} = 12$

$\sin\theta = \dfrac{\text{opp}}{\text{hyp}} = \dfrac{5}{13}$

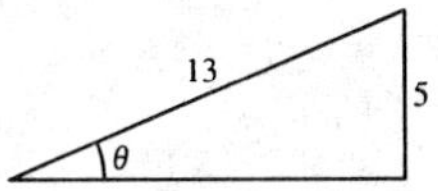

$\cos\theta = \dfrac{\text{adj}}{\text{hyp}} = \dfrac{12}{13}$

$\tan\theta = \dfrac{\text{opp}}{\text{adj}} = \dfrac{5}{12}$

$\csc\theta = \dfrac{\text{hyp}}{\text{opp}} = \dfrac{13}{5}$

$\sec\theta = \dfrac{\text{hyp}}{\text{adj}} = \dfrac{13}{12}$

$\cot\theta = \dfrac{\text{adj}}{\text{opp}} = \dfrac{12}{5}$

4. $c = \sqrt{4^2 + 4^2} = \sqrt{32} = 4\sqrt{2}$

$\sin\theta = \dfrac{\text{opp}}{\text{hyp}} = \dfrac{4}{4\sqrt{2}} = \dfrac{1}{\sqrt{2}} = \dfrac{\sqrt{2}}{2}$

$\cos\theta = \dfrac{\text{adj}}{\text{hyp}} = \dfrac{4}{4\sqrt{2}} = \dfrac{1}{\sqrt{2}} = \dfrac{\sqrt{2}}{2}$

$\tan\theta = \dfrac{\text{opp}}{\text{adj}} = \dfrac{4}{4} = 1$

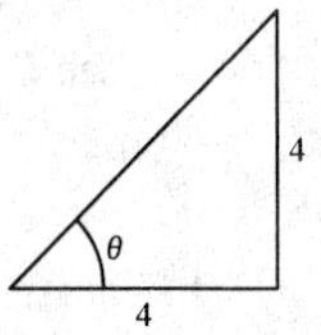

$\csc\theta = \dfrac{\text{hyp}}{\text{opp}} = \dfrac{4\sqrt{2}}{4} = \sqrt{2}$

$\sec\theta = \dfrac{\text{hyp}}{\text{adj}} = \dfrac{4\sqrt{2}}{4} = \sqrt{2}$

$\cot\theta = \dfrac{\text{adj}}{\text{opp}} = \dfrac{4}{4} = 1$

6. $\text{adj} = \sqrt{15^2 - 8^2} = \sqrt{161}$

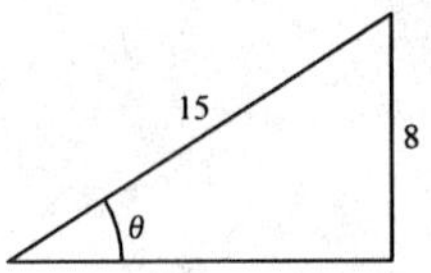

$\sin\theta = \dfrac{\text{opp}}{\text{hyp}} = \dfrac{8}{15}$

$\cos\theta = \dfrac{\text{adj}}{\text{hyp}} = \dfrac{\sqrt{161}}{15}$

$\tan\theta = \dfrac{\text{opp}}{\text{adj}} = \dfrac{8}{\sqrt{161}} = \dfrac{8\sqrt{161}}{161}$

$\csc\theta = \dfrac{\text{hyp}}{\text{opp}} = \dfrac{15}{8}$

$\sec\theta = \dfrac{\text{hyp}}{\text{adj}} = \dfrac{15}{\sqrt{161}} = \dfrac{15\sqrt{161}}{161}$

$\cot\theta = \dfrac{\text{adj}}{\text{opp}} = \dfrac{\sqrt{161}}{8}$

$\text{adj} = \sqrt{7.5^2 - 4^2} = \dfrac{\sqrt{161}}{2}$

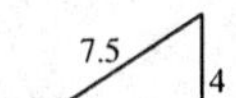

$\sin\theta = \dfrac{\text{opp}}{\text{hyp}} = \dfrac{4}{7.5} = \dfrac{8}{15}$

$\cos\theta = \dfrac{\text{adj}}{\text{hyp}} = \dfrac{\sqrt{161}}{2 \cdot 7.5} = \dfrac{\sqrt{161}}{15}$

$\tan\theta = \dfrac{\text{opp}}{\text{adj}} = \dfrac{4}{(\sqrt{161}/2)} = \dfrac{8}{\sqrt{161}} = \dfrac{8\sqrt{161}}{161}$

$\csc\theta = \dfrac{\text{hyp}}{\text{opp}} = \dfrac{7.5}{4} = \dfrac{15}{8}$

$\sec\theta = \dfrac{\text{hyp}}{\text{adj}} = \dfrac{7.5}{(\sqrt{161}/2)} = \dfrac{15}{\sqrt{161}} = \dfrac{15\sqrt{161}}{161}$

$\cot\theta = \dfrac{\text{adj}}{\text{opp}} = \dfrac{\sqrt{161}}{2 \cdot 4} = \dfrac{\sqrt{161}}{8}$

The function values are the same because the triangles are similar, and corresponding sides are proportional.

8. $\text{hyp} = \sqrt{1^2 + 2^2} = \sqrt{5}$

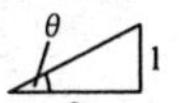

$\sin\theta = \dfrac{\text{opp}}{\text{hyp}} = \dfrac{1}{\sqrt{5}} = \dfrac{\sqrt{5}}{5}$

$\cos\theta = \dfrac{\text{adj}}{\text{hyp}} = \dfrac{2}{\sqrt{5}} = \dfrac{2\sqrt{5}}{5}$

$\tan\theta = \dfrac{\text{opp}}{\text{adj}} = \dfrac{1}{2}$

$\csc\theta = \dfrac{\text{hyp}}{\text{opp}} = \dfrac{\sqrt{5}}{1} = \sqrt{5}$

$\sec\theta = \dfrac{\text{hyp}}{\text{adj}} = \dfrac{\sqrt{5}}{2}$

$\cot\theta = \dfrac{\text{adj}}{\text{opp}} = \dfrac{2}{1} = 2$

$\text{hyp} = \sqrt{3^2 + 6^2} = 3\sqrt{5}$

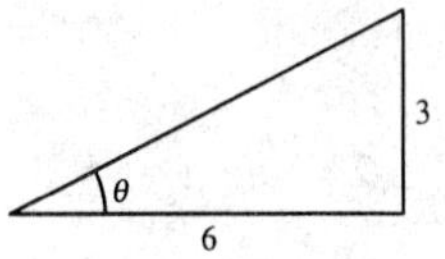

$\sin\theta = \dfrac{3}{3\sqrt{5}} = \dfrac{1}{\sqrt{5}} = \dfrac{\sqrt{5}}{5}$

$\cos\theta = \dfrac{6}{3\sqrt{5}} = \dfrac{2}{\sqrt{5}} = \dfrac{2\sqrt{5}}{5}$

$\tan\theta = \dfrac{3}{6} = \dfrac{1}{2}$

$\csc\theta = \dfrac{3\sqrt{5}}{3} = \sqrt{5}$

$\sec\theta = \dfrac{3\sqrt{5}}{6} = \dfrac{\sqrt{5}}{2}$

$\cot\theta = \dfrac{6}{3} = 2$

The function values are the same because the triangles are similar, and corresponding sides are proportional.

10. opp $= \sqrt{5^2 + 1^2} = \sqrt{26}$

$\sin\theta = \dfrac{\text{opp}}{\text{hyp}} = \dfrac{1}{\sqrt{26}} = \dfrac{\sqrt{26}}{26}$

$\cos\theta = \dfrac{\text{adj}}{\text{hyp}} = \dfrac{5}{\sqrt{26}} = \dfrac{5\sqrt{26}}{26}$

$\tan\theta = \dfrac{\text{opp}}{\text{adj}} = \dfrac{1}{5}$

$\csc\theta = \dfrac{\text{hyp}}{\text{opp}} = \dfrac{\sqrt{26}}{1} = \sqrt{26}$

$\sec\theta = \dfrac{\text{hyp}}{\text{adj}} = \dfrac{\sqrt{26}}{5}$

12. opp $= \sqrt{7^2 - 5^2} = \sqrt{24} = 2\sqrt{6}$

$\sin\theta = \dfrac{\text{opp}}{\text{hyp}} = \dfrac{2\sqrt{6}}{7}$

$\tan\theta = \dfrac{\text{opp}}{\text{adj}} = \dfrac{2\sqrt{6}}{5}$

$\csc\theta = \dfrac{1}{\sin\theta} = \dfrac{7}{2\sqrt{6}} = \dfrac{7\sqrt{6}}{12}$

$\sec\theta = \dfrac{1}{\cos\theta} = \dfrac{7}{5}$

$\cot\theta = \dfrac{1}{\tan\theta} = \dfrac{5}{2\sqrt{6}} = \dfrac{5\sqrt{6}}{12}$

14. adj $= \sqrt{17^2 - 4^2} = \sqrt{237}$

$\sin\theta = \dfrac{\text{opp}}{\text{hyp}} = \dfrac{4}{17}$

$\cos\theta = \dfrac{\text{adj}}{\text{hyp}} = \dfrac{\sqrt{273}}{17}$

$\tan\theta = \dfrac{\text{opp}}{\text{adj}} = \dfrac{4}{\sqrt{273}} = \dfrac{4\sqrt{273}}{273}$

$\sec\theta = \dfrac{1}{\cos\theta} = \dfrac{17}{\sqrt{273}} = \dfrac{17\sqrt{273}}{273}$

$\cot\theta = \dfrac{1}{\tan\theta} = \dfrac{\sqrt{273}}{4}$

16. adj $= \sqrt{8^2 - 3^2} = \sqrt{55}$

$\cos\theta = \dfrac{\text{adj}}{\text{hyp}} = \dfrac{\sqrt{55}}{8}$

$\tan\theta = \dfrac{\text{opp}}{\text{adj}} = \dfrac{3}{\sqrt{55}} = \dfrac{3\sqrt{55}}{55}$

$\csc\theta = \dfrac{1}{\sin\theta} = \dfrac{8}{3}$

$\sec\theta = \dfrac{1}{\cos\theta} = \dfrac{8}{\sqrt{55}} = \dfrac{8\sqrt{55}}{55}$

$\cot\theta = \dfrac{1}{\tan\theta} = \dfrac{\sqrt{55}}{3}$

18. $\sin 30° = \dfrac{1}{2}$, $\tan 30° = \dfrac{\sqrt{3}}{3}$

(a) $\csc 30° = \dfrac{1}{\sin 30°} = 2$

(b) $\cot 60° = \tan(90° - 60°) = \tan 30° = \dfrac{\sqrt{3}}{3}$

(c) $\cos 30° = \dfrac{\sin 30°}{\tan 30°} = \dfrac{(1/2)}{(\sqrt{3}/3)} = \dfrac{3}{2\sqrt{3}} = \dfrac{\sqrt{3}}{2}$

(d) $\cot 30° = \dfrac{1}{\tan 30°} = \dfrac{3}{\sqrt{3}} = \dfrac{3\sqrt{3}}{3} = \sqrt{3}$

20. $\sec\theta = 5$, $\tan\theta = 2\sqrt{6}$

(a) $\cos\theta = \dfrac{1}{\sec\theta} = \dfrac{1}{5}$

(b) $\cot\theta = \dfrac{1}{\tan\theta} = \dfrac{1}{2\sqrt{6}} = \dfrac{\sqrt{6}}{12}$

(c) $\cot(90° - \theta) = \tan\theta = 2\sqrt{6}$

(d) $\sin\theta = \tan\theta\cos\theta = (2\sqrt{6})\left(\dfrac{1}{5}\right) = \dfrac{2\sqrt{6}}{5}$

22. $\tan\beta = 5$

(a) $\cot\beta = \dfrac{1}{\tan\beta} = \dfrac{1}{5}$

(b) $\cos\beta = \dfrac{1}{\sqrt{1 + \tan^2\beta}} = \dfrac{1}{\sqrt{1 + 5^2}}$

$= \dfrac{1}{\sqrt{26}} = \dfrac{\sqrt{26}}{26}$

(c) $\tan(90° - \beta) = \cot\beta = \dfrac{1}{\tan\beta} = \dfrac{1}{5}$

(d) $\csc\beta = \sqrt{1 + \cot^2\beta}$

$= \sqrt{1 + \left(\dfrac{1}{5}\right)^2}$

$= \sqrt{1 + \dfrac{1}{25}} = \sqrt{\dfrac{26}{25}} = \dfrac{\sqrt{26}}{5}$

24. $\cos\theta\sec\theta = \cos\theta\dfrac{1}{\cos\theta} = 1$

26. $\cot\alpha\sin\alpha = \dfrac{\cos\alpha}{\sin\alpha}\sin\alpha = \cos\alpha$

28. $(1 + \sin\theta)(1 - \sin\theta) = 1 - \sin^2\theta = \cos^2\theta$

30. $\sin^2\theta - \cos^2\theta = \sin^2\theta - (1 - \sin^2\theta)$
$= \sin^2\theta - 1 + \sin^2\theta$
$= 2\sin^2\theta - 1$

32. $\dfrac{\tan\beta + \cot\beta}{\tan\beta} = \dfrac{\tan\beta}{\tan\beta} + \dfrac{\cot\beta}{\tan\beta}$
$= 1 + \dfrac{\cot\beta}{(1/\cot\beta)}$
$= 1 + \cot^2\beta = \csc^2\beta$

34. (a) $\csc 30° = \dfrac{1}{\sin 30°} = \dfrac{1}{(1/2)} = 2$

(b) $\sin\dfrac{\pi}{4} = \dfrac{\sqrt{2}}{2}$

36. (a) $\sin\dfrac{\pi}{3} = \dfrac{\sqrt{3}}{2}$

(b) $\csc 45° = \dfrac{1}{\sin 45°} = \dfrac{1}{(\sqrt{2}/2)} = \sqrt{2}$

38. (a) $\tan 23.5° \approx 0.4348$

(b) $\cot 66.5° = \dfrac{1}{\tan 66.5°} \approx 0.4348$

40. (a) $\cos 16°\,18' = \cos\left(16 + \dfrac{18}{60}\right)^{\circ} \approx 0.9598$

(b) $\sin 73°\,56' = \sin\left(73 + \dfrac{56}{60}\right)^{\circ} \approx 0.9609$

42. (a) $\cos 4°\,50'\,15'' = \cos\left(4 + \dfrac{50}{60} + \dfrac{15}{3600}\right)^{\circ}$
≈ 0.9964

(b) $\sec 4°\,50'\,15'' = \dfrac{1}{\cos 4°\,50'\,15''}$
≈ 1.0036

44. (a) $\sec 0.75 = \dfrac{1}{\cos 0.75} \approx 1.3667$

(Note: 0.75 is in radians.)

(b) $\cos 0.75 \approx 0.7317$

46. (a) $\sec\left(\dfrac{\pi}{2} - 1\right) = \dfrac{1}{\cos\left(\dfrac{\pi}{2} - 1\right)} \approx 1.1884$

(b) $\cot\left(\dfrac{\pi}{2} - \dfrac{1}{2}\right) = \dfrac{1}{\tan\left(\dfrac{\pi}{2} - \dfrac{1}{2}\right)} \approx 0.5463$

48. (a) $\cos\theta = \dfrac{\sqrt{2}}{2} \Longrightarrow \theta = 45° = \dfrac{\pi}{4}$

(b) $\tan\theta = 1 \Longrightarrow \theta = 45° = \dfrac{\pi}{4}$

50. (a) $\tan\theta = \sqrt{3} \Longrightarrow \theta = 60° = \dfrac{\pi}{3}$

(b) $\cos\theta = \dfrac{1}{2} \Longrightarrow \theta = 60° = \dfrac{\pi}{3}$

52. (a) $\cot\theta = \dfrac{\sqrt{3}}{3}$

$\tan\theta = \dfrac{3}{\sqrt{3}} = \sqrt{3} \Longrightarrow \theta = 60° = \dfrac{\pi}{3}$

(b) $\sec\theta = \sqrt{2}$

$\cos\theta = \dfrac{1}{\sqrt{2}} = \dfrac{\sqrt{2}}{2} \Longrightarrow \theta = 45° = \dfrac{\pi}{4}$

54. (a) $\cos\theta = 0.9848 \Longrightarrow \theta \approx 10° \approx 0.175$

(b) $\cos\theta = 0.8746 \Longrightarrow \theta \approx 29° \approx 0.506$

56. (a) $\sin\theta = 0.3746 \Longrightarrow \theta \approx 22° \approx 0.384$

(b) $\cos\theta = 0.3746 \Longrightarrow \theta \approx 68° \approx 1.187$

58. $\cos 60° = \dfrac{x}{12}$

$x = 12\cos 60° = 6$

60. $\sin 45° = \dfrac{20}{r}$

$r = \dfrac{20}{\sin 45°} = \dfrac{20}{\sqrt{2}/2} = 20\sqrt{2}$

62. $\tan 20° = \dfrac{25}{x}$

$x = \dfrac{25}{\tan 20°} \approx 68.7$

64. $\cos 75° = \dfrac{25}{r}$

$r = \dfrac{25}{\cos 75°} \approx 96.6$

66. (a)

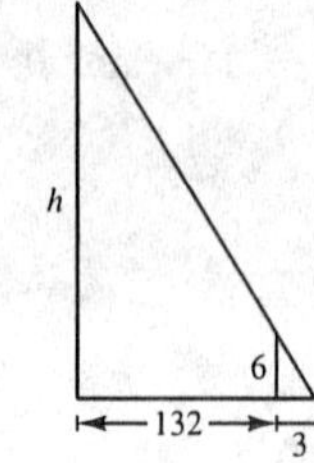

(b) $\tan\theta = \frac{6}{3}$ and $\tan\theta = \frac{h}{135}$

Thus, $\frac{6}{3} = \frac{h}{135}$.

(c) $\frac{135 \cdot 6}{3} = h = 270$ feet

68. $\tan\theta = \frac{\text{opp}}{\text{adj}}$

$\tan 54° = \frac{w}{100}$

$w = 100 \tan 54° \approx 137.6$ feet

70. (a)

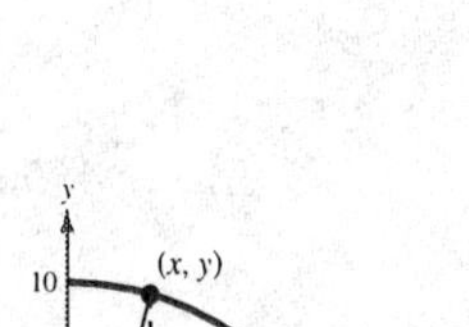

(b) $\sin\theta = \frac{\text{opp}}{\text{hyp}}$

$\sin\theta = \frac{3(1/3)}{20}$

(c) $\sin\theta = \frac{1}{6} \Longrightarrow \theta = 9.59°$

72. $\tan 3° = \frac{x}{15}$

$x = 15 \tan 3°$

$d = 5 + 2x$

$= 5 + 2(15 \tan 3°)$

≈ 6.57 centimeters

74. $x \approx 2.588$, $y \approx 9.659$

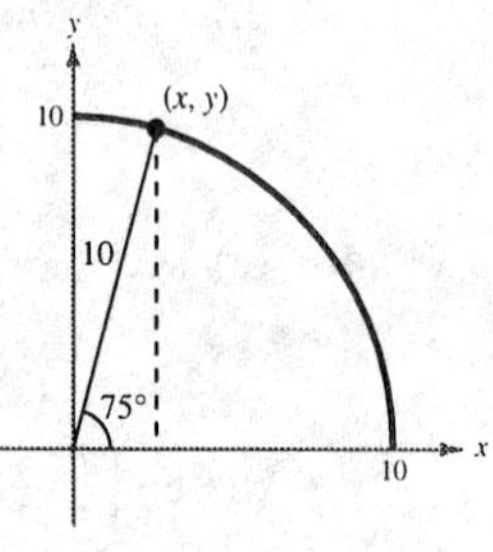

$\sin\theta = \frac{y}{10} \approx 0.97$

$\cos\theta = \frac{x}{10} \approx 0.26$

$\tan\theta = \frac{y}{x} \approx 3.73$

$\csc\theta = \frac{10}{y} \approx 1.04$

$\sec\theta = \frac{10}{x} \approx 3.86$

$\cot\theta = \frac{x}{y} \approx 0.27$

76. (a)

θ	0	0.3	0.6	0.9	1.2	1.5
$\sin\theta$	0	0.2955	0.5646	0.7833	0.9320	0.9975
$\cos\theta$	1	0.9553	0.8253	0.6216	0.3624	0.0707

(b) On [0, 1.5], $\sin\theta$ is an increasing function.

(c) On [0, 1.5], $\cos\theta$ is a decreasing function.

(d) As the angle increases the length of the side opposite the angle increases relative to the length of the hypotenuse and the length of the side adjacent to the angle decreases relative to the length of the hypotenuse. Thus the sine increases and the cosine decreases.

78.

θ	0°	20°	40°	60°	80°
$\cos\theta$	1	0.9397	0.7660	0.5000	0.1736
$\sin(90° - \theta)$	1	0.9397	0.7660	0.5000	0.1736

It seems that $\cos\theta = \sin(90° - \theta)$ for all θ.

θ and $90° - \theta$ are called complementary angles.

80. $\tan\theta = \dfrac{0.672s^2}{3000}$

(a)

s	10	20	30	40	50	60
θ	1.28°	5.12°	11.40°	19.72°	29.25°	38.88°

(b) θ increases at an increasing rate. The function is not linear.

82. $\sec 30° = \csc 60°$

True, because $\sec(90° - \theta) = \csc\theta$.

84. $\cot^2 10° - \csc^2 10° = -1$

True, because

$$1 + \cot^2\theta = \csc^2\theta$$
$$\cot^2\theta = \csc^2\theta - 1$$
$$\cot^2\theta - \csc^2\theta = -1.$$

86. $\tan[(0.8)^2] = \tan^2(0.8)$

False.

$\tan[(0.8)^2] = \tan 0.64 \approx 0.745$

$\tan^2(0.8) = (\tan 0.8)^2 \approx 1.060$

Section 4.4 Trigonometric Functions of Any Angle

Solutions to Even-Numbered Exercises

2. (a) $x = 12, y = -5$

$r = \sqrt{12^2 + (-5)^2} = 13$

$\sin\theta = \frac{y}{r} = \frac{-5}{13} = -\frac{5}{13}$

$\cos\theta = \frac{x}{r} = \frac{12}{13}$

$\tan\theta = \frac{y}{x} = \frac{-5}{12} = -\frac{5}{12}$

$\csc\theta = \frac{r}{y} = \frac{13}{-5} = -\frac{13}{5}$

$\sec\theta = \frac{r}{x} = \frac{13}{12}$

$\cot\theta = \frac{x}{y} = \frac{12}{-5} = -\frac{12}{5}$

(b) $x = -1, y = 1$

$r = \sqrt{(-1)^2 + 1^2} = \sqrt{2}$

$\sin\theta = \frac{y}{r} = \frac{1}{\sqrt{2}} = \frac{\sqrt{2}}{2}$

$\cos\theta = \frac{x}{r} = \frac{-1}{\sqrt{2}} = -\frac{\sqrt{2}}{2}$

$\tan\theta = \frac{y}{x} = \frac{1}{-1} = -1$

$\csc\theta = \frac{r}{y} = \frac{\sqrt{2}}{1} = \sqrt{2}$

$\sec\theta = \frac{r}{x} = \frac{\sqrt{2}}{-1} = -\sqrt{2}$

$\cot\theta = \frac{x}{y} = \frac{-1}{1} = -1$

4. (a) $x = 3, y = 1$

$r = \sqrt{3^2 + 1^2} = \sqrt{10}$

$\sin\theta = \frac{y}{r} = \frac{1}{\sqrt{10}} = \frac{\sqrt{10}}{10}$

$\cos\theta = \frac{x}{r} = \frac{3}{\sqrt{10}} = \frac{3\sqrt{10}}{10}$

$\tan\theta = \frac{y}{x} = \frac{1}{3}$

$\csc\theta = \frac{r}{y} = \frac{\sqrt{10}}{1} = \sqrt{10}$

$\sec\theta = \frac{r}{x} = \frac{\sqrt{10}}{3}$

$\cot\theta = \frac{x}{y} = \frac{3}{1} = 3$

(b) $x = 2, y = -4$

$r = \sqrt{2^2 + (-4)^2} = 2\sqrt{5}$

$\sin\theta = \frac{y}{r} = \frac{-4}{2\sqrt{5}} = -\frac{2\sqrt{5}}{5}$

$\cos\theta = \frac{x}{r} = \frac{2}{2\sqrt{5}} = \frac{\sqrt{5}}{5}$

$\tan\theta = \frac{y}{x} = \frac{-4}{2} = -2$

$\csc\theta = \frac{r}{y} = \frac{2\sqrt{5}}{-4} = -\frac{\sqrt{5}}{2}$

$\sec\theta = \frac{r}{x} = \frac{2\sqrt{5}}{2} = \sqrt{5}$

$\cot\theta = \frac{x}{y} = \frac{2}{-4} = -\frac{1}{2}$

6. (a) $x = 8, y = 15$

$r = \sqrt{8^2 + 15^2} = 17$

$\sin\theta = \frac{y}{r} = \frac{15}{17}$

$\cos\theta = \frac{x}{r} = \frac{8}{17}$

$\tan\theta = \frac{y}{x} = \frac{15}{8}$

$\csc\theta = \frac{r}{y} = \frac{17}{15}$

$\sec\theta = \frac{r}{x} = \frac{17}{8}$

$\cot\theta = \frac{x}{y} = \frac{8}{15}$

(b) $x = -9, y = -40$

$r = \sqrt{(-9)^2 + (-40)^2} = \sqrt{1681} = 41$

$\sin\theta = \frac{y}{r} = \frac{-40}{41} = -\frac{40}{41}$

$\cos\theta = \frac{x}{r} = \frac{-9}{41} = -\frac{9}{41}$

$\tan\theta = \frac{y}{x} = \frac{-40}{-9} = \frac{40}{9}$

$\csc\theta = \frac{r}{y} = \frac{41}{-40} = -\frac{41}{40}$

$\sec\theta = \frac{r}{x} = \frac{41}{-9} = -\frac{41}{9}$

$\cot\theta = \frac{x}{y} = \frac{-9}{-40} = \frac{9}{40}$

8. (a) $x = -5, y = -2$

$r = \sqrt{(-5)^2 + (-2)^2} = \sqrt{29}$

$\sin\theta = \frac{y}{r} = \frac{-2}{\sqrt{29}} = -\frac{2\sqrt{29}}{29}$

$\cos\theta = \frac{x}{r} = \frac{-5}{\sqrt{29}} = -\frac{5\sqrt{29}}{29}$

$\tan\theta = \frac{y}{x} = \frac{-2}{-5} = \frac{2}{5}$

$\csc\theta = \frac{r}{y} = \frac{\sqrt{29}}{-2} = -\frac{\sqrt{29}}{2}$

$\sec\theta = \frac{r}{x} = \frac{\sqrt{29}}{-5} = -\frac{\sqrt{29}}{5}$

$\cot\theta = \frac{x}{y} = \frac{-5}{-2} = \frac{5}{2}$

(b) $x = -\frac{3}{2}, y = 3$

$r = \sqrt{\left(-\frac{3}{2}\right)^2 + 3^2} = \frac{3\sqrt{5}}{2}$

$\sin\theta = \frac{y}{r} = \frac{3}{3\sqrt{5}/2} = \frac{2\sqrt{5}}{5}$

$\cos\theta = \frac{x}{r} = \frac{-3/2}{3\sqrt{5}/2} = -\frac{\sqrt{5}}{5}$

$\tan\theta = \frac{y}{x} = \frac{3}{-3/2} = -2$

$\csc\theta = \frac{r}{y} = \frac{3\sqrt{5}/2}{3} = \frac{\sqrt{5}}{2}$

$\sec\theta = \frac{r}{x} = \frac{3\sqrt{5}/2}{-3/2} = -\sqrt{5}$

$\cot\theta = \frac{x}{y} = \frac{-3/2}{3} = -\frac{1}{2}$

10. (a) $\sin\theta > 0$ and $\cos\theta > 0$

$\frac{y}{r} > 0$ and $\frac{x}{r} > 0$

Quadrant I

(b) $\sin\theta < 0$ and $\cos\theta > 0$

$\frac{y}{r} < 0$ and $\frac{x}{r} > 0$

Quadrant IV

12. (a) $\sec\theta > 0$ and $\cot\theta < 0$

$\frac{r}{x} > 0$ and $\frac{x}{y} < 0$

Quadrant IV

(b) $\csc\theta < 0$ and $\tan\theta > 0$

$\frac{r}{y} < 0$ and $\frac{y}{x} > 0$

Quadrant III

14. $\cos\theta = \frac{x}{r} = \frac{-4}{5} \Rightarrow y = |3|$

θ in Quadrant III $\Rightarrow y = -3$

$\sin\theta = \frac{y}{r} = -\frac{3}{5}$ $\qquad \csc\theta = -\frac{5}{3}$

$\cos\theta = \frac{x}{r} = -\frac{4}{5}$ $\qquad \sec\theta = -\frac{5}{4}$

$\tan\theta = \frac{y}{x} = \frac{3}{4}$ $\qquad \cot\theta = \frac{4}{3}$

16. $\cos\theta = \frac{x}{r} = \frac{8}{17} \Rightarrow y = |15|$

$\tan\theta < 0 \Rightarrow y = -15$

$\sin\theta = \frac{y}{r} = \frac{-15}{17} = -\frac{15}{17}$ $\qquad \csc\theta = -\frac{17}{15}$

$\cos\theta = \frac{x}{r} = \frac{8}{17}$ $\qquad \sec\theta = \frac{17}{8}$

$\tan\theta = \frac{y}{x} = \frac{-15}{8} = -\frac{15}{8}$ $\qquad \tan\theta = -\frac{8}{15}$

18. $\csc\theta = \frac{r}{y} = \frac{4}{1} \Rightarrow x = \left|\sqrt{15}\right|$

$\cot\theta < 0 \Rightarrow x = -\sqrt{15}$

$\sin\theta = \frac{y}{r} = \frac{1}{4}$ $\qquad \csc\theta = 4$

$\cos\theta = \frac{x}{r} = -\frac{\sqrt{15}}{4}$ $\qquad \sec\theta = -\frac{4\sqrt{15}}{15}$

$\tan\theta = \frac{y}{x} = -\frac{\sqrt{15}}{15}$ $\qquad \cot\theta = -\sqrt{15}$

20. $\cot\theta$ is undefined $\Rightarrow \theta = n\pi$

$\frac{\pi}{2} \le \theta \le \frac{3\pi}{2} \Rightarrow \theta = \pi, y = 0, x = -r$

$\sin\theta = \frac{y}{r} = \frac{0}{r} = 0$ $\qquad \csc\theta = \frac{r}{y}$ is undefined.

$\cos\theta = \frac{x}{r} = \frac{-r}{r} = -1$ $\qquad \sec\theta = \frac{r}{x} = -1$

$\tan\theta = \frac{y}{x} = \frac{0}{x} = 0$ $\qquad \cot\theta = \frac{x}{y}$ is undefined.

22. $\tan\theta$ is undefined $\Rightarrow \theta = n\pi + \frac{\pi}{2}$

$\pi \le \theta \le 2\pi \Rightarrow \theta = \frac{3\pi}{2}, x = 0, y = -r$

$\sin\theta = \frac{y}{r} = \frac{-r}{r} = -1$ $\qquad \csc\theta = \frac{r}{y} = -1$

$\cos\theta = \frac{x}{r} = \frac{0}{r} = 0$ $\qquad \sec\theta = \frac{r}{x}$ is undefined.

$\tan\theta = \frac{y}{x}$ is undefined. $\qquad \cot\theta = \frac{x}{y} = \frac{0}{y} = 0$

24. $\left(-x, -\frac{1}{3}x\right)$, Quadrant III

$r = \sqrt{x^2 + \frac{1}{9}x^2} = \frac{\sqrt{10}x}{3}$

$\sin\theta = \frac{y}{r} = \frac{(1/3)-x}{(\sqrt{10}x)/3} = -\frac{\sqrt{10}}{10}$

$\cos\theta = \frac{x}{r} = \frac{-x}{(\sqrt{10}x)/3} = -\frac{3\sqrt{10}}{10}$

$\tan\theta = \frac{y}{x} = \frac{(-1/3)x}{-x} = \frac{1}{3}$

$\csc\theta = \frac{r}{y} = \frac{(\sqrt{10}x)/3}{(-1/3)x} = -\sqrt{10}$

$\sec\theta = \frac{r}{x} = \frac{(\sqrt{10}x)/3}{-x} = -\frac{\sqrt{10}}{3}$

$\cot\theta = \frac{x}{y} = \frac{-x}{(-1/3)x} = 3$

26. $4x + 3y = 0 \Rightarrow y = -\frac{4}{3}x$

$\left(x, -\frac{4}{3}x\right)$, Quadrant IV

$r = \sqrt{x^2 + \frac{16}{5}x^2} = \frac{5}{3}x$

$\sin\theta = \frac{y}{r} = \frac{(-4/3)x}{(5/3)x} = -\frac{4}{5}$

$\cos\theta = \frac{x}{r} = \frac{x}{(5/3)x} = \frac{3}{5}$

$\tan\theta = \frac{y}{x} = \frac{(-4/3)x}{x} = -\frac{4}{3}$

$\csc\theta = -\frac{5}{4}$

$\sec\theta = \frac{5}{3}$

$\tan\theta = -\frac{3}{4}$

28. $\cos \dfrac{3\pi}{2} = \dfrac{x}{r} = \dfrac{0}{1} = 0$

since $(3\pi/2)$ corresponds to $(0, -1)$.

30. $\sec \dfrac{3\pi}{2} = \dfrac{r}{x} = \dfrac{1}{0} \Longrightarrow$ undefined

since $(3\pi/2)$ corresponds to $(0, -1)$.

32. $\tan \pi = \dfrac{y}{x} = \dfrac{0}{-1} = 0$

since π corresponds to $(-1, 0)$.

34. $\csc \pi = \dfrac{r}{y} = \dfrac{1}{0} \Longrightarrow$ undefined

since π corresponds to $(-1, 0)$.

36. (a) $\theta = 309°$

$\theta' = 360° - 309° = 51°$

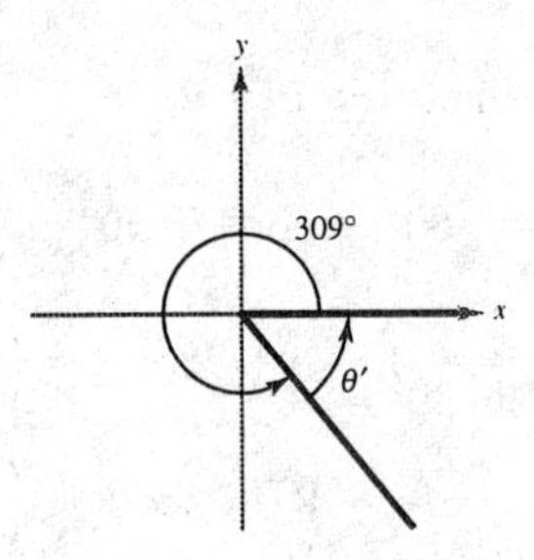

(b) $\theta = 226°$

$\theta' = 226° - 180° = 46°$

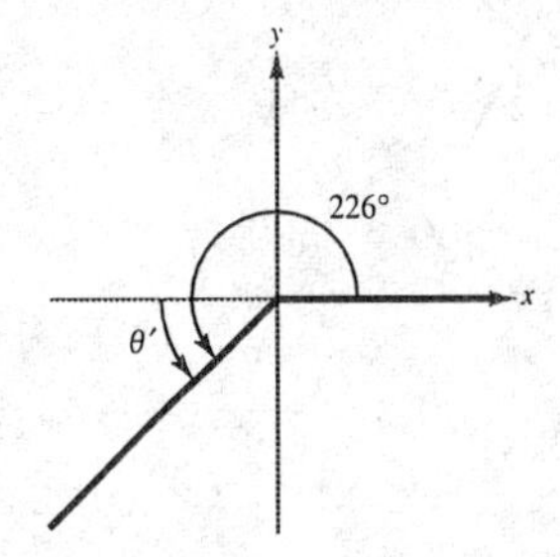

38. (a) $\theta = -145°$ is coterminal with $215°$.

$\theta' = 215° - 180° = 35°$

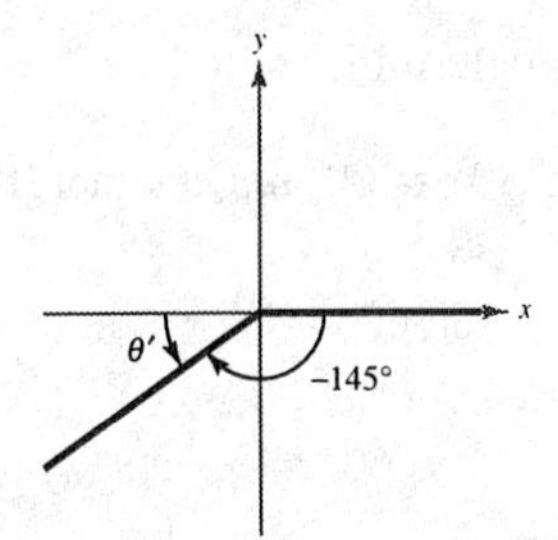

(b) $\theta = -239°$ is coterminal with $121°$.

$\theta' = 180° - 121° = 59°$

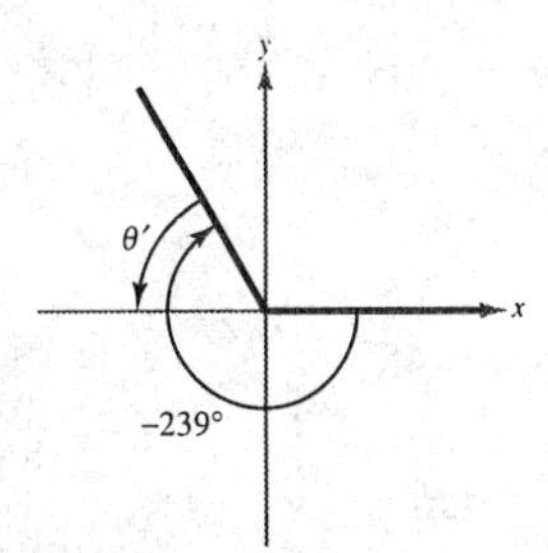

40. (a) $\theta = \dfrac{7\pi}{4}$

$\theta' = 2\pi - \dfrac{7\pi}{4} = \dfrac{\pi}{4}$

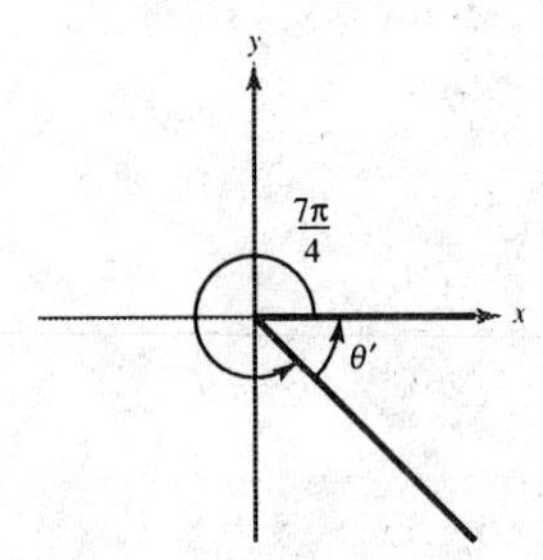

(b) $\theta = \dfrac{8\pi}{9}$

$\theta' = \pi - \dfrac{8\pi}{9} = \dfrac{\pi}{9}$

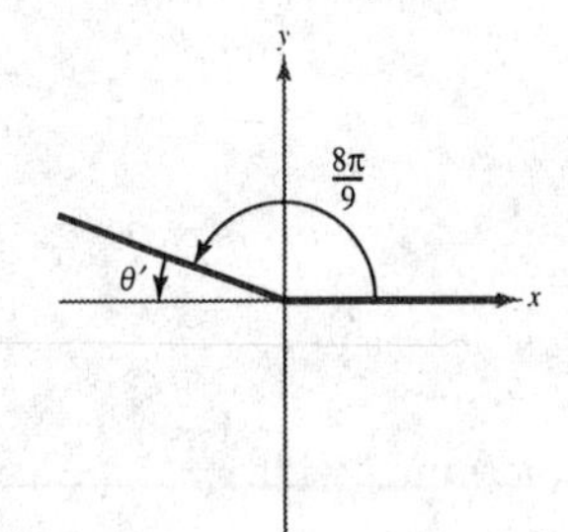

42. (a) $\theta = \frac{11\pi}{3}$ is coterminal with $\frac{5\pi}{3}$.

$\theta' = 2\pi - \frac{5\pi}{3} = \frac{\pi}{3}$

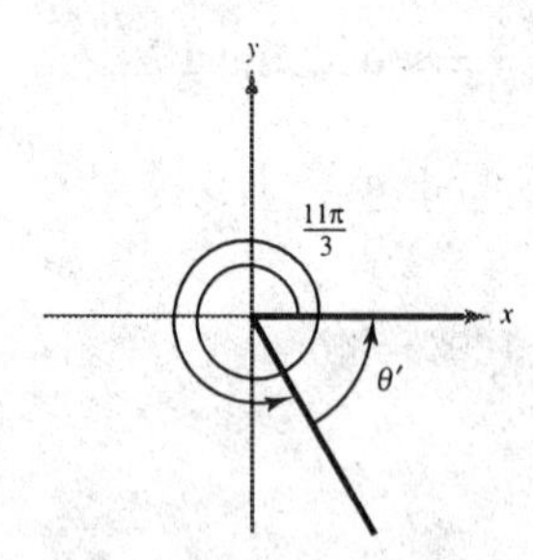

(b) $\theta = -\frac{7\pi}{10}$ is coterminal with $\frac{13\pi}{10}$.

$\theta' = \frac{13\pi}{10} - \pi = \frac{3\pi}{10}$

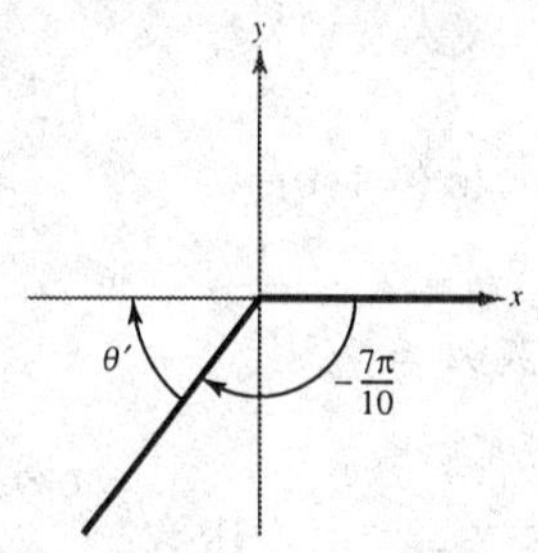

44. (a) $\theta = 300°$, $\theta' = 360° - 300° = 60°$ in Quadrant IV.

$\sin 300° = -\sin 60° = -\frac{\sqrt{3}}{2}$

$\cos 300° = \cos 60° = \frac{1}{2}$

$\tan 300° = -\tan 60° = -\sqrt{3}$

(b) $\theta = 330°$, $\theta' = 360° - 330° = 30°$ in Quadrant IV.

$\sin 330° = -\sin 30° = -\frac{1}{2}$

$\cos 330° = \cos 30° = \frac{\sqrt{3}}{2}$

$\tan 330° = -\tan 30° = -\frac{\sqrt{3}}{3}$

46. (a) $\theta = -405°$, $\theta' = 405° - 360° = 45°$ in Quadrant IV.

$\sin(-405°) = -\sin 45° = -\frac{\sqrt{2}}{2}$

$\cos(-405°) = \cos 45° = \frac{\sqrt{2}}{2}$

$\tan(-405°) = -\tan 45° = -1$

(b) $\theta = -120°$ is coterminal with $240°$.

$\theta' = 240° - 180° = 60°$ in Quadrant III.

$\sin(-120°) = -\sin 60° = -\frac{\sqrt{3}}{2}$

$\cos(-120°) = -\cos 60° = -\frac{1}{2}$

$\tan(-120°) = \tan 60° = \sqrt{3}$

48. (a) $\theta = \frac{\pi}{4}$, $\theta' = \frac{\pi}{4}$ in Quadrant I.

$\sin \frac{\pi}{4} = \frac{\sqrt{2}}{2}$

$\cos \frac{\pi}{4} = \frac{\sqrt{2}}{2}$

$\tan \frac{\pi}{4} = 1$

(b) $\theta = \frac{5\pi}{4}$, $\theta' = \frac{5\pi}{4} - \pi = \frac{\pi}{4}$ in Quadrant III.

$\sin \frac{5\pi}{4} = -\sin \frac{\pi}{4} = -\frac{\sqrt{2}}{2}$

$\cos \frac{5\pi}{4} = -\cos \frac{\pi}{4} = -\frac{\sqrt{2}}{2}$

$\tan \frac{5\pi}{4} = \tan \frac{\pi}{4} = 1$

50. (a) $\theta = -\frac{\pi}{2}$ is coterminal with $\frac{3\pi}{2}$.

$\sin\left(-\frac{\pi}{2}\right) = \sin \frac{3\pi}{2} = -1$

$\cos\left(-\frac{\pi}{2}\right) = \cos \frac{3\pi}{2} = 0$

$\tan\left(-\frac{\pi}{2}\right) = \tan \frac{3\pi}{2}$ is undefined.

(b) $\theta = \frac{\pi}{2}$

$\sin \frac{\pi}{2} = 1$

$\cos \frac{\pi}{2} = 0$

$\tan \frac{\pi}{2}$ is undefined.

52. (a) $\theta = \dfrac{10\pi}{3}$ is coterminal with $\dfrac{4\pi}{3}$.

$\theta' = \dfrac{4\pi}{3} - \pi = \dfrac{\pi}{3}$ in Quadrant III.

$\sin \dfrac{10\pi}{3} = -\sin \dfrac{\pi}{3} = -\dfrac{\sqrt{3}}{2}$

$\cos \dfrac{10\pi}{3} = -\cos \dfrac{\pi}{3} = -\dfrac{1}{2}$

$\tan \dfrac{10\pi}{3} = \tan \dfrac{\pi}{3} = \sqrt{3}$

(b) $\theta = \dfrac{17\pi}{3}$ is coterminal with $\dfrac{5\pi}{3}$.

$\theta' = 2\pi - \dfrac{5\pi}{3} = \dfrac{\pi}{3}$ in Quadrant IV.

$\sin \dfrac{17\pi}{3} = -\sin \dfrac{\pi}{3} = -\dfrac{\sqrt{3}}{2}$

$\cos \dfrac{17\pi}{3} = \cos \dfrac{\pi}{3} = \dfrac{1}{2}$

$\tan \dfrac{17\pi}{3} = \tan \dfrac{\pi}{3} = -\sqrt{3}$

54. (a) $\sec 225° = \dfrac{1}{\cos 225°} \approx -1.4142$

(b) $\sec 135° = \dfrac{1}{\cos 135°} \approx -1.4142$

56. (a) $\csc 330° = \dfrac{1}{\sin 330°} = -2.0000$

(b) $\csc 150° = \dfrac{1}{\sin 150°} = 2.0000$

58. (a) $\cot 1.35 = \dfrac{1}{\tan 1.35} \approx 0.2245$

(b) $\tan 1.35 \approx 4.4552$

60. (a) $\tan\left(-\dfrac{\pi}{9}\right) \approx -0.3640$

(b) $\tan\left(-\dfrac{10\pi}{9}\right) \approx -0.3640$

62. (a) $\sin(-0.65) \approx -0.6052$

(b) $\sin(5.63) \approx -0.6077$

64. (a) $\cos \theta = \dfrac{\sqrt{2}}{2} \Rightarrow$ reference angle is 45° or $\dfrac{\pi}{4}$ and θ is in Quadrant I or IV.

Values in degrees: 45°, 315°

Values in radians: $\dfrac{\pi}{4}, \dfrac{7\pi}{4}$

(b) $\cos \theta = -\dfrac{\sqrt{2}}{2} \Rightarrow$ reference angle is 45° or $\dfrac{\pi}{4}$ and θ is in Quadrant II or III.

Values in degrees: 135°, 225°

Values in radians: $\dfrac{3\pi}{4}, \dfrac{5\pi}{4}$

66. (a) $\sec \theta = 2 \Rightarrow$ reference angle is 60° or $\dfrac{\pi}{3}$ and θ is in Quadrant I or IV.

Values in degrees: 60°, 300°

Values in radians: $\dfrac{\pi}{3}, \dfrac{5\pi}{3}$

(b) $\sec \theta = -2 \Rightarrow$ reference angle is 60° or $\dfrac{\pi}{3}$ and θ is in Quadrant II or III.

Values in degrees: 120°, 240°

Values in radians: $\dfrac{2\pi}{3}, \dfrac{4\pi}{3}$

68. (a) $\sin\theta = \frac{\sqrt{3}}{2} \Rightarrow$ reference angle is 60° or $\frac{\pi}{3}$ and θ is in Quadrant I or II.

Values in degrees: 60°, 120°

Values in radians: $\frac{\pi}{3}, \frac{2\pi}{3}$

(b) $\sin\theta = -\frac{\sqrt{3}}{2} \Rightarrow$ reference angle is 60° or $\frac{\pi}{3}$ and θ is in Quadrant III or IV.

Values in degrees: 240°, 300°

Values in radians: $\frac{4\pi}{3}, \frac{5\pi}{3}$

70. (a) $\cos\theta = 0.8746$

Quadrant I: $\theta = \cos^{-1} 0.8746 \approx 29.00°$

Quadrant IV: $\theta = 360° - 29.00° = 331.00°$

(b) $\cos\theta = -0.2419$

Quadrant II:

$\theta = 180° - \cos^{-1} 0.2419 \approx 104.00°$

Quadrant III:

$\theta = 180° + \cos^{-1} 0.2419 \approx 256.00°$

72. (a) $\sin\theta = 0.0175$

Quadrant I: $\theta = \sin^{-1} 0.0175 \approx 0.018$

Quadrant II: $\theta = \pi - 0.018 = 3.124$

(b) $\sin\theta = -0.6691$

Quadrant III: $\theta = \pi + \sin^{-1} 0.6691 \approx 3.875$

Quadrant IV: $\theta = 2\pi - \sin^{-1} 0.6691 \approx 5.550$

74. (a) $\cot\theta = 5.671 \Rightarrow \frac{1}{\tan\theta} = 5.671$

Quadrant I: $\theta = \tan^{-1}\left(\frac{1}{5.671}\right) \approx 0.175$

Quadrant III: $\theta = \pi + 0.175 = 3.316$

(b) $\cot\theta = -1.280 \Rightarrow \frac{1}{\tan\theta} = -1.280$

Quadrant II: $\theta = \pi - \tan^{-1}\left(\frac{1}{1.280}\right) \approx 2.478$

Quadrant IV: $\theta = 2\pi - \tan^{-1}\left(\frac{1}{1.280}\right) \approx 5.620$

76. $\cot\theta = -3$

$1 + \cot^2\theta = \csc^2\theta$

$1 + (-3)^2 = \csc^2\theta$

$10 = \csc^2\theta$

$\csc\theta > 0$ in Quadrant II.

$\sqrt{10} = \csc\theta$

$\csc\theta = \frac{1}{\sin\theta}$

$\sin\theta = \frac{1}{\csc\theta} = \frac{1}{\sqrt{10}} = \frac{\sqrt{10}}{10}$

78. $\csc\theta = -2$

$1 + \cot^2\theta = \csc^2\theta$

$\cot^2\theta = \csc^2\theta - 1$

$\cot^2\theta = (-2)^2 - 1$

$\cot^2\theta = 3$

$\cot\theta < 0$ in Quadrant IV.

$\cot\theta = -\sqrt{3}$

80. $\sec\theta = -\frac{9}{4}$

$1 + \tan^2\theta = \sec^2\theta$

$\tan^2\theta = \sec^2\theta - 1$

$\tan^2\theta = \left(-\frac{9}{2}\right)^2 - 1$

$\tan^2\theta = \frac{65}{16}$

$\tan\theta > 0$ in Quadrant III.

$\tan\theta = \frac{\sqrt{65}}{4}$

82. $S = 23.1 + 0.442t + 4.3 \sin \frac{\pi t}{6}$

(a) February 1996 $\Rightarrow t = 2$

$S = 23.1 + 0.442(2) + 4.3 \sin \frac{2\pi}{6}$

≈ 27.7 thousand or 27,700 units

(b) February 1997 $\Rightarrow t = 14$

$S = 23.1 + 0.442(14) + 4.3 \sin \frac{14\pi}{6}$

≈ 33.0 thousand or 33,000 units

(c) September 1996 $\Rightarrow t = 9$

$S = 23.1 + 0.442(9) + 4.3 \sin \frac{9\pi}{6}$

≈ 22.8 thousand or 22,800 units

(d) September 1997 $\Rightarrow t = 21$

$S = 23.1 + 0.442(21) + 4.3 \sin \frac{21\pi}{6}$

≈ 28.1 thousand or 28,100 units

84. As θ increases from 0° to 90°, x decreases from 12 cm to 0 cm and y increases from 0 cm to 12 cm. Therefore, $\sin \theta = y/12$ increases from 0 to 1 and $\cos \theta = x/12$ decreases from 1 to 0. Thus, $\tan \theta = y/x$ increases without bound, and when $\theta = 90°$ the tangent is undefined.

86. Selecting the point $(-1.0, 0.8)$ on the terminal side, you obtain $\tan \theta \approx -0.8$.

88. $y = 3^{-x/2}$

x	-4	-2	0	2	4
y	9	3	1	$\frac{1}{3}$	$\frac{1}{9}$

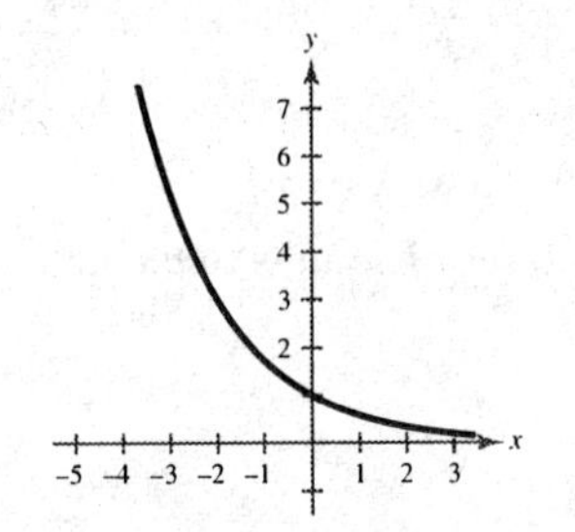

90. $y = \ln x^4$

x	± 6	± 4	± 2	± 1	0
y	7.17	5.55	2.77	0	$-\infty$

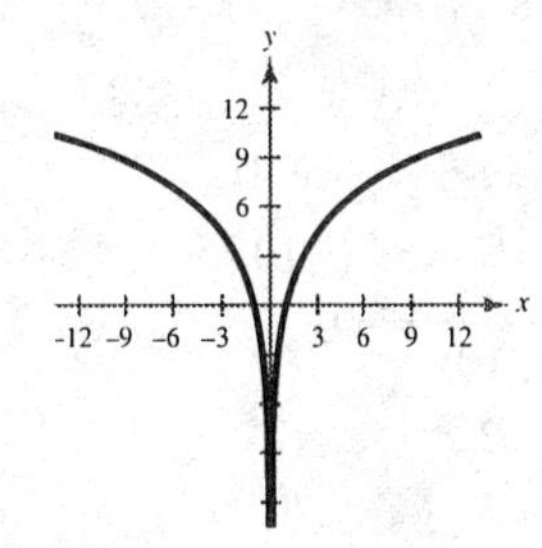

Section 4.5 Graphs of Sine and Cosine Functions

Solutions to Even-Numbered Exercises

2. $y = 2 \cos 3x$

Period $= \dfrac{2\pi}{b} = \dfrac{2\pi}{3}$

Amplitude $= |a| = 2$

4. $y = -3 \sin \dfrac{x}{3}$

Period $= \dfrac{2\pi}{b} = \dfrac{2\pi}{(1/3)} = 6\pi$

Amplitude $= |a| = |-3| = 3$

6. $y = \dfrac{3}{2} \cos \dfrac{\pi x}{2}$

Period $= \dfrac{2\pi}{b} = \dfrac{2\pi}{(\pi/2)} = 4$

Amplitude $= |a| = \dfrac{3}{2}$

8. $y = -\cos \dfrac{2x}{3}$

Period $= \dfrac{2\pi}{b} = \dfrac{2\pi}{(2/3)} = 3\pi$

Amplitude $= |a| = |-1| = 1$

10. $y = \dfrac{1}{3} \sin 8x$

Period $= \dfrac{2\pi}{b} = \dfrac{2\pi}{8} = \dfrac{\pi}{4}$

Amplitude $= |a| = \dfrac{1}{3}$

12. $y = \dfrac{5}{2} \cos \dfrac{x}{4}$

Period $= \dfrac{2\pi}{b} = \dfrac{2\pi}{(1/4)} = 8\pi$

Amplitude $= |a| = \dfrac{5}{2}$

14. $y = \dfrac{2}{3} \cos \dfrac{\pi x}{10}$

Period $= \dfrac{2\pi}{b} = \dfrac{2\pi}{(\pi/10)} = 20$

Amplitude $= |a| = \dfrac{2}{3}$

16. $f(x) = \cos x$, $g(x) = \cos(x + \pi)$

g is a horizontal shift of f π units to the left.

18. $f(x) = \sin 3x$, $g(x) = \sin(-3x)$

g is a reflection of f about the y-axis.

20. $f(x) = \sin x$, $g(x) = \sin 3x$

The period of g is one-third the period of f.

22. $f(x) = \cos 4x$, $g(x) = -2 + \cos 4x$

g is a vertical shift of f two units downward.

24. The period of g is one-half the period of f.

26. Shift the graph of f two units upward to obtain the graph of g.

28. $y = 2 + \sin x$

$y = 3.5 + \sin x$

$y = -2 + \sin x$

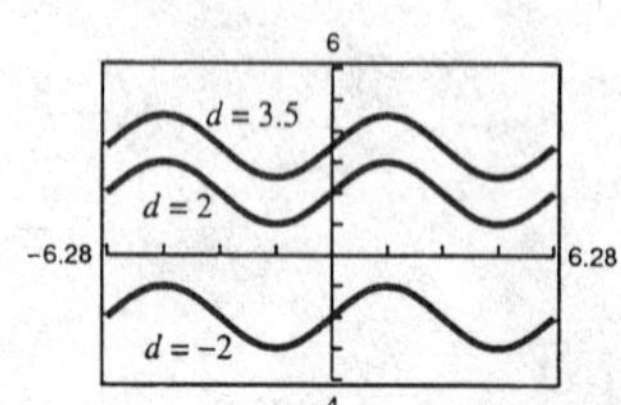

Each value of d produces a vertical shift of $y = \sin x$ upward (or downward) by d units.

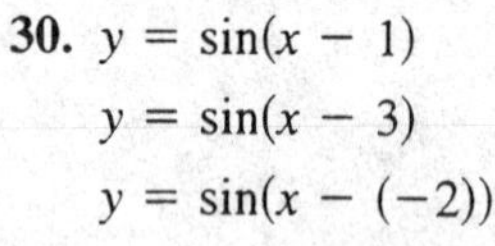

30. $y = \sin(x - 1)$

$y = \sin(x - 3)$

$y = \sin(x - (-2))$

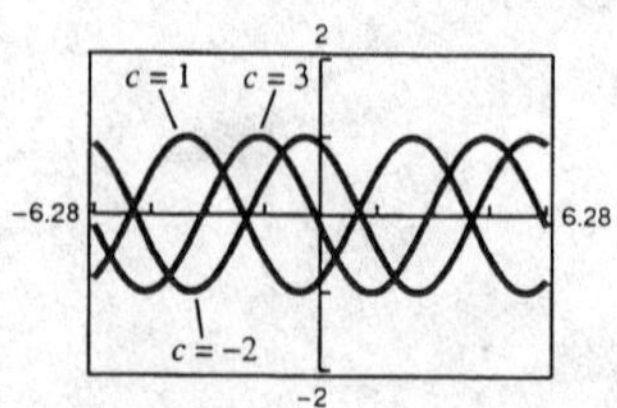

Each value of c produces a horizontal shift of $y = \sin x$ to the left (or right) by c units.

32. $f(x) = \sin x,\ g(x) = \sin \dfrac{x}{3}$

x	0	$\dfrac{\pi}{2}$	π	$\dfrac{3\pi}{2}$	2π
$\sin x$	0	1	0	-1	0
$\sin \dfrac{x}{3}$	0	$\dfrac{1}{2}$	$\dfrac{\sqrt{3}}{2}$	1	$\dfrac{\sqrt{3}}{2}$

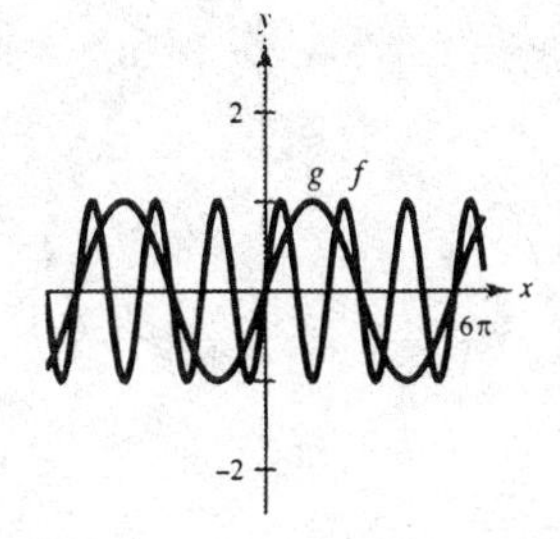

34. $f(x) = 2 \cos 2x,\ g(x) = -\cos 4x$

x	0	$\dfrac{\pi}{4}$	$\dfrac{\pi}{2}$	$\dfrac{3\pi}{4}$	π
$2 \cos 2x$	2	0	-2	0	2
$-\cos 4x$	-1	1	-1	1	-1

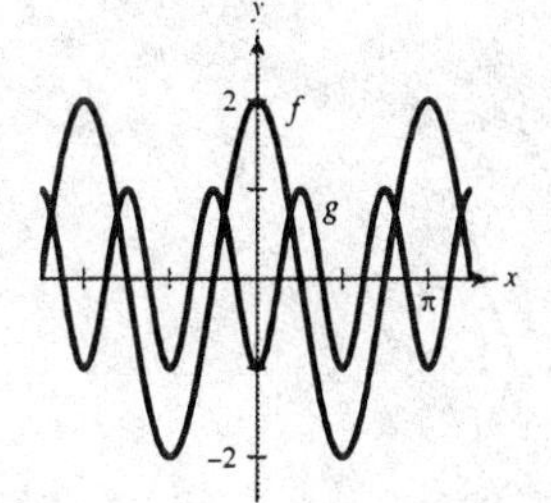

36. $f(x) = 4 \sin \pi x,\ g(x) = 4 \sin \pi x - 3$

x	0	$\dfrac{1}{2}$	1	$\dfrac{3}{2}$	2
$f(x)$	0	4	0	-4	0
$g(x)$	-3	1	-3	-7	-3

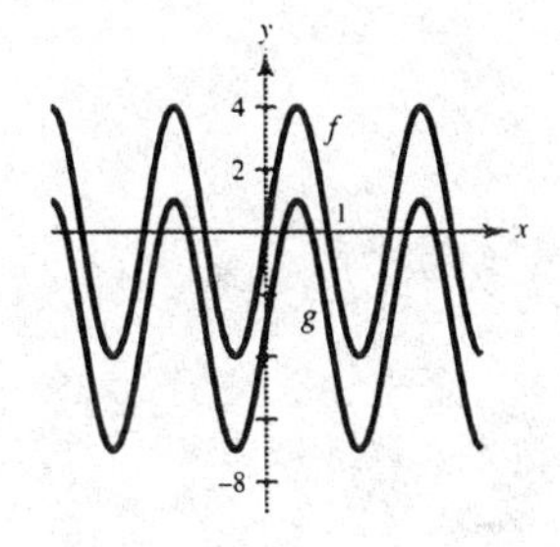

38. $f(x) = -\cos x$
$g(x) = -\cos(x - \pi)$

x	0	$\dfrac{\pi}{2}$	π	$\dfrac{3\pi}{2}$	2π
$-\cos x$	-1	0	1	0	-1
$-\cos(x - \pi)$	1	0	-1	0	1

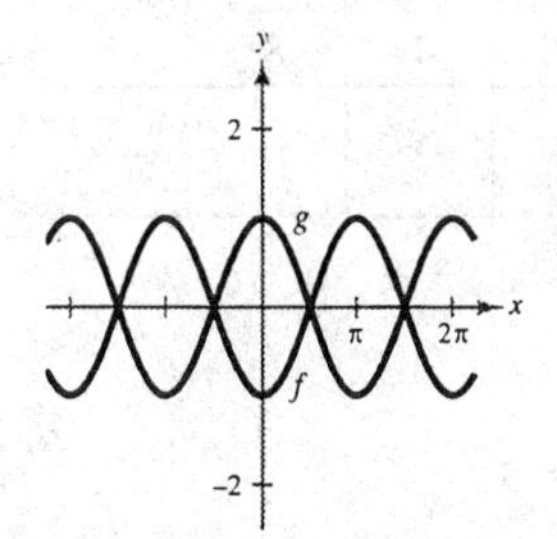

40. $f(x) = \sin x,\ g(x) = -\cos\left(x + \dfrac{\pi}{2}\right)$

x	0	$\dfrac{\pi}{2}$	π	$\dfrac{3\pi}{2}$	2π
$\sin x$	0	1	0	-1	0
$-\cos\left(x - \dfrac{\pi}{2}\right)$	0	1	0	-1	0

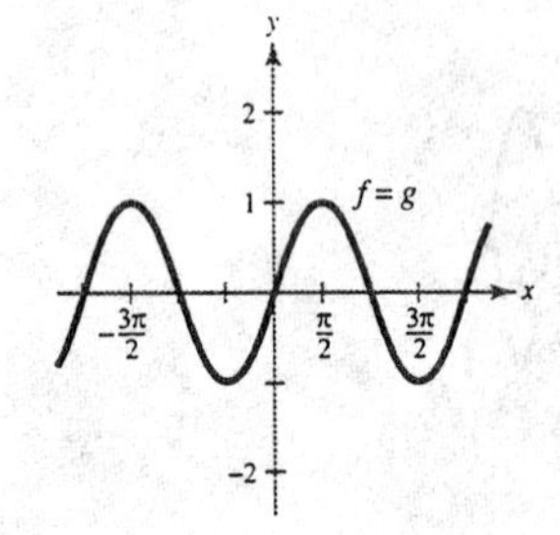

Conjecture: $\sin x = -\cos\left(x + \dfrac{\pi}{2}\right)$

42. $f(x) = \cos x,\ g(x) = -\cos(x - \pi)$

x	0	$\dfrac{\pi}{2}$	π	$\dfrac{3\pi}{2}$	2π
$\cos x$	1	0	-1	0	1
$-\cos(x - \pi)$	1	0	-1	0	1

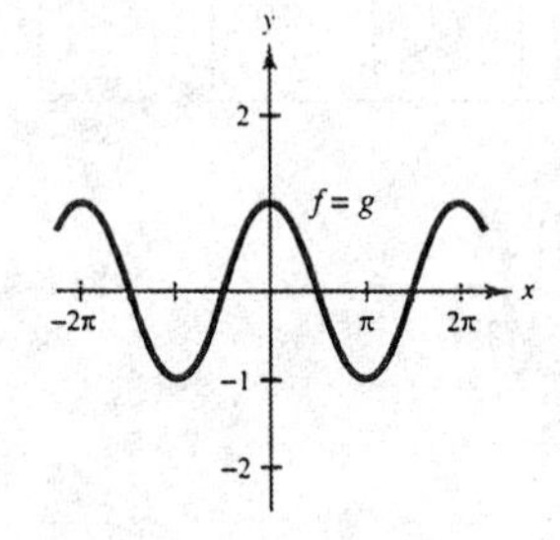

Conjecture: $\cos x = -\cos(x - \pi)$

44. $y = -3\cos 4x$

Period $= \dfrac{2\pi}{4} = \dfrac{\pi}{2}$

Amplitude $= 3$

x	0	$\dfrac{\pi}{8}$	$\dfrac{\pi}{4}$	$\dfrac{3\pi}{8}$	$\dfrac{\pi}{2}$
y	-3	0	3	0	-3

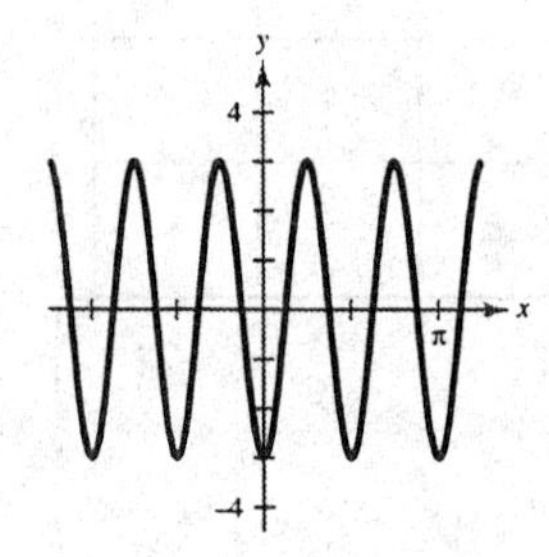

46. $y = \dfrac{3}{2}\sin\dfrac{\pi x}{4}$

Period $= \dfrac{2\pi}{(\pi/4)} = 8$

Amplitude $= \dfrac{3}{2}$

x	0	2	4	6	8
y	0	$\dfrac{3}{2}$	0	$-\dfrac{3}{2}$	0

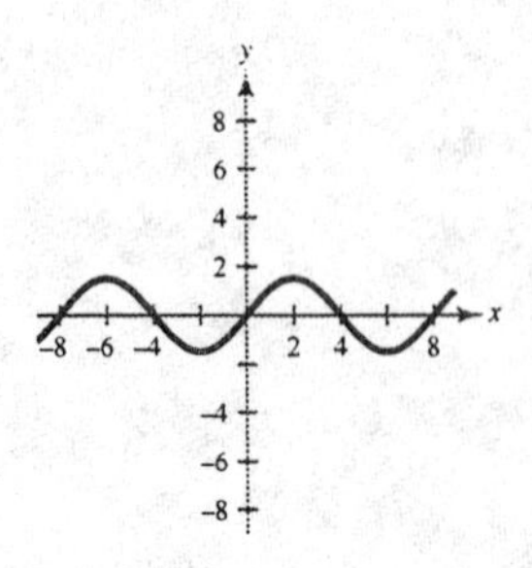

48. $y = 10\cos\dfrac{\pi x}{6}$

Period $= \dfrac{2\pi}{(\pi/6)} = 12$

Amplitude $= 10$

x	0	3	6	9	12
y	10	0	-10	0	10

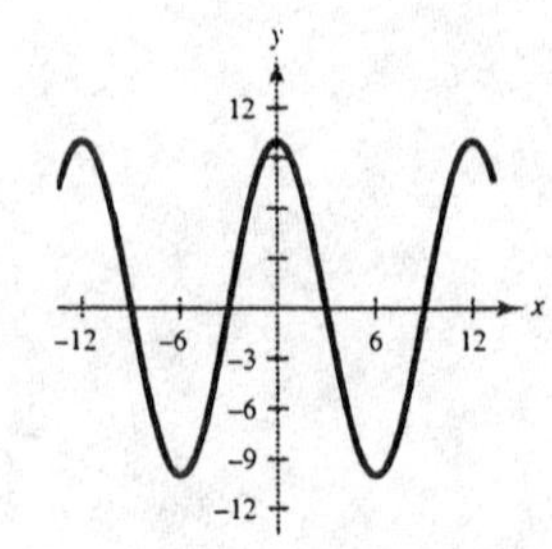

50. $y = \frac{1}{2}\sin(x - \pi)$

Period $= 2\pi$

Amplitude $= \frac{1}{2}$

x	0	$\frac{\pi}{2}$	π	$\frac{3\pi}{2}$	2π
y	0	$-\frac{3}{2}$	0	$\frac{3}{2}$	0

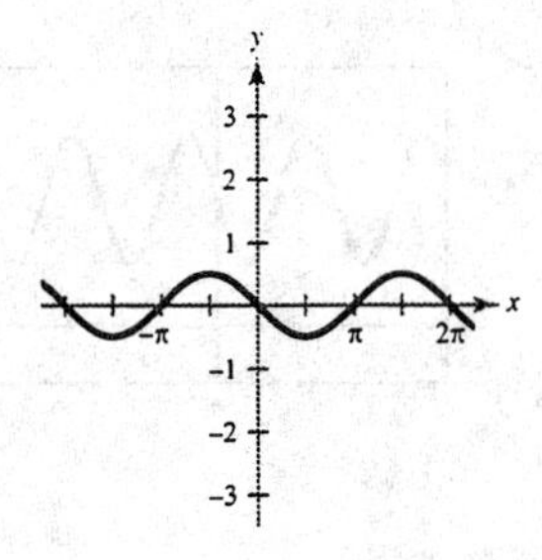

52. $y = 4\cos\left(x + \frac{\pi}{4}\right)$

Period $= 2\pi$

Amplitude $= 4$

x	0	$\frac{\pi}{2}$	π	$\frac{3\pi}{2}$	2π
y	$2\sqrt{2}$	$-2\sqrt{2}$	$-2\sqrt{2}$	$2\sqrt{2}$	$2\sqrt{2}$

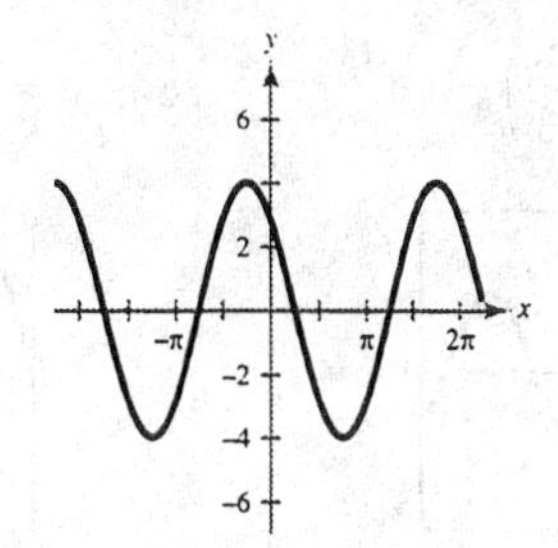

54. $y = -3 + 5\cos\frac{\pi t}{12}$

Period $= \frac{2\pi}{(\pi/12)} = 24$

Amplitude $= 5$

x	0	6	12	18	24
y	2	-3	-8	-3	2

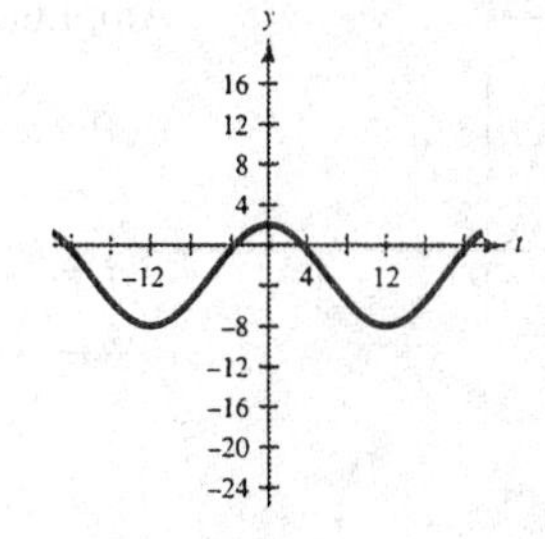

56. $y = 2\cos x - 3$

Period $= 2\pi$

Amplitude $= 2$

x	0	$\frac{\pi}{2}$	π	$\frac{3\pi}{2}$	2π
y	-1	-3	-5	-3	-1

58. $y = 4\cos\left(x + \frac{\pi}{4}\right) + 4$

Period $= 2\pi$

Amplitude $= 4$

x	$-\frac{\pi}{4}$	$\frac{\pi}{4}$	$\frac{3\pi}{4}$	$\frac{5\pi}{4}$	$\frac{7\pi}{4}$
y	8	4	0	4	8

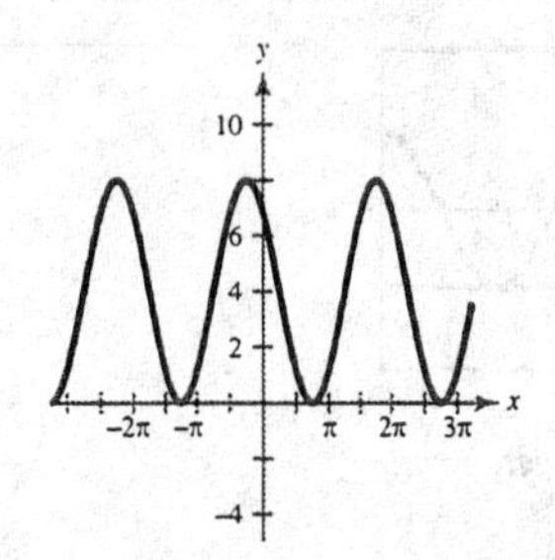

60. $y = -3\cos(6x + \pi)$

Period $= \dfrac{2\pi}{6} = \dfrac{\pi}{3}$

Amplitude $= 3$

x	0	$\frac{\pi}{12}$	$\frac{\pi}{6}$	$\frac{3\pi}{12}$	$\frac{\pi}{3}$
y	3	0	-3	0	3

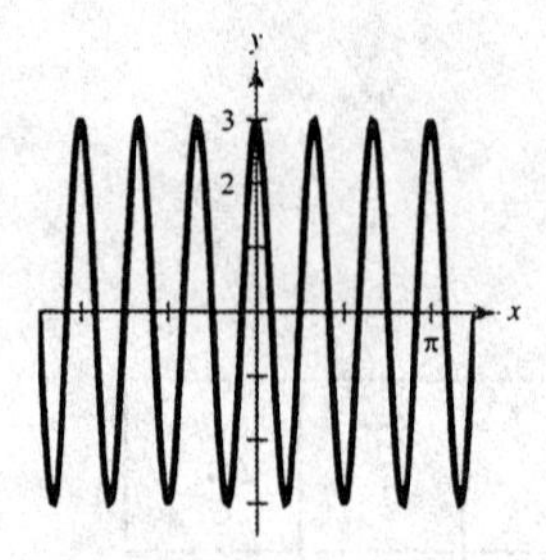

62. $y = -4\sin\left(\dfrac{2}{3}x - \dfrac{\pi}{3}\right)$

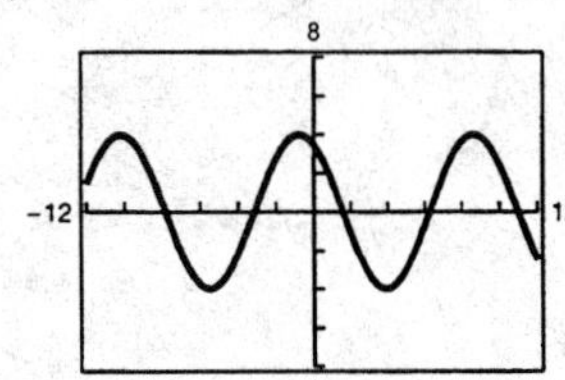

64. $y = 3\cos\left(\dfrac{\pi x}{2} + \dfrac{\pi}{2}\right) - 2$

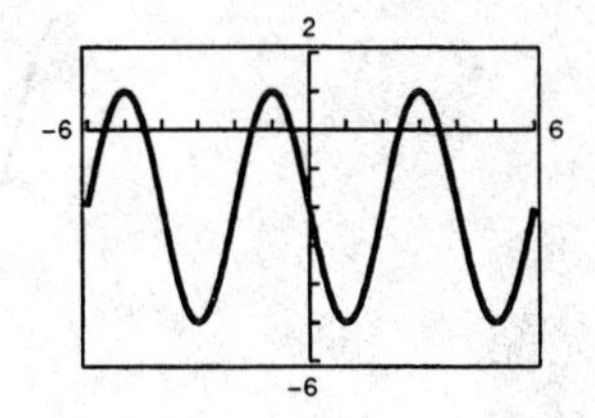

66. $y = 5\sin(\pi - 2x) + 10$

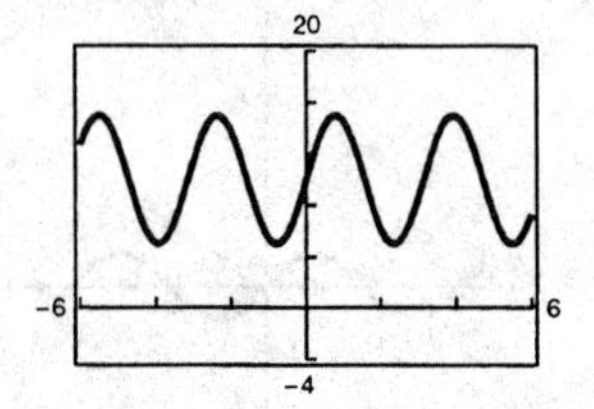

68. $y = \dfrac{1}{100}\sin 120\pi t$

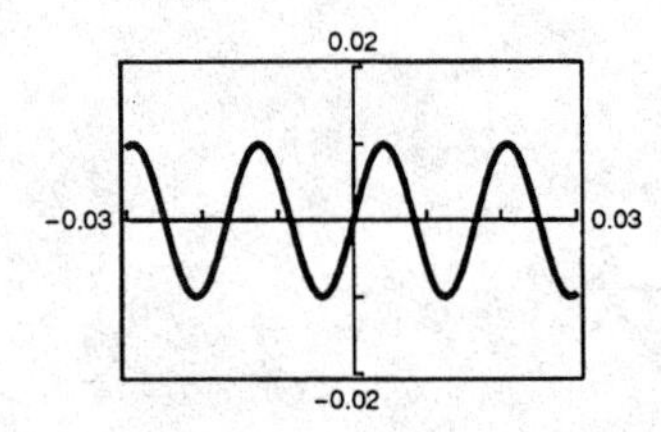

70. $f(x) = a\cos x + d$

Amplitude $= \dfrac{-2 - (-4)}{2} = 1$

Reflected in the x-axis: $a = -1$

$-4 = -1\cos 0 + d$

$d = -3$

$y = -3 - \cos x$

72. $y = a\sin(bx - c)$

Amplitude $= 2 \Longrightarrow a = 2$

Period $= 4\pi$

$\dfrac{2\pi}{b} = 4\pi \Longrightarrow b = \dfrac{1}{2}$

Phase shift: $c = 0$

$y = 2\sin\left(\dfrac{x}{2}\right)$

74. $y = a\sin(bx - c)$

Amplitude $= 2 \Longrightarrow a = 2$

Period $= 4$

$\dfrac{2\pi}{b} = 4 \Longrightarrow b = \dfrac{\pi}{2}$

Phase shift: $\dfrac{c}{b} = -1 \Longrightarrow c = -\dfrac{\pi}{2}$

$y = 2\sin\left(\dfrac{\pi x}{2} + \dfrac{\pi}{2}\right)$

76. $y_1 = \cos x$

$y_2 = -1$

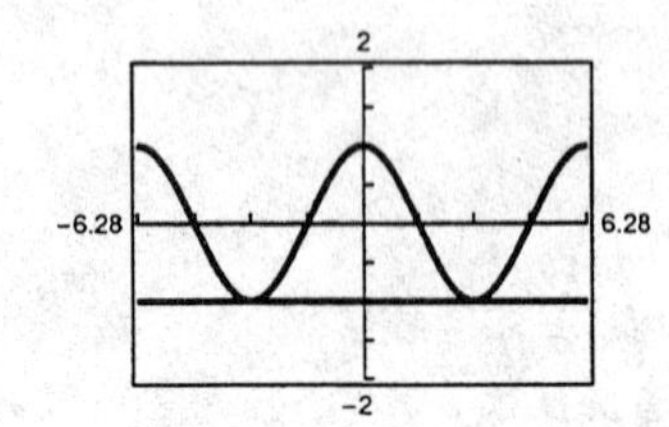

$y_1 = y_2$ when $x = \pi, -\pi$.

78. $y_1 = \sin x$

$y_2 = \dfrac{\sqrt{3}}{2}$

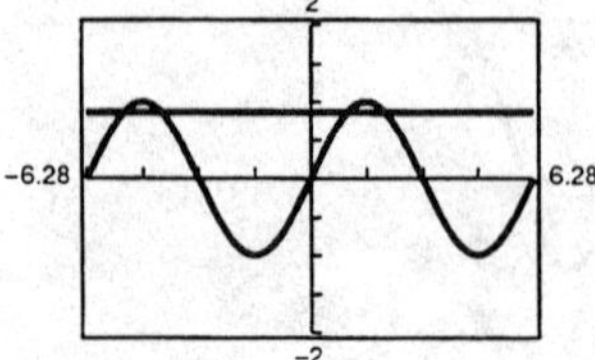

$y_1 = y_2$ when $x = \dfrac{\pi}{3}, \dfrac{2\pi}{3}, -\dfrac{4\pi}{3}, -\dfrac{5\pi}{3}$.

80. (a) In Exercise 79, $f(x) = \cos x$ is even and we saw that $h(x) = \cos^2 x$ is even. Therefore, for $f(x)$ even and $h(x) = [f(x)]^2$, we make the conjecture that $h(x)$ is even.

(b) In Exercise 79, $g(x) = \sin x$ is odd and we saw that $h(x) = \sin^2 x$ is even. Therefore, for $g(x)$ odd and $h(x) = [g(x)]^2$, we make the conjecture that $h(x)$ is even.

(c) From part (c) of 79, we conjecture that the product of an even function and an odd function is odd.

82. The period of the model would change because the time for a respiratory cycle would decrease.

84. (a) $\sin \frac{1}{2} \approx \frac{1}{2} - \frac{(1/2)^3}{3!} + \frac{(1/2)^5}{5!} \approx 0.4794$

$\sin \frac{1}{2} \approx 0.4794$ (by calculator)

(b) $\sin 1 \approx 1 - \frac{1}{3!} + \frac{1}{5!} \approx 0.8417$

$\sin 1 \approx 0.8415$ (by calculator)

(c) $\sin \frac{\pi}{6} \approx 1 - \frac{(\pi/6)^3}{3!} + \frac{(\pi/6)^5}{5!} \approx 0.5000$

$\sin \frac{\pi}{6} = 0.5$ (by calculator)

(d) $\cos(-0.5) \approx 1 - \frac{(-0.5)^2}{2!} + \frac{(-0.5)^4}{4!} \approx 0.8776$

$\cos(-0.5) \approx 0.8776$ (by calculator)

(e) $\cos 1 \approx 1 - \frac{1}{2!} + \frac{1}{4!} \approx 0.5417$

$\cos 1 \approx 0.5403$ (by calculator)

(f) $\cos \frac{\pi}{4} \approx 1 - \frac{(\pi/4)^2}{2!} + \frac{(\pi/4)^4}{4!} = 0.7074$

$\cos \frac{\pi}{4} \approx 0.7071$

The error in the approximation is not the same in each case. The error appears to increase as x moves farther away from 0.

86. $S = 74.50 + 43.75 \sin \frac{\pi t}{6}$

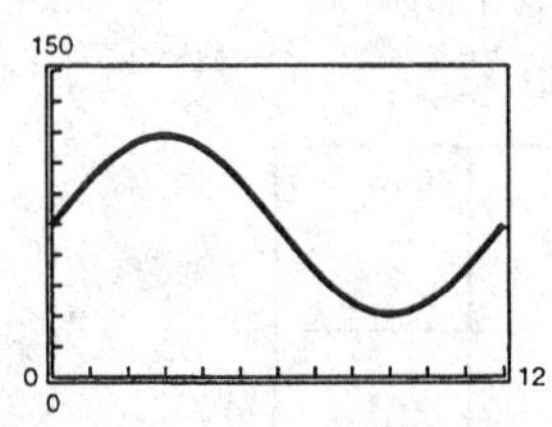

Maximum sales: March ($t = 3$)

Minimum sales: September ($t = 9$)

88. (a) Yes, y is a function of t because for each value of t there corresponds one and only one value of y.

(b) The period is approximately $2(0.375 - 0.125) = 0.5$ seconds.

The amplitude is approximately $\frac{1}{2}(2.35 - 1.65) = 0.35$ centimeters.

(c) One model is $y = 0.35 \sin 4\pi t + 2$.

(d)

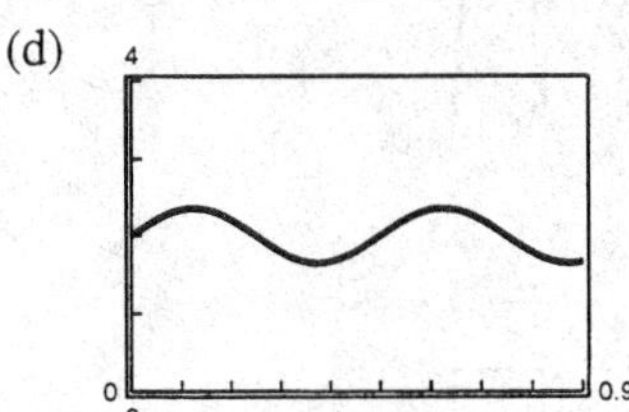

90. $f(x) = 1 - \frac{1}{2}x^2$ is a parabola opening downward.

$g(x) = \cos x$ is periodic.

Section 4.6 Graphs of Other Trigonometric Functions

Solutions to Even-Numbered Exercises

2. $y = \tan \dfrac{x}{2}$

Period $= \dfrac{\pi}{b} = \dfrac{\pi}{(1/2)} = 2\pi$

Asymptotes: $x = -\pi, x = \pi$

Matches graph (d).

4. $y = 2 \csc x$

Period $= \dfrac{2\pi}{b} = \dfrac{2\pi}{1} = 2\pi$

Asymptotes: $x = 0, x = \pi$

Matches graph (a).

6. $y = \dfrac{1}{2} \sec \dfrac{\pi x}{2}$

Period $= \dfrac{2\pi}{b} = \dfrac{2\pi}{(\pi/2)} = 4$

Asymptotes: $x = -1, x = 1$

Matches graph (h).

8. $y = -2 \sec 2\pi x$

Period $= \dfrac{2\pi}{2\pi} = 1$

Asymptotes: $x = -\dfrac{1}{4}, x = \dfrac{1}{4}$

Reflected in x-axis

Matches graph (c).

10. $y = \dfrac{1}{4} \tan x$

Period $= \pi$

Asymptotes: $x = -\dfrac{\pi}{2}, x = \dfrac{\pi}{2}$

x	$-\frac{\pi}{4}$	0	$\frac{\pi}{4}$
y	$-\frac{1}{4}$	0	$\frac{1}{4}$

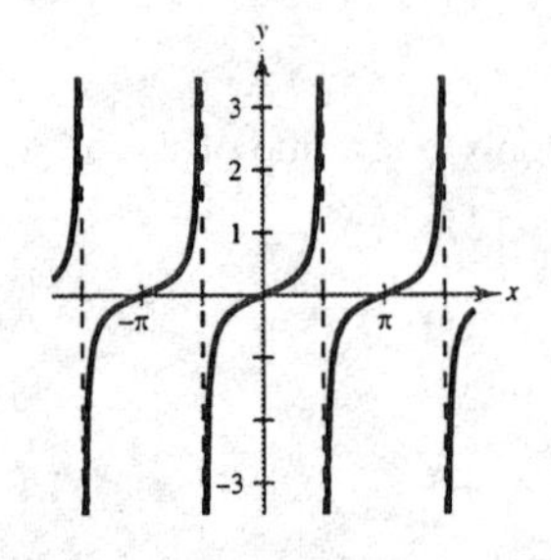

12. $y = -3 \tan \pi x$

Period $= \dfrac{\pi}{\pi} = 1$

Asymptotes: $x = -\dfrac{1}{2}, x = \dfrac{1}{2}$

x	$-\frac{1}{4}$	0	$\frac{1}{4}$
y	3	0	−3

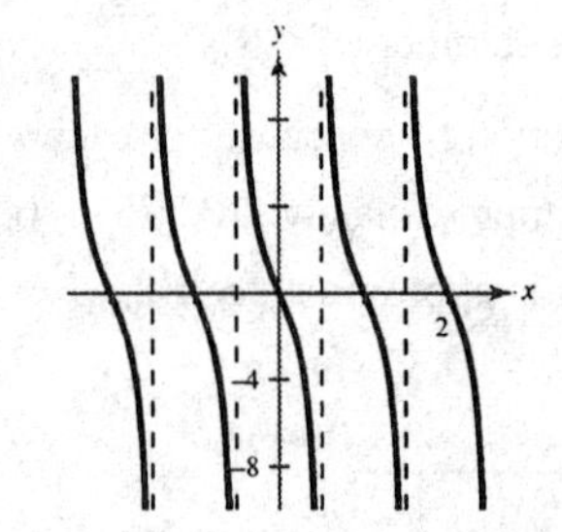

14. $y = \frac{1}{4}\sec x$

Period $= 2\pi$

Asymptotes: $x = -\frac{\pi}{2}, x = \frac{\pi}{2}$

x	$-\frac{\pi}{4}$	0	$\frac{\pi}{4}$
y	0.354	$\frac{1}{4}$	0.354

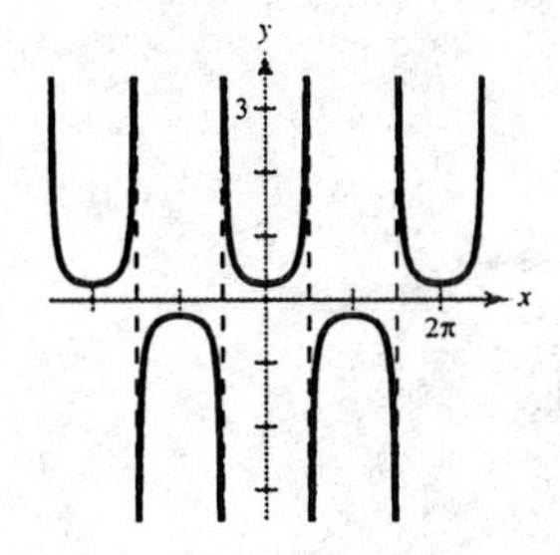

16. $y = 2\sec 4x$

Period $= \frac{2\pi}{4} = \frac{\pi}{2}$

Asymptotes: $x = -\frac{\pi}{8}, x = \frac{\pi}{8}$

x	$-\frac{\pi}{16}$	0	$\frac{\pi}{16}$
y	2.828	2	2.828

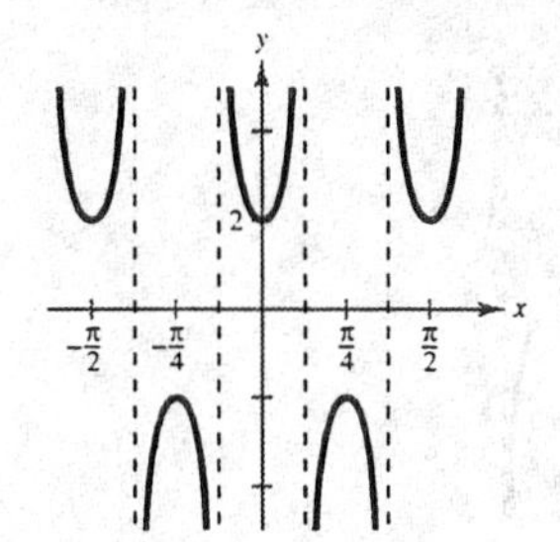

18. $y = -2\sec 4x + 2$

Period $= \frac{2\pi}{4} = \frac{\pi}{2}$

Asymptotes: $x = -\frac{\pi}{8}, x = \frac{\pi}{8}$

x	$-\frac{\pi}{16}$	0	$\frac{\pi}{16}$
y	-0.828	0	-0.828

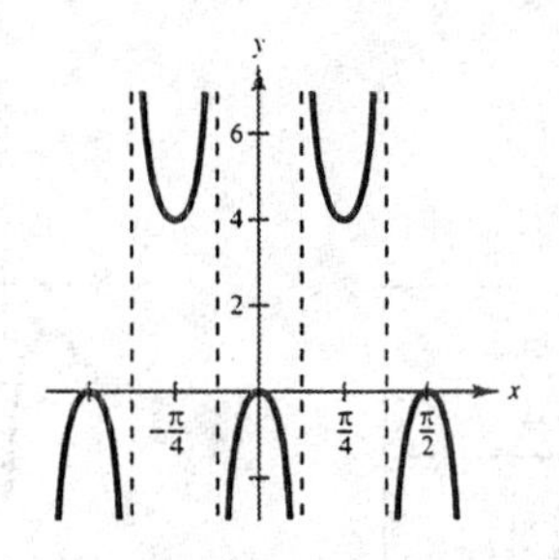

20. $y = \csc \frac{x}{3}$

Period $= \frac{2\pi}{(1/3)} = 6\pi$

Asymptotes: $x = 0, x = 3\pi$

x	π	2π	4π
y	1.155	1.155	-1.155

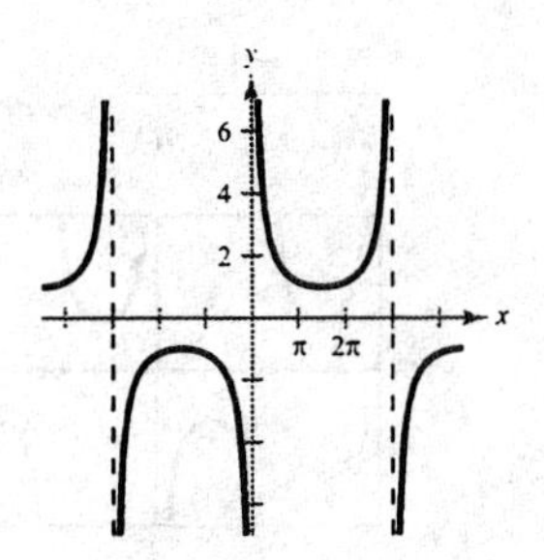

22. $y = 3\cot \frac{\pi x}{2}$

Period $= \frac{\pi}{(\pi/2)} = 2$

Asymptotes: $x = 0, x = 2$

x	$\frac{1}{4}$	1	$\frac{3}{2}$
y	7.243	0	-3

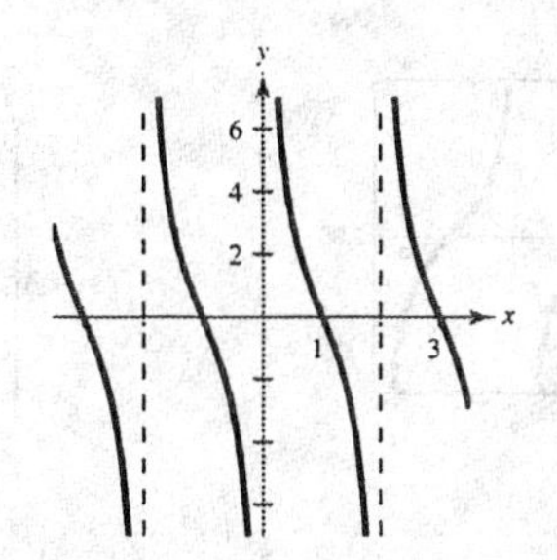

24. $y = -\dfrac{1}{2}\tan x$

Period $= \pi$

Asymptotes: $x = -\dfrac{\pi}{2}, x = \dfrac{\pi}{2}$

x	$-\dfrac{\pi}{4}$	0	$\dfrac{\pi}{4}$
y	$\dfrac{1}{2}$	0	$-\dfrac{1}{2}$

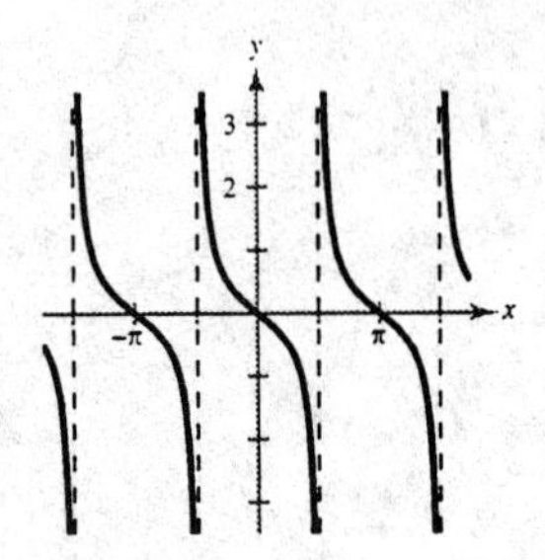

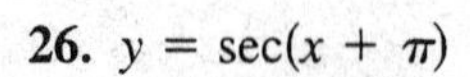

26. $y = \sec(x + \pi)$

Period $= 2\pi$

Asymptotes: $x = -\dfrac{\pi}{2}, x = \dfrac{\pi}{2}$

x	$-\dfrac{\pi}{4}$	0	$\dfrac{\pi}{4}$
y	-1.414	-1	-1.414

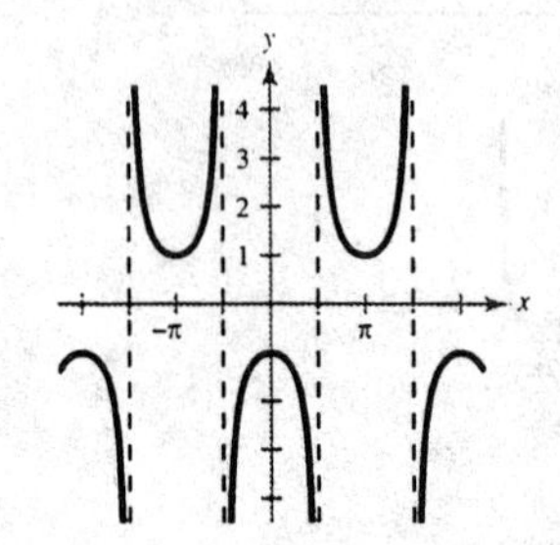

28. $y = \sec(\pi - x)$

Period $= 2\pi$

Asymptotes: $x = -\dfrac{\pi}{2}, x = \dfrac{\pi}{2}$

x	$-\dfrac{\pi}{4}$	0	$\dfrac{\pi}{4}$
y	-1.414	-1	-1.414

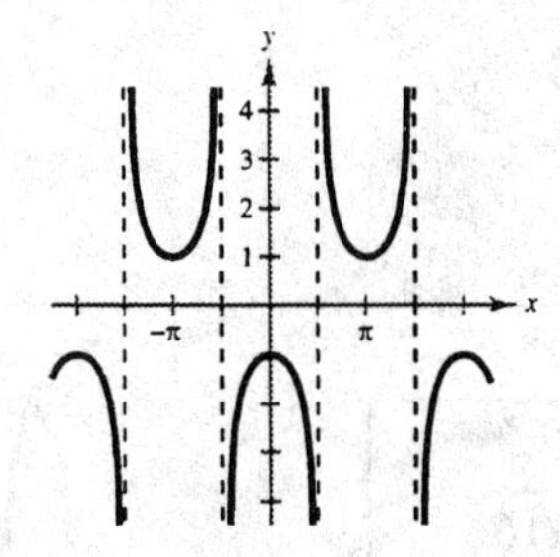

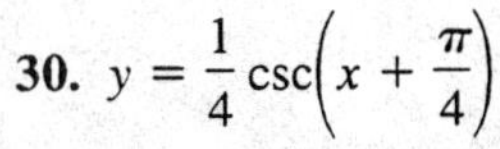

30. $y = \dfrac{1}{4}\csc\left(x + \dfrac{\pi}{4}\right)$

Period: 2π

Asymptotes: $x = -\dfrac{\pi}{4}, \dfrac{3\pi}{4}$

x	0	$\dfrac{\pi}{4}$	$\dfrac{\pi}{2}$
y	$\dfrac{\sqrt{2}}{4}$	$\dfrac{1}{4}$	$\dfrac{\sqrt{2}}{4}$

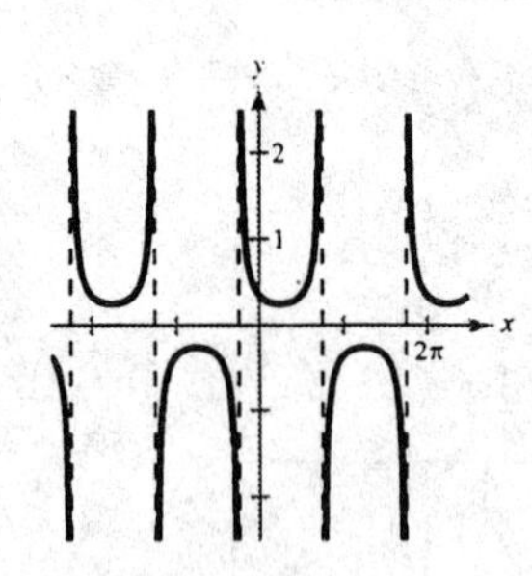

32. $y = -\tan 2x$

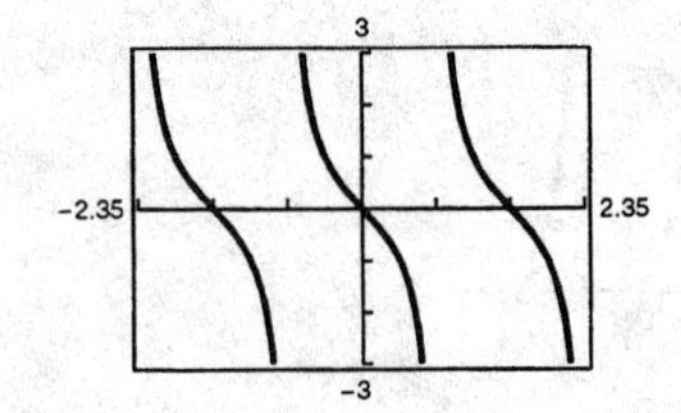

34. $y = \sec \pi x \Longrightarrow y = \dfrac{1}{\cos(\pi x)}$

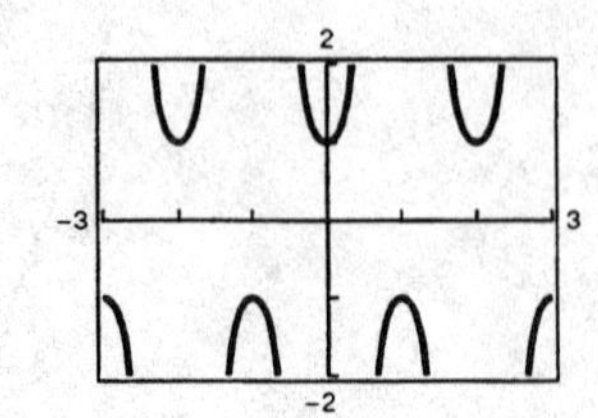

36. $y = -\csc(4x - \pi)$

$y = \dfrac{-1}{\sin(4x - \pi)}$

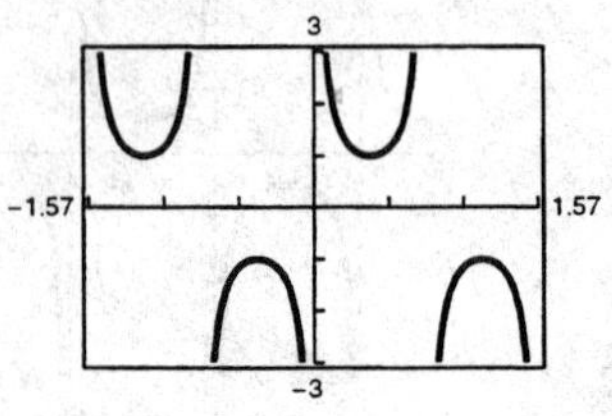

38. $y = 0.1\tan\left(\dfrac{\pi x}{4} + \dfrac{\pi}{4}\right)$

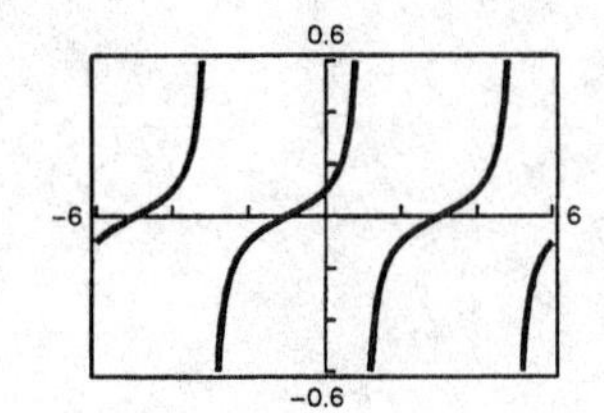

40. $y = \dfrac{1}{3}\sec\left(\dfrac{\pi x}{2} + \dfrac{\pi}{2}\right) \Rightarrow y = \dfrac{1}{3\cos\left(\dfrac{\pi x}{2} + \dfrac{\pi}{2}\right)}$

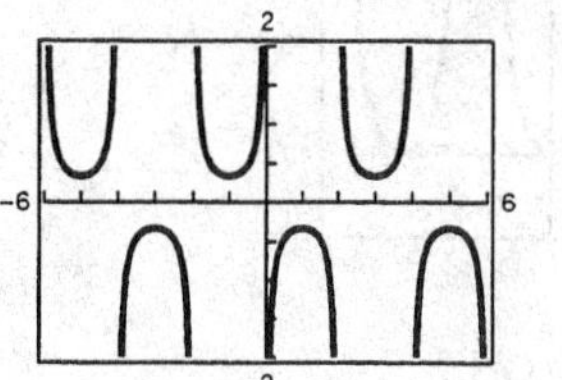

42.

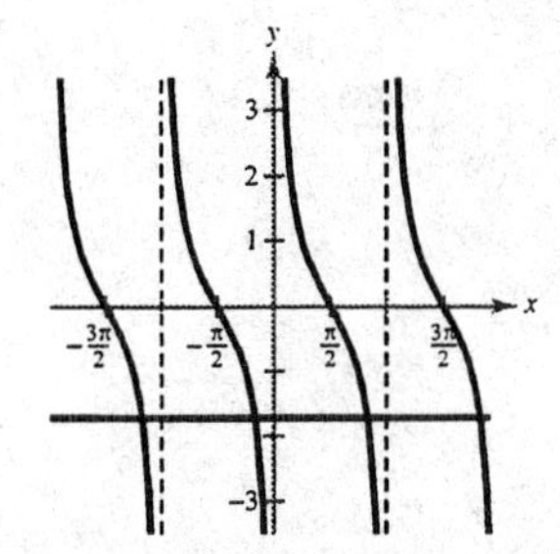

The solutions appear to be:

$x = -\dfrac{7\pi}{6}, -\dfrac{\pi}{6}, \dfrac{5\pi}{6}, \dfrac{11\pi}{6}$

(or in decimal form: $-3.665, -0.524, 2.618, 5.760$)

44.

The solutions appear to be:

$-\dfrac{7\pi}{4}, -\dfrac{5\pi}{4}, \dfrac{\pi}{4}, \dfrac{3\pi}{4}$

(or in decimal form: $-5.498, -3.927, 0.785, 2.356$)

46. $f(x) = \tan x$

$\tan(-x) = -\tan x$

Thus, the function is odd and the graph of $y = \tan x$ is symmetric with the origin.

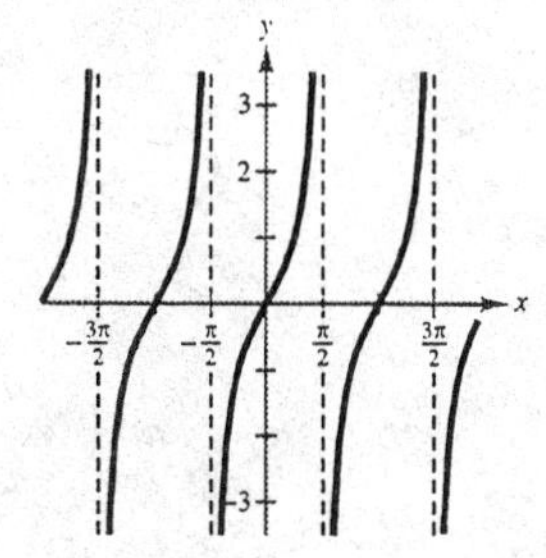

48. For $f(x) = \csc x$, as x approaches π from the left, f approaches ∞. As x approaches π from the right, f approaches $-\infty$.

50. $f(x) = \tan\dfrac{\pi x}{2}$, $g(x) = \dfrac{1}{2}\sec\dfrac{\pi x}{2}$

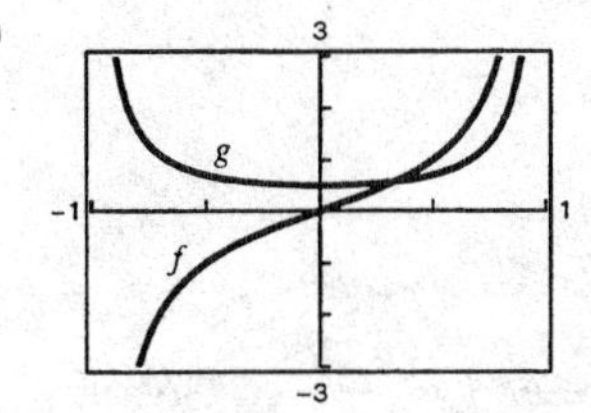

(b) The interval in which $f < g$ is $(-1, 1/3)$.

(c) The interval in which $2f < 2g$ is $(-1, 1/3)$, which is the same interval as part (b).

52. $y_1 = \sin x \sec x$, $y_2 = \tan x$

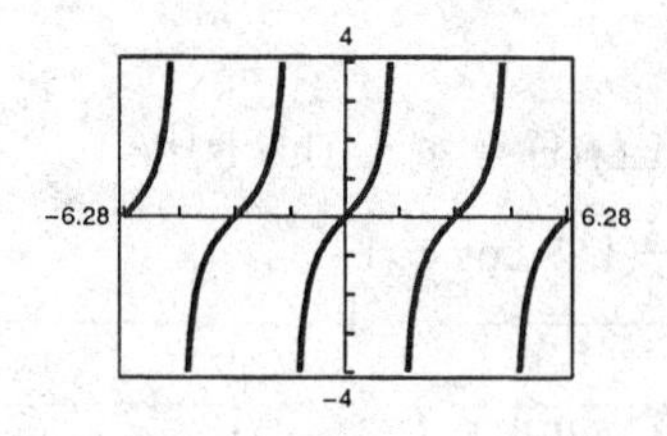

It appears that $y_1 = y_2$.

$\sin x \sec x = \sin x \dfrac{1}{\cos x} = \dfrac{\sin x}{\cos x} = \tan x$

54. $y_1 = \sec^2 x - 1,\ y_2 = \tan^2 x$

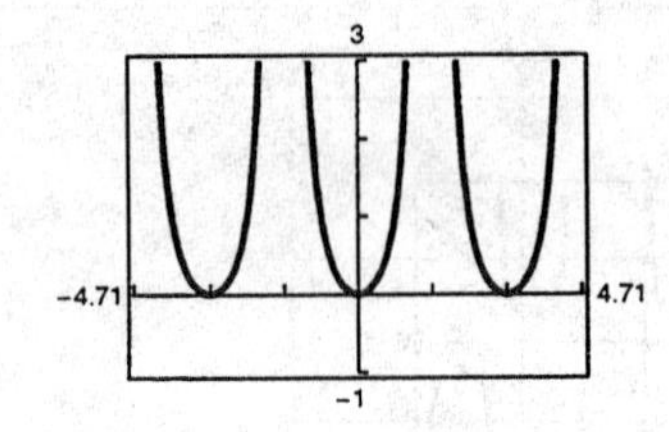

It appears that $y_1 = y_2$.

$1 + \tan^2 x = \sec^2 x$

$\tan^2 x = \sec^2 x - 1$

56. $f(x) = |x \sin x|$

Matches graph (a) as $x \longrightarrow 0, f(x) \longrightarrow 0$.

58. $g(x) = |x| \cos x$

Matches graph (c) as $x \longrightarrow 0, g(x) \longrightarrow 0$.

60. $f(x) = \sin x - \cos\left(x + \dfrac{\pi}{2}\right)$

$g(x) = 2 \sin x$

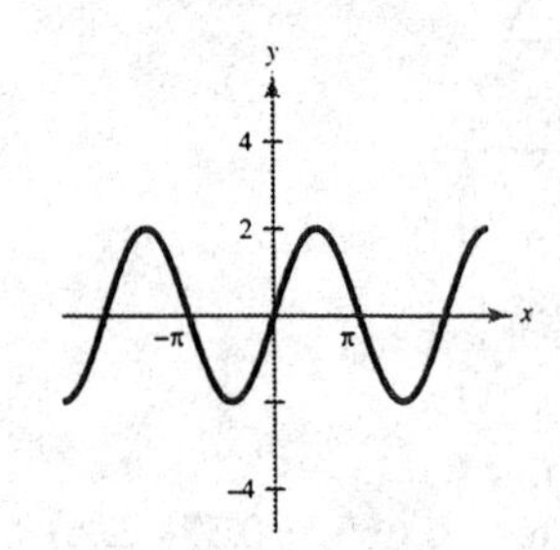

It appears that $f(x) = g(x)$. That is, that

$\sin x - \cos\left(x + \dfrac{\pi}{2}\right) = 2 \sin x$.

62. $f(x) = \cos^2 \dfrac{\pi x}{2}$

$g(x) = \dfrac{1}{2}(1 + \cos \pi x)$

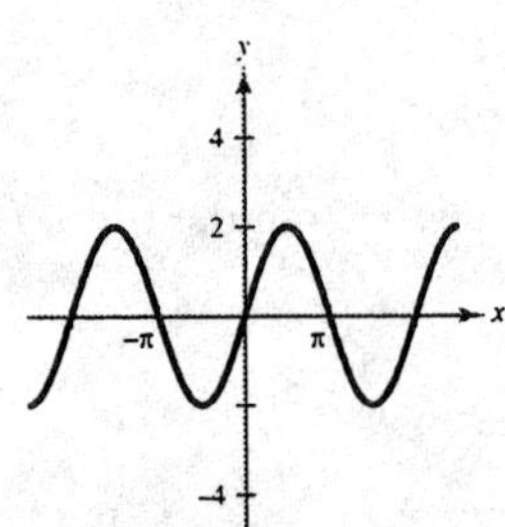

It appears that $f(x) = g(x)$. That is, that

$\cos^2 \dfrac{\pi x}{2} = \dfrac{1}{2}(1 + \cos \pi x)$.

64. $f(x) = e^{-x} \cos x$

Damping factor: e^{-x}

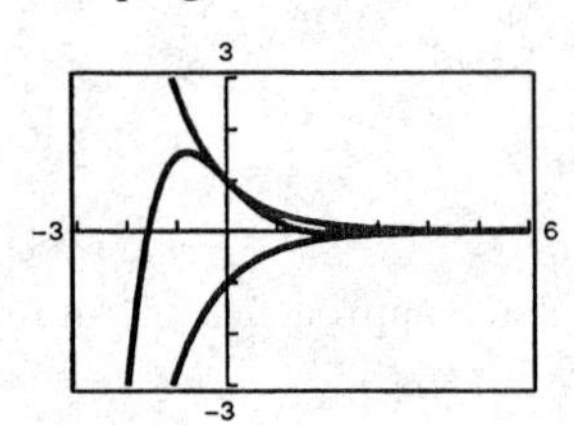

As $x \longrightarrow \infty, f(x) \longrightarrow 0$.

66. $h(x) = 2^{-x^2/4} \sin x$

Damping factor: $2^{-x^2/4}$

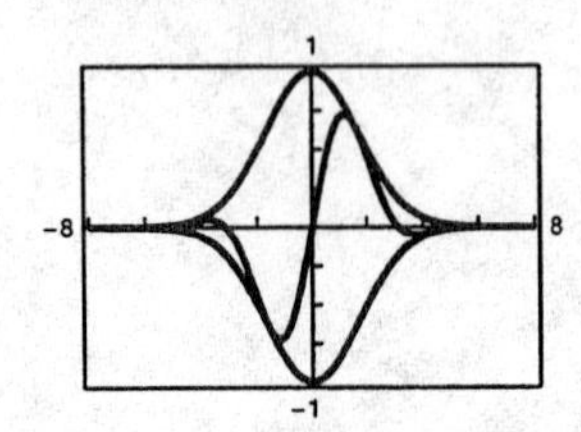

As $x \longrightarrow \infty, h(x) \longrightarrow 0$.

68. $\cos x = \dfrac{36}{d} \Longrightarrow d = \dfrac{36}{\cos x} = 36 \sec x$

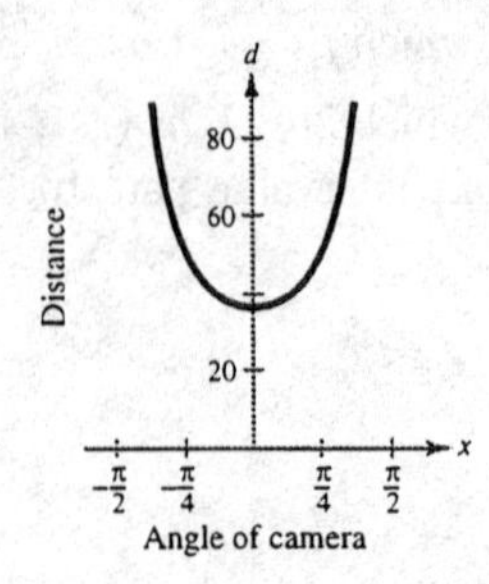

70. $H(t) = 54.33 - 20.38\cos\dfrac{\pi t}{6} - 15.69\sin\dfrac{\pi t}{6}$

$L(t) = 39.36 - 15.70\cos\dfrac{\pi t}{6} - 14.16\sin\dfrac{\pi t}{6}$

(a)

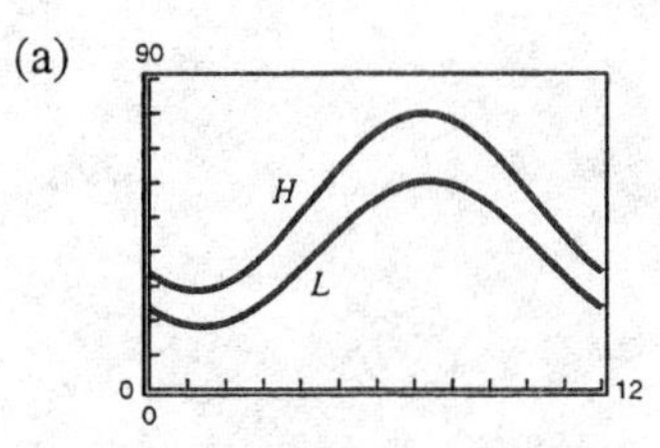

Period of $\cos\dfrac{\pi t}{6}$: $\dfrac{2\pi}{(\pi/6)} = 12$

Period of $\sin\dfrac{\pi t}{6}$: $\dfrac{2\pi}{(\pi/6)} = 12$

Period of $H(t)$: 12

Period of $L(t)$: 12

(b) From the graph, it appears that the greatest difference between high and low temperatures occurs in summer. The smallest difference occurs in winter.

(c) The highest high and low temperatures appear to occur around the middle of July, roughly one month after the time when the sun is northernmost in the sky.

72. $f(x) = x - \cos x$

(a)

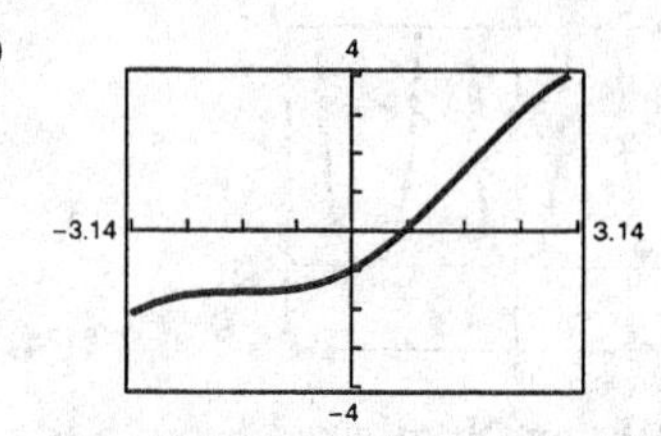

The zero between 0 and 1 appears to occur at $x \approx 0.739$.

(b) $x_n = \cos(x_{n-1})$

$x_0 = 1$

$x_1 = \cos 1 \approx 0.5403$

$x_2 = \cos 0.5403 \approx 0.8576$

$x_3 = \cos 0.8576 \approx 0.6543$

$x_4 = \cos 0.6543 \approx 0.7935$

$x_5 = \cos 0.7935 \approx 0.7014$

$x_6 = \cos 0.7014 \approx 0.7640$

$x_7 = \cos 0.7640 \approx 0.7221$

$x_8 = \cos 0.7221 \approx 0.7504$

$x_9 = \cos 0.7504 \approx 0.7314$

$\vdots$

This sequence appears to be approaching the zero $x = 0.739$.

74. $y_1 = \sec x$

$y_2 = 1 + \dfrac{x^2}{2!} + \dfrac{5x^4}{4!}$

The approximation appears to be good for roughly $[-1.1, 1.1]$.

76. $S = 74 + 3x + 40\sin\dfrac{\pi t}{6}$

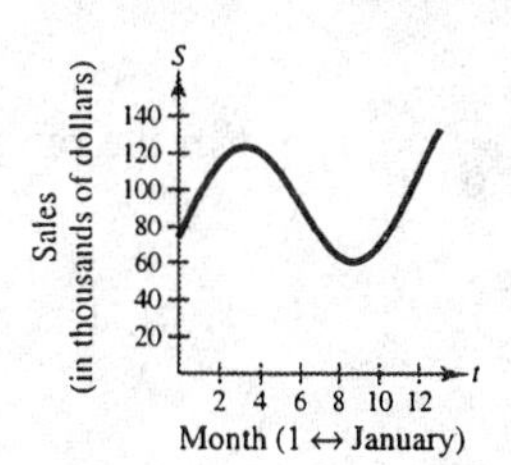

78.

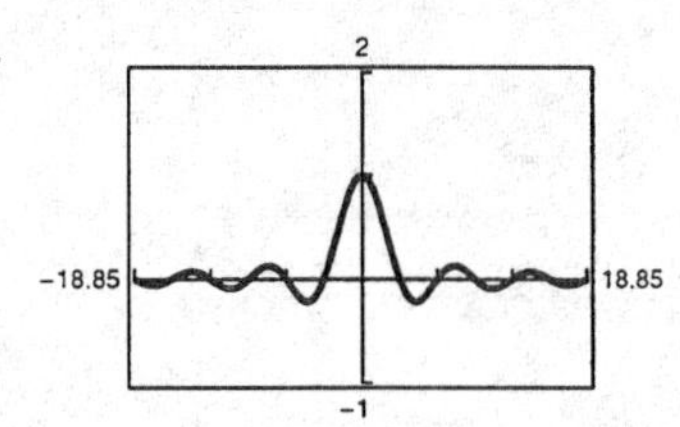

As x tends to 0, $\dfrac{\sin x}{x}$ approaches 1.

80. (a) If a spring of less stiffness is used, then c will be less than 8, $0 < c < 8$.

(b) If the effect of friction is decreased, then b will be less than 0.2231, $0 < b < 0.2231$.

82. $e^{2x} = 54$

$2x = \ln 54$

$x = \frac{1}{2}\ln 54 \approx 1.994$

84. $\ln(x^2 + 1) = 3.2$

$x^2 + 1 = e^{3.2}$

$x^2 = e^{3.2} - 1$

$x = \pm\sqrt{e^{3.2} - 1} \approx \pm 4.851$

Section 4.7 Inverse Trigonometric Functions

Solutions to Even-Numbered Exercises

2. $y = \arccos x$

(a)

x	-1	-0.8	-0.6	-0.4	-0.2
y	3.1416	2.4981	2.2143	1.9823	1.7722

x	0	0.2	0.4	0.6	0.8	1.0
y	1.5708	1.3694	1.1593	0.9273	0.6435	0

(b)

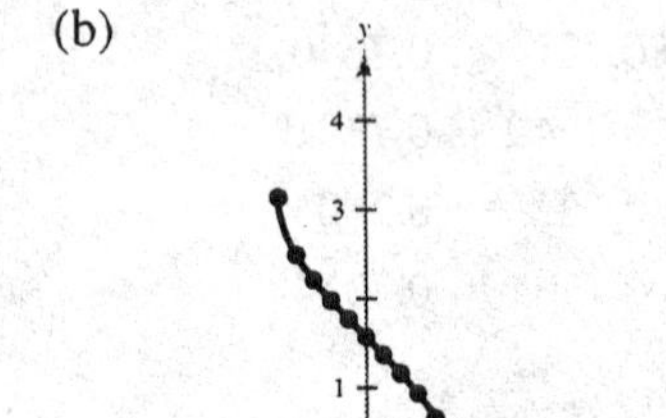

(c)

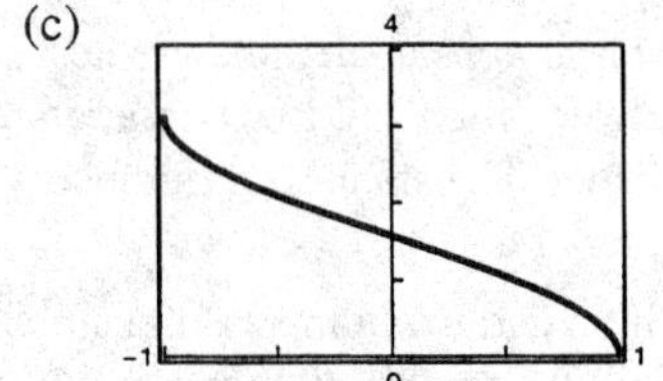

(d) Intercepts are $\left(0, \frac{\pi}{2}\right)$ and $(1, 0)$.

No symmetry

4. False; $\arcsin \frac{1}{2} = \frac{\pi}{6}$. $\frac{5\pi}{6}$ is not in the range of the arcsine function.

6. $y = \arccos x$

$x = \cos y$

$(-1, \pi), \left(-\frac{1}{2}, \frac{2\pi}{3}\right), \left(\frac{\sqrt{3}}{2}, \frac{\pi}{6}\right)$

8. (a) $y = \arccos \frac{1}{2} \Longrightarrow \cos y = \frac{1}{2}$ for $0 \le y \le \pi \Longrightarrow y = \frac{\pi}{3}$

(b) $y = \arccos 0 \Longrightarrow \cos y = 0$ for $0 \le y \le \pi \Longrightarrow y = \frac{\pi}{2}$

10. (a) $y = \arccos\left(-\frac{\sqrt{3}}{2}\right) \Longrightarrow \cos y = -\frac{\sqrt{3}}{2}$ for $0 \le y \le \pi \Longrightarrow y = \frac{5\pi}{6}$

(b) $y = \arcsin\left(-\frac{\sqrt{2}}{2}\right) \Longrightarrow \sin y = -\frac{\sqrt{2}}{2}$ for $-\frac{\pi}{2} \le y \le \frac{\pi}{2} \Longrightarrow y = -\frac{\pi}{4}$

12. (a) $y = \arccos\left(-\frac{1}{2}\right) \Longrightarrow \cos y = -\frac{1}{2}$ for $0 \le y \le \pi \Longrightarrow y = \frac{2\pi}{3}$

(b) $y = \arcsin \frac{\sqrt{2}}{2} \Longrightarrow \sin y = \frac{\sqrt{2}}{2}$ for $-\frac{\pi}{2} \le y \le \frac{\pi}{2} \Longrightarrow y = \frac{\pi}{4}$

14. (a) $y = \arctan 0 \Longrightarrow \tan y = 0 \Longrightarrow y = 0$

(b) $y = \arccos 1 \Longrightarrow \cos y = 1$ for $0 \le y \le \pi \Longrightarrow y = 0$

16. (a) $\arcsin(-0.75) \approx -.85$

(b) $\arccos(-0.7) \approx 2.35$

18. (a) $\arcsin 0.31 \approx 0.32$

(b) $\arccos 0.26 \approx 1.31$

20. (a) $\arctan(0.92) \approx 0.74$

(b) $\arctan 2.8 \approx 1.23$

22. $f(x) = \sin x$

$g(x) = \arcsin x$

$y = x$

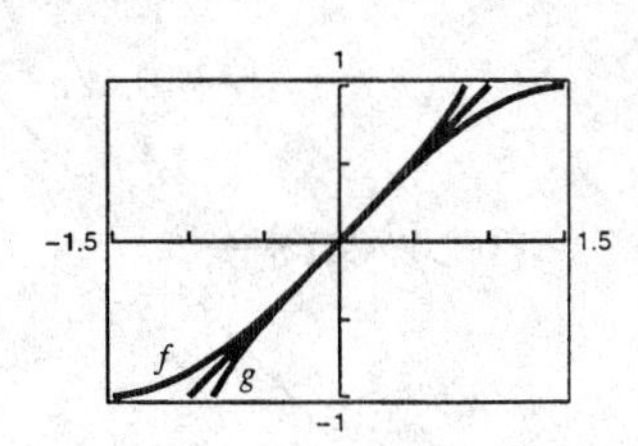

24. $\cos\theta = \dfrac{4}{x}$

$\theta = \arccos \dfrac{4}{x}$

26. $\tan\theta = \dfrac{x+1}{10}$

$\theta = \arctan\left(\dfrac{x+1}{10}\right)$

28. $\tan(\arctan 25) = 25$

30. $\sin[\arcsin(-0.2)] = -0.2$

32. $\arccos\left(\cos\dfrac{7\pi}{2}\right) = \arccos 0 = \dfrac{\pi}{2}$

Note: $(7\pi/2)$ is not in the range of the arccosine function.

34. Let $u = \arcsin\dfrac{4}{5}$,

$\sin u = \dfrac{4}{5}, 0 < u < \dfrac{\pi}{2}$.

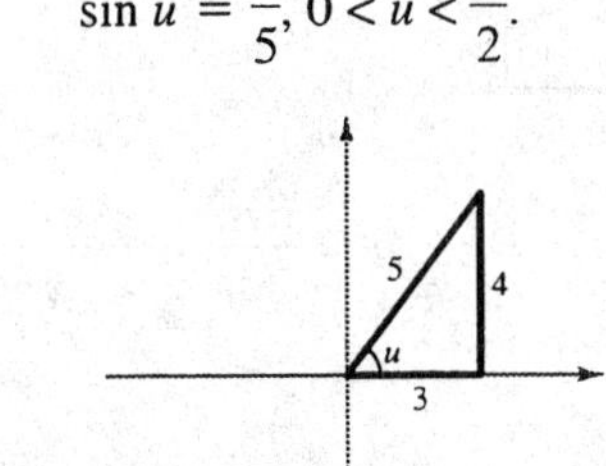

$\sec\left(\arcsin\dfrac{4}{5}\right) = \sec u = \dfrac{\text{hyp}}{\text{adj}} = \dfrac{5}{3}$

36. Let $u = \arccos\dfrac{\sqrt{5}}{5}$,

$\cos u = \dfrac{\sqrt{5}}{5}, 0 < u < \dfrac{\pi}{2}$.

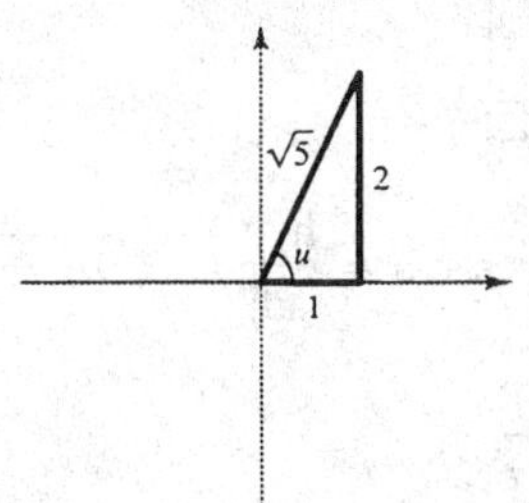

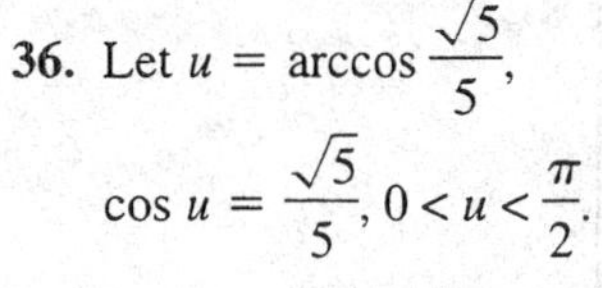

$\sin\left(\arccos\dfrac{\sqrt{5}}{5}\right) = \sin u = \dfrac{2}{\sqrt{5}} = \dfrac{2\sqrt{5}}{5}$

38. Let $u = \arctan\left(-\dfrac{5}{12}\right)$,

$\tan u = -\dfrac{5}{12}, -\dfrac{\pi}{2} < u < 0$.

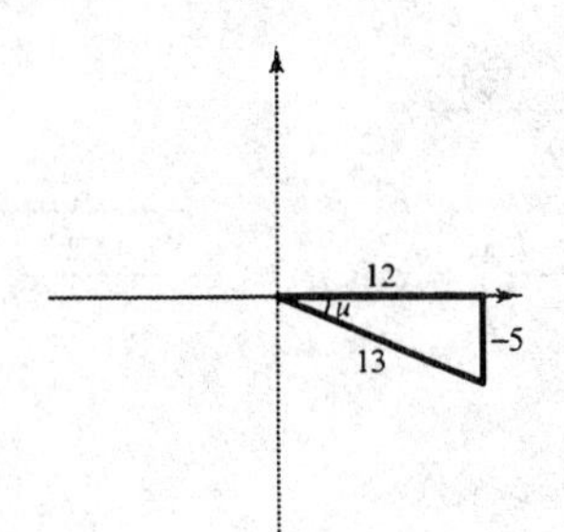

$\csc\left[\arctan\left(-\dfrac{5}{12}\right)\right] = \csc u = \dfrac{\text{hyp}}{\text{opp}} = -\dfrac{13}{5}$

40. Let $u = \arcsin\left(-\dfrac{3}{4}\right)$,

$\sin u = -\dfrac{3}{4}, -\dfrac{\pi}{2} < u < 0$.

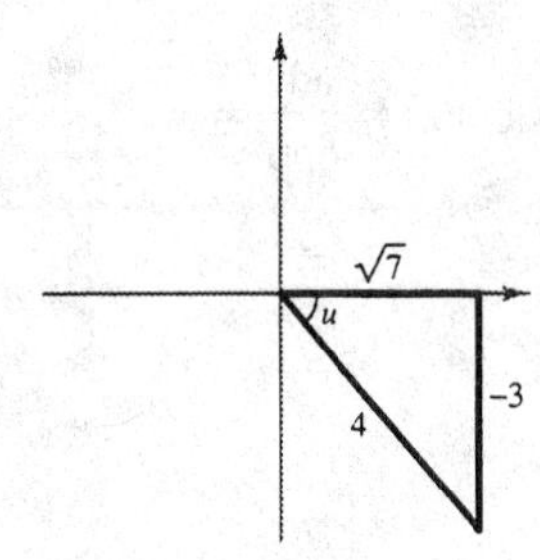

$\tan\left[\arcsin\left(-\dfrac{3}{4}\right)\right] = \tan u = -\dfrac{3}{\sqrt{7}} = -\dfrac{3\sqrt{7}}{7}$

42. Let $u = \arctan \frac{5}{8}$,

$\tan u = \frac{5}{8}, 0 < u < \frac{\pi}{2}$.

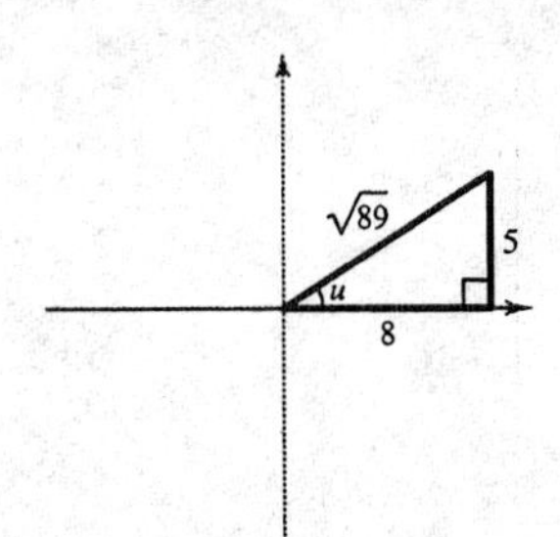

$$\cot\left(\arctan \frac{5}{8}\right) = \cot u = \frac{\text{adj}}{\text{opp}} = \frac{8}{5}$$

46. Let $u = \arctan 3x$,

$\tan u = 3x = \frac{3x}{1}$.

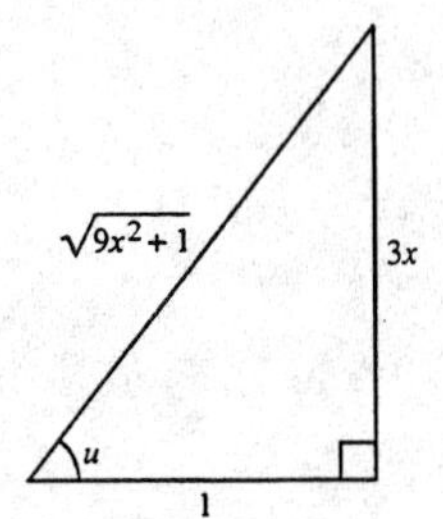

$$\sec(\arctan 3x) = \sec u = \frac{\text{hyp}}{\text{adj}} = \sqrt{9x^2 + 1}$$

50. Let $u = \arctan \frac{1}{x}$,

$\tan u = \frac{1}{x}$.

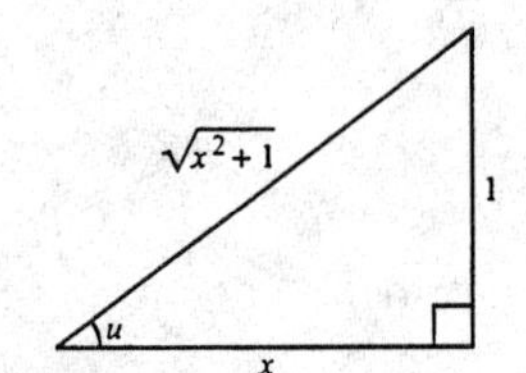

$$\cot\left(\arctan \frac{1}{x}\right) = \cot u = \frac{\text{adj}}{\text{opp}} = x$$

44. Let $u = \arctan x$,

$\tan u = x = \frac{x}{1}$.

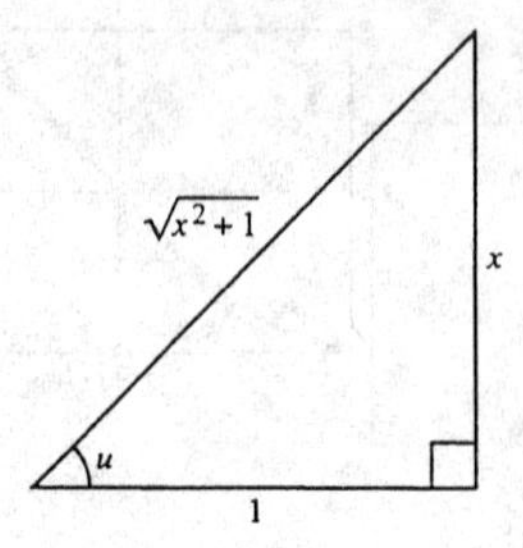

$$\sin(\arctan x) = \sin u = \frac{\text{opp}}{\text{hyp}} = \frac{x}{\sqrt{x^2 + 1}}$$

48. Let $u = \arcsin(x - 1)$,

$\sin u = x - 1 = \frac{x - 1}{1}$.

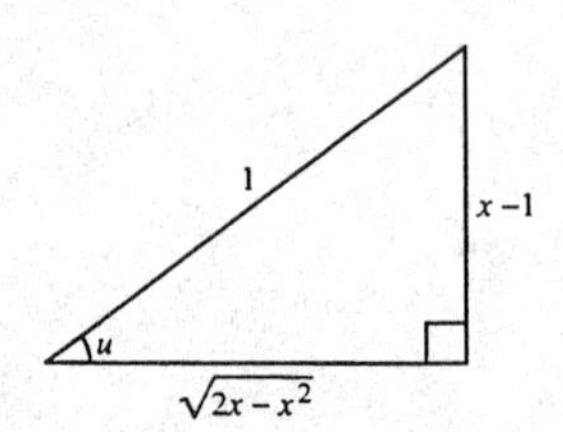

$$\sec[\arcsin (x - 1)] = \sec u = \frac{\text{hyp}}{\text{adj}} = \frac{1}{\sqrt{2x - x^2}}$$

52. Let $u = \arcsin \frac{x - h}{r}$,

$\sin u = \frac{x - h}{r}$.

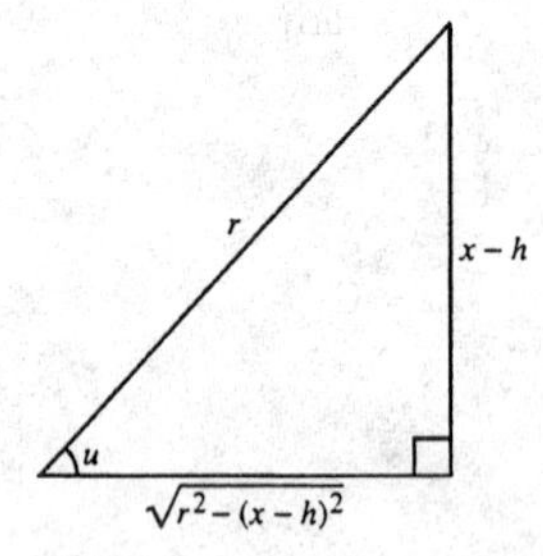

$$\cos\left(\arcsin \frac{x - h}{r}\right) = \cos u = \frac{\sqrt{r^2 - (x - h)^2}}{r}$$

54. $f(x) = \tan\left(\arccos \dfrac{x}{2}\right)$

$g(x) = \dfrac{\sqrt{4 - x^2}}{x}$

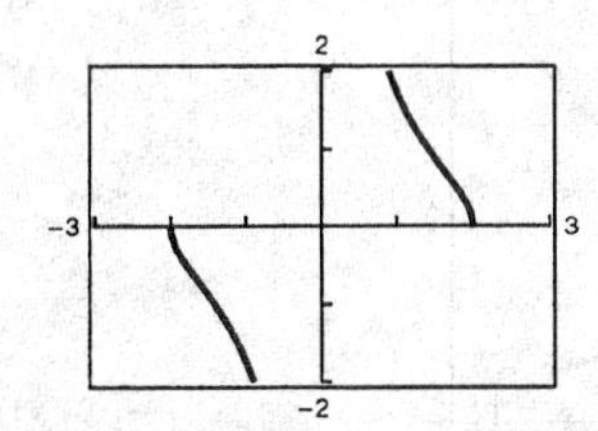

Asymptote: $x = 0$

These are equal because:

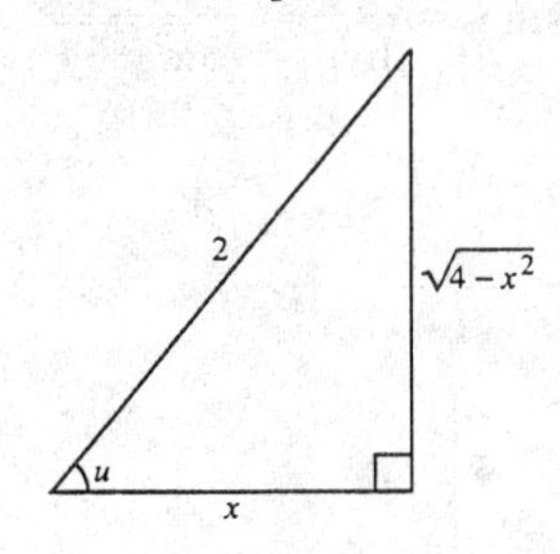

Let $u = \arccos \dfrac{x}{2}$.

$\tan\left(\arccos \dfrac{x}{2}\right) = \tan u = \dfrac{\sqrt{4 - x^2}}{x}$

56. If $\arcsin \dfrac{\sqrt{36 - x^2}}{6} = u$,

then $\sin u = \dfrac{\sqrt{36 - x^2}}{6}$.

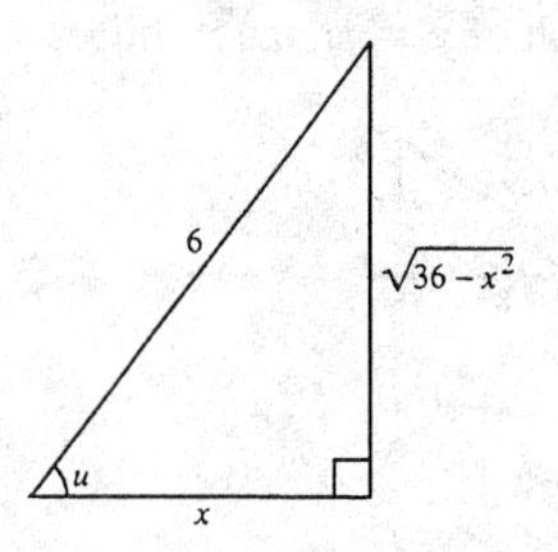

$\arcsin \dfrac{\sqrt{36 - x^2}}{6} = \arccos \dfrac{x}{6}$

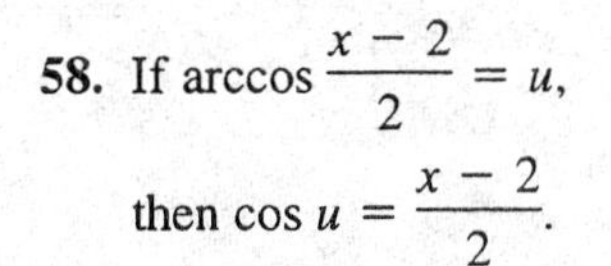

58. If $\arccos \dfrac{x - 2}{2} = u$,

then $\cos u = \dfrac{x - 2}{2}$.

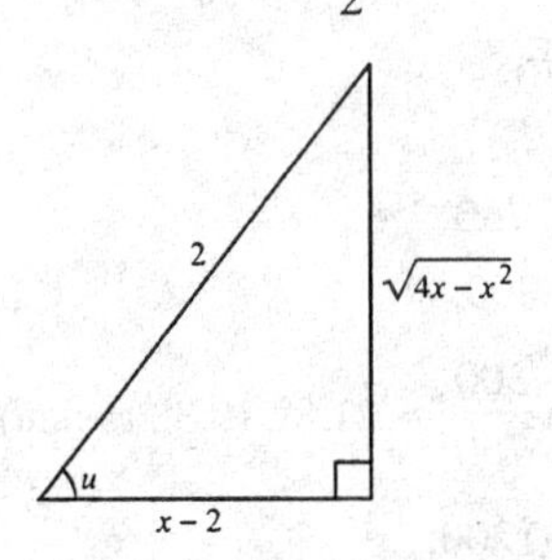

$\arccos \dfrac{x - 2}{2} = \arctan \dfrac{\sqrt{4x - x^2}}{x - 2}$

60. $y = \arcsin \dfrac{x}{2}$

Domain: $-2 \le x \le 2$

Range: $-\dfrac{\pi}{2} \le y \le \dfrac{\pi}{2}$

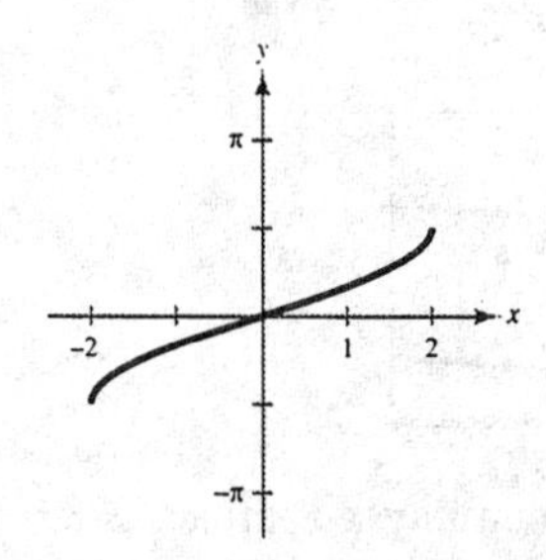

62. $g(t) = \arccos(t + 2)$

Domain: $-3 \le t \le -1$

This is the graph of $y = \arccos t$ shifted two units to the left.

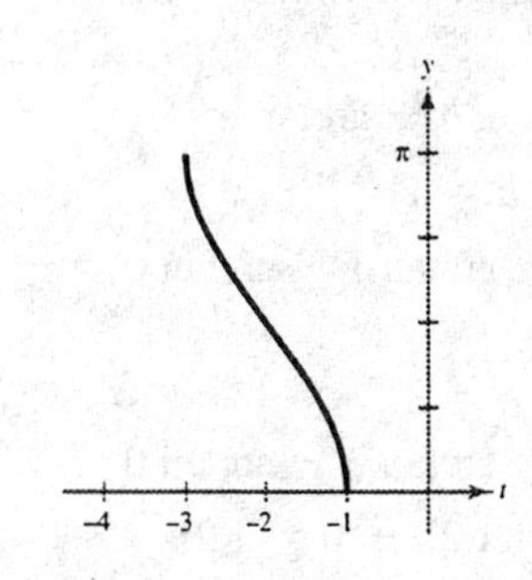

64. $f(x) = \frac{\pi}{2} + \arctan x$

Domain: $(-\infty, \infty)$

Range: $(0, \pi)$

This is the graph of $y = \arctan x$ shifted upward $\frac{\pi}{2}$ units.

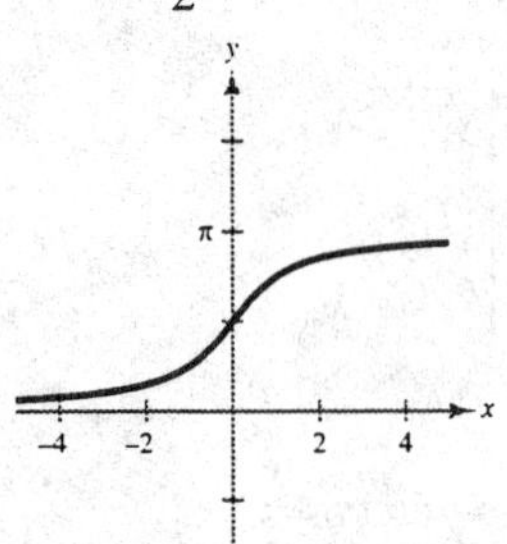

66. $f(x) = \arccos \frac{x}{4}$

Domain: $[-4, 4]$

Range: $[0, \pi]$

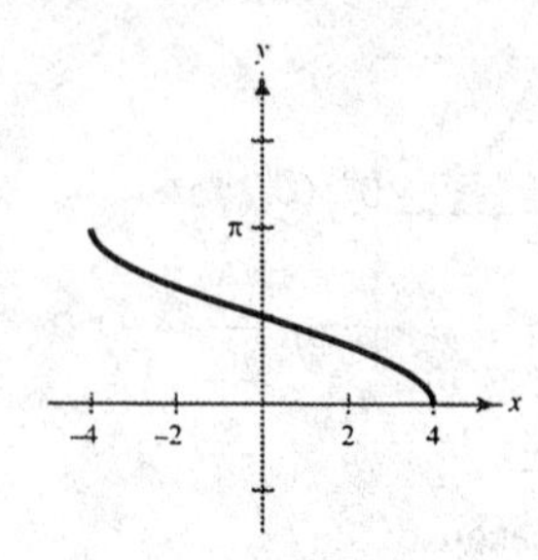

68. $f(t) = 4 \cos \pi t + 3 \sin \pi t$

$$= \sqrt{4^2 + 3^2} \sin\left(\pi t + \arctan \frac{4}{3}\right)$$

$$= 5 \sin\left(\pi t + \arctan \frac{4}{3}\right)$$

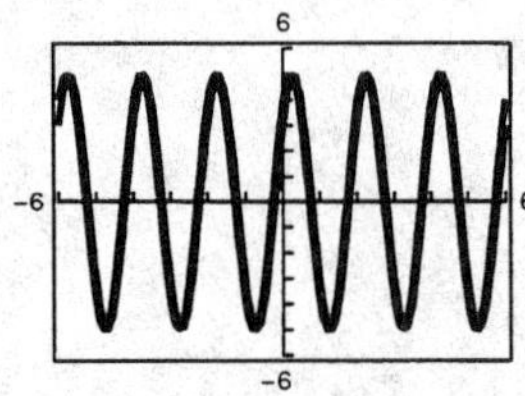

The graph implies that $A \cos \omega t + B \sin \omega t = \sqrt{A^2 + b^2} \sin\left(\omega t + \arctan \frac{A}{B}\right)$ is true.

70. $f(x) = \sqrt{x}$

$g(x) = 6 \arctan x$

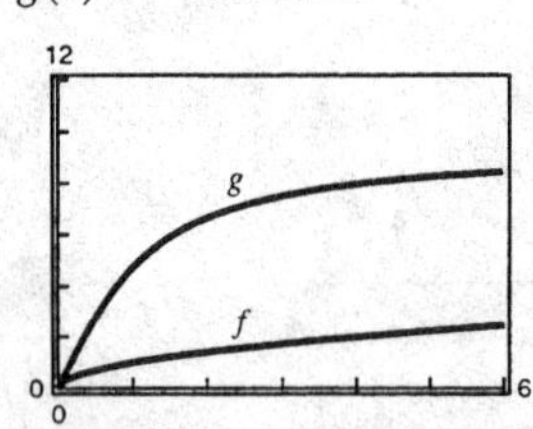

As x increases to infinity, g approaches 3π, but f has no maximum. Using the solve feature of the graphing utility, you find $a \approx 87.54$.

72. (a) $\tan \theta = \frac{s}{750}$

$\theta = \arctan \frac{s}{750}$

(b) When $s = 300$,

$\theta = \arctan \frac{300}{750} \approx 21.8°$, (0.38 radian).

When $s = 1200$,

$\theta = \arctan \frac{1200}{750} \approx 58.0°$, (1.01 radian).

74. Area $= \arctan b - \arctan a$

(a) $a = 0, b = 1$

Area $= \arctan 1 - \arctan 0 = \frac{\pi}{4} - 0 = \frac{\pi}{4}$

(b) $a = -1, b = 1$

Area $= \arctan 1 - \arctan(-1)$

$= \frac{\pi}{4} - \left(-\frac{\pi}{4}\right) = \frac{\pi}{2}$

(c) $a = 0, b = 3$

Area $= \arctan 3 - \arctan 0$

$\approx 1.25 - 0 = 1.25$

$= 1.25$

(d) $a = -1, b = 3$

Area $= \arctan 3 - \arctan(-1)$

$\approx 1.25 - \left(-\frac{\pi}{4}\right) \approx 2.03$

76. (a) $\tan\theta = \frac{x}{20}$

$\theta = \arctan\frac{x}{20}$

(b) When $x = 5$,

$\theta = \arctan\frac{5}{20} \approx 14.0°$, (0.24 rad).

When $x = 12$, $\theta = \arctan\frac{12}{20} \approx 31.0°$.

$\theta = \arctan\frac{12}{20} \approx 31.0°$, (0.54 rad).

78. $y = \operatorname{arcsec} x$ if and only if $\sec y = x$ where $x \le -1 \cup x \ge 1$ and $0 \le y < \pi/2$ and $\pi/2 < y \le \pi$. The domain of y arcsec x is $(-\infty, -1] \cup [1, \infty)$ and the range is $[0, \pi/2) \cup (\pi/2, \pi]$.

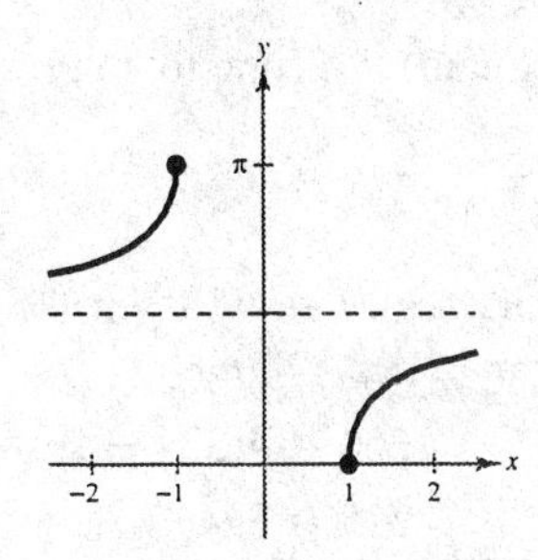

80. (a) $y = \operatorname{arcsec}\sqrt{2} \Rightarrow \sec y = \sqrt{2}$ and $0 \le y < \frac{\pi}{2} \cup \frac{\pi}{2} < y \le \pi \Rightarrow y = \frac{\pi}{4}$

(b) $y = \operatorname{arcsec} 1 \Rightarrow \sec y = 1$ and $0 \le y < \frac{\pi}{2} \cup \frac{\pi}{2} < y \le \pi \Rightarrow y = 0$

(c) $y = \operatorname{arccot}(-\sqrt{3}) \Rightarrow \cot y = -\sqrt{3}$ and $0 < y < \pi \Rightarrow y = \frac{5\pi}{6}$

(d) $y = \operatorname{arccsc} 2 \Rightarrow \csc y = 2$ and $-\frac{\pi}{2} \le y < 0 \cup 0 < y \le \frac{\pi}{2} \Rightarrow y = \frac{\pi}{6}$

82.

$$\begin{aligned} y &= \arctan(-x) \\ \tan y &= -x,\ -\frac{\pi}{2} < y < \frac{\pi}{2} \\ -\tan y &= x \\ \tan(-y) &= x,\ -\frac{\pi}{2} < -y < \frac{\pi}{2} \\ \arctan(\tan(-y)) &= \arctan x \\ -y &= \arctan x \\ y &= -\arctan x \end{aligned}$$

84. $y_2 = \frac{\pi}{2} - y_1$

$$\begin{aligned} \arctan x + \arctan\frac{1}{x} &= y_1 + y_2 \\ &= y_1 + \left(\frac{\pi}{2} - y_1\right) = \frac{\pi}{2} \end{aligned}$$

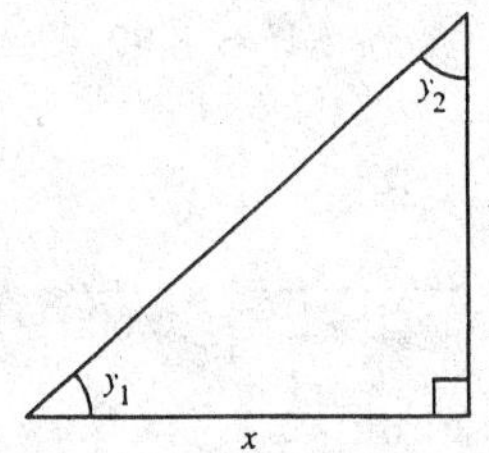

86. $\arcsin x = \operatorname{arsin}\frac{x}{1} = \arctan\frac{x}{\sqrt{1-x^2}}$

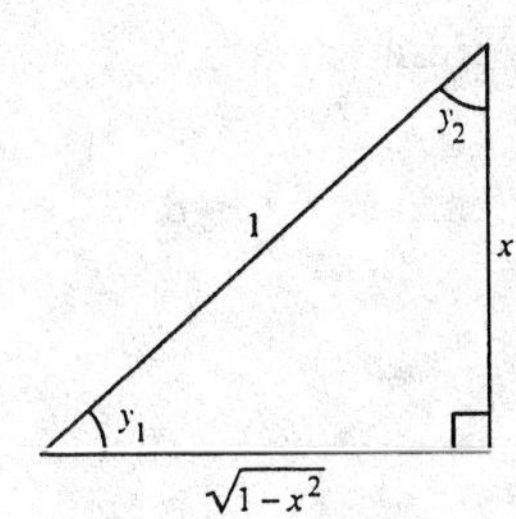

88. $739(1.3) = \$960.70$

90. Rate downstream: $18 + x$

Rate upstream: $18 - x$

$\text{rate} \times \text{time} = \text{distance} \Longrightarrow t = \dfrac{d}{r}$

(Time to go upstream) + (Time to go downstream) = 4

$$\frac{35}{18 - x} + \frac{35}{18 + x} = 4$$

$$35(18 + x) + 35(18 - x) = 4(18 - x)(18 + x)$$

$$630 + 35x + 630 - 35x = 4(324 - x^2)$$

$$1260 = 4(324 - x^2)$$

$$315 = 324 - x^2$$

$$x^2 = 9$$

$$x = \pm 3$$

The speed of the current is 3 miles per hour.

Section 4.8 Applications and Models

Solutions to Even-Numbered Exercises

2. $B = 54°, c = 15$

$A = 90° - B = 90° - 54° = 36°$

$\sin B = \dfrac{b}{c} \Longrightarrow b = c \sin B$

$= 15 \sin 54° \approx 12.14$

$\cos B = \dfrac{a}{c} \Longrightarrow a = c \cos B$

$= 15 \cos 54° \approx 8.82$

4. $A = 8.4°, a = 40.5$

$B = 90° - A = 90° - 8.4° = 81.6°$

$\tan A = \dfrac{a}{b} \Longrightarrow b = \dfrac{a}{\tan A}$

$= \dfrac{40.5}{\tan 8.4°} \approx 274.27$

$\sin A = \dfrac{a}{c} \Longrightarrow c = \dfrac{a}{\sin A}$

$= \dfrac{40.5}{\sin 8.4°} \approx 277.24$

6. $a = 25, c = 35$

$b = \sqrt{c^2 - a^2} = \sqrt{35^2 - 25^2} = \sqrt{600} \approx 24.49$

$\sin A = \dfrac{a}{c} \Longrightarrow A = \arcsin \dfrac{a}{c}$

$= \arcsin \dfrac{25}{35} \approx 45.58$

$\cos B = \dfrac{a}{c} \Longrightarrow B = \arccos \dfrac{a}{c}$

$= \arccos \dfrac{25}{35} \approx 44.42°$

8. $b = 1.32, c = 9.45$

$a = \sqrt{c^2 - b^2} = \sqrt{87.5601} \approx 9.36$

$\cos A = \dfrac{b}{c} \Longrightarrow A = \arccos \dfrac{b}{c}$

$= \arccos \dfrac{1.32}{9.45} \approx 81.97°$

$\sin B = \dfrac{b}{c} \Longrightarrow B = \arcsin \dfrac{b}{c}$

$= \arcsin \dfrac{1.32}{9.45} \approx 8.03°$

10. $B = 65° \, 12', a = 14.2$

$A = 90° - B = 90° - 65° \, 12' = 24° \, 48'$

$\cos B = \dfrac{a}{c} \Longrightarrow c = \dfrac{a}{\cos B}$

$= \dfrac{14.2}{\cos 65° \, 12'} \approx 33.85$

$\tan B = \dfrac{b}{a} \Longrightarrow b = a \tan B$

$= 14.2 \tan 65° \, 12'$

≈ 30.73

12. $\theta = 18°, b = 10$ meters

$\tan \theta = \dfrac{\text{altitude}}{b/2}$

$\text{altitude} = \dfrac{b}{2} \tan \theta$

$= \dfrac{10}{2} \tan 18° \approx 1.62$ meters

14. (a) $\tan \theta = \dfrac{600}{L} \Longrightarrow L = 600 \cot \theta$

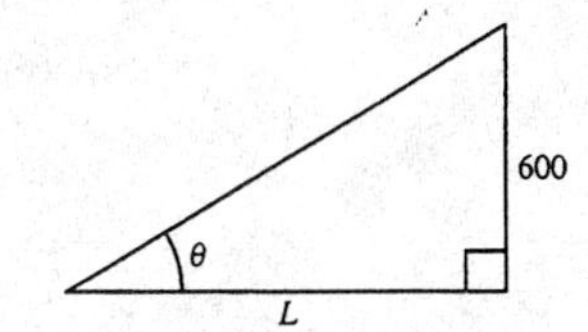

(b)

θ	10°	20°	30°	40°	50°
L	3403	1648	1039	715	503

(c) No, the cotangent function is not a linear function.

16. (a) $\tan \theta = \dfrac{h}{110} \Longrightarrow h = 110 \tan \theta$

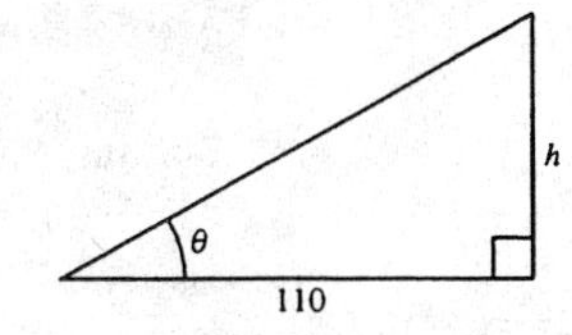

(b)

θ	10°	15°	20°	25°	30°
h	19.4	29.5	40.0	51.3	63.5

18. $\tan 28° = \dfrac{a}{100} \Longrightarrow a = 100 \tan 28°$

$\tan 39.75° = \dfrac{a + s}{100}$

$a + s = 100 \tan 39.75°$

$s = 100 \tan 39.75° - a$

$= 100 \tan 39.75 - 100 \tan 28°$

≈ 30 feet

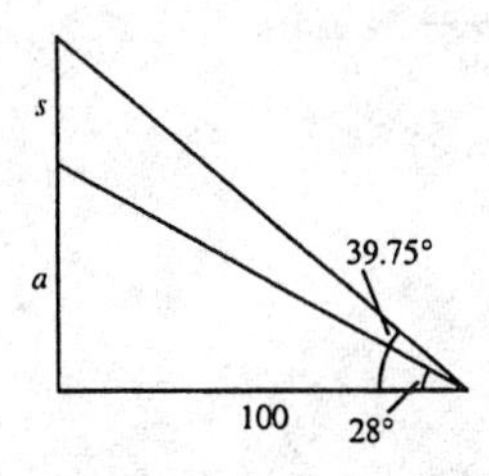

20. $\sin 50° = \dfrac{h}{100}$

$h = 100 \sin 50° \approx 76.6$ feet

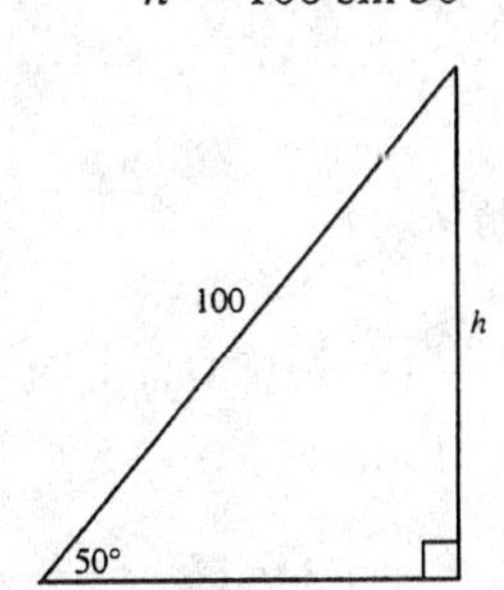

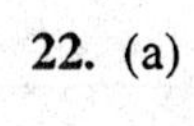

22. (a)

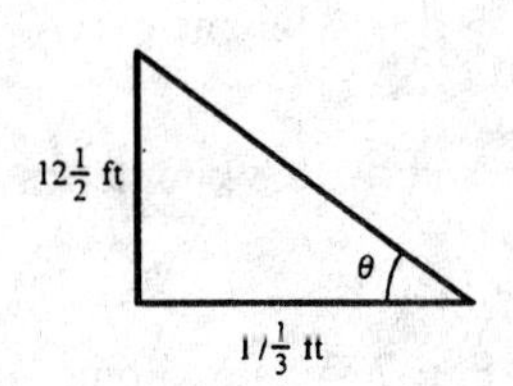

(b) $\tan \theta = \dfrac{12\frac{1}{2}}{17\frac{1}{3}}$

(c) $\theta = \arctan \dfrac{12\frac{1}{2}}{17\frac{1}{3}} \approx 35.8°$

24.

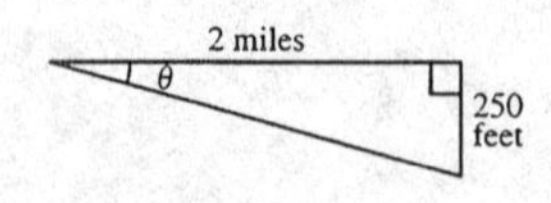

$$\tan \theta = \frac{250}{2(5280)}$$

$$\theta = \arctan \frac{250}{2(5280)} \approx 1.36^\circ$$

26.

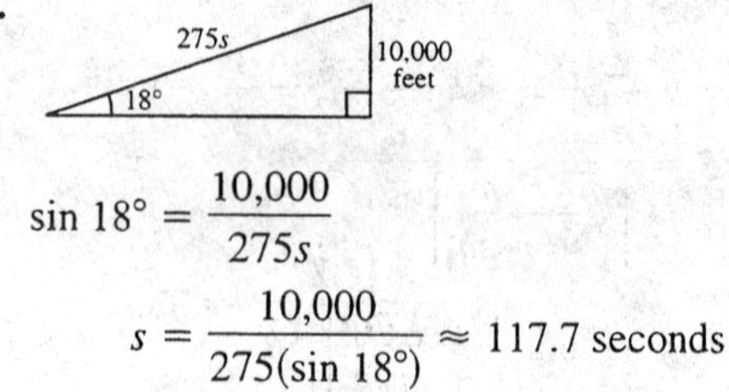

$$\sin 18^\circ = \frac{10{,}000}{275s}$$

$$s = \frac{10{,}000}{275(\sin 18^\circ)} \approx 117.7 \text{ seconds}$$

28.

θ 100x 12x = y 4 miles = 21,120 feet

Angle of grade: $\tan \theta = \dfrac{12x}{100x}$

$\theta = \arctan 0.12 \approx 6.8^\circ$

Change in elevation:

$$\sin \theta = \frac{y}{21{,}120}$$

$$y = 21{,}120 \sin \theta = 21{,}120 \sin(\arctan 0.12)$$

$$\approx 2516.3 \text{ feet}$$

30. $\sin 63^\circ = \dfrac{a}{120} \Longrightarrow a \approx 107$ nautical miles south

$\cos 63^\circ = \dfrac{b}{120} \Longrightarrow b \approx 54.5$ nautical miles west

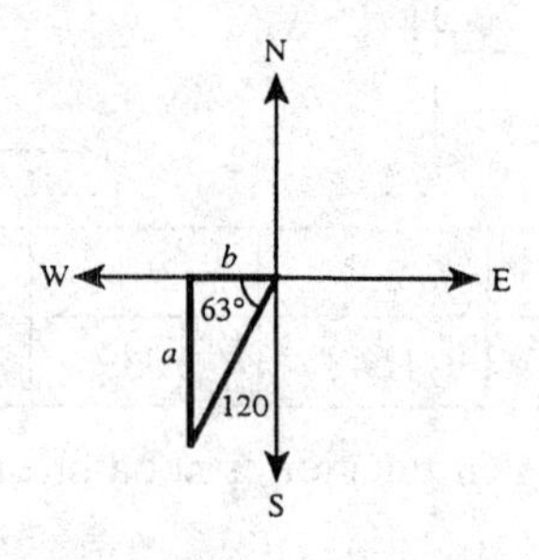

32. $\tan 14^\circ = \dfrac{d}{x} \Longrightarrow x = d \cot 14^\circ$

$$\tan 34^\circ = \frac{d}{y} \Longrightarrow \frac{d}{30 - x} = \frac{d}{30 - d \cot 14^\circ}$$

$$\cot 34^\circ = \frac{30 - d \cot 14^\circ}{d}$$

$$d \cot 34^\circ = 30 - d \cot 14^\circ$$

$$d = \frac{30}{\cot 34^\circ + \cot 14^\circ}$$

$$\approx 5.46 \text{ kilometers}$$

34. $\tan \theta = \dfrac{85}{120} \Longrightarrow \theta = 35.3^\circ$

Bearing: S 35.3° W

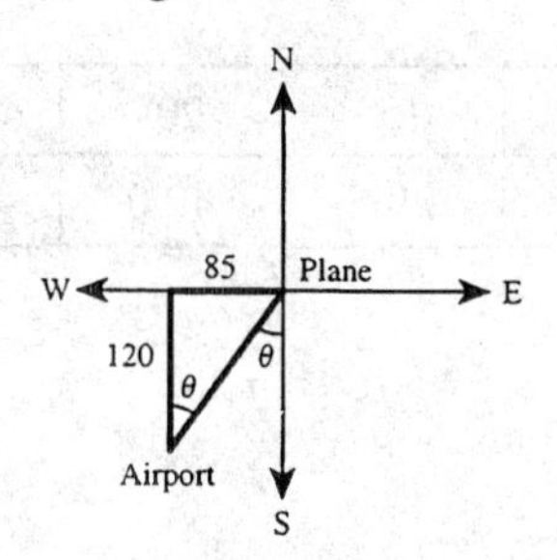

36.

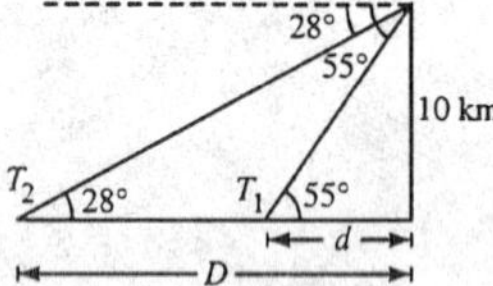

$\cot 55^\circ = \dfrac{d}{10} \Longrightarrow d \approx 7$ kilometers

$\cot 28^\circ = \dfrac{D}{10} \Longrightarrow D \approx 18.8$ kilometers

Distance between towns:

$D - d = 18.8 - 7 = 11.8$ kilometers

38.

3.5° h x

9° h x − 13

$$\tan 3.5^\circ = \frac{h}{x}, \quad \tan 9^\circ = \frac{h}{x - 13}$$

$$x = \frac{h}{\tan 3.5^\circ}, \quad x = \frac{h}{\tan 9^\circ} + 13$$

$$\frac{h}{\tan 3.5^\circ} = \frac{h}{\tan 9^\circ} + 13$$

$$\frac{h}{\tan 3.5^\circ} = \frac{h + 13 \tan 9^\circ}{\tan 9^\circ}$$

$$h \tan 9^\circ = h \tan 3.5^\circ + 13(\tan 9^\circ)(\tan 3.5^\circ)$$

$$h(\tan 9^\circ - \tan 3.5^\circ) = 13(\tan 9^\circ)(\tan 3.5^\circ)$$

$$h = \frac{13(\tan 9^\circ)(\tan 3.5^\circ)}{\tan 9^\circ - \tan 3.5^\circ} \approx 1.3 \text{ miles} \approx 6839 \text{ feet}$$

40. $L_1 = 2x + y = 8 \Longrightarrow m_1 = -2$

$L_2 = x - 5y = -4 \Longrightarrow m_2 = \dfrac{1}{5}$

$$\tan \alpha = \left|\frac{m_2 - m_1}{1 + m_2 m_1}\right|$$

$$\alpha = \arctan\left|\frac{m_2 - m_1}{1 + m_2 m_1}\right|$$

$$= \arctan\left|\frac{\frac{1}{5} - (-2)}{1 + \frac{1}{5}(-2)}\right|$$

$$= \arctan\left(3\tfrac{2}{3}\right) \approx 74.7^\circ$$

42.

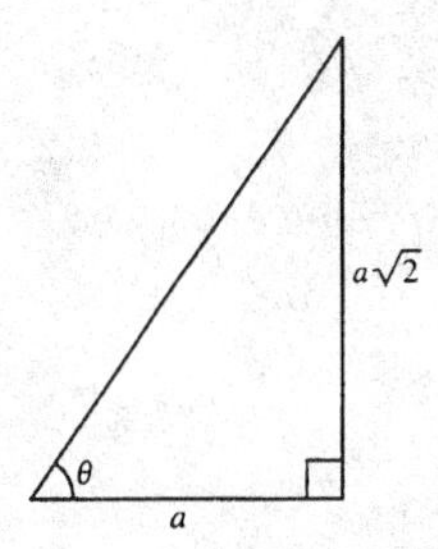

$$\tan \theta = \frac{a\sqrt{2}}{a} = \sqrt{2}$$

$$\theta = \arctan \sqrt{2} \approx 54.7^\circ$$

44.

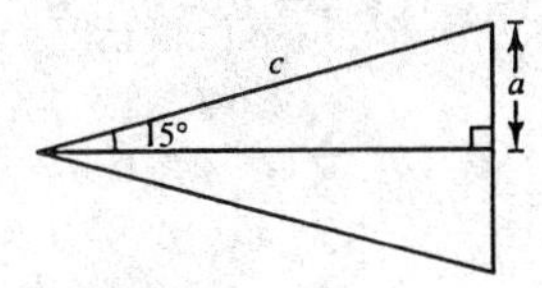

$$c = \frac{35}{2} = 17.5$$

$$\sin 15^\circ = \frac{a}{c}$$

$$a = c \sin 15^\circ = 17.5 \sin 15^\circ \approx 4.53$$

Distance $= 2a \approx 9.06$ centimeters

46.

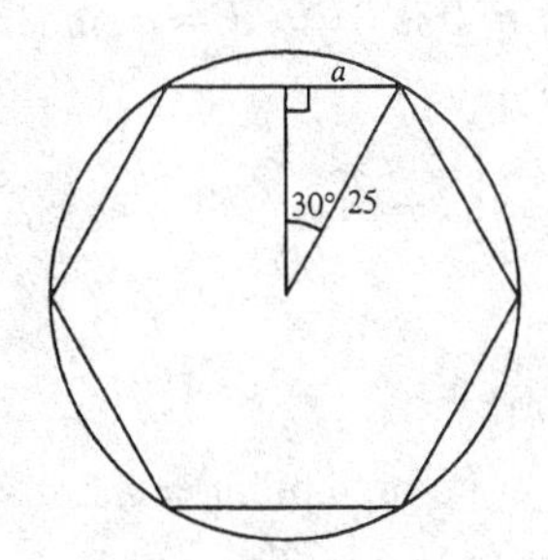

$$\sin 30^\circ = \frac{a}{25}$$

$$a = 25 \sin 30^\circ = 12.5$$

Length of side $= 2a = 2(12.5) = 25$ inches

48.

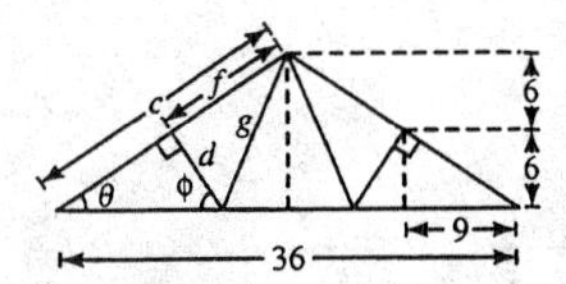

$$\tan \theta = \frac{12}{18}$$

$$\theta = \arctan \frac{2}{3} = 0.588 \text{ rad} \approx 33.7^\circ$$

$$\cos \theta = \frac{18}{c}$$

$$c = \frac{18}{\cos \theta} \approx 21.6$$

$$f \approx \frac{21.6}{2} = 10.8$$

$$\phi \approx 90 - 33.7 = 56.3^\circ$$

$$\sin \phi = \frac{6}{d}$$

$$d = \frac{6}{\sin \phi} \approx 7.2$$

$$g = \sqrt{10.8^2 + 7.2^2} \approx 12.98$$

50. $d = \dfrac{1}{2} \cos 20\pi t$

(a) Maximum displacement: $|a| = \left|\dfrac{1}{2}\right| = \dfrac{1}{2}$

(b) Frequency:

$$\frac{\omega}{2\pi} = \frac{20\pi}{2\pi} = 10$$

(c) Least positive value for t for which $d = 0$:

$$\frac{1}{2} \cos 20\pi t = 0$$

$$\cos 20\pi t = 0$$

$$20\pi t = \arccos 0$$

$$20\pi t = \frac{\pi}{2}$$

$$t = \frac{\pi}{2} \cdot \frac{1}{20\pi} = \frac{1}{40}$$

52. $d = \frac{1}{64}\sin 792\pi t$

(a) Maximum displacement:

$$|a| = \left|\frac{1}{64}\right| = \frac{1}{64}$$

(b) Frequency:

$$\frac{\omega}{2\pi} = \frac{792\pi}{2\pi} = 396$$

(c) Least positive value for t for which $d = 0$:

$$\frac{1}{64}\sin 792\pi t = 0$$
$$\sin 792\pi t = 0$$
$$792\pi t = \arcsin 0$$
$$792\pi t = \pi$$
$$t = \frac{\pi}{792\pi} = \frac{1}{792}$$

54. Displacement at $t = 0$ is $0 \Rightarrow d = a\sin\omega t$

Amplitude: $|a| = 3$

Period: $\frac{2\pi}{\omega} = 6 \Rightarrow \omega = \frac{\pi}{3}$

$$d = 3\sin\left(\frac{\pi t}{3}\right)$$

56. Displacement at $t = 0$ is $2 \Rightarrow d = a\cos\omega t$

Amplitude: $|a| = 2$

Period: $\frac{2\pi}{\omega} = 10 \Rightarrow \omega = \frac{\pi}{5}$

$$d = 2\cos\left(\frac{\pi t}{5}\right)$$

58. At $t = 0$, buoy is at its high point $\Rightarrow d = a\cos\omega t$.

Distance from high to low $= 2|a| = 3.5$

$$|a| = \frac{7}{4}$$

Returns to high point every 10 seconds:

Period $= \frac{2\pi}{\omega} = 10 \Rightarrow \omega = \frac{\pi}{5}$

$$d = \frac{7}{4}\cos\frac{\pi t}{5}$$

60. (a)

θ	L_1	L_2	$L_1 + L_2$
0.1	$\frac{2}{\sin 0.1}$	$\frac{3}{\cos 0.1}$	23.0
0.2	$\frac{2}{\sin 0.2}$	$\frac{3}{\cos 0.2}$	13.1
0.3	$\frac{2}{\sin 0.3}$	$\frac{3}{\cos 0.3}$	9.9
0.4	$\frac{2}{\sin 0.4}$	$\frac{3}{\cos 0.4}$	8.4

(b)

0.5	$\frac{2}{\sin 0.5}$	$\frac{3}{\cos 0.5}$	7.6
0.6	$\frac{2}{\sin 0.6}$	$\frac{3}{\cos 0.6}$	7.2
0.7	$\frac{2}{\sin 0.7}$	$\frac{3}{\cos 0.7}$	7.0
0.8	$\frac{2}{\sin 0.8}$	$\frac{3}{\cos 0.8}$	7.1

The minimum length of the elevator is 7.0 meters.

(c) $L = L_1 + L_2 = \frac{2}{\sin\theta} + \frac{3}{\cos\theta}$

(d)

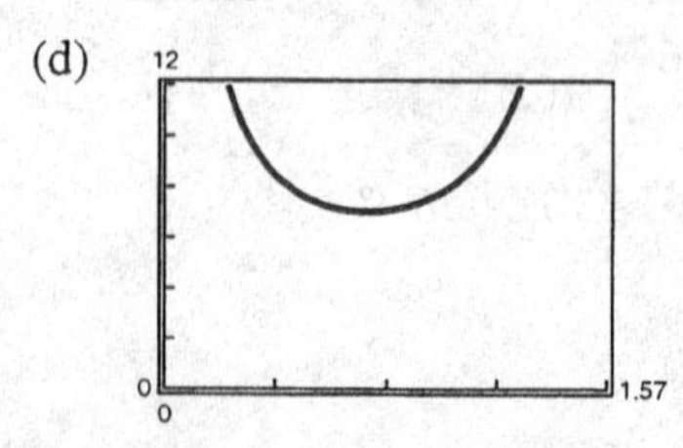

From the graph, it appears that the minimum length is 7.0 meters, which agrees with the estimate of part (b).

62. (a) and (b)

Base 1	*Base2*	*Altitude*	*Area*
8	$8 + 16 \cos 10°$	$8 \sin 10°$	22.1
8	$8 + 16 \cos 20°$	$8 \sin 20°$	42.5
8	$8 + 16 \cos 30°$	$8 \sin 30°$	59.7
8	$8 + 16 \cos 40°$	$8 \sin 40°$	72.7
8	$8 + 16 \cos 50°$	$8 \sin 50°$	80.5
8	$8 + 16 \cos 60°$	$8 \sin 60°$	83.1
8	$8 + 16 \cos 70°$	$8 \sin 70°$	80.7

Maximum is 83.1°

(c) $A = \frac{1}{2}(b_1 + b_2)h = \frac{1}{2}[8 + (8 + 16 \cos \theta)]8 \sin \theta = 64(1 + \cos \theta)(\sin \theta)$

(d)

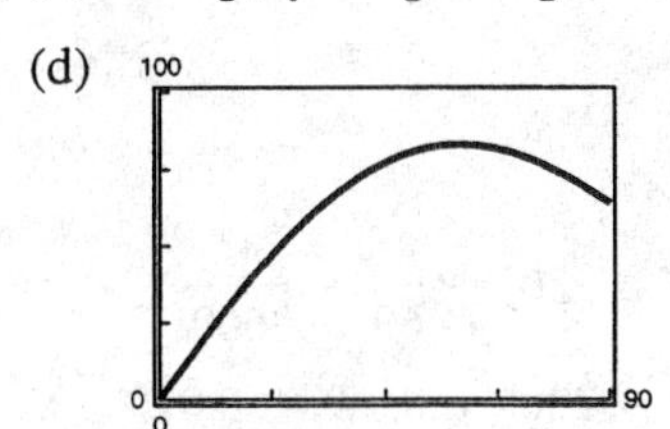

Maximum is 83.1°

64. (a) $A = \text{triangle} - \text{sector}$

$$= \frac{1}{2}10(10 \tan \theta) - \frac{10^2}{2}\theta$$

$$= 50 \tan \theta - 50\theta$$

$$= 50(\tan \theta - \theta)$$

(b)

θ	0	0.3	0.6	0.9	1.2	1.5
A	0	0.47	4.21	18.01	68.61	630.07

(c)

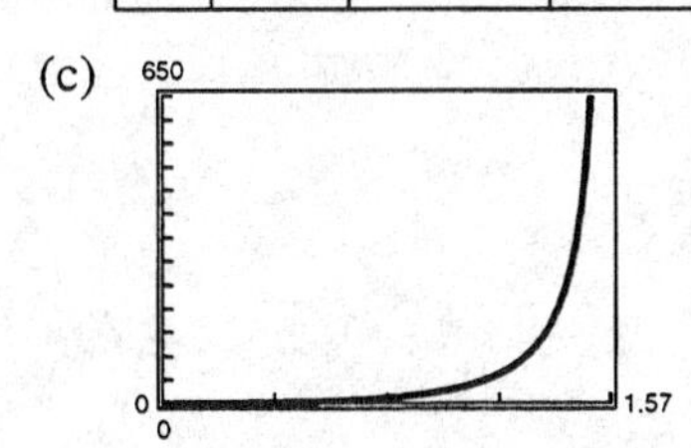

(d) The area approaches ∞.

Review Exercises for Chapter 4

Solutions to Even-Numbered Exercises

2. $\dfrac{2\pi}{9}$

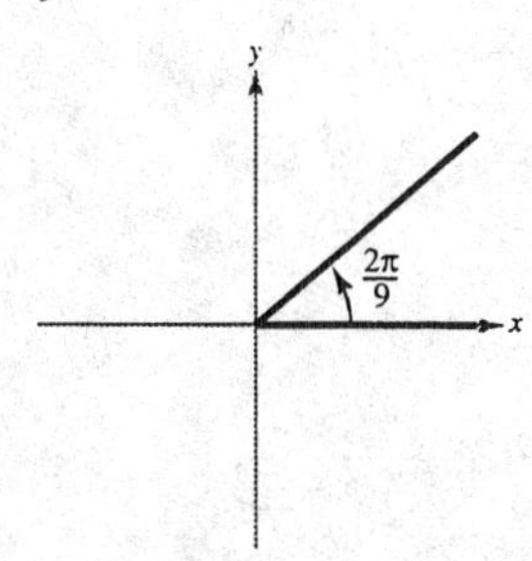

$\dfrac{20\pi}{9}, \dfrac{-16\pi}{9}$

4. $-405°$

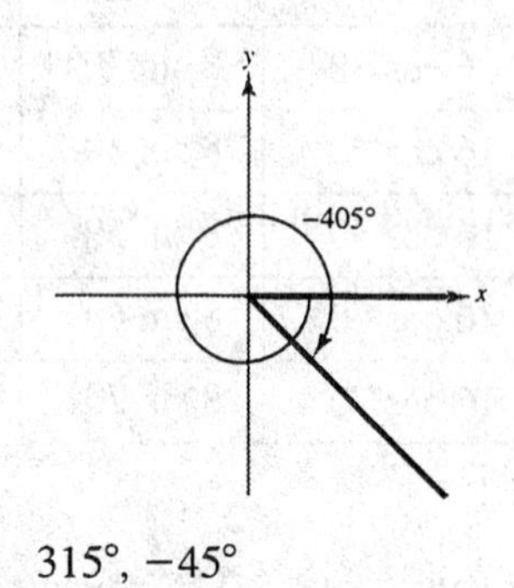

$315°, -45°$

6. $-234° \, 50'' = -\left(234° + \dfrac{50°}{3600}\right)$

$\approx -(234° + 0.01°)$

$= -234.01°$

8. $280° \, 8' \, 50'' = 280° + \dfrac{8°}{60} + \dfrac{50°}{3600}$

$\approx 280° + 0.13° + 0.01°$

$= 280.15°$

10. $25.1° = 25° + (0.1)(60')$

$= 25° \, 6'$

12. $-327.85° = -(327° + (0.85)(60'))$
$= -327° 51'$

14. $-\frac{3\pi}{5} = -\frac{3\pi}{5}\left(\frac{180°}{\pi}\right) = -108°$

16. $1.75 = 1.75\left(\frac{180°}{\pi}\right) \approx 100.27°$

18. $-16.5° = -16.5°\left(\frac{\pi}{180°}\right) \approx -0.2880$

20. $84° 15' = 84° + \frac{15°}{60} = 84.25° = 84.25°\left(\frac{\pi}{180°}\right) \approx 1.4704$

22. $\theta = 640°$ is coterminal with $280°$.
$\theta' = 360° - 280° = 80°$

24. $\theta = \frac{17\pi}{3}$ is coterminal with $\frac{5\pi}{3}$.
$\theta' = 2\pi - \frac{5\pi}{3} = \frac{\pi}{3}$

26. $s = r\theta \Longrightarrow \theta = \frac{s}{r} = \frac{235}{60} = \frac{47}{12}$

28. $s = r\theta \Longrightarrow r = \frac{s}{\theta} = \frac{8.5}{\pi/3} = \frac{25.5}{\pi} \approx 8.1$

30. $\sin\theta = \frac{2}{10} = \frac{1}{5}$

10
2
θ
$\sqrt{96} = 4\sqrt{6}$

$\cos\theta = \frac{4\sqrt{6}}{10} = \frac{2\sqrt{6}}{5}$

$\tan\theta = \frac{2}{4\sqrt{6}} = \frac{\sqrt{6}}{12}$

$\csc\theta = 5$

$\sec\theta = \frac{5}{2\sqrt{6}} = \frac{5\sqrt{6}}{12}$

$\cot\theta = \frac{12}{\sqrt{6}} = 2\sqrt{6}$

32. $\sin\theta = \frac{15}{3\sqrt{41}} = \frac{5\sqrt{41}}{41}$

$3\sqrt{41}$
15
θ
12

$\cos\theta = \frac{12}{3\sqrt{41}} = \frac{4\sqrt{41}}{41}$

$\tan\theta = \frac{15}{12} = \frac{5}{4}$

$\csc\theta = \frac{41}{5\sqrt{41}} = \frac{\sqrt{41}}{5}$

$\sec\theta = \frac{41}{4\sqrt{41}} = \frac{\sqrt{41}}{4}$

$\cot\theta = \frac{4}{5}$

34. $(x, 4x), x > 0$
$x = x, y = 4x$
$r = \sqrt{x^2 + (4x)^2} = \sqrt{17}x$

$\sin\theta = \frac{y}{r} = \frac{4x}{\sqrt{17}x} = \frac{4\sqrt{17}}{17}$

$\cos\theta = \frac{x}{r} = \frac{x}{\sqrt{17}x} = \frac{\sqrt{17}}{17}$

$\tan\theta = \frac{y}{x} = \frac{4x}{x} = 4$

$\csc\theta = \frac{r}{y} = \frac{\sqrt{17}x}{4x} = \frac{\sqrt{17}}{4}$

$\sec\theta = \frac{r}{x} = \frac{\sqrt{17}x}{x} = \sqrt{17}$

$\cot\theta = \frac{x}{y} = \frac{x}{4x} = \frac{1}{4}$

36. $x = 4, y = -8$
$r = \sqrt{4^2 + (-8)^2} = 4\sqrt{5}$

$\sin\theta = \frac{y}{r} = \frac{-8}{4\sqrt{5}} = -\frac{2\sqrt{5}}{5}$

$\cos\theta = \frac{x}{r} = \frac{4}{4\sqrt{5}} = \frac{\sqrt{5}}{5}$

$\tan\theta = \frac{y}{x} = \frac{-8}{4} = -2$

$\csc\theta = \frac{r}{y} = \frac{4\sqrt{5}}{-8} = -\frac{\sqrt{5}}{2}$

$\sec\theta = \frac{r}{x} = \frac{4\sqrt{5}}{4} = \sqrt{5}$

$\cot\theta = \frac{x}{y} = \frac{4}{-8} = -\frac{1}{2}$

38. $x = \frac{2}{3}, y = \frac{5}{2}$

$r = \sqrt{\left(\frac{2}{3}\right)^2 + \left(\frac{5}{2}\right)^2} = \frac{\sqrt{241}}{6}$

$\sin\theta = \frac{y}{r} = \frac{(5/2)}{(\sqrt{241}/6)} = \frac{15}{\sqrt{241}} = \frac{15\sqrt{241}}{241}$

$\cos\theta = \frac{x}{r} = \frac{(2/3)}{(\sqrt{241}/6)} = \frac{4}{\sqrt{241}} = \frac{4\sqrt{241}}{241}$

$\tan\theta = \frac{y}{x} = \frac{(5/2)}{(2/3)} = \frac{15}{4}$

$\csc\theta = \frac{r}{y} = \frac{(\sqrt{241}/6)}{(5/2)} = \frac{2\sqrt{241}}{30} = \frac{\sqrt{241}}{15}$

$\sec\theta = \frac{r}{x} = \frac{(\sqrt{241}/6)}{(2/3)} = \frac{\sqrt{241}}{4}$

$\cot\theta = \frac{x}{y} = \frac{(2/3)}{(5/2)} = \frac{4}{15}$

40. $\tan\theta = \frac{y}{x} = -\frac{12}{5} \Longrightarrow r = 13$

$\sin\theta > 0 \Longrightarrow y = 12, x = -5$

$\sin\theta = \frac{y}{r} = \frac{12}{13}$

$\cos\theta = \frac{x}{r} = -\frac{5}{13}$

$\csc\theta = \frac{r}{y} = \frac{13}{12}$

$\sec\theta = \frac{r}{x} = \frac{13}{-5} = -\frac{13}{5}$

$\cot\theta = \frac{x}{y} = -\frac{5}{12}$

42. $\cos\theta = \frac{x}{r} = \frac{-2}{5} \Longrightarrow y = |\sqrt{21}|$

$\sin\theta > 0 \Longrightarrow y = \sqrt{21}$

$\sin\theta = \frac{y}{r} = \frac{\sqrt{21}}{5}$

$\tan\theta = \frac{y}{x} = -\frac{\sqrt{21}}{2}$

$\csc\theta = \frac{r}{y} = \frac{5}{\sqrt{21}} = \frac{5\sqrt{21}}{21}$

$\sec\theta = \frac{r}{x} = \frac{5}{-2} = -\frac{5}{2}$

$\cot\theta = \frac{x}{y} = \frac{-2}{\sqrt{21}} = -\frac{2\sqrt{21}}{21}$

44. $\sec\frac{\pi}{4} = \frac{1}{\cos(\pi/4)} = \frac{1}{(\sqrt{2}/2)} = \sqrt{2}$

46. $\cot\left(-\frac{5\pi}{6}\right) = \cot\left(\frac{7\pi}{6}\right)$

$= \cot\left(\frac{\pi}{6}\right)$

$= \frac{1}{\tan(\pi/6)} = \sqrt{3}$

48. $\csc 270° = \frac{1}{\sin 270°} = \frac{1}{-1} = -1$

50. $\csc 105° = \frac{1}{\sin 105°} \approx 1.04$

52. $\sin\left(-\frac{\pi}{9}\right) \approx -0.34$

54. $\sec\theta$ is undefined $\Longrightarrow \cos\theta = 0.$

$\theta = 90° = \frac{\pi}{2}, \theta = 270° = \frac{3\pi}{2}$

56. $\tan\theta = \frac{\sqrt{3}}{3}$

$\theta = 30° - \frac{\pi}{6}; \theta = 210° = \frac{7\pi}{6}$

58. $\cot \theta = -1.5399$

$$\frac{1}{\tan \theta} = -1.5399$$

$$\theta = \tan^{-1}\left(\frac{1}{-1.5399}\right) \approx -33^\circ = 327^\circ$$

or 5.7072 radians

$327^\circ - 180^\circ = 147^\circ$ or 2.5656 radians

60. $\csc \theta = 11.4737$

$$\frac{1}{\sin \theta} = 11.4737$$

$$\theta = \sin^{-1}\left(\frac{1}{11.4737}\right) \approx 5^\circ \text{ or } 0.0873 \text{ radians}$$

$180^\circ - 5^\circ = 175^\circ$ or 3.054 radians

62. Period: 4π

Amplitude: $\frac{3}{2}$

64. Period: 4

Amplitude: 4

66. $y = -2 \sin \pi x$

Period $= \frac{2\pi}{\pi} = 2$

Amplitude: $|-2| = 2$

Reflected in x-axis

x	$-\frac{1}{2}$	0	$\frac{1}{2}$
y	2	0	-2

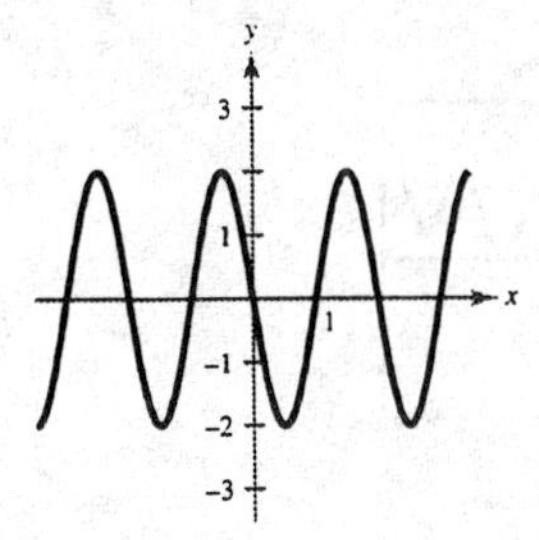

68. $f(x) = 8 \cos\left(-\frac{x}{4}\right)$

Period $= \frac{2\pi}{(1/4)} = 8\pi$

Amplitude: 8

Reflected in y-axis

x	-4π	-2π	0	2π	4π
y	-8	0	8	0	-8

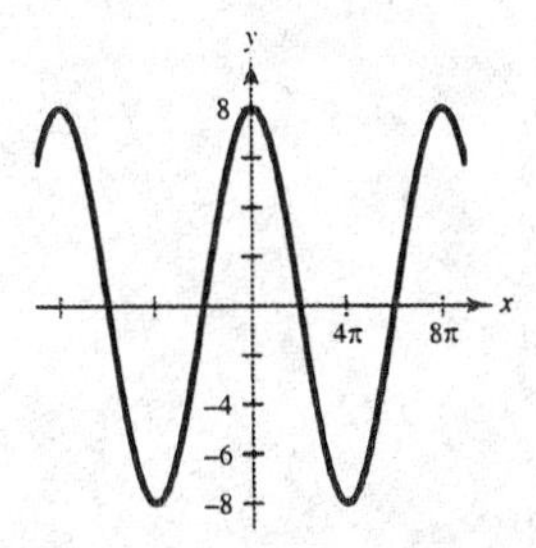

70. $f(x) = -\tan \frac{\pi x}{4}$

Period $= \frac{\pi}{(\pi/4)} = 4$

Asymptotes: $x = -2, x = 2$

Reflected in x-axis

x	-1	0	1
y	1	0	-1

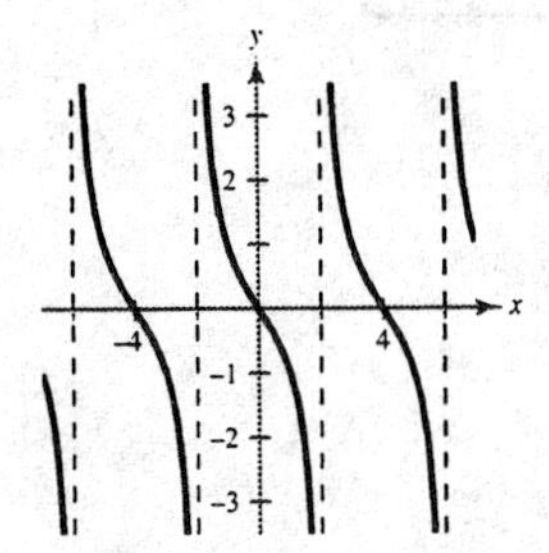

72. $g(t) = 3 \cos(t + \pi)$

Period $= 2\pi$

Amplitude: 3

This is the graph of $y = 3 \cos t$ shifted to the left π units.

t	$-\pi$	$-\frac{\pi}{2}$	0	$\frac{\pi}{2}$	π
$g(t)$	3	0	-3	0	3

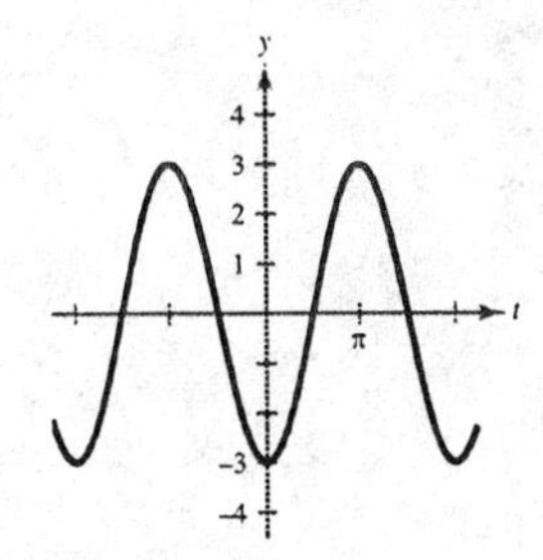

74. $h(t) = \sec\left(t - \dfrac{\pi}{4}\right)$

This is the graph of $y = \sec t$ shifted to the right $(\pi/4)$ units.

Period $= 2\pi$

Asymptotes: $x = -\dfrac{\pi}{4}, x = \dfrac{3\pi}{4}$

t	0	$\dfrac{\pi}{4}$	$\dfrac{\pi}{2}$
$h(t)$	1.414	1	1.414

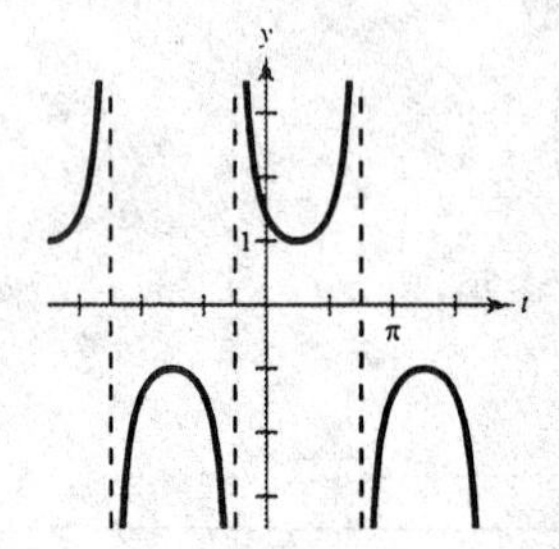

76. $f(t) = 3\csc\left(2t + \dfrac{\pi}{4}\right)$

Period $= \dfrac{2\pi}{2} = \pi$

Asymptotes: $x = -\dfrac{\pi}{8}, x = \dfrac{3\pi}{8}$

t	0	$\dfrac{\pi}{8}$	$\dfrac{\pi}{4}$
$f(t)$	4.243	3	4.243

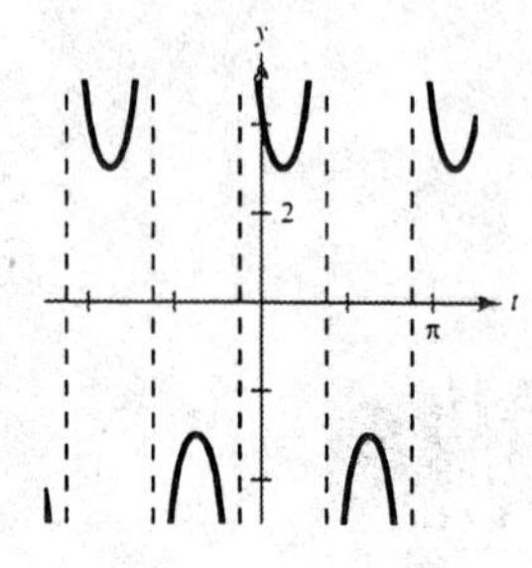

78. $y = 2\arccos x$

Domain: $[-1, 1]$

Range: $[0, 2\pi]$

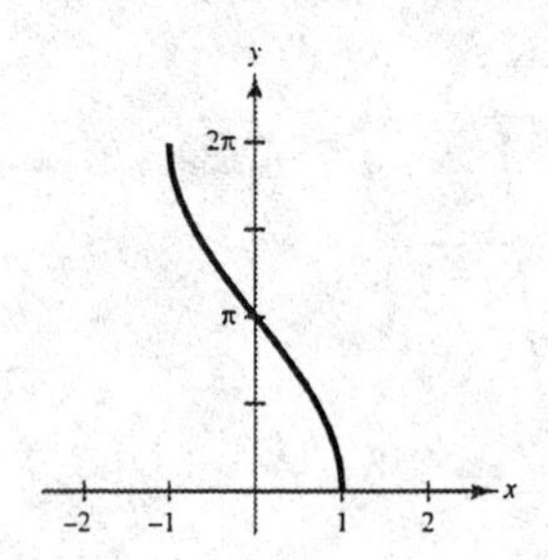

80. $y = \dfrac{x}{3} + \cos \pi x$

Not periodic

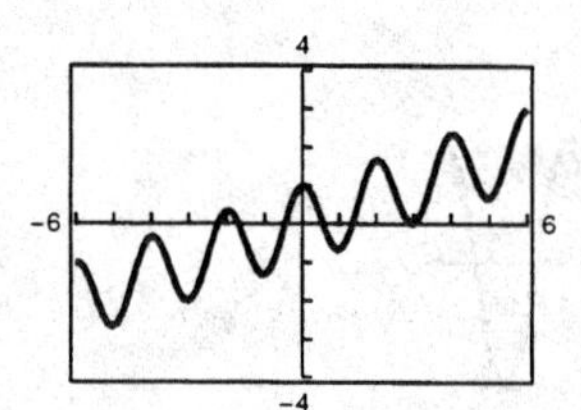

82. $y = 4 - \dfrac{x}{4} + \cos \pi x$

Not periodic

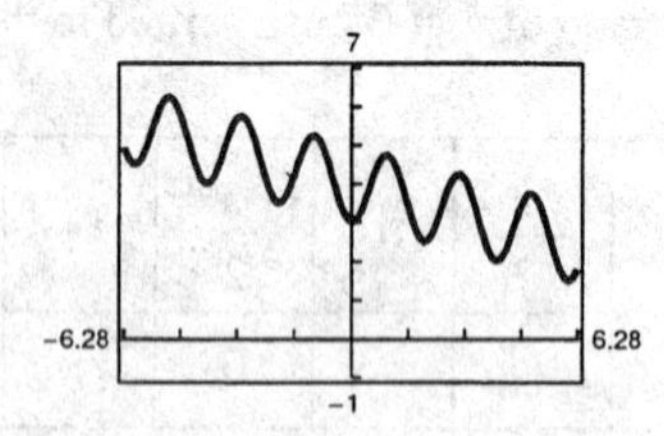

84. $f(\theta) = \cot \dfrac{\pi\theta}{8}$

Period: $\dfrac{\pi}{(\pi/8)} = 8$

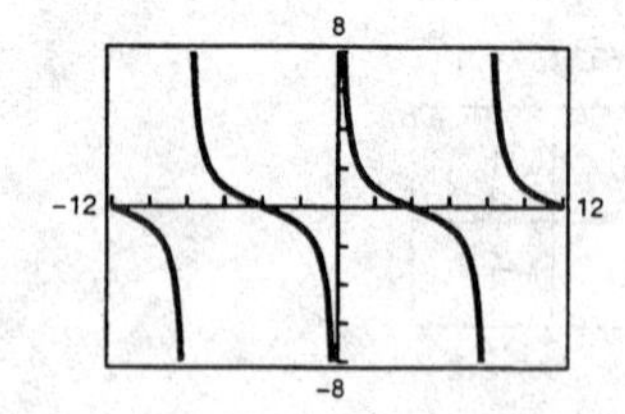

86. $f(x) = \arccos(x - \pi)$

Not periodic

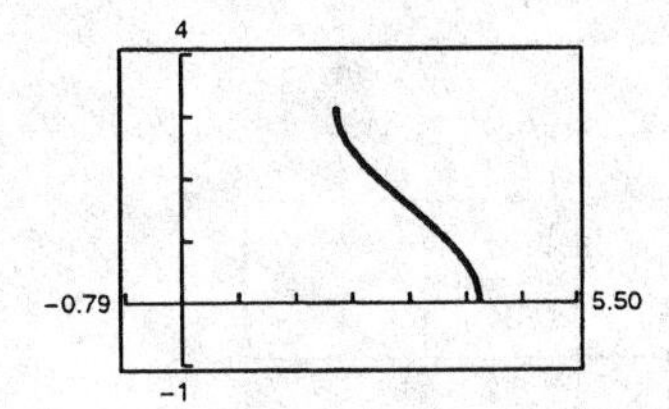

88. $E(t) = 110 \cos\left(120\pi t - \frac{\pi}{3}\right)$

Period: $\frac{2\pi}{120\pi} = \frac{1}{60}$

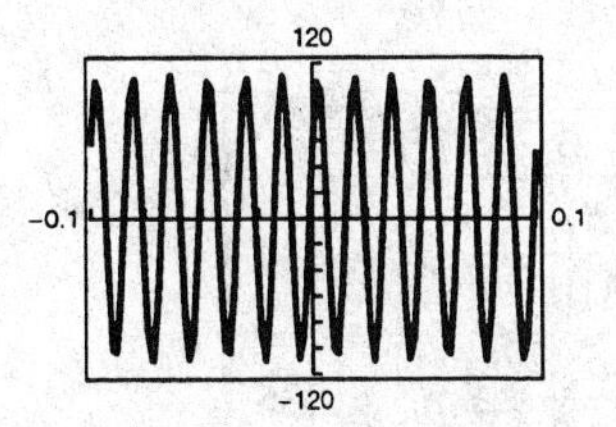

90. $g(x) = \sin e^x$

Not periodic

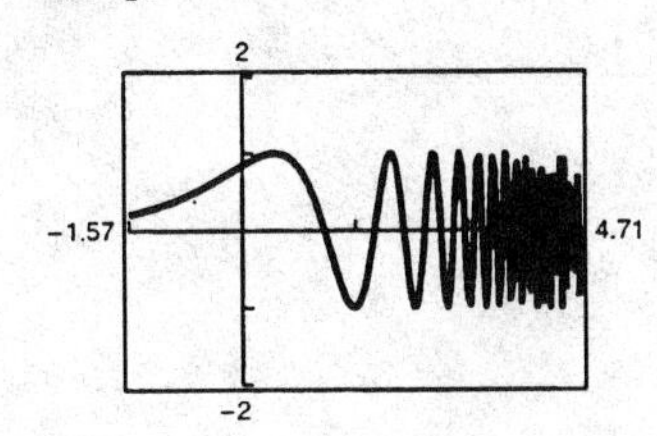

92. $h(x) = 4 \sin^2 x \cos^2 x$

Periodic

Maximum point: $\left(\frac{\pi}{4}, 1\right)$

Minimum point: $(0, 0)$

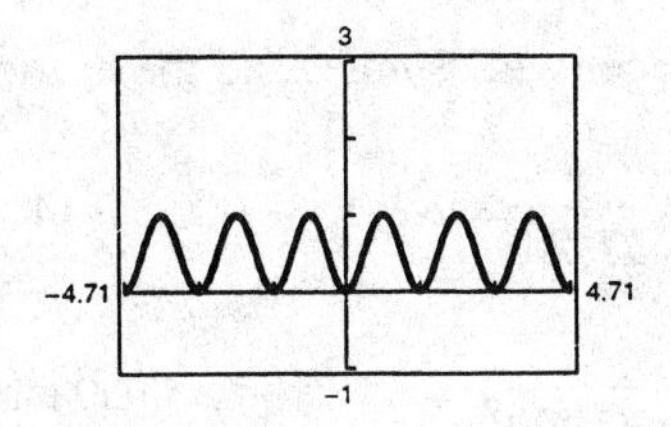

94. $f(x) = a \cos(bx - c)$

Amplitude: 3

Period: $\pi \Longrightarrow f(x) = 3 \cos(2x)$

96. $f(x) = a \cos(bx - c)$

Amplitude: $\frac{1}{2}$

Period: $2 \Longrightarrow f(x) = \frac{1}{2} \cos \pi x$

98. $f(x) = a \tan(bx)$

Period: $\frac{\pi}{2} \Longrightarrow f(x) = a \tan(2x)$.

Passes through point $\left(\frac{\pi}{8}, -2\right) \Longrightarrow -2 = a \tan\left(2\left(\frac{\pi}{8}\right)\right) = a$

Hence, $f(x) = -2 \tan 2x$.

100. Let $u = \arccos \frac{x}{2}$, $\cos u = \frac{x}{2}$.

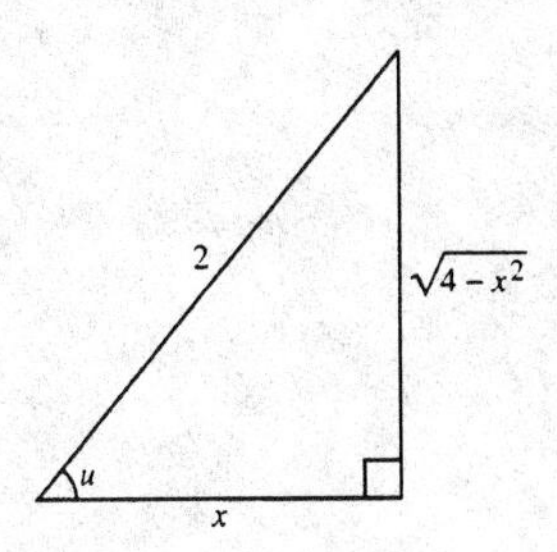

$$\tan\left(\arccos \frac{x}{2}\right) = \tan u = \frac{\sqrt{4 - x^2}}{x}$$

102. Let $u = \arcsin 10x$, $\sin u = 10x$.

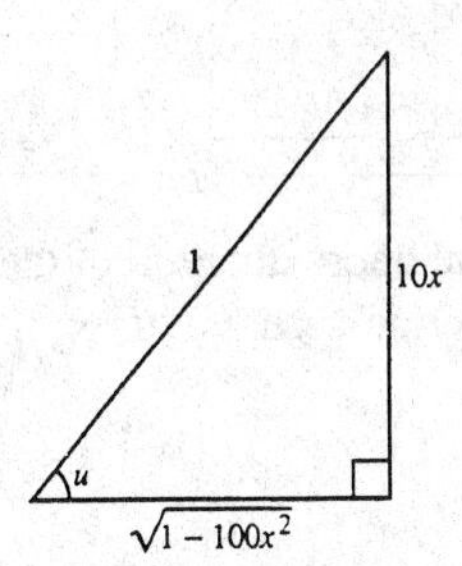

$$\csc(\arcsin 10x) = \csc u = \frac{\text{hyp}}{\text{opp}} = \frac{1}{10x}$$

104. $\tan\theta = \dfrac{70}{30}$

$\theta = \arctan\dfrac{70}{30} \approx 66.8°$

106. $\left.\begin{aligned} \sin 48° &= \frac{d_1}{650} \Longrightarrow d_1 \approx 483 \\ \cos 25° &= \frac{d_2}{810} \Longrightarrow d_2 \approx 734 \end{aligned}\right\} d_1 + d_2 = 1217$

$\left.\begin{aligned} \cos 48° &= \frac{d_3}{650} \Longrightarrow d_3 \approx 435 \\ \sin 25° &= \frac{d_4}{810} \Longrightarrow d_4 \approx 342 \end{aligned}\right\} d_3 - d_4 \approx 93$

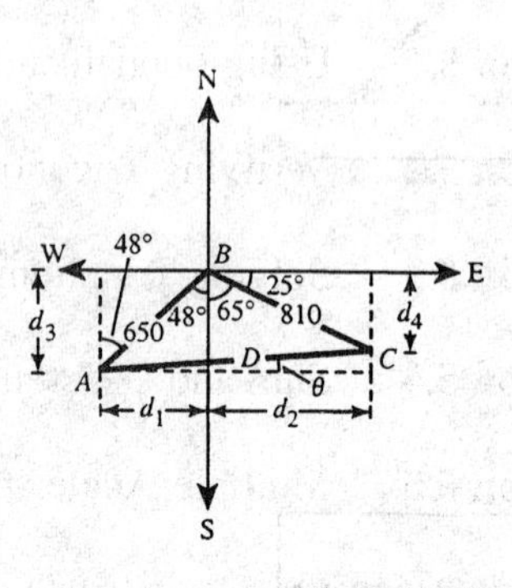

$\tan\theta \approx \dfrac{93}{1217} \Longrightarrow \theta \approx 4.4°$

$\sec 4.4° \approx \dfrac{D}{1217} \Longrightarrow D \approx 1217 \sec 4.4° \approx 1221$

The distance is 1221 miles and the bearing is N 85.6° E.

108. $\tan 14° = \dfrac{y}{37{,}000} \Longrightarrow y = 37{,}000 \tan 14° \approx 9225.1$ feet

$\tan 58° = \dfrac{x + y}{37{,}000} \Longrightarrow x + y = 37{,}000 \tan 58° \approx 59{,}212.4$ feet

$x = 59{,}212.4 - 9225.1 \approx 49{,}987.2$ feet

The towns are approximately 50,000 feet apart.

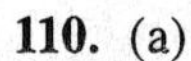

110. (a)

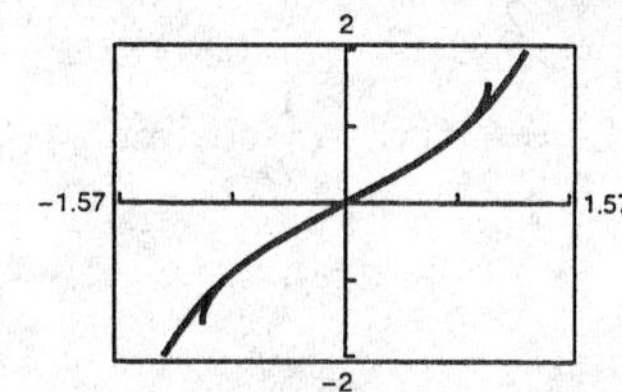

The polynomial approximation of the arcsine function is accurate over $-1 \le x \le 1$.

(b)

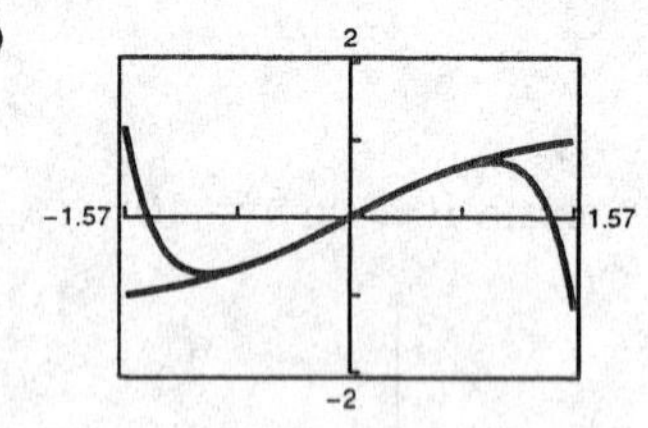

The polynomial approximation of the arctangent function accurate of $-\frac{1}{2} \le x \le \frac{1}{2}$.

(c) The next term appears to be $\dfrac{x^9}{9}$.

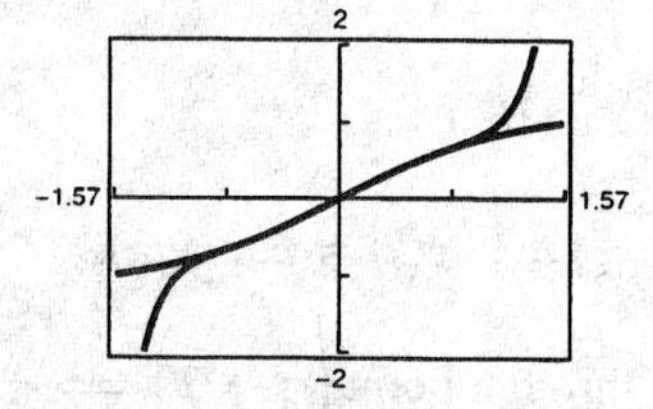

The accuracy of the approximation improved.

CHAPTER 5
Analytic Trigonometry

CHAPTER 5
Analytic Trigonometry

Section 5.1 Using Fundamental Identities

Solutions to Even-Numbered Exercises

2. $\tan x = \dfrac{\sqrt{3}}{3}, \cos x = -\dfrac{\sqrt{3}}{2}$

x is in Quadrant III.

$$\sin x = -\sqrt{1 - \left(-\frac{\sqrt{3}}{2}\right)^2} = -\sqrt{\frac{1}{4}} = -\frac{1}{2}$$

$$\csc x = \frac{1}{\sin x} = -2$$

$$\sec x = \frac{1}{\cos x} = -\frac{2}{\sqrt{3}} = -\frac{2\sqrt{3}}{3}$$

$$\cot x = \frac{1}{\tan x} = \frac{3}{\sqrt{3}} = \sqrt{3}$$

4. $\csc \theta = \dfrac{5}{3}, \tan \theta = \dfrac{3}{4}$

θ is in Quadrant I.

$$\sin \theta = \frac{1}{\csc \theta} = \frac{3}{5}$$

$$\cos \theta = \frac{\sin \theta}{\tan \theta} = \frac{3}{5} \cdot \frac{4}{3} = \frac{4}{5}$$

$$\sec \theta = \frac{1}{\cos \theta} = \frac{5}{4}$$

$$\cot \theta = \frac{1}{\tan \theta} = \frac{4}{3}$$

6. $\cot \phi = -3, \sin \phi = \dfrac{\sqrt{10}}{10}$

ϕ is in Quadrant II.

$$\cos \phi = \cot \phi \sin \phi = -\frac{3\sqrt{10}}{10}$$

$$\tan \phi = \frac{1}{\cot \phi} = -\frac{1}{3}$$

$$\csc \phi = \frac{1}{\sin \phi} = \sqrt{10}$$

$$\sec \phi = \frac{1}{\cos \phi} = -\frac{10}{3\sqrt{10}} = -\frac{\sqrt{10}}{3}$$

8. $\cos\left(\dfrac{\pi}{2} - x\right) = \dfrac{3}{5}, \cos x = \dfrac{4}{5}$

x is in Quadrant I.

$$\sin x = \sqrt{1 - \left(\frac{4}{5}\right)^2} = \frac{3}{5}$$

$$\tan x = \frac{\sin x}{\cos x} = \frac{3}{5} \cdot \frac{5}{4} = \frac{3}{4}$$

$$\csc x = \frac{1}{\sin x} = \frac{5}{3}$$

$$\sec x = \frac{1}{\cos x} = \frac{5}{4}$$

$$\cos x = \frac{1}{\tan x} = \frac{4}{3}$$

10. $\csc x = 5, \cos x > 0$

x is in Quadrant I.

$$\sin x = \frac{1}{\csc x} = \frac{1}{5}$$

$$\cos x = \sqrt{1 - \left(\frac{1}{5}\right)^2} = \frac{2\sqrt{6}}{5}$$

$$\tan x = \frac{\sin x}{\cos x} = \frac{1}{5} \cdot \frac{5}{2\sqrt{6}} = \frac{\sqrt{6}}{12}$$

$$\sec x = \frac{1}{\cos x} = \frac{5}{2\sqrt{6}} = \frac{5\sqrt{6}}{12}$$

$$\cot x = \frac{1}{\tan x} = 2\sqrt{6}$$

12. $\sec \theta = -3, \tan \theta < 0$

θ is in Quadrant II.

$$\cos \theta = \frac{1}{\sec \theta} = -\frac{1}{3}$$

$$\sin \theta = \sqrt{1 - \left(-\frac{1}{3}\right)^2} = \frac{2\sqrt{2}}{3}$$

$$\tan \theta = \frac{\sin \theta}{\cos \theta} = \frac{2\sqrt{2}}{3} \cdot -\frac{3}{1} = -2\sqrt{2}$$

$$\csc \theta = \frac{1}{\sin \theta} = \frac{3}{2\sqrt{2}} = \frac{3\sqrt{2}}{4}$$

$$\cot \theta = \frac{1}{\tan \theta} = -\frac{1}{2\sqrt{2}} = -\frac{\sqrt{2}}{4}$$

14. $\tan\theta$ is undefined, $\sin\theta > 0$.

$$\theta = \frac{\pi}{2}$$

$$\tan\theta = \frac{\sin\theta}{\cos\theta} \text{ is undefined} \Rightarrow \cos\theta = 0$$

$$\sin\theta = \sqrt{1 - 0^2} = 1$$

$$\csc\theta = \frac{1}{\sin\theta} = 1$$

$$\sec\theta = \frac{1}{\cos\theta} \text{ is undefined.}$$

$$\cot\theta = \frac{\cos\theta}{\sin\theta} = \frac{0}{1} = 0$$

16. As $x \to 0^+$,

$$\cos x \to 1 \text{ and } \sec x = \frac{1}{\cos x} \to 1.$$

18. As $x \to \pi^+$,

$$\sin x \to 0 \text{ and } \csc x = \frac{1}{\sin x} \to -\infty.$$

20. $\cot x \sin x = \dfrac{\cos x}{\sin x}\sin x = \cos x$

Matches (b).

22. $(1 - \cos^2 x)(\csc x) = (\sin^2 x)\dfrac{1}{\sin x} = \sin x$

Matches (f).

24. $\dfrac{\sin\left[\left(\frac{\pi}{2}\right) - x\right]}{\cos\left[\left(\frac{\pi}{2}\right) - x\right]} = \dfrac{\cos x}{\sin x} = \cot x$

Matches (c).

26. $\cos^2 x(\sec^2 x - 1) = \cos^2 x(\tan^2 x)$

$$= \cos^2 x\left(\frac{\sin^2 x}{\cos^2 x}\right) = \sin^2 x$$

Matches (c).

28. $\cot x \sec x = \dfrac{\cos x}{\sin x} \cdot \dfrac{1}{\cos x} = \dfrac{1}{\sin x} = \csc x$

Matches (a).

30. $\dfrac{\cos^2\left[\left(\frac{\pi}{2}\right) - x\right]}{\cos x} = \dfrac{\sin^2 x}{\cos x} = \dfrac{\sin x}{\cos x}\sin x$

$$= \tan x \sin x$$

Matches (d).

32. $\sin\phi(\csc\phi - \sin\phi) = \sin\phi\csc\phi - \sin^2\phi$

$$= \sin\phi \cdot \frac{1}{\sin\phi} - \sin^2\phi$$

$$= 1 - \sin^2\phi$$

$$= \cos^2\phi$$

34. $\sec^2 x(1 - \sin^2 x) = \sec^2 x - \sec^2 x \sin^2 x$

$$= \sec^2 x - \frac{1}{\cos^2 x} \cdot \sin^2 x$$

$$= \sec^2 x - \frac{\sin^2 x}{\cos^2 x}$$

$$= \sec^2 x - \tan^2 x$$

$$= 1$$

36. $\dfrac{\csc\theta}{\sec\theta} = \dfrac{\frac{1}{\sin\theta}}{\frac{1}{\cos\theta}} = \dfrac{\cos\theta}{\sin\theta} = \cot\theta$

38. $\dfrac{1}{\tan^2 x + 1} = \dfrac{1}{\sec^2 x} = \dfrac{1}{\frac{1}{\cos^2 x}} = \cos^2 x$

40. $\dfrac{\tan^2\theta}{\sec^2\theta} = \dfrac{\sin^2\theta}{\cos^2\theta} \cdot \dfrac{1}{\sec^2\theta} = \dfrac{\sin^2\theta}{\cos^2\theta} \cdot \dfrac{1}{\frac{1}{\cos^2\theta}} = \dfrac{\sin^2\theta\cos^2\theta}{\cos^2\theta} = \sin^2\theta$

42. $\cot\left(\dfrac{\pi}{2} - x\right)\cos x = \tan x \cos x = \dfrac{\sin x}{\cos x} \cdot \cos x = \sin x$

44. $(\cos t)(1 + \tan^2 t) = (\cos t)(\sec^2 t) = \dfrac{\cos t}{\cos^2 t} = \dfrac{1}{\cos t} = \sec t$

46. $\sec^2 x \tan^2 x + \sec^2 x = \sec^2 x(\tan^2 x + 1) = \sec^2 x(\sec^2 x) = \sec^4 x$

48. $\dfrac{\sec^2 x - 1}{\sec x - 1} = \dfrac{(\sec x + 1)(\sec x - 1)}{\sec x - 1} = \sec x + 1$

50. $1 - 2\cos^2 x + \cos^4 x = (1 - \cos^2 x)(1 - \cos^2 x)$
$= \sin^2 x \sin^2 x$
$= \sin^4 x$

52. $\csc^3 x - \csc^2 x - \csc x - 1 = \csc^2 x(\csc x - 1) - (\csc x - 1)$
$= (\csc^2 x - 1)(\csc x - 1)$
$= \cot^2 x(\csc x - 1)$

54. $(\cot x + \csc x)(\cot x - \csc x) = \cot^2 x - \csc^2 x$
$= -1$

56. $(3 - 3\sin x)(3 + 3\sin x) = 9 - 9\sin^2 x$
$= 9(1 - \sin^2 x)$
$= 9\cos^2 x$

58. $\dfrac{1}{\sec x + 1} - \dfrac{1}{\sec x - 1} = \dfrac{\sec x - 1 - (\sec x + 1)}{(\sec x + 1)(\sec x - 1)}$
$= \dfrac{\sec x - 1 - \sec x - 1}{\sec^2 x - 1}$
$= \dfrac{-2}{\tan^2 x}$
$= -2\left(\dfrac{1}{\tan^2 x}\right)$
$= -2\cot^2 x$

60. $\tan x - \dfrac{\sec^2 x}{\tan x} = \dfrac{\tan^2 x - \sec^2 x}{\tan x}$
$= \dfrac{-1}{\tan x}$
$= -\cot x$

62. $\dfrac{5}{\tan x + \sec x} \cdot \dfrac{\tan x - \sec x}{\tan x - \sec x} = \dfrac{5(\tan x - \sec x)}{\tan^2 x - \sec^2 x}$
$= \dfrac{5(\tan x - \sec x)}{-1}$
$= 5(\sec x - \tan x)$

64. $\dfrac{\tan^2 x}{\csc x + 1} \cdot \dfrac{\csc x - 1}{\csc x - 1} = \dfrac{\tan^2 x(\csc x - 1)}{\csc^2 x - 1}$
$= \dfrac{\tan^2 x(\csc x - 1)}{\cot^2 x}$
$= \tan^2 x(\csc x - 1)\tan^2 x$
$= \tan^4 x(\csc x - 1)$

66. $y_1 = \cos x + \sin x \tan x,\ y_2 = \sec x$

x	0.2	0.4	0.6	0.8	1.0	1.2	1.4
y_1	1.0203	1.0857	1.2116	1.4353	1.8508	2.7597	5.8835
y_2	1.0203	1.0857	1.2116	1.4353	1.8508	2.7597	5.8835

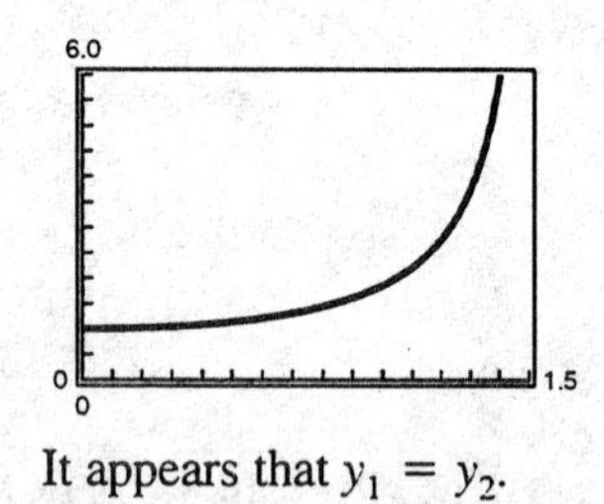

It appears that $y_1 = y_2$.

68. $y_1 = \sec^4 x - \sec^2 - x,\ y_2 = \tan^2 x + \tan^4 x$

x	0.2	0.4	0.6	0.8	1.0	1.2	1.4
y_1	0.0428	0.2107	0.6871	2.1841	8.3087	50.3869	1163.6143
y_2	0.0428	0.2107	0.6871	2.1841	8.3087	50.3869	1163.6143

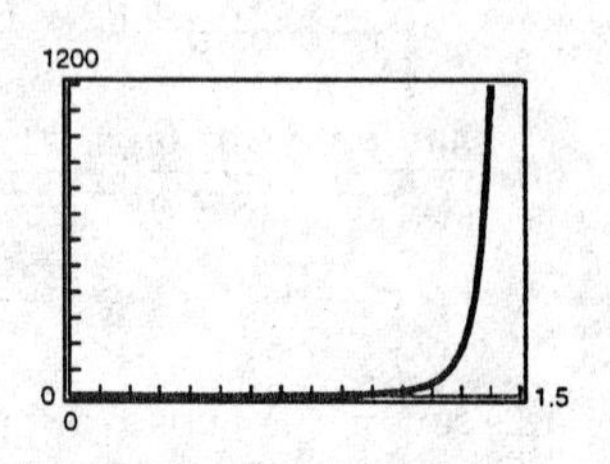

It appears that $y_1 = y_2$.

70. $y_1 = \dfrac{1}{2}\left(\dfrac{1 + \sin\theta}{\cos\theta} + \dfrac{\cos\theta}{1 + \sin\theta}\right)$

y_1 and $y_2 = \sin\theta$

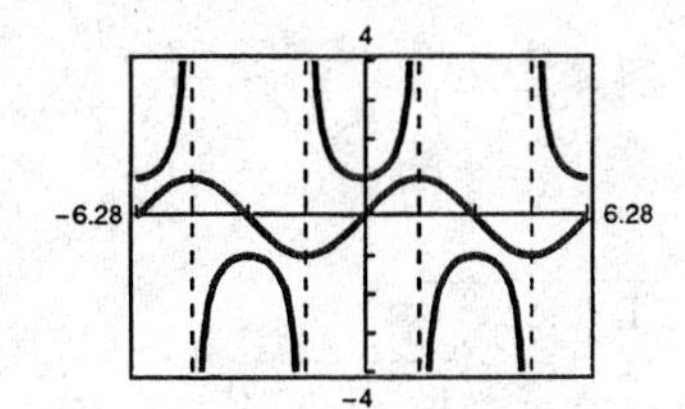

y_1 and $y_2 = \cos\theta$

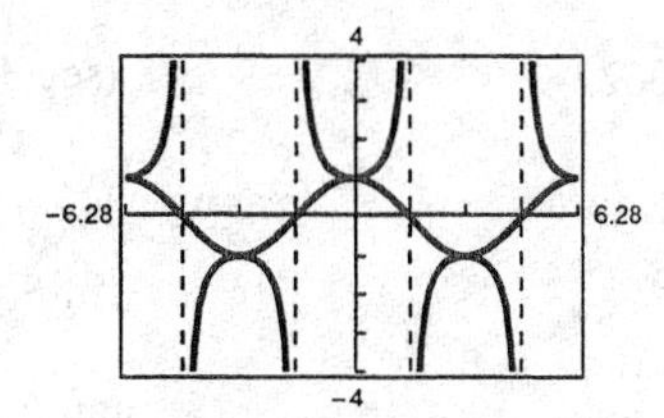

y_1 and $y_2 = \tan\theta$

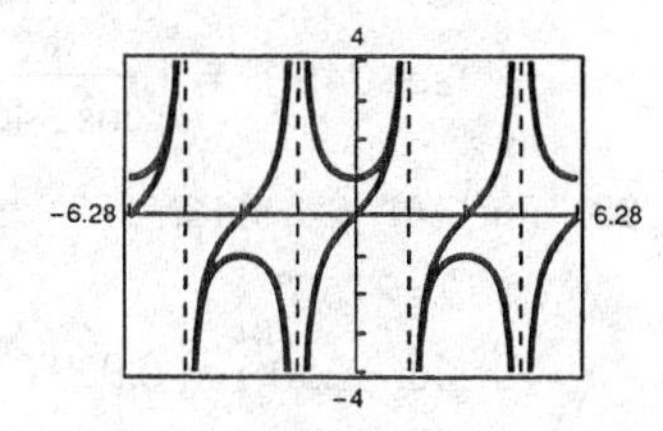

y_1 and $y_2 = \dfrac{1}{\sin\theta} = \csc\theta$

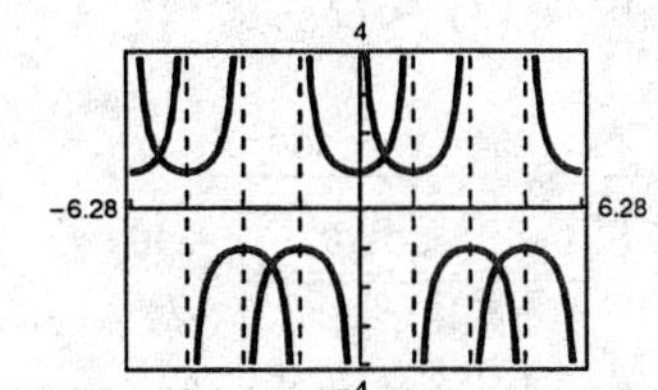

y_1 and $y_2 = \dfrac{1}{\cos\theta} = \sec\theta$

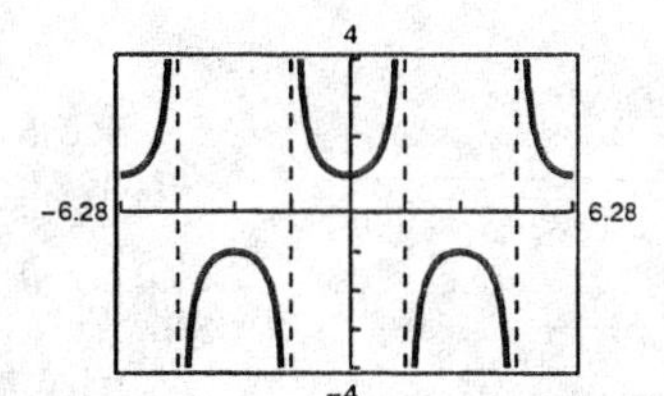

y_1 and $y_2 = \dfrac{1}{\tan\theta} = \cot\theta$

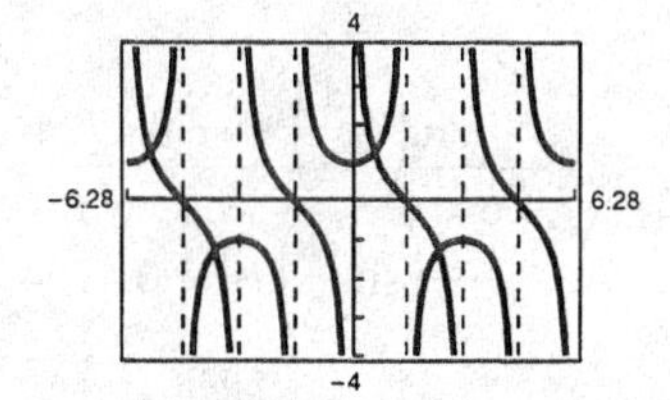

It appears that $\dfrac{1}{2}\left(\dfrac{1 + \sin\theta}{\cos\theta} + \dfrac{\cos\theta}{1 + \sin\theta}\right) = \sec\theta$.

72. Let $x = 2\sin\theta$.

$$\begin{aligned}\sqrt{16 - 4x^2} &= \sqrt{16 - 4(2\sin\theta)^2}\\ &= \sqrt{16(1 - \sin^2\theta)}\\ &= \sqrt{16\cos^2\theta}\\ &= 4\cos\theta\end{aligned}$$

74. Let $x = 2\sec\theta$.

$$\begin{aligned}\sqrt{x^2 - 4} &= \sqrt{(2\sec\theta)^2 - 4}\\ &= \sqrt{4(\sec^2\theta - 1)}\\ &= \sqrt{4\tan^2\theta}\\ &= 2\tan\theta\end{aligned}$$

76. Let $x = 10\tan\theta$.

$$\begin{aligned}\sqrt{x^2 + 100} &= \sqrt{(10\tan\theta)^2 + 100}\\ &= \sqrt{100(\tan^2\theta + 1)}\\ &= \sqrt{100\sec^2\theta}\\ &= 10\sec\theta\end{aligned}$$

78. $\cos\theta = -\sqrt{1 - \sin^2\theta}$

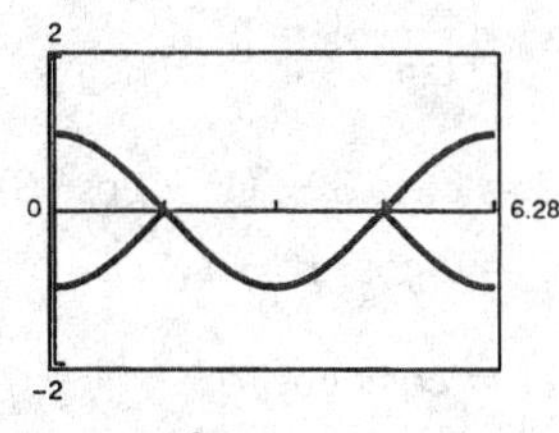

$\dfrac{\pi}{2} \le \theta \le \dfrac{3\pi}{2}$

80. $\tan\theta = \sqrt{\sec^2\theta - 1}$

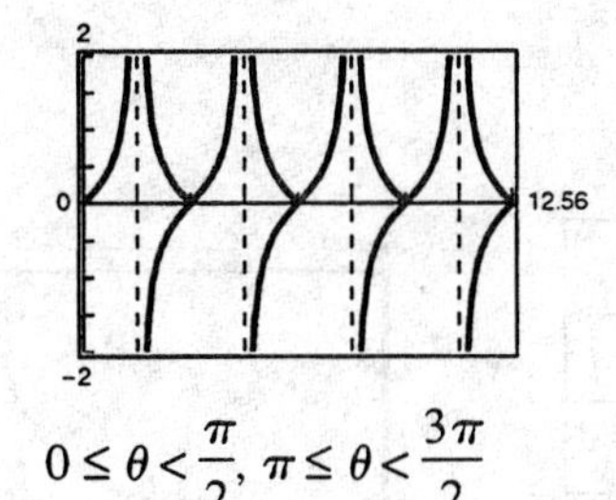

$0 \le \theta < \dfrac{\pi}{2}, \ \pi \le \theta < \dfrac{3\pi}{2}$

82.
$$\begin{aligned}\ln|\cot t| + \ln(1 + \tan^2 t) &= \ln|\cot t|(1 + \tan^2 t)\\ &= \ln\frac{(1 + \tan^2 t)}{|\tan t|}\\ &= \ln\left|\frac{1}{\tan t} + \frac{\tan^2 t}{\tan t}\right|\\ &= \ln|\cot t + \tan t|\end{aligned}$$

84. $\dfrac{1}{5\cos\theta} = \dfrac{1}{5(1/\sec\theta)} = \dfrac{\sec\theta}{5} = \dfrac{1}{5}\sec\theta \ne 5\sec\theta$

Not an identity because $\dfrac{1}{5\cos\theta} \ne \dfrac{5}{\cos\theta}$.

86. $\sin\theta\csc\phi = \sin\theta\dfrac{1}{\sin\theta}$

This may be simplified only if $\theta = \phi$. Thus, $\sin\theta\csc\phi = 1$ is not an identity because θ must be equal to ϕ to be true.

88. $\tan^2\theta + 1 = \sec^2\theta$

(a) $\theta = 346°$

$(\tan 346°)^2 + 1 \approx 1.0622$

$(\sec 346°)^2 = \left(\dfrac{1}{\cos 346°}\right)^2 \approx 1.0622$

(b) $\theta = 3.1$

$(\tan 3.1)^2 + 1 \approx 1.00173$

$(\sec 3.1)^2 = \left(\dfrac{1}{\cos 3.1}\right)^2 \approx 1.00173$

90. $\sin(-\theta) = -\sin\theta$

(a) $\theta = 250°$

$\sin(-250°) \approx 0.9397$

$-(\sin 250°) \approx 0.9397$

(b) $\theta = \dfrac{1}{2}$

$\sin\left(-\dfrac{1}{2}\right) \approx -0.4794$

$-\left(\sin\dfrac{1}{2}\right) \approx -0.4794$

92. $\cos\theta$

$\sin\theta = \pm\sqrt{1-\cos^2\theta}$

$\tan\theta = \dfrac{\sin\theta}{\cos\theta} = \pm\dfrac{\sqrt{1-\cos^2\theta}}{\cos\theta}$

$\csc\theta = \dfrac{1}{\sin\theta} = \pm\dfrac{1}{\sqrt{1-\cos^2\theta}}$

$\sec\theta = \dfrac{1}{\cos\theta}$

$\cot\theta = \dfrac{1}{\tan\theta} = \pm\dfrac{\cos\theta}{\sqrt{1-\cos^2\theta}}$

The sign $+$ or $-$ depends on the choice of θ.

94. The period of the Seward model is $\dfrac{2\pi}{\pi/6} = 12.$

The period of the New Orleans model is $\dfrac{2\pi}{\pi/6} = 12.$

Section 5.2 Verifying Trigonometric Identities

Solutions to Even-Numbered Exercises

2. $\tan y \cot y = \tan y\left(\dfrac{1}{\tan y}\right) = 1$

4. $\cot^2 y(\sec^2 y - 1) = \cot^2 y \tan^2 y = 1$

6. $\cos^2\beta - \sin^2\beta = \cos^2\beta - (1-\cos^2\beta)$
$= 2\cos^2\beta - 1$

8. $2 - \sec^2 z = 2 - (1 + \tan^2 z)$
$= 1 - \tan^2 z$

10. $\cos t(\csc^2 t - 1) = \cos t \cot^2 t$

$= \sin t\left(\dfrac{\cos t}{\sin t}\right)\cot^2 t$

$= \dfrac{1}{\csc t}\cot^3 t$

$= \dfrac{\cot^3 t}{\csc t}$

12.

x	0.2	0.4	0.6	0.8	1.0	1.2	1.4
y_1	1.0203	1.0857	1.2116	1.4353	1.8508	2.7597	5.8835
y_2	1.0203	1.0857	1.2116	1.4353	1.8508	2.7597	5.8835

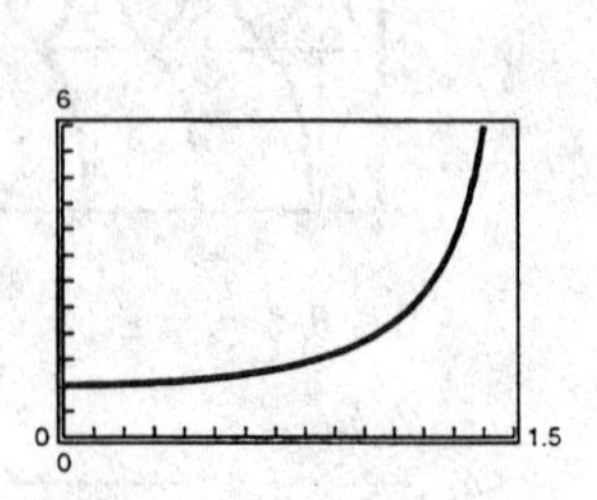

$y_1 = \dfrac{\sec x - 1}{1-\cos x} = \dfrac{\dfrac{1}{\cos x} - 1}{1-\cos x} = \dfrac{1-\cos x}{\cos x(1-\cos x)} = \dfrac{1}{\cos x} = \sec x = y_2$

14.

x	0.2	0.4	0.6	0.8	1.0	1.2	1.4
y_1	0.0403	0.1646	0.3863	0.7386	1.3105	2.3973	5.7135
y_2	0.0403	0.1646	0.3863	0.7386	1.3105	2.3973	5.7135

$$\begin{aligned} y_1 = \sec x - \cos x &= \frac{1}{\cos x} - \cos x \\ &= \frac{1-\cos^2 x}{\cos x} \\ &= \frac{\sin^2 x}{\cos x} \\ &= \sin x\left(\frac{\sin x}{\cos x}\right) \\ &= \sin x \tan x \\ &= y_2 \end{aligned}$$

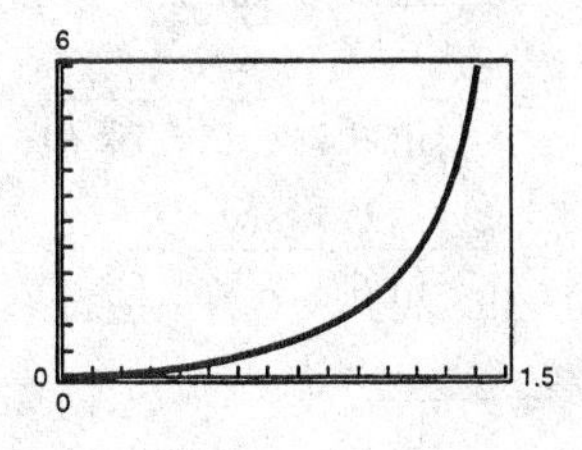

16.

x	0.2	0.4	0.6	0.8	1.0	1.2	1.4
y_1	1.4958	2.2756	3.5939	6.0760	11.6160	28.4287	136.4545
y_2	1.4958	2.2756	3.5939	6.0760	11.6160	28.4287	136.4545

$$\begin{aligned} y_1 = \frac{\sec x + \tan x}{\sec x - \tan x} &= \frac{(\sec x + \tan x)^2}{(\sec x - \tan x)(\sec x + \tan x)} \\ &= \frac{(\sec x + \tan x)^2}{\sec^2 x - \tan^2 x} \\ &= (\sec x + \tan x)^2 \\ &= y_2 \end{aligned}$$

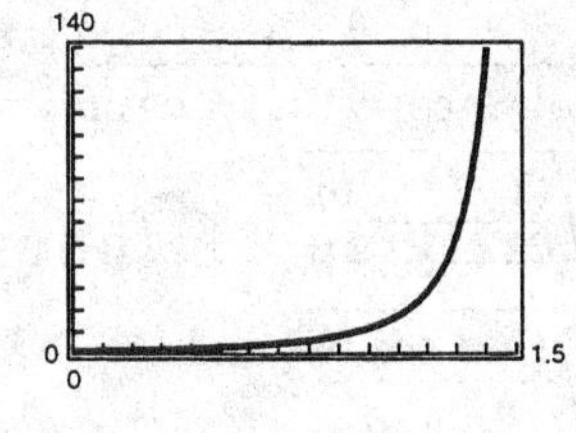

18.

x	0.2	0.4	0.6	0.8	1.0	1.2	1.4
y_1	4.8348	2.1785	1.2064	0.6767	0.3469	0.1409	0.0293
y_2	4.8348	2.1785	1.2064	0.6767	0.3469	0.1409	0.0293

$$\begin{aligned} y_1 &= \frac{1}{\sin x} - \frac{1}{\csc x} \\ &= \csc x - \sin x \\ &= y_2 \end{aligned}$$

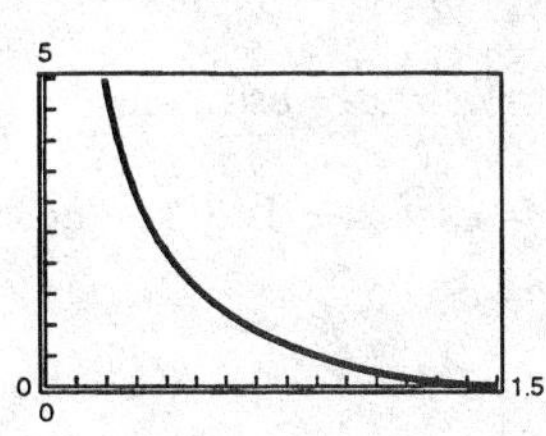

20. $$\begin{aligned} \sec^6 x(\sec x \tan x) - \sec^4 x(\sec x \tan x) &= \sec^4 x(\sec x \tan x)(\sec^2 x - 1) \\ &= \sec^4 x(\sec x \tan x)\tan^2 x \\ &= \sec^5 x \tan^3 x \end{aligned}$$

22. $$\frac{\cos[(\pi/2) - x]}{\sin[(\pi/2) - x]} = \frac{\sin x}{\cos x} = \tan x$$

24. $$\begin{aligned} (1 + \sin y)[1 + \sin(-y)] &= (1 + \sin y)(1 - \sin y) \\ &= 1 - \sin^2 y \\ &= \cos^2 y \end{aligned}$$

26. $$\frac{1+\sec(-\theta)}{\sin(-\theta)+\tan(-\theta)} = \frac{1+\sec\theta}{-\sin\theta-\tan\theta}$$
$$= -\frac{1+\sec\theta}{\sin\theta+\tan\theta}$$
$$= -\frac{1+\sec\theta}{\sin\theta[1+(1/\cos\theta)}$$
$$= -\frac{1+\sec\theta}{\sin\theta(1+\sec\theta)}$$
$$= -\frac{1}{\sin\theta}$$
$$= -\csc\theta$$

28. $$\frac{\tan x+\tan y}{1-\tan x\tan y} = \frac{\dfrac{1}{\cot x}+\dfrac{1}{\cot y}}{1-\dfrac{1}{\cot x}\cdot\dfrac{1}{\cot y}}\cdot\frac{\cot x\cot y}{\cot x\cot y}$$
$$= \frac{\cot y+\cot x}{\cot x\cot y-1}$$

30. $$\frac{\cos x-\cos y}{\sin x+\sin y}+\frac{\sin x-\sin y}{\cos x+\cos y} = \frac{(\cos x-\cos y)(\cos x+\cos y)+(\sin x-\sin y)(\sin x+\sin y)}{(\sin x+\sin y)(\cos x+\cos y)}$$
$$= \frac{\cos^2 x-\cos^2 y+\sin^2 x-\sin^2 y}{(\sin x+\sin y)(\cos x+\cos y)}$$
$$= \frac{(\cos^2 x+\sin^2 x)-(\cos^2 y+\sin^2 y)}{(\sin x+\sin y)(\cos x+\cos y)}$$
$$= 0$$

32. $$\sqrt{\frac{1-\cos\theta}{1+\cos\theta}} = \sqrt{\frac{1-\cos\theta}{1+\cos\theta}\cdot\frac{1-\cos\theta}{1-\cos\theta}}$$
$$= \sqrt{\frac{(1-\cos\theta)^2}{1-\cos^2\theta}}$$
$$= \sqrt{\frac{(1-\cos\theta)^2}{\sin^2\theta}}$$
$$= \frac{1-\cos\theta}{|\sin\theta|}$$

34. $\sec^2 y-\cot^2\left(\frac{\pi}{2}-y\right) = \sec^2 y-\tan^2 y = 1$

36. $\sec^2\left(\frac{\pi}{2}-x\right)-1 = \csc^2 x-1 = \cot^2 x$

38. $$\csc x(\csc x-\sin x)+\frac{\sin x-\cos x}{\sin x}+\cot x = \csc^2 x-\csc x\sin x+1-\frac{\cos x}{\sin x}+\cot x$$
$$= \csc^2 x-1+1-\cot x+\cot x$$
$$= \csc^2 x$$

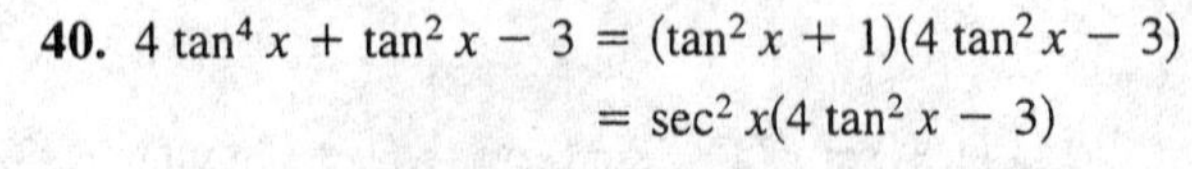

40. $$4\tan^4 x+\tan^2 x-3 = (\tan^2 x+1)(4\tan^2 x-3)$$
$$= \sec^2 x(4\tan^2 x-3)$$

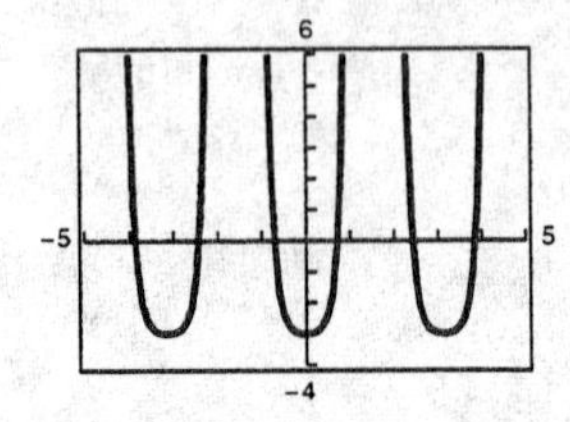

42. $$\sin x(1-2\cos^2 x+\cos^4 x) = \sin x(1-\cos^2 x)^2$$
$$= \sin x(\sin^2 x)^2$$
$$= \sin^5 x$$

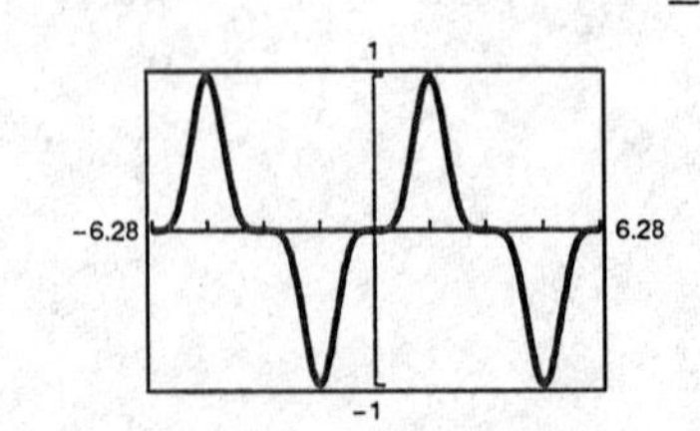

44. $\csc^4 \theta - \cot^4 \theta = (\csc^2 \theta - \cot^2 \theta)(\csc^2 \theta + \cot^2 \theta)$

$= \csc^2 \theta + \cot^2 \theta$

$= \csc^2 \theta + (\csc^2 \theta - 1)$

$= 2 \csc^2 \theta - 1$

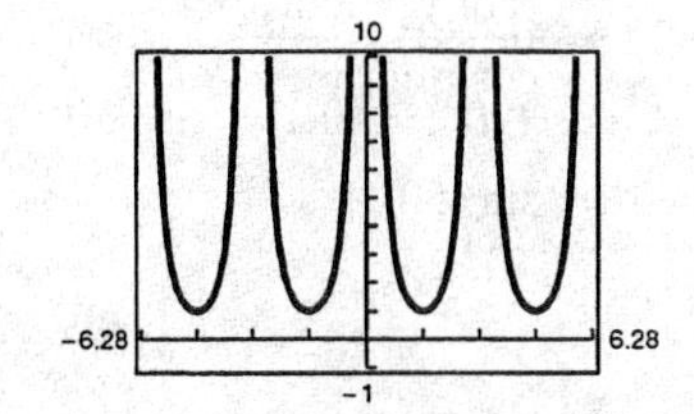

46. $\dfrac{\cot \alpha}{\csc \alpha - 1} \cdot \dfrac{\csc \alpha + 1}{\csc \alpha + 1} = \dfrac{\cot \alpha(\csc \alpha + 1)}{\csc^2 \alpha - 1}$

$= \dfrac{\cot \alpha(\csc \alpha + 1)}{\cot^2 \alpha}$

$= \dfrac{\csc \alpha + 1}{\cot \alpha}$

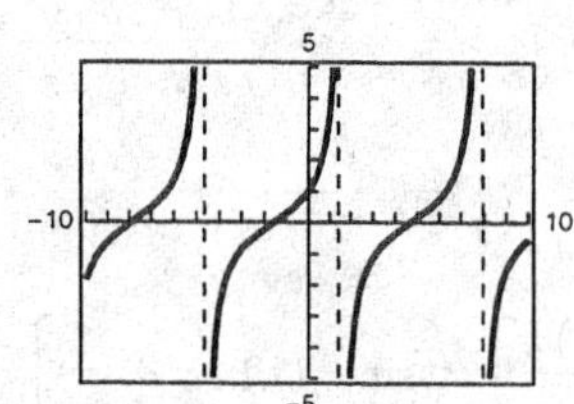

48. $\dfrac{\sin^3 \beta + \cos^3 \beta}{\sin \beta + \cos \beta} = \dfrac{(\sin \beta + \cos \beta)(\sin^2 \beta - \sin \beta \cos \beta + \cos^2 \beta)}{\sin \beta + \cos \beta}$

$= \sin^2 \beta + \cos^2 \beta - \sin \beta \cos \beta$

$= 1 - \sin \beta \cos \beta$

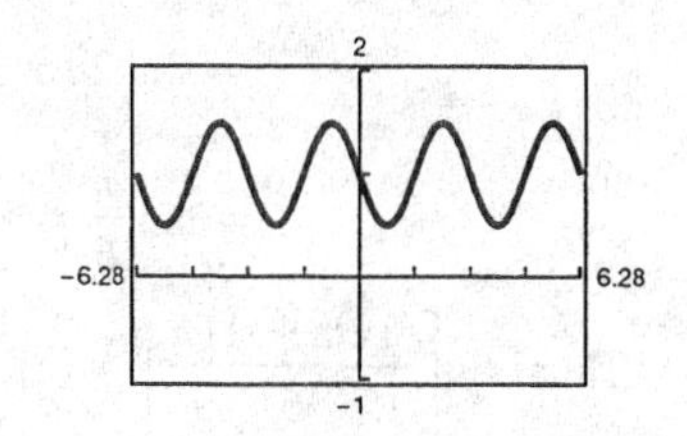

50. The function appears to be $y = \cos x$. Analytically,

$y = \dfrac{\cos x}{1 - \tan x} + \dfrac{\sin x \cdot \cos x}{\sin x - \cos x}$

$= \dfrac{\cos x}{1 - (\sin x/\cos x)} + \dfrac{\sin x \cos x}{\sin x - \cos x}$

$= \dfrac{\cos^2 x}{\cos x - \sin x} - \dfrac{\sin x \cos x}{\cos x - \sin x}$

$= \dfrac{\cos x(\cos x - \sin x)}{\cos x - \sin x} = \cos x.$

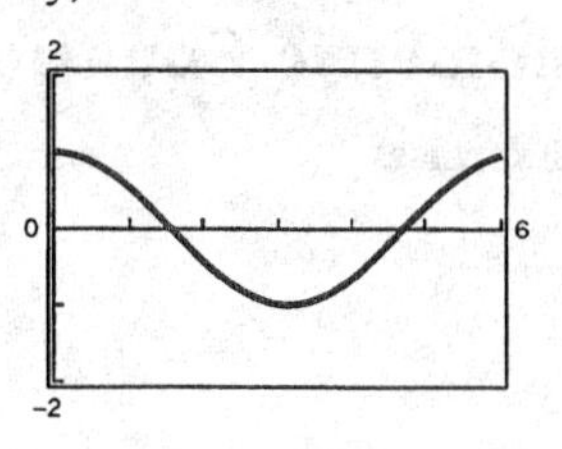

52. The function appears to be $y = \csc t$. Analytically,

$y = \sin t + \dfrac{\cot^2 t}{\csc t}$

$= \dfrac{1 + \cot^2 t}{\csc t}$

$= \dfrac{\csc^2 t}{\csc t} = \csc t.$

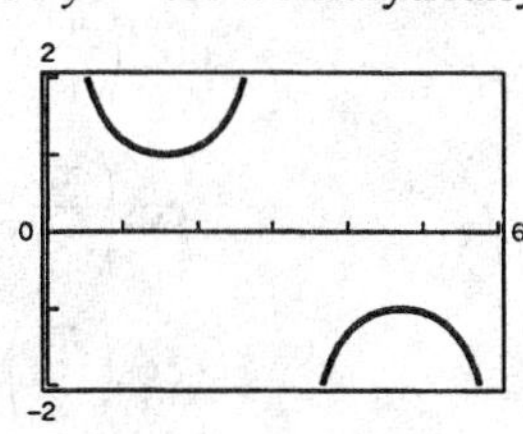

54. $\ln|\sec \theta| = \ln\left|\dfrac{1}{\cos \theta}\right| = \ln|\cos \theta| = -\ln|\cos \theta|$

56. $-\ln|\sec \theta + \tan \theta| = -\ln\left|\dfrac{1}{\cos \theta} + \dfrac{\sin \theta}{\cos \theta}\right|$

$= \ln\left|\dfrac{1 + \sin \theta}{\cos \theta}\right|^{-1}$

$= \ln\left|\dfrac{\cos \theta}{1 + \sin \theta} \cdot \dfrac{1 - \sin \theta}{1 - \sin \theta}\right|$

$= \ln\left|\dfrac{\cos \theta - \cos \theta \sin \theta}{1 - \sin^2 \theta}\right|$

$= \ln\left|\dfrac{\cos \theta - \cos \theta \sin \theta}{\cos^2 \theta}\right|$

$= \ln|\sec \theta - \tan \theta|$

58. $\sin \theta = \sqrt{1 - \cos^2 \theta}$.

True identity is $\sin \theta = \pm\sqrt{1 - \cos^2 \theta}$.

For example, $\sin \theta \neq \sqrt{1 - \cos^2 \theta}$ for $\theta = \dfrac{3\pi}{2}$:

$\sin\left(\dfrac{3\pi}{2}\right) = -1 \neq \sqrt{1 - 0} = 1$

60. $\sqrt{\sin^2 x + \cos^2 x} = \sin x + \cos x$

$\sqrt{\sin^2 x + \cos^2 x} \neq \sin x + \cos x$

The left side is 1 for any x, but the right side is not necessarily 1. The equation is not true for $x = \pi/4$.

62. $\cos^2 18° + \cos^2 72° = \sin^2(90° - 18°) + \cos^2 72°$

$= \sin^2 72° + \cos^2 72°$

$= 1$

64. $\sin^2 12° + \sin^2 40° + \sin^2 50° + \sin^2 78° = \sin^2 12° + \sin^2 78° + \sin^2 40° + \sin^2 50°$

$= \cos^2(90° - 12°) + \sin^2 78° + \cos^2(90° - 40°) + \sin^2 50°$

$= \cos^2 78° + \sin^2 78° + \cos^2 50° + \sin^2 50°$

$= 1 + 1 = 2$

66. $\sin\left[\frac{(12n+1)\pi}{6}\right] = \sin\left[\frac{1}{6}(12n\pi + \pi)\right]$

$= \sin\left(2n\pi + \frac{\pi}{6}\right)$

$= \sin\frac{\pi}{6} = \frac{1}{2}$

Thus, $\sin\left[\frac{(12n+1)\pi}{6}\right] = \frac{1}{2}$ for all integers n.

68. $y = \cos x - \csc x \cdot \cot x$

$= \cos x - \frac{1}{\sin x} \cdot \frac{\cos x}{\sin x}$

$= \cos x(1 - \csc^2 x)$

$= -\cos x \cdot \cot^2 x$

5.3 Solving Trigonometric Equations

Solutions to Even-Numbered Exercises

2. $y = \sin \pi x + \cos \pi x$

$\sin \pi x + \cos \pi x = 0$

$\cos \pi x = -\sin \pi x$

$1 = \frac{-\sin \pi x}{\cos \pi x}$

$1 = -\tan \pi x$

$-1 = \tan \pi x$

$\pi x = -\frac{\pi}{4}, \frac{3\pi}{4}, \frac{7\pi}{4}, \frac{11\pi}{4}$

$x = -\frac{1}{4}, \frac{3}{4}, \frac{7}{4}, \frac{11}{4}$

4. $y = \sec^4\left(\frac{\pi x}{8}\right) - 4$

$\sec^4\left(\frac{\pi x}{8}\right) - 4 = 0$

$\frac{1}{\cos^4(\pi x/8)} + \ = 4$

$\cos^4\left(\frac{\pi x}{8}\right) = \frac{1}{4}$

$\cos\left(\frac{\pi x}{8}\right) = \sqrt[4]{\frac{1}{4}}$

$\cos\left(\frac{\pi x}{8}\right) = \frac{\sqrt{2}}{2}$

$\frac{\pi x}{8} = -\frac{\pi}{4}, \frac{\pi}{4}$

$x = -2, 2$

6. $\csc x - 2 = 0$

(a) $x = \frac{\pi}{6}$

$\csc\frac{\pi}{6} - 2 = \frac{1}{\sin(\pi/6)} - 2$

$= 2 - 2 = 0$

(b) $x = \frac{5\pi}{6}$

$\csc\frac{5\pi}{6} - 2 = \frac{1}{\sin(5\pi/6)} - 2$

$= 2 - 2 = 0$

8. $2\cos^2 4x - 1 = 0$

(a) $x = \frac{\pi}{16}$

$$2\cos^2\left[4\left(\frac{\pi}{16}\right)\right] - 1 = 2\cos^2\frac{\pi}{4} - 1$$
$$= 2\left(\frac{\sqrt{2}}{2}\right)^2 - 1$$
$$= 2\left(\frac{1}{2}\right) - 1 = 1 - 1 = 0$$

(b) $x = \frac{3\pi}{16}$

$$2\cos^2\left[4\left(\frac{3\pi}{16}\right)\right] - 1 = 2\cos^2\frac{3\pi}{4} - 1$$
$$= 2\left(-\frac{\sqrt{2}}{2}\right)^2 - 1$$
$$= 2\left(\frac{1}{2}\right) - 1 = 0$$

10. $\sec^4 x - 4\sec^2 x = 0$

(a) $x = \frac{2\pi}{3}$

$$\sec^4\left(\frac{2\pi}{3}\right) - 4\sec^2\left(\frac{2\pi}{3}\right) = \frac{1}{\cos^4(2\pi/3)} - \frac{4}{\cos^2(2\pi/3)}$$
$$= \frac{1}{(-1/2)^4} - \frac{4}{(-1/2)^2}$$
$$= 16 - 4(4) = 0$$

(b) $x = \frac{5\pi}{3}$

$$\sec^4\left(\frac{5\pi}{3}\right) - 4\sec^2\left(\frac{5\pi}{3}\right) = \frac{1}{\cos^4(5\pi/3)} - \frac{4}{\cos^2(5\pi/3)}$$
$$= \frac{1}{(1/2)^4} - \frac{4}{(1/2)^2}$$
$$= 16 - 4(4) = 0$$

12. $2\sin x - 1 = 0$

$$2\sin x = 1$$
$$\sin x = \frac{1}{2}$$
$$x = \frac{\pi}{6} + 2n\pi$$
$$\text{or } x = \frac{5\pi}{6} + 2n\pi$$

14. $\tan x + 1 = 0$

$$\tan x = -1$$
$$x = \frac{3\pi}{4} + n\pi$$

16. $\csc^2 x - 2 = 0$

$$\csc x = \pm\sqrt{2}$$
$$x = \frac{\pi}{4} + \frac{n\pi}{2}$$

18. $\tan^2 3x = 3$

$$\tan 3x = \pm\sqrt{3}$$
$$3x = \frac{\pi}{3} + n\pi$$
$$x = \frac{\pi}{9} + \frac{n\pi}{3} \quad \text{or} \quad 3x = \frac{2\pi}{9} + \frac{n\pi}{3}$$
$$x = \frac{2\pi}{9} + \frac{n\pi}{3}$$

20. $\sin x(\sin x + 1) = 0$

$$\sin x = 0 \quad \text{or} \quad \sin x = -1$$
$$x = n\pi \qquad x = \frac{3\pi}{2} + 2n\pi$$

22. $\tan 3x(\tan x - 1) = 0$

$$\tan 3x = 0 \quad \text{or} \quad \tan x - 1 = 0$$
$$3x = n\pi \qquad \tan x = 1$$
$$x = \frac{n\pi}{3} \qquad x = \frac{\pi}{4} + n\pi$$

24. $\cos 2x(2\cos x + 1) = 0$

$\cos 2x = 0$ or $2\cos x + 1 = 0$

$2x = \dfrac{\pi}{2} + n\pi \qquad \cos x = -\dfrac{1}{2}$

$x = \dfrac{\pi}{4} + \dfrac{n\pi}{2} \qquad x = \dfrac{2\pi}{3} + 2n\pi$

or $x = \dfrac{4\pi}{3} + 2n\pi$

26. $\tan^2 x - 1 = 0$

$\tan^2 x = 1$

$\tan x = \pm 1$

$x = \dfrac{\pi}{4}, \dfrac{3\pi}{4}, \dfrac{5\pi}{4}, \dfrac{7\pi}{4}$

28. $2\sin^2 x = 2 + \cos x$

$2 - 2\cos^2 x = 2 + \cos x$

$2\cos^2 x + \cos x = 0$

$\cos x(2\cos x + 1) = 0$

$\cos x = 0$ or $2\cos x + 1 = 0$

$x = \dfrac{\pi}{2}, \dfrac{3\pi}{2} \qquad 2\cos x = -1$

$\cos x = -\dfrac{1}{2}$

$x = \dfrac{2\pi}{3}, \dfrac{4\pi}{3}$

30. $\sec x \csc x = 2\csc x$

$\sec x \csc x - 2\csc x = 0$

$\csc x(\sec x - 2) = 0$

$\csc x = 0$ or $\sec x - 2 = 0$

No solution $\qquad \sec x = 2$

$x = \dfrac{\pi}{3}, \dfrac{5\pi}{3}$

32. $\sin 2x = -\dfrac{\sqrt{3}}{2}$

$2x = \dfrac{4\pi}{3} + 2n\pi$ or $2x = \dfrac{5\pi}{3} + 2n\pi$

$x = \dfrac{2\pi}{3} + n\pi \qquad x = \dfrac{5\pi}{6} + n\pi$

$x = \dfrac{2\pi}{3}, \dfrac{5\pi}{3} \qquad x = \dfrac{5\pi}{6}, \dfrac{11\pi}{6}$

34. $\tan 3x = 1$

$3x = \dfrac{\pi}{4} + 2n\pi$ or $3x = \dfrac{5\pi}{4} + 2n\pi$

$x = \dfrac{\pi}{12} + \dfrac{2n\pi}{3} \qquad x = \dfrac{5\pi}{12} + \dfrac{2n\pi}{3}$

$x = \dfrac{\pi}{12}, \dfrac{3\pi}{4}, \dfrac{17\pi}{12} \qquad x = \dfrac{5\pi}{12}, \dfrac{13\pi}{12}, \dfrac{7\pi}{4}$

36. $\sec 4x = 2$

$4x = \dfrac{\pi}{3} + 2n\pi$ or $4x = \dfrac{5\pi}{3} + 2n\pi$

$x = \dfrac{\pi}{12} + \dfrac{n\pi}{2} \qquad x = \dfrac{5\pi}{12} + \dfrac{n\pi}{2}$

$x = \dfrac{\pi}{12}, \dfrac{7\pi}{12}, \dfrac{13\pi}{12}, \dfrac{19\pi}{12} \qquad x = \dfrac{5\pi}{12}, \dfrac{11\pi}{12}, \dfrac{17\pi}{12}, \dfrac{23\pi}{12}$

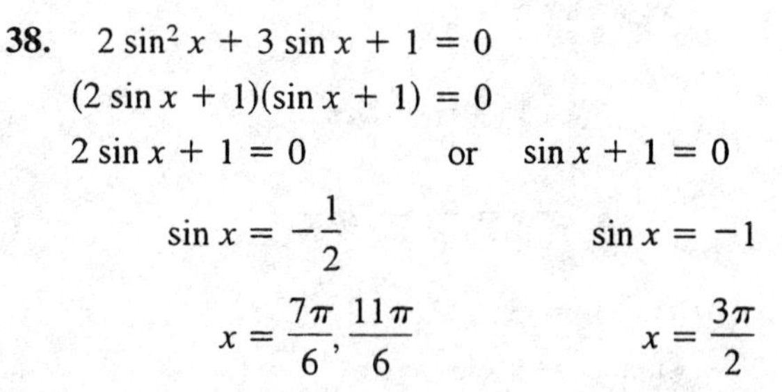

38. $2\sin^2 x + 3\sin x + 1 = 0$

$(2\sin x + 1)(\sin x + 1) = 0$

$2\sin x + 1 = 0$ or $\sin x + 1 = 0$

$\sin x = -\dfrac{1}{2} \qquad \sin x = -1$

$x = \dfrac{7\pi}{6}, \dfrac{11\pi}{6} \qquad x = \dfrac{3\pi}{2}$

40. $\cos x + \sin x \tan x = 2$

$\cos x + \sin x\left(\dfrac{\sin x}{\cos x}\right) = 2$

$\dfrac{\cos^2 x + \sin^2 x}{\cos x} = 2$

$\dfrac{1}{\cos x} = 2$

$\cos x = \dfrac{1}{2}$

$x = \dfrac{\pi}{3}, \dfrac{5\pi}{3}$

42. $2\cos x - \sin x = 0$

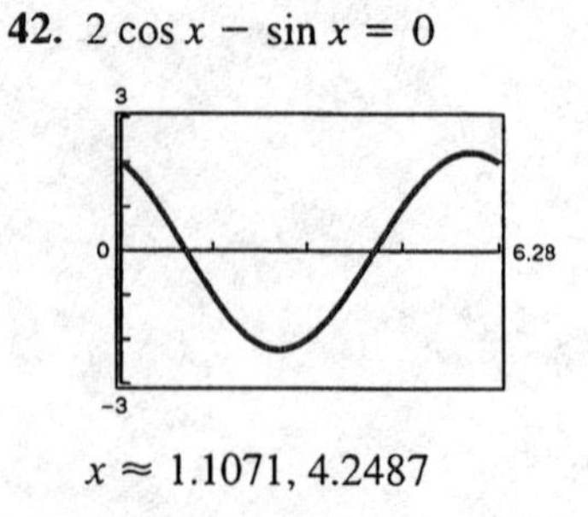

$x \approx 1.1071, 4.2487$

44. $\dfrac{1 + \sin x}{\cos x} + \dfrac{\cos x}{1 + \sin x} - 4 = 0$

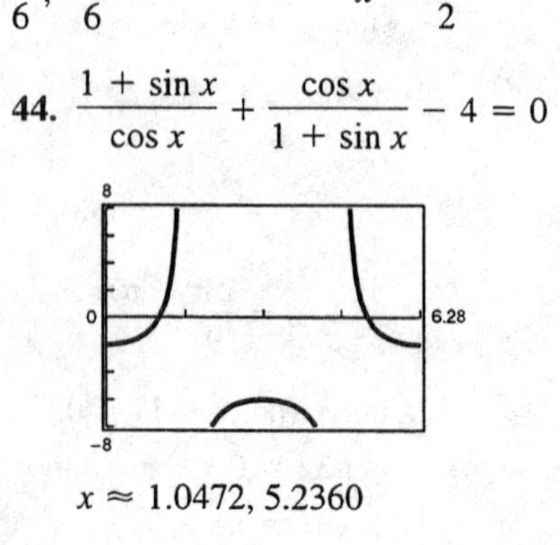

$x \approx 1.0472, 5.2360$

46. $x \cos x - 1 = 0$

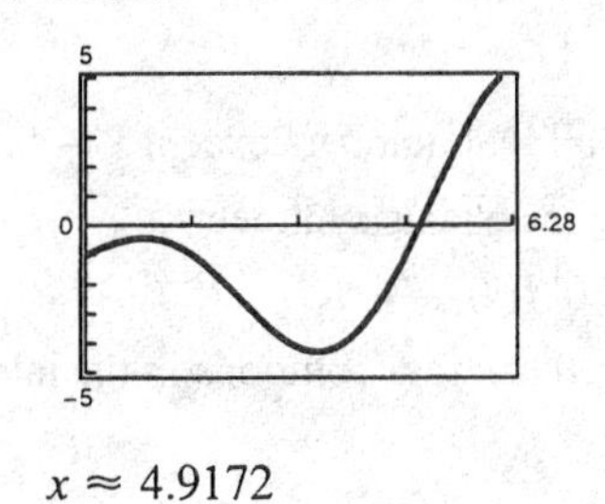

$x \approx 4.9172$

48. $\csc^2 x + 0.5 \cot x - 5 = 0$

$$y_1 = \left(\frac{1}{\sin x}\right)^2 + \frac{1}{2 \tan x} - 5$$

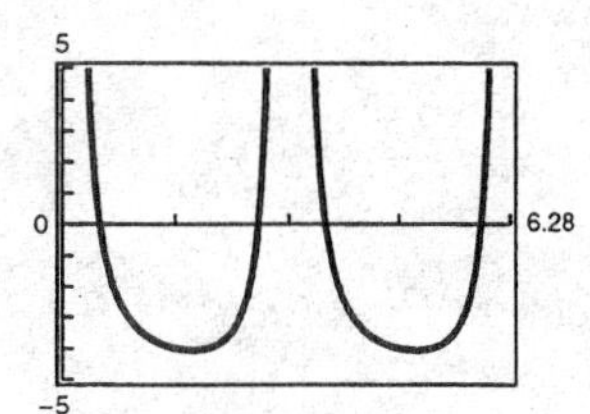

$x \approx 0.5153, 2.7259, 3.6569, 5.8675$

50. $3 \tan^2 x + 4 \tan x - 4 = 0$

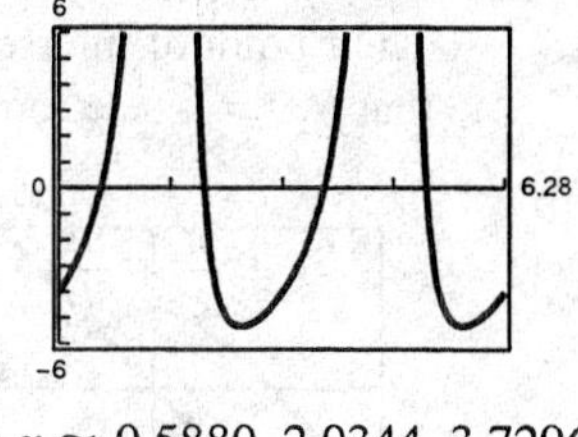

$x \approx 0.5880, 2.0344, 3.7296, 5.1760$

52. (a)

x	0	1	2	3	4	5	6
$f(x)$	-1	0.85	-5.81	-4.03	-7.12	16.63	5.25

Any zeros are in the intervals (0, 1), (1, 2), and (4, 5) because f changes signs in the intervals.

(b)

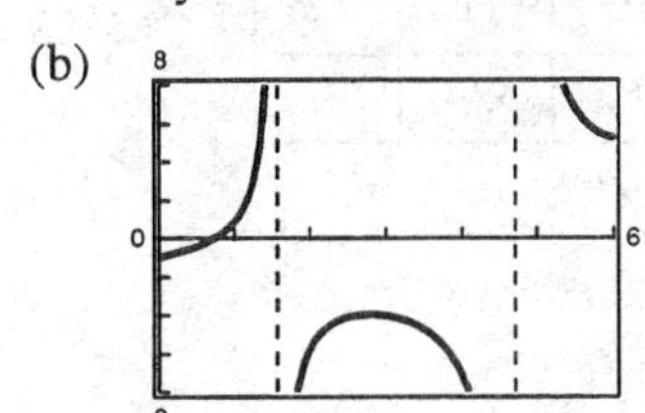

The only interval is (0, 1). It differs from that in part (a) because of the vertical asymptotes.

(c) 0.7391

54. (a)

x	0	1	2	3	4	5	6
$f(x)$	-1	-1.99	1.36	6.88	3.32	-1.81	-1.15

The zeros are in the intervals (1, 2) and (4, 5) because f changes signs in these intervals.

(b)

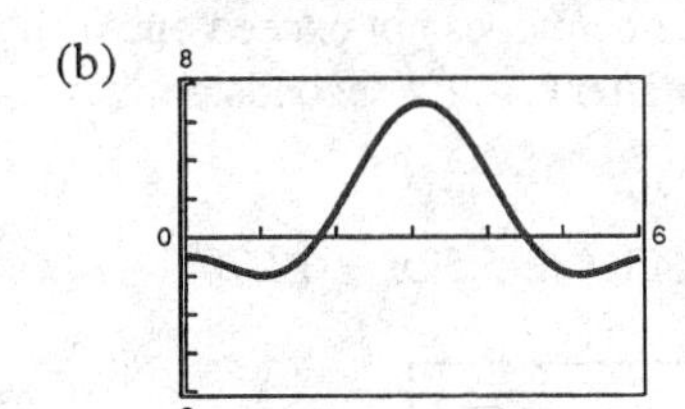

The intervals are the same as in part (a).

(c) 1.7794, 4.5038

56. (a)

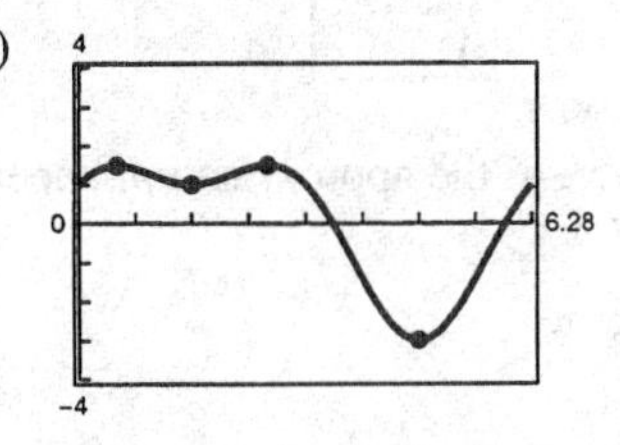

Maximum: (0.5236, 1.5), (2.6180, 1.5)
Minimum: (1.5708, 1.0), (4.7124, -3.0)

(b) $2 \cos x - 4 \sin x \cos x = 0$

$$2 \cos x(1 - 2 \sin x) = 0$$

$$\cos x = 0 \Rightarrow x = \frac{\pi}{2}, \frac{3\pi}{2}$$

$$1 - 2 \sin x = 0 \Rightarrow \sin x = \frac{1}{2} \Rightarrow x = \frac{\pi}{6}, \frac{5\pi}{6}$$

58. Graph $y = \cos x$ and $y = x$ on the same set of axes. Their point of intersection gives the value of c such that $f(c) = c \Rightarrow \cos c = c$.

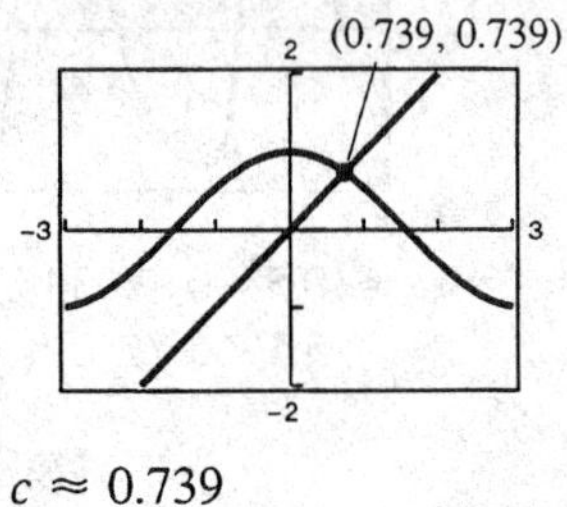

$c \approx 0.739$

60. $f(x) = \dfrac{\sin x}{x}$

(a) Domain: all real numbers except $x = 0$.

(b) The graph has y-axis symmetry.

(c) As $x \to 0, f(x) \to 1$.

(d) $\sin x/x = 0$ has four solutions in the interval $[-8, 8]$.

$$(\sin x)\left(\frac{1}{x}\right) = 0$$

$$\sin x = 0$$

$$x = -2\pi, -\pi, \pi, 2\pi$$

62. $S = 74.50 + 43.75 \sin \dfrac{\pi t}{6}$

t	1	2	3	4	5	6	7	8	9	10	11	12
S	96.4	112.4	118.3	112.4	96.4	74.5	52.6	36.6	30.8	36.6	52.6	74.5

Sales exceed 100,000 units during February, March, and April.

64. Range = 1000 yards = 3000 feet

$v_0 = 1200$ feet per second

$f = \frac{1}{32} v_0^2 \sin 2\theta$

$3000 = \frac{1}{32}(1200)^2 \sin 2\theta$

$\sin 2\theta = 0.066667$

$2\theta \approx 3.8226°$

$\theta \approx 1.9113°$

66. $y_1 = 1.56e^{-0.22t} \cos 4.9t$

$y_2 = 1$

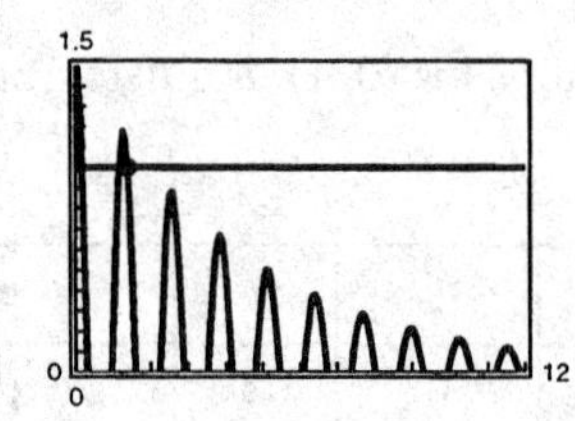

Point of intersection: (1.4, 1)

The displacement does not exceed one inch from equilibrium after $t = 1.4$ seconds.

68. $f(x) = 3 \sin(0.6x - 2)$

(a) Zero: $\sin(0.6x - 2) = 0$

$$0.6x - 2 = 0$$

$$0.6x = 2$$

$$x = \frac{2}{0.6} = \frac{10}{3}$$

(b) $g(x) = -0.45x^2 + 5.52x - 13.70$

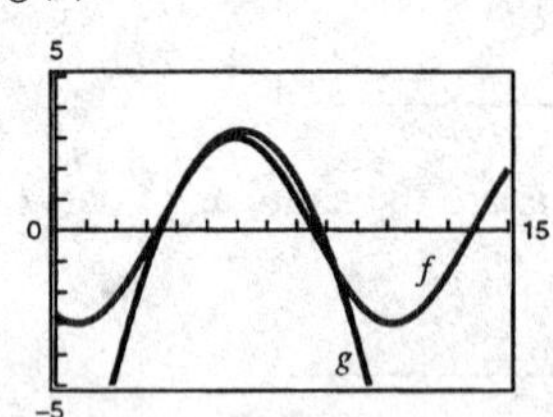

For $3.5 \le x \le 6$ the approximation appears to be good.

(c) $-0.45x^2 + 5.52x - 13.70 = 0$

$$x = \frac{-5.52 \pm \sqrt{(5.52) - 4(-0.45)(-13.70)}}{2(-0.45)}$$

$x \approx 3.46, 8.81$

The zero of g on $[0, 6]$ is 3.46. The zero is close to the zero $\frac{10}{3} \approx 3.33$ of f.

5.4 Sum and Difference Formulas

Solutions to Even-Numbered Exercises

2. (a) $\sin\left(\frac{3\pi}{4} + \frac{5\pi}{6}\right) = \sin\frac{3\pi}{4}\cos\frac{5\pi}{6} + \cos\frac{3\pi}{4}\sin\frac{5\pi}{5}$

$$= \left(\frac{\sqrt{2}}{2}\right)\left(-\frac{\sqrt{3}}{2}\right) + \left(-\frac{\sqrt{2}}{2}\right)\left(\frac{1}{2}\right)$$

$$= -\frac{\sqrt{6} + \sqrt{2}}{4}$$

(b) $\sin\frac{3\pi}{4} + \sin\frac{5\pi}{6} = \frac{\sqrt{2}}{2} + \frac{1}{2} = \frac{\sqrt{2} + 1}{2}$

4. (a) $\cos\left(\frac{2\pi}{3} - \frac{\pi}{6}\right) = \cos\frac{2\pi}{3}\cos\frac{\pi}{6} + \sin\frac{2\pi}{3}\sin\frac{\pi}{6}$

$$= \left(-\frac{1}{2}\right)\left(\frac{\sqrt{3}}{2}\right) + \left(\frac{\sqrt{3}}{2}\right)\left(\frac{1}{2}\right) = 0$$

(b) $\cos\frac{2\pi}{3} + \cos\frac{\pi}{6} = -\frac{1}{2} + \frac{\sqrt{3}}{2} = \frac{\sqrt{3} - 1}{2}$

6. False. It is possible to find the exact value.

$$\sin(75°) = \sin(45° + 30°)$$

$$= \sin 45° \cos 30° + \sin 30° \cos 45°$$

$$= \frac{\sqrt{2}}{2}\frac{\sqrt{3}}{2} + \frac{1}{2}\frac{\sqrt{2}}{2} = \frac{\sqrt{6} + \sqrt{2}}{4}$$

8. $15° = 45° - 30°$

$\sin 15° = \sin(45° - 30°) = \sin 45° \cos 30° - \cos 45° \sin 30°$

$$= \left(\frac{\sqrt{2}}{2}\right)\left(\frac{\sqrt{3}}{2}\right) - \left(\frac{\sqrt{2}}{2}\right)\left(\frac{1}{2}\right) = \frac{\sqrt{2}(\sqrt{3} - 1)}{4} = \frac{\sqrt{2}}{4}(\sqrt{3} - 1)$$

$\cos 15° = \cos(45° - 30°) = \cos 45° \cos 30° + \sin 45° \sin 30°$

$$= \left(\frac{\sqrt{2}}{2}\right)\left(\frac{\sqrt{3}}{2}\right) + \left(\frac{\sqrt{2}}{2}\right)\left(\frac{1}{2}\right) = \frac{\sqrt{2}(\sqrt{3} + 1)}{4} = \frac{\sqrt{2}}{4}(\sqrt{3} + 1)$$

$\tan 15° = \tan(45° - 30°) = \dfrac{\tan 45° - \tan 30°}{1 + \tan 45° \tan 30°}$

$$= \frac{1 - \frac{\sqrt{3}}{3}}{1 + (1)\left(\frac{\sqrt{3}}{3}\right)} = \frac{\frac{3 - \sqrt{3}}{3}}{\frac{3 + \sqrt{3}}{3}} = \frac{3 - \sqrt{3}}{3 + \sqrt{3}} \cdot \frac{3 - \sqrt{3}}{3 - \sqrt{3}} = \frac{12 - 6\sqrt{3}}{6} = 2 - \sqrt{3}$$

10. $165° = 135° + 30°$

$\sin 165° = \sin(135° + 30°) = \sin 135° \cos 30° + \sin 30° \sin 135°$

$$= \sin 45° \cos 30° - \sin 30° \cos 45°$$

$$= \frac{\sqrt{2}}{2} \cdot \frac{\sqrt{3}}{2} - \frac{1}{2} \cdot \frac{\sqrt{2}}{2} = \frac{\sqrt{2}}{4}(\sqrt{3} - 1)$$

$\cos 165° = \cos(135° + 30°) = \cos 135° \cos 30° - \sin 135° \sin 30°$

$$= -\cos 45° \cos 30° - \sin 45° \sin 30°$$

$$= -\frac{\sqrt{2}}{2} \cdot \frac{\sqrt{3}}{2} - \frac{\sqrt{2}}{2} \cdot \frac{1}{2} = -\frac{\sqrt{2}}{4}(\sqrt{3} + 1)$$

$\tan 165° = \tan(135° + 30°) = \dfrac{\tan 135° + \tan 30°}{1 - \tan 135° \tan 30°} = \dfrac{-\tan 45° + \tan 30°}{1 + \tan 45° \tan 30°}$

$$= \frac{-1 + (\sqrt{3}/3)}{1 + (\sqrt{3}/3)} = -2 + \sqrt{3}$$

12. $255° = 300° - 45°$

$$\sin 255° = \sin(300° - 45°) = \sin 300° \cos 45° - \sin 45° \cos 300°$$
$$= -\sin 60° \cos 45° - \sin 45° \cos 60°$$
$$= -\frac{\sqrt{3}}{2} \cdot \frac{\sqrt{2}}{2} - \frac{\sqrt{2}}{2} \cdot \frac{1}{2} = -\frac{\sqrt{2}}{4}(\sqrt{3} + 1)$$

$$\cos 255° = \cos(300° - 45°) = \cos 300° \cos 45° + \sin 300° \sin 45°$$
$$= \cos 60° \cos 45° - \sin 60° \sin 45°$$
$$= \frac{1}{2} \cdot \frac{\sqrt{2}}{2} - \frac{\sqrt{3}}{2} \cdot \frac{\sqrt{2}}{2} = \frac{\sqrt{2}}{4}(1 - \sqrt{3})$$

$$\tan 255° = \tan(300° - 45°) = \frac{\tan 300° - \tan 45°}{1 + \tan 300° \tan 45°}$$
$$= \frac{-\tan 60° + \tan 45°}{1 - \tan 60° \tan 45°} = \frac{-\sqrt{3} - 1}{1 - \sqrt{3}} = 2 + \sqrt{3} = 2 + \sqrt{3}$$

14. $\frac{7\pi}{12} = \frac{\pi}{3} + \frac{\pi}{4}$

$$\sin \frac{7\pi}{12} = \sin\left(\frac{\pi}{3} + \frac{\pi}{4}\right) = \sin \frac{\pi}{3} \cos \frac{\pi}{4} + \sin \frac{\pi}{4} \cos \frac{\pi}{3}$$
$$= \frac{\sqrt{3}}{2} \cdot \frac{\sqrt{2}}{2} + \frac{\sqrt{2}}{2} \cdot \frac{1}{2} = \frac{\sqrt{2}}{3}(\sqrt{3} + 1)$$

$$\cos \frac{7\pi}{12} = \cos\left(\frac{\pi}{3} + \frac{\pi}{4}\right) = \cos \frac{\pi}{3} \cos \frac{\pi}{4} - \sin \frac{\pi}{3} \sin \frac{\pi}{4}$$
$$= \frac{1}{2} \cdot \frac{\sqrt{2}}{2} - \frac{\sqrt{3}}{2} \cdot \frac{\sqrt{2}}{2} = \frac{\sqrt{2}}{4}(1 - \sqrt{3})$$

$$\tan \frac{7\pi}{12} = \tan\left(\frac{\pi}{3} + \frac{\pi}{4}\right) = \frac{\tan(\pi/3) + \tan(\pi/4)}{1 - \tan(\pi/3) \tan(\pi/4)}$$
$$= \frac{\sqrt{3} + 1}{1 - \sqrt{3}} = -2 - \sqrt{3}$$

16. $-\frac{\pi}{12} = \frac{\pi}{6} - \frac{\pi}{4}$

$$\sin\left(-\frac{\pi}{12}\right) = \sin\left(\frac{\pi}{6} - \frac{\pi}{4}\right) = \sin \frac{\pi}{6} \cos \frac{\pi}{4} - \sin \frac{\pi}{4} \cos \frac{\pi}{6}$$
$$= \frac{1}{2} \cdot \frac{\sqrt{2}}{2} - \frac{\sqrt{2}}{2} \cdot \frac{\sqrt{3}}{2} = \frac{\sqrt{2}}{4}(1 - \sqrt{3})$$

$$\cos\left(-\frac{\pi}{12}\right) = \cos\left(\frac{\pi}{6} - \frac{\pi}{4}\right) = \cos \frac{\pi}{6} \cos \frac{\pi}{4} + \sin \frac{\pi}{6} \sin \frac{\pi}{4}$$
$$= \frac{\sqrt{3}}{2} \cdot \frac{\sqrt{2}}{2} + \frac{1}{2} \cdot \frac{\sqrt{2}}{2} = \frac{\sqrt{2}}{4}(\sqrt{3} + 1)$$

$$\tan\left(-\frac{\pi}{12}\right) = \tan\left(\frac{\pi}{6} - \frac{\pi}{4}\right) = \frac{\tan(\pi/6) - \tan(\pi/4)}{1 + \tan(\pi/6) \tan(\pi/4)}$$
$$= \frac{(\sqrt{3}/3) - 1}{1 + (\sqrt{3}/3)} = -2 + \sqrt{3}$$

18. $\sin 140° \cos 50° + \cos 140° \sin 50° = \sin(140° + 50°) = \sin 190°$

20. $\cos 20° \cos 30° + \sin 20° \sin 30° = \cos(30° - 20°)$
$= \cos 10°$

22. $\dfrac{\tan 140° - \tan 60°}{1 + \tan 140° \tan 60°} = \tan(140° - 60°) = \tan 80°$

24. $\cos \dfrac{\pi}{7} \cos \dfrac{\pi}{5} - \sin \dfrac{\pi}{7} \sin \dfrac{\pi}{5} = \cos\left(\dfrac{\pi}{7} + \dfrac{\pi}{5}\right)$
$= \cos \dfrac{12\pi}{35}$

26. $\cos 3x \cos 2y + \sin 3x \sin 2y = \cos(3x - 2y)$

28.

x	0.2	0.4	0.6	0.8	1.0	1.2	1.4
y_1	0.1987	0.3894	0.5646	0.7174	0.8415	0.9320	0.9854
y_2	0.1987	0.3894	0.5646	0.7174	0.8415	0.9320	0.9854

$$\begin{aligned} y_1 &= \sin(3\pi - x) \\ &= \sin 3\pi \cos x - \cos 3\pi \sin x \\ &= 0 - (-1) \sin x \\ &= \sin x \\ &= y_2 \end{aligned}$$

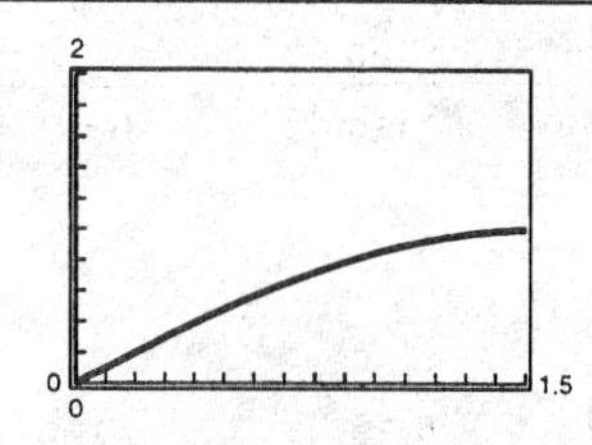

30.

x	0.2	0.4	0.6	0.8	1.0	1.2	1.4
y_1	−0.8335	−0.9266	−0.9829	−0.9999	−0.9771	−0.9153	−0.8170
y_2	−0.8335	−0.9266	−0.9829	−0.9999	−0.9771	−0.9153	−0.8170

$$\begin{aligned} y_1 = \cos\left(\frac{5\pi}{4} - x\right) &= \cos \frac{5\pi}{4} \cos x + \sin \frac{5\pi}{4} \sin x \\ &= -\frac{\sqrt{2}}{2} \cos x - \frac{\sqrt{2}}{2} \sin x \\ &= -\frac{\sqrt{2}}{2}(\cos x + \sin x) \\ &= y_2 \end{aligned}$$

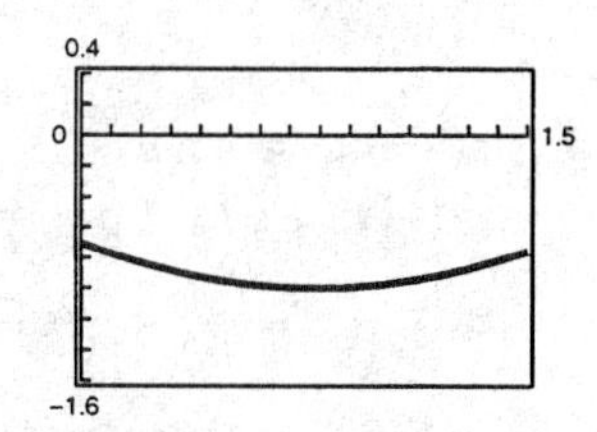

32.

x	0.2	0.4	0.6	0.8	1.0	1.2	1.4
y_1	0.0395	0.1516	0.3188	0.5146	0.7081	0.8687	0.9711
y_2	0.0395	0.1516	0.3188	0.5146	0.7081	0.8687	0.9711

$$\begin{aligned} y_1 &= \sin(x + \pi) \sin(x - \pi) \\ &= [\sin x \cos \pi + \sin \pi \cos x][\sin x \cos \pi - \sin \pi \cos x] \\ &= [-\sin x][-\sin x] \\ &= \sin^2 x \\ &= y_2 \end{aligned}$$

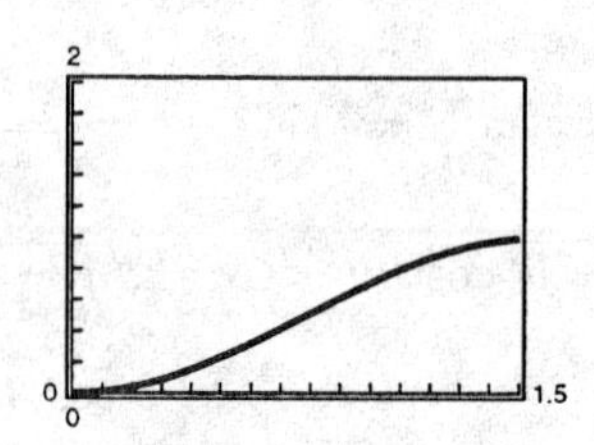

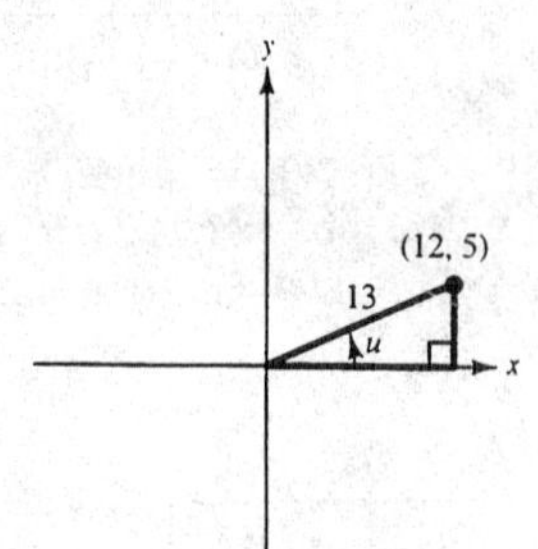

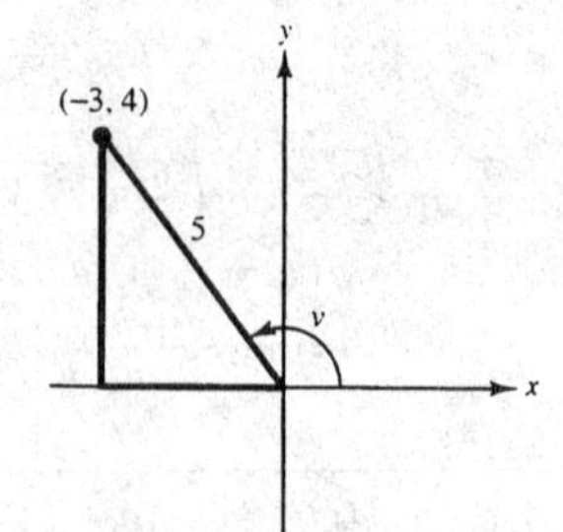

Figures for Exercises 34 and 36

34. $\cos(v - u) = \cos v \cos u + \sin v \sin u$

$= \left(-\frac{3}{5}\right)\left(+\frac{12}{13}\right) + \left(\frac{4}{5}\right)\left(\frac{5}{13}\right)$

$= -\frac{36}{65} + \frac{20}{65} = -\frac{16}{65}$

36. $\sin(u - v) = \sin u \cos v - \cos u \sin v$

$= \left(\frac{5}{13}\right)\left(-\frac{3}{5}\right) - \left(+\frac{12}{13}\right)\left(\frac{4}{5}\right)$

$= -\frac{15}{65} - \frac{48}{65} = -\frac{63}{65}$

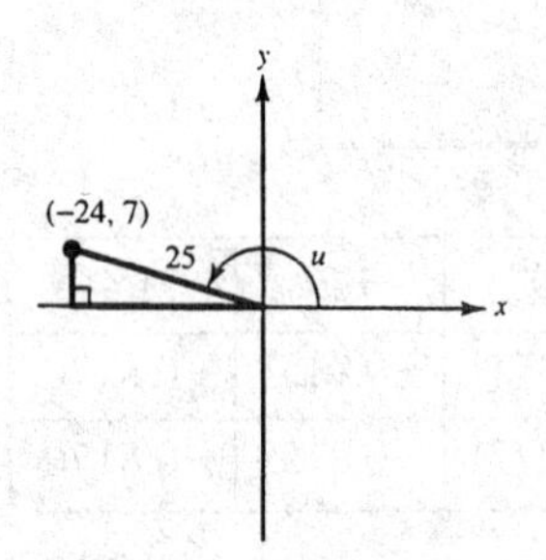

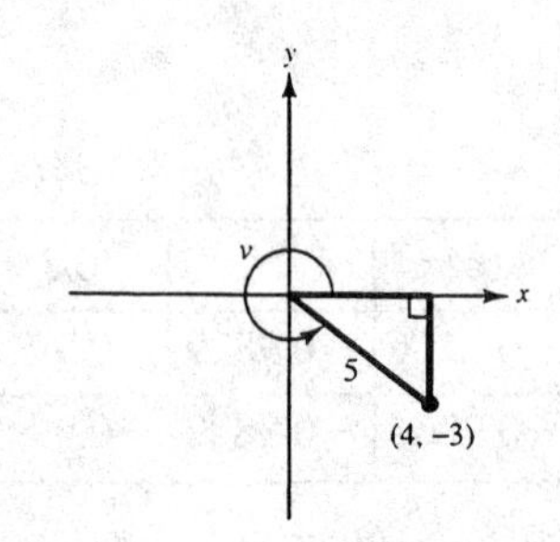

Figures for Exercises 38 and 40

38. $\sin(u + v) = \sin u \cos v + \cos u \sin v$

$= \left(+\frac{7}{25}\right)\left(+\frac{4}{5}\right) + \left(-\frac{24}{25}\right)\left(-\frac{3}{5}\right)$

$= \frac{28}{125} + \frac{72}{125} = \frac{100}{125} = \frac{4}{5}$

40. $\cos(u - v) = \cos u \cos v + \sin u \sin v$

$= \left(-\frac{24}{25}\right)\left(+\frac{4}{5}\right) + \left(+\frac{7}{25}\right)\left(-\frac{3}{25}\right)$

$= -\frac{96}{125} + -\frac{21}{125} = -\frac{117}{125}$

42. $\tan\left(\frac{\pi}{4} - \theta\right) = \frac{\tan(\pi/4) - \tan\theta}{1 + \tan(\pi/4)\tan\theta} = \frac{1 - \tan\theta}{1 + \tan\theta}$

44. $\cos(x + y) + \cos(x - y) = \cos x \cos y - \sin x \sin y + \cos x \cos y + \sin x \sin y$

$= 2 \cos x \cos y$

46. $\sin(n\pi + \theta) = \sin n\pi \cos\theta + \sin\theta \cos n\pi$

$= (0)(\cos\theta) + (\sin\theta)(-1)^n$

$= (-1)^n(\sin\theta)$, where n is an integer.

48. $C = \arctan\frac{a}{b} \Rightarrow \sin C = \frac{a}{\sqrt{a^2 + b^2}}, \cos C = \frac{b}{\sqrt{a^2 + b^2}}$

$\sqrt{a^2 + b^2}\cos(B\theta - C) = \sqrt{a^2 + b^2}\left(\cos B\theta \cdot \frac{b}{\sqrt{a^2 + b^2}} + \sin B\theta \cdot \frac{a}{\sqrt{a^2 + b^2}}\right)$

$= b \cos B\theta + a \sin B\theta$

$= a \sin B\theta + b \cos B\theta$

50. $\tan(\pi + \theta) = \frac{\tan\pi + \tan\theta}{1 - \tan\pi\tan\theta}$

$= \frac{0 + \tan\theta}{1 - (0)\tan\theta}$

$= \tan\theta$

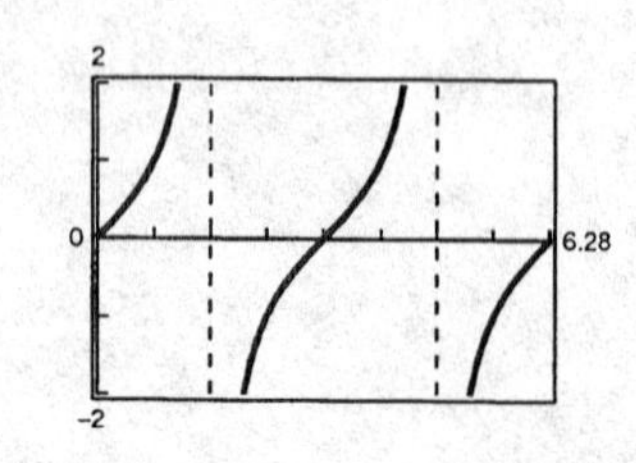

52. $3\sin 2\theta + 4\cos 2\theta$

$a = 3, b = 4, B = 2$

(a) $C = \arctan\frac{b}{a} = \arctan\frac{4}{3} \approx 0.9273$

$$3\sin 2\theta + 4\cos 2\theta = \sqrt{a^2 + b^2}\sin(B\theta + C)$$
$$\approx 5\sin(2\theta + 0.9273)$$

(b) $C = \arctan\frac{a}{b} = \arctan\frac{3}{4} \approx 0.6435$

$$3\sin 2\theta + 4\cos 2\theta = \sqrt{a^2 + b^2}\cos(B\theta - C)$$
$$\approx 5\cos(2\theta - 0.6435)$$

54. $\sin 2\theta - \cos 2\theta$

$a = 1, b = -1, B = 2$

(a) $C = \arctan\frac{b}{a} = \arctan(-1) = -\frac{\pi}{4}$

$$\sin 2\theta - \cos 2\theta = \sqrt{a^2 + b^2}\sin(B\theta + C)$$
$$= \sqrt{2}\sin\left(2\theta - \frac{\pi}{4}\right)$$

(b) Because $b > 0$ in the formula, we write the given expression as:

$-(-\sin 2\theta + \cos 2\theta)$

$a = -1, b = 1, B = 2,$

$C = \arctan\left(\frac{a}{b}\right) = \arctan(-1) = -\frac{\pi}{4}$

Hence,

$$-(-\sin 2\theta + \cos 2\theta) = -\sqrt{a^2 + b^2}\cos(B\theta - C)$$
$$= -\sqrt{2}\cos\left(2\theta + \frac{\pi}{4}\right).$$

56. $C = \arctan\frac{a}{b} = -\frac{3\pi}{4} \Rightarrow a = b, a < 0, b < 0$

$\sqrt{a^2 + b^2} = 5 \Rightarrow a = b = \frac{-5\sqrt{2}}{2}$

$B = 1$

$$5\cos\left(\theta + \frac{3\pi}{4}\right) = -\frac{5\sqrt{2}}{2}\sin\theta - \frac{5\sqrt{2}}{2}\cos\theta$$

58. Let:

$u = \arctan 2x$ and $v = \arccos x$

$\tan u = 2x$ $\cos v = x$

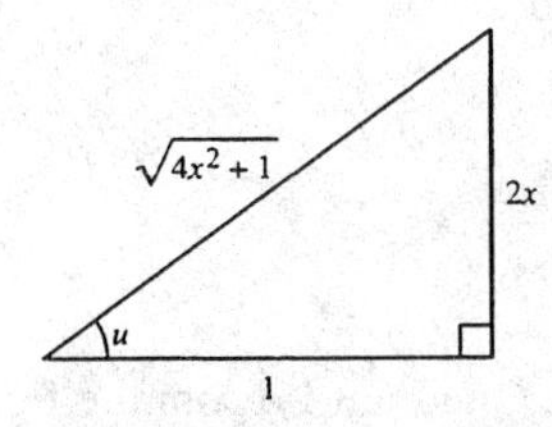

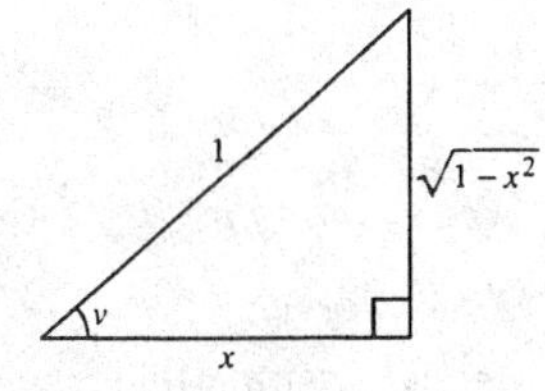

$$\sin(\arctan 2x - \arccos x) = \sin(u - v)$$
$$= \sin u\cos v - \cos u\sin v$$
$$= \frac{2x}{\sqrt{4x^2 + 1}}(x) - \frac{1}{\sqrt{4x^2 + 1}}\left(\sqrt{1 - x^2}\right)$$
$$= \frac{2x^2 - \sqrt{1 - x^2}}{\sqrt{4x^2 + 1}}$$

60.

$$\sin\left(x + \frac{\pi}{6}\right) - \sin\left(x - \frac{\pi}{6}\right) = \frac{1}{2}$$

$$\sin x \cos\frac{\pi}{6} + \cos x \sin\frac{\pi}{6} - \left(\sin x \cos\frac{\pi}{6} - \cos x \sin\frac{\pi}{6}\right) = \frac{1}{2}$$

$$2 \cos x(0.5) = \frac{1}{2}$$

$$\cos x = \frac{1}{2}$$

$$x = \frac{\pi}{3}, \frac{5\pi}{3}$$

62.

$$\tan(x + \pi) + 2 \sin(x + \pi) = 0$$

$$\frac{\tan x + \tan \pi}{1 - \tan x \tan \pi} + 2(\sin x \cos \pi + \cos x \sin \pi) = 0$$

$$\frac{\tan x + 0}{1 - \tan x(0)} + 2[\sin x(-1) + \cos x(0)] = 0$$

$$\frac{\tan x}{1} - 2 \sin x = 0$$

$$\frac{\sin x}{\cos x} = 2 \sin x$$

$$\sin x = 2 \sin x \cos x$$

$$\sin x(1 - 2 \cos x) = 0$$

$$\sin x = 0 \quad \text{or} \quad \cos x = \frac{1}{2}$$

$$x = 0, \pi \qquad x = \frac{\pi}{3}, \frac{5\pi}{3}$$

64. $\tan(x + \pi) - \cos\left(x + \frac{\pi}{2}\right) = 0$

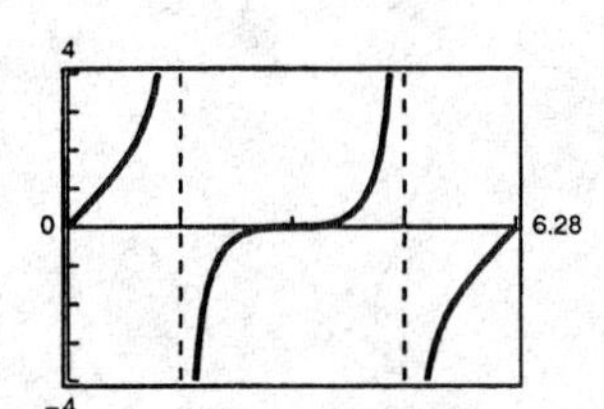

Answers: $(0, 0), (3.14, 0) \Rightarrow x = 0, \pi$

66. From the figure, it appears that $u + v = w$. Assume that u, v, and w are all in Quadrant I. From the figure:

$$\tan u = \frac{s}{3s} = \frac{1}{3}$$

$$\tan v = \frac{s}{2s} = \frac{1}{2}$$

$$\tan s = \frac{s}{s} = 1$$

$$\tan(u + v) = \frac{\tan u + \tan v}{1 - \tan u \tan v} = \frac{1/3 + 1/2}{1 - (1/3)(1/2)} = \frac{5/6}{1 - (1/6)} = 1 = \tan w.$$

Thus, $\tan(u + v) = \tan w$. Because u, v, and w are all in Quadrant I, we have

$$\arctan[\tan(u + v)] = \arctan[\tan w]$$

$$u + v = w.$$

68. $$\frac{\cos(x+h)-\cos x}{h} = \frac{\cos x\cos h - \sin x \sin h - \cos x}{x} = \frac{\cos x(\cos h - 1)}{x} - \frac{\sin x \sin h}{x}$$

70. $$\tan(u+v) = \frac{\sin(u+v)}{\cos(u+v)} = \frac{\sin u \cos v + \sin v \cos u}{\cos u \cos v - \sin u \sin v}$$

$$= \frac{\sin u \cos v + \sin v \cos u}{\cos u \cos v - \sin u \sin v} \cdot \frac{\dfrac{1}{\cos u \cos v}}{\dfrac{1}{\cos u \cos v}} = \frac{\dfrac{\sin u}{\cos u} + \dfrac{\sin v}{\cos v}}{1 - \dfrac{\sin u \sin v}{\cos u \cos v}} = \frac{\tan u + \tan v}{1 - \tan u \tan v}$$

5.5 Multiple-Angle and Product-Sum Formulas

Solutions to Even-Numbered Exercises

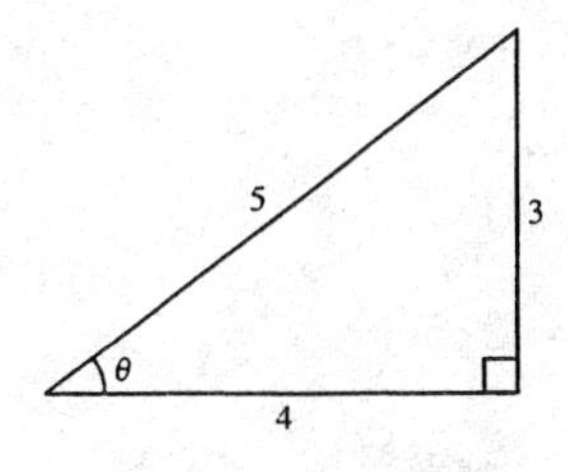

Figure for Exercises 2–8

2. $\tan\theta = \frac{3}{4}$

4. $\sin 2\theta = 2\sin\theta\cos\theta = 2\left(\frac{3}{5}\right)\left(\frac{4}{5}\right) = \frac{24}{25}$

6. $$\sec 2\theta = \frac{1}{\cos 2\theta} = \frac{1}{\cos^2\theta - \sin^2\theta} = \frac{1}{(4/5)^2 - (3/5)^2} = \frac{1}{(16/25) - (9/25)} = \frac{25}{7}$$

8. $$\cot 2\theta = \frac{1}{\tan 2\theta} = \frac{1 - \tan^2\theta}{2\tan\theta} = \frac{1 - (3/4)^2}{2(3/4)} = \frac{7/16}{3/2} = \frac{7}{24}$$

10.
$$\sin 2x + \cos x = 0$$
$$2\sin x\cos x + \cos x = 0$$
$$\cos x(2\sin x + 1) = 0$$
$$\cos x = 0 \quad \text{or} \quad 2\sin x + 1 = 0$$
$$x = \frac{\pi}{2}, \frac{3\pi}{2} \qquad \sin x = -\frac{1}{2}$$
$$x = \frac{7\pi}{6}, \frac{11\pi}{6}$$

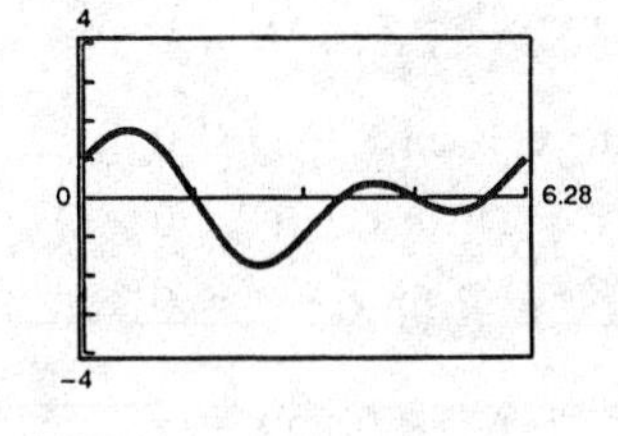

12.

$$\sin 2x \sin x = \cos x$$
$$2 \sin x \cos x \sin x - \cos x = 0$$
$$\cos x(2 \sin^2 x - 1) = 0$$
$$\cos x = 0 \quad \text{or} \quad 2 \sin^2 x - 1 = 0$$
$$x = \frac{\pi}{2}, \frac{3\pi}{2} \qquad \sin^2 x = \frac{1}{2}$$
$$\sin x = \pm\frac{\sqrt{2}}{2}$$
$$x = \frac{\pi}{4}, \frac{3\pi}{4}, \frac{5\pi}{4}, \frac{7\pi}{4}$$

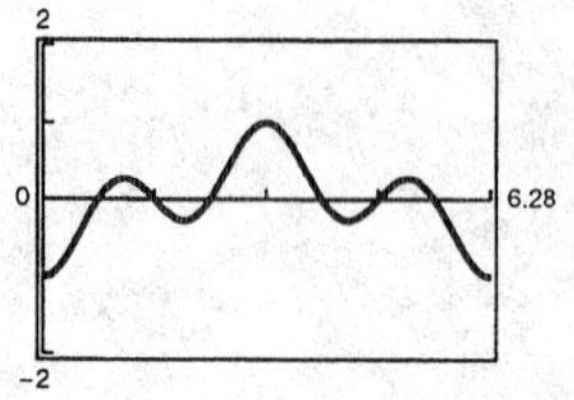

14.

$$\cos 2x + \sin x = 0$$
$$1 - 2 \sin^2 x + \sin x = 0$$
$$2 \sin^2 x - \sin x - 1 = 0$$
$$(2 \sin x + 1)(\sin x - 1) = 0$$
$$2 \sin x + 1 = 0 \quad \text{or} \quad \sin x - 1 = 0$$
$$\sin x = -\frac{1}{2} \qquad \sin x = 1$$
$$x = \frac{7\pi}{6}, \frac{11\pi}{6} \qquad x = \frac{\pi}{2}$$

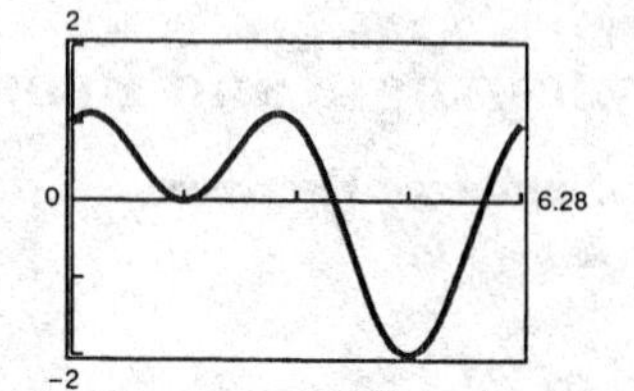

16.

$$\tan 2x - 2 \cos x = 0$$
$$\frac{2 \tan x}{1 - \tan^2 x} = 2 \cos x$$
$$2 \tan x = 2 \cos x(1 - \tan^2 x)$$
$$2 \tan x = 2 \cos x - 2 \cos x \tan^2 x$$
$$2 \tan x = 2 \cos x - 2 \cos x \frac{\sin^2 x}{\cos^2 x}$$
$$2 \tan x = 2 \cos x - 2 \frac{\sin^2 x}{\cos x}$$
$$\tan x = \cos x - \frac{\sin^2 x}{\cos x}$$
$$\frac{\sin x}{\cos x} = \cos x - \frac{\sin^2 x}{\cos x}$$
$$\frac{\sin x}{\cos x} + \frac{\sin^2 x}{\cos x} - \cos x = 0$$
$$\frac{\sin x + \sin^2 x - \cos^2 x}{\cos x} = 0$$
$$\frac{1}{\cos x}[\sin x + \sin^2 x - (1 - \sin^2 x)] = 0$$
$$\sec x[2 \sin^2 x + \sin x - 1] = 0$$
$$\sec x(2 \sin x - 1)(\sin x + 1) = 0$$
$$\sec x = 0 \quad \text{or} \quad 2 \sin x - 1 = 0 \quad \text{or} \quad \sin x + 1 = 0$$
$$x = \frac{\pi}{2}, \frac{3\pi}{2} \qquad \sin x = \frac{1}{2} \qquad \sin x = -1$$
$$x = \frac{\pi}{6}, \frac{5\pi}{6} \qquad x = \frac{3\pi}{2}$$
$$x = \frac{\pi}{6}, \frac{\pi}{2}, \frac{5\pi}{6}, \frac{3\pi}{2}$$

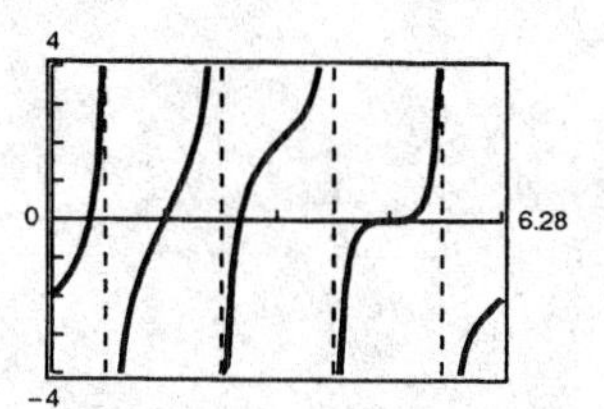

18.
$$\begin{aligned}
(\sin 2x + \cos 2x)^2 &= 1 \\
\sin^2 2x + 2\sin 2x\cos 2x + \cos^2 2x &= 1 \\
2\sin 2x\cos 2x &= 0 \\
\sin 4x &= 0 \\
4x &= n\pi \\
x &= \frac{n\pi}{4} \\
x &= 0, \frac{\pi}{4}, \frac{\pi}{2}, \frac{3\pi}{4}, \pi, \frac{5\pi}{4}, \frac{3\pi}{2}, \frac{7\pi}{4}
\end{aligned}$$

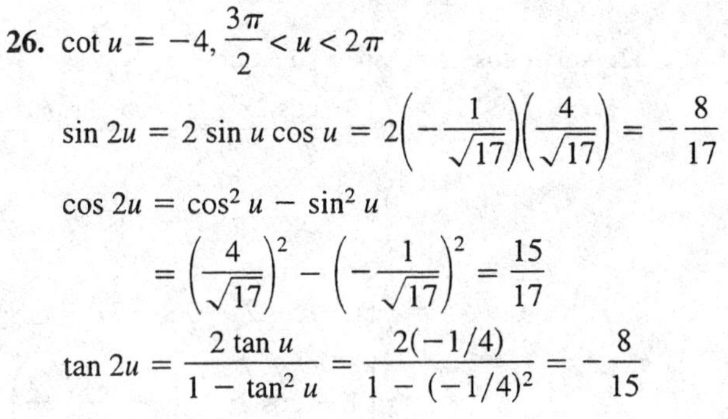

20. $4\sin x\cos x + 2 = 2(2\sin x\cos x) + 2$
$= 2\sin 2x + 2$

22. $(\cos x + \sin x)(\cos x - \sin x) = \cos^2 x - \sin^2 x$
$= \cos 2x$

24. $\cos u = -\frac{2}{3}, \frac{\pi}{2} < u < \pi$

$\sin 2u = 2\sin u\cos u = 2 \cdot \frac{\sqrt{5}}{3}\left(-\frac{2}{3}\right) = -\frac{4\sqrt{5}}{9}$

$\cos 2u = \cos^2 u - \sin^2 u = \frac{4}{9} - \frac{5}{9} = -\frac{1}{9}$

$\tan 2u = \frac{2\tan u}{1 - \tan^2 u} = \frac{2(-\sqrt{5}/2)}{1 - (5/4)} = 4\sqrt{5}$

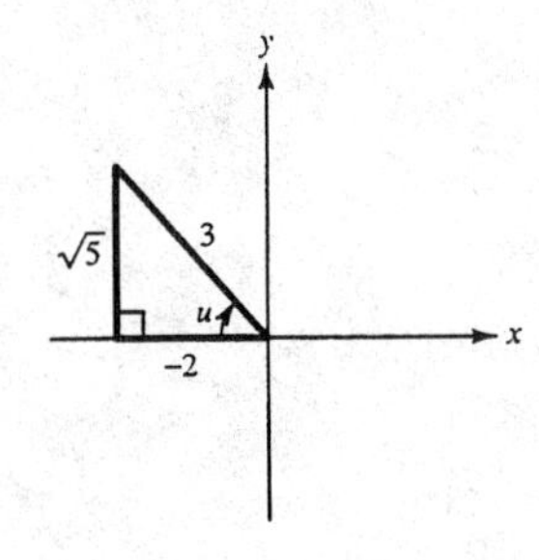

26. $\cot u = -4, \frac{3\pi}{2} < u < 2\pi$

$\sin 2u = 2\sin u\cos u = 2\left(-\frac{1}{\sqrt{17}}\right)\left(\frac{4}{\sqrt{17}}\right) = -\frac{8}{17}$

$\cos 2u = \cos^2 u - \sin^2 u$

$= \left(\frac{4}{\sqrt{17}}\right)^2 - \left(-\frac{1}{\sqrt{17}}\right)^2 = \frac{15}{17}$

$\tan 2u = \frac{2\tan u}{1 - \tan^2 u} = \frac{2(-1/4)}{1 - (-1/4)^2} = -\frac{8}{15}$

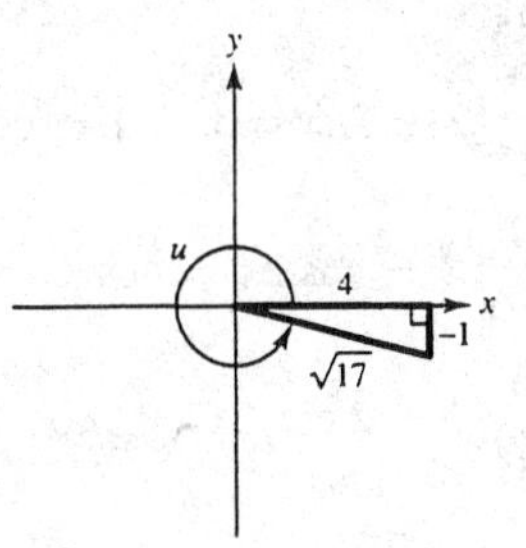

28.
$$\begin{aligned}
\sin^4 x &= (\sin^2 x)(\sin^2 x) \\
&= \left(\frac{1 - \cos 2x}{2}\right)\left(\frac{1 - \cos 2x}{2}\right) \\
&= \frac{1 - 2\cos 2x + \cos^2 2x}{4} \\
&= \frac{1 - 2\cos 2x + \left(\frac{1 + \cos 4x}{2}\right)}{4} \\
&= \frac{2 - 4\cos 2x + 1 + \cos 4x}{8} \\
&= \frac{1}{8}(3 - 4\cos 2x + \cos 4x)
\end{aligned}$$

30. $\cos^6 x = (\cos^2 x)^3 = \left(\dfrac{1 + \cos 2x}{2}\right)^3$

$$= \frac{1}{8}[1 + 3\cos 2x + 3\cos^2 2x + \cos^3 2x]$$

$$= \frac{1}{8}\left[1 + 3\cos 2x + 3 \cdot \frac{1 + \cos 4x}{2} + \cos 2x\left(\frac{1 + \cos 4x}{2}\right)\right]$$

$$= \frac{1}{8}\left[\frac{5}{2} + 3\cos 2x + \frac{3}{2}\cos 4x + \frac{1}{2}\cos 2x + \frac{1}{2}\cos 2x \cdot \cos 4x\right]$$

$$= \frac{1}{8}\left[\frac{5}{2} + \frac{7}{2}\cos 2x + \frac{3}{2}\cos 4x + \frac{1}{2}\frac{1}{2}(\cos 2x + \cos 6x)\right]$$

$$= \frac{1}{32}[10 + 15\cos 2x + 6\cos 4x + \cos 6x]$$

32. $\sin^4 x \cos^2 x = \sin^2 x \sin^2 x \cos^2 x$

$$= \left(\frac{1 - \cos 2x}{2}\right)\left(\frac{1 - \cos 2x}{2}\right)\left(\frac{1 + \cos 2x}{2}\right)$$

$$= \frac{1}{8}(1 - \cos 2x)(1 - \cos^2 2x)$$

$$= \frac{1}{8}(1 - \cos 2x - \cos^2 2x + \cos^3 2x)$$

$$= \frac{1}{8}\left[1 - \cos 2x - \left(\frac{1 + \cos 4x}{2}\right) + \cos 2x\left(\frac{1 + \cos 4x}{2}\right)\right]$$

$$= \frac{1}{16}[2 - 2\cos 2x - 1 - \cos 4x + \cos 2x + \cos 2x \cos 4x]$$

$$= \frac{1}{16}\left[1 - \cos 2x - \cos 4x + \frac{1}{2}\cos 2x + \frac{1}{2}\cos 6x\right]$$

$$= \frac{1}{32}[2 - 2\cos 2x - 2\cos 4x + \cos 2x + \cos 6x]$$

$$= \frac{1}{32}[2 - \cos 2x - 2\cos 4x + \cos 6x]$$

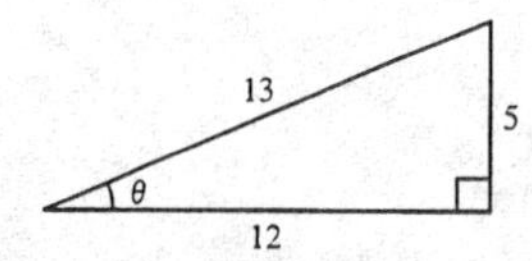

Figure for Exercises 34 and 36

34. $\sin\dfrac{\theta}{2} = \sqrt{\dfrac{1 - \cos\theta}{2}}$

$$= \sqrt{\frac{1 - (12/13)}{2}}$$

$$= \sqrt{\frac{1/13}{2}}$$

$$= \frac{1}{\sqrt{26}}$$

36. $2\sin\dfrac{\theta}{2}\cos\dfrac{\theta}{2} = 2\sqrt{\dfrac{1 - \cos\theta}{2}}\sqrt{\dfrac{1 + \cos\theta}{2}}$

$$= \frac{2\sqrt{1 - \cos^2\theta}}{2}$$

$$= \sqrt{1 - \left(\frac{12}{13}\right)^2}$$

$$= \sqrt{\frac{25}{169}} = \frac{5}{13}$$

38. $\cot\dfrac{\theta}{2} = \dfrac{1}{\tan\dfrac{\theta}{2}} = \dfrac{\sin\theta}{1-\cos\theta} = \dfrac{\frac{5}{13}}{1-\frac{12}{13}} = \dfrac{5}{13}\left(\dfrac{13}{1}\right) = 5$

40. $\sin 165° = \sin\left(\dfrac{1}{2}\cdot 330°\right) = \sqrt{\dfrac{1-\cos 330°}{2}} = \sqrt{\dfrac{1-(\sqrt{3}/2)}{2}} = \dfrac{1}{2}\sqrt{2-\sqrt{3}}$

$\cos 165° = \cos\left(\dfrac{1}{2}\cdot 330°\right) = -\sqrt{\dfrac{1+\cos 330°}{2}} = -\sqrt{\dfrac{1+(\sqrt{3}/2)}{2}} = -\dfrac{1}{2}\sqrt{2+\sqrt{3}}$

$\tan 165° = \tan\left(\dfrac{1}{2}\cdot 330°\right) = \dfrac{\sin 330°}{1+\cos 330°} = \dfrac{-1/2}{1+(\sqrt{3}/2)} = \dfrac{-1}{2+\sqrt{3}} = \sqrt{3}-2$

42. $\sin 67°\,30' = \sin\left(\dfrac{1}{2}\cdot 135°\right) = \sqrt{\dfrac{1-\cos 135°}{2}} = \sqrt{\dfrac{1+(\sqrt{2}/2)}{2}} = \dfrac{1}{2}\sqrt{2+\sqrt{2}}$

$\cos 67°\,30' = \cos\left(\dfrac{1}{2}\cdot 135°\right) = \sqrt{\dfrac{1+\cos 135°}{2}} = \sqrt{\dfrac{1-(\sqrt{2}/2)}{2}} = \dfrac{1}{2}\sqrt{2-\sqrt{2}}$

$\tan 67°\,30' = \tan\left(\dfrac{1}{2}\cdot 135°\right) = \dfrac{\sin 135°}{1+\cos 135°} = \dfrac{\sqrt{2}/2}{1-(\sqrt{2}/2)} = 1+\sqrt{2}$

44. $\sin\dfrac{\pi}{12} = \sin\left[\dfrac{1}{2}\left(\dfrac{\pi}{6}\right)\right] = \sqrt{\dfrac{1-\cos(\pi/6)}{2}} = \sqrt{\dfrac{1-(\sqrt{3}/2)}{2}} = \dfrac{1}{2}\sqrt{2-\sqrt{3}}$

$\cos\dfrac{\pi}{12} = \cos\left[\dfrac{1}{2}\left(\dfrac{\pi}{6}\right)\right] = \sqrt{\dfrac{1+\cos(\pi/6)}{2}} = \dfrac{1}{2}\sqrt{2+\sqrt{3}}$

$\tan\dfrac{\pi}{12} = \tan\left[\dfrac{1}{2}\left(\dfrac{\pi}{6}\right)\right] = \dfrac{\sin(\pi/6)}{1+\cos(\pi/6)} = \dfrac{1/2}{1+(\sqrt{3}/2)} = 2-\sqrt{3}$

46. $\cos u = \dfrac{3}{5},\ 0 < u < \dfrac{\pi}{2}$

$\sin\left(\dfrac{u}{2}\right) = \sqrt{\dfrac{1-\cos u}{2}} = \sqrt{\dfrac{1-(3/5)}{2}} = \dfrac{\sqrt{5}}{5}$

$\cos\left(\dfrac{u}{2}\right) = \sqrt{\dfrac{1+\cos u}{2}} = \sqrt{\dfrac{1+(3/5)}{2}} = \dfrac{2\sqrt{5}}{5}$

$\tan\left(\dfrac{u}{2}\right) = \dfrac{\sin u}{1+\cos u} = \dfrac{4/5}{1+(3/5)} = \dfrac{1}{2}$

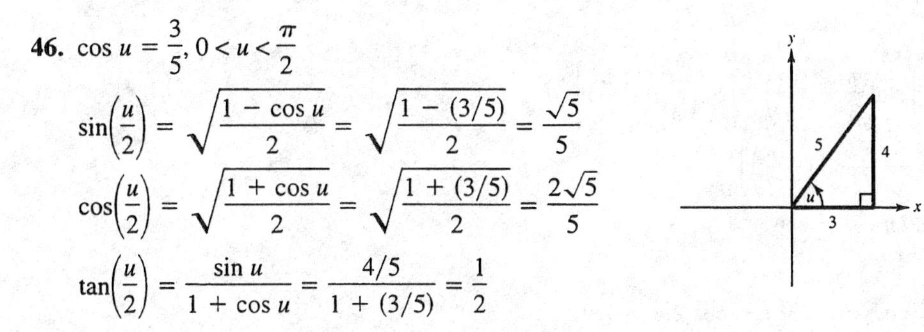

48. $\cot u = 3,\ \pi < u < \dfrac{3\pi}{2}$

$\sin\left(\dfrac{u}{2}\right) = \sqrt{\dfrac{1-\cos u}{2}} = \sqrt{\dfrac{1+(3/\sqrt{10})}{2}} = \sqrt{\dfrac{10+3\sqrt{10}}{20}} = \dfrac{1}{2}\sqrt{\dfrac{10+3\sqrt{10}}{5}}$

$\cos\left(\dfrac{u}{2}\right) = -\sqrt{\dfrac{1+\cos u}{2}} = -\sqrt{\dfrac{1-(3/\sqrt{10})}{2}} = -\sqrt{\dfrac{10-3\sqrt{10}}{20}} = -\dfrac{1}{2}\sqrt{\dfrac{10-3\sqrt{10}}{5}}$

$\tan\left(\dfrac{u}{2}\right) = \dfrac{1-\cos u}{\sin u} = \dfrac{1+(3/\sqrt{10})}{-1/\sqrt{10}} = -3-\sqrt{10}$

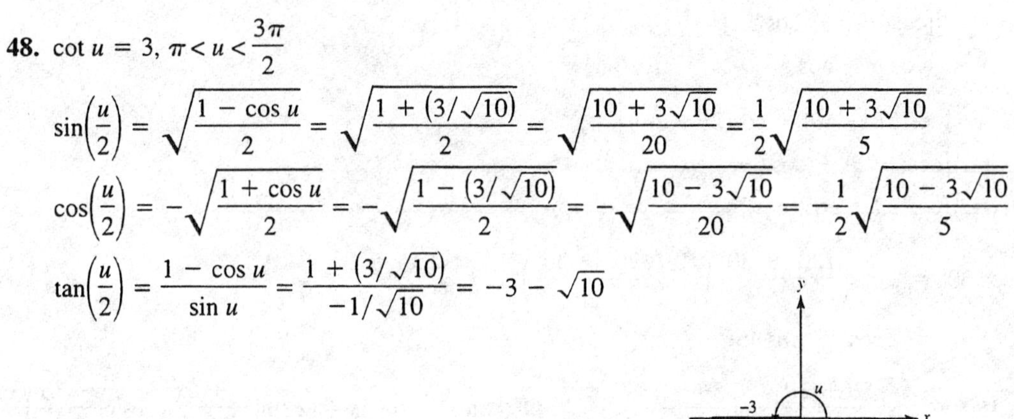

50. $\sqrt{\dfrac{1+\cos 4x}{2}} = \left|\cos\dfrac{4x}{2}\right| = |\cos 2x|$

52. $-\sqrt{\dfrac{1-\cos(x-1)}{2}} = -\left|\sin\left(\dfrac{x-1}{2}\right)\right|$

54. $h(x) = \sin\dfrac{x}{2} + \cos x - 1$

$$\sin\frac{x}{2} + \cos x - 1 = 0$$

$$\pm\sqrt{\frac{1-\cos x}{2}} = 1 - \cos x$$

$$\frac{1-\cos x}{2} = 1 - 2\cos x + \cos^2 x$$

$$1 - \cos x = 2 - 4\cos x + 2\cos^2 x$$

$$2\cos^2 x - 3\cos x + 1 = 0$$

$$(2\cos x - 1)(\cos x - 1) = 0$$

$2\cos x - 1 = 0$ or $\cos x - 1 = 0$

$\cos x = \dfrac{1}{2}$ $\qquad \cos x = 1$

$x = \dfrac{\pi}{3}, \dfrac{5\pi}{3}$ $\qquad x = 0$

0, $\pi/3$, and $5\pi/2$ are all solutions to the equation.

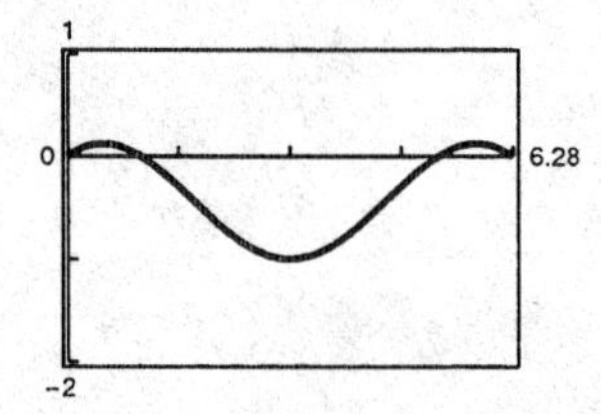

56. $g(x) = \tan\dfrac{x}{2} - \sin x$

$$\tan\frac{x}{2} - \sin x = 0$$

$$\frac{1-\cos x}{\sin x} = \sin x$$

$$1 - \cos x = \sin^2 x$$

$$1 - \cos x = 1 - \cos^2 x$$

$$\cos^2 x - \cos x = 0$$

$$\cos x(\cos x - 1) = 0$$

$\cos x = 0$ or $\cos x - 1 = 0$

$x = \dfrac{\pi}{2}, \dfrac{3\pi}{2}$ $\qquad \cos x = 1$

$x = 0$

0, $\pi/2$, and $3\pi/2$ are all solutions to the equation.

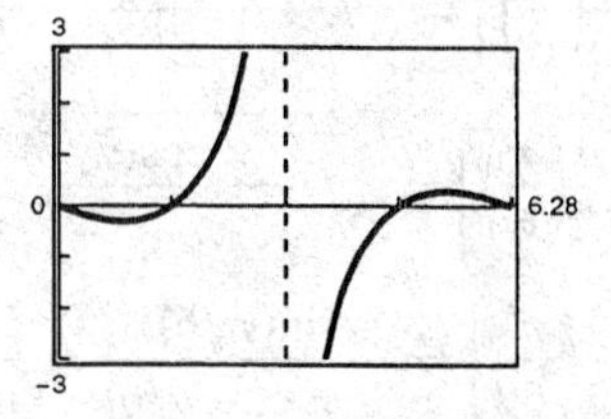

58. $4\sin\dfrac{\pi}{3}\cos\dfrac{5\pi}{6} = 4\cdot\dfrac{1}{2}\left[\sin\left(\dfrac{\pi}{3}+\dfrac{5\pi}{6}\right) + \sin\left(\dfrac{\pi}{3}-\dfrac{5\pi}{6}\right)\right]$

$$= 2\left[\sin\frac{7\pi}{6} + \sin\left(-\frac{\pi}{2}\right)\right]$$

$$= 2\left(\sin\frac{7\pi}{6} - \sin\frac{\pi}{2}\right)$$

60. $3\sin 2\alpha\sin 3\alpha = 3\cdot\frac{1}{2}[\cos(2\alpha - 3\alpha) - \cos(2\alpha + 3\alpha)$

$$= \tfrac{3}{2}[\cos(-\alpha) - \cos 5\alpha]$$

$$= \tfrac{3}{2}(\cos\alpha - \cos 5\alpha)$$

62. $\cos 2\theta\cos 4\theta = \frac{1}{2}[\cos(2\theta - 4\theta) + \cos(2\theta + 4\theta)]$

$$= \tfrac{1}{2}[\cos(-2\theta) + \cos 6\theta]$$

$$= \tfrac{1}{2}(\cos 2\theta + \cos 6\theta)$$

64. $\cos 120^\circ + \cos 30^\circ = 2\cos\left(\dfrac{120^\circ + 30^\circ}{2}\right)\cos\left(\dfrac{120^\circ - 30^\circ}{2}\right)$

$$= 2\cos 75^\circ\cos 45^\circ$$

66. $\sin 5\theta - \sin 3\theta = 2\cos\left(\dfrac{5\theta + 3\theta}{2}\right)\sin\left(\dfrac{5\theta - 3\theta}{2}\right)$

$$= 2\cos 4\theta\sin\theta$$

68. $\sin x + \sin 5x = 2\sin\left(\dfrac{x+5x}{2}\right)\cos\left(\dfrac{x-5x}{2}\right)$

$$= 2\sin 3x\cos(-2x)$$

$$= 2\sin 3x\cos 2x$$

70. $\cos\left(\theta + \frac{\pi}{2}\right) - \cos\left(\theta - \frac{\pi}{2}\right) = -2\sin\left(\frac{\theta + (\pi/2) + \theta - (\pi/2)}{2}\right)\sin\left(\frac{\theta + (\pi/2) - \theta + (\pi/2)}{2}\right)$

$= -2\sin\theta\sin\frac{\pi}{2}$

72. $\cos\left(x + \frac{\pi}{2}\right) + \sin\left(x - \frac{\pi}{2}\right) = 2\sin\left(\frac{x + (\pi/2) + x - (\pi/2)}{2}\right)\cos\left(\frac{x + (\pi/2) - x + (\pi/2)}{2}\right)$

$= 2\sin x\cos\frac{\pi}{2} = 0$

74. $h(x) = \cos 2x - \cos 6x$

$\cos 2x - \cos 6x = 0$

$-2\sin 4x\sin(-2x) = 0$

$2\sin 4x\sin 2x = 0$

$\sin 4x = 0$ or $\sin 2x = 0$

$4x = n\pi$ $2x = n\pi$

$x = \frac{n\pi}{4}$ $x = \frac{n\pi}{2}$

$x = 0, \frac{\pi}{4}, \frac{\pi}{2}, \frac{3\pi}{4}, \pi, \frac{5\pi}{4}, \frac{3\pi}{2}, \frac{7\pi}{4}$ $x = 0, \frac{\pi}{2}, \pi, \frac{3\pi}{2}$

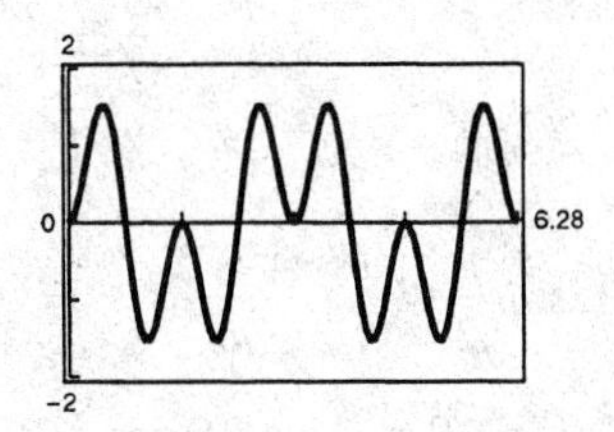

76. $f(x) = \sin^2 3x - \sin^2 x$

$\sin^2 3x - \sin^2 x = 0$

$(\sin 3x + \sin x)(\sin 3x - \sin x) = 0$

$(2\sin 2x\cos x)(2\cos 2x\sin x) = 0$

$\sin 2x = 0 \Rightarrow x = 0, \frac{\pi}{2}, \pi, \frac{3\pi}{2}$ or

$\cos x = 0 \Rightarrow x = \frac{\pi}{2}, \frac{3\pi}{2}$ or

$\cos 2x = 0 \Rightarrow x = \frac{\pi}{4}, \frac{3\pi}{4}, \frac{5\pi}{4}, \frac{7\pi}{4}$ or

$\sin x = 0 \Rightarrow x = 0, \pi$

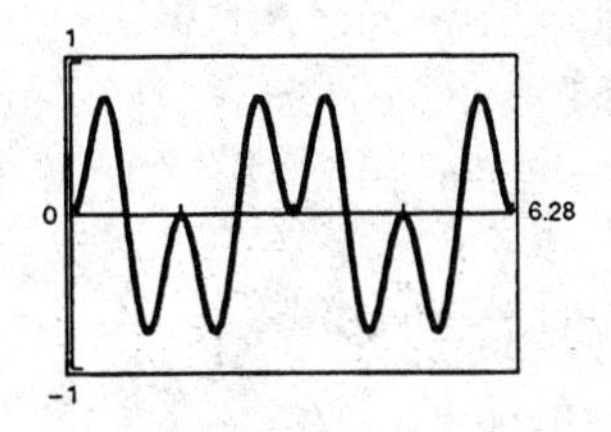

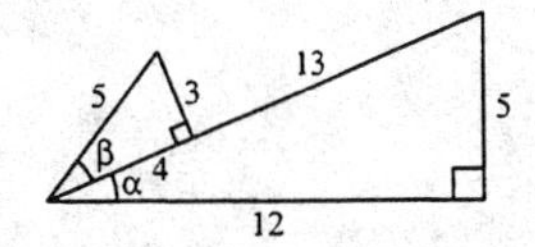

Figure for Exercises 78 and 80

78. $\cos^2\alpha = (\cos\alpha)^2 = \left(\frac{12}{13}\right)^2 = \frac{144}{169}$

$\cos^2\alpha = 1 - \sin^2\alpha$

$= 1 - \left(\frac{5}{13}\right)^2$

$= 1 - \frac{25}{169} = \frac{144}{169}$

80. $\cos\alpha\sin\beta = \left(\frac{12}{13}\right)\left(\frac{3}{5}\right) = \frac{36}{65}$

$\cos\alpha\sin\beta = \sin\left(\frac{\pi}{2} - \alpha\right)\cos\left(\frac{\pi}{2} - \beta\right)$

$= \left(\frac{12}{13}\right)\left(\frac{3}{5}\right) = \frac{36}{65}$

82. $\sec 2\theta = \dfrac{1}{\cos 2\theta} = \dfrac{1}{\cos^2\theta - \sin^2\theta}$

$= \dfrac{1/\cos^2\theta}{1 - (\sin^2\theta/\cos^2\theta)}$

$= \dfrac{\sec^2\theta}{1 - \tan^2\theta}$

$= \dfrac{\sec^2\theta}{1 - (\sec^2\theta - 1)}$

$= \dfrac{\sec^2\theta}{2 - \sec^2\theta}$

84. $\cos^4 x - \sin^4 x = (\cos^2 x - \sin^2 x)(\cos^2 x + \sin^2 x)$

$= (\cos 2x)(1)$

$= \cos 2x$

86. $\sin\left(\dfrac{\alpha}{3}\right)\cos\left(\dfrac{\alpha}{3}\right) = \dfrac{1}{2}\left[2\left(\sin\left(\dfrac{\alpha}{3}\right)\cos\left(\dfrac{\alpha}{3}\right)\right)\right]$

$= \dfrac{1}{2}\sin\dfrac{2\alpha}{3}$

88. $\dfrac{\cos 3\beta}{\cos\beta} = \dfrac{\cos^3\beta - 3\sin^2\beta\cos\beta}{\cos\beta}$

$= \cos^2\beta - 3\sin^2\beta$

$= 1 - \sin^2\beta - 3\sin^2\beta$

$= 1 - 4\sin^2\beta$

90. $\tan\dfrac{u}{2} = \dfrac{1 - \cos u}{\sin u} \quad = \dfrac{1}{\sin u} - \dfrac{\cos u}{\sin u} = \csc u - \cot u$

92. $\sin 4\beta = 2\sin 2\beta\cos 2\beta$

$= 2[2\sin\beta\cos\beta(\cos^2\beta - \sin^2\beta)]$

$= 2[2\sin\beta\cos\beta(1 - \sin^2\beta - \sin^2\beta)]$

$= 4\sin\beta\cos\beta(1 - 2\sin^2\beta)$

Graph: $y_1 = \sin 4\beta$

$y_2 = 4\sin\beta\cos\beta(1 - \sin^2\beta)$

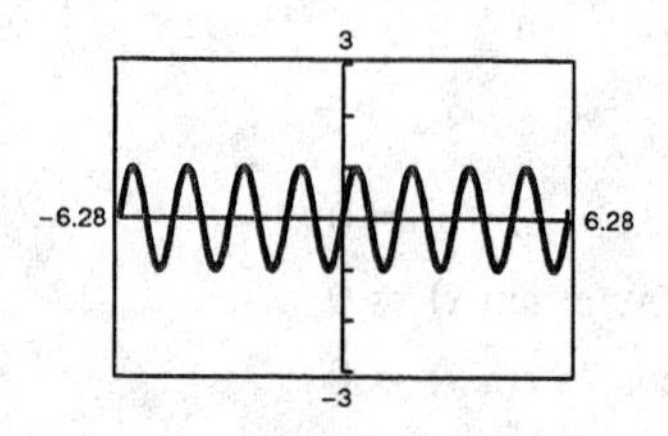

94. $\dfrac{\cos 3x - \cos x}{\sin 3x - \sin x} = \dfrac{-2\sin\left(\dfrac{3x + x}{2}\right)\sin\left(\dfrac{3x - x}{2}\right)}{2\cos\left(\dfrac{3x + x}{2}\right)\sin\left(\dfrac{3x - x}{2}\right)}$

$= \dfrac{-2\sin 2x\sin x}{2\cos 2x\sin x}$

$= -\tan 2x$

96. $f(x) = \cos^2 x = \dfrac{1 + \cos 2x}{2} = \dfrac{1}{2} + \dfrac{\cos 2x}{2}$

Shifted upward by $\dfrac{1}{2}$ unit.

Amplitude: $|a| = \dfrac{1}{2}$

Period: $\dfrac{2\pi}{2} = \pi$

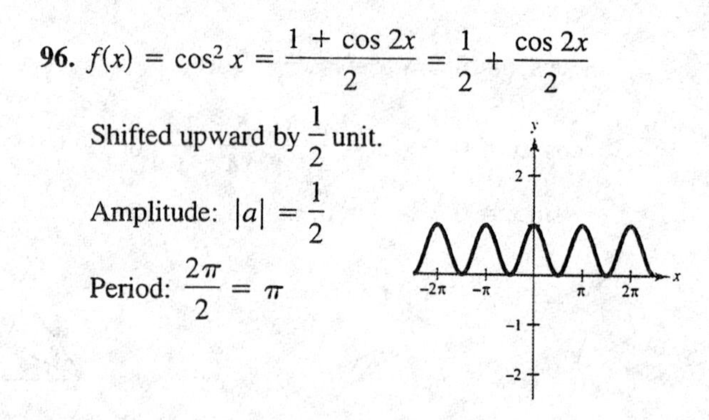

98. $f(x) = \cos 2x - 2\sin x$

(a)

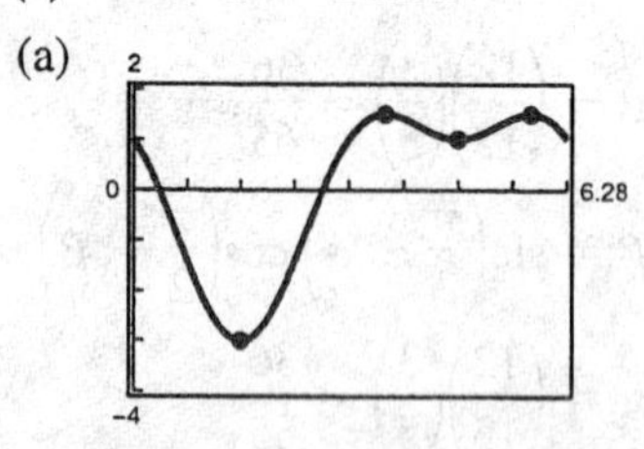

Maximum points: (3.6652, 1.5), (5.7596, 1.5)

Minimum points: (1.5708, −3), (4.7124, 1)

(b) $-2\cos x(2\sin x + 1) = 0$

$-2\cos x = 0$ or $2\sin x + 1 = 0$

$\cos x = 0$ $\qquad \sin x = -\dfrac{1}{2}$

$x = \dfrac{\pi}{2}, \dfrac{3\pi}{2}$ $\qquad x = \dfrac{7\pi}{6}, \dfrac{11\pi}{6}$

$\dfrac{\pi}{2} \approx 1.5708$ $\qquad \dfrac{7\pi}{6} \approx 3.6652$

$\dfrac{3\pi}{2} \approx 4.7124$ $\qquad \dfrac{11\pi}{6} \approx 5.7596$

100. $r = \frac{1}{32} v_0 \sin 2\theta$

$= \frac{1}{32} v_0(\sin \theta \cos \theta)$

$= \frac{1}{16} v_0 \sin \theta \cos \theta$

102. (a) $\sin\left(\frac{\theta}{2}\right) = \frac{b/2}{10} \Rightarrow b = 20 \sin \frac{\theta}{2}$

$\cos\left(\frac{\theta}{2}\right) = \frac{h}{10} \Rightarrow h = 10 \cos \frac{\theta}{2}$

$A = \frac{1}{2}bh = \frac{1}{2}\left(20 \sin \frac{\theta}{2}\right)\left(10 \cos \frac{\theta}{2}\right)$

$= 100 \sin \frac{\theta}{2} \cos \frac{\theta}{2}$

(b) $A = 50\left(2 \sin \frac{\theta}{2} \cos \frac{\theta}{2}\right) = 50 \sin \theta$

(c) The area is maximum when $\theta = \frac{\pi}{2}$, $A = 50$.

104. $\cos(2 \arccos x) = \cos^2(\arccos x) - \sin^2(\arccos x)$

$= x^2 - (1 - x^2) = 2x^2 - 1$

Review Exercises for Chapter 5

Solutions to Even-Numbered Exercises

2. $\dfrac{\sin 2\alpha}{\cos^2 \alpha - \sin^2 \alpha} = \dfrac{\sin 2\alpha}{\cos 2\alpha} = \tan 2\alpha$

4. $\dfrac{\sin^3 \beta + \cos^3 \beta}{\sin \beta + \cos \beta} = \dfrac{(\sin \beta + \cos \beta)(\sin^2 \beta - \sin \beta \cos \beta + \cos^2 \beta)}{\sin \beta + \cos \beta}$

$= 1 - \sin \beta \cos \beta$

$= 1 - \dfrac{1}{2} \sin 2\beta$

6. $1 - 4 \sin^2 x \cos^2 x = 1 - (2 \sin x \cos x)^2$

$= 1 - \sin^2 2x$

$= \cos^2 2x$

8. $\sqrt{\dfrac{1 - \cos^2 x}{1 + \cos x}} = \sqrt{1 - \cos x} = \sqrt{2}\left|\sin \dfrac{x}{2}\right|$

10. $\cos x(\tan^2 x + 1) = \cos x \sec^2 x$

$= \dfrac{1}{\sec x} \sec^2 x$

$= \sec x$

12. $\sin^3 \theta + \sin \theta \cos^2 \theta = \sin \theta(\sin^2 \theta + \cos^2 \theta)$

$= \sin \theta$

14. $\cos^3 x \sin^2 x = \cos x(\cos^2 x) \sin^2 x$

$= \cos x(1 - \sin^2 x) \sin^2 x$

$= (\sin^2 x - \sin^4 x) \cos x$

16. Using a product-sum formula, we have
$\sin 3x \cos 2x = \frac{1}{2}(\sin 5x + \sin x)$.

18. $\sqrt{1 - \cos x} = \sqrt{(1 - \cos x)\dfrac{1 + \cos x}{1 + \cos x}}$

$= \sqrt{\dfrac{\sin^2 x}{1 + \cos x}}$

$= \dfrac{|\sin x|}{\sqrt{1 + \cos x}}$

20. $\cos\left(x + \dfrac{\pi}{2}\right) = \cos x \cos \dfrac{\pi}{2} - \sin x \sin \dfrac{\pi}{2}$

$= (\cos x)(0) - (\sin x)(1)$

$= -\sin x$

22. $\sin(\pi - x) = \sin \pi \cos \pi - \sin x \cos \pi$

$= (0)(\cos x) - (\sin x)(-1)$

$= \sin x$

24. $\dfrac{2 \cos 3x}{\sin 4x - \sin 2x} = \dfrac{2 \cos 3x}{2 \cos 3x \sin x}$

$= \dfrac{1}{\sin x}$

$= \csc x$

26. $\dfrac{\sin(\alpha + \beta)}{\cos \alpha \cos \beta} = \dfrac{\sin \alpha \cos \beta + \cos \alpha \sin \beta}{\cos \alpha \cos \beta}$

$= \dfrac{\sin \alpha \cos \beta}{\cos \alpha \cos \beta} + \dfrac{\cos \alpha \sin \beta}{\cos \alpha \cos \beta}$

$= \tan \alpha + \tan \beta$

28. $\sin 4x = 2 \sin 2x \cos 2x$

$= 2[2 \sin x \cos x(\cos^2 x - \sin^2 x)]$

$= 4 \sin x \cos x(2 \cos^2 x - 1)$

$= 8 \cos^3 x \sin x - 4 \cos x \sin x$

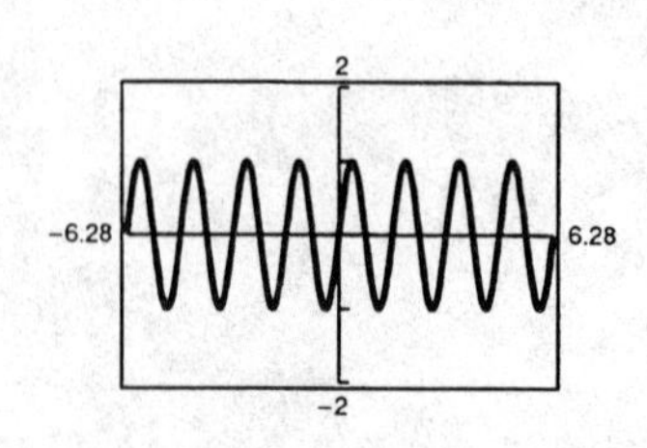

30. $\cos^2 5x - \cos^2 x = (\cos 5x - \cos x)(\cos 5x + \cos x)$

$= (-2 \sin 3x \sin 2x)(2 \cos 3x \cos 2x)$

$= -(2 \sin 2x \cos 2x)(2\sin 3x \cos 3x)$

$= -\sin 4x \sin 6x$

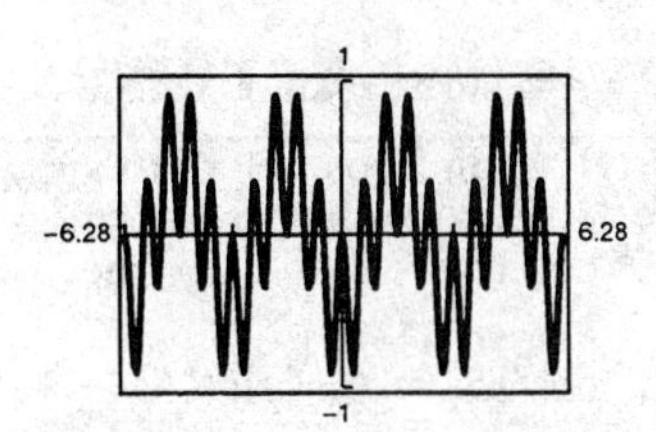

32. $\cos(285°) = \cos(225° + 60°)$

$= \cos 225° \cos 60° - \sin 225° \sin 60°$

$= \left(-\frac{\sqrt{2}}{2}\right)\left(\frac{1}{2}\right) - \left(-\frac{\sqrt{2}}{2}\right)\left(\frac{\sqrt{3}}{2}\right)$

$= \frac{\sqrt{2}}{4}(\sqrt{3} - 1)$

34. $\sin \frac{3\pi}{8} = \sqrt{\frac{1 - \cos(3\pi/4)}{2}}$

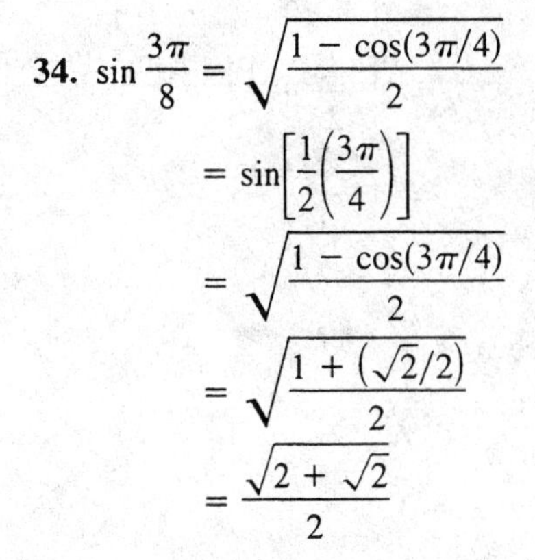

$= \sin\left[\frac{1}{2}\left(\frac{3\pi}{4}\right)\right]$

$= \sqrt{\frac{1 - \cos(3\pi/4)}{2}}$

$= \sqrt{\frac{1 + (\sqrt{2}/2)}{2}}$

$= \frac{\sqrt{2 + \sqrt{2}}}{2}$

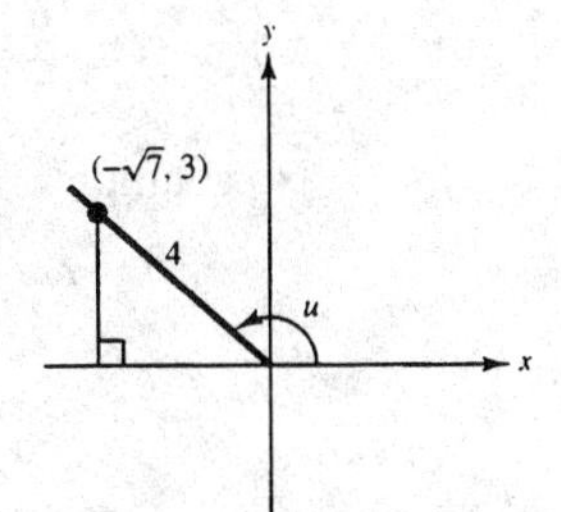

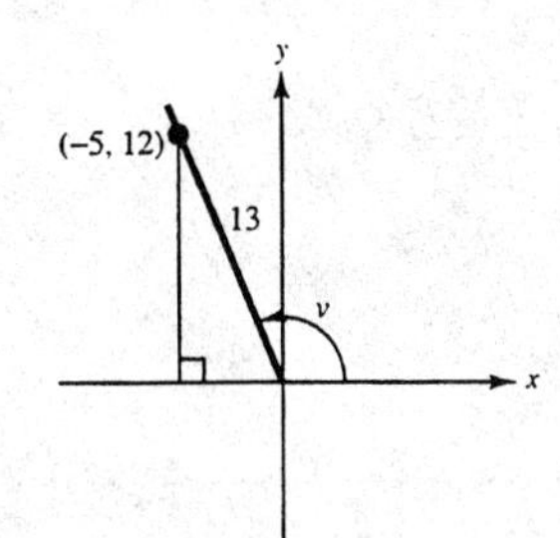

Figures for Exercises 36–40

36. $\tan(u + v) = \frac{\tan u + \tan v}{1 - \tan u \tan v}$

$= \frac{-(3/\sqrt{7}) - (12/5)}{1 - (-3/\sqrt{7})(-12/5)}$

$= \frac{15 + 12\sqrt{7}}{36 - 5\sqrt{7}}$

38. $\sin 2v = 2 \sin v \cos v$

$= 2\left(\frac{12}{13}\right)\left(-\frac{5}{13}\right)$

$= -\frac{120}{169}$

40. $\tan 2v = \frac{\sin 2v}{\cos 2v}$

$= \frac{2 \sin v \cos v}{2 \cos^2 v - 1}$

$= \frac{2(12/13)(-5/13)}{2(-5/13)^2 - 1}$

$= \frac{-120/169}{(50/169) - (169/169)} = \frac{120}{119}$

42. $\sin(x + y) = \sin x + \sin y$. False.

$\sin(x + y) = \sin x \cos y + \cos x \sin y$

44. $4 \sin 45° \cos 15° = 1 + \sqrt{3}$. True.

$$\begin{aligned} 4 \sin 45° \cos 15° &= 4\left(\frac{1}{2}[\sin(45° + 15°) + \sin(45° - 15°)]\right) \\ &= 2[\sin 60° + \sin 30°] \\ &= 2\left[\frac{\sqrt{3}}{2} + \frac{1}{2}\right] \\ &= 2\left(\frac{\sqrt{3} + 1}{2}\right) \\ &= 1 + \sqrt{3} \end{aligned}$$

46. $\csc x - 2 \cot x = 0$

$$\begin{aligned} \frac{1}{\sin x} - \frac{2 \cos x}{\sin x} &= 0 \\ 1 - 2 \cos x &= 0 \\ \cos x &= \frac{1}{2} \\ x &= \frac{\pi}{3}, \frac{5\pi}{3} \end{aligned}$$

48.

$$\begin{aligned} \cos 4x - 7 \cos 2x &= 8 \\ 2 \cos^2 2x - 1 - 7 \cos 2x &= 8 \\ 2 \cos^2 2x - 7 \cos 2x - 9 &= 0 \\ (2 \cos 2x - 9)(\cos 2x + 1) &= 0 \end{aligned}$$

$2 \cos 2x - 9 = 0$ or $\cos 2x + 1 = 0$

$\cos 2x = \frac{9}{2}$ $\qquad$ $\cos 2x = -1$

No solution $\qquad$ $2x = \pi + 2n\pi$

$$x = \frac{\pi}{2} + n\pi$$

$$x = \frac{\pi}{2}, \frac{3\pi}{2}$$

50. $\sin 4x - \sin 2x = 0$

$2 \cos 3x \sin x = 0$

$\cos 3x = 0$ or $\sin x = 0$

$3x = \frac{\pi}{2} + n\pi$ $\qquad$ $x = 0, \pi$

$$x = \frac{\pi}{6} + \frac{n\pi}{3}$$

$$x = \frac{\pi}{6}, \frac{\pi}{2}, \frac{5\pi}{6}, \frac{7\pi}{6}, \frac{3\pi}{2}, \frac{11\pi}{6}$$

52. $y = \cos x - \cos \frac{x}{2}$

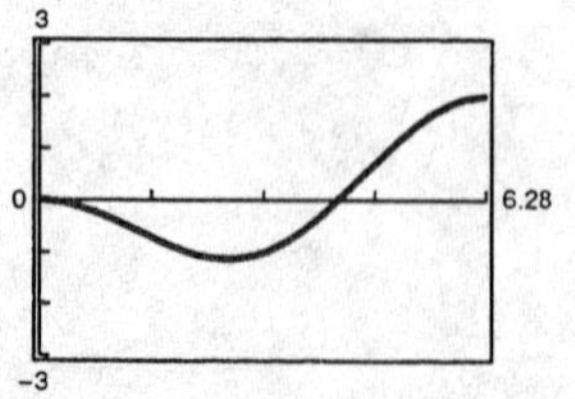

3

0 $\qquad$ 6.28

-3

Zeros: $x \approx 0, 4.1888$

—CONTINUED—

52. —CONTINUED—

$$\cos x - \cos\frac{x}{2} = 0$$

$$\cos x = \cos\frac{x}{2}$$

$$\cos^2 x = \cos^2\frac{x}{2}$$

$$\cos^2 x - \cos^2\frac{x}{2} = 0$$

$$\left(\cos x + \cos\frac{x}{2}\right)\left(\cos x - \cos\frac{x}{2}\right) = 0$$

$$2\cos\frac{3x}{4}\cos\frac{x}{4}\left(-2\sin\frac{3x}{4}\sin\frac{x}{4}\right) = 0$$

$$-4\cos\frac{3x}{4}\cos\frac{x}{4}\sin\frac{3x}{4}\sin\frac{x}{4} = 0$$

$\cos\frac{3x}{4} = 0 \Rightarrow x = \frac{4\pi}{3}$ $\left(x = \frac{2\pi}{3}\text{ is extraneous.}\right)$ or

$\cos\frac{x}{4} = 0 \Rightarrow$ (No solution in $[0, 2\pi)$) or

$\sin\frac{3x}{4} = 0$ or $\sin\frac{x}{4} = 0$

$x = 0, \frac{4\pi}{3}$ $\qquad x = 0$

54. $h(s) = \sin s + \sin 3s + \sin 5s$

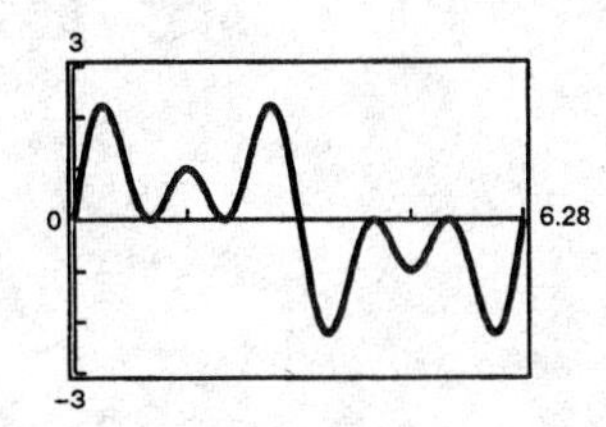

Zeros: $s \approx 0, 1.0472, 2.0944, 3.1416, 4.1888, 5.2360$

$\sin s + \sin 3s + \sin 5s = 0$

$2\sin 3s\cos 2s + \sin 2s = 0$

$\sin 3s(2\cos 2s + 1) = 0$

$\sin 3s = 0$ or $2\cos 2s + 1 = 0$

$3s = n\pi$ $\qquad \cos 2s = -\frac{1}{2}$

$s = \frac{n\pi}{3}$ $\qquad 2s = \frac{2\pi}{3} + 2n\pi$ or $2s = \frac{4\pi}{3} + 2n\pi$

$s = 0, \frac{\pi}{3}, \frac{2\pi}{3}, \pi, \frac{4\pi}{3}, \frac{5\pi}{3}$ $\qquad s = \frac{\pi}{3}, \frac{2\pi}{3}, \frac{4\pi}{3}, \frac{5\pi}{3}$

56. $a\sin x - b = 0$, $\sin x = \frac{b}{a}$

You know that this equation has no solution if $|a| < |b|$ because $-1 \le \sin x \le 1$ for all x. If $|a| < |b|$, then b/a is greater than 1.

58. $\sin\left(x + \frac{\pi}{4}\right) - \sin\left(x - \frac{\pi}{4}\right) = 2 \cos x \sin \frac{\pi}{4}$

$= \sqrt{2} \cos x$

60. $\cos \frac{x}{2} \cos \frac{x}{4} = \frac{1}{2}\left[\cos\left(\frac{x}{2} - \frac{x}{4}\right) + \cos\left(\frac{x}{2} + \frac{x}{4}\right)\right]$

$= \frac{1}{2}\left(\cos \frac{x}{4} + \cos \frac{3x}{4}\right)$

62. $\sin(2 \arctan x) = \sin 2\theta$

$= 2 \sin \theta \cos \theta$

$= 2\left(\frac{x}{\sqrt{x^2 + 1}}\right)\left(\frac{1}{\sqrt{x^2 + 1}}\right)$

$= \frac{2x}{x^2 + 1}$

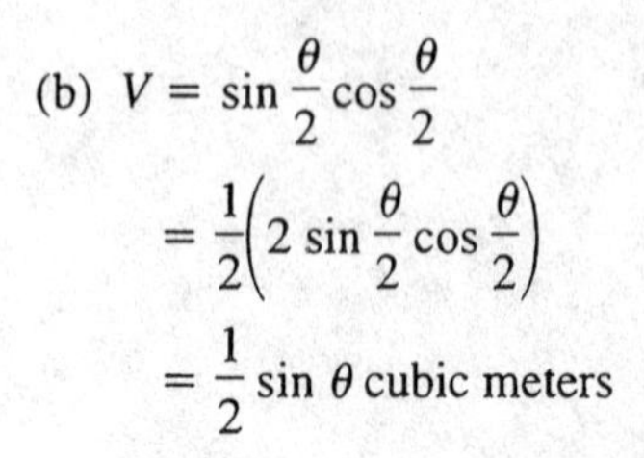

64. $r = \frac{1}{32} v_0^2 \sin 2\theta$

$100 = \frac{1}{32}(80)^2 \sin 2\theta$

$\sin 2\theta = 0.5$

$2\theta = 30°$ or $2\theta = 180° - 30° = 150°$

$\theta = 15°$ $\quad\theta = 75°$

66. Volume V of the trough will be the area A of the isosceles triangle times the length l of the trough.

$V = A \cdot l$

(a) $A = \frac{1}{2}bh$

$\cos \frac{\theta}{2} = \frac{h}{0.5} \Rightarrow h = 0.5 \cos \frac{\theta}{2}$

$\sin \frac{\theta}{2} = \frac{b/2}{0.5} \Rightarrow \frac{b}{2} = 0.5 \sin \frac{\theta}{2}$

$A = 0.5 \sin \frac{\theta}{2} 0.5 \cos \frac{\theta}{2}$

$= (0.5)^2 \sin \frac{\theta}{2} \cos \frac{\theta}{2}$

$= 0.25 \sin \frac{\theta}{2} \cos \frac{\theta}{2}$ square meters

$V = (0.25)(4) \sin \frac{\theta}{2} \cos \frac{\theta}{2}$ cubic meters

$= \sin \frac{\theta}{2} \cos \frac{\theta}{2}$ cubic meters

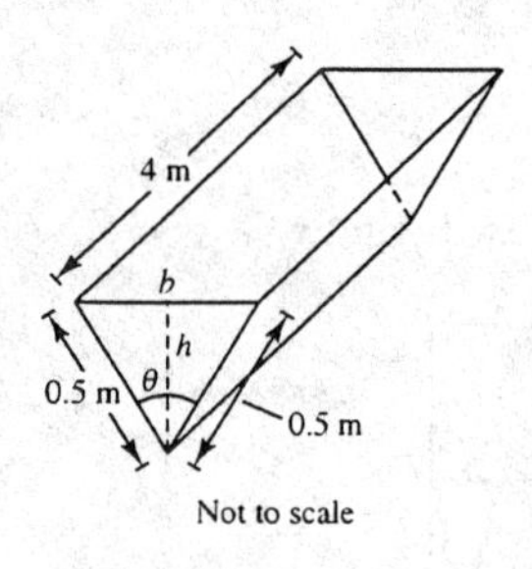

Not to scale

(b) $V = \sin \frac{\theta}{2} \cos \frac{\theta}{2}$

$= \frac{1}{2}\left(2 \sin \frac{\theta}{2} \cos \frac{\theta}{2}\right)$

$= \frac{1}{2} \sin \theta$ cubic meters

Volume is maximum when $\theta = \pi/2$.

CHAPTER 6
Additional Topics in Trigonometry

CHAPTER 6
Additional Topics in Trigonometry

Section 6.1 Law of Sines

Solutions to Even-Numbered Exercises

2.

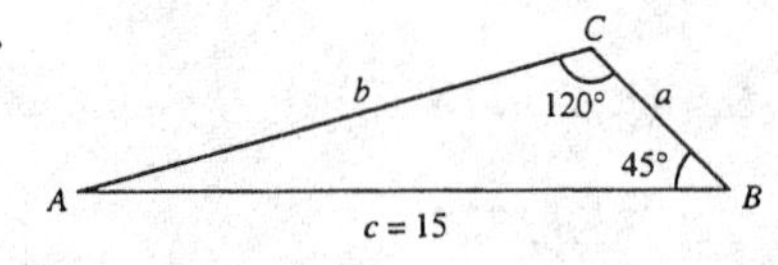

Given: $B = 45°, C = 120°, c = 15$

$A = 180° - B - C = 15°$

$$a = \frac{c}{\sin C}(\sin A) = \frac{15 \sin 15°}{\sin 120°} \approx 4.48$$

$$b = \frac{c}{\sin C}(\sin B) = \frac{15(\sin 45°)}{\sin 120°} \approx 12.25$$

4.

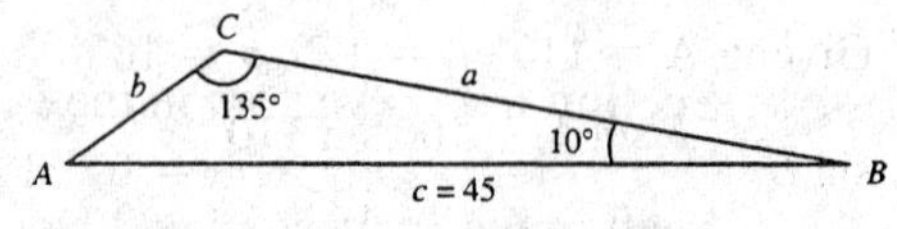

Given: $B = 10°, C = 135°, c = 45$

$A = 180° - B - C = 35°$

$$a = \frac{c}{\sin C}(\sin A) = \frac{45 \sin 35°}{\sin 135°} \approx 36.50$$

$$b = \frac{c}{\sin C}(\sin B) = \frac{45 \sin 35°}{\sin 135°} \approx 11.05$$

6. Given: $A = 60°, a = 9, c = 10$

$$\sin C = \frac{c \sin A}{a} = \frac{10 \sin 60°}{9} \approx 0.9623 \implies C \approx 74.21° \text{ or } C \approx 105.79°$$

<u>Case 1</u>

$C \approx 74.21°$

$B = 180° - A - C \approx 45.79°$

$$b = \frac{a}{\sin A}(\sin B) \approx \frac{9 \sin 45.79°}{\sin 60°} \approx 7.45$$

<u>Case 2</u>

$C \approx 105.79°$

$B = 180° - A - C \approx 14.21°$

$$b = \frac{a}{\sin A}(\sin B) \approx \frac{9 \sin 14.21°}{\sin 60°} \approx 2.55$$

8. Given: $A = 24.3°, C = 54.6°, c = 2.68$

$B = 180° - A - C = 101.1°$

$$a = \frac{c}{\sin C}(\sin A) = \frac{2.68 \sin 24.3°}{\sin 54.6°} \approx 1.35$$

$$b = \frac{c}{\sin C}(\sin B) = \frac{2.68 \sin 101.1°}{\sin 54.6°} \approx 3.23$$

10. Given: $A = 5° 40', B = 8° 15', b = 4.8$

$C = 180° - A - B = 166° 5'$

$$a = \frac{b}{\sin B}(\sin A) = \frac{4.8 \sin 5° 40'}{\sin 8° 15'} \approx 3.30$$

$$c = \frac{b}{\sin B}(\sin C) = \frac{4.8 \sin 166° 5'}{\sin 8° 15'} \approx 8.05$$

12. Given: $C = 85° 20', a = 35, c = 50$

$$\sin A = \frac{a \sin C}{c} = \frac{35 \sin 85° 20'}{50} \approx 0.6977 \implies A \approx 44.24°$$

$B = 180° - A - C \approx 50.43°$

$$b = \frac{c}{\sin C}(\sin B) \approx \frac{50 \sin 50.43°}{\sin 85° 20'} \approx 38.67$$

14. Given: $B = 2° 45', b = 6.2, c = 5.8$

$$\sin C = \frac{c \sin B}{b} = \frac{5.8 \sin 2° 45'}{6.2} \approx 0.04488 \implies C \approx 2.57°$$

$A = 180° - B - C \approx 174.68°$

$$a = \frac{b}{\sin B}(\sin A) \approx \frac{6.2 \sin 174.68°}{\sin 2° 45'} \approx 11.99$$

16. Given: $A = 58°$, $a = 11.4$, $c = 12.8$

$$\sin B = \frac{b \sin A}{a} = \frac{12.8 \sin 58°}{11.4} \approx 0.9522 \Rightarrow B \approx 72.2° \text{ or } B \approx 107.8°$$

Case 1

$B \approx 72.2°$

$C = 180° - A - B \approx 49.8°$

$$c = \frac{a}{\sin A}(\sin C) \approx \frac{11.4 \sin 49.8°}{\sin 58°} \approx 10.27$$

Case 2

$B \approx 107.8°$

$C = 180° - A - B \approx 14.2°$

$$c = \frac{a}{\sin A}(\sin C) \approx \frac{11.4 \sin 14.2°}{\sin 58°} \approx 3.30$$

18. Given: $A = 110°$, $a = 125$, $b = 100$

$$\sin B = \frac{b \sin A}{a} = \frac{100 \sin 110°}{125} \approx 0.75175 \Rightarrow B \approx 48.74°$$

$C = 180° - A - B = 21.26°$

$$c = \frac{a}{\sin A}(\sin C) = \frac{125 \sin 21.26°}{\sin 110°} \approx 48.23$$

20. Given: $A = 60°$, $a = 10$

(a) One solution if $b \leq 10$ or $b = \dfrac{10}{\sin 60°}$.

(b) Two solutions if $10 < b < \dfrac{10}{\sin 60°}$.

(c) No solutions if $b > \dfrac{10}{\sin 60°}$.

22.

$A = 180° - 96° - 22° 50' = 61° 10'$

$$h = \frac{30 \sin 22° 50'}{\sin 61° 10'} \approx 13.3 \text{ meters}$$

24. $$\sin A = \frac{a \sin B}{b} = \frac{500 \sin 46°}{720} \approx 0.4995 \Rightarrow A \approx 30°$$

The bearing from C to A is S 60° W.

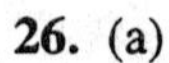

26. (a)

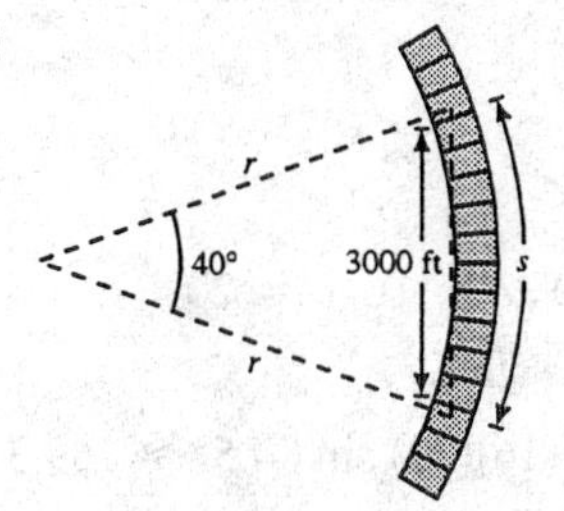

(b) $$r = \frac{3000 \sin[1/2(180° - 40°)]}{\sin 40°} \approx 4385.71 \text{ feet}$$

(c) $$s \approx 40°\left(\frac{\pi}{180°}\right)4385.71 \approx 3061.80 \text{ feet}$$

28. $A = 51°$, $B = 112°$, $c = 2.5$

$C = 180° - 51° - 112° = 17°$

$$a = \frac{c}{\sin C}(\sin A) = \frac{2.5}{\sin 17°}(\sin 51°) \approx 6.65$$

$h \approx 6.65 \sin 68° \approx 6.2$ mi

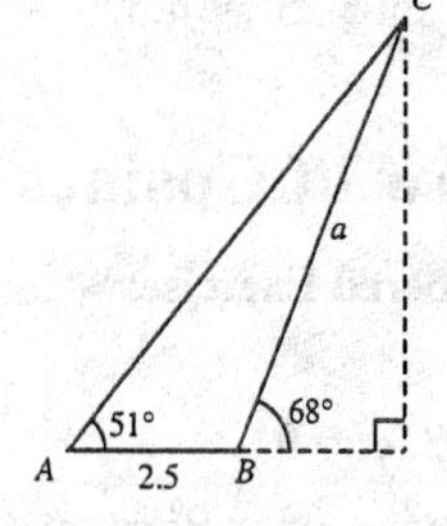

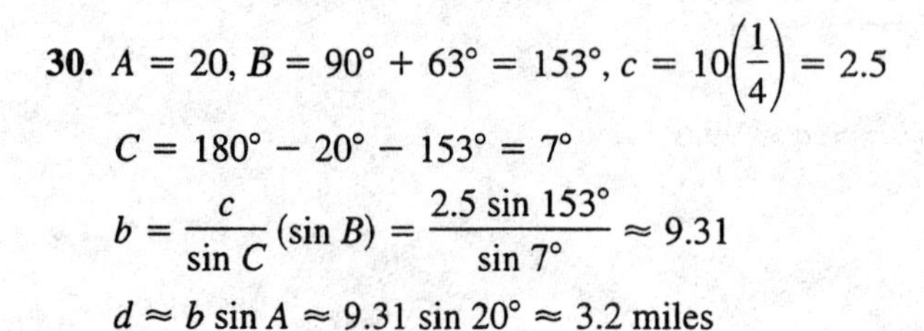

30. $A = 20$, $B = 90° + 63° = 153°$, $c = 10\left(\frac{1}{4}\right) = 2.5$

$C = 180° - 20° - 153° = 7°$

$$b = \frac{c}{\sin C}(\sin B) = \frac{2.5 \sin 153°}{\sin 7°} \approx 9.31$$

$d \approx b \sin A \approx 9.31 \sin 20° \approx 3.2$ miles

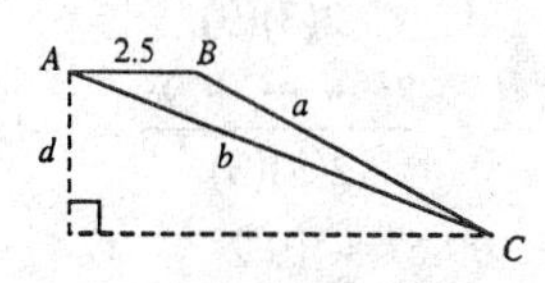

32. (a) $\dfrac{6}{\sin\theta} = \dfrac{1.5}{\sin C}, \sin\theta \neq 0$

$\sin C = \dfrac{1.5\sin\theta}{6} \Rightarrow C = \arcsin\dfrac{1.5\sin\theta}{6}$

$B = 180° - \theta - \arcsin\dfrac{1.5\sin\theta}{6}$

$\dfrac{7.5 - d}{\sin B} = \dfrac{6}{\sin\theta}$

$$d = 7.5 - \frac{6\sin\left(180° - \theta - \arcsin\dfrac{1.5\sin\theta}{6}\right)}{\sin\theta}$$

For $\theta = 0°$, $C = 0°$, $B = 180° \Rightarrow 7.5 - d = 1.5 + 6 \Rightarrow d = 0$.

θ	0°	45°	90°	135°	180°
d	0	0.5338	1.6905	2.6552	3

For $\theta = 180°$, $C = 0°$, $B = 0° \Rightarrow 7.5 - d = 6 - 1.5 \Rightarrow d = 3$.

(b) $\theta = 5°$

$$d = 7.5 - \frac{6\sin\left(180° - 5° - \arcsin\dfrac{1.5\sin 5°}{6}\right)}{\sin 5°} \approx 0.0071 \text{ inch}$$

34. $\alpha = 180 - (\phi + 180 - \theta) = \theta - \phi$

$\dfrac{d}{\sin\phi} = \dfrac{2}{\sin\alpha}$

$d = \dfrac{2\sin\phi}{\sin(\phi - \theta)}$

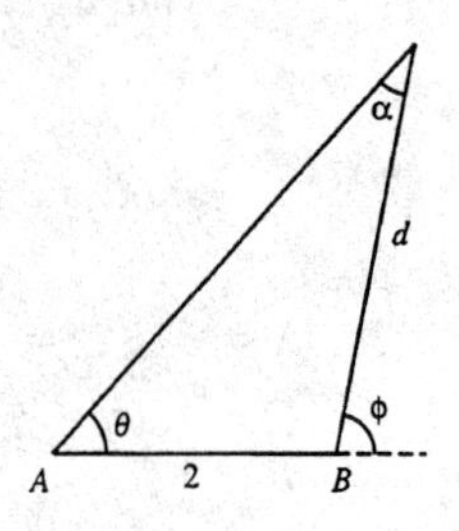

36. $B = 72°\,30'$, $a = 105$, $c = 64$

$\text{Area} = \dfrac{1}{2}ac\sin B = \left(\dfrac{1}{2}\right)(105)(64)\sin 72.5° \approx 3204$

38. $A = 5°\,15'$, $b = 4.5$, $c = 22$

$\text{Area} = \frac{1}{2}bc\sin A$

$= \left(\frac{1}{2}\right)(4.5)(22)\sin 5.25° \approx 4.529$

40. $C = 84°\,30'$, $a = 16$, $b = 20$

$\text{Area} = \frac{1}{2}ab\sin C$

$= \left(\frac{1}{2}\right)(16)(20)\sin 84.5° \approx 159.3$

Section 6.2 Law of Cosines

Solutions to Even-Numbered Exercises

2. Given: $a = 8, b = 3, c = 9$

$\cos A = \dfrac{b^2 + c^2 - a^2}{2bc} = \dfrac{3^2 + 9^2 - 8^2}{2(3)(9)} \approx 0.4815 \Rightarrow A \approx 61.2°$

$\cos c = \dfrac{a^2 + b^2 - c^2}{2ab} = \dfrac{8^2 + 3^2 - 9^2}{2(8)(3)} \approx -0.1667 \Rightarrow c \approx 99.6°$

$B \approx 180° - 61.2° - 99.6° \approx 19.2°$

4. Given: $C = 105°, a = 10, b = 4.5$

$c^2 = a^2 + b^2 - 2ab\cos C = 10^2 + 4.5^2 - 2(10)(4.5)\cos 105° \approx 143.5437 \Rightarrow c \approx 12.0$

$\cos B = \dfrac{a^2 + c^2 - b^2}{2ac} \approx \dfrac{10^2 + (12.0)^2 - (4.5)^2}{2(10)(12.0)} \approx 0.93187 \Rightarrow B \approx 21.3°$

$A = 180° - 105° - 21.3° \approx 53.7°$

6. Given: $a = 55, b = 25, c = 72$

$\cos C = \dfrac{a^2 + b^2 - c^2}{2ab} = \dfrac{55^2 + 25^2 - 72^2}{2(55)(25)} \approx -0.5578 \Rightarrow c \approx 123.91°$

$\cos A = \dfrac{b^2 + c^2 - a^2}{2bc} = \dfrac{25^2 + 72^2 - 55^2}{2(25)(72)} \approx 0.7733 \Rightarrow A \approx 39.35°$

$B = 180° - 123.91° - 39.35° \approx 16.74°$

8. Given: $a = 1.42, b = 0.75, c = 1.25$

$\cos A = \dfrac{b^2 + c^2 - a^2}{2bc} = \dfrac{(0.75)^2 + (1.25)^2 - (1.42)^2}{2(0.75)(1.25)} = 0.05792 \Rightarrow A \approx 86.7°$

$\cos B = \dfrac{a^2 + c^2 - b^2}{2ac} = \dfrac{(1.42)^2 + (1.25)^2 - (0.75)^2}{2(1.42)(1.25)} \approx 0.8497 \Rightarrow B \approx 31.8°$

$180° - 86.7° - 31.8° \approx 61.5°$

10. Given: $B = 75°\,20', a = 6.2, c = 9.5$

$b^2 = a^2 + c^2 - 2ac\cos B = (6.2)^2 + (9.5)^2 - 2(6.2)(9.5)\cos 75°\,20' \approx 98.8636 \Rightarrow b \approx 9.94$

$\sin A = \dfrac{a \sin B}{b} \approx \dfrac{6.2 \sin 75°\,20'}{9.94} \approx 0.6034 \Rightarrow A \approx 37.1°$

$C \approx 180° - 75°\,20' - 37.1° \approx 67.6°$

12.

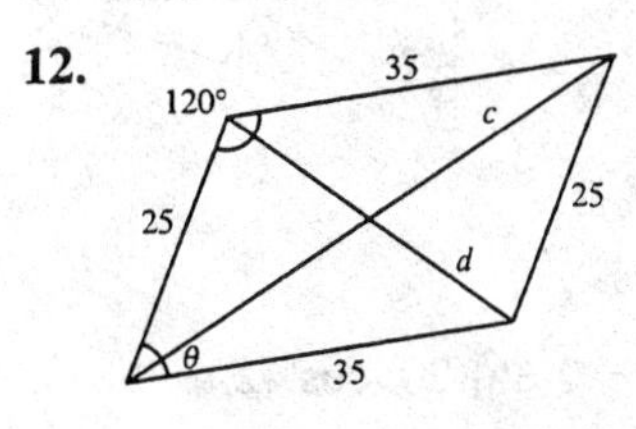

$c^2 = 25^2 + 35^2 - 2(25)(35)\cos 120°$

$= 2725 \Rightarrow c \approx 52.20$

$\theta = 360° - 2(120°) = 60°$

$d^2 = 25^2 + 35^2 - 2(25)(35)\cos 60°$

$= 975 \Rightarrow d \approx 31.22$

14.

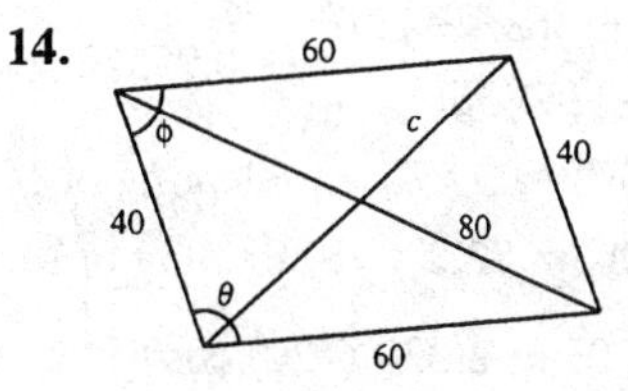

$\cos\theta = \dfrac{40^2 + 60^2 - 80^2}{2(40)(60)} \approx -\dfrac{1}{4} \neq \theta \approx 104.5°$

$\phi \approx 360° - 2(104.5°) \approx 75.5°$

$c^2 \approx 40^2 + 60^2 - 2(40)(60)\cos 75.5° = 4000$

$c \approx 63.25$

16. $\cos\alpha = \dfrac{25^2 + 17.5^2 - 25^2}{2(25)(17.5)}$

$\alpha \approx 69.512°$

$\beta \approx 180 - \alpha \approx 110.488°$

$a^2 = 17.5^2 + 25^2 - 2(17.5)(25)\cos 110.488°$

$a \approx 35.18$

$z = 180 - 2\alpha \approx 40.976$

$\cos\mu = \dfrac{25^2 + 35.18^2 - 17.5^2}{2(25)(35.18)}$

$\mu \approx 27.771°$

$\theta = \mu + z \approx 68.7°$

$\omega = 180° - \mu - \beta \approx 41.741°$

$\phi = \omega + \alpha \approx 111.3°$

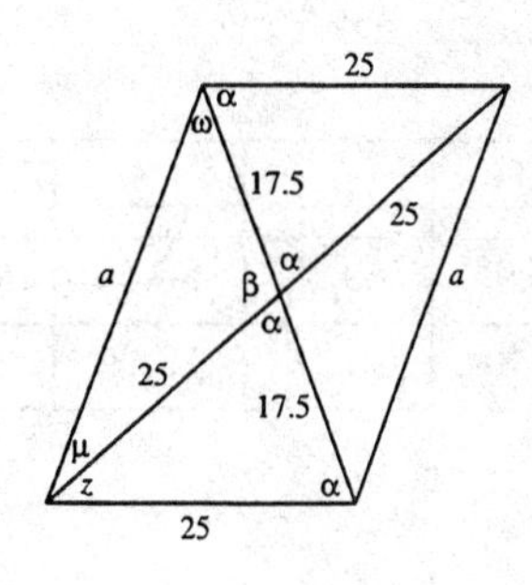

18. $\cos B = \dfrac{1100^2 + 2500^2 - 2000^2}{2(1100)(2500)} = 0.6291$

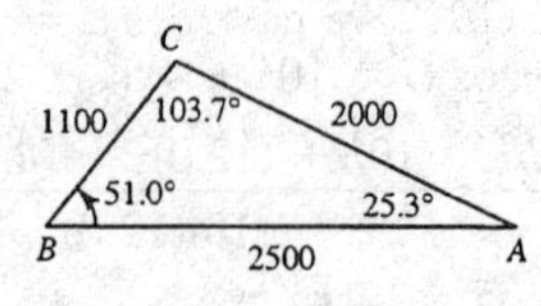

$B \approx 51.0°$

$90 - B \approx 39.$

Bearing at B is approximately N 39° E.

$\cos C = \dfrac{1100^2 + 2000^2 - 2500^2}{2(1100)(2000)} = -0.2364$

$C \approx 103.7°$

$C - (90° - 51.0°) = 64.7°$

Bearing at C is approximately S 64.7° E.

20. $\cos\theta = \dfrac{2^2 + 3^2 - (4.5)^2}{2(2)(3)} \approx -0.60417$

$\theta \approx 127.2°$

22. The angles at the base of the tower are 96° and 84°. The longer guy wire g_1 is given by:

$g_1{}^2 = 75^2 + 100^2 - 2(75)(100)\cos 96° \approx 17{,}192.9 \Rightarrow g_1 \approx 131.1$ feet

The shorter guy wire g_2 is given by:

$g_2{}^2 = 75^2 + 100^2 - 2(75)(100)\cos 84° \approx 14{,}057.1 \Rightarrow g_2 \approx 118.6$ feet

24. Bearing of M from P: N θ E

Bearing of A from P: N ϕ E

Since M is due west of A, it follows that $\theta = M - 90°$ and $\phi = 90° - A$.

$\cos M = \dfrac{165^2 + 216^2 - 368^2}{2(165)(216)} \approx -0.8634 \Rightarrow M \approx 149.7°$

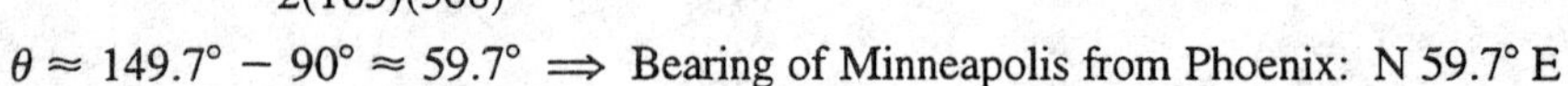

$\cos A = \dfrac{165^2 + 368^2 - 216^2}{2(165)(368)} \approx 0.95515 \Rightarrow A \approx 17.2°$

$\theta \approx 149.7° - 90° \approx 59.7° \Rightarrow$ Bearing of Minneapolis from Phoenix: N 59.7° E

$\phi \approx 90° - 17.2° \approx 72.8° \Rightarrow$ Bearing of Minneapolis from Phoenix: N 72.8° E

26. $x^2 = 330^2 + 420^2 - 2(330)(420)\cos 8°$

$\approx 10{,}797.7$

$x \approx 103.9$ feet

28. $a = 35^2 + 20^2 - 2(35)(20)\cos 42° \approx 584.6$

$a \approx 24$ miles

30. $d^2 = 10^2 + 7^2 - 2(10)(7)\cos\theta$

$\theta = \arccos\left[\dfrac{10^2 + 7^2 - d^2}{2(10)(7)}\right]$

$s = \dfrac{360° - \theta}{360°}(2\pi r) = \dfrac{(360° - \theta)\pi}{45}$

d (inches)	9	10	12	13	14	15	16
θ (degrees)	60.9°	69.5°	88.0°	98.2°	109.6°	122.9°	139.8°
s (inches)	20.88	20.28	18.99	18.28	17.48	16.55	15.37

32. (a) Working with $\triangle OBC$, we have $\cos\alpha = \dfrac{a/2}{R}$.

This implies that $2R = a/\cos\alpha$. Since we know that

$$\frac{a}{\sin A} = \frac{b}{\sin B} = \frac{c}{\sin C},$$

we can complete the proof by showing that $\cos\alpha = \sin A$. The solution of the system

$$A + B + C = 180°$$
$$\alpha - C + A = \beta$$
$$\alpha + \beta = B$$

is $\alpha = 90° - A$. Therefore:

$$2R = \frac{a}{\cos\alpha} = \frac{a}{\cos(90° - A)} = \frac{a}{\sin A}.$$

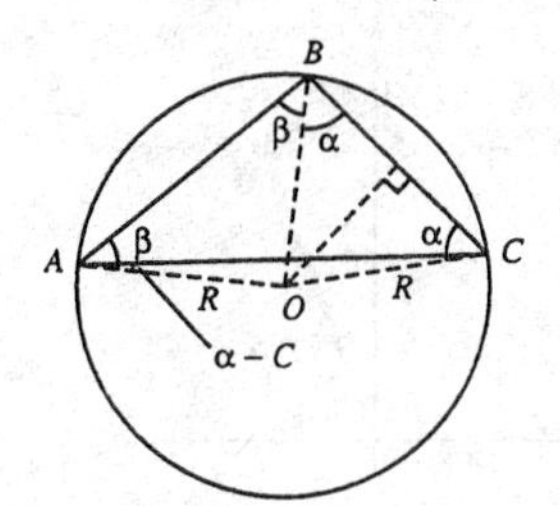

(b) By Heron's Formula, the area of the triangle is

$$\text{Area} = \sqrt{s(s-a)(s-b)(s-c)}.$$

We can also find the area by dividing the area into six triangles and using the fact that the area is 1/2 the base times the height. Using the figure as given, we have

$$\text{Area} = \frac{1}{2}xr + \frac{1}{2}xr + \frac{1}{2}yr + \frac{1}{2}yr + \frac{1}{2}zr + \frac{1}{2}zr$$
$$= r(x + y + z)$$
$$= rs.$$

Therefore: $rs = \sqrt{s(s-a)(s-b)(s-c)} \Rightarrow$

$$r = \sqrt{\frac{(s-a)(s-b)(s-c)}{s}}.$$

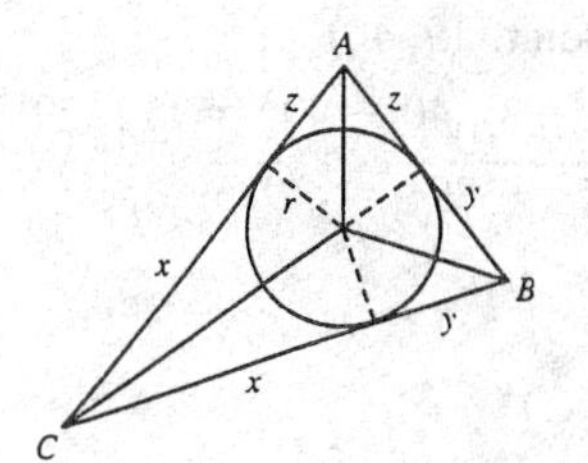

34. Given: $a = 200$ ft, $b = 250$ ft, $c = 325$ ft

$$s = \frac{200 + 250 + 325}{2} \approx 387.5$$

Radius of the inscribed circle: $r = \sqrt{\dfrac{(s-a)(s-b)(s-c)}{s}} = \sqrt{\dfrac{(187.5)(137.5)(62.5)}{387.5}} \approx 64.5$ ft

Circumference of an inscribed circle: $C = 2\pi r \approx 2\pi(64.5) \approx 405$ ft

36. Given: $a = 2.5$, $b = 10.2$, $c = 9$

$$s = \frac{2.5 + 10.2 + 9}{2} = 10.85$$

$$\text{Area} = \sqrt{s(s-a)(s-b)(s-c)}$$
$$= \sqrt{10.85(8.35)(0.65)(1.85)} \approx 10.44$$

38. Given: $a = 75.4$, $b = 52$, $c = 52$

$$s = \frac{75.4 + 52 + 52}{2} = 89.7$$

$$\text{Area} = \sqrt{s(s-a)(s-b)(s-c)}$$
$$= \sqrt{89.7(14.3)(37.7)(37.7)} \approx 1350$$

40. Given: $a = 4.25$, $b = 1.55$, $c = 3.00 \Rightarrow s = \dfrac{4.25 + 1.55 + 3.00}{2} = 4.4$

$\text{Area} = \sqrt{s(s-a)(s-b)(s-c)} = \sqrt{4.4(0.15)(2.85)(1.4)} \approx 1.623$

42.
$$\frac{1}{2}bc(1 - \cos A) = \frac{1}{2}bc\left[1 + \frac{a^2 - (b^2 + c^2)}{2bc}\right]$$
$$= \frac{1}{2}bc\left[\frac{2bc + a^2 - b^2 - c^2}{2bc}\right]$$
$$= \frac{a^2 - (b^2 - 2bc + c^2)}{4}$$
$$= \frac{a^2 - (b - c)^2}{4}$$
$$= \left(\frac{a - (b - c)}{2}\right)\left(\frac{a + (b - c)}{2}\right)$$
$$= \frac{a - b + c}{2} \cdot \frac{a + b - c}{2}$$

Section 6.3 Vectors in the Plane

Solutions to Even-Numbered Exercises

2. Initial point: $(0, 0)$

Terminal point: $(4, -2)$

$\mathbf{v} = \langle 4 - 0, -2 - 0\rangle = \langle 4, -2\rangle$

$\|\mathbf{v}\| = \sqrt{4^2 + (-2)^2} = \sqrt{20} = 2\sqrt{5}$

4. Initial point: $(-1, -1)$

Terminal point: $(3, 5)$

$\mathbf{v} = \langle 3 - (-1), 5 - (-1)\rangle = \langle 4, 6\rangle$

$\|\mathbf{v}\| = \sqrt{4^2 + 6^2} = \sqrt{52} = 2\sqrt{13}$

6. Initial point: $(-4, -1)$

Terminal point: $(3, -1)$

$\mathbf{v} = \langle 3 - (-4), -1 - (-1)\rangle = \langle 7, 0\rangle$

$\|\mathbf{v}\| = \sqrt{7^2 + 0^2} = 7$

8. Initial point: $(3.4, 0)$

Terminal point: $(0, 5.8)$

$\mathbf{v} = \langle 0 - 3.4, 5.8 - 0\rangle = \langle -3.4, 5.8\rangle$

$\|\mathbf{v}\| = \sqrt{(-3.4)^2 + (5.8)^2} \approx 6.7$

10. Initial point: $(-3, 11)$

Terminal point: $(9, 40)$

$\mathbf{v} = \langle 9 - (-3), 40 - 11\rangle = \langle 12, 29\rangle$

$\|\mathbf{v}\| = \sqrt{12^2 + 29^2} = \sqrt{985}$

12. $3\mathbf{v}$

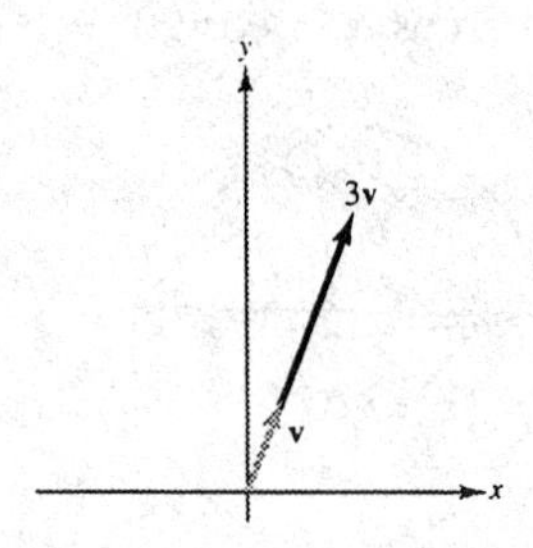

14. $\mathbf{u} + 2\mathbf{v}$

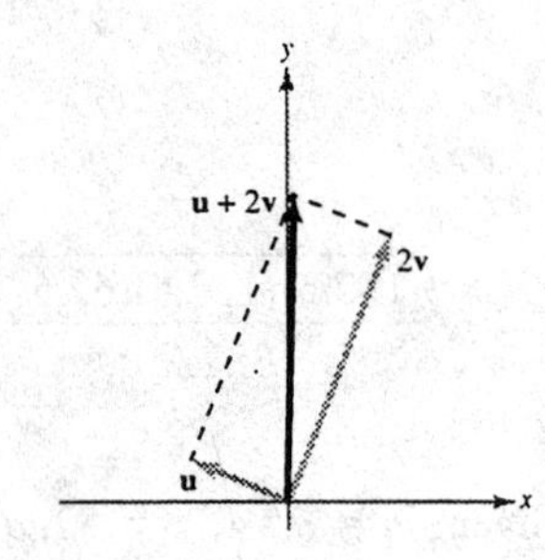

16. $\mathbf{v} - \frac{1}{2}\mathbf{u}$

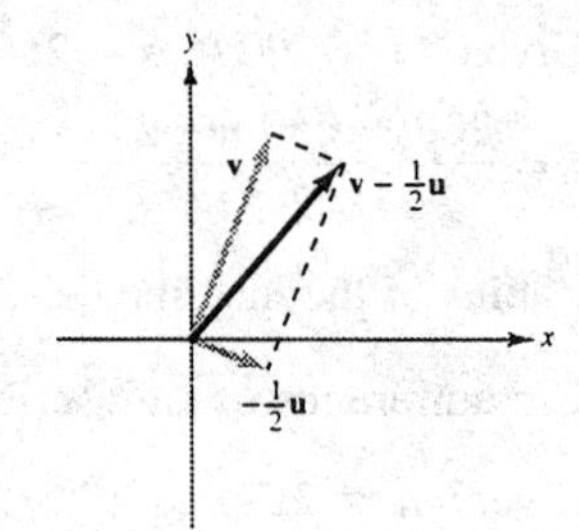

18. $\mathbf{u} = \langle 2, 3\rangle, \mathbf{v} = \langle 4, 0\rangle$

(a) $\mathbf{u} + \mathbf{v} = \langle 6, 3\rangle$

(b) $\mathbf{u} - \mathbf{v} = \langle -2, 3\rangle$

(c) $2\mathbf{u} - 3\mathbf{v} = \langle 4, 6\rangle - \langle 12, 0\rangle = \langle -8, 6\rangle$

20. $\mathbf{u} = 2\mathbf{i} - \mathbf{j}, \mathbf{v} = -\mathbf{i} + \mathbf{j}$

(a) $\mathbf{u} + \mathbf{v} = \mathbf{i}$

(b) $\mathbf{u} - \mathbf{v} = 3\mathbf{i} - 2\mathbf{j}$

(c) $2\mathbf{u} - 3\mathbf{v} = (4\mathbf{i} - 2\mathbf{j}) - (-3\mathbf{i} + 3\mathbf{j}) = 7\mathbf{i} - 5\mathbf{j}$

22. $\mathbf{v} = \langle 0, -3\rangle$

$$\mathbf{u} = \frac{1}{\|\mathbf{v}\|}\mathbf{v} = \frac{1}{\sqrt{0^2 + (-3)^2}}\langle 0, -3\rangle$$

$$= \frac{1}{3}\langle 0, -3\rangle$$

$$= \langle 0, -1\rangle$$

24. $\mathbf{v} = \langle 5, -12\rangle$

$$\mathbf{u} = \frac{1}{\|\mathbf{v}\|}\mathbf{v} = \frac{1}{\sqrt{5^2 + (-12)^2}}\langle 5, -12\rangle$$

$$= \frac{1}{13}\langle 5, -12\rangle$$

$$= \left\langle \frac{5}{13}, -\frac{12}{13}\right\rangle$$

26. $\mathbf{v} = \mathbf{i} + \mathbf{j}$

$$\mathbf{u} = \frac{1}{\|\mathbf{v}\|}\mathbf{v}$$
$$= \frac{1}{\sqrt{1^2 + 1^2}}(\mathbf{i} + \mathbf{j})$$
$$= \frac{1}{\sqrt{2}}(\mathbf{i} + \mathbf{j})$$
$$= \frac{\sqrt{2}}{2}\mathbf{i} + \frac{\sqrt{2}}{2}\mathbf{j}$$

28. $\mathbf{w} = \mathbf{i} - 2\mathbf{j}$

$$\mathbf{u} = \frac{1}{\|\mathbf{w}\|}\mathbf{w}$$
$$= \frac{1}{\sqrt{1^2 + (-2)^2}}(\mathbf{i} - 2\mathbf{j})$$
$$= \frac{1}{\sqrt{5}}(\mathbf{i} - 2\mathbf{j})$$
$$= \frac{\sqrt{5}}{5}\mathbf{i} - \frac{2\sqrt{5}}{5}\mathbf{j}$$

30. $\mathbf{v} = 3\left(\frac{1}{\|\mathbf{u}\|}\mathbf{u}\right)$

$$= 3\left(\frac{1}{\sqrt{4^2 + (-4)^2}}\langle 4, -4\rangle\right)$$
$$= 3\left(\frac{1}{4\sqrt{2}}\langle 4, -4\rangle\right)$$
$$= \left\langle \frac{3}{\sqrt{2}}, -\frac{3}{\sqrt{2}}\right\rangle$$

32. $\mathbf{v} = 10\left(\frac{1}{\|\mathbf{u}\|}\mathbf{u}\right)$

$$= 10\left(\frac{1}{\sqrt{0^2 + (-10)^2}}\langle -10, 0\rangle\right)$$
$$= 10\left(\frac{1}{10}\langle -10, 0\rangle\right)$$
$$= \langle -10, 0\rangle$$

34. $\mathbf{v} = \mathbf{u} + \mathbf{w}$

$$= (2\mathbf{i} - \mathbf{j}) + (\mathbf{i} + 2\mathbf{j})$$
$$= 3\mathbf{i} + \mathbf{j} = \langle 3, 1\rangle$$

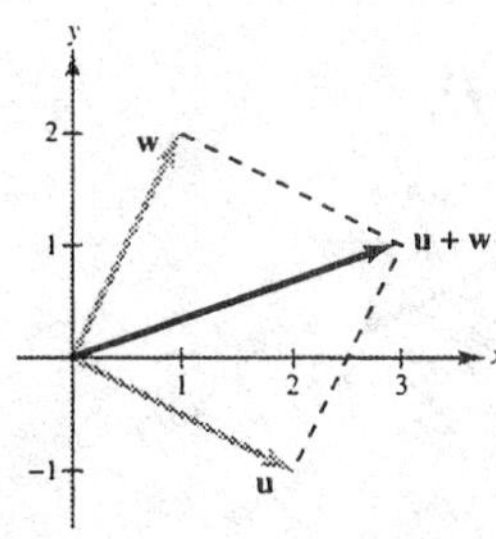

36. $\mathbf{v} = -\mathbf{u} + \mathbf{w}$

$$= -(2\mathbf{i} - \mathbf{j}) + (\mathbf{i} + 2\mathbf{j})$$
$$= -\mathbf{i} + 3\mathbf{j} = \langle -1, 3\rangle$$

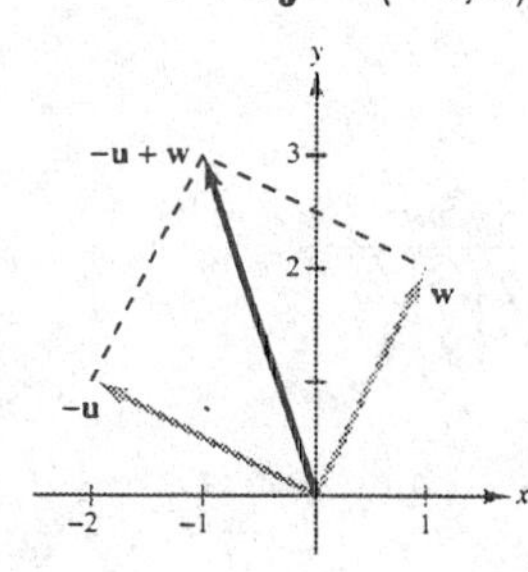

38. $\mathbf{v} = \mathbf{u} - 2\mathbf{w}$

$$= (2\mathbf{i} - \mathbf{j}) - 2(\mathbf{i} + 2\mathbf{j})$$
$$= -5\mathbf{j} = \langle 0, -5\rangle$$

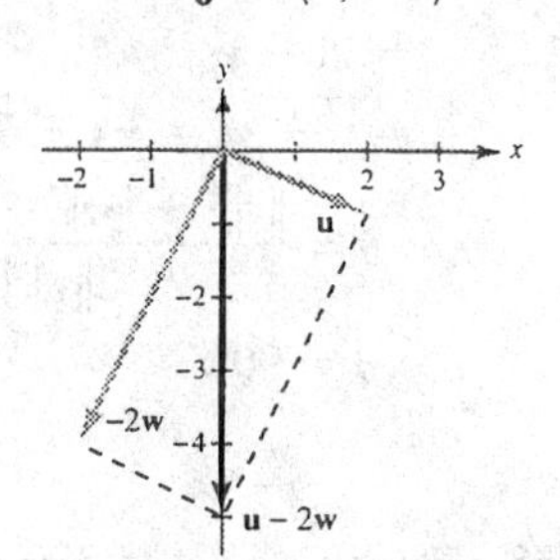

40. $\mathbf{v} = 8(\cos 135°\,\mathbf{i} + \sin 135°\,\mathbf{j})$

$\|\mathbf{v}\| = 8,\ \theta = 135°$

42. $\mathbf{v} = -2\mathbf{i} + 5\mathbf{j}$

$\|\mathbf{v}\| = \sqrt{(-2)^2 + 5^2} = \sqrt{29}$

$\tan\theta = -\frac{5}{2}$

Since $\mathbf{v}$ lies in Quadrant II, $\theta \approx 111.8°$.

44. $\mathbf{v} = \langle \cos 45°, \sin 45°\rangle$

$$= \left\langle \frac{\sqrt{2}}{2}, \frac{\sqrt{2}}{2}\right\rangle$$

46. $\mathbf{v} = \langle 9\cos 90°, 9\sin 90°\rangle$

$$= \langle 0, 9\rangle$$

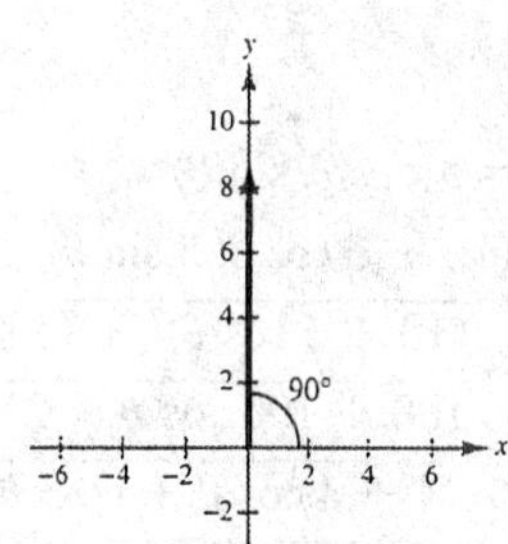

48. $\mathbf{v} = 3\left(\dfrac{1}{\sqrt{3^2 + 4^2}}\right)(3\mathbf{i} + 4\mathbf{j})$

$= \dfrac{3}{5}(3\mathbf{i} + 4\mathbf{j})$

$= \dfrac{9}{5}\mathbf{i} + \dfrac{12}{5}\mathbf{j} = \left\langle \dfrac{9}{5}, \dfrac{12}{5} \right\rangle$

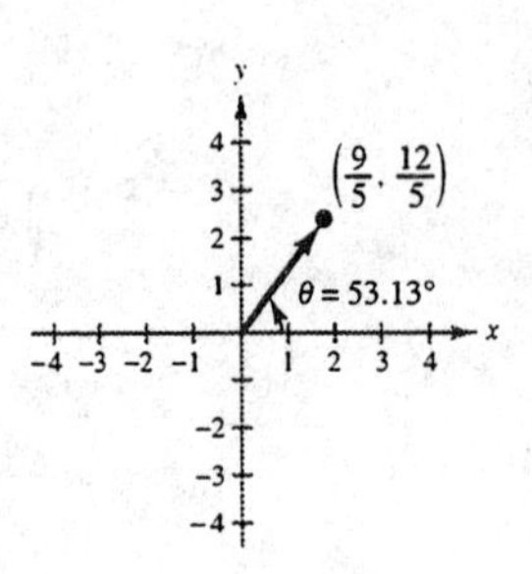

50. $\mathbf{u} = \langle 2\cos 30°, 2\sin 30° \rangle = \langle \sqrt{3}, 1 \rangle$

$\mathbf{v} = \langle 2\cos 90°, 2\sin 90° \rangle = \langle 0, 2 \rangle$

$\mathbf{u} + \mathbf{v} = \langle \sqrt{3}, 3 \rangle$

52. $\mathbf{u} = \langle 35\cos 25°, 35\sin 25° \rangle = \langle 31.72, 14.79 \rangle$

$\mathbf{v} = \langle 50\cos 120°, 50\sin 120° \rangle = \langle -25, 25\sqrt{3} \rangle$

$\mathbf{u} + \mathbf{v} \approx \langle 6.72, 58.09 \rangle$

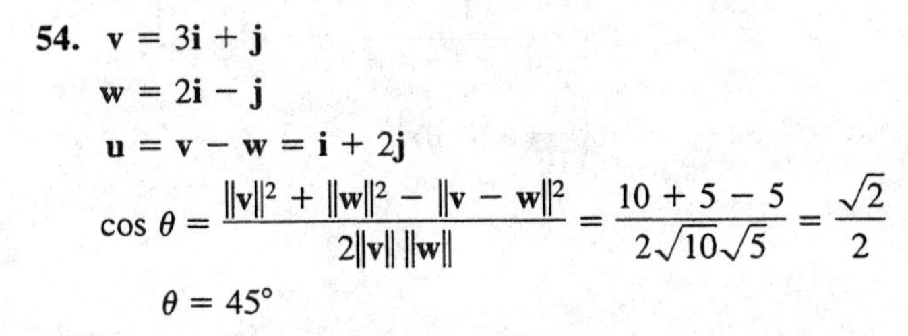

54. $\mathbf{v} = 3\mathbf{i} + \mathbf{j}$

$\mathbf{w} = 2\mathbf{i} - \mathbf{j}$

$\mathbf{u} = \mathbf{v} - \mathbf{w} = \mathbf{i} + 2\mathbf{j}$

$$\cos\theta = \frac{\|\mathbf{v}\|^2 + \|\mathbf{w}\|^2 - \|\mathbf{v} - \mathbf{w}\|^2}{2\|\mathbf{v}\|\,\|\mathbf{w}\|} = \frac{10 + 5 - 5}{2\sqrt{10}\sqrt{5}} = \frac{\sqrt{2}}{2}$$

$\theta = 45°$

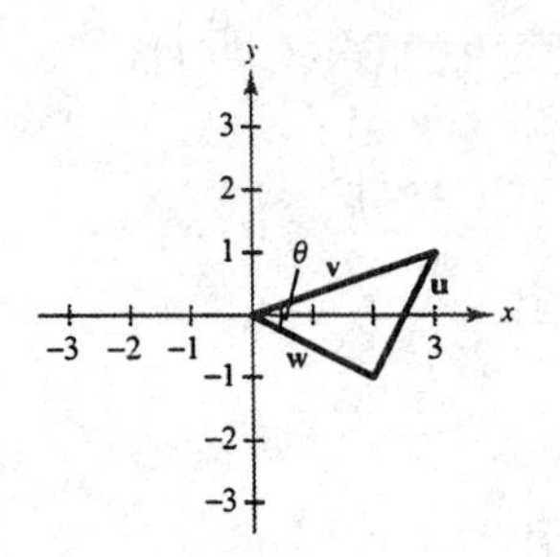

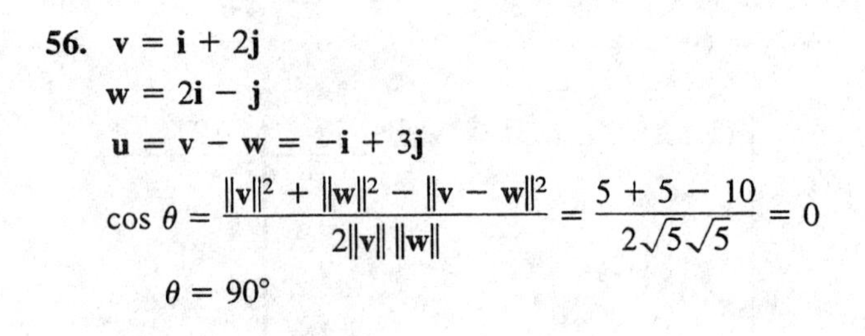

56. $\mathbf{v} = \mathbf{i} + 2\mathbf{j}$

$\mathbf{w} = 2\mathbf{i} - \mathbf{j}$

$\mathbf{u} = \mathbf{v} - \mathbf{w} = -\mathbf{i} + 3\mathbf{j}$

$$\cos\theta = \frac{\|\mathbf{v}\|^2 + \|\mathbf{w}\|^2 - \|\mathbf{v} - \mathbf{w}\|^2}{2\|\mathbf{v}\|\,\|\mathbf{w}\|} = \frac{5 + 5 - 10}{2\sqrt{5}\sqrt{5}} = 0$$

$\theta = 90°$

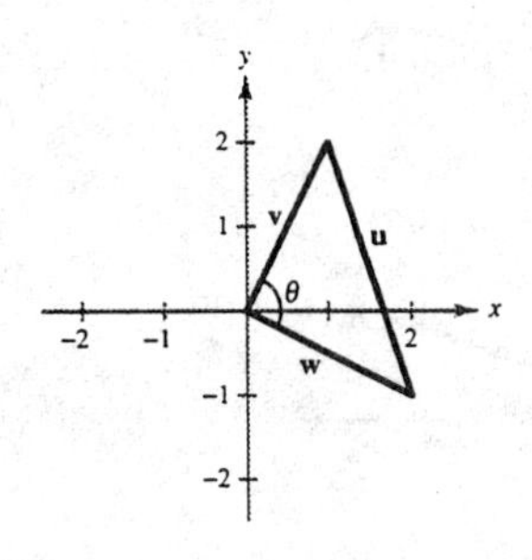

58. Force One: $\mathbf{u} = 3000\mathbf{i}$

Force Two: $\mathbf{v} = 1000\cos\theta\mathbf{i} + 1000\sin\theta\mathbf{j}$

Resultant Force: $\mathbf{u} + \mathbf{v} = (3000 + 1000\cos\theta)\mathbf{i} + 1000\sin\theta\mathbf{j}$

$\|\mathbf{u} + \mathbf{v}\| = \sqrt{(3000 + 1000\cos\theta)^2 + (1000\sin\theta)^2} = 3750$

$$9{,}000{,}000 + 6{,}000{,}000\cos\theta + 1{,}000{,}000 = 14{,}062{,}500$$

$$6{,}000{,}000\cos\theta = 4{,}062{,}500$$

$$\cos\theta = \frac{4{,}062{,}500}{6{,}000{,}000} \approx 0.6771$$

$$\theta \approx 47.4°$$

60. $\mathbf{F}_1 = \langle 10, 0 \rangle$, $\mathbf{F}_2 = 5\langle \cos\theta, \sin\theta \rangle$

(a) $\mathbf{F}_1 + \mathbf{F}_2 = \langle 10 + 5\cos\theta, 5\sin\theta \rangle$

$\|\mathbf{F}_1 + \mathbf{F}_2\| = \sqrt{(10 + 5\cos\theta)^2 + (5\sin\theta)^2}$

$= \sqrt{100 + 100\cos\theta + 25\cos^2\theta + 25\sin^2\theta}$

$= 5\sqrt{4 + 4\cos\theta + \cos^2\theta + \sin^2\theta}$

$= 5\sqrt{4 + 4\cos\theta + 1}$

$= 5\sqrt{5 + 4\cos\theta}$

(b)

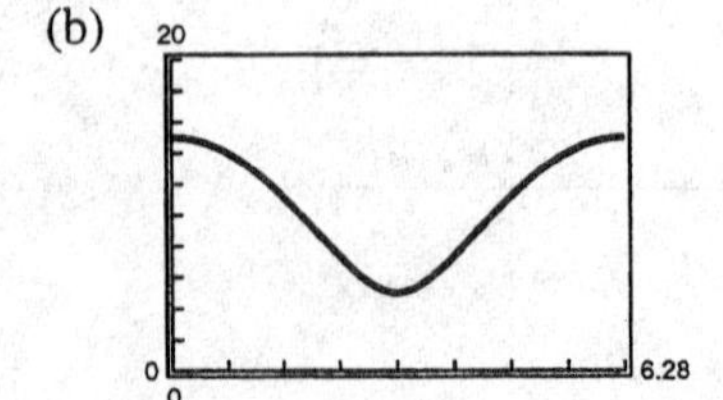

(c) Range: [5, 15]
Maximum is 15 when $\theta = 0$.
Minimum is 5 when $\theta = \pi$.

(d) The magnitude of the resultant is never 0 because the magnitudes of $\mathbf{F}_1$ and $\mathbf{F}_2$ are not the same.

62.

$$\mathbf{u} = (2000\cos 30^\circ)\mathbf{i} + (2000\sin 30^\circ\mathbf{j}) \approx 1732.05\mathbf{i} + 1000\mathbf{j}$$

$$\mathbf{v} = (900\cos(-45^\circ))\mathbf{i} + (900\sin(-45^\circ)\mathbf{j}) \approx 636.4\mathbf{i} + -636.4\mathbf{j}$$

$$\mathbf{u} + \mathbf{v} \approx 2368.4\mathbf{i} + 363.6\mathbf{j}$$

$$\|\mathbf{u} + \mathbf{v}\| \approx \sqrt{(2368.4)^2 + (363.6)^2} \approx 2396.19$$

$$\tan\theta = \frac{363.6}{2368.4} \approx 0.1535 \Rightarrow \theta \approx 8.7^\circ$$

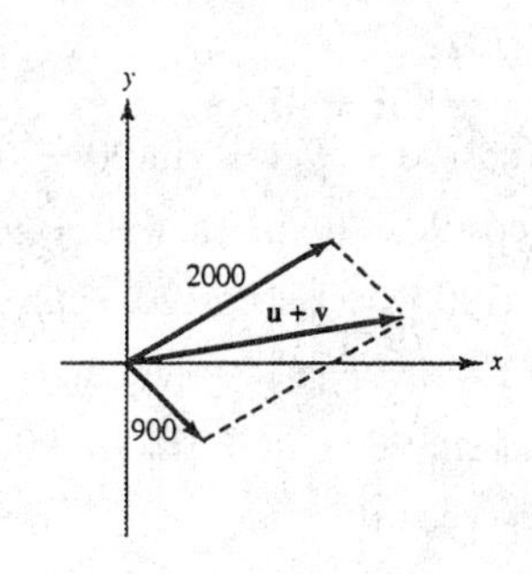

64.

$$\mathbf{u} = (70\cos 30^\circ)\mathbf{i} - (70\sin 30^\circ)\mathbf{j} \approx 60.62\mathbf{i} - 35\mathbf{j}$$

$$\mathbf{v} = (40\cos 45^\circ)\mathbf{i} + (40\sin 45^\circ)\mathbf{j} \approx 28.28\mathbf{i} + 28.28\mathbf{j}$$

$$\mathbf{w} = (60\cos 135^\circ)\mathbf{i} + (60\sin 135^\circ)\mathbf{j} \approx -42.43\mathbf{i} + 42.43\mathbf{j}$$

$$\mathbf{u} + \mathbf{v} + \mathbf{w} = 46.47\mathbf{i} + 35.71\mathbf{j}$$

$$\|\mathbf{u} + \mathbf{v} + \mathbf{w}\| \approx 58.61 \text{ pounds}$$

$$\tan\theta \approx \frac{35.71}{46.47} \approx 0.7684$$

$$\theta \approx 37.5^\circ$$

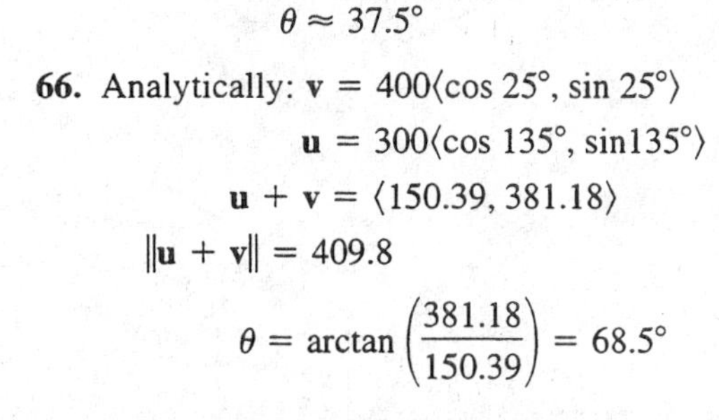

66. Analytically: $\mathbf{v} = 400\langle\cos 25^\circ, \sin 25^\circ\rangle$

$$\mathbf{u} = 300\langle\cos 135^\circ, \sin 135^\circ\rangle$$

$$\mathbf{u} + \mathbf{v} = \langle 150.39, 381.18\rangle$$

$$\|\mathbf{u} + \mathbf{v}\| = 409.8$$

$$\theta = \arctan\left(\frac{381.18}{150.39}\right) = 68.5^\circ$$

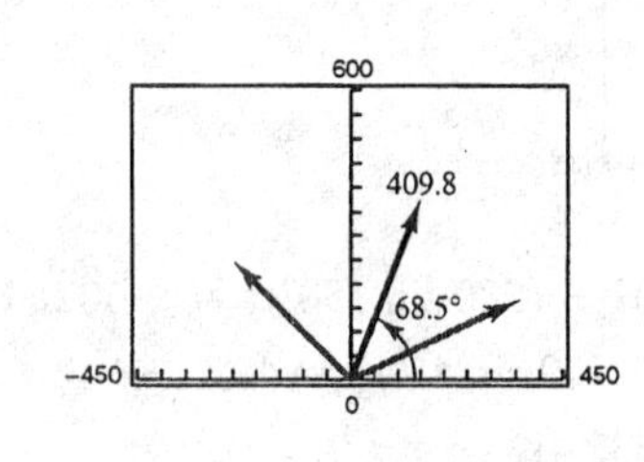

68. Horizontal component of velocity: $1200\cos 6^\circ \approx 1193.4$ ft/sec

Vertical component of velocity: $1200\sin 6^\circ \approx 125.4$ ft/sec

70. Rope $\overrightarrow{AC}$: $\mathbf{u} = 10\mathbf{i} - 24\mathbf{j}$

The vector lies in Quadrant IV and its reference angle is $\arctan\left(\frac{12}{5}\right)$.

$$\mathbf{u} = \|\mathbf{u}\|\left[\cos\left(\arctan\tfrac{12}{5}\right)\mathbf{i} - \sin\left(\arctan\tfrac{12}{5}\right)\mathbf{j}\right]$$

Rope $\overrightarrow{BC}$: $\mathbf{v} = -20\mathbf{i} - 24\mathbf{j}$

The vector lies in Quadrant III and its reference angle is $\arctan\left(\frac{6}{5}\right)$.

$$\mathbf{v} = \|\mathbf{v}\|\left[-\cos\left(\arctan\tfrac{6}{5}\right)\mathbf{i} - \sin\left(\arctan\tfrac{6}{5}\right)\mathbf{j}\right]$$

Resultant: $\mathbf{u} + \mathbf{v} = -5000\mathbf{j}$

$$\|\mathbf{u}\|\cos\left(\arctan\tfrac{12}{5}\right) - \|\mathbf{v}\|\cos\left(\arctan\tfrac{6}{5}\right) = 0$$

$$-\|\mathbf{u}\|\sin\left(\arctan\tfrac{12}{5}\right) - \|\mathbf{v}\|\sin\left(\arctan\tfrac{6}{5}\right) = -5000$$

Solving this system of equations yields: $T_{AC} = \|\mathbf{u}\| \approx 3611.1$ pounds

$T_{BC} = \|\mathbf{v}\| \approx 2169.5$ pounds

72. (a) Rope 1: $\mathbf{u} = \|\mathbf{u}\|(\cos 60^\circ\mathbf{i} + \sin 60^\circ\mathbf{j})$

Rope 2: $\mathbf{v} = \|\mathbf{v}\|(\cos 120^\circ\mathbf{i} + \sin 120^\circ\mathbf{j})$

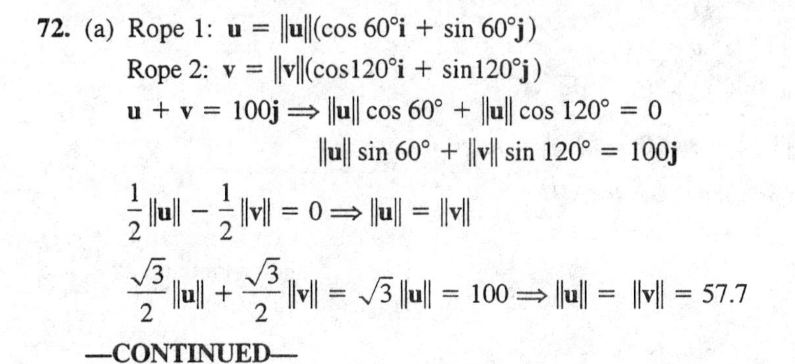

$$\mathbf{u} + \mathbf{v} = 100\mathbf{j} \Rightarrow \|\mathbf{u}\|\cos 60^\circ + \|\mathbf{u}\|\cos 120^\circ = 0$$

$$\|\mathbf{u}\|\sin 60^\circ + \|\mathbf{v}\|\sin 120^\circ = 100\mathbf{j}$$

$$\frac{1}{2}\|\mathbf{u}\| - \frac{1}{2}\|\mathbf{v}\| = 0 \Rightarrow \|\mathbf{u}\| = \|\mathbf{v}\|$$

$$\frac{\sqrt{3}}{2}\|\mathbf{u}\| + \frac{\sqrt{3}}{2}\|\mathbf{v}\| = \sqrt{3}\|\mathbf{u}\| = 100 \Rightarrow \|\mathbf{u}\| = \|\mathbf{v}\| = 57.7$$

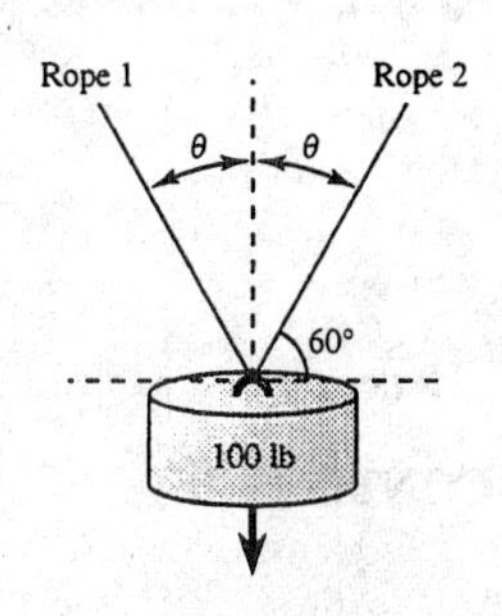

—CONTINUED—

72. —CONTINUED—

(b) $\mathbf{u} = \|\mathbf{u}\|(\cos(90 - \theta)\,\mathbf{i} + \sin(90 - \theta)\,\mathbf{j})$

$\mathbf{v} = \|\mathbf{v}\|(\cos(90 + \theta)\,\mathbf{i} + \sin(90 + \theta)\,\mathbf{j})$

$\mathbf{u} + \mathbf{v} = 100\,\mathbf{j} \Rightarrow \|\mathbf{u}\| \cos(90 - \theta) + \|\mathbf{v}\| \cos(90 + \theta) = 0$

$\Rightarrow \|\mathbf{u}\| = \|\mathbf{v}\|$, and

$100 = \|\mathbf{u}\| \sin(90 - \theta) + \|\mathbf{u}\| \sin(90 + \theta)$

$= 2\|\mathbf{u}\| \cos \theta$

Hence, $\|\mathbf{u}\| = \dfrac{50}{\cos \theta} = 50 \sec \theta$

(c)

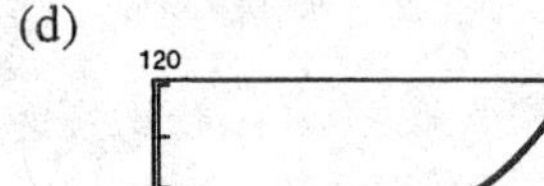

θ	10°	20°	30°	40°	50°	60°
T	50.8	53.2	57.7	65.3	77.8	100

(d)

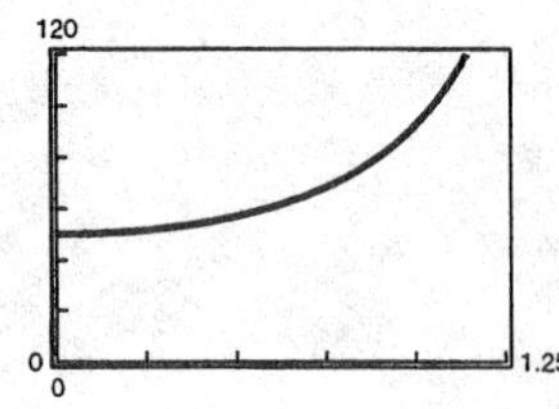

(e) The vertical component of the vectors decreases as θ increases.

74. Plane: $\mathbf{u} = (580 \cos 150°)\mathbf{i} + (580 \sin 150°)\mathbf{j} \approx -502.3\mathbf{i} + 290\mathbf{j}$

Wind: $\mathbf{v} = (60 \cos 45°)\mathbf{i} + (60 \sin 45°)\mathbf{j} \approx 42.4\mathbf{i} + 42.4\mathbf{j}$

$\mathbf{u} + \mathbf{v} \approx -459.9\mathbf{i} + 332.4\mathbf{j}$

$\|\mathbf{u} + \mathbf{v}\| \approx \sqrt{(-459.9)^2 + (332.4)^2} \approx 567.4$

$\tan \theta \approx -\dfrac{332.4}{459.9} \approx -0.7229 \Rightarrow \theta \approx 144.1°$

The ground speed is 567.4 miles per hour and the heading is N 54.1° W.

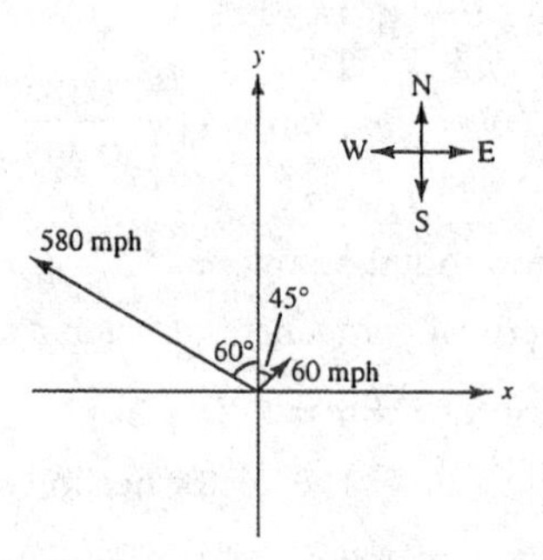

76. (a) Horizontal force: $\mathbf{u} = \|\mathbf{u}\|\,\mathbf{i}$

Weight: $\mathbf{w} = -\mathbf{j}$

Rope Tension: $\mathbf{T} = \|\mathbf{T}\|(\cos 120°\mathbf{i} + \sin 120°\mathbf{j})$

$\mathbf{u} + \mathbf{w} + \mathbf{T} = \mathbf{0} \Rightarrow \|\mathbf{u}\| + \|\mathbf{T}\| \cos 120° = 0$

$-1 + \|\mathbf{T}\| \sin 120° = 0$

Hence, $\|\mathbf{T}\| \dfrac{\sqrt{3}}{2} = 1 \Rightarrow \|\mathbf{T}\| = \dfrac{2}{\sqrt{3}} \approx 1.15$ lbs and $\|\mathbf{u}\| = \|\mathbf{T}\| \left(\dfrac{1}{2}\right) \approx 0.58$ lbs.

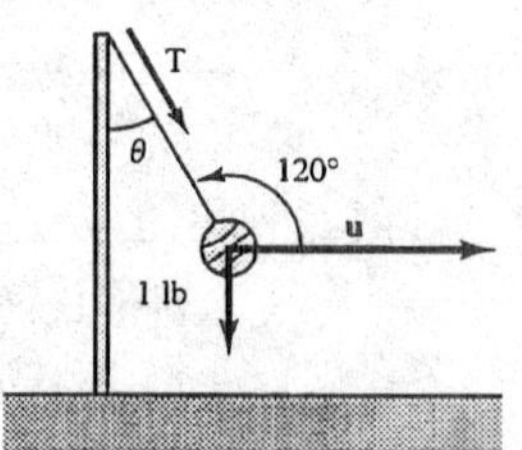

(b) $\mathbf{T} = \|\mathbf{T}\| (\cos(90 + \theta)\mathbf{i} + \sin(90 + \theta)\mathbf{j})$

$\mathbf{u} + \mathbf{w} + \mathbf{T} = \mathbf{0} \Rightarrow \|\mathbf{u}\| + \|\mathbf{T}\| \cos(90 + \theta) = 0$

$\|\mathbf{u}\| - \|\mathbf{T}\| \sin \theta = 0$

$-1 + \|\mathbf{T}\| \sin(90 + \theta) = 0$

$-1 + \|\mathbf{T}\| \cos \theta = 0 \Rightarrow \|\mathbf{T}\| = \sec \theta, 0 < \theta < \dfrac{\pi}{2}$

Hence, $\|\mathbf{u}\| = \|\mathbf{T}\| \sin \theta = \sec \theta \sin \theta = \tan \theta, 0 < \theta < \pi/2$.

—CONTINUED—

76. —CONTINUED—

(c)

θ	0°	10°	20°	30°	40°	50°	60°
T	1	1.02	1.06	1.15	1.31	1.56	2
$\|\mathbf{u}\|$	0	0.18	0.36	0.58	0.84	1.19	1.73

(d)

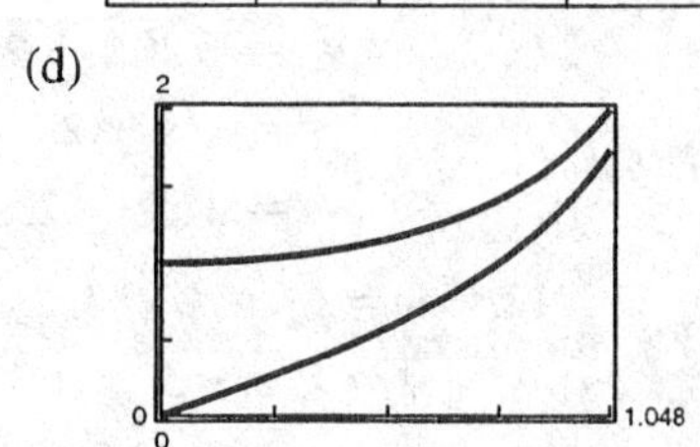

(e) Both **T** and $\|\mathbf{u}\|$ increases as θ increases, and approach each other in magnitude.

78. True, $\mathbf{u} = \dfrac{\mathbf{v}}{\|\mathbf{v}\|}$

80. True

82. The following program is written for a *TI-82* or *TI-83* graphing calculator. The program sketches two vectors $\mathbf{u} = a\mathbf{i} + b\mathbf{j}$ and $\mathbf{v} = c\mathbf{i} + d\mathbf{j}$ in standard position, and then sketches the vector difference $\mathbf{u} - \mathbf{v}$ using the parallelogram law.

```
PROGRAM: SUBVECT
:Input "ENTER A", A
:Input "ENTER B", B
:Input "ENTER C", C
:Input "ENTER D", D
:Line (0, 0, A, B)
:Line (0, 0, C, D)
:Pause
:A-C→E
:B-D→F
:Line (A, B, C, D)
:Line (A, B, E, F)
:Line (0, 0, E, F)
:Pause
:ClrDraw
:Stop
```

84.
$$\mathbf{u} = \langle 80 - 10, 80 - 60\rangle = \langle 70, 20\rangle$$
$$\mathbf{v} = \langle -20 - (-100), 70 - 0\rangle = \langle 80, 70\rangle$$
$$\mathbf{u} - \mathbf{v} = \langle 70 - 80, 20 - 70\rangle = \langle -10, -50\rangle$$
$$\mathbf{v} - \mathbf{u} = \langle 80 - 70, 70 - 20\rangle = \langle 10, 50\rangle$$

Section 6.4 Vectors and Dot Products

Solutions to Even-Numbered Exercises

2. $\mathbf{u} = \langle 5, 12 \rangle, \mathbf{v} = \langle -3, 2 \rangle$

$\mathbf{u} \cdot \mathbf{v} = 5(-3) + 12(2) = 9$

4. $\mathbf{u} = 2\mathbf{i} + 5\mathbf{j}, \mathbf{v} = 9\mathbf{i} - 3\mathbf{j}$

$\mathbf{u} \cdot \mathbf{v} = 2(9) + 5(-3) = 3$

6. $\mathbf{u} = \langle 2, 2 \rangle$

$$\begin{aligned} \|\mathbf{u}\| - 2 &= \sqrt{\mathbf{u} \cdot \mathbf{u}} - 2 \\ &= \sqrt{2(2) + 2(2)} - 2 \\ &= \sqrt{8} - 2 \\ &= 2\sqrt{2} - 2, \text{ scalar} \end{aligned}$$

8. $\mathbf{u} = \langle 2, 2 \rangle, \mathbf{v} = \langle -3, 4 \rangle$

$$\begin{aligned} \mathbf{u} \cdot 2\mathbf{v} = 2\mathbf{u} \cdot \mathbf{v} &= 2[2(-3) + 2(4)] \\ &= 2(2) = 4, \text{ scalar} \end{aligned}$$

10. $\mathbf{u} = \langle 2, -4 \rangle$

$$\begin{aligned} \|\mathbf{u}\| &= \sqrt{\mathbf{u} \cdot \mathbf{u}} \\ &= \sqrt{2(2) + (-4)(-4)} \\ &= \sqrt{20} = 2\sqrt{5} \end{aligned}$$

12. $\mathbf{u} = 6\mathbf{j}$

$$\begin{aligned} \|\mathbf{u}\| &= \sqrt{\mathbf{u} \cdot \mathbf{u}} \\ &= \sqrt{6(6)} = 6 \end{aligned}$$

14. $\mathbf{u} = \langle 1245, 2600 \rangle$

$\mathbf{v} = \langle 12.20, 8.50 \rangle$

Increase prices by 5%: $1.05\mathbf{v}$

$$\begin{aligned} \mathbf{u} \cdot 1.05\mathbf{v} = 1.05\mathbf{u} \cdot \mathbf{v} &= 1.05[1245(12.20) + 2600(8.50)] \\ &= 1.05(37{,}289) \\ &= \$39{,}153.45 \end{aligned}$$

16. $\mathbf{u} = \langle 4, 4 \rangle, \mathbf{v} = \langle 2, 0 \rangle$

$$\begin{aligned} \cos\theta &= \frac{\mathbf{u} \cdot \mathbf{v}}{\|\mathbf{u}\|\,\|\mathbf{v}\|} \\ &= \frac{4(2) + 4(0)}{(4\sqrt{2})(2)} \\ &= \frac{\sqrt{2}}{2} \end{aligned}$$

$\theta = 45°$

18. $\mathbf{u} = 2\mathbf{i} - 3\mathbf{j}, \mathbf{v} = \mathbf{i} - 2\mathbf{j}$

$$\begin{aligned} \cos\theta &= \frac{\mathbf{u} \cdot \mathbf{v}}{\|\mathbf{u}\|\,\|\mathbf{v}\|} \\ &= \frac{2(1) + (-3)(-2)}{\sqrt{2^2 + 3^2}\sqrt{1^2 + 2^2}} \\ &= \frac{8}{\sqrt{65}} \approx 0.992278 \end{aligned}$$

$\theta \approx 7.13°$

20. $\mathbf{u} = \cos\left(\frac{\pi}{4}\right)\mathbf{i} + \sin\left(\frac{\pi}{4}\right)\mathbf{j} = \frac{\sqrt{2}}{2}\mathbf{i} + \frac{\sqrt{2}}{2}\mathbf{j}$

$\mathbf{v} = \cos\left(\frac{\pi}{2}\right)\mathbf{i} + \sin\left(\frac{\pi}{2}\right)\mathbf{j} = \mathbf{j}$

$$\cos\theta = \frac{\mathbf{u} \cdot \mathbf{v}}{\|\mathbf{u}\|\,\|\mathbf{v}\|} = \frac{\frac{\sqrt{2}}{2}(0) + \frac{\sqrt{2}}{2}(1)}{1 \cdot 1} = \frac{\sqrt{2}}{2}$$

$\theta = \frac{\pi}{4}$

22. $\mathbf{u} = -6\mathbf{i} - 3\mathbf{j}, \mathbf{v} = -8\mathbf{i} + 4\mathbf{j}$

$$\begin{aligned} \cos\mathbf{u} &= \frac{\mathbf{u} \cdot \mathbf{v}}{\|\mathbf{u}\|\,\|\mathbf{v}\|} = \frac{-6(-8) + (-3)(-4)}{\sqrt{45}\sqrt{80}} \\ &= \frac{36}{60} = 0.6 \end{aligned}$$

$\theta \approx 53.13°$

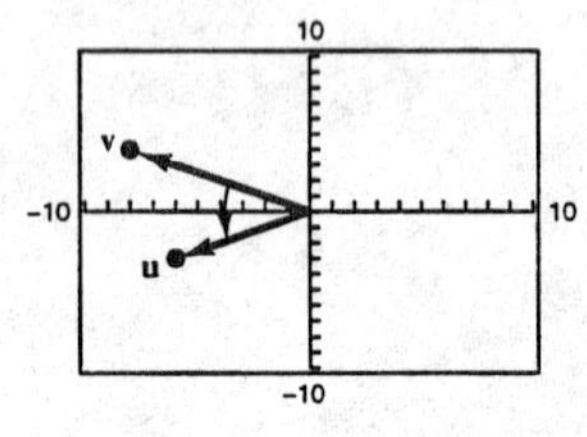

24. $\mathbf{u} = 2\mathbf{i} - 3\mathbf{j}, \mathbf{v} = 4\mathbf{i} + 3\mathbf{j}$

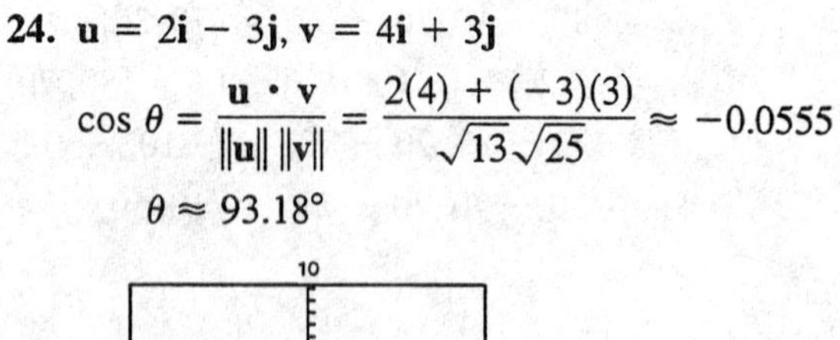

$$\cos\theta = \frac{\mathbf{u} \cdot \mathbf{v}}{\|\mathbf{u}\|\,\|\mathbf{v}\|} = \frac{2(4) + (-3)(3)}{\sqrt{13}\sqrt{25}} \approx -0.0555$$

$\theta \approx 93.18°$

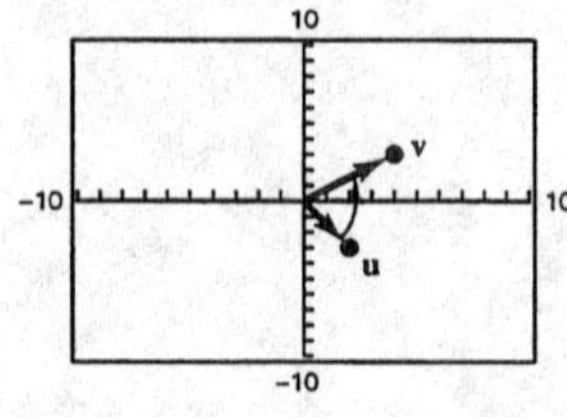

26. $P = (-3, 0), Q = (2, 2), R = (0, 6)$

$\overrightarrow{PQ} = \langle 5, 2\rangle, \overrightarrow{QR} = \langle -2, 4\rangle, \overrightarrow{PR} = \langle 3, 6\rangle,$

$\overrightarrow{QP} = \langle -5, -2\rangle$

$\cos\alpha = \frac{\overrightarrow{PQ}\cdot\overrightarrow{PR}}{\|\overrightarrow{PR}\|\,\|\overrightarrow{PR}\|} = \frac{27}{(\sqrt{29})(\sqrt{45})} \Rightarrow \alpha \approx 41.6°$

$\cos B = \frac{\overrightarrow{QR}\cdot\overrightarrow{QP}}{\|\overrightarrow{QR}\|\,\|\overrightarrow{QP}\|} = \frac{2}{(\sqrt{20})(\sqrt{29})} \Rightarrow \alpha \approx 85.2°$

$\phi = 180° - 41.6° - 85.2° \approx 53.1°$

28. $\|\mathbf{u}\| = 100, \|\mathbf{v}\| = 250, \theta = \frac{\pi}{6}$

$$\begin{aligned}\mathbf{u}\cdot\mathbf{v} &= \|\mathbf{u}\|\,\|\mathbf{v}\|\cos\theta \\ &= (100)(250)\cos\frac{\pi}{6} \\ &= 25{,}000\cdot\frac{\sqrt{3}}{2} \\ &= 12{,}500\sqrt{3}\end{aligned}$$

30. $\mathbf{u} = \langle 15, 45\rangle, \mathbf{v} = \langle -5, 12\rangle$

$\mathbf{u} \neq k\mathbf{v} \Rightarrow$ Not parallel

$\mathbf{u}\cdot\mathbf{v} \neq 0 \Rightarrow$ Not orthogonal

Neither

32. $\mathbf{u} = \mathbf{j}, \mathbf{v} = \mathbf{i} - 2\mathbf{j}$

$\mathbf{u} \neq k\mathbf{v} \Rightarrow$ Not parallel

$\mathbf{u}\cdot\mathbf{v} \neq 0 \Rightarrow$ Not orthogonal

Neither

34. $\mathbf{u} = \langle\cos\theta, \sin\theta\rangle$

$\mathbf{v} = \langle\sin\theta, -\cos\theta\rangle$

$\mathbf{u}\cdot\mathbf{v} = 0 \Rightarrow$ **u** and **v** are orthogonal.

36. $\mathbf{u} = \langle 4, 2\rangle, \mathbf{v} = \langle 1, -2\rangle$

$\mathbf{w}_1 = \text{proj}_{\mathbf{v}}\mathbf{u} = \left(\frac{\mathbf{u}\cdot\mathbf{v}}{\|\mathbf{v}\|^2}\right)\mathbf{v} = 0\langle 1, -2\rangle = \langle 0, 0\rangle$

$\mathbf{w}_2 = \mathbf{u} - \mathbf{w}_1 = \langle 4, 2\rangle - \langle 0, 0\rangle = \langle 4, 2\rangle$

38. $\mathbf{u} = \langle -5, -1\rangle, \mathbf{v} = \langle -1, 1\rangle$

$\mathbf{w}_1 = \text{proj}_{\mathbf{v}}\mathbf{u} = \left(\frac{\mathbf{u}\cdot\mathbf{v}}{\|\mathbf{v}\|^2}\right)\mathbf{v} = \frac{4}{2}\langle -1, 1\rangle = 2\langle -1, 1\rangle$

$$\begin{aligned}\mathbf{w}_2 = \mathbf{u} - \mathbf{w}_1 &= \langle -5, -1\rangle - 2\langle -1, 1\rangle \\ &= \langle -3, -3\rangle = 3\langle -1, -1\rangle\end{aligned}$$

40. Because **u** and **v** are parallel, the projection of **u** onto **v** is **u**.

42. Because **u** and **v** are orthogonal, the projection of **u** onto **v** is **0**.

44. $\mathbf{u} = \langle -8, 3\rangle$

For **v** to be orthogonal to **u**, $\mathbf{u}\cdot\mathbf{v}$ must be equal to 0.

Two possibilities: $\langle 3, 8\rangle, \langle -3, -8\rangle$

46. $\mathbf{u} = -\frac{5}{2}\mathbf{i} - 3\mathbf{j}$

For **v** to be orthogonal to **u**, $\mathbf{u}\cdot\mathbf{v}$ must be equal to 0.

Two possibilities: $\mathbf{v} = 3\mathbf{i} - \frac{5}{2}\mathbf{j}$

$\mathbf{v} = -3\mathbf{i} + \frac{5}{2}\mathbf{j}$

48. (a) $\mathbf{F} = -36{,}000\mathbf{j}$ Gravitational force

$\mathbf{v} = (\cos 12°)\mathbf{i} + (\sin 12°)\mathbf{j}$

$\mathbf{w}_1 = \text{proj}_{\mathbf{v}}\mathbf{F} = \left(\frac{\mathbf{F}\cdot\mathbf{v}}{\|\mathbf{v}\|^2}\right)\mathbf{v} = (\mathbf{F}\cdot\mathbf{v}) \approx -7484.8\mathbf{v}$

The magnitude of this force is 7484.8; therefore, a force of 7484.8 pounds is needed to keep the truck from rolling down the hill.

(b) $\mathbf{w}_2 = \mathbf{F} - \mathbf{w}_1 = -36{,}000\mathbf{j} + 7484.8[(\cos 12°)\mathbf{i} + (\sin 12°)\mathbf{j}]$

$= [(7484.8\cos 12°)\mathbf{i} + (7484.8\sin 12° - 36{,}000)\mathbf{j}]$

$\|\mathbf{w}_2\| \approx 35{,}213.3$ pounds

50. (a) $\text{proj}_{\mathbf{v}}\mathbf{u} = \mathbf{u} \Rightarrow$ **u** and **v** are parallel.

(b) $\text{proj}_{\mathbf{v}}\mathbf{u} = \mathbf{0} \Rightarrow$ **u** and **v** are orthogonal.

52. work $= (2400)(5) = 12{,}000$ foot-pounds

54. work $= (\cos 35°)(15{,}691)(800)$

$\approx 10{,}282{,}651$ newton − meters

56. $P = (1, 3), Q = (-3, 5), \mathbf{v} = -2\mathbf{i} + 2\mathbf{j}$

work $= \|\text{proj}_{\overrightarrow{PQ}}\mathbf{v}\| = \|\overrightarrow{PQ}\|$ where $\overrightarrow{PQ} = \langle -4, 2\rangle$ and $\mathbf{v} = \langle -2, 3\rangle$.

$\text{proj}_{\overrightarrow{PQ}}\mathbf{v} = \left(\frac{\mathbf{v}\cdot\overrightarrow{PQ}}{\|\overrightarrow{PQ}\|}\right)\overrightarrow{PQ} = \left(\frac{14}{20}\right)\langle -4, 2\rangle$

work $= \|\text{proj}_{\overrightarrow{PQ}}\mathbf{v}\|\,\|\overrightarrow{PQ}\| = \left(\frac{4\sqrt{20}}{20}\right)(\sqrt{20}) = 14$

58. Let $\mathbf{u}$ and $\mathbf{v}$ be two sides of the rhombus $\|\mathbf{u}\| = \|\mathbf{v}\|$. The diagonals are $\mathbf{u} + \mathbf{v}$ and $\mathbf{u} - \mathbf{v}$.

$$(\mathbf{u} + \mathbf{v}) \cdot (\mathbf{u} - \mathbf{v}) = \mathbf{u} \cdot \mathbf{u} + \mathbf{v} \cdot \mathbf{u} - \mathbf{u} \cdot \mathbf{v} - \mathbf{v} \cdot \mathbf{v}$$
$$= \|\mathbf{u}\|^2 - \|\mathbf{v}\|^2$$
$$= 0$$

Hence, the diagonals are perpendicular.

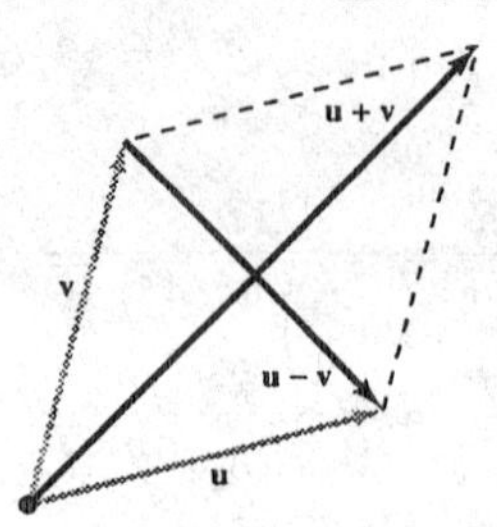

60. (a) $\mathbf{0} \cdot \mathbf{v} = \langle 0, 0\rangle \cdot \langle v_1, v_2\rangle = 0v_1 + 0v_2 = 0$

(b) $\mathbf{u} \cdot (\mathbf{v} + \mathbf{w}) = \langle u_1, u_2\rangle \cdot \langle v_1 + w_1, v_2 + w_2\rangle$
$$= u_1(v_1 + w_1) + u_2(v_2 + w_2)$$
$$= u_1v_1 + u_1w_1 + u_2v_2 + u_2w_2$$
$$= \langle u_1, u_2\rangle \cdot \langle v_1, v_2\rangle + \langle u_1, u_2\rangle \cdot \langle w_1, w_2\rangle$$
$$= \mathbf{u} \cdot \mathbf{v} + \mathbf{u} \cdot \mathbf{w}$$

(c) $c(\mathbf{u} \cdot \mathbf{v}) = c[\langle u_1, u_2\rangle \cdot \langle v_1, v_2\rangle = c(u_1v_1 + u_2v_2)$
$$= (cu_1)v_1 + (cu_2)v_2 = \langle cu_1, cu_2\rangle \cdot \langle v_1, v_2\rangle = c\mathbf{u} \cdot \mathbf{v}$$
$$= u_1(cv_1) + u_2(cv_2) = \langle u_1, u_2\rangle \cdot \langle cv_1, cv_2\rangle = \mathbf{u} \cdot c\mathbf{v}$$

62. (a) $\dfrac{y^2}{x} - x = \dfrac{y^2 - x^2}{x}$

(b) $\dfrac{\csc^2 x}{\cot x} - \cot x = \dfrac{\csc^2 x - \cot^2 x}{\cot x} = \dfrac{1}{\cot x} = \tan x$

64. (a) $\dfrac{y}{z} - \dfrac{z}{1 + y} = \dfrac{y + y^2 - z^2}{z(1 + y)}$

(b) $\dfrac{\tan x}{\sec x} - \dfrac{\sec x}{1 + \tan x} = \dfrac{\tan x + \tan^2 x - \sec^2 x}{\sec x(1 + \tan x)} = \dfrac{\tan x - 1}{\sec x(\tan x + 1)}$

Section 6.5 DeMoivre's Theorem

Solutions to Even-Numbered Exercises

2. $|-5| = \sqrt{5^2 + 0^2}$
$= \sqrt{25} = 5$

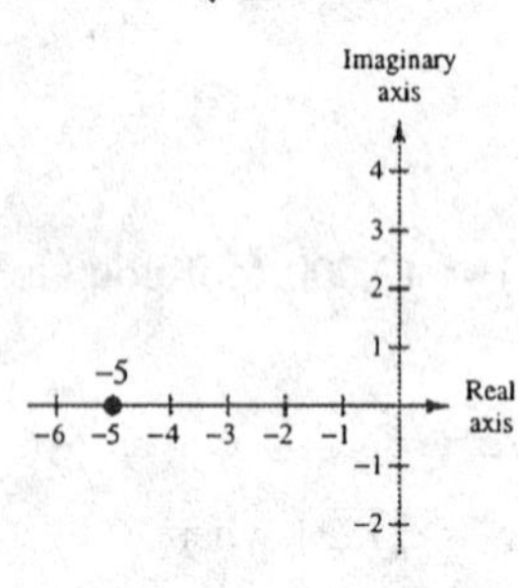

4. $|5 - 12i| = \sqrt{5^2 + (-12)^2}$
$= \sqrt{169} = 13$

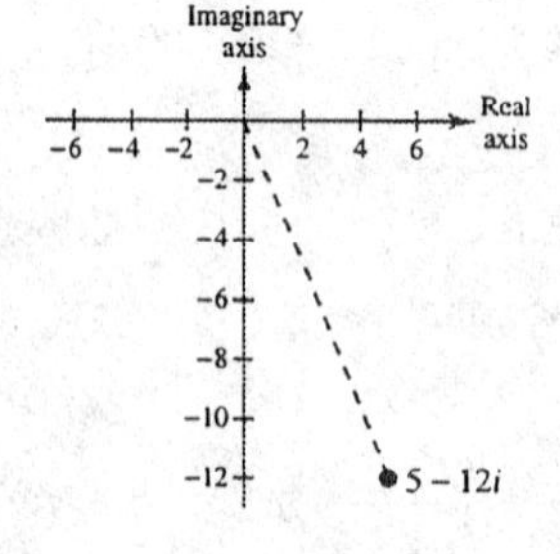

6. $|-8 + 3i| = \sqrt{(-8)^2 + (3)^2}$
$= \sqrt{73}$

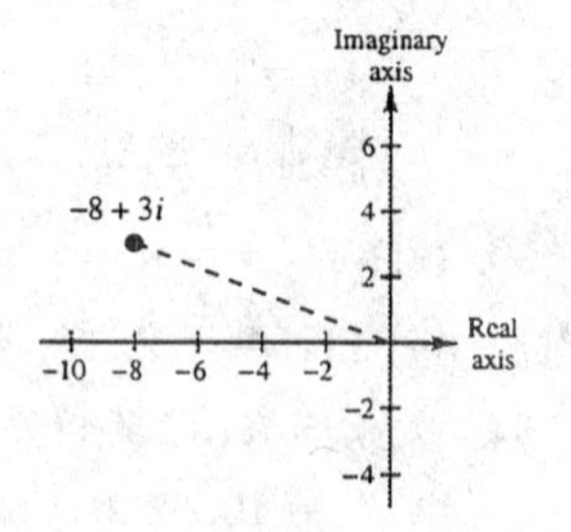

8. $z = 4$

$r = \sqrt{4^2 + 0^2} = \sqrt{16} = 4$

$\tan\theta = \dfrac{0}{4} = 0 \Rightarrow \theta = 0$

$z = 4(\cos 0 + i \sin 0)$

10. $z = -1 + \sqrt{3}i$

$r = \sqrt{(-1)^2 + \left(\sqrt{3}\right)^2} = \sqrt{4} = 2$

$\tan\theta = \dfrac{\sqrt{3}}{-1} = -\sqrt{3} \Rightarrow \theta = \dfrac{2\pi}{3}$

$z = 2\left(\cos\dfrac{2\pi}{3} + i \sin\dfrac{2\pi}{3}\right)$

12. $z = 2 + 2i$

$r = \sqrt{2^2 + 2^2} = \sqrt{8} = 2\sqrt{2}$

$\tan\theta = \frac{2}{2} = 1 \Rightarrow \theta = \frac{\pi}{4}$

$z = 2\sqrt{2}\left(\cos\frac{\pi}{4} + i\sin\frac{\pi}{4}\right)$

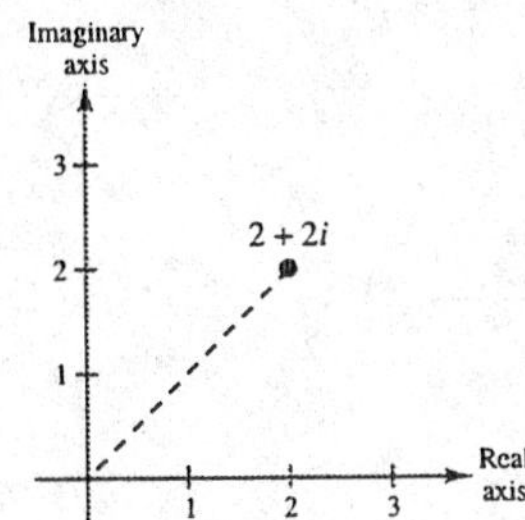

14. $z = -1 + \sqrt{3}i$

$r = \sqrt{(-1)^2 + \left(\sqrt{3}\right)^2} = \sqrt{4} = 2$

$\tan\theta = \frac{\sqrt{3}}{-1} = -\sqrt{3} \Rightarrow 2 = \frac{2\pi}{3}$

$z = 2\left(\cos\frac{2\pi}{3} + i\sin\frac{2\pi}{3}\right)$

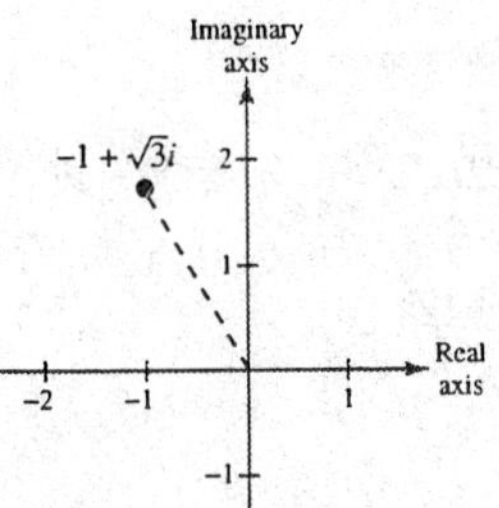

16. $z = \frac{5}{2}\left(\sqrt{3} - i\right)$

$r = \sqrt{\left(\frac{5}{2}\sqrt{3}\right)^2 + \left(\frac{5}{2}(-1)\right)^2} = \sqrt{\frac{100}{4}} = \sqrt{25} = 5$

$\tan\theta = \frac{-1}{\sqrt{3}} = \frac{-\sqrt{3}}{3} \Rightarrow \theta = \frac{11\pi}{6}$

$z = 5\left(\cos\frac{11\pi}{6} + i\sin\frac{11\pi}{6}\right)$

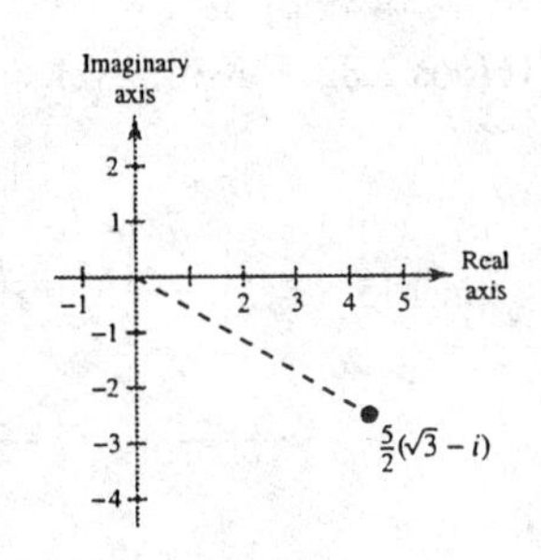

18. $z = 4 + 0i$

$r = \sqrt{4^2 + 0^2} = \sqrt{16} = 4$

$\tan\theta = \frac{0}{4} = 0 \Rightarrow \theta = 0$

$z = 4(\cos 0 + i\sin 0)$

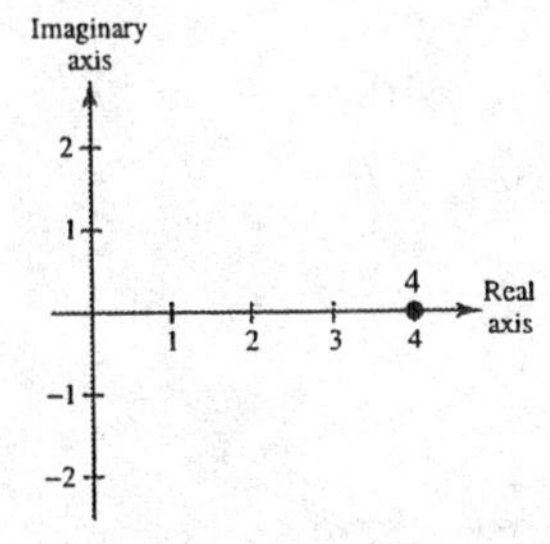

20. $z = 3 - i$

$r = \sqrt{(3)^2 + (-1)^2} = \sqrt{10}$

$\tan\theta = \frac{-1}{3} = \theta \approx -18.4°$

$z = \sqrt{10}(\cos(-18.4°) + i\sin(-18.4°))$

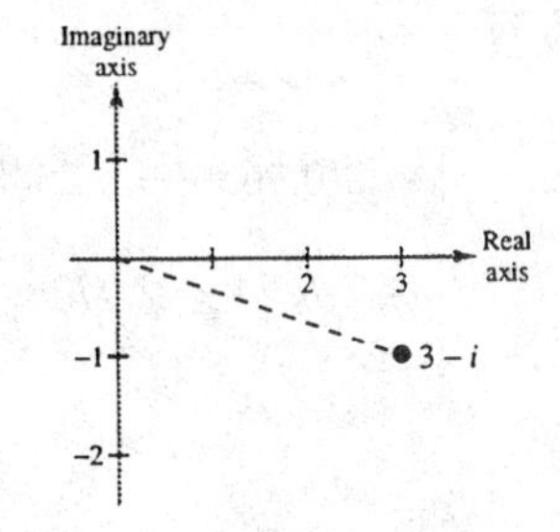

22. $z = 0 - 2i$

$r = \sqrt{0^2 + (-2)^2} = \sqrt{4} = 2$

$\tan\theta = \frac{-2}{0}$, undefined $\Rightarrow \theta = \frac{3\pi}{2}$

$z = 2\left(\cos\frac{3\pi}{2} + i\sin\frac{3\pi}{2}\right)$

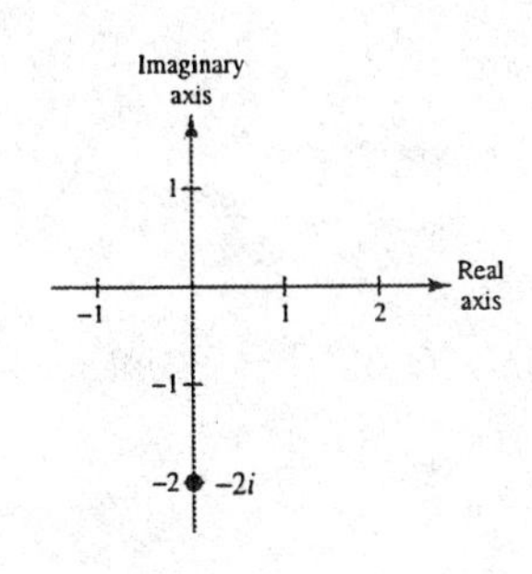

24. $z = 2\sqrt{2} - i$

$r = \sqrt{(2\sqrt{2})^2 + (-1)^2} = \sqrt{9} = 3$

$\tan\theta = \dfrac{-1}{2\sqrt{2}} = -\dfrac{\sqrt{2}}{4} \implies \theta \approx (-19.5°)$

$z = 3(\cos(-19.5°) + i\sin(-19.5°))$

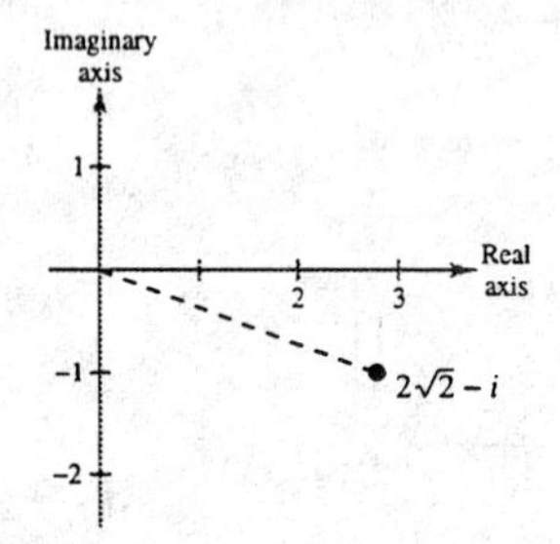

26. $z = 1 + 3i$

$r = \sqrt{1^2 + 3^2} = \sqrt{10}$

$\tan\theta = \frac{3}{1} = 3 \implies \theta \approx 71.6°$

$z \approx \sqrt{10}(\cos 71.6° + i\sin 71.6°)$

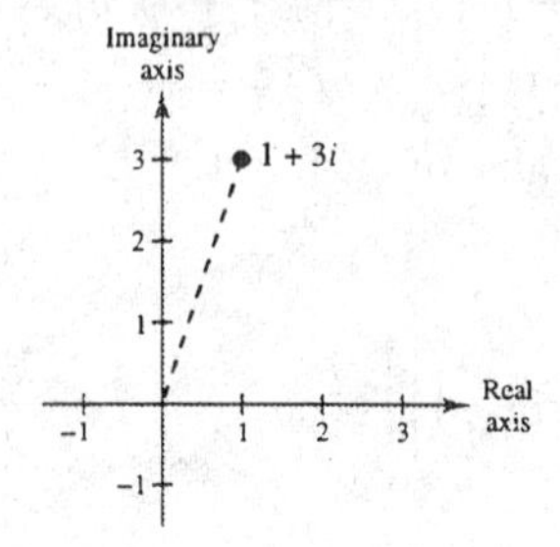

28. $-3 + i \approx 3.16 \angle 2.82$

$= 3.16(\cos 2.82 + i\sin 2.82)$

30. $-8 - 5\sqrt{3}i = 11.79 \angle -2.32$

or $11.79 \angle 3.97$

$-8 - 5\sqrt{3}i = 11.79(\cos 3.97 + i\sin 3.97)$

32. $5(\cos 135° + i\sin 135°) = 5\left[-\dfrac{\sqrt{2}}{2} + i\left(\dfrac{\sqrt{2}}{2}\right)\right]$

$= -\dfrac{5\sqrt{2}}{2} + \dfrac{5\sqrt{2}}{2}i$

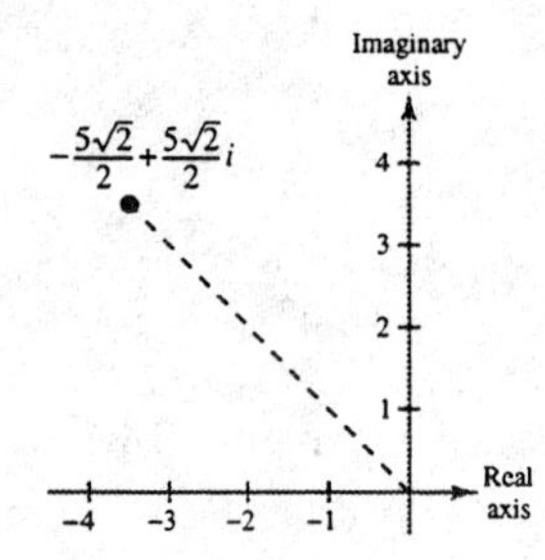

34. $\dfrac{3}{4}(\cos 315° + i\sin 315°) = \dfrac{3}{4}\left[\dfrac{\sqrt{2}}{2} + i\left(-\dfrac{\sqrt{2}}{2}\right)\right]$

$= \dfrac{3\sqrt{2}}{8} - \dfrac{3\sqrt{2}}{8}i$

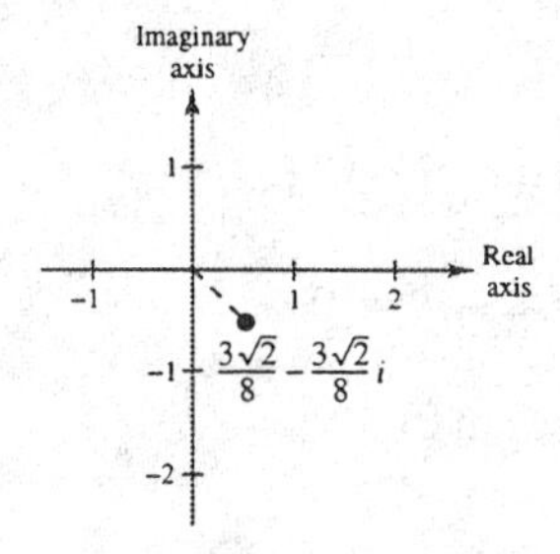

36. $8\left(\cos\dfrac{\pi}{12} + i\sin\dfrac{\pi}{12}\right) = 8(0.9659 + 0.2588i)$

$\approx 7.7274 + 2.0706i$

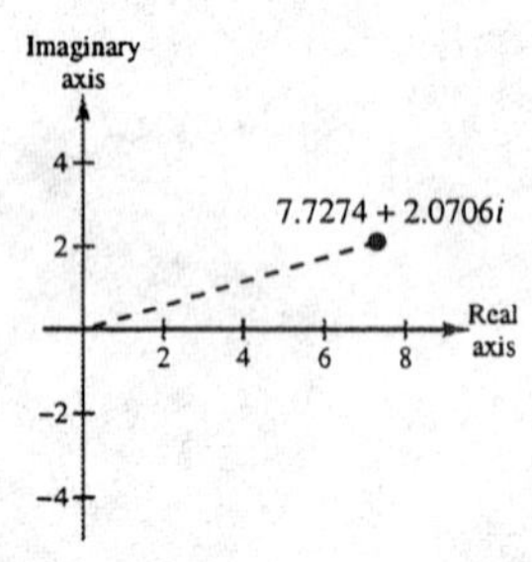

38. $7(\cos 0° + i\sin 0°) = 7$

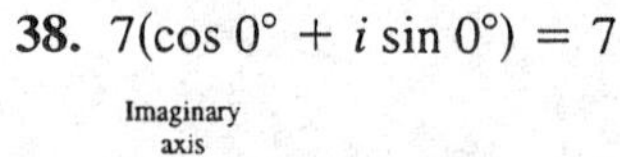

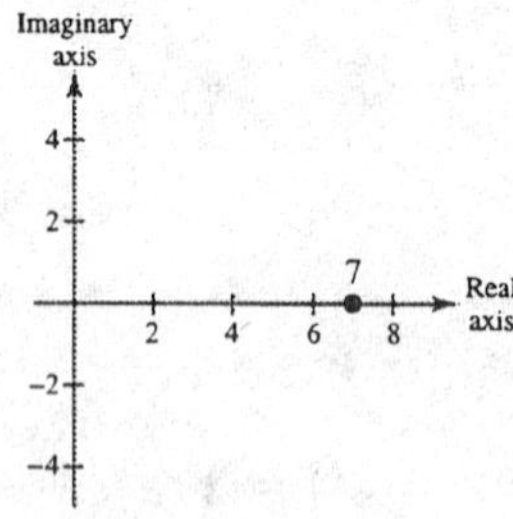

40. $6[\cos(230°\ 30') + i\sin(230°\ 30')] \approx -3.816 - 4.630i$

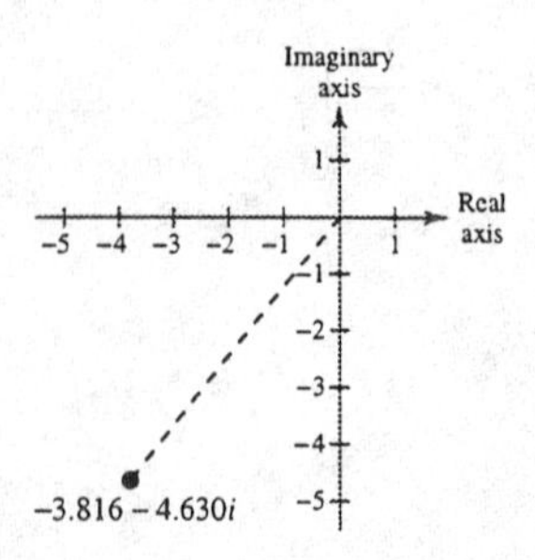

42. $9(\cos 58° + i \sin 58°) \approx 4.77 + 7.63i$

44. $4(\cos 216.5° + i \sin 216.5°) = -3.22 - 2.38i$

46. $z = \frac{1}{2}(1 + \sqrt{3}i)$

$z^2 = \frac{1}{2}(1 + \sqrt{3}i)\frac{1}{2}(1 + \sqrt{3}i) = \frac{1}{2}(-1 + \sqrt{3}i)$

$z^3 = z^2 z = \frac{1}{2}(-1 + \sqrt{3}i)\frac{1}{2}(1 + \sqrt{3}i) = -1$

$z^4 = z^3 z = (-1)\frac{1}{2}(1 + \sqrt{3}i) = \frac{1}{2}(-1 - \sqrt{3}i)$

The absolute value of each is 1.

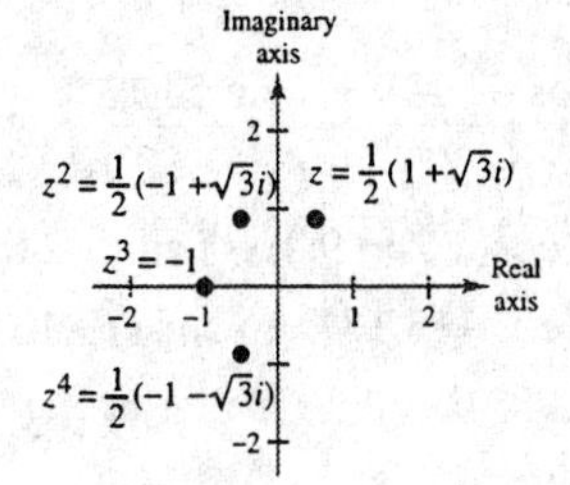

48. $$\left[\frac{3}{2}\left(\cos\frac{\pi}{2} + i\sin\frac{\pi}{2}\right)\right]\left[6\left(\cos\frac{\pi}{4} + i\sin\frac{\pi}{4}\right)\right] = \left(\frac{3}{2}\right)(6)\left[\cos\left(\frac{\pi}{2} + \frac{\pi}{4}\right) + i\sin\left(\frac{\pi}{2} + \frac{\pi}{4}\right)\right]$$
$$= 9\left(\cos\frac{3\pi}{4} + i\sin\frac{3\pi}{4}\right)$$

50. $$[0.5(\cos 100° + i \sin 100°)][0.8(\cos 300° + i \sin 300°)] = (0.5)(0.8)[\cos(100° + 300°) + i \sin(100° + 300°)]$$
$$= 0.4(\cos 400° + i \sin 400°)$$
$$= 0.4(\cos 40° + i \sin 40°)$$

52. $$(\cos 5° + i \sin 5°)(\cos 20° + i \sin 20°) = \cos(5° + 20°) + i \sin(5° + 20°)$$
$$= \cos 25° + i \sin 25°$$

54. $$\frac{5[\cos(4.3) + i\sin(4.3)]}{4[\cos(2.1) + i\sin(2.1)]} = \frac{5}{4}[\cos(4.3 - 2.1) + i\sin(4.3 - 2.1)]$$
$$= \frac{5}{4}[\cos(2.2) + i\sin(2.2)]$$

56. $$\frac{\cos(5\pi/3) + i\sin(5\pi/3)}{\cos(\pi) + i\sin(\pi)} = \cos\frac{2\pi}{3} + i\sin\frac{2\pi}{3}$$

58. $$\frac{9(\cos 20° + i\sin 20°)}{5(\cos 75° + i\sin 75°)} = \frac{9}{5}[\cos(20° - 75°) + i\sin(20° - 75°)]$$
$$= \frac{9}{5}[\cos(-55°) + i\sin(55°)]$$
$$= \frac{9}{5}[\cos 305° + i\sin 305°]$$

60. (a) $\sqrt{3} + i = 2(\cos 30° + i \sin 30°)$

$1 + i = \sqrt{2}(\cos 45° + i \sin 45°)$

(b) $$(\sqrt{3} + i)(1 + i) = [2(\cos 30° + i\sin 30°)][\sqrt{2}(\cos 45° + i\sin 45°)]$$
$$= 2\sqrt{2}(\cos 75° + i\sin 75°)$$
$$= 2\sqrt{2}\left[\left(\frac{\sqrt{6} - \sqrt{2}}{4}\right) + \left(\frac{\sqrt{6} + \sqrt{2}}{4}\right)i\right]$$
$$= (\sqrt{3} - 1) + (\sqrt{3} + 1)i$$

(c) $(\sqrt{3} + i)(1 + i) = \sqrt{3} + (\sqrt{3} + 1)i + i^2 = (\sqrt{3} - 1) + (\sqrt{3} + 1)i$

62. (a) $3 + 4i = 5(\cos 53.13° + i \sin 53.13°)$

$1 - \sqrt{3}i = 2(\cos 300° + i \sin 300°)$

(b) $\dfrac{3 + 4i}{1 - \sqrt{3}i} = \dfrac{5(\cos 53.13° + i \sin 53.13°)}{2(\cos 300° + i \sin 300°)}$

$= 2.5[\cos(-246.9°) + i \sin(-246.9°)]$

$= 2.5(\cos 113.13° + i \sin 113.13°)$

$= -0.9821 + 2.299i$

(c) $\dfrac{3 + 4i}{1 - \sqrt{3}i} = \dfrac{3 + 4i}{1 - \sqrt{3}i} \cdot \dfrac{1 + \sqrt{3}i}{1 + \sqrt{3}i}$

$= \dfrac{3 + (4 + 3\sqrt{3})i^2}{1 + 3}$

$= \dfrac{3 - 4\sqrt{3}}{4} + \dfrac{3 + 4\sqrt{3}}{4}i$

$\approx -0.9821 + 2.299i$

64. (a) $4i = 4(\cos 90° + i \sin 90°)$

$-4 + 2i = 2\sqrt{5}(\cos 153.4° + i \sin 153.4°)$

(b) $\dfrac{4i}{-4 + 2i} = \dfrac{4(\cos 90° + i \sin 90°)}{2\sqrt{5}(\cos 153.4° + i \sin 153.4°)}$

$= 2\sqrt{5}(\cos 296.6° + i \sin 296.6°)$

$\approx 0.400 - 0.800i$

(c) $\dfrac{4i}{-4 + 2i} = \dfrac{4i}{-4 + 2i} \cdot \dfrac{-4 - i}{-4 - 2i}$

$= \dfrac{8 - 16i}{20}$

$= \dfrac{2}{5} - \dfrac{4}{5}i$

$= 0.400 - 0.800i$

66. $z = r(\cos \theta + i \sin \theta)$

$\bar{z} = r(\cos \theta - i \sin \theta)$

$= r[\cos(-\theta) + i \sin(-\theta)]$

68. $z = r(\cos \theta + i \sin \theta)$

$-z = -r(\cos \theta + i \sin \theta)$

$= r(-\cos \theta - i \sin \theta)$

$= r[\cos(\theta + \pi) + i \sin(\theta + \pi)]$

70. Let $\theta = \dfrac{\pi}{6}$.

Let $z = x + iy$ such that: $\tan \dfrac{\pi}{6} = \dfrac{y}{x}$

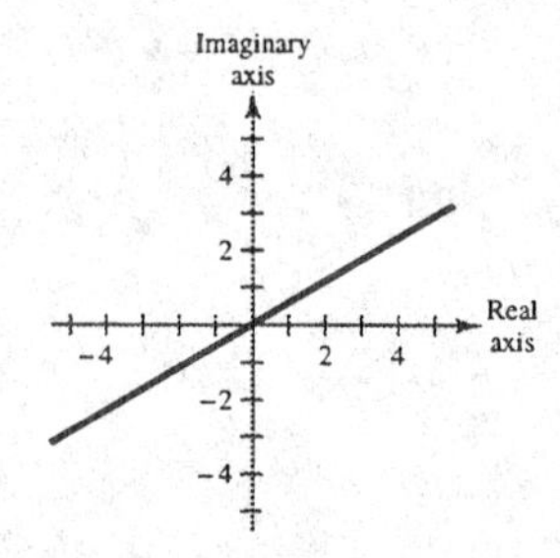

72. $(2 + 2i)^6 = \left[2\sqrt{2}\left(\cos \dfrac{\pi}{4} + i \sin \dfrac{\pi}{4}\right)\right]^6$

$= (2\sqrt{2})^6\left(\cos \dfrac{6\pi}{4} + i \sin \dfrac{6\pi}{4}\right)$

$= 512\left(\cos \dfrac{3\pi}{2} + i \sin \dfrac{3\pi}{2}\right)$

$= -512i$

74. $(1 - i)^{12} = \left[\sqrt{2}\left(\cos \dfrac{7\pi}{4} + i \sin \dfrac{7\pi}{4}\right)\right]^{12}$

$= (\sqrt{2})^{12}(\cos 21\pi + i \sin 21\pi)$

$= 64(\cos \pi + i \sin \pi)$

$= 64(-1)$

$= -64$

76. $4(1 - \sqrt{3}i)^3 = 4\left[2\left(\cos \dfrac{5\pi}{3} + i \sin \dfrac{5\pi}{3}\right)\right]^3$

$= 4[2^3(\cos 5\pi + i \sin 5\pi)]$

$= 32(-1)$

$= -32$

78. $[3(\cos 150° + i \sin 150°)]^4 = 3^4(\cos 600° + i \sin 600°)$

$= 81(\cos 240° + i \sin 240°)$

$= 81(-\cos 60° - i \sin 60°)$

$= -\dfrac{81}{2} - \dfrac{81\sqrt{3}}{2}i$

80. $\left[2\left(\cos \dfrac{\pi}{2} + i \sin \dfrac{\pi}{2}\right)\right]^8 = 2^8(\cos 4\pi + i \sin 4\pi)$

$= 256(\cos 0 + i \sin 0)$

$= 256$

82. $(\cos 0 + i \sin 0)^{20} = \cos 0 + i \sin 0$

$= 1$

84. $(\sqrt{5} - 4i)^3 = -43\sqrt{5} + 4i$

86. $\left[2\left(\cos\dfrac{\pi}{10} + i\sin\dfrac{\pi}{10}\right)\right]^5 = 32\left(\cos\dfrac{\pi}{2}\, i\sin\dfrac{\pi}{2}\right)$
$= 32i$

88. $2^{-1/4}(1 - i)$ is a fourth root of -2 if $-2 = [2^{-1/4}(1 - i)]^4$.

$$\begin{aligned}[2^{-1/4}(1 - i)]^4 &= (2^{-1/4})^4(1 - i)^4\\ &= 2^{-1}(1 - i)^4\\ &= \frac{1}{2}(1 - i)^2(1 - i)^2\\ &= \frac{1}{2}(-2i)(-2i)\\ &= \frac{1}{2}(4i^2)\\ &= \frac{1}{2}(-4) = -2\end{aligned}$$

90. (a) In trigonometric form we have:

$3(\cos 45° + i\sin 45°)$
$3(\cos 135° + i\sin 135°)$
$3(\cos 225° + i\sin 225°)$
$3(\cos 315° + i\sin 315°)$

(b) There are four roots evenly spaced around a circle of radius 3. Therefore, they represent the fourth roots of some number of modulus 81. Raising them to the fourth power shows that they are all fourth roots of -81.

(c) $[3(\cos 45° + i\sin 45°)]^4 = -81$
$[3(\cos 135° + i\sin 135°)]^4 = -81$
$[3(\cos 225° + i\sin 225°)]^4 = -81$
$[3(\cos 315° + i\sin 315°)]^4 = -81$

92. (a) Square roots of $16(\cos 60° + i\sin 60°)$:

$$\sqrt{16}\left[\cos\left(\frac{60° + 360°\,k}{2}\right) + i\sin\left(\frac{60° + 360°\,k}{2}\right)\right],\ k = 0, 1$$

$4(\cos 30° + i\sin 30°)$
$4(\cos 210° + i\sin 210°)$

(b)

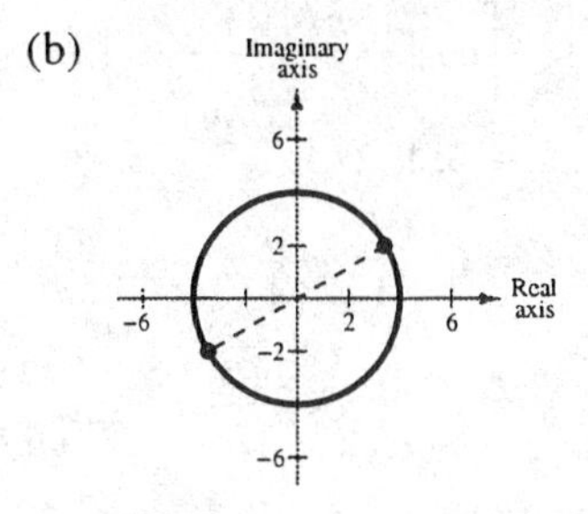

(c) $4\left(\dfrac{\sqrt{3}}{2} + \dfrac{1}{2}i\right) = 2\sqrt{3} + 2i$

$4\left(-\dfrac{\sqrt{3}}{2} - \dfrac{1}{2}i\right) = -2\sqrt{3} - 2i$

94. (a) Fifth roots of $32\left(\cos\dfrac{5\pi}{6} + i\sin\dfrac{5\pi}{6}\right)$:

$$\sqrt[5]{32}\left[\cos\left(\frac{(5\pi/6) + 2k\pi}{5}\right) + i\sin\left(\frac{(5\pi/6) + 2k\pi}{5}\right)\right]$$

$k = 0, 1, 2, 3, 4$

$k = 0$: $2\left(\cos\dfrac{\pi}{6} + i\sin\dfrac{\pi}{6}\right)$

$k = 1$: $2\left(\cos\dfrac{17\pi}{30} + i\sin\dfrac{17\pi}{30}\right)$

$k = 2$: $2\left(\cos\dfrac{29\pi}{30} + i\sin\dfrac{29\pi}{30}\right)$

$k = 3$: $2\left(\cos\dfrac{41\pi}{30} + i\sin\dfrac{41\pi}{30}\right)$

$k = 4$: $2\left(\cos\dfrac{53\pi}{30} + i\sin\dfrac{53\pi}{30}\right)$

(b)

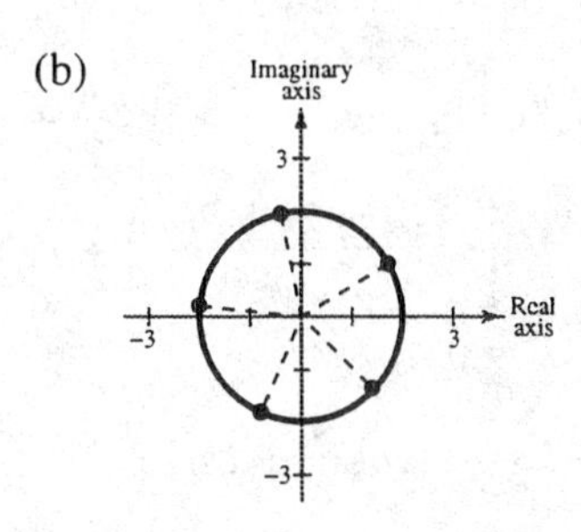

(c) $1.732 + i,\ -0.4158 + 1.956i,\ -1.989 + 0.2091i,$
$-0.8134 - 1.827i,\ 1.486 - 1.338i$

96. (a) Fourth roots of $625i = 625\left(\cos\frac{\pi}{2} + i\sin\frac{\pi}{2}\right)$:

$$\sqrt[4]{625}\left[\cos\left(\frac{(\pi/2) + 2k\pi}{4}\right) + i\sin\left(\frac{(\pi/2) + 2k\pi}{4}\right)\right.$$

$k = 0, 1, 2, 3$

$k = 0$: $5\left(\cos\frac{\pi}{8} + i\sin\frac{\pi}{8}\right)$

$k = 1$: $5\left(\cos\frac{5\pi}{8} + i\sin\frac{5\pi}{8}\right)$

$k = 2$: $5\left(\cos\frac{9\pi}{8} + i\sin\frac{9\pi}{8}\right)$

$k = 3$: $5\left(\cos\frac{13\pi}{8} + i\sin\frac{13\pi}{8}\right)$

(b)

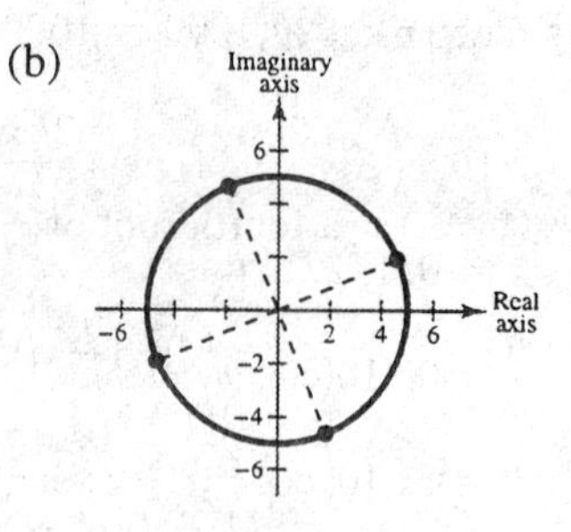

(c) $4.619 + 1.913i, -1.913 + 4.619i, -4.619 - 1.913i, 1.913 - 4.619i$

98. (a) Cube roots of $-4\sqrt{2}(1 - i) = 8\left(\cos\frac{3\pi}{4} + i\sin\frac{3\pi}{4}\right)$:

$$\sqrt[3]{8}\left[\cos\left(\frac{(3\pi/4) + 2k\pi}{3}\right) + i\sin\left(\frac{(3\pi/4) + 2k\pi}{3}\right)\right]$$

$k = 0, 1, 2$

$k = 0$: $2\left(\cos\frac{\pi}{4} + i\sin\frac{\pi}{4}\right)$

$k = 1$: $2\left(\cos\frac{11\pi}{12} + i\sin\frac{11\pi}{12}\right)$

$k = 2$: $2\left(\cos\frac{19\pi}{12} + i\sin\frac{19\pi}{12}\right)$

(b)

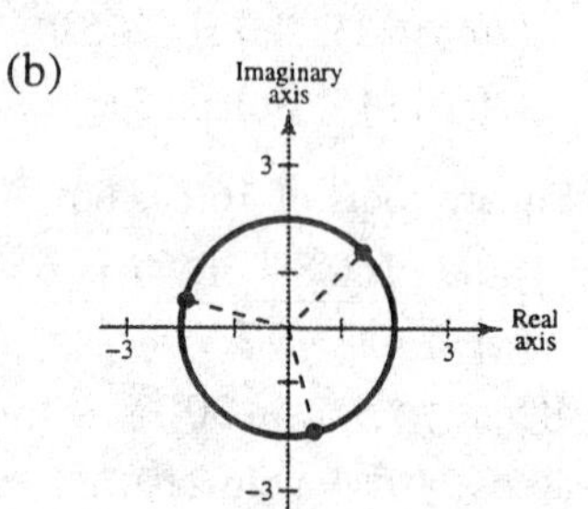

(c) $1.414 + 1.414i, -1.932 + 0.5176i, 0.5176 - 1.9319i$

100. (a) Fourth roots of $i = \cos\frac{\pi}{2} + i\sin\frac{\pi}{2}$:

$$\sqrt[4]{1}\left[\cos\left(\frac{(\pi/2) + 2k\pi}{4}\right) + i\sin\left(\frac{(\pi/2) + 2k\pi}{4}\right)\right.$$

$k = 0, 1, 2, 3$

$k = 0$: $\cos\frac{\pi}{8} + i\sin\frac{\pi}{8}$

$k = 1$: $\cos\frac{5\pi}{8} + i\sin\frac{5\pi}{8}$

$k = 2$: $\cos\frac{9\pi}{8} + i\sin\frac{9\pi}{8}$

$k = 3$: $\cos\frac{13\pi}{8} + i\sin\frac{13\pi}{8}$

(b)

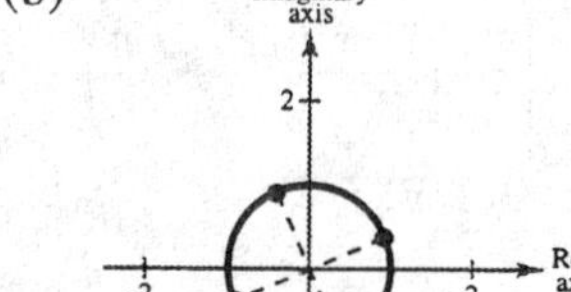

(c) $0.9239 + 0.3827i, -0.3827 + 0.9239i, -0.9239 - 0.3827i, 0.3827 - 0.9239i$

102. (a) Cube roots of $1000 = 1000(\cos 0 + i \sin 0)$:

$$\sqrt[3]{1000}\left(\cos \frac{2k\pi}{3} + i \sin \frac{2k\pi}{3}\right)$$

$k = 0, 1, 2$

$k = 0$: $10(\cos 0 + i \sin 0)$

$k = 1$: $10\left(\cos \frac{2\pi}{3} + i \sin \frac{2\pi}{3}\right)$

$k = 2$: $10\left(\cos \frac{4\pi}{3} + i \sin \frac{4\pi}{3}\right)$

(b)

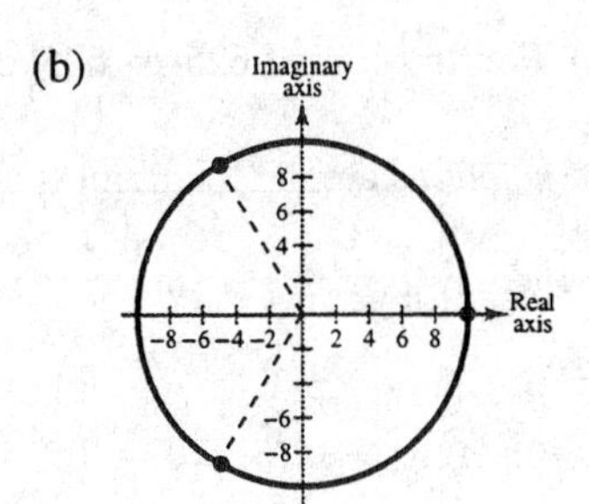

(c) $10, -5 + 5\sqrt{3}i, -5 - 5\sqrt{3}i$

104. (a) The fourth roots of $-4 = 4(\cos 180° + i \sin 180°)$:

$\sqrt{2}(\cos 45° + i \sin 45°)$

$\sqrt{2}(\cos 135° + i \sin 135°)$

$\sqrt{2}(\cos 225° + i \sin 225°)$

$\sqrt{2}(\cos 315° + i \sin 315°)$

(b)

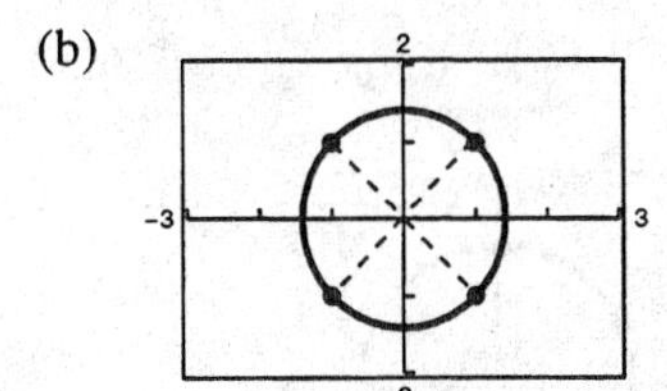

(c) $1 + i, -1 + i, -1 - i, 1 - i$

106. (a) The sixth roots of $64i = 64(\cos 90° + i \sin 90°)$:

$2(\cos 15° + i \sin 15°)$

$2(\cos 75° + i \sin 75°)$

$2(\cos 135° + i \sin 135°)$

$2(\cos 195° + i \sin 195°)$

$2(\cos 255° + i \sin 255°)$

$2(\cos 315° + i \sin 315°)$

(b)

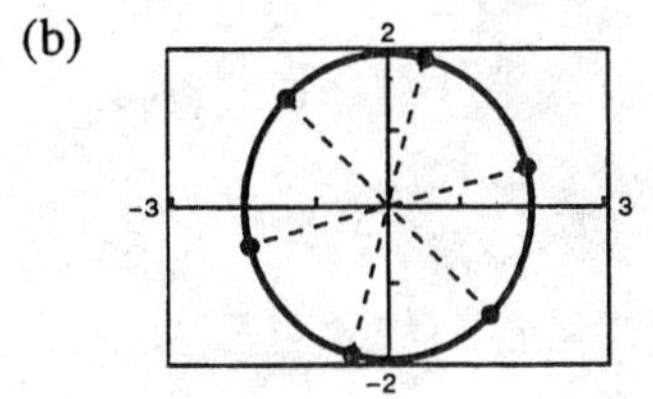

(c) $1.93 + 0.52i, 0.52 + 1.93i, -\sqrt{2} + \sqrt{2}i, -1.93 - 0.52i,$
$-0.52 - -1.93i, \sqrt{2} - \sqrt{2}i$

108. $x^3 + 1 = 0$

$x^3 = -1$

The solutions are the cube roots of $-1 = \cos \pi + i \sin \pi$:

$$\cos\left(\frac{\pi + 2k\pi}{3}\right) + i \sin\left(\frac{\pi + 2k\pi}{3}\right)$$

$k = 0, 1, 2$

$k = 0$: $\cos \frac{\pi}{3} + i \sin \frac{\pi}{3} = \frac{1}{2} + \frac{\sqrt{3}}{2}i$

$k = 1$: $\cos \pi + i \sin \pi = -1$

$k = 2$: $\cos \frac{5\pi}{3} + i \sin \frac{5\pi}{3} = \frac{1}{2} - \frac{\sqrt{3}}{2}i$

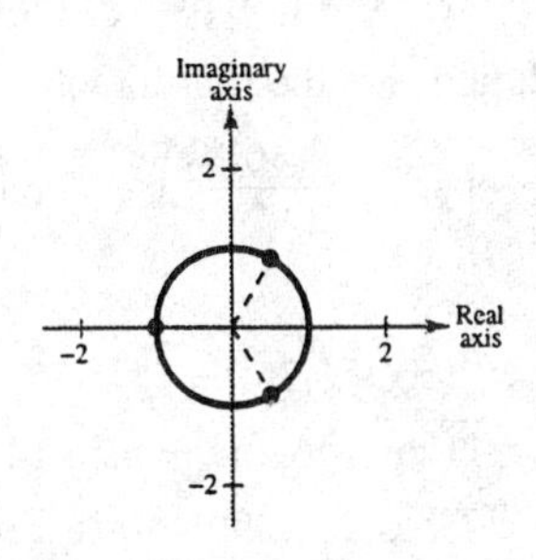

110. $x^4 - 81 = 0$

$x^4 = 81$

The solutions are the fourth roots of 81:

$\sqrt[4]{81}\left(\cos\frac{0 + 2\pi k}{4} + i\sin\frac{0 + 2\pi k}{4}\right)$

$k = 0, 1, 2, 3$

$k = 0$: $3(\cos 0 + i\sin 0) = 3$

$k = 1$: $3\left(\cos\frac{\pi}{2} + i\sin\frac{\pi}{2}\right) = 3i$

$k = 2$: $3(\cos\pi + i\sin\pi) = -3$

$k = 3$: $3\left(\cos\frac{3\pi}{2} + i\sin\frac{3\pi}{2}\right) = -3i$

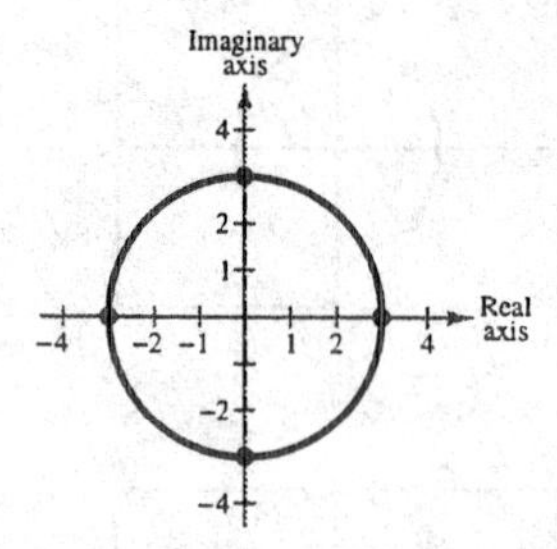

112. $x^6 - 64i = 0$

$x^6 = 64i$

The solutions are the sixth roots of $64i$:

$\sqrt[6]{64}\left[\cos\left(\frac{(\pi/2) + 2k\pi}{6}\right) + i\sin\left(\frac{(\pi/2) + 2k\pi}{6}\right)\right]$

$k = 0, 1, 2, 3, 4, 5$

$k = 0$: $2\left(\cos\frac{\pi}{12} + i\sin\frac{\pi}{12}\right) \approx 1.932 + 0.5176i$

$k = 1$: $2\left(\cos\frac{5\pi}{12} + i\sin\frac{5\pi}{12}\right) \approx 0.5176 + 1.932i$

$k = 2$: $2\left(\cos\frac{3\pi}{4} + i\sin\frac{3\pi}{4}\right) \approx -1.414 + 1.414i$

$k = 3$: $2\left(\cos\frac{13\pi}{12} + i\sin\frac{13\pi}{12}\right) \approx -1.932 - 0.5176i$

$k = 4$: $2\left(\cos\frac{17\pi}{12} + i\sin\frac{17\pi}{12}\right) \approx -0.5176 - 1.932i$

$k = 5$: $2\left(\cos\frac{7\pi}{4} + i\sin\frac{7\pi}{4}\right) \approx 1.414 - 1.414i$

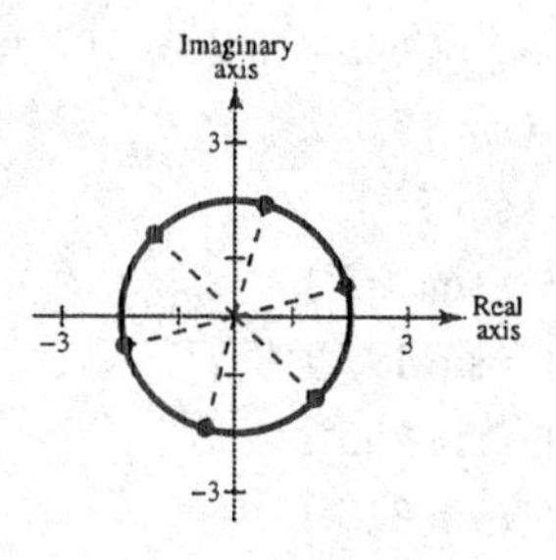

114. $x^4 + (1 + i) = 0$

$x^4 = -1 - i = \sqrt{2}(\cos 225° + i\sin 225°)$

The solutions are the fourth roots of $-1 - i$:

$\sqrt[4]{\sqrt{2}}\left[\cos\left(\frac{225° + 360°k}{4}\right) + i\sin\left(\frac{225° + 360°k}{4}\right)\right]$

$k = 0, 1, 2, 3$

$k = 0$: $\sqrt[8]{2}(\cos 56.25° + i\sin 56.25°) \approx 0.6059 + 0.9067i$

$k = 1$: $\sqrt[8]{2}(\cos 146.25° + i\sin 146.25°) \approx -0.9067 - 0.6059i$

$k = 2$: $\sqrt[8]{2}(\cos 236.25° + i\sin 236.25°) \approx -0.6059 - 0.9067i$

$k = 3$: $\sqrt[8]{2}(\cos 326.25° + i\sin 326.25°) \approx 0.9067 - 0.6059i$

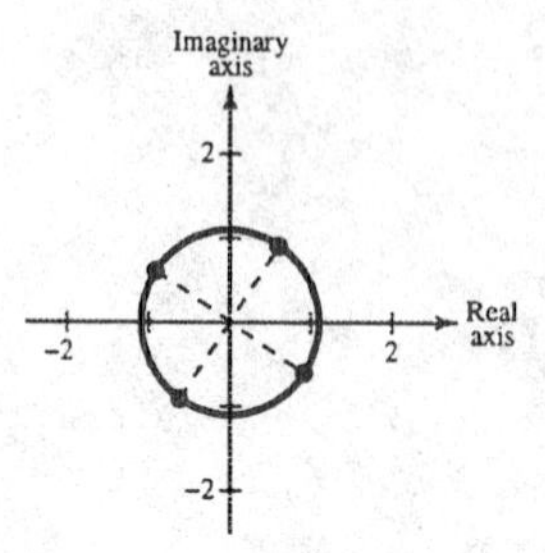

116. $\sin 15° = \frac{h}{50} \Rightarrow h = 50\sin 15° \approx 12.94$

❑ Review Exercises for Chapter 6

Solutions to Even-Numbered Exercises

2. Given: $a = 6, b = 9, c = 45°$

$c^2 = a^2 + b^2 - 2ab\cos 45° \approx 36 + 81 - 2(6)(9)(0.7071) \approx 40.63 \Longrightarrow 6.374$

$\cos B = \dfrac{a^2 + c^2 - b^2}{2ac} \approx \dfrac{36 + 40.63 - 81}{2(6)(6.374)} \approx -0.0571 \Longrightarrow B \approx 93.3°$

$A \approx 180° - 45° - 93.3° = 41.7°$

4. Given: $B = 110°, C = 30°, c = 10.5$

$A = 180° - 110° - 30° = 40°$

$a = \dfrac{c\sin A}{\sin C} = \dfrac{10.5(\sin 40°)}{\sin 30°}$

$\approx \dfrac{10.5(0.6428)}{0.5} \approx 13.5$

$b = \dfrac{c\sin B}{\sin C} = \dfrac{10.5(\sin 110°)}{\sin 30°}$

$\approx \dfrac{10.5(0.9397)}{0.5} \approx 19.7$

6. Given: $a = 80, b = 60, c = 100$

$\cos C = \dfrac{a^2 + b^2 - c^2}{2ab} = \dfrac{6400 + 3600 - 10{,}000}{2(80)(60)}$

$= 0 \Longrightarrow C = 90°$

$\sin A = \dfrac{80}{100} = 0.8 \Longrightarrow A \approx 53.1°$

$\sin B = \dfrac{60}{100} = 0.6 \Longrightarrow B \approx 36.9°$

8. Given: $A = 130°, a = 50, b = 30$

$\sin B = \dfrac{b\sin A}{a} = \dfrac{30\sin 130°}{50} \approx \dfrac{30(0.7660)}{50} \approx 0.4596 \Longrightarrow B \approx 27.4°$

$C \approx 180° - 130° - 27.4° = 22.6°$

$c^2 = a^2 + b^2 - 2ab\cos C \approx 50^2 + 30^2 - 2(50)(30)(0.9232) \approx 630.4 \Longrightarrow c \approx 25.11$

10. Given: $C = 50°, a = 25, c = 22$

$$\sin A = \frac{a \sin C}{c} = \frac{25 \sin 50°}{22} \approx \frac{25(0.7660)}{22} \approx 0.8705 \Longrightarrow A \approx 60.5° \text{ or } 119.5°$$

Case 1:

$A \approx 60.5°$

$B \approx 180° - 50° - 60.5° = 69.5°$

$$b = \frac{c \sin B}{\sin C} \approx \frac{22(0.9367)}{0.7660} \approx 26.90$$

Case 2:

$A \approx 119.5°$

$B \approx 180° - 50° - 119.5° = 10.5°$

$$b = \frac{c \sin B}{\sin C} \approx 5.234$$

12. Given: $B = 150°, a = 64, b = 10$

$$\sin A = \frac{a \sin B}{b} = \frac{64 \sin 150°}{10} \approx 3.2 \Longrightarrow \text{no triangle formed}$$

No solution

14. Given: $a = 2.5, b = 15.0, c = 4.5$

Since $a + c < b$, a triangle in not formed.

No solution

16. Given: $B = 90°, a = 5, c = 12$

$b = \sqrt{12^2 + 5^2} = \sqrt{169} = 13$

$A = \arctan \frac{5}{12} \approx 22.6°$

$C = \arctan \frac{12}{5} \approx 67.4°$

18. $a = 15, b = 8, c = 10$

$$s = \frac{15 + 8 + 10}{2} = 16.5$$

$\text{Area} = \sqrt{16.5(1.5)(8.5)(6.5)} \approx 36.98$

20. $B = 80°, a = 4, c = 8$

$$\text{Area} = \frac{1}{2}ac \sin B = \frac{1}{2}(4)(8)(0.9848) = 15.76$$

22. $a^2 = 5^2 + 8^2 - 2(5)(8) \cos 152°$

$\approx 159.6 \Longrightarrow a \approx 12.63 \text{ ft}$

$b^2 = 5^2 + 8^2 - 2(5)(8) \cos 28°$

$\approx 18.36 \Longrightarrow b \approx 4.285 \text{ ft}$

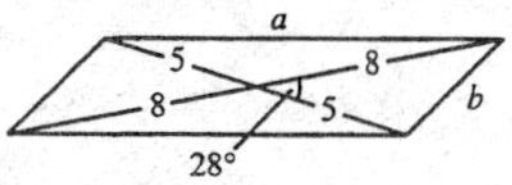

24. $b^2 = a^2 + c^2 - 2ac \cos B$

$= 300^2 + 425^2 - 2(300)(425) \cos(180° - 65°)$

≈ 378392.66

$b \approx 615.1$ meters

26. $$\frac{a}{\sin 75°} = \frac{400}{\sin 37.5°}$$

$$a = \frac{400 \sin 75°}{\sin 37.5°} \approx 634.7 \text{ ft}$$

$$\sin 67.5° = \frac{w}{a}$$

$w = 634.7 \sin 67.5° \approx 586.4 \text{ ft}$

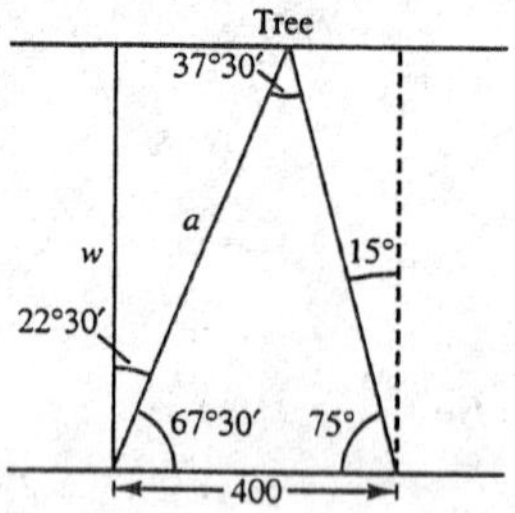

28. By the Law of Sines, $\dfrac{a}{\sin A} = \dfrac{b}{\sin B}$.

$$a = \frac{\sin A}{\sin B} b = \frac{\sin(69°)}{\sin(55°)}(27) \approx 30.8 \text{ meters}$$

30. Initial point: $(0, 1)$

Terminal point: $\left(6, \frac{7}{2}\right)$

$\mathbf{v} = \left\langle 6 - 0, \frac{7}{2} - 1 \right\rangle = \left\langle 6, \frac{5}{2} \right\rangle$

32. Initial point: $(1, 5)$

Terminal point: $(15, 9)$

$\mathbf{v} = \langle 15 - 1, 9 - 5 \rangle = \langle 14, 4 \rangle$

34. $$\left\langle \frac{1}{2} \cos 225°, \frac{1}{2} \sin 225° \right\rangle = \left\langle -\frac{\sqrt{2}}{4}, -\frac{\sqrt{2}}{4} \right\rangle$$

36. $\mathbf{v} = 4\mathbf{i} - \mathbf{j}$

$\|\mathbf{v}\| = \sqrt{4^2 + (-1)^2} = \sqrt{17}$

$\tan\theta = \dfrac{-1}{4} = -\dfrac{1}{4} \Rightarrow \theta \approx 346°$, since θ is in Quadrant IV.

$\mathbf{v} = \sqrt{17}(\mathbf{i}\cos 346° + \mathbf{j}\sin 346°)$

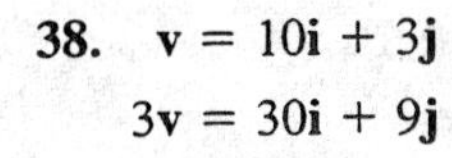

38. $\mathbf{v} = 10\mathbf{i} + 3\mathbf{j}$

$3\mathbf{v} = 30\mathbf{i} + 9\mathbf{j}$

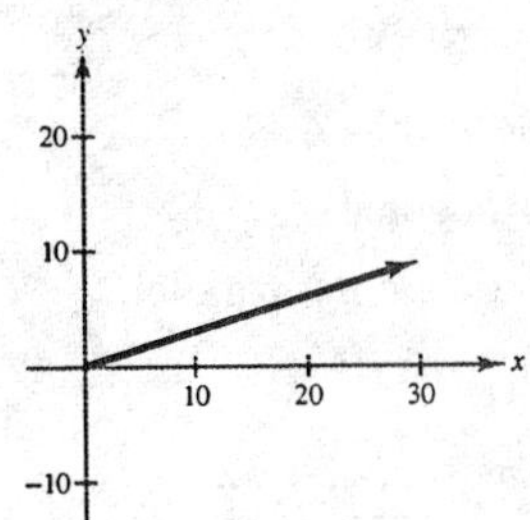

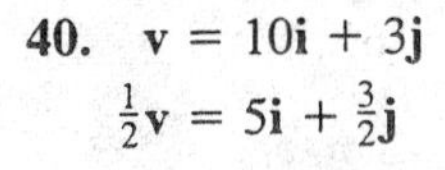

40. $\mathbf{v} = 10\mathbf{i} + 3\mathbf{j}$

$\frac{1}{2}\mathbf{v} = 5\mathbf{i} + \frac{3}{2}\mathbf{j}$

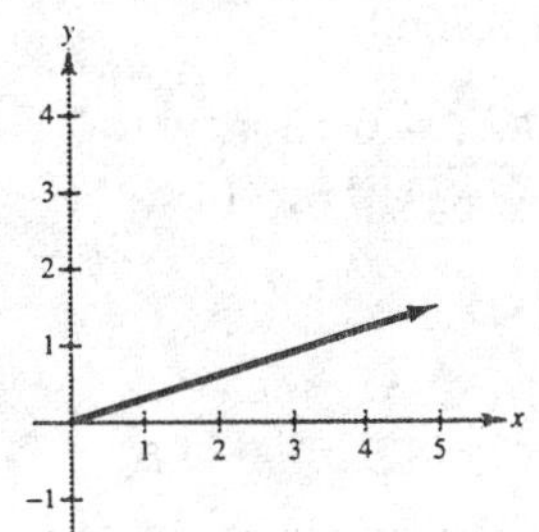

42. $\mathbf{u} = 12[(\cos 82°)\mathbf{i} + (\sin 82°)\mathbf{j}]$

$\mathbf{v} = 8[(\cos(-12°))\mathbf{i} + (\sin(-12°))\mathbf{j}]$

$\mathbf{u} + \mathbf{v} \approx 9.4953\mathbf{i} + 10.2199\mathbf{j}$

$\|\mathbf{u} + \mathbf{v}\| \approx 13.95$

$\tan\theta = \dfrac{10.2199}{9.4953} \Rightarrow \theta \approx 47.11°$

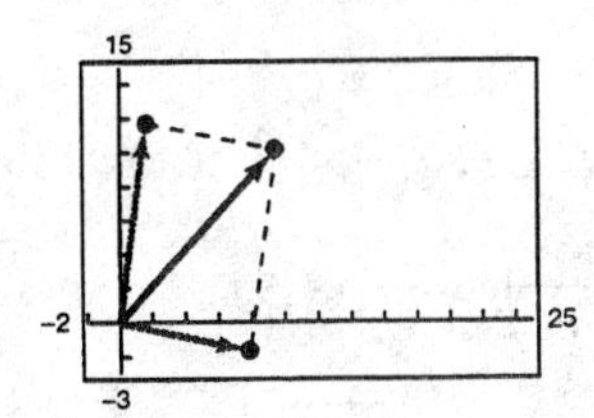

44. Force One: $\mathbf{u} = 85\mathbf{i}$

Force Two: $\mathbf{v} = 50\cos 15°\mathbf{i} + 50\sin 15°\mathbf{j}$

Resultant Force:

$$\mathbf{u} + \mathbf{v} = (85 + 50\cos 15°)\mathbf{i} + (50\sin 15°)\mathbf{j}$$

$$\|\mathbf{u} + \mathbf{v}\| = \sqrt{(85 + 50\cos 15°)^2 + (50\sin 15°)^2}$$

$$= \sqrt{85^2 + 8500\cos 15° + 50^2}$$

$$= 133.92 \text{ lb}$$

$\tan\theta = \dfrac{50\sin 15°}{85 + 50\cos 15°} \Rightarrow \theta \approx 5.5°$ from the 85-pound force.

46. By symmetry, the magnitudes of the tensions are equal.

$\mathbf{T} = \|\mathbf{T}\|(\cos 120°\mathbf{i} + \sin 120°\mathbf{j})$

$\|\mathbf{T}\|\sin 120° = \dfrac{1}{2}(200) \Rightarrow \|\mathbf{T}\| = \dfrac{100}{\sqrt{3}/2} = \dfrac{200}{\sqrt{3}} \approx 115.5$ lbs

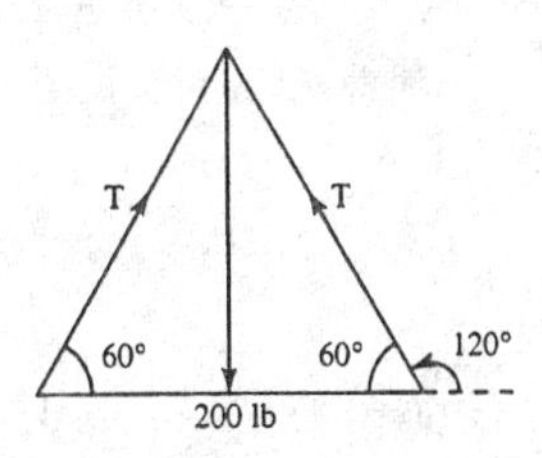

48. Airplane velocity: $\mathbf{u} = 430(\cos(-45°)\mathbf{i} + \sin(-45°)\mathbf{j})$

Wind velocity: $\mathbf{w} = 35(\cos 60°\mathbf{i} + \sin 60°\mathbf{j})$

$$\mathbf{u} + \mathbf{w} = (430\cos(-45) + 35\cos 60)\mathbf{i} + (430\sin(-45) + 35\sin 60)$$

$$= 321.56\mathbf{i} - 273.75\mathbf{j}$$

$\|\mathbf{u} + \mathbf{w}\| = 422.3$ mph

$\theta = \arctan\left(\dfrac{-273.75}{321.56}\right) = -40.4°$

Direction: S 49.6° E

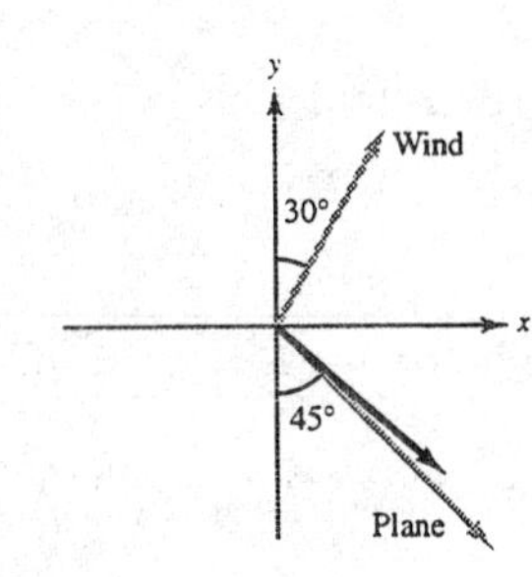

50. Let force one be $\mathbf{u} = 60\mathbf{i}$, and force two be $\mathbf{v} = 100(\cos\theta\mathbf{i} + \sin\theta\mathbf{j})$. The resultant is $125 = \|\mathbf{u} + \mathbf{v}\|$. Hence,

$\mathbf{u} + \mathbf{v} = (60 + 100\cos\theta)\mathbf{i} + (100\sin\theta)\mathbf{j}$.

$$\|\mathbf{u} + \mathbf{v}\| = \sqrt{(60 + 100\cos\theta)^2 + (100\sin\theta)^2} = 125$$

$$3600 + 12000\cos\theta + 10000\cos^2\theta + 10000\sin^2\theta = 125^2$$

$$13600 + 12000\cos\theta = 15{,}625$$

$$\cos\theta = 0.16875$$

$$\theta \approx 80.3^\circ$$

52. $P(0, 3), Q(5, -8)$

$\overrightarrow{PQ} = \langle 5 - 0, -8 - 3\rangle = \langle 5, -11\rangle$

$\|\overrightarrow{PQ}\| = \sqrt{5^2 + (-11)^2} = \sqrt{146}$

$$\frac{\overrightarrow{PQ}}{\|\overrightarrow{PQ}\|} = \frac{1}{\sqrt{146}}\langle 5, -11\rangle$$

54. $\mathbf{u} = \langle 8, 5\rangle, \mathbf{v} = \langle -2, 4\rangle$

$\mathbf{u} \cdot \mathbf{v} = 8(-2) + 5(4) = 4 \neq 0$

$\mathbf{u}$ and $\mathbf{v}$ are not orthogonal.

$\mathbf{u} \neq k\mathbf{v} \Rightarrow \mathbf{u}$ and $\mathbf{v}$ are not parallel.

Neither

56. $\mathbf{u} = \langle -6, -3\rangle, \mathbf{v} = \langle 4, 2\rangle$

$$\cos\theta = \frac{\mathbf{u} \cdot \mathbf{v}}{\|\mathbf{u}\|\,\|\mathbf{v}\|} = \frac{-30}{(\sqrt{45})(\sqrt{20})} = -1$$

$\theta = 180^\circ$

58. $\mathbf{u} = \langle 3, 1\rangle, \mathbf{v} = \langle 4, 5\rangle$

$$\cos\theta = \frac{\mathbf{u} \cdot \mathbf{v}}{\|\mathbf{u}\|\,\|\mathbf{v}\|} = \frac{17}{(\sqrt{10})(\sqrt{41})} \Rightarrow \theta \approx 32.9^\circ$$

60.

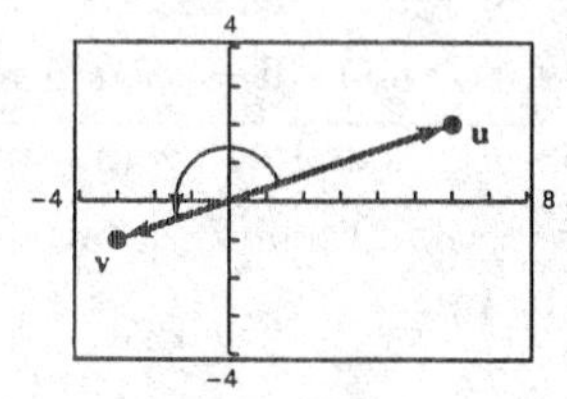

62.

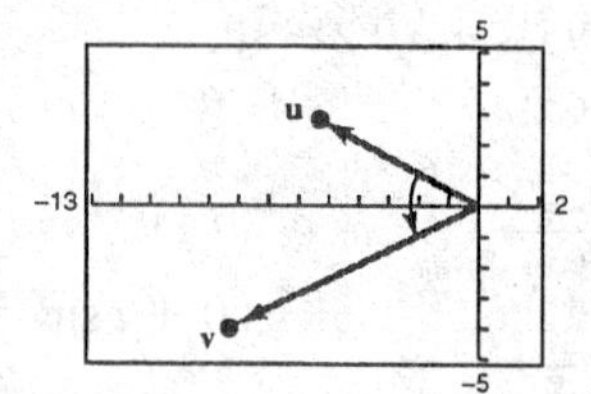

64. $\mathbf{u} = \langle 5, 6\rangle, \mathbf{v} = \langle 10, 0\rangle$

$$\text{proj}_{\mathbf{v}}\mathbf{u} = \left(\frac{\mathbf{u} \cdot \mathbf{v}}{\|\mathbf{v}\|^2}\right)\mathbf{v} = \frac{50}{100}\langle 10, 0\rangle = \langle 5, 0\rangle$$

66. $\mathbf{u} = \langle -3, 5\rangle, \mathbf{v} = \langle -5, 2\rangle$

$$\text{proj}_{\mathbf{v}}\mathbf{u} = \left(\frac{\mathbf{u} \cdot \mathbf{v}}{\|\mathbf{v}\|^2}\right)\mathbf{v} = \frac{25}{29}\langle -5, 2\rangle$$

68. $z = -3 + 3i, \|z\| = \sqrt{9 + 9} = 3\sqrt{2}, \theta = 135^\circ$

$z = 3\sqrt{2}(\cos 135^\circ + i\sin 135^\circ)$

70. $z = 2\sqrt{2} + i, \|z\| = \sqrt{8 + 1} = 3,$

$$\theta = \arctan\left(\frac{1}{2\sqrt{2}}\right) = 19.5^\circ$$

$z = 3(\cos 19.5^\circ + i\sin 19.5^\circ)$

72. $r = 6, \theta = 150^\circ$

$-3\sqrt{3} + 3i = 6(\cos 150^\circ + i\sin 150^\circ)$

74. $r = 7, \theta = 180^\circ$

$-7 = 7(\cos 180^\circ + i\sin 180^\circ)$

76.
$$24(\cos 330^\circ + i\sin 330^\circ) = 24\left(\frac{\sqrt{3}}{2} - \frac{1}{2}i\right) = 12\sqrt{3} - 12i$$

78.
$$8\left(\cos\frac{5\pi}{6} + i\sin\frac{5\pi}{6}\right) = 8\left(-\frac{\sqrt{3}}{2} + \frac{1}{2}i\right) = -4\sqrt{3} + 4i$$

80. (a) $z_1 = -3(1 + i) = 3\sqrt{2}\left(\cos\frac{5\pi}{4} + i\sin\frac{5\pi}{4}\right)$

$z_2 = 2(\sqrt{3} + i) = 4\left(\cos\frac{\pi}{6} + i\sin\frac{\pi}{6}\right)$

(b)
$$z_1z_2 = \left[3\sqrt{2}\left(\cos\frac{5\pi}{4} + i\sin\frac{5\pi}{4}\right)\right]\left[4\left(\cos\frac{\pi}{6} + i\sin\frac{\pi}{6}\right)\right] = 12\sqrt{2}\left(\cos\frac{17\pi}{12} + i\sin\frac{17\pi}{12}\right)$$

$$\frac{z_1}{z_2} = \frac{3\sqrt{2}[\cos(5\pi/4) + i\sin(5\pi/4)]}{4[\cos(\pi/6) + i\sin(\pi/6)]} = \frac{3\sqrt{2}}{4}\left(\cos\frac{13\pi}{12} + i\sin\frac{13\pi}{12}\right)$$

82. $\left[2\left(\cos\frac{4\pi}{15} + i\sin\frac{4\pi}{15}\right)\right]^5 = 2^5\left(\cos\frac{4\pi}{3} + i\sin\frac{4\pi}{3}\right)$

$= 32\left(-\frac{1}{2} - \frac{\sqrt{3}}{2}i\right)$

$= -16 - 16\sqrt{3}i$

84. $(1 - i)^8 = \left[\sqrt{2}(\cos 315° + i\sin 315°)\right]^8$

$= 16(\cos 2520° + i\sin 2520°)$

$= 16(\cos 0° + i\sin 0°)$

$= 16$

86. (a) The trigonometric forms of the four roots shown are:

$4(\cos 60° + i\sin 60°)$

$4(\cos 150° + i\sin 150°)$

$4(\cos 240° + i\sin 240°)$

$4(\cos 330° + i\sin 330°)$

(b) Since there are four evenly spaced roots on the circle of radius 4, they are fourth roots of a complex number of modulus 4^4. In this case, raising them to the fourth power yields $-128 - 128\sqrt{3}$.

(c) $[4(\cos 60° + i\sin 60°)]^4 = -128 - 128\sqrt{3}$

$[4(\cos 150° + i\sin 150°)]^4 = -128 - 128\sqrt{3}$

$[4(\cos 240° + i\sin 240°)]^4 = -128 - 128\sqrt{3}$

$[4(\cos 330° + i\sin 330°)]^4 = -128 - 128\sqrt{3}$

88. (a) The trigonometric forms of the three roots shown are:

$3(\cos 90° + i\sin 90°)$

$3(\cos 210° + i\sin 210°)$

$3(\cos 330° + i\sin 330°)$

(b) Since there are three evenly spaced roots on the circle of radius 3, they are third roots of a complex number of modulus 3^3. In this case, raising them to the fourth power yields $-27i$.

(c) $[3(\cos 90 + i\sin 90)]^3 = -27i$

$[3(\cos 210 + i\sin 210)]^3 = -27i$

$[3(\cos 330 + i\sin 330)]^3 = -27i$

90. Fourth roots of $256 = 256(\cos 0 + i\sin 0)$: $\sqrt[4]{256}\left(\cos\frac{2\pi k}{4} + i\sin\frac{2\pi k}{4}\right)$

$k = 0, 1, 2, 3$

$k = 0$: $4(\cos 0 + i\sin 0) = 4$

$k = 1$: $4(\cos\pi + i\sin\pi) = -4$

$k = 2$: $4\left(\cos\frac{\pi}{2} + i\sin\frac{\pi}{2}\right) = 4i$

$k = 3$: $4\left(\cos\frac{3\pi}{2} + i\sin\frac{3\pi}{2}\right) = -4i$

92. $x^5 - 32 = 0$

$x^5 = 32$

$32 = 32(\cos 0 + i\sin 0)$

$\sqrt[3]{32} = \sqrt[5]{32}\left[\cos\left(0 + \frac{2\pi k}{5}\right) + i\sin\left(0 + \frac{2\pi k}{5}\right)\right]$

$k = 0, 1, 2, 3, 4$

$k = 0$: $2(\cos 0 + i\sin 0) = 2$

$k = 1$: $2\left(\cos\frac{2\pi}{5} + i\sin\frac{2\pi}{5}\right) = 0.6180 + 1.9021i$

$k = 2$: $2\left(\cos\frac{4\pi}{5} + i\sin\frac{4\pi}{5}\right) = -1.6180 + 1.1756i$

$k = 3$: $2\left(\cos\frac{6\pi}{5} + i\sin\frac{6\pi}{5}\right) = -1.6180 - 1.1756i$

$k = 4$: $2\left(\cos\frac{8\pi}{5} + i\sin\frac{8\pi}{5}\right) = 0.6180 - 1.9021i$

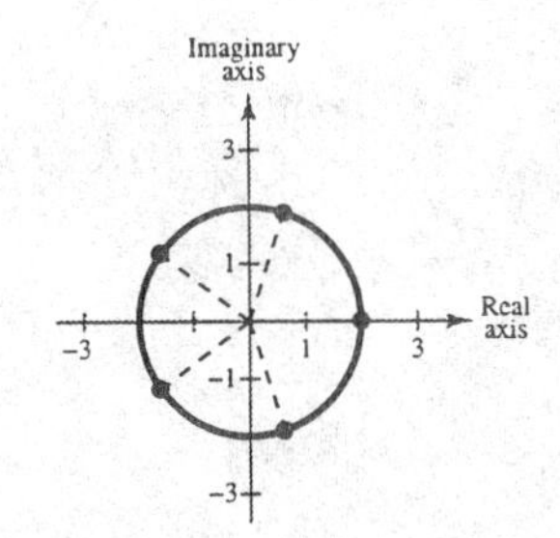

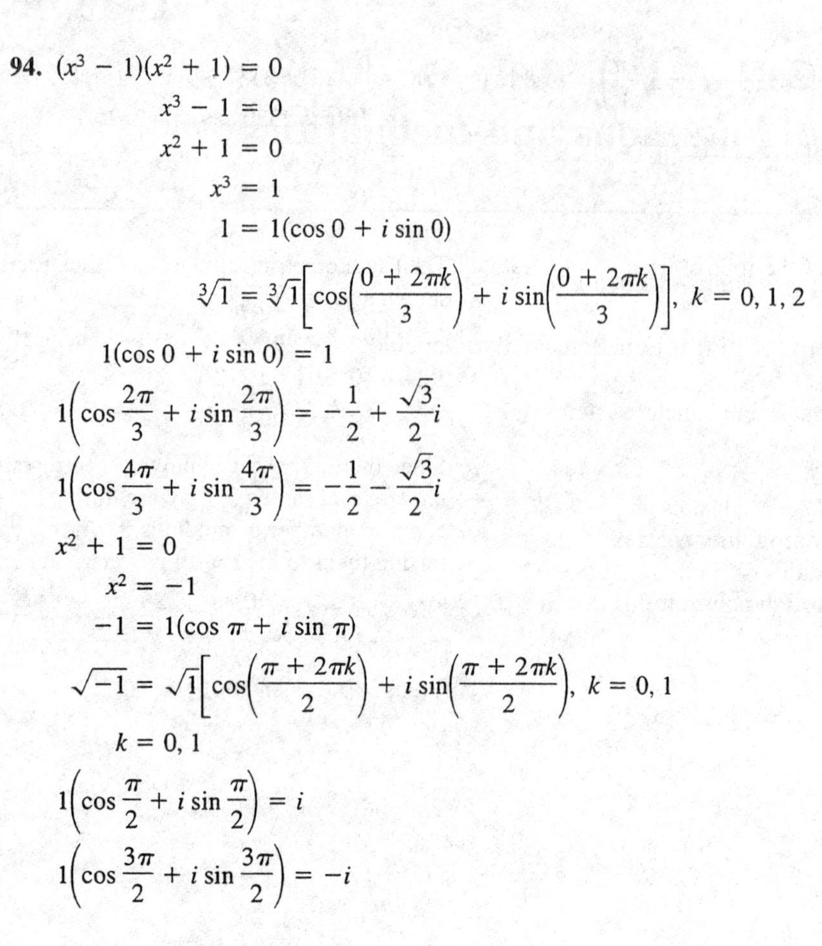

94. $(x^3 - 1)(x^2 + 1) = 0$

$$x^3 - 1 = 0$$

$$x^2 + 1 = 0$$

$$x^3 = 1$$

$$1 = 1(\cos 0 + i \sin 0)$$

$$\sqrt[3]{1} = \sqrt[3]{1}\left[\cos\left(\frac{0 + 2\pi k}{3}\right) + i \sin\left(\frac{0 + 2\pi k}{3}\right)\right], \quad k = 0, 1, 2$$

$$1(\cos 0 + i \sin 0) = 1$$

$$1\left(\cos\frac{2\pi}{3} + i \sin\frac{2\pi}{3}\right) = -\frac{1}{2} + \frac{\sqrt{3}}{2}i$$

$$1\left(\cos\frac{4\pi}{3} + i \sin\frac{4\pi}{3}\right) = -\frac{1}{2} - \frac{\sqrt{3}}{2}i$$

$$x^2 + 1 = 0$$

$$x^2 = -1$$

$$-1 = 1(\cos \pi + i \sin \pi)$$

$$\sqrt{-1} = \sqrt{1}\left[\cos\left(\frac{\pi + 2\pi k}{2}\right) + i \sin\left(\frac{\pi + 2\pi k}{2}\right)\right], \quad k = 0, 1$$

$$k = 0, 1$$

$$1\left(\cos\frac{\pi}{2} + i \sin\frac{\pi}{2}\right) = i$$

$$1\left(\cos\frac{3\pi}{2} + i \sin\frac{3\pi}{2}\right) = -i$$

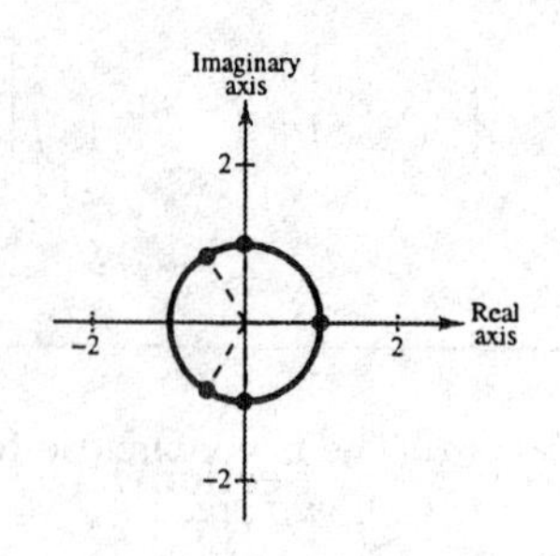

CHAPTER 7
Systems of Equations and Inequalities

CHAPTER 7
Systems of Equations and Inequalities

Section 7.1 Solving Systems of Equations

Solutions to Even-Numbered Exercises

2. $x - y = -4$ Equation 1

$x + 2y = 5$ Equation 2

Solve for x in Equation 1: $x = y - 4$

Substitute for x in Equation 2: $(y - 4) + 2y = 5$

Solve for y: $3y - 4 = 5 \Rightarrow y = 3$

Back-substitute $y = 3$: $x = 3 - 4 = -1$

Answer: $(-1, 3)$

4. $3x + y = 2$ Equation 1

$x^3 + y = 0$ Equation 2

Solve for y in Equation 1: $y = 2 - 3x$

Substitute for y in Equation 2: $x^3 + (2 - 3x) = 0$

Solve for x: $x^3 - 3x + 2 = 0 \Rightarrow (x - 1)^2(x + 2) = 0 \Rightarrow x = 1, -2$

Back-substitute $x = 1$: $y = 2 - 3(1) = -1$

Back-substitute $x = -2$: $y = 2 - 3(-2) = 8$

Answers: $(1, -1), (-2, 8)$

6. $x + y = 0$ Equation 1

$x^3 - 5x - y = 0$ Equation 2

Solve for y in Equation 1: $y = -x$

Substitute for y in Equation 2: $x^3 - 5x - (-x) = 0$

Solve for x: $x^3 - 4x = 0 \Rightarrow x(x^2 - 4) = 0 \Rightarrow x = 0, \pm 2$

Back-substitute $x = 0$: $y = -0 = 0$

Back-substitute $x = 2$: $y = -2$

Back-substitute $x = -2$: $y = -(-2) = 2$

Answers: $(0, 0), (2, -2), (-2, 2)$

8. $y = -2x^2 + 2$ Equation 1

$y = 2(x^4 - 2x^2 + 1)$ Equation 2

Substitute for y in Equation 1: $2(x^4 - 2x^2 + 1) = -2x^2 + 2$

Solve for x: $x^4 - 2x^2 + 1 + x^2 - 1 = 0$

$$x^4 - x^2 = 0$$

$$x^2(x^2 - 1) = 0 \Rightarrow x = 0, \pm 1$$

Back-substitute $x = 0$: $y = -2(0)^2 + 2 = 2$

Back-substitute $x = 1$: $y = -2(1)^2 + 2 = 0$

Back-substitute $x = -1$: $y = -2(-1)^2 + 2 = 0$

Answers: $(0, 2), (1, 0), (-1, 0)$

10. $y = x^3 - 3x^2 + 4$ Equation 1

$y = -2x + 4$ Equation 2

Substitute for y in Equation 1: $-2x + 4 = x^3 - 3x^2 + 4$

Solve for x: $0 = x^3 - 3x^2 + 2x$

$0 = x(x^2 - 3x + 2)$

$0 = x(x - 2)(x - 1) \Rightarrow x = 0, 1, 2$

Back-substitute $x = 0$: $y = -2(0) + 4 = 4$

Back-substitute $x = 1$: $y = -2(1) + 4 = 2$

Back-substitute $x = 2$: $y = -2(2) + 4 = 0$

Answers: $(0, 4), (1, 2), (2, 0)$

12. $x + 2y = 1$ Equation 1

$5x - 4y = -23$ Equation 2

Solve for x in Equation 1: $x = 1 - 2y$

Substitute for x in Equation 2: $5(1 - 2y) - 4y = -23$

Solve for y: $-14y = -28 \Rightarrow y = 2$

Back-substitute $y = 2$: $x = 1 - 2y = 1 - 4 = -3$

Answer: $(-3, 2)$

14. $6x - 3y - 4 = 0$ Equation 1

$x + 2y - 4 = 0$ Equation 2

Solve for x in Equation 2: $x = 4 - 2y$

Substitute for x in Equation 1: $6(4 - 2y) - 3y - 4 = 0$

Solve for y: $24 - 12y - 3y - 4 = 0 \Rightarrow -15y = -20 \Rightarrow y = \frac{4}{3}$

Back-substitute $y = \frac{4}{3}$: $x = 4 - 2y = 4 - 2\left(\frac{4}{3}\right) = \frac{4}{3}$

Answer: $\left(\frac{4}{3}, \frac{4}{3}\right)$

16. $30x - 40y - 33 = 0$ Equation 1

$10x + 20y - 21 = 0$ Equation 2

Solve for x in Equation 2: $x = \frac{21}{10} - 2y$

Substitute for x in Equation 1: $30\left(\frac{21}{10} - 2y\right) - 40y - 33 = 0$

$63 - 60y - 40y - 33 = 0$

$30 = 100y \Rightarrow y = \frac{3}{10}$

Back-substitute: $x = \frac{21}{10} - 2\left(\frac{3}{10}\right) = \frac{15}{10} = \frac{3}{2}$

Answer: $\left(\frac{3}{2}, \frac{3}{10}\right)$

18. $\frac{1}{2}x + \frac{3}{4}y = 10$ Equation 1

$\frac{3}{4}x - y = 4$ Equation 2

Solve for y in Equation 2: $y = \frac{3}{4}x - 4$

Substitute for y in Equation 1: $\frac{1}{2}x + \frac{3}{4}\left(\frac{3}{4}x - 4\right) = 10$

Solve for x: $\frac{1}{2}x + \frac{9}{16}x - 3 = 10 \Rightarrow \frac{17}{16}x = 13 \Rightarrow x = \frac{208}{17}$

Back-substitute $x = \frac{208}{17}$: $y = \frac{3}{4}\left(\frac{208}{17}\right) - 4 = \frac{88}{17}$

Answer: $\left(\frac{208}{17}, \frac{88}{17}\right)$

20. $-\frac{2}{3}x + y = -2$ Equation 1

$2x - 3y = 6$ Equation 2

Solve for y in Equation 1: $y = \frac{2}{3}x - 2$

Substitute for y in Equation 2: $2x - 3\left(\frac{2}{3}x - 2\right) = 6$

Solve for x: $2x - 2x + 6 = 6 \Rightarrow 0 = 0$

Infinite number of solutions, all of form $\left(x, \frac{2}{3}x - 2\right)$. For instance, $(0, -2)$, $(3, 0)$.

22. $x - 2y = 0$ Equation 1

$3x - y = 0$ Equation 2

Solve for x in Equation 1: $x = 2y$

Substitute for x in Equation 2: $3(2y) - y = 0$

Substitute for y: $6y - y = 0 \Rightarrow 5y = 0 \Rightarrow y = 0$

Back-substitute $y = 0$: $x = 2y = 0$

Answer: $(0, 0)$

24. $x + y = 0$

$3x - 2y = 10$

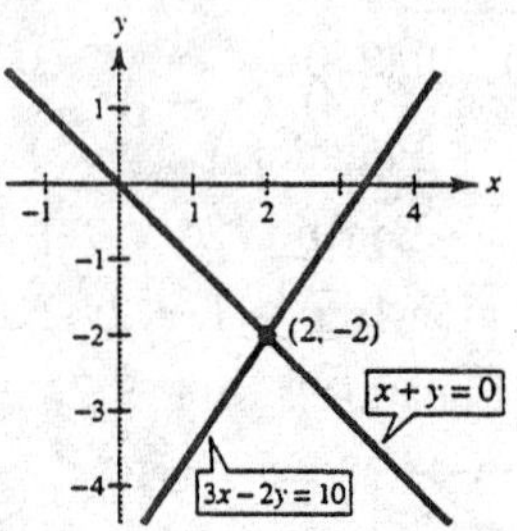

Point of intersection: $(2, -2)$

26. $-x + 2y = 1$

$x - y = 2$

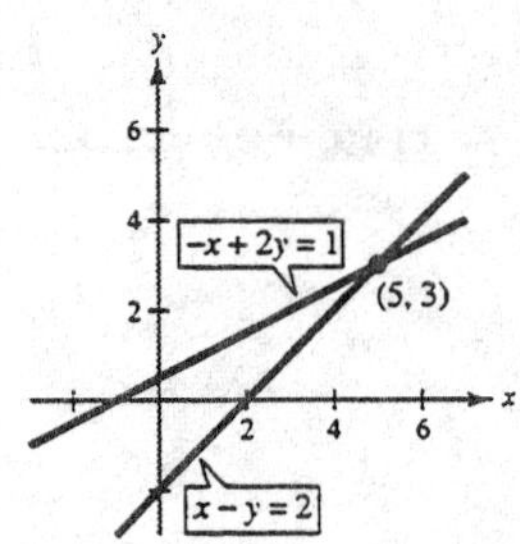

Point of intersection: $(5, 3)$

28. $x - y + 3 = 0$

$x^2 - 4x + 7 = y$

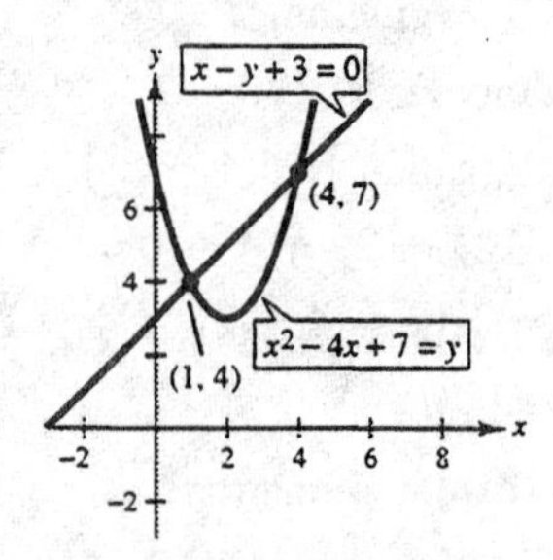

Points of intersection: $(1, 4)$, $(4, 7)$

30. $x - y = 0 \Rightarrow y_1 = x$

$5x - 2y = 6 \Rightarrow y_2 = \frac{5}{2}x - 3$

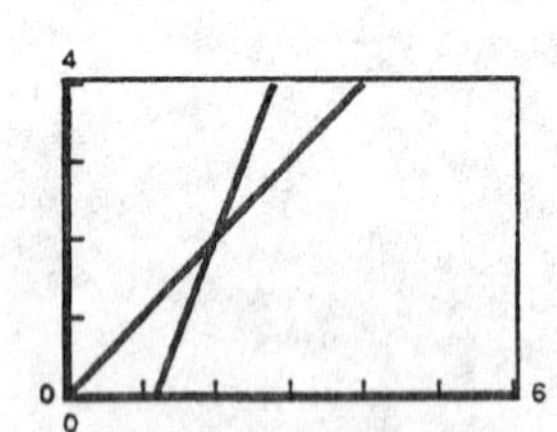

Point of intersection: $(2, 2)$

32. $3x - 2y = 0$

$x^2 + y^2 = 4$

$y_1 = \frac{3}{2}x$

$y_2 = \sqrt{4 - x^2}$

$y_3 = -\sqrt{4 - x^2}$

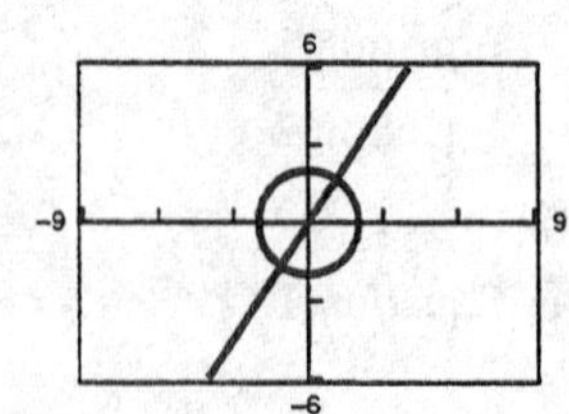

Point of intersection: $\left(\pm\frac{4}{\sqrt{13}}, \pm\frac{6}{\sqrt{13}}\right)$

34. $x^2 + y^2 = 25 \Rightarrow y_1 = \sqrt{25 - x^2}$

$y_2 = -\sqrt{25 - x^2}$

$(x - 8)^2 + y^2 = 41 \Rightarrow y_3 = \sqrt{41 - (x - 8)^2}$

$y_4 = -\sqrt{41 - (x - 8)^2}$

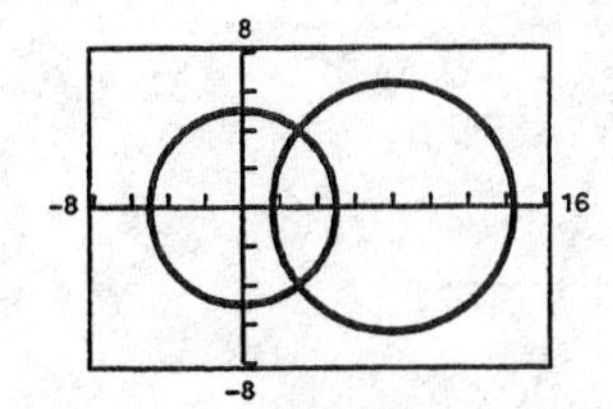

Points of intersection: (3, 4), (3, −4)

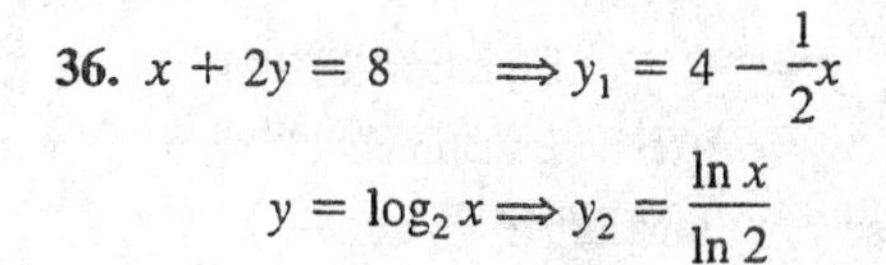

36. $x + 2y = 8 \Rightarrow y_1 = 4 - \frac{1}{2}x$

$y = \log_2 x \Rightarrow y_2 = \frac{\ln x}{\ln 2}$

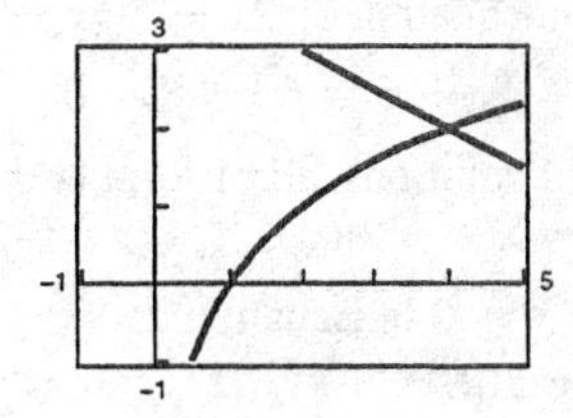

Point of intersection: (4, 2)

38. $x - y = 3 \Rightarrow y_1 = x - 3$

$x - y^2 = 1 \Rightarrow y_2 = \sqrt{x - 1}$

$y_3 = -\sqrt{x - 1}$

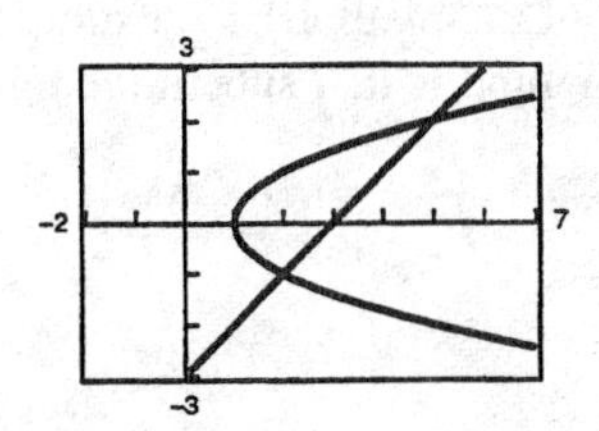

Points of intersection: (5, 2), (2, −1)

40. $x^2 + y^2 = 4 \Rightarrow y_1 = \sqrt{4 - x^2}$

$y_2 = -\sqrt{4 - x^2}$

$2x^2 - y = 2 \Rightarrow y_3 = 2x^2 - 2$

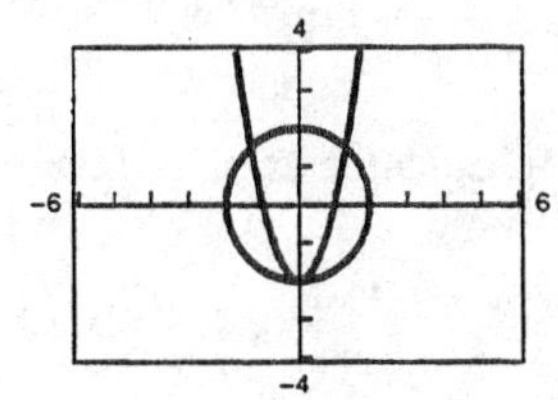

Points of intersection: $(0, -2), \left(\frac{1}{2}\sqrt{7}, \frac{3}{2}\right), \left(-\frac{1}{2}\sqrt{7}, \frac{3}{2}\right)$
or $(0, -2), (\pm 1.323, 1.500)$

42. $x + y = 4$ Equation 1

$x^2 + y = 2$ Equation 2

Solve for y in Equation 1: $y = 4 - x$

Substitute for y in Equation 2: $x^2 + (4 - x) = 2$

Solve for x: $x^2 - x + 2 = 0$

No real solutions because the discriminant in the Quadratic Formula is negative.

Inconsistent. No solution

44. $x^2 + y^2 = 25$ Equation 1

$2x + y = 10$ Equation 2

Solve for y in Equation 2: $y = 10 - 2x$

Substitute for y in Equation 1: $x^2 + (10 - 2x)^2 = 25$

Solve for x: $x^2 + 100 - 40x + 4x^2 = 25 \Rightarrow x^2 - 8x + 15 = 0$

$\Rightarrow (x - 5)(x - 3) = 0 \Rightarrow x = 3, 5$

Back-substitute $x = 3$: $y = 10 - 2(3) = 4$

Back-substitute $x = 5$: $y = 10 - 2(5) = 0$

Answers: (3, 4), (5, 0)

46. $y = (x + 1)^3$

$y = \sqrt{x - 1}$

No points of intersection

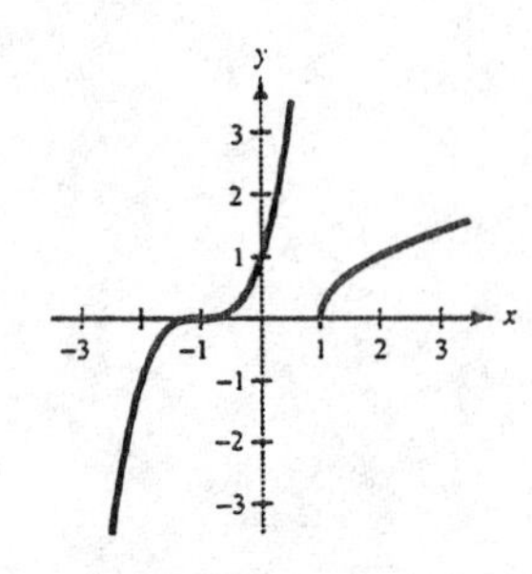

48. $y = x^3 - 2x^2 + x - 1$ Equation 1
$y = -x^2 + 3x - 1$ Equation 2

Substitute for y in Equation 1:
$-x^2 + 3x - 1 = x^3 - 2x^2 + x - 1$

Solve for x: $0 = x^3 - x^2 - 2x$
$0 = x(x^2 - x - 2)$
$0 = x(x - 2)(x + 1) \Rightarrow x = 0, 2, -1$

Back-substitute $x = 0$ in Equation 2:
$y = -0^2 + 3(0) - 1 = -1$

Back-substitute $x = 2$ in Equation 2:
$y = -2^2 + 3(2) - 1 = 1$

Back-substitute $x = -1$ in Equation 2:
$y = -(-1)^2 + 3(-1) - 1 = -5$

Answers: $(0, -1), (2, 1), (-1, -5)$

50. $x^2 + y = 4 \Rightarrow y = 4 - x^2$
$e^x - y = 0 \Rightarrow y = e^x$

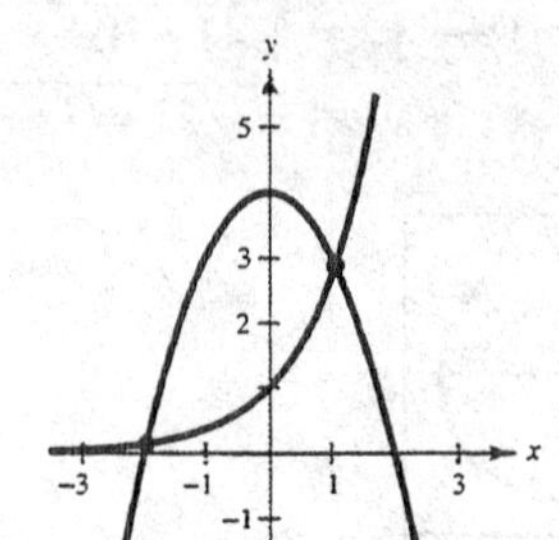

Points of intersection:
Approximately $(-1.96, 0.14), (1.06, 2.88)$

52. $x - 2y = 1$ Equation 1
$y = \sqrt{x - 1}$ Equation 2

Substitute for y in Equation 1: $x - 2\sqrt{x - 1} = 1$

Solve for x:
$x - 1 = 2\sqrt{x - 1}$
$(x - 1)^2 = 4(x - 1)$
$x^2 - 2x + 1 = 4x - 4$
$x^2 - 6x + 5 = 0$
$(x - 1)(x - 5) = 0 \Rightarrow x = 1, 5$

Back-substitute $x = 1$: $y = \sqrt{1 - 1} = 0$
Back-substitute $x = 5$: $y = \sqrt{5 - 1} = 2$

Answers: $(1, 0), (5, 2)$

54. The advantage of the method of substitution of the graphical method is that substitution gives

56. $C = 5.5\sqrt{x} + 10{,}000,\ R = 3.29x$

$$R = C$$
$$3.29x = 5.5\sqrt{x} + 10{,}000$$
$$3.29x - 10{,}000 = 5.5\sqrt{x}$$
$$10.8241x^2 - 65{,}800x + 100{,}000{,}000 = 30.25x$$
$$10.8241x^2 - 65{,}830.25x + 100{,}000{,}000 = 0$$
$$x \approx 3133 \text{ units}$$

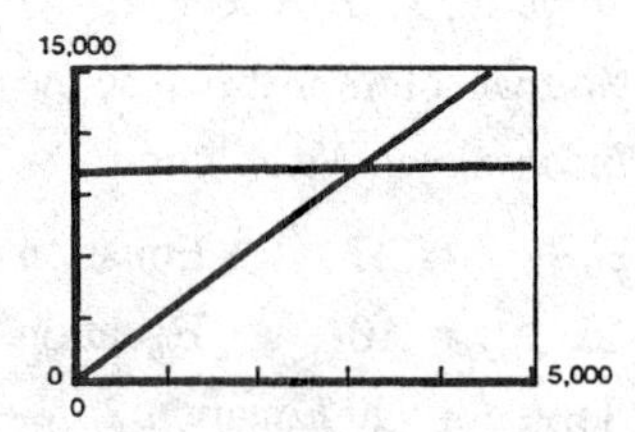

In order for the revenue to break even with the cost, 3133 units must be sold.

58. $C = 0.08x + 50{,}000,\ R = 0.25x$

$R = C$
$0.25x = 0.08x + 50{,}000$
$0.17x = 50{,}000$
$x \approx 294{,}117.6$

In order for the revenue to break even with the cost, 294,118 units must be sold.

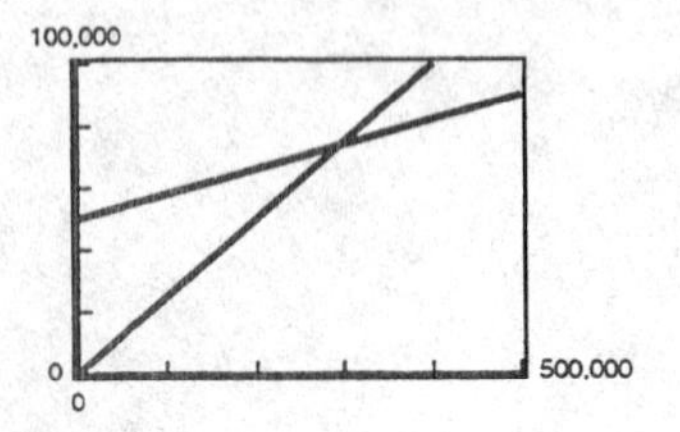

60. $C = 21.60x + 5000,\ R = 34.10x$

$R = C$
$34.10x = 21.60x + 5000$
$12.5x = 5000$
$x = 400$ units

62. (a)

$$\begin{aligned} x + \quad y &= 25{,}000 \quad \text{Equation 1} \\ 0.06x + 0.085y &= 2000 \quad \text{Equation 2} \end{aligned}$$

(b)

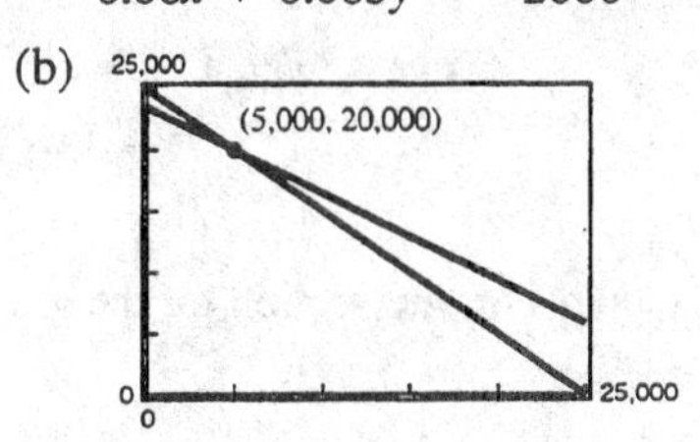

$y_1 = 25{,}000 - x$

$y_2 = \dfrac{1}{0.085}(2000 - 0.06x)$

As x increases, y decreases and the amount of interest decreases.

(c) $\$5000 = x$ yields \$2000 total interest.

64. $20{,}000 + 0.01x = 15{,}000 + 0.02x$

$5000 = 0.01x$

$500{,}000 = x$

For the second offer to be better, you would have to sell more than \$500,000 per year.

66. $V = (D - 4)^2, \; 5 \le D \le 40$

$V = 0.79D^2 - 2D - 4, \; 5 \le D \le 40$

(a)

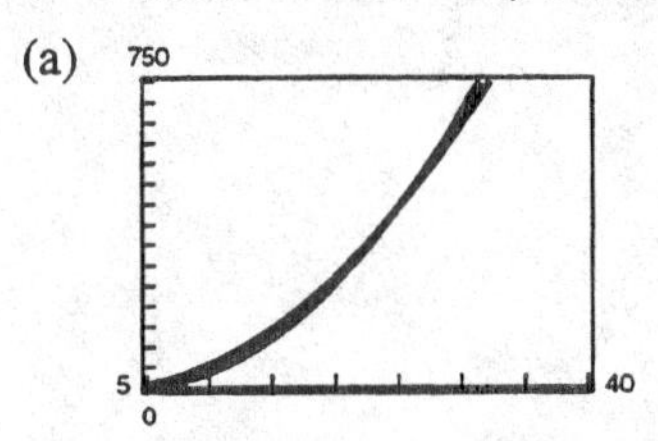

(b) The two graphs intersect at $D = 24.72$.
Algebraically:

$$\begin{aligned} (D - 4)^2 &= 0.79D^2 - 2D - 4 \\ D^2 - 8D + 16 &= 0.79D^2 - 2D - 4 \\ 0.21D^2 - 6D + 20 &= 0 \\ D &\approx 24.72, \; 3.9 \end{aligned}$$

Since $5 \le D \le 40$, the scales agree when $D \approx 24.72$ inches.

(c) V is larger using the Scribner Log Rule when $5 \le D \le 24.7$. V is larger using the Doyle Log Rule when $24.7 \le D \le 40$. Therefore, for large diameters, you would use the Doyle Log Rule.

68. $2l + 2w = 280 \Rightarrow l + w = 140$

$w = l - 20 \Rightarrow l + (l - 20) = 140$

$2l = 160$

$l = 80$

$w = l - 20 = 80 - 20 = 60$

Dimensions: 60×80 centimeters

70. $2l + 2w = 210 \Rightarrow \quad l + 2 = 105$

$l = \frac{3}{2}w \Rightarrow \frac{3}{2}w + w = 105$

$\frac{5}{2}w = 105$

$w = 42$

$l = \frac{3}{2}(42) = 63$

Dimensions: 42×63 feet

72. $A = \frac{1}{2}bh$

$1 = \frac{1}{2}a^2$

$a^2 = 2$

$a = \sqrt{2}$

The dimensions are $\sqrt{2} \times \sqrt{2}$.

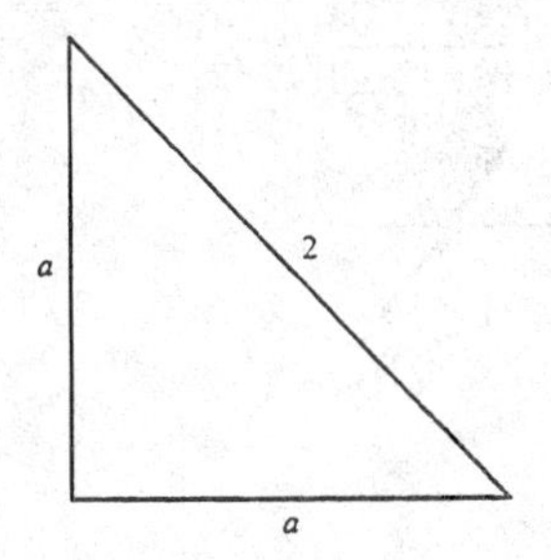

74. The line joining (0, 10) and $(\sqrt{61}, 0)$ intersects the hyperbola at the required point.

$\text{Slope} = \dfrac{-10}{\sqrt{61}}$

$\text{Line: } y = \dfrac{-10}{\sqrt{61}}x + 10$

$10x + \sqrt{61}y = 10\sqrt{61}$

$\dfrac{x^2}{25} - \dfrac{y^2}{36} = 1$

Using a graphing utility, you obtain

$(x, y) = (5.55, 2.89)$

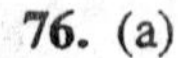

76. (a)

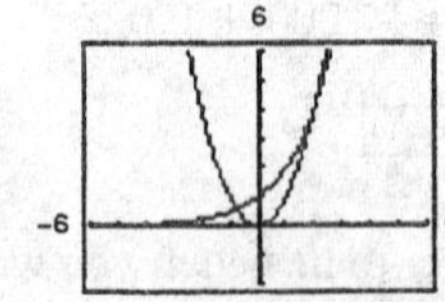

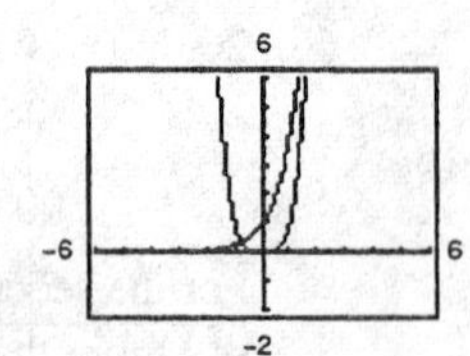

(b) Based on the graphs in part (a) it appears that for $b > 1$, there are three points of intersection for the graphs of $y = b^x$ and $y = x^b$ when b is an even number.

78. $(3.5, 4), (10, 6)$

$$m = \frac{6-4}{10-3.5} = \frac{2}{6.5}$$

$$y - 6 = \frac{2}{6.5}(x - 10)$$

$$6.5y - 39 = 2x - 20$$

$$2x - 6.5y + 19 = 0$$

80. $(4, -2), (4, 5)$

$$x = 4$$

82. $\left(-\frac{7}{3}, 8\right), \left(\frac{5}{2}, \frac{1}{2}\right)$

$$m = \frac{8 - (1/2)}{-(7/3) - (5/2)} = \frac{15/2}{-29/6} = -\frac{45}{29}$$

$$y - \frac{1}{2} = -\frac{45}{29}\left(x - \frac{5}{2}\right)$$

$$29y - \frac{29}{2} = -45x + \frac{225}{2}$$

$$45x + 29y - 127 = 0$$

Section 7.2 Systems of Linear Equations in Two Variables

Solutions to Even-Numbered Exercises

2. $x + 3y = 1$ Equation 1

$-x + 2y = 4$ Equation 2

Add to eliminate x: $5y = 5 \Rightarrow y = 1$

Substitute $y = 1$ in Equation 1: $x + 3(1) = 1 \Rightarrow x = -2$

Answer: $(-2, 1)$

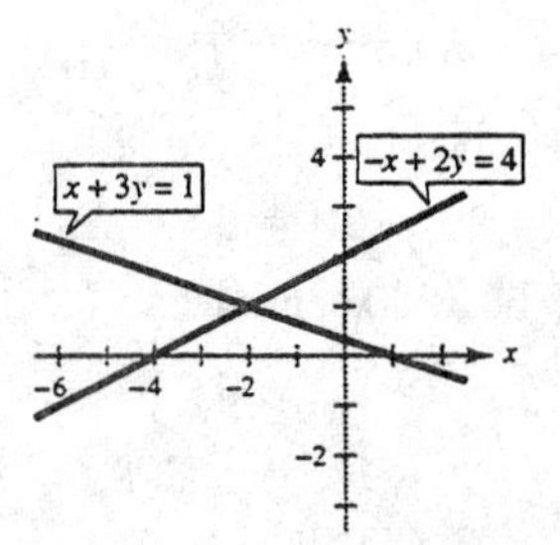

4. $2x - y = 3$ Equation 1

$4x + 3y = 21$ Equation 2

Multiply Equation 1 by 3: $6x - 3y = 9$

Add this to Equation 2 to eliminate y: $10x = 30 \Rightarrow x = 3$

Substitute $x = 3$ in Equation 1: $2(3) - y = 3 \Rightarrow y = 3$

Answer: $(3, 3)$

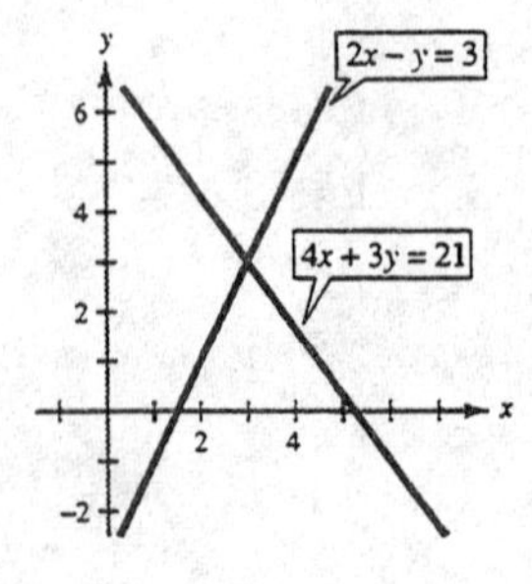

6. $3x + 2y = 3$ Equation 1
$6x + 4y = 14$ Equation 2
Multiply Equation 1 by -2: $-6x - 4y = -6$
Add this to Equation 2: $0 = 8$
There are no solutions.

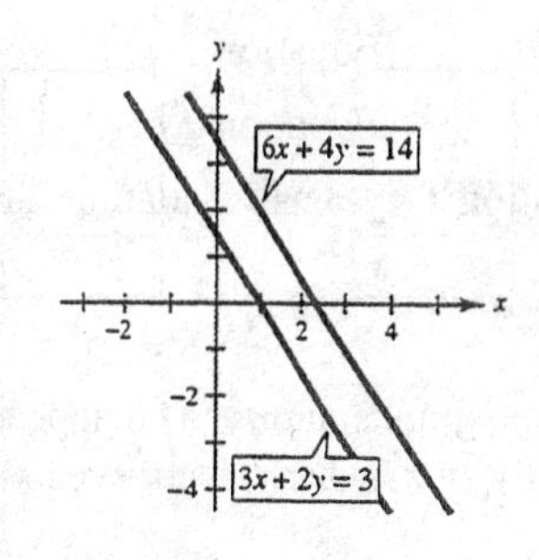

8. $x - 2y = 4$ Equation 1
$6x + 2y = 10$ Equation 2
Add to eliminate y: $7x = 14 \Longrightarrow x = 2$
Substitute $x = 2$ in Equation 1: $2 - 2y = 4 \Longrightarrow y = -1$
Answer: $(2, -1)$

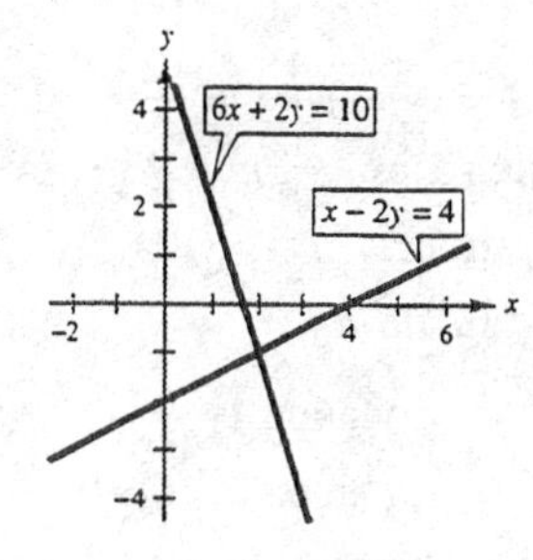

10. $5x + 3y = -18$ Equation 1
$2x - 6y = 1$ Equation 2
Multiply Equation 1 by 2: $10x + 6y = -36$
Add this to Equation 2 to eliminate y: $12x = -35 \Longrightarrow x = -\frac{35}{12}$
Substitute $x = -\frac{35}{12}$ in Equation 2: $2\left(-\frac{35}{12}\right) - 6y = 1 \Longrightarrow y = -\frac{41}{36}$
Answer: $\left(-\frac{35}{12}, -\frac{41}{36}\right)$

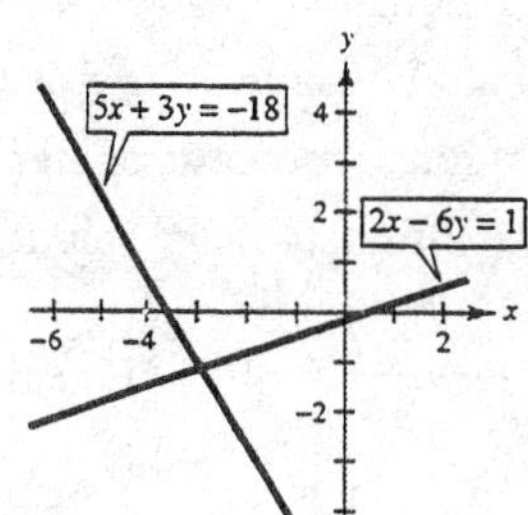

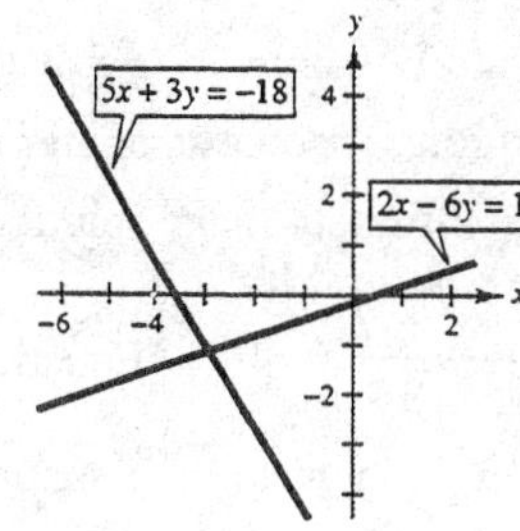

12. $3x - 5y = 2$ Equation 1
$2x + 5y = 13$ Equation 2
Add to eliminate y: $5x = 15$
$x = 3$
Substitute $x = 3$ in Equation 1: $3(3) - 5y = 2 \Longrightarrow y = \frac{7}{5}$
Answer: $\left(3, \frac{7}{5}\right)$

14. $x + 7y = 12$ Equation 1
$3x - 5y = 10$ Equation 2
Multiply Equation 1 by -3: $-3x - 21y = -36$
Add this to Equation 2 to eliminate x: $-26y = -26 \Longrightarrow y = 1$
Substitute $y = 1$ in Equation 1: $x + 7 = 12 \Longrightarrow x = 5$
Answer: $(5, 1)$

16. $8r + 16s = 20$ Equation 1
$16r + 50s = 55$ Equation 2
Multiply Equation 1 by (-2): $-16r - 32s = -40$
Add this to Equation 2 to eliminate r: $18s = 15 \Longrightarrow s = \frac{5}{6}$
Substitute $s = \frac{5}{6}$ in Equation 1: $8r + 16\left(\frac{5}{6}\right) = 20 \Longrightarrow r = \frac{5}{6}$
Answer: $\left(\frac{5}{6}, \frac{5}{6}\right)$

18. $5u + 6v = 24$ Equation 1

$3u + 5v = 18$ Equation 2

Multiply Equation 1 by 3 and Equation 2 by (-5): $15u + 18v = 72$

$-15u - 25v = -90$

Add to eliminate u: $-7v = -18 \Rightarrow v = \frac{18}{7}$

Substitute $v = \frac{18}{7}$ in Equation 2: $3u + 5\left(\frac{18}{7}\right) = 18 \Rightarrow u = \frac{12}{7}$

Answer: $\left(\frac{12}{7}, \frac{18}{7}\right)$

20. $1.8x + 1.2y = 4$ Equation 1

$9x + 6y = 3$ Equation 2

Multiply Equation 1 by (-5): $-9x - 6y = -20$

Add this to Equation 2: $0 = -17$

Inconsistent; no solution

22. $\frac{2}{3}x + \frac{1}{6}y = \frac{2}{3}$ Equation 1

$4x + y = 4$ Equation 2

Multiply Equation 1 by (-6): $-4x - 6y = -4$

Add this to Equation 2: $0 = 0$ (dependent)

The solution set consists of all points lying on the line $4x + y = 4$. Let $x = a$, then $y = 4 - 4a$.

Answer: Infinitely many solutions of the form $(a, 4 - 4a)$, where a is any real number.

24. $\frac{x-1}{2} + \frac{y+2}{3} = 4$ Equation 1

$x - 2y = 5$ Equation 2

Multiply Equation 1 by 6: $3(x - 1) + 2(y + 2) = 24 \Rightarrow 3x + 2y = 23$

Add this to Equation 2 to eliminate y: $4x = 28 \Rightarrow x = 7$

Substitute $x = 7$ in Equation 2: $7 - 2y = 5 \Rightarrow y = 1$

Answer: $(7, 1)$

26. $0.02x - 0.05y = -0.19$ Equation 1

$0.03x + 0.04y = 0.52$ Equation 2

Multiply Equation 1 by 4 and Equation 2 by 5:

$0.08x - 0.2y = -0.76$

$0.15x + 0.2y = 2.6$

Add these to eliminate y: $0.23x = 1.84 \Rightarrow x = 8$

Substitute $x = 8$ in Equation 1: $0.02(8) - 0.05y = -0.19 \Rightarrow y = 7$

Answer: $(8, 7)$

28. $0.2x - 0.5y = -27.8$ Equation 1

$0.3x + 0.4y = 68.7$ Equation 2

Multiply Equation 1 by 4 and Equation 2 by 5:

$0.8x - 2y = -111.2$

$1.5x + 2y = 343.5$

Add these to eliminate y: $2.3x = 232.3 \Rightarrow x = 101$

Substitute $x = 101$ in Equation 1: $0.2(101) - 0.5y = -27.8 \Rightarrow y = 96$

Answer: $(101, 96)$

30. $3b + 3m = 7$ Equation 1

$3b + 5m = 3$ Equation 2

Subtract Equation 2 from Equation 1 to eliminate b: $-2m = 4 \Rightarrow m = -2$

Substitute $m = -2$ in Equation 1: $3b + 3(-2) = 7 \Rightarrow b = \frac{13}{3}$

Answer: $\left(\frac{13}{3}, -2\right)$

32. $2x + y = 5$

$x - 2y = -1$

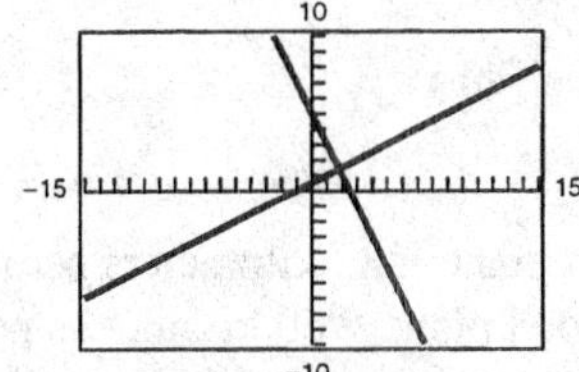
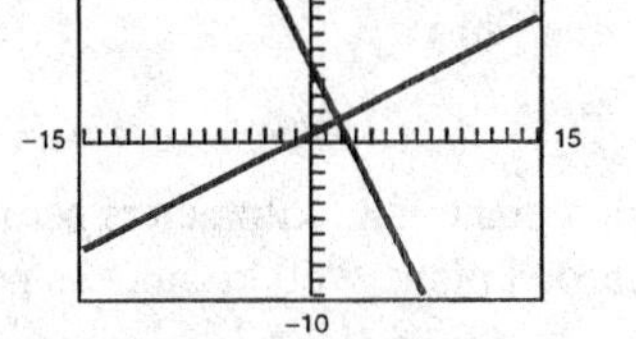

The system is consistent. There is one solution.

34. $4x - 6y = 7$

$2x - 3y = 3.5$

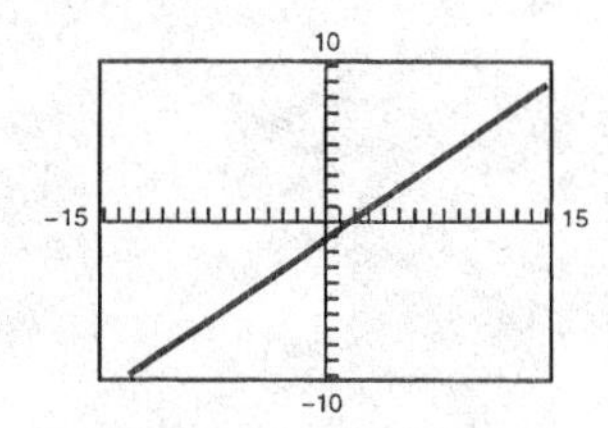

The system is consistent. There are infinitely many solutions.

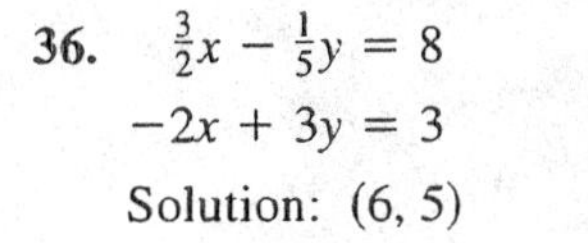

36. $\frac{3}{2}x - \frac{1}{5}y = 8$

$-2x + 3y = 3$

Solution: $(6, 5)$

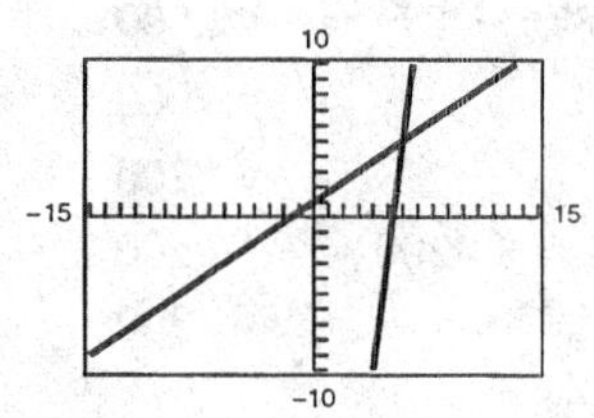

38. $0.5x + 2.2y = 9$

$6x + 0.4y = -22$

Solution: $(-4, 5)$

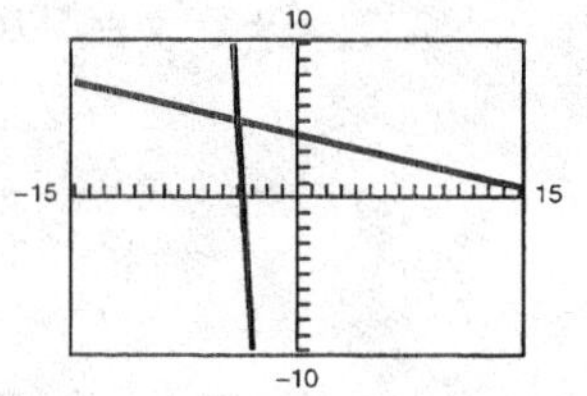

40. $-x + 3y = 17$ Equation 1

$4x + 3y = 7$ Equation 2

Subtract Equation 2 from Equation 1 to eliminate y: $-5x = 10 \Rightarrow x = -2$

Substitute $x = -2$ in Equation 1: $-(-2) + 3y = 17 \Rightarrow y = 5$

Solution: $(-2, 5)$

42. $7x + 3y = 16$ Equation 1

$y = x + 2$ Equation 2

Substitute for y in Equation 1:

$$7x + 3(x + 2) = 16$$
$$7x + 3x + 6 = 16$$
$$10x = 10 \Rightarrow x = 1$$

Substitute $x = 1$ in Equation 2: $y = 1 + 2 = 3$

Solution: $(1, 3)$

44. There are infinitely many systems that have the solution $(8, -2)$. One possible system is:

$$8 - 2 = 6 \Rightarrow x + y = 6$$
$$2(8) - (-2) = 18 \Rightarrow 2x - y = 18$$

46. $21x - 20y = 0$ Equation 1

$13x - 12y = 120$ Equation 2

Multiply Equation 2 by $\left(-\frac{5}{3}\right)$: $-\frac{65}{3}x + 20y = -200$

Add this to Equation 1 to eliminate y: $-\frac{2}{3}x = -200 \Rightarrow x = 300$

Substitute $x = 300$ in Equation 1: $21(300) - 20y = 0 \Rightarrow y = 315$

Solution: $(300, 315)$

The lines are not parallel. The scale on the axes must be changed to see the point of intersection.

48. (a) $x + y = 10$ Equation 1

$x + y = 20$ Equation 2

Subtract Equation 2 from Equation 1: $0 = -10$

System is inconsistent $\Rightarrow$ no solution

(b) $x + y = 3$ Equation 1

$2x + 2y = 6$ Equation 2

Multiply Equation 1 by (-2): $-2x - 2y = -6$

Add this to Equation 2: $0 = 0$ (dependent)

The system has an infinite number of solutions.

50. $15x + 3y = 6$ Equation 1

$-10x + ky = 9$ Equation 2

Multiply Equation 1 by $\frac{2}{3}$: $10x + 2y = 4$

Add this to Equation 2: $ky + 2y = 13$

This system is inconsistent if $ky + 2y = 0$. This occurs when $k = -2$.

52. Let x = the speed of the plane that leaves first and y = the speed of the plane that leaves second.

$y - x = 80$ Equation 1

$2x + \frac{3}{2}y = 3200$ Equation 2

$$\begin{aligned} -2x + 2y &= 160 \\ 2x + \tfrac{3}{2}y &= 3200 \\ \hline \tfrac{7}{2}y &= 3360 \\ y &= 960 \\ 960 - x &= 80 \\ x &= 880 \end{aligned}$$

Answer: First plane: 880 kilometers per hour; Second plane: 960 kilometers per hour

54. Let x = the number of 87 octane gallons; y = the number of 92 octane gallons.

$87x + 92y = 89(500)$ Equation 1

$x + y = 500$ Equation 2

$$\begin{aligned} 87x + 92y &= 44{,}500 \\ -87x - 87y &= -43{,}500 \\ \hline 5y &= 1000 \\ y &= 200 \\ x + 200 &= 500 \\ x &= 300 \end{aligned}$$

Answer: 87 octane: 300 gallons; 92 octane: 200 gallons

56. Let x = the amount invested at 5.75%; y = the amount invested at 6.25%.

$x + y = 32{,}000$ Equation 1

$0.0575x + 0.0625y = 1900$ Equation 2

$$\begin{aligned} -5.75x - 5.75y &= -184{,}000 \\ 5.75x + 6.25y &= 190{,}000 \\ \hline 0.5y &= 6000 \\ y &= 12{,}000 \\ x + 12{,}000 &= 32{,}000 \\ x &= 20{,}000 \end{aligned}$$

The most that can be invested at 5.75% is $20,000.

58. Let x = the number of pairs of $66.95 shoes; y = the number of pairs of $84.95 shoes.

$x + y = 240$ Equation 1

$66.95x + 84.95y = 17{,}652$ Equation 2

$$\begin{aligned} -66.95x - 66.95y &= -16{,}068 \\ 66.95x + 84.95y &= 17{,}652 \\ \hline 18y &= 1584 \\ y &= 88 \\ x + 88 &= 240 \\ x &= 152 \end{aligned}$$

Answer: x = 152 shoes priced at $66.95; y = 88 shoes priced at $84.95

60. Let x = the amount hauled by one company; y = the amount hauled by a second company.

$x + y = 1600$ Equation 1

$y = 4y$ Equation 2

$$\begin{aligned} x + y &= 1600 \\ -x + 4y &= 0 \\ \hline 5y &= 1600 \\ y &= 320 \\ x + 320 &= 1600 \\ x &= 1280 \end{aligned}$$

Answer: One company hauled 320 tons and the other hauled 1280 tons.

62.
$$\begin{aligned} 5b + 10a = 11.7 &\Rightarrow -10b - 20a = -23.4 \\ 10b + 30a = 25.6 &\Rightarrow \quad 10b + 30a = 25.6 \\ &\qquad\qquad 10a = 2.2 \\ &\qquad\qquad a = 0.22 \\ &\quad 5b + 10(0.22) = 11.7 \\ &\qquad\qquad b = 1.9 \end{aligned}$$

Least squares regression line: $y = 0.22x + 1.9$

64.
$$\begin{aligned} 6b + 15a = 23.6 &\Rightarrow -15b - 37.5a = -59 \\ 15b + 55a = 48.8 &\Rightarrow \quad 15b + 55a = 48.8 \\ &\qquad\qquad 17.5a = -10.2 \\ &\qquad\qquad a \approx -0.583 \\ &\qquad\qquad b \approx 5.390 \end{aligned}$$

Least squares regression line: $y = -0.583x + 5.390$

66. $(-3, 0), (-1, 1), (1, 1), (3, 2)$

$4b = 4 \Rightarrow b = 1$

$20a = 6 \Rightarrow a = \frac{3}{10}$

Least squares regression line: $y = \frac{3}{10}x + 1$

68. $(1, 0), (2, 0), (3, 0), (3, 1), (4, 1), (4, 2), (5, 2), (6, 2)$

$$\begin{aligned} 8b + 28a = 8 &\Rightarrow -28b - 98a = -28 \\ 28b + 116a = 37 &\Rightarrow \quad 28b + 116a = 37 \\ &\qquad\qquad 18a = 9 \\ &\qquad\qquad a = \tfrac{1}{2} \\ &\qquad 8b + 14 = 8 \\ &\qquad\qquad b = -\tfrac{3}{4} \end{aligned}$$

Least squares regression line: $y = \frac{1}{2}x - \frac{3}{4}$

70. $(1.00, 450), (1.25, 375), (1.50, 330)$

$n = 3, \quad \sum_{i=1}^{3} x_i = 3.75, \quad \sum_{i=1}^{3} y_i = 1155, \quad \sum_{i=1}^{3} x_i^2 = 4.8125, \quad \sum_{i=1}^{3} x_i y_i = 1413.75$

$$\begin{aligned} 3b + 3.75a = 1155 &\Rightarrow \quad 11.25b + 14.625a = 4331.25 \\ 3.75b + 4.8125a = 1413.75 &\Rightarrow -11.25b - 14.4375a = -4241.25 \\ &\qquad\qquad -0.375a = 90 \\ &\qquad\qquad a = -240 \\ &\qquad\qquad b = 685 \end{aligned}$$

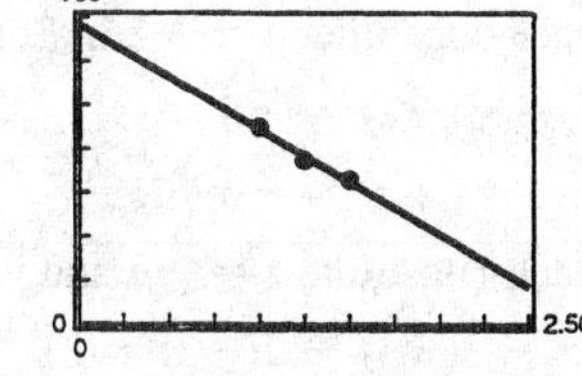

$y = -240x + 685$

When $x = \$1.40$, $y = -240(1.40) + 685 = 349$ units.

72. Supply = Demand

$25 + 0.1x = 100 - 0.05x$

$0.15x = 75$

$x = 500$

$p = 75$

Equilibrium point: $(500, 75)$

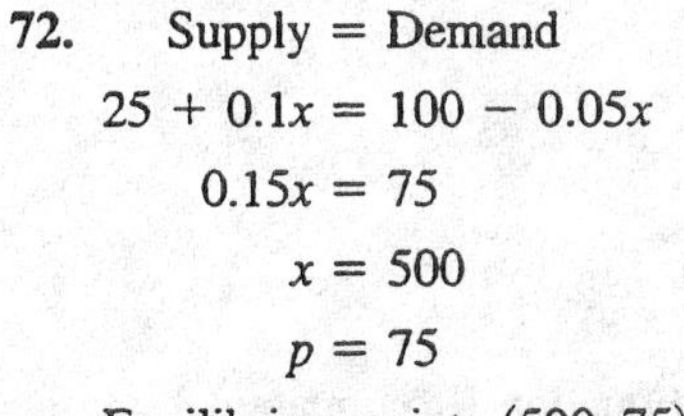

74. Supply = Demand

$225 + 0.0005x = 400 - 0.0002x$

$0.0007x = 175$

$x = 250{,}000$

$p = 350$

Equilibrium point: $(250{,}000, 350)$

76.
$u \cos 2x + v \sin 2x = 0$ Equation 1

$u(-2 \sin 2x) + v(2 \cos 2x) = \csc 2x$ Equation 2

$u = -\dfrac{v \sin 2x}{\cos 2x}$ from Equation 1

$$\left(-v\frac{\sin 2x}{\cos 2x}\right)(-2 \sin 2x) + v(2 \cos 2x) = \csc 2x$$

$$2v^2(\sin^2 2x + \cos^2 2x) = (\csc 2x)(\cos 2x)$$

$$v^2 = \frac{(\csc 2x)\cos 2x}{2(\sin^2 2x + \cos^2 2x)} = \frac{\cot(2x)}{2}$$

$$u = -\frac{(\csc 2x)(\cos 2x)(\sin 2x)}{2(\sin^2 2x + \cos^2 2x)(\cos 2x)} = -\frac{1}{2}$$

78.
$$ue^{2x} + vxe^{2x} = 0$$
$$u(2e^{2x}) + v(2x + 1)e^{2x} = \frac{e^{2x}}{x}$$
$$u = -\frac{vxe^{2x}}{e^{2x}} = -vx \text{ from Equation 1}$$
$$(-vx)2e^{2x} + v(2x + 1)e^{2x} = \frac{e^{2x}}{x}$$
$$ve^{2x} = \frac{e^{2x}}{x} \Rightarrow v = \frac{1}{x}$$
$$\Rightarrow u = -vx = -1$$

80. Domain: all x

82. Domain: all $t \neq \pm 3$

Section 7.3 Multivariable Linear Systems

Solutions to Even-Numbered Exercises

2.
$4x - 3y - 2z = 21$ Equation 1
$6y - 5z = -8$ Equation 2
$z = -2$ Equation 3

Back-substitute $z = -2$ in Equation 2:
$6y - 5(-2) = -8$
$y = -3$

Back-substitute $z = -2$ and $y = -3$ in Equation 1:
$4x - 3(-3) - 2(-2) = 21$
$4x + 13 = 21$
$x = 2$

Answer: $(2, -3, -2)$

4.
$x = 8$ Equation 1
$2x + 3y = 10$ Equation 2
$x = y + 2z = 22$ Equation 3

Back-substitute $x = 8$ in Equation 2:
$2(8) + 3y = 10$
$y = -2$

Back-substitute $x = 8$ and $y = -2$ in Equation 3:
$8 - (-2) + 2z = 22$
$z = 6$

Answer: $(8, -2, 6)$

6.
$5x - 8z = 22$
$3y - 5z = 10$
$z = -4$

Back-substitute $z = -4$ in Equation 2:
$3y - 5(-4) = 10 \Rightarrow y = -\frac{10}{3}$

Back-substitute $z = -4$ in Equation 1:
$5x - 8(-4) = 22 \Rightarrow x = -2$

Answer: $\left(-2, -\frac{10}{3}, -4\right)$

8.
$x - 2y - 3z = 5$
$-x + 3y - 5z = 4$
$2x - 3z = 0$

Add -2 times Equation 1 to Equation 3.
$4y - 9z = -10$

This is the first step in putting the system in row-echelon form.

10.
$$\begin{aligned} x + y + z &= 2 && \text{Equation 1} \\ -x + 3y + 2z &= 8 && \text{Equation 2} \\ 4x + y &= 4 && \text{Equation 3} \end{aligned}$$

$$\begin{aligned} x + y + z &= 2 \\ 4y + 3z &= 10 && \text{Eq.1 + Eq.2} \\ -3y - 4z &= -4 && -4\text{Eq.1 + Eq.3} \end{aligned}$$

$$\begin{aligned} x + y + z &= 2 \\ 12y + 9z &= 30 && 3\text{Eq.2} \\ -12y - 16z &= -16 && 4\text{Eq.3} \end{aligned}$$

$$\begin{aligned} x + y + z &= 2 \\ 12y + 9z &= 30 \\ -7z &= 14 && \text{Eq.2 + Eq.3} \end{aligned}$$

$$\begin{aligned} -7z = 14 &\Rightarrow z = -2 \\ 12y + 9(-2) = 30 &\Rightarrow y = 4 \\ x + 4 - 2 = 2 &\Rightarrow x = 0 \end{aligned}$$

Answer: $(0, 4, -2)$

12.
$$\begin{aligned} 4x + y - 3z &= 11 \\ 2x - 3y + 2z &= 9 \\ x + y + z &= -3 \end{aligned}$$

$$\begin{aligned} x + y + z &= -3 \\ 2x - 3y + 2z &= 9 \\ 4x + y - 3z &= 11 \end{aligned}$$

$$\begin{aligned} x + y + z &= -3 \\ -5y &= 15 \\ -3y + 7z &= 23 \end{aligned}$$

$$\begin{aligned} y = -3 &\Rightarrow -3(-3) - 7z = 23 \\ &\Rightarrow -7z = 14 \\ &\Rightarrow z = -2 \end{aligned}$$

$x + (-3) + (-2) = -3 \Rightarrow x = 2$

Answer: $(2, -3, -2)$

14.
$$\begin{aligned} 2x + 4y + z &= -4 && \text{Equation 1} \\ 2x - 4y + 6z &= 13 && \text{Equation 2} \\ 4x - 2y + z &= 6 && \text{Equation 3} \end{aligned}$$

$$\begin{aligned} 2x + 4y + z &= -4 \\ -8y + 5z &= 17 && -\text{Eq.1 + Eq.2} \\ -10y - z &= 14 && -2\text{Eq. + Eq.3} \end{aligned}$$

$$\begin{aligned} 2x + 4y + z &= -4 \\ -40y + 25z &= 85 && 5\text{Eq.2} \\ -40y - 4z &= 56 && 4\text{Eq.3} \end{aligned}$$

$$\begin{aligned} 2x + 4y + z &= -4 \\ -40y + 25z &= 85 \\ -29z &= -29 && -\text{Eq.2 + Eq.3} \end{aligned}$$

$$\begin{aligned} -29z = -29 &\Rightarrow z = 1 \\ -40y + 25(1) = 85 &\Rightarrow y = -\tfrac{3}{2} \\ 2x + 4\left(-\tfrac{3}{2}\right) + 1 = -4 &\Rightarrow x = \tfrac{1}{2} \end{aligned}$$

Answer: $\left(\frac{1}{2}, -\frac{3}{2}, 1\right)$

16.
$$\begin{aligned} 5x - 3y + 2z &= 3 && \text{Equation 1} \\ 2x + 4y - z &= 7 && \text{Equation 2} \\ x - 11y + 4z &= 3 && \text{Equation 3} \end{aligned}$$

$$\begin{aligned} x - 11y + 4z &= 3 && \text{Interchange} \\ 5x - 3y + 2z &= 3 && \text{Equations 1 and 3} \\ 2x + 4y - z &= 7 \end{aligned}$$

$$\begin{aligned} x - 11y + 4z &= 3 \\ 52y - 18z &= -12 && -5\text{Eq.1 + Eq.2} \\ 26y - 9z &= 1 && -2\text{Eq.1 + Eq.3} \end{aligned}$$

$$\begin{aligned} x - 11y + 4z &= 3 \\ 52y - 18z &= -12 \\ 0 &= 7 && -\tfrac{1}{2}\text{Eq.2 + Eq.3} \end{aligned}$$

Inconsistent; no solution

18.
$$\begin{array}{rcll} 2x + y + 3z &=& 1 & \text{Equation 1} \\ 2x + 6y + 8z &=& 3 & \text{Equation 2} \\ 6x + 8y + 18z &=& 5 & \text{Equation 3} \end{array}$$

$$\begin{array}{rcll} 2x + y + 3z &=& 1 & \\ 5y + 5z &=& 2 & -\text{Eq.1} + \text{Eq.2} \\ 5y + 9z &=& 2 & -3\text{Eq.1} + \text{Eq.3} \end{array}$$

$$\begin{array}{rcll} 2x + y + 3z &=& 1 & \\ 5y + 5z &=& 2 & \\ 4z &=& 0 & -\text{Eq.2} + \text{Eq.3} \end{array}$$

$$4z = 0 \Rightarrow z = 0$$
$$5y + 5(0) = 2 \Rightarrow y = \tfrac{2}{5}$$
$$2x + \tfrac{2}{5} + 3(0) = 1 \Rightarrow x = \tfrac{3}{10}$$

Answer: $\left(\tfrac{3}{10}, \tfrac{2}{5}, 0\right)$

20.
$$\begin{array}{rcll} 2x + y - 3z &=& 4 & \text{Equation 1} \\ 4x + 2z &=& 10 & \text{Equation 2} \\ -2x + 3y - 13z &=& -8 & \text{Equation 3} \end{array}$$

$$\begin{array}{rcll} 2x + y - 3z &=& 4 & -2\text{Eq.1} + \text{Eq.2} \\ -2y + 8z &=& 2 & \text{Eq.1} + \text{Eq.3} \\ 4y - 16z &=& -4 & \end{array}$$

$$\begin{array}{rcll} 2x + y - 3z &=& 4 & \\ y - 4z &=& -1 & -\tfrac{1}{2}\text{Eq.2} \\ 0 &=& 0 & 2\text{Eq.2} + \text{Eq.3} \end{array}$$

$$\begin{array}{rcll} 2x + z &=& 5 & -\text{Eq.2} + \text{Eq.1} \\ y - 4z &=& -1 & \end{array}$$

$z = a$

$y = 4a - 1$

$x = -\tfrac{1}{2}a + \tfrac{5}{2}$

Answer: $\left(-\tfrac{1}{2}a + \tfrac{5}{2}, 4a - 1, a\right)$

22.
$$\begin{array}{rcll} x + 4z &=& 13 & \text{Equation 1} \\ 4x - 2y + z &=& 7 & \text{Equation 2} \\ 2x - 2y - 7z &=& -19 & \text{Equation 3} \end{array}$$

$$\begin{array}{rcll} x + 4z &=& 13 & \\ -2y - 15z &=& -45 & -4\text{Eq.1} + \text{Eq.2} \\ -2y - 15z &=& -45 & -2\text{Eq.1} + \text{Eq.3} \end{array}$$

$$\begin{array}{rcll} x + 4z &=& 13 & \\ -2y - 15z &=& -45 & \\ 0 &=& 0 & -\text{Eq.2} + \text{Eq.3} \end{array}$$

$z = a$

$y = -\tfrac{15}{2}a + \tfrac{45}{2}$

$x = -4a + 13$

Answer: $\left(-4a + 13, -\tfrac{15}{2}a + \tfrac{45}{2}, a\right)$

24.
$$\begin{array}{rcll} x - 3y + 2z &=& 18 & \text{Equation 1} \\ 5x - 13y + 12z &=& 80 & \text{Equation 2} \end{array}$$

$$\begin{array}{rcll} x - 3y + 2z &=& 18 & \\ 2y + 2z &=& -10 & -5\text{Eq.1} + \text{Eq.2} \end{array}$$

$$\begin{array}{rcll} x - 3y + 2z &=& 18 & \\ y + z &=& -5 & \tfrac{1}{2}\text{Eq.2} \end{array}$$

$$\begin{array}{rcll} x + 5z &=& 3 & 3\text{Eq.2} + \text{Eq.1} \\ y + z &=& -5 & \end{array}$$

Let $z = a$,
then $y = -a - 5$, and $x = -5a + 3$.

Answer: $(-5a + 3, -a - 5, a)$

26.
$$\begin{array}{rcll} 2x + 3y + 3z &=& 7 & \text{Equation 1} \\ 4x + 18y + 15z &=& 44 & \text{Equation 2} \end{array}$$

$$\begin{array}{rcll} 2x + 3y + 3z &=& 7 & \\ 12y + 9z &=& 30 & -2\text{Eq.1} + \text{Eq.2} \end{array}$$

$$\begin{array}{rcll} 2x + \tfrac{3}{4}z &=& -\tfrac{1}{2} & -\tfrac{1}{4}\text{Eq.2}+\text{Eq.1} \\ 12y + 9z &=& 30 & \end{array}$$

Let $z = a$, then:

$12y + 9a = 30 \Rightarrow y = -\tfrac{3}{4}a + \tfrac{5}{2}$

$2x + \tfrac{3}{4}a = -\tfrac{1}{2} \Rightarrow x = -\tfrac{3}{8}a - \tfrac{1}{4}$

Answer: $\left(-\tfrac{3}{8}a - \tfrac{1}{4}, -\tfrac{3}{4}a + \tfrac{5}{2}, a\right)$

28.
$$\begin{aligned} x + y + z + w &= 6 && \text{Equation 1}\\ 2x + 3y - w &= 0 && \text{Equation 2}\\ -3x + 4y + z + 2w &= 4 && \text{Equation 3}\\ x + 2y - z + w &= 0 && \text{Equation 4}\end{aligned}$$

$$\begin{aligned} x + y + z + w &= 6 &&\\ y - 2z - 3w &= -12 && -2\text{Eq.}1 + \text{Eq.}2\\ 7y + 4z + 5w &= 22 && 3\text{Eq.}1 + \text{Eq.}3\\ y - 2z &= -6 && -\text{Eq.}1 + \text{Eq.}4\end{aligned}$$

$$\begin{aligned} x + y + z + w &= 6 &&\\ y - 2z - 3w &= -12 &&\\ 18z + 26w &= 106 && -7\text{Eq.}2 + \text{Eq.}3\\ 3w &= 6 && -\text{Eq.}2 + \text{Eq.}4\end{aligned}$$

$$\begin{aligned} 3w &= 6 \Rightarrow w = 2\\ 18z + 26(2) &= 106 \Rightarrow z = 3\\ y - 2(3) - 3(2) &= -12 \Rightarrow y = 0\\ x + 0 + 3 + 2 &= 6 \Rightarrow x = 1\end{aligned}$$

Answer: $(1, 0, 3, 2)$

30.
$$\begin{aligned} 3x - 2y - 6z &= -4 && \text{Equation 1}\\ -3x + 2y + 6z &= 1 && \text{Equation 2}\\ x - y - 5z &= -3 && \text{Equation 3}\end{aligned}$$

$$\begin{aligned} x - y - 5z &= -3 && \text{Interchange the equations}\\ 3x - 2y - 6z &= -4 &&\\ -3x + 2y + 6z &= 1 &&\end{aligned}$$

$$\begin{aligned} x - y - 5z &= -3 &&\\ y + 9z &= 5 && -3\text{Eq.}1 + \text{Eq.}2\\ -y - 9z &= -8 && 3\text{Eq.}1 + \text{Eq.}3\end{aligned}$$

$$\begin{aligned} x - y - 5z &= -3 &&\\ y + 9z &= 5 &&\\ 0 &= -3 && \text{Eq.}2 + \text{Eq.}3\end{aligned}$$

Inconsistent, no solution

32.
$$\begin{aligned} 4x + 3y + 17z &= 0 && \text{Equation 1}\\ 5x + 4y + 22z &= 0 && \text{Equation 2}\\ 4x + 2y + 19z &= 0 && \text{Equation 3}\end{aligned}$$

$$\begin{aligned} 5x + 4y + 22z &= 0 && \text{Interchange the equations}\\ 4x + 3y + 17z &= 0 &&\\ 4x + 2y + 19z &= 0 &&\end{aligned}$$

$$\begin{aligned} x + y + 5z &= 0 && -\text{Eq.}2 + \text{Eq.}1\\ 4x + 3y + 17z &= 0 &&\\ 4x + 2y + 19z &= 0 &&\end{aligned}$$

$$\begin{aligned} x + y + 5z &= 0 &&\\ -y - 3z &= 0 && -4\text{Eq.}1 + \text{Eq.}2\\ -2y - z &= 0 && -4\text{Eq.}1 + \text{Eq.}3\end{aligned}$$

$$\begin{aligned} x + y + 5z &= 0 &&\\ y + 3z &= 0 && -\text{Eq.}2\\ 5z &= 0 && -2\text{Eq.}2 + \text{Eq.}3\end{aligned}$$

$$\begin{aligned} 5z &= 0 \Rightarrow z = 0\\ y + 3(0) &= 0 \Rightarrow y = 0\\ x + 0 + 5(0) &= 0 \Rightarrow x = 0\end{aligned}$$

Answer: $(0, 0, 0)$

34.
$$\begin{aligned} 5x + 5y - z &= 0 && \text{Equation 1}\\ 10x + 5y + 2z &= 0 && \text{Equation 2}\\ 5x + 15y - 9z &= 0 && \text{Equation 3}\end{aligned}$$

$$\begin{aligned} 5x + 5y - z &= 0 &&\\ -5y + 4z &= 0 && -2\text{Eq.}1 + \text{Eq.}2\\ 10y - 8z &= 0 && -\text{Eq.}1 + \text{Eq.}3\end{aligned}$$

$$\begin{aligned} 5x + 3z &= 0 && \text{Eq.}2 + \text{Eq.}1\\ -5y + 4z &= 0 &&\\ 0 &= 0 && 2\text{Eq.}2 + \text{Eq.}3\end{aligned}$$

$$\begin{aligned} z &= a\\ y &= \tfrac{4}{5}a\\ x &= -\tfrac{3}{5}a\end{aligned}$$

Answer: $\left(-\frac{3}{5}a, \frac{4}{5}a, a\right)$

36. When using Gaussian elimination to solve a system of linear equations, a system has no solution when there is a row representing a contradictory equation such as $0 = N$, where N is a nonzero real number.

For instance:
$$\begin{aligned} x + y &= 3 && \text{Equation 1} \\ -x - y &= 3 && \text{Equation 2} \end{aligned}$$

$$\begin{aligned} x + y &= 0 \\ 0 &= 6 && \text{Eq.1 + Eq.2} \end{aligned}$$

No solution

38. There are an infinite number of linear systems that have $\left(-\frac{3}{2}, 4, -7\right)$ as their solution. One such system is:

$$\begin{aligned} 2\left(-\tfrac{3}{2}\right) + 4 - (-7) &= 8 \Rightarrow & 2x + y - z &= 8 \\ 4\left(-\tfrac{3}{2}\right) + 2(4) + (-7) &= -5 \Rightarrow & 4x + 2y + z &= -5 \\ -2\left(-\tfrac{3}{2}\right) + 5(4) - 3(-7) &= 44 \Rightarrow & -2x + 5y - 3z &= 44 \end{aligned}$$

40. $y = ax^2 + bx + c$ passing through

$(0, 3), (1, 4), (2, 3)$

$(0, 3)$: $3 = c$

$(1, 4)$: $4 = a + b + c \Rightarrow 1 = a + b$

$(2, 3)$: $3 = 4a + 2b + c \Rightarrow 0 = 2a + b$

Answer: $a = -1, b = 2, c = 3$

The equation of the parabola is $y = -x^2 + 2x + 3$.

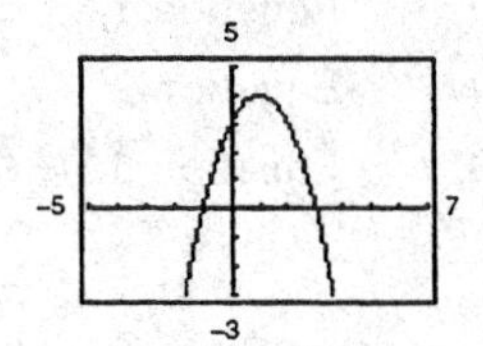

42. $y = ax^2 + bx + c$ passing through

$(1, 3), (2, 2), (3, -3)$

$(1, 3)$: $3 = a + b + c$

$(2, 2)$: $2 = 4a + 2b + c \Rightarrow -1 = 3a + b$

$(3, -3)$: $-3 = 9a + 3b + c \Rightarrow -6 = 8a + 2b$

Answer: $a = -2, b = 5, c = 0$

The equation of the parabola is $y = -2x^2 + 5x$.

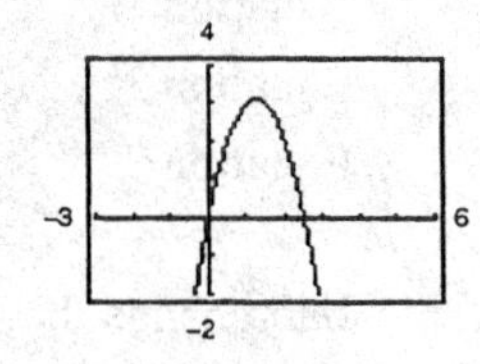

44. $x^2 + y^2 + Dx + Ey + F = 0$ passing through $(0, 0), (0, 6), (3, 3)$

$(0, 0)$: $F = 0$

$(0, 6)$: $36 + 6E + F = 0 \Rightarrow E = -6$

$(3, 3)$: $18 + 3D + 3E + F = 0 \Rightarrow D = 0$

The equation of the circle is $x^2 + y^2 - 6y = 0$.

To graph, complete the square first, then solve for y.

$$\begin{aligned} x^2 + y^2 - 6y + 9 &= 9 \\ x^2 + (y - 3)^2 &= 9 \\ (y - 3)^2 &= 9 - x^2 \\ y - 3 &= \pm\sqrt{9 - x^2} \\ y &= 3 \pm \sqrt{9 - x^2} \end{aligned}$$

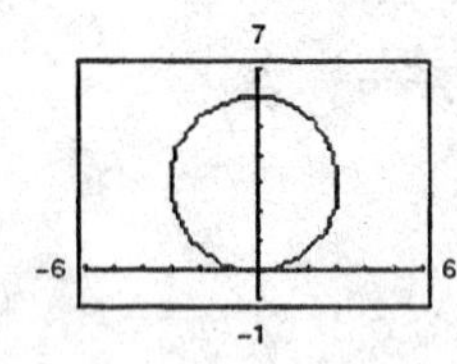

Let $y_1 = 3 + \sqrt{9 - x^2}$ and $y_2 = 3 - \sqrt{9 - x^2}$.

46. $x^2 + y^2 + Dx + Ey + F = 0$ passing through $(0, 0)$, $(0, -2)$, $(3, 0)$

$(0, 0)$: $F = 0$

$(0, -2)$: $4 \quad -2E + F = 0 \Rightarrow E = 2$

$(3, 0)$: $9 + 3D \quad + F = 0 \Rightarrow D = -3$

The equation of the circle is $x^2 + y^2 - 3x + 2y = 0$.

To graph, complete the squares first, then solve for y.

$$x^2 - 3x + \tfrac{9}{4} + y^2 + 2y + 1 = \tfrac{9}{4} + 1$$

$$\left(x - \tfrac{3}{2}\right)^2 + (y + 1)^2 = \tfrac{13}{4}$$

$$(y + 1)^2 = \tfrac{13}{4} - \left(x - \tfrac{3}{2}\right)^2$$

$$y + 1 = \pm\sqrt{\tfrac{13}{4} - \left(x - \tfrac{3}{2}\right)^2}$$

$$y = -1 \pm \sqrt{\tfrac{13}{4} - \left(x - \tfrac{3}{2}\right)^2}$$

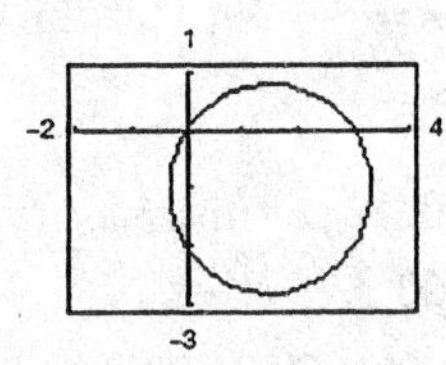

Let $y_1 = -1 + \sqrt{\tfrac{13}{4} - \left(x - \tfrac{3}{2}\right)^2}$ and $y_2 = -1 - \sqrt{\tfrac{13}{4} - \left(x - \tfrac{3}{2}\right)^2}$.

48. $s = \tfrac{1}{2}at^2 + v_0t + s_0$

$(1, 48)$, $(2, 64)$, $(3, 48)$

$48 = \tfrac{1}{2}a + v_0 + s_0 \Rightarrow a + 2v_0 + 2s_0 = 96$

$64 = 2a + 2v_0 + s_0 \Rightarrow 2a + 2v_0 + s_0 = 64$

$48 = \tfrac{9}{2}a + 3v_0 + s_0 \Rightarrow 9a + 6v_0 + 2s_0 = 96$

Solving this system yields $a = -32$, $v_0 = 64$, $s_0 = 0$.

Thus, $s = \tfrac{1}{2}(-32)t^2 + 64t + 0$

$= -16t^2 + 64t$.

50. $s = \tfrac{1}{2}at^2 + v_0t + s_0$

$(1, 132)$, $(2, 100)$, $(3, 36)$

$132 = \tfrac{1}{2}a + v_0 + s_0 \Rightarrow a + 2v_0 + 2s_0 = 264$

$100 = 2a + 2v_0 + s_0 \Rightarrow 2a + 2v_0 + s_0 = 100$

$36 = \tfrac{9}{2}a + 3v_0 + s_0 \Rightarrow 9a + 6v_0 + 2s_0 = 72$

Solving this system yields $a = -32$, $b = 16$, $c = 132$.

Thus, $s = \tfrac{1}{2}(-32)t^2 + 16t + 132$

$= -16t^2 + 16t + 132$.

52. Let x = amount at 5%

Let y = amount at 7%

Let z = amount at 8%

$0.05x + 0.07y + 0.08z = 1520$

$x = 0.5y$

$y = z - 1500$

$5(0.5y) + 7y + 8(y + 1500) = 152{,}000$

$17.5y = 140{,}000$

$y = 8000$

$x = 0.5(8000) = 4000$

$z = 8000 + 1500 = 9500$

Check: $0.05(4000) + 0.07(8000) + 0.08(9500) = 1520$

Answer: x = \$4000 at 5%

y = \$8000 at 7%

z = \$9500 at 8%

54. Let x = amount at 8%
Let y = amount at 9%
Let z = amount at 10%

$$x + y + z = 800{,}000$$
$$0.08z + 0.09y + 0.10z = 67{,}000$$
$$x = 5z$$

$$y + 6z = 800{,}000$$
$$0.09y + 0.5z = 67{,}000$$
$$z = 125{,}000$$
$$y = 800{,}000 - 6(125{,}000) = 50{,}000$$
$$x = 5(125{,}000) = 625{,}000$$

Answer: x = \$625,000 at 8%
y = \$50,000 at 9%
z = \$125,000 at 10%

56. Let C = amount in certificates of deposit
Let M = amount in municipal bonds
Let B = amount in blue-chip stocks
Let G = amount in growth or speculative stocks

$$C + M + B + G = 500{,}000$$
$$0.09C + 0.05M + 0.12B + 0.14G = 0.10(500{,}000)$$
$$B + G = \tfrac{1}{4}(500{,}000)$$

This system has infinitely many solutions.

Let $G = s$, then $B = 125{,}000 - s$
$$M = \tfrac{1}{2}s - 31{,}250$$
$$C = 406{,}250 - \tfrac{1}{2}s.$$

Answer:
$$\left(406{,}250 - \tfrac{1}{2}s, -31{,}250 + \tfrac{1}{2}s, 125{,}000 - s, s\right)$$

One possible solution is to let s = \$100,000.

Certificates of deposit: \$356,250
Municipal bonds: \$18,750
Blue-chip stocks: \$25,000
Growth or speculative stocks: \$100,000

58. (a) To use as little of the 50% solution as possible, the chemist should use no 10% solution.

$$x(0.20) + (10 - x)(0.50) = 10(0.25)$$
$$x(0.20) + 5 - 0.50x = 2.5$$
$$-0.30x = -2.5$$
$$x = 8\tfrac{1}{3} \text{ liters of 20\% solution}$$
$$10 - x = 1\tfrac{2}{3} \text{ liters of 50\% solution}$$

(b) To use as much 50% solution as possible, the chemist should use no 20% solution.

$$x(0.10) + (10 - x)0.50 = 10(0.25)$$
$$0.10x + 5 - 0.50x = 2.5$$
$$-0.40x = -2.5$$
$$x = 6\tfrac{1}{4} \text{ liters of 10\% solution}$$
$$10 - x = 3\tfrac{3}{4} \text{ liters of 50\% solution}$$

(c) To use 2 liters of 50% solution we let x = the number of liters at 10% and y = the number of liters at 20%.

$$0.10x + 0.20y + 2(0.50) = 10(0.25) \qquad \text{Equation 1}$$
$$x + y = 8 \qquad \text{Equation 2}$$

Answer: y = 7 liters of 20% solution; x = 1 liter of 10% solution

60. $I_1 - I_2 + I_3 = 0$ Equation 1
$3I_1 + 2I_2 = 7$ Equation 2
$2I_2 + 4I_3 = 8$ Equation 3

$I_1 - I_2 + I_3 = 0$
$5I_2 - 3I_3 = 7$ -3Eq.1 + Eq.2
$2I_2 + 4I_3 = 8$

$I_1 - I_2 + I_3 = 0$
$10I_2 - 6I_3 = 14$ 2Eq.2
$10I_2 + 20I_3 = 40$ 5Eq.3

$I_1 - I_2 + I_3 = 0$
$10I_2 - 6I_3 = 14$
$26I_3 = 26$ $-$Eq.2 + Eq.3

$26I_3 = 26 \Rightarrow I_3 = 1$
$10I_2 - 6(1) = 14 \Rightarrow I_2 = 2$
$I_1 - 2 + 1 = 0 \Rightarrow I_1 = 1$

Answer: $I_1 = 1$ ampere, $I_2 = 2$ amperes,
$I_3 = 1$ ampere

62. $t_1 - 2t_2 = 0$ Equation 1
$t_1 - 2a = 128$ Equation 2
$t_2 + 2a = 64$ Equation 3

$t_1 - 2t_2 = 0$
$2t_2 - 2a = 128$ $-$Eq.1 + Eq.2
$t_2 + 2a = 64$

$t_1 - 2t_2 = 0$
$2t_2 - 2a = 128$
$3a = 0$ $-\frac{1}{2}$Eq.2 + Eq.3

$3a = 0 \Rightarrow a = 0$
$2t_2 - 2(0) = 128 \Rightarrow t_2 = 64$
$t_1 - 2(64) = 0 \Rightarrow t_1 = 128$

Answer: $a = 0$ ft/sec^2
$t_1 = 128$ lb
$t_2 = 64$ lb

64. $x + y + z = 6$
$(6, 0, 0), (0, 6, 0), (0, 0, 6), (1, 1, 4)$

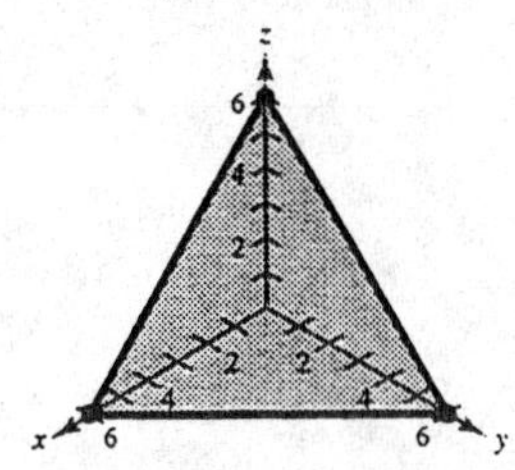

66. $x + 2y + 2z = 6$
$(6, 0, 0), (0, 3, 0), (0, 0, 3), (2, 1, 1)$

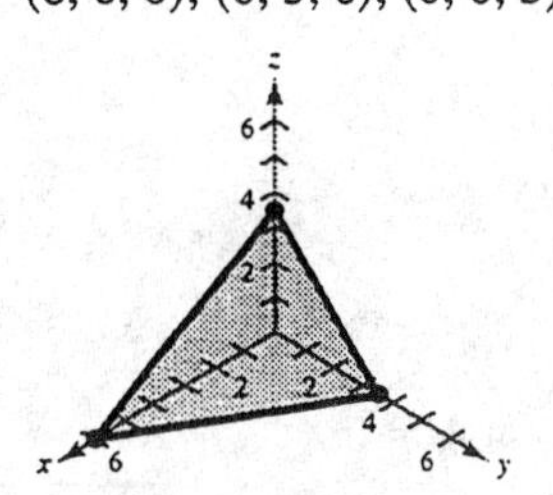

68. Least squares regression parabola through $(-2, 0), (-1, 0), (0, 1), (1, 2), (2, 5)$

$n = 5$

$\sum x_i = 0$ $\sum y_i = 8$ $5c + 10a = 8$
$\sum x_i^2 = 10$ $\sum x_i^3 = 0$ $10b = 12$
$\sum x_i^4 = 34$ $\sum x_i y_i = 12$ $10c + 34a = 22$
$\sum x_i^2 y_i = 22$

Solving this system yields $a = \frac{3}{7}, b = \frac{6}{5}, c = \frac{26}{35}$. Thus, $y = \frac{3}{7}x^2 + \frac{6}{5}x + \frac{26}{35}$.

70. Least squares regression parabola through $(0, 10), (1, 9), (2, 6), (3, 0)$

$n = 4$

$\sum x_i = 6$ $\sum y_i = 25$ $4c + 6b + 14a = 25$
$\sum x_i^2 = 104$ $\sum x_i^3 = 36$ $6c + 14b + 36a = 21$
$\sum x_i^4 = 98$ $\sum x_i y_i = 21$ $14c + 36b + 98a = 33$
$\sum x_i^2 y_i = 33$

Solving this system yields $a = -\frac{5}{4}, b = \frac{9}{20}, c = \frac{199}{20}$. Thus, $y = -\frac{5}{4}x^2 + \frac{9}{20}x + \frac{199}{20}$.

72. (a) Using the quadratic least squares regression feature, we find

$$y = -0.008x^2 + 1.371x + 21.886.$$

(b)

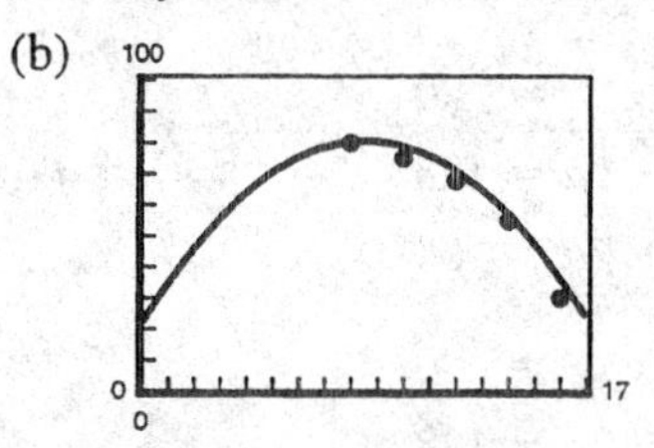

(c) For $x = 170$:

$$y = -0.008(170)^2 + 1.371(170) + 21.886$$
$$= 23.756\%$$

74. $\left.\begin{array}{l} 2x + \lambda = 0 \\ 2y + \lambda = 0 \end{array}\right\} x = y = -\dfrac{\lambda}{2}$

$$x + y - 4 = 0 \Rightarrow 2x - 4 = 0$$
$$2x = 4$$
$$x = 2$$
$$y = 2$$
$$\lambda = -4$$

76. $2 + 2y + 2\lambda = 0$

$2x + 1 + \lambda = 0 \Rightarrow \lambda = -2x - 1$

$2x + y - 100 = 0$

$$2 + 2y + 2(-2x - 1) = 0 \Rightarrow -4x + 2y = 0 \Rightarrow -4x + 2y = 0$$
$$2x + y - 100 = 0 \Rightarrow 2x + y = 100 \Rightarrow 4x + 2y = 200$$
$$4y = 200$$
$$y = 50$$
$$x = 25$$
$$\lambda = -2(25) - 1 = -51$$

78. $P = 47t^2 - 195t + 1109$

$S = 160t + 705$

$P = S$

$47t^2 - 195t + 1109 = 160t + 705$

$47t^2 - 355t + 404 = 0$

$$t = \frac{355 \pm \sqrt{(-355)^2 - 4(47)(404)}}{94} = \frac{355 \pm \sqrt{50,073}}{94} \approx 1.4, 6.2$$

Thus, compact pickup sales will again exceed compact utility sales in 1996.

80. $225 = x(150)$

$x = 1.5$ or 150%

82. $0.48x = 132$

$x = 275$

Section 7.4 Partial Fractions

Solutions to Even-Numbered Exercises

2. $\dfrac{x - 2}{x^2 + 4x + 3} = \dfrac{A}{x + 3} + \dfrac{B}{x + 1}$

4. $\dfrac{4x^2 + 3}{(x - 5)^3} = \dfrac{A}{x - 5} + \dfrac{B}{(x - 5)^2} + \dfrac{C}{(x - 5)^3}$

6. $\dfrac{x - 1}{x(x^2 + 1)^2} = \dfrac{A}{x} + \dfrac{Bx + C}{x^2 + 1} + \dfrac{Dx + E}{(x^2 + 1)^2}$

8. $\dfrac{1}{4x^2-9} = \dfrac{A}{2x+3} + \dfrac{B}{2x-3}$

$1 = A(2x-3) + B(2x+3)$

Let $x = -\dfrac{3}{2}$: $1 = -6A \Rightarrow A = -\dfrac{1}{6}$

Let $x = \dfrac{3}{2}$: $1 = 6B \Rightarrow B = \dfrac{1}{6}$

$\dfrac{1}{4x^2-9} = \dfrac{1}{6}\left[\dfrac{1}{2x-3} - \dfrac{1}{2x+3}\right]$

10. $\dfrac{3}{x^2-3x} = \dfrac{A}{x-3} + \dfrac{B}{x}$

$3 = Ax + B(x-3)$

Let $x = 3$: $3 = 3A \Rightarrow A = 1$

Let $x = 0$: $3 = -3B \Rightarrow B = -1$

$\dfrac{3}{x^2-3x} = \dfrac{1}{x-3} - \dfrac{1}{x}$

12. $\dfrac{5}{x^2+x-6} = \dfrac{A}{x+3} - \dfrac{B}{x-2}$

$5 = A(x-2) + B(x+3)$

Let $x = -3$: $5 = -5A \Rightarrow A = -1$

Let $x = 2$: $5 = 5B \Rightarrow B = 1$

$\dfrac{5}{x^2+x-6} = \dfrac{1}{x-2} - \dfrac{1}{x+3}$

14. $\dfrac{x+1}{x^2+4x+3} = \dfrac{x+1}{(x+3)(x+1)}$

$= \dfrac{1}{x+3}, x \neq -1$

16. $\dfrac{x+2}{x(x-4)} = \dfrac{A}{x} + \dfrac{B}{x-4}$

$x+2 = A(x-4) + Bx$

Let $x = 0$: $2 = -4A \Rightarrow A = -\dfrac{1}{2}$

Let $x = 4$: $6 = 4B \Rightarrow B = \dfrac{3}{2}$

$\dfrac{x+2}{x(x-4)} = \dfrac{1}{2}\left[\dfrac{3}{x-4} - \dfrac{1}{x}\right]$

18. $\dfrac{2x-3}{(x-1)^2} = \dfrac{A}{x-1} + \dfrac{B}{(x-1)^2}$

$2x-3 = A(x-1) + B$

Let $x = 1$: $-1 = B$

Let $x = 0$: $-3 = -A + B$

$-3 = -A - 1$

$2 = A$

$\dfrac{2x-3}{(x-1)^2} = \dfrac{2}{x-1} - \dfrac{1}{(x-1)^2}$

20. $\dfrac{6x^2+1}{x^2(x-1)^3} = \dfrac{A}{x} + \dfrac{B}{x^2} + \dfrac{C}{x-1} + \dfrac{D}{(x-1)^2} + \dfrac{E}{(x-1)^3}$

$6x^2 + 1 = Ax(x-1)^3 + B(x-1)^3 + Cx^2(x-1)^2 + Dx^2(x-1) + Ex^2$

Let $x = 0$: $1 = -B \Rightarrow B = -1$

Let $x = 1$: $7 = E$

Substitute B and E into the equation, expand the binomials, collect like terms, and equate the coefficients of like terms.

$x^3 - 4x^2 + 3x = (A + C)x^4 - (3A + 2C - D)x^3 + (3A + C - D)x^2 - Ax$

$-A = 3 \Rightarrow A = -3$

$A + C = 0 \Rightarrow C = 3$

$-3A - 2C + D = 1$

$9 - 6 + D = 1 \Rightarrow D = -2$

$\dfrac{6x^2+1}{x^2(x-1)^3} = -\dfrac{3}{x} - \dfrac{1}{x^2} + \dfrac{3}{x-1} - \dfrac{2}{(x-1)^2} + \dfrac{7}{(x-1)^3}$

22. $\dfrac{x}{(x-1)(x^2+x+1)} = \dfrac{A}{x-1} + \dfrac{Bx+C}{x^2+x+1}$

$$x = A(x^2+x+1) + (Bx+C)(x-1)$$

Let $x = 1$: $1 = 3A \Rightarrow A = \dfrac{1}{3}$

$$\begin{aligned} x &= \frac{1}{3}(x^2+x+1) + (Bx+C)(x-1) \\ &= \frac{1}{3}x^2 + \frac{1}{3}x + \frac{1}{3} + Bx^2 - Bx + Cx - C \\ &= \left(\frac{1}{3} + B\right)x^2 + \left(\frac{1}{3} - B + C\right)x + \frac{1}{3} - C \end{aligned}$$

Equating coefficients of like powers:

$$0 = \frac{1}{3} + B \Rightarrow B = -\frac{1}{3}$$

$$0 = \frac{1}{3} - C \Rightarrow C = \frac{1}{3}$$

$$\frac{x}{(x-1)(x^2+x+1)} = \frac{1}{3}\left[\frac{1}{x-1} - \frac{x-1}{x^2+x+1}\right]$$

24. $\dfrac{2x^2+x+8}{(x^2+4)^2} = \dfrac{Ax+B}{x^2+4} + \dfrac{Cx+D}{(x^2+4)^2}$

$2x^2 + x + 8 = (Ax+B)(x^2+4) + Cx + D$

$2x^2 + x + 8 = Ax^3 + Bx^2 + (4A+C)x + (4B+D)$

Equating coefficients of like powers:

$0 = A$

$2 = B$

$1 = 4A + C \Rightarrow C = 1$

$8 = 4B + D \Rightarrow D = 0$

$$\frac{2x^2+x+8}{(x^2+4)^2} = \frac{2}{x^2+4} + \frac{x}{(x^2+4)^2}$$

26. $\dfrac{x^2-4x+7}{(x+1)(x^2-2x+3)} = \dfrac{A}{x+1} + \dfrac{Bx+C}{x^2-2x+3}$

$$x^2 - 4x + 7 = A(x^2-2x+3) + Bx(x+1) + C(x+1)$$

Let $x = -1$: $12 = 6A \Rightarrow A = 2$

Let $x = 0$: $7 = 3A + C \Rightarrow C = 1$

Let $x = 1$: $4 = 2A + 2B + 2C$

$$4 = 4 + 2B + 2 \Rightarrow B = -1$$

$$\frac{x^2-4x+7}{(x+1)(x^2-2x+3)} = \frac{2}{x+1} - \frac{x-1}{x^2-2x+3}$$

28. $\dfrac{x+1}{x^3+x} = \dfrac{A}{x} + \dfrac{Bx+C}{x^2+1}$

$x + 1 = A(x^2 + 1) + Bx^2 + Cx$

Let $x = 0$: $1 = A$

$x + 1 = x^2 + 1 + Bx^2 + Cx$

$x + 1 = (1 + B)x^2 + Cx + 1$

Equating coefficients of like powers:

$0 = 1 + B \implies B = -1$

$1 = C$

$\dfrac{x+1}{x^3+x} = \dfrac{1}{x} - \dfrac{x-1}{x^2+1}$

30. $\dfrac{x^2-x}{x^2+x+1} = 1 - \dfrac{2x+1}{x^2+x+1}$

32. $\dfrac{3x^2-7x-2}{x^3-x} = \dfrac{A}{x} + \dfrac{B}{x+1} + \dfrac{C}{x-1}$

$3x^2 - 7x - 2 = A(x^2 - 1) + Bx(x - 1) + Cx(x + 1)$

Let $x = 0$: $-2 = -A \implies A = 2$

Let $x = -1$: $8 = 2B \implies B = 4$

Let $x = 1$: $-6 = 2C \implies C = -3$

$\dfrac{3x^2-7x-2}{x^3-x} = \dfrac{2}{x} + \dfrac{4}{x+1} - \dfrac{3}{x-1}$

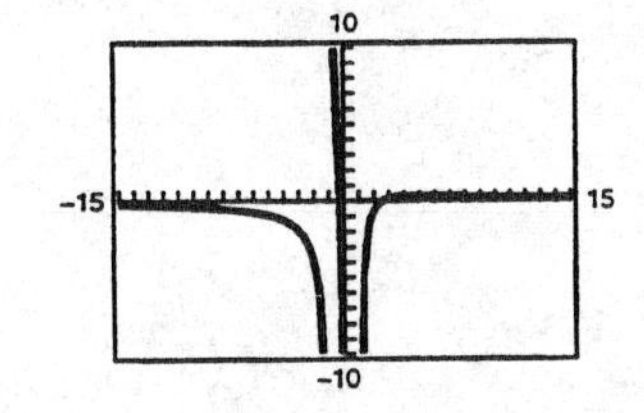

34. $\dfrac{4x^2-1}{2x(x+1)^2} = \dfrac{A}{2x} + \dfrac{B}{x+1} + \dfrac{C}{(x+1)^2}$

$4x^2 - 1 = A(x + 1)^2 + 2Bx(x + 1) + 2Cx$

Let $x = 0$: $-1 = A$

Let $x = -1$: $3 = -2C \implies C = -\dfrac{3}{2}$

Let $x = 1$: $3 = 4A + 4B + 2C$

$3 = -4 + 4B - 3$

$\dfrac{5}{2} = B$

$\dfrac{4x^2-1}{2x(x+1)^2} = \dfrac{1}{2}\left[-\dfrac{1}{x} + \dfrac{5}{x+1} - \dfrac{3}{(x+1)^2}\right]$

36. $\dfrac{x^3}{(x+2)^2(x-2)^2} = \dfrac{A}{x+2} + \dfrac{B}{(x+2)^2} + \dfrac{C}{x-2} + \dfrac{D}{(x-2)^2}$

$x^3 = A(x + 2)(x - 2)^2 + B(x - 2)^2 + C(x + 2)^2(x - 2) + D(x + 2)^2$

Let $x = -2$: $-8 = 16B \implies B = -\dfrac{1}{2}$

Let $x = 2$: $8 = 16D \implies D = \dfrac{1}{2}$

$x^3 = A(x + 2)(x - 2)^2 - \dfrac{1}{2}(x - 2)^2 + C(x + 2)^2(x - 2) + \dfrac{1}{2}(x + 2)^2$

$x^3 - 4x = (A + C)x^3 + (-2A + 2C)x^2 + (-4A - 4C)x + (8A - 8C)$

Equating coefficients of like powers:

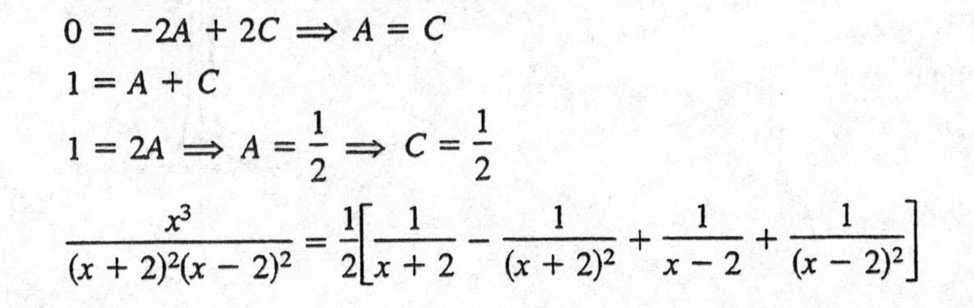

$0 = -2A + 2C \implies A = C$

$1 = A + C$

$1 = 2A \implies A = \dfrac{1}{2} \implies C = \dfrac{1}{2}$

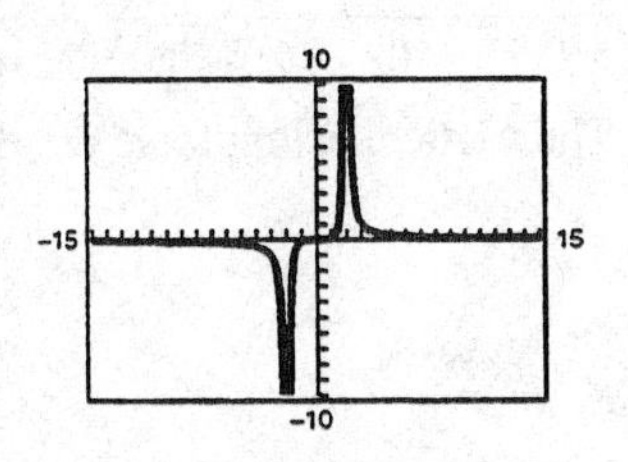

$\dfrac{x^3}{(x+2)^2(x-2)^2} = \dfrac{1}{2}\left[\dfrac{1}{x+2} - \dfrac{1}{(x+2)^2} + \dfrac{1}{x-2} + \dfrac{1}{(x-2)^2}\right]$

38. $\dfrac{x^3 - x + 3}{x^2 + x - 2} = x - 1 + \dfrac{2x + 1}{(x + 2)(x - 1)}$

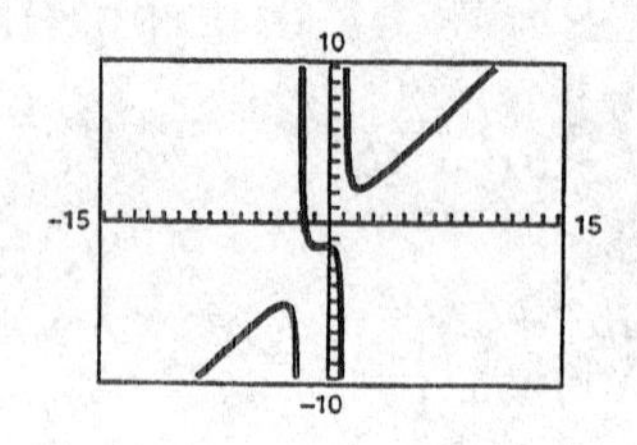

$\dfrac{2x + 1}{(x + 2)(x - 1)} = \dfrac{A}{x + 2} + \dfrac{B}{x - 1}$

$2x + 1 = A(x - 1) + B(x + 2)$

Let $x = -2$: $-3 = -3A \implies A = 1$

Let $x = 1$: $3 = 3B \implies B = 1$

$\dfrac{x^3 - x + 3}{x^2 + x - 2} = x - 1 + \dfrac{1}{x + 2} + \dfrac{1}{x - 1}$

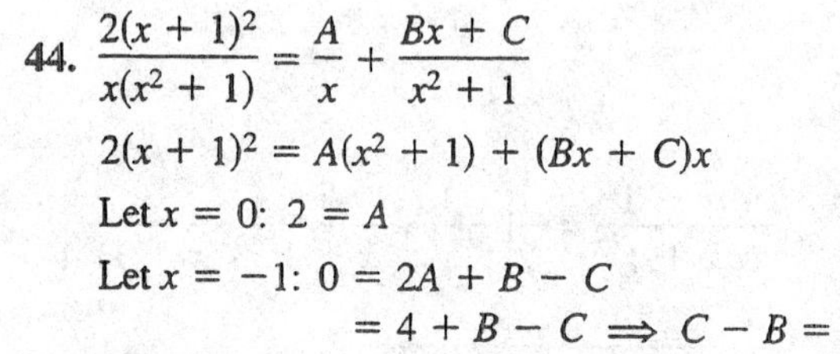

40. $\dfrac{1}{x(x + a)} = \dfrac{A}{x} + \dfrac{B}{x + a}$, a is a constant.

$1 = A(x + a) + Bx$

Let $x = 0$: $1 = aA \implies A = \dfrac{1}{a}$

Let $x = -a$: $1 = -aB \implies B = -\dfrac{1}{a}$

$\dfrac{1}{x(x + a)} = \dfrac{1}{a}\left[\dfrac{1}{x} - \dfrac{1}{x + a}\right]$

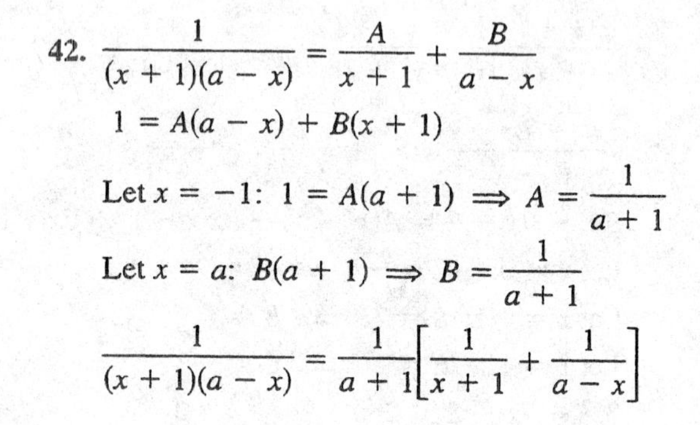

42. $\dfrac{1}{(x + 1)(a - x)} = \dfrac{A}{x + 1} + \dfrac{B}{a - x}$

$1 = A(a - x) + B(x + 1)$

Let $x = -1$: $1 = A(a + 1) \implies A = \dfrac{1}{a + 1}$

Let $x = a$: $B(a + 1) \implies B = \dfrac{1}{a + 1}$

$\dfrac{1}{(x + 1)(a - x)} = \dfrac{1}{a + 1}\left[\dfrac{1}{x + 1} + \dfrac{1}{a - x}\right]$

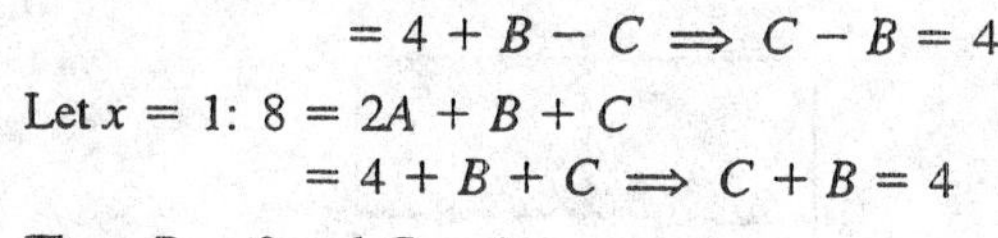

44. $\dfrac{2(x + 1)^2}{x(x^2 + 1)} = \dfrac{A}{x} + \dfrac{Bx + C}{x^2 + 1}$

$2(x + 1)^2 = A(x^2 + 1) + (Bx + C)x$

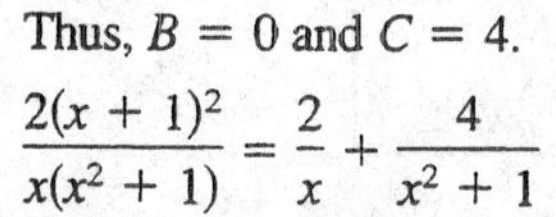

Let $x = 0$: $2 = A$

Let $x = -1$: $0 = 2A + B - C$

$= 4 + B - C \implies C - B = 4$

Let $x = 1$: $8 = 2A + B + C$

$= 4 + B + C \implies C + B = 4$

Thus, $B = 0$ and $C = 4$.

$\dfrac{2(x + 1)^2}{x(x^2 + 1)} = \dfrac{2}{x} + \dfrac{4}{x^2 + 1}$

The vertical asymptotes are the same ($x = 0$).

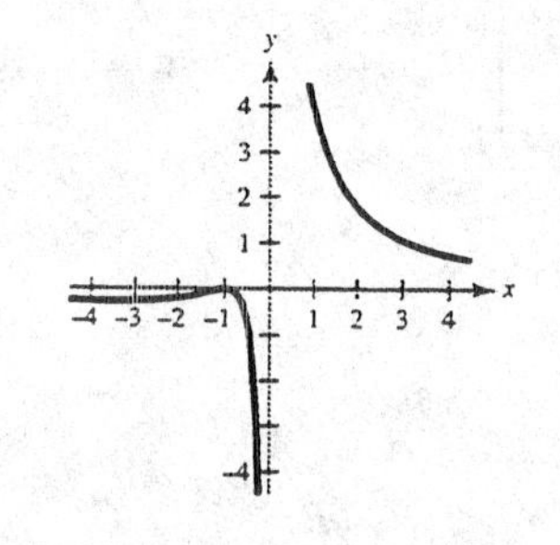

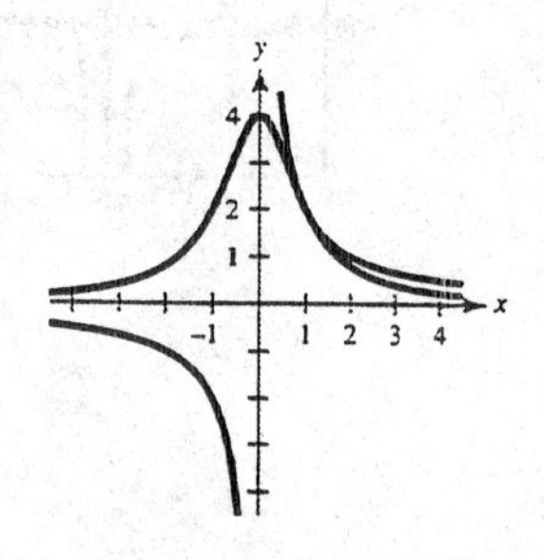

46. $\dfrac{2(4x^2 - 15x + 39)}{x^2(x^2 - 10x + 26)} = \dfrac{A}{x} + \dfrac{B}{x^2} + \dfrac{Cx + D}{x^2 - 10x + 26}$

$2(4x^2 - 15x + 39) = Ax(x^2 - 10x + 26) + B(x^2 - 10x + 26) + (Cx + D)x^2$

Let $x = 0$: $78 = 26B \implies B = 3$

Let $x = 1$: $56 = 17A + 17B + (C + D) \implies 17A + C + D = 5$

Let $x = 2$: $50 = 20A + 10B + (2C + D)4 \implies 20A + 8C + 4D = 20$

Let $x = 3$: $60 = 15A + 5B + (3C + D)9 \implies 15A + 27C + 9D = 45$

Solving for A, C, and D, you obtain $A = C = 0$, $D = 5$. Thus,

$\dfrac{2(4x^2 - 15x + 39)}{x^2(x^2 - 10x + 26)} = \dfrac{3}{x^2} + \dfrac{5}{x^2 - 10x + 26}.$

The vertical asymptotes are the same ($x = 0$).

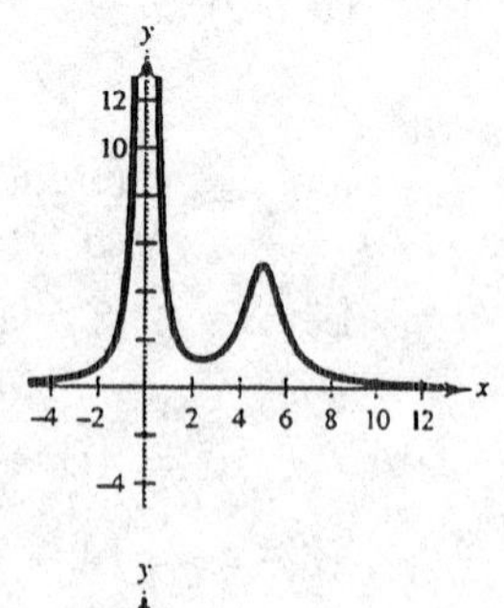

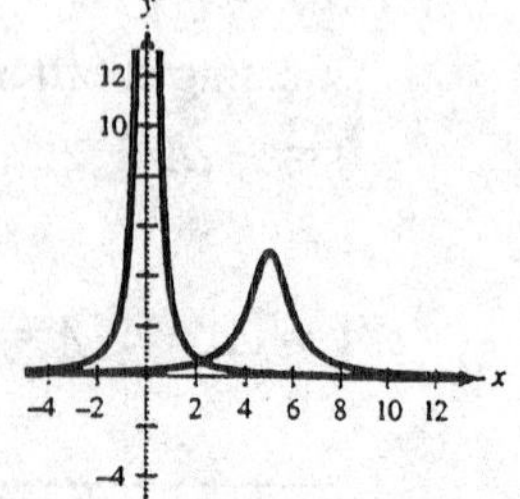

Section 7.5 Systems of Inequalities

Solutions to Even-Numbered Exercises

2. $y \geq 3$

Region above or on horizontal line $y = 3$.
Matches (d).

4. $2x - y \leq -2 \Rightarrow y \geq 2x + 2$

Region above or on line $y = 2x + 2$.
Matches (h).

6. $(x - 2)^2 + (y - 3)^2 > 9$

Region outside circle. Matches (b).

8. $y \leq 1 - x^2$

Region below or on parabola. Matches (c).

10. $x \leq 4$

Using a solid line, graph the vertical line $x = 4$, and shade to the left of this line.

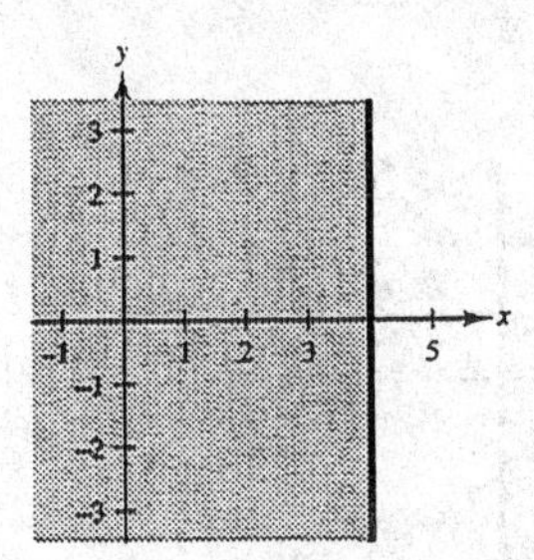

12. $y \leq 3$

Using a solid line, graph the horizontal line $y = 3$, and shade below this line.

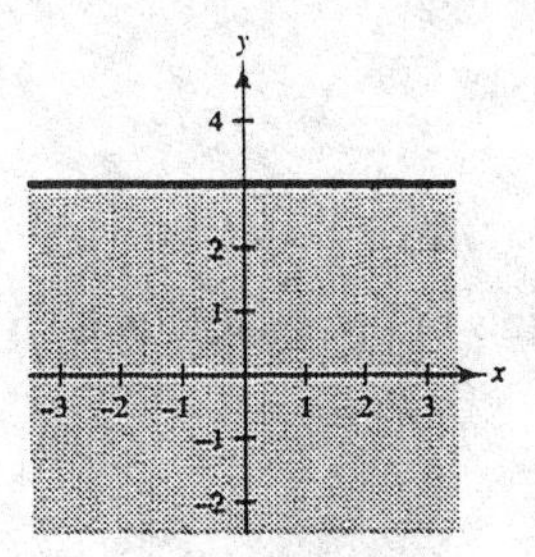

14. $y > 2x - 4$

Using a dashed line, graph $y = 2x - 4$, and shade above the line. (Use (0, 0) as a test point.)

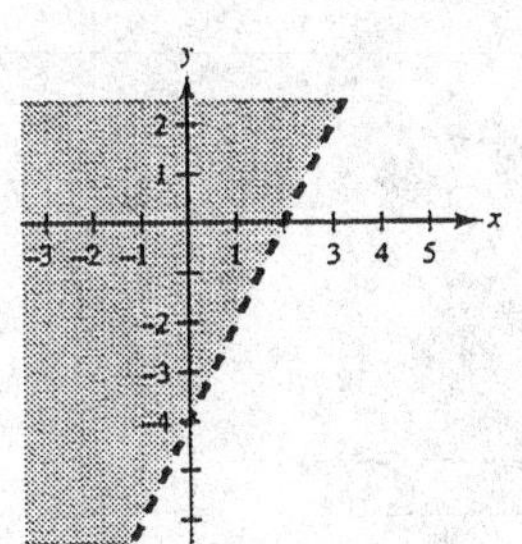

16. $5x + 3y \geq -15$

Using a solid line, graph $5x + 3y = -15$, and shade above the line. (Use (0, 0) as a test point.)

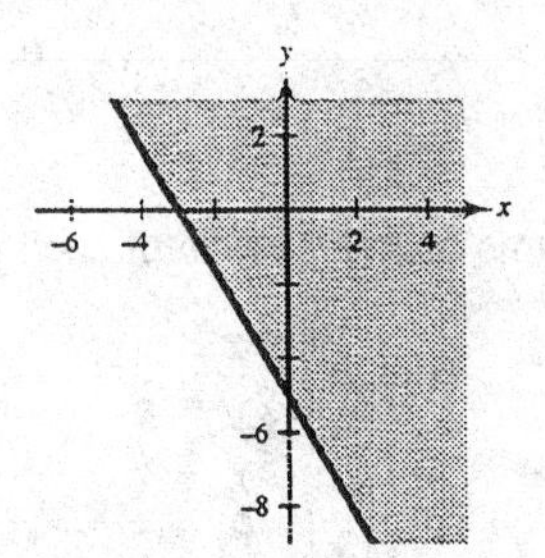

18. $(x + 1)^2 + y^2 < 9$

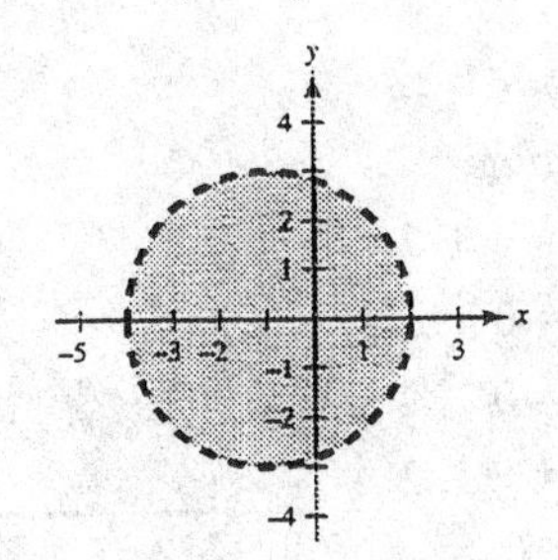

20. $y < \ln x$

Using a dashed line, graph $y = \ln x$, and shade to the right of the curve. (Use (2, 0) as a test point.)

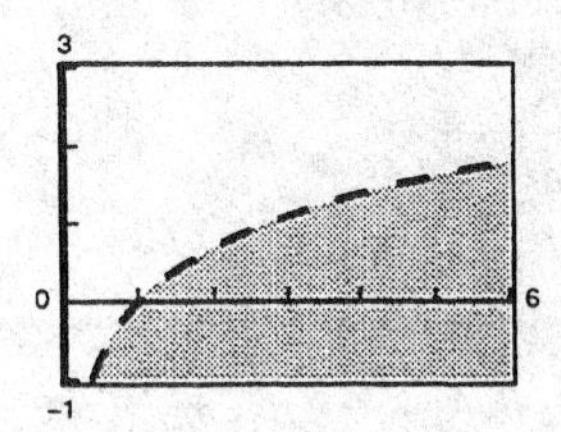

22. $y \leq 6 - \frac{3}{2}x$

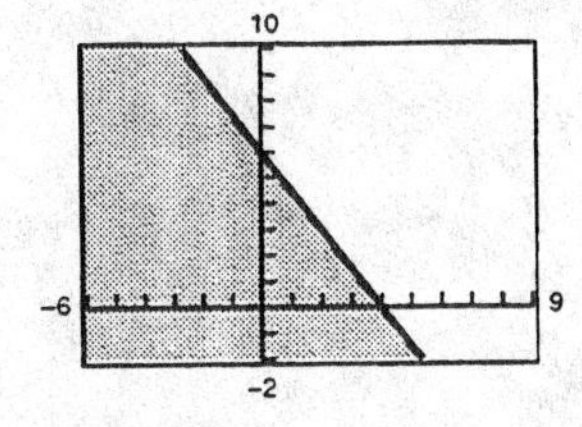

24. $2x^2 - y - 3 > 0$

$y < 2x^2 - 3$

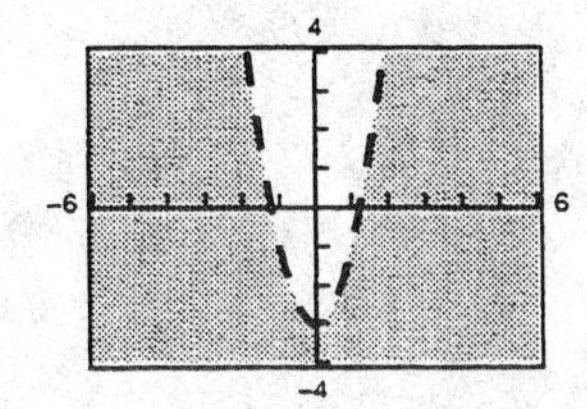

26. The parabola through $(-2, 0)$, $(0, -4)$, $(2, 0)$ is $y = x^2 - 4$. For the shaded region inside the parabola, we have $y \geq x^2 - 4$.

28. The circle shown is $x^2 + y^2 = 9$. For the shaded region inside the circle, we have $x^2 + y^2 \leq 9$.

30. The region to the right of the vertical line $x = 5$. Thus, $x > 5$.

32. $3x + 2y < 6$

$x \quad > 0$

$y > 0$

First, find the points of intersection of each pair of equations.

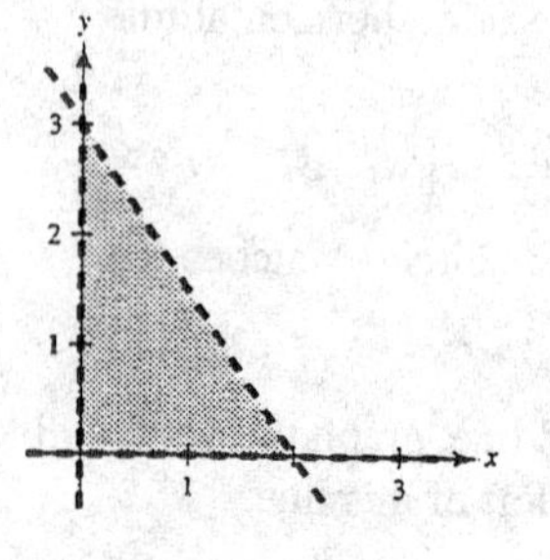

Vertex A	Vertex B	Vertex C
$3x + 2y = 6$	$x = 0$	$3x + 2y = 6$
$x = 0$	$y = 0$	$y = 0$
$(0, 3)$	$(0, 0)$	$(2, 0)$

34. $2x^2 + y \geq 2$

$x \quad \leq 2$

$y \leq 1$

First, find the points of intersection of each pair of equations.

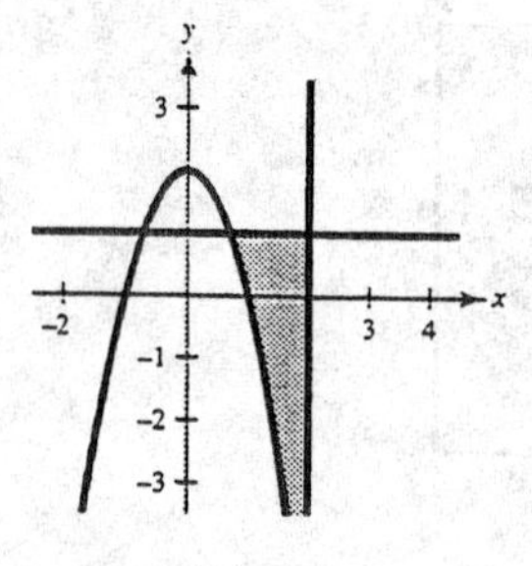

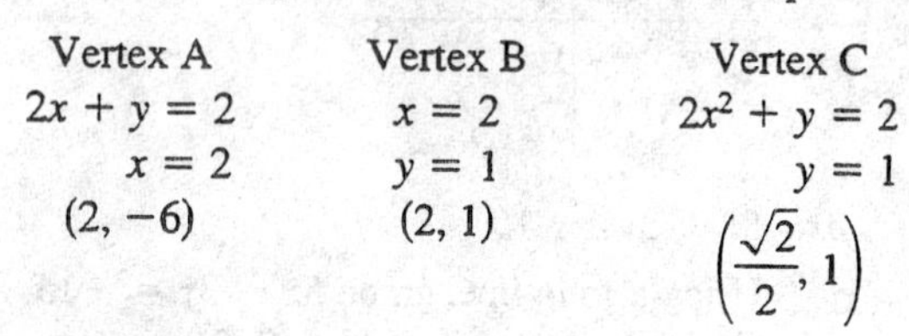

Vertex A	Vertex B	Vertex C
$2x + y = 2$	$x = 2$	$2x^2 + y = 2$
$x = 2$	$y = 1$	$y = 1$
$(2, -6)$	$(2, 1)$	$\left(\frac{\sqrt{2}}{2}, 1\right)$

36. $x - 7y > -36$

$5x + 2y > \quad 5$

$6x - 5y > \quad 6$

First, find the points of intersection of each pair of equations.

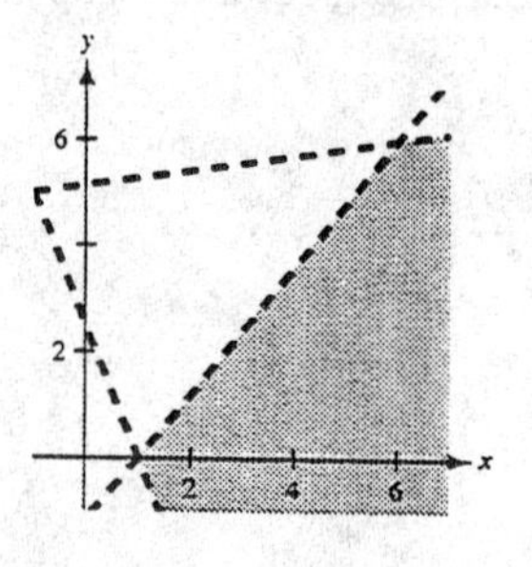

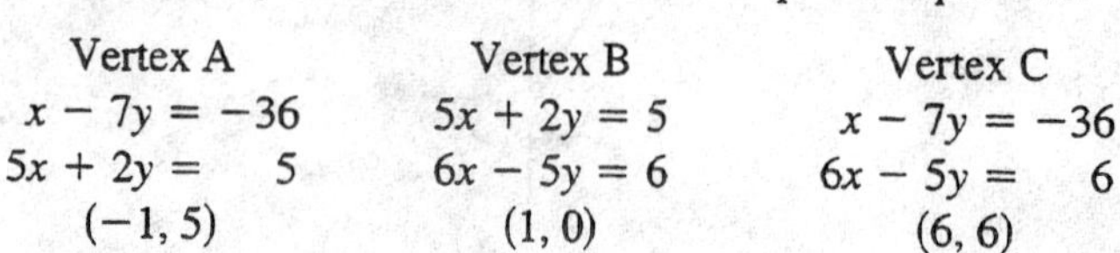

Vertex A	Vertex B	Vertex C
$x - 7y = -36$	$5x + 2y = 5$	$x - 7y = -36$
$5x + 2y = \quad 5$	$6x - 5y = 6$	$6x - 5y = \quad 6$
$(-1, 5)$	$(1, 0)$	$(6, 6)$

38. $x - 2y < -6$

$5x - 3y > -9$

Point of intersection: $(0, 3)$

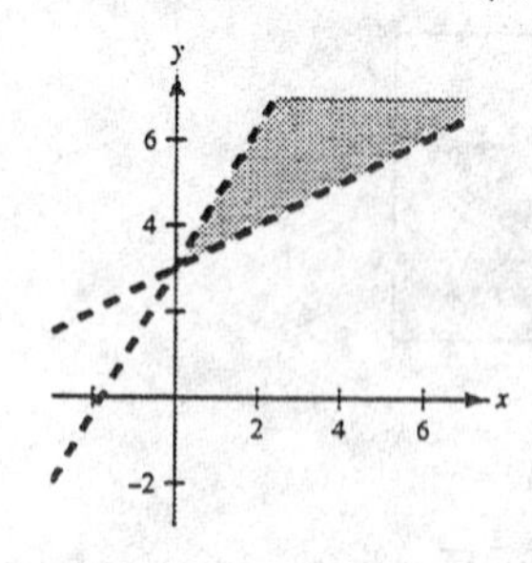

40. $x - y^2 > 0$

$x - y < 2$

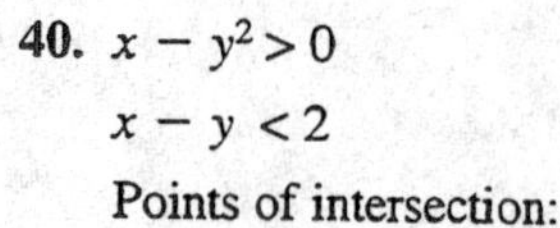

Points of intersection:

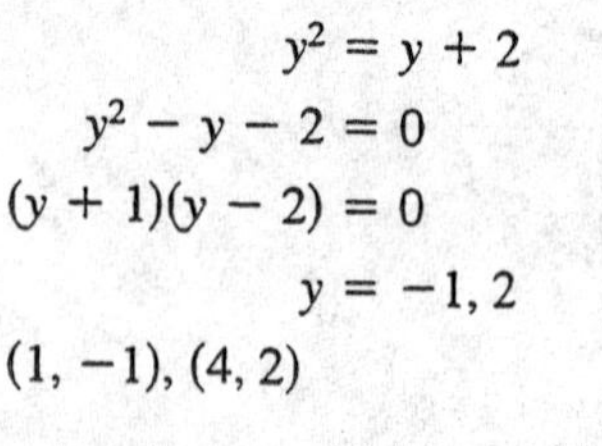

$$y^2 = y + 2$$

$$y^2 - y - 2 = 0$$

$$(y + 1)(y - 2) = 0$$

$$y = -1, 2$$

$(1, -1)$, $(4, 2)$

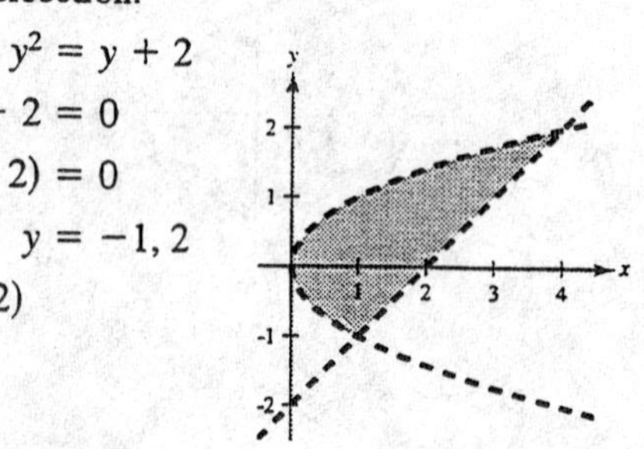

42. $x^2 + y^2 \le 25$

$4x - 3y \le 0$

Points of intersection:

$x^2 + \left(\frac{4}{3}x\right)^2 = 25$

$\frac{25}{9}x^2 = 25$

$x = \pm 3$

$(-3, -4), (3, 4)$

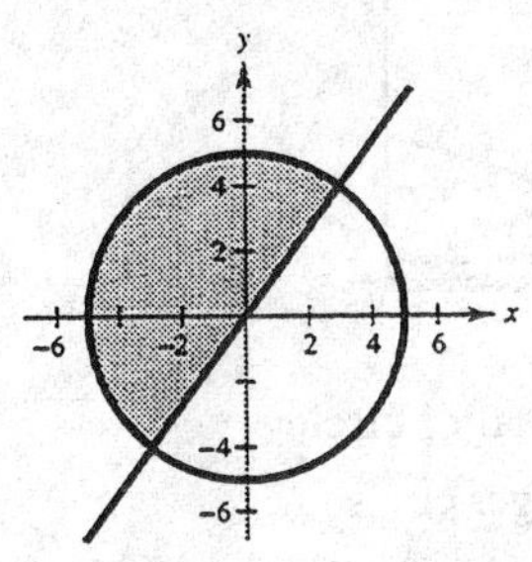

44. $x < 2y - y^2$

$0 < x + y$

Points of intersection:

$-y = 2y - y^2$

$y^2 - 3y = 0$

$y = 0, 3$

$(0, 0), (-3, 3)$

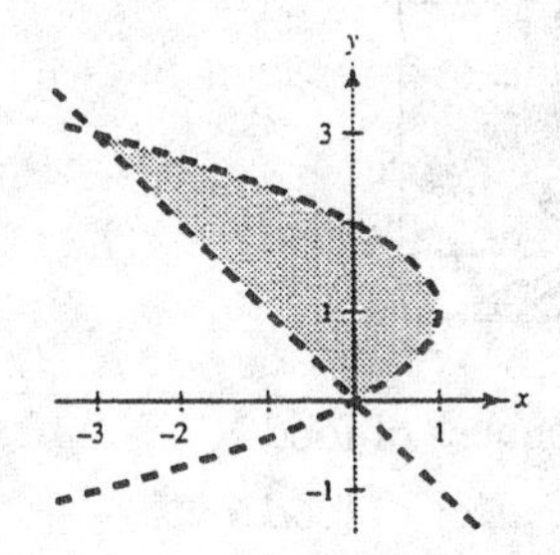

46. $y < -x^2 + 2x + 3$

$y > x^2 - 4x + 3$

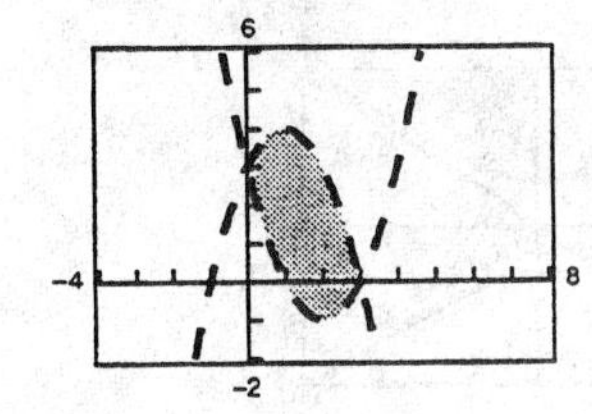

48. $y \ge x^4 - 2x^2 + 1$

$y \le 1 - x^2$

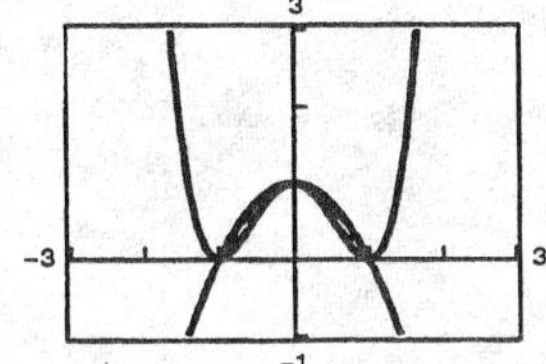

50. $y \le e^{-x^2/2}$

$y \ge 0$

$-2 \le x \le 2$

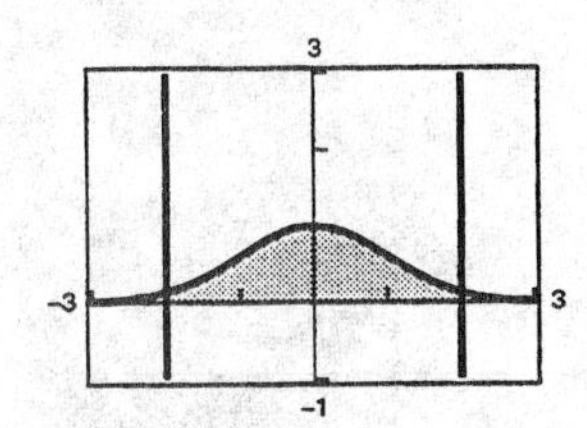

52. $(0, 6), (3, 0)$

Line: $y < 6 - 2x$

$(0, -3), (3, 0)$

Line: $y \ge x - 3$

$x \ge 1$

54. Circle: $x^2 + y^2 > 4$

56. Circle: $x^2 + y^2 \le 16$

$x \ge 0$

$y \ge x$

58. Parallelogram with vertices at $(0, 0), (4, 0), (1, 4), (5, 4)$

$(0, 0), (4, 0)$: $y \ge 0$

$(4, 0), (5, 4)$: $4x - y \le 16$

$(1, 4), (5, 4)$: $y \le 4$

$(0, 0), (1, 4)$: $4x - y \ge 0$

$4x - y \ge 0$

$4x - y \le 16$

$0 \le y \le 4$

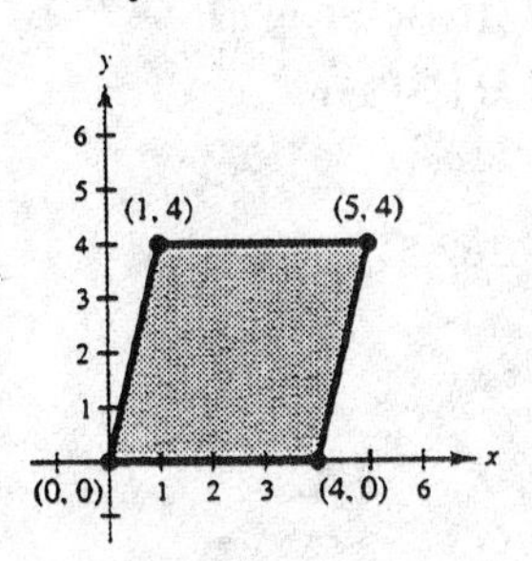

60. Triangle with vertices at $(-1, 0), (1, 0), (0, 1)$

$(-1, 0), (1, 0)$: $y \ge 0$

$(-1, 0), (0, 1)$: $y \le x + 1$

$(0, 1), (1, 0)$: $y \le -x + 1$

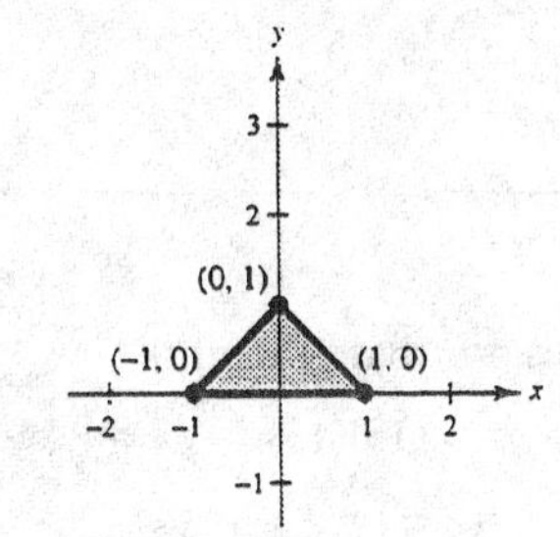

62. x = number of \$15 tickets

y = number of \$25 tickets

$x + y \geq 15{,}000$

$x \geq 8000$

$y \geq 4000$

$15x + 25y \geq 275{,}000$

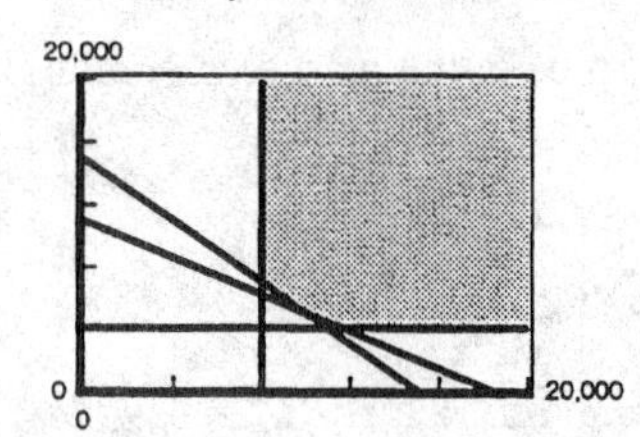

64. x = number of model A

y = number of model B

Demand: $x \geq 2y$

Cost: $8x + 12y \leq 200$

Inventory: $x \geq 4$

$y \geq 2$

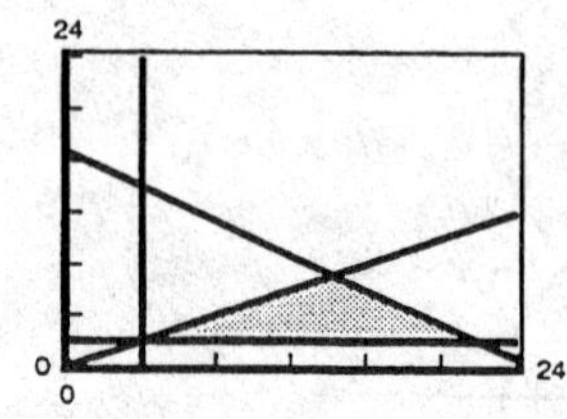

66. x = number of ounces of food X

y = number of ounces of food Y

Calcium: $20x + 10y \geq 300$

Iron: $15x + 10y \geq 150$

Vitamin B: $10x + 20y \geq 200$

$x \geq 0$

$y \geq 0$

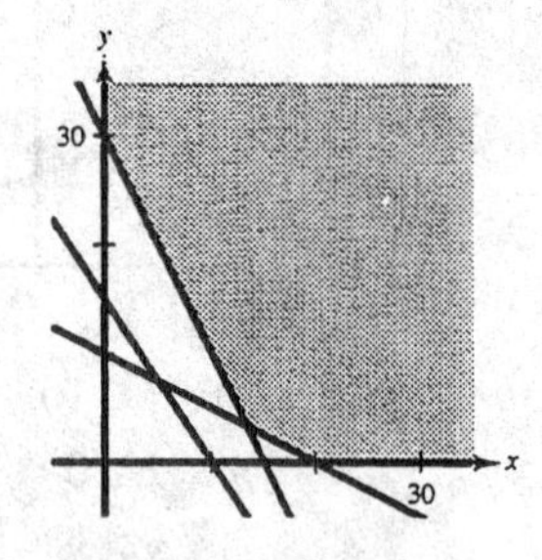

68. x = radius of smaller circle

y = radius of larger circle

(a) Constraints on circles: $\pi y^2 - \pi x^2 \geq 10$

$x > 0$

$y > x$

(b)

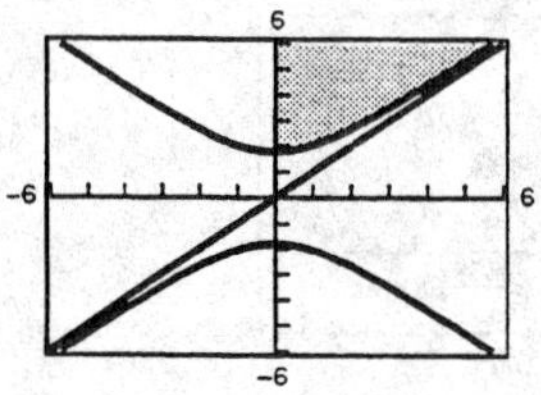

(c) The line is an asymptote to the boundary. The larger the circles, the closer the radii can be and the constraint still be satisfied.

70. Demand = Supply

$60 - x = 10 + \frac{7}{3}x$

$50 = \frac{10}{3}x$

$15 = x$

$45 = p$

Point of equilibrium: $(15, 45)$

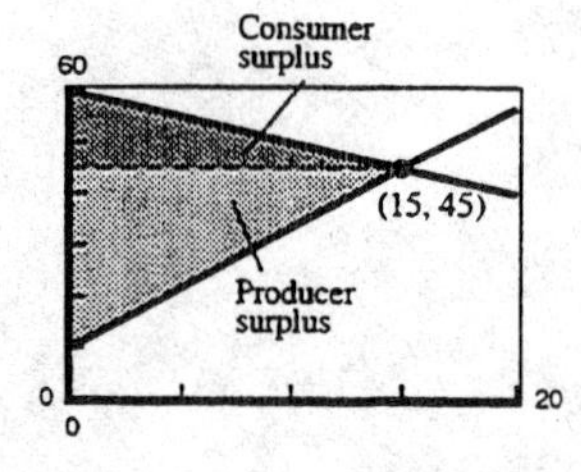

The consumer surplus is the area of the triangle bounded by

$p \leq 60 - x$

$p \geq 45$

$x \geq 0.$

Consumer surplus $= \frac{1}{2}(\text{base})(\text{height})$

$= \frac{1}{2}(15)(15)$

$= \frac{225}{2}$

$= 112.5$

The producer surplus is the area of the triangle bounded by

$p \geq 10 + \frac{7}{3}x$

$p \leq 45$

$x \geq 0.$

Producer surplus $= \frac{1}{2}(\text{base})(\text{height})$

$= \frac{1}{2}(15)(35)$

$= \frac{525}{2}$

$= 262.5$

72.

$$\text{Demand} = \text{Supply}$$
$$140 - 0.00002x = 80 + 0.00001x$$
$$60 = 0.00003x$$
$$2{,}000{,}000 = x$$
$$100 = p$$

Point of equilibrium:(2,000,000, 100)

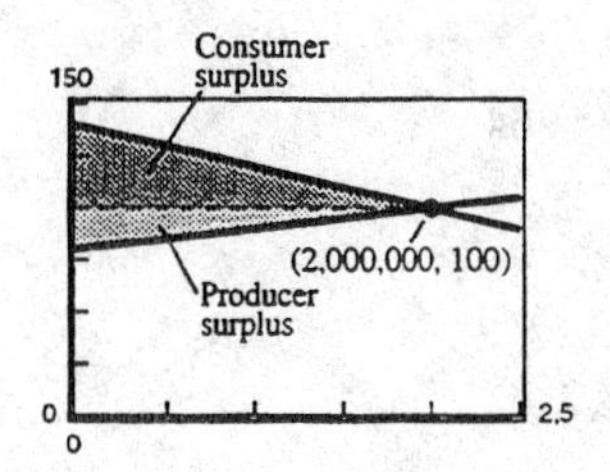

The consumer surplus is the area of the triangle bounded by

$p \leq 140 - 0.00002x$

$p \geq 100$

$x \geq 0.$

$$\text{Consumer surplus} = \tfrac{1}{2}(\text{base})(\text{height})$$
$$= \tfrac{1}{2}(2{,}000{,}000)(40)$$
$$= 40{,}000{,}000 \text{ or } \$40 \text{ million}$$

The producer surplus is the area of the triangle bounded by

$p \geq 80 + 0.00001x$

$p \leq 100$

$x \geq 0.$

$$\text{Producer surplus} = \tfrac{1}{2}(\text{base})(\text{height})$$
$$= \tfrac{1}{2}(2{,}000{,}000)(20)$$
$$= 20{,}000{,}000 \text{ or } \$20 \text{ million}$$

Section 7.6 Linear Programming

Solutions to Even-Numbered Exercises

2. $z = 2x + 8y$

At $(0, 4)$: $z = 2(0) + 8(4) = 32$
At $(0, 0)$: $z = 2(0) + 8(0) = 0$
At $(2, 0)$: $z = 4(0) + 3(2) = 6$

The maximum value is 32 at (0, 4).
The minimum value is 0 at (0, 0).

4. $z = 7x + 3y$

At $(0, 4)$: $z = 7(0) + 3(4) = 12$
At $(0, 0)$: $z = 7(0) + 3(0) = 0$
At $(2, 0)$: $z = 7(2) + 3(0) = 14$

The maximum value is 14 at (2, 0).
The minimum value is 0 at (0, 0).

6. $z = 4x + 3y$

At $(0, 4)$: $z = 4(0) + 3(4) = 12$
At $(3, 0)$: $z = 4(3) + 3(0) = 12$
At $(5, 3)$: $z = 4(5) + 3(3) = 29$
At $(2, 0)$: $z = 2(2) + 8(0) = 4$

The maximum value is 29 at (5, 3).
The minimum value is 6 at (0, 2).

8. $z = x + 6y$

At $(0, 4)$: $z = 0 + 6(4) = 24$
At $(3, 0)$: $z = 3 + 6(0) = 3$
At $(5, 3)$: $z = 5 + 6(3) = 23$
At $(0, 2)$: $z = 0 + 6(2) = 12$

The maximum value is 24 at (0, 4).
The minimum value is 3 at (3, 0).

10. $z = 50x + 35y$

At $(0, 800)$: $z = 50(0) + 35(800) = 28{,}000$
At $(900, 0)$: $z = 50(900) + 35(0) = 45{,}000$
At $(675, 0)$: $z = 50(675) + 35(0) = 33{,}750$
At $(0, 600)$: $z = 50(0) + 35(600) = 21{,}000$

The maximum value is 45,000 at (900, 0).
The minimum value is 21,000 at (0, 600).

12. $z = 16x + 18y$

At $(0, 800)$: $z = 16(0) + 18(800) = 14{,}400$
At $(900, 0)$: $z = 16(900) + 18(0) = 14{,}400$
At $(675, 0)$: $z = 16(675) + 18(0) = 10{,}800$
At $(0, 600)$: $z = 16(0) + 18(600) = 10{,}800$

The maximum value is 14,400 at any point along the line segment connecting (0,800) and (900, 0).
The minimum value is 10,800 at any point along the line segment connecting (645, 0) and (0, 600)

14. $z = 7x + 8y$

At $(0, 8)$: $z = 7(0) + 8(8) = 64$
At $(4, 0)$: $z = 7(4) + 8(0) = 28$
At $(0, 0)$: $z = 7(0) + 8(0) = 0$

The maximum value is 64 at $(0, 8)$.
The minimum value is 0 at $(0, 0)$.

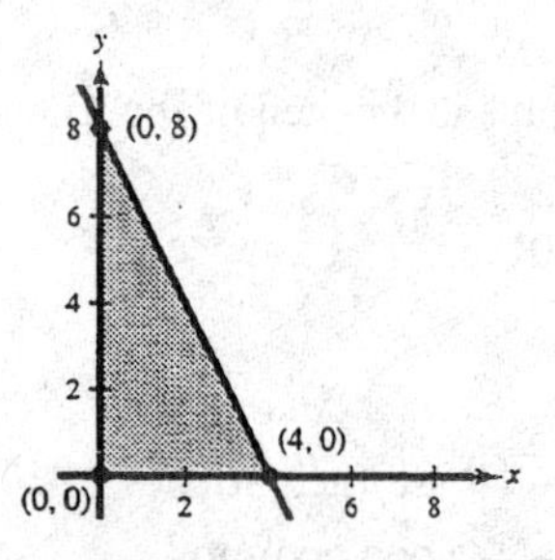

16. $z = 7x + 2y$

At $(0, 8)$: $z = 7(0) + 2(8) = 16$
At $(4, 0)$: $z = 7(4) + 2(0) = 28$
At $(0, 0)$: $z = 7(0) + 2(0) = 0$

The maximum value is 28 at $(4, 0)$.
The minimum value is 0 at $(0, 0)$.

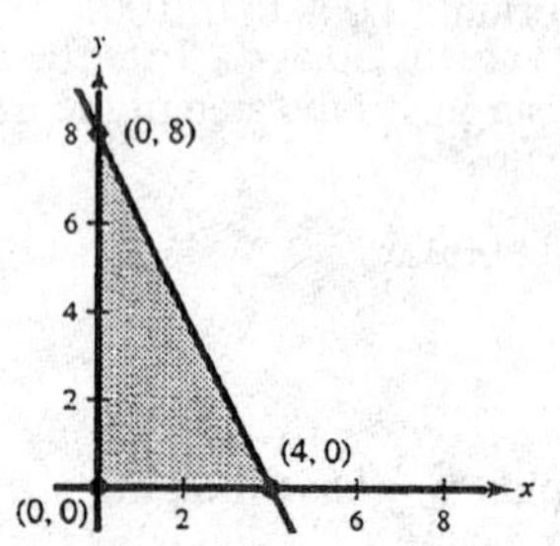

18. $z = 4x + 5y$

At $(0, 0)$: $z = 4(0) + 5(0) = 0$
At $(5, 0)$: $z = 4(5) + 5(0) = 20$
At $(4, 1)$: $z = 4(4) + 5(1) = 21$
At $(0, 3)$: $z = 4(0) + 5(3) = 15$

The maximum value is 21 at $(4, 1)$.
The minimum value is 0 at $(0, 0)$.

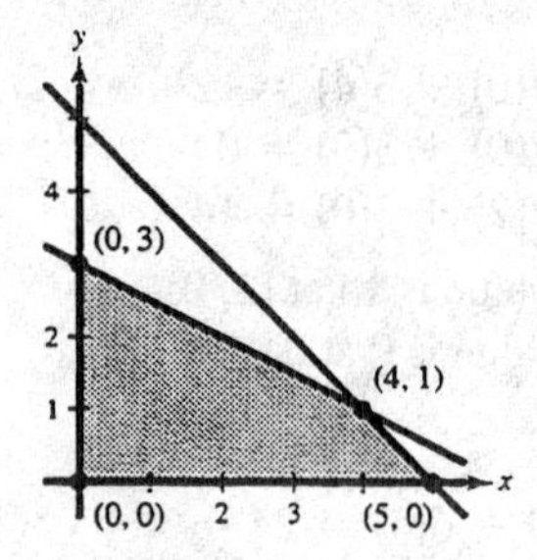

20. $z = 2x - y$

At $(0, 0)$: $z = 2(0) - 0 = 0$
At $(5, 0)$: $z = 2(5) - 0 = 10$
At $(4, 1)$: $z = 2(4) - 0 = 7$
At $(0, 3)$: $z = 2(0) - 3 = -3$

The maximum value is 10 at $(5, 0)$.
The minimum value is -3 at $(0, 3)$.

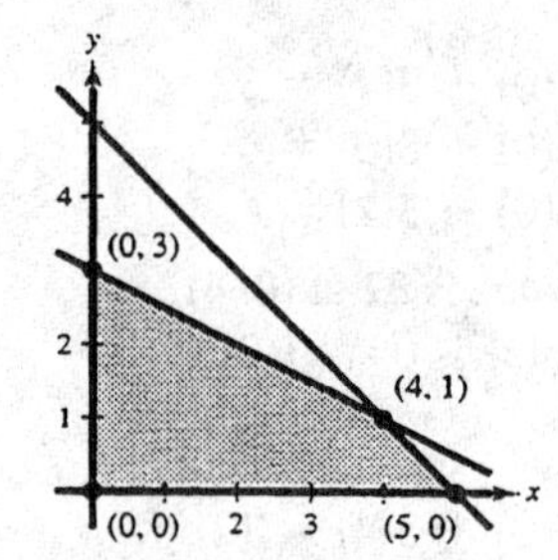

22. $z = x$

At $(0, 0)$: $z = 0$
At $(12, 0)$: $z = 12$
At $(10, 8)$: $z = 10$
At $(6, 16)$: $z = 6$
At $(0, 20)$: $z = 0$

The maximum value is 12 at $(12, 0)$. The minimum value is 0 at any point along the line segment connecting $(0, 0)$ and $(0, 20)$.

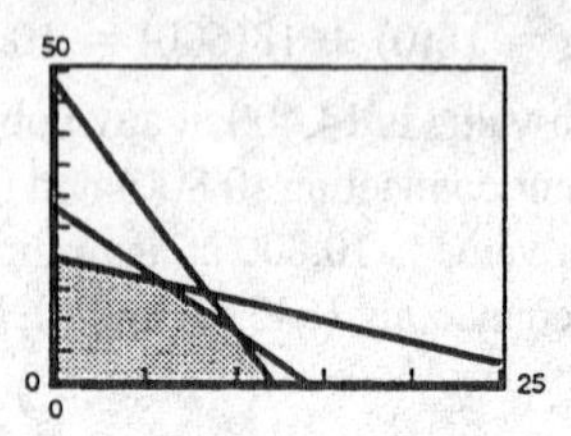

24. $z = y$

At $(0, 0)$: $z = 0$
At $(12, 0)$: $z = 0$
At $(10, 8)$: $z = 8$
At $(6, 16)$: $z = 16$
At $(0, 20)$: $z = 20$

The maximum value is 20 at $(0, 20)$.
The minimum value is 0 at any point along the line segment connecting $(0, 0)$ and $(12, 0)$.

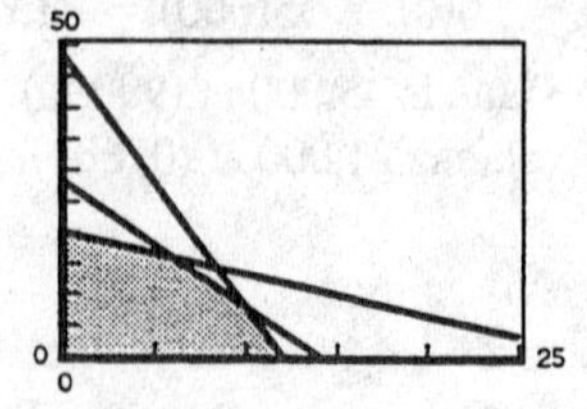

26. $z = 3x + 2y$

At $(0, 0)$: $z = 0$
At $(12, 0)$: $z = 36$
At $(10, 8)$: $z = 46$
At $(6, 16)$: $z = 50$
At $(0, 20)$: $z = 40$

The maximum value is 50 at $(6, 16)$. The minimum value is 0 at $(0, 0)$.

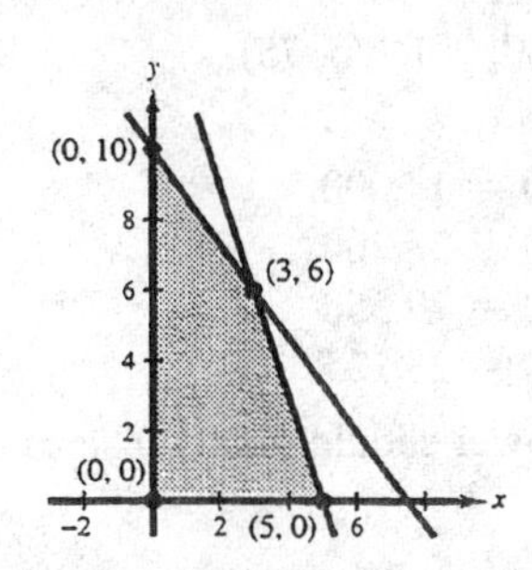

Figure for Exercises 28 and 30

28. $z = 5x + y$

At $(0, 10)$: $z = 5(0) + (10) = 10$
At $(3, 6)$: $z = 5(3) + (6) = 21$
At $(5, 0)$: $z = 5(5) + (0) = 25$
At $(0, 0)$: $z = 5(0) + (0) = 0$

The maximum value is 25 at $(5, 0)$.

30. $z = 3x + y$

At $(0, 10)$: $z = 3(0) + (10) = 10$
At $(3, 6)$: $z = 3(3) + (6) = 15$
At $(5, 0)$: $z = 3(5) + (0) = 15$
At $(0, 0)$: $z = 3(0) + (0) = 0$

The maximum value is 15 at any point along the line segment connecting $(3, 6)$ and $(5, 0)$.

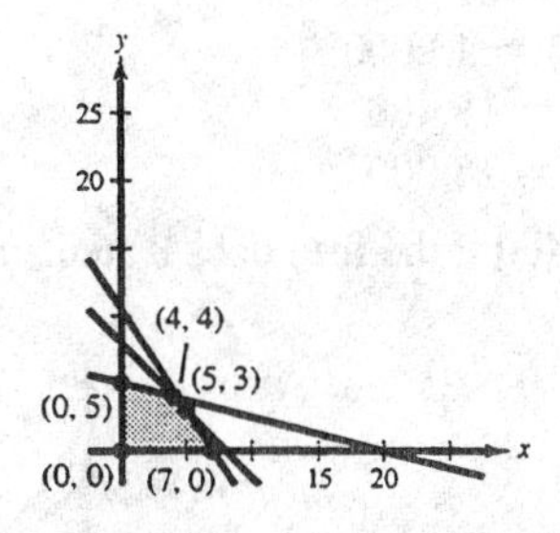

Figure for Exercises 32 and 34

32. $z = 2x + 4y$

At $(0, 5)$: $z = 2(0) + 4(5) = 20$
At $(4, 4)$: $z = 2(4) + 4(4) = 24$
At $(5, 3)$: $z = 2(5) + 4(3) = 22$
At $(7, 0)$: $z = 2(7) + 4(0) = 14$
At $(0, 0)$: $z = 2(0) + 4(0) = 0$

The maximum value is 24 at $(4, 4)$.

34. $z = 4x + y$

At $(0, 5)$: $z = 4(0) + (5) = 5$
At $(4, 4)$: $z = 4(4) + (4) = 20$
At $(5, 3)$: $z = 4(5) + (3) = 23$
At $(7, 0)$: $z = 4(7) + (0) = 28$
At $(0, 0)$: $z = 4(0) + (0) = 0$

The maximum value is 28 at $(7, 0)$.

36. There are an infinite number of objective functions that would have a maximum at $(5, 0)$. One such objective function is $z = x + y$.

38. There are an infinite number of objective functions that would have a minimum at $(5, 0)$. One such objective function is $z = -10x + y$.

40. $x =$ number of acres for crop A; $y =$ number of acres for crop B

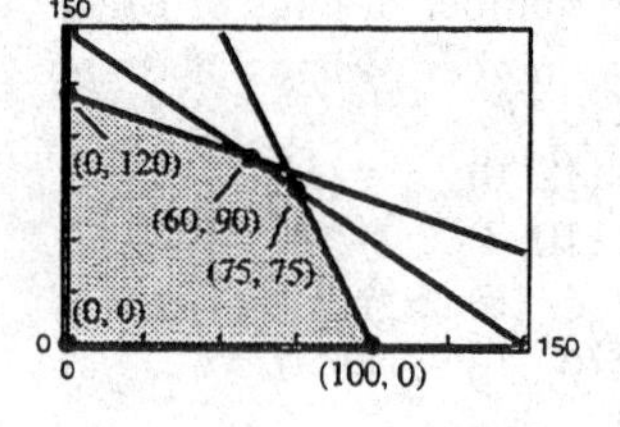

Constraints: $x + y \le 150$

$x + 2y \le 240$

$0.3x + 0.1y \le 30$

$x \ge 0$

$y \ge 0$

Objective function: $P = 140x + 235y$

Vertices: (0, 0), (100, 0), (0, 120), (60, 90), (75, 75)

At (0, 0): $P = 140(0) + 235(0) = 0$
At (100, 0): $P = 140(100) + 235(0) = 14{,}000$
At (0, 120): $P = 140(0) + 235(120) = 28{,}200$
At (60, 90): $P = 140(60) + 235(90) = 29{,}550$
At (75, 75): $P = 140(75) + 235(75) = 28{,}125$

To maximize the profit, the fruit grower should plant 60 acres of crop A and 90 acres of crop B. The maximum profit would be $29,550.

42. Let $x =$ number of audits.
Let $y =$ number of tax returns.

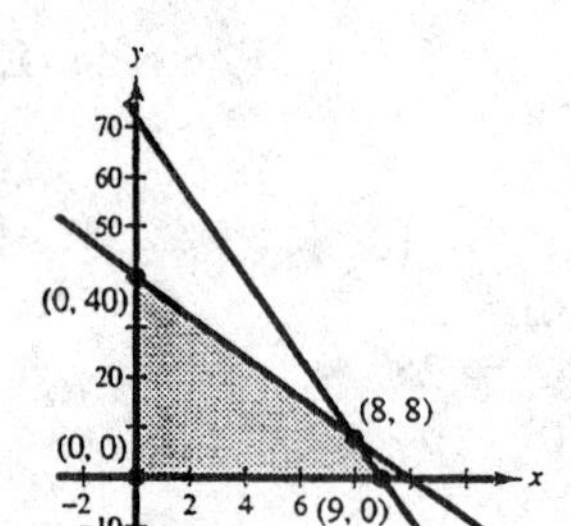

Constraints: $100x + 12.5y \le 900$

$10x + 2.5y \le 100$

$x \ge 0$

$y \ge 0$

Objective function: $R = 2000x + 300y$

Vertices: (0, 0), (0, 40), (8, 8), (9, 0)

At (0, 0): $R = 2000(0) + 300(0) = 0$
At (0, 40): $R = 2000(0) + 300(40) = 12{,}000$
At (8, 8): $R = 2000(8) + 300(8) = 18{,}400$
At (9, 0): $R = 2000(9) + 300(0) = 18{,}000$

The revenue will be maximum $18,400 if the firm does 8 audits and 8 tax returns each week.

44. $x =$ number of Model A
$y =$ number of Model B

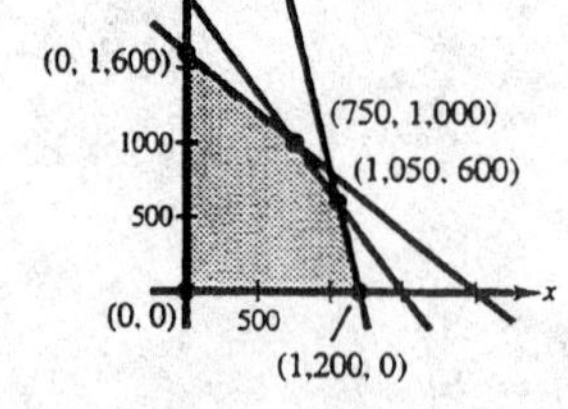

Constraints: $2x + 2.5y \le 4000$

$4x + y \le 4800$

$x + 0.75y \le 1500$

$x \ge 0$

$y \ge 0$

Objective function: $P = 45x + 50y$

Vertices: (0, 0), (0, 1600), (750, 1000), (1050, 600), (1200, 0)

At (0, 0): $P = 45(0) + 50(0) = 0$
At (0, 1600): $P = 45(0) + 50(1600) = 80{,}000$
At (750, 1000): $P = 45(750) + 50(1000) = 83{,}750$
At (1050, 600): $P = 45(1050) + 50(600) = 77{,}250$
At (1200, 0): $P = 45(1200) + 50(0) = 54{,}000$

The maximum profit $83,750 occurs when 750 units of Model A and 1000 units of Model B are produced.

46. x = number of bags of Brand X

y = number of bags of Brand Y

Constraints: $2x + y \geq 12$

$2x + 9y \geq 36$

$2x + 3y \geq 24$

$x \geq 0$

$y \geq 0$

Objective function: $C = 25x + 20y$

Vertices: $(0, 12), (3, 6), (9, 2), (18, 0)$

At $(1, 12)$: $C = 25(0) + 20(12) = 240$
At $(3, 6)$: $C = 25(3) + 20(6) = 195$
At $(9, 2)$: $C = 25(9) + 20(2) = 265$
At $(18, 0)$: $C = 25(18) + 20(0) = 450$

To minimize cost, use three bags of Brand X and six bags of Brand Y for a total cost of \$195 and \$21.67 per bag.

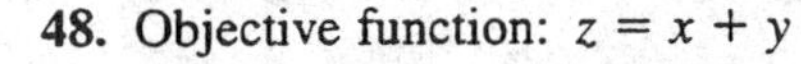

48. Objective function: $z = x + y$

Constraints: $x \geq 0$

$y \geq 0$

$-x + y \leq 1$

$-x + 2y \leq 4$

At $(0, 0)$: $z = 0 + 0 = 0$
At $(0, 1)$: $z = 0 + 1 = 1$
At $(2, 3)$: $z = 2 + 3 = 5$

The constraints do not form a closed set of points. Therefore, $z = x + y$ is unbounded.

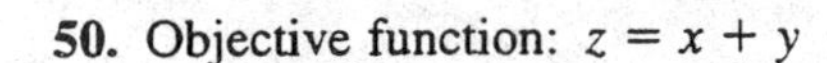

50. Objective function: $z = x + y$

Constraints: $x \geq 0$

$y \geq 0$

$-x + y \leq 0$

$-3x + y \leq 3$

The feasible set is empty.

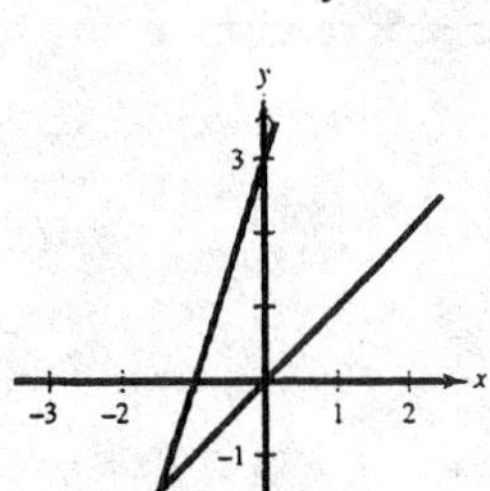

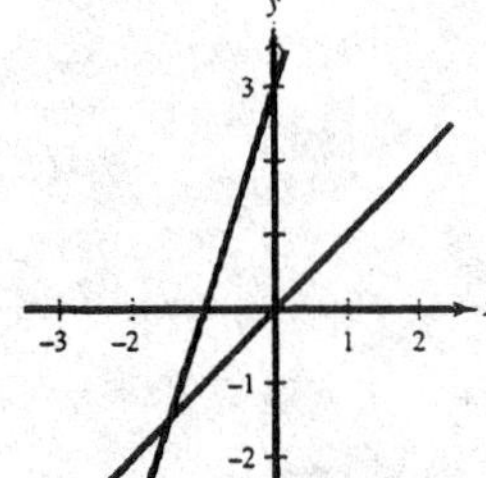

52. Objective function: $z = x + 2y$

Constraints: $x \geq 0$

$y \geq 0$

$x + 2y \leq 4$

$2x + y \leq 4$

At $(0, 0)$: $z = 0 + 2(0) = 0$
At $(0, 2)$: $z = 0 + 2(2) = 4$
At $\left(\frac{4}{3}, \frac{4}{3}\right)$: $z = \frac{4}{3} + 2\left(\frac{4}{3}\right) = 4$
At $(2, 0)$: $z = 2 + 2(0) = 2$

The maximum value is 4 at any point along the line segment connecting $(0, 2)$ and $\left(\frac{4}{3}, \frac{4}{3}\right)$.

54. Constraints: $x \geq 0, y \geq 0, x + 2y \geq 4, x - y \leq 1$

$z = 3x + ty$

At $(0, 0)$: $z = 3(0) + t(0) = 0$
At $(1, 0)$: $z = 3(1) + t(0) = 3$
At $(2, 1)$: $z = 3(2) + t(1) = 6 + t$
At $(0, 2)$: $z = 3(0) + t(2) = 2t$

(a) For the maximum value to be at $(2, 1)$, $z = 6 + t$ must be greater than $z = 2t$ and $z = 3$.

$6 + t > 2t$ and $6 + t > 3$

$6 > t$ $\quad t > -3$

Thus, $-3 < t < 6$.

(b) For maximum value to be at $(0, 2)$, $z = 2t$ must be greater than $z = 6 + t$ and $z = 3$.

$2t > 6 + t$ and $2t > 3$

$6 > t$ $\quad t > \frac{3}{2}$

Thus, $t > 6$.

56. $$\frac{\left(1 + \frac{2}{x}\right)}{x - \frac{4}{x}} = \frac{\frac{x + 2}{x}}{\frac{x^2 - 4}{x}} = \frac{x + 2}{x} \cdot \frac{x}{x^2 - 4} = \frac{x + 2}{x} \cdot \frac{x}{(x + 2)(x - 2)} = \frac{1}{x - 2}$$

58. $$\frac{\left(\frac{1}{x + 1} + \frac{1}{2}\right)}{\left(\frac{3}{2x^2 + 4x + 2}\right)} = \frac{x + 3}{2(x + 1)} \cdot \frac{2(x + 1)^2}{3} = \frac{(x + 3)(x + 1)}{3}$$

❑ Review Exercises for Chapter 7

Solutions to Even-Numbered Exercises

2. $2x = 3(y - 1)$

$y = x$

$2y = 3y - 2$

$3 = y$

$3 = x$

Answer: $(3, 3)$

4. $x^2 + y^2 = 169$

$3x + 2y = 39 \Rightarrow x = \frac{1}{3}(39 - 2y)$

$\left[\frac{1}{3}(39 - 2y)\right]^2 + y^2 = 169$

$\frac{1}{9}(1521 - 156y + 4y^2) + y^2 = 169$

$169 - \frac{52}{3}y + \frac{4}{9}y^2 + y^2 = 169$

$\frac{13}{9}y^2 - \frac{52}{3}y = 0$

$\frac{13}{3}y\left(\frac{1}{3}y - 4\right) = 0 \Rightarrow y = 0, 12$

$y = 0$: $x = \frac{1}{3}(39 - 2(0)) = 13$

$y = 12$: $x = \frac{1}{3}(30 - 2(12)) = 5$

Answer: $(13, 0), (5, 12)$

6. $x = y + 3$

$x = y^2 + 1$

$y + 3 = y^2 + 1$

$0 = y^2 - y - 2$

$0 = (y - 2)(y + 1) \Rightarrow y = 2, -1$

$y = 2$: $x = 2 + 3 = 5$

$y = -1$: $x = -1 + 3 = 2$

Answer: $(5, 2), (2, -1)$

8. $y = 2x^2 - 4x + 1$

$y = x^2 - 4x + 3$

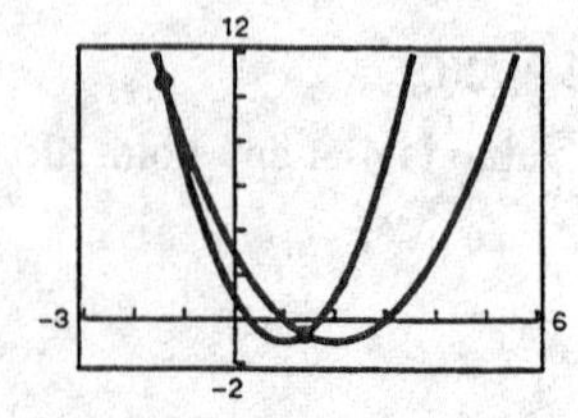

Points of intersection:

$(\sqrt{2}, 5 - 4\sqrt{2}), (-\sqrt{2}, 5 + 4\sqrt{2})$ or

$(1.41, -0.66), (-1.41, 10.66)$

10. $y = \ln(x - 1) - 3$

$y = 4 - \frac{1}{2}x$

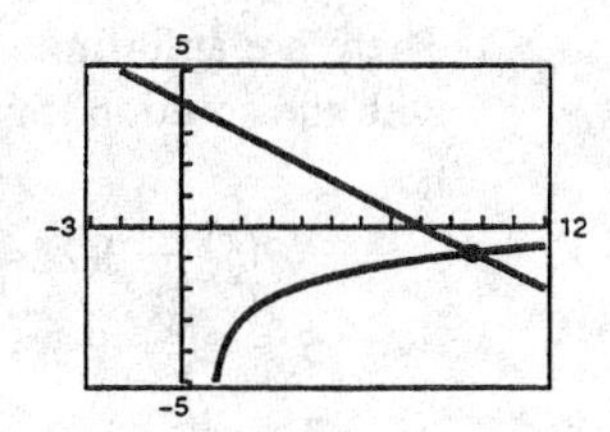

Point of intersection: $(9.68, -0.84)$

12. $40x + 30y = 24 \Rightarrow 40x + 30y = 24$

$20x - 50y = -14 \Rightarrow -40x + 100y = 28$

$130y = 52$

$y = \frac{2}{5}$

$x = \frac{3}{10}$

Answer: $\left(\frac{3}{10}, \frac{2}{5}\right)$

14. $12x + 42y = -17 \Rightarrow 36x + 126y = -51$

$30x - 18y = 19 \Rightarrow 210x + 126y = 133$

$246x = 82$

$x = \frac{1}{3}$

$y = -\frac{1}{2}$

Answer: $\left(\frac{1}{3}, -\frac{1}{2}\right)$

16. $7x + 12y = 63 \Rightarrow -7x - 12y = -63$

$2x + 3y = 15 \Rightarrow 8x + 12y = 60$

$x = -3$

$y = 7$

Answer: $(-3, 7)$

18. $1.5x + 1.5y = 8.5 \Rightarrow 3x + 5y = 17$

$6x + 10y = 24 \Rightarrow -3x - 5y = -12$

$0 = 5$

Inconsistent; no solution

20. There are an infinite number of linear systems with the solution $(-6, 8)$. One possible solution is:

$-6 + 8 = 2 \Rightarrow x + y = 2$

$2(-6) + 3(8) = 12 \Rightarrow 2x + 3y = 12$

22. $y = 22{,}500 + 0.015x$

$y = 20{,}000 + 0.02x$

$22{,}500 + 0.015x = 20{,}000 + 0.02x$

$2500 = 0.005x$

$\$500{,}000 = x$

Answer: \$500,000 or more

24. $2l + 2w = 480$

$l = 1.50w$

$2(1.50w) + 2w = 480$

$5w = 480$

$w = 96$

$l = 144$

The dimensions are 96×144 meters.

26. Supply = Demand

$45 + 0.0002x = 120 - 0.0001x$

$0.0003x = 75$

$x = 250{,}000$ units

$p = \$95.00$

Points of equilibrium: $(250{,}000, 95)$

28.

$$\begin{aligned} x + 2y + 6z &= 4 \\ -3x + 2y - z &= -4 \\ 4x + 2z &= 16 \end{aligned}$$

$$\begin{aligned} x + 2y + 6z &= 4 \\ 8y + 17z &= 8 \\ -8y - 22z &= 0 \end{aligned}$$

$$\begin{aligned} x + 2y + 6z &= 4 \\ 8y + 17z &= 8 \\ -5z &= 8 \end{aligned}$$

$-5z = 8 \Rightarrow z = -\frac{8}{5}$

$8y + 17\left(-\frac{8}{5}\right) = 8 \Rightarrow y = \frac{22}{5}$

$x + 2\left(\frac{22}{5}\right) + 6\left(-\frac{8}{5}\right) = 4 \Rightarrow x = \frac{24}{5}$

Answer: $\left(\frac{24}{5}, \frac{22}{5}, -\frac{8}{5}\right)$ or $(4.8, 4.4, -1.6)$

30. $2x \quad + 6z = -9$ Equation 1
$3x - 2y + 11z = -16$ Equation 2
$3x - y + 7z = -11$ Equation 3

$-x + 2y - 5z = 7$ $-$Eq.2 + Eq.1
$3x - 2y + 11z = -16$
$3x - y + 7z = -11$

$-x + 2y - 5z = 7$
$4y - 4z = 5$ 3Eq.1 + Eq.2
$5y - 8z = 10$ 3Eq.1 + Eq.2

$-x + 2y - 5z = 7$
$4y - 4z = 5$
$-3y \quad = 0$ $-$2Eq.2 + Eq.3

$$-3y = 0 \Rightarrow y = 0$$
$$4(0) - 4z = 5 \Rightarrow z = -\tfrac{5}{4}$$
$$-x + 2(0) - 5\left(-\tfrac{5}{4}\right) = 7 \Rightarrow x = -\tfrac{3}{4}$$

Answer: $\left(-\frac{3}{4}, 0, -\frac{5}{4}\right)$

32. There are an infinite number of linear systems with the solution $\left(5, \frac{3}{2}, 2\right)$. One possible solution is:

$$2(5) + 2\left(\tfrac{3}{2}\right) - 3(2) = 7 \Rightarrow 2x + 2y - 3z = 7$$
$$5 - 2\left(\tfrac{3}{2}\right) + 2 = 4 \Rightarrow x - 2y + z = 4$$
$$-5 + 4\left(\tfrac{3}{2}\right) - 2 = -1 \Rightarrow -x + 4y - z = -1$$

34. $y = ax^2 + bx + c$ through $(-5, 6), (1, 0), (2, 20)$.

$(-5, 6)$: $6 = 25a - 5b + c \Rightarrow 24a - 6b = 6$
$(1, 0)$: $0 = a + b + c \Rightarrow c = -a - b$
$(2, 20)$: $20 = 4a + 2b + c \Rightarrow -8(3a + b = 20)$

$$-14b = -154$$
$$b = 11$$
$$a = 3$$
$$c = -11 - 3 = -14$$

The equation of the parabola is $y = 3x^2 + 11x - 14$.

36. (a) $7b + 156.8a = 169.5$
$156.8b + 3522.78a = 3806.8$

In the first equation, $a = \dfrac{169.5 - 7b}{156.8}$.

Substituting into the second equation:

$$156.8b + 3522.78\left(\frac{169.5 - 7b}{156.8}\right) = 3806.8 \Rightarrow b = 2.799 \text{ and } a = 0.956$$

Thus, $y = 0.956x + 2.799$.

(b)

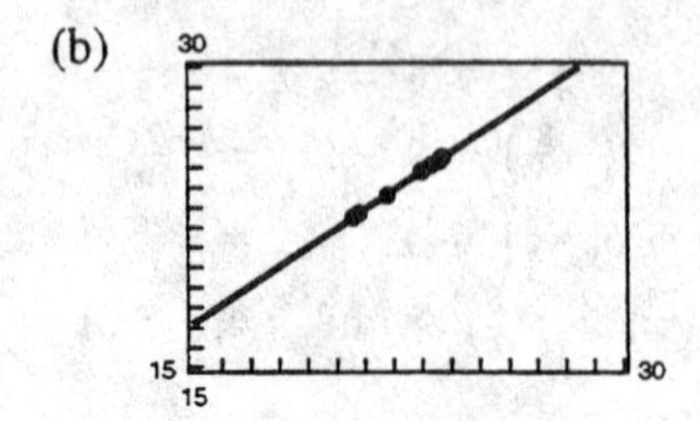

(c) Yes, the data is nearly linear.

(d) The slope shows that a change of one year for the median age for a man's first marriage results in a change of 0.956 in the age of a woman.

38. $\dfrac{-x}{x^2 + 3x + 2} = \dfrac{A}{x + 1} + \dfrac{B}{x + 2}$

$-x = A(x + 2) + B(x + 1)$

Let $x = -1$: $1 = A$

Let $x = -2$: $2 = -B \Rightarrow B = -2$

$\dfrac{-x}{x^2 + 3x + 2} = \dfrac{1}{x + 1} - \dfrac{2}{x + 2}$

40. $\dfrac{9}{x^2 - 9} = \dfrac{A}{x - 3} + \dfrac{B}{x + 3}$

$9 = A(x + 3) + B(x - 3)$

Let $x = 3$: $9 = 6A \Rightarrow A = \dfrac{3}{2}$

Let $x = -3$: $9 = -6B \Rightarrow B = -\dfrac{3}{2}$

$\dfrac{9}{x^2 - 9} = \dfrac{1}{2}\left(\dfrac{3}{x - 3} - \dfrac{3}{x + 3}\right)$

42. $\dfrac{3x^3 + 4x}{(x^2 + 1)^2} = \dfrac{Ax + B}{x^2 + 1} + \dfrac{Cx + D}{(x^2 + 1)^2}$

$3x^3 + 4x = (Ax + B)(x^2 + 1) + Cx + D = Ax^3 + Bx^2 + (A + C)x + B + D$

Equating coefficients of like powers:

$3 = A$

$0 = B$

$4 = 3 + C \Rightarrow C = 1$

$0 = B + D \Rightarrow D = 0$

$\dfrac{3x^3 + 4x}{(x^2 + 1)^2} = \dfrac{3x}{x^2 + 1} + \dfrac{x}{(x^2 + 1)^2}$

44. $2x + 3y \le 24$

$2x + y \le 16$

$x \ge 0$

$y \ge 0$

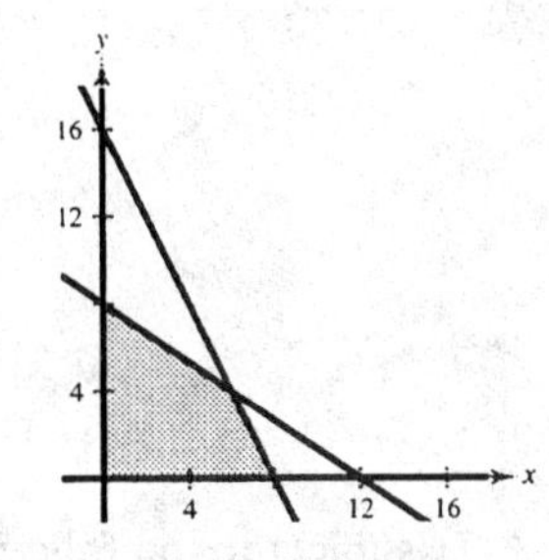

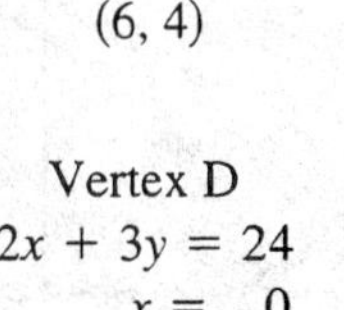

Vertex A
$2x + 3y = 24$
$2x + y = 16$
$(6, 4)$

Vertex B
$2x + y = 16$
$x = 0$
$(0, 16)$
Outside the region

Vertex C
$x = 0$
$y = 0$
$(0, 0)$

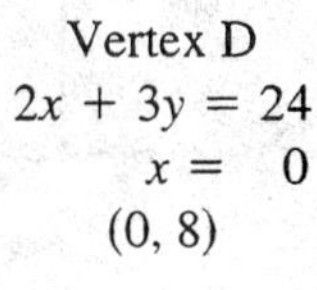

Vertex D
$2x + 3y = 24$
$x = 0$
$(0, 8)$

Vertex E
$2x + 3y = 24$
$y = 0$
$(12, 0)$
Outside the region

Vertex F
$2x + y = 16$
$y = 0$
$(8, 0)$

46. $2x + y \ge 16$

$x + 3y \ge 18$

$0 \le x \le 25$

$0 \le y \le 25$

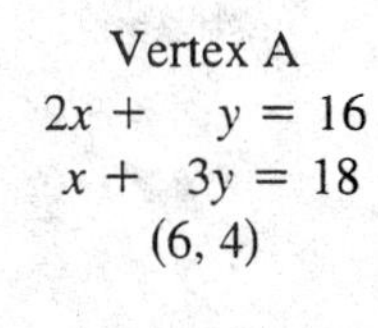

Vertex A
$2x + y = 16$
$x + 3y = 18$
$(6, 4)$

Vertex B
$x + 3y = 18$
$x = 0$
$(0, 6)$
Outside the region

Vertex C
$x + 3y = 18$
$x = 25$
$(25, -7/3)$
Outside the region

Vertex D
$x = 0$
$y = 0$
$(0, 0)$
Outside the region

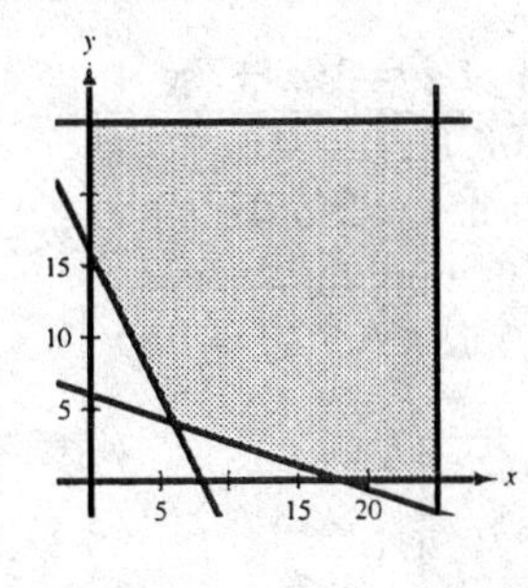

Vertex E
$x = 0$
$y = 25$
$(0, 25)$

Vertex F
$x = 25$
$y = 0$
$(25, 0)$

Vertex G
$x = 25$
$y = 25$
$(25, 25)$

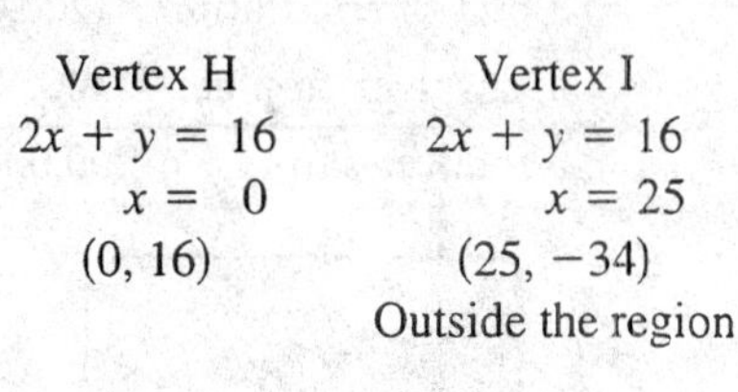

Vertex H
$2x + y = 16$
$x = 0$
$(0, 16)$

Vertex I
$2x + y = 16$
$x = 25$
$(25, -34)$
Outside the region

Vertex J
$2x + y = 16$
$y = 0$
$(8, 0)$
Outside the region

Vertex K
$2x + y = 16$
$y = 25$
$(-4.5, 25)$
Outside the region

Vertex L
$x + 3y = 18$
$y = 0$
$(18, 0)$

Vertex M
$x + 3y = 16$
$y = 25$
$(-57, 25)$
Outside the region

48. $y \le 6 - 2 - x^2$

$y \ge x + 6$

Vertices: $x + 6 = 6 - 2x - x^2$

$x^2 + 3x = 0$

$x(x + 3) = 0 \Rightarrow x - 0, -3$

$(0, 6), (-3, 3)$

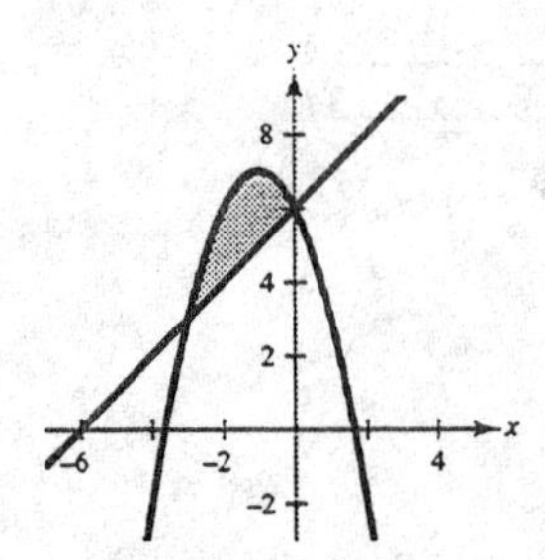

50.

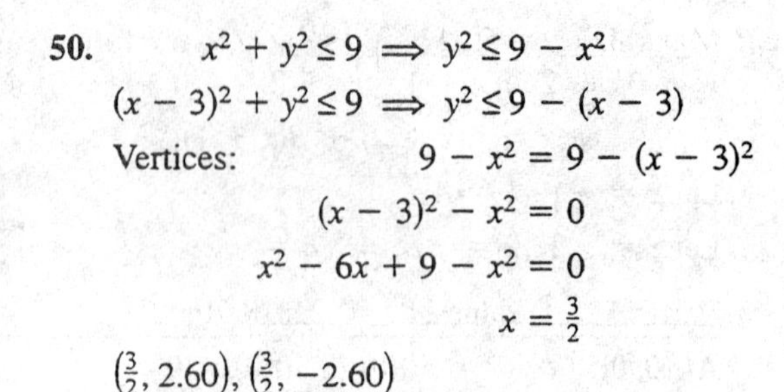

$x^2 + y^2 \le 9 \Rightarrow y^2 \le 9 - x^2$

$(x - 3)^2 + y^2 \le 9 \Rightarrow y^2 \le 9 - (x - 3)$

Vertices: $9 - x^2 = 9 - (x - 3)^2$

$(x - 3)^2 - x^2 = 0$

$x^2 - 6x + 9 - x^2 = 0$

$x = \frac{3}{2}$

$(\frac{3}{2}, 2.60), (\frac{3}{2}, -2.60)$

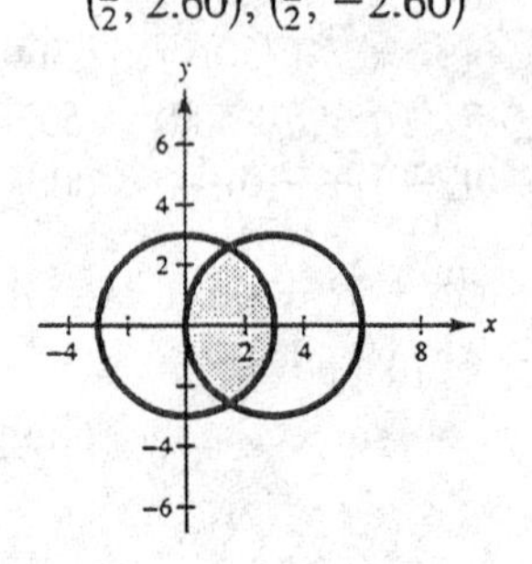

52. Line through (1, 2), (6, 7): $y = x + 1$

Line through (6, 7), (8, 1): $y = -3x + 25$

Line through (1, 2), (8, 1): $y = -\frac{1}{7} + \frac{15}{7} \Rightarrow -x + 15$

System of inequalities: $-x + y \le 1$

$3x + y \le 25$

$x + 7y \ge 15$

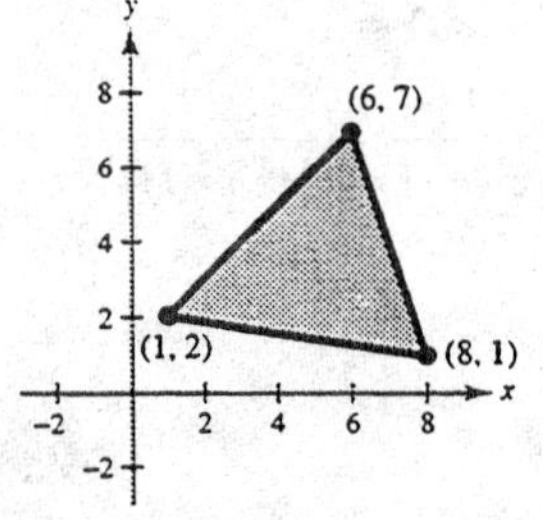

54. x = number of units of Product I

y = number of units of Product II

$20x + 30y \le 24{,}000$

$12x + 8y \le 12{,}400$

$x \ge 0$

$y \ge 0$

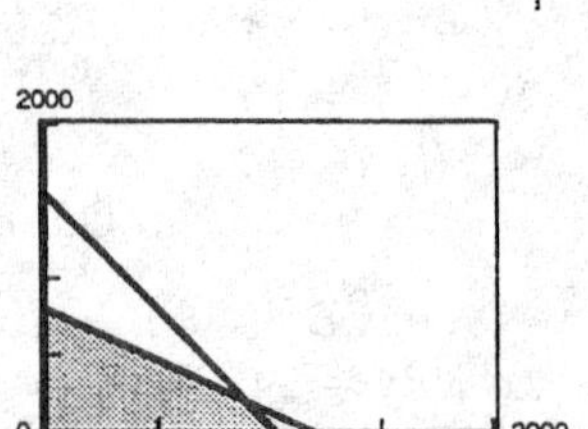

56.

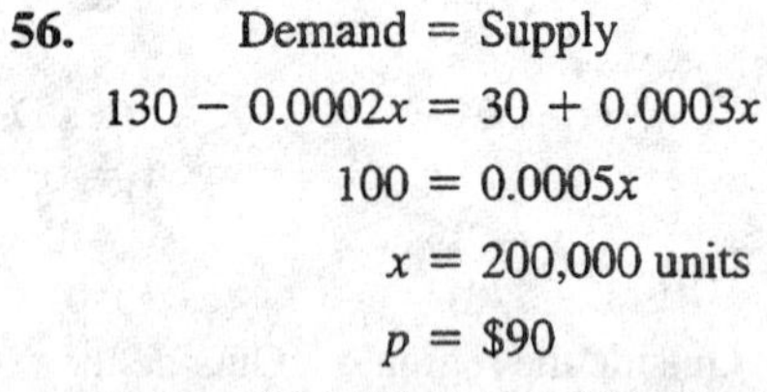

Demand = Supply

$130 - 0.0002x = 30 + 0.0003x$

$100 = 0.0005x$

$x = 200{,}000$ units

$p = \$90$

Point of equilibrium: (200,000, 90)

Consumer surplus: $\frac{1}{2}(200{,}000)(40) = \$4{,}000{,}000$

Producer surplus: $\frac{1}{2}(200{,}000)(60) = \$6{,}000{,}000$

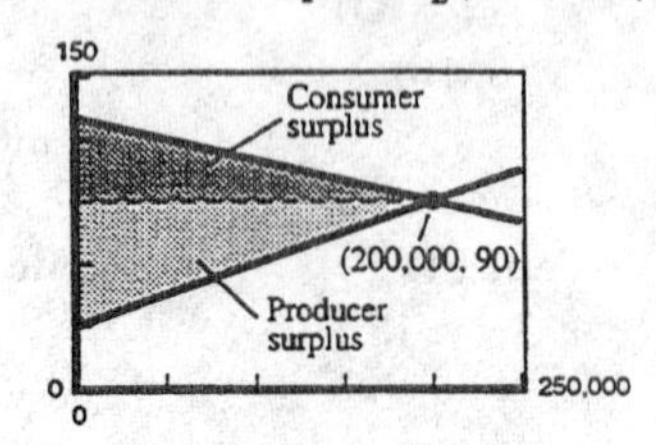

58. Minimize $z = 10x + 7y$ subject to the following constraints:

$x \ge 0$

$y \ge 0$

$2x + y \ge 100$

$x + y \ge 75$

Vertex	Value of $z = 10x + 7y$
At (0, 100):	$z = 10(0) + 7(100) = 700$
At (25, 50):	$z = 10(25) + 7(50)$ $= 600$, minimum value
At (75, 0):	$z = 10(75) + 7(0) = 750$

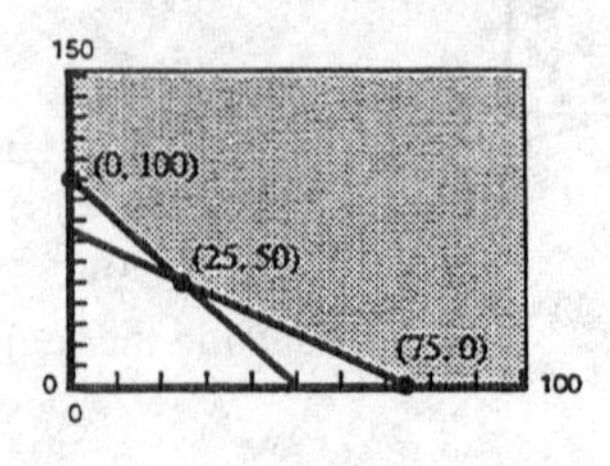

60. Maximize $z = 50x + 70y$ subject to the following constraints:

$$x \geq 0$$
$$y \geq 0$$
$$x + 2y \leq 1500$$
$$5x + 2y \leq 3500$$

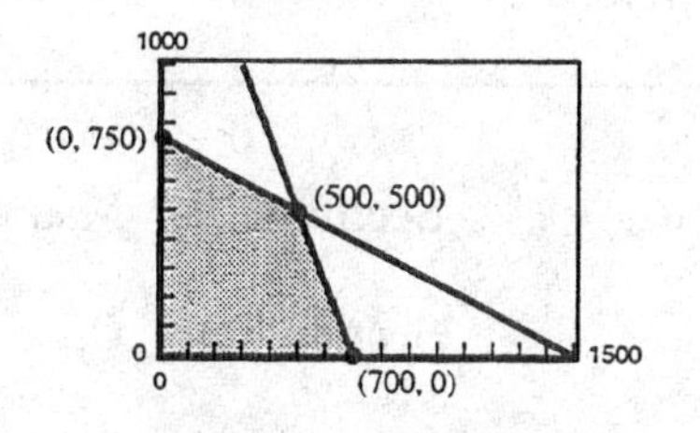

Vertex	Value of $z = 50x + 70y$
At (0, 0):	$z = 50(0) + 70(0) = 0$
At (0, 750):	$z = 50(0) + 77(750) = 52{,}500$
At (500, 500):	$z = 50(500) + 70(500) = 60{,}000$ maximum value
At (700, 0):	$z = 50(700) + 7(0) = 35{,}000$

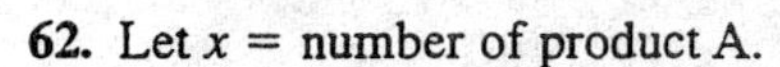

62. Let x = number of product A.

Let y = number of product B.

Maximize $P = 18x + 24y$ subject to the following constraints:

$$4x + 2y \leq 24$$
$$x + 2y \leq 9$$
$$x + y \leq 8$$

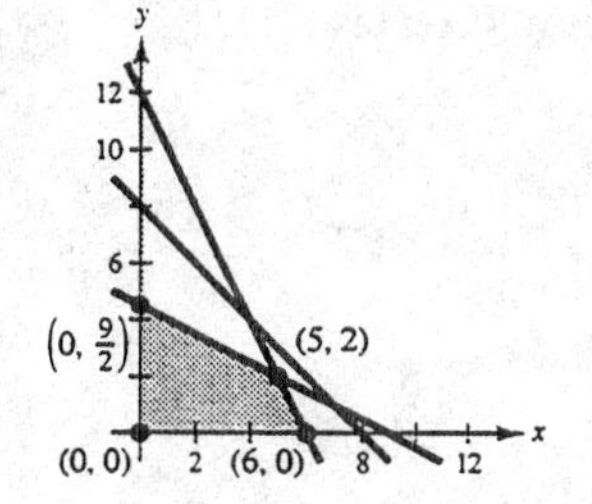

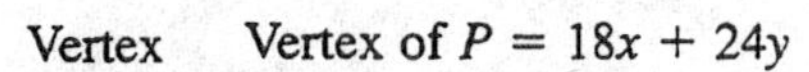

Vertex	Vertex of $P = 18x + 24y$
At (0, 0):	$P = 18(0) + 24(0) = 0$
At (6, 0):	$P = 18(6) + 24(0) = 108$
At (5, 2):	$P = 18(5) + 24(2) = 138$ maximum value
At $\left(0, \frac{9}{2}\right)$:	$P = 19(0) + 24\left(\frac{9}{2}\right) = 108$

The maximum profit is \$138 when we produce 5 units of product A and 2 units of product B.

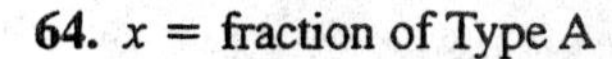

64. x = fraction of Type A

y = fraction of Type B

Constraints: $80x + 92y \geq 88$

$$x + y = 1$$
$$x \geq 0$$
$$y \geq 0$$

Objective function: $C = 1.25x + 1.55y$

At $\left(\frac{1}{3}, \frac{2}{3}\right)$: $C = 1.25\left(\frac{1}{3}\right) + 1.55\left(\frac{2}{3}\right) = 1.45$

The minimum cost is \$1.45 and occurs with a mixture of $\frac{1}{3}A$ and $\frac{2}{3}B$.

CHAPTER 8
Matrices and Determinants

CHAPTER 8
Matrices and Determinants

Section 8.1 Matrices and Systems of Equations

Solutions to Even-Numbered Exercises

2. Since the matrix has one row and four columns, its order is 1×4.

4. Since the matrix has three rows and four columns, its order is 3×4.

6. Since the matrix has one row and one column, its order is 1×1.

8. $7x + 4y = 22$

$5x - 9y = 15$

$$\left[\begin{array}{cc:c} 7 & 4 & 22 \\ 5 & -9 & 15 \end{array}\right]$$

10. $7x - 5y + z = 13$

$19x \qquad -8z = 10$

$$\left[\begin{array}{ccc:c} 7 & -5 & 1 & 13 \\ 19 & 0 & -8 & 10 \end{array}\right]$$

12. $\left[\begin{array}{cc:c} 7 & -5 & 0 \\ 8 & 3 & -2 \end{array}\right]$

$7x - 5y = 0$

$8x + 3y = -2$

14. $\left[\begin{array}{cccc:c} 9 & 12 & 3 & 0 & 0 \\ -2 & 18 & 5 & 2 & 10 \\ 1 & 7 & -8 & 0 & -4 \end{array}\right]$

$$\begin{aligned} 9x + 12y + 3z \qquad &= 0 \\ -2x + 18y + 5z + 2w &= 10 \\ x + 7y - 8z \qquad &= -4 \end{aligned}$$

16. $\begin{bmatrix} 1 & 3 & 0 & 0 \\ 0 & 0 & 1 & 8 \\ 0 & 0 & 0 & 0 \end{bmatrix}$

This matrix is in reduced row-echelon form.

18. $\begin{bmatrix} 1 & 0 & 2 & 1 \\ 0 & 1 & -3 & 10 \\ 0 & 0 & 1 & 0 \end{bmatrix}$

This matrix is in row-echelon form, but not reduced row-echelon form.

20. $\begin{bmatrix} 3 & 6 & 8 \\ 4 & -3 & 6 \end{bmatrix}$

$\frac{1}{3}R_1 \rightarrow \begin{bmatrix} 1 & \boxed{2} & \frac{8}{3} \\ 4 & -3 & 6 \end{bmatrix}$

22. $\begin{bmatrix} 2 & 4 & 8 & 3 \\ 1 & -1 & -3 & 2 \\ 2 & 6 & 4 & 9 \end{bmatrix}$

$\frac{1}{2}R_1 \rightarrow \begin{bmatrix} 1 & \boxed{2} & \boxed{4} & \boxed{\frac{3}{2}} \\ 1 & -1 & -3 & 2 \\ 2 & 6 & 4 & 9 \end{bmatrix}$

$\begin{matrix} \\ -R_1 + R_2 \rightarrow \\ -2R_1 + R_2 \rightarrow \end{matrix} \begin{bmatrix} 1 & 2 & 4 & \frac{3}{2} \\ 0 & \boxed{-3} & -7 & \frac{1}{2} \\ 0 & 2 & \boxed{-4} & \boxed{6} \end{bmatrix}$

24. $\begin{bmatrix} 7 & 1 \\ 0 & 2 \\ -3 & 4 \\ 4 & 1 \end{bmatrix}$

(a) $\begin{bmatrix} 7 & 1 \\ 0 & 2 \\ -3 & 4 \\ 1 & 5 \end{bmatrix}$ (b) $\begin{bmatrix} 1 & 5 \\ 0 & 2 \\ -3 & 4 \\ 7 & 1 \end{bmatrix}$ (c) $\begin{bmatrix} 1 & 5 \\ 0 & 2 \\ 0 & 19 \\ 7 & 1 \end{bmatrix}$

(d) $\begin{bmatrix} 1 & 5 \\ 0 & 2 \\ 0 & 19 \\ 0 & -34 \end{bmatrix}$ (e) $\begin{bmatrix} 1 & 5 \\ 0 & 1 \\ 0 & 19 \\ 0 & -34 \end{bmatrix}$ (f) $\begin{bmatrix} 1 & 0 \\ 0 & 1 \\ 0 & 0 \\ 0 & 0 \end{bmatrix}$ This matrix is in reduced row-echelon form.

26. $\begin{bmatrix} 1 & 2 & -1 & 3 \\ 3 & 7 & -5 & 14 \\ -2 & -1 & -3 & 8 \end{bmatrix}$

$\begin{matrix} \\ -3R_1 + R_2 \rightarrow \\ 2R_1 + R_3 \rightarrow \end{matrix} \begin{bmatrix} 1 & 2 & -1 & 3 \\ 0 & 1 & -2 & 5 \\ 0 & 3 & -5 & 14 \end{bmatrix}$

$\begin{matrix} \\ \\ -3R_2 + R_3 \rightarrow \end{matrix} \begin{bmatrix} 1 & 2 & -1 & 3 \\ 0 & 1 & -2 & 5 \\ 0 & 0 & 1 & -1 \end{bmatrix}$

28. $\begin{bmatrix} 1 & -3 & 0 & -7 \\ -3 & 10 & 1 & 23 \\ 4 & -10 & 2 & -24 \end{bmatrix}$

$\begin{matrix} \\ 3R_1 + R_2 \rightarrow \\ -4R_1 + R_3 \rightarrow \end{matrix} \begin{bmatrix} 1 & -3 & 0 & -7 \\ 0 & 1 & 1 & 2 \\ 0 & 2 & 2 & 4 \end{bmatrix}$

$\begin{matrix} \\ \\ -2R_2 + R_3 \rightarrow \end{matrix} \begin{bmatrix} 1 & -3 & 0 & -7 \\ 0 & 1 & 1 & 2 \\ 0 & 0 & 0 & 0 \end{bmatrix}$

30. $\begin{bmatrix} 1 & 3 & 2 \\ 5 & 15 & 9 \\ 2 & 6 & 10 \end{bmatrix}$

$\begin{matrix} \\ -5R_1 + R_2 \rightarrow \\ -2R_1 + R_3 \rightarrow \end{matrix} \begin{bmatrix} 1 & 3 & 2 \\ 0 & 0 & -1 \\ 0 & 0 & 6 \end{bmatrix}$

$\begin{matrix} \\ \\ 6R_2 + R_3 \rightarrow \end{matrix} \begin{bmatrix} 1 & 3 & 0 \\ 0 & 0 & -1 \\ 0 & 0 & 0 \end{bmatrix}$

$\begin{matrix} \\ -1R_2 \rightarrow \\ \\ \end{matrix} \begin{bmatrix} 1 & 3 & 0 \\ 0 & 0 & 1 \\ 0 & 0 & 0 \end{bmatrix}$

32. $\begin{bmatrix} 1 & -3 \\ -1 & 8 \\ 0 & 4 \\ -2 & 10 \end{bmatrix}$

$\begin{matrix} \\ R_1 + R_2 \rightarrow \\ \\ 2R_1 + R_4 \rightarrow \end{matrix} \begin{bmatrix} 1 & -3 \\ 0 & 5 \\ 0 & 4 \\ 0 & 4 \end{bmatrix}$

$\begin{matrix} \\ \frac{1}{5}R_2 \rightarrow \\ \\ \\ \end{matrix} \begin{bmatrix} 1 & -3 \\ 0 & 1 \\ 0 & 4 \\ 0 & 4 \end{bmatrix}$

$\begin{matrix} \\ \\ -4R_2 + R_3 \rightarrow \\ -4R_2 + R_4 \rightarrow \end{matrix} \begin{bmatrix} 1 & 0 \\ 0 & 1 \\ 0 & 0 \\ 0 & 0 \end{bmatrix}$

34.
$$\begin{aligned} x + 5y &= 0 \\ y &= -1 \\ x + 5(-1) &= 0 \\ x &= 5 \end{aligned}$$

Answer: $(5, -1)$

36.
$$\begin{aligned} x + 2y - 2z &= -1 \\ y + z &= 9 \\ z &= -3 \\ y + (-3) &= 9 \\ y &= 12 \\ x + 2(12) - 2(-3) &= -1 \\ x &= -31 \end{aligned}$$

Answer: $(-31, 12, -3)$

38. $\begin{bmatrix} 1 & 0 & \vdots & -2 \\ 0 & 1 & \vdots & 4 \end{bmatrix}$

$x = -2$

$y = 4$

Answer: $(-2, 4)$

40. $\begin{bmatrix} 1 & 0 & 0 & \vdots & 3 \\ 0 & 1 & 0 & \vdots & -1 \\ 0 & 0 & 1 & \vdots & 0 \end{bmatrix}$

$x = 3$

$y = -1$

$x = 0$

Answer: $(3, -1, 0)$

42. $2x + 6y = 16$

$2x + 3y = 7$

$$\begin{bmatrix} 2 & 6 & \vdots & 16 \\ 2 & 3 & \vdots & 7 \end{bmatrix}$$

$$-R_1 + R_2 \rightarrow \begin{bmatrix} 2 & 6 & \vdots & 16 \\ 0 & -3 & \vdots & -9 \end{bmatrix}$$

$$\begin{matrix} \frac{1}{2}R_1 \rightarrow \\ -\frac{1}{3}R_2 \rightarrow \end{matrix} \begin{bmatrix} 1 & 3 & \vdots & 8 \\ 0 & 1 & \vdots & 3 \end{bmatrix}$$

$y = 3$

$x + 3(3) = 8 \Rightarrow x = -1$

Answer: $(-1, 3)$

44. $x + 2y = 0$

$x + y = 6$

$3x - 2y = 8$

$$\begin{bmatrix} 1 & 2 & \vdots & 0 \\ 1 & 1 & \vdots & 6 \\ 3 & -2 & \vdots & 8 \end{bmatrix}$$

$$\begin{matrix} \\ -R_1 + R_2 \rightarrow \\ -3R_1 + R_3 \rightarrow \end{matrix} \begin{bmatrix} 1 & 2 & \vdots & 0 \\ 0 & -1 & \vdots & 6 \\ 0 & -8 & \vdots & 8 \end{bmatrix}$$

$$\begin{matrix} \\ -R_2 \rightarrow \\ -8R_2 + R_3 \rightarrow \end{matrix} \begin{bmatrix} 1 & 2 & \vdots & 0 \\ 0 & 1 & \vdots & 6 \\ 0 & 0 & \vdots & -40 \end{bmatrix}$$

The system in inconsistent and there is no solution.

46. $2x - y = -0.1$

$3x + 2y = 1.6$

$$\begin{bmatrix} 2 & -1 & \vdots & -0.1 \\ 3 & 2 & \vdots & 1.6 \end{bmatrix}$$

$$-R_2 + R_1 \rightarrow \begin{bmatrix} -1 & -3 & \vdots & -1.7 \\ 3 & 2 & \vdots & 1.6 \end{bmatrix}$$

$$3R_1 + R_2 \rightarrow \begin{bmatrix} -1 & -3 & \vdots & -1.7 \\ 0 & -7 & \vdots & -3.5 \end{bmatrix}$$

$$\begin{matrix} -R_1 \rightarrow \\ -\frac{1}{7}R_2 \rightarrow \end{matrix} \begin{bmatrix} 1 & 3 & \vdots & 1.7 \\ 0 & 1 & \vdots & 0.5 \end{bmatrix}$$

$y = 0.5$

$x + 3(0.5) = 1.7 \Rightarrow x = 0.2$

Answer: $(0.2, 0.5)$

48. $x - 3y = 5$

$-2x + 6y = -10$

$$\begin{bmatrix} 1 & -3 & \vdots & 5 \\ -2 & 6 & \vdots & -10 \end{bmatrix}$$

$$2R_1 + R_2 \rightarrow \begin{bmatrix} 1 & -3 & \vdots & 5 \\ 0 & 0 & \vdots & 0 \end{bmatrix}$$

$y = a$

$x = 3a + 5$

Answer: $(3a + 5, a)$

50. $2x - y + 3z = 24$
$2y - z = 14$
$7x - 5y = 6$

$$\begin{bmatrix} 2 & -1 & 3 & \vdots & 24 \\ 0 & 2 & -1 & \vdots & 14 \\ 7 & -5 & 0 & \vdots & 6 \end{bmatrix}$$

$$\begin{matrix} R_3 - 3R_1 \rightarrow \\ \\ \\ \end{matrix} \begin{bmatrix} 1 & -2 & -9 & \vdots & -66 \\ 0 & 2 & -1 & \vdots & 14 \\ 7 & -5 & 0 & \vdots & 6 \end{bmatrix}$$

$$\begin{matrix} \\ \\ -7R_1 + R_3 \rightarrow \end{matrix} \begin{bmatrix} 1 & -2 & -9 & \vdots & -66 \\ 0 & 2 & -1 & \vdots & 14 \\ 0 & 9 & 63 & \vdots & 468 \end{bmatrix}$$

$$\begin{matrix} \\ 4R_2 \rightarrow \\ \\ \end{matrix} \begin{bmatrix} 1 & -2 & -9 & \vdots & -66 \\ 0 & 8 & -4 & \vdots & 56 \\ 0 & 9 & 63 & \vdots & 468 \end{bmatrix}$$

$$\begin{matrix} \\ -R_3 + R_2 \rightarrow \\ \\ \end{matrix} \begin{bmatrix} 1 & -2 & -9 & \vdots & -66 \\ 0 & -1 & -67 & \vdots & -412 \\ 0 & 9 & 63 & \vdots & 468 \end{bmatrix}$$

$$\begin{matrix} \\ \\ 9R_2 + R_3 \rightarrow \end{matrix} \begin{bmatrix} 1 & -2 & -9 & \vdots & -66 \\ 0 & -1 & -67 & \vdots & -412 \\ 0 & 0 & -540 & \vdots & -3240 \end{bmatrix}$$

$$\begin{matrix} \\ -R_2 \rightarrow \\ -\frac{1}{540}R_3 \rightarrow \end{matrix} \begin{bmatrix} 1 & -2 & -9 & \vdots & -66 \\ 0 & 1 & 67 & \vdots & 412 \\ 0 & 0 & 1 & \vdots & 6 \end{bmatrix}$$

$z = 6$
$y + 67(6) = 412 \implies y = 10$
$x - 2(10) - 9(6) = -66 \implies x = 8$
Answer: $(8, 10, 6)$

52. $2x + 3z = 3$
$4x - 3y + 7z = 5$
$8x - 9y + 15z = 9$

$$\begin{bmatrix} 2 & 0 & 3 & \vdots & 3 \\ 4 & -3 & 7 & \vdots & 5 \\ 8 & -9 & 15 & \vdots & 9 \end{bmatrix}$$

$$\begin{matrix} \\ -2R_1 + R_2 \rightarrow \\ -4R_1 + R_3 \rightarrow \end{matrix} \begin{bmatrix} 2 & 0 & 3 & \vdots & 3 \\ 0 & -3 & 1 & \vdots & -1 \\ 0 & -9 & 3 & \vdots & -3 \end{bmatrix}$$

$$\begin{matrix} \\ \\ -3R_2 + R_3 \rightarrow \end{matrix} \begin{bmatrix} 2 & 0 & 3 & \vdots & 3 \\ 0 & -3 & 1 & \vdots & -1 \\ 0 & 0 & 0 & \vdots & 0 \end{bmatrix}$$

$$\begin{matrix} \\ -\frac{1}{3}R_2 \rightarrow \\ \\ \end{matrix} \begin{bmatrix} 1 & 0 & \frac{3}{2} & \vdots & \frac{3}{2} \\ 0 & 1 & -\frac{1}{3} & \vdots & \frac{1}{3} \\ 0 & 0 & 0 & \vdots & 0 \end{bmatrix}$$

$z = a$
$y = \frac{1}{3}a + \frac{1}{3}$
$x = -\frac{3}{2}a + \frac{3}{2}$
Answer: $\left(-\frac{3}{2}a + \frac{3}{2}, \frac{1}{3}a + \frac{1}{3}, a\right)$

54. $4x + 12y - 7z - 20w = 22$
$3x + 9y - 5z - 28w = 30$

$$\begin{bmatrix} 4 & 12 & -7 & -20 & \vdots & 22 \\ 3 & 9 & -5 & -28 & \vdots & 30 \end{bmatrix}$$

$$\begin{matrix} -R_2 + R_1 \rightarrow \\ \\ \end{matrix} \begin{bmatrix} 1 & 3 & -2 & 8 & \vdots & -8 \\ 3 & 9 & -5 & -28 & \vdots & 30 \end{bmatrix}$$

$$\begin{matrix} \\ -3R_1 + R_2 \rightarrow \end{matrix} \begin{bmatrix} 1 & 3 & -2 & 8 & \vdots & -8 \\ 0 & 0 & 1 & -52 & \vdots & 54 \end{bmatrix}$$

$$\begin{matrix} 2R_2 + R_1 \rightarrow \\ \\ \end{matrix} \begin{bmatrix} 1 & 3 & 0 & -96 & \vdots & 100 \\ 0 & 0 & 1 & -52 & \vdots & 54 \end{bmatrix}$$

$w = a$
$z = 52a + 54$
$y = b$
$x = -3b + 96a + 100$
Answer: $(-3b + 96a + 100, b, 52a + 54, a)$

56. $x + 2y = 0$
$2x + 4y = 0$

$$\begin{bmatrix} 1 & 2 & \vdots & 0 \\ 2 & 4 & \vdots & 0 \end{bmatrix}$$

$$\begin{matrix} \\ -2R_1 + R_2 \rightarrow \end{matrix} \begin{bmatrix} 1 & 2 & \vdots & 0 \\ 0 & 0 & \vdots & 0 \end{bmatrix}$$

$y = a$
$x = -2a$
Answer: $(-2a, a)$

58. $\begin{aligned} 2x + 10y + 2z &= 6 \\ x + 5y + 2z &= 6 \\ x + 5y + z &= 3 \\ -3x - 15y - 3z &= -9 \end{aligned}$

$$\begin{bmatrix} 2 & 10 & 2 & \vdots & 6 \\ 1 & 5 & 2 & \vdots & 6 \\ 1 & 5 & 1 & \vdots & 3 \\ -3 & -15 & -3 & \vdots & -9 \end{bmatrix}$$

$$\tfrac{1}{2}R_1 \rightarrow \begin{bmatrix} 1 & 5 & 1 & \vdots & 3 \\ 1 & 5 & 2 & \vdots & 6 \\ 1 & 5 & 1 & \vdots & 3 \\ -3 & -15 & -3 & \vdots & -9 \end{bmatrix}$$

$$\begin{matrix} \\ -R_1 + R_2 \rightarrow \\ -R_1 + R_3 \rightarrow \\ 3R_1 + R_4 \rightarrow \end{matrix} \begin{bmatrix} 1 & 5 & 1 & \vdots & 3 \\ 0 & 0 & 1 & \vdots & 3 \\ 0 & 0 & 0 & \vdots & 0 \\ 0 & 0 & 0 & \vdots & 0 \end{bmatrix}$$

$$-R_2 + R_1 \rightarrow \begin{bmatrix} 1 & 5 & 0 & \vdots & 0 \\ 0 & 0 & 1 & \vdots & 3 \\ 0 & 0 & 0 & \vdots & 0 \\ 0 & 0 & 0 & \vdots & 0 \end{bmatrix}$$

$z = 3, y = a, x = -5a$

Answer: $(-5a, a, 3)$

60. $\begin{aligned} x + 2y + 2z + 4w &= 11 \\ 3x + 6y + 5z + 12w &= 30 \end{aligned}$

$$\begin{bmatrix} 1 & 2 & 2 & 4 & \vdots & 11 \\ 3 & 6 & 5 & 12 & \vdots & 30 \end{bmatrix}$$

$$-3R_1 + R_2 \rightarrow \begin{bmatrix} 1 & 2 & 2 & 4 & \vdots & 11 \\ 0 & 0 & -1 & 0 & \vdots & -3 \end{bmatrix}$$

$$\begin{matrix} 2R_2 + R_1 \rightarrow \\ -R_2 \rightarrow \end{matrix} \begin{bmatrix} 1 & 2 & 0 & 4 & \vdots & 5 \\ 0 & 0 & 1 & 0 & \vdots & 3 \end{bmatrix}$$

$w = a, z = 3, y = b, x = -2b - 4a + 5$

Answer: $(-2b - 4a + 5, b, 3, a)$

62. $\begin{aligned} x + 2y + z + 3w &= 0 \\ x - y + w &= 0 \\ y - z + 2w &= 0 \end{aligned}$

$$\begin{bmatrix} 1 & 2 & 1 & 3 & \vdots & 0 \\ 1 & -1 & 0 & 1 & \vdots & 0 \\ 0 & 1 & -1 & 2 & \vdots & 0 \end{bmatrix}$$

$$-R_1 + R_2 \rightarrow \begin{bmatrix} 1 & 2 & 1 & 3 & \vdots & 0 \\ 0 & -3 & -1 & -2 & \vdots & 0 \\ 0 & 1 & -1 & 2 & \vdots & 0 \end{bmatrix}$$

$$4R_3 + R_2 \rightarrow \begin{bmatrix} 1 & 2 & 1 & 3 & \vdots & 0 \\ 0 & 1 & -5 & 6 & \vdots & 0 \\ 0 & 1 & -1 & 2 & \vdots & 0 \end{bmatrix}$$

$$\begin{matrix} -2R_2 + R_1 \rightarrow \\ \\ -R_2 + R_3 \rightarrow \end{matrix} \begin{bmatrix} 1 & 0 & 11 & -9 & \vdots & 0 \\ 0 & 1 & -5 & 6 & \vdots & 0 \\ 0 & 0 & 4 & -4 & \vdots & 0 \end{bmatrix}$$

$$\tfrac{1}{4}R_3 \rightarrow \begin{bmatrix} 1 & 0 & 11 & -9 & \vdots & 0 \\ 0 & 1 & -5 & 6 & \vdots & 0 \\ 0 & 0 & 1 & -1 & \vdots & 0 \end{bmatrix}$$

$$\begin{matrix} -11R_3 + R_1 \rightarrow \\ 5R_3 + R_2 \rightarrow \\ \\ \end{matrix} \begin{bmatrix} 1 & 0 & 0 & 2 & \vdots & 0 \\ 0 & 1 & 0 & 1 & \vdots & 0 \\ 0 & 0 & 1 & -1 & \vdots & 0 \end{bmatrix}$$

$w = a, z = a, y = -a, x = -2a$

Answer: $(-2a, -a, a, a)$

64. (a) In the row-echelon form of an augmented matrix that corresponds to an inconsistent system of linear equations, there exists a row consisting of all zeros except for the entry in the last column.

(b) In the row-echelon form of an augmented matrix that corresponds to a system with an infinite number of solutions, there are fewer rows with nonzero entries than there are variables. Nor does the last row consist of all zeros except for the entry in the last column.

66. x = amount at 9%, y = amount at 10%, z = amount at 12%

$$\begin{aligned} x + y + z &= 500{,}000 \\ 0.09x + 0.010y + 0.12z &= 52{,}000 \\ 2.5x - y &= 0 \end{aligned}$$

$$\left[\begin{array}{ccc:c} 1 & 1 & 1 & 500{,}000 \\ 0.09 & 0.01 & 0.12 & 52{,}000 \\ 2.5 & -1 & 0 & 0 \end{array}\right]$$

$$\begin{array}{r} \\ -0.09R_1 + R_2 \rightarrow \\ -2.5R_1 + R_3 \rightarrow \end{array} \left[\begin{array}{ccc:c} 1 & 1 & 1 & 500{,}000 \\ 0 & 0.10 & 0.03 & 7{,}000 \\ 0 & -3.5 & -2.5 & -1{,}250{,}000 \end{array}\right]$$

$$\begin{array}{r} \\ 100R_2 \rightarrow \\ 2R_3 \rightarrow \end{array} \left[\begin{array}{ccc:c} 1 & 1 & 1 & 500{,}000 \\ 0 & 1 & 3 & 700{,}000 \\ 0 & -7 & -5 & -2{,}500{,}000 \end{array}\right]$$

$$\begin{array}{r} -R_2 + R_1 \rightarrow \\ \\ 7R_2 + R_3 \rightarrow \end{array} \left[\begin{array}{ccc:c} 1 & 0 & -2 & -200{,}000 \\ 0 & 1 & 3 & 700{,}000 \\ 0 & 0 & 16 & 2{,}400{,}000 \end{array}\right]$$

$$\begin{array}{r} \\ \\ \frac{1}{16}R_3 \rightarrow \end{array} \left[\begin{array}{ccc:c} 1 & 0 & -2 & -200{,}000 \\ 0 & 1 & 3 & 700{,}000 \\ 0 & 0 & 1 & 150{,}000 \end{array}\right]$$

$z = 150{,}000$, $y = 250{,}000$, $x = 100{,}000$

Answer: \$100,000 at 9%, \$250,000 at 10%, \$150,000 at 12%

68.

$$\begin{aligned} I_1 - I_2 + I_3 &= 0 \\ 2I_1 + 2I_2 &= 7 \\ 2I_2 + 4I_3 &= 8 \end{aligned}$$

$$\left[\begin{array}{ccc:c} 1 & -1 & 1 & 0 \\ 2 & 2 & 0 & 7 \\ 0 & 2 & 4 & 8 \end{array}\right]$$

$$\begin{array}{r} \\ -2R_1 + R_2 \rightarrow \\ \\ \end{array} \left[\begin{array}{ccc:c} 1 & -1 & 1 & 0 \\ 0 & 4 & -2 & 7 \\ 0 & 2 & 4 & 8 \end{array}\right]$$

$$\begin{array}{r} \\ R_3 \rightarrow \\ R_2 \rightarrow \end{array} \left[\begin{array}{ccc:c} 1 & -1 & 1 & 0 \\ 0 & 2 & 4 & 8 \\ 0 & 4 & -2 & 7 \end{array}\right]$$

$$\begin{array}{r} \\ \frac{1}{2}R_2 \rightarrow \\ \\ \end{array} \left[\begin{array}{ccc:c} 1 & -1 & 1 & 0 \\ 0 & 1 & 2 & 4 \\ 0 & 4 & -2 & 7 \end{array}\right]$$

$$\begin{array}{r} \\ \\ -4R_2 + R_3 \rightarrow \end{array} \left[\begin{array}{ccc:c} 1 & -1 & 1 & 0 \\ 0 & 1 & 2 & 4 \\ 0 & 0 & -10 & -9 \end{array}\right]$$

$$\begin{array}{r} \\ \\ -\frac{1}{10}R_3 \rightarrow \end{array} \left[\begin{array}{ccc:c} 1 & -1 & 1 & 0 \\ 0 & 1 & 2 & 4 \\ 0 & 0 & 1 & \frac{9}{10} \end{array}\right]$$

$I_3 = \frac{9}{10}$ amperes

$I_2 + 2\left(\frac{9}{10}\right) = 4 \Rightarrow I_2 = \frac{11}{5}$ amperes

$I_1 - \frac{11}{5} + \frac{9}{10} = 0 \Rightarrow I_1 = \frac{13}{10}$ amperes

70. $f(x) = ax^2 + bx + c$

$f(1) = a + b + c = 9$

$f(2) = 4a + 2b + c = 8$

$f(3) = 9a + 3b + c = 5$

$$\left[\begin{array}{ccc:c} 1 & 1 & 1 & 9 \\ 4 & 2 & 1 & 8 \\ 9 & 3 & 1 & 5 \end{array}\right]$$

$$\begin{array}{r} \\ -4R_1 + R_2 \rightarrow \\ -9R_1 + R_3 \rightarrow \end{array} \left[\begin{array}{ccc:c} 1 & 1 & 1 & 9 \\ 0 & -2 & -3 & -28 \\ 0 & -6 & -8 & -76 \end{array}\right]$$

$$\begin{array}{r} \\ -\frac{1}{2}R_2 \rightarrow \\ -3R_2 + R_3 \rightarrow \end{array} \left[\begin{array}{ccc:c} 1 & 1 & 1 & 9 \\ 0 & 1 & \frac{3}{2} & 14 \\ 0 & 0 & 1 & 8 \end{array}\right]$$

$c = 8$

$b + \frac{3}{2}(8) = 14 \Rightarrow b = 2$

$a + 8 + 2 = 9 \Rightarrow a = -1$

Answer: $y = -x^2 + 2x + 8$

72. $f(x) = ax^3 + bx^2 + cx + d$

$f(1) = a + b + c + d = 0.875$

$f(4) = 64a + 16b + 4c + d = -1$

$f(-2) = -8a + 4b - 2c + d = -4$

$f(-4) = -64a + 16b - 4c + d = -1$

$$\begin{bmatrix} 1 & 1 & 1 & 1 & \vdots & 0.875 \\ 64 & 16 & 4 & 1 & \vdots & -1 \\ -8 & 4 & -2 & 1 & \vdots & -4 \\ -64 & 16 & -4 & 1 & \vdots & -1 \end{bmatrix} \Rightarrow \begin{bmatrix} 1 & 0 & 0 & 0 & \vdots & -\frac{1}{8} \\ 0 & 1 & 0 & 0 & \vdots & 0 \\ 0 & 0 & 1 & 0 & \vdots & 2 \\ 0 & 0 & 0 & 1 & \vdots & -1 \end{bmatrix}$$

$y = -\frac{1}{8}x^3 + 2x - 1$

74. $f(x) = ax^4 + bx^3 + cx^2 + dx + e$

$f(-2) = 16a - 8b + 4c - 2d + e = 10$

$f(-1) = a - b + c - d + e = 1.5$

$f(1) = a + b + c + d + e = -0.5$

$f(2) = 16a + 8b + 4c + 2d + e = -6$

$f(3) = 81a + 27b + 9c + 3d + e = -2.5$

$$\begin{bmatrix} 16 & -8 & 4 & -2 & 1 & \vdots & 10 \\ 1 & -1 & 1 & -1 & 1 & \vdots & 1.5 \\ 1 & 1 & 1 & 1 & 1 & \vdots & -0.5 \\ 16 & 8 & 4 & 2 & 1 & \vdots & -6 \\ 81 & 27 & 9 & 3 & 1 & \vdots & -2.5 \end{bmatrix} \Rightarrow \begin{bmatrix} 1 & 0 & 0 & 0 & 0 & \vdots & 0.5 \\ 0 & 1 & 0 & 0 & 0 & \vdots & -1 \\ 0 & 0 & 1 & 0 & 0 & \vdots & -2 \\ 0 & 0 & 0 & 1 & 0 & \vdots & 0 \\ 0 & 0 & 0 & 0 & 1 & \vdots & 2 \end{bmatrix}$$

$y = \frac{1}{2}x^4 - x^3 - 2x^2 + 2$

76. (a) $f(x) = ax^2 + bx + c$

$f(0) = c = 5.0$

$f(15) = 225a + 15b + c = 9.6$

$f(30) = 900a + 30b + c = 12.4$

$$\begin{bmatrix} 0 & 0 & 1 & \vdots & 5.0 \\ 225 & 15 & 1 & \vdots & 9.6 \\ 900 & 30 & 1 & \vdots & 12.4 \end{bmatrix} \rightarrow \begin{bmatrix} 1 & 0 & 0 & \vdots & -0.004 \\ 0 & 1 & 0 & \vdots & 0.367 \\ 0 & 0 & 1 & \vdots & 5 \end{bmatrix}$$

$y = -0.004x^2 + 0.367x + 5$

(b)

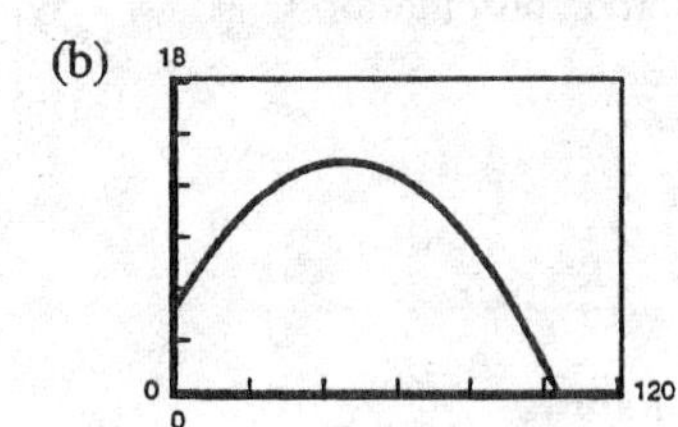

Maximum height $\approx$ 13 feet

Strikes ground ($y = 0$) when $x \approx$ 104 feet.

(c) Complete the square:

$-0.004(x^2 - 91.75x + 2104.5) + 5 + 8.418$

Maximum height = 13.418 feet

$$\text{Range: } y = 0 \Rightarrow x = \frac{-0.367 \pm \sqrt{0.367^2 + 4(0.004)5}}{-0.008}$$

$$\approx 103.793 \text{ feet}$$

78. (a) $x_1 + x_2 = 300$

$x_1 + x_3 = 150 + x_4 \Rightarrow x_1 + x_3 - x_4 = 150$

$x_2 + 200 = x_3 + x_5 \Rightarrow x_2 - x_3 - x_5 = -200$

$x_4 + x_5 = 350$

$$\begin{bmatrix} 1 & 1 & 0 & 0 & 0 & \vdots & 300 \\ 1 & 0 & 1 & -1 & 0 & \vdots & 150 \\ 0 & 1 & -1 & 0 & -1 & \vdots & -200 \\ 0 & 0 & 0 & 1 & 1 & \vdots & 350 \end{bmatrix}$$

$$\begin{matrix} \\ -R_1 + R_2 \rightarrow \\ \\ \\ \end{matrix} \begin{bmatrix} 1 & 1 & 0 & 0 & 0 & \vdots & 300 \\ 0 & -1 & 1 & -1 & 0 & \vdots & -150 \\ 0 & 1 & -1 & 0 & -1 & \vdots & -200 \\ 0 & 0 & 0 & 1 & 1 & \vdots & 350 \end{bmatrix}$$

$$\begin{matrix} \\ \\ R_2 + R_3 \rightarrow \\ \\ \end{matrix} \begin{bmatrix} 1 & 1 & 0 & 0 & 0 & \vdots & 300 \\ 0 & -1 & 1 & -1 & 0 & \vdots & -150 \\ 0 & 0 & 0 & -1 & -1 & \vdots & -350 \\ 0 & 0 & 0 & 1 & 1 & \vdots & 350 \end{bmatrix}$$

$$\begin{matrix} \\ -R_2 \rightarrow \\ -R_3 \rightarrow \\ R_3 + R_4 \rightarrow \end{matrix} \begin{bmatrix} 1 & 1 & 0 & 0 & 0 & \vdots & 300 \\ 0 & 1 & -1 & 1 & 0 & \vdots & 150 \\ 0 & 0 & 0 & 1 & 1 & \vdots & 350 \\ 0 & 0 & 0 & 0 & 0 & \vdots & 0 \end{bmatrix}$$

Let $x_5 = t$.

$x_4 + t = 350 \Rightarrow x_4 = 350 - t$

Let $x_3 = s$.

$x_2 - s + 350 - t = 150 \Rightarrow x_2 = -200 + s + t$

$x_1 - 200 + s + t = 300 \Rightarrow x_1 = 500 - s - t$

(b) When $x_2 = 200$ and $x_3 = 50$:

$x_2 = -200 + s + t$

$200 = -200 + 50 + t \Rightarrow t = 350.$

$x_5 = 350, x_4 = 0, x_3 = 50, x_2 = 200,$

$x_1 = 100$

(c) When $x_2 = 150$ and $x_3 = 0$:

$x_2 = -200 + s + t$

$150 = -200 + 0 + t \Rightarrow t = 350.$

$x_5 = 350, x_4 = 0, x_3 = 0, x_2 = 150,$

$x_1 = 150$

80. (a) Each of the network's four junctions gives rise to a linear equation as follows.

input = output

$400 + x_2 = x_1$

$x_1 + x_3 = x_4 + 600$

$300 = x_2 + x_3 + x_5$

$x_4 + x_5 = 100$

We reorganize these equations, form the augmented matrix, and use Gauss-Jordan elimination.

$$\begin{bmatrix} 1 & -1 & 0 & 0 & 0 & 400 \\ 1 & 0 & 1 & -1 & 0 & 600 \\ 0 & 1 & 1 & 0 & 1 & 300 \\ 0 & 0 & 0 & 1 & 1 & 100 \end{bmatrix} \Rightarrow \begin{bmatrix} 1 & 0 & 1 & 0 & 1 & 700 \\ 0 & 1 & 1 & 0 & 1 & 300 \\ 0 & 0 & 0 & 1 & 1 & 100 \\ 0 & 0 & 0 & 0 & 0 & 0 \end{bmatrix}$$

Letting $x_5 = t$ and $x_3 = s$ be the free variables, we can write the solution as follows.

$x_1 = 700 - s - t$

$x_2 = 300 - s - t$

$x_3 = s$

$x_4 = 100 - t$

$x_5 = t$

(b) If $x_3 = 0$ and $x_5 = 100$, then the solution is $x_1 = 600, x_2 = 200, x_3 = 0, x_4 = 0, x_5 = 100.$

(c) If $x_3 = x_5 = 100$, then the solution is $x_1 = 500, x_2 = 100, x_3 = 100, x_4 = 0, x_5 = 100.$

82. $1.279 - 0.0049t = 1.411 - 0.0078t$

$$0.0029t = 0.132$$

$$t = 45.5, \text{ or the year 2026.}$$

t is rounded up because the Winter Olympics will take place in the year 2026.

84. $g(x) = 3^{-x/2}$

x	-2	0	2	4	6
y	3	1	$\frac{1}{3}$	$\frac{1}{9}$	$\frac{1}{27}$

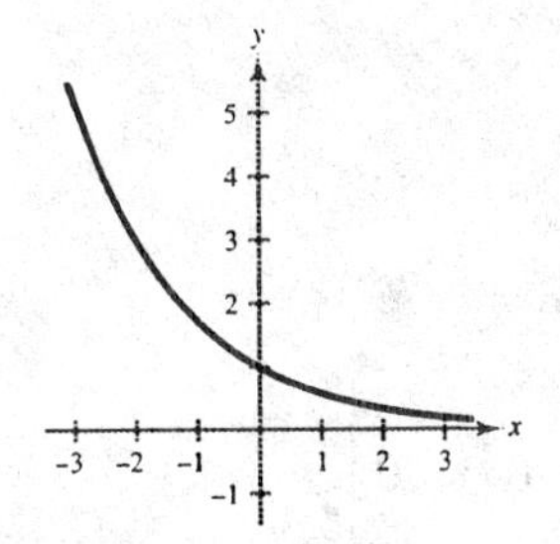

86. $f(x) = 3 + \ln x \Rightarrow y - 3 \ln x \Rightarrow e^{y-3} = x$

x	0.05	0.14	0.37	1	2.72
y	0	1	2	3	4

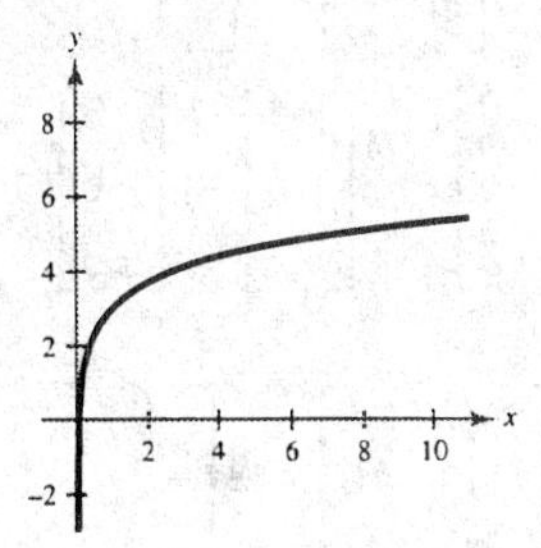

Section 8.2 Operations with Matrices

Solutions to Even-Numbered Exercises

2. $x = 13, y = 12$

4. $x + 2 = 2x + 6, \quad y + 2 = 11$

$$-4 = x \qquad y = 9$$

6. (a) $A + B = \begin{bmatrix} 1 & 2 \\ 2 & 1 \end{bmatrix} + \begin{bmatrix} -3 & -2 \\ 4 & 2 \end{bmatrix} = \begin{bmatrix} 1-3 & 2-2 \\ 2+4 & 1+2 \end{bmatrix} = \begin{bmatrix} -2 & 0 \\ 6 & 3 \end{bmatrix}$

(b) $A - B = \begin{bmatrix} 1 & 2 \\ 2 & 1 \end{bmatrix} - \begin{bmatrix} -3 & -2 \\ 4 & 2 \end{bmatrix} = \begin{bmatrix} 1+3 & 2+2 \\ 2-4 & 1-2 \end{bmatrix} = \begin{bmatrix} 4 & 4 \\ -2 & -1 \end{bmatrix}$

(c) $3A = 3\begin{bmatrix} 1 & 2 \\ 2 & 1 \end{bmatrix} = \begin{bmatrix} 3(1) & 3(2) \\ 3(2) & 3(1) \end{bmatrix} = \begin{bmatrix} 3 & 6 \\ 6 & 3 \end{bmatrix}$

(d) $3A - 2B = \begin{bmatrix} 3 & 6 \\ 6 & 3 \end{bmatrix} - 2\begin{bmatrix} -3 & -2 \\ 4 & 2 \end{bmatrix} = \begin{bmatrix} 3+6 & 6+4 \\ 6-8 & 3-4 \end{bmatrix} = \begin{bmatrix} 9 & 10 \\ -2 & -1 \end{bmatrix}$

8. (a) $A + B = \begin{bmatrix} 2 & 1 & 1 \\ -1 & -1 & 4 \end{bmatrix} + \begin{bmatrix} 2 & -3 & 4 \\ -3 & 1 & -2 \end{bmatrix} = \begin{bmatrix} 2+2 & 1-3 & 1+4 \\ -1-3 & -1+1 & 4-2 \end{bmatrix} = \begin{bmatrix} 4 & -2 & 5 \\ -4 & 0 & 2 \end{bmatrix}$

(b) $A - B = \begin{bmatrix} 2 & 1 & 1 \\ -1 & -1 & 4 \end{bmatrix} + \begin{bmatrix} 2 & -3 & 4 \\ -3 & 1 & -2 \end{bmatrix} = \begin{bmatrix} 2-2 & 1-(-3) & 1-4 \\ -1-(-3) & -1-1 & 4-(-2) \end{bmatrix} = \begin{bmatrix} 0 & 4 & -3 \\ 2 & -2 & 6 \end{bmatrix}$

(c) $3A = 3\begin{bmatrix} 2 & 1 & 1 \\ -1 & -1 & 4 \end{bmatrix} = \begin{bmatrix} 3(2) & 3(1) & 3(1) \\ 3(-1) & 3(-1) & 3(4) \end{bmatrix} = \begin{bmatrix} 6 & 3 & 3 \\ -3 & -3 & 12 \end{bmatrix}$

(d) $3A - 2B = \begin{bmatrix} 6 & 3 & 3 \\ -3 & -3 & 12 \end{bmatrix} - 2\begin{bmatrix} 2 & -3 & 4 \\ -3 & 1 & -2 \end{bmatrix} = \begin{bmatrix} 6 & 3 & 3 \\ -3 & -3 & 12 \end{bmatrix} + \begin{bmatrix} -4 & 6 & -8 \\ 6 & -2 & 4 \end{bmatrix}$

$$= \begin{bmatrix} 2 & 9 & -5 \\ 3 & -5 & 16 \end{bmatrix}$$

10. (a) $A + B = \begin{bmatrix} 3 \\ 2 \\ -1 \end{bmatrix} + \begin{bmatrix} -4 \\ 6 \\ 2 \end{bmatrix} = \begin{bmatrix} 3-4 \\ 2+6 \\ -1+2 \end{bmatrix} = \begin{bmatrix} -1 \\ 8 \\ 1 \end{bmatrix}$

(b) $A - B = \begin{bmatrix} 3 \\ 2 \\ -1 \end{bmatrix} - \begin{bmatrix} -4 \\ 6 \\ 2 \end{bmatrix} = \begin{bmatrix} 3+4 \\ 2-6 \\ -1-2 \end{bmatrix} = \begin{bmatrix} 7 \\ -4 \\ -3 \end{bmatrix}$

(c) $3A = 3\begin{bmatrix} 3 \\ 2 \\ -1 \end{bmatrix} = \begin{bmatrix} 3(3) \\ 3(2) \\ 3(-1) \end{bmatrix} = \begin{bmatrix} 9 \\ 6 \\ -3 \end{bmatrix}$

(d) $3A - 2B = \begin{bmatrix} 9 \\ 6 \\ -3 \end{bmatrix} - 2\begin{bmatrix} -4 \\ 6 \\ 2 \end{bmatrix} = \begin{bmatrix} 9 \\ 6 \\ -3 \end{bmatrix} + \begin{bmatrix} 8 \\ -12 \\ -4 \end{bmatrix} = \begin{bmatrix} 17 \\ -6 \\ -7 \end{bmatrix}$

12. $2X = 2A - B$

$X = A - \frac{1}{2}B = \begin{bmatrix} -2 & -1 \\ 1 & 0 \\ 3 & -4 \end{bmatrix} - \frac{1}{2}\begin{bmatrix} 0 & 3 \\ 2 & 0 \\ -4 & -1 \end{bmatrix} = \begin{bmatrix} -2 & -1 \\ 1 & 0 \\ 3 & -4 \end{bmatrix} - \begin{bmatrix} 0 & \frac{3}{2} \\ 1 & 0 \\ -2 & -\frac{1}{2} \end{bmatrix} = \begin{bmatrix} -2 & -\frac{5}{2} \\ 0 & 0 \\ 5 & -\frac{7}{2} \end{bmatrix}$

14. $2A + 4B = -2X$

$X = -A - 2B = -1\begin{bmatrix} -2 & -1 \\ 1 & 0 \\ 3 & -4 \end{bmatrix} - 2\begin{bmatrix} 0 & 3 \\ 2 & 0 \\ -4 & -1 \end{bmatrix} = \begin{bmatrix} 2 & 1 \\ -1 & 0 \\ -3 & 4 \end{bmatrix} + \begin{bmatrix} 0 & -6 \\ -4 & 0 \\ 8 & 2 \end{bmatrix} = \begin{bmatrix} 2 & -5 \\ -5 & 0 \\ 5 & 6 \end{bmatrix}$

16. (a) $AB = \begin{bmatrix} 2 & -1 \\ 1 & 4 \end{bmatrix}\begin{bmatrix} 0 & 0 \\ 3 & -3 \end{bmatrix} = \begin{bmatrix} 2(0) + (-1)3 & 2(0) + (-1)(-3) \\ 1(0) + 4(3) & 1(0) + 4(-3) \end{bmatrix} = \begin{bmatrix} -3 & 3 \\ 12 & -12 \end{bmatrix}$

(b) $BA = \begin{bmatrix} 0 & 0 \\ 3 & -3 \end{bmatrix}\begin{bmatrix} 2 & -1 \\ 1 & 4 \end{bmatrix} = \begin{bmatrix} 0(2) + (0)1 & 0(-1) + (0)(4) \\ 3(2) + (-3)(1) & 3(-1) + (-3)4 \end{bmatrix} = \begin{bmatrix} 0 & 0 \\ 3 & -15 \end{bmatrix}$

(c) $A^2 = \begin{bmatrix} 2 & -1 \\ 1 & 4 \end{bmatrix}\begin{bmatrix} 2 & -1 \\ 1 & 4 \end{bmatrix} = \begin{bmatrix} 2(2) + (-1)(1) & 2(-1) + (-1)4 \\ 1(2) + 4(1) & 1(-1) + 4(4) \end{bmatrix} = \begin{bmatrix} 3 & -6 \\ 6 & 15 \end{bmatrix}$

18. (a) $AB = \begin{bmatrix} 1 & -1 \\ 1 & 1 \end{bmatrix}\begin{bmatrix} 1 & 3 \\ -3 & 1 \end{bmatrix} = \begin{bmatrix} 1(1) + (-1)(-3) & 1(3) + (-1)(1) \\ 1(1) + 1(-3) & 1(3) + 1(1) \end{bmatrix} = \begin{bmatrix} 4 & 2 \\ -2 & 4 \end{bmatrix}$

(b) $BA = \begin{bmatrix} 1 & 3 \\ -3 & 1 \end{bmatrix}\begin{bmatrix} 1 & -1 \\ 1 & 1 \end{bmatrix} = \begin{bmatrix} 1(1) + (3)1 & 1(-1) + 3(1) \\ -3(1) + (1)(1) & -3(-1) + 1(1) \end{bmatrix} = \begin{bmatrix} 4 & 2 \\ -2 & 4 \end{bmatrix}$

(c) $A^2 = \begin{bmatrix} 1 & -1 \\ 1 & 1 \end{bmatrix}\begin{bmatrix} 1 & 1 \\ 1 & 1 \end{bmatrix} = \begin{bmatrix} 1(1) + (-1)(1) & 1(-1) + (-1)(1) \\ 1(1) + (1)(1) & 1(-1) + 1(1) \end{bmatrix} = \begin{bmatrix} 0 & -2 \\ 2 & 0 \end{bmatrix}$

20. (a) $AB = \begin{bmatrix} 3 & 2 & 1 \end{bmatrix}\begin{bmatrix} 2 \\ 3 \\ 0 \end{bmatrix} = [3(2) + 2(3) + 1(0)] = [12]$

(b) $BA = \begin{bmatrix} 2 \\ 3 \\ 0 \end{bmatrix}\begin{bmatrix} 3 & 2 & 1 \end{bmatrix} = \begin{bmatrix} 2(3) & 2(2) & 2(1) \\ 3(3) & 3(2) & 3(1) \\ 0(3) & 0(2) & 0(1) \end{bmatrix} = \begin{bmatrix} 6 & 4 & 2 \\ 9 & 6 & 3 \\ 0 & 0 & 0 \end{bmatrix}$

(c) The number of columns of A does not equal the number of rows of A; the multiplication is not possible.

22. A is 3×3, B is $3 \times 2 \Rightarrow AB$ is 3×2.

$$\begin{bmatrix} 0 & -1 & 0 \\ 4 & 0 & 2 \\ 8 & -1 & 7 \end{bmatrix}\begin{bmatrix} 2 & 1 \\ -3 & 4 \\ 1 & 6 \end{bmatrix} = \begin{bmatrix} 3 & -4 \\ 10 & 16 \\ 26 & 46 \end{bmatrix}$$

24. A is 3×3, B is $3 \times 3 \Rightarrow AB$ is 3×3.

$$\begin{bmatrix} 1 & 0 & 0 \\ 0 & 4 & 0 \\ 0 & 0 & -2 \end{bmatrix}\begin{bmatrix} 3 & 0 & 0 \\ 0 & -1 & 0 \\ 0 & 0 & 5 \end{bmatrix} = \begin{bmatrix} 3 & 0 & 0 \\ 0 & -4 & 0 \\ 0 & 0 & -10 \end{bmatrix}$$

26. A is 2×1, B is $1 \times 4 \Rightarrow AB$ is 2×4.

$$\begin{bmatrix} 10 \\ 12 \end{bmatrix}\begin{bmatrix} 6 & -2 & 1 & 6 \end{bmatrix} = \begin{bmatrix} 60 & -20 & 10 & 60 \\ 72 & -24 & 12 & 72 \end{bmatrix}$$

28. A is 2×4, B is $2 \times 2 \Rightarrow AB$ is not defined.

30. $$\begin{bmatrix} 11 & -12 & 4 \\ 14 & 10 & 12 \\ 6 & -2 & 9 \end{bmatrix}\begin{bmatrix} 12 & 10 \\ -5 & 12 \\ 15 & 16 \end{bmatrix} = \begin{bmatrix} 252 & 30 \\ 298 & 452 \\ 217 & 180 \end{bmatrix}$$

32. A is 3×3, B is $4 \times 2 \Rightarrow AB$ is not defined.

34. $$\begin{bmatrix} 15 & -18 \\ -4 & 12 \\ -8 & 22 \end{bmatrix}\begin{bmatrix} -7 & 22 & 1 \\ 8 & 16 & 24 \end{bmatrix} = \begin{bmatrix} -249 & 42 & -417 \\ 124 & 104 & 284 \\ 232 & 176 & 520 \end{bmatrix}$$

36. $A = \begin{bmatrix} 1 & -2 & 3 \\ -1 & 3 & -1 \\ 2 & -5 & 5 \end{bmatrix}, X = \begin{bmatrix} x \\ y \\ z \end{bmatrix}, B = \begin{bmatrix} 9 \\ -6 \\ 17 \end{bmatrix}$

$$\left[\begin{array}{ccc|c} 1 & -2 & 3 & 9 \\ -1 & 3 & -1 & -6 \\ 2 & -5 & 5 & 17 \end{array}\right]$$

$$\begin{array}{r} \\ R_1 + R_2 \rightarrow \\ -2R_1 + R_3 \rightarrow \end{array}\left[\begin{array}{ccc|c} 1 & -2 & 3 & 9 \\ 0 & 1 & 2 & 3 \\ 0 & -1 & -1 & -1 \end{array}\right]$$

$$\begin{array}{r} 2R_2 + R_1 \rightarrow \\ \\ R_2 + R_3 \rightarrow \end{array}\left[\begin{array}{ccc|c} 1 & 0 & 7 & 15 \\ 0 & 1 & 2 & 3 \\ 0 & 0 & 1 & 2 \end{array}\right]$$

$$\begin{array}{r} -7R_3 + R_1 \rightarrow \\ -2R_3 + R_2 \rightarrow \\ \end{array}\left[\begin{array}{ccc|c} 1 & 0 & 0 & 1 \\ 0 & 1 & 0 & -1 \\ 0 & 0 & 1 & 2 \end{array}\right]$$

$x = 1, y = -1, z = 2$

38. $A = \begin{bmatrix} 1 & 1 & -3 \\ -1 & 2 & 0 \\ 0 & -1 & 1 \end{bmatrix}, X = \begin{bmatrix} x \\ y \\ z \end{bmatrix}, B = \begin{bmatrix} -1 \\ 1 \\ 0 \end{bmatrix}$

$$\left[\begin{array}{ccc|c} 1 & 1 & -3 & -1 \\ -1 & 2 & 0 & 1 \\ 0 & -1 & 1 & 0 \end{array}\right]$$

$$\begin{array}{r} \\ R_1 + R_2 \rightarrow \\ \\ \end{array}\left[\begin{array}{ccc|c} 1 & 1 & -3 & -1 \\ 0 & 3 & -3 & 0 \\ 0 & -1 & 1 & 0 \end{array}\right]$$

$$\begin{array}{r} \\ \frac{1}{3}R_2 \rightarrow \\ \frac{1}{3}R_2 + R_3 \rightarrow \end{array}\left[\begin{array}{ccc|c} 1 & 1 & -3 & -1 \\ 0 & 1 & -1 & 0 \\ 0 & 0 & 0 & 0 \end{array}\right]$$

$$\begin{array}{r} -R_2 + R_1 \rightarrow \\ \\ \\ \end{array}\left[\begin{array}{ccc|c} 1 & 0 & -2 & -1 \\ 0 & 1 & -1 & 0 \\ 0 & 0 & 0 & 0 \end{array}\right]$$

Let $z = a$, then $y = a, x = 2a - 1$.

40. $A = \begin{bmatrix} 5 & 4 \\ 1 & 2 \end{bmatrix}$

$$f(A) = A^2 - 7A + 6 = \begin{bmatrix} 5 & 4 \\ 1 & 2 \end{bmatrix}\begin{bmatrix} 5 & 4 \\ 1 & 2 \end{bmatrix} - 7\begin{bmatrix} 5 & 4 \\ 1 & 2 \end{bmatrix} + 6\begin{bmatrix} 1 & 0 \\ 0 & 1 \end{bmatrix} = \begin{bmatrix} 0 & 0 \\ 0 & 0 \end{bmatrix}$$

42. $A = \begin{bmatrix} 8 & -4 \\ 2 & 2 \end{bmatrix}$

$$f(A) = A^2 - 10A + 24 = \begin{bmatrix} 8 & -4 \\ 2 & 2 \end{bmatrix}\begin{bmatrix} 8 & -4 \\ 2 & 2 \end{bmatrix} - 10\begin{bmatrix} 8 & -4 \\ 2 & 2 \end{bmatrix} + 24\begin{bmatrix} 1 & 0 \\ 0 & 1 \end{bmatrix} = \begin{bmatrix} 0 & 0 \\ 0 & 0 \end{bmatrix}$$

44. $A = \begin{bmatrix} 3 & 3 \\ 4 & 4 \end{bmatrix}_1 B = \begin{bmatrix} 1 & -1 \\ -1 & 1 \end{bmatrix}$

$$AB = \begin{bmatrix} 3 & 3 \\ 4 & 4 \end{bmatrix}\begin{bmatrix} 1 & -1 \\ -1 & 1 \end{bmatrix} = \begin{bmatrix} 0 & 0 \\ 0 & 0 \end{bmatrix}$$

$AB = O$ but $A \neq O$ and $B \neq O$.

For 46–54, A is of order 2×3, B is of order 2×3, C is of order 3×2 and D is of order 2×2.

46. $B - 3C$ is not possible. B and C are not of the same order.

48. BC is possible. The resulting order is 2×2.

50. $CB - D$ is not possible. The order of CB is 3×3, but the order of D is 2×2.

52. $(BC)D$ is possible. The resulting order is 2×2.

54. $(BC - D)A$ is possible. The resulting order is 2×3.

56. $$1.10\begin{bmatrix} 100 & 90 & 70 & 30 \\ 40 & 20 & 60 & 60 \end{bmatrix} = \begin{bmatrix} 110 & 99 & 77 & 33 \\ 44 & 22 & 66 & 66 \end{bmatrix}$$

58. $$BA = [\$20.50 \quad \$26.50 \quad \$29.50] \begin{bmatrix} 5{,}000 & 4{,}000 \\ 6{,}000 & 10{,}000 \\ 8{,}000 & 5{,}000 \end{bmatrix} = [\$497{,}500 \quad \$494{,}500]$$

The entries represent the costs of the three models of the product at the two warehouses.

60. $$A = \begin{bmatrix} 0 & -i \\ i & 0 \end{bmatrix}$$

$$A^2 = \begin{bmatrix} 0 & -i \\ i & 0 \end{bmatrix}\begin{bmatrix} 0 & -i \\ i & 0 \end{bmatrix} = \begin{bmatrix} 1 & 0 \\ 0 & 1 \end{bmatrix} = I, \text{ the identity matrix.}$$

62. $$ST = \begin{bmatrix} 1.0 & 0.5 & 0.2 \\ 1.6 & 1.0 & 0.2 \\ 2.5 & 2.0 & 0.2 \end{bmatrix}\begin{bmatrix} 12 & 10 \\ 9 & 8 \\ 6 & 5 \end{bmatrix}$$

$$= \begin{bmatrix} \$17.70 & \$15.00 \\ \$29.40 & \$25.00 \\ \$50.40 & \$43.00 \end{bmatrix}$$

This represents the labor cost for each boat size at each plant.

64. $$P^3 = P^2P = \begin{bmatrix} 0.4 & 0.15 & 0.15 \\ 0.28 & 0.53 & 0.17 \\ 0.32 & 0.32 & 0.68 \end{bmatrix}\begin{bmatrix} 0.5 & 0.1 & 0.1 \\ 0.2 & 0.7 & 0.1 \\ 0.2 & 0.2 & 0.8 \end{bmatrix} = \begin{bmatrix} 0.300 & 0.175 & 0.175 \\ 0.308 & 0.433 & 0.217 \\ 0.392 & 0.392 & 0.608 \end{bmatrix}$$

$$P^4 = P^3P = \begin{bmatrix} 0.300 & 0.175 & 0.175 \\ 0.308 & 0.433 & 0.217 \\ 0.392 & 0.392 & 0.608 \end{bmatrix}\begin{bmatrix} 0.6 & 0.1 & 0.1 \\ 0.2 & 0.7 & 0.1 \\ 0.2 & 0.2 & 0.8 \end{bmatrix} = \begin{bmatrix} 0.250 & 0.188 & 0.188 \\ 0.315 & 0.377 & 0.248 \\ 0.435 & 0.435 & 0.565 \end{bmatrix}$$

$$P^5 = P^4P = \begin{bmatrix} 0.250 & 0.188 & 0.188 \\ 0.315 & 0.377 & 0.248 \\ 0.435 & 0.435 & 0.565 \end{bmatrix}\begin{bmatrix} 0.6 & 0.1 & 0.1 \\ 0.2 & 0.7 & 0.1 \\ 0.2 & 0.2 & 0.8 \end{bmatrix} = \begin{bmatrix} 0.225 & 0.194 & 0.194 \\ 0.314 & 0.345 & 0.267 \\ 0.461 & 0.461 & 0.539 \end{bmatrix}$$

$$P^6 = \begin{bmatrix} 0.213 & 0.197 & 0.197 \\ 0.311 & 0.326 & 0.280 \\ 0.477 & 0.477 & 0.523 \end{bmatrix}$$

$$P^7 = \begin{bmatrix} 0.206 & 0.198 & 0.198 \\ 0.308 & 0.316 & 0.288 \\ 0.486 & 0.486 & 0.514 \end{bmatrix}$$

$$P^8 = \begin{bmatrix} 0.203 & 0.199 & 0.199 \\ 0.305 & 0.309 & 0.292 \\ 0.492 & 0.492 & 0.508 \end{bmatrix}$$

—CONTINUED—

64. —CONTINUED—

As P is raised to higher and higher powers, the resulting matrices appear to be approaching the matrix

$$\begin{bmatrix} 0.2 & 0.2 & 0.2 \\ 0.3 & 0.3 & 0.3 \\ 0.5 & 0.5 & 0.5 \end{bmatrix}.$$

66. (a) $A = \begin{bmatrix} 0 & 2 \\ 0 & 0 \end{bmatrix}, B = \begin{bmatrix} 0 & 2 & 3 \\ 0 & 0 & 4 \\ 0 & 0 & 0 \end{bmatrix}$

(b) A^2 and B^3 are both zero matrices.

(c) If A is 4×4, then A^4 will be the zero matrix.

(d) If A is $n \times n$, then A^n is the zero matrix.

68. $\log_5(5^3 \cdot 4) = 3 \log_5 5 + \log_5 4 = 3 + \log_5 4$

70. $\ln \frac{100}{e^2} = \ln 100 - \ln e^2 = \ln 100 - 2$

Section 8.3 The Inverse of a Square Matrix

Solutions to Even-Numbered Exercises

2. $AB = \begin{bmatrix} 1 & -1 \\ -1 & 2 \end{bmatrix}\begin{bmatrix} 2 & 1 \\ 1 & 1 \end{bmatrix} = \begin{bmatrix} 2-1 & 1-1 \\ -2+2 & -1=2 \end{bmatrix} = \begin{bmatrix} 1 & 0 \\ 0 & 1 \end{bmatrix}$

$BA = \begin{bmatrix} 2 & 1 \\ 1 & 1 \end{bmatrix}\begin{bmatrix} 1 & -1 \\ -1 & 2 \end{bmatrix} = \begin{bmatrix} 2-1 & -2+2 \\ 1-1 & -1+2 \end{bmatrix} = \begin{bmatrix} 1 & 0 \\ 0 & 1 \end{bmatrix}$

4. $AB = \begin{bmatrix} 1 & -1 \\ 2 & 3 \end{bmatrix}\begin{bmatrix} \frac{3}{5} & \frac{1}{5} \\ -\frac{2}{5} & \frac{1}{5} \end{bmatrix} = \begin{bmatrix} \frac{3}{5}+\frac{2}{5} & \frac{1}{5}-\frac{1}{5} \\ \frac{6}{5}-\frac{6}{5} & \frac{2}{5}+\frac{3}{5} \end{bmatrix} = \begin{bmatrix} 1 & 0 \\ 0 & 1 \end{bmatrix}$

$AB = \begin{bmatrix} \frac{3}{5} & \frac{1}{5} \\ -\frac{2}{5} & \frac{1}{5} \end{bmatrix}\begin{bmatrix} 1 & -1 \\ 2 & 3 \end{bmatrix} = \begin{bmatrix} \frac{3}{5}+\frac{2}{5} & -\frac{3}{5}+\frac{3}{5} \\ -\frac{2}{5}+\frac{2}{5} & \frac{2}{5}+\frac{3}{5} \end{bmatrix} = \begin{bmatrix} 1 & 0 \\ 0 & 1 \end{bmatrix}$

6. $AB = \begin{bmatrix} 2 & -17 & 11 \\ -1 & 11 & -7 \\ 0 & 3 & -2 \end{bmatrix}\begin{bmatrix} 1 & 1 & 2 \\ 2 & 4 & -3 \\ 3 & 6 & -5 \end{bmatrix}$

$= \begin{bmatrix} 2-34+33 & 2-68+66 & 4+51-55 \\ -1+22-21 & -1+44-42 & -2-33+35 \\ 6-6 & 12-12 & -9+10 \end{bmatrix} = \begin{bmatrix} 1 & 0 & 0 \\ 0 & 1 & 0 \\ 0 & 0 & 1 \end{bmatrix}$

$BA = \begin{bmatrix} 1 & 1 & 2 \\ 2 & 4 & -3 \\ 3 & 6 & -5 \end{bmatrix}\begin{bmatrix} 2 & -17 & 11 \\ -1 & 11 & -7 \\ 0 & 3 & -2 \end{bmatrix} = \begin{bmatrix} 2-1 & -17+11+6 & 11-7-4 \\ 4-4 & -34+44-9 & 22-28+6 \\ 6-6 & -51+66-15 & 33-42+10 \end{bmatrix} = \begin{bmatrix} 1 & 0 & 0 \\ 0 & 1 & 0 \\ 0 & 0 & 1 \end{bmatrix}$

8. $AB = \frac{1}{3}\begin{bmatrix} -1 & 1 & 0 & -1 \\ 1 & -1 & 1 & 0 \\ -1 & 1 & 2 & 0 \\ 0 & -1 & 1 & 1 \end{bmatrix}\begin{bmatrix} -3 & 1 & 1 & -3 \\ -3 & -1 & 2 & -3 \\ 0 & 1 & 1 & 0 \\ -3 & -2 & 1 & 0 \end{bmatrix}$

$= \frac{1}{3}\begin{bmatrix} 3-3+0+3 & -1-1+0+2 & -1+2+0-1 & 3-3+0+0 \\ -3+3+0+0 & 1+1+1+0 & 1-2+1+0 & -3+3+0+0 \\ 3-3+0+0 & -1-1+2+0 & -1+2+2+0 & 3-3+0+0 \\ 0+3+0-3 & 0+1+1-2 & 0-2+1+1 & 0+3+0+0 \end{bmatrix}$

$= \frac{1}{3}\begin{bmatrix} 3 & 0 & 0 & 0 \\ 0 & 3 & 0 & 0 \\ 0 & 0 & 3 & 0 \\ 0 & 0 & 0 & 3 \end{bmatrix} = I_4$

$BA = \frac{1}{3}\begin{bmatrix} -3 & 1 & 1 & -3 \\ -3 & -1 & 2 & -3 \\ 0 & 1 & 1 & 0 \\ -3 & -2 & 1 & 0 \end{bmatrix}\begin{bmatrix} -1 & 1 & 0 & -1 \\ 1 & -1 & 1 & 0 \\ -1 & 1 & 2 & 0 \\ 0 & -1 & 1 & 1 \end{bmatrix}$

$= \frac{1}{3}\begin{bmatrix} 3+3-0+3 & -3-1+1+3 & 0+1+2-3 & 3+0+0-3 \\ 3-1-2+0 & -3+1+2+3 & 0-1+4-3 & 3+0+0-3 \\ 0+1-1+0 & 0-1+1+0 & 0+1+2+0 & 0+0+0+0 \\ 3-2-1+0 & -3+2+1+0 & 0-2+2+0 & 3+0+0+0 \end{bmatrix}$

$= \frac{1}{3}\begin{bmatrix} 3 & 0 & 0 & 0 \\ 0 & 3 & 0 & 0 \\ 0 & 0 & 3 & 0 \\ 0 & 0 & 0 & 3 \end{bmatrix} = I_4$

10. $[A \ \vdots \ I] = \left[\begin{array}{cc|cc} 1 & 2 & 1 & 0 \\ 3 & 7 & 0 & 1 \end{array}\right]$

$-3R_1 + R_2 \rightarrow \left[\begin{array}{cc|cc} 1 & 2 & 1 & 0 \\ 0 & 1 & -3 & 1 \end{array}\right]$

$-2R_2 + R_1 \rightarrow \left[\begin{array}{cc|cc} 1 & 0 & 7 & -2 \\ 0 & 1 & -3 & 1 \end{array}\right] = [I \ \vdots \ A^{-1}]$

$A^{-1} = \begin{bmatrix} 7 & -2 \\ -3 & 1 \end{bmatrix}$

12. $[A \ \vdots \ I] = \left[\begin{array}{cc|cc} -7 & 33 & 1 & 0 \\ 4 & -19 & 0 & 1 \end{array}\right]$

$2R_2 + R_1 \rightarrow \left[\begin{array}{cc|cc} 1 & -5 & 1 & 2 \\ 4 & -19 & 0 & 1 \end{array}\right]$

$-4R_1 + R_2 \rightarrow \left[\begin{array}{cc|cc} 1 & -5 & 1 & 2 \\ 0 & 1 & -4 & -7 \end{array}\right]$

$5R_2 + R_1 \rightarrow \left[\begin{array}{cc|cc} 1 & 0 & -19 & -33 \\ 0 & 1 & -4 & -7 \end{array}\right] = [I \ \vdots \ A^{-1}]$

$A^{-1} = \begin{bmatrix} -19 & -33 \\ -4 & -7 \end{bmatrix}$

14. $[A \ \vdots \ I] = \begin{bmatrix} 11 & 1 & \vdots & 1 & 0 \\ -1 & 0 & \vdots & 0 & 1 \end{bmatrix}$

$10R_2 + R_1 \rightarrow \begin{bmatrix} 1 & 1 & \vdots & 1 & 10 \\ -1 & 0 & \vdots & 0 & 1 \end{bmatrix}$

$R_1 + R_2 \rightarrow \begin{bmatrix} 1 & 1 & \vdots & 1 & 10 \\ 0 & 1 & \vdots & 1 & 11 \end{bmatrix}$

$-R_2 + R_1 \rightarrow \begin{bmatrix} 1 & 0 & \vdots & 0 & -1 \\ 0 & 1 & \vdots & 1 & 11 \end{bmatrix} = [I \ \vdots \ A^{-1}]$

$A^{-1} = \begin{bmatrix} 0 & -1 \\ 1 & 11 \end{bmatrix}$

16. $[A \ \vdots \ I] = \begin{bmatrix} 2 & 3 & \vdots & 1 & 0 \\ 1 & 4 & \vdots & 0 & 1 \end{bmatrix}$

$\begin{matrix} R_2 \rightarrow \\ R_1 \rightarrow \end{matrix} \begin{bmatrix} 1 & 3 & \vdots & 0 & 1 \\ 2 & 4 & \vdots & 1 & 0 \end{bmatrix}$

$-2R_1 + R_2 \rightarrow \begin{bmatrix} 1 & 4 & \vdots & 0 & 1 \\ 0 & -5 & \vdots & 1 & -2 \end{bmatrix}$

$-\frac{1}{5}R_2 \rightarrow \begin{bmatrix} 1 & 4 & \vdots & 0 & 1 \\ 0 & 1 & \vdots & -\frac{1}{5} & \frac{2}{5} \end{bmatrix}$

$-4R_2 + R_1 \rightarrow \begin{bmatrix} 1 & 0 & \vdots & \frac{4}{5} & -\frac{3}{5} \\ 0 & 1 & \vdots & -\frac{1}{5} & \frac{2}{5} \end{bmatrix} = [I \ \vdots \ A^{-1}]$

$A^{-1} = \frac{1}{5}\begin{bmatrix} 4 & -3 \\ -1 & 2 \end{bmatrix}$

18. $A = \begin{bmatrix} -2 & 5 \\ 6 & -15 \\ 0 & 1 \end{bmatrix}$ A has no inverse because it is not square.

20. $[A \ \vdots \ I] = \begin{bmatrix} 1 & 2 & 2 & \vdots & 1 & 0 & 0 \\ 3 & 7 & 9 & \vdots & 0 & 1 & 0 \\ -1 & -4 & -7 & \vdots & 0 & 0 & 1 \end{bmatrix}$

$\begin{matrix} \\ -3R_1 + R_2 \rightarrow \\ R_1 + R_3 \rightarrow \end{matrix} \begin{bmatrix} 1 & 2 & 2 & \vdots & 1 & 0 & 0 \\ 0 & 1 & 3 & \vdots & -3 & 1 & 0 \\ 0 & -2 & -5 & \vdots & 1 & 0 & 1 \end{bmatrix}$

$\begin{matrix} -2R_2 + R_1 \rightarrow \\ \\ 2R_2 + R_3 \rightarrow \end{matrix} \begin{bmatrix} 1 & 0 & -4 & \vdots & 7 & -2 & 0 \\ 0 & 1 & 3 & \vdots & -3 & 1 & 0 \\ 0 & 0 & 1 & \vdots & -5 & 2 & 1 \end{bmatrix}$

$\begin{matrix} 4R_3 + R_1 \rightarrow \\ -3R_3 + R_2 \rightarrow \\ \\ \end{matrix} \begin{bmatrix} 1 & 0 & 0 & \vdots & -13 & 6 & 4 \\ 0 & 1 & 0 & \vdots & 12 & -5 & -3 \\ 0 & 0 & 1 & \vdots & -5 & 2 & 1 \end{bmatrix} = [I \ \vdots \ A^{-1}]$

$A^{-1} = \begin{bmatrix} -13 & 6 & 4 \\ 12 & -5 & -3 \\ -5 & 2 & 1 \end{bmatrix}$

22. $[A \ \vdots \ I] = \left[\begin{array}{ccc:ccc} 1 & 0 & 0 & 1 & 0 & 0 \\ 3 & 0 & 0 & 0 & 1 & 0 \\ 2 & 5 & 5 & 0 & 0 & 1 \end{array}\right] \begin{array}{l} \\ -3R_1 + R_2 \rightarrow \\ -2R_1 + R_3 \rightarrow \end{array} \left[\begin{array}{ccc:ccc} 1 & 0 & 0 & 1 & 0 & 0 \\ 0 & 0 & 0 & -3 & 1 & 0 \\ 0 & 5 & 5 & -2 & 0 & 1 \end{array}\right]$

Since the first three entries of row 2 are all zeros, the inverse of A does not exist.

24.

$$[A \ \vdots \ I] = \left[\begin{array}{cccc:cccc} 1 & 3 & -2 & 0 & 1 & 0 & 0 & 0 \\ 0 & 2 & 4 & 6 & 0 & 1 & 0 & 0 \\ 0 & 0 & -2 & 1 & 0 & 0 & 1 & 0 \\ 0 & 0 & 0 & 5 & 0 & 0 & 0 & 1 \end{array}\right]$$

$$\begin{array}{r} \\ \frac{1}{2}R_2 \rightarrow \\ \\ \frac{1}{5}R_4 \rightarrow \end{array} \left[\begin{array}{cccc:cccc} 1 & 3 & -2 & 0 & 1 & 0 & 0 & 0 \\ 0 & 1 & 2 & 3 & 0 & \frac{1}{2} & 0 & 0 \\ 0 & 0 & -2 & 1 & 0 & 0 & 1 & 0 \\ 0 & 0 & 0 & 1 & 0 & 0 & 0 & \frac{1}{5} \end{array}\right]$$

$$\begin{array}{r} -3R_2 + R_1 \rightarrow \\ R_3 + R_2 \rightarrow \\ -R_4 + R_3 \rightarrow \\ \\ \end{array} \left[\begin{array}{cccc:cccc} 1 & 0 & -8 & -9 & 1 & -\frac{3}{2} & 0 & 0 \\ 0 & 1 & 0 & 4 & 0 & \frac{1}{2} & 1 & 0 \\ 0 & 0 & -2 & 0 & 0 & 0 & 1 & -\frac{1}{5} \\ 0 & 0 & 0 & 1 & 0 & 0 & 0 & \frac{1}{5} \end{array}\right]$$

$$\begin{array}{r} -4R_3 + R_1 \rightarrow \\ -4R_4 + R_2 \rightarrow \\ \frac{1}{2}R_3 \rightarrow \\ \\ \end{array} \left[\begin{array}{cccc:cccc} 1 & 0 & 0 & 0 & 1 & -\frac{3}{2} & -4 & \frac{4}{5} \\ 0 & 1 & 0 & 0 & 0 & \frac{1}{2} & 1 & -\frac{4}{5} \\ 0 & 0 & 1 & 0 & 0 & 0 & -\frac{1}{2} & \frac{1}{10} \\ 0 & 0 & 0 & 1 & 0 & 0 & 0 & \frac{1}{5} \end{array}\right]$$

$$\begin{array}{r} 9R_4 + R_1 \rightarrow \\ \\ \\ \\ \end{array} \left[\begin{array}{cccc:cccc} 1 & 0 & 0 & 0 & 1 & -\frac{3}{2} & -4 & \frac{13}{5} \\ 0 & 1 & 0 & 0 & 0 & \frac{1}{2} & 1 & -\frac{4}{5} \\ 0 & 0 & 1 & 0 & 0 & 0 & -\frac{1}{2} & \frac{1}{10} \\ 0 & 0 & 0 & 1 & 0 & 0 & 0 & \frac{1}{5} \end{array}\right] = [I \ \vdots \ A^{-1}]$$

$$A^{-1} = \frac{1}{10}\begin{bmatrix} 10 & -15 & -40 & 26 \\ 0 & 5 & 10 & -8 \\ 0 & 0 & -5 & 1 \\ 0 & 0 & 0 & 2 \end{bmatrix}$$

26. $A = \begin{bmatrix} 10 & 5 & -7 \\ -5 & 1 & 4 \\ 3 & 2 & -2 \end{bmatrix}$

$A^{-1} = \begin{bmatrix} -10 & -4 & 27 \\ 2 & 1 & -5 \\ -13 & -5 & 35 \end{bmatrix}$

28. $A = \begin{bmatrix} 3 & 2 & 2 \\ 2 & 2 & 2 \\ -4 & 4 & 3 \end{bmatrix}$

$A^{-1} = \frac{1}{2}\begin{bmatrix} 2 & -2 & 0 \\ 14 & -17 & 2 \\ -16 & 20 & -2 \end{bmatrix}$

30. $A = \begin{bmatrix} 2 & 0 & 0 \\ 0 & 3 & 0 \\ 0 & 0 & 5 \end{bmatrix}$

$A^{-1} = \frac{1}{30}\begin{bmatrix} 15 & 0 & 0 \\ 0 & 10 & 0 \\ 0 & 0 & 6 \end{bmatrix}$

32. $A = \begin{bmatrix} -1 & 0 & 1 & 0 \\ 0 & 2 & 0 & -1 \\ 2 & 0 & -1 & 0 \\ 0 & -1 & 0 & 1 \end{bmatrix}$

$A^{-1} = \begin{bmatrix} 1 & 0 & 1 & 0 \\ 0 & 1 & 0 & 1 \\ 2 & 0 & 1 & 0 \\ 0 & 1 & 0 & 2 \end{bmatrix}$

34. $A = \begin{bmatrix} 4 & 8 & -7 & 14 \\ 2 & 5 & -4 & 6 \\ 0 & 2 & 1 & -7 \\ 3 & 6 & -5 & 10 \end{bmatrix}$

$A^{-1} = \begin{bmatrix} 27 & -10 & 4 & -29 \\ -16 & 5 & -2 & 18 \\ -17 & 4 & -2 & 20 \\ -7 & 2 & -1 & 8 \end{bmatrix}$

36. $A = \begin{bmatrix} a & b \\ c & d \end{bmatrix}, A^{-1} = \dfrac{1}{ad - bc}\begin{bmatrix} d & -b \\ -c & a \end{bmatrix}$

(a) $A = \begin{bmatrix} 5 & -2 \\ 2 & 3 \end{bmatrix}$

$A^{-1} = \dfrac{1}{15 + 4}\begin{bmatrix} 3 & 2 \\ -2 & 5 \end{bmatrix} = \dfrac{1}{19}\begin{bmatrix} 3 & 2 \\ -2 & 5 \end{bmatrix}$

(b) $A = \begin{bmatrix} 7 & 12 \\ -8 & -5 \end{bmatrix}$

$A^{-1} = \dfrac{1}{-35 + 96}\begin{bmatrix} -5 & -12 \\ 8 & 7 \end{bmatrix} = \dfrac{1}{61}\begin{bmatrix} -5 & -12 \\ 8 & 7 \end{bmatrix}$

38. $\begin{bmatrix} x \\ y \end{bmatrix} = \begin{bmatrix} -3 & 2 \\ -2 & 1 \end{bmatrix}\begin{bmatrix} 0 \\ 3 \end{bmatrix} = \begin{bmatrix} 6 \\ 3 \end{bmatrix}$

Answer: (6, 3)

40. $\begin{bmatrix} x \\ y \end{bmatrix} = \begin{bmatrix} -3 & 2 \\ -2 & 1 \end{bmatrix}\begin{bmatrix} 1 \\ -2 \end{bmatrix} = \begin{bmatrix} -7 \\ -4 \end{bmatrix}$

Answer: $(-7, -4)$

42. $\begin{bmatrix} x \\ y \\ z \end{bmatrix} = \begin{bmatrix} 1 & 1 & -1 \\ -3 & 2 & -1 \\ 3 & -3 & 2 \end{bmatrix}\begin{bmatrix} -1 \\ 2 \\ 0 \end{bmatrix} = \begin{bmatrix} 1 \\ 7 \\ -9 \end{bmatrix}$

Answer: $(1, 7, -9)$

44. $\begin{bmatrix} x \\ y \\ z \\ w \end{bmatrix} = \begin{bmatrix} -24 & 7 & 1 & -2 \\ -10 & 3 & 0 & -1 \\ -29 & 7 & 3 & -2 \\ 12 & -3 & -1 & 1 \end{bmatrix}\begin{bmatrix} 1 \\ -2 \\ 0 \\ -3 \end{bmatrix} = \begin{bmatrix} -32 \\ -13 \\ -37 \\ 15 \end{bmatrix}$

Answer: $(-32, -13, -37, 15)$

46. $A = \begin{bmatrix} 18 & 12 \\ 30 & 24 \end{bmatrix}$

$A^{-1} = \dfrac{1}{423 - 360}\begin{bmatrix} 24 & -12 \\ -30 & 18 \end{bmatrix}$

Answer: $\left(\frac{1}{2}, \frac{1}{3}\right)$

48. $A = \begin{bmatrix} 13 & -6 \\ 26 & -12 \end{bmatrix}$

$A^{-1} = \dfrac{1}{-156 + 156}\begin{bmatrix} -12 & 6 \\ -26 & 13 \end{bmatrix} \Rightarrow A^{-1}$ does not exist.

No solution

50. $A = \begin{bmatrix} 3 & 2 \\ 2 & 10 \end{bmatrix}$

$A^{-1} = \dfrac{1}{30 - 4}\begin{bmatrix} 10 & -2 \\ -2 & 3 \end{bmatrix}$

$\begin{bmatrix} x \\ y \end{bmatrix} = \dfrac{1}{26}\begin{bmatrix} 10 & -2 \\ -2 & 3 \end{bmatrix}\begin{bmatrix} 1 \\ 6 \end{bmatrix} = \begin{bmatrix} -\frac{1}{13} \\ \frac{8}{13} \end{bmatrix}$

Answer: $\left(-\frac{1}{13}, \frac{8}{13}\right)$

52. $A = \begin{bmatrix} 4 & -2 & 3 \\ 2 & 2 & 5 \\ 8 & -5 & -2 \end{bmatrix}$

$A^{-1} = \dfrac{1}{82}\begin{bmatrix} -21 & 19 & 16 \\ -44 & 32 & 14 \\ 26 & -4 & -12 \end{bmatrix}$

$\begin{bmatrix} x \\ y \\ z \end{bmatrix} = \dfrac{1}{82}\begin{bmatrix} -21 & 19 & 16 \\ -44 & 32 & 14 \\ 26 & -4 & -12 \end{bmatrix}\begin{bmatrix} -2 \\ 16 \\ 4 \end{bmatrix} = \begin{bmatrix} 5 \\ 8 \\ -2 \end{bmatrix}$

Answer: $(5, 8, -2)$

54. $A = \begin{bmatrix} 2 & 3 & 5 \\ 3 & 5 & 9 \\ 5 & 9 & 17 \end{bmatrix}$ A^{-1} does not exist.

No solution

56. $A = \begin{bmatrix} 2 & 5 & 0 & 1 \\ 1 & 4 & 2 & -2 \\ 2 & -2 & 5 & 1 \\ 1 & 0 & 0 & -3 \end{bmatrix}$

$$A^{-1} \approx \begin{bmatrix} 0.338 & -0.352 & 0.141 & 0.394 \\ 0.042 & 0.164 & -0.066 & -0.117 \\ -0.141 & 0.230 & 0.108 & -0.164 \\ 0.113 & -0.117 & 0.047 & -0.202 \end{bmatrix}$$

$$\begin{bmatrix} x \\ y \\ z \\ w \end{bmatrix} = \begin{bmatrix} 0.338 & -0.352 & 0.141 & 0.394 \\ 0.042 & 0.164 & -0.066 & -0.117 \\ -0.141 & 0.230 & 0.108 & -0.164 \\ 0.113 & -0.117 & 0.047 & -0.202 \end{bmatrix} \begin{bmatrix} 11 \\ -7 \\ 3 \\ -1 \end{bmatrix} = \begin{bmatrix} 6.21 \\ -0.77 \\ -2.67 \\ 2.40 \end{bmatrix}$$

Answer: (6.21, −0.77, −2.67, 2.40)

For 58 and 60 use $A = \begin{bmatrix} 1 & 1 & 1 \\ 0.065 & 0.07 & 0.09 \\ 0 & 2 & -1 \end{bmatrix}$. **Using the methods of this section, we have** $A^{-1} = \frac{1}{11}\begin{bmatrix} 50 & -600 & -4 \\ -13 & 200 & 5 \\ -26 & 400 & -1 \end{bmatrix}$.

58. $X = A^{-1}B = \frac{1}{11}\begin{bmatrix} 50 & -600 & -4 \\ -13 & 200 & 5 \\ -26 & 400 & -1 \end{bmatrix} \begin{bmatrix} 45{,}000 \\ 3750 \\ 0 \end{bmatrix} = \begin{bmatrix} 0 \\ 15{,}000 \\ 30{,}000 \end{bmatrix}$

Answer: $0 in AAA bonds, $15,000 in A bonds, and $30,000 in B bonds.

60. $X = A^{-1}B = \frac{1}{11}\begin{bmatrix} 50 & -600 & -4 \\ -13 & 200 & 5 \\ -26 & 400 & -1 \end{bmatrix} \begin{bmatrix} 500{,}000 \\ 38{,}000 \\ 0 \end{bmatrix} = \begin{bmatrix} 200{,}000 \\ 100{,}000 \\ 200{,}000 \end{bmatrix}$

Answer: $200,000 in AAA bonds, $100,000 in A bonds, and $200,000 in B bonds.

62. True. The definition of the inverse A^{-1} of an $n \times n$ matrix A is an $n \times n$ matrix such that $AA^{-1} = A = I$. Thus, the multiplication of an invertible matrix and its inverse is commutative. One example is:

$$\begin{bmatrix} 2 & 1 \\ 5 & 0 \end{bmatrix} \begin{bmatrix} 0 & \frac{1}{5} \\ 1 & -\frac{2}{5} \end{bmatrix} = \begin{bmatrix} 1 & 0 \\ 0 & 1 \end{bmatrix} = \begin{bmatrix} 0 & \frac{1}{5} \\ 1 & -\frac{2}{5} \end{bmatrix} \begin{bmatrix} 2 & 1 \\ 5 & 0 \end{bmatrix}$$

64. $A = \begin{bmatrix} 2 & 0 & 4 \\ 0 & 1 & 4 \\ 1 & 1 & -1 \end{bmatrix}$

$$A^{-1} = \frac{1}{14}\begin{bmatrix} 5 & -4 & 4 \\ -4 & 6 & 8 \\ 1 & 2 & -2 \end{bmatrix}$$

$$\begin{bmatrix} I_1 \\ I_2 \\ I_3 \end{bmatrix} = \frac{1}{14}\begin{bmatrix} 5 & -4 & 4 \\ -4 & 6 & 8 \\ 1 & 2 & -2 \end{bmatrix} \begin{bmatrix} 10 \\ 10 \\ 0 \end{bmatrix} = \begin{bmatrix} \frac{5}{7} \\ \frac{10}{7} \\ \frac{15}{7} \end{bmatrix}$$

Answer: $I_1 = \frac{5}{7}$ amps, $I_2 = \frac{10}{7}$ amps, $I_3 = \frac{15}{7}$ amps

Section 8.4 The Determinant of a Square Matrix

Solutions to Even-Numbered Exercises

2. -8

4. $\begin{vmatrix} -3 & 1 \\ 5 & 2 \end{vmatrix} = (-3)(2) - (5)(1) = -11$

6. $\begin{vmatrix} 2 & -2 \\ 4 & 3 \end{vmatrix} = (2)(3) - (4)(-2) = 14$

8. $\begin{vmatrix} 4 & -3 \\ 0 & 0 \end{vmatrix} = (4)(0) - (0)(-3) = 0$

10. $\begin{vmatrix} 2 & -3 \\ -6 & 9 \end{vmatrix} = (2)(9) - (-6)(-3) = 0$

12. $\begin{vmatrix} -2 & 2 & 3 \\ 1 & -1 & 0 \\ 0 & 1 & 4 \end{vmatrix} = 0\begin{vmatrix} 2 & 3 \\ -1 & 0 \end{vmatrix} - 1\begin{vmatrix} -2 & 3 \\ 1 & 0 \end{vmatrix} + 4\begin{vmatrix} -2 & 2 \\ 1 & -1 \end{vmatrix} = 0(3) - 1(-3) + 4(0) = 3$

14. $\begin{vmatrix} 1 & 1 & 2 \\ 3 & 1 & 0 \\ -2 & 0 & 3 \end{vmatrix} = -2\begin{vmatrix} 1 & 2 \\ 1 & 0 \end{vmatrix} - 0\begin{vmatrix} 1 & 2 \\ 3 & 0 \end{vmatrix} + 3\begin{vmatrix} 1 & 1 \\ 3 & 1 \end{vmatrix} = -2$

16. $\begin{vmatrix} 1 & 0 & 0 \\ -4 & -1 & 0 \\ 5 & 1 & 5 \end{vmatrix} = (1)(-1)(5) = -5$ (Lower Triangular)

18. $\begin{vmatrix} 0.1 & 0.2 & 0.3 \\ -0.3 & 0.2 & 0.2 \\ 0.5 & 0.4 & 0.4 \end{vmatrix} = -0.022$

20. $\begin{vmatrix} 2 & 3 & 1 \\ 0 & 5 & -2 \\ 0 & 0 & -2 \end{vmatrix} = -20$

22. $\begin{bmatrix} 11 & 0 \\ -3 & 2 \end{bmatrix}$

(a) $M_{11} = 2$
$M_{12} = -3$
$M_{21} = 0$
$M_{22} = 11$

(b) $C_{11} = M_{11} = 2$
$C_{12} = M_{12} = 3$
$C_{21} = M_{21} = 0$
$C_{22} = M_{22} = 11$

24. $\begin{bmatrix} -2 & 9 & 4 \\ 7 & -6 & 0 \\ 6 & 7 & -6 \end{bmatrix}$

(a) $M_{11} = \begin{vmatrix} -6 & 0 \\ 7 & -6 \end{vmatrix} = 36$

$M_{12} = \begin{vmatrix} 7 & 0 \\ 6 & -6 \end{vmatrix} = -42$

$M_{13} = \begin{vmatrix} 7 & -6 \\ 6 & 7 \end{vmatrix} = 85$

$M_{21} = \begin{vmatrix} 9 & 4 \\ 7 & -6 \end{vmatrix} = -82$

$M_{22} = \begin{vmatrix} -2 & 4 \\ 6 & -6 \end{vmatrix} = -12$

$M_{23} = \begin{vmatrix} -2 & 9 \\ 6 & 7 \end{vmatrix} = -68$

$M_{31} = \begin{vmatrix} 9 & 4 \\ -6 & 0 \end{vmatrix} = 24$

$M_{32} = \begin{vmatrix} -2 & 4 \\ 7 & 0 \end{vmatrix} = -28$

$M_{33} = \begin{vmatrix} -2 & 9 \\ 7 & -6 \end{vmatrix} = -51$

(b) $C_{11} = (-1)^2 M_{11} = 36$
$C_{12} = (-1)^3 M_{12} = 42$
$C_{13} = (-1)^4 M_{13} = 85$
$C_{21} = (-1)^3 M_{21} = 82$
$C_{22} = (-1)^4 M_{22} = -12$
$C_{23} = (-1)^5 M_{23} = 68$
$C_{31} = (-1)^4 M_{31} = 24$
$C_{32} = (-1)^5 M_{32} = 28$
$C_{33} = (-1)^6 M_{33} = -51$

26. (a) $\begin{vmatrix} -3 & 4 & 2 \\ 6 & 3 & 1 \\ 4 & -7 & -8 \end{vmatrix} = -6\begin{vmatrix} 4 & 2 \\ -7 & -8 \end{vmatrix} + 3\begin{vmatrix} -3 & 2 \\ 4 & -8 \end{vmatrix} - 1\begin{vmatrix} -3 & 4 \\ 4 & -7 \end{vmatrix} = -6(-18) + 3(16) - (5) = 151$

(b) $\begin{vmatrix} -3 & 4 & 2 \\ 6 & 3 & 1 \\ 4 & -7 & -8 \end{vmatrix} = 2\begin{vmatrix} 6 & 3 \\ 4 & -7 \end{vmatrix} - \begin{vmatrix} -3 & 4 \\ 4 & -7 \end{vmatrix} - 8\begin{vmatrix} -3 & 4 \\ 6 & 3 \end{vmatrix} = 2(-54) - (5) - 8(-33) = 151$

28. (a) $\begin{vmatrix} 10 & -5 & 5 \\ 30 & 0 & 10 \\ 0 & 10 & 1 \end{vmatrix} = 0\begin{vmatrix} -5 & 5 \\ 0 & 10 \end{vmatrix} - 10\begin{vmatrix} 10 & 5 \\ 30 & 10 \end{vmatrix} + \begin{vmatrix} 10 & -5 \\ 30 & 0 \end{vmatrix} = 0(-50) - 10(-50) + 150 = 650$

(b) $\begin{vmatrix} 10 & -5 & 5 \\ 30 & 0 & 10 \\ 0 & 10 & 1 \end{vmatrix} = 10\begin{vmatrix} 0 & 10 \\ 10 & 1 \end{vmatrix} - 30\begin{vmatrix} -5 & 5 \\ 10 & 1 \end{vmatrix} + 0\begin{vmatrix} -5 & 5 \\ 0 & 10 \end{vmatrix} = 10(-100) - 30(-55) + 0(-50) = 650$

30. (a) $\begin{vmatrix} 10 & 8 & 3 & -7 \\ 4 & 0 & 5 & -6 \\ 0 & 3 & 2 & 7 \\ 1 & 0 & -3 & 2 \end{vmatrix} = 0\begin{vmatrix} 8 & 3 & -7 \\ 0 & 5 & -6 \\ 0 & -3 & 2 \end{vmatrix} - 3\begin{vmatrix} 10 & 3 & -7 \\ 4 & 5 & -6 \\ 1 & -3 & 2 \end{vmatrix} + 2\begin{vmatrix} 10 & 8 & -7 \\ 4 & 0 & -6 \\ 1 & 0 & 2 \end{vmatrix} - 7\begin{vmatrix} 10 & 8 & 3 \\ 4 & 0 & 5 \\ 1 & 0 & -3 \end{vmatrix}$

$= 0(-64) - 3(-3) + 2(-112) - 7(136) = -1167$

(b) $\begin{vmatrix} 10 & 8 & 3 & -7 \\ 4 & 0 & 5 & -6 \\ 0 & 3 & 2 & 7 \\ 1 & 0 & -3 & 2 \end{vmatrix} = 10\begin{vmatrix} 0 & 5 & -6 \\ 3 & 2 & 7 \\ 0 & -3 & 2 \end{vmatrix} - 4\begin{vmatrix} 8 & 3 & -7 \\ 3 & 2 & 7 \\ 0 & -3 & 2 \end{vmatrix} + 0\begin{vmatrix} 8 & 3 & -7 \\ 0 & 5 & -6 \\ 0 & -3 & 2 \end{vmatrix} - 1\begin{vmatrix} 8 & 3 & -7 \\ 0 & 5 & -6 \\ 3 & 2 & 7 \end{vmatrix}$

$= 10(24) - 4(245) + 0(-64) - 1(427) = -1167$

32. Expand by Row 3.

$\begin{vmatrix} 2 & -1 & 3 \\ 1 & 4 & 4 \\ 1 & 0 & 2 \end{vmatrix} = 1\begin{vmatrix} -1 & 3 \\ 4 & 4 \end{vmatrix} + 2\begin{vmatrix} 2 & -1 \\ 1 & 4 \end{vmatrix} = 1(-16) + 2(9) = 2$

34. $\begin{vmatrix} -3 & 0 & 0 \\ 7 & 11 & 4 \\ 1 & 2 & 2 \end{vmatrix} = (-3)(11)(2) = -66$ (Lower Triangular)

36. Expand by Row 2.

$\begin{vmatrix} 3 & 6 & -5 & 4 \\ -2 & 0 & 6 & 0 \\ 1 & 1 & 2 & 2 \\ 0 & 3 & -1 & -1 \end{vmatrix} = -(-2)\begin{vmatrix} 6 & -5 & 4 \\ 1 & 2 & 2 \\ 3 & -1 & -1 \end{vmatrix} - 6\begin{vmatrix} 3 & 6 & 4 \\ 1 & 1 & 2 \\ 0 & 3 & -1 \end{vmatrix} = 2(-63) - 6(-3) = -108$

38. Expand by Row 3.

$\begin{vmatrix} 1 & 4 & 3 & 2 \\ -5 & 6 & 2 & 1 \\ 0 & 0 & 0 & 2 \\ 3 & -2 & 1 & 5 \end{vmatrix} = 0$

40. Expand by Column 1.

$$\begin{vmatrix} 5 & 2 & 0 & 0 & -2 \\ 0 & 1 & 4 & 3 & 2 \\ 0 & 0 & 2 & 6 & 3 \\ 0 & 0 & 3 & 4 & 1 \\ 0 & 0 & 0 & 0 & 2 \end{vmatrix} = 5\begin{vmatrix} 1 & 4 & 3 & 2 \\ 0 & 2 & 6 & 3 \\ 0 & 3 & 4 & 1 \\ 0 & 0 & 0 & 2 \end{vmatrix} = 5 \cdot 1 \begin{vmatrix} 2 & 6 & 3 \\ 3 & 4 & 1 \\ 0 & 0 & 2 \end{vmatrix} = 5(-20) = -100$$

42. $\begin{vmatrix} 5 & -8 & 0 \\ 9 & 7 & 4 \\ -8 & 7 & 1 \end{vmatrix} = 223$

44. $\begin{vmatrix} 3 & 0 & 0 \\ -2 & 5 & 0 \\ 12 & 5 & 7 \end{vmatrix} = 105$

46. $\begin{vmatrix} 0 & -3 & 8 & 2 \\ 8 & 1 & -1 & 6 \\ -4 & 6 & 0 & 9 \\ -7 & 0 & 0 & 14 \end{vmatrix} = 7441$

48. $\begin{vmatrix} -2 & 0 & 0 & 0 & 0 \\ 0 & 3 & 0 & 0 & 0 \\ 0 & 0 & -1 & 0 & 0 \\ 0 & 0 & 0 & 2 & 0 \\ 0 & 0 & 0 & 0 & -4 \end{vmatrix} = -48$

50. $\begin{vmatrix} w & cx \\ y & cz \end{vmatrix} = cwz - cxy = c(wz - xy)$

$c\begin{vmatrix} w & x \\ y & z \end{vmatrix} = c(wz - xy)$

Thus, $\begin{vmatrix} w & cx \\ y & cz \end{vmatrix} = c\begin{vmatrix} w & x \\ y & z \end{vmatrix}$.

52. $\begin{vmatrix} w & x \\ cw & cx \end{vmatrix} = cxw - cxw = cxw = 0$

Thus, $\begin{vmatrix} w & x \\ cw & cx \end{vmatrix} = 0.$

54.
$$\begin{aligned}
\begin{vmatrix} a+b & a & a \\ a & a+b & a \\ a & a & a+b \end{vmatrix} &= (a+b)\begin{vmatrix} a+b & a \\ a & a+b \end{vmatrix} - a\begin{vmatrix} a & a \\ a & a+b \end{vmatrix} + a\begin{vmatrix} a & a \\ a+b & a \end{vmatrix} \\
&= (a+b)[(a+b)^2 - a^2] - a[a(a+b) - a^2] + a[a^2 - a(a+b)] \\
&= (a+b)^3 - a^2(a+b) - a^2(a+b) + a^3 + a^3 - a^2(a+b) \\
&= (a+b)^3 - 3a^2(a+b)2a^3 \\
&= a^3 + 3a^2b + 3ab^2 + b^3 - 3a^3 - 3a^2b + 2a^3 \\
&= 3ab^2 + b^3 = b^2(3a+b)
\end{aligned}$$

56.
$$\begin{vmatrix} x-2 & -1 \\ -3 & x \end{vmatrix} = 0$$

$$\begin{aligned}
x(x-2) - (-3)(-1) &= 0 \\
x^2 - 2x - 3 &= 0 \\
(x+1)(x-3) &= 0 \\
x = -1 \text{ or } x &= 3
\end{aligned}$$

58. $\begin{vmatrix} 3x^2 & -3y^2 \\ 1 & 1 \end{vmatrix} = 3x^2 - (-3y^2) = 3x^2 + 3y^2$

60. $\begin{vmatrix} e^{-x} & xe^{-x} \\ -e^{-x} & (1-x)e^{-x} \end{vmatrix} = (1-x)e^{-2x} - (-xe^{-2x}) = e^{-2x} - xe^{-2x} + xe^{-2x} = e^{-2x}$

62. $\begin{vmatrix} x & x\ln x \\ 1 & 1+\ln x \end{vmatrix} = x(1 + \ln x) - x\ln x = x + x\ln x - x\ln x = x$

64. (a) $|A| = \begin{vmatrix} -2 & 1 \\ 4 & -2 \end{vmatrix} = 0$

(b) $|B| = \begin{vmatrix} 1 & 2 \\ 0 & -2 \end{vmatrix} = -1$

(c) $AB = \begin{bmatrix} -2 & 1 \\ 4 & -2 \end{bmatrix}\begin{bmatrix} 1 & 2 \\ 0 & -1 \end{bmatrix} = \begin{bmatrix} -2 & -5 \\ 4 & 10 \end{bmatrix}$

(d) $|AB| = \begin{vmatrix} -2 & -5 \\ 4 & 10 \end{vmatrix} = 0$

66. (a) $|A| = \begin{vmatrix} 2 & 0 & 1 \\ 1 & -1 & 2 \\ 3 & 1 & 0 \end{vmatrix} = 0$ (b) $|B| = \begin{vmatrix} 2 & -1 & 4 \\ 0 & 1 & 3 \\ 3 & -2 & 1 \end{vmatrix} = -7$

(c) $AB = \begin{bmatrix} 2 & 0 & 1 \\ 1 & -1 & 2 \\ 3 & 1 & 0 \end{bmatrix}\begin{bmatrix} 2 & -1 & 4 \\ 0 & 1 & 3 \\ 3 & -2 & 1 \end{bmatrix} = \begin{bmatrix} 7 & -4 & 9 \\ 8 & -6 & 3 \\ 6 & -2 & 15 \end{bmatrix}$

(d) $|AB| = \begin{vmatrix} 7 & -4 & 9 \\ 8 & -6 & 3 \\ 6 & -2 & 15 \end{vmatrix} = 0$

68. (a) $\begin{vmatrix} 4 & 5 & 6 \\ 7 & 8 & 9 \\ 10 & 11 & 12 \end{vmatrix} = 0$ $\begin{vmatrix} 10 & 11 & 12 \\ 13 & 14 & 15 \\ 16 & 17 & 18 \end{vmatrix} = 0$

$\begin{vmatrix} 33 & 34 & 35 \\ 36 & 37 & 38 \\ 39 & 40 & 41 \end{vmatrix} = 0$ $\begin{vmatrix} -5 & -4 & -3 \\ -2 & -1 & 0 \\ 1 & 2 & 3 \end{vmatrix} = 0$

$\begin{vmatrix} 19 & 20 & 21 & 22 \\ 23 & 24 & 25 & 26 \\ 27 & 28 & 29 & 30 \\ 31 & 32 & 33 & 34 \end{vmatrix} = 0$ $\begin{vmatrix} 57 & 58 & 59 & 60 \\ 61 & 62 & 63 & 64 \\ 65 & 66 & 67 & 68 \\ 69 & 70 & 71 & 72 \end{vmatrix} = 0$

For an $n \times n$ matrix $(n > 2)$ with consecutive integer entries, the determinant appears to be 0.

(b) $\begin{vmatrix} x & x+1 & x+2 \\ x+3 & x+4 & x+5 \\ x+6 & x+7 & x+8 \end{vmatrix} = x\begin{vmatrix} x+4 & x+5 \\ x+7 & x+8 \end{vmatrix} - (x+1)\begin{vmatrix} x+3 & x+5 \\ x+6 & x+8 \end{vmatrix} + (x+2)\begin{vmatrix} x+3 & x+4 \\ x+6 & x+7 \end{vmatrix}$

$$\begin{aligned} &= x[(x+4)(x+8) - (x+7)(x+5)] - (x+1)[(x+3)(x+8) \\ &\quad - (x+6)(x+5)] + (x+2)[(x+3)(x+7) - (x+6)(x+4)] \\ &= x[(x^2+12x+32) - (x^2+12x+35)] - (x+1)[(x^2+11x+24) \\ &\quad - (x^2+11x+30)] + (x+2)[(x^2+10x+21) - (x^2+10x+24)] \\ &= -3x - (x+1)(-6) + (x+2)(-3) \\ &= -3x + 6x + 6 - 3x - 6 = 0 \end{aligned}$$

70. Let $A = \begin{bmatrix} x_{11} & x_{12} & x_{13} \\ x_{21} & x_{22} & x_{23} \\ x_{31} & x_{32} & x_{33} \end{bmatrix}$ and $|A| = 5$.

$$2A = \begin{bmatrix} 2x_{11} & 2x_{12} & 2x_{13} \\ 2x_{21} & 2x_{22} & 2x_{23} \\ 2x_{31} & 2x_{32} & 2x_{33} \end{bmatrix}$$

$$\begin{aligned} |2A| &= 2x_{11}\begin{vmatrix} 2x_{22} & 2x_{23} \\ 2x_{32} & 2x_{33} \end{vmatrix} - 2x_{12}\begin{vmatrix} 2x_{21} & 2x_{23} \\ 2x_{31} & 2x_{33} \end{vmatrix} + 2x_{13}\begin{vmatrix} 2x_{21} & 2x_{22} \\ 2x_{31} & 2x_{32} \end{vmatrix} \\ &= 2[x_{11}(4x_{22}x_{33} - 4x_{32}x_{23}) - x_{12}(4x_{21}x_{33} - 4x_{31}x_{23}) + x_{13}(4x_{21}x_{32} - 4x_{31}x_{22})] \\ &= 8[x_{11}(x_{22}x_{33} - x_{32}x_{23}) - x_{12}(x_{21}x_{33} - x_{31}x_{23}) + x_{13}(x_{21}x_{32} - x_{31}x_{22})] \\ &= 8|A| \end{aligned}$$

Thus, $|2A| = 8|A| = 8(5) = 40$.

72. (a) (-5) times Row 1 is added to Row 2.

(b) (-2) times Row 2 is added to Row 1.

Section 8.5 Applications of Matrices and Determinants

Solutions to Even-Numbered Exercises

2. Vertices: $(-2, 1), (1, 6), (3, -1)$

$$\text{Area} = -\frac{1}{2}\begin{vmatrix} -2 & 1 & 1 \\ 1 & 6 & 1 \\ 3 & -1 & 1 \end{vmatrix} = -\frac{1}{2}(-19 + 1 - 13) = \frac{31}{2} \text{ square units}$$

4. Vertices: $(0, 0), (4, 5), (5, -2)$

$$\text{Area} = -\frac{1}{2}\begin{vmatrix} 0 & 0 & 1 \\ 4 & 5 & 1 \\ 5 & -2 & 1 \end{vmatrix} = -\frac{1}{2}\begin{vmatrix} 4 & 5 \\ 5 & -2 \end{vmatrix} = \frac{33}{2} \text{ square units}$$

6. Vertices: $(0, 4), (2, 3), (5, 0)$

$$\text{Area} = -\frac{1}{2}\begin{vmatrix} 0 & 4 & 1 \\ 2 & 3 & 1 \\ 5 & 0 & 1 \end{vmatrix} = -\frac{1}{2}(12 - 15) = \frac{3}{2} \text{ square units}$$

8. Vertices: $(0, -2), (-1, 4), (3, 5)$

$$\text{Area} = -\frac{1}{2}\begin{vmatrix} 0 & -2 & 1 \\ -1 & 4 & 1 \\ 3 & 5 & 1 \end{vmatrix} = -\frac{1}{2}(-17 - 6 - 2) = \frac{25}{2} \text{ square units}$$

10. Vertices: $(-2, 4), (1, 5), (3, -2)$

$$\text{Area} = -\frac{1}{2}\begin{vmatrix} -2 & 4 & 1 \\ 1 & 5 & 1 \\ 3 & -2 & 1 \end{vmatrix} = -\frac{1}{2}(-17 + 8 - 14) = \frac{23}{2} \text{ square units}$$

12. $$4 = \pm\frac{1}{2}\begin{vmatrix} -4 & 2 & 1 \\ -3 & 5 & 1 \\ -1 & x & 1 \end{vmatrix}$$

$$\pm 8 = \begin{vmatrix} -3 & 5 \\ -1 & x \end{vmatrix} - \begin{vmatrix} -4 & 2 \\ -1 & x \end{vmatrix} + \begin{vmatrix} -4 & 2 \\ -3 & 5 \end{vmatrix}$$

$$\pm 8 = -3x + 5 - (-4x + 2) - 20 + 6$$

$$\pm 8 = -3x + 5 + 4x - 2 - 20 + 6$$

$$\pm 8 = x - 11$$

$$x = 11 \pm 8$$

$$x = 19 \text{ or } x = 3$$

14. $-0.4x + 0.8y = 1.6$

$0.2x + 0.3y = 2.2$

$$D = \begin{vmatrix} -0.4 & 0.8 \\ 0.2 & 0.3 \end{vmatrix} = -0.28$$

$$x = \frac{\begin{vmatrix} 1.6 & 0.8 \\ 2.2 & 0.3 \end{vmatrix}}{-0.28} = \frac{-1.18}{-0.28} = \frac{32}{7}$$

$$y = \frac{\begin{vmatrix} -0.4 & 1.6 \\ 0.2 & 2.2 \end{vmatrix}}{-0.28} = \frac{-1.20}{-0.28} = \frac{30}{7}$$

Answer: $\left(\frac{32}{7}, \frac{30}{7}\right)$

16. $4x - 2y + 3z = -2$
$2x + 2y + 5z = 16$
$8x - 5y - 2z = 4$

$$D = \begin{vmatrix} 4 & -2 & 3 \\ 2 & 2 & 5 \\ 8 & -5 & -2 \end{vmatrix} = -82$$

$$x = \frac{\begin{vmatrix} -2 & -2 & 3 \\ 16 & 2 & 5 \\ 4 & -5 & -2 \end{vmatrix}}{-82} = \frac{-401}{-82} = 5$$

$$y = \frac{\begin{vmatrix} 4 & -2 & 3 \\ 2 & 16 & 5 \\ 8 & 4 & -2 \end{vmatrix}}{-82} = \frac{-656}{-82} = 8$$

$$z = \frac{\begin{vmatrix} 4 & -2 & -2 \\ 2 & 2 & 16 \\ 8 & -5 & 4 \end{vmatrix}}{-82} = \frac{164}{-82} = -2$$

Answer: $(5, 8, -2)$

18. $2x + 3y + 5z = 4$
$3x + 5y + 9z = 7$
$5x + 9y + 17z = 13$

$$D = \begin{vmatrix} 2 & 3 & 5 \\ 3 & 5 & 9 \\ 5 & 9 & 17 \end{vmatrix} = 0$$

Cramer's Rule does not apply.

20. Vertices: $(0, 30), (85, 0), (20, -50)$

$$\text{Area} = -\frac{1}{2}\begin{vmatrix} 0 & 30 & 1 \\ 85 & 0 & 1 \\ 20 & -50 & 1 \end{vmatrix} = 3100 \text{ square units}$$

22. Points: $(-3, -5), (6, 1), (10, 2)$

$$\begin{vmatrix} -3 & -5 & 1 \\ 6 & 1 & 1 \\ 10 & 2 & 1 \end{vmatrix} = \begin{vmatrix} 6 & 1 \\ 10 & 2 \end{vmatrix} - \begin{vmatrix} -3 & -5 \\ 10 & 2 \end{vmatrix} + \begin{vmatrix} -3 & -5 \\ 6 & 1 \end{vmatrix} = -15 \neq 0$$

The points are not collinear.

24. Points: $(0, 1), (4, -2), (-8, 7)$

$$\begin{vmatrix} 0 & 1 & 1 \\ 4 & -2 & 1 \\ -8 & 7 & 1 \end{vmatrix} = \begin{vmatrix} 4 & -2 \\ -8 & 7 \end{vmatrix} - \begin{vmatrix} 0 & 1 \\ -8 & 7 \end{vmatrix} + \begin{vmatrix} 0 & 1 \\ 4 & -2 \end{vmatrix} = 0$$

The points are collinear.

26. Points: $(2, 3), (3, 3.5), (-1, 2)$

$$\begin{vmatrix} 2 & 3 & 1 \\ 3 & 3.5 & 1 \\ -1 & 2 & 1 \end{vmatrix} = \begin{vmatrix} 3 & 3.5 \\ -1 & 2 \end{vmatrix} - \begin{vmatrix} 2 & 3 \\ -1 & 2 \end{vmatrix} + \begin{vmatrix} 2 & 3 \\ 3 & 3.5 \end{vmatrix} = \frac{1}{2} \neq 0$$

The points are not collinear.

28. Points: $(0, 0), (-2, 2)$

$$\text{Equation: } \begin{vmatrix} x & y & 1 \\ 0 & 0 & 1 \\ -2 & 2 & 1 \end{vmatrix} = -(2x + 2y) = 0 \text{ or } x + y = 0$$

30. Points: $(10, 7), (-2, -7)$

$$\text{Equation: } \begin{vmatrix} x & y & 1 \\ 10 & 7 & 1 \\ -2 & -7 & 1 \end{vmatrix} = -70 + 14 - (-7x + 2y) + 7x - 10y = 0 \text{ or } 7x - 6y - 28 = 0$$

32. Points: $\left(\frac{2}{3}, 4\right)$, $(6, 12)$

Equation: $\begin{vmatrix} x & y & 1 \\ \frac{2}{3} & 4 & 1 \\ 6 & 12 & 1 \end{vmatrix} = -16 - (12x - 6y) + 4x - \frac{2}{3}y = 0$ or $3x - 2y + 6 = 0$

34. $\begin{vmatrix} -6 & 2 & 1 \\ -5 & x & 1 \\ -3 & 5 & 1 \end{vmatrix} = 0$

$$\begin{vmatrix} -5 & x \\ -3 & 5 \end{vmatrix} - \begin{vmatrix} -6 & 2 \\ -3 & 5 \end{vmatrix} + \begin{vmatrix} -6 & 2 \\ -5 & x \end{vmatrix} = 0$$

$$-25 + 3x + 24 - 6x + 10 = 0$$

$$-3x = -9$$

$$x = 3$$

36. The uncoded row matrices are the rows of the 6×3 matrix on the left.

$$\begin{matrix} \text{P} & \text{L} & \text{E} \\ \text{A} & \text{S} & \text{E} \\ & \text{S} & \text{E} \\ \text{N} & \text{D} & \\ \text{M} & \text{O} & \text{N} \\ \text{E} & \text{Y} & \end{matrix} \quad \begin{bmatrix} 16 & 12 & 5 \\ 1 & 19 & 5 \\ 0 & 19 & 5 \\ 14 & 4 & 0 \\ 13 & 15 & 14 \\ 5 & 25 & 0 \end{bmatrix} \begin{bmatrix} 4 & 2 & 1 \\ -3 & -3 & -1 \\ 3 & 2 & 1 \end{bmatrix} = \begin{bmatrix} 43 & 6 & 9 \\ -38 & -45 & -13 \\ -42 & -47 & -14 \\ 44 & 16 & 10 \\ 49 & 9 & 12 \\ -55 & -65 & -20 \end{bmatrix}$$

Answer: [43 6. 9], [−38 −45 −13], [−42 −47 −14], [44 16 10], [49 9 12], [−55 −65 −20]

38. B E A M _ M E _ U P _ S C O T T Y _

[2 5 1] [13 0 13] [5 0 21][16 0 19] [3 15 20][20 25 0]

$[2\ 5\ 1]A = [16\ 35\ 42]$

$[13\ 0\ 13]A = [0\ -26\ -65]$

$[5\ 0\ 21]A = [-16\ -74\ -137]$

$[16\ 0\ 19]A = [-3\ -44\ -101]$

$[3\ 15\ 20]A = [28\ 31\ 1]$

$[20\ 25\ 0]A = [95\ 215\ 265]$

Cryptogram: 16 35 42 0 −26 −65 −16 −74 −137 −3 −44 −101 28 31 1 95 215 265

40. O P E R A T I O N _ O V E R L O R D

[15 16 5] [18 1 20] [9 15 14] [0 15 22] [5 18 12][15 18 4]

$[15\ 16\ 5]A = [58\ 122\ 139]$

$[18\ 1\ 20]A = [1\ -37\ -95]$

$[9\ 15\ 14]A = [40\ 67\ 55]$

$[0\ 15\ 22]A = [23\ 17\ -19]$

$[5\ 18\ 12]A = [47\ 88\ 88]$

$[15\ 18\ 4]A = [65\ 140\ 164]$

Cryptogram: 58 122 139 1 −37 −95 40 67 55 23 17 −19 47 88 88 65 140 164

42. $A^{-1} = \begin{bmatrix} -2 & -3 & -1 \\ -3 & -3 & -1 \\ -2 & -4 & -1 \end{bmatrix}$

$$\begin{bmatrix} 9 & -1 & -9 \\ 38 & -19 & -19 \\ 28 & -9 & -19 \\ -80 & 25 & 41 \\ -64 & 21 & 31 \\ 9 & -5 & -4 \end{bmatrix} \begin{bmatrix} -2 & -3 & -1 \\ -3 & -3 & -1 \\ -2 & -4 & -1 \end{bmatrix} = \begin{bmatrix} 3 & 12 & 1 \\ 19 & 19 & 0 \\ 9 & 19 & 0 \\ 3 & 1 & 14 \\ 3 & 5 & 12 \\ 5 & 4 & 0 \end{bmatrix} \begin{matrix} C & L & A \\ S & S & \_ \\ I & S & \_ \\ C & A & N \\ C & E & L \\ E & D & \_ \end{matrix}$$

Message: CLASS IS CANCELED

44. $A^{-1} = \begin{bmatrix} -13 & 6 & 4 \\ 12 & -5 & -3 \\ -5 & 2 & 1 \end{bmatrix}$

$$\begin{bmatrix} 13 & -9 & -59 \\ 61 & 112 & 106 \\ -17 & -73 & -131 \\ 11 & 24 & 29 \\ 65 & 144 & 172 \end{bmatrix} \begin{bmatrix} -13 & 6 & 4 \\ 12 & -5 & -3 \\ -5 & 2 & 1 \end{bmatrix} = \begin{bmatrix} 18 & 5 & 20 \\ 21 & 18 & 14 \\ 0 & 1 & 20 \\ 0 & 4 & 1 \\ 23 & 14 & 0 \end{bmatrix} \begin{matrix} R & E & T \\ U & R & N \\ \_ & A & T \\ \_ & D & A \\ W & N & \_ \end{matrix}$$

Message: RETURN AT DAWN

46. Let A be the 2×2 matrix needed to decode the message.

$$\begin{bmatrix} -19 & -19 \\ 37 & 16 \end{bmatrix} A = \begin{bmatrix} 0 & 19 \\ 21 & 5 \end{bmatrix} \begin{matrix} \_ & S \\ U & E \end{matrix}$$

$$A = \begin{bmatrix} -19 & -19 \\ 37 & 16 \end{bmatrix}^{-1} \begin{bmatrix} 0 & 19 \\ 21 & 5 \end{bmatrix} = \begin{bmatrix} \frac{16}{399} & \frac{19}{399} \\ -\frac{37}{399} & -\frac{19}{399} \end{bmatrix} \begin{bmatrix} 0 & 19 \\ 21 & 5 \end{bmatrix} = \begin{bmatrix} 1 & 1 \\ -1 & -2 \end{bmatrix}$$

$$\begin{bmatrix} 5 & 2 \\ 25 & 11 \\ -2 & -7 \\ -15 & -15 \\ 32 & 14 \\ -8 & -13 \\ 38 & 19 \\ -19 & -19 \\ 37 & 16 \end{bmatrix} \begin{bmatrix} 1 & 1 \\ -1 & -2 \end{bmatrix} = \begin{bmatrix} 3 & 1 \\ 14 & 3 \\ 5 & 12 \\ 0 & 15 \\ 18 & 4 \\ 5 & 18 \\ 19 & 0 \\ 0 & 19 \\ 21 & 5 \end{bmatrix} \begin{matrix} C & A \\ N & C \\ E & L \\ \_ & O \\ R & D \\ E & R \\ S & \_ \\ \_ & S \\ U & E \end{matrix}$$

Message: CANCEL ORDERS SUE

Review Exercises for Chapter 8

Solutions to Even-Numbered Exercises

2. $\left[\begin{array}{rrr|r} 8 & -7 & 4 & 12 \\ 3 & -5 & 2 & 20 \\ 5 & 3 & -3 & 26 \end{array}\right]$

4. $\left[\begin{array}{rrrr|r} 13 & 16 & 7 & 3 & 2 \\ 1 & 21 & 8 & 5 & 12 \\ 4 & 10 & -4 & 3 & -1 \end{array}\right]$

$$\begin{aligned} 13x + 16y + 7z + 3w &= 2 \\ x + 21y + 8z + 5w &= 12 \\ 4x + 10y - 4z + 3w &= -1 \end{aligned}$$

6. $\begin{bmatrix} 1 & 1 & 1 & 0 \\ 1 & 1 & 0 & 1 \\ 1 & 0 & 1 & 1 \\ 0 & 1 & 1 & 1 \end{bmatrix}$

$\begin{matrix} \\ -R_1 + R_2 \rightarrow \\ -R_1 + R_3 \rightarrow \\ \\ \end{matrix} \begin{bmatrix} 1 & 1 & 1 & 0 \\ 0 & 0 & -1 & 1 \\ 0 & -1 & 0 & 1 \\ 0 & 1 & 1 & 1 \end{bmatrix}$

$\begin{matrix} R_3 + R_1 \rightarrow \\ \\ \\ R_3 + R_4 \rightarrow \end{matrix} \begin{bmatrix} 1 & 0 & 1 & 1 \\ 0 & 0 & -1 & 1 \\ 0 & -1 & 0 & 1 \\ 0 & 0 & 1 & 2 \end{bmatrix}$

$\begin{matrix} \\ -R_2 \rightarrow \\ -R_3 \rightarrow \\ R_2 + R_4 \rightarrow \end{matrix} \begin{bmatrix} 1 & 0 & 1 & 1 \\ 0 & 0 & 1 & -1 \\ 0 & 1 & 0 & -1 \\ 0 & 0 & 0 & 3 \end{bmatrix}$

$\begin{matrix} R_3 \rightarrow \\ R_2 \rightarrow \\ \frac{1}{3}R_4 \rightarrow \\ \\ \end{matrix} \begin{bmatrix} 1 & 0 & 1 & 1 \\ 0 & 1 & 0 & -1 \\ 0 & 0 & 1 & -1 \\ 0 & 0 & 0 & 1 \end{bmatrix}$

$\begin{matrix} -R_4 + R_1 \rightarrow \\ R_4 + R_2 \rightarrow \\ R_4 + R_3 \rightarrow \\ \\ \end{matrix} \begin{bmatrix} 1 & 0 & 1 & 0 \\ 0 & 1 & 0 & 0 \\ 0 & 0 & 1 & 0 \\ 0 & 0 & 0 & 1 \end{bmatrix}$

$\begin{matrix} -R_3 + R_1 \rightarrow \\ \\ \\ \\ \end{matrix} \begin{bmatrix} 1 & 0 & 0 & 0 \\ 0 & 1 & 0 & 0 \\ 0 & 0 & 1 & 0 \\ 0 & 0 & 0 & 1 \end{bmatrix}$

8. $\begin{bmatrix} 1 & 0 & 0 & -6 & -4 & 3 \\ 0 & 1 & 0 & 11 & 6 & -5 \\ 0 & 0 & 1 & -2 & -1 & 1 \end{bmatrix}$

10. $\begin{bmatrix} 1 & 0 & \frac{8}{7} \\ 0 & 1 & \frac{10}{7} \\ 0 & 0 & 0 \end{bmatrix}$

12. $\left[\begin{array}{rr|r} 2 & -5 & 2 \\ 3 & -7 & 1 \end{array}\right]$

$R_2 - R_1 \rightarrow \left[\begin{array}{rr|r} 1 & -2 & -1 \\ 3 & -7 & 1 \end{array}\right]$

$-3R_1 + R_2 \rightarrow \left[\begin{array}{rr|r} 1 & -2 & -1 \\ 0 & -1 & 4 \end{array}\right]$

$\begin{matrix} -2R_2 + R_1 \rightarrow \\ -R_2 \rightarrow \end{matrix} \left[\begin{array}{rr|r} 1 & 0 & -9 \\ 0 & 1 & -4 \end{array}\right]$

$x = -9$

$y = -4$

Answer: $(-9, -4)$

14. $\left[\begin{array}{rr|r} 0.2 & -0.1 & 0.07 \\ 0.4 & -0.5 & -0.01 \end{array}\right]$

$\begin{matrix} 5R_1 \rightarrow \\ -2R_1 + R_2 \rightarrow \end{matrix} \left[\begin{array}{rr|r} 1 & -0.5 & 0.35 \\ 0 & -0.3 & -0.15 \end{array}\right]$

$\begin{matrix} -\frac{5}{3}R_2 + R_1 \rightarrow \\ -\frac{10}{3}R_2 \rightarrow \end{matrix} \left[\begin{array}{rr|r} 1 & 0 & 0.6 \\ 0 & 1 & 0.5 \end{array}\right]$

$x = 0.6$

$y = 0.5$

Answer: $(0.6, 0.5)$

16.

$$\left[\begin{array}{ccc|c} 2 & 3 & 1 & 10 \\ 2 & -3 & -3 & 22 \\ 4 & -2 & 3 & -2 \end{array}\right]$$

$$\begin{matrix} R_3 \to \\ \\ R_1 \to \end{matrix}\left[\begin{array}{ccc|c} 4 & -2 & 3 & -2 \\ 2 & -3 & -3 & 22 \\ 2 & 3 & 1 & 10 \end{array}\right]$$

$$\begin{matrix} \\ R_1 - 2R_2 \to \\ R_1 - 2R_3 \to \end{matrix}\left[\begin{array}{ccc|c} 4 & -2 & 3 & -2 \\ 0 & 4 & 9 & -46 \\ 0 & -8 & 1 & -22 \end{array}\right]$$

$$\begin{matrix} \\ \\ 2R_2 + R_3 \to \end{matrix}\left[\begin{array}{ccc|c} 4 & -2 & 3 & -2 \\ 0 & 4 & 9 & -46 \\ 0 & 0 & 19 & -114 \end{array}\right]$$

$19z = -114 \Rightarrow z = -6$

$4y + 9(-6) = -46 \Rightarrow y = 2$

$4x - 2(2) + 3(-6) = -2 \Rightarrow x = 5$

Answer: $(5, 2, -6)$

18.

$$\left[\begin{array}{ccc|c} 2 & 3 & 3 & 3 \\ 6 & 6 & 12 & 13 \\ 12 & 9 & -1 & 2 \end{array}\right]$$

$$\begin{matrix} \\ -3R_1 + R_2 \to \\ -6R_1 + R_3 \to \end{matrix}\left[\begin{array}{ccc|c} 2 & 3 & 3 & 3 \\ 0 & -3 & 3 & 4 \\ 0 & -3 & -25 & -24 \end{array}\right]$$

$$\begin{matrix} R_2 + R_1 \to \\ \\ -R_2 + R_3 \to \end{matrix}\left[\begin{array}{ccc|c} 2 & 0 & 6 & 7 \\ 0 & -3 & 3 & 4 \\ 0 & 0 & -28 & -28 \end{array}\right]$$

$$\begin{matrix} \frac{1}{2}R_1 \to \\ -\frac{1}{3}R_2 \to \\ -\frac{1}{28}R_3 \to \end{matrix}\left[\begin{array}{ccc|c} 1 & 0 & 3 & \frac{7}{2} \\ 0 & 1 & -1 & -\frac{4}{3} \\ 0 & 0 & 1 & 1 \end{array}\right]$$

$z = 1$

$y - 1 = -\frac{4}{3} \Rightarrow y = -\frac{1}{3}$

$x + 3(1) = \frac{7}{2} \Rightarrow x = \frac{1}{2}$

Answer: $\left(\frac{1}{2}, -\frac{1}{3}, 1\right)$

20.

$$\left[\begin{array}{ccc|c} 3 & 21 & -29 & -1 \\ 2 & 15 & -21 & 0 \end{array}\right]$$

$$\begin{matrix} -R_2 + R_1 \to \\ \\ \end{matrix}\left[\begin{array}{ccc|c} 1 & 6 & -8 & -1 \\ 2 & 15 & -21 & 0 \end{array}\right]$$

$$\begin{matrix} \\ -2R_1 + R_2 \to \end{matrix}\left[\begin{array}{ccc|c} 1 & 6 & -8 & -1 \\ 0 & 3 & -5 & 2 \end{array}\right]$$

$$\begin{matrix} -2R_2 + R_1 \to \\ \\ \end{matrix}\left[\begin{array}{ccc|c} 1 & 0 & 2 & -5 \\ 0 & 3 & -5 & 2 \end{array}\right]$$

Let $z = a$.

$3y - 5a = 2 \Rightarrow \frac{5}{3}a + \frac{2}{3}$

$x + 2a = -5 \Rightarrow -2a - 5$

Answer: $\left(-2a - 5, \frac{5}{3}a + \frac{2}{3}, a\right)$

22.

$$\left[\begin{array}{cccc|c} 1 & 2 & 0 & 1 & 3 \\ 0 & -3 & 3 & 0 & 0 \\ 4 & 4 & 1 & 2 & 0 \\ 2 & 0 & 1 & 0 & 3 \end{array}\right]$$

$$\begin{matrix} \\ -\frac{1}{3}R_2 \to \\ -4R_1 + R_3 \to \\ -2R_1 + R_3 \to \end{matrix}\left[\begin{array}{cccc|c} 1 & 2 & 0 & 1 & 3 \\ 0 & 1 & -1 & 0 & 0 \\ 0 & -4 & 1 & -2 & -12 \\ 0 & -4 & 1 & -2 & -3 \end{array}\right]$$

$$\begin{matrix} \\ \\ \\ -R_3 + R_4 \to \end{matrix}\left[\begin{array}{cccc|c} 1 & 2 & 0 & 1 & 3 \\ 0 & 1 & -1 & 0 & 0 \\ 0 & -4 & 1 & -2 & -12 \\ 0 & 0 & 0 & 0 & 9 \end{array}\right]$$

$$\begin{aligned} x + 2y \quad + w &= 3 \\ y - z \quad &= 0 \\ -4y + z - 2w &= -12 \\ 0 &= 9 \end{aligned}$$

Inconsistent, no solution

24. $x + 9 = A(x + 2)^2 + B(x + 1)(x + 2) + C(x + 1)$

$x + 9 = A(x^2 + 4x + 4) + B(x^2 + 3x + 2) + Cx + C$

$x + 9 = Ax^2 + 4Ax + 4A + Bx^2 + 3Bx + 2B + Cx + C$

$x + 9 = (A + B)x^2 + (4A + 3B + C)x + 4A + 2B + C$

Equating coefficients of corresponding terms:

$0 = A + B$

$1 = 4A + 3B + C$

$9 = 4A + 2B + C$

$$\begin{bmatrix} 1 & 1 & 0 & \vdots & 0 \\ 4 & 3 & 1 & \vdots & 1 \\ 4 & 2 & 1 & \vdots & 9 \end{bmatrix}$$

$$\begin{matrix} \\ -4R_1 + R_2 \rightarrow \\ -4R_1 + R_3 \rightarrow \end{matrix} \begin{bmatrix} 1 & 1 & 0 & \vdots & 0 \\ 0 & -1 & 1 & \vdots & 1 \\ 0 & -2 & 1 & \vdots & 9 \end{bmatrix}$$

$$\begin{matrix} \\ -R_2 \rightarrow \\ -2R_2 + R_3 \rightarrow \end{matrix} \begin{bmatrix} 1 & 1 & 0 & \vdots & 0 \\ 0 & 1 & -1 & \vdots & -1 \\ 0 & 0 & -1 & \vdots & 7 \end{bmatrix}$$

$-C = 7 \Rightarrow C = -7$

$B - (-7) = -1 \Rightarrow B = -8$

$A - 8 = 0 \Rightarrow A = 8$

$$\frac{x + 9}{(x + 1)(x + 2)^2} = \frac{8}{x + 1} - \frac{8}{x + 2} - \frac{7}{(x + 2)^2}$$

26. $\begin{bmatrix} -1 & 3 & \vdots & 5 \\ 4 & -1 & \vdots & 2 \end{bmatrix} \rightarrow \begin{bmatrix} 1 & 0 & \vdots & 1 \\ 0 & 1 & \vdots & 2 \end{bmatrix}$

$(x, y) = (1, 2)$

28. $\begin{bmatrix} 3 & 0 & 6 & \vdots & 0 \\ -2 & 1 & 0 & \vdots & 5 \\ 0 & 1 & 2 & \vdots & 3 \end{bmatrix} \rightarrow \begin{bmatrix} 1 & 0 & 0 & \vdots & -2 \\ 0 & 1 & 0 & \vdots & 1 \\ 0 & 0 & 1 & \vdots & 1 \end{bmatrix}$

$(x, y, z) = (-2, 1, 1)$

30. $-2\begin{bmatrix} 1 & 2 \\ 5 & -4 \\ 6 & 0 \end{bmatrix} + 8\begin{bmatrix} 7 & 1 \\ 1 & 2 \\ 1 & 4 \end{bmatrix} = \begin{bmatrix} -2 & -4 \\ -10 & 8 \\ -12 & 0 \end{bmatrix} + \begin{bmatrix} 56 & 8 \\ 8 & 16 \\ 8 & 32 \end{bmatrix} = \begin{bmatrix} 54 & 4 \\ -2 & 24 \\ -4 & 32 \end{bmatrix}$

32. $\begin{bmatrix} 1 & 5 & 6 \\ 2 & -4 & 0 \end{bmatrix}\begin{bmatrix} 6 & -2 & 8 \\ 4 & 0 & 0 \end{bmatrix}$ is undefined.

34. $\begin{bmatrix} 4 \\ 6 \end{bmatrix}\begin{bmatrix} 6 & -2 \end{bmatrix} = \begin{bmatrix} 24 & -8 \\ 36 & -12 \end{bmatrix}$

36. $\begin{bmatrix} 2 & 1 \\ 6 & 0 \end{bmatrix}\left(\begin{bmatrix} 4 & 2 \\ -3 & 1 \end{bmatrix} + \begin{bmatrix} -2 & 4 \\ 0 & 4 \end{bmatrix}\right) = \begin{bmatrix} 2 & 1 \\ 6 & 0 \end{bmatrix}\begin{bmatrix} 2 & 6 \\ -3 & 5 \end{bmatrix}$

$$= \begin{bmatrix} 2(2) + 1(-3) & 2(6) + 1(5) \\ 6(2) + 0 & 6(6) + 0 \end{bmatrix}$$

$$= \begin{bmatrix} 1 & 17 \\ 12 & 36 \end{bmatrix}$$

38. $-5\begin{bmatrix} 2 & 0 \\ 7 & -2 \\ 8 & 2 \end{bmatrix} + 4\begin{bmatrix} 4 & -2 \\ 6 & 11 \\ -1 & 3 \end{bmatrix} = \begin{bmatrix} 6 & -8 \\ -11 & 54 \\ -44 & 2 \end{bmatrix}$

40. $\begin{bmatrix} -2 & 3 & 10 \\ 4 & -2 & 2 \end{bmatrix}\begin{bmatrix} 1 & 1 \\ -5 & 2 \\ 3 & 2 \end{bmatrix} = \begin{bmatrix} 13 & 24 \\ 20 & 4 \end{bmatrix}$

42. $X = \frac{1}{6}(4A + 3B) = \frac{1}{6}\left(4\begin{bmatrix} -4 & 0 \\ 1 & -5 \\ -3 & 2 \end{bmatrix} + 3\begin{bmatrix} 1 & 2 \\ -2 & 1 \\ 4 & 4 \end{bmatrix}\right)$

$= \frac{1}{6}\begin{bmatrix} -13 & 6 \\ -2 & -17 \\ 0 & 20 \end{bmatrix}$

44. $X = \frac{1}{3}(2A - 5B) = \frac{1}{3}\left(2\begin{bmatrix} -4 & 0 \\ 1 & -5 \\ -3 & 2 \end{bmatrix} - 5\begin{bmatrix} 1 & 2 \\ -2 & 1 \\ 4 & 4 \end{bmatrix}\right)$

$= \frac{1}{3}\begin{bmatrix} -13 & -10 \\ 12 & -15 \\ -26 & -16 \end{bmatrix}$

46. $2x + 3y + z = 10$

$2x - 3y - 3z = 22$

$4x - 2y + 3z = -2$

$\begin{bmatrix} 2 & 3 & 1 \\ 2 & -3 & -3 \\ 4 & -2 & 3 \end{bmatrix}\begin{bmatrix} x \\ y \\ z \end{bmatrix} = \begin{bmatrix} 10 \\ 22 \\ -2 \end{bmatrix}$

48. $\begin{bmatrix} 3 & -10 \\ 4 & 2 \end{bmatrix}^{-1} = \frac{1}{46}\begin{bmatrix} 2 & 10 \\ -4 & 3 \end{bmatrix}$

50. $A = \begin{bmatrix} 1 & 4 & 6 \\ 2 & -3 & 1 \\ -1 & 18 & 16 \end{bmatrix}$

A^{-1} does not exist because $|A| = 0$.

52. $\begin{vmatrix} 8 & 5 \\ 2 & -4 \end{vmatrix} = 8(-4) - 2(5) = -42$

54. $\begin{vmatrix} x & x^2 \\ 1 & 2x \end{vmatrix} = x(2x) - 1(x^2) = 2x^2 - x^2 = x^2$

56. $\begin{vmatrix} 0 & 3 & 1 \\ 5 & -2 & 1 \\ 1 & 6 & 1 \end{vmatrix} = -3(5 - 1) + 1(30 + 2)$

$= -12 + 32 = 20$

58. $\begin{vmatrix} -5 & 6 & 0 & 0 \\ 0 & 1 & -1 & 2 \\ -3 & 4 & -5 & 1 \\ 1 & 6 & 0 & 3 \end{vmatrix} = -5\begin{vmatrix} 1 & -1 & 2 \\ 4 & -5 & 1 \\ 6 & 0 & 3 \end{vmatrix} - 6\begin{vmatrix} 0 & -1 & 2 \\ -3 & -5 & 1 \\ 1 & 0 & 3 \end{vmatrix}$ (Expansion along Row 1.)

$= -5[6(-1 + 10) + 3(-5 + 4)] - 6[(-1 + 10) + 3(0 - 3)]$

$= -5[54 \quad -3] - 6[9 \quad -9]$

$= -255$

60. $x + 3y = 23$

$-x + 2y = -18$

$\begin{bmatrix} 1 & 3 \\ -6 & 2 \end{bmatrix}^{-1} \Rightarrow \begin{bmatrix} 0.1 & -0.15 \\ 0.3 & 0.05 \end{bmatrix}\begin{bmatrix} x \\ y \end{bmatrix} = \begin{bmatrix} 0.1 & -0.15 \\ 0.3 & 0.05 \end{bmatrix}\begin{bmatrix} 23 \\ -18 \end{bmatrix} = \begin{bmatrix} 5 \\ 6 \end{bmatrix}$

$x = 5, y = 6$

Answer: $(5, 6)$

62. $$\begin{aligned} x - 3y - 2z &= 8 \\ -2x + 7y + 3z &= -19 \\ x - y - 3z &= 3 \end{aligned}$$

$$\begin{bmatrix} 1 & -3 & -2 \\ -2 & 7 & 3 \\ 1 & -1 & -3 \end{bmatrix}^{-1} \Rightarrow \begin{bmatrix} -18 & -7 & 5 \\ -1 & -1 & 1 \\ -5 & -2 & 1 \end{bmatrix}\begin{bmatrix} x \\ y \\ z \end{bmatrix} = \begin{bmatrix} -18 & -7 & 5 \\ -1 & -1 & 1 \\ -5 & -2 & 1 \end{bmatrix}\begin{bmatrix} 8 \\ -19 \\ 3 \end{bmatrix} = \begin{bmatrix} 4 \\ -2 \\ 1 \end{bmatrix}$$

$x = 4, y = -2, z = 1$

Answer: $(4, -2, 1)$

64. $$\begin{aligned} 2x + 4y &= -12 \\ 3x + 4y - 2z &= -14 \\ -x + y + 2z &= -6 \end{aligned}$$

$$\begin{bmatrix} 2 & 4 & 0 \\ 3 & 4 & -2 \\ -1 & 1 & 2 \end{bmatrix}^{-1} \Rightarrow \begin{bmatrix} 2.5 & -2 & -2 \\ -3 & 1 & 1 \\ 1.75 & -1.5 & -1 \end{bmatrix}\begin{bmatrix} x \\ y \\ z \end{bmatrix} = \begin{bmatrix} 2.5 & -2 & -2 \\ -3 & 1 & 1 \\ 1.75 & -1.5 & -1 \end{bmatrix}\begin{bmatrix} -12 \\ -14 \\ -6 \end{bmatrix} = \begin{bmatrix} 10 \\ -8 \\ 6 \end{bmatrix}$$

$x = 10, y = -8, z = 6$

Answer: $(10, -8, 6)$

66. $$\begin{aligned} 2x + 3y - 4z &= 1 \\ x - y + 2z &= -4 \\ 3x + 7y - 10z &= 0 \end{aligned}$$

$$\begin{bmatrix} 2 & 3 & -4 \\ 1 & -1 & 2 \\ 3 & 7 & -10 \end{bmatrix}^{-1} \text{ does not exist.}$$

The system is inconsistent. No solution.

68. $$x = \frac{\begin{vmatrix} -10 & -1 \\ -1 & 2 \end{vmatrix}}{\begin{vmatrix} 2 & -1 \\ 3 & 2 \end{vmatrix}} = \frac{-21}{7} = -3$$

$$y = \frac{\begin{vmatrix} 2 & -10 \\ 3 & -1 \end{vmatrix}}{\begin{vmatrix} 2 & -1 \\ 3 & 2 \end{vmatrix}} = \frac{28}{7} = 4$$

$(x, y) = (-3, 4)$

70. Cramer's Rule does not apply.

$$\begin{vmatrix} 13 & -6 \\ 26 & -12 \end{vmatrix} = 0$$

72. $$x = \frac{\begin{vmatrix} 1.6 & 0.8 \\ 2.2 & 0.3 \end{vmatrix}}{\begin{vmatrix} -0.4 & 0.8 \\ 0.2 & 0.3 \end{vmatrix}} = \frac{-1.28}{-0.28} = \frac{32}{7}$$

$$y = \frac{\begin{vmatrix} -0.4 & 1.6 \\ 0.2 & 2.2 \end{vmatrix}}{\begin{vmatrix} -0.4 & 0.8 \\ 0.2 & 0.3 \end{vmatrix}} = \frac{-1.2}{-0.28} = \frac{30}{7}$$

$$(x, y) = \left(\frac{32}{7}, \frac{30}{7}\right)$$

74. $x = \dfrac{\begin{vmatrix} 10 & -21 & -7 \\ 4 & 2 & -2 \\ 5 & -21 & 7 \end{vmatrix}}{\begin{vmatrix} 14 & -21 & -7 \\ -4 & 2 & -2 \\ 56 & -21 & 7 \end{vmatrix}} = \dfrac{1176}{1568} = \dfrac{3}{4}$

$y = \dfrac{\begin{vmatrix} 14 & 10 & -7 \\ -4 & 4 & -2 \\ 56 & 5 & 7 \end{vmatrix}}{\begin{vmatrix} 14 & -21 & -7 \\ -4 & 2 & -2 \\ 56 & -21 & 7 \end{vmatrix}} = \dfrac{1400}{1568} = \dfrac{25}{28}$

$z = \dfrac{\begin{vmatrix} 14 & -21 & 10 \\ -4 & 2 & 4 \\ 56 & -21 & 5 \end{vmatrix}}{\begin{vmatrix} 14 & -21 & -7 \\ -4 & 2 & -2 \\ 56 & -21 & 7 \end{vmatrix}} = \dfrac{-4088}{1568} = -\dfrac{73}{28}$

$(x, y, z) = \left(\dfrac{3}{4}, \dfrac{25}{28}, -\dfrac{73}{28}\right)$

76. x = number of liters of 75% acid

y = number of liters of 50% acid

$x + y = 100$

$0.75x + 0.50y = 60$

$\begin{bmatrix} 1 & 1 \\ 0.75 & 0.50 \end{bmatrix} \begin{bmatrix} x \\ y \end{bmatrix} = \begin{bmatrix} 100 \\ 60 \end{bmatrix}$

$D = \begin{vmatrix} 1 & 1 \\ 0.75 & 0.50 \end{vmatrix} = -0.25$

$x = \dfrac{\begin{vmatrix} 100 & 1 \\ 60 & 0.50 \end{vmatrix}}{-0.25} = \dfrac{-10}{-0.25} = 40$

$y = \dfrac{\begin{vmatrix} 1 & 100 \\ 0.75 & 60 \end{vmatrix}}{-0.25} = \dfrac{-15}{-0.25} = 60$

Answer: 40 liters of 75% acid; 60 liters of 50% acid

78. x = number of units produced

y = number of units produced

$x - y = 0$

$-3.75x + 5.25y = 25{,}000$

$\begin{bmatrix} 1 & -1 \\ -3.75 & 5.25 \end{bmatrix} \begin{bmatrix} x \\ y \end{bmatrix} = \begin{bmatrix} 0 \\ 25{,}000 \end{bmatrix}$

$D = \begin{vmatrix} 1 & -1 \\ -3.75 & 5.25 \end{vmatrix} = 1.5$

$y = \dfrac{\begin{vmatrix} 1 & 0 \\ -3.75 & 25{,}000 \end{vmatrix}}{1.5} \approx 16{,}667$ units must be sold.

80. $\begin{vmatrix} 2 - \lambda & 5 \\ 3 & -8 - \lambda \end{vmatrix} = 0$

$(2 - \lambda)(-8 - \lambda) - 15 = 0$

$-16 - 6\lambda + \lambda^2 - 15 = 0$

$\lambda^2 + 6\lambda - 31 = 0$

$\lambda = \dfrac{-6 \pm \sqrt{36 - 4)(-31)}}{2}$

$\lambda = -3 \pm 2\sqrt{10}$

82. $(-4, 0), (4, 0), (0, 6)$

$\text{Area} = \dfrac{1}{2}\begin{vmatrix} -4 & 0 & 1 \\ 4 & 0 & 1 \\ 0 & 6 & 1 \end{vmatrix} = \dfrac{1}{2}(48) = 24$ square units

84. $\left(\dfrac{3}{2}, 1\right), \left(4, -\dfrac{1}{2}\right), (4, 2)$

$\text{Area} = \dfrac{1}{2}\begin{vmatrix} \frac{3}{2} & 1 & 1 \\ 4 & -\frac{1}{2} & 1 \\ 4 & 2 & 1 \end{vmatrix} = \dfrac{1}{2}\left(\dfrac{25}{4}\right) = \dfrac{25}{8}$ square units

86. $(2, 5), (6, -1)$

$$\begin{vmatrix} x & y & 1 \\ 2 & 5 & 1 \\ 6 & -1 & 1 \end{vmatrix} = 0$$

$$6x + 4y - 32 = 0$$
$$3x + 2y - 16 = 0$$

88. $(-0.8, 0.2), (0.7, 3.2)$

$$\begin{vmatrix} x & y & 1 \\ -0.8 & 0.2 & 1 \\ 0.7 & 3.2 & 1 \end{vmatrix} = 0$$

$$-3x + 1.5y - 2.7 = 0$$
$$10x - 5y + 9 = 0$$

90. $\begin{bmatrix} 1 & 1 & 1 \\ 4 & -10 & 0 \\ 0 & 10 & -2 \end{bmatrix}^{-1} \begin{bmatrix} 0 \\ 12 \\ -6 \end{bmatrix} = \begin{bmatrix} 1.2353 \\ -0.7059 \\ -0.5294 \end{bmatrix} = \begin{bmatrix} \frac{21}{17} \\ -\frac{12}{17} \\ -\frac{9}{17} \end{bmatrix}$

CHAPTER 9
Sequences, Probability, and Statistics

CHAPTER 9
Sequences, Probability, and Statistics

Section 9.1 Sequences and Summation Notation

Solutions to Even-Numbered Exercises

2. $a_n = 4n - 3$

$a_1 = 4(1) - 3 = 1$

$a_2 = 4(2) - 3 = 5$

$a_3 = 4(3) - 3 = 9$

$a_4 = 4(4) - 3 = 13$

$a_5 = 4(5) - 3 = 17$

4. $a_n = \left(\frac{1}{2}\right)^n$

$a_1 = \left(\frac{1}{2}\right)^1 = \frac{1}{2}$

$a_2 = \left(\frac{1}{2}\right)^2 = \frac{1}{4}$

$a_3 = \left(\frac{1}{2}\right)^3 = \frac{1}{8}$

$a_4 = \left(\frac{1}{2}\right)^4 = \frac{1}{16}$

$a_5 = \left(\frac{1}{2}\right)^5 = \frac{1}{32}$

6. $a_n = \left(-\frac{1}{2}\right)^n$

$a_1 = \left(-\frac{1}{2}\right)^1 = -\frac{1}{2}$

$a_2 = \left(-\frac{1}{2}\right)^2 = \frac{1}{4}$

$a_3 = \left(-\frac{1}{2}\right)^3 = -\frac{1}{8}$

$a_4 = \left(-\frac{1}{2}\right)^4 = \frac{1}{16}$

$a_5 = \left(-\frac{1}{2}\right)^5 = -\frac{1}{32}$

8. $a_n = \dfrac{n}{n+1}$

$a_1 = \dfrac{1}{1+1} = \dfrac{1}{2}$

$a_2 = \dfrac{2}{2+1} = \dfrac{2}{3}$

$a_3 = \dfrac{3}{3+1} = \dfrac{3}{4}$

$a_4 = \dfrac{4}{4+1} = \dfrac{4}{5}$

$a_5 = \dfrac{5}{5+1} = \dfrac{5}{6}$

10. $a_n = \dfrac{3n^2 - n + 4}{2n^2 + 1}$

$a_1 = \dfrac{3(1)^2 - 1 + 4}{2(1)^2 + 1} = 2$

$a_2 = \dfrac{3(2)^2 - 2 + 4}{2(2)^2 + 1} = \dfrac{14}{9}$

$a_3 = \dfrac{3(3) - 3 + 4}{2(3)^2 + 1} = \dfrac{28}{19}$

$a_4 = \dfrac{3(4) - 4 + 4}{2(4)^2 + 1} = \dfrac{16}{11}$

$a_5 = \dfrac{3(5)^2 - 5 + 4}{2(5)^2 + 1} = \dfrac{74}{51}$

12. $a_n = 1 + (-1)^n$

$a_1 = 1 + (-1)^1 = 0$

$a_2 = 1 + (-1)^2 = 2$

$a_3 = 1 + (-1)^3 = 0$

$a_4 = 1 + (-1)^4 = 2$

$a_5 = 1 + (-1)^5 = 0$

14. $a_n = \dfrac{3^n}{4^n}$

$a_1 = \dfrac{3^1}{4^1} = \dfrac{3}{4}$

$a_2 = \dfrac{3^2}{4^2} = \dfrac{9}{16}$

$a_3 = \dfrac{3^3}{4^3} = \dfrac{27}{64}$

$a_4 = \dfrac{3^4}{4^4} = \dfrac{81}{256}$

$a_5 = \dfrac{3^5}{4^5} = \dfrac{243}{1024}$

16. $a_n = \dfrac{10}{n^{2/3}} = \dfrac{10}{\sqrt[3]{n^2}}$

$a_1 = \dfrac{10}{1} = 10$

$a_2 = \dfrac{10}{\sqrt[3]{2^2}} = \dfrac{10}{\sqrt[3]{4}}$

$a_3 = \dfrac{10}{\sqrt[3]{3^2}} = \dfrac{10}{\sqrt[3]{9}}$

$a_4 = \dfrac{10}{\sqrt[3]{4^2}} = \dfrac{10}{\sqrt[3]{16}}$

$a_5 = \dfrac{10}{\sqrt[3]{5^2}} = \dfrac{10}{\sqrt[3]{25}}$

18. $a_n = \dfrac{n!}{n}$

$a_1 = \dfrac{1!}{1} = 1$

$a_2 = \dfrac{2!}{2} = 1$

$a_3 = \dfrac{3!}{3} = 2$

$a_4 = \dfrac{4!}{4} = 6$

$a_5 = \dfrac{5!}{5} = 24$

20. $a_n = (-1)^n\left(\dfrac{n}{n+1}\right)$

$a_1 = (-1)^1\dfrac{1}{1+1} = -\dfrac{1}{2}$

$a_2 = (-1)^2\dfrac{2}{1+2} = \dfrac{2}{3}$

$a_3 = (-1)^3\dfrac{3}{3+1} = -\dfrac{3}{4}$

$a_4 = (-1)^4\dfrac{4}{4+1} = \dfrac{4}{5}$

$a_5 = (-1)^5\dfrac{5}{5+1} = -\dfrac{5}{6}$

22. $a_n = n(n-1)(n-2)$

$a_1 = 1(1-1)(1-2) = 0$

$a_2 = 2(2-1)(2-2) = 0$

$a_3 = 3(3-1)(3-2) = 6$

$a_4 = 4(4-1)(4-2) = 24$

$a_5 = 5(5-1)(5-2) = 60$

24. $a_n = \dfrac{2^n}{n!}$

$a_{10} = \dfrac{2^{10}}{10!} = \dfrac{1024}{3{,}628{,}800}$

$= \dfrac{4}{14{,}175}$

26. $a_1 = 15,\ a_{k+1} = a_k + 3$

$a_1 = 15$

$a_2 = a_1 + 3 = 15 + 3 = 18$

$a_3 = a_2 + 3 = 18 + 3 = 21$

$a_4 = a_3 + 3 = 21 + 3 = 24$

$a_5 = a_4 + 3 = 24 + 3 = 27$

28. $a_1 = 32,\ a_{k+1} = \dfrac{1}{2}a_k$

$a_1 = 32$

$a_2 = \dfrac{1}{2}a_1 = \dfrac{1}{2}(32) = 16$

$a_3 = \dfrac{1}{2}a_2 = \dfrac{1}{2}(16) = 8$

$a_4 = \dfrac{1}{2}a_3 = \dfrac{1}{2}(8) = 4$

$a_5 = \dfrac{1}{2}a_4 = \dfrac{1}{2}(4) = 2$

30. $a_n = 2 - \dfrac{4}{n}$

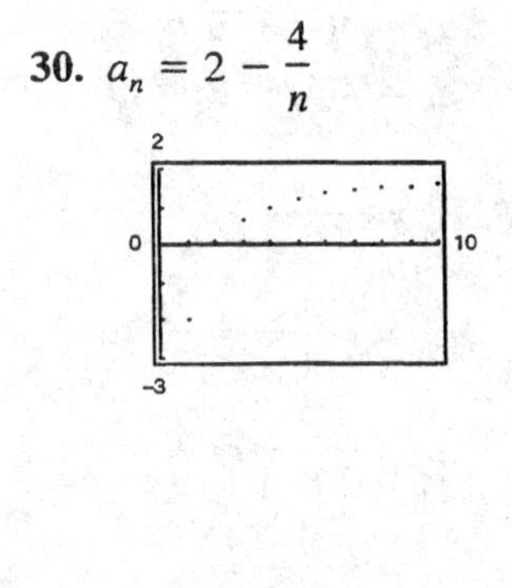

32. $a_n = 8(0.75)^{n-1}$

34. $a_n = \dfrac{3n^2}{n^2+1}$

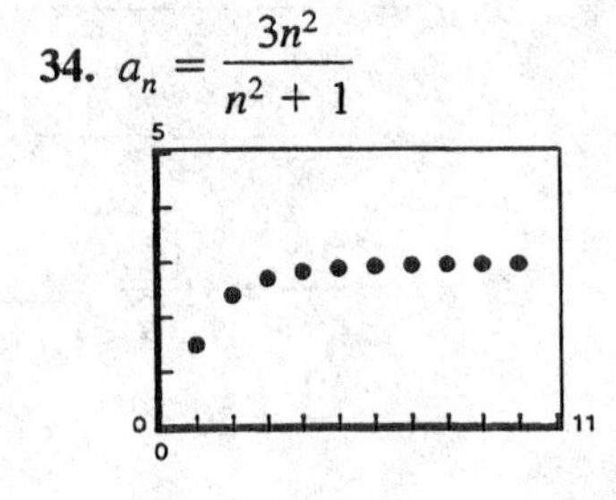

36. $a_n = \dfrac{8n}{n+1}$

$a_n \to 8$ as $n \to \infty$

$a_1 = 4,\ a_4 = \dfrac{8(4)}{9} = \dfrac{32}{9}$

Matches graph (b).

38. $a_n = \dfrac{4^n}{n!}$

$a_n \to 0$ as $n \to \infty$

$a_1 = 4,\ a_4 = \dfrac{4^4}{4!} = \dfrac{256}{24} = 10\dfrac{2}{3}$

Matches graph (a).

40. $\dfrac{4!}{7!} = \dfrac{4!}{7\cdot 6\cdot 5\cdot 4!} = \dfrac{1}{210}$

42. $\dfrac{25!}{23!} = \dfrac{25\cdot 24\cdot 23!}{23!} = 600$

44. $\dfrac{(n+2)!}{n!} = \dfrac{(n+2)(n+1)n!}{n!} = (n+2)(n+1)$

46. $\dfrac{(2n+2)!}{(2n)!} = \dfrac{(2n+2)(2n+1)(2n)!}{(2n)!}$

$= (2n+2)(2n+1)$

48. $3, 7, 11, 15, 19, \ldots$

$a_n = 4n - 1$

50. $1, \dfrac{1}{4}, \dfrac{1}{9}, \dfrac{1}{16}, \dfrac{1}{25}, \ldots$

$a_n = \dfrac{1}{n^2}$

52. $\dfrac{2}{1}, \dfrac{3}{3}, \dfrac{4}{5}, \dfrac{5}{7}, \dfrac{6}{9}, \ldots$

$a_n = \dfrac{n+1}{2n-1}$

54. $\dfrac{1}{3}, \dfrac{2}{9}, \dfrac{4}{27}, \dfrac{8}{81}, \ldots$

$a_n = \dfrac{2^{n-1}}{3^n}$

56. $1 + \dfrac{1}{2}, 1 + \dfrac{3}{4}, 1 + \dfrac{7}{8}, 1 + \dfrac{15}{16}, 1 + \dfrac{31}{32}, \ldots$

$a_n = 1 + \dfrac{2^n - 1}{2^n}$

58. $2, -4, 6, -8, 10, \ldots$

$a_n = (-1)^{n+1}(2n)$

60. $1, 2, \frac{2^2}{2}, \frac{2^3}{6}, \frac{2^4}{24}, \frac{2^5}{120}, \ldots$

$a_n = \frac{2^{n-1}}{(n-1)!}$

62. $a_1 = 25, \quad a_{k+1} = a_k - 5$

$a_1 = 25$

$a_2 = a_1 - 5 = 25 - 5 = 20$

$a_3 = a_2 - 5 = 20 - 5 = 15$

$a_4 = a_3 - 5 = 15 - 5 = 10$

$a_5 = a_4 - 5 = 10 - 5 = 5$

In general, $a_n = 30 - 5n$.

64. $a_1 = 14, \quad a_{k+1} = (-2)a_k$

$a_1 = 14$

$a_2 = (-2)a_1 = (-2)(14) = -28$

$a_3 = (-2)a_2 = (-2)(-28) = 56$

$a_4 = (-2)a_3 = (-2)(56) = -112$

$a_5 = (-2)(a_4) = (-2)(-112) = 224$

In general, $a_n = 14(-2)^{n-1}$.

66. $\sum_{i=1}^{6} (3i - 1) = (3 \cdot 1 - 1) + (3 \cdot 2 - 1) + (3 \cdot 3 - 1) + (3 \cdot 4 - 1) + (3 \cdot 5 - 1) + (3 \cdot 6 - 1) = 57$

68. $\sum_{k=1}^{5} 6 = 6 + 6 + 6 + 6 + 6 = 30$

70. $\sum_{k=0}^{5} 3i^2 = 3\sum_{i=0}^{5} i^2 = 3(0^2 + 1^2 + 2^2 + 3^2 + 4^2 + 5^2) = 165$

72. $\sum_{j=3}^{5} \frac{1}{j} = \frac{1}{3} + \frac{1}{4} + \frac{1}{5} = \frac{47}{60}$

74. $\sum_{k=2}^{5} (k + 1)(k - 3) = (2 + 1)(2 - 3) + (3 + 1)(3 - 3) + (4 + 1)(4 - 3) + (5 + 1)(5 - 3) = 14$

76. $\sum_{j=0}^{4} (-2)^j = (-2)^0 + (-2)^1 + (-2)^2 + (-2)^3 + (-2)^4 = 11$

78. $\sum_{j=1}^{10} \frac{3}{j + 1} \approx 6.06$

80. $\sum_{k=0}^{4} \frac{(-1)^k}{k!} = \frac{3}{8} = 0.375$

82. $\frac{5}{1+1} + \frac{5}{1+2} + \frac{5}{1+3} + \cdots + \frac{5}{1+15} = \sum_{i=1}^{15} \frac{5}{1+i}$

84. $\left[1 - \left(\frac{1}{6}\right)^2\right] + \left[1 - \left(\frac{2}{6}\right)^2\right] + \cdots + \left[1 - \left(\frac{6}{6}\right)^2\right] = \sum_{k=1}^{6} \left[1 - \left(\frac{k}{6}\right)^2\right]$

86. $1 - \frac{1}{2} + \frac{1}{4} - \frac{1}{8} + \cdots - \frac{1}{128} = \frac{1}{2^0} - \frac{1}{2^1} + \frac{1}{2^2} - \frac{1}{2^3} + \cdots - \frac{1}{2^7} = \sum_{n=0}^{7} \left(-\frac{1}{2}\right)^2$

88. $\frac{1}{1 \cdot 3} + \frac{1}{2 \cdot 4} + \frac{1}{3 \cdot 5} + \cdots + \frac{1}{10 \cdot 12} = \sum_{k=1}^{10} \frac{1}{k(k+2)}$

90. $\frac{1}{2} + \frac{2}{4} + \frac{6}{8} + \frac{24}{16} + \frac{120}{32} + \frac{720}{64} = \sum_{k=1}^{6} \frac{k!}{2^k}$

92. (a) $A_1 = 100(101)[(1.01)^1 - 1] = \$101.00$
$A_2 = 100(101)[(1.01)^2 - 1] = \$203.01$
$A_3 = 100(101)[(1.01)^3 - 1] \approx \$306.04$
$A_4 = 100(101)[(1.01)^4 - 1] \approx \$410.10$
$A_5 = 100(101)[(1.01)^5 - 1] \approx \$515.20$
$A_6 = 100(101)[(1.01)^6 - 1] \approx \$621.35$
(b) $A_{60} = 100(101)[(1.01)^{60} - 1] \approx \$8248.64$
(c) $A_{240} = 100(101)[(1.01)^{240} - 1] \approx \$99{,}914.79$

94. $a_0 = 0.1\sqrt{82 + 9 \cdot 0^2} \approx 0.91$
$a_1 = 0.1\sqrt{82 + 9 \cdot 1^2} \approx 0.95$
$a_2 = 0.1\sqrt{82 + 9 \cdot 2^2} \approx 1.09$
$a_3 = 0.1\sqrt{82 + 9 \cdot 3^2} \approx 1.28$
$a_4 = 0.1\sqrt{82 + 9 \cdot 4^2} \approx 1.50$
$a_5 = 0.1\sqrt{82 + 9 \cdot 5^2} \approx 1.75$
$a_6 = 0.1\sqrt{82 + 9 \cdot 6^2} \approx 2.01$
$a_7 = 0.1\sqrt{82 + 9 \cdot 7^2} \approx 2.29$
$a_8 = 0.1\sqrt{82 + 9 \cdot 8^2} \approx 2.57$
$a_9 = 0.1\sqrt{82 + 9 \cdot 9^2} \approx 2.85$
$a_{10} = 0.1\sqrt{82 + 9 \cdot 10^2} \approx 3.13$

96. $\sum_{n=5}^{14} (1.39 + 0.18n - 1.02 \ln n) \approx \$8.55$

98. $b_n = \frac{a_{n+1}}{a_n}$; $b_1 = 1, b_2 = 2, b_3 = \frac{3}{2}, b_4 = \frac{5}{3}, \ldots$

$b_2 = 1 + \frac{1}{b_1} = 1 + \frac{1}{1} = 2$

$b_3 = 1 + \frac{1}{b_2} = 1 + \frac{1}{2} = \frac{3}{2}$

$b_4 = 1 + \frac{1}{b_3} = 1 + \frac{2}{3} = \frac{5}{3}$

$b_5 = 1 + \frac{1}{b_4} = 1 + \frac{3}{5} = \frac{8}{5}$

$b_n = 1 + \frac{1}{b_{n-1}}$

Section 9.2 Arithmetic Sequences

Solutions to Even-Numbered Exercises

2. $4, 7, 10, 13, 16, \ldots$
Arthimetic sequence, $d = 3$

4. $3, \frac{5}{2}, 2, \frac{3}{2}, 1, \ldots$
Arthimetic sequence, $d = -\frac{1}{2}$

6. $\frac{1}{3}, \frac{2}{3}, \frac{4}{3}, \frac{8}{3}, \frac{16}{3}, \ldots$
Not an arithmetic sequence

8. $\ln 1, \ln 2, \ln 3, \ln 4, \ln 5, \ldots$
Not an arithmetic sequence

10. $1^2, 2^2, 3^2, 4^2, 5^2, \ldots$
Not an arithmetic sequence

12. $a_n = (2^n)n$
$2, 8, 24, 64, 160$
Not an arithmetic sequence

14. $a_n = 1 + (n - 1)4$
$1, 5, 9, 13, 17$
Arithmetic sequence, $d = 4$

16. $a_n = 2^{n-1}$
$1, 2, 4, 8, 16$
Not an arithmetic sequence

18. $a_n = (-1)^n$
$-1, 1, -1, 1, -1$
Not an arithmetic sequence

20. $a_1 = 2, a_{k+1} = a_k + \frac{2}{3}$
$a_2 = 2 + \frac{2}{3} = \frac{8}{3}$
$a_3 = \frac{8}{3} + \frac{2}{3} = \frac{10}{3}$
$a_4 = \frac{10}{3} + \frac{2}{3} = \frac{12}{3} = 4$
$a_5 = 4 + \frac{2}{3} = \frac{14}{3}$
$a_n = \frac{4}{3} + \frac{2}{3}n$

22. $a_1 = 72, a_{k+1} = a_k - 6$
$a_2 = 72 - 6 = 66$
$a_3 = 66 - 6 = 60$
$a_4 = 60 - 6 = 54$
$a_5 = 54 - 6 = 48$
$a_n = 78 - 6n$

24. $a_1 = 0.375, a_{k+1} = a_k + 0.25$
$a_2 = 0.375 + 0.25 = 0.625$
$a_3 = 0.625 + 0.25 = 0.875$
$a_4 = 0.875 + 0.25 = 1.125$
$a_5 = 1.125 + 0.25 = 1.375$
$a_n = 0.125 + 0.25n$

26. $a_1 = 5, d = -\frac{3}{4}$
$a_1 = 5$
$a_2 = 5 - \frac{3}{4} = \frac{17}{4}$
$a_3 = \frac{17}{4} - \frac{3}{4} = \frac{14}{4} = \frac{7}{2}$
$a_4 = \frac{7}{2} - \frac{3}{4} = \frac{11}{4}$
$a_5 = \frac{11}{4} - \frac{3}{4} = \frac{8}{4} = 2$

28. $a_1 = 16.5, d = 0.25$
$a_1 = 16.5$
$a_2 = 16.5 + 0.25 = 16.75$
$a_3 = 16.75 + 0.25 = 17$
$a_4 = 17 + 0.25 = 17.25$
$a_5 = 17.25 + 0.25 = 17.5$

30. $a_4 = 16, a_{10} = 46$
$16 = a_4 = a_1 + (n - 1)d = a_1 + 3d$
$46 = a_{10} = a_1 + (n - 1)d = a_1 + 9d$
Answer: $a_1 = 1, d = 5$
$a_1 = 1$
$a_2 = 1 + 5 = 6$
$a_3 = 6 + 5 = 11$
$a_4 = 11 + 5 = 16$
$a_5 = 16 + 5 = 21$

32. $a_3 = 19, a_{15} = -1.7$
$19 = a_3 = a_1 + (n - 1)\,d = a_1 + 2d$
$-1.7 = a_{15} = a_1(n - 1)\,d = a_1 + 14d$
Answer: $a_1 = 22.45, d = -1.725$
$a_1 = 22.45$
$a_2 = 22.45 - 1.725 = 20.725$
$a_3 = 20.725 - 1.725 = 19$
$a_4 = 19 - 1.725 = 17.275$
$a_5 = 17.275 - 1.725 = 15.55$

34. $a_1 = 15, d = 4$
$a_n = a_1 + (n - 1)\,d = 15 + (n - 1)\,4$

36. $a_1 = 0, d = -\frac{2}{3}$
$a_n = a_1 + (n - 1)\,d = (n - 1)\left(-\frac{2}{3}\right)$

38. $a_1 = -y, d = 5y$
$a_n = a_1 + (n - 1)\,d = -y + (n - 1)\,(5y)$

40. $10, 5, 0, -5, -10, \ldots$
$d = -5$
$a_n = a_1 + (n - 1)\,d = 10 + (n - 1)\,(-5)$

42. $a_1 = -4, a_5 = 16$
$a_n = a_1 + (n - 1)\,d$
$16 = -4 + 4d$
$d = 5$
$a_n = a_1 + (n - 1)\,d = -4 + (n - 1)\,5$

44. $a_5 = 190, a_{10} = 115$
$a_{10} = a_5 + 5d \Rightarrow 115 = 190 + 5d \Rightarrow d = -15$
$a_1 = a_5 - 4d \Rightarrow a_1 = 190 - 4(-15) = 250$
$a_n = a_1 + (n - 1)\,d = 250 + (n - 1)\,(-15)$

46. $a_n = 3n - 5$
$d = 3$ so the sequence is increasing.
and $a_1 = -2$.
Matches (d).

48. $a_n = 25 - 3n$
$d = -3$ so the sequence is decreasing.
and $a_1 = 22$.
Matches (a).

50. $a_n = -5 + 2n$

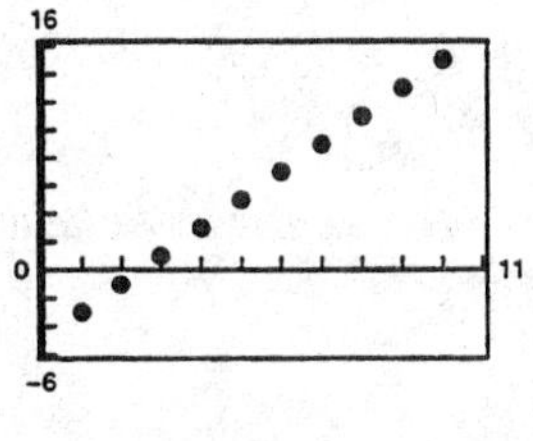

16
0
11
-6

52. $a_n = -0.3n + 8$

10
0
11
0

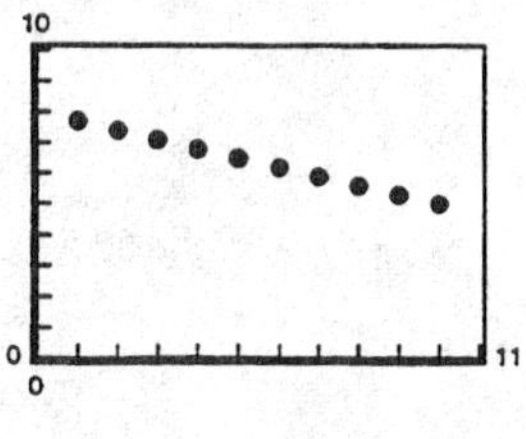

54. You can use the first two terms of an arithmetic sequence to find the nth term of the sequence by subtracting the first term from the second term to find the common difference and then find the nth term by adding $(n - 1)$ times the common difference to the first term.

56. $2, 8, 14, 20, \ldots, n = 25$

$a_n = 6n - 4$

$a_1 = 2$ and $a_{25} = 146$

$S_{25} = \frac{25}{2}(2 + 146) = 1850$

58. $0.5, 0.9, 1.3, 1.7, \ldots, n = 10$

$a_n = 0.4n + 0.1$

$a_1 = 0.5$ and $a_{10} = 4.1$

$S_{10} = \frac{10}{2}(0.5 + 4.1) = 23$

60. $1.50, 1.45, 1.40, 1.35, \ldots, n = 20$

$a_n = -0.05n + 1.55$

$a_1 = 1.50$ and $a_{20} = 0.55$

$S_{20} = \frac{20}{2}(1.50 + 0.55) = 20.5$

62. $a_1 = 15, a_{100} = 307, n = 100$

$S_{100} = \frac{100}{2}(15 + 307) = 16{,}100$

64. $a_n = 2n$

$a_1 = 2, a_{100} = 200, n = 100$

$$\sum_{n=1}^{100} 2n = \frac{100}{2}(2 + 200) = 10{,}100$$

66. $a_n = 7n$

$a_{51} = 357, a_{100} = 700$

$$\sum_{n=51}^{100} 7n = \frac{50}{2}(357 + 700) = 26{,}425$$

68. $$\sum_{n=51}^{100} n - \sum_{n=1}^{50} n = \frac{50}{2}(51 + 100) - \frac{50}{2}(1 + 50) = 3775 - 1275 = 2500$$

70. $a_n = 1000 - n$

$a_1 = 999, a_{250} = 750, n = 250$

$$\sum_{n=1}^{250} (1000 - n) = \frac{250}{2}(999 + 750) = 218{,}625$$

72. $a_1 = \frac{5}{2}, a_{100} = 52, n = 100$

$$\sum_{n=1}^{100} \frac{n+4}{2} = \frac{100}{2}\left(\frac{5}{2} + 52\right) = 2725$$

74. $a_0 = \frac{1}{2}, a_{100} = -18\frac{1}{4}, n = 101$

$$\sum_{n=0}^{100} \frac{8 - 3n}{16} = \frac{101}{2}\left(\frac{1}{2} - 18\frac{1}{4}\right) = -896.375$$

76. $a_1 = 4.525, a_{200} = 9.5, n = 200$

$$\sum_{j=1}^{200} (4.5 + 0.025j) = \frac{200}{2}(4.525 + 9.5) = 1402.5$$

78. $a_1 = -10, a_{61} = 50, n = 61$

$$\sum_{i=0}^{61} (i - 10) = \frac{61}{2}(-10 + 50) = 1220$$

80. (a) $a_1 = 36{,}800, d = 1750$

$a_6 = a_1 + 5d = 36{,}800 + 5(1750) = \$45{,}550$

(b) $S_6 = \frac{6}{2}[36{,}800 + 45{,}550] = \$247{,}050$

82. $a_1 = 15, d = 3, n = 36$

$a_{36} = 15 + 35(3) = 120$

$S_{36} = \frac{36}{2}(15 + 120) = 2430$ seats

84. $a_1 = 15, a_7 = 21, d = 1, n = 7$

$S_7 = \frac{7}{2}(15 + 21) = \frac{7}{2}(36) = 126$ logs

86. $a_1 = 93, d = 89 - 93 = -4, n = 8$

$a_8 = -4(8) + 97 = 65$

$S_8 = \frac{8}{2}(93 + 65) = 4(158) = 632$ bales

88. $a_1 = 4.9, a_2 = 14.7, a_3 = 24.5,$

$a_4 = 34.3 \Rightarrow d = 9.8$

$a_1 = 4.9 = 9.8(1) + c \Rightarrow c = -4.9$

$a_n = 9.8n - 4.9$

$a_{10} = 9.8(10) - 4.9 = 93.1$

$S_{10} = \frac{10}{2}(4.9 + 93.1) = 490$ meters

90. If an arithmetic sequence is defined by $a_1, a_2, a_3, a_4, \ldots$ and the common difference is $a_2 - a_1 = d$.

(a) A constant C is added to each term: $a_1 + C, a_2 + C, a_3 + C, a_4 + C, \ldots$

The resulting sequence is arithmetic, and the common difference is the original common difference:

$$a_2 + C - (a_1 + C) = a_2 + C - a_1 - C$$
$$= a_2 - a_1 = d.$$

(b) Each term is multiplied by a nonzero constant C: $Ca_1, Ca_2, Ca_3, Ca_4, \ldots$

The resulting sequence is arithmetic, and the common difference is C times the original common difference:

$$Ca_2 - Ca_1 = C(a_2 - a_1) = Cd.$$

(c) If each term is squared, the sequence is not arithmetic.

92. Let $S_n = \frac{n}{2}(a_1 + a_n)$ be the sum of the first n terms of the original sequence.

$$S_n' = \frac{n}{2}(a_1 + 5 + a_n + 5) = \frac{n}{2}(a_1 + a_n + 10) = \frac{n}{2}(a_1 + a_n) + \frac{n}{2}(10)$$
$$= \frac{n}{2}(a_1 + a_n) + 5n$$
$$= S_n + 5n$$
$$= S + 5$$

Section 9.3 Geometric Sequences

Solutions to Even-Numbered Exercises

2. $3, 12, 48, 192, \ldots$

Geometric sequence, $r = 4$

4. $1, -2, 4, -8, \ldots$

Geometric sequence, $r = -2$

6. $5, 1, 0.2, 0.04, \ldots$

Geometric sequence, $r = \frac{1}{5} = 0.2$

8. $9, -6, 4, -\frac{8}{3}, \ldots$

Geometric sequence, $r = -\frac{2}{3}$

10. $\frac{1}{5}, \frac{2}{7}, \frac{3}{9}, \frac{4}{11}, \ldots$

Not a geometric sequence

12. $a_1 = 6, r = 2$

$a_1 = 6$
$a_2 = 6(2)^1 = 12$
$a_3 = 6(2)^2 = 24$
$a_4 = 6(2)^3 = 48$
$a_5 = 6(2)^4 = 96$

14. $a_1 = 1, r = \frac{1}{3}$

$a_1 = 1$
$a_2 = 1\left(\frac{1}{3}\right)^1 = \frac{1}{3}$
$a_3 = 1\left(\frac{1}{3}\right)^2 = \frac{1}{9}$
$a_4 = 1\left(\frac{1}{3}\right)^3 = \frac{1}{27}$
$a_5 = 1\left(\frac{1}{3}\right)^4 = \frac{1}{81}$

16. $a_1 = 6, r = -\frac{1}{4}$

$a_1 = 6$
$a_2 = 6\left(-\frac{1}{4}\right)^1 = -\frac{3}{2}$
$a_3 = 6\left(-\frac{1}{4}\right)^2 = \frac{3}{8}$
$a_4 = 6\left(-\frac{1}{4}\right)^3 = -\frac{3}{32}$
$a_5 = 6\left(-\frac{1}{4}\right)^4 = \frac{3}{128}$

18. $a_1 = 2, r = \sqrt{3}$

$a_1 = 2$
$a_2 = 2(\sqrt{3})^1 = 2\sqrt{3}$
$a_3 = 2(\sqrt{3})^2 = 6$
$a_4 = 2(\sqrt{3})^3 = 6\sqrt{3}$
$a_5 = 2(\sqrt{3})^4 = 18$

20. $a_1 = 5, r = 2x$

$a_1 = 5$
$a_2 = 5(2x)^1 = 10x$
$a_3 = 5(2x)^2 = 20x^2$
$a_4 = 5(2x)^3 = 40x^3$
$a_5 = 5(2x)^4 = 80x^4$

22. $a_1 = 81, a_{k+1} = \frac{1}{3}a_k$

$a_1 = 81$
$a_2 = \frac{1}{3}(81) = 27$
$a_3 = \frac{1}{3}(27) = 9$
$a_4 = \frac{1}{3}(9) = 3$
$a_5 = \frac{1}{3}(3) = 1$
$a_n = 243\left(\frac{1}{3}\right)^n$

24. $a_1 = 5, a_{k+1} = -2a_k$

$a_1 = 5$
$a_2 = -2(5) = -10$
$a_3 = -2(-10) = 20$
$a_4 = -2(20) = -40$
$a_5 = -2(-40) = 80$
$a_n = 5(-2)^{n-1}$

26. $a_1 = 36,\ a_{k+1} = -\dfrac{2}{3}a_k$

$a_1 = 36$

$a_2 = -\dfrac{2}{3}(36) = -24$

$a_3 = -\dfrac{2}{3}(-24) = 16$

$a_4 = -\dfrac{2}{3}(16) = -\dfrac{32}{3}$

$a_5 = -\dfrac{2}{3}\left(-\dfrac{32}{3}\right) = \dfrac{64}{9}$

$a_n = 36\left(-\dfrac{2}{3}\right)^{n-1}$

28. $a_1 = 5,\ r = \dfrac{3}{2},\ n = 8$

$a_n = a_1 r^{n-1}$

$a_8 = 5\left(\dfrac{3}{2}\right)^7 = \dfrac{10{,}935}{128}$

30. $a_1 = 8,\ r = \sqrt{5},\ n = 9$

$a_n = a_1 r^{n-1}$

$a_9 = 8(\sqrt{5})^8 = 5000$

32. $a_1 = 1,\ r = -\dfrac{x}{3},\ n = 7$

$a_n = a_1 r^{n-1}$

$a_7 = 1\left(-\dfrac{x}{3}\right)^6 = \dfrac{x^6}{729}$

34. $a_1 = 1000,\ r = 1.005,$

$n = 60$

$a_n = a_1 r^{n-1}$

$a_6 = 1000(1.005)^{59}$

36. $a_2 = 3,\ a_5 = \frac{3}{64},\ n = 1$

$a_2 r^3 = a_5$

$3r^3 = \frac{3}{64}$

$r^3 = \frac{1}{64}$

$r = \frac{1}{4}$

$a_2 = a_1 r$

$3 = a_1\left(\frac{1}{4}\right)$

$a_1 = 12$

38. $a_3 = \frac{16}{3},\ a_5 = \frac{64}{27},\ n = 7$

$a_3 r^2 = a_5$

$\frac{16}{3} r^2 = \frac{64}{27}$

$r^2 = \frac{4}{9}$

$r = \pm\frac{2}{3}$

$a_7 = a_5 r^2 = \frac{64}{27}\left(\pm\frac{2}{3}\right)^2 = \frac{256}{243}$

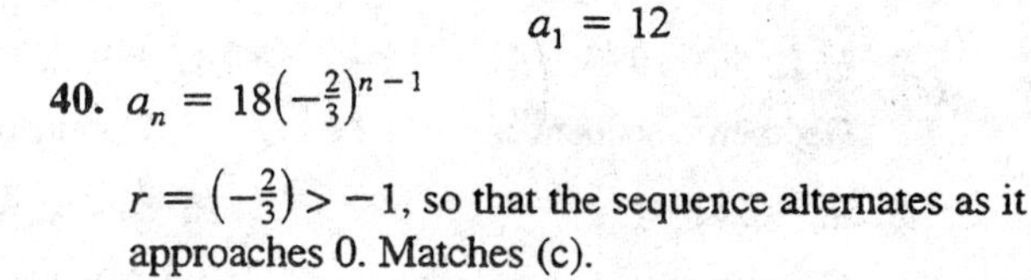

40. $a_n = 18\left(-\frac{2}{3}\right)^{n-1}$

$r = \left(-\frac{2}{3}\right) > -1$, so that the sequence alternates as it approaches 0. Matches (c).

42. $a_n = 18\left(-\frac{3}{2}\right)^{n-1}$

$r = \left(-\frac{3}{2}\right) < -1$, so the sequence alternates as it approaches ∞. Matches (d).

44. $a_n = 12(-0.4)^{n-1}$

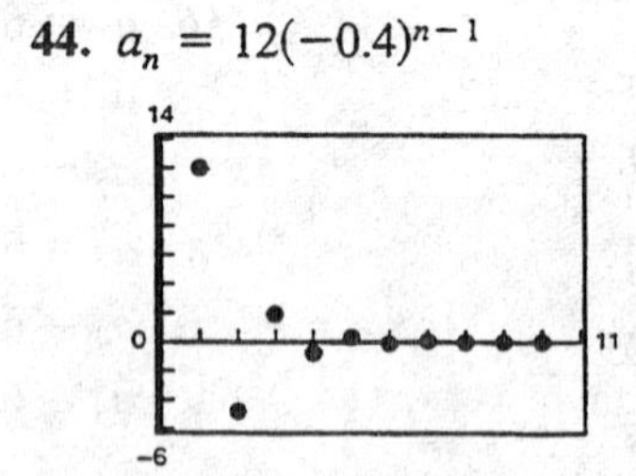

46. $a_n = 2(-1.4)^{n-1}$

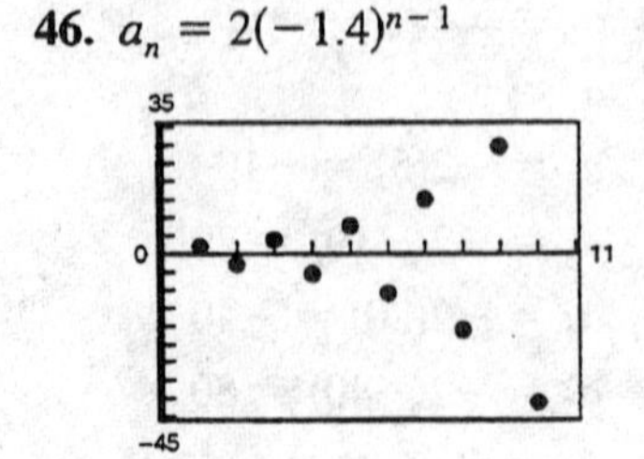

48. To use the first two terms of a geometric sequence to find the nth term, first divide the second term by the first term to obtain the common ratio. The nth term is the first term multiplied by the common ratio raised to the $(n - 1)$ power.

50. $A = P\left(1 + \frac{r}{n}\right)^{nt} = 2500\left(1 + \frac{0.12}{n}\right)^{n(20)}$

(a) $n = 1$, $A = 2500(1 + 0.12)^{20} \approx \$24,115.73$

(b) $n = 2$, $A = 2500\left(1 + \frac{0.12}{2}\right)^{2(20)} \approx \$25,714.29$

(c) $n = 4$, $A = 2500\left(1 + \frac{0.12}{4}\right)^{4(20)} \approx \$26,602.23$

(d) $n = 12$, $A = 2500\left(1 + \frac{0.12}{12}\right)^{12(20)} \approx \$27,231.38$

(e) $n = 365$, $A = 2500\left(1 + \frac{0.12}{365}\right)^{365(20)} \approx \$27,547.07$

52. P = population after n years

P_0 = initial population = 250,000

r = rate of increase = 1.3%

n = number of years = 30

$P = P_0(1 + r)^n = 250,000(1.013)^{30} \approx 368,318$

54. $8, 12, 18, 27, \frac{81}{2}, \ldots$

$S_1 = 8$

$S_2 = 8 + 12 = 20$

$S_3 = 8 + 12 + 18 = 38$

$S_4 = 8 + 12 + 18 + 27 = 65$

56. $\sum_{n=1}^{9} (-2)^{n-1} \Rightarrow a_1 = 1, r = -2$

$S_a = \frac{1(1 - (-2)^9)}{1 - (-2)} = 171$

58. $\sum_{i=1}^{6} 32\left(\frac{1}{4}\right)^{i-1} \Rightarrow a_1 = 32, r = \frac{1}{4}$

$S_6 = 32\frac{(1 - (1/4)^6)}{1 - (1/4)} = \frac{1365}{32}$

60. $\sum_{n=0}^{15} 2\left(\frac{4}{3}\right)^n = \sum_{n=1}^{16} 2\left(\frac{4}{3}\right)^{n-1} \Rightarrow a_1 = 2, r = \frac{4}{3}$

$S_{16} = 2\left(\frac{1 - (4/3)^{16}}{1 - (4/3)}\right) \approx 592.65$

62. $\sum_{i=1}^{10} 5\left(-\frac{1}{3}\right)^{i-1} \Rightarrow a_1 = 5, r = -\frac{1}{3}$

$S_{10} = 5\left(\frac{1 - (-1/3)^{10}}{1 - (-1/3)}\right) \approx 3.75$

64. $\sum_{n=0}^{6} 500(1.04)^n = \sum_{n=1}^{7} 500(1.04)^{n-1} \Rightarrow a_1 = 500, r = 1.04$

$S_7 = 500\left(\frac{1 - (1.04)^7}{1 - 1.04}\right) \approx 3949.15$

66. $2 - \frac{1}{2} + \frac{1}{8} - \cdots + \frac{1}{2048}$

$r = -\frac{1}{4}$ and $\frac{1}{2048} = 2\left(-\frac{1}{4}\right)^{n-1} \Rightarrow n = 7$

Thus, the sum can be written as $\sum_{n=1}^{7} 2\left(-\frac{1}{4}\right)^{n-1}$.

68. $A = \sum_{n=1}^{60} 50\left(1 + \frac{0.12}{12}\right)^n$

$= 50(1.01) \cdot \frac{(1 - (1.01)^{60})}{(1 - 1.01)} \approx \$4124.32$

70. Let $N = 12t$ be the total number of deposits.

$A = Pe^{r/12} + Pe^{2r/12} + \cdots + Pe^{Nr/12} = \sum_{n=1}^{N} Pe^{r/12 \cdot n}$

$= Pe^{r/12}\frac{(1 - (e^{r/12})^N)}{(1 - e^{r/12})}$

$= Pe^{r/12}\frac{(1 - (e^{r/12})^{12t})}{1 - e^{r/12}}$

$= \frac{Pe^{r/12}(e^{rt} - 1)}{(e^{r/12} - 1)}$

72. $P = \$75, r = 9\%, t = 25$ years

(a) Compounded monthly: $A = 75\left[\left(1 + \dfrac{0.09}{12}\right)^{12(25)} - 1\right]\left(1 + \dfrac{12}{0.09}\right) \approx \$84{,}714.78$

(b) Compounded continuously: $A = \dfrac{75e^{0.09/12}(e^{0.09(25)} - 1)}{e^{0.09/12} - 1} \approx \$85{,}196.05$

74. $P = \$20, r = 6\%, t = 50$ years

(a) Compounded monthly: $A = 20\left[\left(1 + \dfrac{0.06}{12}\right)^{12(50)} - 1\right]\left(1 + \dfrac{12}{0.06}\right) \approx \$76{,}122.54$

(b) Compounded continuously: $A = \dfrac{20e^{0.06/12}(e^{0.06(50)} - 1)}{e^{0.06/12} - 1} \approx \$76{,}533.16$

76. $W = \$2000, t = 20, r = 9\%$

$$P = W\left(\frac{12}{r}\right)\left[1 - \left(1 + \frac{r}{12}\right)^{-12t}\right]$$

$$P = 2000\left(\frac{12}{0.09}\right)\left[1 - \left(1 + \frac{0.09}{12}\right)^{-12(20)}\right] \approx \$222{,}289.91$$

78. $\displaystyle\sum_{n=5}^{14} 3.49e^{0.108n} = \sum_{n=6}^{15} 3.49(e^{0.108})^{n-1} = \sum_{n=1}^{15} 3.49(e^{0.108})^{n-1} - \sum_{n=1}^{5} 3.49(e^{0.108})^{n-1}$

$$= 3.49\left(\frac{1 - (e^{0.108})^{15}}{1 - e^{0.108}}\right) - 3.49\left(\frac{1 - (e^{0.108})^{5}}{1 - e^{0.108}}\right) \approx \$102.1 \text{ billion}$$

80. $a_n = 30{,}000(1.05)^{n-1}$

$$T = \sum_{n=1}^{40} 30{,}000(1.05)^{n-1} = 30{,}000\frac{(1 - 1.05^{40})}{(1 - 1.05)} \approx \$3{,}623{,}993.23$$

82. $a_1 = 2, r = \dfrac{2}{3}$

$$\sum_{n=0}^{\infty} 2\left(\frac{2}{3}\right)^n = \frac{a_1}{1 - r} = \frac{2}{1 - (2/3)} = 6$$

84. $a_1 = 2, r = -\dfrac{2}{3}$

$$\sum_{n=0}^{\infty} 2\left(-\frac{2}{3}\right)^n = \frac{a_1}{1 - r} = \frac{2}{1 - (-2/3)} = \frac{6}{5}$$

86. $a_1 = 1, r = \dfrac{1}{10}$

$$\sum_{n=0}^{\infty} \left(\frac{1}{10}\right)^n = \frac{a_1}{1 - r} = \frac{1}{1 - (1/10)} = \frac{10}{9}$$

88. $3 - 1 + \dfrac{1}{3} - \dfrac{1}{9} + \cdots = \displaystyle\sum_{n=0}^{\infty} 3\left(-\frac{1}{3}\right)^n = \frac{a_1}{1 - r} = \frac{3}{1 - (-1/3)} = \frac{9}{4}$

90. $0.\overline{297} = \displaystyle\sum_{n=0}^{\infty} 0.297(0.001)^n = \frac{0.297}{1 - 0.001} = \frac{0.297}{0.999} = \frac{297}{999} = \frac{11}{37}$

92. $1.3\overline{8} = 1.3 + \displaystyle\sum_{n=0}^{\infty} 0.08(0.1)^n = 1.3 + \frac{0.08}{1 - 0.1} = 1.3 + \frac{0.08}{0.9} = 1\frac{3}{10} + \frac{4}{45} = 1\frac{7}{18} = \frac{25}{18}$

94. $f(x) = 2\left[\dfrac{1 - (0.8)^x}{1 - (0.8)}\right], \displaystyle\sum_{n=0}^{\infty} 2\left(\frac{4}{5}\right)^n = \frac{2}{1 - (4/5)} = 10$

The horizontal asymptote of $f(x)$ is $y = 10$. This corresponds to the sum of the series.

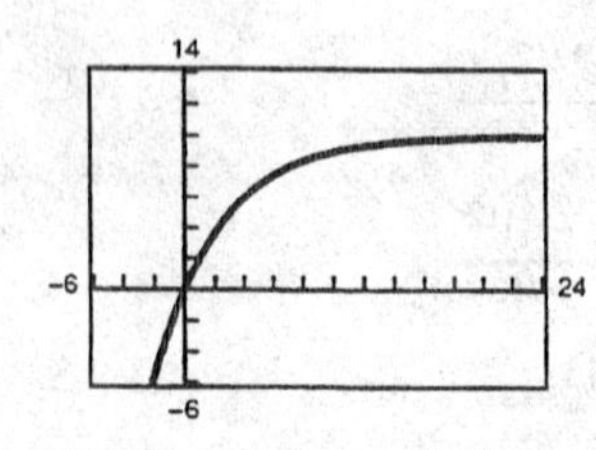

96. Let x = amount of cement. Then $90 - x$ is the amount of sand. Thus,

$$\frac{x}{90 - x} = \frac{1}{4}$$
$$4x = 90 - x$$
$$5x = 90$$
$$x = 18 \text{ pounds of cement}$$
$$\Rightarrow 90 - 18 = 72 \text{ pounds of sand.}$$

98. Let $2n$ and $2n + 2$ be the two consecutive even integers.

$$(2n)(2n + 2) = 624$$
$$4n^2 + 4n - 624 = 0$$
$$n^2 + n - 156 = 0$$
$$(n - 12)(n + 13) = 0$$

Since the integers are positive, $n = 12$, and the two integers are 24 and 26.

Section 9.4 Mathematical Induction

Solutions to Even-Numbered Exercises

2. $P_k = \dfrac{1}{(k + 1)(k + 3)}$

$$P_{k+1} = \frac{1}{((k + 1) + 1)((k + 1) + 3)} = \frac{1}{(k + 2)(k + 4)}$$

4. $P_k = \dfrac{k}{2}(3k - 1)$

$$P_{k+1} = \frac{k + 1}{2}(3(k + 1) - 1) = \frac{k + 1}{2}(3k + 2)$$

6. 1. When $n = 1$, $S_1 = 3 = 1(2 \cdot 1 + 1)$.

2. Assume that

$$S_k = 3 + 7 + 11 + 15 + \cdots + (4k - 1) = k(2k + 1).$$

Then,

$$\begin{aligned} S_{k+1} &= 3 + 7 + 11 + 15 + \cdots + (4k - 1) + [4(k + 1) - 1] \\ &= S_k + [4(k + 1) - 1] \\ &= k(2k + 1) + (4k + 3) \\ &= 2k^2 + 5k + 3 \\ &= (k + 1)(2k + 3) \\ &= (k + 1)[2(k + 1) + 1]. \end{aligned}$$

We conclude by mathematical induction that the formula is valid for all positive interger values of n.

8. 1. When $n = 1$,

$$S_1 = 1 = \frac{1}{2}(3 \cdot 1 - 1).$$

2. Assume that $S_k = 1 + 4 + 7 + 10 + \cdots + (3k - 2) = \dfrac{k}{2}(3k - 1)$.

Then,

$$\begin{aligned} S_{k+1} &= 1 + 4 + 7 + 10 + \cdots + (3k - 2) + (3(k + 1) - 2) \\ &= S_k + (3(k + 1) - 2) \\ &= \frac{k}{2}(3k - 1) + (3k + 1) \\ &= \frac{3k^2 - k + 6k + 2}{2} \\ &= \frac{3k^2 + 5k + 2}{2} \\ &= \frac{(k + 1)(3k + 2)}{2} \\ &= \frac{k + 1}{2}[3(k + 1) - 1]. \end{aligned}$$

Therefore, we conclude that this formula holds for all positive integer values of n.

10. 1. When $n = 1$, $S_1 = 2 = 3^1 - 1$.

2. Assume that

$$S_k = 2(1 + 3 + 3^2 + 3^3 + \cdots + 3^{k+1}) = 3^k - 1.$$

Then,

$$\begin{aligned} S_{k+1} &= 2(1 + 3 + 3^2 + 3^3 + \cdots + 3^{k-1}) + 2 \cdot 3^{k+1-1} \\ &= S_k + 2 \cdot 3^k \\ &= 3^k - 1 + 2 \cdot 3^k \\ &= 3 \cdot 3^k - 1 \\ &= 3^{k+1} - 1. \end{aligned}$$

Therefore, we conclude that this formula holds for all positive integer values of n.

12. 1. When $n = 1$,

$$S_1 = 1 = \frac{1(1 + 1)(2 \cdot 1 + 1)}{6}.$$

2. Assume that $S_k = 1^2 + 2^2 + 3^2 + 4^2 + \cdots + k^2 = \dfrac{k(k + 1)(2k + 1)}{6}$.

Then,

$$\begin{aligned} S_{k+1} &= 1^2 + 2^2 + 3^2 + 4^2 + \cdots + k^2 + (k + 1)^2 \\ &= S_k + (k + 1)^2 \\ &= \frac{k(k + 1)(2k + 1)}{6} + (k + 1)^2 \\ &= \frac{k(k + 1)(2k + 1) + 6(k + 1)^2}{6} \\ &= \frac{(k + 1)[2k^2 + k + 6k + 6]}{6} \\ &= \frac{(k + 1)(k + 2)(2k + 3)}{6}. \end{aligned}$$

Therefore, we conclude that this formula holds for all positive integer values of n.

14. 1. When $n = 1$, $S_1 = 2 = 1 + 1$.

2. Assume that

$$S_k = \left(1 + \frac{1}{1}\right)\left(1 + \frac{1}{2}\right)\left(1 + \frac{1}{3}\right) \cdots \left(1 + \frac{1}{k}\right) = k + 1.$$

Then,

$$\begin{aligned} S_{k+1} &= \left(1 + \frac{1}{1}\right)\left(1 + \frac{1}{2}\right)\left(1 + \frac{1}{3}\right) \cdots \left(1 + \frac{1}{k}\right)\left(1 + \frac{1}{k + 1}\right) \\ &= (S_k)\left(1 + \frac{1}{k + 1}\right) \\ &= (k + 1)\left(1 + \frac{1}{k + 1}\right) \\ &= k + 1 + 1 \\ &= k + 2. \end{aligned}$$

Therefore, we conclude that this formula holds for all positive integer values of n.

16. 1. When $n = 1$,

$$S_1 = 1^4 = \frac{1(1+1)(2 \cdot 1 + 1)(3 \cdot 1^2 + 3 \cdot 1 - 1)}{30}.$$

2. Assume that

$$S_k = \sum_{i=1}^{k} i^4 = \frac{k(k+1)(2k+1)(3k^2+3k-1)}{30}.$$

Then,

$$\begin{aligned}
S_{k+1} &= S_k + (k+1)^4 \\
&= \frac{k(k+1)(2k+1)(3k^2+3k-1)}{30} + (k+1)^4 \\
&= \frac{k(k+1)(2k+1)(3k^2+3k-1) + 30(k+1)^4}{30} \\
&= \frac{(k+1)[k(2k+1)(3k^2+3k-1) + 30(k+1)^3]}{30} \\
&= \frac{(k+1)(6k^4 + 39k^3 + 91k^2 + 89k + 30)}{30} \\
&= \frac{(k+1)(k+2)(2k+3)(3k^2+9k+5)}{30} \\
&= \frac{(k+1)(k+2)(2(k+1)+1)(3(k+1)^2 + 3(k+1) - 1)}{30}.
\end{aligned}$$

Therefore, we conclude that this formula holds for all positive integer values of n.

18. 1. When $n = 1$,

$$S_1 = \frac{1}{2} = \frac{1}{2 \cdot 1 + 1}.$$

2. Assume that

$$S_k = \sum_{i=0}^{k} \frac{1}{(2i-1)(2i+1)} = \frac{k}{2k+1}.$$

Then,

$$\begin{aligned}
S_{k+1} &= S_k + \frac{1}{(2(k+1)-1)(2(k+1)+1)} \\
&= \frac{k}{2k+1} + \frac{1}{(2k+1)(2k+3)} \\
&= \frac{k(2k+3)+1}{(2k+1)(2k+3)} \\
&= \frac{2k^2+3k+1}{(2k+1)(2k+3)} \\
&= \frac{(2k+1)(k+1)}{(2k+1)(2k+3)} \\
&= \frac{k+1}{2(k+1)+1}.
\end{aligned}$$

Therefore, we conclude that this formula holds for all positive integer values of n.

20. $\displaystyle\sum_{n=1}^{50} n = \frac{50(50+1)}{2} = 1275$

22. $\displaystyle\sum_{n=1}^{10} n^3 = \frac{10^2(10+1)^2}{4} = 3025$

24. $\sum_{n=1}^{8} n^5 = \dfrac{8^2(8+1)^2(2(8)^2+2(8)-1)}{12} = 61{,}776$

26. $\sum_{n=1}^{10}(n^3-n^2) = \sum_{n=1}^{10} n^3 - \sum_{n=1}^{10} n^2 = \dfrac{10^2(10+1)^2}{4} - \dfrac{10(10+1)(2(10)+1)}{6}$

$= 3025 - 385 = 2640$

28. $\sum_{j=1}^{4}\left(2+\frac{5}{2}j-\frac{3}{2}j^2\right) = \sum_{j=1}^{4} 2 + \frac{5}{2}\sum_{j=1}^{4} j - \frac{3}{2}\sum_{j=1}^{4} j^2$

$= 8 + \frac{5}{2}\left(\frac{4(4+1)}{2}\right) - \frac{3}{2}\left(\frac{4(4+1)(2(4)+1)}{6}\right)$

$= 8 + 25 - 45 = -12$

30. $25 + 22 + 19 + 16 + \cdots = \sum_{n=0}^{\infty}(25-3n)$

32. $3 - \frac{9}{2} + \frac{27}{4} - \frac{81}{8} + \cdots = \sum_{n=1}^{\infty} 3\left(-\frac{3}{2}\right)^{n-1}$

34. $\frac{1}{2\cdot 3} + \frac{1}{3\cdot 4} + \frac{1}{4\cdot 5} + \frac{1}{5\cdot 6} + \cdots + \frac{1}{(n+1)(n+2)} + \cdots = \sum_{n=1}^{\infty}\frac{1}{(n+1)(n+2)}$

36. 1. When $n = 7$, $\left(\frac{4}{3}\right)^7 \approx 7.4915 > 7$.

2. Assume that $\left(\frac{4}{3}\right)^k > k, k > 7$.

Then, $\left(\frac{4}{3}\right)^{k+1} = \left(\frac{4}{3}\right)^k\left(\frac{4}{3}\right) > k\left(\frac{4}{3}\right) = k + \frac{k}{3} > k + 1$ for $k > 7$.

Thus, $\left(\frac{4}{3}\right)^{k+1} > k + 1$.

Therefore, $\left(\frac{4}{3}\right)^n > n$.

38. 1. When $n = 1$, $\left(\frac{x}{y}\right)^2 < \left(\frac{x}{y}\right)$ and $(0 < x < y)$.

2. Assume that

$\left(\frac{x}{y}\right)^{k+1} < \left(\frac{x}{y}\right)^k$

$\left(\frac{x}{y}\right)^{k+1} < \left(\frac{x}{y}\right)^k \Rightarrow \left(\frac{x}{y}\right)\left(\frac{x}{y}\right)^{k+1} < \left(\frac{x}{y}\right)\left(\frac{x}{y}\right)^k \Rightarrow \left(\frac{x}{y}\right)^{k+2} < \left(\frac{x}{y}\right)^{k+1}$.

Therefore, $\left(\frac{x}{y}\right)^{n+1} < \left(\frac{x}{y}\right)^n$ for all integers $n \geq 1$.

40. 1. When $n = 1$, $\left(\frac{a}{b}\right)^1 = \frac{a^1}{b^1}$.

2. Assume that $\left(\frac{a}{b}\right)^k = \frac{a^k}{b^k}$.

Then, $\left(\frac{a}{b}\right)^{k+1} = \left(\frac{a}{b}\right)^k\left(\frac{a}{b}\right) = \frac{a^k}{b^k}\cdot\frac{a}{b} = \frac{a^{k+1}}{b^{k+1}}$.

Thus, $\left(\frac{a}{b}\right)^n = \frac{a^n}{b^n}$.

42. 1. When $n = 1$, $\ln x_1 = \ln x_1$.

2. Assume that

$$\ln(x_1 x_2 x_3 \ldots x_k) = \ln x_1 + \ln x_2 + \ln x_3 + \cdots + \ln x_k.$$

$$\begin{aligned}\text{Then, } \ln(x_1 x_2 x_3 \ldots x_k\, x_{k+1}) &= \ln[(x_1 x_2 x_3 \ldots x_k)x_{k+1}] \\ &= \ln(x_1 x_2 x_3 \ldots x_k) + \ln x_{k+1} \\ &= \ln x_1 + \ln x_2 + \ln x_3 + \cdots + \ln x_k + \ln x_{k+1}.\end{aligned}$$

Thus, $\ln(x_1 x_2 x_3 \ldots x_n) = \ln x_1 + \ln x_2 + \ln x_3 + \cdots + \ln x_n$.

44. 1. When $n = 1$, $a + bi$ and $a - bi$ are complex conjugates by definition.

2. Assume that $(a + bi)^k$ and $(a - bi)^k$ are complex conjugates.

That is, if $(a + bi)^k = c + di$, then $(a - bi)^k = c - di$.

Then,

$$\begin{aligned}(a + bi)^{k+1} &= (a + bi)^k(a + bi) = (c + di)(a + bi) \\ &= (ac - bd) + i(bc + ad) \\ \text{and } (a - bi)^{k+1} &= (a - bi)^k\,(a - bi) = (c - di)(a - bi) \\ &= (ac - bd) - i(bc + ad).\end{aligned}$$

This implies that $(a + bi)^{k+1}$ and $(a - bi)^{k+1}$ are complex conjugates. Therefore, $(a + bi)^n$ and $(a - bi)^n$ are complex conjugates for $n \geq 1$.

46. 1. When $n = 1$, $\tan(x + \pi) = \dfrac{\tan x + \tan \pi}{1 - \tan x \tan \pi} = \dfrac{\tan x + 0}{1 - 0} = \tan x$.

2. Assume $\tan(x + k\pi) = \tan x$.

$$\begin{aligned}\text{Then, } \tan(x + (k + 1)\pi) &= \tan[(x + \pi) + k\pi] \\ &= \frac{\tan(x + \pi) + \tan k\pi}{1 - \tan(x + \pi)\tan k\pi} \\ &= \frac{\tan x + \tan k\pi}{1 - \tan x \tan k\pi} \\ &= \frac{\tan x + 0}{1 - (\tan x)(0)} = \tan x.\end{aligned}$$

Thus, $\tan(x + n\pi) = \tan x$.

48. 1. When $n = 1$, $(2^{2(1)-1} + 3^{2(1)-1}) = 2 + 3 = 5$ and 5 is a factor.

2. Assume that 5 is a factor of $(2^{2k-1} + 3^{2k-1})$.

$$\begin{aligned}\text{Then, } (2^{2(k+1)-1} + 3^{2(k+1)-1}) &= (2^{2k+2-1} + 3^{2k+2-1}) \\ &= (2^{2k-1}2^2 + 3^{2k-1}3^2) \\ &= (4 \cdot 2^{2k-1} + 9 \cdot 3^{2k-1}) \\ &= (2^{2k-1} + 3^{2k-1}) + (2^{2k-1} + 3^{2k-1}) \\ &\quad + (2^{2k-1} + 3^{2k-1}) + (2^{2k-1} + 3^{2k-1}) + 5 \cdot 3^{2k-1}.\end{aligned}$$

Since 5 is a factor of each set of parenthesis and 5 is a factor of $5 \cdot 3^{2k-1}$, then 5 is a factor of the whole sum. Thus, 5 is a factor of $(2^{2n-1} + 3^{2n-1})$ for every positive integer n.

50. (a) If P_3 is true and P_k implies P_{k+1}, then P_n is true for integers $n \geq 3$.

(b) If $P_1, P_2, P_3, \ldots, P_{50}$ are all true, then P_n is true for integers $1 \leq n \leq 50$.

(c) If P_1, P_2, and P_3 are all true, but the truth of P_k does not imply that P_{k+1} is true, then you may only conclude that P_1, P_2, and P_3 are true.

(d) If P_2 is true and P_{2k} implies P_{2k+2}, then P_{2n} is true for any positive integer n.

52. $a_0 = 10, \ a_n = 4a_{n-1}$

$a_0 = 10$

$a_1 = 4(10) = 40$

$a_2 = 4(40) = 160$

$a_3 = 4(160) = 640$

$a_4 = 4(640) = 2560$

54. $a_0 = 0, \ a_1 = 2, \ a_n = a_{n-1} + 2a_{n-2}$

$a_0 = 0$

$a_1 = 2$

$a_2 = 2 + 2(0) = 2$

$a_3 = 2 + 2(2) = 6$

$a_4 = 6 + 2(2) = 10$

56. $f(1) = 2, \ a_n = n - a_{n-1}$

$a_1 = f(1) = 2$

$a_2 = n - a_1 = 2 - 2 = 0$

$a_3 = n - a_2 = 3 - 0 = 3$

$a_4 = n - a_3 = 4 - 3 = 1$

$a_5 = n - a_5 = 5 - 1 = 4$

a_n: 2 0 3 1 4

First differences: −2 3 −2 3

Second differences: 5 −5 5

Since neither the first differences nor the second differences are equal, the sequence does not have a linear or quadratic model.

58. $f(2) = -3, \ a_n = -2a_{n-1}$

$a_2 = f(2) = -3 \Longrightarrow -3 = -2a_1$

$a_1 = \frac{3}{2}$

$a_3 = -2a_2 = -2(-3) = 6$

$a_4 = -2a_3 = -2(6) = -12$

$a_5 = -2a_4 = -2(-12) = 24$

a_n: $\frac{3}{2}$ −3 6 −12 24

First differences: $-\frac{9}{2}$ 9 −18 36

Second differences: $\frac{27}{2}$ −27 54

Since neither the first differences nor the second differences are equal, the sequence does not have a linear or quadratic model.

60. $a_0 = 2, \ a_n = (a_{n-1})^2$

$a_0 = 2$

$a_1 = a_0^2 = 2^2 = 4$

$a_2 = a_1^2 = 4^2 = 16$

$a_3 = a_2^2 = 16^2 = 256$

$a_4 = a_3^2 = 256^2 = 65{,}536$

a_n: 2 4 16 256 65,536

First differences: 2 12 240 65,280

Second differences: 10 228 65,040

Since neither the first differences nor the second differences are equal, the sequence does not have a linear or quadratic model.

62. $f(1) = 0, \ a_n = a_{n-1} + 2n$

$a_1 = 0$

$a_2 = a_1 + 2(2) = 0 + 4 = 4$

$a_3 = a_2 + 2(3) = 4 + 6 = 10$

$a_4 = a_3 + 2(4) = 10 + 8 = 18$

$a_5 = a_4 + 2(5) = 18 + 10 = 28$

a_n: 0 4 10 18 28

First differences: 4 6 8 10

Second differences: 2 2 2

Since the second differences are equal, the sequence has a quadratic model.

64. $a_0 = 0, \ a_n = a_{n-1} - 1$

$a_0 = 0$

$a_1 = a_0 - 1 = 0 - 1 = -1$

$a_2 = a_1 - 1 = -1 - 1 = -2$

$a_3 = a_2 - 1 = -2 - 1 = -3$

$a_4 = a_3 - 1 = -3 - 1 = -4$

a_n: 0 −1 −2 −3 −4

First differences: −1 −1 −1 −1

Second differences: 0 0 0

Since the first differences are equal, the sequence has a linear model.

66. $a_0 = 7,\ a_1 = 6,\ a_3 = 10$

Let $a_n = an^2 + bn + c$. Thus,

$$\begin{aligned} a_0 &= a(0)^2 + b(0) + c = 7 \Longrightarrow & c &= 7 \\ a_1 &= a(1)^2 + b(1) + c = 6 \Longrightarrow & a + b + c &= 6 \\ & & a + b &= -1 \\ a_3 &= {}_a(3)^2 + b(3) + c = 10 \Longrightarrow & 9a + 3b + c &= 10 \\ & & 9a + 3b &= 3 \\ & & 3a + b &= 1 \end{aligned}$$

By elimination: $-a - b = 1$

$$\begin{aligned} 3a + b &= 1 \\ \hline 2a &= 2 \\ a &= 1 \Longrightarrow b = -2 \end{aligned}$$

Thus, $a_n = n^2 - 2n + 7$.

68. $a_0 = 3,\ a_2 = 0,\ a_6 = 36$

Let $a_n = an^2 + bn + c$. Thus,

$$\begin{aligned} a_0 &= a(0)^2 + b(0) + c = 3 \Longrightarrow & c &= 3 \\ a_2 &= a(2)^2 + b(2) + c = 0 \Longrightarrow & 4a + 2b + c &= 0 \\ & & 4a + 2b &= -3 \\ a_6 &= a(6)^2 + b(6) + c = 36 \Longrightarrow & 36a + 6b + c &= 36 \\ & & 36a + 6b &= 33 \\ & & 12a + 2b &= 11 \end{aligned}$$

By elimination: $-4a - 2b = 3$

$$\begin{aligned} 12a + 2b &= 11 \\ 8a &= 14 \\ a &= \tfrac{7}{4} \Longrightarrow b = -5 \end{aligned}$$

Thus, $a_n = \frac{7}{4}n^2 - 5n + 3$.

70. $x - y^3 = 0 \Longrightarrow x = y^3$

$x - 2y^2 = 0$

$y^3 - 2y^2 = 0$

$y^2(y - 2) = 0 \Longrightarrow y = 0, 2$

When $y = 0$: $x = 0^3 = 0$.

When $y = 2$: $x = 2^3 = 8$.

Points of intersection: $(0, 0)$ and $(8, 2)$

72.
$$\begin{aligned} 2x + y - 2z &= 1 \\ x \qquad - z &= 1 \\ 3x + 3y + z &= 12 \end{aligned}$$

$$A = \begin{bmatrix} 2 & 1 & -2 \\ 1 & 0 & -1 \\ 3 & 3 & 1 \end{bmatrix},\quad A^{-1} = \frac{1}{4}\begin{bmatrix} -3 & 7 & 1 \\ 4 & -8 & 0 \\ -3 & 3 & 1 \end{bmatrix}$$

$$\begin{bmatrix} x \\ y \\ z \end{bmatrix} = \frac{1}{4}\begin{bmatrix} -3 & 7 & 1 \\ 4 & -8 & 0 \\ -3 & 3 & 1 \end{bmatrix}\begin{bmatrix} 1 \\ 1 \\ 12 \end{bmatrix} = \begin{bmatrix} 4 \\ -1 \\ 3 \end{bmatrix}$$

Thus, $x = 4, y = -1, z = 3$.

Answer: $(4, -1, 3)$

Section 9.5 The Binomial Theorem

Solutions to Even-Numbered Exercises

2. ${}_8C_6 = \dfrac{8!}{6!2!} = \dfrac{8 \cdot 7}{2 \cdot 1} = 28$

4. ${}_{20}C_{20} = \dfrac{20!}{20!0!} = 1$

6. ${}_{12}C_5 = \dfrac{12!}{5!7!} = \dfrac{12 \cdot 11 \cdot 10 \cdot 9 \cdot 8}{5 \cdot 4 \cdot 3 \cdot 2 \cdot 1} = 792$

8. ${}_{10}C_4 = \dfrac{10!}{6!4!} = \dfrac{10 \cdot 9 \cdot 8 \cdot 7}{4 \cdot 3 \cdot 2 \cdot 1} = 210$

10. ${}_{10}C_6 = \dfrac{10!}{6!4!} = 210$

12.

```
              1
             1 1
            1 2 1
           1 3 3 1
          1 4 6 4 1
        1 5 10 10 5 1
       1 6 15 20 15 6 1
     1 7 21 35 35 21 7 1
   1 8 28 56 70 56 28 8 1
```

14.

```
            1
           1 1
          1 2 1
         1 3 3 1
        1 4 6 4 1
      1 5 10 10 5 1
    1 6 15 (20) 15 6 1
```

${}_6C_3 = 20$, the 4th entry in the 7th row.

16.

```
              1
             1 1
            1 2 1
           1 3 3 1
          1 4 6 4 1
        1 5 10 10 5 1
       1 6 15 20 15 6 1
     1 7 21 35 35 21 7 1
   1 8 28 56 70 56 28 (8) 1
```

${}_8C_7 = 8$, the 8th entry in the 9th row.

18. $(x + 1)^6 = {}_6C_0x^6 + {}_6C_1x^5(1) + {}_6C_2x^4(1)^2 + {}_6C_3x^3(1)^3 + {}_6C_4x^2(1)^4 + {}_6C_5x(1)^5 + {}_6C_6(1)^6$
$= x^6 + 6x^5 + 15x^4 + 20x^3 + 15x^2 + 6x + 1$

20. $(a + 3)^4 = {}_4C_0a^4 + {}_4C_1a^3(3) + {}_4C_2a^2(3)^2 + {}_4C_3a(3)^3 + {}_4C_4(3)^4$
$= a^4 + 12a^3 + 54a^2 + 108a + 81$

22. $(y - 2)^5 = {}_5C_0y^5 - {}_5C_1y^4(2) + {}_5C_2y^3(2)^2 - {}_5C_3y^2(2)^3 + {}_5C_4y(2)^4 - {}_5C_5(2)^5$
$= y^5 - 10y^4 + 40y^3 - 80y^2 + 80y - 32$

24. $(x + y)^6 = {}_6C_0x^6 + {}_6C_1x^5y + {}_6C_2x^4y^2 + {}_6C_3x^3y^3 + {}_6C_4x^2y^4 + {}_6C_5xy^5 + {}_6C_6y^6$
$= x^6 + 6x^5y + 15x^4y^2 + 20x^3y^3 + 15x^2y^4 + 6xy^5 + y^6$

26. $(x + 2y)^4 = {}_4C_0x^4 + {}_4C_1x^3(2y) + {}_4C_2x^2(2y)^2 + {}_4C_3x(2y)^3 + {}_4C_4(2y)^4$
$= x^4 + 4x^3(2y) + 6x^2(4y^2) + 4x(8y^3) + 16y^4$
$= x^4 + 8x^3y + 24x^2y^2 + 32xy^3 + 16y^4$

28. $(2x - y)^5 = {}_5C_0(2x)^5 - {}_5C_1(2x)^4y + {}_5C_2(2x)^3y^2 - {}_5C_3(2x)^2y^3 + {}_5C_4(2x)y^4 - {}_5C_5(2x)y^5$
$= 32x^5 - 5(16x^4)y + 10(8x^3)y^2 - 10(4x^2)y^3 + 5(2x)y^4 - y^5$
$= 32x^5 - 80x^4y + 80x^3y^2 - 40x^2y^3 + 10xy^4 - y^5$

30. $(5 - 3y)^3 = 5^3 - 3(5)^2 3y + 3(5)(3y)^2 - (3y)^3$
$= 125 - 225y + 135y^2 - 27y^3$

32. $(x^2 + y^2)^6 = {}_6C_0(x^2)^6 + {}_6C_1(x^2)^5(y^2) + {}_6C_2(x^2)^4(y^2)^2 + {}_6C_3(x^2)^3(y^2)^3 + {}_6C_4(x^2)^2(y^2)^4$
$+ {}_6C_5(x^2)(y^2)^5 + {}_6C_6(y^2)^6$
$= x^{12} + 6x^{10}y^2 + 15x^8y^4 + 20x^6y^6 + 15x^4y^8 + 6x^2y^{10} + y^{12}$

34. $\left(\frac{1}{x}+2y\right)^6 = {}_6C_0\left(\frac{1}{x}\right)^6 + {}_6C_1\left(\frac{1}{x}\right)^5(2y) + {}_6C_2\left(\frac{1}{x}\right)^4(2y)^2 + {}_6C_3\left(\frac{1}{x}\right)^3(2y)^3 + {}_6C_4\left(\frac{1}{x}\right)^2(2y)^4 + {}_6C_5\left(\frac{1}{x}\right)(2y)^5 + {}_6C_6(2y)^6$

$= 1\left(\frac{1}{x}\right)^6 + 6(2)\left(\frac{1}{x}\right)^5 y + 15(4)\left(\frac{1}{x}\right)^4 y^2 + 20(8)\left(\frac{1}{x}\right)^3 y^3 + 15(16)\left(\frac{1}{x}\right)^2 + y^4 + 6(32)\left(\frac{1}{x}\right)y^5 + 1(64)y^6$

$= \frac{1}{x^6} + \frac{12y}{x^5} + \frac{60y^2}{x^4} + \frac{160y^3}{x^3} + \frac{240y^4}{x^2} + \frac{192y^5}{x} + 64y^6$

36. $3(x+1)^5 - 4(x+1)^3 = 3[{}_5C_0x^5 + {}_5C_1x^4(1) + {}_5C_2x^3(1)^2 + {}_5C_3x^2(1)^3 + {}_5C_4x(1)^4 + {}_5C_5(1)^5]$

$- 4[{}_3C_0x^3 + {}_3C_1x^2(1) + {}_3C_2x(1)^2 + {}_3C_3(1)^3]$

$= 3[(1)x^5 + 5x^4 + 10x^3 + 10x^2 + 5x + 1] - 4[(1)x^3 + 3x^2 + 3x + 1]$

$= 3x^5 + 15x^4 + 26x^3 + 18x^2 + 3x - 1$

38. 5th row of Pascal's Triangle: 1 5 10 5 1

$(x+2y)^5 = (1)x^5 + 5x^4 2y + 10x^3(2y)^2 + 10x^2(2y)^3 + 5x(2y)^4 + (2y)^5$

$= x^5 + 10x^4y + 40x^3y^2 + 80x^2y^3 + 80xy^4 + 32y^5$

40. 5th row of Pascal's Triangle: 1 5 10 10 5 1

$(3y+2)^5 = (3y)^5 + 5(3y)^4(2) + 10(3y)^3(2)^2 + 10(3y)^2(2)^3 + 5(3y)(2)^4 + (2)^5$

$= 243y^5 + 810y^4 + 1080y^3 + 720y^2 + 240y + 32$

42. The term involving x^8 in the expansion of $(x^2+3)^{12}$ is ${}_{12}C_8(x^2)^4(3)^8 = \frac{12!}{(12-8)!8!} \cdot 3^8x^8 = 3{,}247{,}695x^8$. The coefficient is 3,247,695.

44. The term involving x^2y^8 in the expansion of $(4x-y)^{10}$ is ${}_{10}C_8(4x)^2(-y)^8 = \frac{10!}{(10-8)!8!} \cdot 16x^2y^8 = 720x^2y^8$. The coefficient is 720.

46. The term involving x^6y^2 in the expansion of $(2x-3y)^8$ is ${}_8C_2(2x)^6(-3y)^2 = \frac{8!}{(8-6)!2!}(64x^6)(9y^2) = 16{,}128x^6y^2$. The coefficient is 16,128.

48. The term involving z^6 in the expansion of $(z^2-1)^{12}$ is ${}_{12}C_9(z^2)^3(-1)^9 = \frac{12}{(12-9)!9!}z^6(-1) = -220z^6$. The coefficient is -220.

50. The expansions of $(x+y)^n$ and $(x-y)^n$ are almost the same except that the signs of the terms in the expansion of $(x-y)^n$ alternate from positive to negative.

52. $(2\sqrt{t}-1)^3 = (2\sqrt{t})^3 + 3(2\sqrt{t})^2(-1) + 3(2\sqrt{t})(-1) + (-1)^3$

$= 8t^{3/2} - 12t + 6t^{1/2} - 1$

54. $(u^{3/5}+2)^5 = (u^{3/5})^5 + 5(u^{3/5})^4(2) + 10(u^{3/5})^3(2)^2 + 10(u^{3/5})^2(2)^3 + 5(u^{3/5})(2)^4 + 2^5$

$= u^3 + 10u^{12/5} + 40u^{9/5} + 80u^{6/5} + 80u^{3/5} + 32$

56. $\frac{f(x+h)-f(x)}{h} = \frac{(x+h)^4 - x^4}{h}$

$= \frac{x^4 + 4x^3h + 6x^2h^2 + 4xh^3 + h^4 + x^4}{h}$

$= \frac{h(4x^3 + 6x^2h + 4xh^2 + h^3)}{h}$

$= 4x^3 + 6x^2h + 4xh^2 + h^3, h \neq 0$

58. $\dfrac{f(x+h)-f(x)}{h} = \dfrac{\dfrac{1}{x+h} - \dfrac{1}{x}}{h}$

$= \dfrac{\dfrac{x-(x+h)}{x(x+h)}}{h}$

$= \dfrac{\dfrac{-h}{x(x+h)}}{h}$

$= -\dfrac{1}{x(x+h)}, h \neq 0$

60. $(2-i)^5 = {}_5C_0 2^5 - {}_5C_1 2^4 i + {}_5C_2 2^3 i^2 - {}_5C_3 2^2 i^3 + {}_5C_4 2 i^4 - {}_5C_5 i^5$

$= 32 - 80i - 80 + 40i + 10 - i$

$= -38 - 41i$

62. $(5+\sqrt{-9})^3 = (5+3i)^3$

$= 5^3 + 3 \cdot 5^2(3i) + 3 \cdot 5(3i)^2 + (3i)^3$

$= 125 + 225i - 135 - 27i$

$= -10 + 198i$

64. $(5-\sqrt{3}i)^4 = 5^4 - 4 \cdot 5^3(\sqrt{3}i) + 6 \cdot 5^2(\sqrt{3}i)^2 - 4 \cdot 5(\sqrt{3}i)^3 + (\sqrt{3}i)^4$

$= 625 - 500\sqrt{3}i - 450 + 60\sqrt{3}i + 9$

$= 184 - 440\sqrt{3}i$

66. ${}_{10}C_3\left(\frac{1}{4}\right)^3\left(\frac{3}{4}\right)^7 = 120\left(\frac{1}{64}\right)\left(\frac{2187}{16{,}384}\right) \approx 0.2503$

68. ${}_8C_4\left(\frac{1}{2}\right)^4\left(\frac{1}{2}\right)^4 = 70\left(\frac{1}{16}\right)\left(\frac{1}{16}\right) \approx 0.273$

70. $(2.005)^{10} = (2+0.005)^{10} = 2^{10} + 10(2)^9(0.005) + 45(2)^8(0.005)^2 + 120(2)^7(0.005)^3 + 210(2)^6(0.005)^4$

$+ 252(2)^5(0.005)^5 + 210(2)^4(0.005)^6 + 120(2)^3(0.005)^7 + 45(2)^8(0.005)^2$

$+ 10(2)(0.005)^9 + (0.005)^{10}$

$= 1024 + 25.6 + 0.288 + 0.00192 + 0.0000084 + \cdots$

≈ 1049.890

72. $(1.98)^9 = (2-0.02)^9 = 2^9 - 9(2)^8(0.02) + 36(2)^7(0.02)^2 - 84(2)^6(0.02)^3 + 126(2)^5(0.02)^4$

$- 126(2)^4(0.02)^5 + 84(2)^3(0.02)^6 - 36(2)^2(0.02)^7 + 9(2)(0.02)^8 - (0.02)^9$

$= 512 - 46.08 + 1.8432 - 0.043008 + 0.00064512$

≈ 467.721

74. $f(x) = -x^4 + 4x^2 - 1,\ g(x) = f(x-3)$

$g(x) = f(x-3)$

$= -(x-3)^4 + 4(x-3)^2 - 1$

$= -(x^4 + 4x^3(-3) + 6x^2(-3)^2 + 4x(-3)^3 + (-3)^4) + 4(x^2 - 6x + 9) - 1$

$= -x^4 + 12x^3 - 54x^2 + 108x - 81 + 4x^2 - 24x + 36 - 1$

$= -x^4 + 12x^3 - 50x^2 + 84x - 46$

The graph of g is the same as the graph of f shifted 3 units to the right.

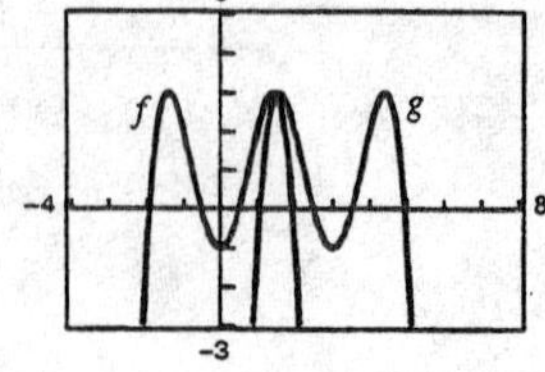

76. $f(x) = 2x^2 - 4x + 1, g(x) = f(x + 3)$

$$\begin{aligned} g(x) &= f(x + 3) \\ &= 2(x + 3)^2 - 4(x + 3) + 1 \\ &= 2[x^2 + 6x + 9] - 4x - 12 + 1 \\ &= 2x^2 + 8x + 7 \end{aligned}$$

$g(x)$ if $f(x)$ shifted 3 units to the left.

78. $0 = (1 - 1)^n = {}_nC_0 - {}_nC_1 + {}_nC_2 - {}_nC_3 + \cdots + (\pm {}_nC_n) = 0$

80. ${}_nC_0 + {}_nC_1 + {}_nC_2 + {}_nC_3 + \cdots + {}_nC_n = (1 + 1)^n = 2^n$

82. (a) ${}_{25}C_6 = 177{,}100$

(b) $2({}_{25}C_2 + {}_{25}C_4) = 25{,}900$

(c) $\sum_{k=0}^{5}[({}_{10}C_k)({}_8C_{5-k})] = 8568$

(d) ${}_{18}C_5 = 8568$

(c) and (d) are equal.

84. $p(x) = 1 - 2x + \frac{3}{2}x^2 - \frac{1}{2}x^3 + \frac{1}{16}x^4 = f(x)$

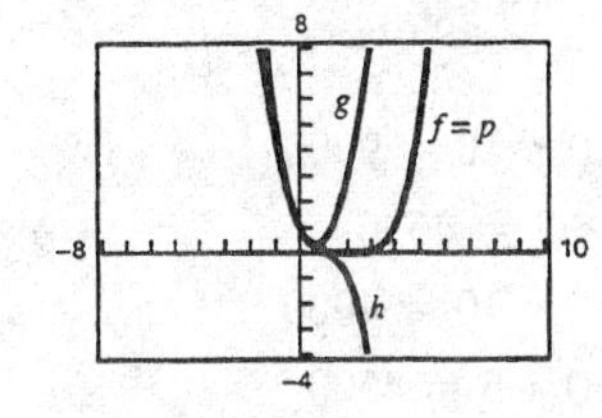

$p(x)$ is the expansion of $f(x)$.

86. (a)

$$\begin{aligned} g(t) &= f(t + 4) \\ &= 0.1043(t + 4)^2 + 0.7100(t + 4) + 4.6852 \\ &= 0.1043t^2 + 1.5444t + 9.1940 \end{aligned}$$

(b)

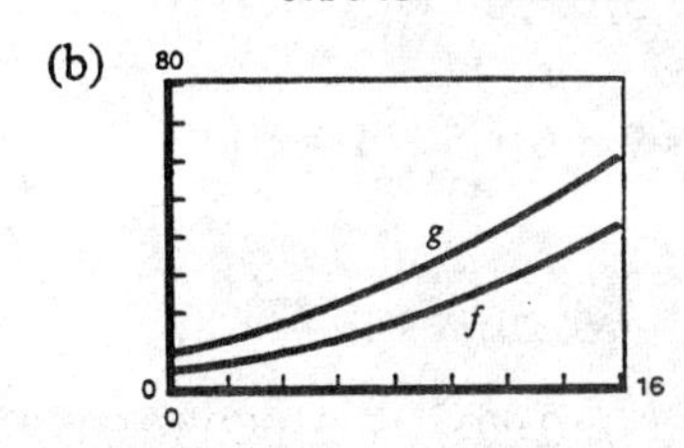

88. $g(x) = f(x - 3)$

$g(x)$ is shifted three units to the right of $f(x)$.

90. $g(x) = -f(x)$

$g(x)$ is the reflection of $f(x)$ in the x-axis.

Section 9.6 Counting Principles

Solutions to Even-Numbered Exercises

2. Even integers: 2, 4, 6, 8, 10, 12

6 ways

4. Greater than 9: 10, 11, 12

3 ways

6. Divisible by 7: 7

1 way

8. Two *distinct* integers whose sum is 8:

$1 + 7, 2 + 6, 3 + 5$

3 ways

10. Monitors: 3

Keyboards: 2

Computers: 4

Total: $3 \cdot 2 \cdot 4 = 24$ ways

12. Math courses: 2

Science courses: 3

Social sciences and humanities courses: 5

Total: $2 \cdot 3 \cdot 5 = 30$ ways

14. $2^{10} = 1024$

16. 1st position: 4 choices
2nd position: 3 choices
3rd position: 2 choices
4th position: 1 choice
5th position: 6 choices
6th position: 5 choices
7th position: 4 choices
8th position: 3 choices
9th position: 2 choices
10th position: 1 choice
Total: $(4!)(6!) = 17{,}280$

18. $24 \cdot 24 \cdot 10 \cdot 10 \cdot 10 \cdot 10 = 5{,}760{,}000$

20. (a) $9 \cdot 10 \cdot 10 \cdot 10 = 9000$
(b) $9 \cdot 9 \cdot 8 \cdot 7 = 4536$
(c) $4 \cdot 10 \cdot 10 \cdot 10 = 4000$
(d) $9 \cdot 10 \cdot 10 \cdot 5 = 4500$

22. $50^3 = 125{,}000$

24. (a) $5! = 5 \cdot 4 \cdot 3 \cdot 2 \cdot 1 = 120$
(b) $(3!)(2!) = 3 \cdot 2 \cdot 1 \cdot 2 \cdot 1 = 12$

26. ${}_nP_r = \dfrac{n!}{(n-r)!}$

$${}_5P_5 = \frac{5!}{(5-5)!} = \frac{5!}{0!} = 120$$

28. ${}_{20}P_2 = \dfrac{20!}{18!} = 20.19 = 380$

30. ${}_7P_4 = \dfrac{7!}{3!} = 7 \cdot 6 \cdot 5 \cdot 4 = 840$

32. ${}_nP_5 = 18 \cdot {}_{n-2}P_4$

Note: $n \geq 6$ for this to be defined.

$$\frac{n!}{(n-5)!} = 18\left(\frac{(n-2)!}{(n-6)!}\right)$$

$$n(n-1)(n-2)(n-3)(n-4) = 18(n-2)(n-3)(n-4)(n-5)$$

(We can divide by $(n-2)$, $(n-3)$, $(n-4)$ since $n \neq 2$, $n \neq 3$, and $n \neq 4$.)

$$n^2 - n = 18n - 90$$

$$n^2 - 19n + 90 = 0$$

$$(n-1)(n-10) = 0$$

$$n = 9 \text{ or } n = 10$$

34. ${}_{100}P_5 = 9{,}034{,}502{,}400$

36. ${}_{10}P_8 = 1{,}814{,}400$

38. ${}_{10}C_7 = 120$

40. The symbol ${}_nP_r$ means to choose and order r elements out of a collection of n elements.

42. A B C D
A C B D
D B C A
D C B A

44. $6! = 720$ ways

46. $4! = 24$ ways

48. $\dfrac{8!}{3!5!} = 56$

50. $\dfrac{11!}{1!4!4!2!} = \dfrac{11!}{4!4!2!} = 34{,}650$

52. ${}_6C_3 = \dfrac{6!}{3!3!} = 20$

ABC, ABD, ABE, ABF, ACD, ACE, ACF, ADE, ADF, AEF, BCD, BCE, BCF, BDE, BDF, BEF, CDE, CDF, CEF, DEF

54. ${}_{12}C_{10} = 66$ ways

56. ${}_{50}C_6 = 15{,}890{,}700$ ways

58. ${}_{80}C_5 = 24{,}040{,}016$ subsets

60. There are 9 good units and 3 defective units.

(a) ${}_9C_4 = 126$ ways

(b) ${}_9C_2 \cdot {}_3C_2 = 36 \cdot 3 = 108$ ways

(c) $${}_9C_4 + {}_9C_3 \cdot {}_3C_1 + {}_9C_2 \cdot {}_3C_2 = \frac{9!}{5!4!} + \frac{9!}{6!3!} \cdot \frac{3!}{2!1!} + \frac{9!}{7!2!} \cdot \frac{3!}{1!2!}$$
$$= 126 + 84 \cdot 3 + 36 \cdot 3$$
$$= 486$$

62. Select type of card for three of a kind: ${}_{13}C_1$

Select three of four cards for three of a kind: ${}_4C_3$

Select type of card for pair: ${}_{12}C_1$

Select two of four cards for pair: ${}_4C_2$

${}_{13}C_1 \cdot {}_4C_3 \cdot {}_{12}C_1 \cdot {}_4C_2 = 13 \cdot 4 \cdot 12 \cdot 6 = 3744$ ways to get a full house

64. (a) ${}_3C_2 = \dfrac{3!}{2!1!} = 3$ relationships

(b) ${}_8C_2 = \dfrac{8!}{2!6!} = \dfrac{8 \cdot 7}{2} = 28$ relationships

(c) ${}_{12}C_2 = \dfrac{12!}{2!10!} = \dfrac{12 \cdot 11}{2} = 66$ relationships

(d) ${}_{20}C_2 = \dfrac{20!}{2!18!} = \dfrac{20 \cdot 19}{2} = 190$ relationships

66. ${}_6C_2 - 6 = 15 - 6 = 9$ diagonals

68. ${}_{10}C_2 - 10 = 45 - 10 = 35$ diagonals

70. $${}_nC_n = \frac{n!}{(n-n)!n!} = \frac{n!}{0!n!} = \frac{n!}{n!0!} = \frac{n!}{(n-0)!0!} = {}_nC_0$$

72. $${}_nC_r = \frac{n!}{(n-r)!r!}$$
$$= \frac{n(n-1)(n-2)\cdots(n-r+1)(n-r)!}{(n-r)!r!}$$
$$= \frac{n(n-1)(n-2)\cdots(n-r+1)}{r!}$$
$$= \frac{{}_nP_r}{r!}$$

74. $$\frac{4}{t} + \frac{3}{2t} = 1$$
$$\frac{8+3}{2t} = 1$$
$$11 = 2t$$
$$t = \frac{11}{2} = 5.5$$

76. $$e^{x/3} = 16$$
$$\frac{x}{3} = \ln 16$$
$$x = 3 \ln 16 \approx 8.32$$

Section 9.7 Probability

Solutions to Even-Numbered Exercises

2. {2, 3, 4, 5, 6, 7, 8, 9, 10, 11, 12}

4. {(red, red), (red, blue), (red, black), (blue, blue), (blue, black)}

6. {SSS, SSF, SFS, FSS, SFF, FFS, FSF, FFF}

8. $E = \{HHH, HHT, HTH, HTT\}$

$$P(E) = \frac{n(E)}{n(s)} = \frac{4}{8} = \frac{1}{2}$$

10. $E = \{HHH, HHT, HTH, THH\}$

$$P(E) = \frac{n(E)}{n(s)} = \frac{4}{8} = \frac{1}{2}$$

12. The probability that the card is *not* a face card is the complement of getting a face card. (See Exercise 11.)

$$P(E') = 1 - P(E) = 1 - \frac{3}{13} = \frac{10}{13}$$

14. There are six possible cards in each of 4 suits: $6 \cdot 4 = 24$

$$P(E) = \frac{n(E)}{n(s)} = \frac{24}{52} = \frac{6}{13}$$

16. $E = \{(1, 6), (2, 5), (2, 6), (3, 4), (3, 5), (3, 6), (4, 3), (4, 4), (4, 5), (4, 6), (5, 2), (5, 3), (5, 4), (5, 5), (5, 6), (6, 1), (6, 2), (6, 3), (6, 4), (6, 5), (6, 6)\}$

$$P(E) = \frac{n(E)}{n(s)} = \frac{21}{36} = \frac{7}{12}$$

18. $E = \{(1, 1), (1, 2), (2, 1), (6, 6)\}$

$$P(E) = \frac{n(E)}{n(s)} = \frac{4}{36} = \frac{1}{9}$$

20. $E = \{(1, 1), (1, 2), (1, 4), (1, 6), (2, 1), (2, 3), (2, 5), (3, 2), (3, 4), (3, 6), (4, 1), (4, 3), (4, 5), (5, 2), (5, 4), (5, 6), (6, 1), (6, 3), (6, 5)\}$

$$P(E) = \frac{n(E)}{n(s)} = \frac{19}{36}$$

22. $P(E) = \frac{{}_2C_2}{{}_6C_2} = \frac{1}{15}$

24. $$P(E) = \frac{{}_1C_1 \cdot {}_2C_1 + {}_1C_1 \cdot {}_3C_1 + {}_2C_1 \cdot {}_3C_1}{{}_6C_2} = \frac{2 + 3 + 6}{15} = \frac{11}{15}$$

26. $1 - p = 1 - 0.36 = 0.64$

28. $1 - p = 1 - 0.84 = 0.16$

30. (a) $0.392(101) = 39.592$ million $= 39{,}592{,}000$

(b) 0.213

(c) $0.213 + 0.267 = 0.48$

32. (a) $\frac{34}{100} = 0.34$

(b) $\frac{45}{100} = 0.45$

(c) $\frac{23}{100} = 0.23$

34. (a) $\frac{18 + 12}{72} = \frac{30}{72} = \frac{5}{12}$

(b) $1 - \frac{5}{12} = \frac{7}{12}$

(c) $\frac{10}{72} = \frac{5}{36}$

36. $1 - 0.37 - 0.44 = 0.19$

38. (a) $\frac{{}_6C_5}{{}_8C_5} = \frac{6}{56} = \frac{3}{28}$

(b) $\frac{{}_6C_4 \cdot {}_2C_1}{{}_8C_5} = \frac{15 \cdot 2}{56} = \frac{15}{28}$

(c) $\frac{3}{28} + \frac{15}{28} = \frac{18}{28} = \frac{9}{14}$

40. Total ways to insert paychecks: $5! = 120$ ways

5 correct: 1 way

4 correct: not possible

3 correct: 10 ways

2 correct: 20 ways

1 correct: 45 ways

0 correct: 44 ways

(a) $\frac{45}{120} = \frac{3}{8}$

(b) $\frac{45 + 20 + 10 + 1}{120} = \frac{19}{30}$

42. (a) $\frac{1}{{}_4P_4} = \frac{1}{24}$

(b) $\frac{1}{{}_3P_3} = \frac{1}{6}$

44. $\frac{{}_{13}C_1 \cdot {}_4C_3 \cdot {}_{12}C_1 \cdot {}_4C_2}{{}_{52}C_5} = \frac{13 \cdot 4 \cdot 12 \cdot 6}{2{,}598{,}960}$

$= \frac{3744}{2{,}598{,}960}$

$= \frac{6}{4165}$

46. (a) $\frac{{}_{16}C_5}{{}_{20}C_5} = \frac{4368}{15{,}504} = \frac{91}{323} \approx 0.282$ (5 good units)

(b) $\frac{{}_{16}C_4 \cdot {}_4C_1}{{}_{20}C_5} = \frac{1820 \cdot 4}{15{,}504} = \frac{455}{969} \approx 0.470$ (4 good units)

(c) Probability of at least one defective unit = 1 − (Probability of no defective units.)
Probability of no defective units = Probability of 5 good units

P (at least one defective unit) $= 1 - \frac{91}{323} = \frac{232}{323} \approx 0.718$

48. (a) $P(EE) = \frac{20}{40} \cdot \frac{20}{40} = \frac{1}{4}$

(b) $P(EO \text{ or } OE) = 2\left(\frac{20}{40}\right)\left(\frac{20}{40}\right) = \frac{1}{2}$

(c) $P(N_1 < 30, N_2 < 30) = \frac{29}{40} \cdot \frac{29}{40} = \frac{841}{1600}$

(d) $P(N_1N_1) = \frac{40}{40} \cdot \frac{1}{40} = \frac{1}{40}$

50. (a) $P(AA) = (0.90)^2 = 0.81$

(b) $P(NN) = (0.10)^2 = 0.01$

(c) $P(A) = 1 - P(NN) = 1 - 0.01 = 0.99$

52. (a) $P(BBBB) = \left(\frac{1}{2}\right)^4 = \frac{1}{16}$

(b) $P(BBBB) + P(GGGG) = \left(\frac{1}{2}\right)^4 + \left(\frac{1}{2}\right)^4 = \frac{1}{8}$

(c) $P(\text{at least one boy}) = 1 - P(\text{no boys})$

$= 1 - P(GGGG) = 1 - \frac{1}{16} = \frac{15}{16}$

54. $(0.78)^3 = 0.474552$

56. (a) If the *center* of the coin falls within the circle of radius $d/2$ around a vertex, the coin will cover the vertex.

$$P(\text{coin covers a vertex}) = \frac{\text{Area in which coin may fall so that it covers a vertex}}{\text{Total area}}$$

$$= \frac{n\left[\pi\left(\frac{d}{2}\right)^2\right]}{nd^2} = \frac{1}{4}\pi$$

(b) Experimental results will vary.

58. If a weather forecast indicates that the probability of rain is 40%, this means the meteorological records indicate that over an extended period of time with similar weather conditions it will rain 40% of the time.

60. $8.6\% + 11.4\% + 14.5\% + 9.2\% = 43.7\%$

Section 9.8 Exploring Data: Measures of Central Tendency

Solutions to Even-Numbered Exercises

2. Mean: $\dfrac{30 + 32 + 32 + 33 + 34 + 37 + 39}{7} = \dfrac{237}{2} \approx 33.86$

Median: 33

Mode: 32

4. Mean: $\dfrac{20 + 32 + 32 + 33 + 34 + 37 + 39}{7} = \dfrac{227}{2} \approx 32.43$

Median: 33

Mode: 32

6. Mean: $\dfrac{30 + 32 + 32 + 34 + 37 + 39}{6} = \dfrac{204}{6} = 34$

Median: $\dfrac{32 + 34}{2} = 33$

Mode: 32

8. (a) Mean: $\dfrac{11 + 13 + 13 + 14 + 15 + 18 + 20}{7} = \dfrac{104}{7} \approx 14.86$

Median: 14

Mode: 13

The mean, median and mode are each increased by 6.

(b) The mean, median and mode will each be increased by k. If $a_1, \ldots a_n$ are the original measurements, then the new mean is

$$\frac{(a_1 + k) + \ldots + (a_n + k)}{n} = \frac{a_1 + \ldots + a_n}{n} + \frac{nk}{n}.$$

$$= (\text{original mean}) + k$$

10. Mean: 10.7

Median: 10.5

Modes: 6 and 10

12. Mean: 48.39

Median: 49

Mode: None

14. Mean: $\dfrac{150 + 260 + 320 + 320 + 410 + 460 + 385}{7} = \dfrac{2305}{7} = 329.3$

Median: 320

Mode: 320

16. (a) average $= \dfrac{0(14) + 1(26) + 2(7) + 3(2) + 4(1)}{50} = \dfrac{50}{50} = 1$

(b) batting average $= \dfrac{50}{200} = .250$

18. Yes, at least half fo the 254 employees earn at or above the median hourly wage.

20. 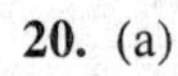
(a)

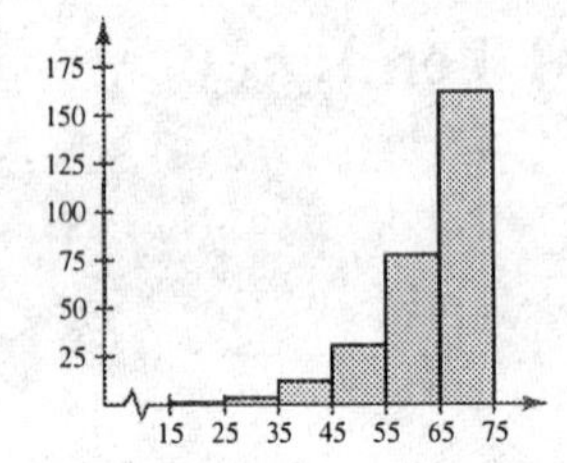

The mean is approximately 63.3.

(b)

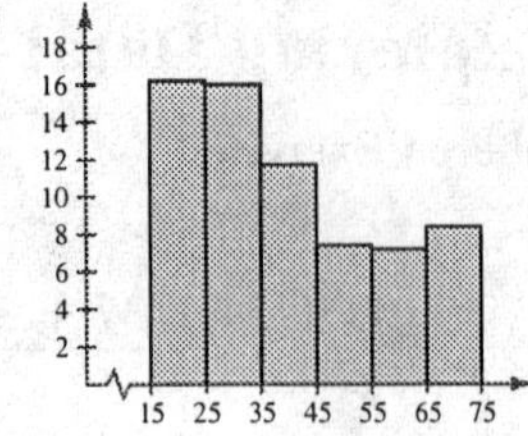

The mean is approximately 39.8.

22. (a)

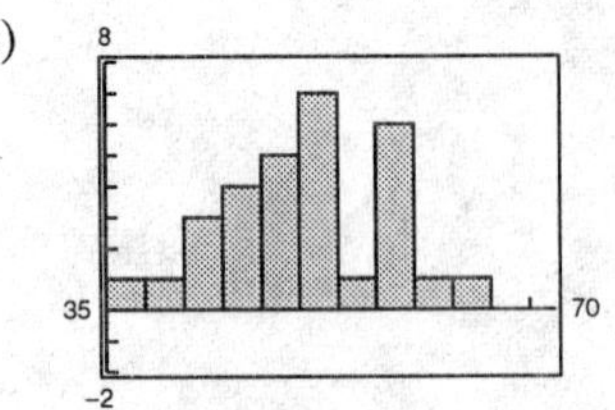

(b)

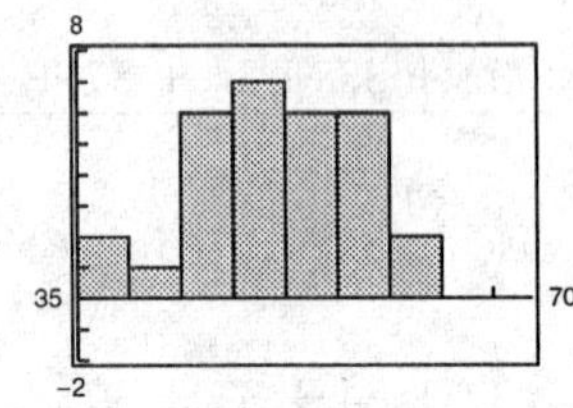

The width of the intervals is changed.

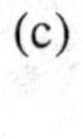
(c)

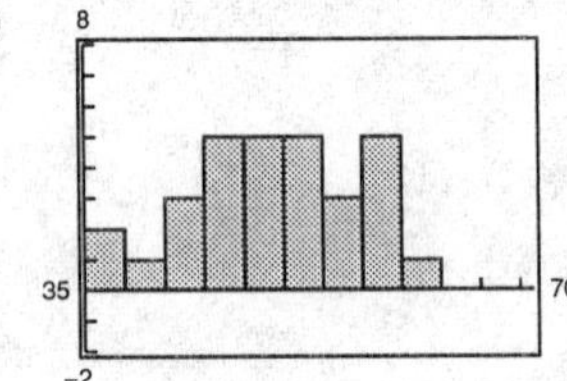

The endpoints of the intervals are changed.

(d) The median is 50.

(e) The mean is 49.9.

Section 9.9 Exploring Data: Measures of Dispersion

Solutions to Even-Numbered Exercises

2. $\bar{x} = \dfrac{2 + 3 + 6 + 9 + 15}{5} = \dfrac{35}{5} = 7$

$v = \dfrac{(2 - 7)^2 + (3 - 7)^2 + (6 - 7)^2 + (9 - 7)^2 + (15 - 7)^2}{5} = \dfrac{110}{5} = 22$

$\sigma = \sqrt{v} = \sqrt{22} \approx 4.69$

4. $\bar{x} = 2$

$v = 0$ (see the discussion preceeding Example 1)

$\sigma = 0$

6. $\bar{x} = \dfrac{1 + 1 + 1 + 5 + 5 + 5}{6} = \dfrac{18}{6} = 3$

$v = \dfrac{(1 - 3)^2 + (1 - 3)^2 + (1 - 3) + (5 - 3)^2 + (5 - 3)^2 + (5 - 3)^2}{6} = \dfrac{24}{6} = 4$

$\sigma = \sqrt{v} = \sqrt{4} = 2$

8. The mean is $\bar{x} = \dfrac{192}{8} = 24$. Then,

$$\sigma = \sqrt{\frac{10^2 + 25^2 + 50^2 + 26^2 + 15^2 + 33^2 + 29^2 + 4^2}{8} - 24^2}$$

$$= \sqrt{\frac{6072}{8} - 576} = \sqrt{183} \approx 13.53$$

10. $\sigma = \sqrt{\dfrac{x_1^2 + \ldots x_{10}^2}{10} - \bar{x}^2}$ $\quad \bar{x} = \dfrac{79}{10} = 7.9$

$\sigma = \sqrt{\dfrac{763}{10} - (7.9)^2} = \sqrt{13.89} \approx 3.73$

12. $\bar{x} = 300$

$\sigma = 101.55$

$v = \sigma^2 = 10{,}312$

14. $\bar{x} = 7.72$

$\sigma = 2.19$

$v = \sigma^2 \approx 4.81$

16. $\bar{x} = 6.64$

$\sigma = 1.92$

$v = \sigma^2 \approx 3.69$

18. (a) $\bar{x} = 15;\ \sigma = 3.19$

(b) $\bar{x} = 15;\ \sigma = 2.83$

(c) $\bar{x} = 25;\ \sigma = 2.83$

(d) $\bar{x} = 5;\ \sigma = 3.19$

20. Each measurement is 8 units from the mean of 12.

22. The first histogram has the smaller standard deviation because it is more bunched about the mean.

24. At least $\frac{3}{4}$ of the heights: [64.5, 77.7]
At least $\frac{8}{9}$ of the heights: [61.2, 81.0]

26. (a) $\bar{x} \approx 40.62,\ \sigma \approx 6.81$

(b) The number of data that lie within the $[40.62 - 2(6.81), 40.62 + 2(6.81)]$ or $[27, 54.24]$ is 34. The percent of data within two standard deviations is $\frac{34}{36}$ or $\approx 94\%$.

28. (a) $\bar{x} \approx 49.9,\ \sigma \approx 6.17$

(b) 1 bale has a weight (37) that differs by more than 2 standard deviations from the mean.

30.

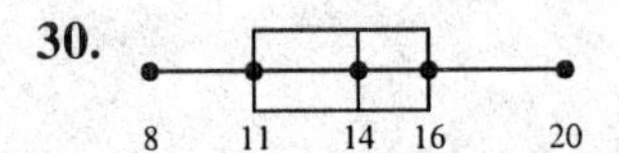

32.

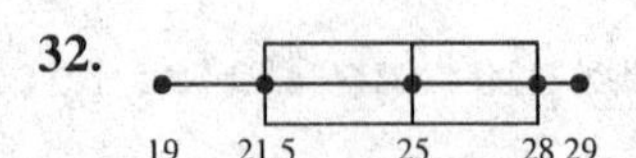

34.

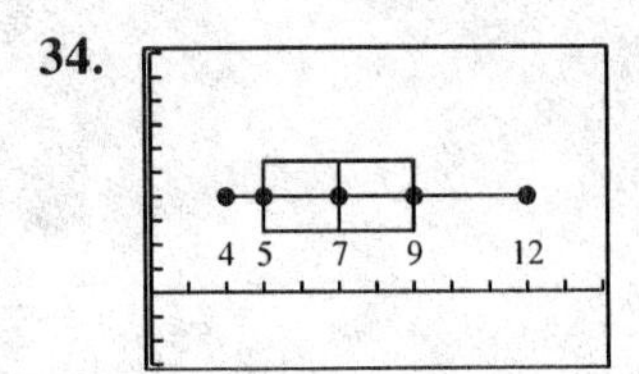

36.

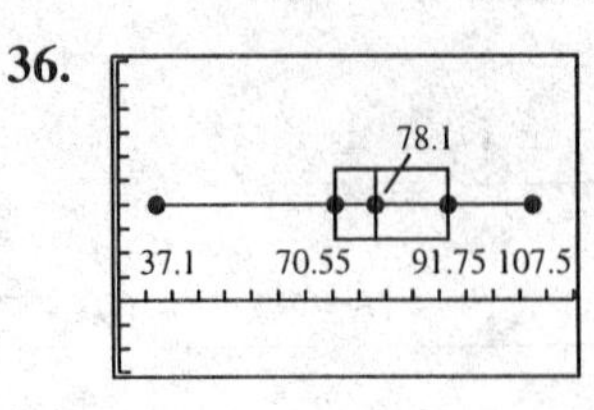

Review Exercises for Chapter 9

Solutions to Even-Numbered Exercises

2. $a_n = \dfrac{5n}{2n-1}$

$a_1 = \dfrac{5(1)}{2(1)-1} = 5$

$a_2 = \dfrac{5(2)}{2(2)-1} = \dfrac{10}{3}$

$a_3 = \dfrac{5(3)}{2(3)-1} = 3$

$a_4 = \dfrac{5(4)}{2(4)-1} = \dfrac{20}{7}$

$a_5 = \dfrac{5(5)}{2(5)-1} = \dfrac{25}{9}$

4. $a_n = n(n-1)$

$a_1 = 1(1-1) = 0$

$a_2 = 2(2-1) = 2$

$a_3 = 3(3-1) = 6$

$a_4 = 4(4-1) = 12$

$a_5 = 5(5-1) = 20$

6. $a_n = 4(0.4)^{n-1}$

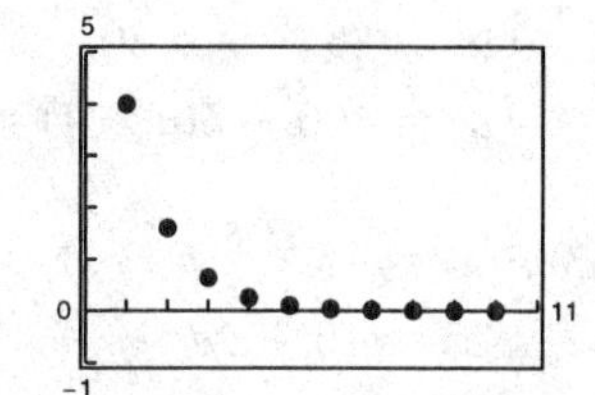

8. $a_n = 5 - \dfrac{3}{n}$

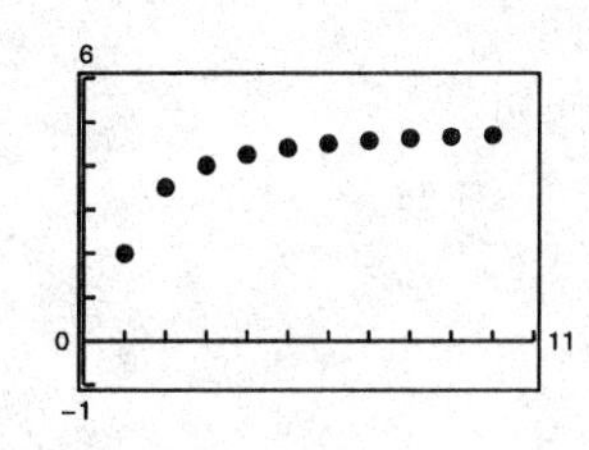

10. $2(1^2) + 2(2^2) + 2(3^2) + \cdots + 2(9^2) = \sum_{k=1}^{9} 2k^2$

12. $1 - \dfrac{1}{3} + \dfrac{1}{9} - \dfrac{1}{27} + \ldots = \sum_{k=0}^{\infty}\left(-\dfrac{1}{3}\right)^k$

14. $\sum_{k=2}^{5} 4k = 8 + 12 + 16 + 20 = 56$

16. $\sum_{i=1}^{8} \dfrac{i}{i+1} = \dfrac{1}{2} + \dfrac{2}{3} + \dfrac{3}{4} + \dfrac{4}{5} + \dfrac{5}{6} + \dfrac{6}{7} + \dfrac{7}{8} + \dfrac{8}{9} \approx 6.17$

18. $\sum_{j=0}^{4}(j^2+1) = \sum_{j=0}^{4} j^2 + \sum_{j=0}^{4} 1 = \dfrac{4(5)(9)}{6} + 5(1) = 35$

20. $\sum_{n=1}^{100}\left(\dfrac{1}{n} - \dfrac{1}{n+1}\right) = \left(\dfrac{1}{1} - \dfrac{1}{2}\right) + \left(\dfrac{1}{2} - \dfrac{1}{3}\right) + \left(\dfrac{1}{3} - \dfrac{1}{4}\right) + \cdots + \left(\dfrac{1}{99} - \dfrac{1}{100}\right) + \left(\dfrac{1}{100} - \dfrac{1}{101}\right)$

$= \dfrac{1}{1} - \dfrac{1}{101} = \dfrac{100}{101}$

22. $a_1 = 8, d = -2$

$a_1 = 8$

$a_2 = 8 - 2 = 6$

$a_3 = 6 - 2 = 4$

$a_4 = 4 - 2 = 2$

$a_5 = 2 - 2 = 0$

24. $a_2 = 14, a_6 = 22$

$a_6 = a_2 + 4d$

$22 = 14 + 4d$

$8 = 4d$

$2 = d$

$a_1 = a_2 - d$

$a_1 = 14 - 2 = 12$

$a_2 = 12 + 2 = 14$

$a_3 = 14 + 2 = 16$

$a_4 = 16 + 2 = 18$

$a_5 = 18 + 2 = 20$

26. $a_1 = 15, a_{k+1} = a_k + \frac{5}{2}$

$a_1 = 15$

$a_2 = 15 + \frac{5}{2} = \frac{35}{2}$

$a_3 = \frac{35}{2} + \frac{5}{2} = \frac{40}{2} = 20$

$a_4 = 20 + \frac{5}{2} = \frac{45}{2}$

$a_5 = \frac{45}{2} + \frac{5}{2} = \frac{50}{2} = 25$

$a_n = 15 + \frac{5}{2}(n-1) = \frac{25}{2} + \frac{5}{2}n$

28. $a_1 = 100, a_{k+1} = a_k - 5$

$a_1 = 100$

$a_2 = 100 - 5 = 95$

$a_3 = 95 - 5 = 90$

$a_4 = 90 - 5 = 85$

$a_5 = 85 - 5 = 80$

$a_n = 100 - 5(n - 1) = 105 - 5n$

30. $a_3 = 15 + \frac{5}{2}(n - 1) = \frac{25}{2} + \frac{5}{2}n$

$28 = 10 + 2d$

$18 = 2d$

$9 = d$

$a_n = 10 + (n - 1)9 = 1 + 9n$

$$\sum_{n=1}^{20}(1 + 9n) = \sum_{n=1}^{20} 1 + 9\sum_{n=1}^{20} n = 20(1) + 9\left[\frac{(20)(21)}{2}\right] = 1910$$

32. $\sum_{j=1}^{8}(20 - 3j) = \sum_{j=1}^{8} 20 - 3\sum_{j=1}^{8} j = 8(20) - 3\left[\frac{(8)(9)}{2}\right] = 52$

34. $\sum_{k=1}^{25}\left(\frac{3k + 1}{4}\right) = \frac{3}{4}\sum_{k=1}^{25} k + \sum_{k=1}^{25}\frac{1}{4} = \frac{3}{4}\left[\frac{(25)(26)}{2}\right] + 25\left(\frac{1}{4}\right) = 250$

36. $\sum_{n=20}^{80} n = \sum_{n=1}^{80} n - \sum_{n=1}^{19} n = \frac{(80)(81)}{2} - \frac{(19)(20)}{2} = 3050$

38. $a_1 = 123, d = 112 - 123 = -11$

$n = 8$

$a_8 = (-11)8 + 134 = 46$

$S_8 = \frac{8}{2}(123 + 46) = 676$

40. $a_1 = 2, r = 2$

$a_1 = 2$

$a_2 = 2(2) = 4$

$a_3 = 4(2) = 8$

$a_4 = 8(2) = 16$

$a_5 = 16(2) = 32$

42. $a_1 = 2, a_3 = 12$

$a_3 = a_1 r^2$

$12 = 2r^2$

$6 = r^2$

$\pm\sqrt{6} = r$

$a_1 = 2$

$a_2 = 2(\sqrt{6}) = 2\sqrt{6}$

$a_3 = 2\sqrt{6}(\sqrt{6}) = 12$

$a_4 = 12(\sqrt{6}) = 12\sqrt{6}$

$a_5 = 12\sqrt{6}(6) = 72$

or

$a_1 = 2$

$a_2 = 2(-\sqrt{6}) = -2\sqrt{6}$

$a_3 = -2\sqrt{6}(-\sqrt{6}) = 12$

$a_4 = 12(-\sqrt{6}) = -12\sqrt{6}$

$a_5 = -12\sqrt{6}(-\sqrt{6}) = 72$

44. $a_1 = 200, a_{k+1} = 0.1a_k$

$a_1 = 200$

$a_2 = 0.1(200) = 20$

$a_3 = 0.1(20) = 2$

$a_4 = 0.1(2) = 0.2$

$a_5 = 0.1(0.2) = 0.02$

$a_n = 200(0.1)^{n-1}$

46. $a_1 = 18, a_{k+1} = \frac{5}{3}a_k$

$a_1 = 18$

$a_2 = \frac{5}{3}(18) = 30$

$a_3 = \frac{5}{3}(30) = 50$

$a_4 = \frac{5}{3}(50) = \frac{250}{3}$

$a_5 = \frac{5}{3}\left(\frac{250}{3}\right) = \frac{1250}{9}$

$a_n = 18\left(\frac{5}{3}\right)^{n-1}$

48. $a_1 = 100, r = 1.05$

$a_n = 100(1.05)^{n-1}$

$$\sum_{n=1}^{20} 100(1.05)^{n-1} = 100\left[\frac{1-(1.05)^{20}}{1-1.05}\right] \approx 3306.60$$

50. $\displaystyle\sum_{i=1}^{5} 3^{i-1} = \frac{1-3^5}{1-3} = 121$

52. $\displaystyle\sum_{i=1}^{\infty}\left(\frac{1}{3}\right)^{i-1} = \frac{1}{1-(1/3)} = \frac{3}{2}$

54. $\displaystyle\sum_{k=1}^{\infty} 1.3\left(\frac{1}{10}\right)^{k-1} = \frac{1.3}{1-(1/10)} = \frac{13}{9}$

56. $\displaystyle\sum_{i=1}^{25} 100(1.06)^{i-1} = 100\left[\frac{1-(1.06)^{25}}{1-1.06}\right] \approx 5486.45$

58. $\displaystyle\sum_{i=1}^{40} 32{,}000(1.055)^{i-1} = 32{,}000\left[\frac{1-(1.055)^{40}}{1-1.055}\right]$

60. $$A = \sum_{i=1}^{120} 100\left(1+\frac{0.065}{12}\right)^i = 100\left(1+\frac{0.065}{12}\right)\left(\frac{1-\left(1+\frac{0.065}{12}\right)^{120}}{-\frac{0.0065}{12}}\right) \approx \$16{,}931.53$$

62. 1. When $n = 1, 1 = \frac{1}{4}(1+3) = 1$.

2. Assume that $1 + \frac{3}{2} + 2 + \frac{5}{2} + \cdots + \frac{1}{2}(k+1) = \frac{k}{4}(k+3)$. Then

$$\begin{aligned}1 + \frac{3}{2} + 2 + \frac{5}{2} + \cdots + \frac{1}{2}(k+1) + \frac{1}{2}(k+2) &= \frac{k}{4}(k+3) + \frac{1}{2}(k+2)\\ &= \frac{k(k+3)+2(k+2)}{4}\\ &= \frac{k^2+5k+4}{4}\\ &= \frac{(k+1)(k+4)}{4}\\ &= \frac{k+1}{4}[(k+1)+3].\end{aligned}$$

Thus, the formula holds for all positive intergers n.

64. 1. When $n = 1, a + 0 \cdot d = a = \frac{1}{2}[2a + (1-1)d] = a$.

2. Assume that $\displaystyle\sum_{k=0}^{i-1}(a+kd) = \frac{i}{2}[2a+(i-1)d]$. Then,

$$\begin{aligned}\sum_{k=0}^{i+1-1}(a+kd) &= \frac{i}{2}[2a+(i-1)d] + [a+id]\\ &= \frac{2ia + i(i-1)d + 2a + 2id}{2} = \frac{2a(i+1) + id(i+1)}{2} = \left(\frac{i+1}{2}\right)[2a+id].\end{aligned}$$

Thus, the formula holds for all positive integers n.

66. $_{10}C_7 = \dfrac{10!}{3!7!} = 120$

68. $_{12}P_3 = \dfrac{12!}{9!} = 1320$

70. $(a - 3b)^5 = a^5 - 5a^4(3b) + 10a^3(3b)^2 - 10a^2(3b)^3 + 5a(3b)^4 - (3b)^5$
$= a^5 - 15a^4b + 90a^3b^2 - 270a^2b^3 + 405ab^4 - 243b^5$

72. $(3x + y^2)^7 = (3x)^7 + 7(3x)^6y^2 + 21(3x)^5(y^2)^2 + 35(3x)^4(y^2)^3 + 35(3x)^3(y^2)^4 + 21(3x)(y^2)^5 + 7(3x)(y^2)^6 + (y^2)^7$
$= 2187x^7 + 5103x^6y^2 + 5103x^5y^4 + 2835x^4y^6 + 945x^3y^8 + 189x^2y^{10} + 21xy^{12} + y^{14}$

74. $(4 - 5i)^3 = 4^3 - 3(4)^2(5i) + 3(4)(5i)^2 - (5i)^3$
$= 64 - 240i - 300 + 125i$
$= -236 - 115i$

76. $2^3 = 8$

78.

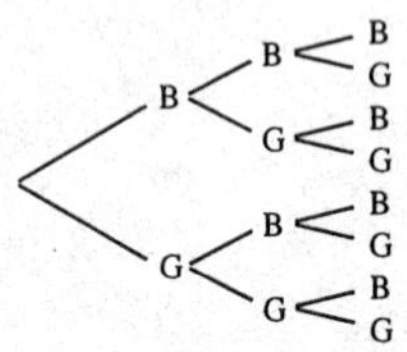

B — B — B
B — B — G
B — G — B
B — G — G
G — B — B
G — B — G
G — G — B
G — G — G

80. $P(E) = \dfrac{n(E)}{n(S)} = \dfrac{1}{5!} = \dfrac{1}{120}$

82. $\left(\dfrac{6}{6}\right)\left(\dfrac{5}{6}\right)\left(\dfrac{4}{6}\right)\left(\dfrac{3}{6}\right)\left(\dfrac{2}{6}\right)\left(\dfrac{1}{6}\right) = \dfrac{6!}{6^6} = \dfrac{720}{46,656} = \dfrac{5}{324}$

84. $(0.8)^3 = 0.512$

86. (a) $\frac{208}{500} = 0.416$

(b) $\frac{400}{500} = 0.8$

(c) $\frac{37}{500} = 0.074$

88. (a) $\bar{x} = 12, \sigma = 2.27$

(b) $\bar{x} = 30, \sigma = 1.08$

(c) $\bar{x} = 60, \sigma = 1.08$

(d) $\bar{x} = 4, \sigma = 2.27$

90. Mean: 13, Median: 12, $\sigma = 4.78$

92. Mean: 320.23, Median: 322.15, $\sigma = 51.70$

94.

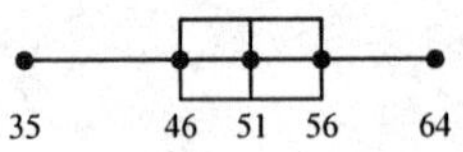

35 46 51 56 64

CHAPTER 10
Topics in Analytic Geometry

CHAPTER 10
Topics in Analytic Geometry

Section 10.1 Introduction to Conics: Parabolas

Solutions to Even-Numbered Exercises

2. $x^2 = 2y$
Vertex: $(0, 0)$
$p = \frac{1}{2} > 0$
Opens upward.
Matches graph (b).

4. $y^2 = 12x$
Vertex: $(0, 0)$
$p = 3 > 0$
Opens to the right.
Matches graph (f).

6. $(x + 3)^2 = -2(y - 1)$
Vertex: $(-3, 1)$
$p = -\frac{1}{2} < 0$
Opens downward.
Matches graph (c).

8. $y = 2x^2 \Rightarrow x^2 = 4\left(\frac{1}{8}\right)y$
Vertex: $(0, 0)$
Focus: $\left(0, \frac{1}{8}\right)$
Directrix: $y = -\frac{1}{8}$

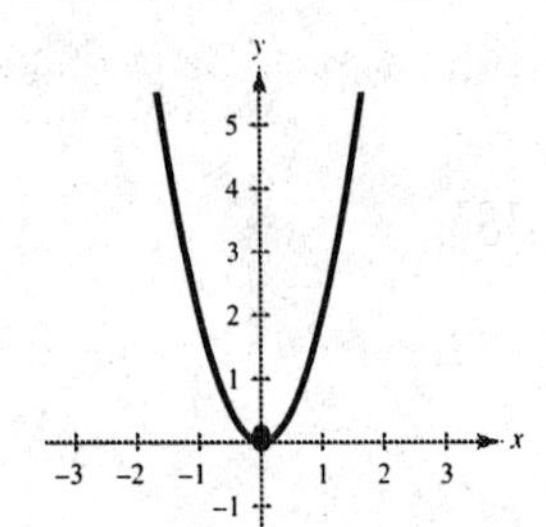

10. $y^2 = 3x \Rightarrow 4\left(\frac{3}{4}\right)x$
Vertex: $(0, 0)$
Focus: $\left(\frac{3}{4}, 0\right)$
Directrix: $x = -\frac{3}{4}$

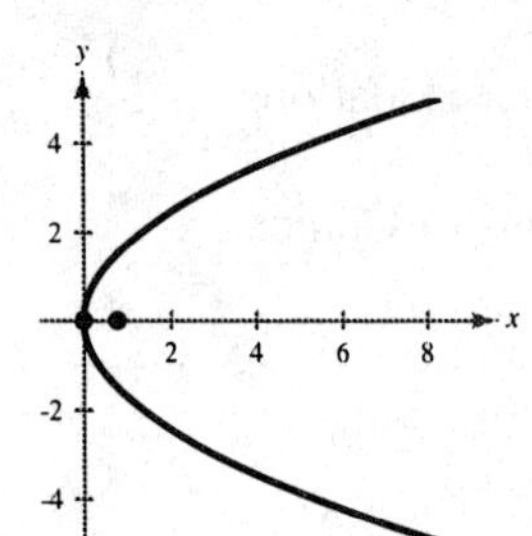

12. $x + y^2 = 0$
$y^2 = -x = 4\left(-\frac{1}{4}\right)x$
Vertex: $(0, 0)$
Focus: $\left(-\frac{1}{4}, 0\right)$
Directrix: $x = \frac{1}{4}$

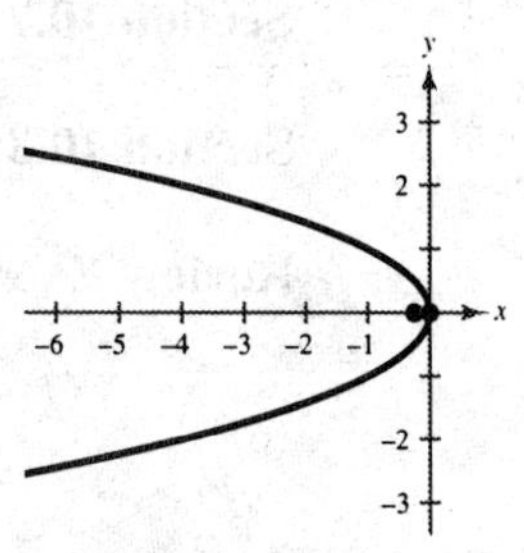

14. $(x + 3) + (y - 2)^2 = 0$
$(y - 2)^2 = 4\left(-\frac{1}{4}\right)(x + 3)$
Vertex: $(-3, 2)$
Focus: $\left(-3 + \left(-\frac{1}{4}\right), 2\right) \Rightarrow \left(-\frac{13}{4}, 2\right)$
Directrix: $x = -3 - \left(-\frac{1}{4}\right) = -\frac{11}{4}$

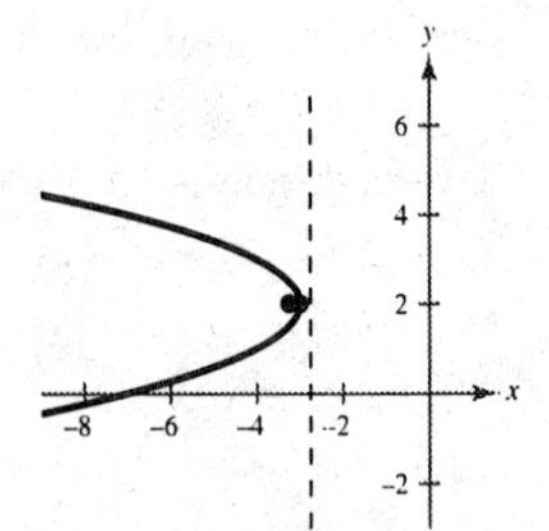

16. $\left(x + \frac{1}{2}\right)^2 = 4(y - 3) = 4(1)(y - 3)$
Vertex: $\left(-\frac{1}{2}, 3\right)$
Focus: $\left(-\frac{1}{2}, 3 + 1\right) \Rightarrow \left(-\frac{1}{2}, 4\right)$
Directrix: $y = 3 - 1 = 2$

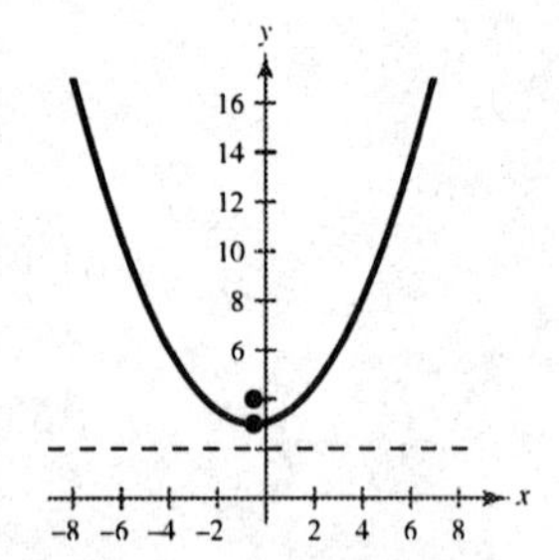

18. $4x - y^2 - 2y - 33 = 0$
$y^2 + 2y + 1 = 4x - 33 + 1$
$(y + 1)^2 = 4(1)(x - 8)$
Vertex: $(8, -1)$
Focus: $(9, -1)$
Directrix: $x = 7$

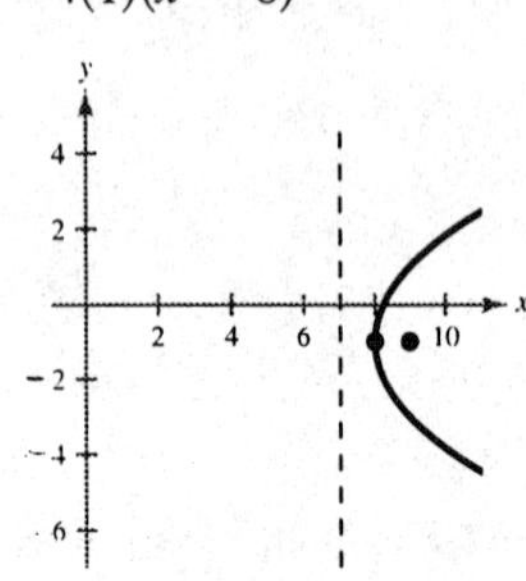

20. $y^2 - 4y - 4x = 0$
$y^2 - 4y + 4 = 4x + 4$
$(y - 2)^2 = 4(1)(x + 1)$
Vertex: $(-1, 2)$
Focus: $(0, 2)$
Directrix: $x = -2$

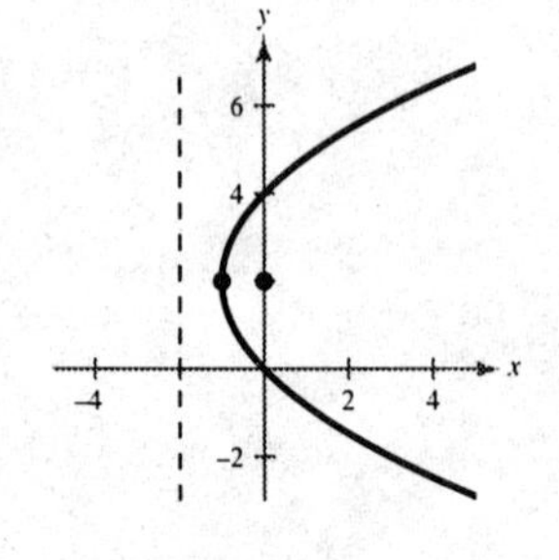

22. $x^2 - 2x + 8y + 9 = 0$

$x^2 - 2x + 1 = -8y - 9 + 1$

$(x - 1)^2 = -8(y + 1) = 4(-2)(y + 1)$

Vertex: $(1, -1)$
Focus: $(1, -3)$
Directrix: $y = 1$

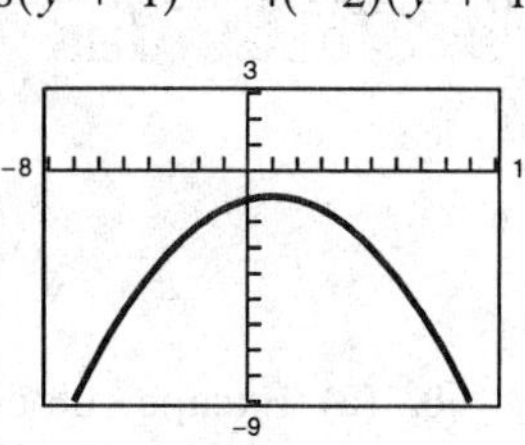

24. $y^2 - 4x - 4 = 0$

$y^2 = 4x + 4 = 4(1)(x + 1)$

Vertex: $(-1, 0)$
Focus: $(0, 0)$
Directrix: $x = -2$

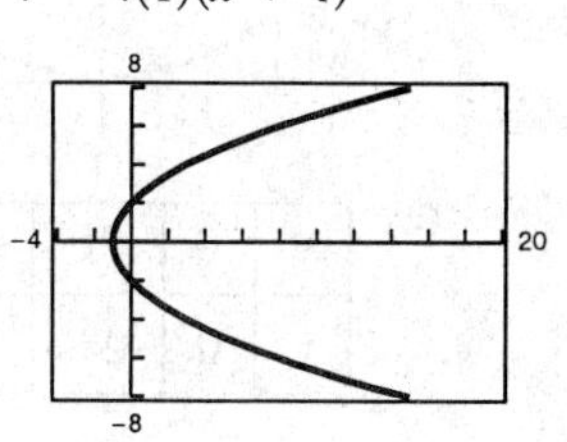

26. $x^2 + 12y = 0 \Rightarrow y_1 = -\frac{1}{12}x^2$

$x + y - 3 = 0 \Rightarrow y_2 = 3 - x$

Using the trace or intersect feature, the point of tangency is $(6, -3)$.

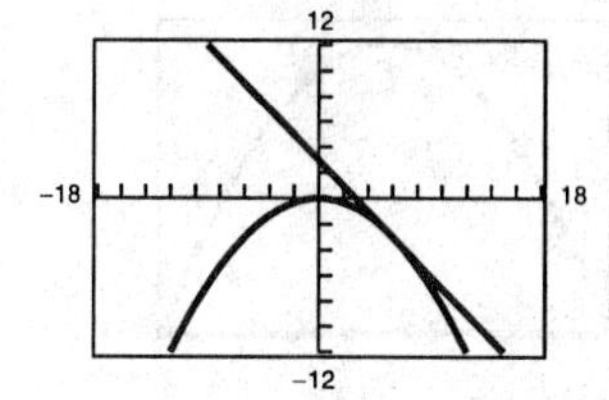

28. $y^2 = 9x,\ y \geq 0$

$y = \sqrt{9x} = 3\sqrt{x}$

30. No, it is not possible for a parabola to intersect its directrix. If the graph crossed the directrix there would exist points nearer the directrix than the focus.

32. Point: $(-2, 6)$

$x = ay^2$

$-2 = a(6)^2$

$-\frac{1}{18} = a$

$x = -\frac{1}{18}y^2$

34. Focus: $(2, 0) \Rightarrow p = 2$

$y^2 = 4px$

$y^2 = 8x$

36. Focus: $(0, -2) \Rightarrow p = -2$

$x^2 = 4py$

$x^2 = -8y$

38. Directrix: $x = 3 \Rightarrow p = -3$

$(y - k)^2 = 4p(x - h)$

$y^2 = 4(-3)x$

$y^2 = -12x$

40. Directrix: $x = -2 \Rightarrow p = 2$

$y^2 = 4px$

$y^2 = 8x$

42. Vertical axis
Passes through: $(-2, -2)$

$x^2 = 4py$

$(-2)^2 = 4p(-2)$

$4 = -8p$

$p = -\frac{1}{2}$

$x^2 = -2y$

44. Vertex: $(5, 3) \Rightarrow h - 5,\ k = 3$

Passes through: $(4.5, 4)$

$(y - k)^2 = 4p(x - h)$

$(y - 3)^2 = 4p(x - 5)$

$1 = 4p(4.5 - 5)$

$p = -\frac{1}{2}$

$(y - 3)^2 = -2(x - 5)$

46. Vertex: $(3, -3) \Rightarrow h = 3,\ k = -3$

Passes through: $(0, 0)$

$(x - h)^2 = 4p(y - k)$

$(x - 3)^3 = 4p(y + 3)$

$9 = 12p$

$p = \frac{3}{4}$

$(x - 3)^2 = 3(y + 3)$

48. Vertex: $(-1, 2) \Rightarrow h = -1, k = 2$
Focus: $(-1, 0) \Rightarrow p = -2$

$(x - h)^2 = 4p(y - k)$

$(x + 1)^2 = 4(-2)(y - 2)$

$(x + 1)^2 = -8(y - 2)$

50. Vertex: $(-2, 1) \Rightarrow h = -2, k = 1$
Directrix: $x = 1 \Rightarrow p = -3$

$(y - k)^2 = 4p(x - h)$

$(y - 1)^2 = 4(-3)(x - (-2))$

$(y - 1)^2 = -12(x + 2)$

52. Focus: $(0, 0)$
Directrix: $y = 4 \Rightarrow p = -2 \Rightarrow h = 0, k = 2$

$(x - h)^2 = 4p(y - k)$

$x^2 = 4(-2)(y - 2)$

$x^2 = -8(y - 2)$

54. $(y + 1)^2 = 2(x - 2)$

$y + 1 = \pm\sqrt{2(x - 2)}$

$y = -1 \pm \sqrt{2(x - 2)}$

Lower half of parabola: $y = -1 - \sqrt{2(x - 2)}$

56. 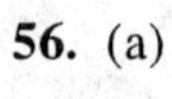 (a)

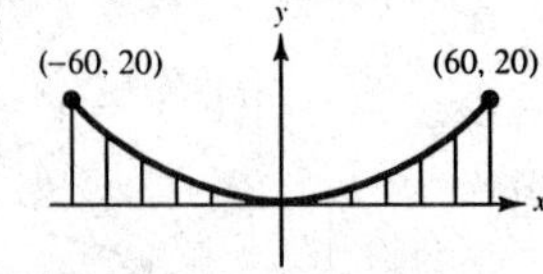

(c)

x	0	20	40	60
y	0	$2\frac{2}{9}$	$8\frac{8}{9}$	20

(b) $(x - 0)^2 = 4p(y - 0)$

$x^2 = 4py$

At (60, 20): $60^2 = 4p(20) \Rightarrow p = 45$

$x^2 = 4(45)y$

$y = \dfrac{x^2}{180}$

58. $R = 375x - \frac{3}{2}x^2$

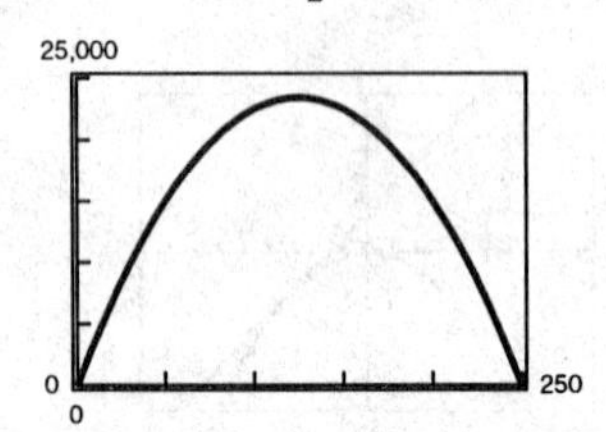

$x = 125$ maximizes R.

60. (a) Excape velocity: $17{,}500\sqrt{2}$

(b) $x^2 = 4p(y - 4100)$ and $p = -4100$.

$x^2 = -16{,}400(y - 4100)$

62. $y = -0.08x^2 + x + 4$

(a)

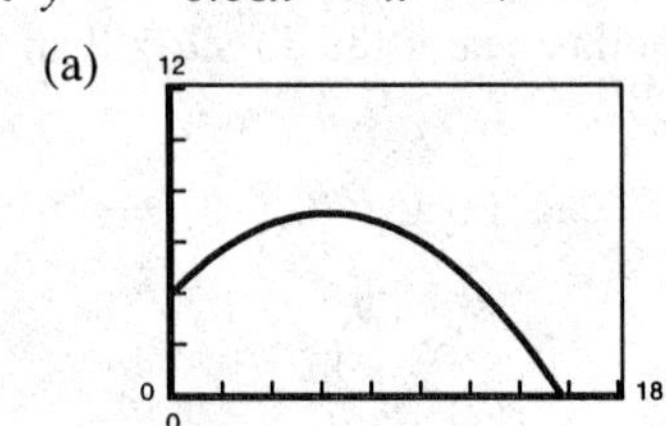

(b) The highest point is approximately (6.25, 7.125). The range is approximately 15.69 feet.

64. $A = \dfrac{4}{3}pb^{3/2}$

(a) For $p = 2$ and $b = 4$: $A = \dfrac{4}{3}(2)(4)^{3/2} = \dfrac{64}{3}$

(b) The parabola becomes narrower for $0 \le y \le b$.

66.

$$2y = x^2$$

$$4\left(\frac{1}{2}\right)y = x^2$$

$$p = \frac{1}{2}$$

Focus: $\left(0, \dfrac{1}{2}\right)$

$$d_1 = \frac{1}{2} - b$$

$$d_2 = \sqrt{(-3 - 0)^2 + \left(\frac{9}{2} - \frac{1}{2}\right)^2} = 5$$

$$\frac{1}{2} - b = 5$$

$$b = -\frac{9}{2}$$

$$m = \frac{-(9/2) - (9/2)}{0 + 3} = -3$$

Tangent line: $y = -3x - \dfrac{9}{2} \Rightarrow 6x + 2y + 9 = 0$

x-intercept: $\left(-\dfrac{3}{2}, 0\right)$

68.

$$y = -2x^2$$

$$-\frac{1}{2}y = x^2$$

$$4\left(-\frac{1}{8}\right)y = x^2$$

$$p = -\frac{1}{8}$$

Focus: $\left(0, -\dfrac{1}{8}\right)$

$$d_1 = \frac{1}{8} + b$$

$$d_2 = \sqrt{(3 - 0)^2 + \left(-18 - \left(\frac{1}{8}\right)\right)^2} = \frac{145}{8}$$

$$\frac{1}{8} + b = \frac{145}{8}$$

$$b = \frac{144}{8} = 18$$

$$m = \frac{-18 - 18}{3 - 0} = -12$$

Tangent line: $y = -12x + 18 \Rightarrow 12x + y - 18 = 0$

x-intercept: $\left(\dfrac{3}{2}, 0\right)$

70. $y = -\dfrac{16}{v^2}x^2 + s$

550 miles per hour $=$ 806.67 feet per second.

$$y = -\frac{16}{806/67^2}x^2 + 42{,}000$$

$$y = 0 \Rightarrow \frac{16}{806.67^2}x^2 = 42{,}000$$

$$x^2 \approx 1{,}708{,}115{,}666.67 \Rightarrow x \approx 41{,}329.37 \text{ feet}$$

72. $f(x) = 2x^3 - 3x^2 + 50x - 75$

$$\begin{array}{r|rrrr} \frac{3}{2} & 2 & -3 & 50 & -75 \\ & & 3 & 0 & 75 \\ \hline & 2 & 0 & 50 & 0 \end{array}$$

$$2x^3 - 3x^2 + 50x - 75 = \left(x - \tfrac{3}{2}\right)(2x^2 + 50)$$
$$= \left(x - \tfrac{3}{2}\right)(x - 5i)(x - 5i)$$

Zeros: $\frac{3}{2}, \pm 5i$

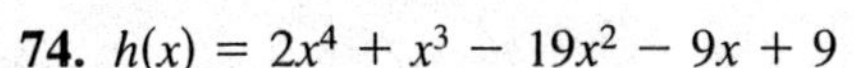

74. $h(x) = 2x^4 + x^3 - 19x^2 - 9x + 9$

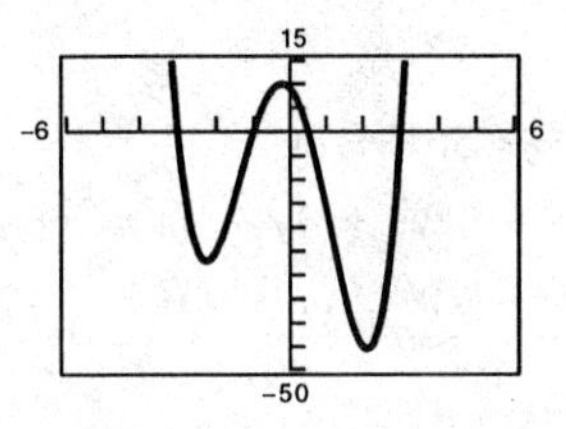

Zeros: $\pm 3, -1, \frac{1}{2}$

Section 10.2 Ellipses

Solutions to Even-Numbered Exercises

2. $\dfrac{x^2}{9} + \dfrac{y^2}{4} = 1$

Center: $(0, 0)$

$a = 3, b = 2$

Horizontal major axis

Matches graph (c).

4. $\dfrac{y^2}{4} + \dfrac{x^2}{4} = 1$

Center: $(0, 0)$

Circle of radius: 2

Matches graph (f).

6. $\dfrac{(x + 2)^2}{4} + \dfrac{(y + 2)^2}{16} = 1$

Center: $(-2, -2)$

$a = 4, b = 2$

Vertical major axis

Matches graph (e).

8. $\dfrac{x^2}{144} + \dfrac{y^2}{169} = 1$

$a^2 = 169, b^2 = 144, c^2 = 25$

Center: $(0, 0)$

Foci: $(0, \pm 5)$

Vertices: $(0, \pm 13)$

$e = \dfrac{5}{13}$

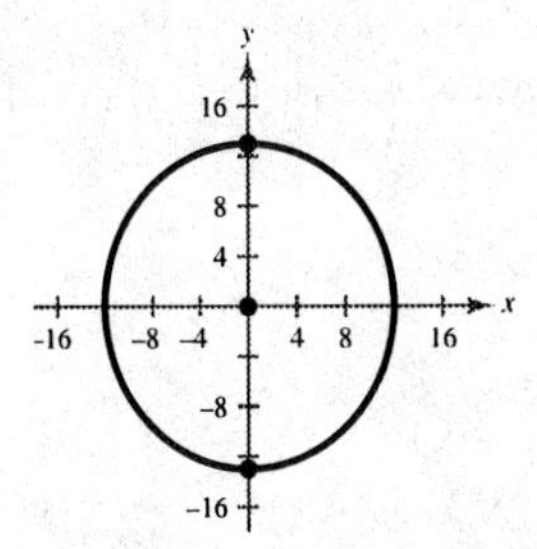

10. $\dfrac{x^2}{169} + \dfrac{y^2}{144} = 1$

$a^2 = 169, b^2 = 144, c^2 = 25$

Center: $(0, 0)$

Foci: $(\pm 5, 0)$

Vertices: $(\pm 13, 0)$

$e = \dfrac{c}{a} = \dfrac{5}{13}$

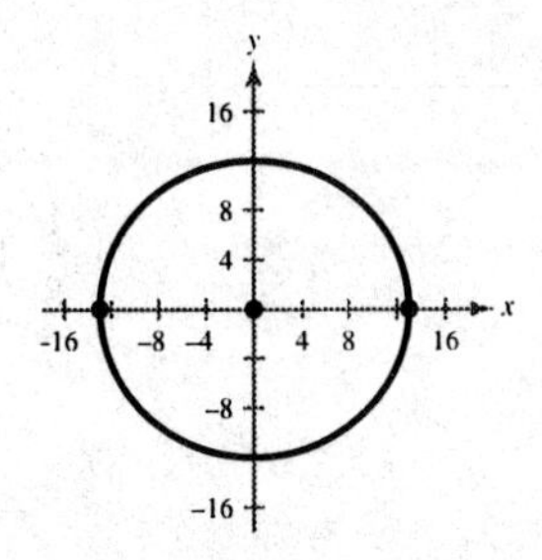

12. $\dfrac{x^2}{28} + \dfrac{y^2}{64} = 1$

$a^2 = 64, b^2 = 28, c^2 = 36$

Center: $(0, 0)$

Foci: $(0, \pm 6)$

Vertices: $(0, \pm 8)$

$e = \dfrac{c}{a} = \dfrac{6}{8} = \dfrac{3}{4}$

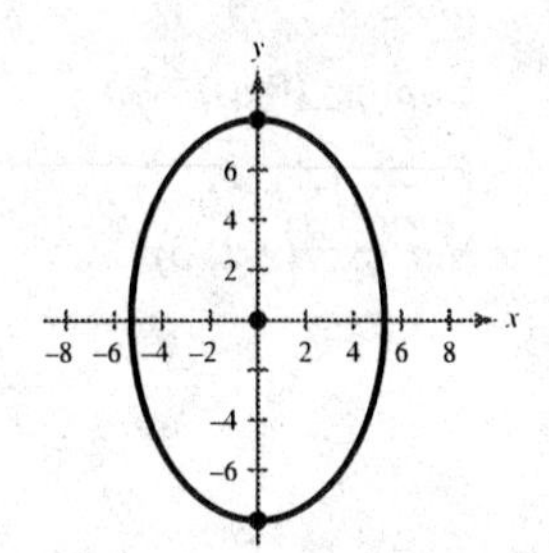

14. $\dfrac{(x+2)^2}{1} + \dfrac{(y+4)^2}{1/4} = 1$

$a^2 = 1, b^2 = \dfrac{1}{4}, c^2 = \dfrac{3}{4}$

Center: $(-2, -4)$

Foci: $\left(-2 \pm \dfrac{\sqrt{3}}{2}, -4\right)$

Vertices: $(-1, -4), (-3, -4)$

$e = \dfrac{\sqrt{3}}{2}$

16. $9x^2 + 4y^2 - 36x + 8y + 31 = 0$

$9(x^2 - 4x + 4) + 4(y^2 + 2y + 1) = -31 + 36 + 4$

$\dfrac{(x-2)^2}{1} + \dfrac{(y+1)^2}{9/4} = 1$

$a^2 = \dfrac{9}{4}, b^2 = 1, c^2 = \dfrac{5}{4}$

Center: $(2, -1)$

Foci: $\left(2, -1 \pm \dfrac{\sqrt{5}}{2}\right)$

Vertices: $\left(2, \dfrac{1}{2}\right), \left(2, -\dfrac{5}{2}\right)$

$e = \dfrac{\sqrt{5}}{3}$

18. $9x^2 + 25y^2 - 36x - 50y + 61 = 0$

$9(x^2 - 4x + 4) + 25(y^2 - 2y + 1) = -61 + 36 + 25$

$9(x-2)^2 + 25(y-1)^2 = 0$

Degenerate ellipse with center $(2, 1)$ as the only point

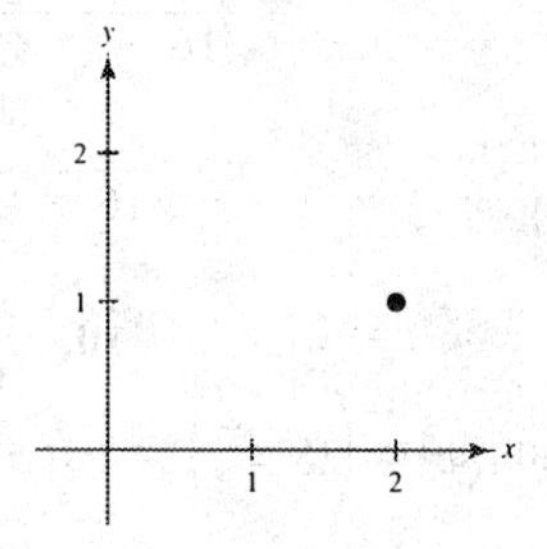

20. $x^2 + 4y^2 = 4$

$\dfrac{x^2}{4} + \dfrac{y^2}{1} = 1$

$a^2 = 4, b^2 = 1, c^2 = 3$

Center: $(0, 0)$

Foci: $(\pm\sqrt{3}, 0)$

Vertex: $(\pm 2, 0)$

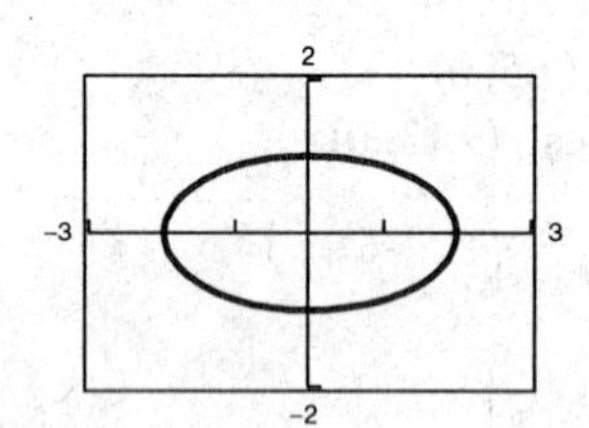

22. $$36x^2 + 9y^2 + 48x - 36y + 43 = 0$$

$$36\left(x^2 + \frac{4}{3}x + \frac{4}{9}\right) + 9(y^2 - 4y + 4) = -43 + 16 + 36$$

$$\frac{[x + (2/3)]^2}{1/4} + \frac{(y - 2)^2}{1} = 1$$

$a^2 = 1, b^2 = \frac{1}{4}, c^2 = \frac{3}{4}$

Center: $\left(-\frac{2}{3}, 2\right)$

Foci: $\left(-\frac{2}{3}, 2 \pm \frac{\sqrt{3}}{2}\right)$

Vertices: $\left(-\frac{2}{3}, 3\right), \left(-\frac{2}{3}, 1\right)$

24. $$\frac{(x + 1)^2}{16} + \frac{(y - 2)^2}{25} = 1$$

$$\frac{y - 2}{25} = 1 - \frac{(x + 1)^2}{16}$$

$$\frac{y - 2}{5} = \pm\sqrt{1 - \frac{(x + 1)^2}{16}}$$

$$y - 2 = \pm 5\sqrt{\frac{16}{16} - \frac{(x + 1)^2}{16}}$$

$$y - 2 = \pm\frac{5}{4}\sqrt{16 - (x + 1)^2}$$

$$y = 2 \pm \frac{5}{4}\sqrt{16 - (x + 1)^2}$$

Bottom half of ellipse: $y = 2 - \frac{5}{4}\sqrt{16 - (x + 1)^2}$

26. Vertices: $(0, \pm 8) \Rightarrow a = 8$

Foci: $(0, \pm 4) \Rightarrow c = 4$

$b^2 = a^2 - c^2 = 64 - 16 = 48$

Center: $(0, 0) = (h, k)$

$$\frac{(y - k)^2}{a^2} + \frac{(x - h)^2}{b^2} = 1$$

$$\frac{y^2}{64} + \frac{x^2}{48} = 1$$

28. Vertices: $(\pm 2, 0) \Rightarrow a = 2$

Endpoints of minor axis: $\left(0, \pm\frac{3}{2}\right) \Rightarrow b = \frac{3}{2}$

$$\frac{x^2}{a^2} + \frac{y^2}{b^2} = 1$$

$$\frac{x^2}{2^2} + \frac{y^2}{(3/2)^2} = 1$$

$$\frac{x^2}{4} + \frac{4y^2}{9} = 1$$

30. Foci: $(\pm 2, 0) \Rightarrow c = 2$

Major axis length: $8 \Rightarrow a = 4$

$b^2 = a^2 - c^2 = 16 - 4 = 12$

$$\frac{x^2}{a^2} + \frac{y^2}{b^2} = 1$$

$$\frac{x^2}{16} + \frac{y^2}{12} = 1$$

32. Major axis vertical

Passes through: $(0, 4)$ and $(2, 0)$

$a = 4, b = 2$

$$\frac{x^2}{b^2} + \frac{y^2}{a^2} = 1$$

$$\frac{x^2}{4} + \frac{y^2}{16} = 1$$

34. Vertices: $(4, \pm 4) \Rightarrow a = 4$

Center: $(4, 0) \Rightarrow h = 4, k = 0$

Endpoints of minor axis: $(1, 0), (7, 0) \Rightarrow b = 3$

$$\frac{(x - h)^2}{b^2} + \frac{(y - k)^2}{a^2} = 1$$

$$\frac{(x - 4)^2}{9} + \frac{y^2}{16} = 1$$

36. Vertices: $(0, -1), (4, -1) \Rightarrow a = 2$
Center: $(2, -1) \Rightarrow h = 2, k = -1$
Endpoints of minor axis: $(2, 0), (2, -2) \Rightarrow b = 1$

$$\frac{(x-h)^2}{a^2} + \frac{(y-k)^2}{b^2} = 1$$

$$\frac{(x-2)^2}{4} + \frac{(y+1)^2}{1} = 1$$

38. Foci: $(0, 0), (4, 0) \Rightarrow c = 2, h = 2, k = 0$
Major axis length: $8 \Rightarrow a = 4$

$b^2 = a^2 - c^2 = 16 - 4 = 12$

$$\frac{(x-h)^2}{a^2} + \frac{(y-k)^2}{b^2} = 1$$

$$\frac{(x-2)^2}{16} + \frac{y^2}{12} = 1$$

40. Center: $(2, -1) \Rightarrow h = 2, k = -1$

Vertex: $\left(2, \frac{1}{2}\right) \Rightarrow a = \frac{3}{2}$

Minor axis length: $2 \Rightarrow b = 1$

$$\frac{(x-h)}{b^2} + \frac{(y-k)^2}{a^2} = 1$$

$$\frac{(x-2)^2}{1} + \frac{(y+1)^2}{(3/2)^2} = 1$$

$$(x-2)^2 + \frac{4(y+1)^2}{9} = 1$$

42. Center: $(3, 2) = (h, k)$

$a = 3c$

Foci: $(1, 2), (5, 2) \Rightarrow c = 2, a = 6$

$b^2 = a^2 - c^2 = 36 - 4 = 32$

$$\frac{(x-h)^2}{a^2} + \frac{(y-k)^2}{b^2} = 1$$

$$\frac{(x-3)^2}{36} + \frac{(y-2)^2}{32} = 1$$

44. Vertices: $(5, 0), (5, 12) \Rightarrow a = 6$
Endpoints of the minor axis:
$(0, 6), (10, 6) \Rightarrow b = 5$
Center: $(5, 6) \Rightarrow h = 5, k = 6$

$$\frac{(x-h)^2}{b^2} + \frac{(y-k)^2}{a^2} = 1$$

$$\frac{(x-5)^2}{25} + \frac{(y-6)^2}{36} = 1$$

46. Vertices: $(\pm 3, 0) \Rightarrow a = 3$
Half of minor axis length: $2 \Rightarrow b = 2$
$c^2 = a^2 - b^2 = 9 - 4 = 5 \Rightarrow c = \sqrt{5}$
Place the tacks $\sqrt{5}$ feet from the center: $(\pm\sqrt{5}, 0)$
Length of string: $2a = 2(3) = 6$ feet

48. Area of ellipse = 2 (area of circle)

$$\pi ab = 2\pi r^2$$

$$\pi a(10) = 2\pi(10)^2$$

$$\pi a(10) = 200$$

$$a = 20$$

Length of major axis: $2a = 2(20) = 40$ units

50. Vertices: $(0, \pm 8) \Rightarrow a = 8, h = 0, k = 0$

Eccentricity: $e = \frac{1}{2} = \frac{c}{a}$

$$\frac{1}{2} = \frac{c}{8}$$

$$c = 4$$

$b^2 = c^2 - a^2 = 64 - 16 = 48$

$$\frac{x^2}{b^2} + \frac{y^2}{a^2} = 1$$

$$\frac{x^2}{48} + \frac{y^2}{64} = 1$$

52. Center: $(0, 0) \Rightarrow h = 0, k = 0$

$2a = 0.34 + 4.08 = 4.42$

$a = 2.21$

$c = 2.21 - 0.34 = 1.87$

$b^2 = a^2 - c^2 = 4.8841 - 3.4969 = 1.3872$

$$\frac{x^2}{a^2} + \frac{y^2}{b^2} = 1$$

$$\frac{x^2}{4.88} + \frac{y^2}{1.39} = 1$$

54. For $\frac{x^2}{a^2} + \frac{y^2}{b^2} = 1$, we have $c^2 = a^2 - b^2$.

When $x = c$:

$$\frac{c^2}{a^2} + \frac{y^2}{b^2} = 1 \Rightarrow y^2 = b^2\left(1 - \frac{a^2 - b^2}{a^2}\right) \Rightarrow y^2 = \frac{b^4}{a^2} \Rightarrow 2y = \frac{2b^2}{a}.$$

56. $\dfrac{x^2}{9} + \dfrac{y^2}{16} = 1$

$a = 4, b = 3, c = \sqrt{7}$

Points on the ellipse: $(\pm 3, 0), (0, \pm 4)$

Length of latus recta: $\dfrac{2b^2}{a} = \dfrac{2(3)^2}{4} = \dfrac{9}{2}$

Additional points: $\left(\pm\dfrac{9}{4}, -\sqrt{7}\right), \left(\pm\dfrac{9}{4}, \sqrt{7}\right)$

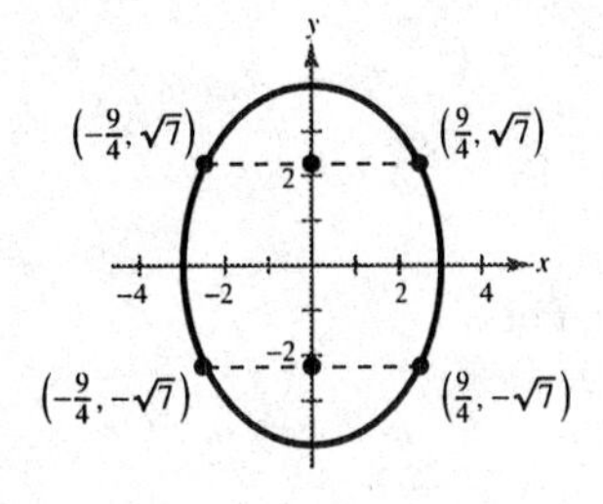

58. $5x^2 + 3y^2 = 15$

$\dfrac{x^3}{3} + \dfrac{y^2}{5} = 1$

$a = \sqrt{5}, b = \sqrt{3}, c = \sqrt{2}$

Points on the ellipse: $(\pm\sqrt{3}, 0), (0, \pm\sqrt{5})$

Length of latus recta: $\dfrac{2b^2}{a} = \dfrac{2 \cdot 3}{\sqrt{5}} = \dfrac{6\sqrt{5}}{5}$

Additional points: $\left(\pm\dfrac{3\sqrt{5}}{5}, -\sqrt{2}\right), \left(\pm\dfrac{3\sqrt{5}}{5}, \sqrt{2}\right)$

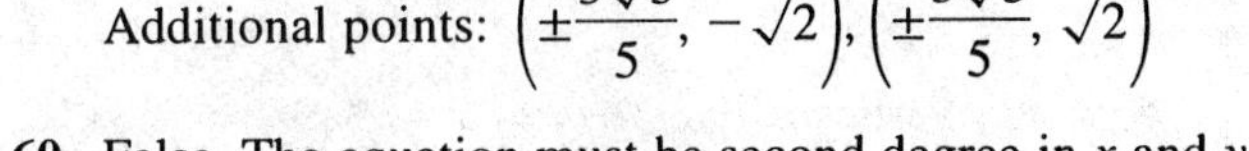

60. False. The equation must be second degree in x and y.

Section 10.3 Hyperbolas

Solutions to Even-Numbered Exercises

2. $\dfrac{y^2}{16} - \dfrac{x^2}{4} = 1$

Center: $(0, 0)$

$a = 4, b = 2$

Vertical transverse axis

Matches graph (c).

4. $\dfrac{y^2}{16} - \dfrac{x^2}{9} = 1$

Center: $(0, 0)$

$a = 4, b = 3$

Vertical transverse axis

Matches graph (b).

6. $\dfrac{x^2}{9} - \dfrac{y^2}{16} = 1$

$a = 3, b = 4,$

$c = \sqrt{4^2 + 3^2} = 5$

Center: $(0, 0)$

Vertices: $(\pm 3, 0)$

Foci: $(\pm 5, 0)$

Asymptotes: $y = \pm\dfrac{4}{3}x$

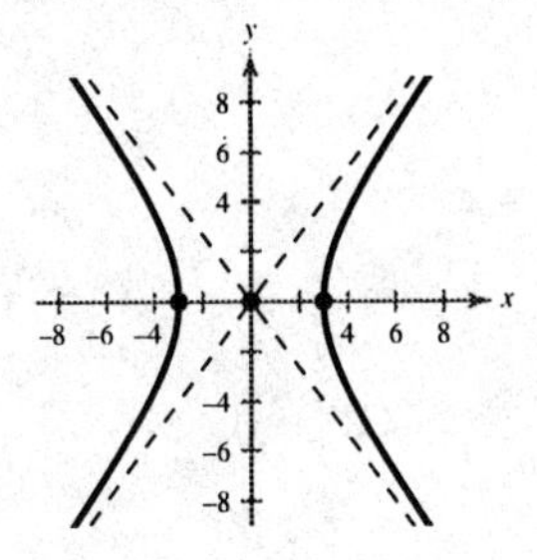

8. $\dfrac{y^2}{9} - \dfrac{x^2}{1} = 1$

$a = 3, b = 1,$

$c = \sqrt{3^2 + 1^2} = \sqrt{10}$

Center: $(0, 0)$

Vertices: $(0, \pm 3)$

Foci: $(0, \pm\sqrt{10})$

Asymptotes: $y = \pm 3x$

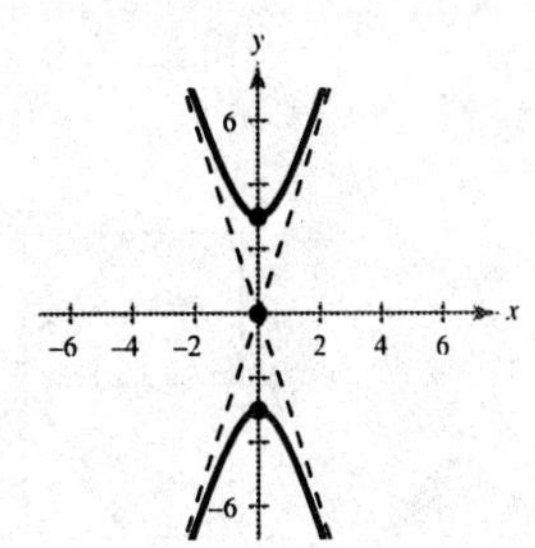

10. $\dfrac{x^2}{36} - \dfrac{y^2}{4} = 1$

$a = 6, b = 2,$

$c = \sqrt{36 + 4} = 2\sqrt{10}$

Center: $(0, 0)$

Vertices: $(\pm 6, 0)$

Foci: $(\pm 2\sqrt{10}, 0)$

Asymptotes: $y = \pm\dfrac{1}{3}x$

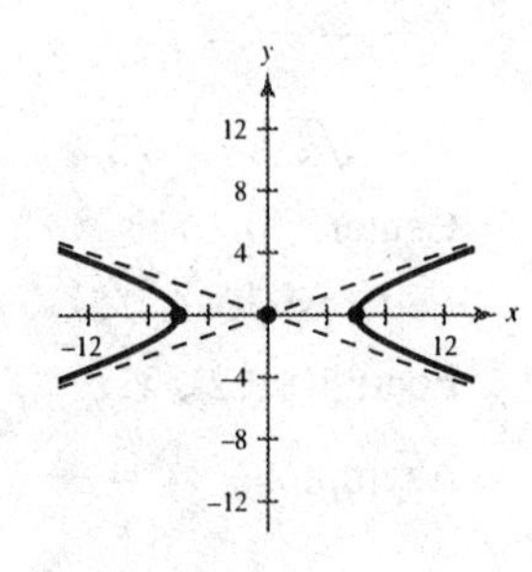

12. $\dfrac{(x+1)^2}{144} - \dfrac{(y-4)^2}{25} = 1$

$a = 12, b = 5, c = 13$

Center: $(-1, 4)$

Vertices: $(-13, 4), (11, 4)$

Foci: $(-14, 4), (12, 4)$

Asymptotes: $y = 4 \pm \dfrac{5}{12}(x+1)$

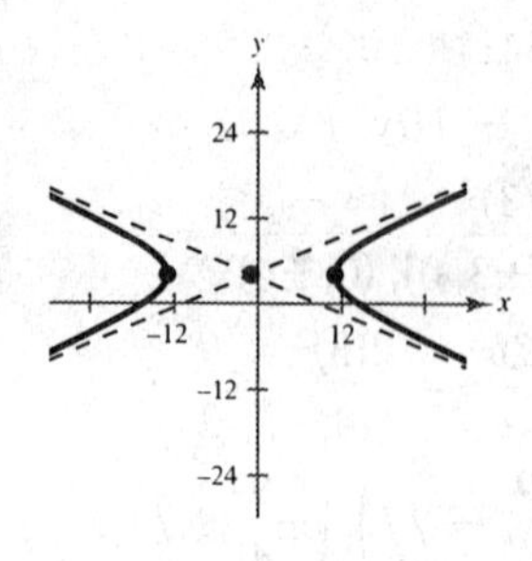

14. $\dfrac{(y-1)^2}{1/4} - \dfrac{(x+3)^2}{1/9} = 1$

$a = \dfrac{1}{2}, b = \dfrac{1}{3}, c = \dfrac{\sqrt{13}}{6}$

Center: $(-3, 1)$

Vertices: $\left(-3, \dfrac{1}{2}\right), \left(-3, \dfrac{3}{2}\right)$

Foci: $\left(-3, 1 \pm \dfrac{1}{6}\sqrt{13}\right)$

Asymptotes: $y = 1 \pm \dfrac{3}{2}(x+3)$

16. $x^2 - 9y^2 + 36y - 72 = 0$

$x^2 - 9(y^2 - 4y + 4) = 72 - 36$

$x^2 - 9(y-2)^2 = 36$

$\dfrac{x^2}{36} - \dfrac{(y-2)^2}{4} = 1$

$a = 6, b = 2, c = \sqrt{36+4} = 2\sqrt{10}$

Center: $(0, 2)$

Vertices: $(\pm 6, 2)$

Foci: $(\pm 2\sqrt{10}, 2)$

Asymptotes: $y = 2 \pm \dfrac{1}{3}x$

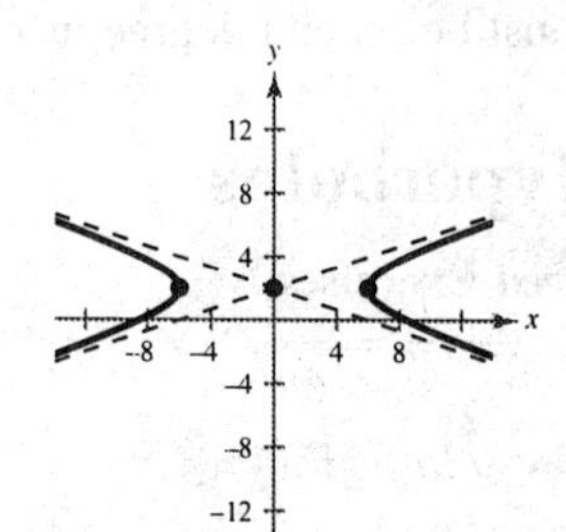

18. $16y^2 - x^2 + 2x + 64y + 63 = 0$

$16(y^2 + 4y + 4) - (x^2 - 2x + 1) = -63 + 64 - 1$

$16(y+2)^2 - (x-1) = 0$

$y + 2 = \pm\frac{1}{4}(x-1)$

Degenerate hyperbola is two intersecting lines.

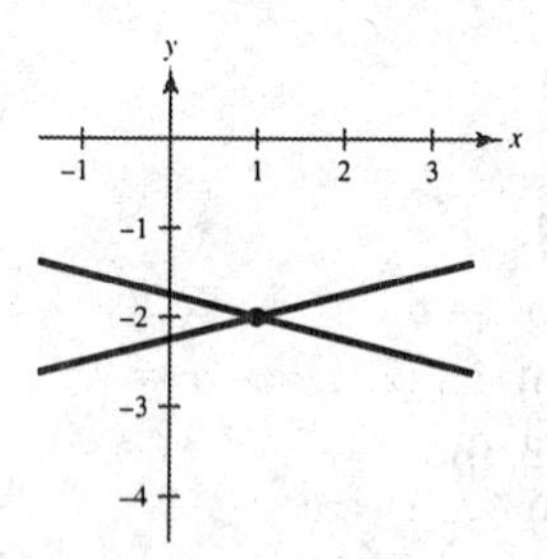

20. $3y^2 - 5x^2 = 15$

$\dfrac{y^2}{5} - \dfrac{x^2}{3} = 1$

$a = \sqrt{5}, b = \sqrt{3}, c = \sqrt{5+3} = 2\sqrt{2}$

Center: $(0, 0)$

Vertices: $(0, \pm\sqrt{5})$

Foci: $(0, \pm 2\sqrt{2})$

Asymptotes: $y = \pm\dfrac{\sqrt{5}}{\sqrt{3}}x$

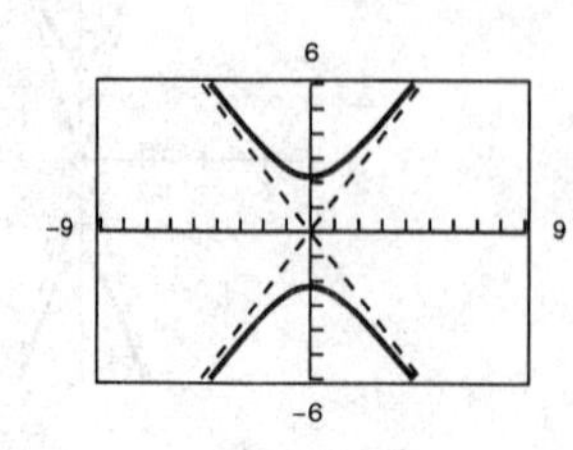

22. $$9x^2 - y^2 + 54x + 10y + 55 = 0$$
$$9(x^2 + 6x + 9) - (y^2 - 10y + 25) = -55 + 81 - 25$$
$$\frac{(x+3)^2}{1/9} - \frac{(y-5)^2}{1} = 1$$
$$a = \frac{1}{3},\ b = 1,\ c = \frac{\sqrt{10}}{3}$$

Center: $(-3, 5)$

Vertices: $\left(-3 \pm \frac{1}{3}, 5\right)$

Foci: $\left(-3 \pm \frac{\sqrt{10}}{3}, 5\right)$

Asymptotes: $y = 5 \pm 3(x + 3)$

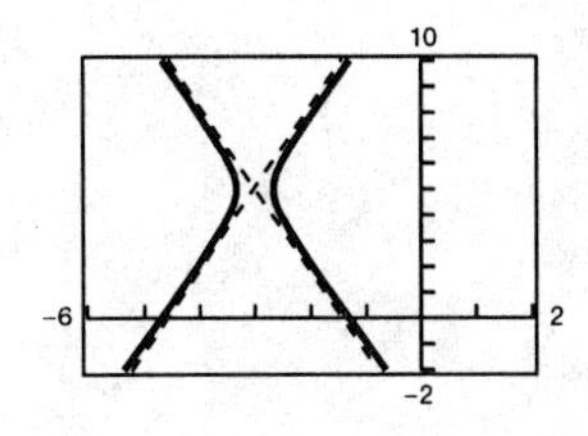

24. Vertices: $(\pm 3, 0) \Rightarrow a = 3$

Foci: $(\pm 5, 0) \Rightarrow c = 5$

$b^2 = c^2 - a^2 = 25 - 9 = 16 \Rightarrow b = 4$

$$\frac{x^2}{a^2} - \frac{y^2}{b^2} = 1$$
$$\frac{x^2}{9} - \frac{y^2}{16} = 1$$

26. Vertices: $(0, \pm 3) \Rightarrow a = 3$

Asymptotes: $y = \pm 3x \Rightarrow \frac{a}{b} = 3, b = 1$

Center: $(0, 0) = (h, k)$

$$\frac{(y-k)^2}{a^2} - \frac{(x-h)^2}{b^2} = 1$$
$$\frac{y^2}{9} - x^2 = 1$$

28. Foci: $(\pm 10, 0) \Rightarrow c = 10$

Asymptotes: $y = \pm\frac{3}{4}x \Rightarrow \frac{b}{a} = \frac{3m}{4m}$

$$c^2 = a^2 + b^2 \Rightarrow 100 = (3m)^2 + (4m)^2$$
$$100 = 25m^2$$
$$2 = m$$

$a = 3(2) = 6$

$b = 4(2) = 8$

$$\frac{x^2}{a^2} - \frac{y^2}{b^2} = 1$$
$$\frac{x^2}{36} - \frac{y^2}{64} = 1$$

30. Vertices: $(2, 3), (2, -3) \Rightarrow a = 3$

Center: $(2, 0)$

Foci: $(2, 5), (2, -5) \Rightarrow c = 5$

$b^2 = c^2 - a^2 = 25 - 9 = 16$

$$\frac{(y-k)^2}{a^2} - \frac{(x-h)^2}{b^2} = 1$$
$$\frac{y^2}{9} - \frac{(x-2)^2}{16} = 1$$

32. Vertices: $(-2, 1), (2, 1) \Rightarrow a = 2$

Center: $(0, 1)$

Foci: $(-3, 1), (3, 1) \Rightarrow c = 3$

$b^2 = c^2 - a^2 = 9 - 4 = 5$

$$\frac{(x-h)^2}{a^2} - \frac{(y-k)^2}{b^2} = 1$$
$$\frac{x^2}{4} - \frac{(y-1)^2}{5} = 1$$

34. Vertices: $(-2, 1), (2, 1) \Rightarrow a = 2$

Solution point: $(4, 3)$

Center: $(0, 1) = (h, k)$

$$\frac{(x-h)^2}{a^2} - \frac{(y-k)^2}{b^2} = 1$$
$$\frac{x^2}{4} - \frac{(y-1)^2}{b^2} = 1 \Rightarrow b^2 = \frac{4(y-1)^2}{x^2 - 4}$$
$$= \frac{4(2^2)}{16 - 4} = \frac{16}{12}$$
$$= \frac{4}{3}$$
$$\frac{x^2}{4} - \frac{(y-1)^2}{4/3} = 1$$

36. Vertices: $(3, 0), (3, 4) \Rightarrow a = 2$

Asymptotes: $y = \frac{2}{3}x,\ y = 4 - \frac{2}{3}x$

$\frac{a}{b} = \frac{2}{3} \Rightarrow b = 3$

Center: $(3, 2) = (h, k)$

$$\frac{(y-k)^2}{a^2} - \frac{(x-h)^2}{b^2} = 1$$

$$\frac{(y-2)^2}{4} - \frac{(x-3)^2}{9} = 1$$

38. $\frac{(x-3)^2}{4} - \frac{(y-1)^2}{9} = 1$

$$\frac{(x-3)^2}{4} - 1 = \frac{(y-1)^2}{9}$$

$$\frac{(x-3)^2}{4} - \frac{4}{4} = \frac{(y-1)^2}{9}$$

$$\pm\frac{1}{2}\sqrt{(x-3)^2 - 4} = \frac{y-1}{3}$$

$$\pm\frac{3}{2}\sqrt{(x-3)^2 - 4} = y - 1$$

$$1 \pm \frac{3}{2}\sqrt{(x-3)^2 - 4} = y$$

$1 + \left(\frac{3}{2}\right)\sqrt{(x-3)^2 - 4} = y$ is the top half of the graph of the hyperbola.

40. Since $\overline{AB} = 100$ feet and the sound takes one second longer to reach B than A, the explosion must occur on the vertical line through A and B below A.

Foci: $(\pm 4400, 0) \Rightarrow c = 4400$
Center: $(0, 0) = (h, k)$

$\frac{\overline{CD}}{1100} - \frac{\overline{AE}}{1100} = 5 \Rightarrow 2a = 5500,\ a = \frac{5500}{2} = 2750$

$b^2 = c^2 - a^2 = (4400)^2 - (2750)^2 = 11{,}797{,}500$

$$\frac{x^2}{(2750)^2} - \frac{y^2}{11{,}797{,}500} = 1$$

$$y^2 = 11{,}797{,}500\left(\frac{x^2}{(2750)^2} - 1\right)$$

$$y^2 = 11{,}797{,}500\left(\frac{(4400)^2}{(2750)^2} - 1\right) = 18{,}404{,}100$$

$$y = -4290$$

The explosion occurs at $(4400, -4290)$.

42. Let (x, y) be such that the difference of the distances from $(c, 0)$ and $(-c, 0)$ is $2a$ (again only deriving one of the forms).

$$2a = \left|\sqrt{(x+c)^2 + y^2} - \sqrt{(x-c) + y^2}\right|$$

$$2a + \sqrt{(x-c)^2 + y^2} = \sqrt{(x+c)^2 + y^2}$$

$$4a^2 + 4a\sqrt{(x-c)^2 + y^2} + (x-c)^2 + y^2 = (x+c)^2 + y^2$$

$$4a\sqrt{(x-c)^2 + y^2} = 4cx - 4a^2$$

$$a\sqrt{(x-c)^2 + y^2} = cx - a^2$$

$$a^2(x^2 - 2cx + c^2 + y^2) = c^2x^2 - 2a^2cx + a^4$$

$$a^2(c^2 - a^2) = (c^2 - a^2)x^2 - a^2y^2$$

Let $b^2 = c^2 - a^2$. Then $a^2b^2 = b^2x^2 - a^2y^2 \Rightarrow 1 = \frac{x^2}{a^2} - \frac{y^2}{b^2}$.

44. $x^2 + 4y^2 - 6x + 16y + 21 = 0$

$A = 1, C = 4$

$AC = 1(4) = 4 > 0 \Rightarrow$ Ellipse

46. $y^2 - 4y - 4x = 0$

$A = 0, C = 1$

$AC = 0(1) = 0 \Rightarrow$ Parabola

48. $4y^2 - 2x^2 - 4y - 8x - 15 = 0$

$A = -2, C = 4$

$AC = (-2)(4) = -8 < 0 \Rightarrow$ Hyperbola

50. $4x^2 + 4y^2 - 16y + 15 = 0$

$A = 4, C = 4$

$A = C \Rightarrow$ Circle

Section 10.4 Rotation and Systems of Quadratic Equations

Solutions to Even-Numbered Exercises

2. $\theta = 45°$; Point: $(3, 3)$

$x' = x\cos\theta - y\sin\theta = 3\cos 45° - 3\sin 45° = 0$

$y' = x\sin\theta + y\cos\theta = 3\sin 45° + 3\cos 45° = 3\sqrt{2}$

Thus, $(x', y') = \left(0, 3\sqrt{2}\right)$.

4. $\theta = 60°$; Point: $(3, 1)$

$x' = x\cos\theta - y\sin\theta = 3\cos 60° - 1\sin 60° = \frac{3}{2} - \frac{\sqrt{3}}{2}$

$y' = x\sin\theta + y\cos\theta = 3\sin 60° + 1\cos 60° = \frac{3\sqrt{3}}{2} + \frac{1}{2}$

Thus, $(x', y') = \left(\frac{1}{2}(3 - \sqrt{3}), \frac{1}{2}(3\sqrt{3} + 1)\right)$.

6. $xy - 4 = 0$

$A = 0, B = 1, C = 0$

$\cot 2\theta = \frac{A - C}{B} = 0 \Rightarrow 2\theta = \frac{\pi}{2} \Rightarrow \theta = \frac{\pi}{4}$

$$x = x'\cos\frac{\pi}{4} - y'\sin\frac{\pi}{4} = x'\left(\frac{\sqrt{2}}{2}\right) - y'\left(\frac{\sqrt{2}}{2}\right) = \frac{x' - y'}{\sqrt{2}}$$

$$y = x'\sin\frac{\pi}{4} + y'\cos\frac{\pi}{4} = x'\left(\frac{\sqrt{2}}{2}\right) + y'\left(\frac{\sqrt{2}}{2}\right) = \frac{x' + y'}{\sqrt{2}}$$

$$xy - 4 = 0$$

$$\left(\frac{x' - y'}{\sqrt{2}}\right)\left(\frac{x' + y'}{\sqrt{2}}\right) - 4 = 0$$

$$\frac{(x')^2 - (y')^2}{2} = 4$$

$$\frac{(x')^2}{8} - \frac{(y')^2}{8} = 1$$

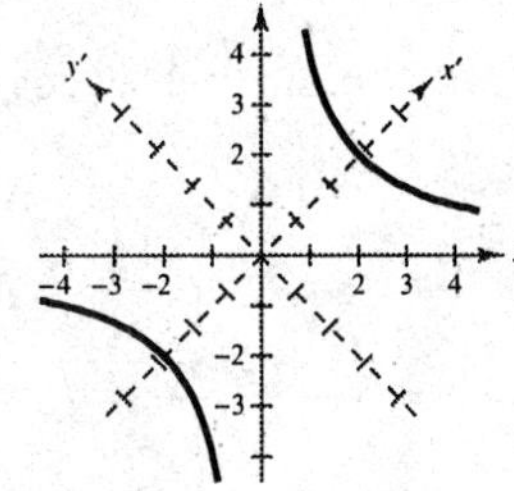

8. $xy + x - 2y + 3 = 0$

$A = 0, B = 1, C = 0$

$$\cot 2\theta = \frac{A - C}{B} = 0 \Rightarrow 2\theta = \frac{\pi}{2} \Rightarrow \theta = \frac{\pi}{4}$$

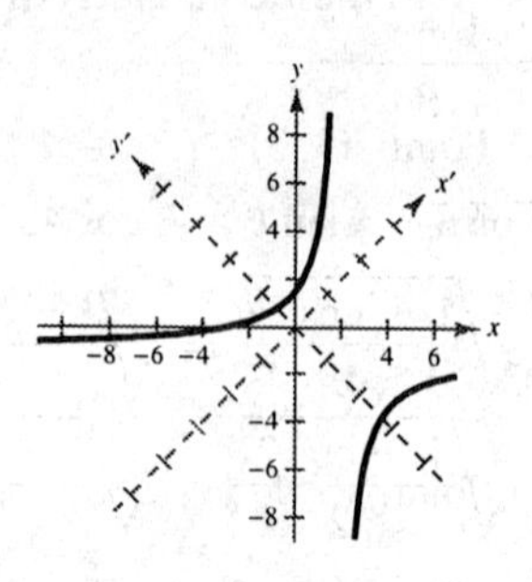

$$x = x'\cos\frac{\pi}{4} - y'\sin\frac{\pi}{4} = x'\left(\frac{\sqrt{2}}{2}\right) - y'\left(\frac{\sqrt{2}}{2}\right) = \frac{x' - y'}{\sqrt{2}}$$

$$y = x'\sin\frac{\pi}{4} + y'\cos\frac{\pi}{4} = x'\left(\frac{\sqrt{2}}{2}\right) + y'\left(\frac{\sqrt{2}}{2}\right) = \frac{x' + y'}{\sqrt{2}}$$

$$xy + x - 2y + 3 = 0$$

$$\left(\frac{x' - y'}{\sqrt{2}}\right)\left(\frac{x' + y'}{\sqrt{2}}\right) + \left(\frac{x' - y'}{\sqrt{2}}\right) - 2\left(\frac{x' + y'}{\sqrt{2}}\right) + 3 = 0$$

$$\frac{(x')^2}{2} - \frac{(y')^2}{2} + \frac{x'}{\sqrt{2}} - \frac{y'}{\sqrt{2}} - \frac{2x'}{\sqrt{2}} - \frac{2y'}{\sqrt{2}} + 3 = 0$$

$$\left[(x')^2 - \sqrt{2}x' + \left(\frac{\sqrt{2}}{2}\right)^2\right] - \left[(y')^2 + 3\sqrt{2}y' + \left(\frac{3\sqrt{2}}{2}\right)^2\right] = -6 + \left(\frac{\sqrt{2}}{2}\right)^2 - \left(\frac{3\sqrt{2}}{2}\right)^2$$

$$\left(x' - \frac{\sqrt{2}}{2}\right)^2 - \left(y' + \frac{3\sqrt{2}}{2}\right)^2 = -10$$

$$\frac{\left(y' + \frac{3\sqrt{2}}{2}\right)^2}{10} - \frac{\left(x' - \frac{\sqrt{2}}{2}\right)^2}{10} = 1$$

10. $13x^2 + 6\sqrt{3}xy + 7y^2 - 16 = 0$

$A = 13, B = 6\sqrt{3}, C = 7$

$$\cot 2\theta = \frac{A - C}{B} = \frac{1}{\sqrt{3}} \Rightarrow 2\theta = \frac{\pi}{3} \Rightarrow \theta = \frac{\pi}{6}$$

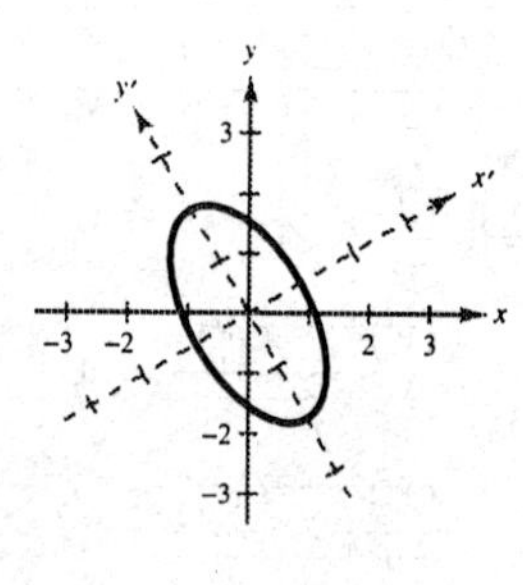

$$x = x'\cos\frac{\pi}{6} - y'\sin\frac{\pi}{6} = x'\left(\frac{\sqrt{3}}{2}\right) - y'\left(\frac{1}{2}\right) = \frac{\sqrt{3}x' - y'}{2}$$

$$y = x'\sin\frac{\pi}{6} + y'\cos\frac{\pi}{6} = x'\left(\frac{1}{2}\right) + y'\left(\frac{\sqrt{3}}{2}\right) = \frac{x' + \sqrt{3}y'}{2}$$

$$13x^2 + 6\sqrt{3}xy + 7y^2 - 16 = 0$$

$$13\left(\frac{\sqrt{3}x' - y'}{2}\right)^2 + 6\sqrt{3}\left(\frac{\sqrt{3}x' - y'}{2}\right)\left(\frac{x' + \sqrt{3}y'}{2}\right) + 7\left(\frac{x' + \sqrt{3}y'}{2}\right)^2 - 16 = 0$$

$$\frac{39(x')^2}{4} - \frac{13\sqrt{3}x'y'}{2} + \frac{13(y')^2}{4} + \frac{18(x')^2}{4} + \frac{18\sqrt{3}x'y'}{4} - \frac{6\sqrt{3}x'y'}{4}$$

$$-\frac{18(y')^2}{4} + \frac{7(x')^2}{4} + \frac{7\sqrt{3}x'y'}{2} + \frac{21(y')^2}{4} - 16 = 0$$

$$16(x')^2 + 4(y')^2 = 16$$

$$\frac{(x')^2}{1} + \frac{(y')^2}{4} = 1$$

12. $2x^2 - 3xy - 2y^2 + 10 = 0$

$A = 2, B = -3, C = -2$

$\cot 2\theta = \dfrac{A - C}{B} = -\dfrac{4}{3} \Rightarrow \theta \approx 71.57°$

$\cos 2\theta = -\dfrac{4}{5}$

$\sin\theta = \sqrt{\dfrac{1 - \cos 2\theta}{2}} = \sqrt{\dfrac{1 - (-4/5)}{2}} = \dfrac{3}{\sqrt{10}}$

$\cos\theta = \sqrt{\dfrac{1 + \cos 2\theta}{2}} = \sqrt{\dfrac{1 + (-4/5)}{2}} = \dfrac{1}{\sqrt{10}}$

$x = x'\cos\theta - y'\sin\theta = x'\left(\dfrac{1}{\sqrt{10}}\right) - y'\left(\dfrac{3}{\sqrt{10}}\right) = \dfrac{x' - 3y'}{\sqrt{10}}$

$y = x'\sin\theta + y'\cos\theta = x'\left(\dfrac{3}{\sqrt{10}}\right) + y'\left(\dfrac{1}{\sqrt{10}}\right) = \dfrac{3x' + y'}{\sqrt{10}}$

$$2x^2 - 3xy - 2y^2 + 10 = 0$$

$$2\left(\frac{x' - 3y'}{\sqrt{10}}\right)^2 - 3\left(\frac{x' - 3y'}{\sqrt{10}}\right)\left(\frac{3x' + y'}{\sqrt{10}}\right) - 2\left(\frac{3x' + y'}{\sqrt{10}}\right)^2 + 10 = 0$$

$$\frac{(x')^2}{5} - \frac{6x'y'}{5} + \frac{9(y')^2}{5} - \frac{9(x')^2}{10} + \frac{24x'y'}{10} + \frac{9(y')^2}{10} - \frac{9(x')^2}{5} - \frac{6x'y'}{5} - \frac{(y')^2}{5} + 10 = 0$$

$$-\frac{5}{2}(x')^2 + \frac{5}{2}(y')^2 = -10$$

$$\frac{(x')^2}{4} - \frac{(y')^2}{4} = 1$$

14. $16x^2 - 24xy + 9y^2 - 60x - 80y + 100 = 0$

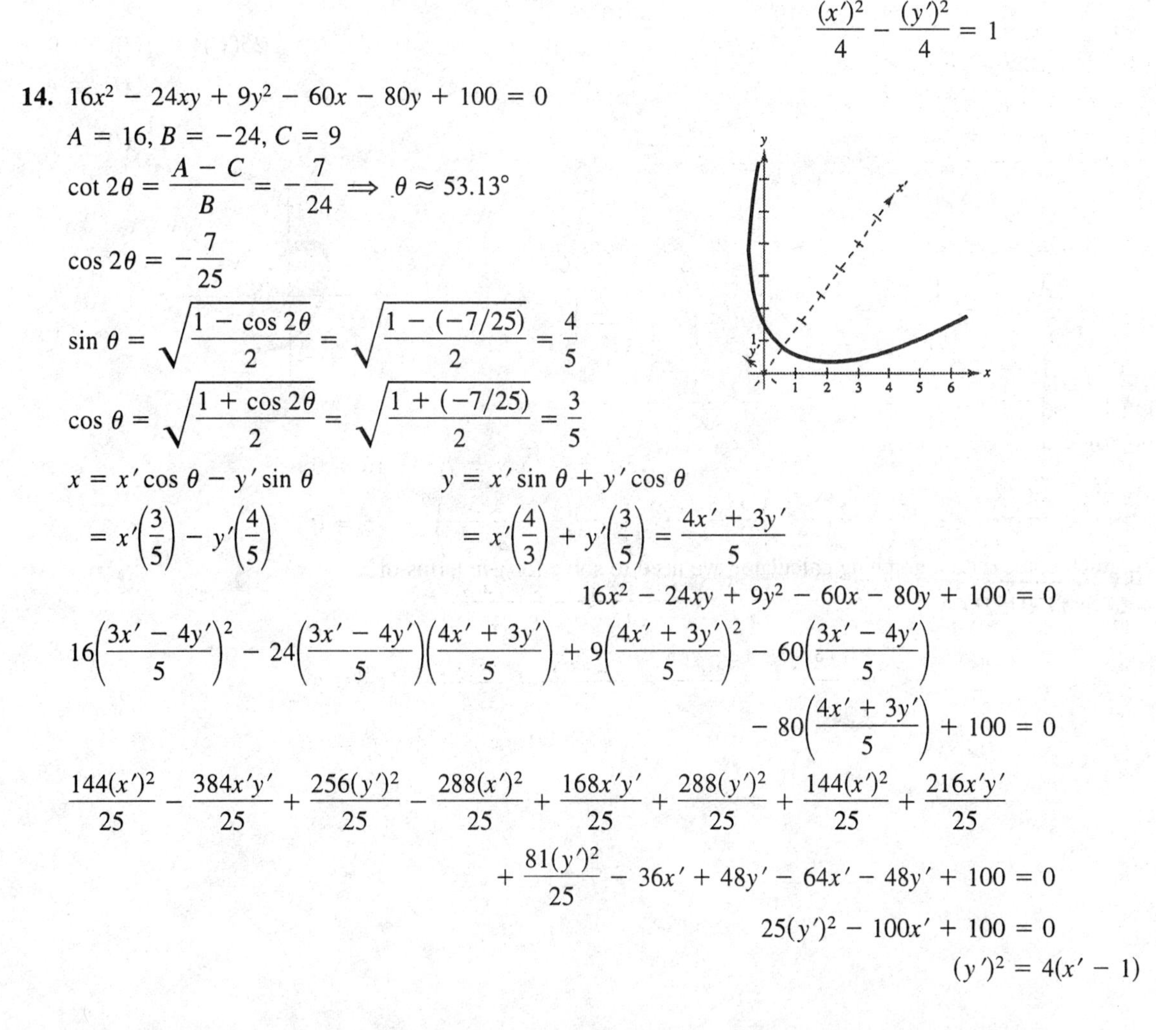

$A = 16, B = -24, C = 9$

$\cot 2\theta = \dfrac{A - C}{B} = -\dfrac{7}{24} \Rightarrow \theta \approx 53.13°$

$\cos 2\theta = -\dfrac{7}{25}$

$\sin\theta = \sqrt{\dfrac{1 - \cos 2\theta}{2}} = \sqrt{\dfrac{1 - (-7/25)}{2}} = \dfrac{4}{5}$

$\cos\theta = \sqrt{\dfrac{1 + \cos 2\theta}{2}} = \sqrt{\dfrac{1 + (-7/25)}{2}} = \dfrac{3}{5}$

$x = x'\cos\theta - y'\sin\theta = x'\left(\dfrac{3}{5}\right) - y'\left(\dfrac{4}{5}\right)$

$y = x'\sin\theta + y'\cos\theta = x'\left(\dfrac{4}{3}\right) + y'\left(\dfrac{3}{5}\right) = \dfrac{4x' + 3y'}{5}$

$$16x^2 - 24xy + 9y^2 - 60x - 80y + 100 = 0$$

$$16\left(\frac{3x' - 4y'}{5}\right)^2 - 24\left(\frac{3x' - 4y'}{5}\right)\left(\frac{4x' + 3y'}{5}\right) + 9\left(\frac{4x' + 3y'}{5}\right)^2 - 60\left(\frac{3x' - 4y'}{5}\right) - 80\left(\frac{4x' + 3y'}{5}\right) + 100 = 0$$

$$\frac{144(x')^2}{25} - \frac{384x'y'}{25} + \frac{256(y')^2}{25} - \frac{288(x')^2}{25} + \frac{168x'y'}{25} + \frac{288(y')^2}{25} + \frac{144(x')^2}{25} + \frac{216x'y'}{25} + \frac{81(y')^2}{25} - 36x' + 48y' - 64x' - 48y' + 100 = 0$$

$$25(y')^2 - 100x' + 100 = 0$$

$$(y')^2 = 4(x' - 1)$$

16. $9x^2 + 24xy + 16y^2 + 80x - 60y = 0$

$A = 9, B = 24, C = 16$

$\cot 2\theta = \dfrac{A - C}{B} = -\dfrac{7}{24} \Rightarrow \theta \approx 53.13°$

$\cos 2\theta = -\dfrac{7}{25}$

$\sin \theta = \sqrt{\dfrac{1 - \cos 2\theta}{2}} = \sqrt{\dfrac{1 - (-7/25)}{2}} = \dfrac{4}{5}$

$\cos \theta = \sqrt{\dfrac{1 + \cos 2\theta}{2}} = \sqrt{\dfrac{1 + (-7/25)}{2}} = \dfrac{3}{5}$

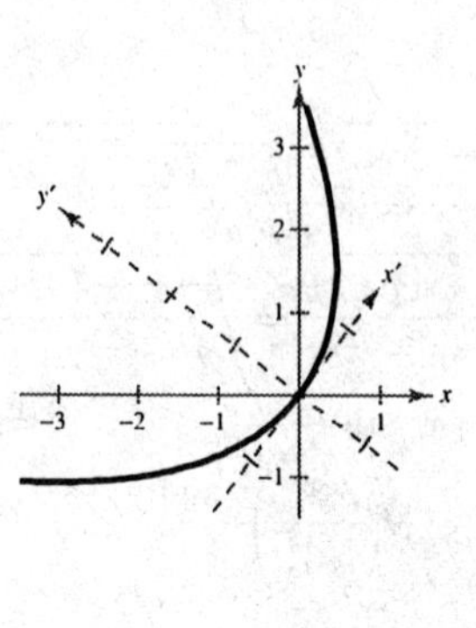

$x = x' \cos \theta - y' \sin \theta$

$= x'\left(\dfrac{3}{5}\right) - y'\left(\dfrac{4}{5}\right)$

$= \dfrac{3x' - 4y'}{5}$

$y = x' \sin \theta + y' \cos \theta$

$= x'\left(\dfrac{4}{5}\right) + y'\left(\dfrac{3}{5}\right)$

$= \dfrac{4x' + 3y}{5}$

$$9x^2 + 24xy + 16y^2 + 80x - 60y = 0$$

$$9\left(\frac{3x' - 4y'}{5}\right)^2 + 24\left(\frac{3xy - 4y'}{5}\right)\left(\frac{4x' + 3y'}{5}\right) + 16\left(\frac{4x' + 3y'}{5}\right)^2 + 80\left(\frac{3x' - 4y'}{5}\right) - 60\left(\frac{4x' + 3y'}{5}\right) = 0$$

$$\frac{81(x')^2}{25} - \frac{216x'y'}{25} + \frac{144(y')^2}{25} + \frac{288(x')^2}{25} - \frac{168x'y'}{25} - \frac{288(y')^2}{25} + \frac{256(x')^2}{25} + \frac{384x'y'}{25}$$

$$+ \frac{144(y')^2}{25} + 48x' - 64x' - 48x' - 36x' = 0$$

$$25(x')^2 - 100y' = 0$$

$$(x')^2 = 4y'$$

$$\frac{1}{4}(x')^2 = y'$$

18. $x^2 - 4xy + 2y^2 = 6$

$A = 1, B = -4, C = 2$

$\cot 2\theta = \dfrac{A - C}{B} = \dfrac{1 - 2}{-4} = \dfrac{1}{4}$

$\dfrac{1}{\tan 2\theta} = \dfrac{1}{4}$

$\tan 2\theta = 4$

$2\theta \approx 75.96$

$\theta \approx 37.98°$

To graph conic with a graphing calculator, we need to solve for y in terms of x.

—CONTINUED—

18. —CONTINUED—

$x^2 - 4xy + 2y^2 = 6$

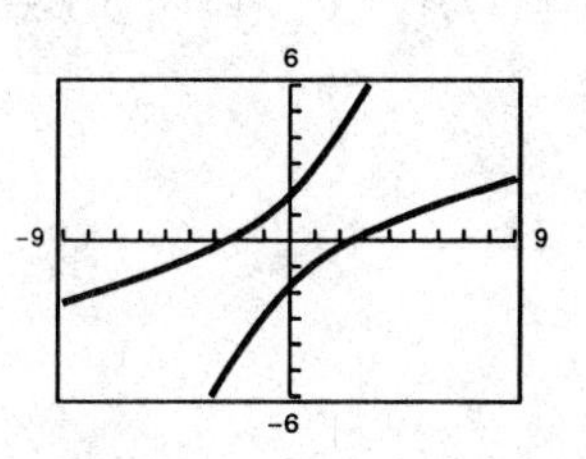

$$y^2 - 2xy + x^2 = 3 - \frac{x^2}{2} + x^2$$

$$(y - x)^2 = 3 + \frac{x^2}{2}$$

$$y - x = \pm\sqrt{3 + \frac{x^2}{2}}$$

$$y = x \pm \sqrt{3 + \frac{x^2}{2}}$$

Enter $y_1 = x + \sqrt{3 + \frac{x^2}{2}}$ and $y_2 = x - \sqrt{3 + \frac{x^2}{2}}$.

20. $40x^2 + 36xy + 25y^2 = 52$

$A = 40, B = 36, C = 25$

$$\cot 2\theta = \frac{A - C}{B} = \frac{40 - 25}{36} = \frac{5}{12}$$

$$\frac{1}{\tan 2\theta} = \frac{5}{12}$$

$$\tan 2\theta = \frac{12}{5}$$

$$2\theta \approx 67.38°$$

$$\theta \approx 33.69°$$

Solve for y in terms of x by completing the square:

$$25y^2 + 36xy = 52 - 40x^2$$

$$y^2 + \frac{36}{25}xy = \frac{52}{25} - \frac{40}{25}x^2$$

$$y^2 + \frac{36}{25}xy + \frac{324}{625}x^2 = \frac{52}{25} - \frac{40}{25}x^2 + \frac{324}{625}x^2$$

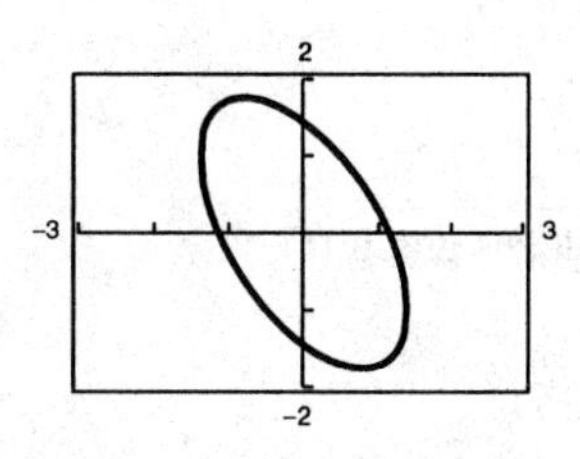

$$\left(y + \frac{18}{25}x\right)^2 = \frac{1300 - 676x^2}{625}$$

$$y + \frac{18}{25}x = \pm\sqrt{\frac{1300 - 676x^2}{625}}$$

$$y = \frac{-18x \pm \sqrt{1300 - 676x^2}}{25}$$

Enter $y_1 = \dfrac{-18x + \sqrt{1300 - 676x^2}}{25}$ and $y_2 = \dfrac{-18x - \sqrt{1300 - 676x^2}}{25}$.

22. $4x^2 - 12xy + 9y^2 + (4\sqrt{13} - 12)x - (6\sqrt{13} + 8)y = 91$

$A = 4, B = -12, C = 9$

$\cot 2\theta = \dfrac{A - C}{B} = \dfrac{4 - 9}{-12} = \dfrac{5}{12}$

$\dfrac{1}{\tan 2\theta} = \dfrac{5}{12}$

$\tan 2\theta = \dfrac{12}{5}$

$2\theta \approx 67.38°$

$\theta \approx 33.69°$

Solve for y in terms of x with the quadratic formula:

$$4x^2 - 12xy + 9y^2 + (4\sqrt{13} - 12)x - (6\sqrt{13} + 8)y = 91$$

$$9y^2 - (12x + 6\sqrt{13} + 8)y + (4x^2 + 4\sqrt{13}x - 12x - 91) = 0$$

$$a = 9, b = -(12x + 6\sqrt{13} + 8), c = 4x^2 + 4\sqrt{13}x - 12x - 91$$

$$y = \frac{-b \pm \sqrt{b^2 - 4ac}}{2a}$$

$$y = \frac{(12x + 6\sqrt{13} + 8) \pm \sqrt{(12x + 6\sqrt{13} + 8)^2 - 4(9)(4x^2 + 4\sqrt{13}x - 12x - 91)}}{18}$$

$$= \frac{(12x + 6\sqrt{13} + 8) \pm \sqrt{624x + 3808 + 96\sqrt{13}}}{18}$$

Enter $y_1 = \dfrac{12x + 6\sqrt{13} + 8 + \sqrt{624x + 3808 + 96\sqrt{13}}}{18}$

and $y_2 = \dfrac{12x + 6\sqrt{13} + 8 - \sqrt{624x + 3808 + 96\sqrt{13}}}{18}$.

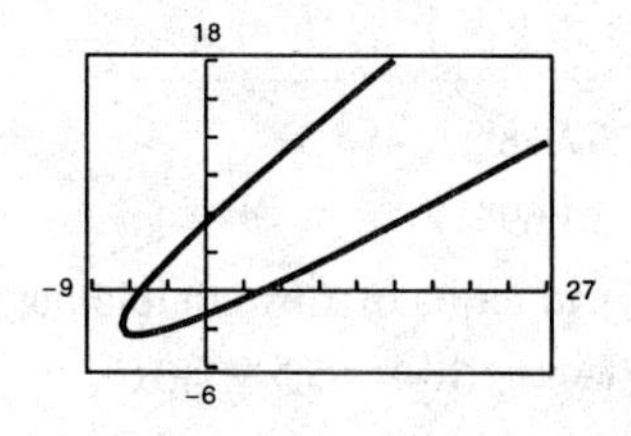

24. $x^2 + 2xy + y^2 = 0$

$(x + y)^2 = 0$

$x + y = 0$

$y = -x$

The graph is a line. Matches graph (f).

26. $x^2 - xy + 3y^2 - 5 = 0$

$A = 1, B = -1, C = 3$

$B^2 - 4AC = (-1)^2 - 4(1)(3) = -11$

The graph is an ellipse.

$\cot 2\theta = \dfrac{A - C}{B} = \dfrac{1 - 3}{-1} = 2 \Rightarrow \theta \approx 13.28°$

Matches graph (a).

28. $x^2 - 4xy + 4y^2 + 10x - 30 = 0$

$A = 1, B = -4, C = 4$

$B^2 - 4AC = (-4)^2 - 4(1)(4) = 0$

The graph is a parabola.

$\cot 2\theta = \dfrac{A - C}{B} = \dfrac{1 - 4}{-4} = \dfrac{3}{4} \Rightarrow \theta \approx 12.66°$

Matches graph (c).

30. $x^2 - 4xy - 2y^2 - 6 = 0$

$A = 1, B = -4, C = -2$

$B^2 - 4AC = (-4)^2 - 4(1)(-2) = 24 > 0$

Hyperbola

32. $2x^2 + 4xy + 5y^2 + 3x - 4y - 20 = 0$

$A = 2, B = 4, C = 5$

$B^2 - 4AC = 4^2 - 4(2)(5) = 16 - 40 = -24 < 0$

Ellipse

34. $36x^2 - 60xy + 25y^2 + 9y = 0$

$A = 36, B = -60, C = 25$

$B^2 - 4AC = (-60)^2 - 4(36)(25)$

$= 3600 - 3600 = 0$

Parabola

36. $x^2 + xy + 4y^2 + x + y - 4 = 0$

$A = 1, B = 1, C = 4$

$B^2 - 4AC = 1^2 - 4(1)(4) = -15 < 0$

Ellipse

38.
$$x^2 + y^2 - 2x + 6y + 10 = 0$$
$$(x^2 - 2x + 1) + (y^2 + 6y + 9) = -10 + 1 + 9$$
$$(x - 1)^2 + (y + 3)^2 = 0$$

Point at $(1, -3)$

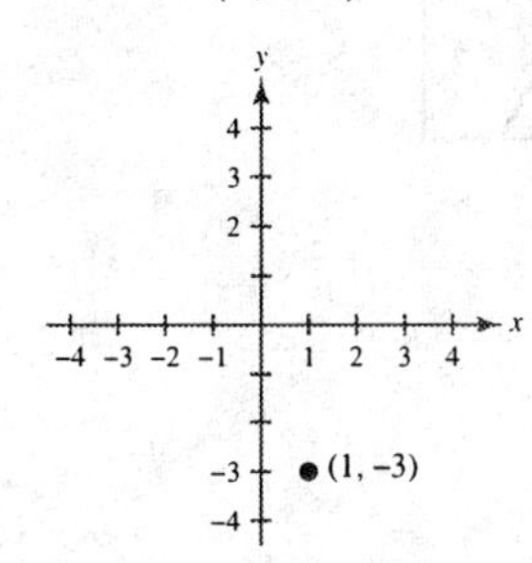

40.
$$x^2 - 10xy + y^2 = 0$$
$$y^2 - 10xy + 25x^2 = 25x^2 - x^2$$
$$(y - 5x)^2 = 24x^2$$
$$y - 5x = \pm\sqrt{24x^2}$$
$$y = 5x \pm 2\sqrt{6}x$$
$$y = \left(5 \pm 2\sqrt{6}\right)x$$

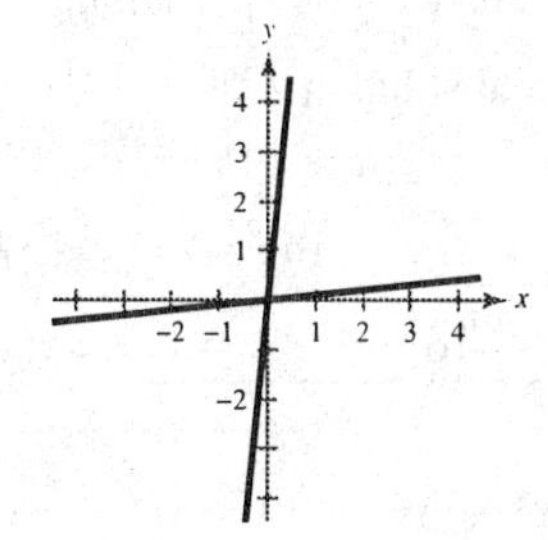

42.
$$-x^2 - y^2 - 8x + 20y - 7 = 0 \Rightarrow (x + 4)^2 + (y - 10)^2 = 109$$
$$x^2 + 9y^2 + 8x + 4y + 7 = 0 \Rightarrow (x + 4)^2 + 9\left(y + \tfrac{2}{9}\right)^2 = \tfrac{85}{9}$$
$$8y^2 + 24y = 0$$
$$8y(y + 3) = 0$$
$$y = 0 \text{ or } y = -3$$

For $y = 0$: $x^2 + 9(0)^2 + 8x + 4(0) + 7 = 0$
$$(x + 7)(x + 1) = 0$$
$$x = -7, -1$$

For $y = -3$: $x^2 + 9(-3)^2 + 8x + 4(-3) + 7 = 0$
$$x^2 + 8x + 76 = 0$$

No real solution

Points of intersection: $(-7, 0), (-1, 0)$

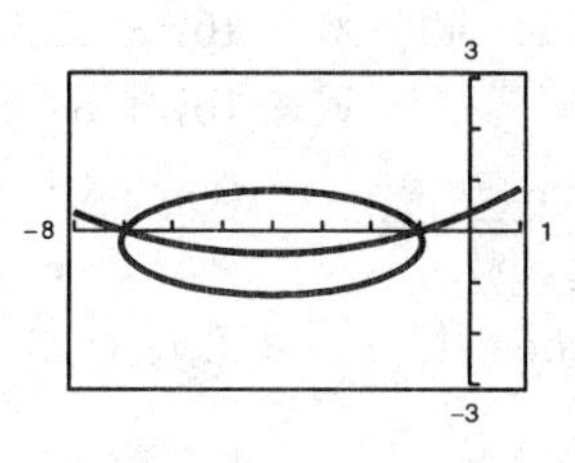

44.
$$x^2 - 4y^2 - 20x - 64y - 172 = 0 \Rightarrow (x - 10)^2 - 4(y + 8)^2 = 16$$
$$16x^2 + 4y^2 - 320x + 64y - 1600 = 0 \Rightarrow 16(x - 10)^2 + 4(y + 8)^2 = 256$$
$$17x^2 - 340x + 1428 = 0$$
$$(17x - 238)(x - 6) = 0$$
$$x = 6 \text{ or } x = 14$$

When $x = 6$: $6^2 - 4y^2 - 20(6) - 64y - 172 = 0$
$$-4y^2 - 64y - 256 = 0$$
$$y^2 + 16y + 64 = 0$$
$$(y + 8)^2 = 0$$
$$y = -8$$

Points of intersection: $(6, -8), (14, -8)$

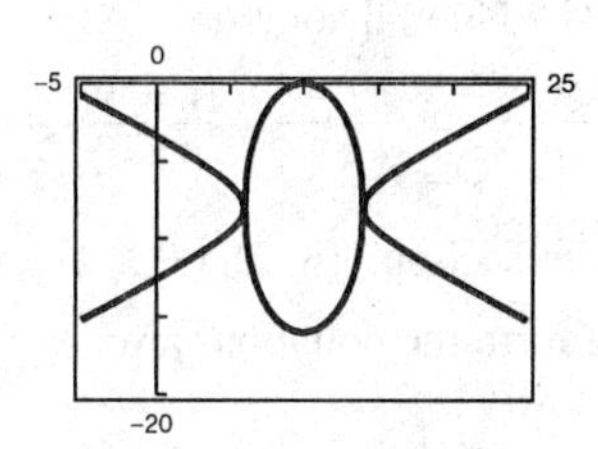

46. $x^2 + 4y^2 - 2x - 8y + 1 = 0 \Rightarrow (x-1)^2 + 4(y-1)^2 = 4$

$\underline{-x^2 \qquad + 2x - 4y - 1 = 0} \Rightarrow y = -\frac{1}{4}(x-1)^2$

$$\begin{aligned} 4y^2 \qquad -12y &= 0 \\ 4y(y-3) &= 0 \\ y = 0 \text{ or } y &= 3 \end{aligned}$$

When $y = 0$: $x^2 + 4(0)^2 - 2x - 8(0) + 1 = 0$

$$\begin{aligned} x^2 - 2x + 1 &= 0 \\ (x-1)^2 &= 0 \\ x &= 1 \end{aligned}$$

When $y = 3$: $-x^2 + 2x - 4(3) - 1 = 0$

$$x^2 - 2x + 13 = 0$$

No real solution

Point of intersection: $(1, 0)$

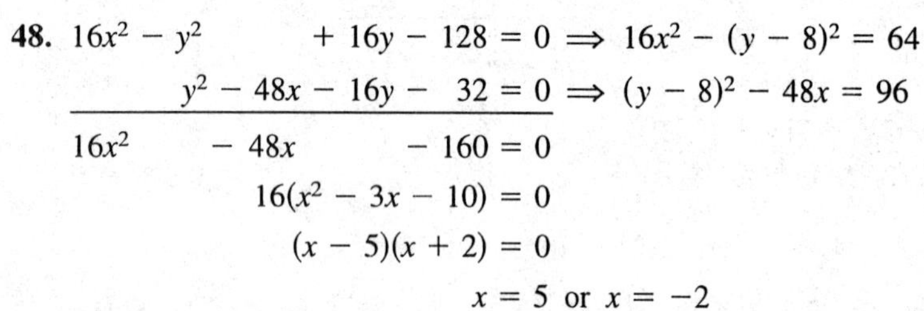

48. $16x^2 - y^2 \qquad + 16y - 128 = 0 \Rightarrow 16x^2 - (y-8)^2 = 64$

$\underline{\qquad y^2 - 48x - 16y - 32 = 0} \Rightarrow (y-8)^2 - 48x = 96$

$$\begin{aligned} 16x^2 \qquad - 48x \qquad - 160 &= 0 \\ 16(x^2 - 3x - 10) &= 0 \\ (x-5)(x+2) &= 0 \\ x = 5 \text{ or } x &= -2 \end{aligned}$$

When $x = 5$: $y^2 - 48(5) - 16y - 32 = 0$

$$\begin{aligned} y^2 - 16y - 272 &= 0 \\ y &= 8 \pm 4\sqrt{21} \end{aligned}$$

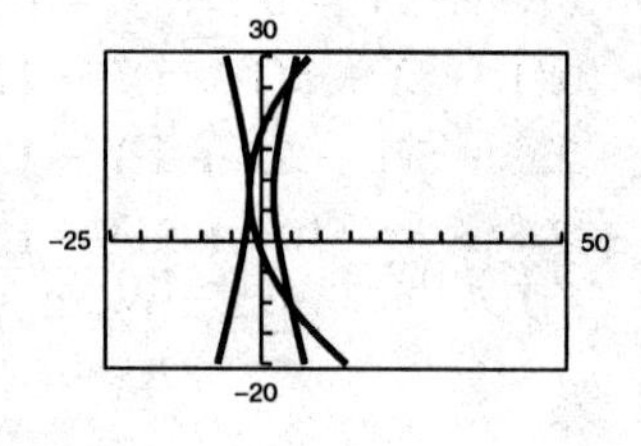

When $x = -2$: $y^2 - 48(-2) - 16y - 32 = 0$

$$\begin{aligned} y^2 - 16y + 64 &= 0 \\ (y-8)^2 &= 0 \\ y &= 8 \end{aligned}$$

Points of intersection: $(5, 8 + 4\sqrt{21}), (5, 8 - 4\sqrt{21}), (-2, 8)$

50. $4x^2 + 9y^2 - 36y = 0$

$x^2 + 9y - 27 = 0 \Rightarrow x^2 = 27 - 9y$

$$\begin{aligned} 4(27 - 9y) + 9y^2 - 36y &= 0 \\ 9y^2 - 72y + 108 &= 0 \\ 9(y-6)(y-2) &= 0 \\ y = 6 \text{ or } y &= 2 \end{aligned}$$

When $y = 6$: $x^2 = 27 - 9(6) = -27$

No real solution

When $y = 2$: $x^2 = 27 - 9(2) = 9$

$$x = \pm 3$$

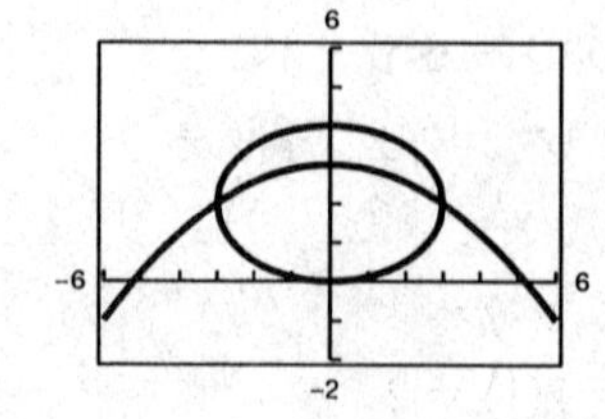

Points of intersection: $(3, 2), (-3, 2)$

In standard form the equations are:

$$\frac{x^2}{9} + \frac{(y-2)^2}{4} = 1$$

$$y = -\frac{x^2}{9} + 3$$

52. $x^2 + 2y^2 - 4x + 6y - 5 = 0$

$x - 4x - y + 4 = 0 \Rightarrow x^2 - 4x = y - 4$

$y - 4 + 2y^2 + 6y - 5 = 0$

$2y^2 + 7y - 9 = 0$

$(2y + 9)(y - 1) = 0$

$y = -\dfrac{9}{2}$ or $y = 1$

When $y = 1$: $x^2 - 4x - 1 + 4 = 0$

$(x - 3)(x - 1) = 0$

$x = 1$ or $x = 3$

When $y = -\dfrac{9}{2}$: $x^2 - 4x - \left(-\dfrac{9}{2}\right) + 4 = 0$

$x^2 - 4x + \dfrac{17}{2} = 0$

No real solution

Points of intersection: $(1, 1), (3, 1)$

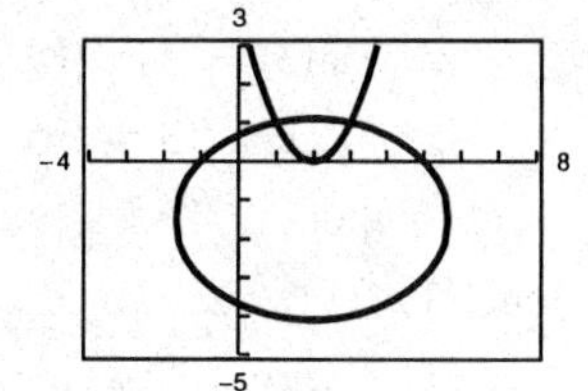

In standard form the equations are: $\dfrac{(x - 2)^2}{27/2} + \dfrac{2[y + (3/2)]^2}{27/2} = 1$

$y = x^2 - 4x + 4$

54. $5x^2 - 2xy + 5y^2 - 12 = 0$

$x + y - 1 = 0 \Rightarrow y = 1 - x$

$5x^2 - 2x(1 - x) + 5(1 - x)^2 - 12 = 0$

$5x^2 - 2x + 2x^2 + 5(1 - 2x + x^2) - 12 = 0$

$5x^2 - 2x + 2x^2 + 5 - 10x + 5x^2 - 12 = 0$

$12x^2 - 12x - 7 = 0$

$x = \dfrac{3 \pm \sqrt{30}}{6}$

When $x = \dfrac{3 + \sqrt{30}}{6}$: $y = 1 - \dfrac{3 + \sqrt{30}}{6} = \dfrac{3 - \sqrt{30}}{6}$

When $x = \dfrac{3 - \sqrt{30}}{6}$: $y = 1 - \dfrac{3 - \sqrt{30}}{6} = \dfrac{3 + \sqrt{30}}{6}$

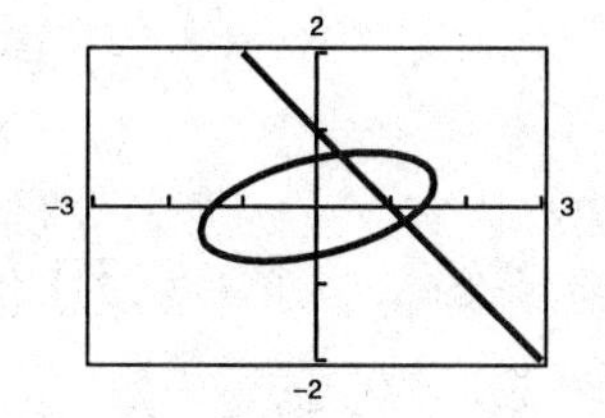

Points of intersection: $\left(\dfrac{1}{6}(3 + \sqrt{30}), \dfrac{1}{6}(3 - \sqrt{30})\right), \left(\dfrac{1}{6}(3 - \sqrt{30}), \dfrac{1}{6}(3 + \sqrt{30})\right)$

56. In Exercise 10, the equation of the rotated ellipse is:

$\dfrac{(x')^2}{1} + \dfrac{(y')^2}{4} = 1$

$a^2 = 4 \Rightarrow a = 2$

$b^2 = 1 \Rightarrow b = 1$

Length of major axis is $2a = 2(2) = 4$.

Length of minor axis is $2b = 2(1) = 2$.

58. $f(x) = \dfrac{2x}{2 - x}$

y-intercept: $(0, 0)$

Vertical asymptote: $x = 2$

Horizontal asymptote: $y = -2$

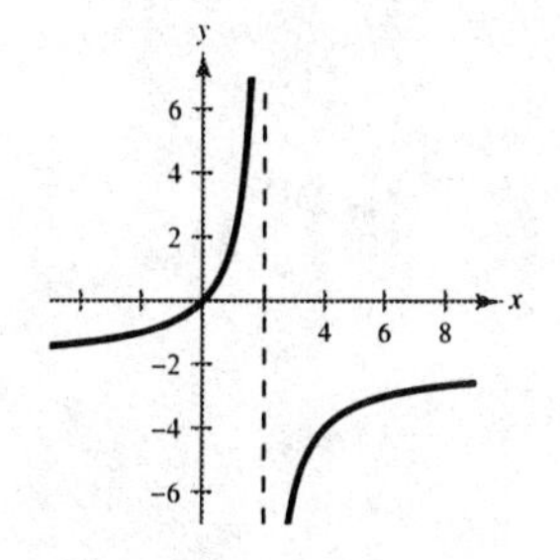

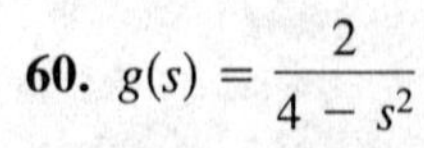

60. $g(s) = \dfrac{2}{4 - s^2}$

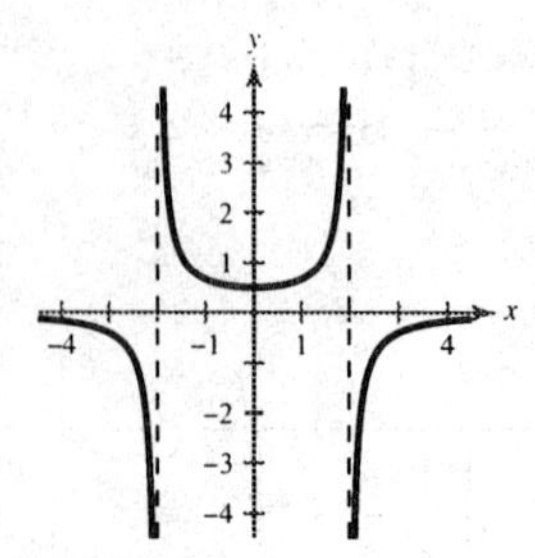

Intercept: $\left(0, \dfrac{1}{2}\right)$

Vertical asymptotes: $s = 2, s = -2$

Horizontal asymptote: $y = 0$

Section 10.5 Parametric Equations

Solutions to Even-Numbered Exercises

2. $x = 4\cos^2\theta,\ y = 2\sin\theta$

(a)

θ	$-\frac{\pi}{2}$	$-\frac{\pi}{4}$	0	$\frac{\pi}{4}$	$\frac{\pi}{2}$
x	0	2	4	2	0
y	-2	$-\sqrt{2}$	0	$\sqrt{2}$	2

(b)

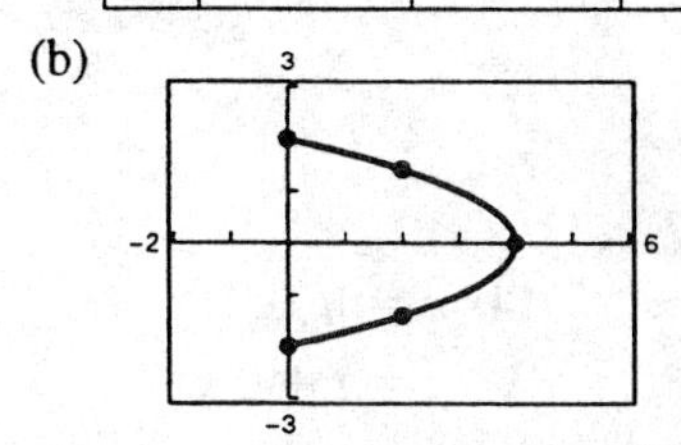

(c) $\frac{x}{4} = \cos^2\theta,\ \frac{y}{2} = \sin^2\theta$

$$\frac{x}{4} + \frac{y^2}{4} = 1$$

Parabola

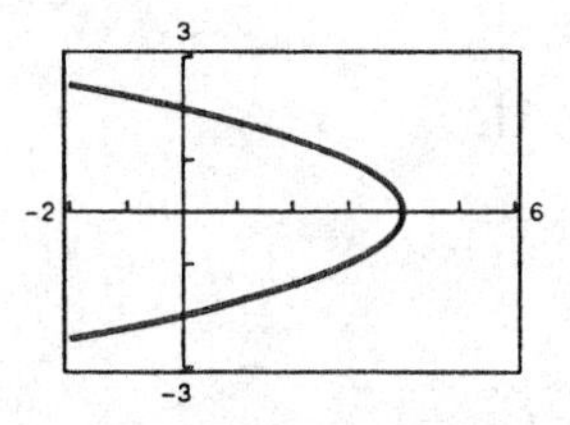

The rectangular version continues the graph into the second and third quadrants.

4. $x = t,\ y = \frac{1}{2}t$

$y = \frac{1}{2}x$ or $x - 2y = 0$

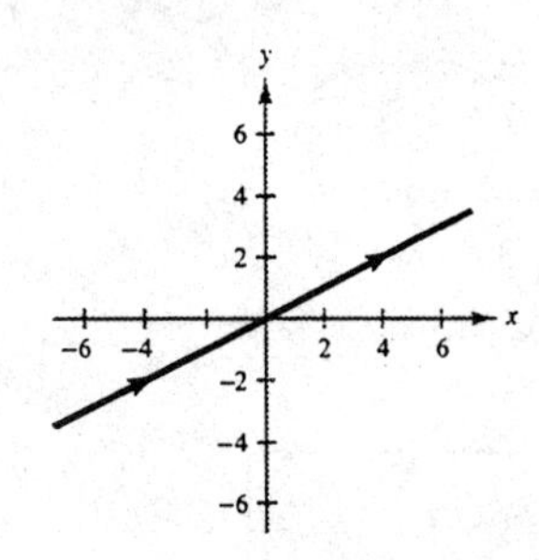

6. $x = 3 - 2t,\ y = 2 + 3t$

$$y = 2 + 3\left(\frac{3 - x}{2}\right)$$

$3x + 2y - 13 = 0$

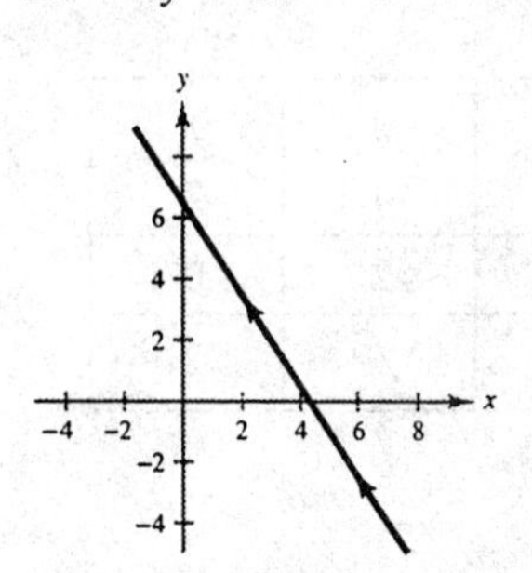

8. $x = t,\ y = t^3$

$y = x^3$

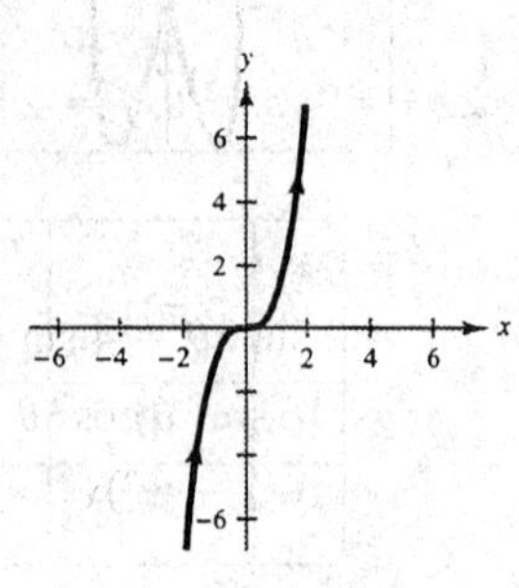

10. $x = \sqrt{t}$

$y = 1 - t$

$y = 1 - x^2,\ x \geq 0$

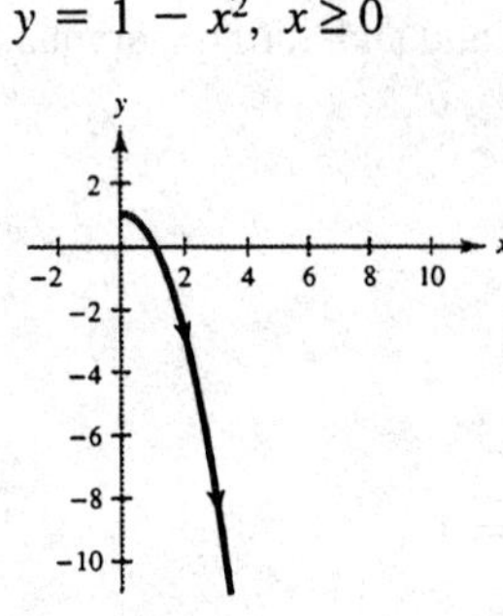

12. $x = t - 1,\ y = \frac{t}{t - 1}$

$$y = \frac{x + 1}{x + 1 - 1}$$

$$y = \frac{x + 1}{x}$$

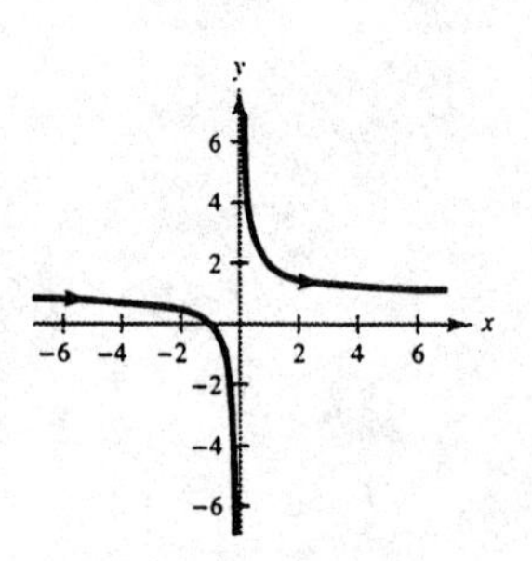

14. $x = |t - 1|$

$y = t + 2$

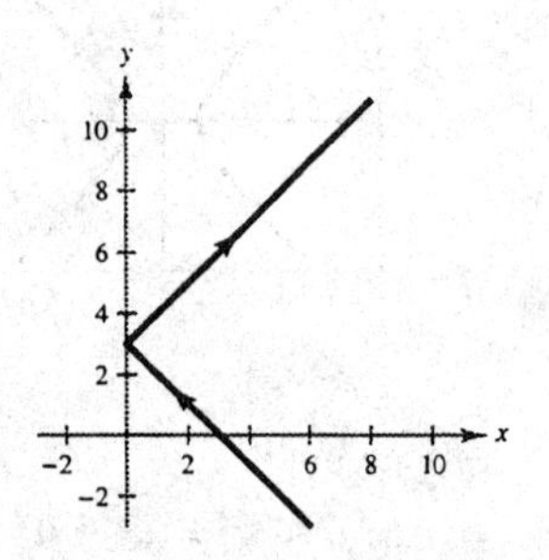

Eliminating the parameter t, $t = y - 2$ and

$$x = |t - 1|$$
$$= |(y - 2) - 1|$$
$$= |y - 3|.$$

16.

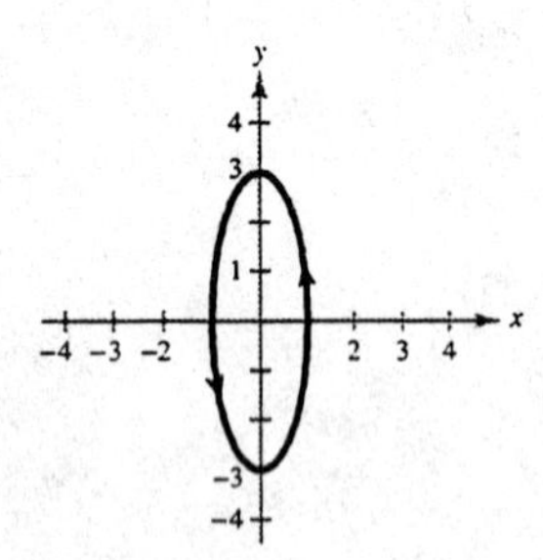

$$\left.\begin{aligned} x = \cos\theta &\Longrightarrow x^2 = \cos^2\theta \\ y = 3\sin\theta &\Longrightarrow \frac{y^2}{9} = \sin^2\theta \end{aligned}\right\} \sin^2\theta + \cos^2\theta = 1 = x^2 + \frac{y^2}{9}$$

Ellipse

18.

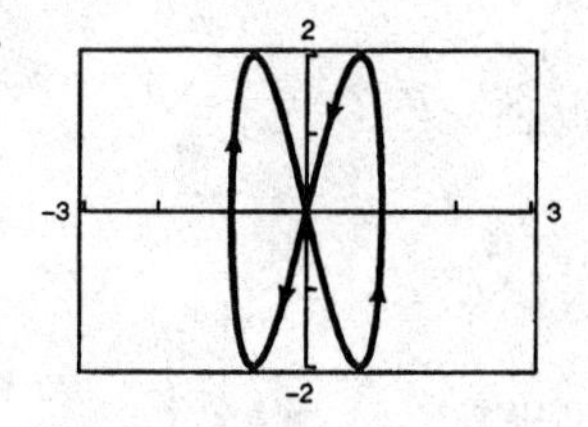

$x = \cos\theta$

$y = 2\sin 2\theta = 4\sin\theta\cos\theta$

$y^2 = 16\sin^2\theta\cos^2\theta = 16(1 - \cos^2\theta)\cos^2\theta$

$y^2 = 16(1 - x^2)x^2 = 16x^2(1 - x^2)$

20.

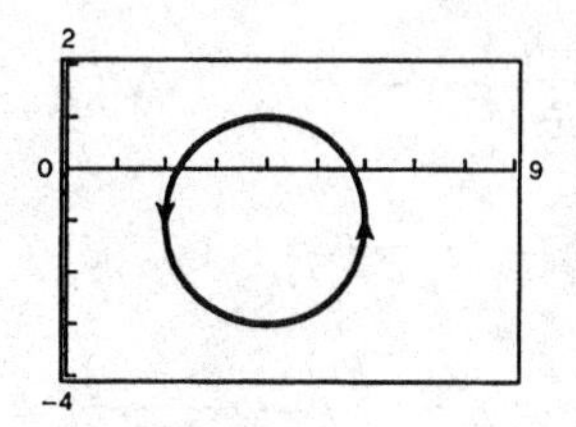

$x = 4 + 2\cos\theta \Rightarrow \cos\theta = \dfrac{x-4}{2}$

$y = -1 + 2\sin\theta \Rightarrow \sin\theta = \dfrac{y+1}{2}$

$\left(\dfrac{x-4}{2}\right)^2 + \left(\dfrac{y+1}{2}\right)^2 = \cos^2\theta + \sin^2\theta = 1$

$(x-4)^2 + (y+1)^2 = 4$

Circle

22.

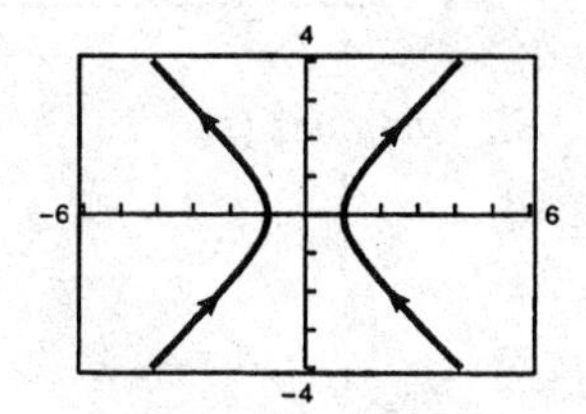

$x = \sec\theta$

$y = \tan\theta$

$\tan^2\theta + 1 = \sec^2\theta$

$y^2 + 1 = x^2$

$x^2 - y^2 = 1$

Hyperbola

24.

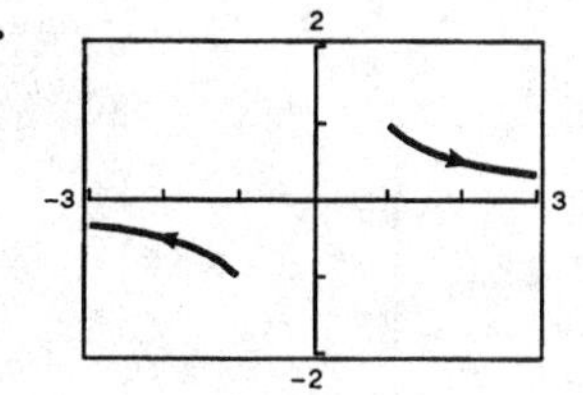

$x = \sec\theta$

$y = \cos\theta$

$xy = \sec\theta\cos\theta = 1$

$y = \dfrac{1}{x}$

26. $x = e^{2t}$

$y = e^t \Rightarrow y^2 = e^{2t}$

$y^2 = x,\ y > 0$

$y = \sqrt{x}$

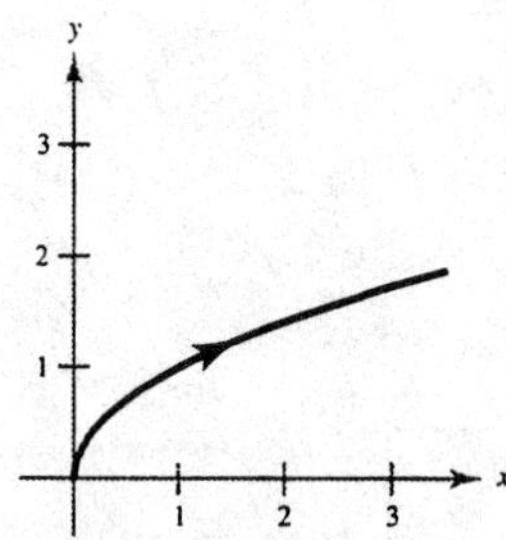

28. $x = \ln 2t \Rightarrow e^x = 2t$

$y = t^2$

$y = \left(\dfrac{e^x}{2}\right)^2 = \dfrac{e^{2x}}{4} = \dfrac{1}{4}e^{2x}$

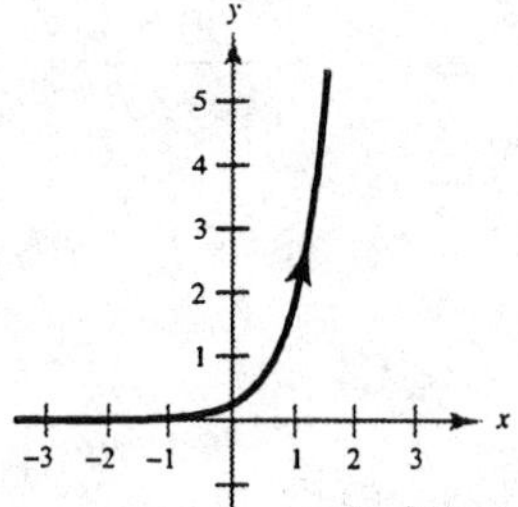

30. Each curve represents a portion of the line $y = x$.

(a) Domain: $-\infty < x < \infty$
Orientation: Left to right

(b) Domain: $x \geq 0$
Orientation: Left to right

(c) Domain: $-\infty < x < \infty$
Orientation: Right to left

(d) Domain: $-\infty < x < \infty$
Orientation: Left to right

32. $x = h + r\cos\theta$

$y = k + r\sin\theta$

$\dfrac{(x-h)}{r} = \cos\theta, \quad \dfrac{y-k}{r} = \sin\theta$

$\cos^2\theta + \sin^2\theta = \dfrac{(x-h)^2}{r^2} + \dfrac{(y-k)^2}{r^2} = 1$

$(x-h)^2 + (y-k)^2 = r^2$

34. $x = h + a\sec\theta$

$y = k + b\tan\theta$

$$\frac{x-h}{a} = \sec\theta, \qquad \frac{y-k}{b} = \tan\theta$$

$$\sec^2\theta - \tan^2\theta = \frac{(x-h)^2}{a^2} - \frac{(y-k)^2}{b^2} = 1$$

36. $x = 1 + t(5 - 1)$

$y = 4 + t(-2 - 4)$

$x = 1 + 4t$

$y = 4 - 6t$

(Solution not unique.)

38. From Exercise 32:

$x = -3 + 3\cos\theta, \quad h = -3, r = 3$

$y = 1 + 3\sin\theta, \quad k = 1, r = 3$

40. $a = 1, c = 2, b = \sqrt{c^2 - a^2} = \sqrt{3}$

From Exercise 34, $x = \sqrt{3}\tan\theta, y = \sec\theta$.

42. $y = x^2$

Examples:

$x = t, \; y = t^2$

$x = \frac{1}{2}t, \; y = \frac{1}{4}t^2$

44.

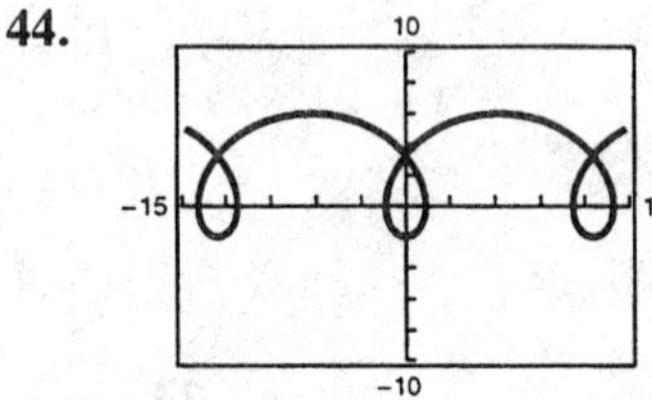

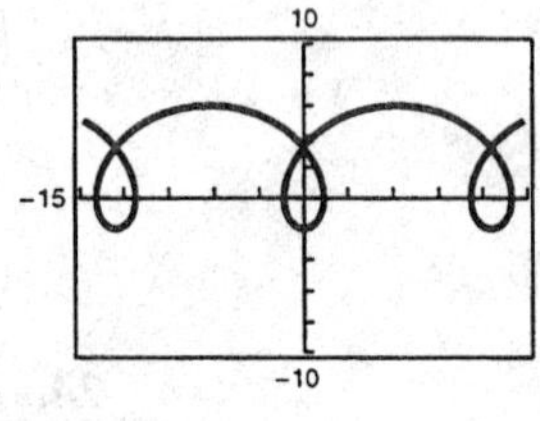

46.

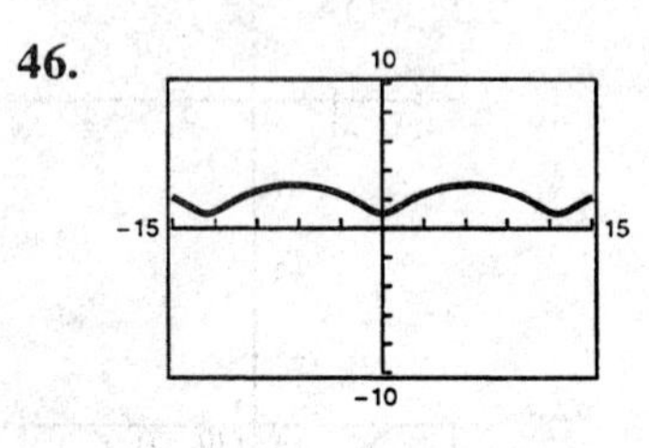

48.

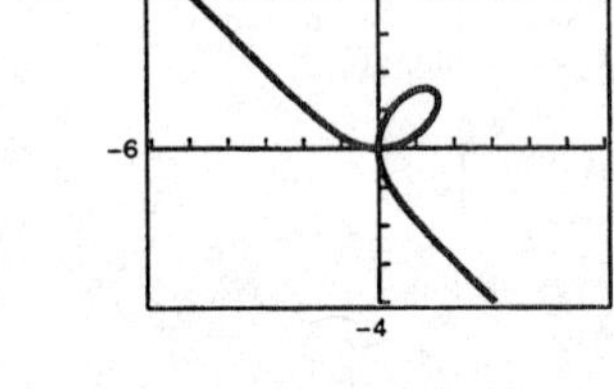

50. Matches (c).

52. Matches (a).

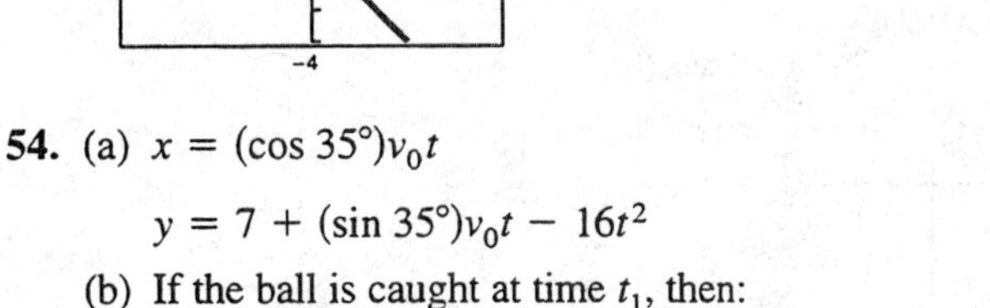

54. (a) $x = (\cos 35°)v_0 t$

$y = 7 + (\sin 35°)v_0 t - 16t^2$

(b) If the ball is caught at time t_1, then:

$90 = (\cos 35°)v_0 t_1$

$4 = 7 + (\sin 35°)v_0 t_1 - 16t_1^2.$

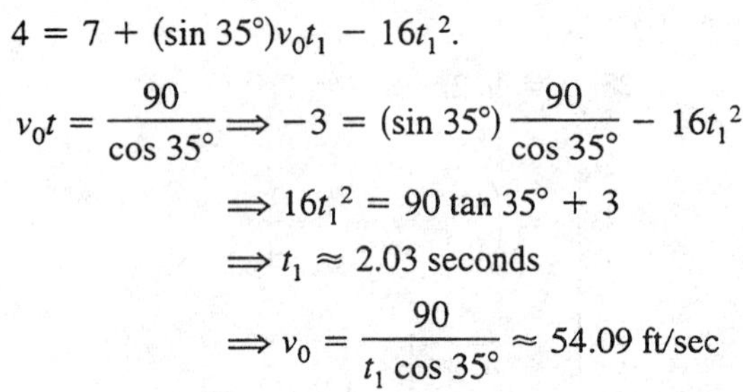

$$v_0 t = \frac{90}{\cos 35°} \Longrightarrow -3 = (\sin 35°)\frac{90}{\cos 35°} - 16t_1^2$$

$$\Longrightarrow 16t_1^2 = 90\tan 35° + 3$$

$$\Longrightarrow t_1 \approx 2.03 \text{ seconds}$$

$$\Longrightarrow v_0 = \frac{90}{t_1 \cos 35°} \approx 54.09 \text{ ft/sec}$$

(c)

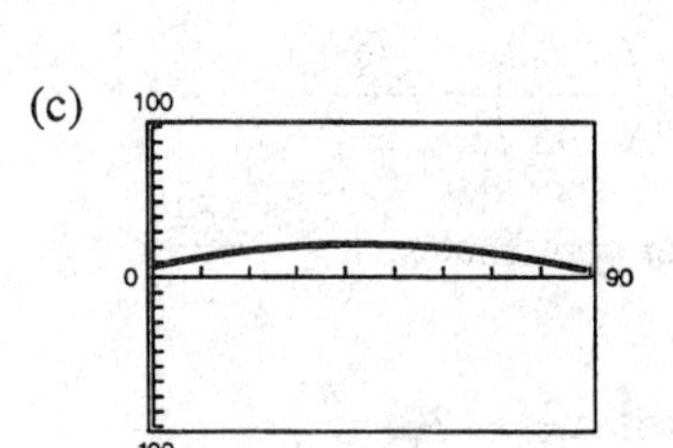

Maximum height $\approx$ 55 feet

(d) From part (b), $t_1 \approx 2.03$ seconds.

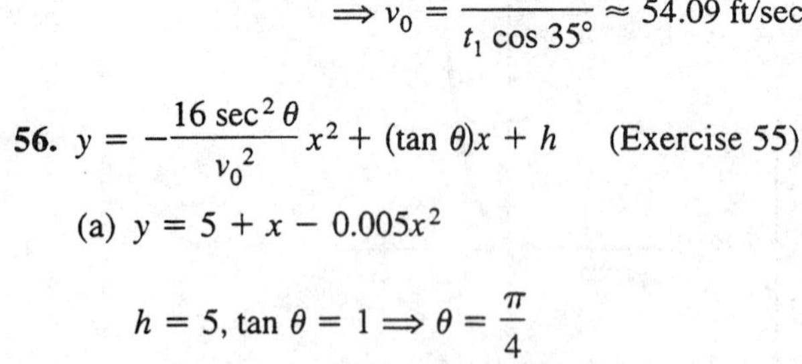

56. $y = -\frac{16\sec^2\theta}{v_0^2}x^2 + (\tan\theta)x + h$ (Exercise 55)

(a) $y = 5 + x - 0.005x^2$

$$h = 5, \tan\theta = 1 \Longrightarrow \theta = \frac{\pi}{4}$$

$$-0.005 = \frac{-16\sec^2\theta}{v_0^2} = \frac{-16(2)}{v_0^2}$$

$$v_0^2 = \frac{-32}{-0.005}$$

$$v_0^2 = 6400$$

$$v_0 = 80$$

(b)

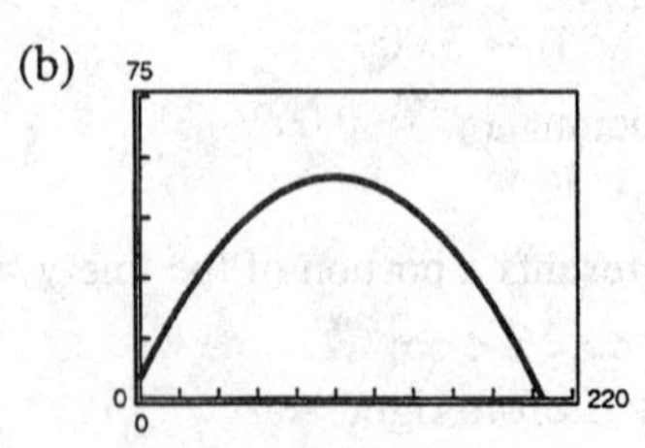

(c) Maximum height: 55 feet

Range $\approx$ 205 feet

Section 10.6 Polar Coordinates

Solutions to Even-Numbered Exercises

2. Polar coordinates: $\left(4, \dfrac{3\pi}{2}\right)$

$x = 4\cos\left(\dfrac{3\pi}{2}\right) = 0,\ y = 4\sin\left(\dfrac{3\pi}{2}\right) = -4$

Rectangular coordinates: $(0, -4)$

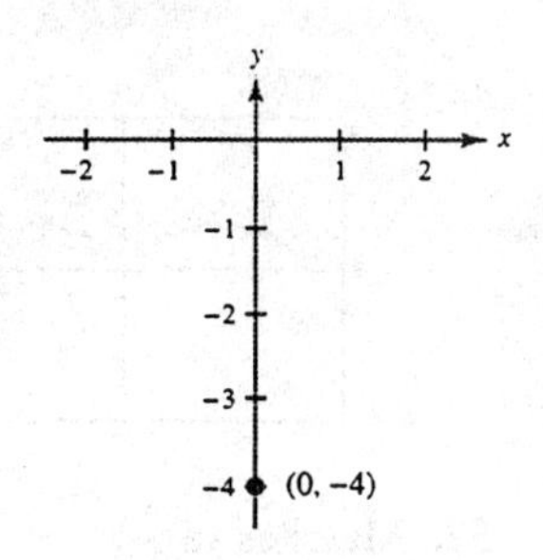

4. Polar coordinates: $(0, -\pi)$

$x = 0\cos(-\pi) = 0,\ y = 0\sin(-\pi) = 0$

Rectangular coordinates: $(0, 0)$

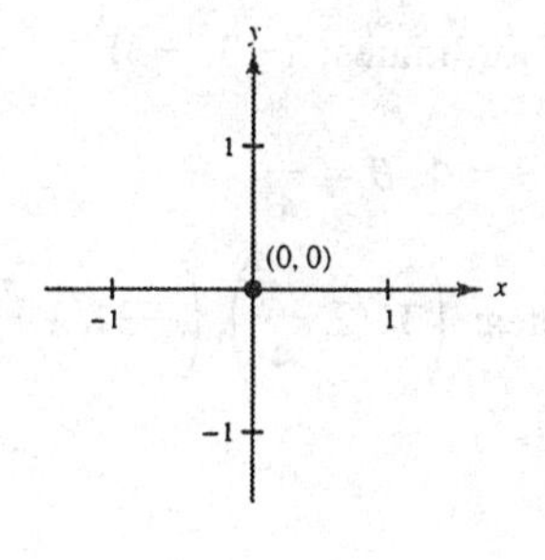

6. Polar coordinates: $\left(-1, \dfrac{-3\pi}{4}\right)$

$x = -1\cos\left(\dfrac{-3\pi}{4}\right) = \dfrac{\sqrt{2}}{2}$

$y = -1\sin\left(\dfrac{-3\pi}{4}\right) = \dfrac{\sqrt{2}}{2}$

Rectangular coordinates: $\left(\dfrac{\sqrt{2}}{2}, \dfrac{\sqrt{2}}{2}\right)$

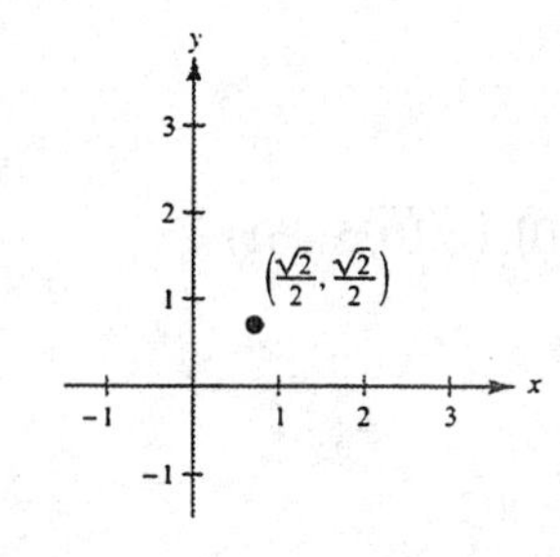

8. Polar coordinates: $\left(32, \dfrac{5\pi}{2}\right)$

$x = 32\cos\left(\dfrac{5\pi}{2}\right) = 0,\ y = 32\sin\left(\dfrac{5\pi}{2}\right) = 32$

Rectangular coordinates: $(0, 32)$

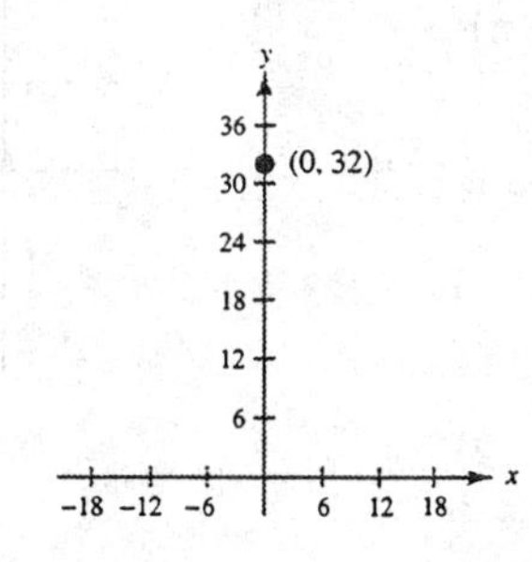

10. Polar coordinates: $(-3, -1.57)$

$x = -3\cos(-1.57) \approx -0.0024$

$y = -3\sin(-1.57) \approx 3.000$

Rectangular coordinates: $(-0.0024, 3)$

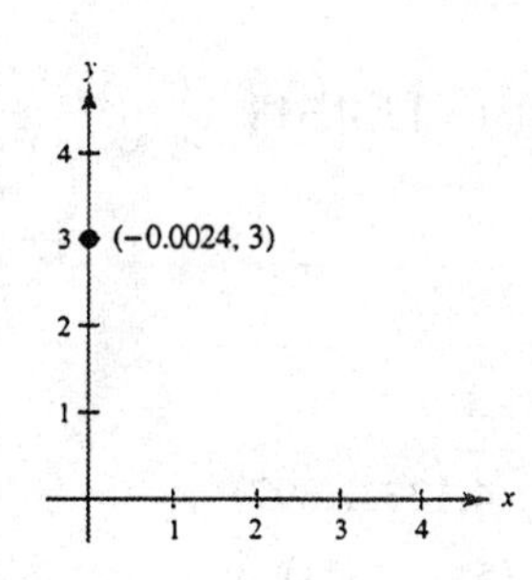

12. $(r, \theta) = \left(-2, \dfrac{7\pi}{6}\right) \Rightarrow (x, y) = (1.732, 1.0)$

14. $(r, \theta) = (8.25, 3.5) \Rightarrow (x, y) = (-7.726, -2.894)$

16. Rectangular coordinates: $(0, -5)$

$r = 5$, $\tan\theta$ undefined, $\theta = \dfrac{\pi}{2}$

Polar coordinates: $\left(5, \dfrac{3\pi}{2}\right), \left(-5, \dfrac{\pi}{2}\right)$

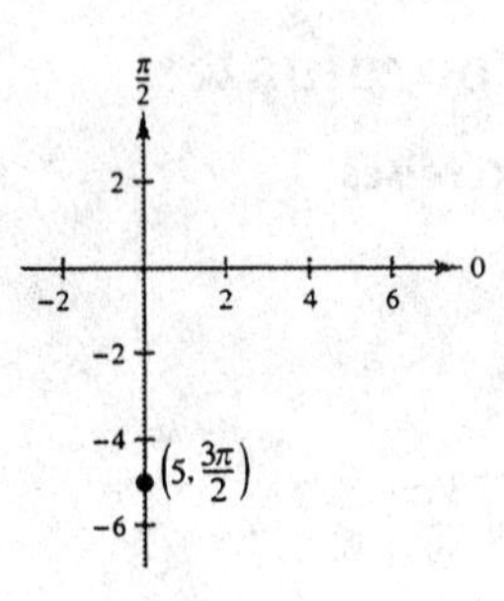

18. Rectangular coordinates: $(-3, -3)$

$r = 3\sqrt{2}$, $\tan\theta = 1$, $\theta = \dfrac{\pi}{4}$

Polar coordinates: $\left(3\sqrt{2}, \dfrac{5\pi}{4}\right), \left(-3\sqrt{2}, \dfrac{\pi}{4}\right)$

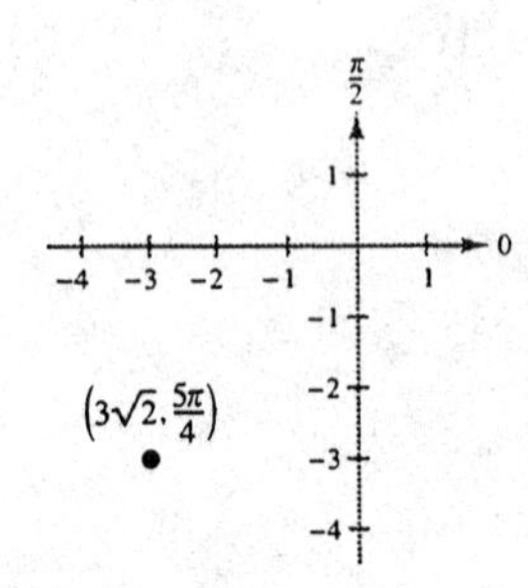

20. Rectangular coordinates: $(3, -1)$

$r = \sqrt{9+1} = \sqrt{10}$,

$\tan\theta = -\dfrac{1}{3}$, $\theta = \dfrac{\pi}{4}$

Polar coordinates: $\left(-\sqrt{10}, 2.820\right), \left(\sqrt{10}, 5.961\right)$

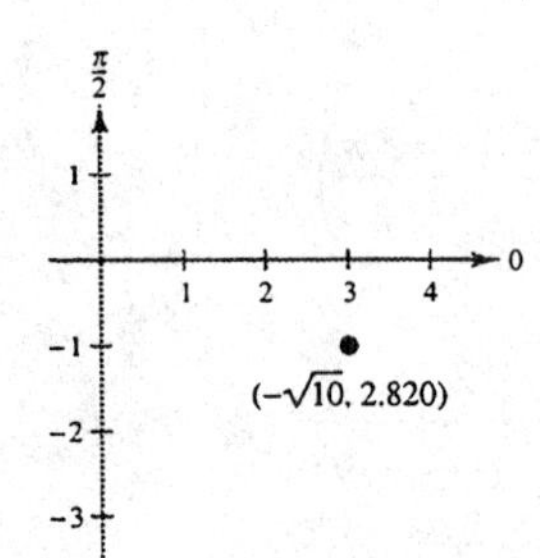

22. Rectangular coordinates: $(2, 0)$

$r = 2$, $\tan\theta = 0$, $\theta = 0$

Polar coordinates: $(2, 0), (-2, \pi)$

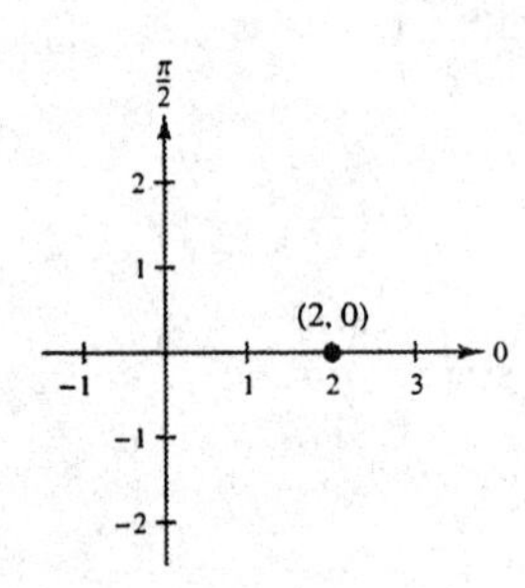

24. Rectangular coordinates: $(5, 12)$

$r = \sqrt{25 + 144} = 13$, $\tan\theta = \frac{12}{5}$,

$\theta \approx 1.176$

Polar coordinates: $(13, 1.176), (-13, 4.318)$

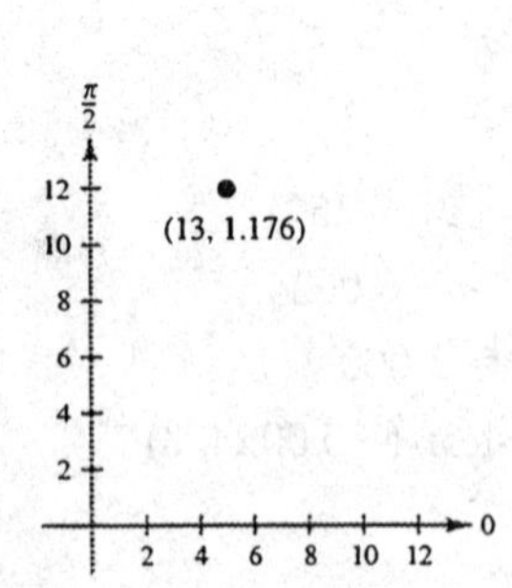

26. $(x, y) = (-4, 1) \Rightarrow (r, \theta) = (4.123, 2.897)$

28. $(x, y) = \left(3\sqrt{2}, 3\sqrt{2}\right) \Rightarrow (r, \theta) = (6.0, 0.785)$

30. $(x, y) = (0, -5) \Rightarrow (r, \theta) = (5, -1.571)$

32. False. For instance when $r = 0$ any value of θ gives the same point.

34. $x^2 + y^2 = a^2$

$r = a$

36. $x^2 + y^2 - 2ay = 0$

$r^2 - 2ar \sin \theta = 0$

$r(r - 2a \sin \theta) = 0$

$r = 2a \sin \theta$

38. $y = b$

$r \sin \theta = b$

$r = b \csc \theta$

40. $x = a$

$r \cos \theta = a$

$r = a \sec \theta$

42. $4x + 7y - 2 = 0$

$4r \cos \theta + 7r \sin \theta - 2 = 0$

$r(4 \cos \theta + 7 \sin \theta) = 2$

$$r = \frac{2}{4 \cos \theta + 7 \sin \theta}$$

44. $y = x$

$r \cos \theta = r \sin \theta$

$1 = \tan \theta$

$$\theta = \frac{\pi}{4}$$

46. $y^2 - 8x - 16 = 0$

$r^2 \sin^2 \theta - 8r \cos \theta = 16$

$r^2 - r^2 \cos^2 \theta - 8r \cos \theta - 16 = 0$

$r^2 \cos^2 \theta + 8r \cos \theta + 16 = r^2$

$(r \cos \theta + 4)^2 = r^2$

$r = \pm(r \cos \theta + 4)$

$$r = \frac{4}{1 - \cos \theta}$$

$$\text{or } r = \frac{-4}{1 + \cos \theta}$$

48. $r = 4 \cos \theta$

$r^2 = 4r \cos \theta$

$x^2 + y^2 = 4x$

$x^2 + y^2 - 4x = 0$

50. $r = 4$

$r^2 = 16$

$x^2 + y^2 = 16$

52. $r^2 = \sin 2\theta = 2 \sin \theta \cos \theta$

$$r^2 = 2\left(\frac{y}{r}\right)\left(\frac{x}{r}\right) = \frac{2xy}{r^2}$$

$r^4 = 2xy$

$(x^2 + y^2)^2 = 2xy$

54.

$$r = \frac{1}{1 - \cos \theta}$$

$r - r \cos \theta = 1$

$\sqrt{x^2 + y^2} - x = 1$

$x^2 + y^2 = 1 + 2x + x^2$

$y^2 = 2x + 1$

56.

$$r = \frac{6}{2 \cos \theta - 3 \sin \theta}$$

$$r = \frac{6}{2(x/r) - 3(y/r)}$$

$$r = \frac{6r}{2x - 3y}$$

$$1 = \frac{6}{2x - 3y}$$

$2x - 3y = 6$

58. $r = 8$

$r^2 = 64$

$x^2 + y^2 = 64$

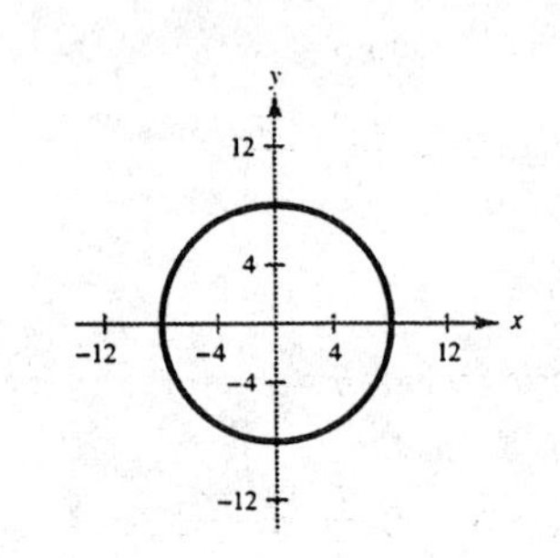

60. $\theta = \dfrac{5\pi}{6}$

$\tan\theta = \tan\dfrac{5\pi}{6}$

$\dfrac{y}{x} = -\dfrac{1}{\sqrt{3}}$

$\sqrt{3}y = -x$

$x + \sqrt{3}y = 0$

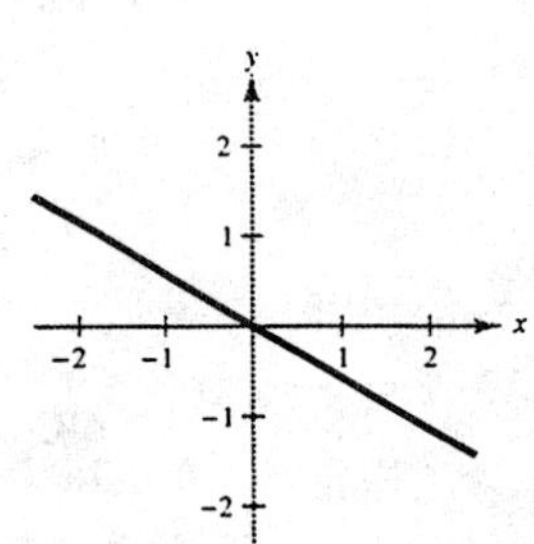

62. $r = 2\csc\theta$

$r\sin\theta = 2$

$y = 2$

$y - 2 = 0$

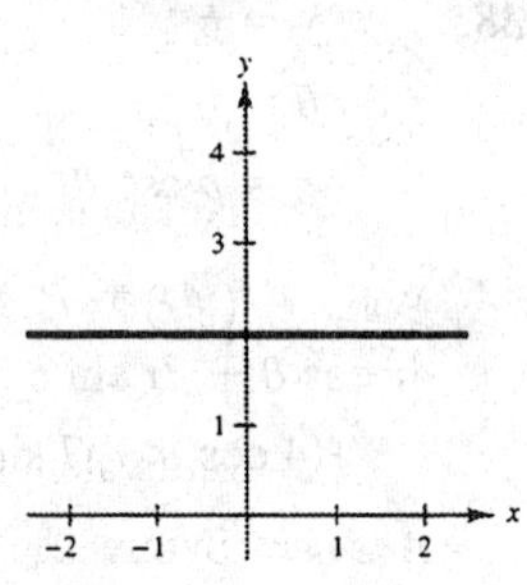

64. $r = \cos\theta + 3\sin\theta = 2\left(\frac{1}{2}\cos\theta + \frac{3}{2}\sin\theta\right)$

Using Exercise 63, we see that this is the equation of a circle of radius

$\sqrt{h^2 + k^2} = \sqrt{\left(\frac{1}{2}\right)^2 + \left(\frac{3}{2}\right)^2} = \sqrt{\frac{1}{4} + \frac{9}{4}} = \frac{1}{2}\sqrt{10}$ and center $(h, k) = \left(\frac{1}{2}, \frac{3}{2}\right)$.

$\left(x - \frac{1}{2}\right)^2 + \left(y - \frac{3}{2}\right)^2 = \frac{5}{2}$

66. (a) Horizontal movement: x-coordinate changes.
Vertical movement: y-coordinate changes.

(b) Horizontal movement: r and θ both change.
Vertical movement: r and θ both change.

(c) Unlike r and θ, x and y measure horizontal and vertical change, respectively.

68. By Cramer's Rule, $x = -\frac{5}{26}$, $y = \frac{55}{26}$.

70. By Cramer's Rule, $u = \frac{295}{89}$, $v = \frac{844}{89}$, $w = \frac{-672}{89}$.

72. By Cramer's Rule, $x = -2$, $y = \frac{3}{2}$, $z = 1$, $w = 4$.

Section 10.7 Graphs of Polar Equations

Solutions to Even-Numbered Exercises

2. Cardioid

4. Lemniscate

6. Limaçon

8. $r = 16\cos 3\theta$

$\theta = \dfrac{\pi}{2}$: $-r = 16\cos(3(-\theta))$

$-r = 16\cos(-3\theta)$

$-r = 16\cos 3\theta$

Not an equivalent equation

Polar axis: $r = 16\cos(3(-\theta))$

$r = 16\cos(-3\theta)$

$r = 16\cos 3\theta$

Equivalent equation

Pole: $-r = 16\cos 3\theta$

Not an equivalent equation

Answer: Symmetric with respect to polar axis

10. $r = 6\sin\theta$

$\theta = \dfrac{\pi}{2}$: $-r = 6\sin(-\theta)$

$-r = -6\sin\theta$

$r = 6\sin\theta$

Equivalent equation

Polar axis: $r = 6\sin(-\theta)$

$r = -6\sin\theta$

Not an equivalent equation

Pole: $-r = 6\sin\theta$

Not an equivalent equation

Answer: Symmetric with respect to $\theta = \pi/2$

12. $r^2 = 25 \sin 2\theta$

$\theta = \dfrac{\pi}{2}$: $(-r)^2 = 25 \sin(2(\theta))$

$r^2 = -25 \sin 2\theta$

Not an equivalent equation

Polar axis: $r^2 = 25 \sin(2(-\theta))$

$r^2 = -25 \sin 2\theta$

Not an equivalent equation

Pole: $(-r)^2 = 25 \sin 2\theta$

$r^2 = 25 \sin 2\theta$

Equivalent equation

Answer: Symmetric with respect to pole

14. $|r| = |6 + 12 \cos \theta| \leq |6| + |12 \cos \theta|$

$= 6 + 12|\cos \theta| \leq 18$

$\cos \theta = 1$

$\theta = 0$

Maximum: $|r| = 18$ when $\theta = 0$

Zero: $r = 0$ when $\theta = \dfrac{2\pi}{3}, \dfrac{4\pi}{3}$

16. $|r| = |5 \sin 2\theta|$

Maximum: $|r| = 5$ when $\theta = \dfrac{\pi}{4}, \dfrac{3\pi}{4}, \dfrac{5\pi}{4}, \dfrac{7\pi}{4}$

Zero: $r = 0$ when $\theta = 0, \dfrac{\pi}{2}, \pi, \dfrac{3\pi}{2}, 2\pi$

18. Circle: $r = 2$

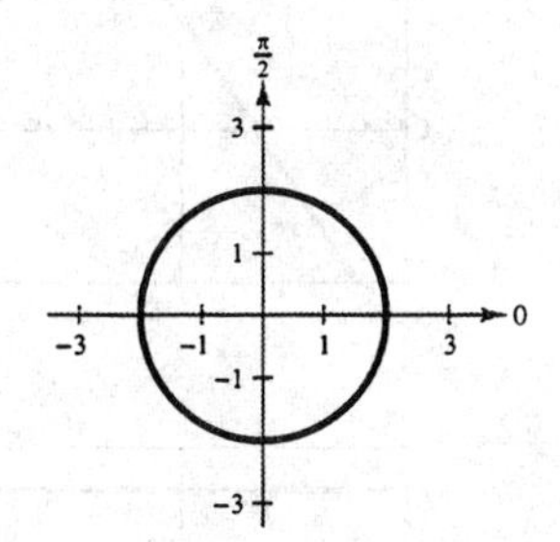

20. $r = -\dfrac{\pi}{4}$

Circle

Radius: $\dfrac{\pi}{4}$

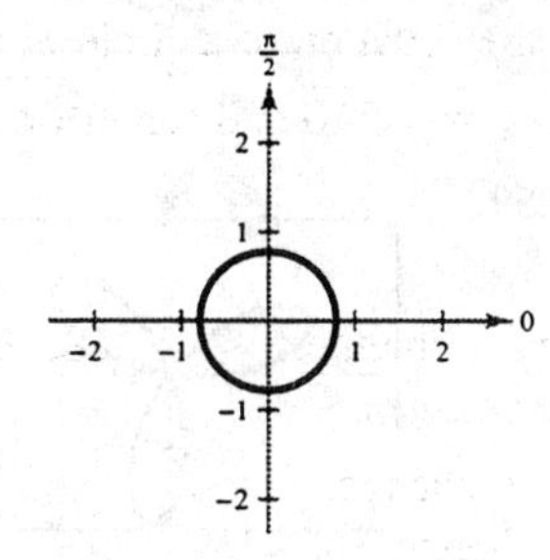

22. $r = 3 \cos \theta$

Circle

Radius: $\dfrac{3}{2}$

Center: $\left(\dfrac{3}{2}, 0\right)$

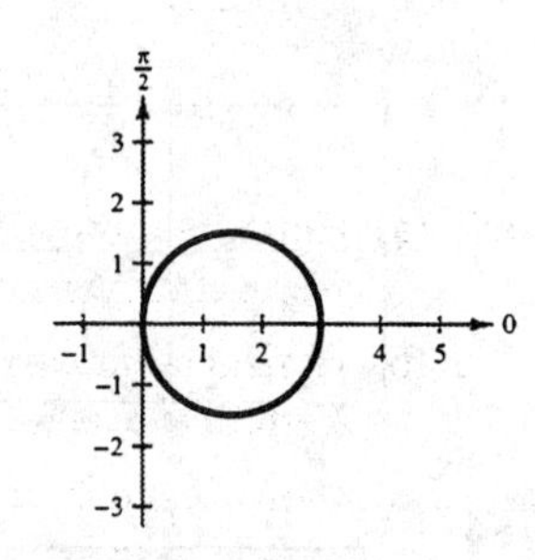

24. $r = 2 - 2 \sin \theta$

Cardioid

26. $r = 1 + \cos \theta$

Cardioid

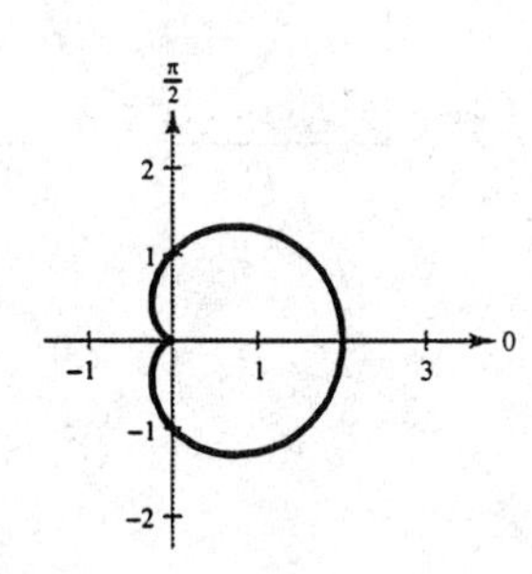

28. $r = 5 - 4 \sin \theta$

Dimpled limaçon

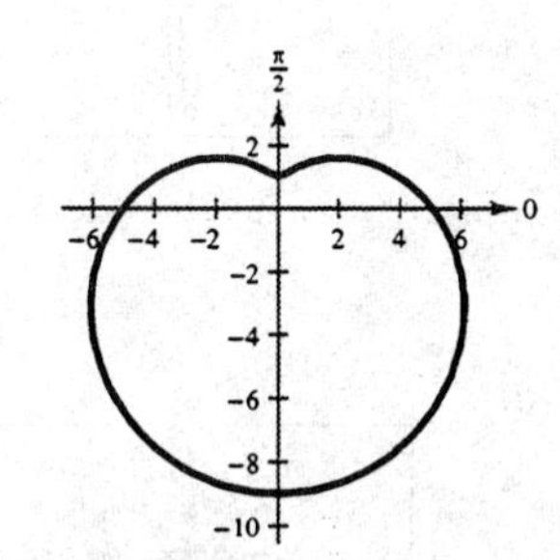

30. $r = 4 + 3 \cos \theta$

Dimpled limaçon

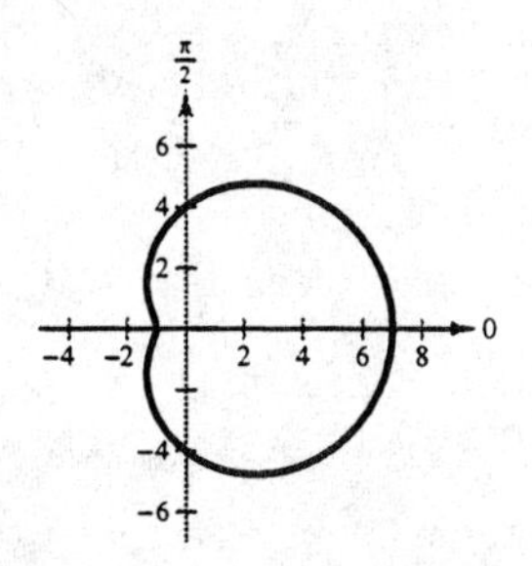

32. $r = -\sin 5\theta$

Rose curve

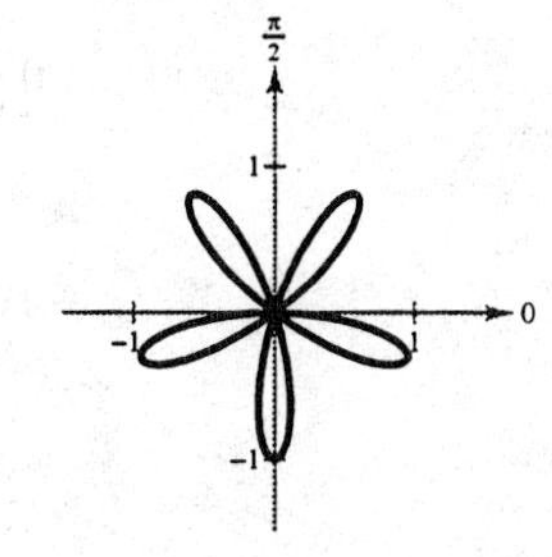

34. $r = 3 \cos 2\theta$

Rose curve

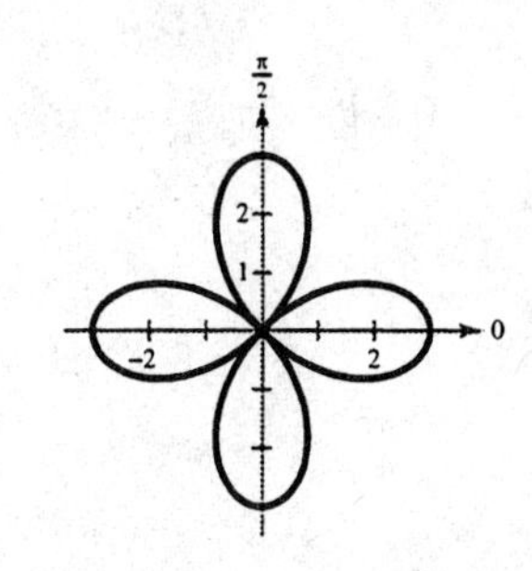

36. $r = \theta$

Symmetric with respect to

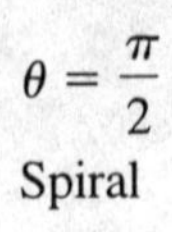

$\theta = \dfrac{\pi}{2}$

Spiral

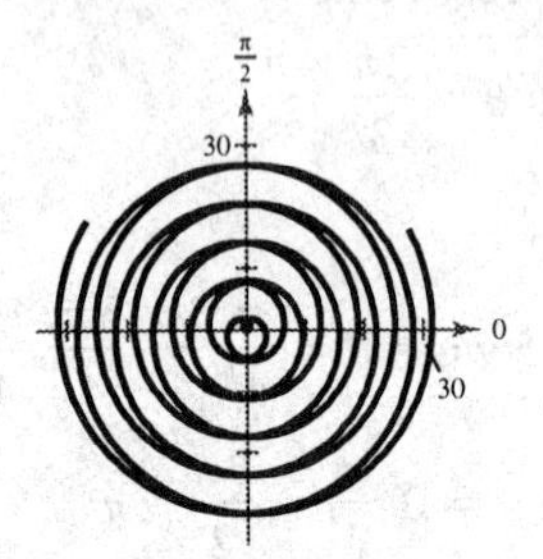

38.

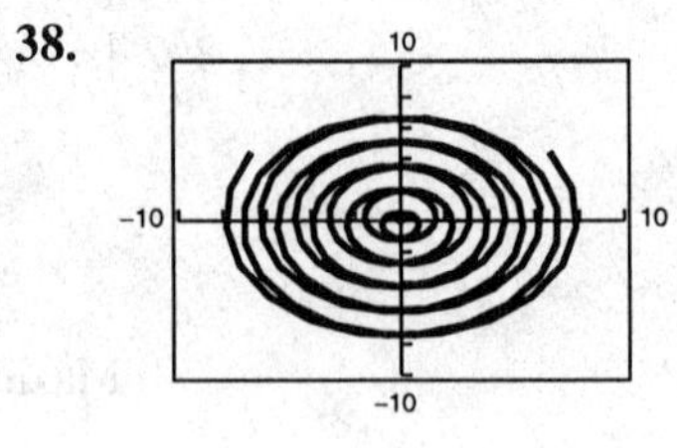

40.

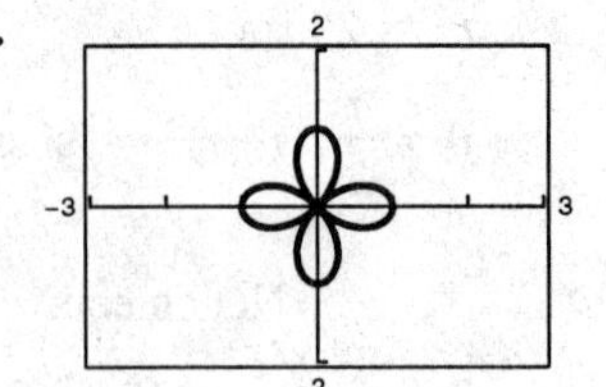

42.

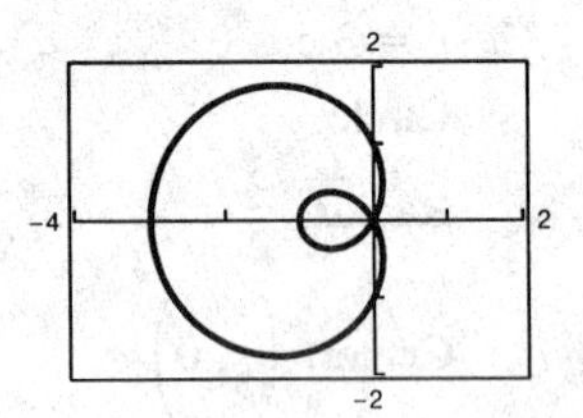

44.

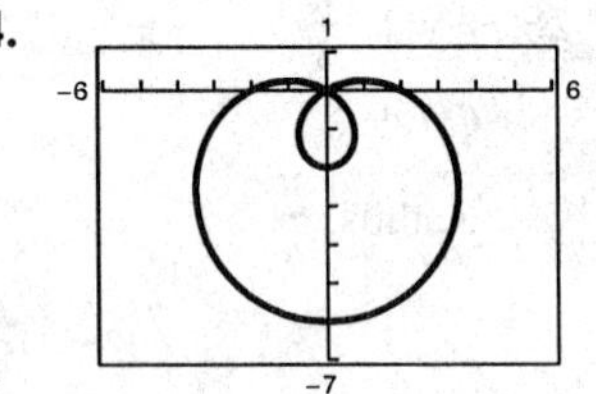

46.

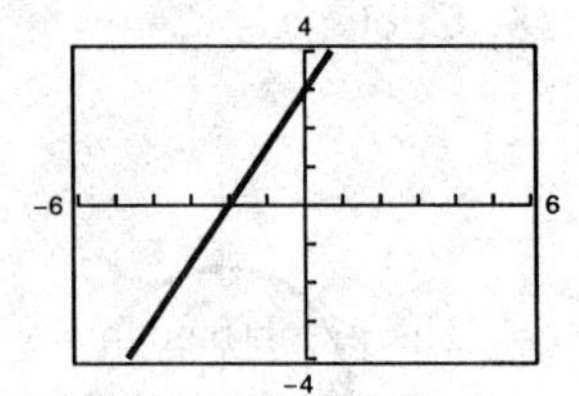

48.

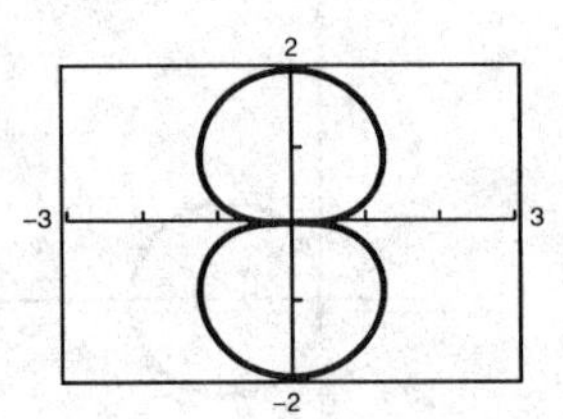

50.

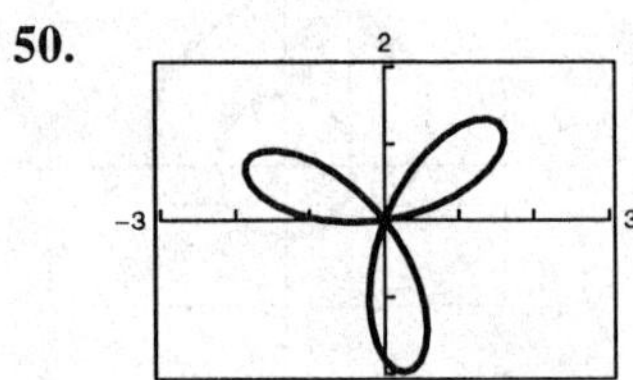

52.

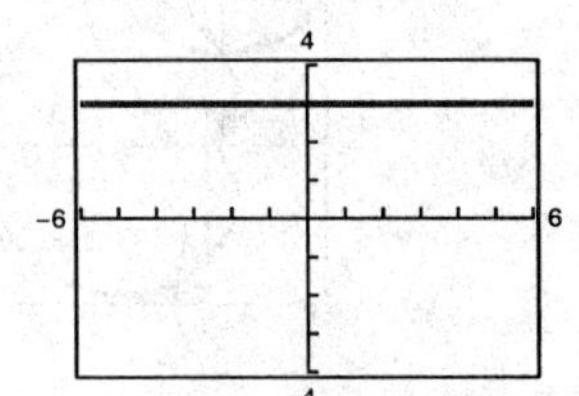

54.

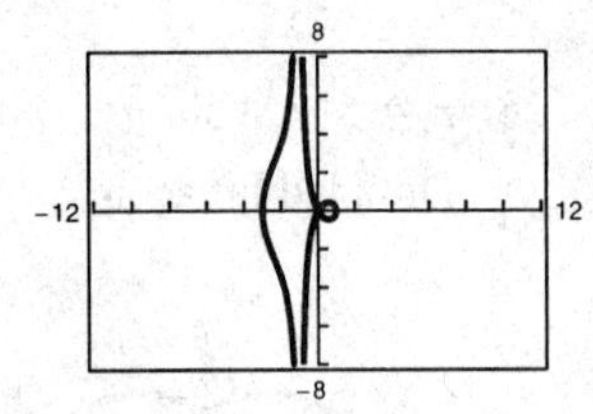

56.

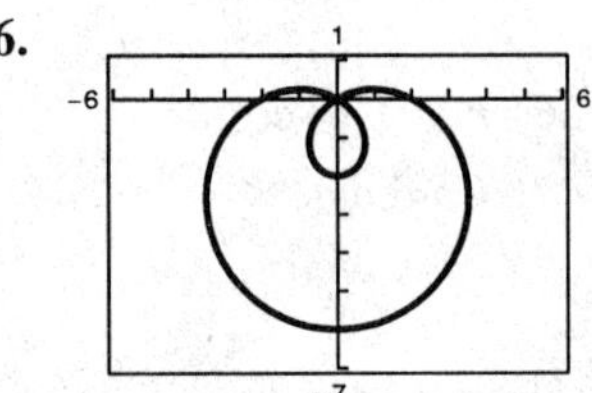

$0 \le \theta < 2\pi$

58.

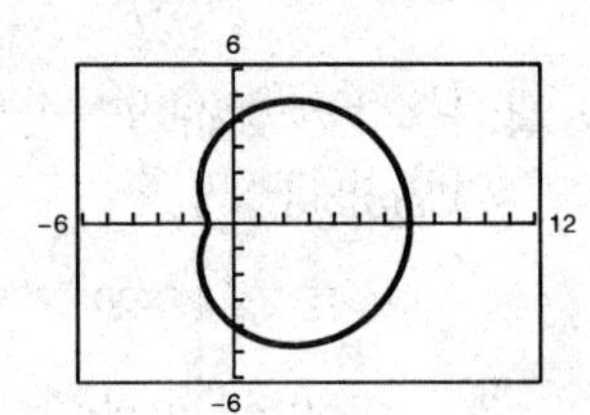

$0 \le \theta < 2\pi$

60.

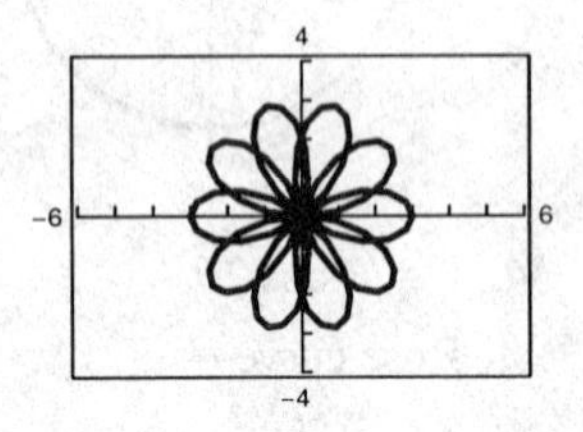

$0 \le \theta < 4\pi$

62.

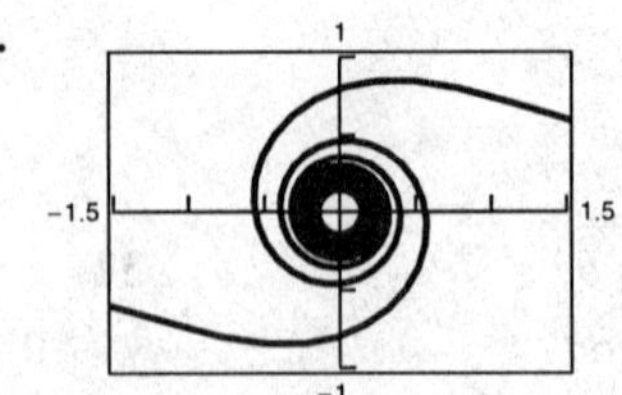

$0 < \theta < \infty$

64. $$r = 2 + \csc\theta = 2 + \frac{1}{\sin\theta}$$

$$r\sin\theta = 2\sin\theta + 1$$

$$r(r\sin\theta) = 2r\sin\theta + r$$

$$\left(\pm\sqrt{x^2+y^2}\right)(y) = 2y + \left(\pm\sqrt{x^2+y^2}\right)$$

$$\left(\pm\sqrt{x^2+y^2}\right)(y-1) = 2y$$

$$\left(\pm\sqrt{x^2+y^2}\right) = \frac{2y}{y-1}$$

$$x^2 + y^2 = \frac{4y^2}{(y-1)^2}$$

$$x^2 = \frac{y^2(3+2y-y^2)}{(y-1)^2}$$

$$x = \pm\sqrt{\frac{y^2(3+2y-y^2)}{(y-1)^2}}$$

$$= \pm\left|\frac{y}{y-1}\right|\sqrt{3+2y-y^2}$$

The graph has an asymptote at $y = 1$.

66.

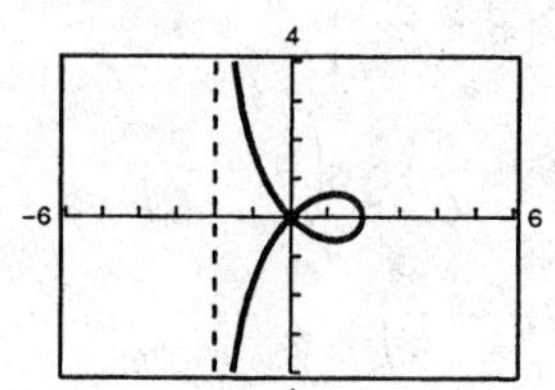

68. (a)

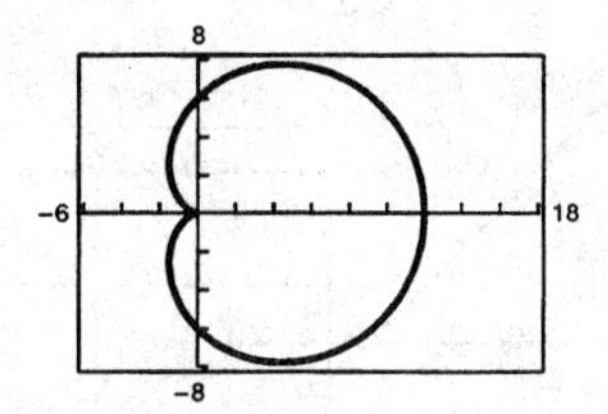

(b)

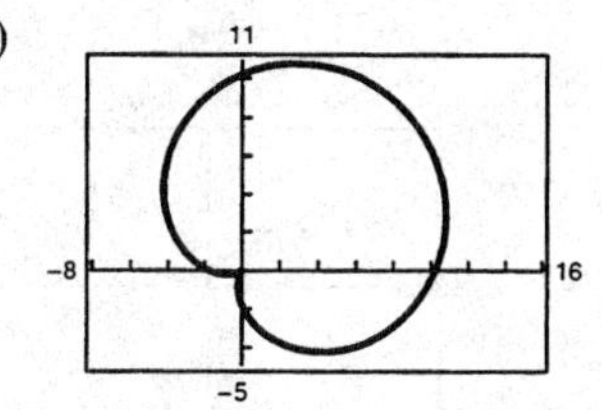

(c)

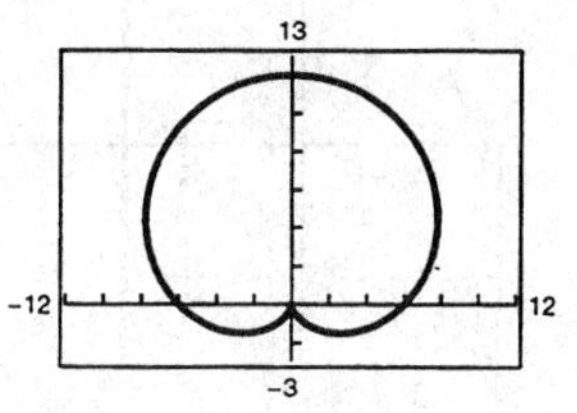

The angle ϕ rotates the graph around the pole. In part (c), $r = 6\left[1 + \cos\left(\theta - \frac{\pi}{2}\right)\right] = 6[1 + \sin\theta]$.

70. Use the result of Exercise 69.

(a) Rotation: $\phi = \frac{\pi}{2}$

Orginal graph: $r = f(\sin\theta)$

Rotated graph: $r = f\left(\sin\left(\theta - \frac{\pi}{2}\right)\right) = f(-\cos\theta)$

(b) Rotation: $\phi = \pi$

Orginal graph: $r = f(\sin\theta)$

Rotated graph: $r = f(\sin(\theta - \pi)) = f(-\sin\theta)$

(c) Rotation: $\phi = \pi$

Orginal graph: $r = f(\sin\theta)$

Rotated graph: $r = f\left(\sin\left(\theta - \frac{3\pi}{2}\right)\right) = f(\cos\theta)$

72. (a) $r = 2\sin\left[2\left(\theta - \dfrac{\pi}{6}\right)\right]$

$= 4\sin\left(\theta - \dfrac{\pi}{6}\right)\cos\left(\theta - \dfrac{\pi}{6}\right)$

$= 2\sin\left(2\theta - \dfrac{\pi}{3}\right)$

$= \sin 2\theta - \sqrt{3}\cos 2\theta$

(b) $r = 2\sin\left[2\left(\theta - \dfrac{\pi}{2}\right)\right]$

$= 2\sin(2\theta - \pi)$

$= -2\sin 2\theta$

$= -4\sin\theta\cos\theta$

(c) $r = 2\sin\left[2\left(\theta - \dfrac{2\pi}{3}\right)\right]$

$= 4\sin\left(\theta - \dfrac{2\pi}{3}\right)\cos\left(\theta - \dfrac{2\pi}{3}\right)$

$= 2\sin\left(2\theta - \dfrac{4\pi}{3}\right)$

$= \sqrt{3}\cos 2\theta - \sin 2\theta$

(d) $r = 2\sin[2(\theta - \pi)]$

$= 2\sin(2\theta - 2\pi)$

$= 2\sin 2\theta$

$= 4\sin\theta\cos\theta$

74. (a) $r = 3\sec\theta$

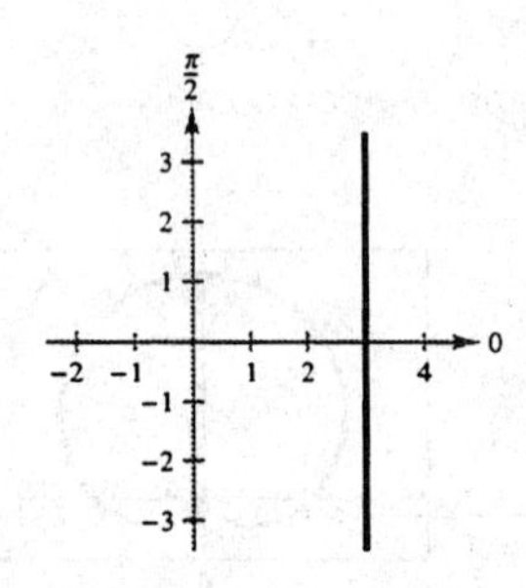

(b) $r = 3\sec\left(\theta - \dfrac{\pi}{4}\right)$

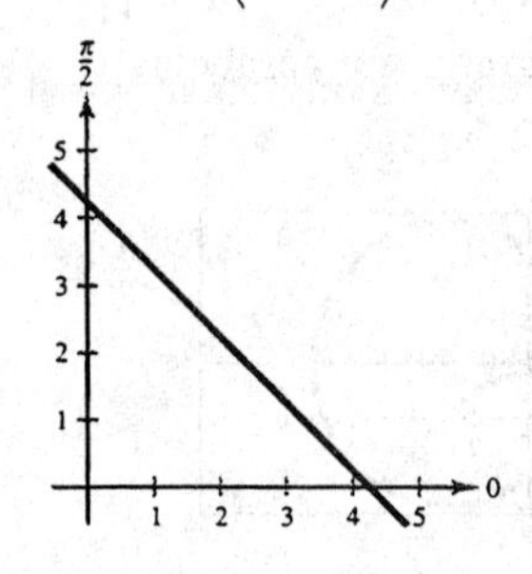

(c) $r = 3\sec\left(\theta + \dfrac{\pi}{3}\right)$

(d) $r = 3\sec\left(\theta - \dfrac{\pi}{2}\right)$

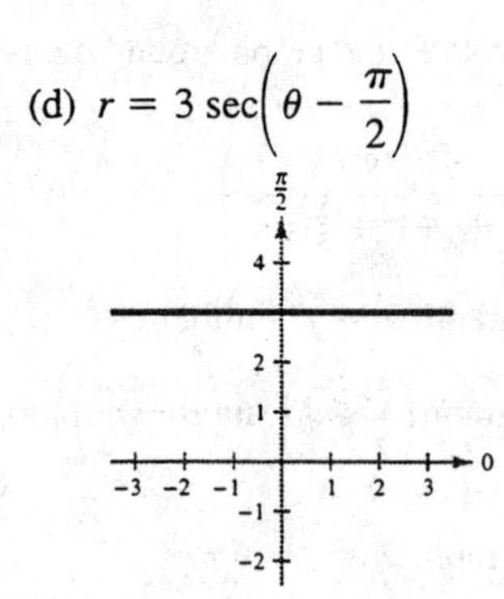

76. $r = 3\sin k\theta$

(a)

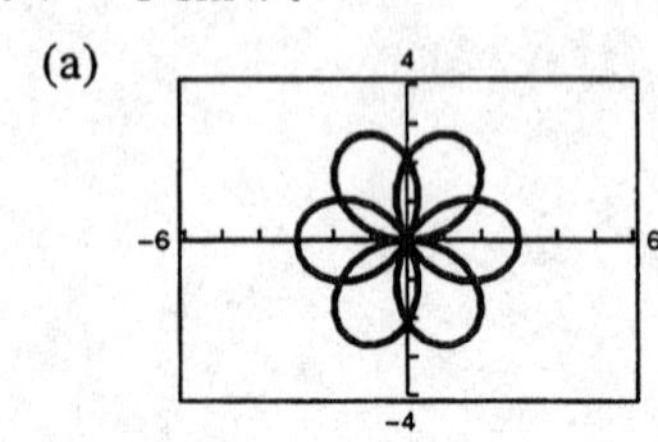

(b)

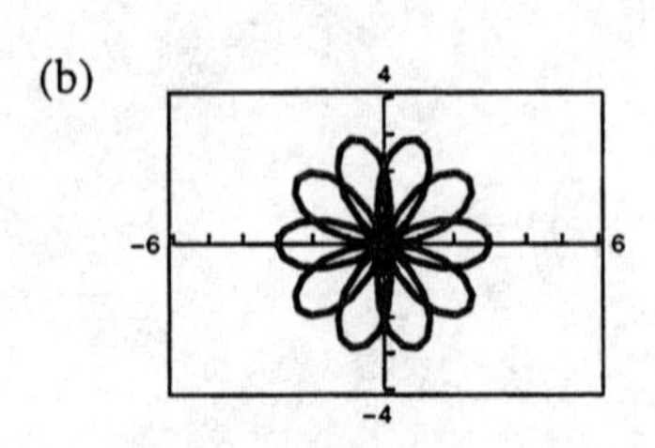

(c) Yes

Section 10.8 Polar Equations of Conics

Solutions to Even-Numbered Exercises

2.

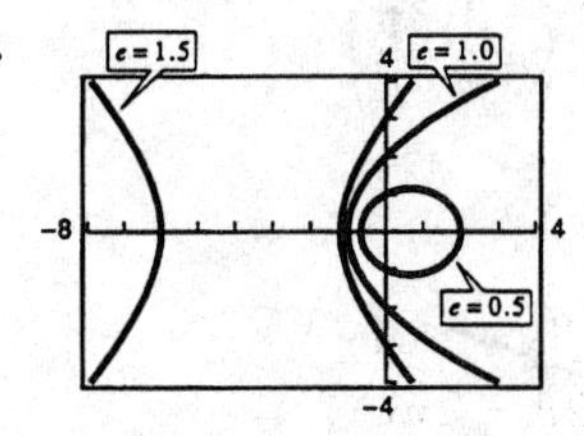

4.

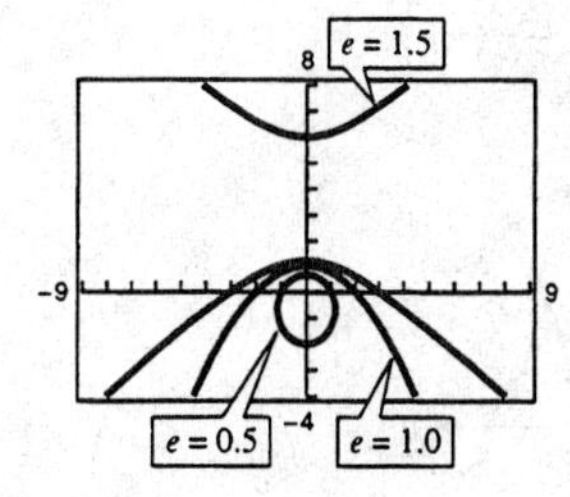

6. $r = \dfrac{3}{2 - \cos \theta}$

Ellipse

Matches (c).

8. $r = \dfrac{4}{1 + \sin \theta}$

Parabola opening downward

Matches (a).

10. $r = \dfrac{3}{1 + \sin \theta}$

$e = 1$, the graph is a parabola.

Vertex: $(3/2, \pi/2)$

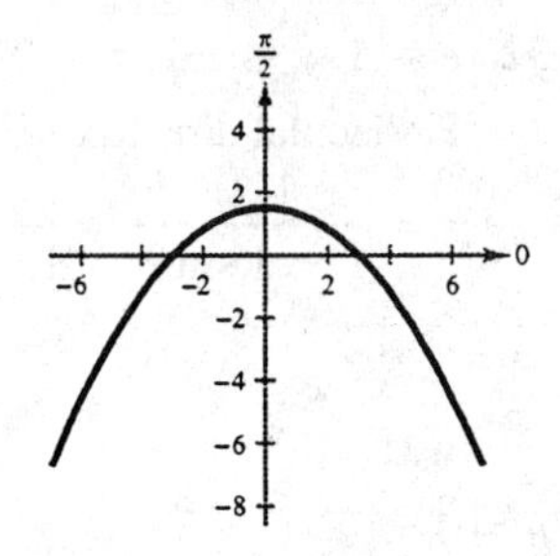

12. $r = \dfrac{6}{1 + \cos \theta}$

$e = 1$, the graph is a parabola.

Vertex: $(3, 0)$

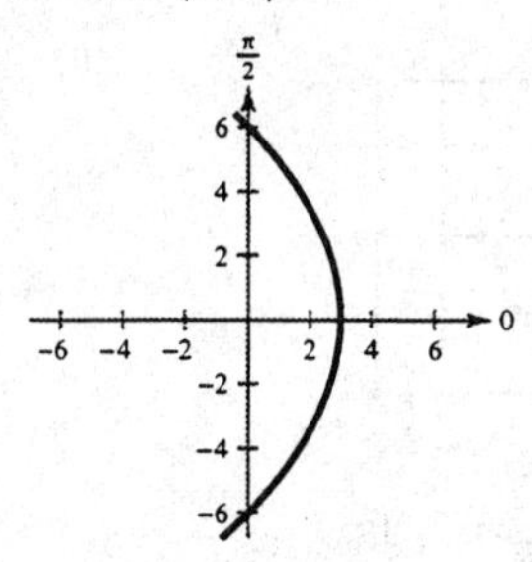

14. $r = \dfrac{3}{3 + \sin \theta} = \dfrac{1}{1 + \frac{1}{3} \sin \theta}$

$e = \dfrac{1}{3} < 1$, the graph is an ellipse.

Vertices: $\left(\dfrac{3}{4}, \dfrac{\pi}{2}\right), \left(\dfrac{3}{2}, \dfrac{3\pi}{2}\right)$

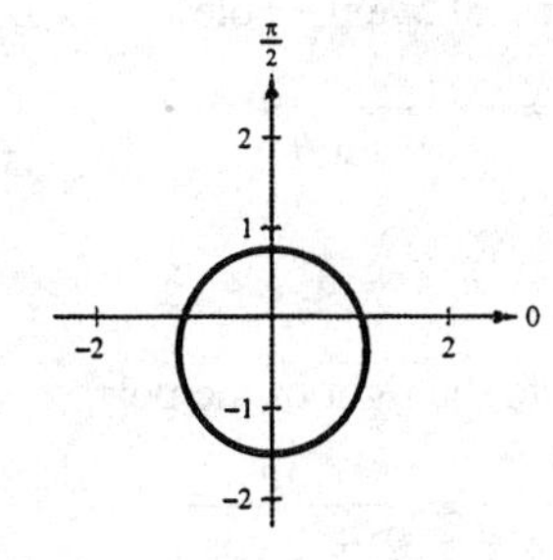

16. $r = \dfrac{6}{3 - 2 \cos \theta} = \dfrac{2}{1 - \frac{2}{3} \cos \theta}$

$e = \dfrac{2}{3} < 1$, the graph is an ellipse.

Vertices: $(6, 0), \left(\dfrac{6}{5}, \pi\right)$

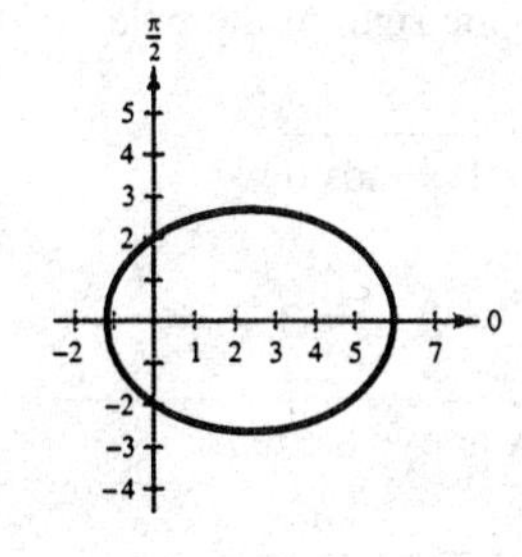

18. $r = \dfrac{5}{-1 + 2\cos\theta} = \dfrac{-5}{1 - 2\cos\theta}$

$e = 2 > 1$, the graph is a hyperbola.

Vertices: $(5, 0), \left(-\dfrac{5}{3}, \pi\right)$

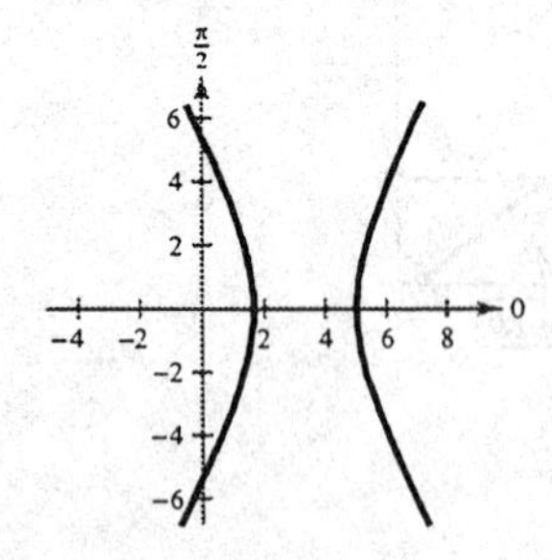

20. $r = \dfrac{3}{2 + 6\sin\theta} = \dfrac{3/2}{1 + 3\sin\theta}$

$e = 3 > 1$, the graph is a hyperbola.

Vertices: $\left(\dfrac{3}{8}, \dfrac{\pi}{2}\right), \left(-\dfrac{3}{4}, \dfrac{3\pi}{2}\right)$

22.

24.

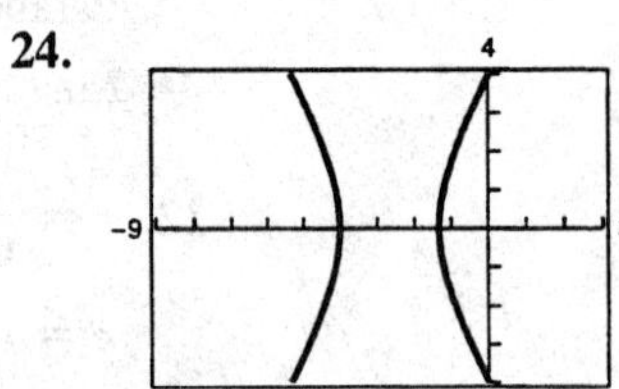

26.

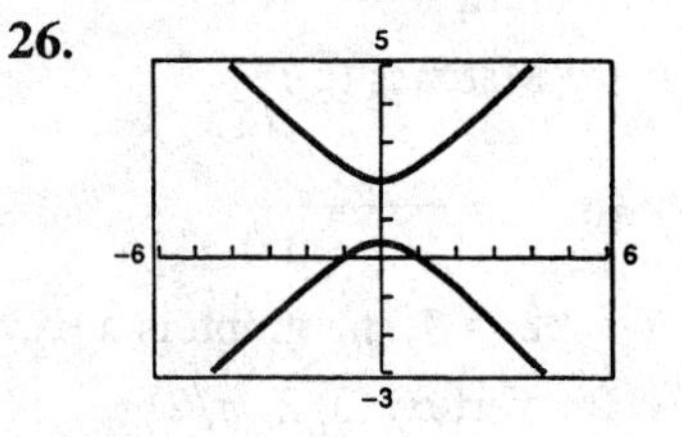

28.

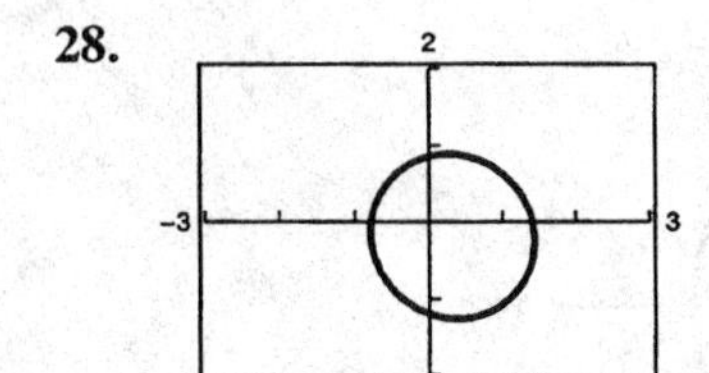

30.

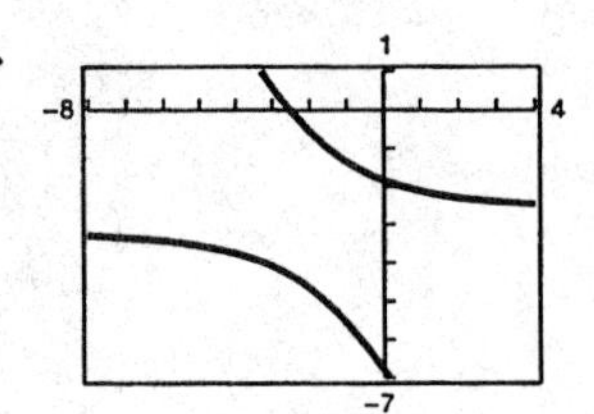

32. $e = 1, y = -2, p = 2$

Horizontal directrix below the pole

$r = \dfrac{1(2)}{1 - 1\sin\theta} = \dfrac{2}{1 - \sin\theta}$

34. $e = \dfrac{3}{4}, y = -2, p = 2$

Horizontal directrix below the pole

$r = \dfrac{3/4(2)}{1 - (3/4)\sin\theta} = \dfrac{6}{4 - 3\sin\theta}$

36. $e = \dfrac{3}{2}, x = -1, p = 1$

Vertical directrix to the left of the pole

$r = \dfrac{3/2(1)}{1 - (3/2)\cos\theta} = \dfrac{3}{2 - 3\cos\theta}$

38. Vertex: $(4, 0) \Rightarrow e = 1, p = 8$

Vertical directrix to the right of the pole

$r = \dfrac{1(8)}{1 + 1\cos\theta} = \dfrac{8}{1 + \cos\theta}$

40. Vertex: $\left(10, \dfrac{\pi}{2}\right) \Rightarrow e = 1, p = 20$

Horizontal directrix above the pole

$r = \dfrac{1(20)}{1 + 1\sin\theta} = \dfrac{20}{1 + \sin\theta}$

42. Center: $\left(1, \dfrac{3\pi}{2}\right)$; $c = 1, a = 3, e = \dfrac{1}{3}$

Horizontal directrix above the axis

$r = \dfrac{(1/3)p}{1 + (1/3)\sin\theta} = \dfrac{p}{3 + \sin\theta}$

$2 = \dfrac{p}{3 + \sin(\pi/2)}$

$p = 8$

$r = \dfrac{8}{3 + \sin\theta}$

44. Center: $(6, 0)$; $c = 6, a = 4, e = \dfrac{3}{2}$

Vertical directrix to the right of the pole

$r = \dfrac{(3/2)p}{1 + (3/2)\cos\theta} = \dfrac{3p}{2 + 3\cos\theta}$

$2 = \dfrac{3p}{2 + 3\cos\theta}$

$p = \dfrac{10}{3}$

$r = \dfrac{3(10/3)}{2 + 3\cos\theta} = \dfrac{10}{2 + 3\cos\theta}$

46. Center: $\left(\frac{5}{2}, \frac{\pi}{2}\right)$; $c = \frac{5}{2}$, $a = \frac{3}{2}$, $e = \frac{5}{3}$

Horizontal directrix above the pole

$$r = \frac{(5/3)p}{1 + (5/3)\sin\theta} = \frac{5p}{3 + 5\sin\theta}$$

$$1 = \frac{5p}{3 + 5\sin(-3\pi/2)}$$

$$p = \frac{8}{5}$$

$$r = \frac{5(8/5)}{3 + 5\sin\theta} = \frac{8}{3 + 5\sin\theta}$$

48.

$$\frac{x^2}{a^2} - \frac{y^2}{b^2} = 1$$

$$\frac{r^2\cos^2\theta}{a^2} - \frac{r^2\sin^2\theta}{b^2} = 1$$

$$\frac{r^2\cos^2\theta}{a^2} - \frac{r^2(1 - \cos^2\theta)}{b^2} = 1$$

$$r^2b^2\cos^2\theta - r^2a^2 + r^2a^2\cos^2\theta = a^2b^2$$

$$r^2(b^2 + a^2)\cos^2\theta - r^2a^2 = a^2b^2$$

$$a^2 + b^2 = c^2$$

$$r^2c^2\cos^2\theta - r^2a^2 = a^2b^2$$

$$r^2\left(\frac{c}{a}\right)^2\cos^2\theta - r^2 = b^2,\ e = \frac{c}{a}$$

$$r^2e^2\cos^2\theta - r^2 = b^2$$

$$r^2(e^2\cos^2\theta - 1) = b^2$$

$$r^2 = \frac{b^2}{e^2\cos^2\theta - 1}$$

$$= \frac{-b^2}{1 - e^2\cos^2\theta}$$

50. $\frac{x^2}{25} + \frac{y^2}{16} = 1$

$a = 5, b = 4, c = 3, e = \frac{3}{5}$

$$r^2 = \frac{400}{25 - 9\cos^2\theta}$$

52. $\frac{x^2}{36} - \frac{y^2}{4} = 1$

$a = 6, b = 2, c = 2\sqrt{10}, e = \frac{\sqrt{10}}{3}$

$$r^2 = \frac{-4}{1 - (10/9)\cos^2\theta} = \frac{-36}{9 - 10\cos^2\theta}$$

54. Ellipse

One focus: $(4, 0)$

Vertices: $(5, 0)$, $(5, \pi)$

$a = 5, c = 4, b = 3, e = \frac{4}{5}$

$$r^2 = \frac{9}{1 - (16/25)\cos^2\theta} = \frac{225}{25 - 16\cos^2\theta}$$

56. Minimum distance occurs when $\theta = \pi$.

$$r = \frac{(1 - e^2)}{1 - e\cos\pi} = \frac{(1 - e)(1 + e)a}{1 + e} = a(1 - e)$$

Maximum distance occurs when $\theta = 0$.

$$r = \frac{(1 - e^2)a}{1 - e\cos 0} = \frac{(1 - e)(1 + e)a}{1 - e} = a(1 + e)$$

58. Perihelion distance $= a(1 - e) = 4.4362 \times 10^9$ kilometers

Aphelion distance $= a(1 + e) = 7.3638 \times 10^9$ kilometers

$$r = \frac{(1 - e^2)a}{1 - e\cos\theta} = \frac{5.5368 \times 10^9}{1 - 0.2481\cos\theta} \quad \text{(Exercise 55)}$$

Review Exercises for Chapter 10

Solutions to Even-Numbered Exercises

2. $4x^2 - y^2 = 4$

$$\frac{x^2}{1} - \frac{y^2}{4} = 1$$

Hyperbola with center (0, 0) and a horizontal transverse axis. Matches graph (c).

4. $y^2 - 4x^2 = 4$

$$\frac{y^2}{4} - \frac{x^2}{1} = 1$$

Hyperbola with center (0, 0) and a vertical transverse axis. Matches graph (b).

6. $8y + x^2 = 0$

$$-8y = x^2$$

The graph is a parabola.

Vertex: (0, 0)

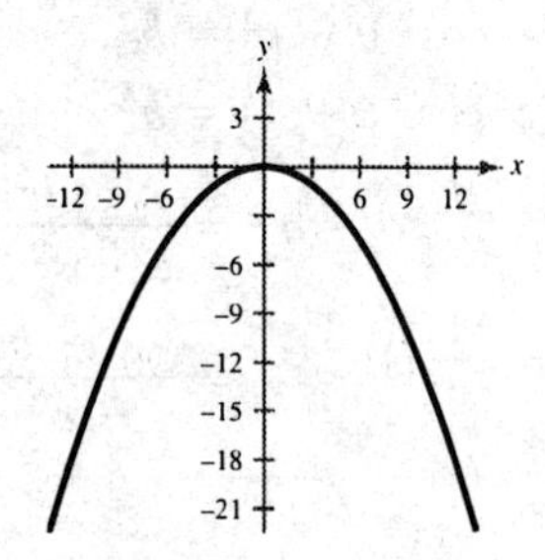

8. $y^2 - 12y - 8x + 20 = 0$

$AC = 0(1) = 0$

The graph is a parabola.

$$y^2 - 12y - 8x + 20 = 0$$

$$(y - 6)^2 = 8(x + 2)$$

Vertex: $(-2, 6)$

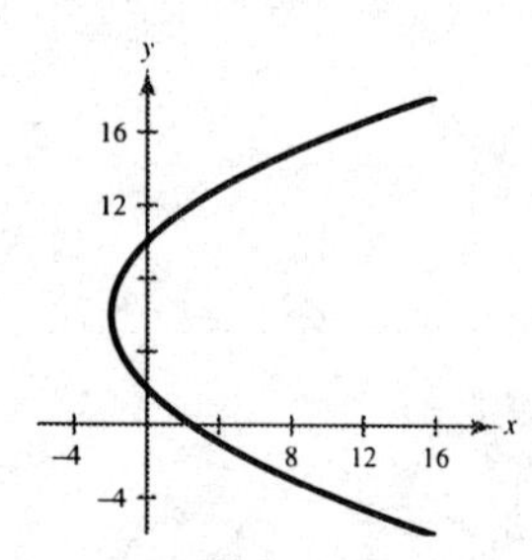

10. $16x^2 + 16y^2 - 16x + 24y - 3 = 0$

$A = C = 16$

The graph is a circle.

$$16x^2 + 16y^2 - 16x + 24y - 3 = 0$$

$$\left(x - \frac{1}{2}\right)^2 + \left(y + \frac{3}{4}\right)^2 = 1$$

Center: $\left(\frac{1}{2}, -\frac{3}{4}\right)$

Radius: 1

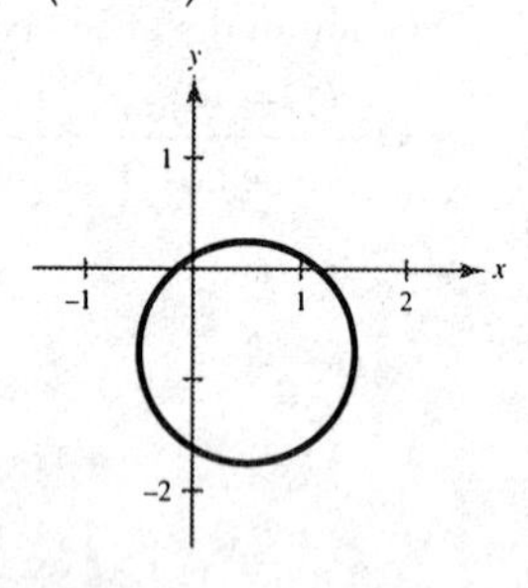

12. $2x^2 + 6y^2 = 18$

$AC = 2(6) = 12 > 0$

The graph is an ellipse.

$$\frac{2x^2}{18} + \frac{6y^2}{18} = \frac{18}{18}$$

$$\frac{x^2}{9} + \frac{y^2}{3} = 1$$

Center: (0, 0)

Vertex: $(\pm 3, 0)$

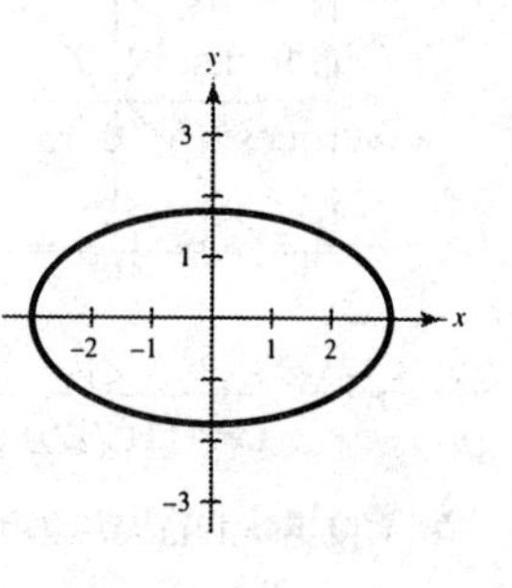

14. $4x^2 + y^2 - 16x + 15 = 0$

$AC = 4(1) > 0$

The graph is an ellipse.

$$4x^2 + y^2 - 16x + 15 = 0$$

$$\frac{(x - 2)^2}{1/4} + y^2 = 1$$

Center: (2, 0)

Vertices: $(2, \pm 1)$

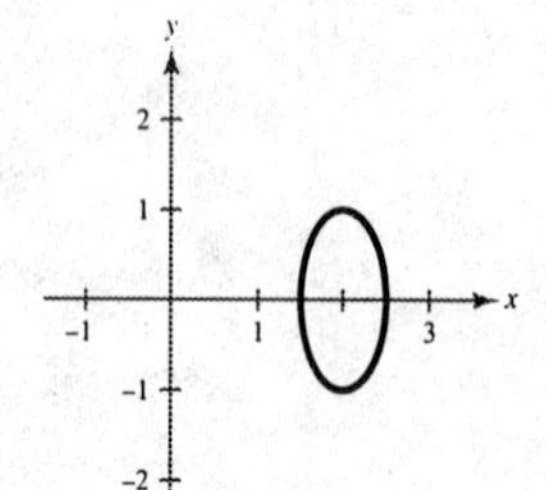

16. $x^2 - 9y^2 + 10x + 18y + 7 = 0$

$AC = 1(-9) = -9 < 0$

The graph is a hyperbola.

$x^2 + 10x - 9y^2 + 18y = -7$

$$\frac{(x + 5)^2}{9} - \frac{9(y - 1)^2}{9} = 1$$

Center: $(-5, 1)$

Vertices: $(-8, 1), (-2, 1)$

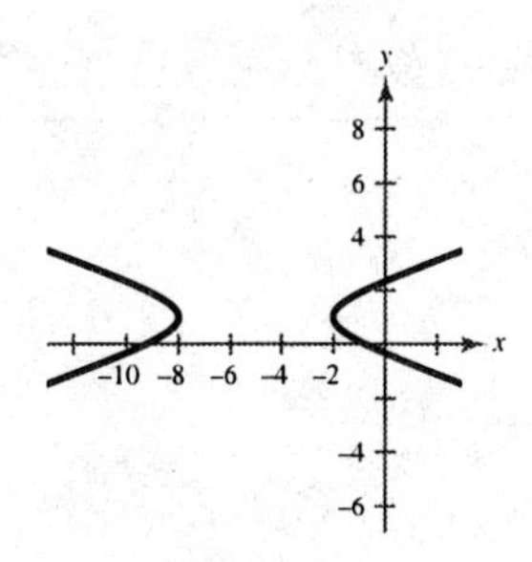

18. $4x^2 - 4y^2 - 4x + 8y - 11 = 0$

$AC = 4(-4) = -16 < 0$

The graph is a hyperbola.

$4x^2 - 4y^2 - 4x + 8y - 11 = 0$

$$\frac{(x - 1/2)^2}{2} - \frac{(y - 1)^2}{2} = 1$$

Center: $\left(\frac{1}{2}, 1\right)$

Vertices: $\left(\frac{1}{2} \pm \sqrt{2}, 1\right)$

To use a graphing calculator, we need to solve for y in terms of x.

$$(y - 1)^2 = \left(x - \frac{1}{2}\right)^2 - 2$$

$$y = 1 \pm \sqrt{\left(x - \frac{1}{2}\right)^2 - 2}$$

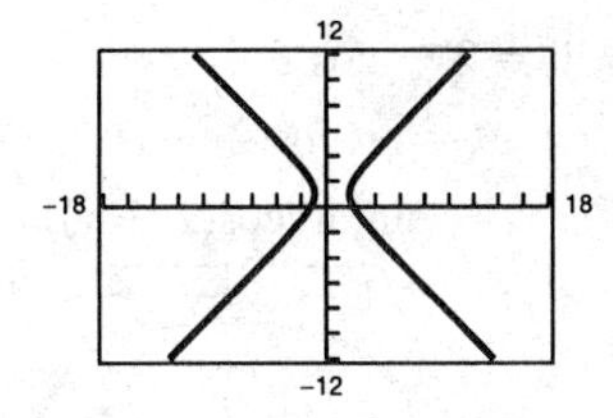

20. $40x^2 + 36xy + 25y^2 - 52 = 0$

$B^2 - 4AC = 36^2 - 4(40)(25) = -2704 < 0$

The graph is an ellipse.

To use a graphing calculator, we need to solve for y in terms of x.

$$25y^2 + 36xy = 52 - 40x^2$$

$$y^2 + \frac{36}{25}xy + \frac{324}{625} = \frac{1}{625}(1624 - 1000x^2)$$

$$\left(y + \frac{18}{25}x\right)^2 = \frac{1}{625}(1624 - 1000x^2)$$

$$y + \frac{18}{25}x = \pm\frac{1}{25}\sqrt{1624 - 1000x^2}$$

$$y = \frac{18}{25}x \pm \frac{1}{25}\sqrt{1624 - 1000x^2}$$

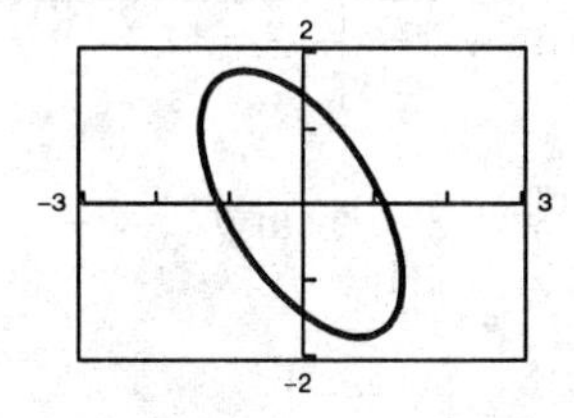

22. $9x^2 + 6y^2 + 4xy - 20 = 0$

$B^2 - 4AC = 4^2 - 4(9)(6) = -220 < 0$

The graph is an ellipse.

To use a graphing utility, we need to solve for y in terms of x.

$6y^2 + 4xy + 9x^2 - 20 = 0$

$$y = \frac{4x \pm \sqrt{16x^2 - 4(6)(9x^2 - 20)}}{2(6)}$$

$$= -\frac{2x \pm \sqrt{120 - 50x^2}}{6}$$

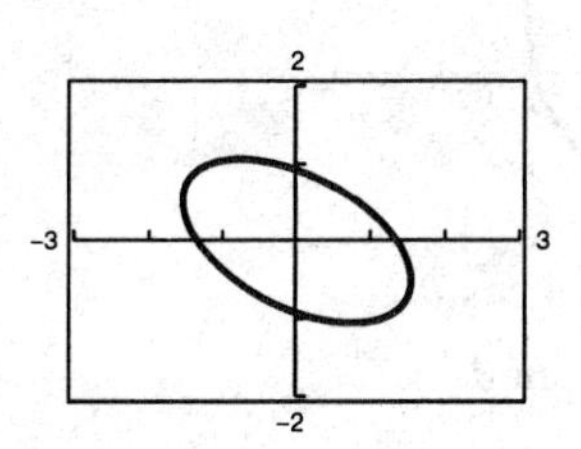

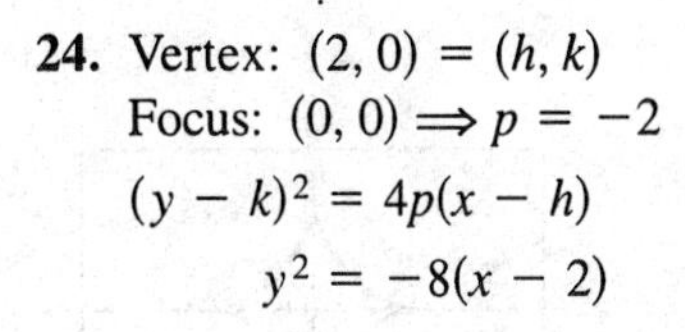

24. Vertex: $(2, 0) = (h, k)$

Focus: $(0, 0) \Longrightarrow p = -2$

$(y - k)^2 = 4p(x - h)$

$y^2 = -8(x - 2)$

26. Vertex: $(2, 2) = (h, k)$

Directrix: $y = 0 \Longrightarrow p = 2$

$(x - h)^2 = 4p(y - k)$

$(x - 2)^2 = 8(y - 2)$

28. Vertices: $(2, 0), (2, 4) \Rightarrow a = 2, (h, k) = (2, 2)$

Foci: $(2, 1), (2, 3) \Rightarrow c = 1$

$b^2 = a^2 - c^2 = 4 - 1 = 3$

$$\frac{(x-h)^2}{b^2} + \frac{(y-k)^2}{a^2} = 1$$

$$\frac{(x-2)^2}{3} + \frac{(y-2)^2}{4} = 1$$

30. Vertices: $(0, 1), (4, 1) \Rightarrow a = 2, (h, k) = (2, 1)$

Endpoints of minor axis: $(2, 0), (2, 2) \Rightarrow b = 1$

$$\frac{(x-h)^2}{a^2} + \frac{(y-k)^2}{b^2} = 1$$

$$\frac{(x-2)^2}{4} + (y-1)^2 = 1$$

32. Vertices: $(2, 2), (-2, 2) \Rightarrow a = 2, (h, k) = (0, 2)$

Foci: $(4, 2), (-4, 2) \Rightarrow c = 4$

$b^2 = c^2 - a^2 = 16 - 4 = 12$

$$\frac{(x-h)^2}{a^2} - \frac{(y-k)^2}{b^2} = 1$$

$$\frac{x^2}{4} - \frac{(y-2)^2}{12} = 1$$

34. Foci: $(3, \pm 2) \Rightarrow c = 2, (h, k) = (3, 0)$

Asymptotes: $y = \pm 2(x-3) \Rightarrow \frac{a}{b} = 2, a = 2b$

$b^2 = c^2 - a^2 = 4 - 4b^2 \Rightarrow b^2 = \frac{4}{5}, a^2 = \frac{16}{5}$

$$\frac{(y-k)^2}{a^2} - \frac{(x-h)^2}{b^2} = 1$$

$$\frac{y^2}{16/5} - \frac{(x-3)^2}{4/5} = 1 \Rightarrow \frac{5y^2}{16} - \frac{5(x-3)^2}{4} = 1$$

36. $x = t + 4, y = t^2$

$t = x - 4$

$y = (x-4)^2$

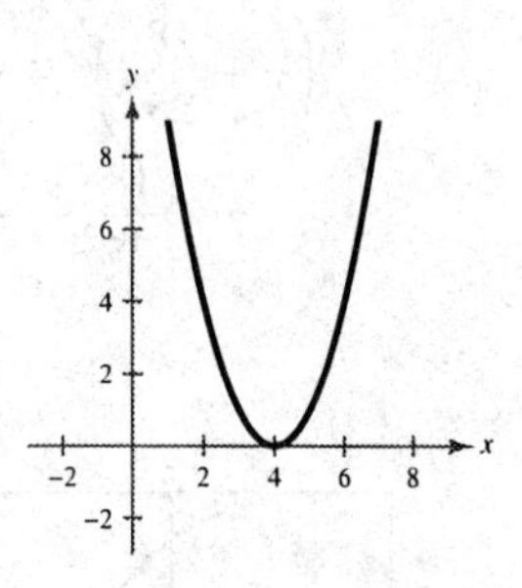

38. $x = \frac{1}{t}, y = 2t + 3$

$t = \frac{1}{x}$

$y = \frac{2}{x} + 3$

$y = \frac{2 + 3x}{x}$

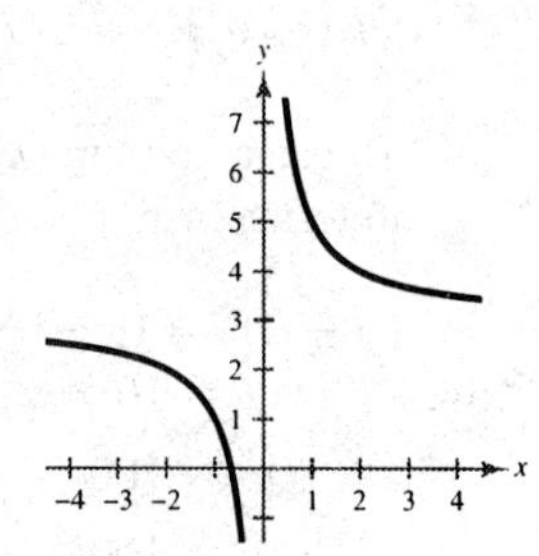

40. $x = 3 + 3\cos\theta, y = 2 + 5\sin\theta$

$\cos\theta = \frac{x-3}{3}, \sin\theta = \frac{y-2}{5}$

$$\frac{(x-3)^2}{9} + \frac{(y-2)^2}{25} = 1$$

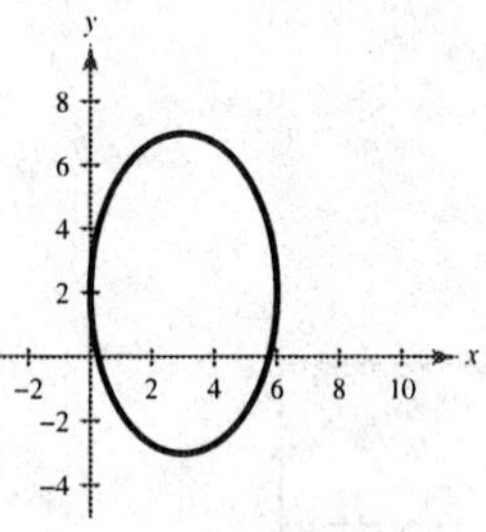

42. $x = 2\theta - \sin\theta, y = 2 - \cos\theta$

$\cos\theta = 2 - y \Rightarrow \theta = \arccos(2 - y)$

$x = 2\arccos(2-y) - \sin(\arccos(2-y))$

$x = 2\arccos(2-y) - \sqrt{1-(2-y)^2} + 4\pi n$

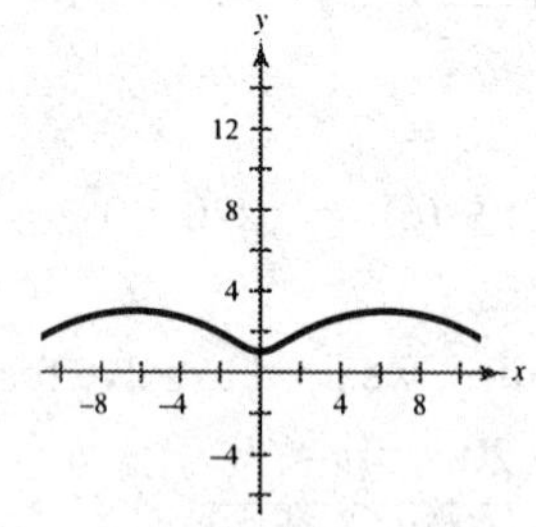

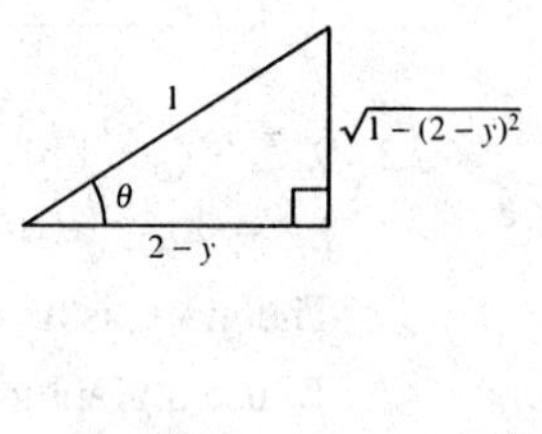

44. $x = -3 + 4\cos\theta$

$y = 4 + 3\sin\theta$

$$\frac{(x+3)^2}{4^2} + \frac{(y-4)^2}{3^2} = 1$$

46. (a)

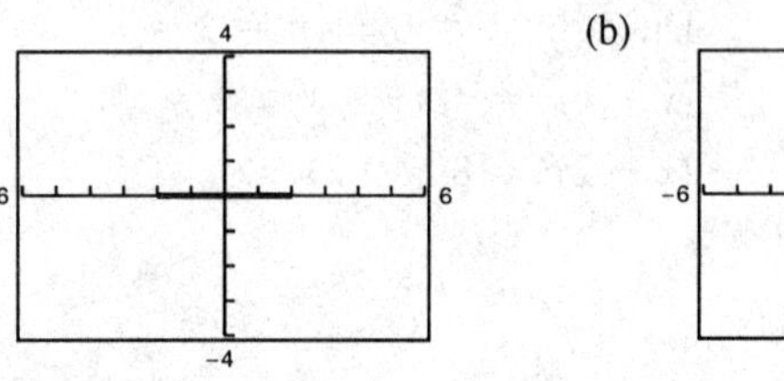

(b)

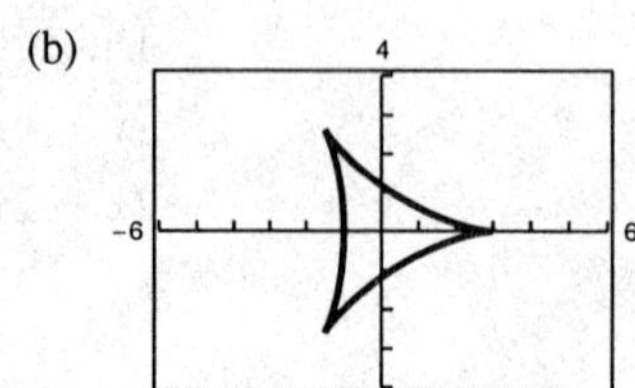

(c)

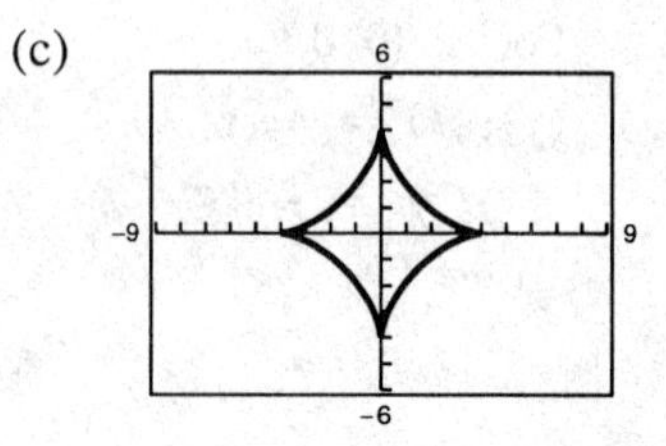

–CONTINUED–

46. –CONTINUED–

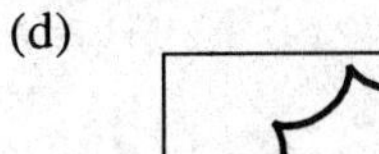

(d)

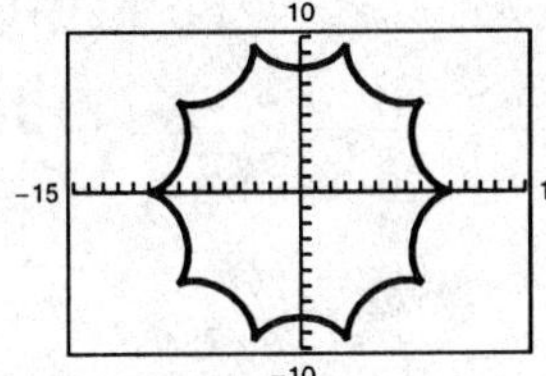

(e)

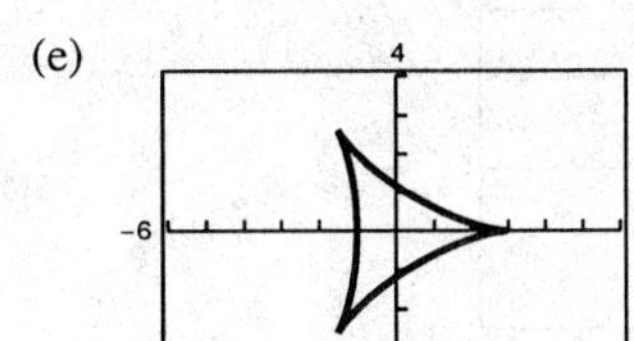

(f)

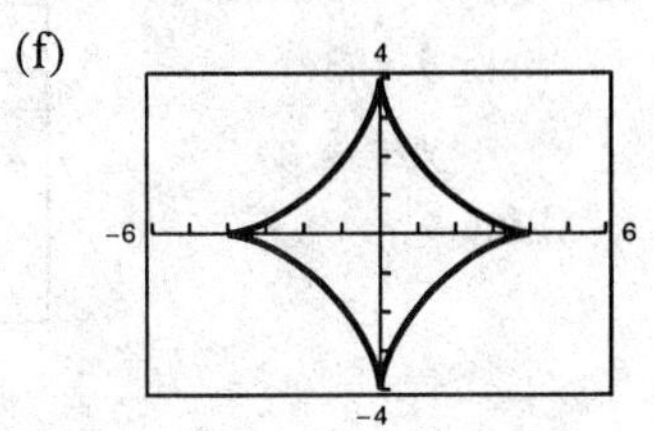

48. $x^2 + y^2 - 4x = 0$

$$r^2 - 4r\cos\theta = 0$$

$$r = 4\cos\theta$$

50. $r = 4\sec\left(\theta - \dfrac{\pi}{3}\right)$

$$r\cos\left(\theta - \frac{\pi}{3}\right) = 4$$

$$r\left(\cos\theta\cos\frac{\pi}{3} + \sin\theta\sin\frac{\pi}{3}\right) = 4$$

$$\frac{1}{2}r\cos\theta + \frac{\sqrt{3}}{2}r\sin\theta = 4$$

$$\frac{1}{2}x + \frac{\sqrt{3}}{2}y = 4$$

$$x + \sqrt{3}y = 8$$

52. $r = \dfrac{1}{2 - \cos\theta}$

$$2r - r\cos\theta = 1$$

$$2\sqrt{x^2 + y^2} = x + 1$$

$$2\sqrt{x^2 + y^2} - x = 1$$

$$4(x^2 + y^2) = x^2 + 2x + 1$$

$$3x^2 + 4y^2 - 2x - 1 = 0$$

54. $r = 10$

$$r^2 = 100$$

$$x^2 + y^2 = 100$$

56. $\theta = \dfrac{\pi}{12}$

Line

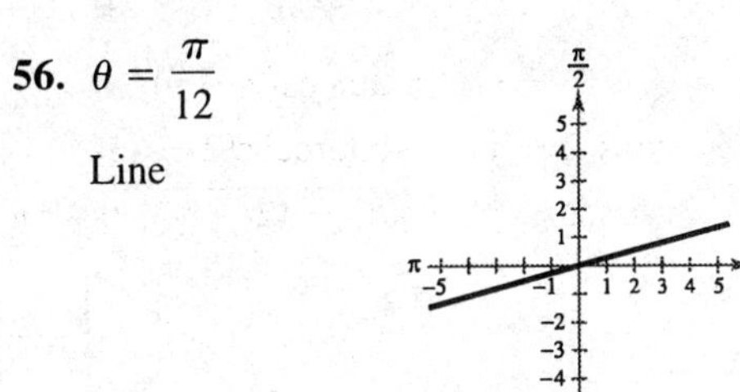

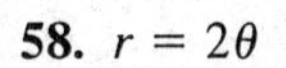

58. $r = 2\theta$

Symmetric with respect to $\theta = \pi/2$

Spiral

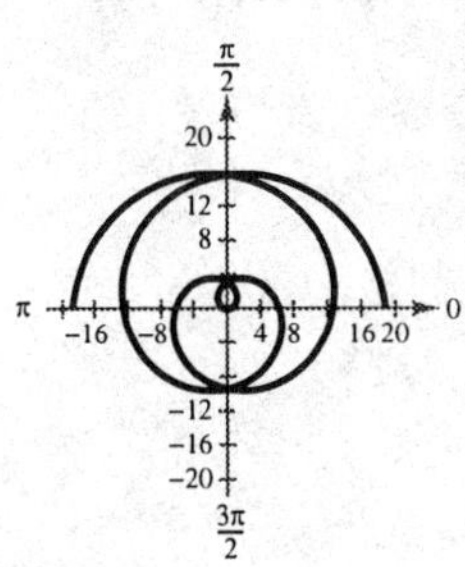

60. $r = 3 - 4\cos\theta$

Symmetric with respect to polar axis

$\dfrac{a}{b} = \dfrac{3}{4} < 0 \Rightarrow$ Limaçon with inner loop

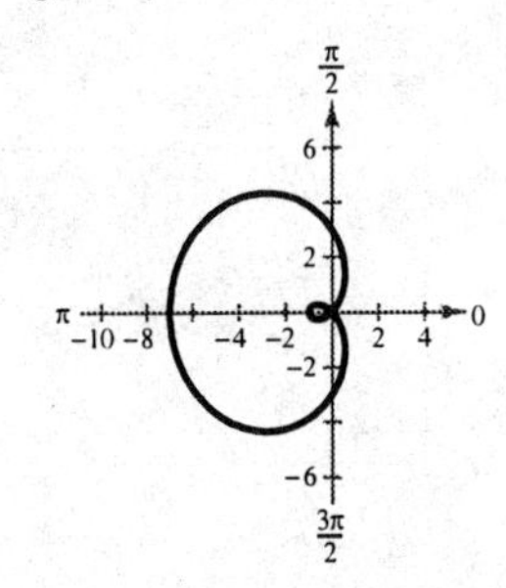

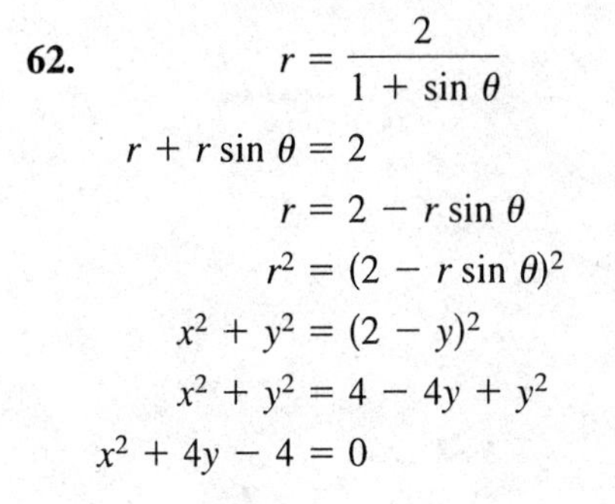

62. $r = \dfrac{2}{1 + \sin\theta}$

$$r + r\sin\theta = 2$$

$$r = 2 - r\sin\theta$$

$$r^2 = (2 - r\sin\theta)^2$$

$$x^2 + y^2 = (2 - y)^2$$

$$x^2 + y^2 = 4 - 4y + y^2$$

$$x^2 + 4y - 4 = 0$$

64. $r = 3 \csc \theta$

$r \sin \theta = 3$

$y = 3$

Line

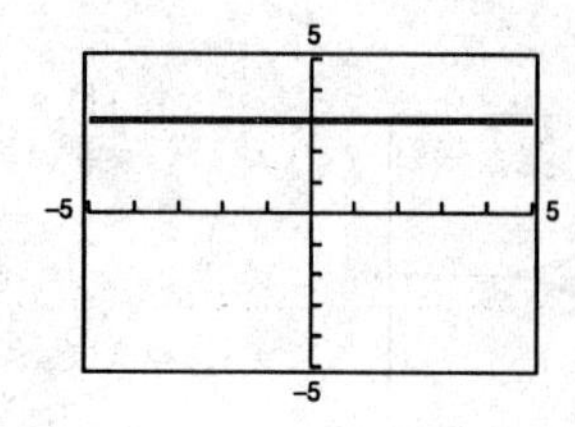

66. $r = \dfrac{4}{5 - 3 \cos \theta}$

$r = \dfrac{4/5}{1 - (3/5) \cos \theta}, \; e = \dfrac{3}{5}$

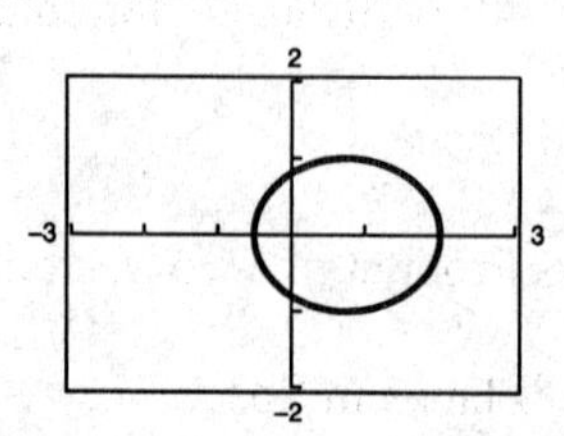

Ellipse symmetric with polar axis and having vertices at $(2, 0)$ and $(1/2, \pi)$.

68. $m = \sqrt{3}$

$\tan \theta = \sqrt{3}$

$\theta = \dfrac{\pi}{3}$

70. Parabola: $r = \dfrac{ep}{1 + e \sin \theta}, \; e = 1$

Vertex: $\left(2, \dfrac{\pi}{2}\right)$

Focus: $(0, 0) \implies p = 4$

$r = \dfrac{4}{1 + \sin \theta}$

72. Hyperbola: $r = \dfrac{ep}{1 + e \cos \theta}$

Vertices: $(1, 0), (7, 0) \implies a = 3$

One focus: $(0, 0) \implies c = 4$

$e = \dfrac{c}{a} = \dfrac{4}{3}, \; p = \dfrac{7}{4}$

$r = \dfrac{(4/3)(7/4)}{1 + (4/3) \cos \theta} = \dfrac{7/3}{1 + (4/3) \cos \theta} = \dfrac{7}{3 + 4 \cos \theta}$

CHAPTER 11
Analytic Geometry in Three Dimensions

CHAPTER 11

Analytic Geometry in Three Dimensions

Section 11.1 The Three-Dimensional Coordinate System

Solutions to Even-Numbered Exercises

2. $A(-1, 4, 3), B(1, 3, -2)$

4. $A(6, 2, -3), B(2, -1, 2)$

6.

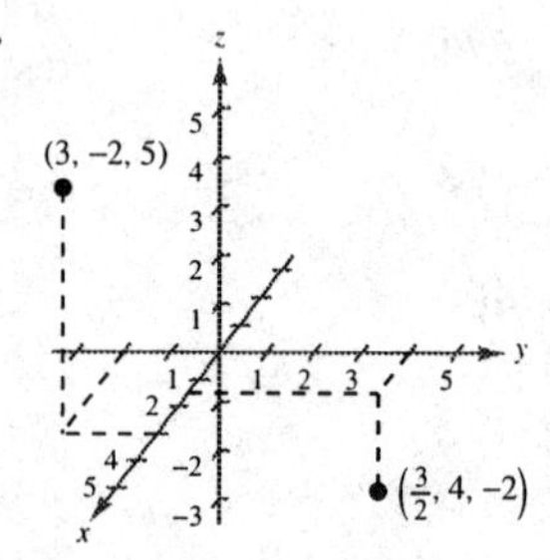

8.

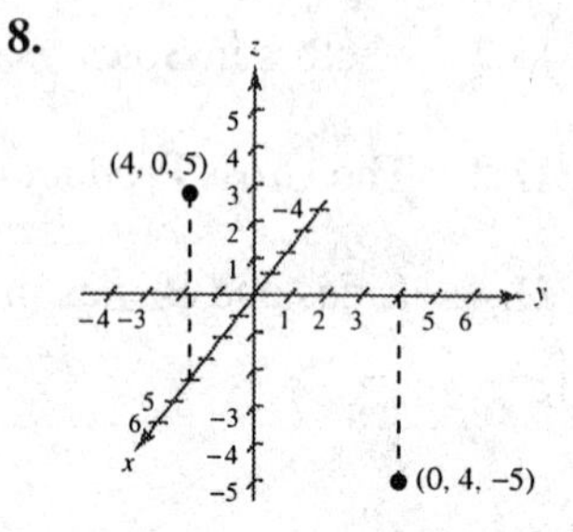

10. $x = 7, y = -2, z = -1 \Rightarrow (7, -2, -1)$

12. $x = 0, y = 3, z = 2 \Rightarrow (0, 3, 2)$

14. No. The scales depend on the magnitudes of the measurements recorded for each coordinate.

16. Octant VI

18. Octants III, IV, VII, or VIII

20. Octants I, II, VII, or VIII

22. $d_1 = \sqrt{(-2-0)^2 + (3-0)^2 + (4-0)^2} = \sqrt{4+9+16} = \sqrt{29}$

$d_2 = \sqrt{(-2-0)^2 + (3-0)^2 + (0-0)^2} = \sqrt{4+9} = \sqrt{13}$

$d_3 = \sqrt{(-2+2)^2 + (3-3)^2 + (4-0)^2} = \sqrt{16} = 4$

$d_1^2 = d_2^2 + d_3^2 = 29$

24. $d_1 = \sqrt{(-4-2)^2 + (4+1)^2 + (1-2)^2} = \sqrt{36+25+1} = \sqrt{62}$

$d_2 = \sqrt{(-4+2)^2 + (4-5)^2 + (1-0)^2} = \sqrt{4+1+1} = \sqrt{6}$

$d_3 = \sqrt{(2+2)^2 + (-1-5)^2 + (2-0)^2} = \sqrt{16+36+4} = \sqrt{56}$

$d_1^2 = d_2^2 + d_3^2 = 62$

26. $d_1 = \sqrt{(7-5)^2 + (1-3)^2 + (3-4)^2} = \sqrt{4+4+1} = \sqrt{9} = 3$

$d_2 = \sqrt{(3-7)^2 + (5-1)^2 + (3-3)^2} = \sqrt{16+16} = \sqrt{32} = 4\sqrt{2}$

$d_3 = \sqrt{(3-5)^2 + (5-3)^2 + (3-4)^2} = \sqrt{4+4+1} = \sqrt{9} = 3$

$d_1 = d_3 = 3$. Isosceles triangle

28. $d_1 = \sqrt{(5-0)^2 + (0-2)^2 + (0-0)^2} = \sqrt{25+4} = \sqrt{29}$

$d_2 = \sqrt{(0-0)^2 + (0-2)^2 + (-3-0)^2} = \sqrt{4+9} = \sqrt{13}$

$d_3 = \sqrt{(0-5)^2 + (0-0)^2 + (-3-0)^2} = \sqrt{25+9} = \sqrt{34}$

Neither.

30. Midpoint: $\left(\frac{2+6}{2}, \frac{-2+6}{2}, \frac{-8+18}{2}\right) = (4, 2, 5)$

32. Midpoint: $\left(\frac{-3-6}{2}, \frac{5+4}{2}, \frac{7+10}{2}\right) = \left(-\frac{9}{2}, \frac{9}{2}, \frac{17}{2}\right)$

34. $x_2 = 2x_m - x_1 = 2(5) - 3 = 7$

$y_2 = 2y_m - y_1 = 2(8) - 0 = 16$

$z_2 = 2z_m - z_1 = 2(7) - 2 = 12$

$(7, 16, 12)$

36. $(x+3)^2 + (y-4)^2 + (z-3)^2 = 4$

38. $(x-1)^2 + (y+2)^2 + (z-3)^2 = 5^2 = 25$

40. radius $= \dfrac{\text{diameter}}{2} = 3$: $(x-0)^2 + (y-5)^2 + (z+9)^2 = 3^2 = 9$

42. Center: $\left(\frac{2-1}{2}, \frac{-2+4}{2}, \frac{2+6}{2}\right) = \left(\frac{1}{2}, 1, 4\right)$

Radius: $\sqrt{\left(2-\frac{1}{2}\right)^2 + (-2-1)^2 + (2-4)^2} = \sqrt{\frac{9}{4} + 9 + 4} = \sqrt{\frac{61}{4}}$

Sphere: $\left(x - \frac{1}{2}\right)^2 + (y-1)^2 + (z-4)^2 = \frac{61}{4}$

44. $(x^2 - 6x + 9) + (y^2 + 4y + 4) + z^2 = -9 + 9 + 4$

$(x-3)^2 + (y+2)^2 + z^2 = 4$

Center: $(3, -2, 0)$
Radius: 2

46. $x^2 + (y^2 - 8y + 16) + (z^2 - 6z + 9) = -13 + 16 + 9$

$x^2 + (y-4)^2 + (z-3)^2 = 12$

Center: $(0, 4, 3)$
Radius: $\sqrt{12} = 2\sqrt{3}$

48. $x^2 + y^2 + z^2 - x - 3y - 2z = -\frac{5}{2}$

$\left(x^2 - x + \frac{1}{4}\right) + \left(y^2 - 3y + \frac{9}{4}\right) + (z^2 - 2z + 1) = -\frac{5}{2} + \frac{1}{4} + \frac{9}{4} + 1$

$\left(x - \frac{1}{2}\right)^2 + \left(y - \frac{3}{2}\right)^2 + (z-1)^2 = \perp$

Center: $\left(\frac{1}{2}, \frac{3}{2}, 1\right)$

Radius: 1

50. (a)

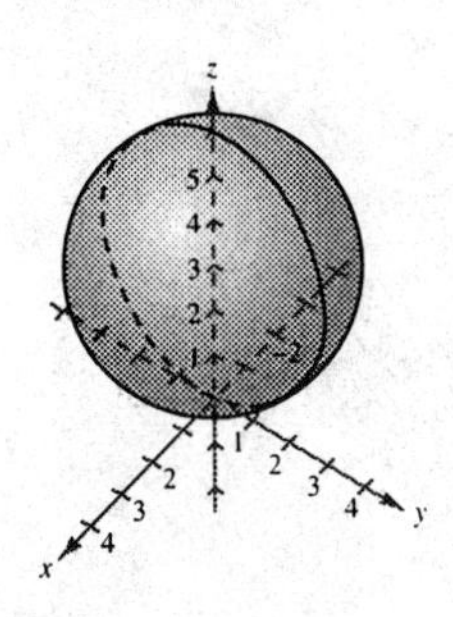

(b)

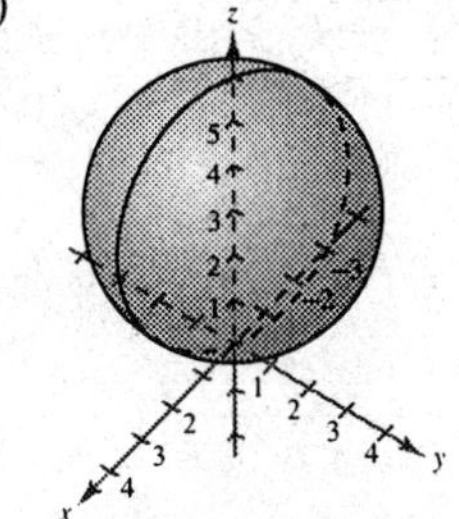

52. $z^2 = -x^2 - y^2 + 4y + 4 \Rightarrow$

$\begin{cases} z_1 = \sqrt{-x^2 - y^2 + 4y + 4} \\ z_2 = -\sqrt{-x^2 - y^2 + 4y + 4} \end{cases}$

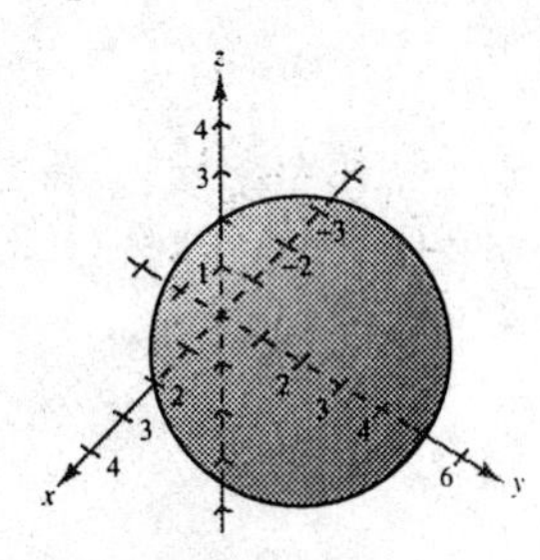

54. $x^2 + y^2 + 6y + (z^2 - 8z + 16) = -21 + 16$

$x^2 + y^2 + 6y + (z-4)^2 = -5$

$z_1 = 4 + \sqrt{-5 - x^2 - y^2 - 6y}$

$z_2 = 4 - \sqrt{-5 - x^2 - y^2 - 6y}$

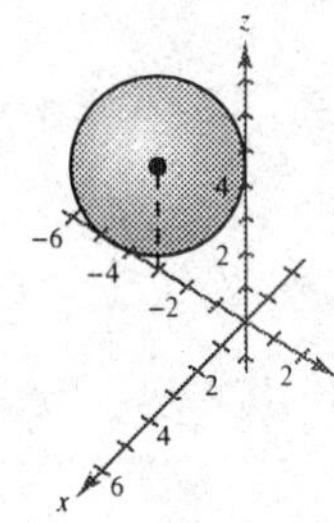

56. $d_1(t) = 0.07t^2 - 2.43t + 12 = 0$ when $t = 5.96$.

$d_2(5.96) \approx 0.16(5.96)^2 - 3.93(5.96) + 20 \approx 2.26$

Yes, the aircraft will be less than 3 miles apart when the first aircraft lands.

$d_2 - d_1 = .09t^2 - 1.5t + 8 = 3$ when $t \approx 4.6$.

Section 11.2 Vectors in Space

Solutions to Even-Numbered Exercises

2. $\mathbf{v} = \langle 1 - 1, 4 - 4, 0 - 4 \rangle = \langle 0, 0, -4 \rangle$

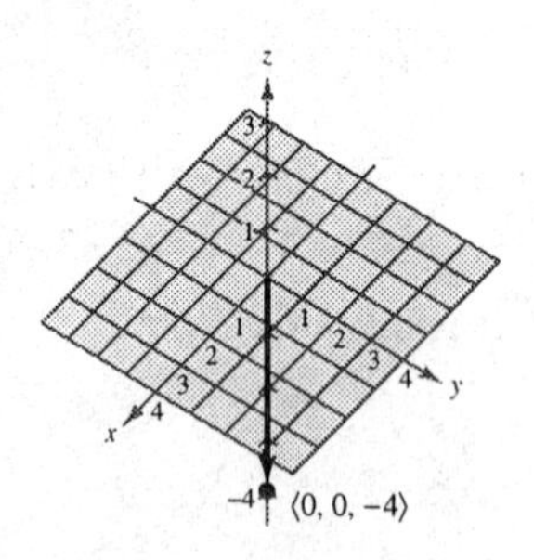

4. $\mathbf{v} = \langle 4 - (-4), 0 - 5, 0 - 5 \rangle = \langle 8, -5, -5 \rangle$

any vector of for $c\langle 8, -5, -5 \rangle$, $c > 0$, is parallel to $\mathbf{v}$.

any vector of form $c\langle 8, -5, -5 \rangle$, $c < 0$, is opposite direction.

6. $\mathbf{v} = \langle q_1, q_2, q_3 \rangle$. Since $\mathbf{v}$ lies in the xz-plane, $q_z = 0$. Since $\mathbf{v}$ makes an angle of 60°, $q_1 = c \cdot \sqrt{3}/2$, and $q_3 = c\frac{1}{2}$, for some constant c. Finally $\|\mathbf{v}\| = 10$ implies that $\sqrt{c^2\frac{3}{4} + c\frac{2}{4}} = 10 \Rightarrow c = 10$. Thus, $\mathbf{v} = \langle 10\sqrt{3}/2, 0, 5 \rangle = \langle 5\sqrt{3}, 0, 5 \rangle$.

8. (a)

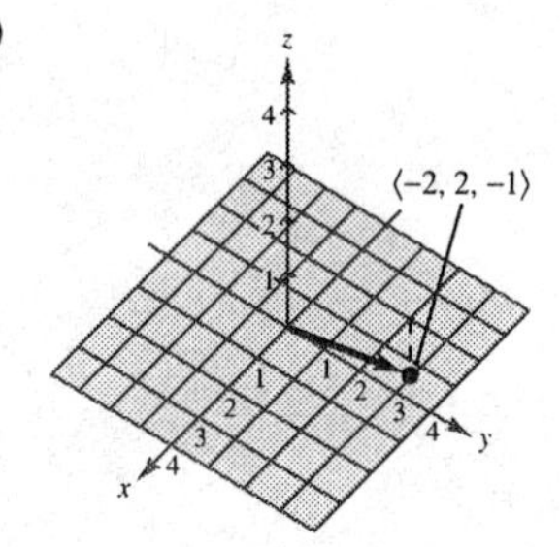

(b)

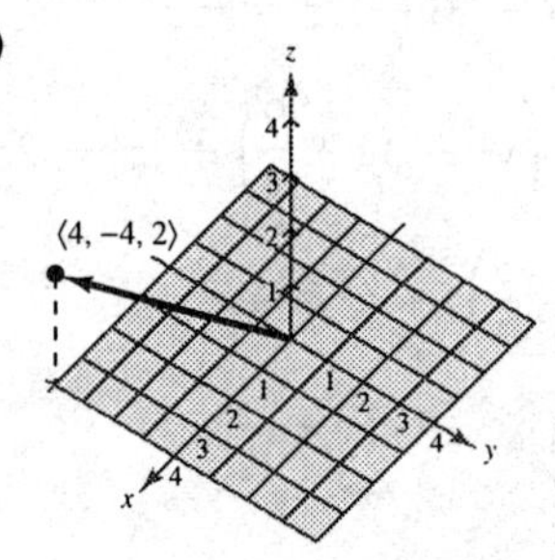

(c)

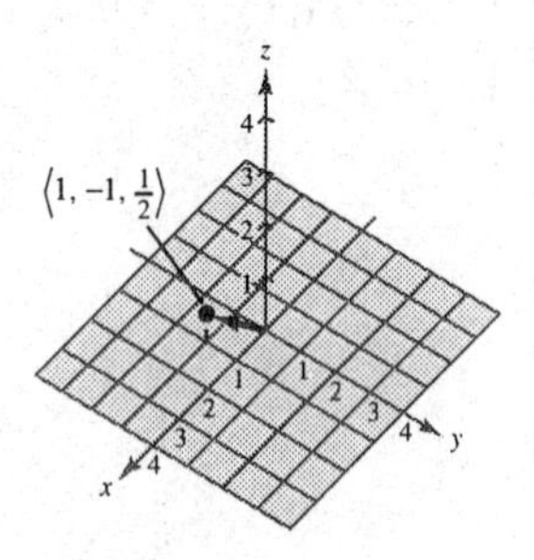

(d)

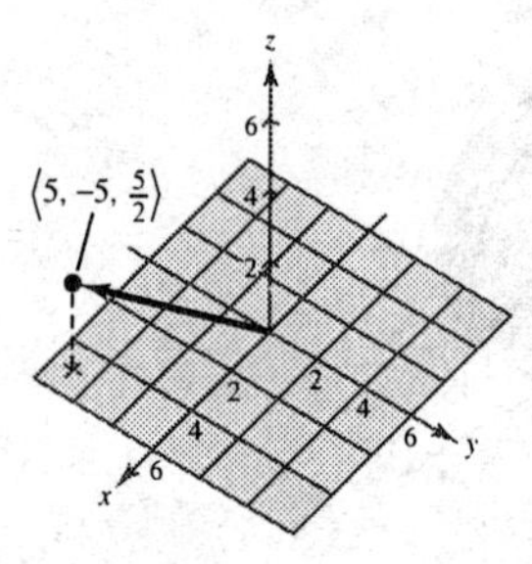

10. $\mathbf{z} = -7\mathbf{u} + \mathbf{v} - \frac{1}{5}\mathbf{w} = -7\langle -1, 3, 2 \rangle + \langle 1, -2, -2 \rangle = \frac{1}{5}\langle 5, 0, -5 \rangle$
$= \langle 7, -23, -15 \rangle$

12. $\mathbf{u} + \mathbf{v} - 2\mathbf{w} + \mathbf{z} = \mathbf{o} \Rightarrow \mathbf{z} = 2\mathbf{w} - \mathbf{u} - \mathbf{v}$
$= 2\langle 5, 0, -5 \rangle - \langle -1, 3, 2 \rangle - \langle 1, -2, -2 \rangle$
$= \langle 10, -1, -10 \rangle$

14. $\|\mathbf{v}\| = \sqrt{4^2 + (-3)^2 + (-7)^2} = \sqrt{74}$

16. $\mathbf{v} = \langle 1 - 0, 2 - (-1), -2 - 0 \rangle = \langle 1, 3, -2 \rangle$
$\|\mathbf{v}\| = \sqrt{1 + 9 + 4} = \sqrt{14}$

18. (a) $\dfrac{\mathbf{u}}{\|\mathbf{u}\|} = \dfrac{\langle -3, 5, 10 \rangle}{\sqrt{134}} = \dfrac{1}{\sqrt{134}}(-3\mathbf{i} + 5\mathbf{j} + 10\mathbf{k})$

(b) $\dfrac{-1}{\sqrt{134}}(-3\mathbf{i} + 5\mathbf{j} + 10\mathbf{k})$

20. $2\mathbf{u} + \frac{5}{2}\mathbf{v} = 2\langle -1, 3, 4 \rangle + \frac{5}{2}\langle 5, 4.5, -6 \rangle = \langle \frac{21}{2}, \frac{69}{4}, -7 \rangle$

22. $\dfrac{\mathbf{v}}{\|\mathbf{v}\|} = \dfrac{\langle 5, 4.5, -6\rangle}{\sqrt{5^2 + 4.5^2 + 6^2}} = \left\langle \dfrac{2\sqrt{13}}{13}, \dfrac{9\sqrt{13}}{65}, \dfrac{-12\sqrt{13}}{65} \right\rangle$

$\approx \langle 0.5547, 0.4992, 0.6656\rangle$

24. $\mathbf{u} \cdot \mathbf{v} = 2(9) + 5(-3) + (-3)(1) = 0$

26. $\cos\theta = \dfrac{\mathbf{u} \cdot \mathbf{v}}{\|\mathbf{u}\|\,\|\mathbf{v}\|} = \dfrac{-120}{\sqrt{1700}\sqrt{73}} \Rightarrow \theta \approx 109.92°$

28. $\mathbf{u} \cdot \mathbf{v} = 3 - 5 + 2 = 0$

Orthogonal

30. $\mathbf{v} = \langle -4 - (-2), 8 - 7, 1 - 4\rangle = \langle -2, 1, -3\rangle$

$\mathbf{u} = \langle 0 - (-4), 6 - 8, 7 - 1\rangle = \langle 4, -2, 6\rangle$

Since $\mathbf{u} = -2\mathbf{v}$, the points are collinear.

32. $\mathbf{v} = \langle -1 - 0, 5 - 4, 6 - 4\rangle = \langle -1, 1, 2\rangle$

$\mathbf{u} = \langle -2 - (-1), 6 - 5, 7 - 6\rangle = \langle -1, 1, 1\rangle$

Since $\mathbf{u}$ and $\mathbf{v}$ are not parallel, the points are not collinear.

34. Let the points be P_1, P_2, P_3, and P_4.

$\overrightarrow{P_1P_2} = \langle 8, -2, 5\rangle$, $\overrightarrow{P_4P_3} = \langle 8, -2, 5\rangle$ parallel and same length

$\overrightarrow{P_1P_4} = \langle 2, 3, -1\rangle$, $\overrightarrow{P_2P_3} = \langle 2, 3, -1\rangle$ parallel and equal length

36. $\langle 4, \frac{3}{2}, -\frac{1}{4}\rangle = \langle x - 2, y - 1, z + \frac{3}{2}\rangle \Rightarrow \langle 6, \frac{5}{2}, -\frac{7}{4}\rangle$

38. (a)

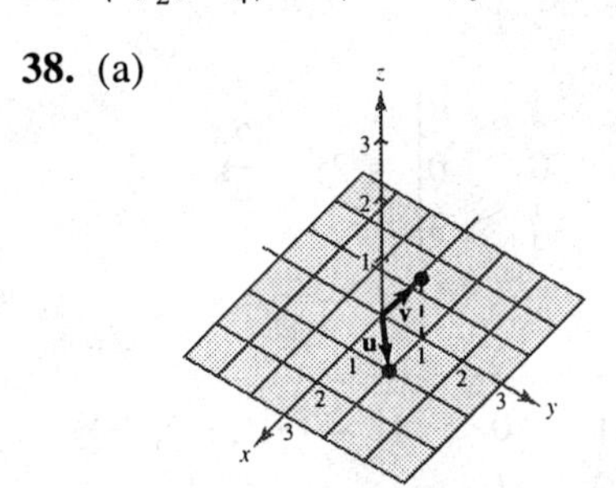

(b) $\mathbf{w} = a\mathbf{u} + b\mathbf{v} = a\langle 1, 1, 0\rangle + b\langle 0, 1, 1\rangle$

$\mathbf{0} = \langle a, a + b, b\rangle \Rightarrow a = b = 0$

(c) $\mathbf{w} = \langle 1, 2, 1\rangle = a\langle 1, 1, 0\rangle + b\langle 0, 1, 1\rangle$

$1 = a$

$2 = a + b$

$1 = b$

Hence $a = b = 1$

(d) $\mathbf{w} = \langle 1, 2, 3\rangle = a\langle 1, 1, 0\rangle + b\langle 0, 1, 1\rangle$

$1 = a$

$2 = a + b$

$3 = b$

Impossible

40. $\overrightarrow{PQ_1} = \langle 0, -24, -12\sqrt{21}\rangle$

$\overrightarrow{PQ_2} = \langle 12\sqrt{3}, 12, -12\sqrt{21}\rangle$

$\overrightarrow{PQ_3} = \langle -12\sqrt{3}, 12, -12\sqrt{21}\rangle\langle 0, -24, 0\rangle$

Let $\mathbf{F}_1$, $\mathbf{F}_2$, and $\mathbf{F}_3$ be the tension on each wire. Since $\|\mathbf{F}_1\| = \|\mathbf{F}_2\| = \|\mathbf{F}_3\|$, there exists a constant c such that

$\mathbf{F}_1 = C\langle 0, -24, -12\sqrt{21}\rangle$

$\mathbf{F}_2 = C\langle 12\sqrt{3}, 12, -12\sqrt{21}\rangle$

$\mathbf{F}_3 = C\langle -12\sqrt{3}, 12, -12\sqrt{21}\rangle$

The total force is $-30\mathbf{k} = \mathbf{F}_1 + \mathbf{F}_2 + \mathbf{F}_3 \Rightarrow$ the vertical ($\mathbf{k}$) component satisfies $-10 = -12\sqrt{21}c \Rightarrow c = \dfrac{5}{6\sqrt{21}}$. Hence,

$\mathbf{F}_1 = \left\langle 0, \dfrac{-20}{\sqrt{21}}, -10 \right\rangle$

$\mathbf{F}_2 = \left\langle \dfrac{10}{\sqrt{7}}, \dfrac{10}{\sqrt{21}}, -10 \right\rangle$

$\mathbf{F} = \left\langle \dfrac{-10}{\sqrt{7}}, \dfrac{10}{\sqrt{21}}, -10 \right\rangle$

$\|\mathbf{F}_1\| = \|\mathbf{F}_2\| = \|\mathbf{F}_3\| \approx 10.91$ pounds

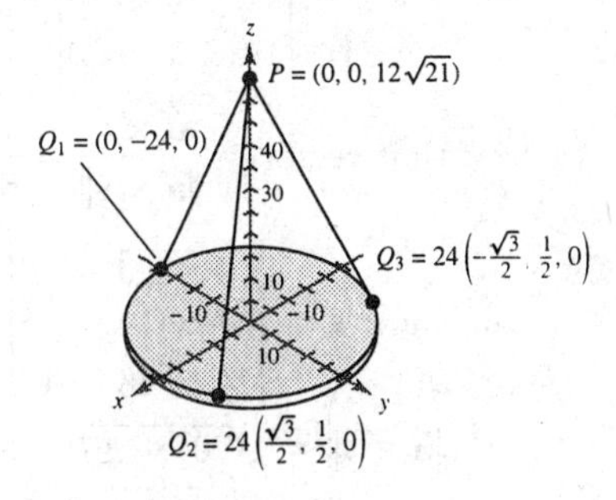

$Q_1 = (0, -24, 0)$

$Q_2 = (20.8, 12, 0)$

$Q_3 = (-20.8, 12, 0)$

$P = (0, 0, 55)$

Section 11.3 The Cross Product of Two Vectors

Solutions to Even-Numbered Exercises

2. $\mathbf{j} \times \mathbf{k} = \begin{vmatrix} \mathbf{i} & \mathbf{j} & \mathbf{k} \\ 0 & 1 & 0 \\ 0 & 0 & 1 \end{vmatrix} = \mathbf{i}$

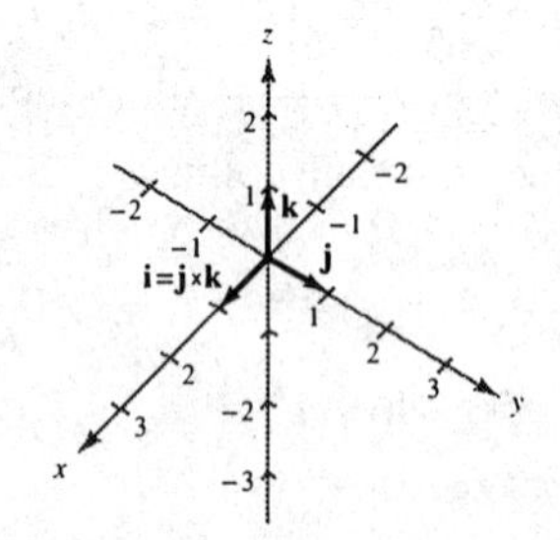

4. $\mathbf{k} \times \mathbf{i} = \begin{vmatrix} \mathbf{i} & \mathbf{j} & \mathbf{k} \\ 0 & 0 & 1 \\ 1 & 0 & 0 \end{vmatrix} = \mathbf{j}$

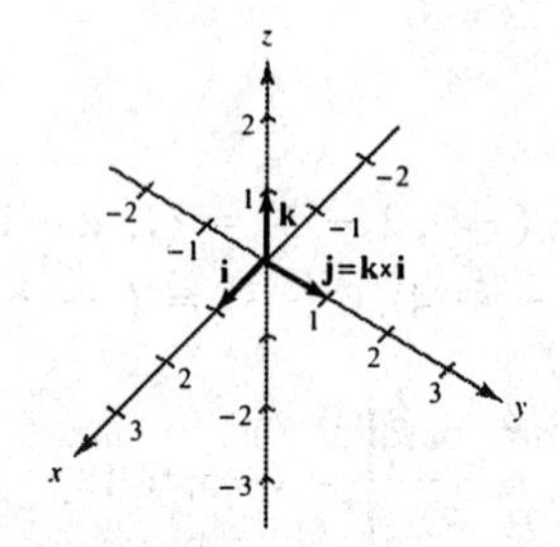

6. $\mathbf{u} \times \mathbf{v} = \begin{vmatrix} \mathbf{i} & \mathbf{j} & \mathbf{k} \\ -3 & 2 & 3 \\ 0 & 1 & 0 \end{vmatrix} = -3\mathbf{i} - 3\mathbf{k}$

$= \langle -3, 0, -3 \rangle$

8. $\mathbf{u} \times \mathbf{v} = \begin{vmatrix} \mathbf{i} & \mathbf{j} & \mathbf{k} \\ -5 & 5 & 11 \\ 2 & 2 & 3 \end{vmatrix} = \langle -7, 37, -20 \rangle$

10. $\mathbf{u} \times \mathbf{v} = \begin{vmatrix} \mathbf{i} & \mathbf{j} & \mathbf{k} \\ 0 & 0 & 6 \\ -1 & 3 & 1 \end{vmatrix} = \langle -18, -6, 0 \rangle$

$= -18\mathbf{i} - 6\mathbf{j}$

12. $\mathbf{u} \times \mathbf{v} = \begin{vmatrix} \mathbf{i} & \mathbf{j} & \mathbf{k} \\ \frac{2}{3} & 0 & 0 \\ 0 & \frac{1}{3} & -3 \end{vmatrix} = 2\mathbf{j} + \frac{2}{9}\mathbf{k}$

14. $\mathbf{u} \times \mathbf{v} = \begin{vmatrix} \mathbf{i} & \mathbf{j} & \mathbf{k} \\ 4 & -2 & 6 \\ -1 & 5 & 7 \end{vmatrix} = \langle -44, -34, 18 \rangle$

16. $\mathbf{u} \times \mathbf{v} = \begin{vmatrix} \mathbf{i} & \mathbf{j} & \mathbf{k} \\ -1 & 0 & 1 \\ 0 & 1 & -2 \end{vmatrix} = \langle -1, -2, -1 \rangle$

$= -\mathbf{i} - 2\mathbf{j} - \mathbf{k}$

18. $\mathbf{u} \times \mathbf{v} = \begin{vmatrix} \mathbf{i} & \mathbf{j} & \mathbf{k} \\ 1 & 2 & 0 \\ 1 & -3 & 0 \end{vmatrix} = -5\mathbf{k};\ \|\mathbf{u} \times \mathbf{v}\| = 5$

Unit vector $= \dfrac{\mathbf{u} \times \mathbf{v}}{\|\mathbf{u} \times \mathbf{v}\|} = \dfrac{1}{5}(-5\mathbf{k}) = -\mathbf{k}$

20. $\mathbf{u} \times \mathbf{v} = \begin{vmatrix} \mathbf{i} & \mathbf{j} & \mathbf{k} \\ 7 & -14 & 5 \\ 14 & 28 & -15 \end{vmatrix} = 70\mathbf{i} + 175\mathbf{j} + 392\mathbf{k}$

$\|\mathbf{u} \times \mathbf{v}\| = \sqrt{70^2 + 175^2 + 392^2}$

$= \sqrt{189{,}189} = 21\sqrt{429}$

Unit vector $= \dfrac{\mathbf{u} \times \mathbf{v}}{\|\mathbf{u} \times \mathbf{v}\|} = \dfrac{1}{21\sqrt{249}}\langle 70, 175, 392 \rangle$

$= \dfrac{1}{3\sqrt{429}}\langle 10, 25, 56 \rangle$

22. $\mathbf{u} \times \mathbf{v} = \begin{vmatrix} \mathbf{i} & \mathbf{j} & \mathbf{k} \\ 1 & -2 & 2 \\ 2 & -1 & -2 \end{vmatrix} = 6\mathbf{i} + 6\mathbf{j} + 3\mathbf{k}$

$\|\mathbf{u} \times \mathbf{v}\| = \sqrt{36 + 36 + 9} = 9$

Unit vector $= \dfrac{\mathbf{u} \times \mathbf{v}}{\|\mathbf{u} \times \mathbf{v}\|} = \dfrac{1}{9}(6\mathbf{i} + 6\mathbf{j} + 3\mathbf{k})$

$= \dfrac{2}{3}\mathbf{i} + \dfrac{2}{3}\mathbf{j} + \dfrac{1}{3}\mathbf{k}$

24. $\mathbf{u} \times \mathbf{v} = \begin{vmatrix} \mathbf{i} & \mathbf{j} & \mathbf{k} \\ 1 & 2 & 2 \\ 1 & 0 & 1 \end{vmatrix} = 2\mathbf{i} + \mathbf{j} - 2\mathbf{k}$

Area $= \|\mathbf{u} \times \mathbf{v}\| = \|2\mathbf{i} + \mathbf{j} - 2\mathbf{k}\|$

$= \sqrt{4 + 1 + 4} = 3$

26. $\mathbf{u} \times \mathbf{v} = \begin{vmatrix} \mathbf{i} & \mathbf{j} & \mathbf{k} \\ -2 & 3 & 2 \\ 1 & 2 & 4 \end{vmatrix} = \langle 8, 10, -7 \rangle$

Area $= \|\mathbf{u} \times \mathbf{v}\| = \sqrt{8^2 + 10^2 + (-7)^2} = \sqrt{213}$

28. $\mathbf{u} \times \mathbf{v} = \begin{vmatrix} \mathbf{i} & \mathbf{j} & \mathbf{k} \\ 4 & -3 & 2 \\ 5 & 0 & 1 \end{vmatrix} = \langle -3, 6, 15 \rangle$

Area $= \|\mathbf{u} \times \mathbf{v}\| = \sqrt{(-3)^2 + 6^2 + 15^2}$

$= \sqrt{270} = 3\sqrt{30}$

30. $\overrightarrow{AB} = \langle -1 - 3, 8 - 5, 5 - 0 \rangle = \langle -4, 3, 5 \rangle$ is parallel to

$\overrightarrow{DC} = \rangle 1 - 5, 3 - 0, 11 - 6 \rangle = \langle -4, 3, 5 \rangle$

$\overrightarrow{AD} = \langle 2, -5, 6 \rangle$ is parallel to $\overrightarrow{BC} = \langle 2, -5, 6 \rangle$.

$\overrightarrow{AB} \times \overrightarrow{AD} = \begin{vmatrix} \mathbf{i} & \mathbf{j} & \mathbf{k} \\ -4 & 3 & 5 \\ 2 & -5 & 6 \end{vmatrix} = \langle 43, 34, 14 \rangle$

Area $= \|\overrightarrow{AB} \times \overrightarrow{AD}\| = \sqrt{43^2 + 34^2 + 14^2} = \sqrt{3201}$

32. $\mathbf{u} = \langle 2 - 1, 0 - (-4), 2 - 3 \rangle = \langle 1, 4, -1 \rangle$

$\mathbf{v} = \langle -2 - 1, 2 - (-4), 0 - 3 \rangle = \langle -3, 6, -3 \rangle$

$\mathbf{u} \times \mathbf{v} = \begin{vmatrix} \mathbf{i} & \mathbf{j} & \mathbf{k} \\ 1 & 4 & -1 \\ -3 & 6 & -3 \end{vmatrix} = \langle -6, 6, 18 \rangle$

Area $= \frac{1}{2}\|\mathbf{u} \times \mathbf{v}\| = \frac{1}{2}\sqrt{(-6)^2 + 6^2 + 18^2}$

$= \frac{1}{2}\sqrt{396} = 3\sqrt{11}$

34. $\mathbf{u} = \langle -2 - 2, -4 - 4, 0 - 0 \rangle = \langle -4, -8, 0 \rangle$

$\mathbf{v} = \langle 0 - 2, 0 - 4, 4 - 0 \rangle = \langle -2, -4, 4 \rangle$

$\mathbf{u} \times \mathbf{v} = \begin{vmatrix} \mathbf{i} & \mathbf{j} & \mathbf{k} \\ -4 & -8 & 0 \\ -2 & -4 & 4 \end{vmatrix} = \langle -32, 16, 0 \rangle$

Area $= \frac{1}{2}\|\mathbf{u} \times \mathbf{v}\| = \frac{1}{2}\sqrt{(-32)^2 + 16^2}$

$= \frac{1}{2}\sqrt{1280} = 8\sqrt{5}$

36. $\mathbf{u} \cdot (\mathbf{v} \times \mathbf{w}) = \begin{vmatrix} 20 & 10 & 10 \\ 1 & 4 & 4 \\ 0 & 2 & 2 \end{vmatrix} = 20(0) - 10(2) + 10(2) = 0$

38. $\mathbf{u} \cdot (\mathbf{v} \times \mathbf{w}) = \begin{vmatrix} 1 & 1 & 10 \\ 0 & 3 & 3 \\ 3 & 0 & 3 \end{vmatrix} = 1(9) - 1(-9) + 3(-9) = -9$

Volume $= |\mathbf{u} \cdot (\mathbf{v} \times \mathbf{w})| = |-9| = 9$

40. $\mathbf{u} = \langle 2 - 3, -2 - 0, 2 - 0 \rangle = \langle -1, -2, 2 \rangle$

$\mathbf{v} = \langle 4 - 3, 1 - 0, 2 - 0 \rangle = \langle 1, 1, 2 \rangle$

$\mathbf{w} = \langle -1 - 3, 5 - 0, 4 - 0 \rangle = \langle -4, 5, 4 \rangle$

$\mathbf{u} \cdot (\mathbf{v} \times \mathbf{w}) = \begin{vmatrix} -1 & -2 & 2 \\ 1 & 1 & 2 \\ -4 & 5 & 4 \end{vmatrix}$

$= -1(-6) + 2(12) + 2(9) = 48$

Volume $= 48$

42. $\overrightarrow{PQ} = 0.16\mathbf{k}$

$\overrightarrow{PQ} \times \mathbf{F} = \begin{vmatrix} \mathbf{i} & \mathbf{j} & \mathbf{k} \\ 0 & 0 & 0.16 \\ 0 & -1000\sqrt{3} & -1000 \end{vmatrix} = 160\sqrt{3}\mathbf{i}$

$\|\overrightarrow{PQ} \times \mathbf{F}\| = 160\sqrt{3}$ ft • lb

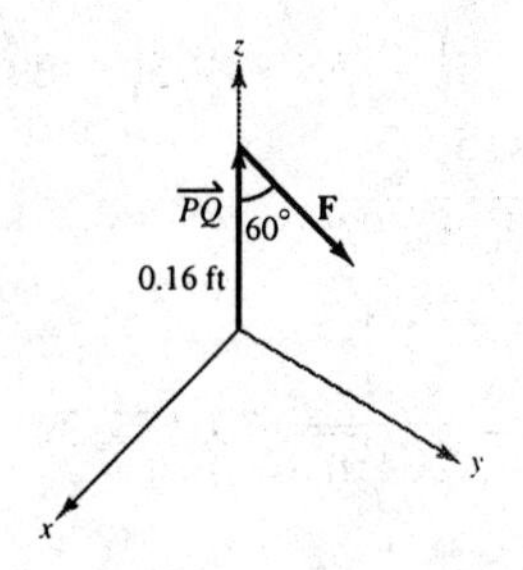

44. $\mathbf{u} \times \mathbf{v} = \begin{vmatrix} \mathbf{i} & \mathbf{j} & \mathbf{k} \\ u_1 & u_2 & u_3 \\ v_1 & v_2 & v_3 \end{vmatrix} = (u_2v_3 - v_2u_3)\mathbf{i} - (u_1v_3 - v_1u_3)\mathbf{j} + (u_1v_2 - v_1u_2)\mathbf{k}$

$\mathbf{u} \cdot (\mathbf{u} \times \mathbf{v}) = u_1(u_2v_3 - v_2u_3) - u_2(u_1v_3 - v_1u_3) + u_3(u_1v_2 - v_1u_2)$

$= 0$

Similarly, $\mathbf{v} \cdot (\mathbf{u} \times \mathbf{v}) = 0$.

Section 11.4 Lines and Planes in Space

Solutions to Even-Numbered Exercises

2. $P(1, 2, 2) \Rightarrow (t = 0)$

$Q(7, 0, 12) \Rightarrow (t = 2)$

$\overrightarrow{PQ} = \langle 6, -2, 10\rangle$ which is proportional to $\langle 3, -1, 5\rangle$.

4. The lines are parallels because they have the same direction numbers.

6. $x = x_1 + at = 0 + 3t = 3t$

$y = y_1 + bt = 0 - 7t = -7t$

$z = z_1 + ct = 0 - 10t = -10t$

(a) Parametric equations: $x = 3t$, $y = -7t$, $z = -10t$

(b) Symmetric equations: $\frac{x}{3} = \frac{y}{-7} = \frac{z}{-10}$

8. $x = x_1 + at = 5 + 4t$

$y = y_1 + bt = 0 + 0t$

$z = z_1 + ct = 10 + 3t$

(a) Parametric equations: $x = 5 + 4t$, $y = 0$, $z = 10 + 3t$

(b) Symmetric equations: $\frac{x - 5}{4} = \frac{z - 10}{3}$, $y = 0$

10. $x = x_1 + at = 10 + 3t$

$y = y_1 + bt = -18 - 3t$

$z = z_1 + ct = 36 + t$

(a) Parametric equations: $x = 10 + 3t$, $y = -18 - 3t$, $z = 36 + t$

(b) Symmetric equations: $\frac{x - 10}{3} = \frac{y + 18}{-3} = z - 36$

12. The line is parallel to the z-axis: $x = -3$, $y = 8$, $z = t$ (or $z = 15 + t$)

14.

$$6 + t = 7 + 5s \Rightarrow t - 5s = 1$$

$$-5 - 2t = 5 + 2s \Rightarrow 2t + 2s = -10$$

$$1 + 3t = -8 + 3s \Rightarrow 3t - 3s = -9$$

$$\left.\begin{aligned} t - 5s &= 1 \\ t + s &= -5 \end{aligned}\right\} 6s = -6 \Rightarrow s = -1, t = -4$$

Point: $(2, 3, -11)$

$$\cos\theta = \frac{\langle 1, -2, 3\rangle \cdot \langle 5, 2, 3\rangle}{\sqrt{14}\sqrt{38}} = \frac{10}{\sqrt{532}} \Rightarrow \theta \approx 1.1224 = 64.31°$$

16.

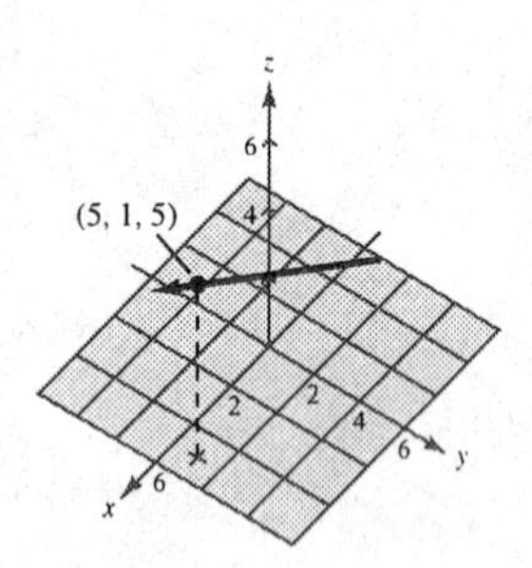

18. $a(x - x_1) + b(y - y_1) + c(z - z_1) = 0$

$0(x - 2) + 0(y - 3) + 1(z - 5) = 0$

$z - 5 = 0$

20. $0(x - 0) - 3(y - 0) + 5(z - 0) = 0$

$-3y + 5z = 0$

22. $\mathbf{n} = \langle 3, 1, -2\rangle$

$3(x - 1) + 1(y - 3) - 2(z - 1) = 0$

$3x + y - 2z - 4 = 0$

24. $\mathbf{u} = \langle 2, -6, 2 \rangle$, $\mathbf{v} = \langle -3, -3, 0 \rangle$

$$\mathbf{u} \times \mathbf{v} = \begin{vmatrix} \mathbf{i} & \mathbf{j} & \mathbf{k} \\ 2 & -6 & 2 \\ -3 & -3 & 0 \end{vmatrix} = \langle 6, -6, -24 \rangle$$

$\mathbf{n} = \langle -1, 1, 4 \rangle$

Plane: $-1(x - 4) + 1(y + 1) + 4(z - 3) = 0$

$$-x + y + 4z - 7 = 0$$

26. $\mathbf{n} = \mathbf{k}$ is the normal vector: $z - 3 = 0$

28. $\langle 1, 2, 2 \rangle$ and $\langle -4, 2, 0 \rangle$ are parallel to plane

$$\mathbf{n} = \langle 1, 2, 2 \rangle \times \langle -4, 2, 0 \rangle = \begin{vmatrix} \mathbf{i} & \mathbf{j} & \mathbf{k} \\ 1 & 2 & 2 \\ -4 & 2 & 0 \end{vmatrix} = \langle -4, -8, 10 \rangle$$

$$-4(x - 4) - 8(y - 0) + 10(z - 0) = 0$$

$$-4x - 8y + 10z + 16 = 0$$

$$-2x - 4y + 5z + 8 = 0$$

30. $\mathbf{n}_1 = \langle 3, 2, -1 \rangle$, $\mathbf{n}_2 = \langle 1, -4, 2 \rangle$

$$\cos\theta = \frac{\mathbf{n}_1 \cdot \mathbf{n}_2}{\|\mathbf{n}_1\|\,\|\mathbf{n}_2\|} = \frac{-7}{\sqrt{14}\sqrt{21}} = \frac{-1}{\sqrt{6}}$$

$\theta = 65.9°$

32. $\mathbf{n}_1 = \langle 1, -5, -1 \rangle$

$\mathbf{n}_2 = \langle 5, -25, -5 \rangle = 5\mathbf{n}_1 \Rightarrow$ parallel

34.

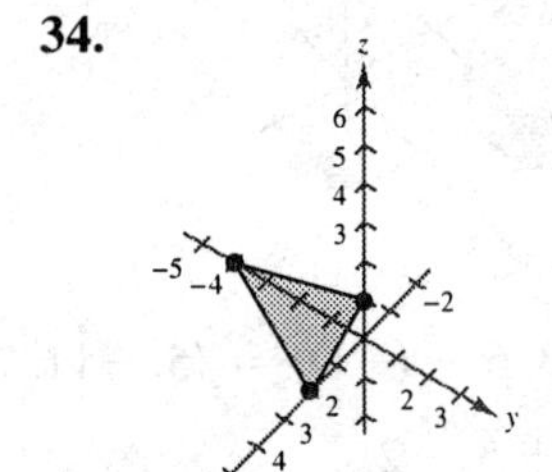

36.

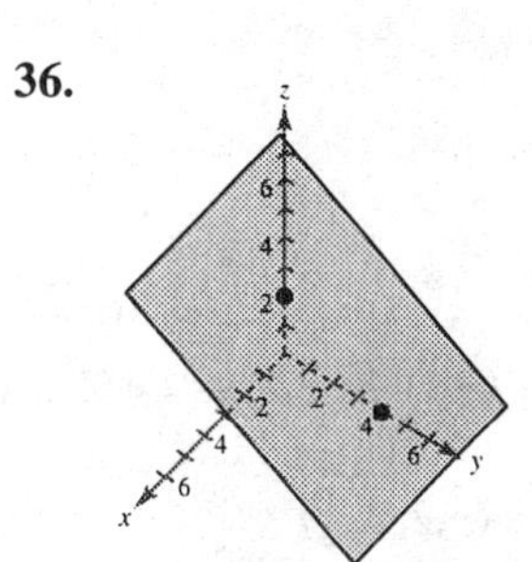

38.

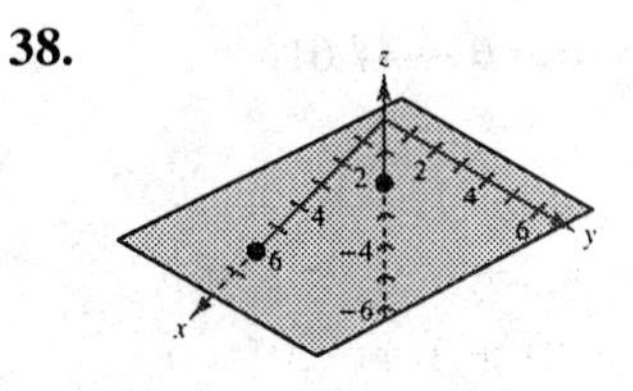

40.

42. $P = (2, 0, 0)$ on plane, $Q = (1, 2, 3)$, $\mathbf{n} = \langle 2, -1, 1 \rangle$, $\overrightarrow{PQ} = \langle -1, 2, 3 \rangle$.

$$D = \frac{|\overrightarrow{PQ} \cdot \mathbf{n}|}{\|\mathbf{n}\|} = \frac{|-1|}{\sqrt{6}} = \frac{1}{\sqrt{6}}$$

44. False, the line could be skew lines.

❑ Review Exercises for Chapter 11

Solutions to Even-Numbered Exercises

2. $A(3, 4, 0)$, $B(-1, -3, 5)$

4. y-axis $\Rightarrow x = z = 0$

$(0, -7, 0)$

6. Octants II and III

8. $d_1 = \sqrt{(4-0)^2 + (3-0)^2 + (2-4)^2} = \sqrt{16 + 9 + 4} = \sqrt{29}$

$d_2 = \sqrt{(4-4)^2 + (5-3)^2 + (5-2)^2} = \sqrt{4 + 9} = \sqrt{13}$

$d_3 = \sqrt{(4-0)^2 + (5-0)^2 + (5-4)^2} = \sqrt{16 + 25 + 1} = \sqrt{42}$

$d_1^2 + d_2^2 + d_3^2 = 42$

10. $\left(\dfrac{6.2 + 3.6}{2}, \dfrac{4.5 - 3.5}{2}, \dfrac{-3 + 10}{2}\right) = (4.9, 0.5, 3.5)$

12. $(x - 3)^2 + (y + 2)^2 + (z - 4)^2 = 4^2 = 16$

14. Center: $(2, 3, 2)$

Radius: $\sqrt{2^2 + 3^2 + 2^2} = \sqrt{17}$

$(x - 2)^2 + (y - 3)^2 + (z - 2)^2 = 17$

16. $(x^2 - 10x + 25) + (y^2 + 6y + 9) + (z^2 - 4z + 4) = -34 + 25 + 9 + 4$

$(x - 5)^2 + (y + 3)^2 + (z - 2)^2 = 4$

Center: $(5, -3, 2)$

Radius: 2

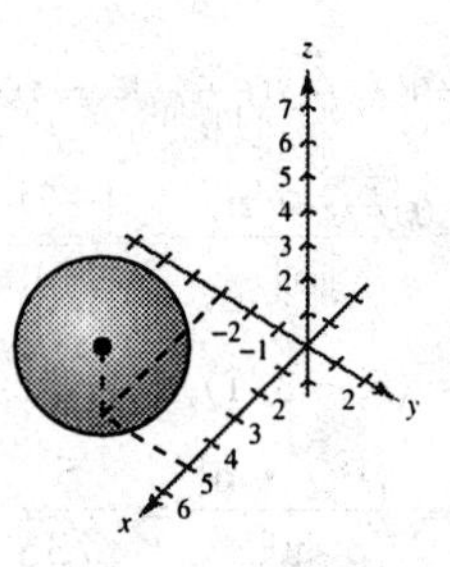

18. $\overrightarrow{PQ} = \langle -3, -2, 2 - (-1), 3 - 2\rangle = \langle -5, 3, 1\rangle$

20. $\overrightarrow{PQ} = \langle 5 - 0, -8 - 3, 6 - (-1)\rangle = \langle 5, -11, 7\rangle$

22. $\mathbf{u} \cdot \mathbf{v} = \langle 8, 5, -8\rangle \cdot \langle -2, 4, \frac{1}{2}\rangle$

$= -16 + 20 - 4 = 0$

Orthogonal

24. $\cos\theta = \dfrac{\mathbf{u} \cdot \mathbf{v}}{\|\mathbf{u}\|\,\|\mathbf{v}\|} = \dfrac{12 + 5 - 2}{\sqrt{11}\sqrt{45}}$

$= \dfrac{5}{\sqrt{11}\sqrt{45}} \Rightarrow \theta \approx 47.61°$

26. $\mathbf{u} \times \mathbf{v} = \begin{vmatrix} \mathbf{i} & \mathbf{j} & \mathbf{k} \\ 10 & 15 & 5 \\ 5 & -3 & 0 \end{vmatrix} = \langle 15, 25, -105\rangle$

28. $2\mathbf{u} + 3\mathbf{v} - 5\mathbf{w} = 2\langle 2, -3, 3\rangle + 3\langle 1, 4, -2\rangle - 5\langle -2, 1, 2\rangle$

$= \langle 17, 1, -10\rangle$

30. $\cos\theta = \dfrac{\mathbf{u} \cdot \mathbf{v}}{\|\mathbf{u}\|\,\|\mathbf{v}\|} = \dfrac{2 - 12 - 6}{\sqrt{22}\sqrt{21}} = \dfrac{-16}{\sqrt{462}} \Rightarrow \theta \approx 138.10°$

32. $\mathbf{v} \times \mathbf{w} = \begin{vmatrix} \mathbf{i} & \mathbf{j} & \mathbf{k} \\ 1 & 4 & -2 \\ -2 & 1 & 2 \end{vmatrix} = \langle 10, 2, 9\rangle$

34. $\mathbf{u} \cdot (\mathbf{v} + \mathbf{w}) = \langle 2, -3, 3\rangle \cdot \langle -1, 5, 0\rangle = -17$

$\mathbf{u} \cdot \mathbf{v} + \mathbf{u} \cdot \mathbf{w} = (2 - 12 - 6) + (-4 - 3 + 6)$

$= -17$

36. $\mathbf{u} \cdot (\mathbf{v} \times \mathbf{w}) = \begin{vmatrix} 2 & -3 & 3 \\ 1 & 4 & -2 \\ -2 & 1 & 2 \end{vmatrix} = 2(10) + 3(-2) + 3(9) = 41$

Volume $= |\mathbf{u} \cdot (\mathbf{u} \times \mathbf{w})| = 41$

38. $\mathbf{n} = \langle 1, -3, 4\rangle$ is perpendicular to plane

$$\mathbf{u} = 3\frac{\mathbf{n}}{\|\mathbf{n}\|} = \frac{3}{\sqrt{26}}\langle 1, -3, 4\rangle$$

40. $\mathbf{v} = \langle 3 - (-1), 6 - 3, -1 - 5\rangle = \langle 4, 3, -6\rangle$

$x = -1 + 4t, y = 3 + 3t, z = 5 - 6t$

42. $\mathbf{v} = \mathbf{j}$

$x = 2$

$y = 1 + t$

$z = 4$

44. The line is parallel to

$$\mathbf{u} \times \mathbf{v} = \begin{vmatrix} \mathbf{i} & \mathbf{j} & \mathbf{k} \\ 0 & 2 & 2 \\ 2 & 0 & 2 \end{vmatrix} = \langle 4, 4, -4\rangle,$$

or $\langle 1, 1, -1\rangle$. Hence, $x = -5 + t, y = 6 + t, z = 7 - t$.

46. $\mathbf{u} = \langle 5, 0, 2\rangle, \mathbf{v} = \langle 2, 3, 8\rangle$

$$\mathbf{n} = \mathbf{u} \times \mathbf{v} = \begin{vmatrix} \mathbf{i} & \mathbf{j} & \mathbf{k} \\ 5 & 0 & 2 \\ 2 & 3 & 8 \end{vmatrix} = \langle -6, -36, 15\rangle$$

Plane: $6(x - 0) - 36(y - 0) + 15(z - 0) = 0$

$$-6x - 36y + 15z = 0$$

$$-2x - 12y + 5z = 0$$

48. $\mathbf{n} = \mathbf{k}$ normal vector

Plane: $0(x - 5) + 0(y - 3) + 1(z - 2) = 0$

$$z - 2 = 0$$

50. $\mathbf{n} = \langle 2, -20, 6\rangle, P = (0, 0, 1)$ in plane, $Q = (2, 3, 10), \overrightarrow{PQ} = \langle 2, 3, 9\rangle$.

$$D = \frac{|\overrightarrow{PQ} \cdot \mathbf{n}|}{\|\mathbf{n}\|} = \frac{|-2|}{\sqrt{440}} = \frac{1}{\sqrt{110}} \approx 0.0953$$

52. $\mathbf{n} = \langle 5, -3, 1\rangle, P = (0, 1, 0)$ in second plane, $Q = (0, 0, 2)$ in top plane, $\overrightarrow{PQ} = \langle 0, -1, 2\rangle$.

$$D = \frac{|\overrightarrow{PQ} \cdot \mathbf{n}|}{\|\mathbf{n}\|} = \frac{5}{\sqrt{35}}$$

54.

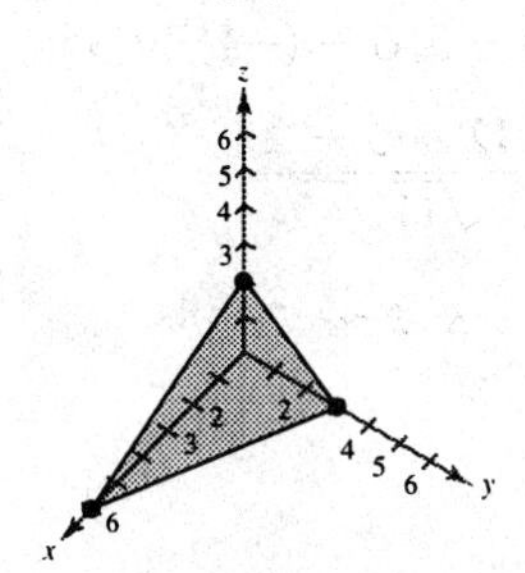

CHAPTER 12
Limits and an Introduction to Calculus

CHAPTER 12
Limits and an Introduction to Calculus

Section 12.1 Introduction to Limits

Solutions to Even-Numbered Exercises

2. $\lim_{x\to 5} f(x) = 12$ means that the value of $f(x)$ approaches 12 as x approaches 5.

4. $\lim_{x\to 4}\left(\frac{1}{2}x^2 - 2x + 3\right) = 3$ The limit is reached.

x	3.9	3.99	3.999	4.0	4.001	4.01	4.1
$f(x)$	2.805	2.980	2.998	3	3.002	3.020	3.205

6. $\lim_{x\to -1} \dfrac{x+1}{x^2 - x - 2} = -\dfrac{1}{3}$ The limit is not reached.

x	−1.1	−1.01	−1.001	−1.0	−0.999	−0.99	−0.9
$f(x)$	−0.3226	−0.3322	−0.3332	?	−0.3334	−0.3344	−0.3348

8. In general you cannot use a graphing utility to determine whether a limit can be reached. It is important to analyze a function analytically.

10. $\lim_{x\to -3} \dfrac{x+3}{x^2 + x - 6} = -\dfrac{1}{5}$

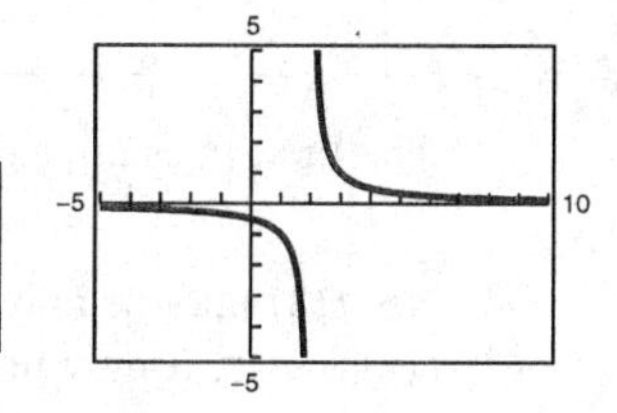

x	−3.1	−3.01	−3.001	−3.0	−2.999	−2.99	−2.9
$f(x)$	−0.1961	−0.1996	−0.19996	?	−0.20004	−0.2004	−0.2041

12. $\lim_{x\to -3} \dfrac{\sqrt{1-x} - 2}{x+3} = -\dfrac{1}{4}$

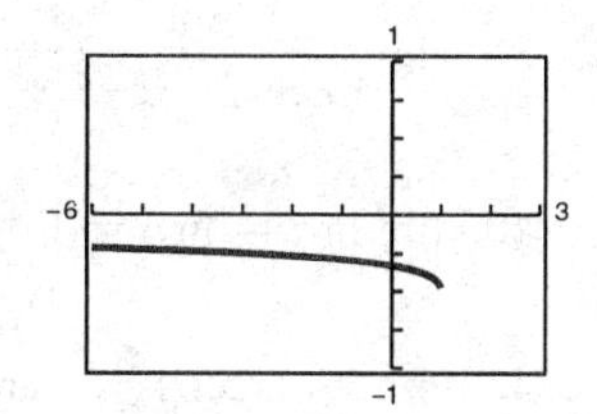

x	−3.1	−3.01	−3.001	−3.0	−2.999	−2.99	−2.9
$f(x)$	−0.2485	−0.2498	−0.25	?	−0.25	−0.2502	−0.2516

14. $\lim_{x\to 2} \dfrac{\dfrac{1}{x+2} - \dfrac{1}{4}}{x-2} = -\dfrac{1}{16}$

x	1.9	1.99	1.999	2.0	2.001	2.01	2.1
$f(x)$	−0.0641	−0.0627	−0.0625	?	−0.0625	−0.0623	−0.0610

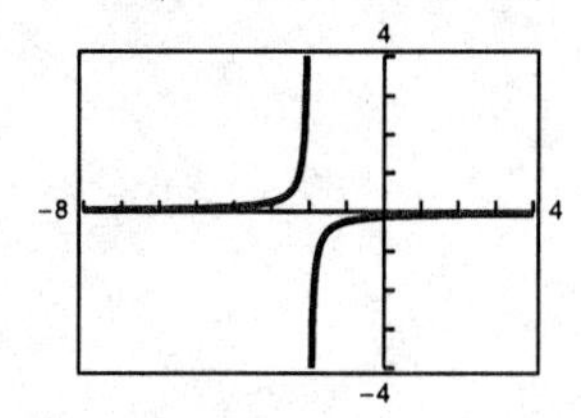

16. $\lim_{x \to 0} \dfrac{\cos x - 1}{x} = 0$

x	-0.1	-0.01	-0.001	0	0.001	0.01	0.1
$f(x)$	-0.050	-0.005	-0.0005	?	-0.0005	-0.005	-0.050

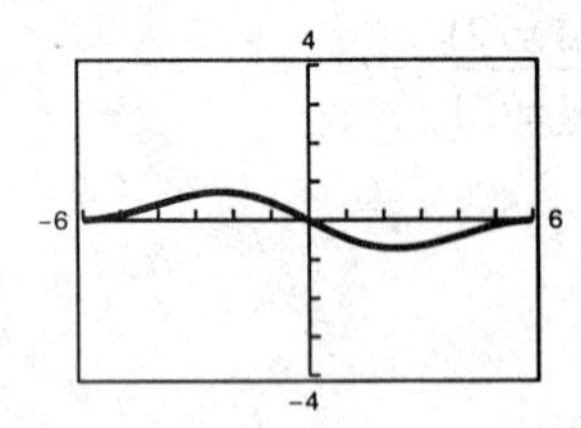

18. $\lim_{x \to -4} (x^2 - 3) = 13$

20. $\lim_{x \to 2} \dfrac{3x^2 - 12}{x - 2} = 12$

22. The limit does not exist because $f(x)$ does not approach a real number as x approaches 1.

24. $\lim_{x \to \pi/3} \sec x = 2$

26.

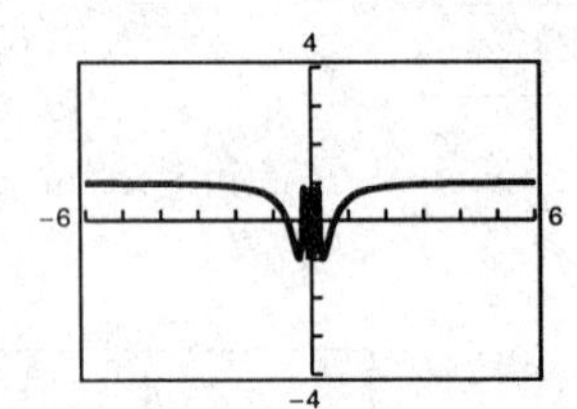

The limit does not exist.

28.

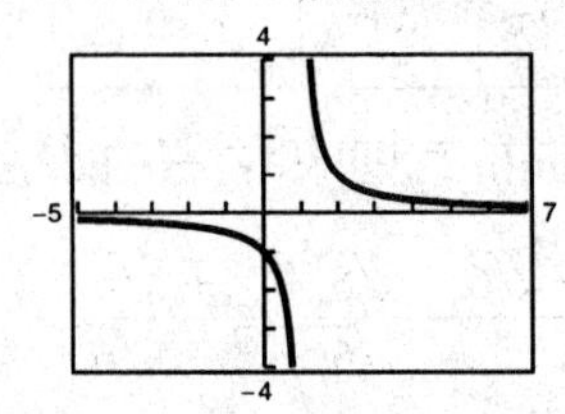

$\lim_{x \to 3} f(x) = \frac{1}{2}$

30.

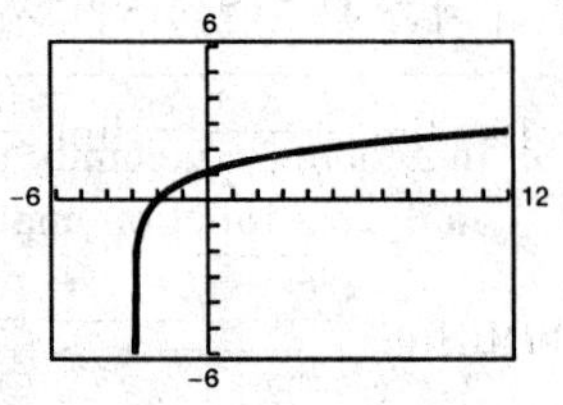

$\lim_{x \to 4} f(x) = \ln 7$

32. No. $f(2)$ may or may not exist. Furthermore, if $f(2)$ exists, it may not be equal to 4.

34. $\lim_{x \to 2} \sqrt[3]{10x + 7} = \sqrt[3]{10(2) + 7} = \sqrt[3]{27} = 3$

36. $\lim_{x \to -5} \dfrac{4}{x + 2} = \dfrac{4}{-5 + 2} = -\dfrac{4}{3}$

38. $\lim_{x \to 8} \dfrac{\sqrt{x + 1}}{x - 4} = \dfrac{\sqrt{8 + 1}}{x - 4} = \dfrac{3}{4}$

40. $\lim_{x \to e} \ln x = \ln e = 1$

42. $\lim_{x \to 1/4} \dfrac{\tan \pi x}{2} = \dfrac{\tan(\pi/4)}{2} = \dfrac{1}{2}$

44. $\lim_{x \to 1} \arccos \dfrac{x}{2} = \arccos \dfrac{1}{2} = \dfrac{\pi}{3} \approx 1.0472$

46. $\lim_{x \to 2} f(x) = 0$. As x approaches 2 from both sides, $f(x)$ approaches 0.

48. (a) $\lim_{x \to c} [f(x) + g(x)]^2 = \left[\dfrac{3}{2} - \dfrac{1}{2}\right]^2 = 1$

(b) $\lim_{x \to c} [6f(x)\,g(x)] = 6\left(\dfrac{3}{2}\right)\left(-\dfrac{1}{2}\right) = -\dfrac{9}{2}$

(c) $\lim_{x \to c} \dfrac{5\,g(x)}{4\,f(x)} = \dfrac{5(-1/2)}{4(3/2)} = -\dfrac{5}{12}$

(d) $\lim_{x \to c} \dfrac{1}{\sqrt{f(x)}} = \dfrac{1}{\sqrt{3/2}} = \sqrt{\dfrac{2}{3}}$

50. $\lim_{x \to 3} f(x) = \frac{1}{6}$

Domain: $x \neq \pm 3$ or $(-\infty, -3), (-3, 3), (3, \infty)$

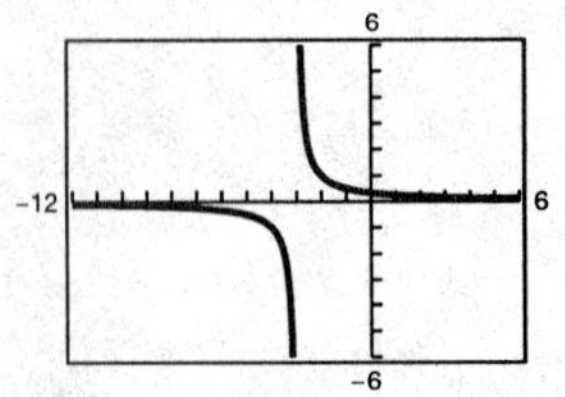

It is difficult to determine the domain solely by the graph because it is not obvious that the function is undefined at $x = 3$.

Section 12.2 Techniques for Evaluating Limits

Solutions to Even-Numbered Exercises

2. $\lim_{x\to-2}\left(\frac{1}{4}x^3 - 3x\right) = \frac{1}{4}(-2)^3 - 3(-2) = -2 + 6 = 4$

4. $\lim_{x\to4}\frac{x-1}{x^2+2x+3} = \frac{4-1}{4^2+2(4)+3} = \frac{3}{27} = \frac{1}{9}$

6. $\lim_{x\to-1} e^{x^2-1} = e^0 = 1$

8. $\lim_{t\to\pi/2}\cot t = \frac{\cos(\pi/2)}{\sin(\pi/2)} = 0$

10. $h(x) = \frac{x^2-3x}{x}$

$h_2(x) = x - 3$

(a) $\lim_{x\to-2} h(x) = -5$

(b) $\lim_{x\to0} h(x) = -3$

(c) $\lim_{x\to3} h(x) = 0$

12. $f(x) = \frac{x^2-1}{x+1}$

$f_2(x) = x - 1$

(a) $\lim_{x\to1} f(x) = 0$

(b) $\lim_{x\to2} f(x) = 1$

(c) $\lim_{x\to3} h(x) = -2$

14. $\lim_{x\to3}\frac{3-x}{x^2-9} = \lim_{x\to3}\frac{3-x}{(x-3)(x+3)} = \lim_{x\to3} -\frac{1}{x+3} = -\frac{1}{6}$

16. $\lim_{x\to-2}\frac{t^3+8}{t+2} = \lim_{x\to-2}\frac{(t+2)(t^2-2t+4)}{(t+2)} = \lim_{t\to-2}(t^2-2t+4) = 12$

18.
$$\lim_{z\to0}\frac{\sqrt{6-z}-\sqrt{6}}{z} = \lim_{z\to0}\frac{\sqrt{6-z}-\sqrt{6}}{z}\cdot\frac{\sqrt{6-z}+\sqrt{6}}{\sqrt{6-z}+\sqrt{6}}$$
$$= \lim_{z\to0}\frac{(6-z)-6}{z\left(\sqrt{6-z}+\sqrt{6}\right)}$$
$$= \lim_{z\to0}\frac{-1}{\sqrt{6-z}+\sqrt{6}} = \frac{-1}{2\sqrt{6}}$$

20.
$$\lim_{x\to2}\frac{4-\sqrt{18-x}}{x-2} = \lim_{x\to2}\frac{4-\sqrt{18-x}}{x-2}\cdot\frac{4+\sqrt{18-x}}{4+\sqrt{18-x}}$$
$$= \lim_{x\to2}\frac{16-(18-x)}{(x-2)\left(4+\sqrt{18-x}\right)}$$
$$= \lim_{x\to2}\frac{1}{4+\sqrt{18-x}} = \frac{1}{8}$$

22.
$$\lim_{x\to0}\frac{\frac{1}{5-x}-\frac{1}{5}}{x} = \lim_{x\to0}\frac{5-(5-x)}{(5-x)5x} = \lim_{x\to0}\frac{1}{5(5-x)} = \frac{1}{25}$$

24.
$$\lim_{x\to\pi/2}\frac{1-\sin x}{x} = \lim_{x\to\pi/2}\frac{1-\sin x}{\cos x}\cdot\frac{1+\sin x}{1+\sin x}$$
$$= \lim_{x\to\pi/2}\frac{1-\sin^2 x}{\cos x(1+\sin x)} = \lim_{x\to\pi/2}\frac{\cos^2}{\cos x(1+\sin x)}$$
$$= \lim_{x\to\pi/2}\frac{\cos x}{1+\sin x} = 0$$

26. $\lim\limits_{x\to 2} \dfrac{\sqrt{x+14}-4}{x-2} \cdot \dfrac{\sqrt{x+14}+4}{\sqrt{x+14}+4} = \lim\limits_{x\to 2} \dfrac{(x+14)-16}{(x-2)(\sqrt{x+14}+4)}$

$= \lim\limits_{x\to 2} \dfrac{x-2}{(x-2)(\sqrt{x+14}+4)}$

$= \lim\limits_{x\to 2} \dfrac{1}{\sqrt{x+14}+4} = \dfrac{1}{\sqrt{16}+4} = \dfrac{1}{8}$

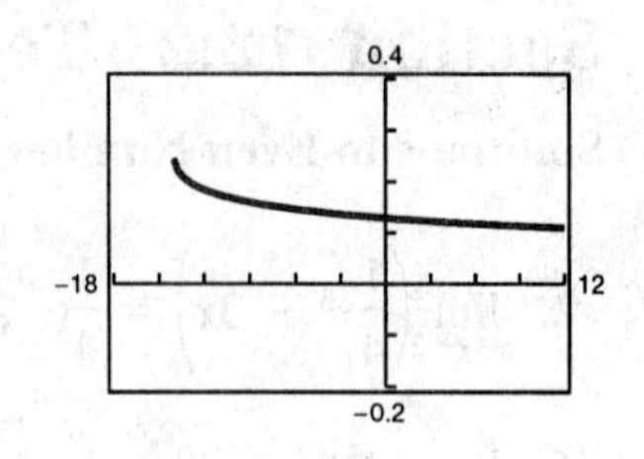

x	1.9	1.99	1.99	2	2.001	2.01	2.1
$f(x)$	0.1252	0.1250	0.1250	?	0.1250	0.1250	0.1248

28. $\lim\limits_{x\to 0} \dfrac{\dfrac{1}{2x-5}+\dfrac{1}{5}}{x} = \lim\limits_{x\to 0} \dfrac{5+(2x-5)}{x(2x-5)5} = \lim\limits_{x\to 0} \dfrac{2}{5(2x-5)} = \dfrac{-2}{25}$

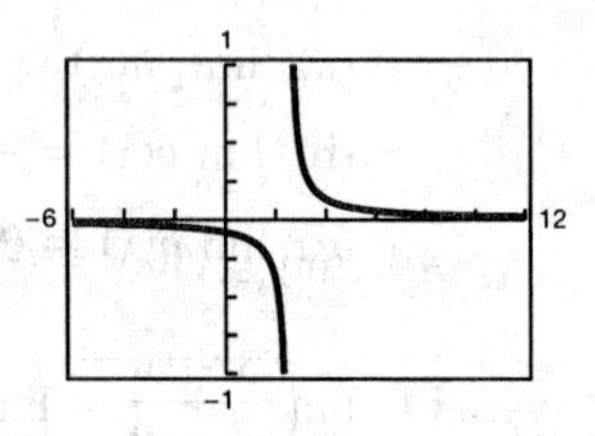

x	−0.1	−0.01	−0.001	0	0.001	0.01	0.1
$f(x)$	−0.0796	−0.0797	−0.0800	?	−0.0800	−0.0803	−0.0833

30. $\lim\limits_{x\to 1} f(x) = -0.2$

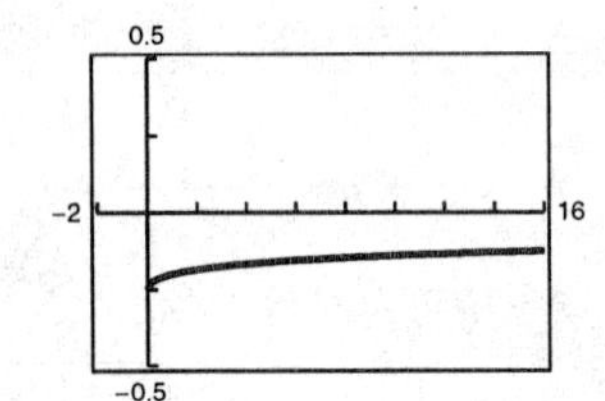

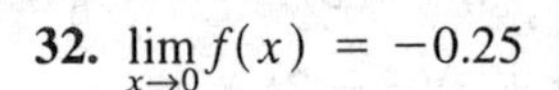

32. $\lim\limits_{x\to 0} f(x) = -0.25$

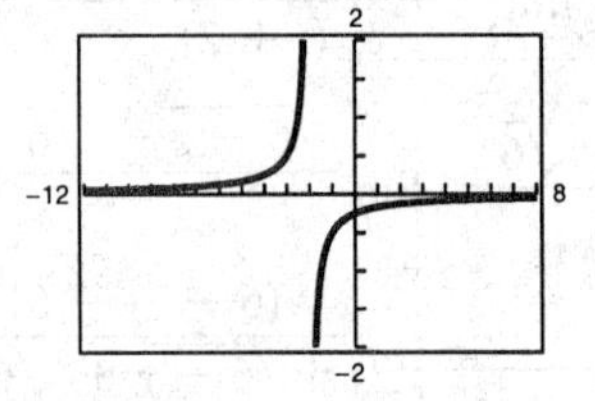

34. $\lim\limits_{x\to 3^+} |x-3| = \lim\limits_{x\to 3^-} |x-3| = \lim\limits_{x\to 3} |x-3| = 0$

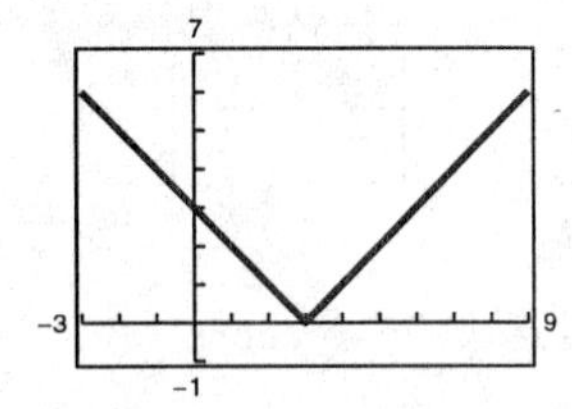

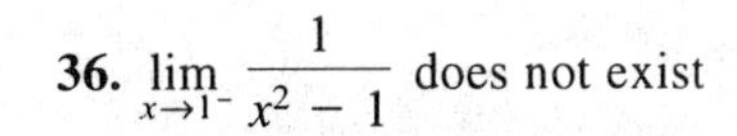

36. $\lim\limits_{x\to 1^-} \dfrac{1}{x^2-1}$ does not exist

$\lim\limits_{x\to 1^+} \dfrac{1}{x^2-1}$ does not exist

$\lim\limits_{x\to 1} \dfrac{1}{x^2-1}$ does not exist

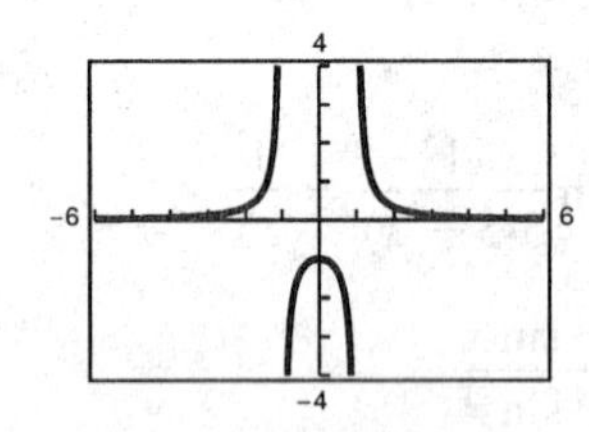

38. $\lim\limits_{x\to 1^-} f(x) = 4 - 1 = 3$

$\lim\limits_{x\to 1^+} f(x) = 3 - 1 = 2$

$\lim\limits_{x\to 1} f(x)$ does not exist

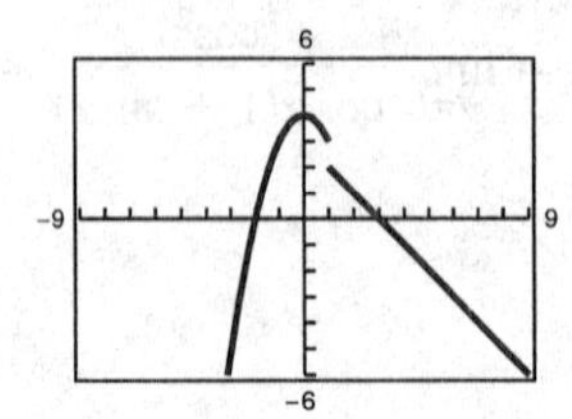

40. $\lim\limits_{h\to 0} \dfrac{f(x+h)-f(x)}{h} = \lim\limits_{h\to 0} \dfrac{\sqrt{x+h}-\sqrt{x}}{h} \cdot \dfrac{\sqrt{x+h}+\sqrt{x}}{\sqrt{x+h}+\sqrt{x}}$

$= \lim\limits_{h\to 0} \dfrac{(x+h)-x}{h\left(\sqrt{x+h}+\sqrt{x}\right)}$

$= \lim\limits_{h\to 0} \dfrac{1}{\sqrt{x+h}+\sqrt{x}} = \dfrac{1}{2\sqrt{x}}$

42. $\lim\limits_{h\to 0} \dfrac{f(x+h)-f(x)}{h} = \lim\limits_{h\to 0} \dfrac{[4-2(x+h)-(x+h)^2]-[4-2x-x^2]}{h}$

$= \lim\limits_{h\to 0} \dfrac{4-2x-2h-x^2-2xh-h^2-4+2x+x^2}{h}$

$= \lim\limits_{h\to 0} \dfrac{-2h-2xh-h^2}{h} = \lim\limits_{h\to 0}(-2-2x-h) = -2-2x$

44. $\lim\limits_{x\to 0^+} x^2 \ln x = 0$

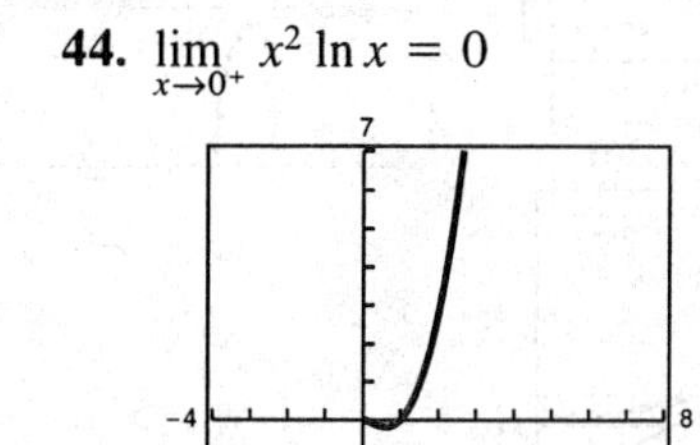

46. $\lim\limits_{x\to 0} \dfrac{\sin 3x}{x} = 3$

48. $\lim\limits_{x\to 0} \dfrac{1-\cos 2x}{x} = 0$

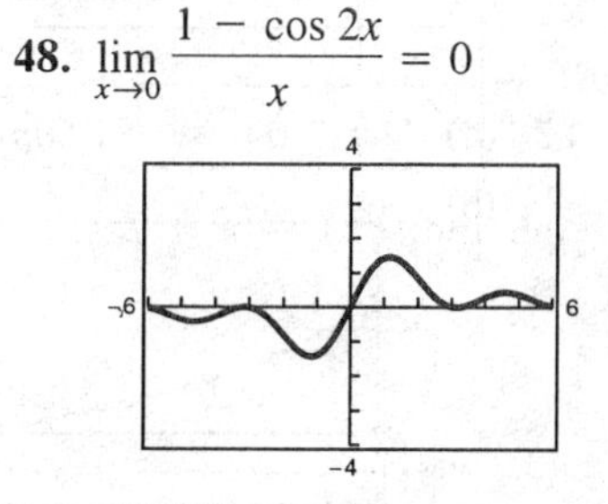

50. $\lim\limits_{x\to 0} (1+2x)^{1/x} \approx 7.389$

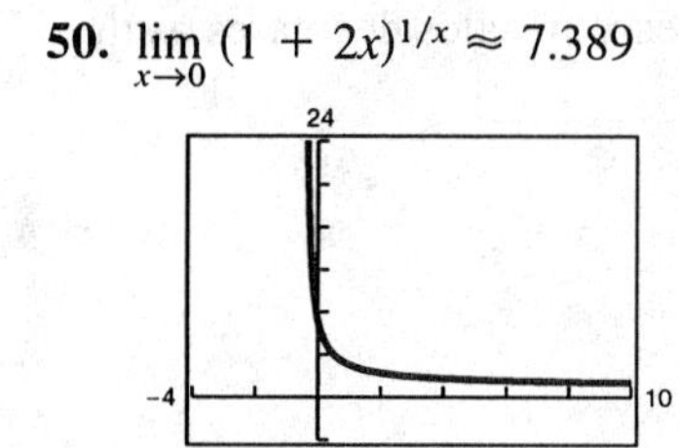

52. $\lim\limits_{x\to 0} f(x) = 0$

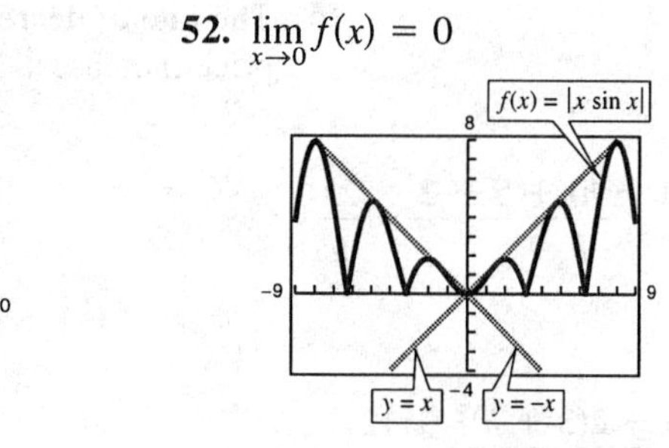

54. $\lim\limits_{x\to 0} f(x) = 0$

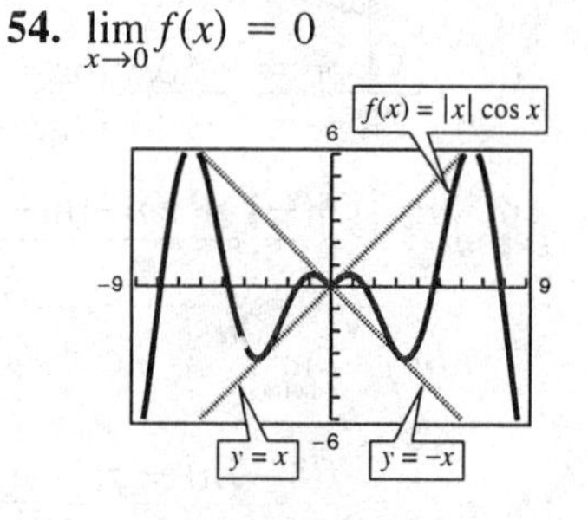

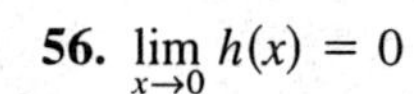

56. $\lim\limits_{x\to 0} h(x) = 0$

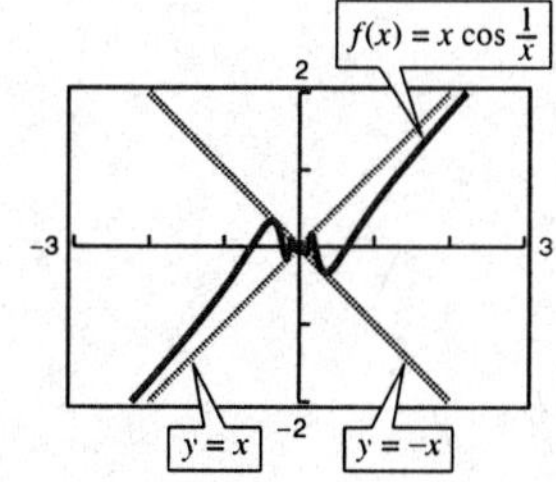

58. (a) Can be evaluated by direct substitution.

$\lim\limits_{x\to 0} \dfrac{x}{\cos x} = \dfrac{0}{\cos 0} = 0$

(b) Cannot be evaluated by direct substitution.

$\lim\limits_{x\to 0} \dfrac{1-\cos x}{x} = 0$

(See Section 12.1, Exercise 16)

60. $v(2) = \lim\limits_{t\to 2} \dfrac{s(2)-s(t)}{2-t} = \lim\limits_{t\to 2} \dfrac{(-64+128)-(-16t^2+128)}{2-t}$

$= \lim\limits_{t\to 2} \dfrac{16t^2-64}{2-t} = \lim\limits_{t\to 2} \dfrac{16(t+2)(t-2)}{2-t}$

$= \lim\limits_{t\to 2} -16(t+2) = -64$ feet per second

Section 12.3 The Tangent Line Problem

Solutions to Even-Numbered Exercises

2. Slope is -1 at (x, y).

4. Slope is -2 at (x, y).

6. (a) $\dfrac{f(3) - f(1)}{3 - 1} > \dfrac{f(3) - f(2)}{3 - 2}$

(b) $\dfrac{f(3) - f(1)}{3 - 1} < f'(1)$

8. Slope ≈ 0

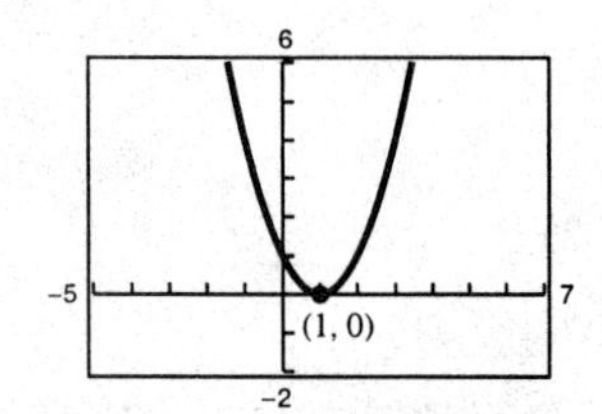

10. Slope ≈ 3

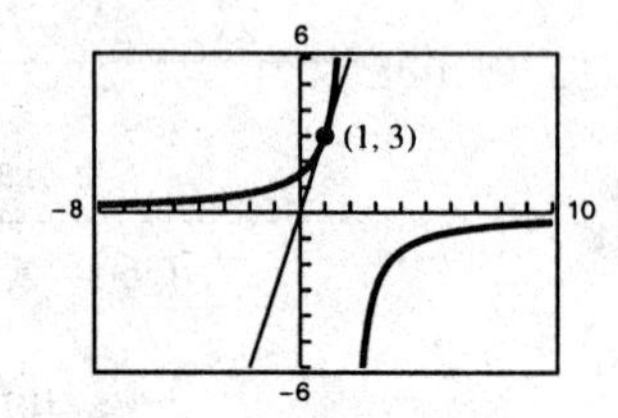

12. (a) $N = 1.04p^2 - 81.50p + 1613.31$

(b)

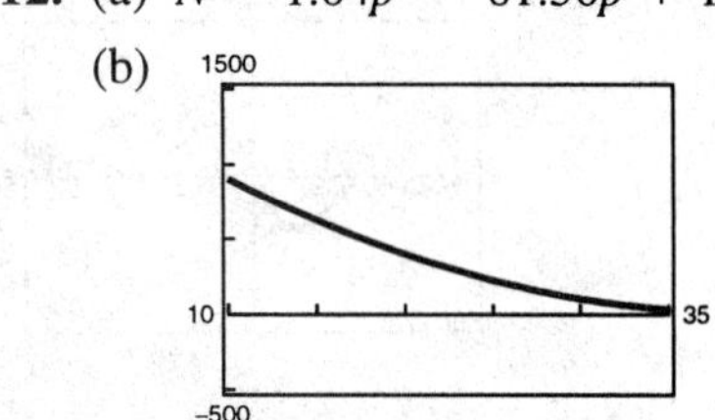

Slope $= -50.3$ for $p = 15$
Slope $= -19.1$ for $p = 30$

(c) 1500 10 35 -500

(d) The rate of decrease in sales decreases as the price increases.

14. $m_{\text{sec}} = \dfrac{h(-1 + h) - h(-1)}{h} = \dfrac{2(-1 + h) + 5 - 3}{h} = \dfrac{2h}{h}$

$m = \lim\limits_{h \to 0} \dfrac{2h}{h} = 2$

16. $m_{\text{sec}} = \dfrac{f(3 + h) - f(3)}{h} = \dfrac{10(3 + h) - 2(3 + h)^2 - 12}{h}$

$= \dfrac{-2h - 2h^2}{h} = -2 - 4h,\ h \neq 0$

$m = \lim\limits_{h \to 0} (-2 - 4h) = -2$

18. $m_{\text{sec}} = \dfrac{g(4 + h) - g(4)}{h} = \dfrac{\dfrac{1}{4 + h - 2} - \dfrac{1}{2}}{h} = \dfrac{\dfrac{1}{2 + h} - \dfrac{1}{2}}{h}$

$= \dfrac{-h}{(2 + h)2h} = \dfrac{1}{2(2 + h)},\ h \neq 0$

$m = \lim\limits_{h \to 0} \left(\dfrac{-1}{2(2 + h)}\right) = -\dfrac{1}{4}$

20. $m_{\text{sec}} = \dfrac{h(-1 + h) - h(-1)}{h} = \dfrac{\sqrt{-1 + h + 10} - 3}{h} \cdot \dfrac{\sqrt{h + 9} + 3}{\sqrt{h + 9} + 3}$

$= \dfrac{(h + 9) - 9}{h\left[\sqrt{h + 9} + 3\right]} = \dfrac{1}{\sqrt{h + 9} + 3},\ h \neq 0$

$m = \lim\limits_{h \to 0} \dfrac{1}{\sqrt{h + 9} + 3} = \dfrac{1}{6}$

22. $m_{\text{sec}} = \dfrac{g(x + h) - g(x)}{h} = \dfrac{(x + h)^3 - x^3}{h} = \dfrac{3x^2h + 3xh^2 + h^3}{h}$

$= 3x^2 + 3xh + h^2, h \neq 0$

$m = \lim_{h \to 0} (3x^2 + 3xh + h^2) = 3x^2$

At $(1, 1)$, $m = 3(1)^2 = 3$

At $(-2, -8)$, $m = 3(-2)^2 = 12$

24. $m_{\text{sec}} = \dfrac{g(x + h) - g(x)}{h} = \dfrac{\sqrt{x + h - 1} - \sqrt{x - 1}}{h} \cdot \dfrac{\sqrt{x + h - 1} + \sqrt{x - 1}}{\sqrt{x + h - 1} + \sqrt{x - 1}}$

$= \dfrac{(x + h - 1) - (x - 1)}{h[\sqrt{x + h - 1} + \sqrt{x - 1}]} = \dfrac{1}{\sqrt{x + h - 1} + \sqrt{x - 1}}, h \neq 0$

$m = \lim_{h \to 0} \dfrac{1}{\sqrt{x + h - 1} + \sqrt{x - 1}} = \dfrac{1}{2\sqrt{x - 1}}$

At $(5, 2)$, $m = \dfrac{1}{2\sqrt{5 - 1}} = \dfrac{1}{4}$

At $(10, 3)$, $m = \dfrac{1}{2\sqrt{10 - 1}} = \dfrac{1}{6}$

26. $f'(x) = \lim_{h \to 0} \dfrac{f(x + h) - f(x)}{h} = \lim_{h \to 0} \dfrac{(-4(x + h) + 2) - (-4x + 2)}{h} = \lim_{h \to 0} \dfrac{-4h}{h} = -4$

28. $f'(x) = \lim_{h \to 0} \dfrac{f(x + h) - f(x)}{h} = \lim_{h \to 0} \dfrac{((x + h)^2 - 2(x + h) + 3) - (x^2 - 2x + 3)}{h}$

$= \lim_{h \to 0} \dfrac{2xh + h^2 - 2h}{h} = \lim_{h \to 0} (2x + h - 2) = 2x - 2$

30. $h'(s) = \lim_{h \to 0} \dfrac{h(s + h) - h(s)}{h} = \lim_{h \to 0} \dfrac{\dfrac{1}{\sqrt{s + h + 1}} - \dfrac{1}{x + 1}}{h}$

$= \lim_{h \to 0} \dfrac{\sqrt{s + 1} - \sqrt{s + h + 1}}{h\sqrt{s + h + 1}\sqrt{s + 1}} \cdot \dfrac{\sqrt{s + 1} + \sqrt{s + h + 1}}{\sqrt{s + 1} + \sqrt{s + h + 1}}$

$= \lim_{h \to 0} \dfrac{(s + 1) - (s + h + 1)}{h\sqrt{s + h + 1}\sqrt{s + 1}[\sqrt{s + 1} + \sqrt{s + h + 1}]}$

$= \lim_{h \to 0} \dfrac{-1}{h\sqrt{s + h + 1}\sqrt{s + 1}[\sqrt{s + 1} + \sqrt{s + h + 1}]}$

$= \dfrac{-1}{(s + 1)2\sqrt{s + 1}} = \dfrac{-1}{2(s + 1)^{3/2}}$

32. $m_{\text{sec}} = \dfrac{f(2 + h) - f(2)}{h} = \dfrac{(2 + h)^3 - (2 + h) - 6}{h}$

$= \dfrac{h^3 + 6h^2 + 11h}{h} = h^2 + 6h + 11, h \neq 0$

$m = \lim_{h \to 0} (h^2 + 6h + 11) = 11$

Tangent line: $y - 6 = 11(x - 2)$

$y = 11x - 16$

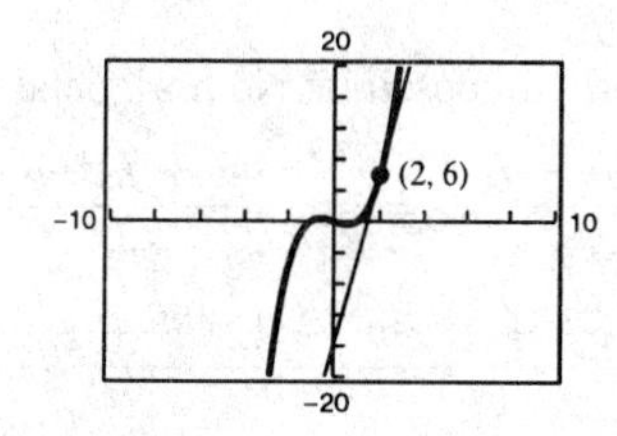

34. $m_{\text{sec}} = \dfrac{f(2+h) - f(2)}{h} = \dfrac{\left(2(2+h) + \dfrac{4}{2+h}\right) - 6}{h}$

$= \dfrac{(4+2h)(2+h) + 4 - 6(2+h)}{(2+h)h} = \dfrac{2h + 2h^2}{(2+h)h} = \dfrac{2+2h}{2+h}, h \neq 0$

$m = \lim_{h \to 0} \left(\dfrac{2+2h}{2+h}\right) = 1$

Tangent line: $y - 6 = 1(x - 2)$

$y = x + 4$

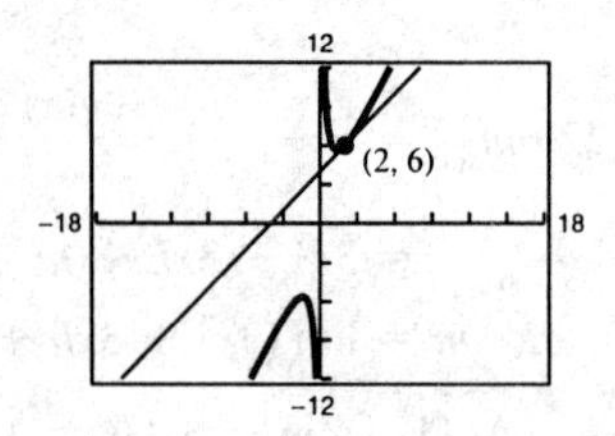

36.

x	-2	-1.5	-1	-0.5	0	0.5	1	1.5	2
$f(x)$	1	1.225	1.414	1.581	1.732	1.871	2	2.121	2.236
$f(x)$	0.5	0.408	0.354	0.316	0.289	0.267	0.25	0.236	0.224

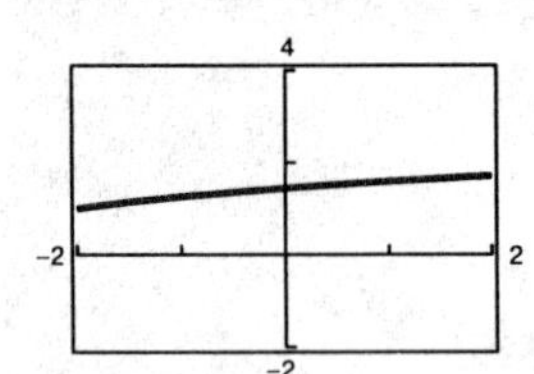

$f(x) = \sqrt{x+3}$

$f'(x) = \dfrac{1}{2\sqrt{x+3}}$

38.

x	-2	-1.5	-1	-0.5	0	0.5	1	1.5	2
$f(x)$	0	-0.7	-1	-1.071	-1	-0.833	-0.6	-0.318	0
$f(x)$	-2	-0.92	-0.333	-0.020	0.25	0.407	0.52	0.603	0.667

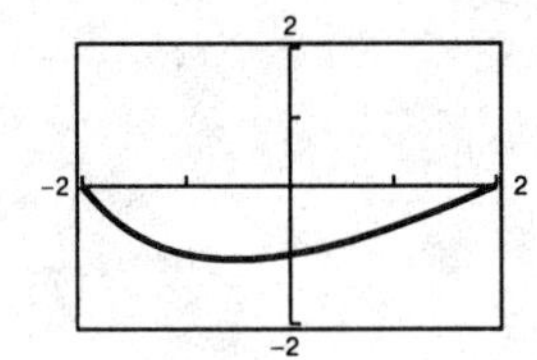

$f(x) = \dfrac{x^2 - 4}{x + 4}$

$f'(x) = \dfrac{x^2 + 8x + 4}{(x+4)^2}$

40. $f'(x) = \lim_{h \to 0} \dfrac{f(x+h) - f(x)}{h} = \lim_{h \to 0} \dfrac{(x+h)^3 + 3(x+h) - (x^3 + 3x)}{h}$

$= \lim_{h \to 0} \dfrac{x^3 + 3x^2h + 3xh^2 + h^3 + 3x + 3h - x^3 - 3x}{h}$

$= \lim_{h \to 0} \dfrac{3x^2h + 3xh^2 + h^3 + 3h}{h}$

$= \lim_{h \to 0} (3x^2 + 3xh + h^2 + 3) = 3x^2 + 3$

$f'(x) = 3x^2 + 3 = 0$ Impossible; No horizontal tangents.

42. $f'(x) = \lim_{h \to 0} \dfrac{f(x+h) - f(x)}{h} = \lim_{h \to 0} \dfrac{3(x+h)^4 + 4(x+h)^3 - (3x^4 + 4x^3)}{h}$

$= \lim_{h \to 0} \dfrac{(12x^3h + 18x^2h^2 + 12xh^3 + 3h^4) + (12x^2h + 12xh^2 + 4h^3)}{h}$

$= 12x^3 + 12x^2$

$f'(x) = 0 = 12x^3 + 12x^2 = 12x^2(x+1) \Rightarrow 0, -1$

f has horizontal tangents at $(0, 0)$ and $(-1, -1)$.

44. Matches (a).
(Derivative approaches $-\infty$ when x approaches 0.)

46. Matches (c).
(Derivative decreases until origin, then increases.)

Section 12.4 Limits at Infinity and Limits of Sequences

Solutions to Even-Numbered Exercises

2. Horizontal asymptote: $y = -2$
Matches (a).

4. Horizontal asymptote: $y = 0$
Domain all x. Matches (c).

6. No horizontal asymptote.
Matches (d).

8. $\lim_{x\to\infty} \dfrac{4}{2x+3} = 0$

10. $\lim_{x\to\infty} \dfrac{1-5x}{1+2x} = -\dfrac{5}{2}$

12. $\lim_{x\to-\infty} \dfrac{x^2+1}{x^2} = 1$

14. $\lim_{y\to\infty} \dfrac{4y^4}{y^2+1}$ Does not exist.

16. $\lim_{x\to\infty} \dfrac{2x^2-4}{(x-1)^2} = \lim_{x\to\infty} \dfrac{2x^2-6}{x^2 - {}^2x + 1} = 2$

18. $\lim_{x\to\infty} \left[3 + \dfrac{2x^2}{(x+5)^2}\right] = 3 + 2 = 5$

20. $\lim_{x\to\infty} \left[\dfrac{x}{2x+1} + \dfrac{3x^2}{(x-3)^2}\right] = \dfrac{1}{2} + 3 = \dfrac{7}{2}$

22.

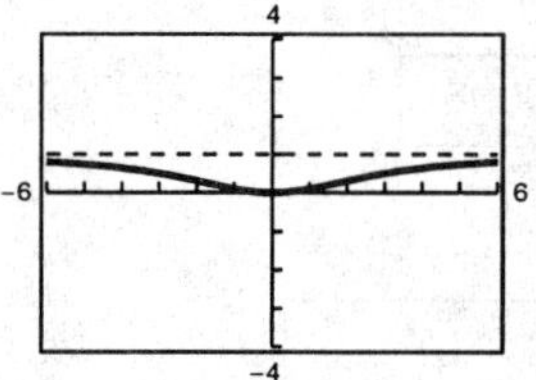

Horizontal asymptote: $y = 1$

24.

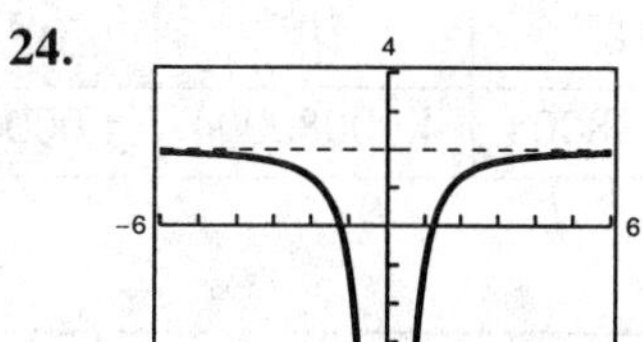

Horizontal asymptote: $y = 2$

26.

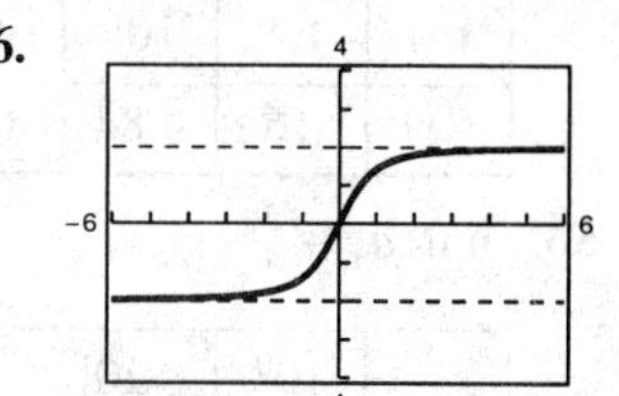

Horizontal asymptotes: $y = 2$, $y = -2$

28.

x	10^0	10^1	10^2	10^3	10^4	10^5	10^6
$f(x)$	-0.162	-0.0167	-0.00167	-1.67×10^{-4}	-1.7×10^{-5}	-1.7×10^{-6}	-2×10^{-7}

$\lim_{x\to\infty} \left(3x - \sqrt{9x^2+1}\right) = 0$

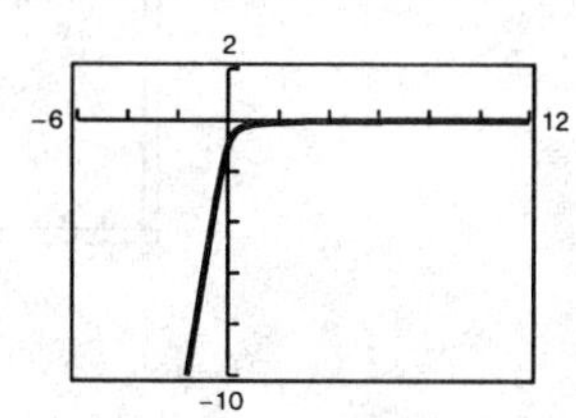

30.

x	10^0	10^1	10^2	10^3	10^4	10^5	10^6
$f(x)$	0.508	0.5008	0.50008	0.5	0.5	0.5	0.5

$\lim_{x\to\infty} 4\left(4x - \sqrt{16x^2 - x}\right) = \frac{1}{2}$

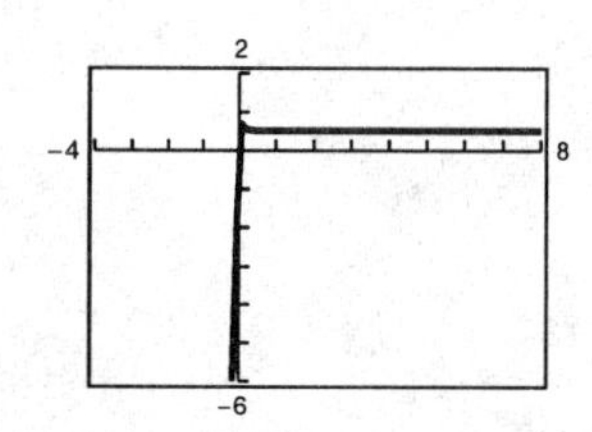

32. (a)

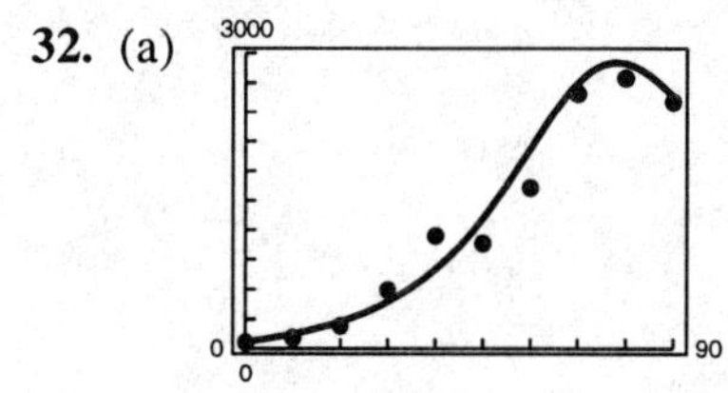

The model is a good fit.

(b) For 1975, $t = 75$ and $N(75) \approx 2875$ thousand

(c) The model approaches 0 as $t \to \infty$.

34. $\dfrac{2}{3}, \dfrac{5}{4}, \dfrac{8}{5}, \dfrac{11}{6}, 2$ • $\lim_{n\to\infty} \dfrac{3n-1}{n+2} = 3$

36. $\dfrac{2}{3}, \dfrac{8}{19}, \dfrac{18}{85}, \dfrac{32}{261}, \dfrac{50}{631}$ • $\lim_{n\to\infty} \dfrac{2n^2}{n^4+n+1} = 0$

38. $-3, -3, 1, 3, 5$ • $\lim_{n\to\infty} (2n-5)$ Does not exist.

40. $\frac{1}{6}, \frac{1}{20}, \frac{1}{42}, \frac{1}{72}, \frac{1}{110}$ • $\lim_{n\to\infty} \frac{(2n-1)!}{(2n+1)} = \lim_{n\to\infty} \frac{1}{(2n+1)(2n)} = 0$

42. $1, -\frac{1}{4}, \frac{1}{9}, -\frac{1}{16}, \frac{1}{25}$ • $\lim_{n\to\infty} \frac{(-1)^{n+1}}{n^2} = 0$

44. $\frac{1}{5}, -\frac{1}{2}, \frac{5}{7}, -\frac{7}{8}, 1$ • $\lim_{n\to\infty} (-1)^{n+1}\left(\frac{2n-1}{n+4}\right)$ does not exist.

46. $\lim_{n\to\infty} a_n = 12$

x	10^0	10^1	10^2	10^3	10^4	10^5	10^6
$f(x)$	20	12.8	12.08	12.008	12.0008	12.00008	12.000008

48. $\lim_{n\to\infty} a_n = 4$

x	10^0	10^1	10^2	10^3	10^4	10^5	10^6
$f(x)$	16	4.84	4.0804	4.0088004	4.00080004	4.00008	4.000008

50. $\lim_{n\to\infty} a_n = \frac{3}{4}$

x	10^0	10^1	10^2	10^3	10^4	10^5	10^6
$f(x)$	1	0.7975	0.754975	0.75049975	0.75005	0.750005	0.7500005

52. Diverges

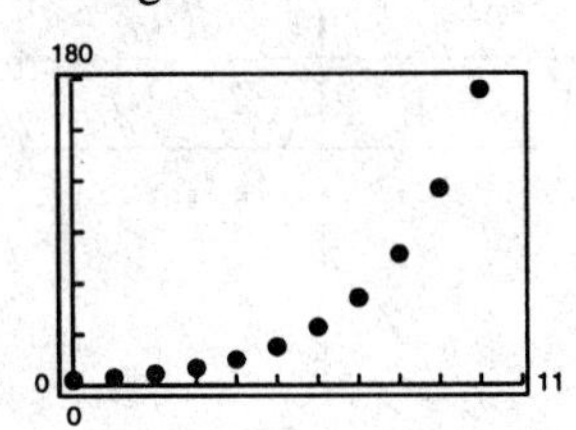

54. Converges to 6

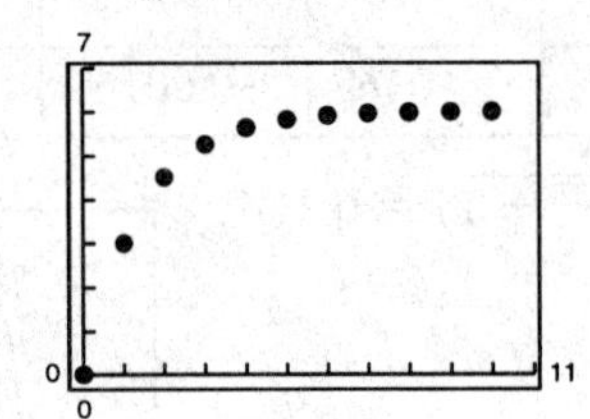

Section 12.5 The Area Problem

Solutions to Even-Numbered Exercises

2. $\sum_{i=1}^{30} i^2 = \frac{n(n+1)(2n+1)}{6} = \frac{30(31)(61)}{6} = 9455$

4. $\sum_{k=1}^{50} (2k+1) = 2\sum_{k=1}^{50} k + \sum_{k=1}^{50} 1 = 2\frac{50(51)}{2} + 50 = 2600$

6. $\sum_{j=1}^{10} (j^3 - 3j^2) = \frac{10^2(11)^2}{4} - 3\left(\frac{10(11)(21)}{6}\right) = 1870$

8. $S(n) = \sum_{i=1}^{n} \frac{i}{n^2} = \frac{1}{n^2}\frac{n(n+1)}{2} = \frac{n+1}{2n}$

n	10^0	10^1	10^2	10^3
$S(n)$	1	0.55	0.505	0.5005

$\lim_{n\to\infty} S(n) = \frac{1}{2}$

10. $S(n) = \sum_{i=1}^{n} \frac{i^3}{n^4} = \frac{1}{n^4}\left[\frac{n^2(n+1)^2}{4}\right] = \frac{n^2+2n+1}{4n^2}$

n	10^0	10^1	10^2	10^3
$S(n)$	1	0.3025	0.255025	0.25050025

$\lim_{n\to\infty} S(n) = \frac{1}{4}$

12. $S(n) = \sum_{i=1}^{n} \frac{24i^2}{n^3} = \frac{24}{n^3}\left[\frac{n(n+1)(2n+1)}{6}\right] = \frac{4}{n^2}(2n^2+3n+1) = \frac{8n^2+12n-4}{n^2}$

n	10^0	10^1	10^2	10^3
$S(n)$	24	9.24	8.1204	8.012004

$\lim_{n\to\infty} S(n) = 8$

14. $S(n) = \sum_{i=1}^{n}\left[3 - 2\left(\frac{i}{n}\right)\right]\frac{1}{n} = \frac{1}{n}\left[3n - \frac{2}{n}\frac{n(n+1)}{2}\right] = 3 - \frac{n+1}{n}$

n	10^0	10^1	10^2	10^3
$S(n)$	1	1.9	1.99	1.999

$\lim_{n\to\infty} S(n) = 2$

16. $S(n) = \sum_{i=1}^{n}\left(\frac{4}{n} + \frac{2i}{n^2}\right)\left(\frac{2i}{n}\right) = \frac{2}{n}\left[\frac{4}{n}\frac{n(n+1)}{2} + \frac{2}{n^2}\frac{n(n+1)(2n+1)}{6}\right]$

$$= \frac{2}{n}\left[\frac{4n^2+4n}{2n} + \frac{2(2n^2+3n+1)}{6n}\right] = \frac{16n^2+18n+2}{3n^2}$$

n	10^0	10^1	10^2	10^3
$S(n)$	12.0	5.94	5.3934	5.3393

$\lim_{n\to\infty} S(n) = \frac{16}{3}$

18. The width of each rectangle is 1. The height is obtained by evaluating f at the right hand endpoint of each interval.

$$A \approx \sum_{i=1}^{5} f(-2+i)(1) = \sum_{i=1}^{5}[(-2+i)^2 + 1] = 20$$

20. The width of each rectangle is $\frac{1}{2}$. The height is obtained by evaluating f at the right hand endpoint of each interval.

$$A \approx \sum_{i=1}^{4} f\left(2 + \frac{i}{2}\right)\left(\frac{1}{2}\right) = \sum_{i=1}^{4} \frac{1}{2}\left(2 + \frac{i}{2} - 2\right)^3\left(\frac{1}{2}\right)$$

$$= \sum_{i=1}^{4} \frac{1}{4}\left(\frac{i}{2}\right)^3 = 3.125$$

22. The width of each rectangle is $3/n$. The height is

$$f\left(\frac{3i}{n}\right) = 9 - \left(\frac{3i}{n}\right)^2.$$

$$A \approx \sum_{i=1}^{n} \left(9 - \left(\frac{3i}{n}\right)^2\right)\frac{3}{n}$$

(Note: exact area is 18)

n	4	8	20	50
Approximate area	14.344	16.242	17.314	17.7282

24. The width of each rectangle is $(2 - (-1))/n = 3/n$. The height is

$$f\left(-1 + \frac{3i}{n}\right) = 3 - \frac{1}{4}\left(-1 + \frac{3i}{n}\right)^3.$$

$$A \approx \sum_{i=1}^{n} \left[3 - \frac{1}{4}\left(-1 + \frac{3i}{n}\right)^3\right]\frac{3}{n}$$

(Note: exacth area is $8\frac{1}{16} \approx 8.0625$)

n	4	8	20	50
Approximate area	7.113	7.614	7.8895	7.994

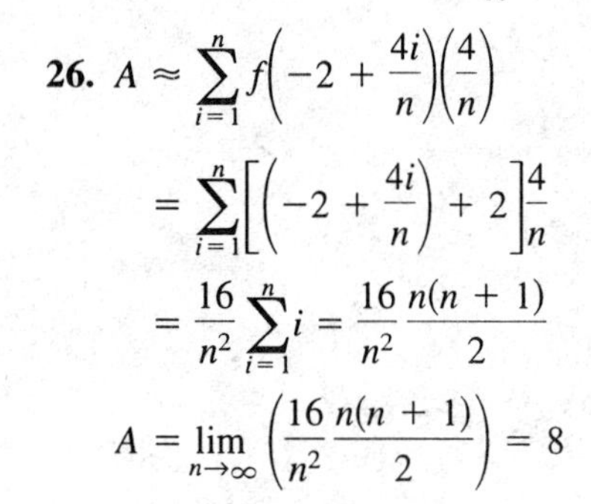

26. $A \approx \sum_{i=1}^{n} f\left(-2 + \frac{4i}{n}\right)\left(\frac{4}{n}\right)$

$$= \sum_{i=1}^{n} \left[\left(-2 + \frac{4i}{n}\right) + 2\right]\frac{4}{n}$$

$$= \frac{16}{n^2}\sum_{i=1}^{n} i = \frac{16}{n^2}\,\frac{n(n+1)}{2}$$

$$A = \lim_{n\to\infty}\left(\frac{16}{n^2}\,\frac{n(n+1)}{2}\right) = 8$$

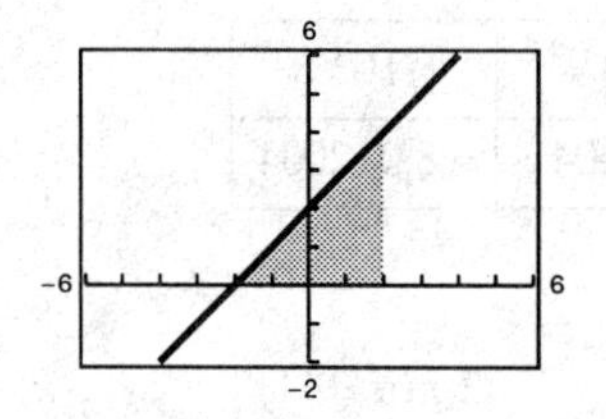

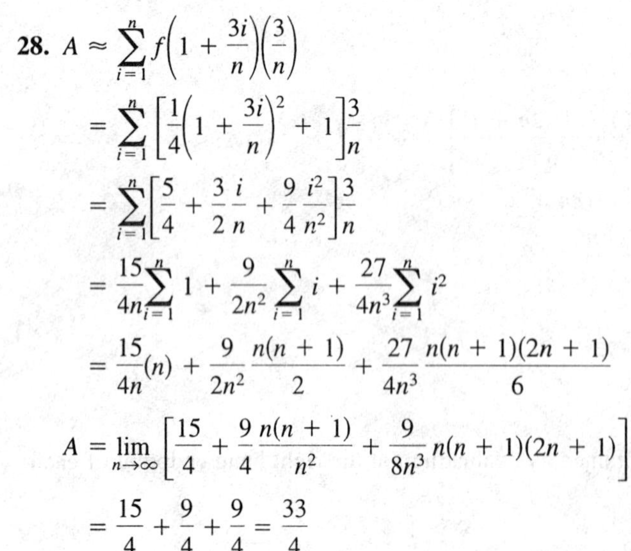

28. $A \approx \sum_{i=1}^{n} f\left(1 + \frac{3i}{n}\right)\left(\frac{3}{n}\right)$

$$= \sum_{i=1}^{n} \left[\frac{1}{4}\left(1 + \frac{3i}{n}\right)^2 + 1\right]\frac{3}{n}$$

$$= \sum_{i=1}^{n} \left[\frac{5}{4} + \frac{3}{2}\frac{i}{n} + \frac{9}{4}\frac{i^2}{n^2}\right]\frac{3}{n}$$

$$= \frac{15}{4n}\sum_{i=1}^{n} 1 + \frac{9}{2n^2}\sum_{i=1}^{n} i + \frac{27}{4n^3}\sum_{i=1}^{n} i^2$$

$$= \frac{15}{4n}(n) + \frac{9}{2n^2}\,\frac{n(n+1)}{2} + \frac{27}{4n^3}\,\frac{n(n+1)(2n+1)}{6}$$

$$A = \lim_{n\to\infty}\left[\frac{15}{4} + \frac{9}{4}\,\frac{n(n+1)}{n^2} + \frac{9}{8n^3}\,n(n+1)(2n+1)\right]$$

$$= \frac{15}{4} + \frac{9}{4} + \frac{9}{4} = \frac{33}{4}$$

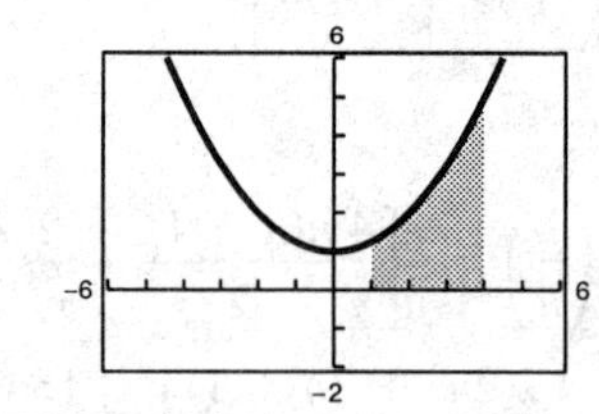

30. $A \approx \sum_{i=1}^{n} f\left(2\frac{i}{n}\right)\left(\frac{2}{n}\right)$

$$= \sum_{i=1}^{n}\left[4\left(\frac{2i}{n}\right) - \left(\frac{2i}{n}\right)^3\right]\left(\frac{2}{n}\right)$$

$$= \frac{16}{n^2}\sum_{i=1}^{n} i = \frac{16}{n^4}\sum_{i=1}^{n} i^3$$

$$= \frac{16}{n^2}\frac{n(n+1)}{n^2} - \frac{16}{n^4}\frac{n^2(n+1)^2}{4}$$

$$A = \lim_{n\to\infty}\left[8\frac{n(n+1)}{2} - 4\frac{n^2(n+1)^2}{n^4}\right] = 8 - 4 = 4$$

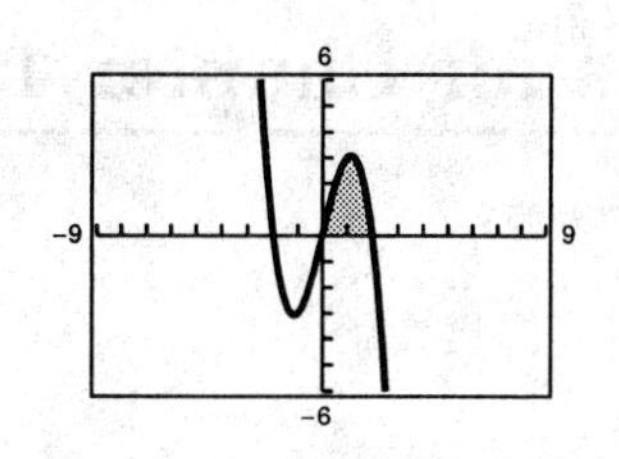

32. $A \approx \sum_{i=1}^{n} f\left(-1 + \frac{2i}{n}\right)\left(\frac{2}{n}\right)$

$$= \sum_{i=1}^{n}\left[\left(-1 + \frac{2i}{n}\right)^2 - \left(-1 + \frac{2i}{n}\right)^3\right]\left(\frac{2}{n}\right)$$

$$= \sum_{i=1}^{n}\left[\left(1 - \frac{4i}{n} + \frac{4i^2}{n^2}\right) - \left(-1 + \frac{6i}{n} - \frac{12i^2}{n^2} + \frac{8i^3}{n^3}\right)\right]\frac{2}{n}$$

$$= \sum_{i=1}^{n}\left[2 - \frac{10i}{n} + \frac{16i^2}{n^2} - \frac{8i^3}{n^3}\right]\frac{2}{n}$$

$$= \frac{4}{n}\sum_{i=1}^{n} 1 - \frac{20}{n^2}\sum_{i=1}^{n} i + \frac{32}{n^3}\sum_{i=1}^{n} i^2 - \frac{16}{n^4}\sum_{i=1}^{n} i^3$$

$$= \frac{4}{n}(n) - \frac{20}{n^2}\frac{n(n+1)}{2} + \frac{32}{n^3}\frac{n(n+1)(2n+1)}{6} - \frac{16}{n^4}\frac{n^2(n+1)}{4}$$

$$\lim_{n\to\infty} A = 4 - 10 + \frac{32}{3} - 4 = \frac{2}{3}$$

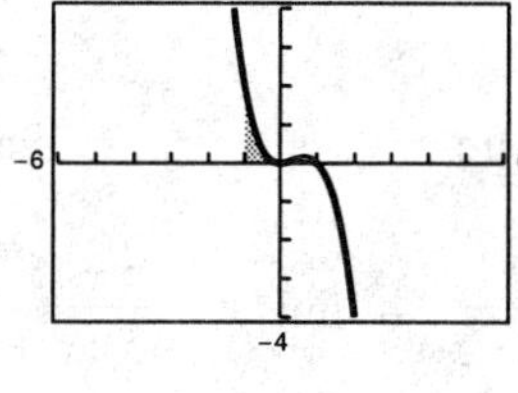

34.

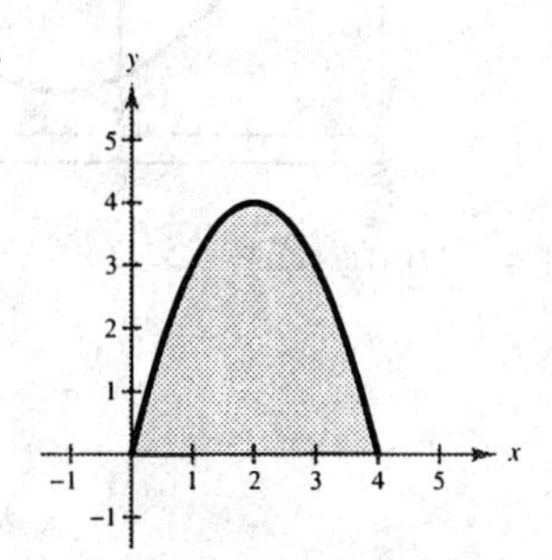

$5 < \text{Area} < 12 \Rightarrow$ (a)

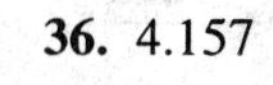

36. 4.157

38. 5.188

Review Exercises for Chapter 12

Solutions to Even-Numbered Exercises

2. $\lim_{x\to3}\dfrac{\dfrac{x}{x+1}-\dfrac{3}{4}}{(x-3)} = 0.0625 = \dfrac{1}{16}$

x	2.9	2.99	2.999	3	3.001	3.01	3.1
$f(x)$	0.0641	0.0627	0.0625	?	0.0625	0.0623	0.0610

4. $\lim_{x\to-2}(5-2x-x^2) = 5-2(-2)-(-2)^2 = 5$

6. $\lim_{x\to2}\dfrac{3x+5}{5x-3} = \dfrac{3(2)+5}{5(2)-3} = \dfrac{11}{7}$

8. $\lim_{t\to3}\dfrac{t^2-9}{t-3} = \lim_{t\to3}\dfrac{(t-3)(t+3)}{t-3} = \lim_{t\to3}(t+3) = 6$

10. $\lim_{x\to5}\dfrac{x-5}{x^2+5x-50} = \lim_{x\to5}\dfrac{x-5}{(x-5)(x+10)} = \lim_{x\to5}\dfrac{1}{x+10} = \dfrac{1}{15}$

12. $\lim_{x\to4}\dfrac{x^3-64}{x^2-16} = \lim_{x\to4}\dfrac{(x-4)(x^2+4x+16)}{(x-4)(x+4)} = \lim_{x\to4}\dfrac{x^2+4x+16}{x+4} = \dfrac{16+16+16}{8} = 6$

14. $\lim_{v\to0}\dfrac{\sqrt{v+7}-3}{v}$

$\dfrac{\sqrt{7}-3}{0}$ does not exist.

16. $\lim_{x\to0}\dfrac{(1/1+x)-1}{x} = \lim_{x\to0}\dfrac{1-(1+x)}{x(1+x)}$

$= \lim_{x\to0}\dfrac{-x}{x(1+x)} = -1$

18. $\lim_{x\to2^+}f(x) = \lim_{x\to2^+}(x^2-3)$

$= 4-3 = 1$

20. 

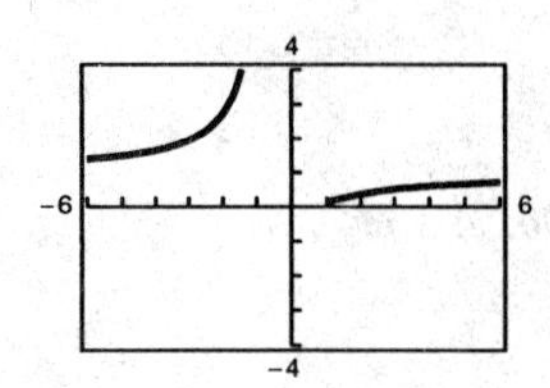

$\lim_{x\to0}e^{-2/x}$

Limit does not exist.

22.

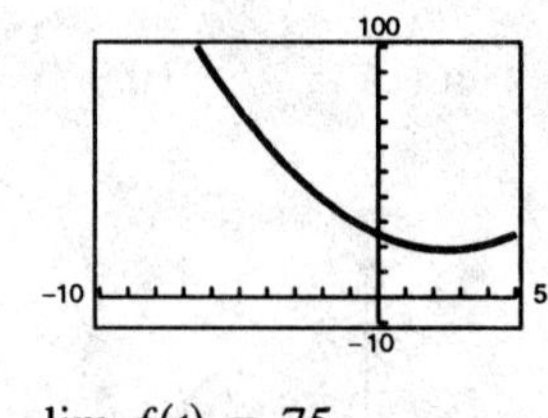

$\lim_{t\to-5}f(t) = 75$

24. (a)

x	1.1	1.01	1.001	1.0001
$f(x)$	−0.4881	−0.4988	−0.4999	−0.5000

$\lim_{x\to1^+}\dfrac{1-\sqrt{x}}{x-1} = -\dfrac{1}{2}$

(b) $\lim_{x\to1^+}\dfrac{1-\sqrt{x}}{x-1}\cdot\dfrac{1+\sqrt{x}}{1+\sqrt{x}} = \lim_{x\to1^+}\dfrac{1-x}{(x-1)(x-\sqrt{x})} = \lim_{x\to1^+}\dfrac{-1}{1+\sqrt{x}} = -\dfrac{1}{2}$

26. (a) $\lim_{x\to c}\sqrt[3]{f(x)} = \sqrt[3]{27} = 3$

(b) $\lim_{x\to c}\dfrac{f(x)}{18} = \dfrac{27}{18} = \dfrac{3}{2}$

(c) $\lim_{x\to c}[f(x)\,g(x)] = (27)(12) = 324$

(d) $\lim_{x\to c}[f(x)-2g(x)] = 27-2(12) = 3$

28. $\lim_{h\to0}\dfrac{f(x+h)-f(x)}{h} = \lim_{h\to0}\dfrac{(x+h)^3+5-(x^3+5)}{h}$

$= \lim_{h\to0}\dfrac{x^3+3x^2h+3xh^2+h^3-x^3}{h}$

$= \lim_{h\to0}(3x^2+3xh+h^2) = 3x^2$

30.

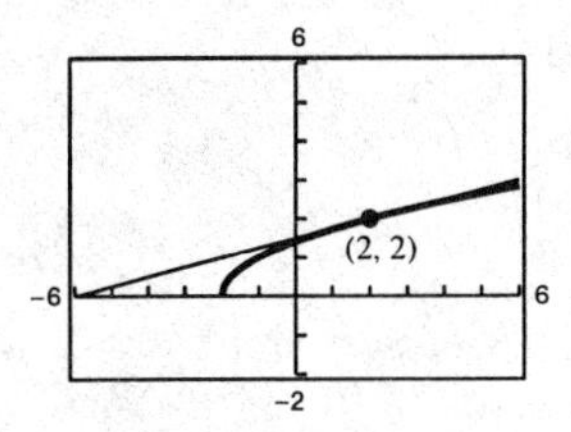

Slope is $\frac{1}{4}$ at (2, 2).

32. $m = \lim_{h\to 0} \dfrac{g(x+h) - g(x)}{h} = \lim_{h\to 0} \dfrac{\sqrt{x+h} - \sqrt{x}}{h} \cdot \dfrac{\sqrt{x+h} + \sqrt{x}}{\sqrt{x+h} + \sqrt{x}}$

$= \lim_{h\to 0} \dfrac{(x+h) - x}{h\left[\sqrt{x+h} + \sqrt{x}\right]} = \lim_{h\to 0} \dfrac{1}{\sqrt{x+h} + \sqrt{x}} = \dfrac{1}{2\sqrt{x}}$

(a) At (1, 1), $m = \dfrac{1}{2\sqrt{1}} = \dfrac{1}{2}$

(b) At (4, 2), $m = \dfrac{1}{2\sqrt{4}} = \dfrac{1}{4}$

34. $m = \lim_{h\to 0} \dfrac{f(x+h) - f(x)}{h} = \lim_{h\to 0} \dfrac{(1/4)(x+h)^4 - (1/4)x^4}{h}$

$= \lim_{h\to 0} \dfrac{(1/4)[x^4 + 4x^3h + 6x^2h^2 + 4xh^3 + h^4 - x^4]}{h}$

$= \lim_{h\to 0} \dfrac{1}{4}[4x^3 + 6x^2h + 4xh^2 + h^3] = x^3$

(a) At $(-2, 4)$, $m = (-^3) = -8$

(b) At $\left(1, \frac{1}{4}\right)$, $m = (1)^3 = 1$

36. $f'(x) = \lim_{h\to 0} \dfrac{f(x+h) - f(x)}{h} = \lim_{h\to 0} \dfrac{3(x+h) - 3x}{h} = 3$

38. $f'(x) = \lim_{h\to 0} \dfrac{f(x+h) - f(x)}{h} = \lim_{h\to 0} \dfrac{((1/2)(x+h)^2 + 3) - ((1/2)x^2 + 3)}{h}$

$= \lim_{h\to 0} \dfrac{(1/2)(x^2 + 2xh + h^2)}{h} = \lim_{h\to 0} \left(xh + \dfrac{1}{2}h^2\right) = x$

40. $g'(s) = \lim_{h\to 0} \dfrac{g(s+h) - g(s)}{h} = \lim_{h\to 0} \dfrac{\dfrac{4}{s+h+5} - \dfrac{4}{s+5}}{h}$

$= \lim_{h\to 0} \dfrac{4s + 20 - 4s - 4h - 20}{(s+h+5)(s+5)h} = \lim_{h\to 0} \dfrac{-4}{(s+h+5)(s+5)}$

$= \dfrac{-4}{(s+5)^2}$

42. From Exercise 41, $y = 8.73t^2 - 6.23t + 0.54t$. The derivative is $y' = 17.46t - 6.23$. At $t = 5$, $y' \approx 81.07$ feet/second.

44. Matches (a). Derivative is negative for $x < 0$, positive for $x > 0$.

46. Matches (c). Derivative is average negative.

48. $\lim_{x\to\infty} \dfrac{7x}{14x + 2} = \dfrac{7}{14} = \dfrac{1}{2}$

50. $\lim_{x\to\infty} \dfrac{x^2}{2x + 3}$ Does not exist.

52. $\lim_{x\to\infty} (x-2)^{-3} = \lim_{x\to\infty} \dfrac{1}{(x-2)^3} = 0$

54. $\lim_{n\to\infty} \left(\dfrac{2}{n}\right)\left[n + \dfrac{2}{n}\left(\dfrac{n(n-1)}{2} - n\right)\right] = \lim_{n\to\infty} \left[2 + 2\dfrac{(n-1)}{n} - \dfrac{4}{n}\right] = 2 + 2 = 4$

56. $\sum_{i=1}^{n}\left[4-\left(\frac{3i}{n}\right)^2\right]\left(\frac{3i}{n^2}\right)=\frac{12}{n^2}\sum_{i=1}^{n} i-\frac{27}{n^4}\sum_{i=1}^{n} i^3=\frac{12}{n^2}\frac{n(n+1)}{2}-\frac{27}{n^4}\frac{n^2(n+1)^2}{4}$

$$=\frac{24n^2+24n-27(n^2+2n+1)}{4n^2}$$

$$=\frac{-3n^2-30n-27}{4n^2}=\frac{-3}{4n^2}(n^2+10n+9)$$

$$=\frac{-3(n+1)(n+9)}{4n^2}$$

n	10^0	10^1	10^2	10^3
$S(n)$	-15	-1.5675	-0.8257	-0.7575

$\lim_{n\to\infty} Sn=-\frac{3}{4}$

58. Width of rectangle: $\frac{1}{4}$; Height is f evaluated at right endpoint.

Area $\approx \frac{1}{4}\left[f\left(\frac{1}{4}\right)+f\left(\frac{1}{2}\right)+f\left(\frac{3}{4}\right)+f(1)\right]$

$=\frac{1}{4}\left[4-\left(\frac{1}{4}\right)^2+4-\left(\frac{1}{2}\right)^2+4-\left(\frac{3}{4}\right)^2+4-1\right]$

$=\frac{1}{4}\left[15-\frac{14}{16}\right]=\frac{113}{32}=3.53125$

60. $f(x)=4x-x^2$

n	4	8	20	50
Approximate Area	10	10.5	10.64	10.6624

(Exact area is $10\frac{2}{3}$)

62. $A\approx\sum_{i=1}^{6}\left[2\left(3+\frac{i}{2}\right)-6\right]\left(\frac{1}{2}\right)$

$=\sum_{i=1}^{6}(i)\frac{1}{2}=\frac{21}{2}$ approximate area

$A=\lim_{n\to\infty}\sum_{i=1}^{n}\left[2\left(3+\frac{3i}{n}\right)-6\right]\left(\frac{3}{n}\right)$

$=\lim_{n\to\infty}\sum_{i=1}^{n}\frac{18i}{n^2}=\lim_{n\to\infty}\frac{18}{n^2}\sum_{i=1}^{n} i$

$=\lim_{n\to\infty}\frac{18}{n^2}\frac{n(n+1)}{2}=9$ exact area

64. $A\approx\sum_{i=1}^{4}8\left(\frac{i}{4}-\left(\frac{i}{4}\right)^2\right)\frac{1}{4}$

$=2\sum_{i=1}^{4}\left(\frac{i}{4}-\frac{i^2}{16}\right)=5-\frac{15}{4}=\frac{5}{4}$ approximate area

$A=\lim_{n\to\infty}\sum_{i=1}^{n}8\left(\left(\frac{i}{n}\right)-\left(\frac{i}{n}\right)^2\right)\frac{i}{n}$

$=\lim_{n\to\infty}\left[\frac{8}{n^2}\sum_{i=1}^{n} i-\frac{8}{n^3}\sum i^2\right]$

$=\lim_{n\to\infty}\left[\frac{8}{n^2}\frac{n(n+1)}{2}-\frac{8}{n^3}\frac{n(n+1)(2n+1)}{6}\right]=4-\frac{8}{3}=\frac{4}{3}$ exact area

66. $A \approx \sum_{i=1}^{6}\left[4 - \left[\left(1 + \frac{i}{2}\right) - 2\right]^2\right]\left(\frac{1}{2}\right)$

$= \sum_{i=1}^{6}\left[4 - \left(\frac{i}{2} - 1\right)^2\right]\left(\frac{1}{2}\right)$

$= \sum_{i=1}^{6}\left[4 - \frac{i^2}{4} + i - 1\right]\left(\frac{1}{2}\right)$

$= \sum_{i=1}^{6}\left(3 + i - \frac{i^2}{4}\right)\frac{1}{2} = 8.125$

$A = \lim_{n\to\infty} \sum_{i=1}^{n}\left[4 - \left[\left(1 + \frac{3i}{n}\right) - 2\right]^2\right]\left(\frac{3}{n}\right)$

$= \lim_{n\to\infty} \sum_{i=1}^{n}\left[4 - \left(\frac{3i}{n} - 1\right)^2\right]\left(\frac{3}{n}\right)$

$= \lim_{n\to\infty} \frac{3}{n}\sum_{i=1}^{n}\left(3 + \frac{6i}{n} - \frac{9i^2}{n^2}\right)$

$= \lim_{n\to\infty} \left[\frac{9}{n}\sum_{i=1}^{n} 1 + \frac{18}{n^2}\sum_{i=1}^{n} i - \frac{27}{n^3}\sum_{i=1}^{n} i^2\right]$

$= \lim_{n\to\infty} \left[\frac{9}{n}(n) + \frac{18}{n^2}\frac{n(n+1)}{2} - \frac{27}{n^3}\frac{n(n+1)(2n+1)}{6}\right]$

$= 9 + 9 - 9 = 9$ exact area

PART II
Chapter Project Solutions

Chapter P Project Modeling the Volume of a Box

(a) Area of base = $x \cdot x = x^2$, area of top = $x \cdot x = x^2$, Area of one side = $x \cdot h$, area of four sides = $4xh$

(b) Surface Area = area of base + area of top + 4· (area of side)

Surface Area $= x^2 + x^2 + 4xh$

Surface Area $= 2x^2 + 4xh$

(c) From part (b), start with $216 = 2x^2 + 4xh$.

$$216 = 2x^2 + 4xh$$
$$216 - 2x^2 = 4xh$$
$$\frac{216 - 2x^2}{4x} = h$$
$$\frac{108 - x^2}{2x} = h$$

(d) Volume of the box = length of box × width of box × height of box. The length and width of the box are given by x, and the height of the box is given by the expression for h from part (c).

$$V = x \cdot x \cdot \left(\frac{108 - x^2}{2x} \right)$$
$$= \frac{108x^2 - x^4}{2x}$$
$$= 54x - \frac{1}{2}x^3, \quad 0 \le x \le \sqrt{108}$$

(e)

Base, x	*Height*	*Surface Area*	*Volume*
1.0	53.5	216.0	53.5
1.5	32.3	216.0	79.3
2.0	26.0	216.0	104.0
2.5	20.4	216.0	127.2
3.0	16.5	216.0	148.5
3.5	13.7	216.0	167.6
4.0	11.5	216.0	184.0
4.5	9.8	216.0	197.4
5.0	8.3	216.0	207.5
5.5	7.1	216.0	213.8
6.0	6.0	216.0	216.0
6.5	5.1	216.0	213.7
7.0	4.2	216.0	206.5
7.5	3.5	216.0	194.1
8.0	2.8	216.0	176.0

8.5	2.1	216.0	151.9
9.0	1.5	216.0	121.5
9.5	0.9	216.0	84.3
10.0	0.4	216.0	40.0

It is clear from the table that the box with greatest volume is a cube with 6-inch sides.

Questions for Further Exploration

1. The height of the boxes increases without bound as x gets closer and closer to 0. Of all boxes with square bases and surface areas of 216 square inches, there is not a tallest box because the height of the boxes increases without bound as the length x of the base gets smaller and smaller.

2. The height of the boxes gets closer and closer (but does not become) 0 as x gets closer and closer to $\sqrt{108}$. Of all boxes with square bases and surface areas of 216 square inches, there is not a shortest box because the height of the boxes becomes infinitely small as the length x of the approaches $\sqrt{108}$.

3.

Base x	5.9	5.99	5.999	6.001	6.01	6.1
Volume V	215.9105	215.9991005	215.999991	215.999991	215.9990995	215.9095

The table lends further support to the answer in part (e) because as x gets closer and closer to 6, the volume V gets closer and closer to 216.

4. Begin by writing a model for the volume of a rectangular box with a surface area of 216 square inches and base x inches by $2x$ inches.

Surface area = area of base + area of top + 2 · area of short side + 2 · area of long side

$$216 = 2x^2 + 2x^2 + 2(xh) + 2(2xh)$$

$$216 = 4x^2 + 6xh$$

$$216 - 4x^2 = 6xh$$

$$\frac{216 - 4x^2}{6x} = h$$

$$\frac{36}{x} - \frac{2x}{3} = h$$

Having written the height in terms of x, we can write the volume in terms of x.

Volume of box = length of box · width of box · height of box

$$V = 2x \cdot x \cdot \left(\frac{36}{x} - \frac{2x}{3}\right)$$

$$V = 72x - \frac{4x^3}{3}, \quad 0 \le x \le \sqrt{54}$$

Using this model for volume, we can make create a table like the one in part (e) to find the maximum volume of this type of box.

Width of base, x	Length of base, $2x$	Height	Surface Area	Volume
1.0	2.0	35.333	216	70.667
1.5	3.0	23.000	216	103.500
2.0	4.0	16.667	216	133.333
2.5	5.0	12.733	216	159.167
3.0	6.0	10.000	216	180.000
3.5	7.0	7.952	216	194.833
4.0	8.0	6.333	216	202.667
4.5	9.0	5.000	216	202.500
5.0	10.0	3.867	216	193.333
5.5	11.0	2.879	216	174.167
6.0	12.0	2.000	216	144.000
6.5	13.0	1.205	216	101.833
7.0	14.0	0.476	216	46.667

From the results of the experiment, it appears that the box of greatest volume occurs when $3.5 < x < 4.5$. We can construct another table to narrow down the value of x giving the greatest volume.

Width of base, x	Length of base, $2x$	Height	Surface Area	Volume
3.6	7.2	7.600	216	196.992
3.7	7.4	7.263	216	198.863
3.8	7.6	6.940	216	200.44
3.9	7.9	6.631	216	201.708
4.0	8.0	6.333	216	202.667
4.1	8.2	6.047	216	203.305
4.2	8.4	5.771	216	203.616
4.3	8.6	5.505	216	203.591
4.4	8.8	5.248	216	203.221

From this table, it appears that the box of greatest volume occurs when $4.1 < x < 4.3$. We can construct yet another table to narrow down the value of x giving the greatest volume even more.

Width of base, x	Length of base, $2x$	Height	Surface Area	Volume
4.15	8.30	5.908	216	203.500
4.16	8.32	5.881	216	203.532
4.17	8.34	5.853	216	203.558
4.18	8.36	5.826	216	203.580
4.19	8.38	5.799	216	203.600
4.20	8.40	5.771	216	203.616
4.21	8.42	5.744	216	203.629
4.22	8.44	5.717	216	203.638
4.23	8.46	5.691	216	203.644
4.24	8.48	5.664	216	203.647
4.25	8.50	5.637	216	203.646
4.26	8.52	5.611	216	203.642

From the table, it appears that the box of the greatest volume occurs when $x \approx 4.24$ and $h \approx 5.66$. Better approximations may be obtained by creating additional tables (for example, creating a table covering values from 4.23 to 4.25 with an increment of 0.001).

Chapter 1 Project Modeling the Area of a Plot

(a) The sum of the length and width of the plot must equal 50 meters. Thus, if the length is x meters, the width is $(x - 50)$ meters.

$$\begin{aligned}\text{Area } A(x) &= \text{length} \times \text{width} \\ &= x(50 - x)\end{aligned}$$

(b)

x	0	5	10	15	20	25	30	35	40	45	50
$A(x)$	0	225	400	525	600	625	600	525	400	225	0

According to the table, it appears that the dimensions of 25 meters and $50 - 25 = 25$ meters should enclose the maximum area.

(c) The domain of the function is $0 \le x \le 50$. The graph of the function, with window settings of $0 \le x \le 50$ and $0 \le y \le 700$, is shown below. Using the graphing utility's trace feature, it appears that $x = 25$ gives the maximum area.

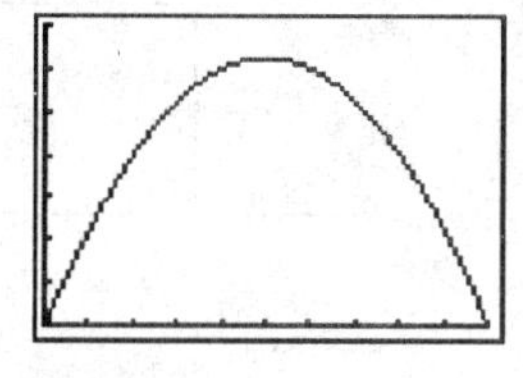

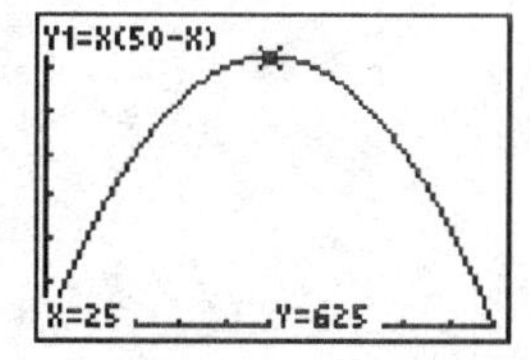

(d) Begin with the area function $A(x) = x(50 - x)$ and complete the square as follows.

$$\begin{aligned}A(x) &= x(50 - x) \\ &= 50x - x^2 \\ &= -(x^2 - 50x) \\ &= -(x^2 - 50x + 25^2 - 25^2) \\ &= -(x^2 - 50x + 625) + 625 \\ &= 625 - (x - 25)^2\end{aligned}$$

The value of this expression is at its largest when $(x - 25)^2 = 0$, which happens only when $x = 25$. Therefore, the maximum area is given when $x = 25$.

(e) Answers will vary. For instance for part (b), the numerical approach could be time consuming when pinpointing the accuracy of an estimate but involves rather simple computations.

Questions for Further Exploration

1. A greater area could be enclosed if the plot were circular in shape. For instance, for a circle with circumference 100 meters,

$$\begin{aligned}100 &= 2\pi r \\ 100/(2\pi) &= r \\ 15.9155 &\approx r\end{aligned}$$

Thus, the radius would be approximately 15.9 meters. The area of a circle with this radius is

$$\begin{aligned}\text{Area} &= \pi r^2 \\ &= \pi\left(\frac{100}{2\pi}\right)^2 \\ &= \frac{50^2}{\pi} \approx 795.8 \text{ square meters.}\end{aligned}$$

2. Using an analytical approach, the maximum area that could be enclosed with double the amount of fencing is found by completing the square of the expression $A(x) = x[2(50) - x] = x(100 - x)$.

$$\begin{aligned}A(x) &= x(100 - x) \\ &= 100x - x^2 \\ &= -(x^2 - 100x) \\ &= -(x^2 - 100x + 50^2 - 50^2) \\ &= -(x^2 - 100x + 2500) + 2500 \\ &= 2500 - (x - 25)^2\end{aligned}$$

Thus, the maximum area that can be enclosed in a rectangular plot with 200 meters of fencing (twice the original amount of fencing) is 2500 square meters, or four times the maximum area enclosed by 100 meters of fencing. When the amount of fencing is doubled we can enclose at least double the area and can, in fact, enclose quadruple the area.

3. If the rectangular plot runs along the side of a building and there are 100 meters of fencing, an expression for the area of the plot is $A(x) = x(100 - 2x)$. By graphing this function with a graphing utility, we can graphically find the dimensions that yield a maximum area, as shown below.

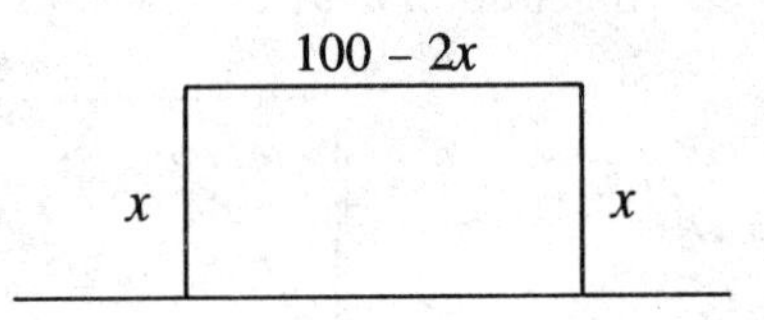

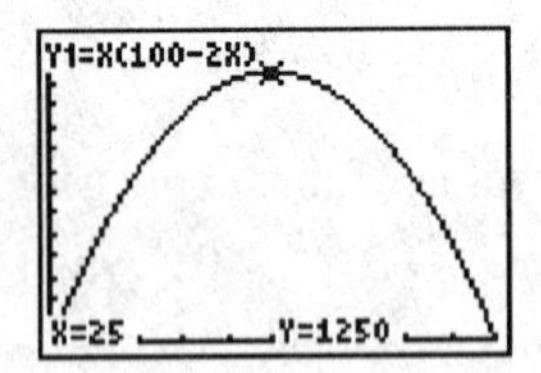

From the graph, we can see that the maximum area of 1250 square meters is given when $x = 25$ and $100 - 2(25) = 50$. The required dimensions are 25 meters by 50 meters.

Chapter 2 Project Graphs of Polynomial and Rational Functions

(a) From the graph, we can see that the zeros of f are the vertical asymptotes of g.

(b) From the graph, it appears that the minimum point of the graph of f is $(0, -4)$. The graph of g has a relative maximum at $x = 0$, the same value of x at which f has its minimum. In general, it seems that the relative minimum of a function is located at the same x-value as the relative maximum of the reciprocal of the function. Also, the relative maximum of a function is located at the same x-value as the relative minimum of the reciprocal function.

(c) Because the reciprocal function g has a horizontal asymptote at $y = 0$, this tells us that for very large and very small values of x the function f goes to infinity.

(d) Using the graph of P, we know that the graph of $q(x) = 1/P(x)$ will have vertical asymptotes at $x = -1$ and $x = 5$ and will have a relative minimum at $x = 2$. A sketch of the graph of $q(x) = 1/P(x)$ is shown at the right.

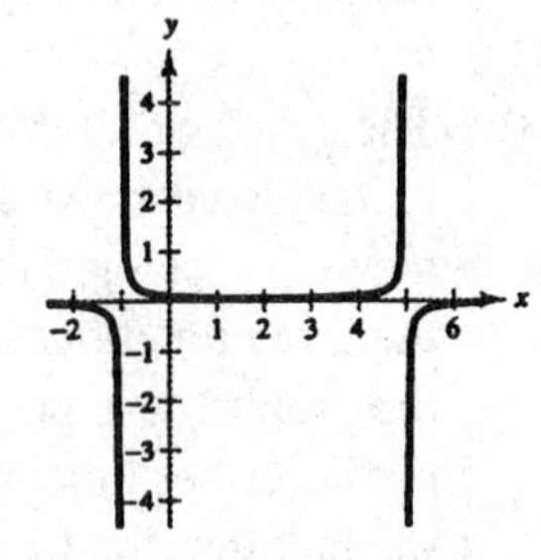

Questions for Further Exploration

1. Because the graph of the polynomial function $f(x) = x^2 + 9$ has no real zeros, this indicates that the graph of the reciprocal function $g(x) = 1/f(x)$ has no vertical asymptotes.

2. The graph of $f(x) = (x-1)(x-2)(x-3)$ and its reciprocal function $g(x) = 1/f(x)$ in the window $-1 \le x \le 5$ and $-10 \le y \le 10$ is shown at the right. At $x = 1$, 2, and 3, f has a zero and g has a vertical asymptote. At $x \approx 1.4$, f has a relative maximum and g has a relative minimum. At $x \approx 2.6$, f has a relative minimum and g has a relative maximum.

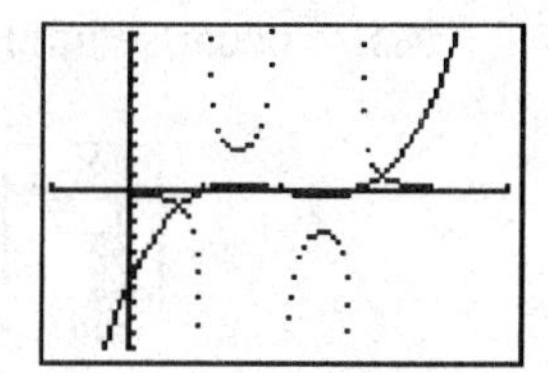

3. (a) Because the given graph has a relative maximum at $x = -1$ and vertical asymptotes at $x = -4$ and $x = 2$, the graph of $f(x)$ has a relative minimum at $x = -1$ and zeros at $x = -4$ and $x = 2$. A sketch of the graph of $f(x)$ is given at the right.

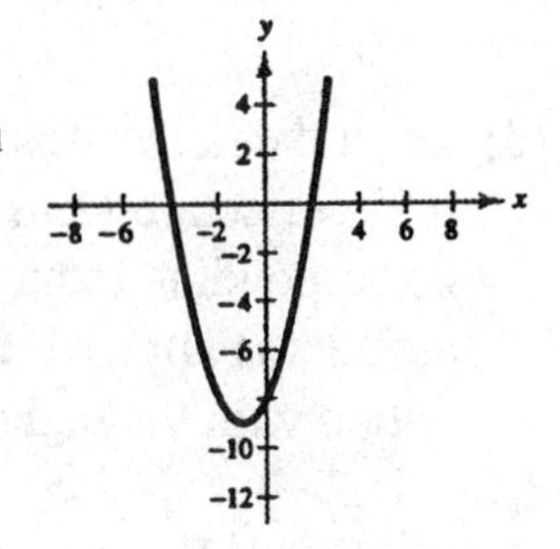

(b) Because the given graph has a relative minimum at $x = -3$, a relative maximum at $x = 1.5$, and vertical asymptotes at $x = -5$, $x = -1$, and $x = 4$, the graph of $f(x)$ has a relative maximum at $x = -3$, a relative minimum at $x = 1.5$, and zeros at $x = -5$, $x = -1$, and $x = 4$. A sketch of the graph of $f(x)$ is given at the right.

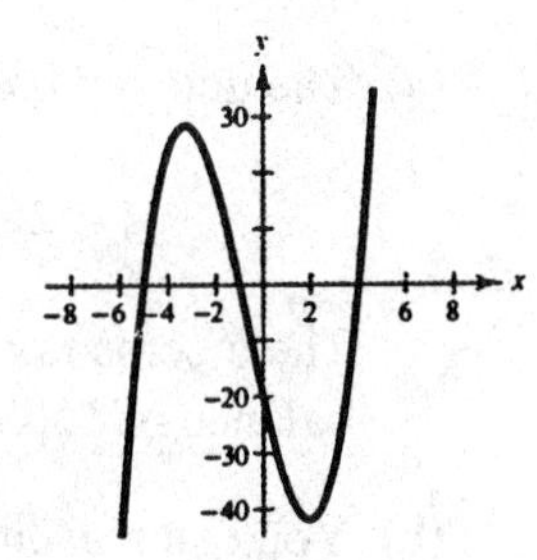

Chapter 3 Project A Graphical Approach to Compound Interest

(a) For \$1000 invested at the 6.2% annual interest rate, compounded annually, the function that gives the balance as a function of the time t in years is $A = P\left(1 + \frac{r}{n}\right)^{nt}$, where $P = \$1000$, $r = 0.062$, and $n = 1$, or $A = 1000(1+0.062)^t$. For \$1000 invested at the 6.1% annual interest rate, compounded quarterly, $P = \$1000$, $r = 0.061$, and $n = 4$, and the function is $A = 1000(1+0.061/4)^{4t}$. For \$1000 invested at the 6.0% annual interest rate, compounded continuously, the function that gives the balance as a function of time t in years is $A = Pe^{rt}$, where $P = 1000$ and $r = 0.06$, or $A = 1000e^{0.06t}$.

(b) The graph of all three functions in the same viewing rectangle is shown at the right. It is difficult to find a viewing rectangle that distinguishes among the graphs of the three functions and also conveys the overall shapes of the graphs. It is possible to find a viewing rectangle that distinguishes among the graphs, but these often obscure the fact that the functions are exponential functions.

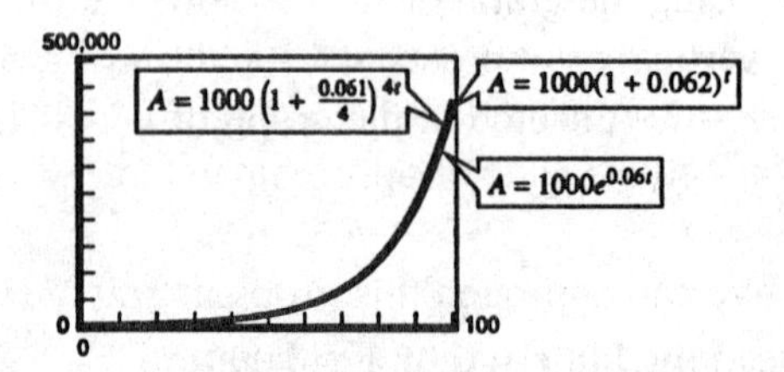

(c) Using the table feature of a graphing utility, we can create tables of the balances in each account after 25, 50, 75, and 100 years. We see from the tables that the best choice after 25 years is 6.1% interest compounded quarterly. This is the same option that yields the greatest balances after 50, 75, and 100 years, because interest is earned on the principal and the previously earned interest.

X	Y1
1	1062
25	4499
50	20241
75	91062
100	409687

Y1=1000(1.062)^X

X	Y2
1	1062.4
25	4542.6
50	20635
75	93734
100	425794

Y2=1000(1+.061/...

X	Y3
1	1061.8
25	4481.7
50	20086
75	90017
100	403429

Y3=1000e^(.06X)

(d) From the tables above, the balance after one year in the first account is \$1062, corresponding to a 6.2% effective yield. The balance after one year in the second account is approximately \$1062.4, corresponding to an approximately 6.24% effective yield. The balance after one year in the third account is approximately \$1061.8, corresponding to an approximately 6.18% effective yield. The option with the highest yield is the best option.

Questions for Further Exploration

1. (a) The untaxed balance at the end of 40 years is

$$A = 25{,}000(1+0.08)^{40} \approx \$543{,}113.04.$$

The income tax due on the interest alone is 0.3(518,113.04) ≈ \$155,433.91, so you are left with a balance of \$543,113.04 – \$155,433.91 = \$387,679.13.

(b) You can reason that only 70% of the earned interest will remain in the account each year. The taxed balance at the end of 40 years is

$$A = 25{,}000[1+0.08(0.7)]^{40} \approx \$221{,}053.16.$$

Thus the tax-deferred plan will produce a significantly greater balance at the end of 40 years.

2. To compare these three options, compute the effective yield for each.

(a) 8.02% compounded monthly:

$$\text{Effective yield} = \left(1+\frac{0.0802}{12}\right)^{12(1)} \approx 1.0832$$

(b) 8.03% compounded daily:

$$\text{Effective yield} = \left(1+\frac{0.0803}{365}\right)^{365(1)} \approx 1.0836$$

(c) 8% compounded continuously:

$$\text{Effective yield} = e^{0.08(1)} \approx 1.0833$$

Thus, an annual interest rate of 8.03% compounded daily in option (b) would produce a larger balance because this option has a slightly higher effective yield, approximately 8.36%, as compared to the effective yields of approximately 8.32% in option (a) and 8.33% in option (c).

3. We can approach this problem graphically. Use a graphing utility to define $y_1 = (1000(1+0.08)^x - 1000)\cdot 0.7 + 1000$ for the tax-deferred plan and $y_2 = 1000[1+0.08(0.7)]^x$ for the plan that is not tax deferred. Then graph $y_3 = y_1 - y_2$ and $y_4 = 100$ on the same screen. By finding the point of intersection of these two graphs, we can find how long it would take for the balance of one account to exceed the balance in the other account by \$100. We see that it would take approximately 10.56 years to obtain a difference of at least \$100.

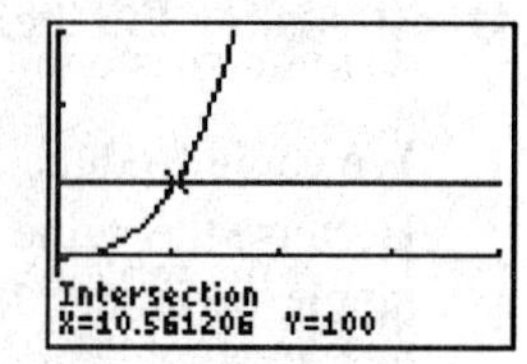

To find how long it would take for the balance of one account to exceed the balance in the other account by \$100,000, graph $y_3 = y_1 - y_2$ and $y_4 = 100{,}000$ on the same screen. We see that it would take approximately 69.11 years to obtain a difference of at least \$100,000.

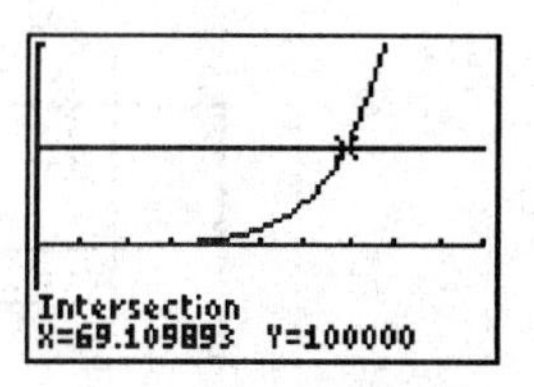

4. The answer will vary depending on how many years there are until retirement. This can be seen graphically by using a graphing utility to graph option (a) as $y_1 = 25{,}000(1.05)^x$ and option (b) as $y_2 = (25{,}000(1.07)^x - 25{,}000)\cdot 0.6 +$ 25,000. You can see from the graph that the point of intersection of these two graphs is approximately (16.80, 56,734.58).

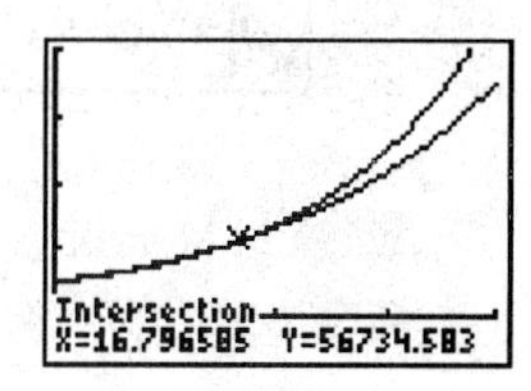

Let's compare each case at 12 years and at 40 years.

(a) For $x = 12$, $A = 25{,}000(1+0.05)^{12} \approx 44{,}896.41$.

For $x = 40$, $A = 25{,}000(1+0.05)^{40} \approx 175{,}999.72$.

(b) For $x = 12$, $A = 25{,}000(1+0.07)^{12} \approx 56{,}304.79$. After taxes on the earned interest, there would be $(56{,}304.79 - 25{,}000)\cdot 0.6 + 25{,}000 = 43{,}782.87$ left.

For $x = 40$, $A = 25{,}000(1+0.07)^{40} \approx 374{,}361.45$. After taxes on the earned interest, there would be $(374{,}361.45 - 25{,}000)\cdot 0.6 + 25{,}000 = 234{,}616.87$ left.

These calculations lead us to believe that if the length of the investment (i.e., time until retirement) is 16 years or less, the tax-free plan in option (a) is better. However, if the length of the investment is 17 years or more, the tax-deferred plan in option (b) is better.

Chapter 4 Project Fitting a Model to Data

(a) The data appears to be best modeled by a quadratic model.

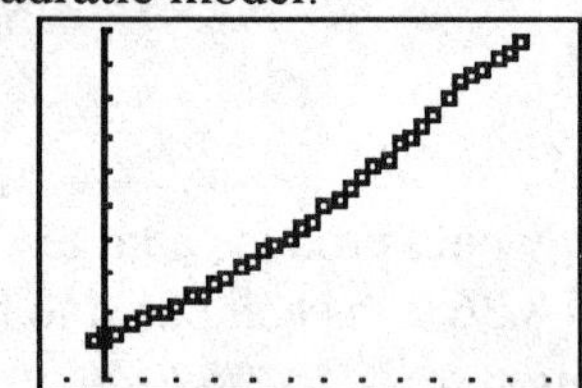

(b) Using a graphing utility, the best fitting quadratic model (where $x = 0$ represents 1960) is $y = 0.015x^2 + 0.764x + 315.838$.

Questions for Further Exploration

1. We can introduce an oscillation by adding a term involving a trigonometric function to the model. Because the value of the trigonometric portion varies by at most 2.5 parts per million from the quadratic model, the amplitude is 2.5. Because the trigonometric portion is 0 in January, 2.5 in April, 0 in July, and – 2.5 in October, using the sine function will work. Because January of each year is given by whole number values, the period is 1. Thus, the trigonometric term needed in the model is $2.5\sin(2\pi x)$, and the new model is $y = 0.015x^2 + 0.764x + 315.838 + 2.5\sin(2\pi x)$.

2.

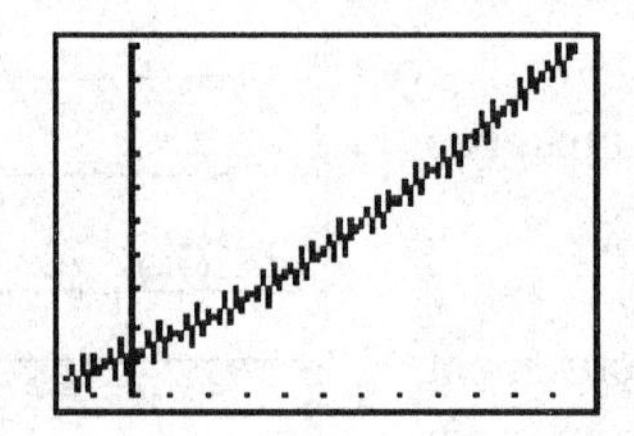

3. Answers will vary. Earth's seasons could be one factor.

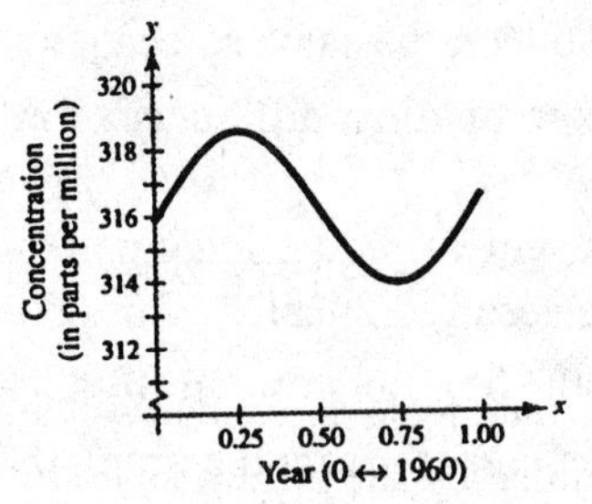

4. No, the model found in Question 1 is not periodic because the model increases each year.

5. (a) In January 2000, $x = 40$:

$$y = 0.015x^2 + 0.764x + 315.838 + 2.5\sin(2\pi x)$$
$$y = 0.015(40)^2 + 0.764(40) + 315.838 + 2.5\sin(2\pi(40))$$
$$\approx 370.40$$

(b) In January 2010, $x = 50$:

$$y = 0.015x^2 + 0.764x + 315.838 + 2.5\sin(2\pi x)$$
$$y = 0.015(50)^2 + 0.764(50) + 315.838 + 2.5\sin(2\pi(50))$$
$$\approx 391.54$$

(c) In January 2010, $x = 60$:

$$y = 0.015x^2 + 0.764x + 315.838 + 2.5\sin(2\pi x)$$
$$y = 0.015(60)^2 + 0.764(60) + 315.838 + 2.5\sin(2\pi(60))$$
$$\approx 415.68$$

Chapter 5 Project Projectile Motion

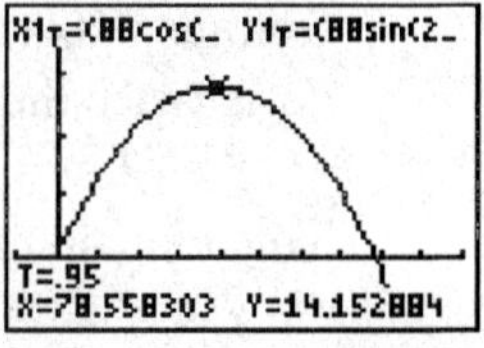

(a) Using the trace feature, the maximum height attained by the projectile is approximately 14 feet (see figure at the right).

(b) From the figure at the right, we can see that the maximum height is attained at approximately 0.95 second.

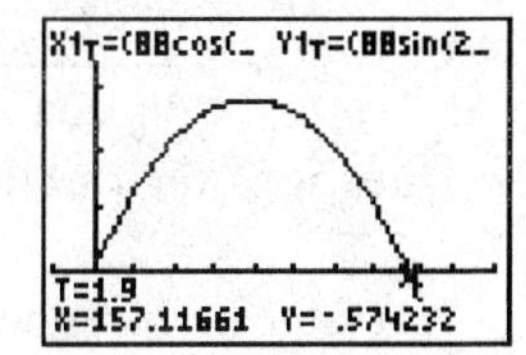

(c) Using the trace feature to locate the x-intercept on the right, we can see that the projectile is in the air for approximately 1.9 seconds (see figure at the right).

(d) From the figure at the right, we can see that the range of the projectile is approximately 157 feet.

Questions for Further Exploration

1.

$$(88\sin 20^\circ)t - 16t^2 = 0$$

$$t(88\sin 20^\circ - 16t) = 0$$

$$t = 0$$

$$88\sin 20^\circ - 16t = 0 \Rightarrow t = \frac{88\sin 20^\circ}{16} \approx 1.88 \text{ seconds}$$

$$x(t) = (88\cos 20^\circ)\left(\frac{88\sin 20^\circ}{16}\right)$$

$$\approx 155.6 \text{ feet}$$

2. Using the trace feature, we can see that the maximum height attained by the projectile in this situation is approximately 68 feet, and the range of the projectile is approximately 472 feet. (See figures at the right.)

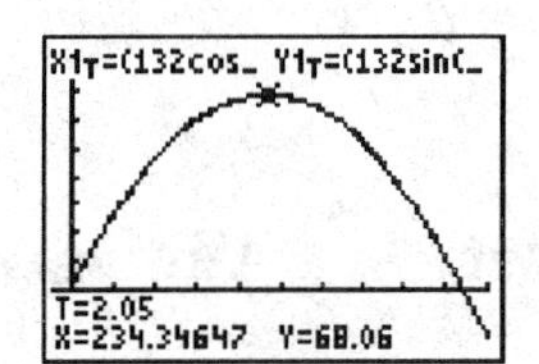

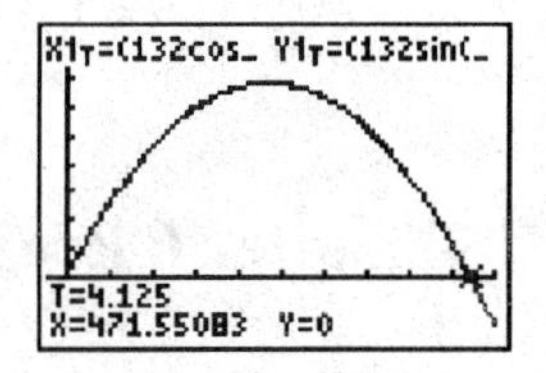

3. For $\theta = 20^\circ$, range is approximately 72 feet.

For $\theta = 30^\circ$, range is approximately 98 feet.

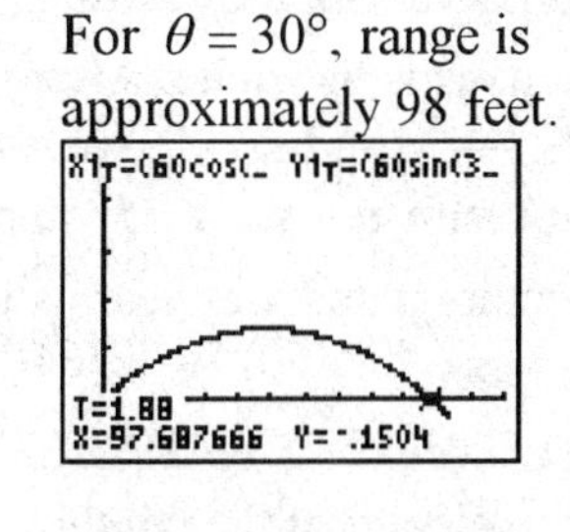

For $\theta = 40^\circ$, range is approximately 111 feet.

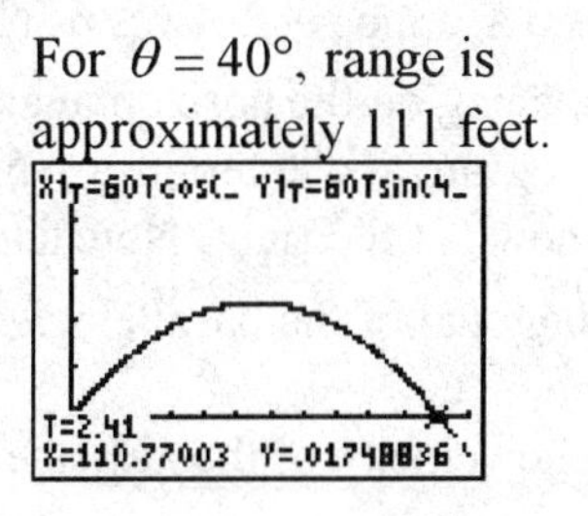

For $\theta = 50^\circ$, range is approximately 111 feet.

For $\theta = 60^\circ$, range is approximately 98 feet.

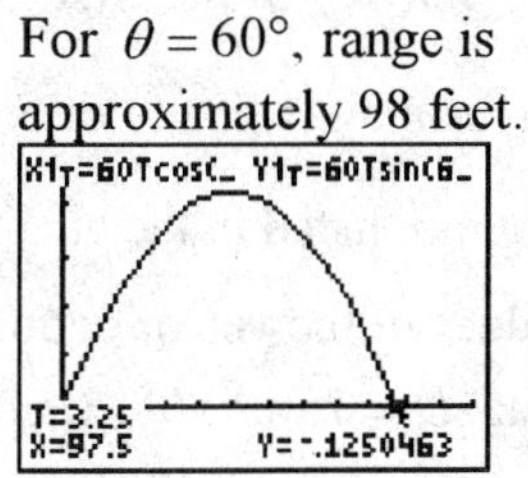

Because the range for $\theta = 40^\circ$ and $\theta = 50^\circ$ is the same, it appears that the angle that would produce the maximum range is $\theta = 45^\circ$.

4. Because the shape of the path of the projectile is a parabola, we know that the time that it takes the projectile to reach its maximum height and the time it takes for the projectile to return to the ground from its maximum height are equal. Or, in other words, the total length of time the projectile is in the air is twice the time it takes the projectile to reach its maximum height.

5. To eliminate t from the parametric equations:

$$x = (v_0 \cos\theta)t \qquad y = (v_0 \sin\theta)t - 16t^2$$

$$t = \frac{x}{v_0 \cos\theta} \quad \Rightarrow \quad = \frac{(v_0 \sin\theta)x}{v_0 \cos\theta} - 16\left(\frac{x}{v_0 \cos\theta}\right)^2$$

$$= (\tan\theta)x - \frac{16\sec^2\theta}{v_0^2}x^2$$

For $x = 200$ and $v_0 = 80$, we find θ as follows:

$$y = (\tan\theta)x - \frac{16\sec^2\theta}{v_0^2}x^2$$

$$0 = 200\tan\theta - \frac{16\sec^2\theta}{80^2}(200)^2$$

$$0 = 200\tan\theta - 100\sec^2\theta$$

$$0 = 200\tan\theta - 100(1+\tan^2\theta)$$

$$0 = \tan^2\theta - 2\tan\theta + 1$$

$$0 = (\tan\theta - 1)^2$$

$$\tan\theta = 1 \quad \Rightarrow \quad \theta = 45°$$

Chapter 6 Project Adding Vectors Graphically

(a) To show both vectors and their sum, we can use the viewing rectangle $-6 \le x \le 6$, $-2 \le y \le 6$. Note that this is a "square" setting. This is, the spacing on the horizontal and vertical axes is the same. After running the program and entering $A = 5$, $B = 2$, $C = -4$, and $D = 3$, we obtain the screen shown at the right. Note that the vector sum $\mathbf{u} + \mathbf{v} = \mathbf{i} + 5\mathbf{j}$ appears as the diagonal of the parallelogram. The vectors **u** and **v** appear as two of the sides of the parallelogram.

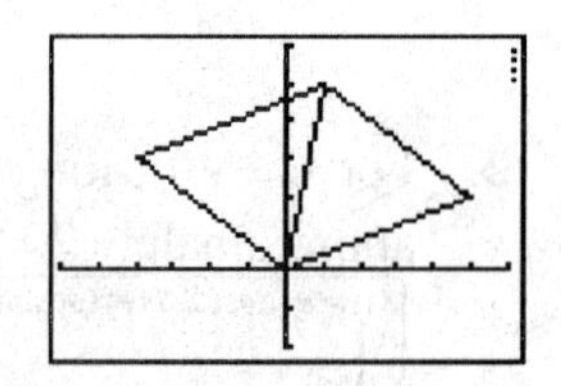

(b) The velocity of the airplane can be represented by the vector $\mathbf{v}_1 = 400\langle\cos 150°, \sin 150°\rangle$ and the velocity of the wind by the vector $\mathbf{v}_2 = 75\langle\cos 50°, \sin 50°\rangle$. Thus the resultant velocity of the airplane can be represented by $\mathbf{v}_1 + \mathbf{v}_2$. With the program listed in the chapter project, we do not need to evaluate the numerical values of the vector coordinates. Simply enter: $A = 400\cos 150°$, $B = 400\sin 150°$, $C = 75\cos 50°$, and $D = 75\sin 50°$. Using $-400 \le x \le 200$ and $-100 \le y \le 300$, we obtain the screen at the right.

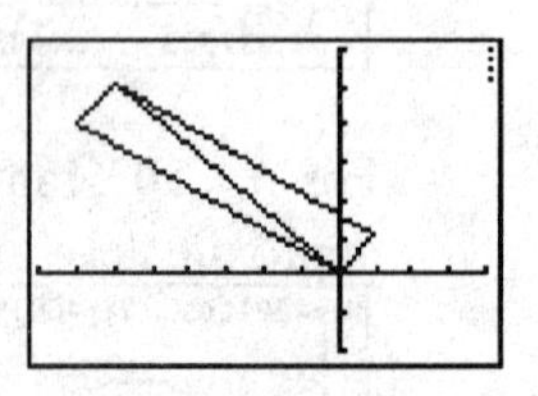

1. Run the ADDVECT program with $A = 3$, $B = 4$, $C = -5$, and $D = 1$. The resulting graphing utility display is shown below.

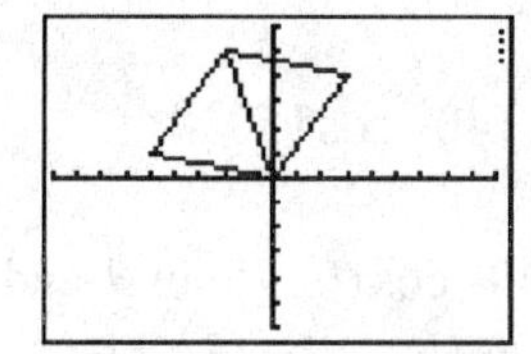

2. Run the ADDVECT program with $A = 5$, $B = -4$, $C = 3$, and $D = 2$. The resulting graphing utility display is shown below.

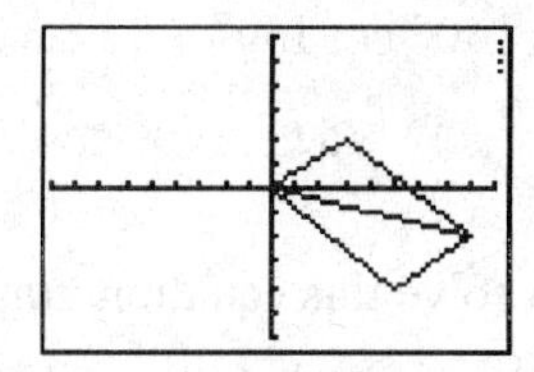

3. Run the ADDVECT program with $A = -4$, $B = 4$, $C = -2$, and $D = -6$. The resulting graphing utility display is shown below.

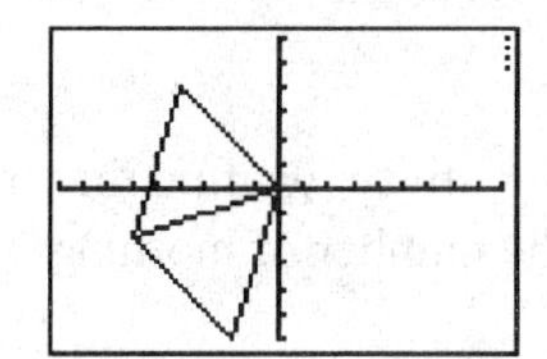

4. Run the ADDVECT program with $A = 7$, $B = 3$, $C = -2$, and $D = -6$. The resulting graphing utility display is shown below.

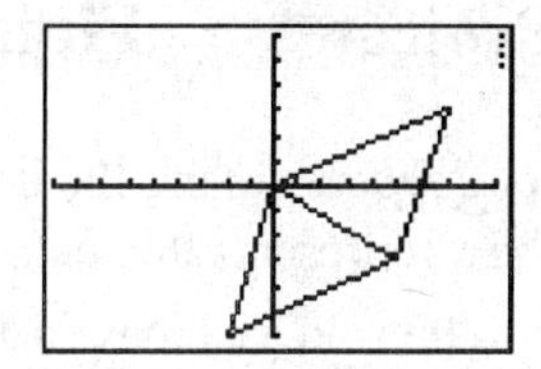

5. To find the speed of the airplane after encountering the wind, find the length of the resultant vector.

$$\|\mathbf{v}_1 + \mathbf{v}_2\| = \sqrt{(400\cos 150° + 75\cos 50°)^2 + (400\sin 150° + 75\sin 50°)^2}$$
$$\approx 394 \text{ miles per hour}$$

Because the original speed of the airplane was 400 miles per hour, the plane is now traveling slower.

6. To obtain the answer experimentally, try substituting different values for the wind's velocity into the program ADDVECT and approximate the components of the resultant vector. Then use the approximations to find the velocity. To obtain the answer analytically, solve the following equation for w, the wind's velocity.

$$\tan 140° = \frac{400\sin 150° + w\sin 50°}{400\cos 150° + w\cos 50°}$$
$$\tan 140°\,(400\cos 150° + w\cos 50°) = 400\sin 150° + w\sin 50°$$
$$400\tan 140°\,\cos 150° + w\tan 140°\,\cos 50° = 400\sin 150° + w\sin 50°$$
$$w(\tan 140°\,\cos 50° - \sin 50°) = 400\sin 150° - 400\tan 140°\,\cos 150°$$
$$w = \frac{400\sin 150° - 400\tan 140°\,\cos 150°}{\tan 140°\,\cos 50° - \sin 50°}$$
$$w \approx 69.5 \text{ miles per hour}$$

7. To obtain the answer experimentally, try substituting different values for the wind's direction into the program ADDVECT and approximate the components of the resultant vector. Then use the approximations to find the direction. To obtain the answer analytically, solve the following equation for θ, the wind's direction.

$$\tan 140° = \frac{400 \sin 150° + 75 \sin \theta}{400 \cos 150° + 75 \cos \theta}$$

$$\tan 140° (400 \cos 150° + 75 \cos \theta) = 400 \sin 150° + 75 \sin \theta$$

$$400 \tan 140° \cos 150° + 75 \tan 140° \cos \theta = 400 \sin 150° + 75 \sin \theta$$

$$75 \tan 140° \cos \theta - 75 \sin \theta = 400 \sin 150° - 400 \tan 140° \cos 150°$$

We can solve this equation graphically by graphing $y_1 = 75 \tan 140° \cos \theta - 75 \sin \theta$ and $y_2 = 400 \sin 150° - 400 \tan 140° \cos 150°$ in the same window, as at the right. These two equation intersect at approximately (27.8. – 90.7). Thus, the direction of the wind is about N 62.2 °E.

Chapter 7 Project Fitting Models to Data

(a) *Morning Newspapers* Using the linear regression feature of a graphing utility, we find that the linear model that best represents this data $y = 1.039t + 30.929$. To project the number of morning newspapers to be sold each day in 1998, evaluate the model at $t = 18$.

$$\begin{aligned} y &= 1.039t + 30.929 \\ &= 1.039(18) + 30.929 \\ &= 49.631 \end{aligned}$$

In 1998, approximately 49.6 million morning newspapers will be sold each day.

Evening Newspapers Using the linear regression feature of a graphing utility, we find that the linear model that best represents this data $y = -1.165t + 32.062$. To project the number of evening newspapers to be sold each day in 1998, evaluate the model at $t = 18$.

$$\begin{aligned} y &= -1.165t + 32.062 \\ &= -1.165(18) + 32.062 \\ &= 11.092 \end{aligned}$$

In 1998, approximately 11.1 million evening newspapers will be sold each day.

Sunday Newspapers Using the linear regression feature of a graphing utility, we find that the linear model that best represents this data is $y = 0.707t + 54.894$. To project the number of Sunday newspapers to be sold each week in 1998, evaluate the model at $t = 18$.

$$\begin{aligned} y &= 0.707t + 54.894 \\ &= 0.707(18) + 54.894 \\ &= 67.62 \end{aligned}$$

In 1998, approximately 67.6 million Sunday newspapers will be sold each week.

(b) Using the linear regression feature of a graphing utility, we find that the linear model that best represents the data for morning newspaper companies is $y = 16.101t + 397.258$. The linear model for the evening newspaper company data is $y = -30.234t + 1376.939$. The linear model for the Sunday newspaper company data is $y = 12.448t + 736.924$.

(c) In 1981, the average circulation per morning newspaper company was

$$\frac{30{,}600{,}000}{408} = 75{,}000 \text{ newspapers per day}$$

In 1992, the average circulation per morning newspaper company was

$$\frac{42{,}400{,}000}{596} \approx 71{,}141 \text{ newspapers per day}$$

Thus, the average circulation per morning newspaper company decreased from 1981 to 1992.

(d) The best newspaper investment choice would be a morning newspaper company because there are fewer morning newspaper companies to compete with.

Chapter 8 Project Row Operations and Graphing

(a) To solve the system of equations $2x - 4y = 9$ and $x + 5y = 15$ by Gauss-Jordan elimination, perform the following steps.

$$\left[\begin{array}{cc|c} 2 & -4 & 9 \\ 1 & 5 & 15 \end{array}\right] \quad \begin{array}{c} R_2 \rightarrow \\ R_1 \rightarrow \end{array} \left[\begin{array}{cc|c} 1 & 5 & 15 \\ 2 & -4 & 9 \end{array}\right] \quad -2R_1 + R_2 \rightarrow \left[\begin{array}{cc|c} 1 & 5 & 15 \\ 0 & -14 & -21 \end{array}\right]$$

$$-\tfrac{1}{14}R_2 \rightarrow \left[\begin{array}{cc|c} 1 & 5 & 15 \\ 0 & 1 & 1.5 \end{array}\right] \quad -5R_2 + R_1 \rightarrow \left[\begin{array}{cc|c} 1 & 0 & 7.5 \\ 0 & 1 & 1.5 \end{array}\right]$$

The solution is (7.5, 1.5).

To solve the system of equations $6x + 2y = -19$ and $3x - y = -5$ by Gauss-Jordan elimination, perform the following steps.

$$\left[\begin{array}{cc|c} 6 & 2 & -19 \\ 3 & -1 & -5 \end{array}\right] \quad \tfrac{1}{6}R_1 \rightarrow \left[\begin{array}{cc|c} 1 & 1/3 & -19/6 \\ 3 & -1 & -5 \end{array}\right] \quad -3R_1 + R_2 \rightarrow \left[\begin{array}{cc|c} 1 & 1/3 & -19/6 \\ 0 & -2 & 9/2 \end{array}\right]$$

$$-\tfrac{1}{2}R_2 \rightarrow \left[\begin{array}{cc|c} 1 & 1/3 & -19/6 \\ 0 & 1 & -9/4 \end{array}\right] \quad -\tfrac{1}{3}R_2 + R_1 \rightarrow \left[\begin{array}{cc|c} 1 & 0 & -29/12 \\ 0 & 1 & -9/4 \end{array}\right]$$

The solution is $\left(-\frac{29}{12}, -\frac{9}{4}\right) \approx (-2.4167, -2.25)$.

(b) The last two screens of results from running the program are given below for each system of equations (results for the first system are given on the left, and results for the second system are given on the right).

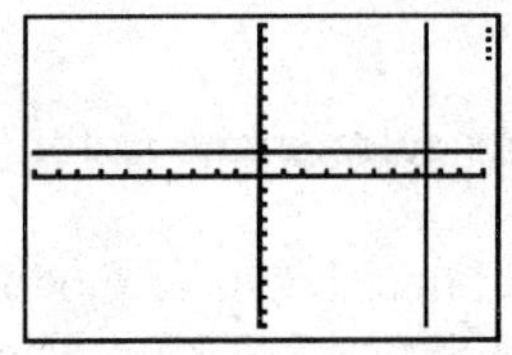

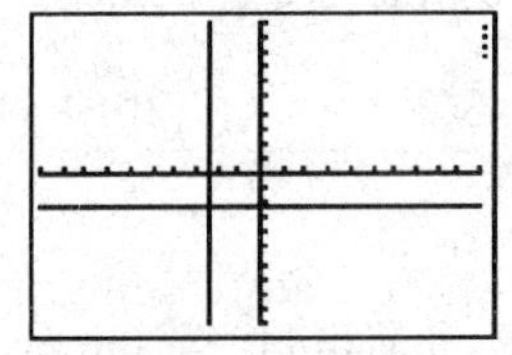

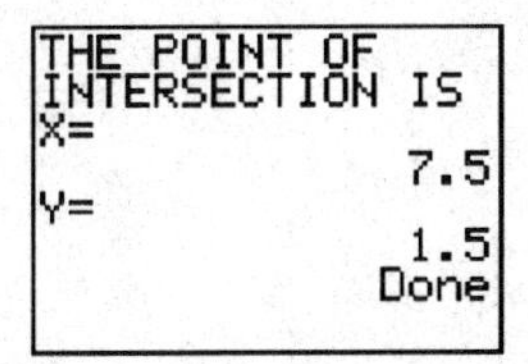

THE POINT OF
INTERSECTION IS
X=
-2.416666667
Y=
-2.25
Done

Notice that these are the same results as were obtained in part (a).

(c) In the running of the program, multiplying a row of the matrix by a constant has no effect on the graph of the corresponding linear equation.

(d) In the running of the program, adding a multiple of the first row of the matrix to the second row to obtain a 0 below the leading 1 results in the graph of a horizontal line.

(e) As the matrix is transformed, the point of intersection remains the same on the graphs of the corresponding linear equations.

Questions for Further Exploration

1. When finding a point of intersection, the program is more accurate than using zoom and trace because it uses a mathematical formula to obtain the solution. Examples will vary.

2. For this solution, the program gives the final result shown in the screen at the right. The program draws only one line in all but the last screen because the graph of the first equation in the system is a vertical line and the first equation does not represent a function. The solution of this system by hand is shown below.

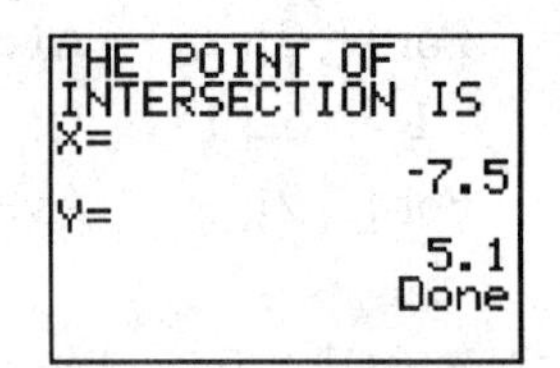

$$\left[\begin{array}{cc|c} 2 & 0 & -15 \\ 3 & 5 & 3 \end{array}\right] \quad \tfrac{1}{2}R_1 \rightarrow \left[\begin{array}{cc|c} 1 & 0 & -7.5 \\ 3 & 5 & 3 \end{array}\right] \quad -3R_1 + R_2 \rightarrow \left[\begin{array}{cc|c} 1 & 0 & -7.5 \\ 0 & 5 & 25.5 \end{array}\right] \quad \tfrac{1}{5}R_2 \rightarrow \left[\begin{array}{cc|c} 1 & 0 & -7.5 \\ 0 & 1 & 5.1 \end{array}\right]$$

3. An error message appeared because the program tried to divide the first equation by the coefficient of x, which for this system is 0.

4. Programs will vary. Note that some Texas Instruments graphing calculators have a reduced row-echelon form command that could be used for this task rather than writing a program to do it.

Chapter 9 Project   Exploring Difference Quotients

(a) For $f(x) = 3x + 4$ on the interval [2, 6], the average rate of change is

$$\frac{f(6) - f(2)}{6 - 2} = \frac{3(6) + 4 - (3(2) + 4)}{4} = \frac{12}{4} = 3$$

On the interval [20, 30], the average rate of change is

$$\frac{f(30) - f(20)}{30 - 20} = \frac{3(30) + 4 - (3(20) + 4)}{10} = \frac{30}{10} = 3,$$ the same average rate as for [2, 6].

(b) For $f(x) = x^2$ on the interval [1, 3], the average rate of change is

$$\frac{f(3) - f(1)}{3 - 1} = \frac{3^2 - 1^2}{2} = \frac{8}{2} = 4$$

On the interval [4, 6], the average rate of change is

$$\frac{f(6)-f(4)}{6-4}=\frac{6^2-4^2}{2}=\frac{20}{2}=10$$, which is a different average rate than for [1, 3].

No, these rates of change are not equal because $f(x)=x^2$ is not a linear function.

(c) Answers will vary. For instance, the speeds shown on the car's speedometer might range from 40 miles per hour to 60 miles per hour over the 1-hour trip but the distance traveled divided by the time to make the trip will equal 50 miles per hour.

Questions for Further Exploration

1. $f(x)=x^2-2$

For [1, 3]: Average rate of change $=\frac{f(3)-f(1)}{3-1}=\frac{3^2-2-(1^2-2)}{3-1}=\frac{8}{2}=4$

For [1, 2]: Average rate of change $=\frac{f(2)-f(1)}{2-1}=\frac{2^2-2-(1^2-2)}{2-1}=\frac{3}{1}=3$

For [1, 1.5]: Average rate of change $=\frac{f(1.5)-f(1)}{1.5-1}=\frac{1.5^2-2-(1^2-2)}{1.5-1}=\frac{1.25}{0.5}=2.5$

For [1, 1.1]: Average rate of change $=\frac{f(1.1)-f(1)}{1.1-1}=\frac{1.1^2-2-(1^2-2)}{1.1-1}=\frac{0.21}{0.1}=2.1$

(a)
$$\frac{f(1+h)-f(1)}{h+1-1}=\frac{(1+h)^2-2-(1^2-2)}{h}$$
$$=\frac{(1+h)^2-2-1+2}{h}$$
$$=\frac{(h^2+2h+1)-1^2}{h}$$
$$=\frac{h^2+2h}{h}=h+2$$

(b) Tables will vary. One example is shown below

h	[1, 1 + h]	Average rate of change
2.0	[1, 3]	4
1.0	[1, 2]	3
0.5	[1, 1.5]	2.5
0.3	[1. 1.3]	2.3
0.1	[1, 1.1]	2.1
0.05	[1, 1.05]	2.05
0.01	[1, 1.01]	2.01
0.001	[1, 1.001]	2.001

The average rate of change of f approaches 2 as $h\to 0$.

(c) The slope of this tangent line is 2 and the tangent line passes through $(1, f(1)) = (1, -1)$. The y-intercept of the tangent line is

$$y = mx + b$$
$$-1 = 2(1)x + b$$
$$b = -3$$

The equation of the tangent line is $y = 2x - 3$. The graph of $f(x)$ and the tangent line at (1 – 1) is given in the graph at the right. The graphs intersect at one point: (1 – 1).

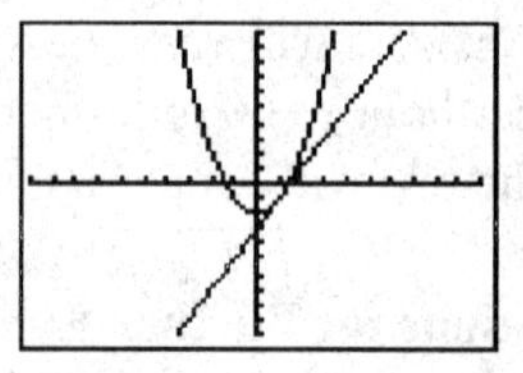

2. (a) For [68, 71]: Average rate of change $= \dfrac{0.08 - 0.06}{71 - 68} = \dfrac{0.02}{3} \approx 0.00667$

For [71, 74]: Average rate of change $= \dfrac{0.10 - 0.08}{74 - 71} = \dfrac{0.02}{3} \approx 0.00667$

For [74, 75]: Average rate of change $= \dfrac{0.13 - 0.10}{75 - 74} = \dfrac{0.03}{1} = 0.03$

For [75, 78]: Average rate of change $= \dfrac{0.15 - 0.13}{78 - 75} = \dfrac{0.02}{3} \approx 0.00667$

For [78, 81]: Average rate of change $= \dfrac{0.20 - 0.15}{81 - 78} = \dfrac{0.05}{3} \approx 0.01667$

For [81, 85]: Average rate of change $= \dfrac{0.22 - 0.20}{85 - 81} = \dfrac{0.02}{4} = 0.005$

For [85, 88]: Average rate of change $= \dfrac{0.25 - 0.22}{88 - 85} = \dfrac{0.03}{3} = 0.01$

For [88, 91]: Average rate of change $= \dfrac{0.29 - 0.25}{91 - 88} = \dfrac{0.04}{3} \approx 0.01333$

For [91, 95]: Average rate of change $= \dfrac{0.32 - 0.29}{95 - 91} = \dfrac{0.03}{4} \approx 0.0075$

(b) The average rate of change was the largest between 1974 and 1975. The average rate of change was the smallest between 1981 and 1985.

(c) There is no interval in the table that has an average rate of change of zero. For an interval such as 1971 to 1972 the average rate of change is zero. This means there was no change in the cost of first-class postage during that interval.

Chapter 10 Project Polar, Rectangular, and Parametric Forms

(a)

$$r = \frac{15}{3 - 2\cos\theta}$$

$$3r - 2r\cos\theta = 15$$

$$3\sqrt{x^2 + y^2} - 2x = 15$$

$$3\sqrt{x^2 + y^2} = 2x + 15$$

$$9x^2 + 9y^2 = 4x^2 + 60x + 225$$

$$5x^2 + 9y^2 - 60x = 225$$

$$5(x^2 - 12x + 36) + 9y^2 = 225 + 180$$

$$5(x-6)^2 + 9y^2 = 405$$

$$\frac{5(x-6)^2}{405} + \frac{9y^2}{405} = 1$$

$$\frac{(x-6)^2}{81} + \frac{y^2}{45} = 1$$

$$\frac{(x-6)^2}{9^2} + \frac{y^2}{(3\sqrt{5})^2} = 1$$

(b) From the standard form of the ellipse in part (a), we see that the center of the ellipse is at (6, 0). Because $a = 9$ and $b = 3\sqrt{5}$, $c^2 = a^2 - b^2 = 81 - 45 = 36$. Because in general the foci of an ellipse with a horizontal major axis are at $(h - c, 0)$ and $(h + c, 0)$, we know that the foci for this ellipse are at (0, 0) and (12, 0). The eccentricity is given by $e = c/a = 6/9 \approx 0.667$. The results are the same as in Example 1 in Section 10.8. The graph in function mode is given below.

(c)

$$\frac{(x-6)^2}{9^2} + \frac{y^2}{(3\sqrt{5})^2} = 1$$

$$\frac{(x-6)^2}{9^2} + \frac{y^2}{(3\sqrt{5})^2} = \cos^2\theta + \sin^2\theta$$

Let $t = \theta$,

$$\frac{(x-6)^2}{9^2} + \frac{y^2}{(3\sqrt{5})^2} = \cos^2 t + \sin^2 t$$

Thus,

$$\frac{(x-6)^2}{9^2} = \cos^2 t \qquad \text{and} \qquad \frac{y^2}{(3\sqrt{5})^2} = \sin^2 t$$

$$\frac{x-6}{9} = \cos t \qquad \frac{y}{3\sqrt{5}} = \sin t$$

$$x = 6 + 9\cos t \qquad y = 3\sqrt{5}\sin t$$

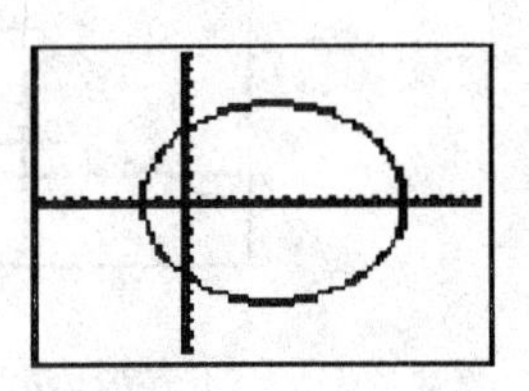

Questions for Further Exploration

1. The conic section is a hyperbola.

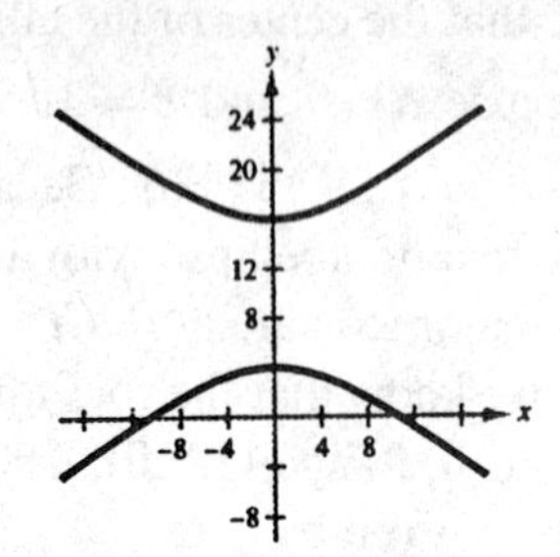

2.
$$r = \frac{32}{3+5\sin\theta}$$
$$3r + 5r\sin\theta = 32$$
$$3\sqrt{x^2+y^2} + 5y = 32$$
$$3\sqrt{x^2+y^2} = 32 - 5y$$
$$9x^2 + 9y^2 = 1024 - 320y + 25y^2$$
$$9x^2 - 16y^2 + 320y = 1024$$
$$9x^2 - 16(y^2 - 20y + 100) = 1024 - 1600$$
$$9x^2 - 16(y-10)^2 = -576$$
$$\frac{9x^2}{-576} - \frac{16(y-10)^2}{-576} = 1$$
$$\frac{(y-10)^2}{36} - \frac{x^2}{64} = 1$$
$$\frac{(y-10)^2}{6^2} - \frac{x^2}{8^2} = 1$$

Yes, the graph obtained in function mode agrees with the graph obtained in Question 1. The polar equation is easier to obtain because only one equation must be entered into the graphing utility.

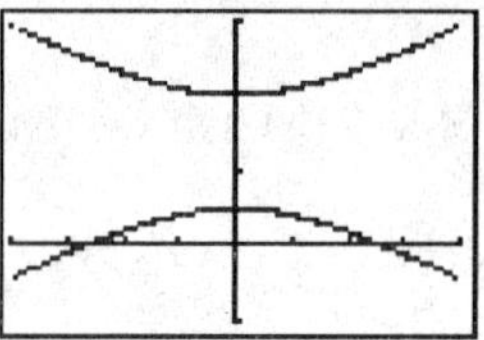

3.

$$\frac{(y-10)^2}{6^2} - \frac{x^2}{8^2} = 1$$
$$\frac{(y-10)^2}{6^2} - \frac{x^2}{8^2} = \sec^2 t - \tan^2 t$$

$$\Rightarrow \quad \frac{(y-10)^2}{6^2} = \sec^2 t \qquad -\frac{x^2}{8^2} = -\tan^2 t$$
$$\frac{y-10}{6} = \sec t \quad \text{and} \quad \frac{x}{8} = \tan t$$
$$y = 10 + 6\sec t \qquad x = 8\tan t$$

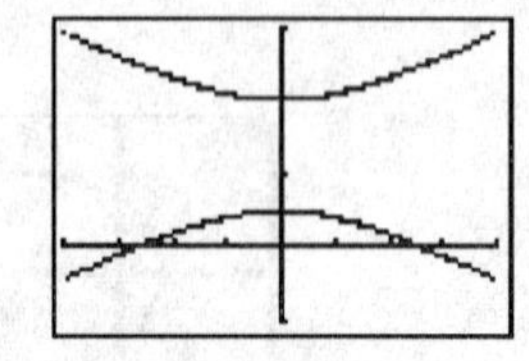

Yes, the graph in parametric mode agrees with the graph obtained in Question 1.

4. From the graph at the right, we can see that the comet would travel in an elliptical orbit about the sun. So, yes, with appropriate scaling of the coordinate system, the given parametric equations could represent the motion of a comet about the sun.

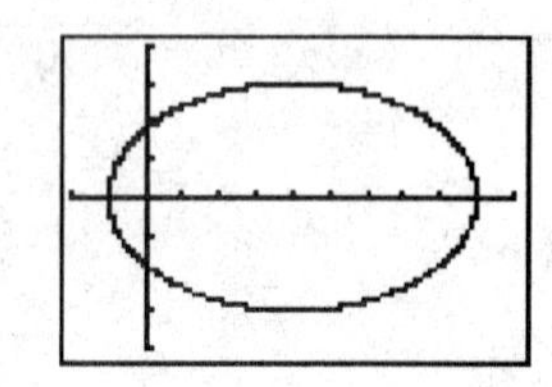

Chapter 11 Project Precalculus with Limits

Questions for Further Exploration

1. Using the direction numbers 3, – 2, and 4, you know that the direction vector for the line is $\mathbf{v} = \langle 3, -2, 4 \rangle$. To find a point on the line, let $t = 0$ and obtain $P = (-2, 0, 1)$. Because $Q = (3, -1, 4)$, we have

$$\overrightarrow{PQ} = \langle 3-(-2), -1-0, 4-1 \rangle = \langle 5, -1, 3 \rangle$$

and we can form the cross product

$$\mathbf{v} \times \overrightarrow{PQ} = \begin{vmatrix} \mathbf{i} & \mathbf{j} & \mathbf{k} \\ 3 & -2 & 4 \\ 5 & -1 & 3 \end{vmatrix} = -2\mathbf{i} + 11\mathbf{j} + 7\mathbf{k} = \langle -2, 11, 7 \rangle$$

Finally, the distance is

$$D = \frac{\left\| \mathbf{v} \times \overrightarrow{PQ} \right\|}{\|\mathbf{v}\|} = \frac{\sqrt{(-2)^2 + 11^2 + 7^2}}{\sqrt{3^2 + (-2)^2 + 4^2}} = \frac{\sqrt{174}}{\sqrt{29}} = \sqrt{6}.$$

2. Find a set of direction numbers for the line that passes through the points. Begin by letting $P = (-2, 3, 1)$ and $R = (2, 3, 0)$. Then a direction vector for the line passing through P and R is given by

$$\mathbf{v} = \overrightarrow{PR} = \langle 2-(-2), 3-3, 0-1 \rangle = \langle 4, 0, -1 \rangle$$

Let $Q = (10, 3, -2)$. Then we have

$$\overrightarrow{PQ} = \langle 10-(-2), 3-3, -2-1 \rangle = \langle 12, 0, -3 \rangle$$

and we can form the cross product

$$\mathbf{v} \times \overrightarrow{PQ} = \begin{vmatrix} \mathbf{i} & \mathbf{j} & \mathbf{k} \\ 4 & 0 & -1 \\ 12 & 0 & -3 \end{vmatrix} = 0\mathbf{i} + 0\mathbf{j} + 0\mathbf{k} = \langle 0, 0, 0 \rangle$$

Finally, the distance is

$$D = \frac{\left\| \mathbf{v} \times \overrightarrow{PQ} \right\|}{\|\mathbf{v}\|} = \frac{\sqrt{0^2 + 0^2 + 0^2}}{\sqrt{4^2 + 0^2 + (-1)^2}} = \frac{0}{\sqrt{17}} = 0.$$

3. Using the direction numbers 3, – 1, and 4 for the first line, you know that the direction vector for the line is $\mathbf{v} = \langle 3, -1, 4 \rangle$. To find a point on this first line, let $t = 0$ and obtain $P = (-1, 2, 0)$. To find a point Q on the second line, let $t = 0$ and obtain $Q = (2, 0, -3)$. Thus,

$$\overrightarrow{PQ} = \langle 2-(-1), 0-2, -3-0 \rangle = \langle 3, -2, -3 \rangle$$

and we can form the cross product

$$\mathbf{v} \times \overrightarrow{PQ} = \begin{vmatrix} \mathbf{i} & \mathbf{j} & \mathbf{k} \\ 3 & -1 & 4 \\ 3 & -2 & -3 \end{vmatrix} = 11\mathbf{i} + 21\mathbf{j} - 3\mathbf{k} = \langle 11, 21, -3 \rangle$$

Finally, the distance is

$$D = \frac{\left\| \mathbf{v} \times \overrightarrow{PQ} \right\|}{\|\mathbf{v}\|} = \frac{\sqrt{11^2 + 21^2 + (-3)^2}}{\sqrt{3^2 + (-1)^2 + 4^2}} = \frac{\sqrt{571}}{\sqrt{26}}.$$

4. Rewrite the symmetric equations of the lines as:

Line 1: $x = t, y = 2t, z = 3t$

Line 2: $x = 1 - t, y = 4 + t, z = t - 1$; Direction vector: $\langle -1, 1, 1 \rangle$

Begin by finding two points on Line 1. Let $t = 0$ and obtain $A = (0, 0, 0)$. Let $t = 1$ and obtain $B = (1, 2, 3)$. Use the direction vector for Line 2 with point A to find a third point C to form a plane that contains Line 1 and is parallel to Line 2.

$C = (0 - 1, 0 + 1, 0 + 1) = (-1, 1, 1)$

Now find the distance between a point on Line 2 and this plane formed by points A, B, and C. To do so, first find a normal to the plane by finding the cross product of two vectors in the plane. Two such vectors are

$$\overrightarrow{AB} = \langle 1-0, 2-0, 3-0 \rangle = \langle 1, 2, 3 \rangle$$

$$\overrightarrow{AC} = \langle -1-0, 1-0, 1-0 \rangle = \langle -1, 1, 1 \rangle$$

$$\mathbf{n} = \overrightarrow{AB} \times \overrightarrow{AC} = \begin{vmatrix} \mathbf{i} & \mathbf{j} & \mathbf{k} \\ 1 & 2 & 3 \\ -1 & 1 & 1 \end{vmatrix} = -\mathbf{i} - 4\mathbf{j} + 3\mathbf{k} = \langle -1, -4, 3 \rangle$$

Now find a point Q on Line 2. Let $t = 0$ and obtain $(1, 4, -1)$. The vector from point A on the plane to point Q on Line 2 is given by

$$\overrightarrow{AQ} = \langle 1-0, 4-0, -1-0 \rangle = \langle 1, 4, -1 \rangle$$

Using the formula for distance between a plane and a point produces

$$D = \frac{\left| \overrightarrow{AQ} \cdot \mathbf{n} \right|}{\|\mathbf{n}\|} = \frac{\left| \langle 1, 4, -1 \rangle \cdot \langle -1, -4, 3 \rangle \right|}{\sqrt{(-1)^2 + (-4)^2 + 3^2}}$$

$$= \frac{|-1-16-3|}{\sqrt{1+16+9}}$$

$$= \frac{20}{\sqrt{26}} = \frac{20}{\sqrt{26}} \cdot \frac{\sqrt{26}}{\sqrt{26}} = \frac{10\sqrt{26}}{13}$$

Chapter 12 Project Tangent Lines to Sine Curves

(a) By approximating the slope of the sine curve at 17 points and connecting them with a smooth curve, we get the following graph.

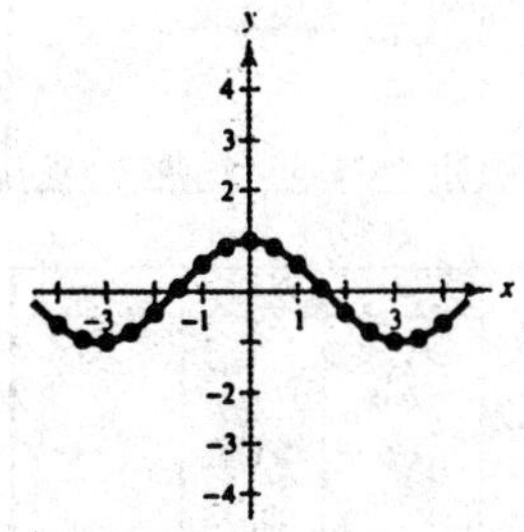

The curve appears to be the graph of the cosine function.

(b) Using the table feature of a graphing utility, we obtain the following table.

x	0	0.5	1.0	1.5	2.0	2.5	3.0	3.5	4.0
m	≈ 1	0.875	0.536	0.066	− 0.421	− 0.804	− 0.991	− 0.935	− 0.650

Using a graphing utility, a graph of these points is shown below. The results of this numerical approach is similar to the results in part (a). The graph again appears to be the cosine function.

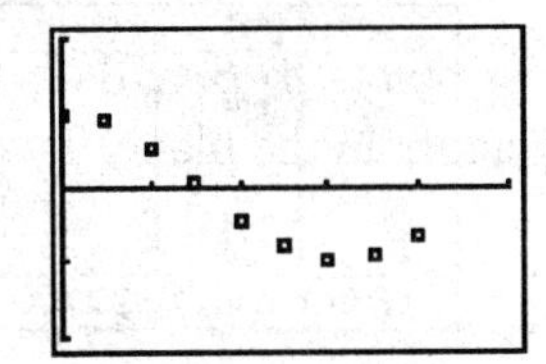

(c) To calculate the slope of the tangent line to $f(x) = \sin x$, start with the definition.

$$\begin{aligned}
\lim_{h\to 0} \frac{f(x+h)-f(x)}{h} &= \lim_{h\to 0} \frac{\sin(x+h)-\sin x}{h} \\
&= \lim_{h\to 0} \frac{\sin x \cos h + \cos x \sin h - \sin x}{h} \\
&= \lim_{h\to 0} \frac{\cos x \sin h - (\sin x)(1-\cos h)}{h} \\
&= \lim_{h\to 0}\left[(\cos x)\left(\frac{\sin h}{h}\right) - (\sin x)\left(\frac{1-\cos h}{h}\right)\right] \\
&= \cos x\left[\lim_{h\to 0}\left(\frac{\sin h}{h}\right)\right] - \sin x\left[\lim_{h\to 0}\left(\frac{1-\cos h}{h}\right)\right] \\
&= (\cos x)(1) - (\sin x)(0) \\
&= \cos x
\end{aligned}$$

Questions for Further Exploration

1. *Graphically:* Use a graphing utility to graph $f(x) = \dfrac{\sin x}{x}$ and use the trace feature to investigate values of the function to the left and right of $x = 0$ (note that the function is undefined *at* $x = 0$). The graph at the right provides support that the value of $\lim\limits_{h \to 0} \dfrac{\sin h}{h}$ is 1.

   Numerically: Use a graphing utility to create a table of values for $f(x) = \dfrac{\sin x}{x}$ at values of x near 0. Notice from the table that as x approaches 0 from the left and from the right, the value of $f(x)$ gets closer and closer to 1.

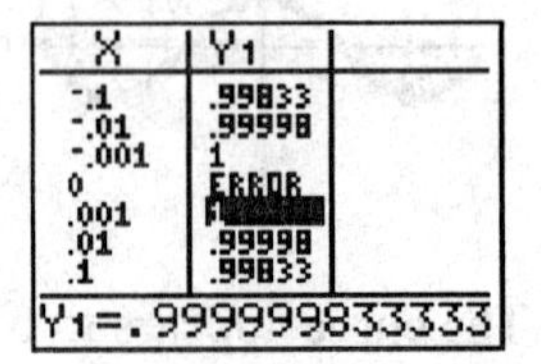

2. *Graphically:* Use a graphing utility to graph $f(x) = \dfrac{\cos x - 1}{x}$ and use the trace feature to investigate values of the function to the left and right of $x = 0$ (note that the function is undefined *at* $x = 0$). The graph at the right provides support that the value of $\lim\limits_{h \to 0} \dfrac{\cos h - 1}{h}$ is 0.

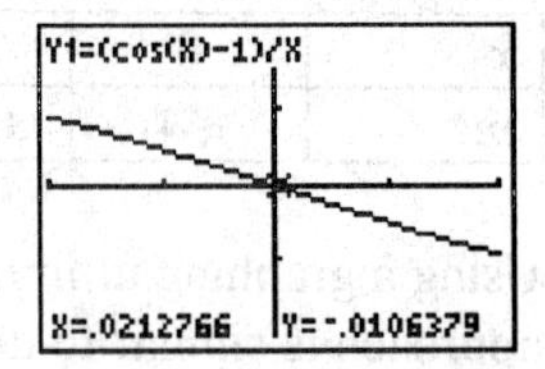

   Numerically: Use a graphing utility to create a table of values for $f(x) = \dfrac{\cos x - 1}{x}$ at values of x near 0. Notice from the table that as x approaches 0 from the left and from the right, the value of $f(x)$ gets closer and closer to 0.

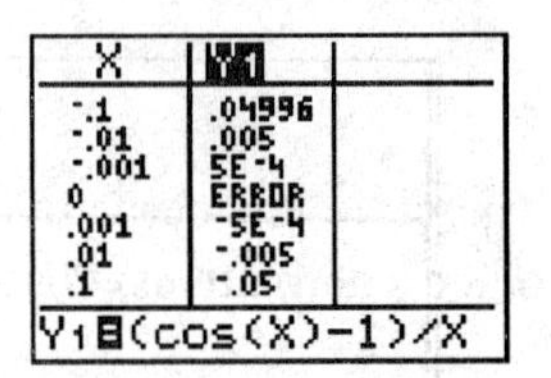

3. *Graphical Approach:* Begin by graphing $g(x) = \cos x$ with a graphing utility as shown at the right and estimating the slope of this curve at various values of x. For instance, the slope is approximately 0 at $x = 0$, approximately -0.5 at $x = 0.5$, and so on. A graph of the slope estimates connected by a smooth curve is shown below. The graph appears to be the reflection of the sine curve in the x-axis.

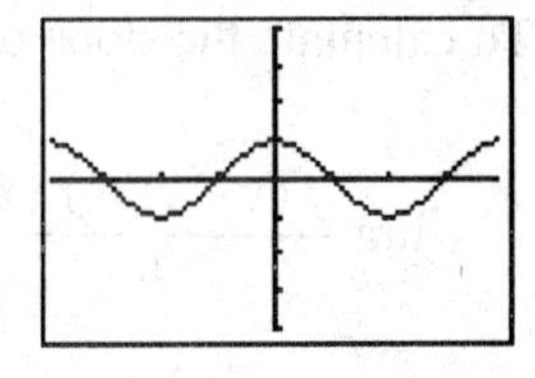

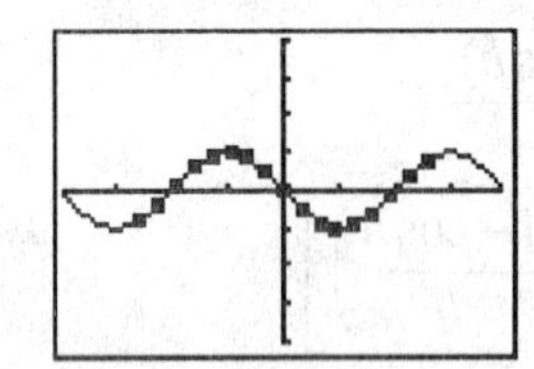

   Numerical Approach: Using the table feature of a graphing utility, we obtain the following table of approximations for the slope of the cosine curve at x.

x	0	0.5	1.0	1.5	2.0	2.5	3.0	3.5	4.0
m	-0.005	-0.484	-0.844	-0.998	-0.907	-0.594	-0.136	0.355	0.760

   Using a graphing utility, a graph of these points is shown below. The results of this numerical approach is similar to the results of the graphical approach. The graph again appears to be the reflection of the sine curve in the x-axis.

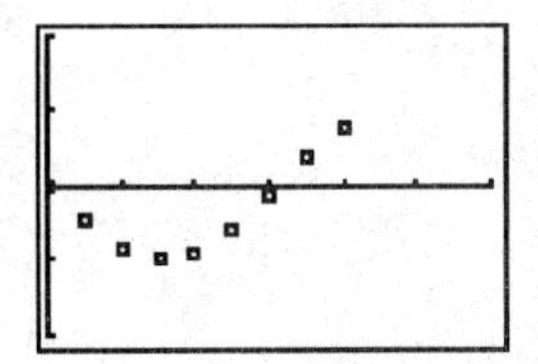

Analytical Approach: Use the definition of the slope of the tangent line to $f(x) = \sin x$.

$$\lim_{h\to 0}\frac{f(x+h)-f(x)}{h} = \lim_{h\to 0}\frac{\cos(x+h)-\cos x}{h}$$

$$= \lim_{h\to 0}\frac{\cos x\cos h-\sin x\sin h-\cos x}{h}$$

$$= \lim_{h\to 0}\frac{\cos x(\cos h-1)-\sin x\sin h}{h}$$

$$= \lim_{h\to 0}\left[(\cos x)\left(\frac{\cos h-1}{h}\right)-(\sin x)\left(\frac{\sin h}{h}\right)\right]$$

$$= \cos x\left[\lim_{h\to 0}\left(\frac{\cos h-1}{h}\right)\right]-\sin x\left[\lim_{h\to 0}\left(\frac{\sin h}{h}\right)\right]$$

$$= (\cos x)(0)-(\sin x)(-)$$

$$= -\sin x$$

Therefore, the derivative of cos x is – sin x. It is not true that the sine and cosine functions are each other's derivatives; cos x is the derivative of sin x.

PART III
Focus on Concepts Solutions

Chapter P

1. (b) **2.** (c) **3.** (d) **4.** (a)

5. An identity is true for all values of the variable, and a conditional equation is true for only some values of the variable.

6. (a) Negative (b) Positive

7. (a) Because a linear equation can have no more than one solution and a quadratic equation can have no more than two solutions, these are solutions of neither.
(b) Because a linear equation can have 0 or 1 solution and a quadratic can have 0, 1, or 2 solutions, this could be the solution of both.
(c) Quadratic
(d) Neither

8. Dividing by x does not yield an equivalent equation. $x = 0$ is also a solution.

9. (b)

10. (a) $x = a$, $x = b$

(b)

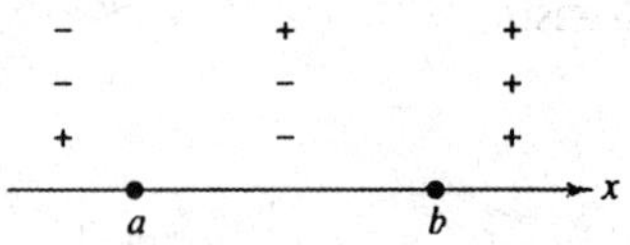

(c) The real zeros of the polynomial.

Chapter 1

1. No, this relationship is not a function. The element 3 in the domain corresponds to two different elements in the range.

2. (a)

Xmin = – 15
Xmax = 6
Xscl = 3
Ymin = – 18
Ymax = 6
Yscl = 3

(b)

Xmin = – 24
Xmax = 36
Xscl = 6
Ymin = – 54
Ymax = 12
Yscl = 6

3. (a) Even function. The graph is a reflection in the x-axis.
(b) Even function. The graph is a reflection in the y-axis.
(c) Even function. The graph is a vertical translation of f.
(d) Neither even nor odd. The graph is a horizontal translation of f.

4. (a) $g(t) = \frac{3}{4} f(t)$ (b) $g(t) = f(t) + 10{,}000$

(c) $g(t) = f(t-2)$ (d) $g(t) = f\left(\frac{t}{2}\right)$

5. False. A correlation coefficient close to -1 implies that the model fits the data well and that the slope of the model is negative.

6. True

Chapter 2

1. The president of the company would prefer the conditions (a) and (b) because profits would be increasing.

2. (a) Degree: 3; leading coefficient: positive
(b) Degree: 2; leading coefficient: positive
(c) Degree: 4; leading coefficient: positive
(d) Degree: 5; leading coefficient: positive

3. (a) No, because a second-degree equation has at most 2 real solutions.
(b) Yes, because a third-degree equation has at most 3 real solutions.

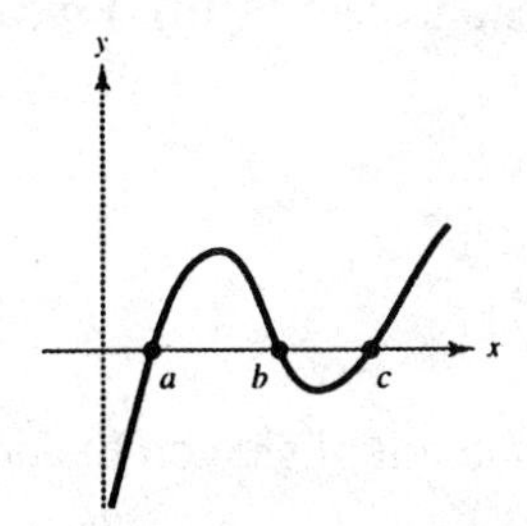

(c) Yes, because a fourth-degree equation has at most 4 real solutions—this would contain a set of repeated real solutions.

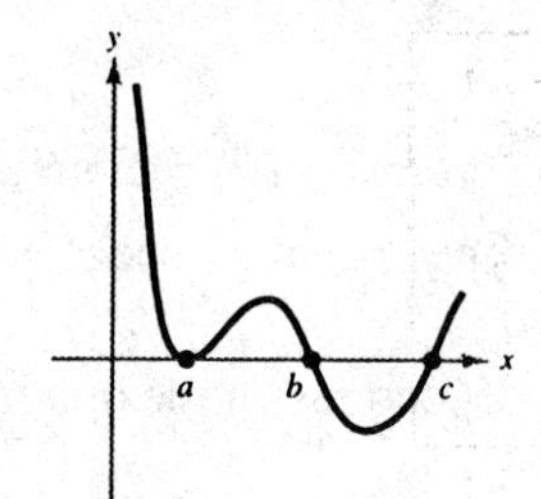

(d) Yes, because a fifth-degree equation has at most 5 real solutions—this could contain two sets of repeated real solutions.

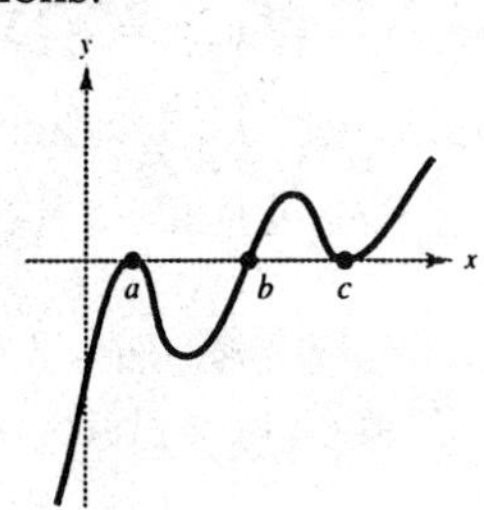

4. Because this has a vertical asymptote at $x = 2$, a horizontal asymptote at $y = 2$, and a zero at $x = 3$, this graph resembles the graph of the function $f(x) = \frac{2(x-3)}{x-2}$.

5. Because this has vertical asymptotes at $x \approx \pm\sqrt{2}$, a horizontal asymptote at $y = 0$, and no real zeros, this graph resembles the graph of the function $f(x) = \frac{2}{x^2 - 2}$.

6. Because this has a vertical asymptote at $x = 2$, a horizontal asymptote at $y = 0$, and no real zeros, this graph resembles the graph of the function $f(x) = \frac{2}{x-2}$.

7. Because this has a vertical asymptote at $x = 2$, a horizontal asymptote at $y = 2$, and a zero at $x = 0$, this graph resembles the graph of the function $f(x) = \frac{2x}{x-2}$.

Chapter 3

1. $b < d < a < c$; b and d are negative.

2. (a) $\log_b uv = \log_b u + \log_b v$. True
   (b) $\log_b (u+v) = (\log_b u)(\log_b v)$. False. For example, $2.04 \approx \log(10+100) \neq (\log 10)(\log 100) = 2$.
   (c) $\log_b (u-v) = \log_b u - \log_b v$. False. For example, $1.95 \approx \log(100-10) \neq \log 100 - \log 10 = 1$.
   (d) $\log_b \frac{u}{v} = \log_b u - \log_b v$. True

3. It would be most advantageous to double the interest rate or time because that would double the exponent in the exponential function.

4. (a) Logarithmic (b) Logistic (c) Exponential (d) Linear
   (e) None of the above (f) Exponential

Chapter 4

1. (a) An angle in standard position has its vertex at the origin and the initial side on the positive x-axis.
   (b) A negative angle is a clockwise rotation of the terminal side.
   (c) Coterminal angles are two angles in standard position where the terminal sides coincide.
   (d) An obtuse angle has a magnitude between 90° and 180°.

2. If a fan of greater diameter is installed on the motor, the speed of the tips of the blade increases. The linear velocity is proportional to the radius.

3. This is false because for each θ there corresponds exactly one value of y.

4. At the zeros of $g(\theta) = \cos\theta$, $f(\theta) = \sec\theta$ is undefined because $\sec\theta = 1/\cos\theta$.

5. Determine the trigonometric function of the reference angle and, depending on the quadrant in which the obtuse angle lies, prefix the appropriate sign.

6. matches graph (d) because the period is 2π and the amplitude is 3.

7. matches graph (a) because the period is 2π and, because $a < 0$, the graph is reflected about the x-axis.

8. matches graph (b) because the period is 2 and the amplitude is 2.

9. matches graph (c) because the period is 4π and the amplitude is 2.

10. (a) $f(t-2c) = f(t)$ because the left side represents a two-period shift
(b) $f\left(t+\frac{1}{2}c\right) \neq f\left(\frac{1}{2}t\right)$ because the left side represents a horizontal translation and the right side represents a period change.
(c) $f\left(\frac{1}{2}[t+c]\right) = f\left(\frac{1}{2}t\right)$ because both sides represent the same period change.

11. Finding the amplitude is not part of the analysis of the other four trigonometric functions because their range is $(-\infty, \infty)$.

12. (a) When A is changed from $\frac{1}{5}$ to $\frac{1}{3}$, the displacement is increased.
(b) When k is changed from $\frac{1}{10}$ to $\frac{1}{3}$, the friction damps the oscillations more quickly.
(c) When b is changed from 6 to 9, the frequency of the oscillations increases.

13. False, because $3\pi/4$ is not in the range of the arctangent function.

Chapter 5

1. An identity is true for all values of the variable and a conditional equation is true for some values of the variable.

2. When proving an identity, use the fundamental identities and rules of algebra to transform one expression into another. To solve a trigonometric equation, use standard algebraic techniques and identities to isolate a trigonometric function involved in the equation. Find the value of the variable by using the inverse of the trigonometric function.

3. Reciprocal identities: $\csc\theta = \dfrac{1}{\sin\theta}$, $\sec\theta = \dfrac{1}{\cos\theta}$, $\cot\theta = \dfrac{1}{\tan\theta}$

Quotient identities: $\tan\theta = \dfrac{\sin\theta}{\cos\theta}$, $\cot\theta = \dfrac{\cos\theta}{\sin\theta}$

Pythagorean identities: $\sin^2\theta + \cos^2\theta = 1$, $\tan^2\theta + 1 = \sec^2\theta$, $1 + \cot^2\theta = \csc^2\theta$

4. No, $\cos\theta = \pm\sqrt{1-\sin^2\theta}$.

5. False. The order in which algebraic operations and fundamental identities are used to verify an identity may vary.

6. (a) True. The period of tangent is π.
(b) False. The period of cosine is 2π.
(c) False. $\sec\theta\cos\theta = 1$.
(d) True
(e) True. $\sin(-\alpha) = -\sin\alpha$

7. $y_1 = y_2 + 1$

8. $y_1 = 1 - y_2$

9. One point of intersection

10. Three points of intersection

11. The graph shows 5 zeros.

12. The graph shows 4 zeros.

13. The most likely value of d is 2π.

Chapter 6

1. $\dfrac{a}{\sin A} = \dfrac{b}{\sin B} = \dfrac{c}{\sin C}$

2. $a^2 = b^2 + c^2 - 2bc\cos A$, $b^2 = a^2 + c^2 - 2ac\cos B$, $c^2 = a^2 + b^2 - 2ab\cos C$

3. True

4. When one of the angles in a triangle is a right triangle, the Law of Cosines simplifies to the Pythagorean Theorem.

5. False. There may be no solution, one solution, or two solutions.

6. Direction and magnitude characterize a vector in the plane.

7. Vectors **A** and **C** appear to be equivalent (same direction and magnitude).

8. The magnitude of the resultant will be greater in (a) because the angle between the vectors is acute.

9. If $k > 0$, the direction is the same and the magnitude is k times as great. If $k < 0$, the result is a vector in the opposite direction and the magnitude is k times as great.

10. The sum of the vectors **u** and **v** is the diagonal of the parallelogram with **u** and **v** as its adjacent sides.

11. Graph (b) gives the difference **u** – **v**. Visualize the sum of **u** and – **v**.

12. $z_1 z_2 = -4$ and $z_1 / z_2 = (-1/4)z_1^{\,2}$

13. (a) Three roots are not shown.
(b) The modulus of each is 2, and the arguments are 120°, 210°, and 300°.

Chapter 7

1. A solution of a system is an ordered pair that satisfies each equation in the system.

2. For a linear system the result will be a contradictory equation such as $0 = N$, where N is a nonzero real number. For nonlinear systems there may be an equation with imaginary roots.

3. There will be a contradictory equation of the form $0 = N$, where N is a nonzero real number.

4. The algebraic methods yield exact solutions.

5. (a) The maximum number of solutions when both equations are linear is one.
   (b) The maximum number of solutions when one equation is linear and other is quadratic is two.
   (c) The maximum number of solutions when both equations are quadratics is four.

6. An inconsistent system of equations has no solution.

7. The graph of a system of linear equations in two variables that has no solution two lines that are distinct and parallel. An example of such a system is given at the right.

$$x + 2y = 3$$
$$2x + 4y = 9$$

8. The operations on a system of linear equations that produce equivalent systems of equations are:
   (a) Interchange any two equations.
   (b) Multiply an equation by a nonzero constant.
   (c) Add a multiple of one equation to any other equation in the system.

9. No, the given systems of equations are not equivalent because in the first system when -2 times Equation 1 is added Equation 2, the constant is -11. In the second system, this same equation (the second equation in the system) has a constant of 1.

10. The first system is inconsistent because 4 times Equation 1 added to Equation 2 yields $0 = 20$.

11. (d) 12. (b) 13. (c) 14. (a)

15. (a) The boundary would be included in the solution.
    (b) The solution would be the half-plane on the opposite side of the boundary.

Chapter 8

1. The three elementary row operations are:
   (a) Interchange any two rows.
   (b) Multiply a row by a nonzero constant.
   (c) Add a multiple of one row to any other row.

2. They are the same.

3. A matrix in row-echelon form is in reduced row-echelon form if every column that has a leading 1 has zeros in every position above and below its leading 1.

4. Consistent—an infinite number of solutions (because the last row of the matrix is equivalent to $0 = 0$)

5. Inconsistent (because the last row of the matrix is equivalent to $0 = 8$)

6. Consistent—a unique solution (because the last row is equivalent to $z = -3$)

7. Consistent—an infinite number of solutions (because the last row of the matrix is equivalent to $y + z = 0$)

8. (a) The operation can be performed because A and $3B$ have the same dimensions.
(b) The operation can be performed because the number of columns in A is equal to the number of rows in B.

9. (a) The operation can be performed because A and $3B$ have the same dimensions.
(b) The operation cannot be performed because the number of columns in A does not equal the number of rows in B.

10. (a) The operation cannot be performed because A and $3B$ do not have the same dimensions.
(b) The operation can be performed because the number of columns in A is equal to the number of rows in B.

11. The matrix must be square and its determinant nonzero.

12. A square matrix is a square array of numbers and a determinant is a real number associated with a square matrix.

13. No. The matrix must be square.

14. If A is a square matrix, the cofactor C_{ij} of the entry a_{ij} is $(-1)^{i+j} M_{ij}$, where M_{ij} is the determinant obtained by deleting the ith row and the jth column of A. The determinant of A is the sum of the entries of any row or column of A multiplied by their respective cofactors.

15. No. The first two yield a unique solution to the system and the third yields an infinite number of solutions.

Chapter 9

1. Natural numbers

2. (a) Odd-numbered terms are negative.
(b) Even-numbered terms are negative.

3. True by the definition of factorial.

4. True by the properties of sums.

5. True by the properties of sums.

6. True because $\sum_{j=1}^{6} 2^j = 2+4+8+16+32+64$ and $\sum_{j=3}^{8} 2^{j-2} = 2+4+8+16+32+64$.

7. (a) Each term is obtained by adding the same constant (common difference) to the previous term.
(b) Each term is obtained by multiplying the same constant (common ratio) by the previous term.

8. (a) Arithmetic. There is a constant difference between consecutive terms.
(b) Geometric. Each term is a constant multiple of the previous term. In this case the common ratio is greater than 1.

9. Each term of the sequence is defined in terms of the previous term.

10. Increased powers of real numbers between 0 and 1 approach zero.

11. (d)

12. (a)

13. (b)

14. (c)

15. The signs of the terms alternate in the expansion of $(x-y)^n$.

16. They are the same.

17. ${}_{10}P_6 > {}_{10}C_6$. Changing the order of any of the six elements selected results in a different permutation but the same combination.

18. Meteorological records indicate that over an extended period of time with similar weather conditions it will rain 60% of the time.

19. Of the different measures of central tendency, only the mean is affected by an extreme measure in the data list.

20. If the standard deviation of a set of measures is 0, then the numbers in the set are identical.

Chapter 10

1. (a) Vertical translation
(b) Horizontal translation
(c) Reflection in the y-axis
(d) The parabola is narrower.

2. (a) Major axis horizontal
(b) Circle
(c) Ellipse is flatter.
(d) Horizontal translation

3. The extended diagonals of the central rectangle are asymptotes of the hyperbola.

4. The number b must be less than 5. The ellipse becomes more circular and approaches a circle of radius 5.

5. The orientation of the graph would be reversed.

6. (a) The speed would double.
(b) The elliptical orbit would be flatter. The length of the major axis is greater.

7. False. The following are two sets of parametric equations for the line.

$$x = t, \quad y = 3 - 2t \qquad\qquad x = 3t, \quad y = 3 - 6t$$

8. False. $(2, \pi/4)$, $(-2, 5\pi/4)$, and $(2, 9\pi/4)$ all represent the same point.

9. (a) Symmetric to the pole
(b) Symmetric to the polar axis
(c) Symmetric to the line $\theta = \pi/2$

10. (a) The graphs are the same.
(b) The graphs are the same.

Chapter 11

1. x = directed distance from yz-plane to P
   y = directed distance from xz-plane to P
   z = directed distance from xy-plane to P

2. This surface is a sphere with the equation: $(x-h)^2+(y-k)^2+(z-j)^2=r^2$

3. The xz-trace of a surface consists of all points that are common to both the surface and the xz-plane. To find the equation for this trace, substitute $y=0$ into the equation for the surface.

4. The trace is a circle.

5. The trace is a line.

6. Direction and magnitude characterize a vector in space.

7. Vectors **B** and **D** appear to be equivalent (same direction and magnitude).

8. The vector has the same direction or opposite direction depending on whether k is positive or negative, respectively. The magnitude is changed by the factor $|k|$.

9. The angle θ between the two vectors is in the interval $90° < \theta < 180°$.

10. The vectors must be in three-space because the cross product is orthogonal to the two given vectors.

11. False. From the properties of cross products we know that $\mathbf{u} \times \mathbf{v} = -(\mathbf{v} \times \mathbf{u})$.

12. (c)

13. The lines are parallel because their direction numbers are proportional.

14. False. The two lines could be skew—not parallel but do not intersect.

15. True

16. (a) Line (b) Line (c) Plane

Chapter 12

1. From the graph, it appears that the limit exists as x approaches 2.

2. From the graph, it appears that the limit does not exist as x approaches 2.

3. From the graph, it appears that the limit does not exist as x approaches 2.

4. From the graph, it appears that the limit exists as x approaches 2.

5. No, $f(3)$ may or may not exist because the limit only explains the behavior of the function as x *approaches* 3.

6. When direct substitution in a fractional function produces 0 in both the numerator and denominator, the resulting fraction $\frac{0}{0}$ is called an indeterminate form. (There are other indeterminate forms studied in calculus.)

7. The functions agree at all real numbers x except $x = -2$.

8. False. The tangent line $y = 3x - 2$ intersects the graph of $y = x^3$ at more than one point.

9. The tangent line is used to determine the rate at which a graph rises or falls at a single point.

10. The slope at the point can be estimated as 2.

11. The slope at the point appears to be 0.

12. False. Not every rational function has a horizontal asymptote. For instance, $y = \dfrac{x^2}{x+1}$ does not.

13. When the degree of the numerator of a rational function is less than the degree of the denominator, the limit at infinity is 0. If the degree of the numerator is equal to the degree of the denominator, the limit at infinity is the ratio of the coefficients of the highest-powered terms.

14. Divide the interval from a to b into n subintervals of equal width. Form a collection of rectangles over the subintervals whose heights are given by the function. The exact area is the limit of the sum of the areas of the n rectangles as n approaches ∞.

15. (c)